A TEXTBOOK OF ENGINEERING MATHEMATICS

A TEXTBOOK OF

ENGINEERING MATHEMATICS

For

B.E., B.TECH. SEMESTER I
ACCORDING TO THE LATEST SYLLABUS OF
PUNJAB TECHNICAL UNIVERSITY, JALANDHAR

By

N.P. BALI
Former Principal
S.B. College, Gurgaon
Haryana

USHA PAUL
Formerly Head, Dept. of Mathematics,
Kanya Maha Vidyalaya, Jalandhar
Ex-Member Board of Studies
Ex-Faculty of Sciences
Guru Nanak Dev University, Amritsar
Punjab

LAXMI PUBLICATIONS (P) LTD
(An ISO 9001:2008 Company)
BENGALURU • CHENNAI • COCHIN • GUWAHATI • HYDERABAD
JALANDHAR • KOLKATA • LUCKNOW • MUMBAI • RANCHI • NEW DELHI
BOSTON (USA) • ACCRA (GHANA) • NAIROBI (KENYA)

A TEXTBOOK OF ENGINEERING MATHEMATICS

Printed and bound in India

Typeset at Goswami Associates, Delhi

First Edition: 2006; Second Edition : 2007; Third Edition: 2008; Fourth Edition: 2009; Fifth Edition : 2010;
Sixth Edition: 2011; Seventh Edition: 2013; Eighth Edition: 2014; Ninth Edition: 2015, Tenth Edition : 2016;

ISBN 978-81-318-0807-8

Published in India by

Laxmi Publications (P) Ltd.
(An ISO 9001:2008 Company)
113, GOLDEN HOUSE, DARYAGANJ,
NEW DELHI - 110002, INDIA
Telephone : 91-11-4353 2500, 4353 2501
Fax : 91-11-2325 2572, 4353 2528
www.laxmipublications.com info@laxmipublications.com

Branches

✆	Bengaluru	080-26 75 69 30	
✆	Chennai	044-24 34 47 26,	24 35 95 07
✆	Cochin	0484-237 70 04,	405 13 03
✆	Guwahati	0361-254 36 69,	251 38 81
✆	Hyderabad	040-27 55 53 83,	27 55 53 93
✆	Jalandhar	0181-222 12 72	
✆	Kolkata	033-22 27 43 84	
✆	Lucknow	0522-220 99 16	
✆	Mumbai	022-24 91 54 15,	24 92 78 69
✆	Ranchi	0651-220 44 64	

C—

CONTENTS

PART A

PART B

Preface to the Eleventh Edition

It is with the grace of God and over whelming response given by learned professors and discerning students that our book **'A Textbook of Engineering Mathematics-I'** is gaining increasing popularity. We thank the almighty that our hardwork paid off and also place on record our sense of gratitude to the esteemed readers for their so much appreciation of our work.

Every year we revise the book as per university and our readers' requirements and bring the necessary alterations in the subject matter of the book. This edition of the book is strictly according to the revised (new) syllabus of Engineering Mathematics-I. Authors observed that these days the trend of paper setting of various universities is to test the general understanding of the subject. So keeping in view this trend, this new edition is revised with lot of care, dedication and patience.

The following are the salient features of this edition:

(*i*) Many new solved examples and problems are added in each chapter. Also hints are given alongside typical unsolved problems.

(*ii*) Where needed, recapitulation of the topic is given in the beginning of the chapter. Working rules for lengthy formulae are also given.

(*iii*) At the end of each chapter **'Review of the Chapter'** is introduced so that students can revise the chapter at a glance. Also at the end of each chapter **'Short Answer Type Questions'** are given which form compulsory section (containing ten questions each of two marks) of the paper.

The present edition includes all the questions set in the last ten years university papers. These are mostly in the form of solved examples. These will certainly make students familiar with university pattern.

We have tried our best to make the book 'mistake-free' but inspite of our best efforts some errors might have crept in the book. Report of any such error and all suggestions for improving the future edition of the book are welcome and will be gratefully acknowledged.

It is hoped that book in its new form will attract more readers and will be found to be of much more utility.

We wish our readers the very best of luck for their brilliant success in life.

—Authors

Preface to the First Edition

This book of Mathematics has been specially written to meet the requirements of B.E./B. Tech. first year students of various institutions, universities and Engineering Courses.

The salient features of the book are:

(*i*) The book presents the subject matter in a very systematic, simple and lucid style, so that students themselves will be able to understand the solutions of the problems.

(*ii*) Each chapter starts with necessary definitions and complete proofs of the standard theorems followed by solved examples. For convenience of students, working rules for the applications of theorems in questions are given. Also lists of important results are given at the end of chapters, where needed.

(*iii*) For convenience of students, lengthy chapters are divided into small units.

(*iv*) In the beginning of some chapters, some reference topics are discussed in detail inspite of the fact that these topics are not in the syllabus of certain universities. It is done because without the knowledge of these topics students cannot understand the main topic of the syllabus *e.g.*, in second chapter of 'Analytical solid geometry' unit I of 'planes and lines' is discussed because without the knowledge of plane and line, next topics of sphere, cone and cylinder cannot be understood. Similarly in chapter fourth the topics of asymptotes multiple points etc., are discussed because without the detailed knowledge of these topics, 'curve tracing' is not possible.

(*v*) *The most distinguished and outstanding feature* of this book is that each topic contains a large number of solved examples (Simple as well as typical). Many examples have been selected from various university papers so as to make students familiar with university pattern.

This book serves triple purpose *viz. textbook, help book and solved university* papers and authors are sure that the study of this book will instill confidence in students.

The authors of the book possess more than three decades of rich experience of teaching Mathematics to Graduate as well as Postgraduate classes and have first hand experience of the problems and difficulties faced by students.

Suggestions for improvement of the book will be most gratefully received.

—Authors

SYLLABUS

PUNJAB TECHNICAL UNIVERSITY, JALANDHAR

BTAM-101, ENGINEERING MATHEMATICS–I

Objective/s and Expected Outcome

"Math and basic science are certainly the foundations of any engineering program. This fact will not change in the foreseeable future" said by Ellis et al. Engineering Mathematics is an essential tool for describing and analyzing engineering processes and systems. Mathematics also enables precise representation and communication of knowledge. Core mathematics courses have broader objectives than just supporting engineering programs. The learning objectives of core mathematics courses can be put into three categories: (1) Content Objectives: Students should learn fundamental mathematical concepts and how to apply them. (2) Skill Objectives: Students should learn critical thinking, modeling/problem solving and effective uses of technology. (3) Communication Objectives: Students should learn how to read mathematics and use it to communicate knowledge. The students are expected to understand the fundamentals of the mathematics to apply while designing technology and creating innovations.

PART A

1. **Differential Calculus:** Curve tracing: Tracing of Standard Cartesian; Parametric and Polar curves; Curvature of Cartesian, Parametric and Polar curves. **(6)**
2. **Integral Calculus:** Rectification of standard curves; Areas bounded by standard curves; Volumes and surfaces of revolution of curves; Applications of integral calculus to find centre of gravity and moment of inertia. **(6)**
3. **Partial Derivatives:** Function of two or more variables; Partial differentiation; Homogeneous functions and Euler's theorem; Composite functions; Total derivative; Derivative of an implicit function; Change of variable; Jacobians. **(6)**
4. **Applications of Partial Differentiation:** Tangent and normal to a surface; Taylor's and Maclaurin's series for a function of two variables; Errors and approximations; Maxima and minima of function of several variables; Lagrange's method of undetermined multipliers. **(6)**

PART B

5. **Multiple Integrals:** A brief introduction of cylinder, cone and standard conicoids. Double and triple integral and their evaluation, change of order of integration, change of variable, Application of double and triple integration to find areas and volumes. **(6)**
6. **Vector Calculus:** Scalar and vector fields, differentiation of vectors, velocity and acceleration. Vector differential operators: Del, Gradient, Divergence and Curl, their physical interpretations. Formulae involving Del applied to point functions and their products. Line, surface and volume integrals. **(8)**
7. **Application of Vector Calculus:** Flux, Solenoidal and Irrotational vectors. Gauss Divergence theorem. Green's theorem in plane, Stoke's theorem (without proofs) and their applications. **(4)**

PART A

- **Differential Calculus**
- **Integral Calculus**
- **Partial Differentiation**
- **Application of Partial Differentiation**

1

Differential Calculus

The Chapter of Differential Calculus is divided into two Sections—Section A : Curve Tracing and Section B : Curvature.

SECTION A: CURVE TRACING

1.1. INTRODUCTION

Students are already familiar with the idea of tracing a curve by giving different values to x and finding corresponding values of y and joining these points by free hand curves. But this type of tracing gives us only a rough idea of what the graph looks like. But to have more accuracy in sketching the curves we will now use the techniques of domains, intervals of increase and decrease of the curve, maxima, minima, concepts of concavity and convexity, asymptotes etc. Many of these techniques are already studied in details by the students in lower classes but some of these like concavity convexity, asymptotes, multiple points are new to them. So first of all we will have discussion about these new topics.

1.2. CONCAVITY, CONVEXITY, POINT OF INFLEXION

Def. (a) Concave upward or convex downward

Let $y = f(x)$ be any continuous curve in (a, b).

A curve is said to concave upward (or convex downward) on (a, b) if all the points of the curve lie above any tangent to it on that interval [See Fig. (*i*)].

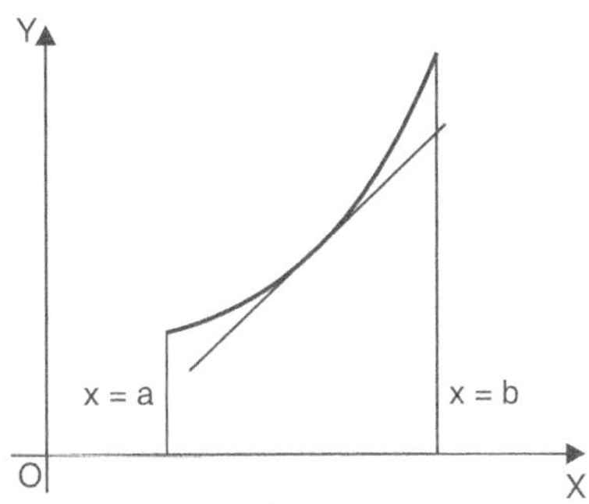

Fig. (*i*)

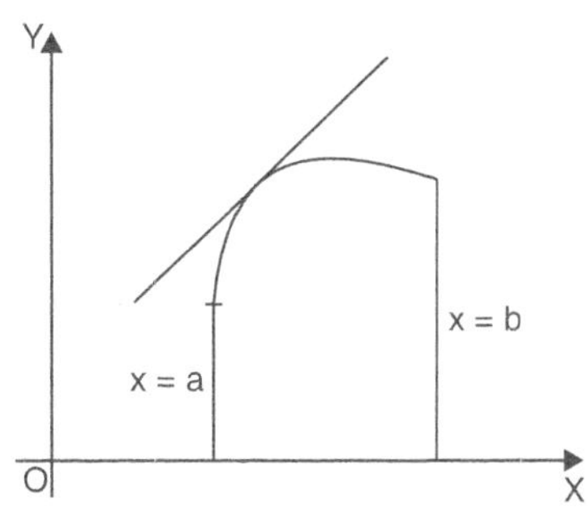

Fig. (*ii*)

(b) Concave downward or Convex upward

A curve is said to be concave downward (or convex upward) on (a, b) if all the points of the curve lies below any tangent to it on that interval [See Fig. (*ii*)].

(c) Point of Inflexion *(P.T.U., Dec. 2005, Dec. 2007)*

A point on the curve is said to be a point of inflexion if the curve crosses the tangent at that point *i.e.*, two portions of the curve on the two sides of P lie on different sides of the tangent at that point [See Fig. (*iii*)].

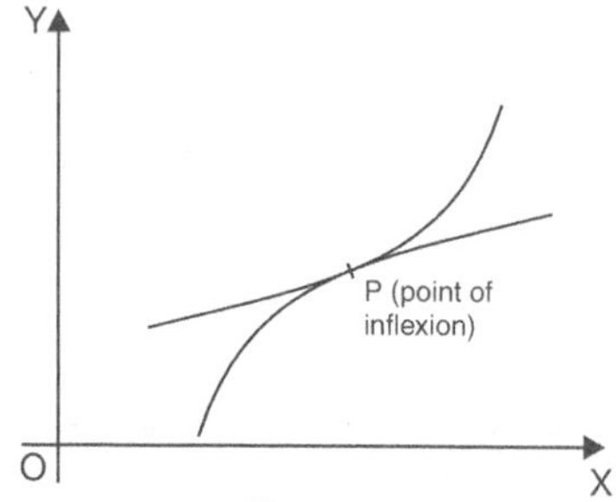

Fig. (*iii*)

1.3. CRITERIA FOR CONCAVITY, CONVEXITY AND INFLEXION

To determine whether a curve $y = f(x)$ is concave upward, concave downward or has a point of inflexion; we take P$(c, f(c))$ any point on the curve such that $f''(c)$ is finite then

(*i*) Curve is concave upward or convex downward at P if $f''(c) > 0$

(*ii*) Curve is concave downward or convex upward at P if $f''(c) < 0$

(*iii*) Curve has a point of inflexion at P if $f''(c) = 0$ but $f'''(c) \neq 0$.

Proof. Let Q$(c + h, f(c + h))$ be another point on the curve very near to P. Q will lie on the right hand side of P if $h > 0$ and will be on the left hand side of P if $h < 0$ (See Figs.).

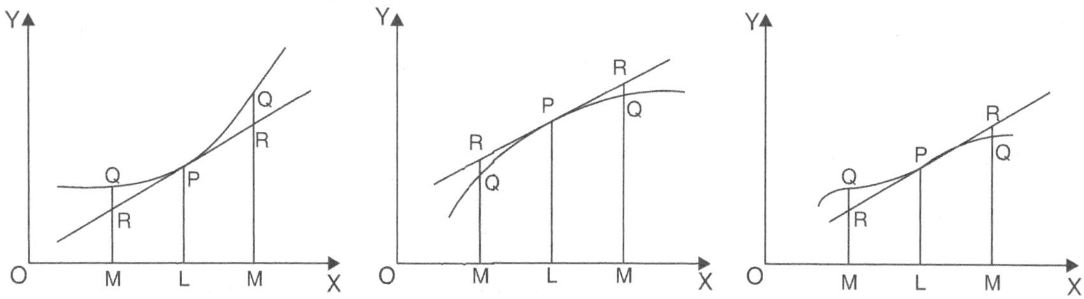

Draw tangent at P and perpendiculars from P and Q on OX ; Let QM ; the perpendicular from Q meets tangent at R.

(Find co-ordinates of R) abscissa of R is same as that of Q *i.e.*, $c + h$ and ordinate is MR. Equation of tangent at P$(c, f(c))$ is $y - f(c) = f'(c)(x - c)$ by using formula $y - y_1 = m(x - x_1)$.

It intersects MQ at R$(c + h, \text{MR})$

$$\therefore \quad \text{MR} - f(c) = f'(c)(c + h - c)$$

$$\text{MR} = f(c) + f'(c) \cdot h = f(c) + h\, f'(c)$$

Now $$\text{QR} = \text{MQ} - \text{MR} = f(c + h) - f(c) - hf'(c) \qquad ...(1)$$

By Taylor's theorem with remainder after two terms

$$f(c + h) = f(c) + hf'(c) + \frac{h^2}{2!} f''(c + \theta h) \qquad \text{where } 0 < \theta < 1$$

$\therefore$ From (1),

$$\text{QR} = \frac{h^2}{2!} f''(c + \theta h) \qquad 0 < \theta < 1 \quad ...(2)$$

Case I. Let $f''(c) > 0 \quad \because \quad f''(x)$ is continuous at c

$\therefore \quad f''(x) > 0 \quad \forall \ x$ in the neighbourhood of c

$c + \theta h \in$ neighbourhood of c

$\therefore \qquad f''(c + \theta h) > 0, \quad \dfrac{h^2}{2!}$ is +ve

$\therefore$ From (2) QR is +ve

$\therefore$ Hence curve is concave upward at P if $f''(c) > 0$.

Case II. If $f''(c) < 0$

Then $f''(c + \theta h) < 0, \dfrac{h^2}{2!}$ is positive

$\therefore$ QR is negative

Hence curve is concave downward at P.

Case III. Let $f''(c) = 0$ but $f'''(c) \neq 0$

Then again by Taylor's theorem with remainder after three terms.

$$f(c + h) = f(c) + h f'(c) + \frac{h^2}{2!} f''(c) + \frac{h^3}{3!} f'''(a + \theta_1 h) \ ; \ 0 < \theta_1 < 1$$

$\because \qquad f''(c) = 0$

$$\therefore \qquad f(c + h) = f(c) + h f'(c) + \frac{h^3}{3!} f'''(a + \theta_1 h)$$

From (1) $\qquad \text{QR} = \dfrac{h^3}{3!} f'''(a + \theta_1 h)$ where $0 < \theta_1 < 1$

Suppose $f'''(x)$ is also continuous at c

$\therefore \quad f'''(x)$ and $f'''(c)$ have the same sign

$\therefore \quad f'''(c + \theta_1 h)$ has the same sign as that of $f'''(c)$.

But $\dfrac{h^3}{3!}$ is +ve when h is +ve and is –ve when h is –ve

i.e., $\dfrac{h^3}{3!}$ changes sign with the change in the sign of h *i.e.,* QR changes sign with the change in sign of h.

Hence the curve has point of inflexion at P if $f''(c) = 0$ and $f'''(c) \neq 0$.

1.4. (a) WORKING RULE TO FIND CONCAVITY AND CONVEXITY OF A CURVE IN AN INTERVAL (*a*, *b*)

(1) Evaluate $\dfrac{d^2y}{dx^2}$.

(2) Find the interval (a, b) for which $\dfrac{d^2y}{dx^2} > 0$; then (a, b) is the interval for the curve being concave upward or convex downward.

(3) Find the interval (a, b) for which $\dfrac{d^2y}{dx^2} < 0$; then (a, b) is the interval for the curve being concave downward or convex upward.

1.4. (b) WORKING RULE TO FIND THE POINTS OF INFLEXION

(1) Evaluate $\frac{d^2y}{dx^2}$.

(2) Find the values of x for which $\frac{d^2y}{dx^2} = 0$ and also the values of x (if any) where $\frac{d^2y}{dx^2}$ does not exist.

(3) All these values of x obtained in step (2) will be expected points of inflexion.

(4) Let $x = a$ be any one of the values.

It will be the point of inflexion if

$$\frac{d^3y}{dx^3} \neq 0 \quad \text{at} \quad x = a.$$

or
$$\frac{d^2y}{dx^2} \text{ changes sign at } x = a.$$

Note. For point of inflexion $\frac{d^2y}{dx^2} = 0$ but converse is not true *i.e.*, if $\frac{d^2y}{dx^2} = 0$ at $x = a$ then $x = a$ may not be the point of inflexion.

1.5. INCREASING AND DECREASING FUNCTION

A function $y = f(x)$ is said to increasing in an interval (a, b) if $f'(x) > 0 \ \forall \ x \in (a, b)$ and decreasing in an interval (a, b) if $f'(x) < 0 \ \forall \ x \in (a, b)$

e.g., if
$$y = x^3 - 3x + 2$$

then
$$\frac{dy}{dx} = 3x^2 - 3 = 3(x^2 - 1)$$

$y = f(x)$ will be increasing in the interval where

$$f'(x) = \frac{dy}{dx} > 0 \ i.e., \ x^2 - 1 > 0 \quad \text{or} \quad x^2 > 1 \text{ or } |x| > 1$$

i.e., either $x > 1$ or $x < -1$

i.e., function increases in the interval $(-\infty, -1) \cup (1, \infty)$ and function decreases when $x^2 - 1 < 0$

i.e., $x^2 < 1$ or $|x| < 1$ or $-1 < x < 1$

i.e., function decreases in the interval $(-1, 1)$.

ILLUSTRATIVE EXAMPLES

Example 1. *Find the values of x for which* $y = x^4 - 6x^3 + 12x^2 + 5x + 7$ *is concave upwards or downwards. Also determine the points of inflexion.*

Sol.
$$y = x^4 - 6x^3 + 12x^2 + 5x + 7$$

$$\frac{dy}{dx} = 4x^3 - 18x^2 + 24x + 5$$

$$\frac{d^2y}{dx^2} = 12x^2 - 36x + 24$$

Curve will concave upward in the interval for which

$$\frac{d^2y}{dx^2} > 0$$

i.e., $12x^2 - 36x + 24 > 0$

or $x^2 - 3x + 2 > 0$

or $\left(x - \frac{3}{2}\right)^2 > \frac{9}{4} - 2$

or $\left(x - \frac{3}{2}\right)^2 > \frac{1}{4}$ or $\left|x - \frac{3}{2}\right| > \frac{1}{2}$

i.e., either $x - \frac{3}{2} > \frac{1}{2}$ or $x - \frac{3}{2} < -\frac{1}{2}$

or $x > 2$ or $x < 1$

i.e., $x \in (-\infty, 1) \cup (2, \infty)$

$\therefore$ Curve is concave upward in $(-\infty, 1) \cup (2, \infty)$

Curve will concave down if $\frac{d^2y}{dx^2} < 0$

i.e., $x^2 - 3x + 2 < 0$

or $\left|x - \frac{3}{2}\right| < \frac{1}{2}$

or $\frac{3}{2} - \frac{1}{2} < x < \frac{3}{2} + \frac{1}{2}$ or $1 < x < 2$

$\therefore$ $x \in (1, 2)$

$\therefore$ Curve is concave downward in (1, 2).

For points of inflexion ;

$$\frac{d^2y}{dx^2} = 0 \quad \text{or} \quad \frac{d^2y}{dx^2} \text{ does not exit.}$$

i.e., $x^2 - 3x + 2 = 0$

or $x = 1, 2$

$$\frac{d^3y}{dx^3} = 24x - 36$$

At $x = 1$, $\frac{d^3y}{dx^3} = -12 \neq 0$; for $x = 1, y = 19$ | from (1)

$\therefore$ (1, 19) is the point of inflexion.

At $x = 2$, $\frac{d^3y}{dx^3} = 48 - 36 = 12 \neq 0$ for $x = 2, y = 23$

$\therefore$ (2, 23) is also the point of inflexion.

Hence (1, 19) and (2, 23) are two points of inflexion.

Example 2. *Find the points of inflexion of the following curves:*

(*i*) $y = x^3 + 8x^2 - 270x$ **(P.T.U., Dec. 2005)**

(*ii*) $x = 3y^4 - 4y^3 + 5$.

Sol. (*i*) $$y = x^3 + 8x^2 - 270x \quad \ldots(1)$$

$$\frac{dy}{dx} = 3x^2 + 16x - 270$$

$$\frac{d^2y}{dx^2} = 6x + 16$$

$$\frac{d^3y}{dx^3} = 6$$

$$\frac{d^2y}{dx^2} = 0 \text{ gives } x = -\frac{16}{6} = -\frac{8}{3}$$

and at $x = -\frac{8}{3}$, $\frac{d^3y}{dx^3} \neq 0$

$\therefore$ $x = -\frac{8}{3}$ gives the point of inflexion.

From (1), $$y = -\frac{512}{27} + \frac{512}{9} + \frac{2160}{3} = \frac{20464}{27}$$

$\therefore$ Point of inflexion is $\left(-\frac{8}{3}, \frac{20464}{27}\right)$.

(*ii*) $$x = 3y^4 - 4y^3 + 5$$

$$\frac{dx}{dy} = 12y^3 - 12y^2 = 12\,(y^3 - y^2)$$

$$\frac{d^2x}{dy^2} = 12\,(3y^2 - 2y)$$

$$\frac{d^3x}{dy^3} = 12\,(6y - 2) = 24\,(3y - 1)$$

for points of inflexion $\frac{d^2x}{dy^2} = 0 \Rightarrow 3y^2 - 2y = 0$ which gives $y = 0, y = \frac{2}{3}$

for both the values of y; $\frac{d^3x}{dy^3} \neq 0$ $\therefore$ $y = 0, y = \frac{2}{3}$ gives points of inflexion.

$\therefore$ Points of inflexion are (5, 0) and $\left(\frac{119}{27}, \frac{2}{3}\right)$.

Example 3. *Find points of inflexion of the curve*

$$x = a(2\theta - \sin\theta), \quad y = a(2 - \cos\theta).$$

Sol. $$x = a(2\theta - \sin\theta), \quad y = a(2 - \cos\theta)$$

$$\frac{dx}{d\theta} = a(2 - \cos\theta), \quad \frac{dy}{d\theta} = a\sin\theta$$

$$\frac{dy}{dx} = \frac{\frac{dy}{d\theta}}{\frac{dx}{d\theta}} = \frac{a\sin\theta}{a(2-\cos\theta)} = \frac{\sin\theta}{2-\cos\theta}$$

$$\frac{d^2y}{dx^2} = \frac{(2-\cos\theta)\cos\theta - \sin\theta\,(\sin\theta)}{(2-\cos\theta)^2}\,\frac{d\theta}{dx}$$

$$= \frac{2\cos\theta - \cos^2\theta - \sin^2\theta}{(2-\cos\theta)^2} \cdot \frac{1}{a(2-\cos\theta)} = \frac{2\cos\theta - 1}{a(2-\cos\theta)^3}$$

$$\frac{d^2y}{dx^2} = 0 \quad \text{when} \quad 2\cos\theta - 1 = 0 \quad \therefore \quad \cos\theta = \frac{1}{2} = \cos\frac{\pi}{3}$$

$\therefore$ $$\theta = 2n\pi \pm \frac{\pi}{3}\ ;\ n \in \mathrm{N}$$

Now $$\frac{d^2y}{dx^2} < 0 \quad \text{if} \quad \cos\theta < \frac{1}{2} \qquad \mid \because \ 2-\cos\theta \text{ is +ve as } \cos\theta < 1$$

and $$\frac{d^2y}{dx^2} > 0 \quad \text{if} \quad \cos\theta > \frac{1}{2}$$

$\therefore$ $\dfrac{d^2y}{dx^2}$ changes sign at $\theta = 2n\pi \pm \dfrac{\pi}{3}$

$\therefore$ $$x = a\left[4n\pi \pm \frac{2\pi}{3} \mp \frac{\sqrt{3}}{2}\right],\ y = \frac{3a}{2}$$

$\therefore$ $\left\{a\left(4n\pi \pm \dfrac{2\pi}{3} \mp \dfrac{\sqrt{3}}{2}\right), \dfrac{3a}{2}\right\}$ are points of inflexion.

Example 4. *Determine a, b so that the curve $y = ax^3 + 3bx^2$ has a point of inflexion at (–1, 2).*

Sol. $$y = ax^3 + 3bx^2 \qquad \text{...(1)}$$

$$\frac{dy}{dx} = 3ax^2 + 6bx$$

$$\frac{d^2y}{dx^2} = 6ax + 6b \qquad \text{...(2)}$$

$$\frac{d^3y}{dx^3} = 6a \qquad \text{...(3)}$$

$\because$ P(–1, 2) is the point of inflexion.

$\therefore$ (*i*) P(–1, 2) lies on the curve (1) (*ii*) at (–1, 2) $\dfrac{d^2y}{dx^2} = 0$ (*iii*) $\dfrac{d^3y}{dx^3}$ at $(-1, 2) \neq 0$

From (1) $\quad 2 = -a + 3b$

From (2) $\quad -6a + 6b = 0 \quad \therefore \quad a = b$

$\therefore \quad 2 = -a + 3a \quad \therefore \quad 2a = 2 \quad \therefore \quad a = 1$

$\therefore \quad a = 1,\ b = 1$

$$\frac{d^3y}{dx^3} = 6a = 6 \neq 0$$

$\therefore$ We have $a = 1, b = 1$

Example 5. *Find the points of inflexion of the curve* $y = \dfrac{x^2+1}{x^2-1}$. *Also find the values of x for which function is concave upward and concave downward.*

Sol.

$$y = \frac{x^2+1}{x^2-1}$$

$$\frac{dy}{dx} = \frac{(x^2-1)\,2x-(x^2+1)\,2x}{(x^2-1)^2} = \frac{-4x}{(x^2-1)^2}$$

$$\frac{d^2y}{dx^2} = -4\,\frac{(x^2-1)^2 - x\cdot 2(x^2-1)2x}{(x^2-1)^4}$$

$$= \frac{-4\,[x^4-2x^2+1-4x^4+4x^2]}{(x^2-1)^4} = \frac{4(3x^4-2x^2-1)}{(x^2-1)^4}$$

$$= \frac{4(x^2-1)(3x^2+1)}{(x^2-1)^4} = \frac{4(3x^2+1)}{(x^2-1)^3}$$

The expected points of inflexion can be the values of x for which $\dfrac{d^2y}{dx^2} = 0$ but $\dfrac{d^3y}{dx^3} \neq 0$ and also the values of x for which $\dfrac{d^2y}{dx^2}$ does not exist. $\dfrac{d^2y}{dx^2} = 0$ gives $3x^2+1=0$ which does not give any real value of x. Secondly the points at which $\dfrac{d^2y}{dx^2}$ does not exist are given by $x = \pm 1$ but these two points do not belong to the domain of the given function $\therefore$ these points cannot be considered for points of inflexion.

$\therefore$ $f(x)$ curve has no points of inflexion.

Now curve will concave upward if $\dfrac{d^2y}{dx^2} > 0$

i.e., $$\frac{4(3x^2+1)}{(x^2-1)^3} > 0 \quad i.e., \quad x^2-1>0$$

or $$x^2 > 1 \quad \text{or} \quad |x| > 1 \quad i.e., \quad \text{either } x > 1 \text{ or } x < -1$$

$\therefore$ Curve concave upwards in the interval $(-\infty, -1) \cup (1, \infty)$.

Curve will concave downward if $\dfrac{d^2y}{dx^2} < 0$

i.e., $$x^2-1<0 \quad \therefore \quad x^2<1 \quad \text{or} \quad |x|<1$$

i.e., $$-1 < x < 1$$

$\therefore$ Curve concaves downward is the interval $(-1, 1)$.

Example 6. *Find the points of inflexion (if any) of the curve* $y = (x^3 - 6x^2 + 9x + 6)/6$.

(P.T.U., May 2005)

Sol. Equation of the curve is

$$y = \frac{1}{6}(x^3 - 6x^2 + 9x + 6)$$

$$\frac{dy}{dx} = \frac{1}{6}(3x^2 - 12x + 9)$$

$$\frac{d^2y}{dx^2} = \frac{1}{6}(6x - 12) = x - 2$$

$$\frac{d^3y}{dx^3} = 1$$

For points of inflexion $\frac{d^2y}{dx^2} = 0$ but $\frac{d^3y}{dx^3} \neq 0$ at those points.

$$\frac{d^2y}{dx^2} = 0 \quad \text{gives} \quad x - 2 = 0 \quad i.e., \quad x = 2 \quad \text{and} \quad \frac{d^3y}{dx^3} \neq 0$$

$\therefore$ $x = 2$ is the point of inflexion and for $x = 2, y = \frac{1}{6}(8 - 24 + 18 + 6) = \frac{4}{3}$

$\therefore$ Coordinates of the point of inflexion are $\left(2, \frac{4}{3}\right)$.

Example 7. *As x moves from left to right through the point c = 2, is the graph of $f(x) = x^3 - 3x + 2$ rising or is it falling ? Give reason for your answer.* **(P.T.U., Dec. 2004)**

Sol. Given $f(x) = x^3 - 3x + 2$

$$f'(x) = 3x^2 - 3$$

For graph of the curve find its intersection with x-axis and y-axis

$$y = f(x) = x^3 - 3x + 2 = (x - 1)^2 (x + 2)$$

Curve intersects x-axis at (1, 0) and (– 2, 0) and y-axis at (0, 2)

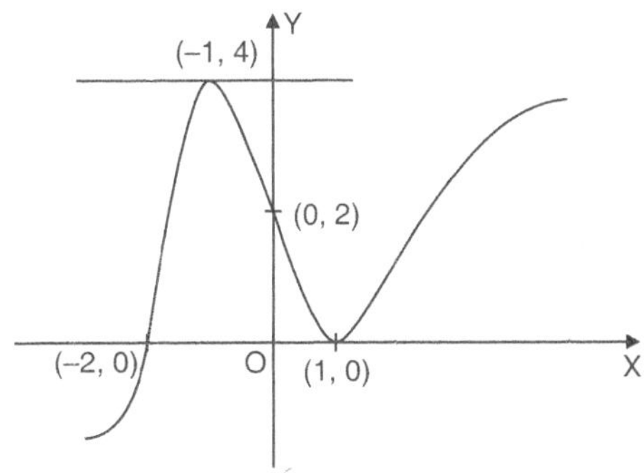

Tangent to the curve is parallel to x-axis where

$$\frac{dy}{dx} = 0 \quad i.e., \quad f'(x) = 0 \quad i.e., \quad 3x^2 - 3 = 0$$

i.e., $x^2 = 1$ or $x = \pm 1$.

$\therefore$ Tangent is parallel to x-axis at (1, 0) and at (–1, 4).

Curve increases when $f'(x) > 0$ *i.e.* $x^2 - 1 > 0$

or $x^2 > 1$ or $|x| > 1$

i.e., either $x > 1$ or $x < -1$ and decreases when $|x| < 1$ *i.e.,* $-1 < x < 1$. Graph of the curve is shown in the figure.

$\therefore$ As x moves from left to right $f(x)$ is rising in the interval $(-\infty, -1) \cup (1, \infty)$ and falling in the interval (–1, 1).

i.e., graph of $f(x)$ is falling at $c = 2$.

TEST YOUR KNOWLEDGE

1. Examine the curve $y = x^3 - 18x^2 + 5x + 7$ for concavity upward and concavity downward.
2. Determine the concavity and convexity of the curve $y = x + \frac{4}{x}$.
3. Find the points of inflexion of the following curves

   (*i*) $y = \frac{x(x^2 - 1)}{3x^2 + 1}$ (*ii*) $a^2y^2 = x^2(a^2 - x^2)$

   (*iii*) $54y = (x + 5)^2(x^3 - 10)$

4. Show that the line joining the two points of inflexion of the curve $y^2(x - a) = x^2(x + a)$; $x \neq \pm a$ subtends an angle $\frac{\pi}{3}$ at the origin.

5. Show that abscissa of the points of inflexion on the curve $x = a - b \cos \theta$, $y = a\theta - b \sin \theta$ is $\frac{a^2 - b^2}{a}$.

Answers

1. Concave upward in $(6, \infty)$, concave downward in $(-\infty, 6)$
2. Concave upward in $(0, \infty)$, concave downward in $(-\infty, 0)$
3. (*i*) (0, 0), (1, 0), (–1, 0) (*ii*) (0, 0) (*iii*) $x = -2, -4 \pm \sqrt{18}$.

1.6. ASYMPTOTES

Definition. *A straight line 'l' is called an asymptote of an infinite branch of a curve iff the line lies on one side of branch of curve and the perpendicular distance of a point on it from any point P on that branch of the curve tends to zero as P approaches infinity along that branch.*

Rectangular Asymptotes. If an asymptote to a curve is either parallel to x-axis or y-axis then it is called a **rectangular asymptote**. Asymptote parallel to x-axis is called **horizontal asymptote** and asymptote parallel to y-axis is called **vertical asymptote**.

Oblique Asymptote. An asymptote which is neither parallel to x-axis nor parallel to y-axis is called an **oblique asymptote** (See Figs. below).

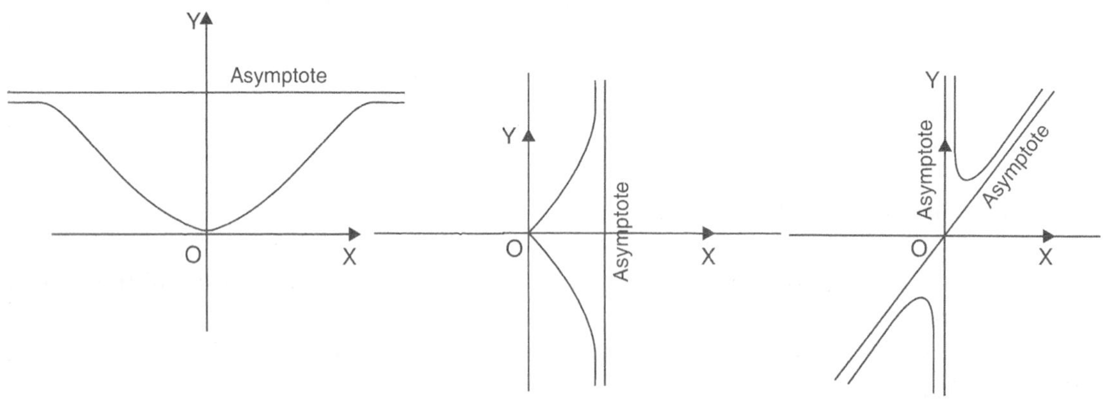

1.7. (a) RULE TO FIND VERTICAL ASYMPTOTES

Case I. If the equation of the curve can be put into the form $y = \frac{N(x)}{D(x)}$ when $N(x)$, $D(x)$ are polynomials without any common factor then *vertical asymptotes* are obtained by equating to zero the real linear factors of $D(x)$.

Case II. If the equation of the curve cannot be put into the form $y = \frac{N(x)}{D(x)}$ then *vertical Asymptotes* are obtained by equating to zero the real linear factors in the coefficient of the highest power of y in the equation of the given curve.

1.7. (b) RULE TO FIND HORIZONTAL ASYMPTOTES

Case I. If the equation of the curve can be put into the form $x = \dfrac{N(y)}{D(y)}$ where N(*y*) and D(*y*) are two polynomials in *y* without any common factor then horizontal asymptotes are obtained by equating to zero the real linear factors of D(*y*).

Case II. If the equation of the curve cannot be put into the form $x = \dfrac{N(y)}{D(y)}$ then horizontal asymptotes are obtained by equating to zero the real linear factors in the coefficient of highest power of *x* in the equation of the given curve.

1.8. CONDITION THAT A LINE *y* = *mx* + *c* MAY BE AN ASYMPTOTE OF THE CURVE *y* = *f* (*x*); *m* AND *c* BOTH BEING REAL

Proof. Let $y = mx + c$ be any oblique asymptote of the curve $y = f(x)$ and P(*x*, *y*) be any point on the curve. Draw PM ⊥ on $y = mx + c$ then

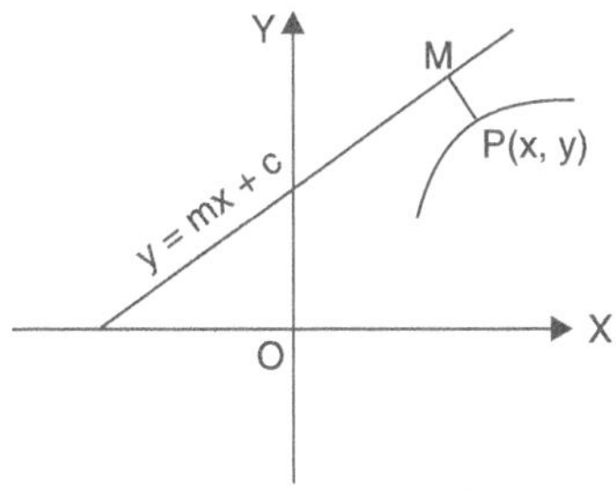

$$PM = \frac{|y - mx - c|}{\sqrt{1+m^2}} \quad ...(1)$$

As $x \to \infty$ or $-\infty$

$| PM | \to 0$ $\quad [\because \; y = mx + c$ is an asymptote]

$$\therefore \quad \underset{x \to \infty}{\text{Lt}} \frac{|y - mx - c|}{\sqrt{1+m^2}} = 0$$

or $$\underset{x \to \infty}{\text{Lt}} | y - mx - c | = 0 \quad \text{or} \quad \underset{x \to \infty}{\text{Lt}} (y - mx - c) = 0$$

or $$\underset{x \to \infty}{\text{Lt}} (y - mx) = c$$

Also $$\underset{x \to \infty}{\text{Lt}} (y - mx - c) = 0 \text{ gives us } \underset{x \to \infty}{\text{Lt}} \left(\frac{y}{x} - m - \frac{c}{x}\right) = 0$$

i.e., $$m = \underset{x \to \infty}{\text{Lt}} \left(\frac{y}{x} - \frac{c}{x}\right) = \underset{x \to \infty}{\text{Lt}} \frac{y}{x}$$

Hence $y = mx + c$ will be an asymptote of the curve if

$$m = \underset{x \to \infty}{\text{Lt}} \frac{y}{x} \quad \text{and} \quad c = \underset{\substack{x \to \infty \\ \text{or } -\infty}}{\text{Lt}} (y - mx).$$

1.9. OBLIQUE ASYMPTOTES OF THE CURVE IN GENERAL FORM i.e., *f*(*x*, *y*) = 0

Proof. Given curve is $f(x, y) = 0$

Let this equation be of *n*th order

Given equation can be put in the form

$$x^n \phi_n\left(\frac{y}{x}\right) + x^{n-1}\phi_{n-1}\left(\frac{y}{x}\right) + x^{n-2}\phi_{n-2}\left(\frac{y}{x}\right) + \cdots + x\phi_1\left(\frac{y}{x}\right) + \phi_0\left(\frac{y}{x}\right) = 0 \quad ...(1)$$

Divide by x^n

$$\phi_n\left(\frac{y}{x}\right)+\frac{1}{x}\phi_{n-1}\left(\frac{y}{x}\right)+\frac{1}{x^2}\phi_{n-2}\left(\frac{y}{x}\right)-\cdots-\frac{1}{x^{n-1}}\phi_1\left(\frac{y}{x}\right)+\frac{1}{x^n}\phi_0\left(\frac{y}{x}\right)=0 \quad \ldots(2)$$

Let $y = mx + c$ be any one of its asymptotes then $\underset{x\to\infty}{\text{Lt}} \dfrac{y}{x} = m$

$\therefore$ Let $x \to \infty$ in (1), we get

$$\phi_n(m) = 0 \quad \ldots(3)$$

$\therefore$ Roots of the equation (3) will give us the values of m and the real values of m will represent the slopes of the different asymptotes.

Let m_1 be any one of the real roots of (3) $\therefore$ $\phi_n(m_1) = 0$ ...(4)
and the asymptote becomes

$$y = m_1 x + c_1$$

$$\frac{y}{x} = m_1 + \frac{c_1}{x}$$

Substituting the value of $\dfrac{y}{x}$ in (1)

$$x^n\phi_n\left(m_1+\frac{c_1}{x}\right)+x^{n-1}\phi_{n-1}\left(m_1+\frac{c_1}{x}\right)+x^{n-2}\phi_{n-2}\left(m_1+\frac{c_1}{x}\right)$$
$$+\cdots+x\phi_1\left(m_1+\frac{c_1}{x}\right)+\phi_0\left(m_1+\frac{c_1}{x}\right)x=0$$

Expand by Taylor's theorem, we get

$$x^n\left\{\phi_n(m_1)+\frac{c_1}{x}\phi_n'(m_1)+\frac{c_1^2}{2x^2}\phi_n''(m_1)+\cdots\right\}$$
$$+x^{n-1}\left\{\phi_{n-1}(m_1)+\frac{c_1}{x}\phi'_{n-1}(m_1)+\frac{c_1^2}{2x^2}\phi''_{n-1}(m_1)\right\}$$
$$+x^{n-2}\left\{\phi_{n-2}(m_1)+\frac{c_1}{x}\phi'_{n-2}(m_1)+\frac{c_1^2}{2x^2}\phi''_{n-2}(m_1)\right\}+\cdots=0$$

or

$$x^n\{\phi_n(m_1)\}+x^{n-1}\{c_1\phi'_n(m_1)+\phi_{n-1}(m_1)\}+x^{n-2}\left\{\frac{c_1^2}{2}\phi''_n(m_1)+c_1\phi'_{n-1}(m_1)+\phi_{n-2}(m)\right\}+\cdots=0$$

Put $\phi_n(m_1) = 0$ and divide by x^{n-1}

$$[c_1\phi'_n(m_1)+\phi_{n-1}(m_1)]+\frac{1}{x}\left[\frac{c_1^2}{2}\phi_n''(m_1)+c_1\phi_{n-1}'(m_1)+\phi_{n-2}(m_1)\right]+\cdots=0 \quad \ldots(5)$$

Let $x \to \infty$, we get

$$c_1\phi_n'(m_1) + \phi_{n-1}(m_1) = 0$$

$\therefore$

$$c_1 = -\frac{\phi_{n-1}(m_1)}{\phi_n'(m_1)} \quad \text{provided } \phi_n'(m_1) \neq 0$$

$\therefore$ Asymptote corresponding to the slope m_1 is

$$y = m_1 x - \frac{\phi_{n-1}(m_1)}{\phi_n'(m_1)} \text{ when } \phi_n'(m_1) \neq 0$$

Similarly, if $m_2, m_3, \ldots$ are the roots of (3)

Then we have corresponding asymptotes

$$y = m_2 x - \frac{\phi_{n-1}(m_2)}{\phi_n'(m_2)}, \; y = m_3 x - \frac{\phi_{n-1}(m_3)}{\phi_n'(m_3)}$$

provided $\phi_n'(m_2) \neq 0,\ \phi_n'(m_3) \neq 0 \ldots$

Cor 1. If $\phi_n'(m_1) = 0$ suppose $\phi_{n-1}(m_1) \neq 0$, then

$$c_1 \phi_n'(m_1) + \phi_{n-1}(m_1) = 0$$

does not determine any finite value of c_1 $\therefore$ There is no asymptote corresponding to m_1.

Cor 2. If $\phi_n'(m_1) = 0$ and $\phi_{n-1}(m_1) = 0$. Then from (5) we have when $x \to \infty$

$$\frac{c_1^2}{2} \phi_n''(m_1) + c_1 \phi_{n-1}(m_1) + \phi_{n-2}(m_1) = 0$$

which is quadratic in c_1 provided $\phi_n''(m_1) \neq 0$

$\therefore$ When $\phi_n'(m_1) = 0$, $\phi_{n-1}(m_1) = 0$ but $\phi_n''(m_1) \neq 0$

i.e., m_1 is a repeated root of $\phi_n(m) = 0$ twice then there are two paralled asymptotes to the curve corresponding to the slope m_1, *i.e.*, $y = m_1 x + c'$, and $y = m_1 x + c''$, are two asymptotes.

1.10. WORKING RULE TO FIND OBLIQUE ASYMPTOTES

Step 1. Put $x = 1, y = m$ in nth degree terms and $(n-1)$th degree terms in $f(x, y) = 0$

Step 2. Obtain $\phi_n(m), \phi_{n-1}(m)$

Step 3. Find all the real roots of $\phi_n(m) = 0$

Step 4. For every non repeated real root of $\phi_n(m) = 0$, corresponding value of c is given by

$$c\phi_n'(m) + \phi_{n-1}(m) = 0 \quad \text{provided } \phi_n'(m) \neq 0$$

The ansymptote $y = mx + c$ is obtained by substituing the values of m and c

$\phi_n'(m) = 0 \Rightarrow$ there is no asymptote corresponding to this value of m.

Step 5. For every repeated real root of $\phi_n(m) = 0$ occuring twice, corresponding values of c are given by

$$\frac{c^2}{2!} \phi_n''(m) + c\phi_{n-1}'(m) + \phi_{n-2}(m) = 0$$

provided $\phi_n''(m) \neq 0$. In this case we get two parallel asymptotes to the curve.

$\because$ for one value of m we get two values of c.

Note 1. For every root of $\phi_n(m) = 0$ repeated thrice or more we can proceed similarly.

Note 2. A rational algebraic curve of degree n cannot have more than n asymptotes.

ILLUSTRATIVE EXAMPLES

Example 1. *Find the asymptotes (if any) parallel to the co-ordinate axes of the following curves*

(i) $x^2y^2 - xy^2 - x^2y + x + y + 1 = 0$ *(ii)* $(x + y)^2 = x^2y^2 - 7$

(P.T.U., Dec. 2012)

(iii) $x^2y^3 + x^2 + 3y^2 - 9xy + 8x - 25 = 0$ **(P.T.U., Dec. 2013)**

Sol. *(i)* Equation of the curve is $x^2y^2 - xy^2 - x^2y + x + y + 1 = 0$

For asymptotes parallel to x-axis equate coefficient of highest power of x to zero *i.e.,* coefficient of x^2 to zero which gives $y^2 - y = 0$ or $y(y - 1) = 0$ or $y = 0, y = 1$

$\therefore$ Asymptotes parallel to x-axis are $y = 0, y = 1$

For asymptotes parallel to y-axis equate coefficient of highest power of y to zero *i.e.,* coefficient of y^2 to zero which gives $x^2 - x = 0$ *i.e.,* $x(x - 1) = 0$ or $x = 0, x = 1$

$\therefore$ Asymptotes parallel to y-axis are $x = 0, x = 1$

(ii)
$$(x + y)^2 = x^2y^2 - 7$$
$$x^2 + 2xy + y^2 = x^2y^2 - 7$$
or
$$x^2 + 2xy + y^2 - x^2y^2 + 7 = 0$$

Highest power of x is 2

$\therefore$ Asymptotes parallel to x-axis is

$$1 - y^2 = 0 \quad \therefore \quad y = \pm 1$$

Highest power of y is 2

Asymptote parallel to y-axis is

$$1 - x^2 = 0 \quad \therefore \quad x = \pm 1$$

$\therefore$ Asymptotes parallel to co-ordinate axes are

$$y - 1 = 0, y + 1 = 0, x - 1 = 0, x + 1 = 0$$

(iii) Equation of the curve is $x^2y^3 + x^2 + 3y^2 - 9xy + 8x - 25 = 0$

For asymptotes parallel to x-axis equate coefficient of highest power of x to zero *i.e.,* equate coefficient of x^2 to zero *i.e.,* $y^3 + 1 = 0$ or $(y + 1)(y^2 - y + 1) = 0$ which gives $y = -1$, the only real asymptote parallel to x-axis. For asymptote parallel to y-axis equate coefficient of highest power of y to zero *i.e.,* equate coefficient of y^3 to zero *i.e.,* $x^2 = 0$ which gives $x = 0$ as asymptote parallel to y-axis.

Hence the asymptotes parallel to coordinate axes are $x = 0$ and $y = -1$.

Example 2. *Find the asymptotes of the curve*

$$x^3 - x^2y - xy^2 + y^3 + 2x^2 - 4y^2 + 2xy + x + y + 1 = 0.$$

Sol. Coefficient of highest power of $x = 1$

$\therefore$ no asymptote parallel to x-axis.

Coefficient of highest power of $y = 1$

$\therefore$ no asymptote parallel to y-axis.

For oblique asymptotes.

Put $x = 1,\ y = m$ in 3rd degree terms.

We get $\phi_3(m) = 1 - m - m^2 + m^3$

Put $x = 1,\ y = m$ in 2nd degree terms, we get

$$\phi_2(m) = 2 - 4m^2 + 2m$$

Put $x = 1$, $y = m$ in the 1st degree terms

$$\phi_1(m) = 1 + m$$
$$\phi_0(m) = 1$$

$\therefore$

$$\phi_3(m) = 1 - m - m^2 + m^3$$
$$\phi_3'(m) = -1 - 2m + 3m^2$$
$$\phi_3''(m) = -2 + 6m$$
$$\phi_3'''(m) = 6$$
$$\phi_2(m) = 2 - 4m^2 + 2m \; ; \; \phi_2'(m) = -8m + 2$$
$$\phi_1(m) = 1 + m \; ; \; \phi_1'(m) = 1$$
$$\phi_0(m) = 1$$

$\phi_3(m) = 0$ gives $1 - m - m^2 + m^3 = 0$

$$(1 - m)(1 - m^2) = 0$$
$$(1 - m)^2(1 + m) = 0$$

$\therefore$ $m = -1, m = 1, 1$

For $m = -1$, c is given by

$$\phi_3'(m)c + \phi_2(m) = 0$$
$$(-1 + 2 + 3)c + (2 - 4 - 2) = 0$$
$$4c - 4 = 0 \quad \therefore \quad c = 1$$

$\therefore$ Corresponding asymptote is

$$y = -x + 1 \quad \text{or} \quad x + y - 1 = 0$$

when $m = 1, 1$ *i.e.,* 1 is the repeated root of $\phi_3(m) = 0$

$\therefore$ The two values of c are given by

$$\frac{c^2}{2!}\phi_3''(m) + c \,.\, \phi_2'(m) + \phi_1(m) = 0$$

or

$$\frac{c^2}{2}(-2 + 6) + c(-8 + 2) + 2 = 0$$
$$2c^2 - 6c + 2 = 0 \quad \text{or} \quad c^2 - 3c + 1 = 0$$
$$c = \frac{3 \pm \sqrt{9 - 4}}{2} = \frac{3 \pm \sqrt{5}}{2}$$

$\therefore$ Asymptotes corresponding to $m = 1$ are

$$y = x + \frac{3 \pm \sqrt{5}}{2} \quad \therefore \quad x - y + \frac{3 + \sqrt{5}}{2} = 0$$

Hence Asymptotes of (1) are and $x - y + \dfrac{3 - \sqrt{5}}{2} = 0$

$$x + y = 1, \quad x - y + \frac{3 \pm \sqrt{5}}{2} = 0.$$

1.11. ASYMPTOTES FOR POLAR CURVES

Prove that the asymptotes of the polar curve $\dfrac{1}{r} = f(\theta)$ *are given by* $r \sin(\theta - \alpha) = \dfrac{1}{f'(\alpha)}$ *where* α *is a root of the equation* $f(\theta) = 0$.

Proof. Equation of the curve is $\frac{1}{r} = f(\theta)$ ....(1)

Let $P(r, \theta)$ be any point on the curve we know that in cartesian system we have

$$\left.\begin{aligned} x &= r\cos\theta \\ y &= r\sin\theta \end{aligned}\right] \quad \text{...(2)}$$

Substituting the value of r from (1) in (2)

$$x = \frac{\cos\theta}{f(\theta)}, \quad y = \frac{\sin\theta}{f(\theta)}$$

$$\therefore \qquad \frac{y}{x} = \tan\theta$$

If $P(r, \theta) \to \infty$ along the curve then $r \to \infty$ and hence $\frac{1}{r} \to 0$ and from (1) : $f(\theta) \to 0$

Let α be the root of $f(\theta) = 0$ $\therefore$ $f(\alpha) = 0$

$$\therefore \qquad \theta \to \alpha \quad \text{as} \quad r \to \infty$$

In cartesian system, the equation of the oblique asymptote to the curve $y = f(x)$ is $y = mx + c$

where $$m = \underset{P\to\infty}{\text{Lt}} \frac{y}{x} \quad \text{and} \quad c = \underset{P\to\infty}{\text{Lt}} (y - mx) \qquad \text{(art. 1.8)}$$

In polar form $m = \underset{\theta\to\alpha}{\text{Lt}} \tan\theta = \tan\alpha$

$$c = \underset{\theta\to\alpha}{\text{Lt}} [y - mx] = \underset{\theta\to\alpha}{\text{Lt}} [r\sin\theta - \tan\alpha \,.\, r\cos\theta]$$

$$c = \underset{\theta\to\alpha}{\text{Lt}} \left[\frac{1}{f(\theta)}\sin\theta - \frac{\tan\alpha \,.\, \cos\theta}{f(\theta)}\right]$$

$$= \underset{\theta\to\alpha}{\text{Lt}} \frac{1}{f(\theta)}\left[\sin\theta - \frac{\sin\alpha \,.\, \cos\theta}{\cos\alpha}\right] = \underset{\theta\to\alpha}{\text{Lt}} \frac{\sin(\theta-\alpha)}{f(\theta)\cos\alpha} \text{ which is } \left(\frac{0}{0}\right) \text{ form}$$

Apply L' Hospital Rule (differentiate numerator and denominator separately w.r.t. θ)

$$= \underset{\theta\to\alpha}{\text{Lt}} \frac{\cos(\theta-\alpha)}{f'(\theta)\cos\alpha} = \frac{1}{f'(\alpha)\cos\alpha}$$

$\therefore$ Equation of the asymptote is

$$y = mx + c$$

i.e., $$r\sin\theta = \tan\alpha.\, r\cos\theta + \frac{1}{f'(\alpha)\cos\alpha}$$

or $$r[\sin\theta - \tan\alpha\cos\theta] = \frac{1}{f'(\alpha)\cos\alpha}$$

or $$r\left[\sin\theta - \frac{\sin\alpha\cos\theta}{\cos\alpha}\right] = \frac{1}{f'(\alpha)\cos\alpha}$$

or $$r\sin(\theta-\alpha) = \frac{1}{f'(\alpha)}$$

or $$r\sin(\theta-\alpha) = \frac{1}{f'(\alpha)}$$

or $$r = \frac{1}{f'(\alpha)\sin(\theta-\alpha)}$$

Note 1. Only those polar curves will have asymptotes which possess an infinite branch or in other words the polar curve for which r remains finite has no asymptote.

Note 2. If $r \to \infty$ when $\theta \to \alpha$ it is not necessary that $\theta = \alpha$ is an asymptote of the curve. Asymptote might not exist and even if it exists it may be parallel to the radius vector along $\theta = \alpha$.

Suggestion. For finding asymptotes of the polar curves it is always better that, if convenient, convert polar curve to cartesian by the transformation $x = r\cos\theta$, $y = r\sin\theta$.

Example 3. *Find asymptote of the curve $r\sin\theta = a\cos 2\theta$.*

Sol. Equation of the curve is

$$r\sin\theta = a\cos 2\theta \quad \text{or} \quad \frac{1}{r} = \frac{\sin\theta}{a\cos 2\theta}$$

Compare it with $$\frac{1}{r} = f(\theta) \quad \therefore \quad f(\theta) = \frac{\sin\theta}{a\cos 2\theta}$$

$$f(\theta) = 0 \quad \text{gives} \quad \frac{\sin\theta}{a\cos 2\theta} = 0 \quad \text{or} \quad \sin\theta = 0 \quad \text{or} \quad \theta = n\pi$$

$$f'(\theta) = \frac{1}{a}\frac{\cos 2\theta\cos\theta - \sin\theta\,.\,2(-\sin 2\theta)}{\cos^2 2\theta} = \frac{1}{a}\frac{\cos 2\theta\cos\theta + 2\sin\theta\sin 2\theta}{\cos^2 2\theta}$$

Put $\theta = n\pi$

$$f'(n\pi) = \frac{1}{a\cos^2 2n\pi}[\cos 2n\pi\cos n\pi + 2\sin 2n\pi\sin n\pi]$$

$$= \frac{(-1)^n}{a}$$

Hence the asymptote to (1) is

$$r\sin(\theta - n\pi) = \frac{1}{(-1)^n/a} \qquad \left[r\sin(\theta-\alpha) = \frac{1}{f'(\alpha)}\right]$$

or $$-r\sin(n\pi - \theta) = \frac{a}{(-1)^n}$$

(*b*) Symmetric about y-axis if it is unchanged on changing x to $-x$

i.e., $f(-x, y) = f(x, y)$

(*c*) Symmetric about origin if it is unchanged on changing x to $-x$ and y to $-y$

If $f(-x, -y) = f(x, y)$

(*d*) Symmetric about $y = x$

If $f(y, x) = f(x, y)$ *i.e.*, on interchanging x and y, $f(x, y)$ is unchanged

(*e*) Symmetric about $y = -x$ if it is unchanged on changing x to $-y$ and y to $-x$

i.e., $f(-y, -x) = f(x, y)$

(2) Origin. Check whether origin lies on the curve. If curve passes through the origin, find tangents at the origin and also check whether origin is node, cusp or conjugate point.

[Rule to find tangent at the origin (if it exists) ; Tangent at the origin to any curve is obtained by equating to zero the lowest degree terms of $f(x, y)$ *e.g.*, if the curve is $x^2 + 3axy + y^2 + x + y = 0$ then tangent at the origin is $x + y = 0$].

(3) Domain and Range. Find domain and range of the curve

[Rule to find domain and range ; If curve is put in the form $y = f(x)$ then $D_f = \{x ; f(x)$ is defined$\}$ and $R_f = \{y = f(x); x \in D_f\}$ *i.e.*, Find all those possible values of x for which $f(x)$ is real and defined. Also find values of y for all the values of $x \in D_f$ then values of y will form R_f] Knowledge of D_f, R_f gives us an idea of the extent of the curve in the xy-plane.

(4) Points of Intersection. Find the points of intersection of the curve with co-ordinate axes and the line of symmetry by putting $x = 0$, $y = 0$ and $y = x$.

(5) Tangents. Find the slopes of the tangents at all the points obtained in step 4. (note that slope of the tangent at any point of the curve $y = f(x)$ is the value of $\frac{dy}{dx}$ at that point)

Equation of the tangent at (x_1, y_1) is

$$y - y_1 = \left(\frac{dy}{dx}\right)_{(x_1, y_1)} (x - x_1)\,.$$

(6) Asymptotes. Find all the asymptotes of the curve and position of the curve relative to its asymptotes [for asymptotes consult articles 1.6 to 1.9].

(7) Maxima and Minima. Find the points where the function has maximum or minimum values .

[Rule to find maxima, minima ; put the curve in form $y = f(x)$ find $\frac{dy}{dx}, \frac{d^2y}{dx^2}$; for extreme values $\frac{dy}{dx} = 0$; for maximum value $\frac{dy}{dx} = 0, \frac{d^2y}{dx^2} < 0$; for minimum value $\frac{dy}{dx} = 0, \frac{d^2y}{dx^2} > 0$].

(8) Intervals of Increase and Decrease. Find the intervals in which y increases or decreases. Consult art 1.5.

(9) Intervals of concavity convexity. Consult art. 1.4(*a*).

(10) Points of inflexion. Consult art. 1.4(*b*).

(11) Additional Points. Find the points on the curve at which $\frac{dy}{dx} = 0$ or $\frac{dy}{dx} = \infty$

Because where $\frac{dy}{dx} = 0$, at that point tangent to the curve is parallel to x-axis and where $\frac{dy}{dx} = \infty$; tangent is parallel to y-axis.

Note 1. It is obvious that it is not possible for the students to verify all the above properties of a curve before tracing the curve as there is limitation of time in examination. So while tracing, students can easily omit inconvenient steps but they can do so only if they have sufficient practice of curve tracing before the examination. So students should have as much practice as they can have.

Note 2. In paper, students must mention the headings of all the salient features of curve tracing mentioned above.

ILLUSTRATIVE EXAMPLES

Example 1. *Trace the curve* $x^3 + y^3 = 3axy, \quad a \geq 0.$

(P.T.U., Dec. 2012, May 2015, Dec. 2015)

Sol. The equation of the curve is

$$x^3 + y^3 - 3axy = 0 \qquad \text{...(1)}$$

1. Symmetry. The curve is neither symmetric about x-axis, nor y-axis but about $y = x$.

2. Origin. The curve passes through the origin (0, 0) and the tangents at the origin are given by $3axy = 0$

i.e., $x = 0, y = 0$ *i.e.,* x-axis and y-axis.

3. Domain and Range. From (1) it is clear that x and y both cannot be negative $\because$ then LHS will be negative but RHS will be positive which is impossible $\therefore$ no portion of the curve will lie in 3rd quadrant.

4. Points of Intersection. Curve meets x-axis (put $y = 0$ in (1) we get $x = 0$) *i.e.,* at (0, 0) Curve meets y-axis (put $x = 0$ we get $y = 0$,) *i.e.,* at (0, 0) $\therefore$ curve only passes through (0, 0); Curve intersects $y = x$ where $x^3 + x^3 = 3ax^2$ or $2x^3 = 3ax^2$ or $x = \frac{3a}{2}$ $\therefore$ Point of intersection with $y = x$ is $\left(\frac{3a}{2}, \frac{3a}{2}\right)$

5. Tangents. To find slope of the tangent at $\left(\frac{3a}{2}, \frac{3a}{2}\right)$ differentiate (1) w.r.t. x

$$3x^2 + 3y^2\frac{dy}{dx} = 3ax\frac{dy}{dx} + 3ay$$

$$(y^2 - ax)\frac{dy}{dx} = ay - x^2$$

$$\frac{dy}{dx} = \frac{ay - x^2}{y^2 - ax}; \quad \left(\frac{dy}{dx}\right)_{\left(\frac{3a}{2}, \frac{3a}{2}\right)} = \frac{\frac{3a^2}{2} - \frac{9a^2}{4}}{\frac{9a^2}{4} - \frac{3a^2}{2}} = -1$$

$\therefore$ At $\left(\frac{3a}{2}, \frac{3a}{2}\right)$ slope of the tangent = –1

$\therefore$ Tangent makes an angle of 135° with x-axis at $\left(\frac{3a}{2}, \frac{3a}{2}\right)$

6. Asymptotes. Asymptote of $x^3 + y^3 = 3axy$ are given by putting $x = 1, y = m$

$\phi_3(m) = 1 + m^3$, $\phi_3'(m) = 3m^2$, $\phi_3''(m) = 6m$

$\phi_2(m) = -3am$, $\phi_2'(m) = -3a$

$\phi_3(m) = 0$ given $1 + m^3 = 0$

or $(1 + m)(1 - m + m^2) = 0$ only real value of m is – 1.

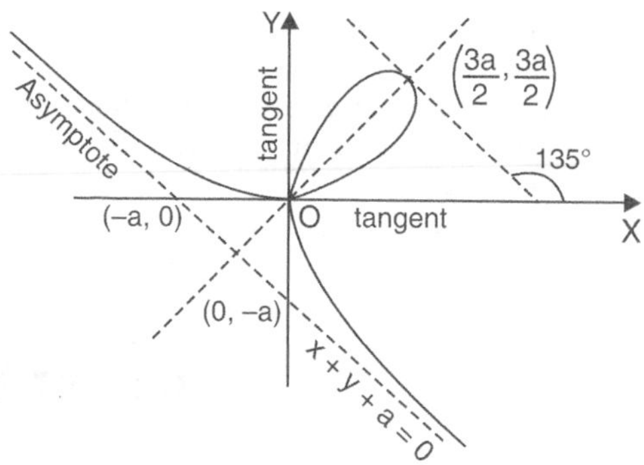

$\therefore$ Asymptote with slope $m = -1$ is $y = -x + c$ where c is given by

$$c\phi_3'(m) + \phi_2(m) = 0 \text{ at } m = -1$$

$$c(3) + (-3a)(-1) = 0$$

or $$c + a = 0 \quad \therefore \quad c = -a$$

$\therefore$ Asymptote is $y = -x - a$ or $x + y + a = 0$

Sketch of the curve is shown in the figure.

Example 2. *Trace the curve* $a^2y^2 = x^2(a^2 - x^2)$. **(P.T.U., May 2009)**

Sol. Equation of the curve is

$$a^2y^2 = x^2(a^2 - x^2) \quad \text{...(1)}$$

1. Symmetry. Curve is symmetrical about x-axis, y-axis and origin.

2. Domain. $$y^2 = \frac{x^2(a^2 - x^2)}{a^2}$$

y is defined when $x^2 < a^2$ *i.e.,* $|x| < a$ *i.e.,* $-a < x < a$

$\therefore$ $D_f = (-a, a)$ $\therefore$ curve lies between $x = a$ and $x = -a$.

3. Origin. The curve passes through the origin and tangent at the origin

$$a^2x^2 - a^2y^2 = 0 \quad \text{or} \quad y = \pm x.$$

4. Points of Intersection. The curve intersects x-axis where $y = 0$

i.e., at $x = 0, x = \pm a$ *i.e.,* at $(0, 0), (a, 0)$; $(-a, 0)$.

It intersects y-axis where $x = 0$ $\therefore$ $y = 0$ $\therefore$ it intersects y-axis at $(0, 0)$.

5. Tangents. Now slope of the tangents at $(a, 0)$ and $(-a, 0)$ are given by

Differentiate (1) w.r.t. x

$$2a^2y\frac{dy}{dx} = x^2(-2x) + (a^2 - x^2) \,.\, 2x$$

$$= 2x(-x^2 + a^2 - x^2)$$

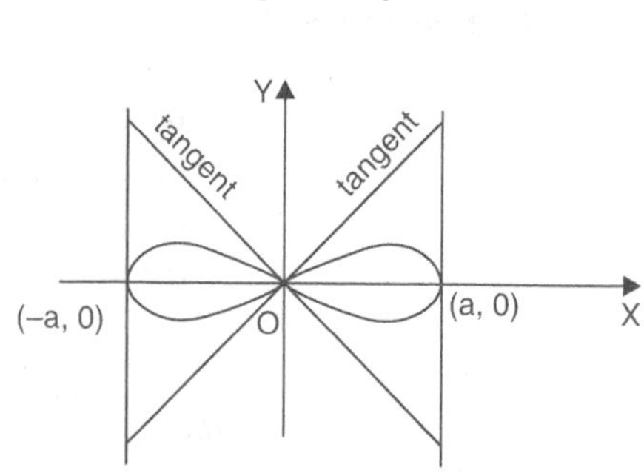

$$\therefore \quad \frac{dy}{dx} = \frac{x(a^2 - 2x^2)}{a^2y}$$

$$\left(\frac{dy}{dx}\right)_{(a,0)} = \frac{a(-a^2)}{2} = \infty$$

$\therefore$ Tangent at $(a, 0)$ is parallel to y-axis.

Similarly, $$\left(\frac{dy}{dx}\right)_{(-a, 0)} = \infty$$

$\therefore$ Tangent at $(-a, 0)$ is also parallel to y-axis.

6. **Asymptotes.** Asymptotes to $x^4 - a^2y^2 - a^2x^2 = 0$; curve has no asymptote.

Sketch of the curve is shown in figure.

Example 3. *Sketch the curve*

$$x(x^2 + y^2) = a(3x^2 - y^2)$$

or $$y^2(a + x) = x^2(3a - x)$$ **(P.T.U. May 2014)**

Sol. Equation of the curve is

$$y^2 (a + x) = x^2(3a - x) \qquad ...(1)$$

1. **Symmetry.** Curve is symmetric about x-axis (changing y to $-y$ does not change the curve).

2. **Domain.** From (1) $y^2 = \dfrac{x^2(3a - x)}{a + x}$ $\quad x \neq -a \quad$ or $\quad y = \pm x\sqrt{\dfrac{3a - x}{a + x}}$

y will exist when $3a - x > 0$ and $a + x > 0$ *i.e.,* $3a > x$ and $x > -a$

i.e., $$-a < x < 3a.$$

$\therefore$ Curve lies between $x = -a$ and $x = 3a$.

$3a - x$ and $a + x$ both cannot be negative $\because$ $3a < -a$ is impossible.

3. **Origin.** Curve passes through the origin and tangents at the origin are given by $3x^2 - y^2 = 0$ (equating the least degree terms in the equation to zero)

i.e., $$y = \sqrt{3}x \text{ and } y = -\sqrt{3}x.$$

4. **Intersection with axes.** Curve intersects x-axis (put $y = 0$) at $x = 0$ and $x = 3a$ *i.e.,* at $(0, 0)$ and $(3a, 0)$

It intersects y-axis (put $x = 0$) at $(0, 0)$ only.

5. **Tangents.** From (1)

$$y^2 (a + x) = x^2 (3a - x)$$

Differentiate w.r.t. x

$$2y\frac{dy}{dx}(a + x) + y^2 = x^2 (-1) + (3a - x)\, 2x = -3x^2 + 6ax$$

$$2y\frac{dy}{dx}(a + x) = -3x^2 + 6ax - y^2$$

$$= -3x^2 + 6ax - \frac{x^2(3a - x)}{a + x}$$

$$\frac{dy}{dx} = \pm\frac{3x(2a - x) - \dfrac{x^2(3a - x)}{a + x}}{2x\sqrt{\dfrac{3a - x}{a + x}}\,(a + x)}$$

$$= \pm \frac{3x(2a-x)(a+x) - x^2(3a-x)}{2x\sqrt{\dfrac{3a-x}{a+x}}\,(a+x)^2}$$

$$= \pm \frac{6a^2x - 2x^3}{2x\sqrt{3a-x}\,(a+x)^{3/2}}$$

$$\frac{dy}{dx} = \frac{3a^2 - x^2}{\sqrt{3a-x}\,(a+x)^{3/2}}$$

At $(3a, 0)$, $\quad \dfrac{dy}{dx} = \infty$

$\therefore$ At $(3a, 0)$ tangent is parallel to y-axis.

Now, tangent is parallel to x-axis where

$$\frac{dy}{dx} = 0$$

i.e., $\quad 3a^2 - x^2 = 0$

or $\quad x = \pm\sqrt{3}\,a.$

$\quad x = \sqrt{3}\,a$

But $-\sqrt{3}\,a$ does not belong to the domain $(-a, 3a)$

$\therefore$ Tangent is parallel to x-axis at $(\sqrt{3}\,a, 0)$ only.

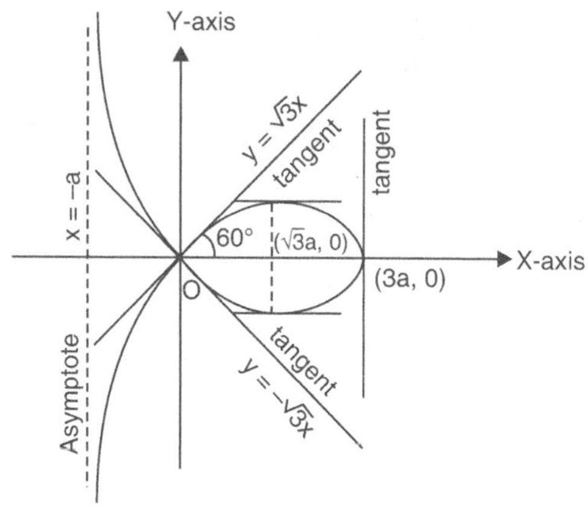

6. Asymptotes. Asymptote parallel to y-axis is (equate coefficient of highest power of y to zero) $a + x = 0$ *i.e.*, $x = -a$.

Sketch of the curve is shown in the figure.

Example 4. *Sketch the curve* $y = x + \dfrac{1}{x}$.

Sol. Equation of the curve is

$$y = x + \frac{1}{x} \qquad \text{...(1)}$$

1. Symmetry. Curve is only symmetric about origin *i.e.*, x changed to $-x$; y changed to $-y$ does not change the curve.

Also when x is positive, y is +ve

When x is negative ; y is –ve

$\therefore$ Curve lies in Ist and 3rd quadrants only.

2. Domain and Range. y is not defined at $x = 0$ $\therefore$ $D_f = R$ except 0

$\therefore$ Curve is discontinuous at $x = 0$

For range $\quad xy = x^2 + 1 \quad$ or $\quad x^2 - xy + 1 = 0$

$\therefore \quad x = \dfrac{y \pm \sqrt{y^2 - 4}}{2}$; x is defined when

$$y^2 - 4 \geq 0 \quad i.e., \quad y^2 \geq 4 \quad \text{or} \quad |y| \geq 2$$

$\therefore$ either $\quad y \geq 2 \quad$ or $\quad y \leq -2$

$\therefore \quad R_f = (-\infty, -2] \cup [2, \infty)$

3. Intersection with Co-ordinate Axes. Curve intersects x-axis (put $y = 0$); $x^2 + 1 = 0$ which does not give any real value of x $\therefore$ curve does not intersect x-axis.

It intersects y-axis where $x = 0$ but $x = 0 \notin D_f$ $\therefore$ Curve does not intersect any of the axes.

4. Asymptotes. Equation of the curve is

$$xy = x^2 + 1 \quad \text{or} \quad xy - x^2 - 1 = 0$$

Asymptote parallel to y-axis is $x = 0$

There is no asymptote parallel to x-axis

For oblique asymptotes

$$\phi_2(m) = m - 1 \qquad \text{(put } x = 1, y = m \text{ in second degree terms)}$$

$$\phi_1(m) = 0 \quad \phi_2'(m) = 1$$

$$\phi_0(m) = -1$$

$$\phi_2(m) = 0 \quad \text{gives } m = 1$$

c is given by

$$c\phi_2'(m) + \phi_1(m) = 0$$

$\therefore$ $$c \cdot 1 + 0 = 0 \quad \therefore \quad c = 0$$

$\therefore$ Asymptote is $$y = mx + c \quad i.e., \ y = x$$

5. Tangents. $$\frac{dy}{dx} = 1 - \frac{1}{x^2}$$

Tangent is parallel to x-axis where $\dfrac{dy}{dx} = 0$

i.e., $$1 - \frac{1}{x^2} = 0 \quad \text{or} \quad x^2 = 1 \quad \text{or} \quad x = \pm 1$$

when $x = 1, y = 2$; when $x = -1, y = -2$

$\therefore$ Tangent is $\|$ to x-axis at $(1, 2)$ and $(-1, -2)$

6. Intervals of Increase and Decrease

$$\frac{dy}{dx} > 0 \qquad \text{or} \quad 1 - \frac{1}{x^2} > 0 \text{ or } \ x^2 > 1$$

or $$|x| > 1 \quad \Rightarrow \text{ either } x > 1 \qquad \text{or} \quad x < -1$$

$\therefore$ Curve increases in $(-\infty, -1) \cup (1, \infty)$ and decreases in $(-1, 1)$ when $\dfrac{dy}{dx} < 0$

7. Intervals of Concavity and Convexity

$$\frac{d^2y}{dx^2} = \frac{2}{x^3}$$

$$\frac{d^3y}{dx^3} = -\frac{6}{x^4}$$

Curve concaves upward when $\dfrac{2}{x^3} > 0$ or $x^3 > 0$ *i.e.,* $x > 0$ *i.e.,* $x \in (0, \infty)$ concaves downward when $x < 0$ *i.e.,* $x \in (-\infty, 0)$. There is no point of inflexion.

Sketch of the curve is shown in the figure.

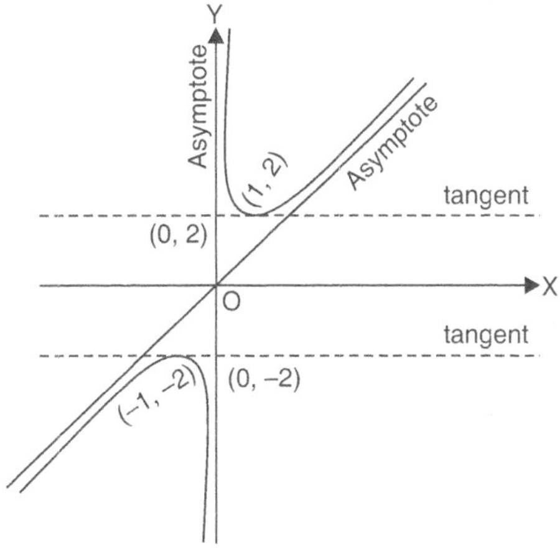

Example 5. *Trace the curve* $y^2 (a - x) = x^3$, $a > 0$.

Sol. Equation of the curve is $y^2(a - x) = x^3$

1. Symmetry. The curve is symmetric about x-axis $\because$ If we change y to $-y$ equation remains unchanged.

2. Domain and Range. $y^2 = \dfrac{x^3}{a-x}$ $\quad\therefore\quad y = \pm\sqrt{\dfrac{x^3}{a-x}}$, $\quad x \neq a$

$\therefore$ y is defined when $\dfrac{x^3}{a-x} \geq 0$

When $x > 0$, then $\dfrac{x^3}{a-x} \geq 0$ iff $a - x > 0$ or $x < a$ *i.e.*, $0 < x < a$

When $x < 0$, then $\dfrac{x^3}{a-x} \geq 0$ iff $a - x < 0$ or $x > a$ *i.e.*, $a < x < 0$ which is impossible $\because$ $a > 0$

$\therefore$ $D_f = [0, a)$ $\therefore$ curve lies between 0 and a.

As y can have both +ve as well as –ve values $\therefore$ Range is the set of all real numbers.

3. Origin. Curve passes through the origin and the tangent at the origin is $ay^2 = 0$ *i.e.*, $y = 0$ [obtained by equating the lowest degree terms in the equation to zero.]

4. Points of Intersection with Axes. Curve intersects x-axis put $(y = 0)$ at $x^3 = 0$ *i.e.*, at (0, 0). It intersects y-axis (put $x = 0$) at (0, 0) origin.

5. Asymptotes. In the equation of the curve

coeff. of $x^3 = 1$ $\quad\therefore\quad$ no horizontal asymptotes

coeff. of y^2 is $a - x$ $\quad\therefore\quad$ vertical asymptote is $x = a$

For oblique asymptote write the equation of the curve in the form $x^3 + xy^2 - ay^2 = 0$

Put $x = 1, y = m$ in the highest degree terms, we get $1 + m^2 = 0$ it gives no real value of m

$\therefore$ There is no oblique asymptote.

6. Intervals of Increase and Decrease

$$y = \pm\frac{x^{3/2}}{\sqrt{a-x}} \text{ take only +ve sign, we get } y = \frac{x^{3/2}}{\sqrt{a-x}}$$

$$\frac{dy}{dx} = \frac{\sqrt{x}\,(3a - 2x)}{2(a-x)^{3/2}}$$

$$\frac{dy}{dx} > 0 \quad \text{when} \quad 3a - 2x > 0 \quad i.e., \quad x < \frac{3a}{2}$$

As $\quad x = \dfrac{3a}{2}$ does not belong to the domain

$$\therefore \quad \frac{dy}{dx} > 0 \quad \text{when} \quad 0 < x < a$$

$\therefore$ y increases in $0 < x < a$.

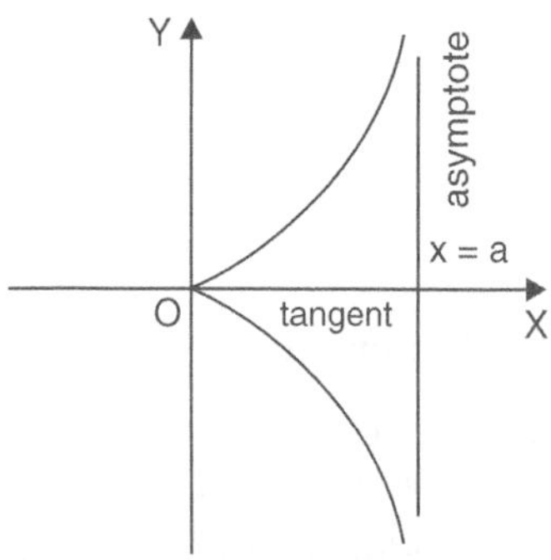

7. Additional Points

Tangents parallel to x-axis is given by $\frac{dy}{dx} = 0$ *i.e.*, at $x = \frac{3a}{2}$ which is outside the domain.

$\therefore$ Within the domain there is no tangent parallel to x-axis.

Tangent parallel to y-axis is given by $\frac{dy}{dx} = \infty$, *i.e.*, at $x = a$. Which again does not belong to the domain.

Now the sketch of the curve is given in the figure

Example 6. *Trace the curve* $y = \frac{x^2 + 1}{x + 1}$

Sol. Equation of the curve is

$$y = \frac{x^2 + 1}{x + 1} \quad ...(1)$$

1. Symmetry. Curve is not symmetrical about any line.

2. Domain. y exists for all real values of x except $x = -1$

$\therefore$ Curve is discontinuous at $x = -1$

$\therefore$ $D_f = R$ except -1

for Range $yx + y = x^2 + 1$

or $x^2 - y \,.\, x + (1 - y) = 0$

or $$x = \frac{y \pm \sqrt{y^2 - 4(1 - y)}}{2} = \frac{y \pm \sqrt{y^2 + 4y - y}}{2}$$

$$x = \frac{y \pm \sqrt{(y + 2)^2 - 8}}{2}$$

x exists if $(y + 2)^2 - 8 \geq 0$ or $(y + 2)^2 \geq 8$

$$|y + 2| \geq 2\sqrt{2} \quad \therefore \quad \text{either } y + 2 \geq 2\sqrt{2} \quad \text{or } y + 2 \leq -2\sqrt{2}$$

$\therefore$ $$y \geq 2\sqrt{2} - 2 \quad \text{or} \quad y \leq -2\sqrt{2} - 2$$

$\therefore$ $$y \in (-\infty, -2\sqrt{2} - 2] \cup [2\sqrt{2} - 2, \infty)$$

$\therefore$ Curve lies outside $y = 2(\sqrt{2} - 1)$ and $y = -2(\sqrt{2} + 1)$

3. Origin. Curve does not pass through the origin.

4. Intersection with Co-ordinate Axes

Curve intersects x-axis (put $y = 0$) ; $x^2 + 1 = 0$ no real point

Curve intersect y-axis (put $x = 0$) ; $y = 1$

$\therefore$ Point of intersection with y-axis is $(0, 1)$

5. Tangents. Slope of the tangent at $(0, 1)$ is obtained from $\frac{dy}{dx}$

$$\frac{dy}{dx} = \frac{(x+1)2x - (x^2+1)}{(x+1)^2} = \frac{2x^2 + 2x - x^2 - 1}{(x+1)^2}$$

$$\frac{dy}{dx} = \frac{x^2 + 2x - 1}{(x+1)^2} = \frac{(x+1)^2 - 2}{(x+1)^2}$$

$$\left(\frac{dy}{dx}\right)_{(0,\,1)} = \frac{1-2}{1} = -1$$

$\therefore$ Tangent to the curve makes an angle of 135° at (0, 1) with +ve direction of x-axis.

6. Asymptotes. Equation of the curve is

$$y(x + 1) = x^2 + 1$$

Asymptote parallel to y-axis is $x + 1 = 0$; *i.e.*, $x = -1$

Now equation of the curve can be rewritten as

$$(xy - x^2) + y - 1 = 0$$

For oblique asymptotes $\phi_2(m) = m - 1$ [Put $x = 1, y = m$ in 2nd degree terms]

$$\phi_2'(m) = 1$$

$$\phi_1(m) = m$$

$$\phi_2(m) = 0 \text{ gives } m = 1$$

c is given by $c\phi_2'(m) + \phi_1(m) = 0$

$$c \,.\, 1 + 1 = 0 \quad \therefore \quad c = -1$$

$\therefore$ Asymptote is $y = mx + c$ *i.e.*, $y = x - 1$

7. Maxima, Minima

For maxima, minima $\dfrac{dy}{dx} = 0$ *i.e.*, $(x + 1)^2 - 2 = 0$

i.e., $x + 1 = \pm\sqrt{2} \quad \therefore \quad x = \sqrt{2} - 1, -\sqrt{2} - 1$

$\therefore$ Tangent is || to x-axis at $x = \sqrt{2} - 1$ and at $x = -\sqrt{2} - 1$

When $x = \sqrt{2} - 1$, $y = \dfrac{(\sqrt{2}-1)^2 + 1}{\sqrt{2} - 1 + 1} = \dfrac{2 + 1 - 2\sqrt{2} + 1}{\sqrt{2}} = \dfrac{4 - 2\sqrt{2}}{\sqrt{2}} = 2\sqrt{2} - 2 = 2(\sqrt{2} - 1)$

When $x = -\sqrt{2} - 1$, $y = \dfrac{(\sqrt{2}+1)^2 + 1}{-\sqrt{2} - 1 + 1} = \dfrac{2 + 2\sqrt{2} + 1 + 1}{-\sqrt{2}} = \dfrac{4 + 2\sqrt{2}}{-\sqrt{2}}$

$$= -2\sqrt{2} - 2 = -2(\sqrt{2} + 1)$$

$$\frac{d^2y}{dx^2} = \frac{(x+1)^2 \,.\, 2(x+1) - \{(x+1)^2 - 2\}\{2(x+1)\}}{(x+1)^4}$$

$$= 2\,\frac{(x+1)^2 - (x+1)^2 + 2}{(x+1)^3} = \frac{4}{(x+1)^3}$$

At $\quad x = \sqrt{2} - 1, \quad \dfrac{d^2y}{dx^2} > 0$

$\therefore$ at $(\sqrt{2} - 1, 2\sqrt{2} - 2)$ curve has minimum value

At $\quad x = -\sqrt{2} - 1, \quad \dfrac{d^2y}{dx^2} < 0$

$\therefore$ at $(-\sqrt{2} - 1, -2\sqrt{2} - 2)$ curve has maximum value

8. Intervals of Increase and Decrease.

$$\frac{dy}{dx} > 0 \quad \text{when} \quad (x + 1)^2 > 2$$

$$|x + 1| > \sqrt{2} \quad i.e., \text{ either } \quad x + 1 > \sqrt{2} \;;\; i.e., x > \sqrt{2} - 1$$

$$\text{or} \quad x + 1 < -\sqrt{2} \;;\; x < -\sqrt{2} - 1$$

$\therefore$ Curve increases in $(-\infty, -\sqrt{2} - 1) \cup (\sqrt{2} - 1, \infty)$

and decreases in $(-\sqrt{2} - 1, \sqrt{2} - 1)$

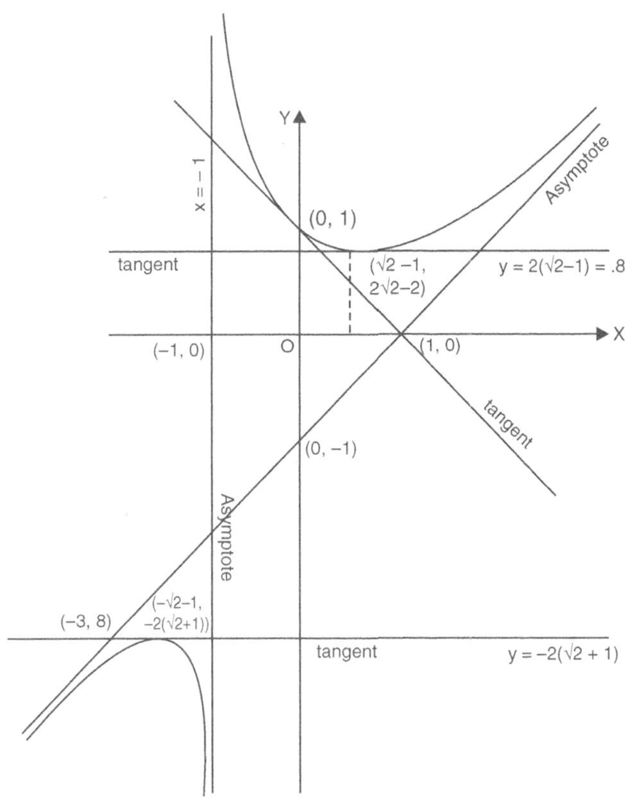

9. Concavity, Convexity

$$\frac{d^2y}{dx^2} = \frac{(x+1)^2 \cdot 2(x+1) - \{(x+1)^2 - 2\}\, 2(x+1)}{(x+1)^4}$$

$$= 2 \cdot \frac{(x+1)^2 - (x+1)^2 + 2}{(x+1)^3} = \frac{4}{(x+1)^3}$$

Curve concaves upward if $x + 1 > 0$ *i.e.*, $x > -1$ and concaves downward when $x < -1$

$\frac{d^2y}{dx^2} \neq 0$ $\therefore$ There is no point of inflexion.

Sketch of the curve is shown in figure.

Example 7. *Trace the curve* $y = \frac{x^3 + 1}{x}$. **(P.T.U., Dec. 2004)**

Sol. Equation of the curve is

$$y = \frac{x^3 + 1}{x} \qquad ...(1)$$

1. Symmetry. Curve is not symmetric about any line.

2. Domain. y does not exist at $x = 0$

$\therefore$ D_f = all real values of x except 0.

$\therefore$ Curve is discontinuous at $x = 0$

3. Intersection with Co-ordinate Axes. Curve intersects x-axis at (put $y = 0$)

$$x^3 + 1 = 0 \quad \therefore \quad x = -1 \text{ i.e., at } (-1, 0)$$

Curve does not intersect y-axis

4. Tangent. Slope of the tangent at (–1, 0) is obtained by differentiating (1) w.r.t. x

$$\frac{dy}{dx} = 2x - \frac{1}{x^2}; \left(\frac{dy}{dx}\right)_{(-1, 0)} = -2 - 1 = -3$$

$\therefore$ Tangent at (–1, 0) makes as an $\pi - \tan^{-1} 3$ with +ve direction of x-axis.

5. Asymptote

$$xy = x^3 + 1$$

Curve has asymptote $x = 0$, parallel to y-axis *i.e.*, y-axis is the only asymptote to the curve.

6. Maxima, Minima

$$\frac{dy}{dx} = 2x - \frac{1}{x^2}; \frac{dy}{dx} = 0 \text{ gives } x^3 = \frac{1}{2}. \quad i.e., \quad x = \frac{1}{2^{1/3}} = 0.79$$

$$x = \frac{1}{2^{1/3}}, y = \frac{\frac{1}{2} + 1}{\frac{1}{2^{1/3}}} = \frac{3}{2} \times 2^{1/3} = 1.19$$

$$\frac{d^2y}{dx^2} = 2 + \frac{2}{x^3}$$

At $x = \frac{1}{2^{1/3}}$; $\frac{d^2y}{dx^2}$ is positive

$\therefore$ Curve has minimum value at $\left(\frac{1}{2^{1/3}}, \frac{3}{2}2^{1/3}\right)$ and tangent at this point is parallel to x-axis.

7. Concavity, Convexity, Point of Inflexion

$$\frac{d^2y}{dx^2} > 0 \text{ for } 2 + \frac{2}{x^3} > 0$$

For $(x > 0)$ $\quad x^3 + 1 > 0 \quad \therefore \quad x + 1 > 0, x > -1$; But $x > 0 \quad \therefore \quad x \in (0, \infty)$

for $x < 0, x^3 + 1 < 0 \quad \therefore \quad x < -1 \quad \Rightarrow \quad x \in (-\infty, -1)$

$\therefore$ Curve is concave upward in $(-\infty, -1) \cup (0, \infty)$

Concave downward for $\frac{d^2y}{dx^2} < 0$

$$2 + \frac{2}{x^3} < 0 \quad \text{or} \quad 1 + \frac{1}{x^3} < 0$$

When $x > 0 \quad x^3 + 1 < 0 \quad$ or $\quad x + 1 < 0 \quad$ *i.e.,* $\quad x < -1$ impossible

When $x < 0 \quad x^3 + 1 > 0 \quad$ or $\quad x + 1 > 0 \quad$ or $\quad x > -1$

$\therefore \quad -1 < x < 0 \quad \therefore \quad$ Concave downward in the interval $(-1, 0)$

$$\frac{d^2y}{dx^2} = 0, \quad x^3 + 1 = 0 \quad \Rightarrow \quad x + 1 = 0 \quad \Rightarrow \quad x = -1;$$

$$\frac{d^3y}{dx^3} = \frac{-6}{x^4} \neq 0 \text{ at } x = -1$$

When $x = -1, y = 0$

$\therefore \quad (-1, 0)$ is the point of Inflexion of the curve.

8. Intervals of Increase and Decrease

$$\frac{dy}{dx} > 0 ;\ 2x - \frac{1}{x^2} > 0 \quad \text{or} \quad 2x^3 > 1 \quad \text{or} \quad x > \frac{1}{2^{1/3}}$$

$\therefore$ Curve increases (rises up) in the interval $\left(\frac{1}{2^{1/3}}, \infty\right)$ *i.e.,* $(0.79, \infty)$. Curve decreases when $\frac{dy}{dx} < 0$ *i.e.,* $\frac{2x^3 - 1}{x^2} < 0$ *i.e.,* $2x^3 - 1 < 0 \quad \because \quad x^2$ is always +ve *i.e.,* $x < \frac{1}{2^{1/3}} = 0.79$

But at $x = 0$, curve is discontinuous

$\therefore$ Curve decreases in the interval $(0, 0.79)$

Also when $x < 0$, $\frac{dy}{dx} < 0$.

$\therefore$ Curve also decreases in the interval $(-\infty, 0)$

$\therefore$ Curve decreases in the interval $(-\infty, 0) \cup (0, 0.79)$

9. Additional Points

As the above conditions are not sufficient to trace the accurate curve.

We will find some additional points on the curve

x	–3	–2	–1.5	–1	–.5	.5	1	2
y	8.6	3.5	1.6	0	–1.75	2.25	2	4.5

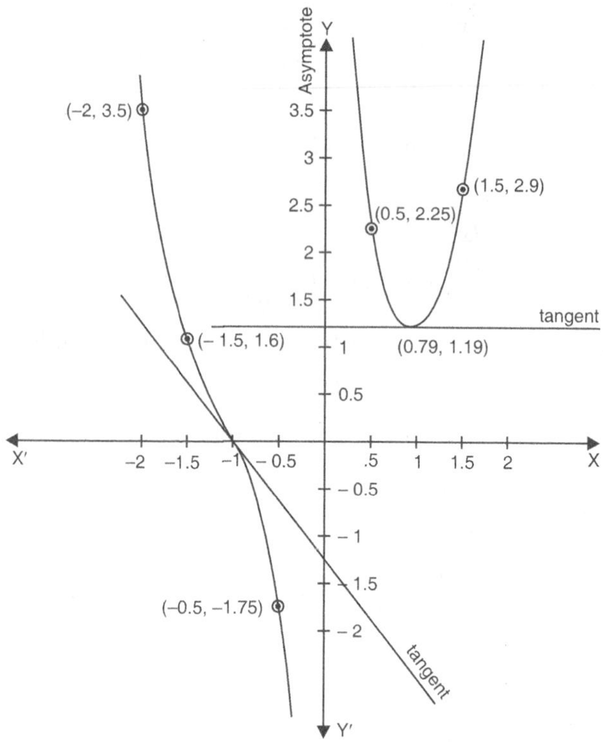

Example 8. *Trace the curve* $y = \dfrac{x^3}{1+x^2}$.

Sol. Equation of the curve is

$$y = \frac{x^3}{1+x^2} \qquad ...(1)$$

1. Symmetry. Curve is not symmetric about any line.

When x is negative y is negative when x is positive y is +ve

$\therefore$ Curve lies in first and third quadrants.

2. Origin. Curve passes through the origin and the equation of the tangent at the origin is (obtained by equating the lowest degree terms to zero) *i.e.*, $y = 0$ *i.e.*, x-axis is tangent at the origin

3. Domain. y exists for all real values of x $\therefore$ $D_f = R$.

4. Intersection with Co-ordinate Axes. For intersection with x-axis put ($y = 0$), we get $x = 0$. For intersection with y-axis put $x = 0$ we get $y = 0$ $\therefore$ Curve only passes through origin.

5. Asymptotes. Curve is $(1 + x^2)y = x^3$ or $x^2y - x^3 + y = 0$

There is no real rectangular asymptote.

For oblique asymptote

$$\phi_3(m) = m - 1 \qquad \text{(Put } x = 1, y = m \text{ in } x^2y - x^3\text{)}$$

$$\phi_2(m) = 0 \qquad \phi_3'(m) = 1$$

$$\phi_3(m) = 0 \Rightarrow \qquad m = 1$$

$$\therefore \qquad c\phi_3'(m) + \phi_2(m) = 0$$

$$c \cdot 1 + 0 = 0 \quad \therefore \qquad c = 0$$

$\therefore$ Asymptote is $y = x$

6. Maxima, Minima.

$$y = \frac{x^3}{1+x^2}$$

$$\frac{dy}{dx} = \frac{(1+x^2)\,3x^2 - x^3 \cdot 2x}{(1+x^2)^2} = \frac{3x^2 + x^4}{(1+x^2)^2} \qquad ...(2)$$

For maximum, minimum values

$$\frac{dy}{dx} = 0 \Rightarrow \frac{3x^2 + x^4}{(1+x^2)^2} = 0$$

$$x^2(3 + x^2) = 0 \Rightarrow x = 0$$

$$\frac{d^2y}{dx^2} = \frac{(1+x^2)^2 . (6x + 4x^3) - (3x^2 + x^4)2(1+x^2)2x}{(1+x^2)^4}$$

$$= \frac{(1+x^2)(6x + 4x^3) - 4x(3x^2 + x^4)}{(1+x^2)^3}$$

$$= \frac{6x + 6x^3 + 4x^3 + 4x^5 - 12x^3 - 4x^5}{(1+x^2)^3} = \frac{6x - 2x^3}{(1+x^2)^3} = \frac{2x(3 - x^2)}{(1+x^2)^3}$$

At $x = 0$; $\quad \dfrac{d^2y}{dx^2} = 0$

So (0, 0) is a saddle point. There is no point of maxima, minima on the curve.

7. Concavity, Convexity, Point of Inflexion.

$$\frac{d^2y}{dx^2} > 0 \Rightarrow x(3 - x^2) > 0$$

i.e., when $x > 0$, $3 > x^2$ or $x^2 < 3$ or $|x| < \sqrt{3}$ or $-\sqrt{3} < x < \sqrt{3}$

But $x > 0$ $\therefore$ Curve concaves upward in the interval $(0, \sqrt{3})$

When $x < 0$; $3 - x^2 < 0$ $\therefore$ $3 < x^2$ or $x^2 > 3$

or $|x| > \sqrt{3}$ $\therefore$ either $x > \sqrt{3}$ or $x < -\sqrt{3}$

But $x \not> \sqrt{3}$ $\because$ $x < 0$

$\therefore$ Curve concaves upward in $(-\infty, -\sqrt{3})$

$\therefore$ Combining the two, the curve concaves upward in $(-\infty, -\sqrt{3}) \cup (0, \sqrt{3})$

If $\frac{d^2y}{dx^2} < 0$ then $x(3 - x^2) < 0$

$\therefore$ when $x > 0$, $3 - x^2 < 0$ or $x^2 > 3$

$\therefore$ either $x > \sqrt{3}$ or $x < -\sqrt{3}$

But $x \nless -\sqrt{3}$ $\because$ $x > 0$ $\therefore$ $x > \sqrt{3}$ $\therefore$ $x \in (\sqrt{3}, \infty)$

$\therefore$ Curve concaves downward in $(\sqrt{3}, \infty)$

when $x < 0$, $3 - x^2 > 0$, *i.e.*, $-\sqrt{3} < x < \sqrt{3}$

But $x < 0$ $\therefore$ Curve concaves downward in $(-\sqrt{3}, 0)$

$\therefore$ Combining the two curve concaves downward in $(-\sqrt{3}, 0) \cup (\sqrt{3}, \infty)$

Points of Inflexion.

$$\frac{d^2y}{dx^2} = 0 \text{ for } x = 0, x = \sqrt{3}, x = -\sqrt{3}$$

$$\frac{d^3y}{dx^3} = 2\,\frac{(1+x^2)^3(3-3x^2) - (3x - x^3)\cdot 3(1+x^2)^2\cdot 2x}{(1+x^2)^6}$$

$$= 2\,\frac{3(1+x^2)(1-x^2) - 6x(3x - x^3)}{(1+x^2)^4}$$

$$= 2\cdot\frac{3 - 3x^4 - 18x^2 + 6x^4}{(1+x^2)^4} = 2\cdot\frac{3 - 18x^2 + 3x^4}{(1+x^2)^4} = \frac{6(1 - 6x^2 + x^4)}{(1+x^2)^4}$$

At $x = 0$, $\frac{d^3y}{dx^3} \neq 0$ $\therefore$ $x = 0$ is a point of inflexion

At $x = \pm\sqrt{3}$, $\frac{d^3y}{dx^3} = \frac{6(1 - 6.3 + 9)}{(1+3)^4} = \frac{6(-8)}{4^4} \neq 0$

$\therefore$ $x = \sqrt{3}$, $y = \frac{3\sqrt{3}}{4}$; $x = -\sqrt{3}$, $y = \frac{-3\sqrt{3}}{4}$

$\therefore$ Points of inflexion are $(0, 0)$, $\left(\sqrt{3}, \frac{3\sqrt{3}}{4}\right)$, $\left(-\sqrt{3}, \frac{-3\sqrt{3}}{4}\right)$

Slope of the tangents at $\left(\sqrt{3}, \frac{3\sqrt{3}}{4}\right)$ and $\left(-\sqrt{3}, \frac{-3\sqrt{3}}{4}\right)$ are given by

$$\tan\theta = \frac{9+9}{4^2}, \frac{9+9}{4^2} \quad \textit{i.e.}, \tan\theta = \frac{9}{8} \text{ at each point.} \qquad \text{(Using 2)}$$

$\therefore$ Tangent at both the points of inflexion make $\tan^{-1}\frac{9}{8}$ angle with +ve direction of x-axis.

8. Intervals of Increase and Decrease. $\frac{dy}{dx} > 0$ for all values of x.

$\therefore$ Curve is increasing and never decreases.

9. Additional Points

We can choose some additional points also

x	–3	–2	–1	0	1	2	3
y	–2.7	–1.6	–.5	0	.5	1.6	2.7

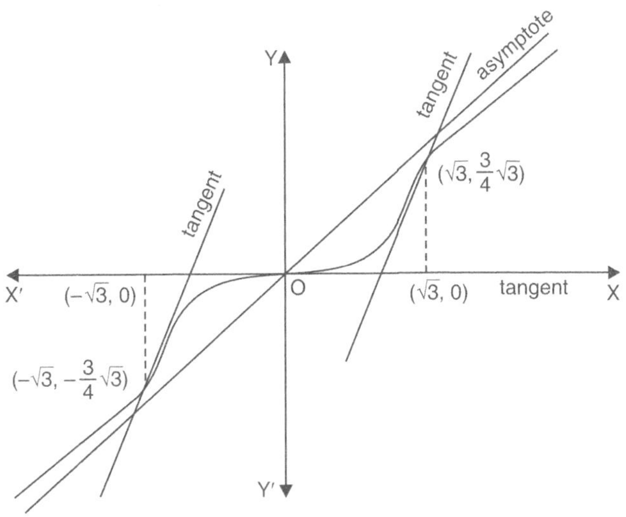

Example 9. *Trace the curve* $y = \dfrac{x^3 - 3}{2x - 4}$. **(P.T.U., May 2005)**

Sol. Equation of the curve is $y = \dfrac{x^3 - 3}{2x - 4} = \dfrac{x^3 - 3}{2(x - 2)}$

1. Symmetry. The curve is not symmetrical about any line.

2. Domain. y exists $\forall$ x except $x = 2$, but when x is negative y is positive and when x is positive y may be positive or negative except at $x = 2$ $\therefore$ curve does not lie in 3rd quadrant.

3. Origin. Curve does not pass through the origin.

4. Intersection with Axes. Curve intersects x-axis at $(3^{1/3}, 0) = (1.4, 0)$ and y-axis at $\left(0, \dfrac{3}{4}\right) = (0, 0.75)$.

5. Asymptotes. Equation of the curve can be put into the form $2y(x - 2) = x^3 - 3$.

It has only one asymptote parallel to y-axis which is $x - 2 = 0$ *i.e.*, $x = 2$.

6. Tangents. $\dfrac{dy}{dx} = \dfrac{1}{2}\dfrac{(x - 2) . 3x^2 - (x^3 - 3)}{(x - 2)^2} = \dfrac{1}{2}\dfrac{2x^3 - 6x^2 + 3}{(x - 2)^2}$

At (1.4, 0); $\dfrac{dy}{dx}$ is –ve $\therefore$ tangent at (1.4, 0) makes obtuse angle with the +ve direction of x-axis.

At (0, 0.75); $\dfrac{dy}{dx}$ is +ve $\therefore$ tangent at (0, 0.75) makes acute angle with the +ve direction of x-axis.

7. Additional points. We can choose some additional points on the curve

We have $$y = \frac{x^3 - 3}{2(x-2)} = \frac{1 - \frac{3}{x^3}}{2\left(\frac{1}{x^2} - \frac{2}{x^3}\right)}$$

When $x \to -\infty$ $y \to +\infty$ [$\because$ when x is –ve, y is +ve]

x	–1.5	–1	–0. 5	0	0.5	1	1.5	1.8
y	$\frac{51}{56} = .9$	$\frac{2}{3} = .66$	$\frac{5}{8} = .62$	$\frac{3}{4} = .75$	$\frac{23}{24} = .9$	1	$-\frac{3}{8} = -.37$	–7.08
$\frac{dy}{dx}$	–ve	–ve	+ve	+ve	+ve	–ve	–ve	–ve

Sketch of the curve is given below:

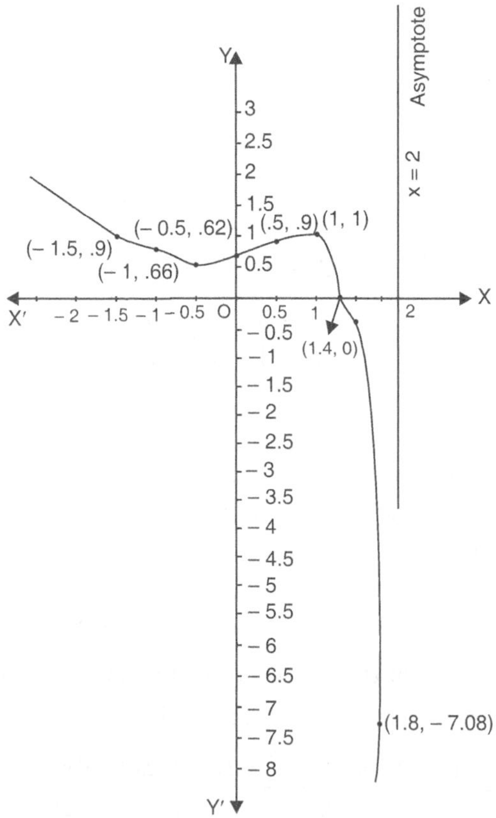

Example 10. *Trace the curve* $y = x^3 - 12x - 16$. **(P.T.U., May 2008)**

Sol. Equation of the curve is

$$y = x^3 - 12x - 16 \qquad ...(1)$$

1. Symmetry. Curve is not symmetric about any line

2. Origin. Curve does not pass through the origin

3. Domain. Curve exists $\forall$ real x.

4. Intersection with Coordinate Axes. Curve intersects x-axis (put $y = 0$)

i.e., $x^3 - 12x - 16 = 0$

or $(x + 2)^2 (x - 4) = 0$ or $x = -2, x = 4$ *i.e.,* at $(-2, 0), (4, 0)$

It intersects y-axis (put $x = 0$) at $(0, -16)$

5. Asymptote. Curve has no asymptote

6. Tangents. $\frac{dy}{dx} = 3x^2 - 12$

At $(-2, 0)$; $\frac{dy}{dx} = 0$ $\therefore$ Tangent is parallel to x-axis at $(-2, 0)$

At $(4, 0)$; $\frac{dy}{dx} = 36$ *i.e.,* $\tan\theta = 36$

$\therefore$ at $(4, 0)$ tangent makes an acute angle $\tan^{-1} 36$ with x-axis

Also $\frac{dy}{dx} = 0$ at $3x^2 - 12 = 0$ *i.e.,* $x^2 = 4$ *i.e.,* $x = \pm 2$

$\therefore$ Tangent is also parallel to x-axis at $(2, -32)$

7. Concavity. $y = x^3 - 12x - 16$

$$\frac{dy}{dx} = 3x^2 - 12$$

$$\frac{d^2y}{dx^2} = 6x$$

$\therefore$ Curve concaves upward for $x > 0$

and downward for $x < 0$

$x = 0$ is the point of inflexion

8. Intervals of Increase and Decrease . $\frac{dy}{dx} = 3(x^2 - 4)$

Curve increases for $\frac{dy}{dx} > 0$ *i.e.,* $x^2 > 4$ or $|x| > 2$ *i.e.,* increases in the interval $(-\infty, -2) \cup (2, \infty)$ and decreases in the interval $(-2, 2)$

9. Additional Points. $y \to \infty$ as $x \to \infty$

and $y \to -\infty$ as $x \to -\infty$

y is +ve for $x > 4$ and –ve for $x < 4$

Sketch of the curve is shown in the figure:

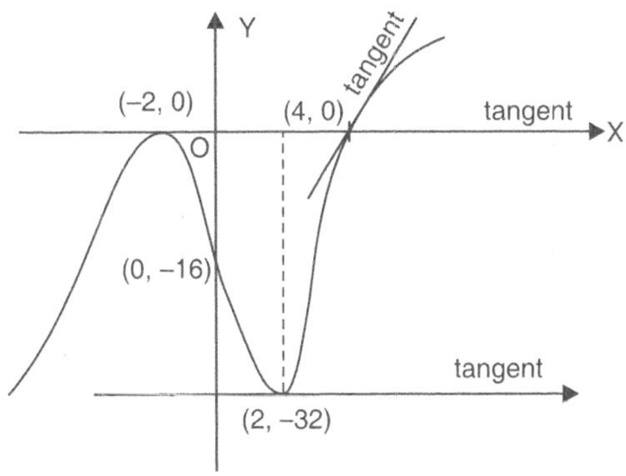

Example 11. *Sketch the curve* $y = \frac{1}{6}(x^3 - 6x^2 + 9x + 6)$.

Sol. 1. Symmetry. No symmetry

2. Domain. $D_f = R$

3. Origin. Does not pass through the origin

4. Intersection with Co-ordinate Axes. Curve intersects x-axis (put $y = 0$)

$$x^3 - 6x^2 + 9x + 6 = 0$$

It is not convenient to find points of intersection. $\therefore$ we can only guess the location of points of intersection by taking

$$f(0) = 6 = +\text{ve}$$
$$f(1) = 1 - 6 + 9 + 6 = 10 = +\text{ve}$$
$$f(2) = 8 - 24 + 18 + 6 = 8 = +\text{ve}$$

which shows that on R.H.S. of origin curve does not meet x-axis.

Now $\quad f(-1) = -1 - 6 - 9 + 6 = -10 = -\text{ve}, f(0) = 1 = +\text{ve}$

$$f(-1) < 0, f(0) > 0$$

$\therefore$ One point of intersection lies between $(-1, 0)$.

5. Asymptote. No asymptote of the curve.

6. Intervals of Increase and Decrease.

$$\frac{dy}{dx} = \frac{1}{6}(3x^2 - 12x + 9) = \frac{1}{2}(x^2 - 4x + 3)$$

$$\frac{dy}{dx} > 0 \quad \text{when} \quad x^2 - 4x + 3 > 0$$

or $\quad (x - 2)^2 > 4 - 3 = 1$

or $\quad |x - 2| > 1$

i.e., either $x - 2 > 1$ or $x - 2 < -1$.

or $\quad x > 3$ or $x < 1$

$\therefore$ Curve increases in $(3, \infty)$ or $(-\infty, 1)$ *i.e.*, $(-\infty, 1) \cup (3, \infty)$

$$\frac{dy}{dx} < 0 \quad \text{when} \quad (x - 2)^2 < 1$$

or $\quad |x - 2| < 1$ or $2 - 1 < x < 2 + 1$

$$1 < x < 3$$

$\therefore$ Curve decreases in $(1, 3)$.

7. Maxima, Minima. Curve has extreme value where

$$\frac{dy}{dx} = 0 \quad i.e., \quad x^2 - 4x + 3 = 0$$

$\therefore \quad x = 1, 3$

$$\frac{d^2y}{dx^2} = \frac{1}{2}(2x - 4) = x - 2$$

At $x = 1$, $\quad \frac{d^2y}{dx^2} = -\text{ve}$

$\therefore$ Curve has maximum value at $x = 1$ *i.e.*, at $\left(1, \frac{5}{3}\right)$

At $x = 3$; $\dfrac{d^2y}{dx^2} = 1 = +\text{ve}$

$\therefore$ Curve has minimum value at (3, 1).

$\because$ Tangents to the curve are parallel to x-axis at (1, 5/3) and (3, 1)

8. Concavity, Convexity, Point of Inflexion

$$\frac{d^2y}{dx^2} = x - 2;\quad \frac{d^3y}{dx^3} = 1$$

Curve concaves upward for $x > 2$ *i.e,* in $(2, \infty)$

Curve concaves downward of $x < 2$ *i.e.,* $(-\infty, 2)$

Point of inflexion at $x = 2$ $\because$ at $x = 2$ $\dfrac{d^2y}{dx^2} = 0$ but $\dfrac{d^3y}{dx^3} \neq 0$ $\therefore$ $\left(2, \dfrac{4}{3}\right)$ is point of inflexion

Slope of tangent at $x = 2$ is given by

$$\left(\frac{dy}{dx}\right)_{x=2} = \frac{1}{2}(4 - 8 + 3) = -\frac{1}{2}.$$

$\therefore$ Tangent at (2, 4/3) makes an angle of $\pi - \dfrac{1}{2}$ with +ve direction of x-axis.

9. Additional Points. We can find some more points on the curve as given below:

x	-3	-2	-1	0	1	2	3
y	-17	$-\frac{22}{3}$	$-\frac{5}{3}$	1	$\frac{5}{3}$	$\frac{4}{3}$	1

When $x \to \infty, y \to \infty$

When $x \to -\infty, y \to -\infty$

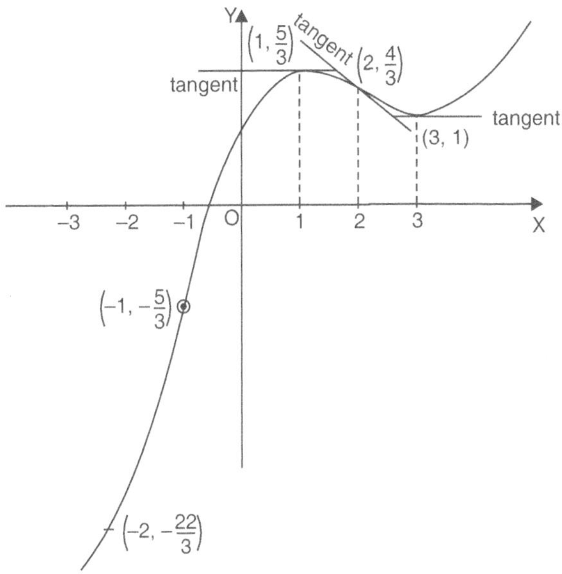

Example 12. *Trace the curve* $x = (y-1)(y-2)(y-3)$. **(P.T.U., May 2011)**

Sol. 1. Symmetry. Nil or $x = y^3 - 6y^2 + 11y - 6$

2. Origin. Does not pass through the origin

3. Domain Range. All real values of y

4. Intersection with Co-ordinate Axes. Curve intersects y-axis (put $x = 0$) at $y = 1$, $y = 2$, $y = 3$ *i.e.*, at (0, 1), (0, 2), (0, 3) curve intersects x-axis (put $y = 0$) ; $x = -6$ *i.e.*, $(-6, 0)$.

5. Asymptotes. Nil

6. Tangents.

$$\frac{dx}{dy} = (y-1)(y-2) + (y-2)(y-3) + (y-3)(y-1) = 3y^2 - 12y + 11$$

$$\frac{dx}{dy} = 0 \text{ when } y = \frac{12 \pm \sqrt{144 - 132}}{6} = \frac{12 \pm 2\sqrt{3}}{6} = 2 \pm \frac{1}{\sqrt{3}}$$

$$\therefore \quad \frac{dx}{dy} = 0 \text{ at } y = 2.6,\ 1.4$$

At $y = 2.6$; $x = -3.84$

and at $y = 1.4$; $x = 3.84$

$\therefore$ Tangents to the curve are parallel to y-axis at $(-3.84, 2.6)$ and $(3.84, 1.4)$

7. Additional Points. when $y < 0$ then $x < 0$

when y lies between 0 and 1; $x < 0$

when y lies between 1 and 2; $x > 0$

when y lies between 2 and 3; $x < 0$

when y is greater than 3 $x > 0$

when $x \to \infty$, $y \to \infty$

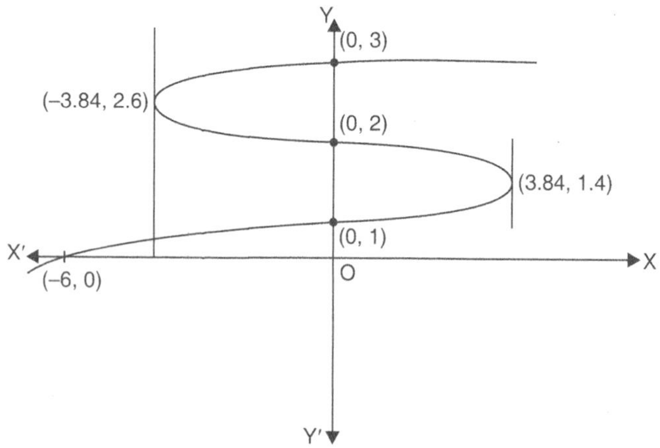

Example 13. *Trace the curve* $\left(\frac{x}{a}\right)^{2/3} + \left(\frac{y}{b}\right)^{2/3} = 1$. **(P.T.U., Jan 2010)**

Sol. Equation of the curve is

$$\left(\frac{x}{a}\right)^{2/3} + \left(\frac{y}{b}\right)^{2/3} = 1 \qquad \text{...(1)}$$

1. Symmetry. Curve is symmetrical about both x-axis , y-axis and origin.

2. Origin. Curve does not pass through origin.

3. Domain and Range.

$$\left(\frac{y}{b}\right)^{2/3} = 1 - \left(\frac{x}{a}\right)^{2/3}$$

or $$y = b\left\{1 - \left(\frac{x}{a}\right)^{2/3}\right\}^{3/2}$$

y exists if $\quad 1 - \left(\frac{x}{a}\right)^{2/3} > 0 \quad \text{or} \quad 1 > \left(\frac{x}{a}\right)^{2/3}$

or $$\left(\frac{x}{a}\right)^{2/3} < 1 \quad \text{or} \quad \left(\frac{x}{a}\right)^{2} < 1 \quad \text{or} \quad \left|\frac{x}{a}\right| < 1$$

or $$-1 < \frac{x}{a} < 1 \quad \text{or} \quad -a < x < a$$

$\therefore$ Domain of the curve $(-a, a)$ and similarly range of the curve $(-b, b)$.

4. Intersection with Co-ordinate Axes. Curve intersects x-axis (put $y = 0$)

$$\left(\frac{x}{a}\right)^{2/3} = 1 \quad \text{or} \quad \left(\frac{x}{a}\right)^{2} = 1 \quad \text{or} \quad x = \pm a$$

$\therefore$ Points of intersection with x-axis $(a, 0)$; $(-a, 0)$

Points of intersection with y-axis $(0, b)$; $(0, -b)$

5. Tangents. Slopes of the Tangents at $(\pm a, 0)$ and $(0, \pm b)$ are obtained from the value of $\frac{dy}{dx}$

$$\frac{2}{3}\left(\frac{x}{a}\right)^{-\frac{1}{3}}\left(\frac{1}{a}\right) + \frac{2}{3}\left(\frac{y}{b}\right)^{-\frac{1}{3}}\left(\frac{1}{b}\right)\frac{dy}{dx} = 0$$

$\therefore$ $$\frac{dy}{dx} = -\frac{\left(\frac{x}{a}\right)^{-1/3}\frac{1}{a}}{\left(\frac{y}{b}\right)^{-1/3}\left(\frac{1}{b}\right)} = -\frac{y^{1/3}b^{2/3}}{x^{1/3}a^{2/3}}$$

$$\left(\frac{dy}{dx}\right)_{(a,0)} = 0 \,;\, \left(\frac{dy}{dx}\right)_{(-a,0)} = 0$$

$\therefore$ At $(a, 0)$, $(-a, 0)$ tangents are parallel to x-axis

$$\left(\frac{dy}{dx}\right)_{(0,b)} = -\infty, \left(\frac{dy}{dx}\right)_{(0,-b)} = -\infty$$

$\therefore$ Tangents are parallel to y-axis at $(0, b)$ and $(0, -b)$.

6. Asymptote. Nil.

7. Maxima, Minima. Curve has extreme value where

$$\frac{dy}{dx} = 0 \quad i.e., \quad y = 0 \quad \therefore \quad x = \pm a$$

extreme values at $(\pm a, 0)$

$$\frac{d^2y}{dx^2} = -\left(\frac{b}{a}\right)^{2/3} \frac{x^{1/3} \cdot \frac{1}{3} y^{-2/3} \frac{d}{dx} y - y^{1/3} \frac{1}{3} x^{-2/3}}{x^{2/3}}$$

$$= -\left(\frac{b}{a}\right)^{2/3} \frac{1}{3x^{2/3}} \left\{ -x^{1/3} y^{-2/3} \frac{y^{1/3}}{x^{1/3}} \left(\frac{b}{a}\right)^{2/3} - y^{1/3} x^{-2/3} \right\}$$

$$= +\left(\frac{b}{a}\right)^{2/3} \frac{1}{3x^{2/3}} \cdot b^{2/3} x^{-2/3} y^{-1/3} \left[\frac{x^{2/3}}{a^{2/3}} + \frac{y^{2/3}}{b^{2/3}} \right]$$

$$= \frac{1}{3}\left(\frac{b^2}{a}\right)^{2/3} \cdot \frac{1}{x^{4/3} y^{1/3}} \qquad | \because \text{ of } (1)$$

$$= \frac{1}{3}\left[\frac{b^2}{ax^2}\right]^{2/3} \cdot \frac{1}{y^{1/3}}$$

$$\frac{d^2y}{dx^2} > 0 \text{ if } y > 0$$

And $$\frac{d^2y}{dx^2} < 0 \text{ if } y < 0$$

$\therefore$ Curve has maximum values at $(\pm a, 0)$ for $y < 0$ and curve has minimum value at $(\pm a, 0)$ for $y > 0$

Sketch of the curve is shown in the figure.

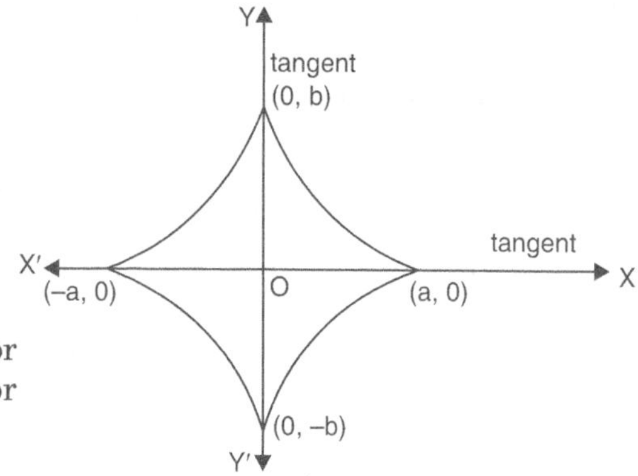

Example 14. *Trace the curve* $y^2 = \dfrac{x-3}{x^2-6x-7}$.

Sol. Equation of the curve is

$$y^2 = \frac{x-3}{x^2-6x-7} = \frac{x-3}{(x+1)(x-7)}$$

1. Symmetry. Curve is symmetric about x-axis (changing y to $-y$ does not change the equation)

2. Origin. Curve does not pass through the origin

3. Domain and Range. y is not defined at $x = -1$, $x = 7$ also when $x > 3$, $x < 7$

Curve is not defined in (3, 7) also when $x < -1$, y is not defined

$\therefore$ Curve is not defined in $(-\infty, -1) \cup (3, 7)$

For range; solve the equation for x

$$x^2y^2 - 6xy^2 - 7y^2 = x - 3$$

or $$(y^2)x^2 - (6y^2 + 1)x + (-7y^2 + 3) = 0$$

$$x = \frac{6y^2 + 1 \pm \sqrt{(6y^2+1)^2 - 4y^2(-7y^2+3)}}{2y^2}$$

x exists when

$$(6y^2 + 1)^2 - 4y^2(-7y^2 + 3) \geq 0$$

$$36y^4 + 12y^2 + 1 + 28y^4 - 12y^2 \geq 0$$

$$64y^4 + 1 \geq 0$$

$\therefore$ x exists for all values of y $\quad\therefore$ Range = R

4. Intersection with Co-ordinate Axes.

Curve intersects x-axis where $y = 0$ *i.e.*, at $x = 3$ *i.e.*, at (3, 0)

and y-axis (put $x = 0$) $\qquad y^2 = \dfrac{-3}{-7} = \dfrac{3}{7}$

$\therefore$ At $\left(0, \sqrt{\dfrac{3}{7}}\right)$ and $\left(0, -\sqrt{\dfrac{3}{7}}\right)$

i.e., At (0, 0.65) and (0, –0.65)

5. Tangents. Slopes of the tangent at these points are obtained by evaluating $\dfrac{dy}{dx}$ at these points

$$2y\,\frac{dy}{dx} = \frac{(x^2 - 6x - 7)(1) - (x - 3)(2x - 6)}{(x^2 - 6x - 7)^2}$$

$$= \frac{x^2 - 6x - 7 - 2x^2 + 12x - 18}{(x^2 - 6x - 7)^2} = \frac{-x^2 + 6x - 25}{(x^2 - 6x - 7)^2}$$

$$\frac{dy}{dx} = \frac{\pm 1}{2 \cdot \dfrac{\sqrt{x-3}}{\sqrt{x^2 - 6x - 7}}} \times \frac{-(x^2 - 6x + 25)}{(x^2 - 6x - 7)^2}$$

$$\frac{dy}{dx} = \pm \frac{x^2 - 6x + 25}{2\sqrt{x-3}\,(x^2 - 6x - 7)^{3/2}}$$

$$\left(\frac{dy}{dx}\right)_{(3,0)} = \infty \quad \therefore \text{ At (3, 0) tangent is parallel to } y\text{-axis.}$$

At $\left(0, \pm\sqrt{\dfrac{3}{7}}\right)$, $\qquad \dfrac{dy}{dx} = \pm \dfrac{25}{2\sqrt{-3}(-7)^{3/2}}$ imaginary

$\therefore$ No tangent at $\left(0, \pm\sqrt{\dfrac{3}{7}}\right)$

6. Asymptotes. Equation of the curve is $(x^2 - 6x - 7)\,y^2 = x - 3$

Asymptotes || to x-axis are $y^2 = 0$ *i.e.*, $y = 0$, *i.e.*, x-axis itself.

[Equating coefficient of highest power of x to zero]

Asymptotes || to y-axis are

$$x^2 - 6x - 2 = 0 \quad i.e., \quad x = -1, x = 7$$

No oblique asymptotes.

7. Maxima, Minima. Curve attains the extreme value at $\frac{dy}{dx} = 0$

i.e., at $$x^2 - 6x + 25 = 0$$

$$x = \frac{6 \pm \sqrt{36 - 100}}{2} = 3 \pm i4 \quad \text{(imaginary)}$$

no extreme point.

8. Additional Points. When $x > 7$ or $-1 < x < 3$ we get real values of y

x	7	8	9	10
y	∞	$\frac{5}{9}$	$\frac{3}{10}$	$\frac{7}{33}$

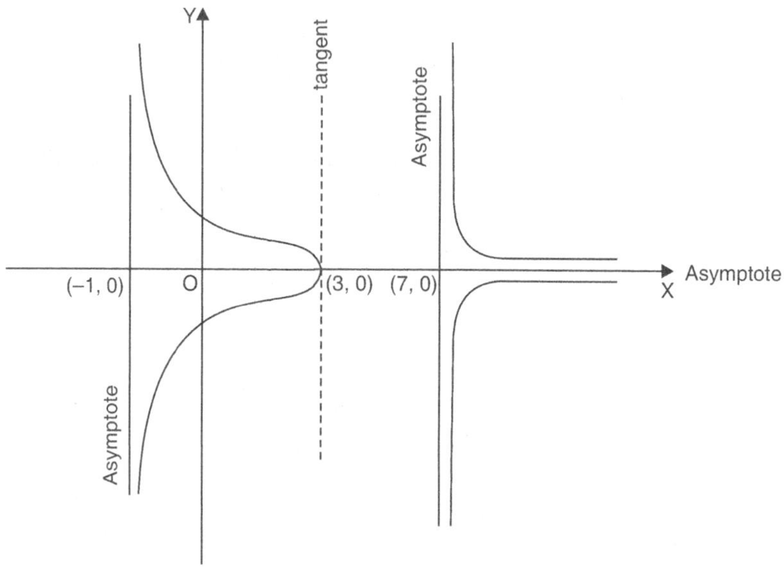

Example 15. *Trace the curve* $y = c \cosh \frac{x}{c}$. **(P.T.U., Dec. 2003)**

Sol. 1. Symmetry. The curve is symmetric about y-axis

[$\because$ Changing of x to $-x$ does not change the equation]

2. Origin. Curve does not pass through the origin.

3. Domain and Range. y is defined for all real values of x is $D_f = R$. But y always has +ve values $\forall\ x$ $\therefore$ Range is the set of +ve values of y only $\therefore$ curve lies in the upper half plane.

4. Intersection with Axes. Curve intersects x-axis (put $y = 0$) *i.e.*, $c \cosh \frac{x}{c} = 0$ but $c \neq 0$ $\therefore$ $\cosh \frac{x}{c} = 0$ *i.e.*, $\frac{e^{x/c} + e^{-x/c}}{2} = 0$ *i.e.*, $e^{\frac{2x}{c}} = -1$ which is impossible $\therefore$ curve does not intersect x-axis.

5. Tangents. $\frac{dy}{dx} = \sinh \frac{x}{c}$

Tangent is parallel to x-axis when $\frac{dy}{dx} = 0$ *i.e.*, $\sinh \frac{x}{c} = 0$ *i.e.*, $x = 0$ *i.e.*, at $(0, c)$; $\frac{dy}{dx} = \infty$ at no real point. $\therefore$ no where tangent is parallel to y-axis.

6. Asymptotes. Curve does not have any asymptote.

7. Maxima, Minima. The curve has maximum or minimum values where ; $\frac{dy}{dx} = 0$ *i.e.*, at $x = 0$ *i.e.*, at $(0, c)$ and at $(0, c)$ $\frac{d^2y}{dx^2} = \frac{1}{c} \cosh \frac{x}{c} > 0$

$\therefore$ curve has minimum value at $(0, c)$

Now sketch of the curve is given in the figure:

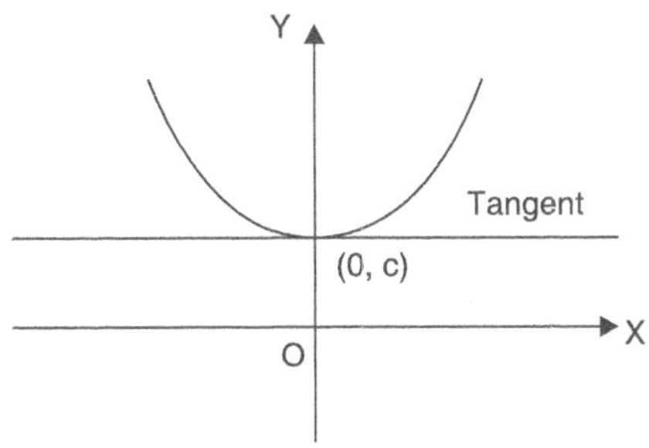

8. Additional Points. We see that when $x \to \infty$ y also tends to $+\infty$ and when $x \to -\infty$ again y tends to $+\infty$.

TEST YOUR KNOWLEDGE

Trace the following curves from 1 to 19:

1. $y^2 = x(x - a)^2$; $a > 0$

2. $(a^2 + x^2)\, y^2 = x^2(a^2 - x^2)$

3. $y^2 = (x - 2)(x - 5)^2$

4. $ay^2 = x^2(a - x)$

5. $y = \frac{x^2}{1 + x^2}$

6. $y = \frac{x}{x^2 - 1}$

7. $y = \frac{x^2}{x^2 - 1}$

8. $y = x^3 - 3x^2 + 3$

9. $y = x(x + 1)(x - 2)$ or $y = x^3 - x^2 - 2x$ **(*P.T.U., May 2006*)**

10. $y^2(a + x) = x^2(a - x)$ or $x\,(x^2 + y^2) = a\,(x^2 - y^2)$
[**Hint.** Similar to solved examples 3]

11. $y^2(2a - x) = x^3$

12. $xy^2 = 4a^2(2a - x)$

13. $y^2 = x^3$ **(*P.T.U., Jan. 2009*)**

14. $y^2 = (x + 1)^3$

15. $y^2(x^2 - 9) = -9x^2$

16. $a^2\, x^2 = y^3(2a - y)$

17. $x^3 + y^3 = ax^2$

18. $x^4 - y^4 + xy = 0$

19. $y = \frac{(x - 1)(x - 3)}{x^2}$. **(*P.T.U., Dec. 2007*)**

Answers

1.

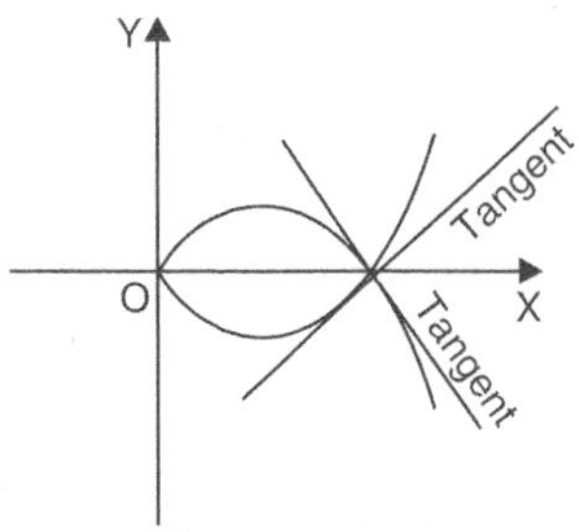

2.

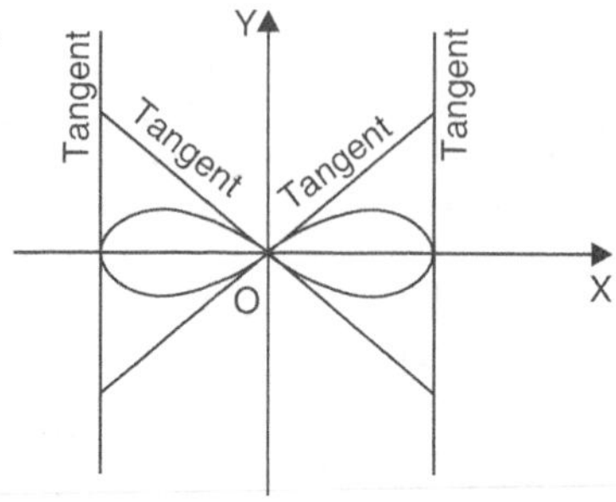

3.

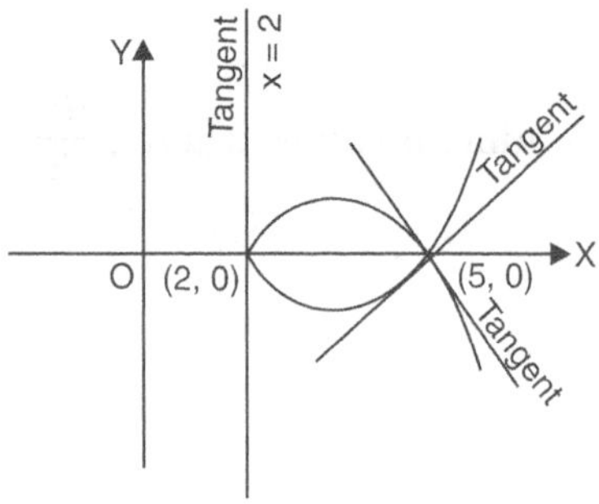

4.

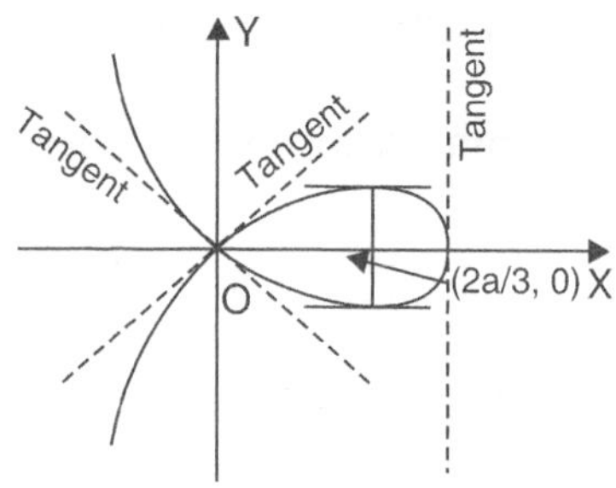

5.

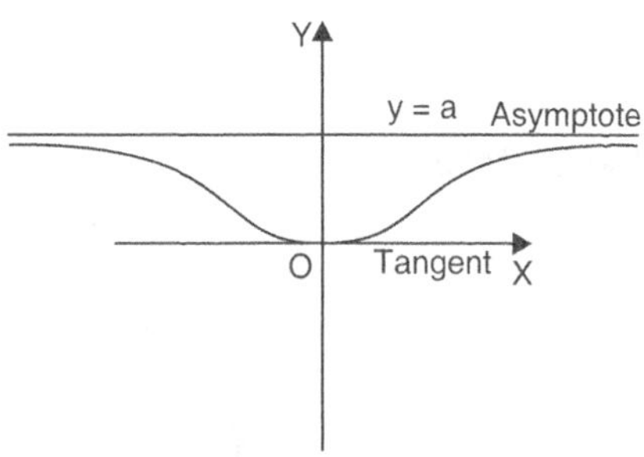

6.

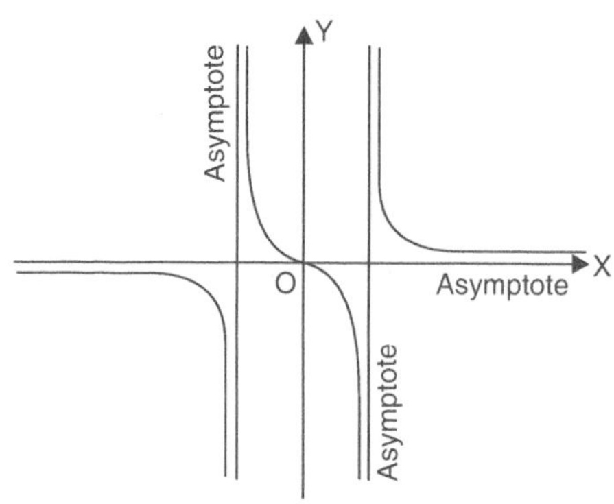

7.

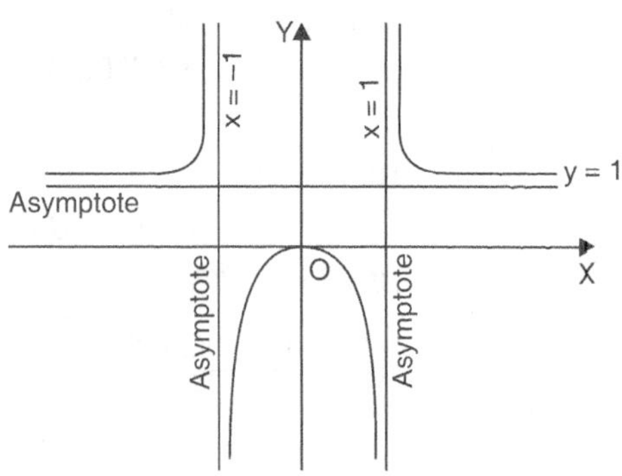

8.

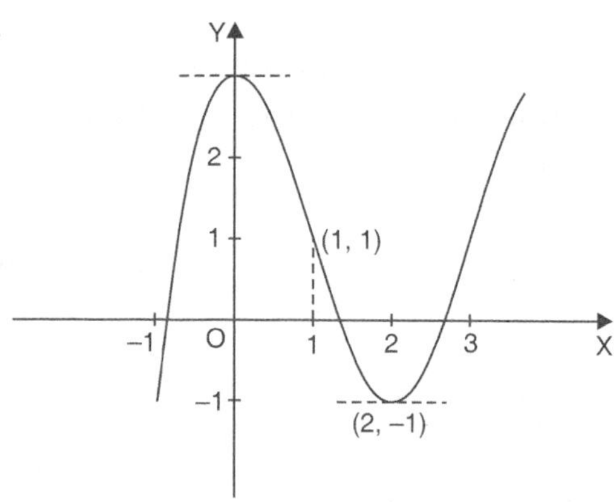

9.

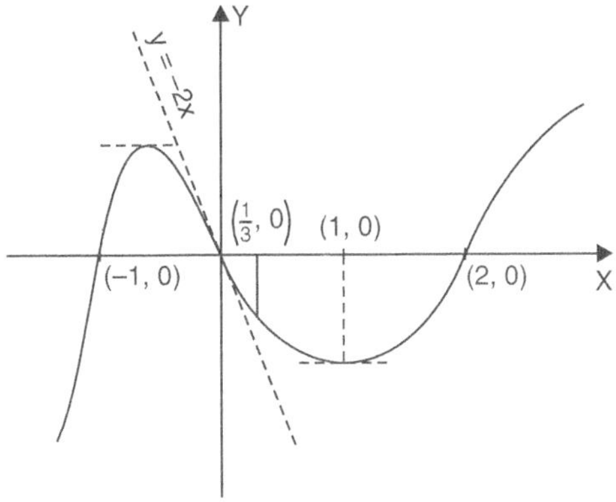

10.

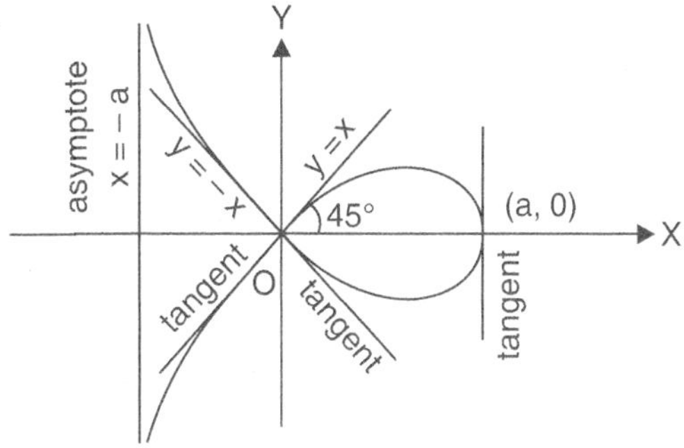

11.

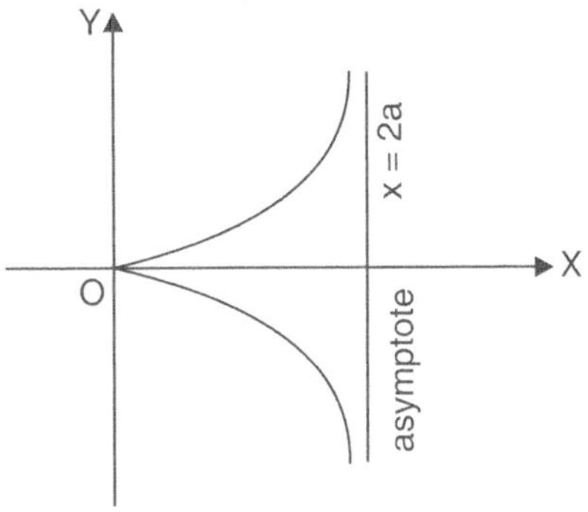

12.

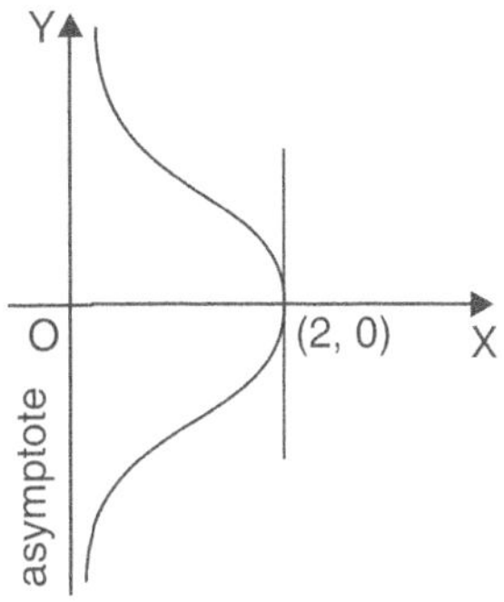

13.

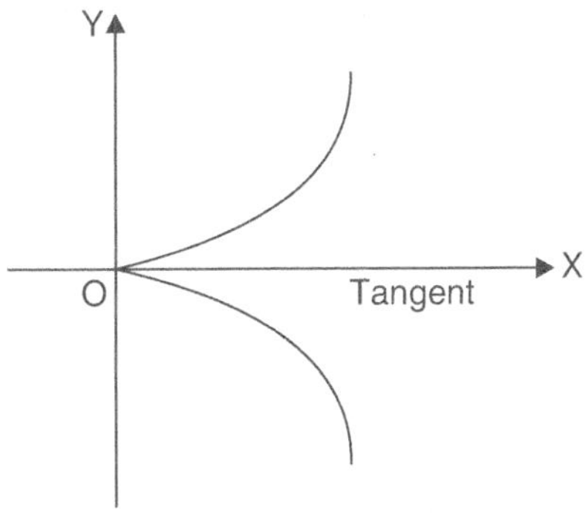

14.

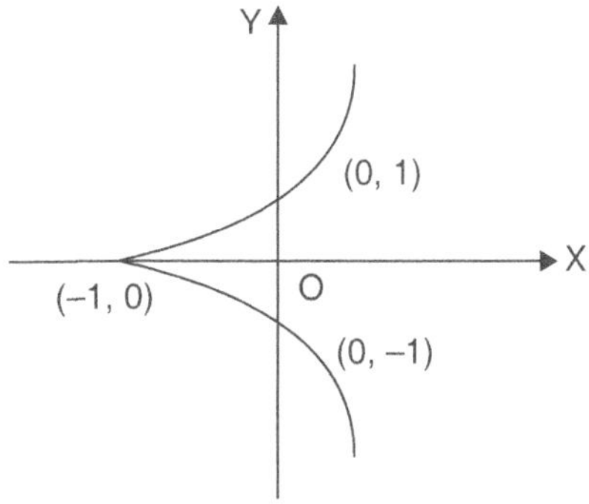

15.

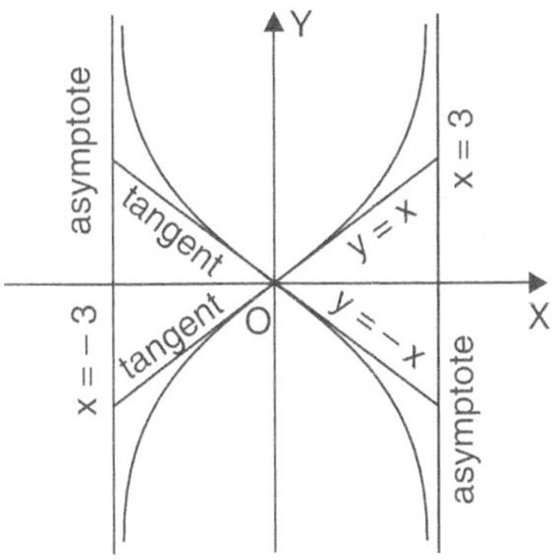

16.

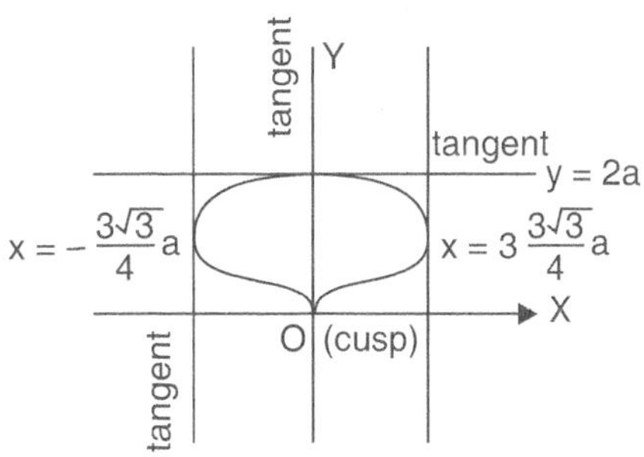

17.

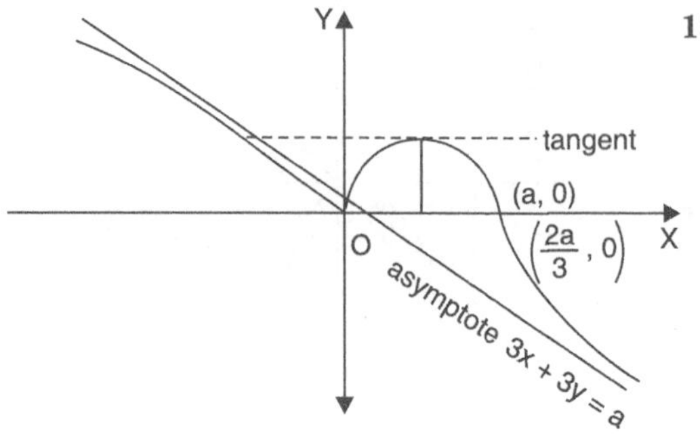

18.

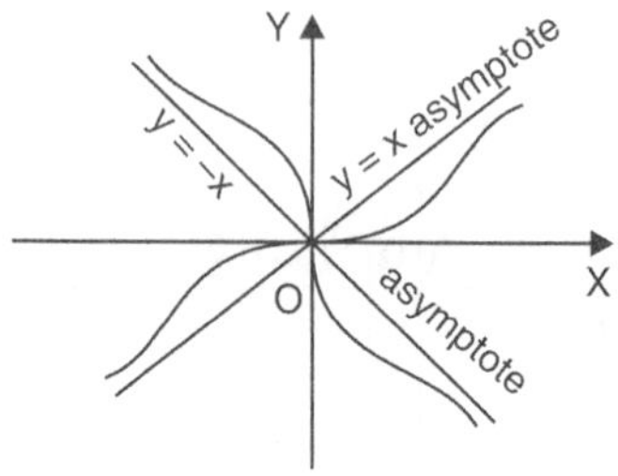

19.

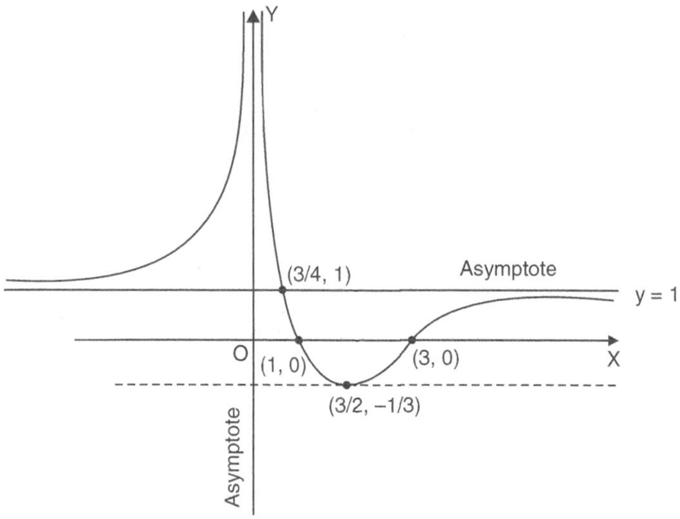

1.14. PARAMETRIC CURVES

If $x = f(t)$ and $y = g(t)$ are two continuous function of t, then the set of points $(x, y) = \{f(t), g(t)\}$ describes a curve in co-ordinate plane. Then $x = f(t), y = g(t)$ are known as parametric equations for the curve and t is known as the parameter *e.g.*, $x = a \cos t$, $y = a \sin t$ is the parametric equation of the circle $x^2 + y^2 = a^2$.

1.15. WORKING RULE TO TRACE A PARAMETRIC CURVE

Note. [If the parametric equation can easily be changed to cartesian by eliminating the parameter, then prefer to change it and trace the curve as done earlier].

In case elimination of parameter is not easy then proceed as follows:

Let $x = f(t)$ and $y = g(t)$ be the parameter equation of the curve

(1) Symmetry.

(*a*) If $x = f(t)$ is an even function of t and $y = g(t)$ is an odd function of t then curve is **symmetric about *x*-axis.**

(b) If $x = f(t)$ is an odd function of t and $y = g(t)$ is an even function of t then curve is **symmetric about y-axis.**

(c) If $x = f(t)$ and $y = g(t)$ are both odd functions of t, then the curve is symmetric in opposite quadrants.

2. Origin. If we get a real value of t when we put $x = 0$ and y becomes zero for that value of t then curve passes through the origin.

3. Intersection with Co-ordinate Axes. Find intersection of the curve with x-axis by putting $y = 0$ and with y-axis by putting $x = 0$.

4. Tangents. Find the points where tangents to the curve are parallel to co-ordinate axes *i.e.*, where

$$\frac{dy}{dx} = 0 \quad \text{and} \quad \frac{dy}{dx} \to \infty$$

5. Region.

(a) Find the regions in which curve does not lie.

(b) Consider the signs of $\frac{dx}{dt}, \frac{dy}{dt}$.

(c) Consider the values of $x, y, \frac{dx}{dt}, \frac{dy}{dt}, \frac{dy}{dx}$.

6. Asymptotes. Find asymptotes ; if any.

Method to find Asymptotes of the parametric curves

Let the equation of the parametric curve be $x = f(t), y = g(t)$

(a) If $x \to \infty, y \to y_1$ when $t \to t_1$ (say) then $y = y_1$ is an asymptote parallel to x-axis.

(b) If $y \to \infty, x \to x_1$ when $t \to t_2$ (say) then $x = x_1$ is an asymptote parallel to y-axis.

(c) If $x \to \infty, y \to \infty$ when $t \to t_3$ (say) then slope of the oblique asymptote $y = mx + c$ is given by $\underset{t \to t_3}{\text{Lt}} \frac{y}{x} = m$ and $c = \underset{t \to t_3}{\text{Lt}} (y - mx) = \underset{t \to t_3}{\text{Lt}} [g(t) - mf(t)]$

ILLUSTRATIVE EXAMPLES

Example 1. *Trace the curve $x = a(\theta - \sin\theta)$, $y = a(1 - \cos\theta)$.*

Sol. Parametric equation of the curve

$$x = a(\theta - \sin\theta); \quad y = a(1 - \cos\theta) \qquad ...(1)$$

1. Symmetry. x is an odd function of θ and y is an even function of θ

$\therefore$ Curve is symmetric about y-axis.

2. Origin. $x = 0 \Rightarrow \theta - \sin\theta = 0 \Rightarrow \theta = 0$

$\therefore$ $y = a(1 - 1) = 0$

$\therefore$ Curve passes through (0, 0)

3. Intersection with Co-ordinate Axes. Curve intersects x-axis (put $y = 0$)

$$a(1 - \cos\theta) = 0 \quad \therefore \quad \cos\theta = 1 \quad \therefore \quad \theta = 0, \quad \theta = 2\pi$$

Put $\theta = 0, 2\pi$ in (1), we get $x = 0, \quad x = 2\pi a$

$\therefore$ Curve intersects x-axis at (0, 0), $(2a\pi, 0)$

Curve intersects y-axis where $x = 0$

$\therefore$ $\theta - \sin\theta = 0 \quad \therefore \quad \theta = 0$

Put $\theta = 0$ in (1), we get $x = 0, y = 0$

$\therefore$ Curve intersects y-axis at (0, 0)

4. Tangents. $\dfrac{dx}{d\theta} = a(1 - \cos\theta)$

$$\frac{dy}{d\theta} = a \sin\theta$$

$$\therefore \quad \frac{dy}{dx} = \frac{a \sin\theta}{a(1-\cos\theta)} = \frac{\sin\theta}{1-\cos\theta}$$

$$\frac{dy}{dx} = \frac{2 \sin\theta/2 \cos\theta/2}{2\sin^2\theta/2} = \cot\frac{\theta}{2}$$

$$\frac{dy}{dx} = 0 \quad \text{where } \cot\frac{\theta}{2} = 0 \quad \text{or} \quad \tan\frac{\theta}{2} = \infty \ \therefore\ \theta = \pi$$

$\therefore$ Tangent to the curve is parallel to x-axis at $(a\pi, 2a)$

$$\frac{dy}{dx} = \infty \quad \text{when } \cot\frac{\theta}{2} = \infty \ \therefore\ \tan\frac{\theta}{2} = 0 = \tan 0, \tan\pi, \ \therefore\ \theta = 0, \theta = 2\pi$$

$\therefore$ Tangent to the curve is parallel to y-axis at $(0, 0)$, $(2a\pi, 0)$.

5. Region. $y = a \,.\, 2\sin^2\dfrac{\theta}{2}$

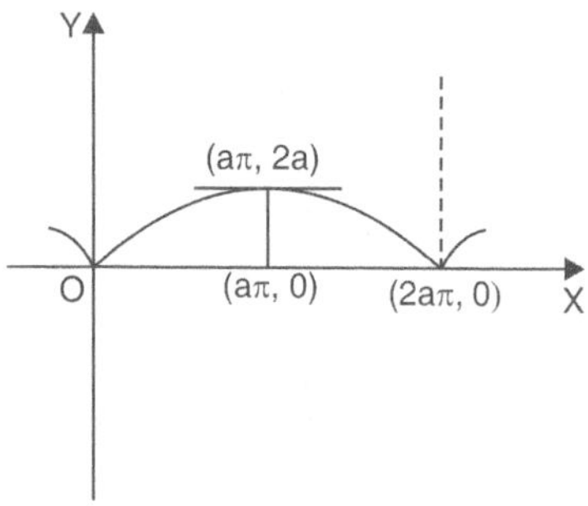

We know that $0 \le \sin^2\dfrac{\theta}{2} \le 1$

$$\therefore \quad 0 \le \frac{y}{2a} \le 1 \quad \text{or} \quad 0 \le y \le 2a$$

$\therefore$ Curve lies between $y = 0$ and $y = 2a$

As x increases from 0 to $a\pi$, y also increases from 0 to $2a$ and as x increases from $a\pi$ to $2a\pi$, y decreases from $2a$ to 0.

6. Asymptotes. Curve has no asymptotes

Sketch of the curve is shown in the adjoining figure.

Example 2. *Trace the curve*

$$x = a(\theta + \sin\theta),\ y = a(1 + \cos\theta) \quad -\pi \le \theta \le \pi.$$

Sol. 1. Symmetry. x is an odd function of θ and y is an even function of θ.

$\therefore$ Curve is symmetric about y-axis.

2. Origin. Curve does not pass through the origin.

3. Intersection with Co-ordinate Axes. Curve intersects x-axis (put $y = 0$)

$$a(1 + \cos\theta) = 0 \quad \therefore \quad \cos\theta = -1 = \cos\pi, \cos(-\pi)$$

$$\therefore \quad \theta = \pi, -\pi \quad i.e., \quad \text{at } (a\pi, 0) \text{ and } (-a\pi, 0)$$

Curve intersects y-axis (put $x = 0$)

$$0 = a(\theta + \sin\theta) \ \Rightarrow\ \theta = 0$$

$\therefore$ Curve intersects y-axis at $(0, 2a)$.

4. Tangents. $\dfrac{dx}{d\theta} = a(1 + \cos\theta)$, $\dfrac{dy}{d\theta} = -a\sin\theta$.

$$\frac{dy}{dx} = \frac{-a \sin \theta}{a(1+\cos \theta)} = -\frac{2 \sin \frac{\theta}{2} \cos \frac{\theta}{2}}{2 \cos^2 \frac{\theta}{2}} = -\tan \frac{\theta}{2}$$

$$\frac{dy}{dx} = 0 \quad \Rightarrow \quad \tan \frac{\theta}{2} = 0 \quad \Rightarrow \quad \frac{\theta}{2} = 0, \pi, -\pi$$

$\therefore$ $\theta = 0, 2\pi, -2\pi \quad \therefore \quad \theta = 0$ only $\because \quad 2\pi, -2\pi \notin -\pi < \theta < \pi$

$\therefore$ $x = 0, y = 2a$

$\therefore$ Tangent is parallel to x-axis at $(0, 2a)$

$$\frac{dy}{dx} = \infty \quad \text{When} \quad -\tan \frac{\theta}{2} = \tan \frac{\pi}{2} \text{ or } \tan \frac{-\pi}{2}$$

$$\theta = -\pi, +\pi$$

$\therefore$ When $\theta = -\pi$

$\therefore$ $x = -a\pi, y = 0$ *i.e.*, $(-a\pi, 0)$

When $\theta = \pi, x = a\pi, y = 0$ *i.e.*, $(a\pi, 0)$

$\therefore$ Tangent is parallel to y-axis at $(a\pi, 0), (-a\pi, 0)$

5. Asymptotes. No asymptote

6. Region. $\dfrac{dx}{d\theta} = a(1 + \cos \theta), \quad \dfrac{dy}{d\theta} = -a \sin \theta$

$\dfrac{dx}{d\theta}$ is always positive for all values of θ

Now when θ lies between $-\pi$ to 0

Then $\dfrac{dy}{d\theta}$ is +ve $\therefore$ y increases between $(-\pi, 0)$

When θ lies between 0 to π

$\dfrac{dy}{d\theta}$ is negative $\therefore$ y decreases in $(0, \pi)$

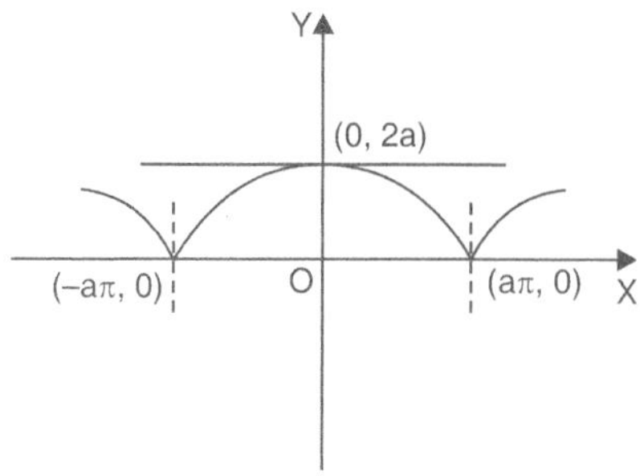

Sketch of the curve is shown in the adjoining figure.

Example 3. *Trace the curve*

$$x = a \cos t + \frac{1}{2} a \log \tan^2 \frac{t}{2}$$

$$y = a \sin t \quad a > 0.$$

Sol. 1. Symmetry. The curve is symmetric about x-axis $\because$ on changing t to $-t$; x remains unchanged and y changes to $-y$.

2. Origin. For no real value of t, x and y become zero $\therefore$ curve does not pass through the origin.

3. Intersection with Axes. Curve does not meet x-axis $\because$ $y = 0 \Rightarrow t = 0 \Rightarrow x = \infty$

The curve meets y-axis at $(0, \pm a)$ $\because$ when $x = 0 \Rightarrow t = \pm\frac{\pi}{2} \Rightarrow y = \pm a$.

4. Region. As $-1 \le \sin t \le 1 \quad \therefore \quad -a \le a \sin t \le a \quad \therefore \quad -a \le y \le a$

$\therefore$ Entire curve lies within $y = -a$ and $y = a$.

5. Asymptotes. $\underset{t \to 0}{\text{Lt}}\ x = -\infty$ and $\underset{t \to 0}{\text{Lt}}\ y = 0$

$\therefore \quad y = 0$ *i.e.*, x-axis is an asymptote of the curve

6. Tangent. $\dfrac{dx}{dt} = -a \sin t + \dfrac{a}{2} \cdot \dfrac{1}{\tan^2 \frac{t}{2}} \cdot 2 \tan \dfrac{t}{2} \sec^2 \dfrac{t}{2} \cdot \dfrac{1}{2}$

$$= -a \sin t + \frac{a}{2} \frac{\sec^2 t/2}{\tan t/2} = a\left[-\sin t + \frac{1}{2} \frac{1}{\sin \frac{t}{2} \cos \frac{t}{2}}\right]$$

$$= a\left[-\sin t + \frac{1}{\sin t}\right] = a\,\frac{\cos^2 t}{\sin t}$$

$$\frac{dy}{dt} = a \cos t \quad \therefore \quad \frac{dy}{dx} = \frac{a \cos t}{a \cos^2 t} \sin t = \tan t$$

$$\frac{dy}{dx} = 0 \;\Rightarrow\; \tan t = 0 \;\Rightarrow\; t = 0 \;\Rightarrow\; x \to -\infty \text{ and } y = 0$$

$\Rightarrow \quad y = 0$ or x-axis is an asymptote.

Also $\dfrac{dy}{dx} = \infty \;\Rightarrow\; t = \pm\dfrac{\pi}{2} \;\Rightarrow\; x = 0, y = \pm a$

$\therefore$ at $(0, \pm a)$ tangents are parallel y-axis.

7. Additional points

t	$-\pi$	$-\frac{\pi}{2}$	0	$\frac{\pi}{2}$	π
x	∞	0	$-\infty$	0	∞
y	0	$-a$	0	a	0
$\frac{dy}{dx}$	0	$-\infty$	0	∞	0

8. Intervals of Increase and Decrease. In the interval $0 < t < \dfrac{\pi}{2}$; $\dfrac{dx}{dt} > 0$

$\therefore$ x increases

$\dfrac{dy}{dt} > 0 \qquad \therefore$ y increases

In the interval $\dfrac{\pi}{2} < t < \pi \quad \dfrac{dx}{dt} > 0 \qquad \therefore$ x increases

$\dfrac{dy}{dt} < 0 \qquad \therefore$ y decreases

$\therefore \quad 0 < t < \dfrac{\pi}{2} \quad \dfrac{dy}{dx}$ is +ve $\qquad \therefore$ curve increases and in $\frac{\pi}{2} < t < \pi$ $\frac{dy}{dx}$ is negative $\therefore$ curve decreases.

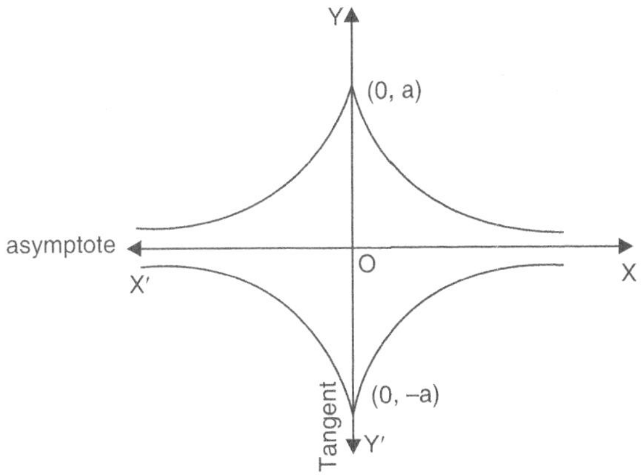

TEST YOUR KNOWLEDGE

Trace the following curves:

1. $x = a(t + \sin t), y = a(1 - \cos t)$ **2.** $x = a(t - \sin t), y = a\,(1 + \cos t)$

3. $x = a \cos^3 \theta \,;\, y = a \sin^3 \theta$

[**Hint.** Better change this equation to cartesian equation and see example (13) art. 1.13]

4. $x = a \sin^2 t,\ y = \dfrac{a \sin^3 t}{\cos t}$

[**Hint.** Eliminate t we get $y^2 = \dfrac{x^3}{a - x}$ same as example (5) art 1.13]

5. $x = \dfrac{3at}{1+t^3},\ y = \dfrac{3at^2}{1+t^2}$

[**Hint.** Eliminate t we get $x^3 + y^3 = 3axy$ same as example (1) art 1.13]

6. $x = \dfrac{1-t^2}{1+t^2},\ y = \dfrac{2t}{1+t^2}$

[**Hint.** Eliminate t ; $x^2 + y^2 = 1$; a circle with centre (0, 0) radius 1]

7. $x = a\ [\cos \theta - \log (1 + \cos \theta)]$

$y = a \sin \theta$

8. $x = a \sin 2\theta\,(1 + \cos 2\,\theta),\ y = a \cos 2\,\theta\,(1 - \cos 2\theta)$

9. $x = t^2,\ y = t - \dfrac{1}{3}t^3$

[**Hint.** Eliminate t we get $y^2 = t^2\left(1 - \dfrac{1}{3}t^2\right)^2$ *i.e.*, $9y^2 = x(3 - x)^2$

10. $x = \dfrac{a(t + t^3)}{1+t^4},\ y = \dfrac{a(t - t^3)}{1+t^4}\quad a > 0$

[**Hint.** Eliminate t ; $x^2 = \dfrac{a^2\, t^2(1+t^2)^2}{(1+t^4)^2},\ y^2 = \dfrac{a^2t^2(1-t^2)^2}{(1+t^4)^2};\ \dfrac{x}{y} = \dfrac{1+t^2}{1-t^2}\quad \therefore\ t^2 = \dfrac{x-y}{x+y}$ substituting in the value of x^2 or y^2, we get $(x^2 + y^2)^2 = a(x^2 - y^2)$]

Answers

1.

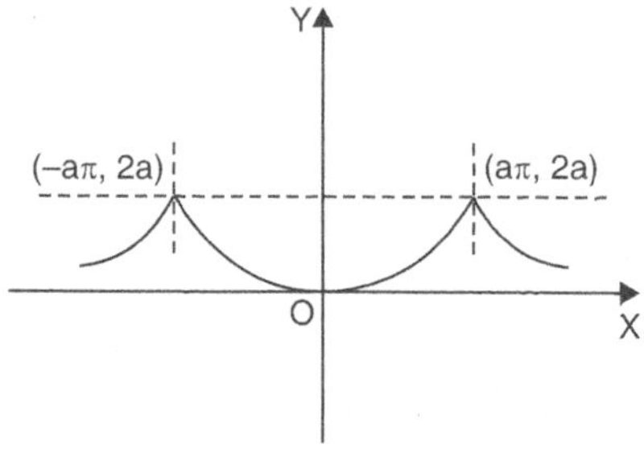

2.

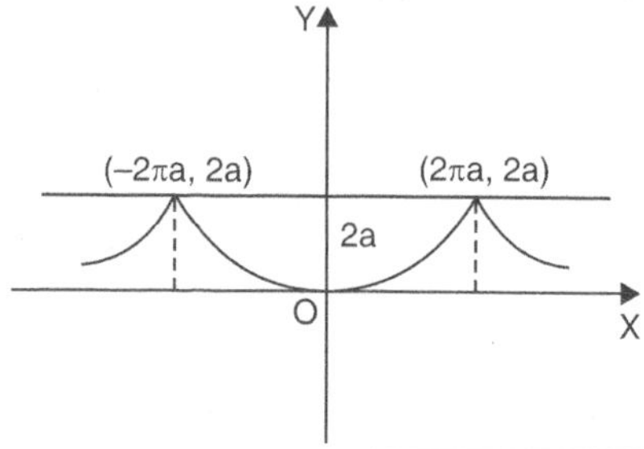

3.

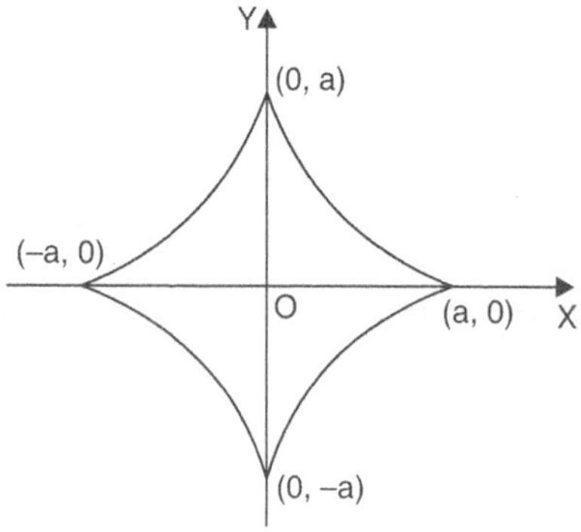

4.

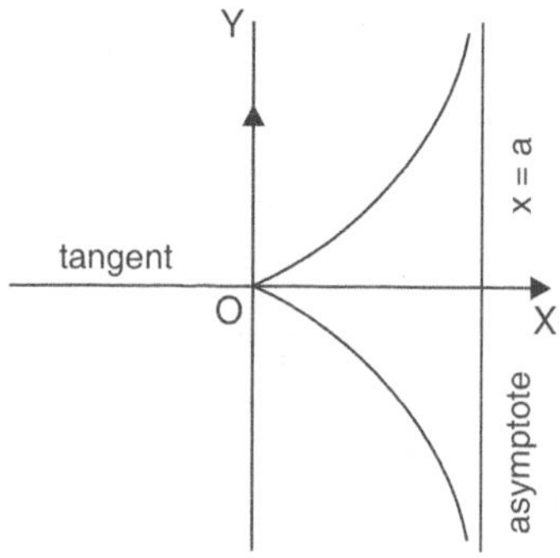

5.

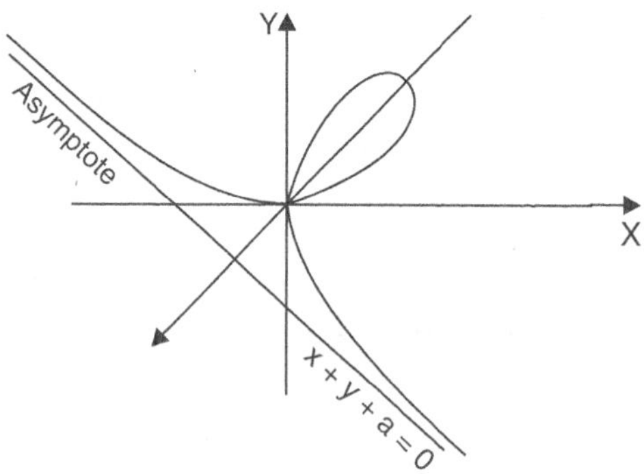

6.

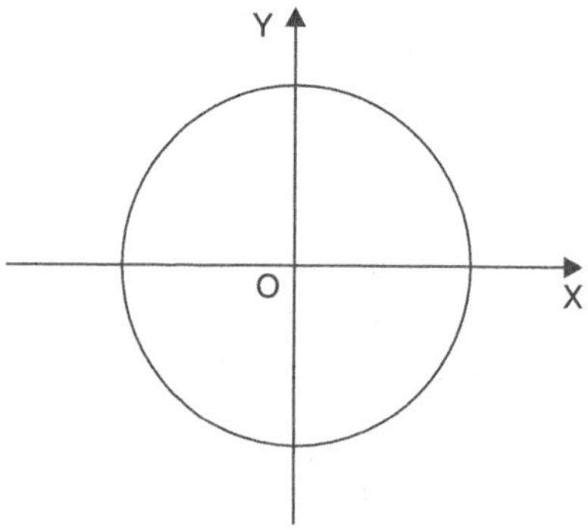

7.

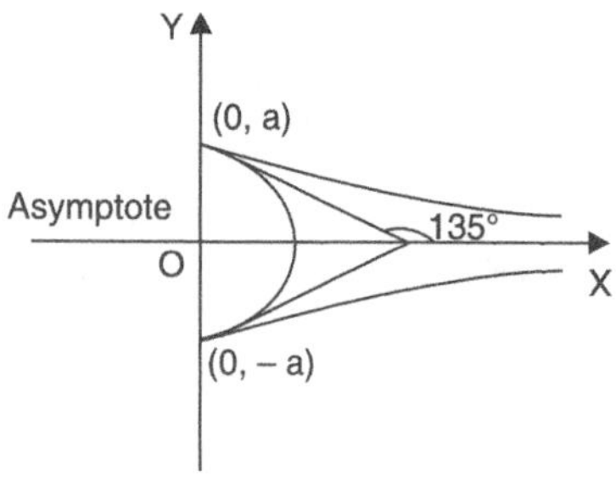

8.

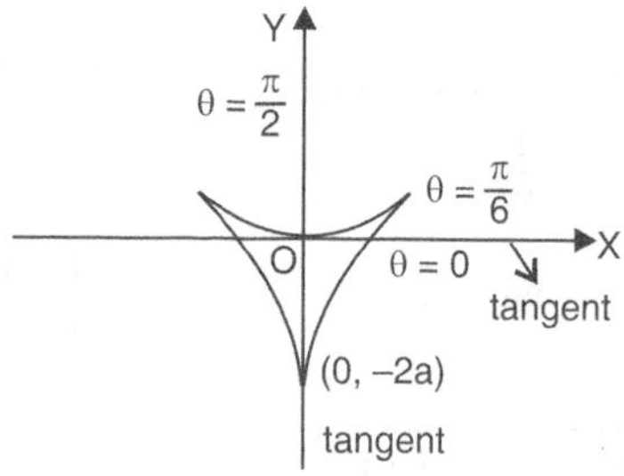

9.

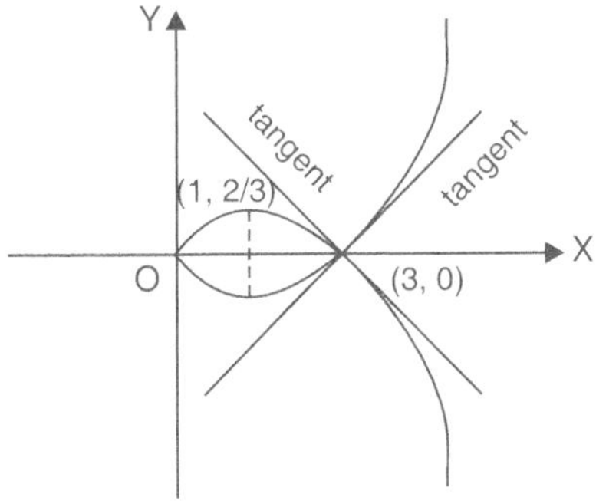

10.

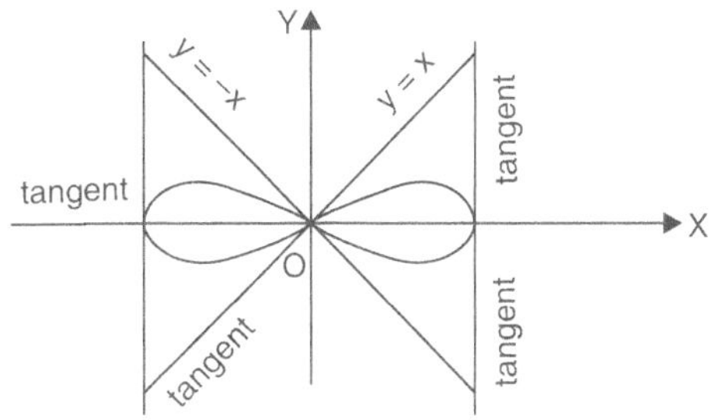

1.16. WORKING RULE TO TRACE A CURVE IN POLAR CO-ORDINATES

Salient features to trace a curve in polar co-ordinates are same as in the case of cartesian co-ordinates but first of all we will discuss how to apply these features when curve is given in polar co-ordinates.

In polar co-ordinates we have one point called the pole (generally taken as origin) and one line called the initial line (generally taken as x-axis).

Let the equation of the curve in polar co-ordinates be $f(r, \theta) = 0$

1. **Symmetry.** Curve is symmetric

(*a*) w.r.t. initial line (*i.e.*, x-axis) if equation remains unchanged when θ is changed to $-\theta$.

(*b*) w.r.t. line $\theta = \frac{\pi}{2}$ (*i.e.*, y-axis) if equation remains unchanged when θ is changed to $\pi - \theta$ or when θ is changed to $-\theta$ and r is changed to $-r$.

(*c*) w.r.t. line $\theta = \frac{\pi}{4}$ (*i.e.*, $y = x$) if equation remains unchanged when θ is changed to $\frac{\pi}{2} - \theta$.

(*d*) w.r.t. line $\theta = \frac{3\pi}{4}$ (*i.e.*, $y = -x$) if equation remains unchanged when θ is changed to $\frac{3\pi}{2} - \theta$.

(*e*) w.r.t. pole (origin) if equation remains unchanged when θ is changed to $\pi + \theta$ and when r is changed to $-r$.

2. **Pole or Origin.** Curve passes through the pole if $r = 0$ for some real values of θ and if $r = 0$ gives no real value of θ then curve does not pass through the pole.

3. **Tangent at the Pole (Origin).** The real value of θ for which $r = 0$ gives tangent at the pole.

4. **Points of Intersection with Initial Line and the Line $\theta = \frac{\pi}{2}$.** Find the points where the curve meets the initial line $\theta = 0$ (put $\theta = 0$ in curve get the value of r say r_1, then

3. Asymptote. The curve has no asymptote $\because$ r does not tend to ∞ for any real value of θ.

4. Region. As $-1 \le \sin\theta \le 1 \ \forall\, \theta$

$\therefore\ 0 \le 1 + \sin\theta \le 2 \quad \therefore\quad r \le 2a$

$\therefore$ given curve lies entirely within the circle $r = 2a$.

5. Points of Intersection. Curve intersects the initial line *i.e.*, $\theta = 0$ at $(a, 0)$ and the line $\theta = \dfrac{\pi}{2}$ at $\left(2a, \dfrac{\pi}{2}\right)$.

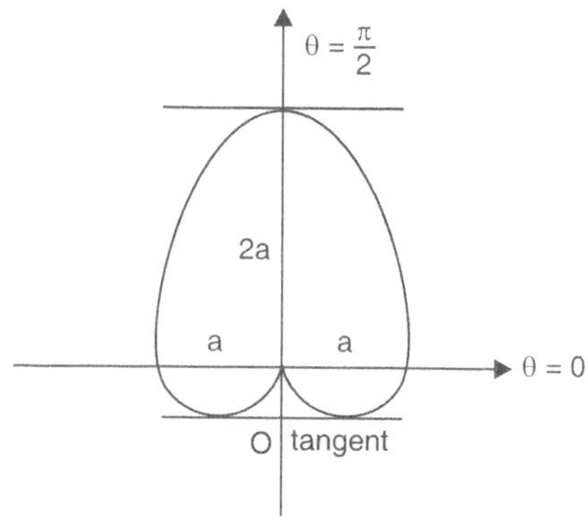

6. Additional Points. $\tan\phi = r\,\dfrac{d\theta}{dr}$

$$= a(1 + \sin\theta) \cdot \frac{1}{a\cos\theta}$$

$$\tan\phi = \frac{1+\sin\theta}{\cos\theta} = \frac{(\cos\theta/2 + \sin\theta/2)^2}{\cos^2\theta/2 - \sin^2\theta/2} = \frac{\cos\dfrac{\theta}{2} + \sin\dfrac{\theta}{2}}{\cos\dfrac{\theta}{2} - \sin\dfrac{\theta}{2}} = \frac{1+\tan\theta/2}{1-\tan\theta/2}$$

$$= \tan\left(\frac{\pi}{4} + \frac{\theta}{2}\right)$$

$$\therefore \qquad \phi = \frac{\pi}{4} + \frac{\theta}{2}$$

θ	0	$\frac{\pi}{6}$	$\frac{\pi}{3}$	$\frac{\pi}{2}$	π	2π
r	a	$1.5a$	$1.87a$	$2a$	a	a
ϕ	$\frac{\pi}{4}$	$\frac{\pi}{3}$	$\frac{5}{12}\pi$	$\frac{\pi}{2}$	$\frac{3\pi}{4}$	$\frac{5\pi}{4}$.

Example 3. *Trace the curve $r = a \sin 3\theta$, $a > 0$.* (Called three leaved rose)

Sol. 1. Symmetry. The curve is symmetrical about the line $\theta = \dfrac{\pi}{2}$

2. Pole (origin). $r = 0$ for $\sin 3\theta = 0 = \sin 0, \pi, 2\pi, 3\pi,$

i.e., when $\theta = 0, \dfrac{\pi}{3}, \dfrac{2\pi}{3}, \pi, \ldots$

$\therefore$ Curve passes through the pole and tangents at the pole are $\theta = 0$, $\theta = \dfrac{\pi}{3}$, $\theta = \dfrac{2\pi}{3}$, $\theta = \pi$.

3. Asymptote. Curve has no asymptote.

4. Region. We know that $-1 \le \sin 3\theta \le 1$ $\therefore -a \le r \le a$ $\therefore$ curve lies within the circle $r = a$.

5. Additional Points.

θ	0	$\frac{\pi}{6}$	$\frac{\pi}{3}$	$\frac{\pi}{2}$	$\frac{2\pi}{3}$	$\frac{5\pi}{6}$	π
r	0	a	0	$-a$	0	a	0

$$\tan\phi = r\frac{d\theta}{dr} = \frac{a\sin 3\theta}{3a\cos 3\theta}$$

$$\tan\phi = \frac{1}{3}\tan 3\theta$$

Plot all these points and trace the curve.

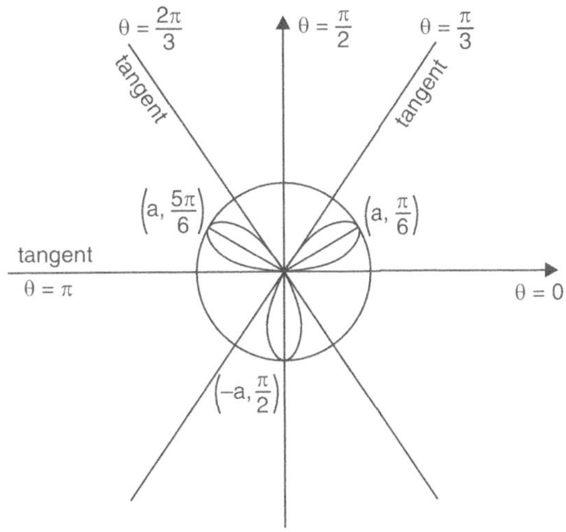

Example 4. *Trace the curve $r = a + b\cos\theta$ $(a > b)$.*

Sol. 1. Symmetry. The curve is symmetric about the initial line $\because$ on changing θ to $-\theta$ curve remains unchanged.

2. Pole or Origin. Put $r = 0$ we get $a + b\cos\theta = 0$

i.e., $\cos\theta = -\frac{a}{b}$ which is impossible

$\because$ $\frac{a}{b} > 1$ and $\cos\theta < -1$ $\therefore$ curve does not pass through the pole.

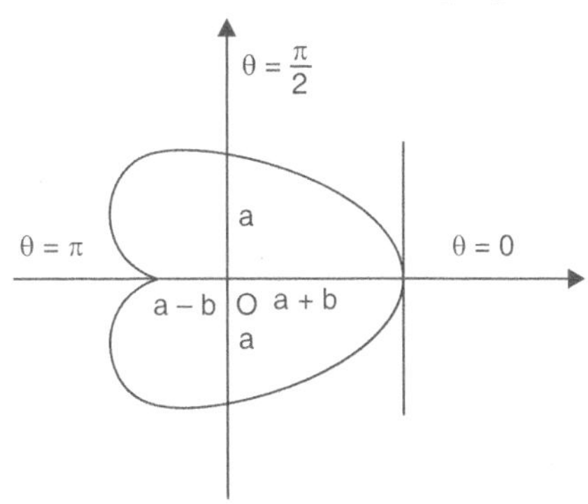

3. Asymptote. The given curve does not have any asymptote $\because$ r remains always finite.

4. Region. as $|\cos\theta| \le 1$ and $a > b$ $\therefore$ r is always positive.

5. Points of Intersection. Curve intersects initial line where $\theta = 0$ at $(a + b, 0)$ and line $\theta = \frac{\pi}{2}$ at $\left(a, \frac{\pi}{2}\right)$

Intersection of the curve with $\theta = \pi$ is $(a - b, \pi)$.

6. Additional Points. $\tan\phi = r\dfrac{d\theta}{dr} = (a + b\cos\theta)\dfrac{1}{-b\sin\theta} = -\dfrac{a + b\cos\theta}{b\sin\theta}$

θ	0	$\frac{\pi}{3}$	$\frac{\pi}{2}$	$\frac{2\pi}{3}$	π
r	$a+b$	$a+\frac{b}{2}$	a	$a-\frac{b}{2}$	$a-b$
ϕ	$-\frac{\pi}{2}$	—	$\pi - \tan^{-1}\frac{a}{b}$	—	$-\frac{\pi}{2}$

Example 5. *Trace the curve* $r^2\theta = a^2$.

Sol. 1. Symmetry. The curve is symmetric about the pole (r changed to $-r$ does not change the equation)

2. Pole. $r \neq 0$ for any real value of θ. $\therefore$ Curve does not pass through the pole.

3. Asymptotes. $r^2 = \dfrac{a^2}{\theta}$; $r \to \infty$ as $\theta \to 0$

$\therefore$ $\theta = 0$ may be tested for being asymptote

$$\frac{1}{r} = \pm\frac{\sqrt{\theta}}{a} = f(\theta) \text{ say}$$

$$f'(\theta) = \pm\frac{1}{a}\cdot\frac{1}{2\sqrt{\theta}} \Rightarrow f'(0) = \infty$$

$\therefore$ Asymptote is given by

$$r\sin(\theta - 0) = \frac{1}{f'(0)} \Rightarrow r\sin\theta = 0$$

$\Rightarrow$ $\theta = 0$ is the asymptote $\because$ $r \neq 0$

4. Region. Curve does not exist for –ve values of θ.

5. Additional Points.

θ	0	$\frac{\pi}{6}$	$\frac{\pi}{3}$	$\frac{\pi}{2}$	π
r	∞	$\pm a\sqrt{\frac{6}{\pi}}$	$\pm a\sqrt{\frac{3}{\pi}}$	$\pm a\sqrt{\frac{2}{\pi}}$	$\pm a\sqrt{\frac{1}{\pi}}$

Note that as θ increases from 0 to π, r decreases from ∞ to $\dfrac{a}{\sqrt{\pi}}$; r has equal and opposite values.

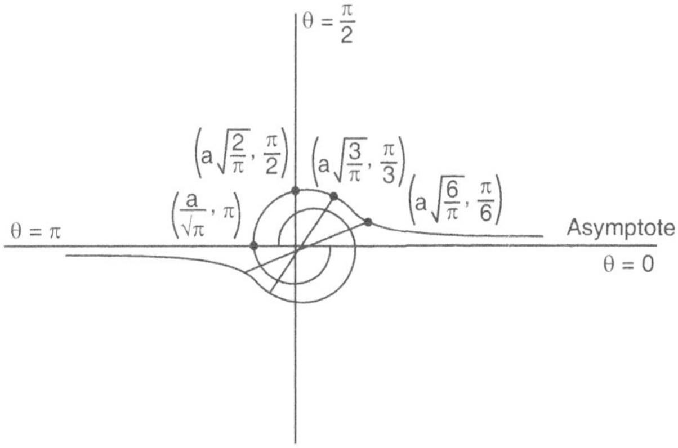

Example 6. *Trace the curve* $r^2 = a^2 \cos 2\theta$ *(known as lamniscate).* **(P.T.U., Dec. 2013)**

Sol. Equation of the curve is $r^2 = a^2 \cos 2\theta$ ...(1)

1. Symmetry. On changing θ to $-\theta$ and r to $-r$ curve remains unchanged $\therefore$ Curve is symmetric about initial line $\theta = 0$ and line $\theta = \dfrac{\pi}{2}$.

2. Pole or Origin. Put $r = 0$ in (1), we get $\cos 2\theta = 0$ *i.e.,* $\cos 2\theta = \cos \dfrac{\pi}{2} = \cos\left(2n\pi \pm \dfrac{\pi}{2}\right)$

$$\therefore \qquad \theta = n\pi \pm \frac{\pi}{4} \text{ for } n = 0, 1, 2, 3, \ldots$$

$$\therefore \quad \theta \text{ has the values} \qquad -\frac{\pi}{4}, \frac{\pi}{4}, \frac{3\pi}{4}, \frac{5\pi}{4}, \frac{7\pi}{4}, \ldots$$

$$\therefore \quad \text{Tangents at the pole are;} \quad \theta = \frac{\pi}{4}, \frac{3\pi}{4}, \frac{5\pi}{4}, \frac{7\pi}{4}, \ldots$$

3. Region. $\cos 2\theta$ is –ve in 2nd and 3rd quadrants.

i.e.,
$$\frac{\pi}{2} < 2\theta < \pi \quad \text{and} \quad \pi < 2\theta < \frac{3\pi}{2}$$

i.e.,
$$\frac{\pi}{4} < \theta < \frac{\pi}{2} \quad \text{and} \quad \frac{\pi}{2} < \theta < \frac{3\pi}{4}$$

$\therefore$ Curve does not lie in the portion between $\left(\dfrac{\pi}{4} \text{ and } \dfrac{\pi}{2}\right)$ and $\left(\dfrac{\pi}{2}, \dfrac{3\pi}{4}\right)$.

4. Points of Intersection. Curve intersects $\theta = 0$ at $r^2 = a^2$ *i.e.,* $r = \pm a$ *i.e.,* at $(-a, 0)$ and $(a, 0)$.

5. Asymptotes. Since r is finite $\therefore$ There is no asymptote.

6. Additional Points. $\tan \phi = r \dfrac{d\theta}{dr}$

$$\text{From (1);} \qquad 2r \frac{dr}{d\theta} = a^2 (-\sin 2\theta) \cdot 2$$

$$\therefore \qquad \tan\phi = r\frac{r}{-a^2\sin 2\theta} = \frac{a^2\cos 2\theta}{-a^2\sin 2\theta} = -\cot 2\theta$$

$$\therefore \qquad \phi = \frac{\pi}{2} + 2\theta$$

θ	0	$\frac{\pi}{4}$	$\frac{3\pi}{4}$	π
r	a	0	0	$-a$
ϕ	$\frac{\pi}{2}$	π	2π	$\frac{5\pi}{2}$

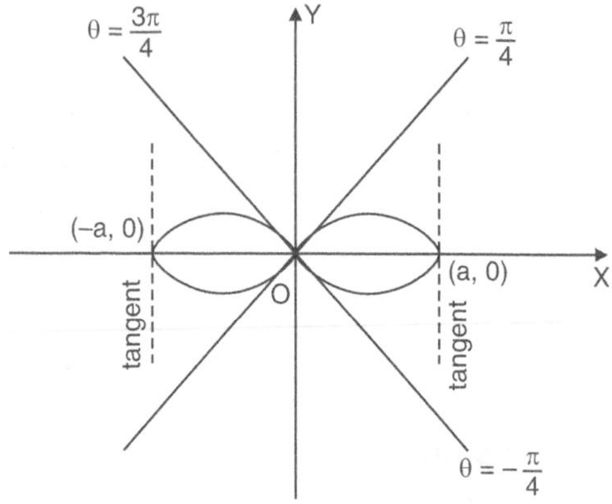

Sketch of the curve is shown in adjoining figure.

TEST YOUR KNOWLEDGE

Trace the following curves:

1. $r = a(1 + \cos\theta)$ **(P.T.U., Dec. 2011; Dec. 2012)**
2. $r = 2a\cos\theta$
3. $r = a + b\cos\theta\ (a < b)$
4. $r = a\cos 3\theta$
5. $r = a\sin 2\theta$
6. $r^2 = a^2\sin 2\theta$
7. $r = ae^{m\theta}\ (a > 0, m > 0)$
8. $r^{-1/2} = a^{-1/2}\cos\frac{\theta}{2}$ or $a = r\cos^2\frac{\theta}{2}$
9. $r\cos\theta = a\sin^2\theta$

   [**Hint.** Change to cartesian coordinates $(x^2 + y^2)x = ay^2$]
10. $x^5 + y^5 = 5ax^2y^2$.

Answers

1.

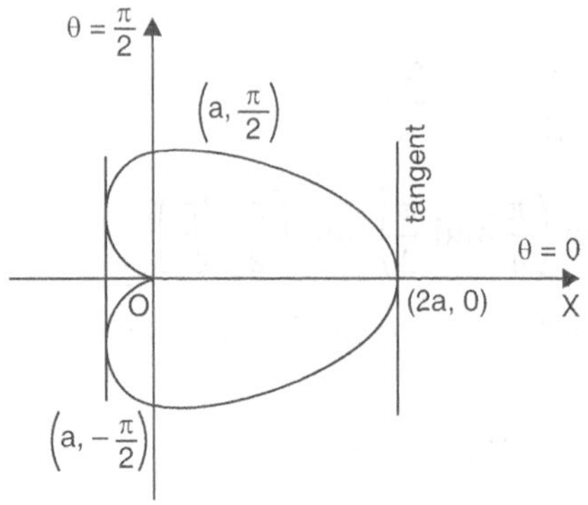

2.

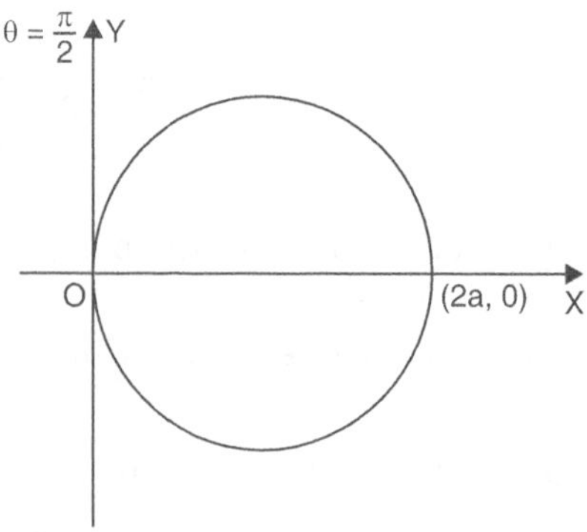

3.

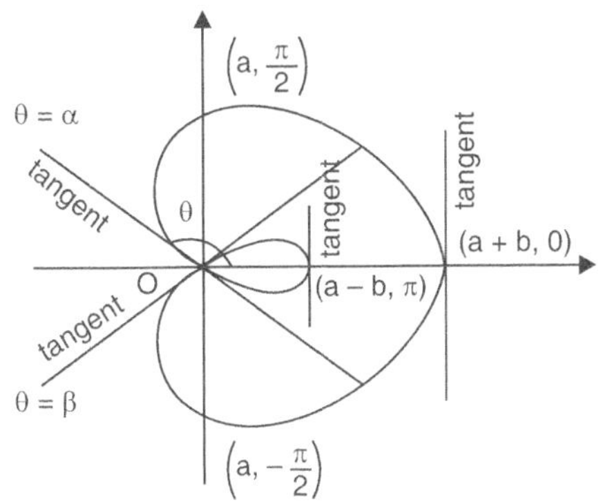

$\theta = \cos^{-1}\left(-\frac{a}{b}\right)$ has two values in the interval $(0, 2\pi)$ which lie in 2nd and 3rd quadrant (say α, β)

4.

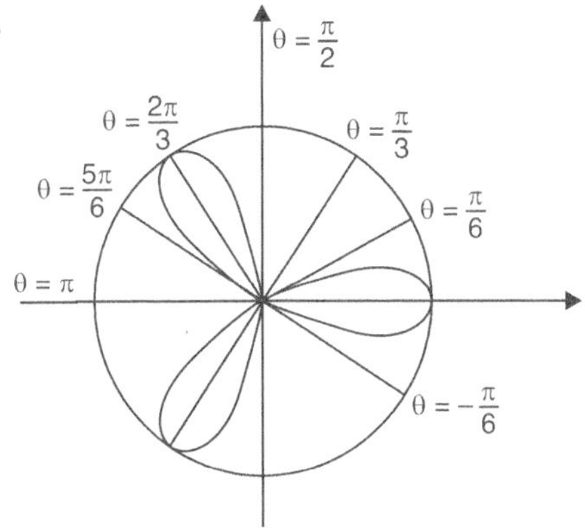

5.

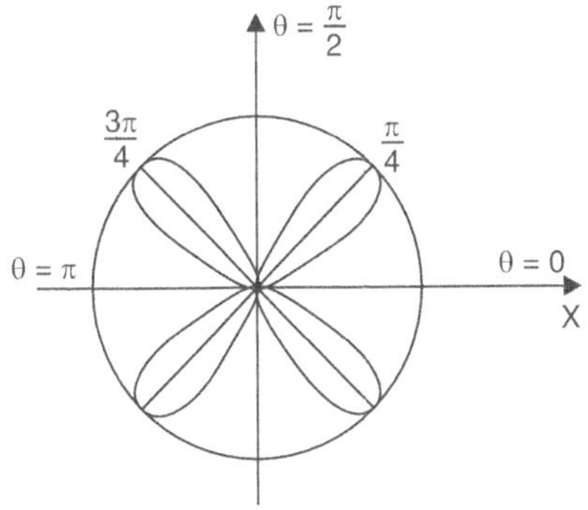

6.

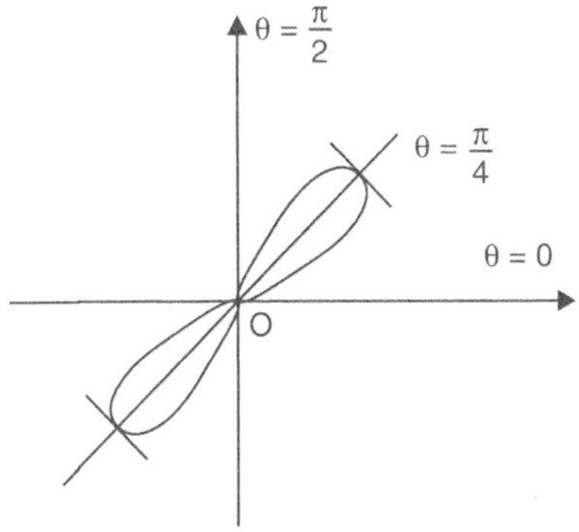

7.

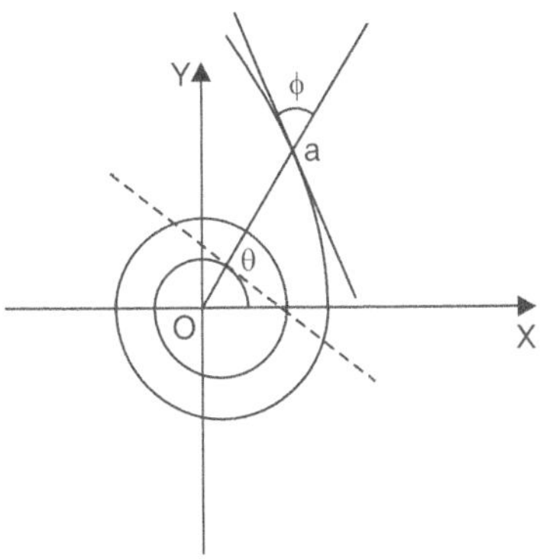

8.

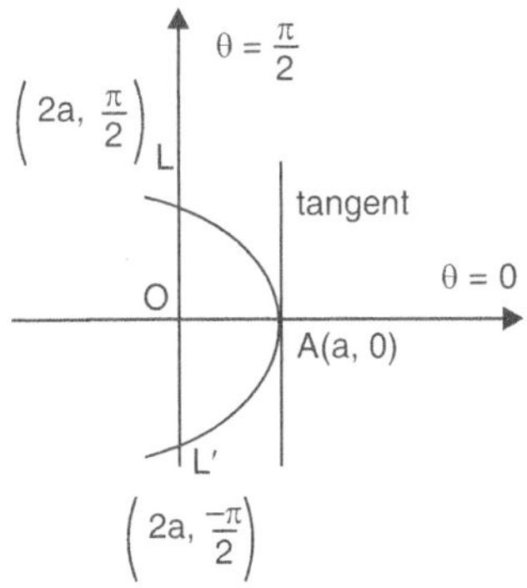

9.

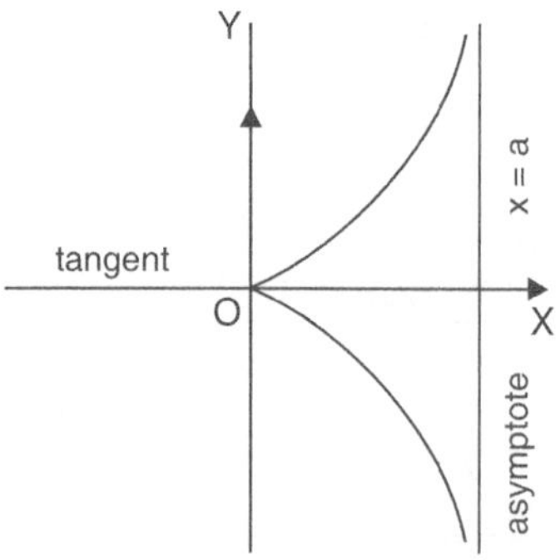

10.

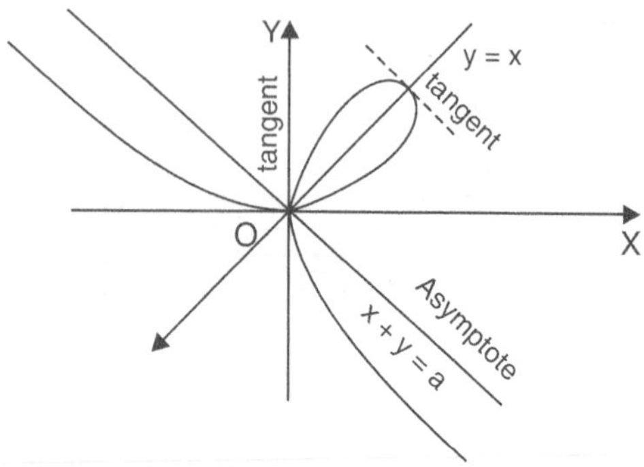

REVIEW OF THE CHAPTER

1. **Concavity, Convexity**—A curve $y = f(x)$ is said to concave upward (or convex downward) on (a, b) if all the points of the curve lie **above** any tangent to it on that interval.
   A curve $y = f(x)$ is said to concave downward (or convex upward) on (a, b) if all the points of the curve lie **below** any tangent to it on that interval.
2. **Point of Inflexion**—A point on the curve is said to be a point of inflexion if the curve crosses the tangent at that point *i.e.*, two portions of the curve on the two sides of the point lie on different sides of the tangent at that point.
3. **Criteria for Concavity, Convexity and Inflexion**—Find $\frac{d^2y}{dx^2}$, the interval for which $\frac{d^2y}{dx^2} > 0$ is called interval of concave upward or convex downward. The interval for which $\frac{d^2y}{dx^2} < 0$ is called interval of concave downward or convex upward. If $\frac{d^2y}{dx^2} = 0$ for certain value of x (say a) but $\frac{d^3y}{dx^3} \neq 0$ at $x = a$ then $x = a$ is the point of inflexion (or if $\frac{d^2y}{dx^2}$ changes sign at $x = a$).

   Note: For point of inflexion $\frac{d^2y}{dx^2} = 0$ but converse is not true.

   For points of inflexion we consider those points also (belonging to the domain of $y = f(x)$) for which $\frac{d^2y}{dx^2}$ does not exist.
4. **Increasing and Decreasing Function**—A function $y = f(x)$ is said to be increasing in an interval (a, b) if $f'(x) > 0 \ \forall \ x \in (a, b)$ and decreasing in an interval (a, b) if $f'(x) < 0 \ \forall \ x \in (a, b)$.
5. **Asymptote**—A straight line is called an asymptote of an infinite branch of a curve iff the line lies on one side of the branch of the curve and $\perp$ distance from any point on curve on it tends to zero as the point approaches infinity along that branch.
6. **How to Find Asymptotes**—Asymptotes parallel to x-axis and y-axis are called rectangular asymptotes.
   Coefficient of the highest power of x equated to zero gives horizontal asymptote and coefficient of the highest power of y equated to zero gives the vertical asymptote.

To find oblique asymptote put $x = 1$, $y = m$ in the nth degree (highest degree) and $(n-1)$th degree terms and obtain $\phi_n(m)$, $\phi_{n-1}(m)$. Find the real values of m from $\phi_n(m) = 0$ then substitute in $\phi_n'(m) + \phi_{n-1}(m) = 0$; $\phi_n'(m) \neq 0$ and get the value of c. Then asymptote will be $y = mx + c$. If $\phi_n'(m) = 0$ then c will have two values given by $\frac{c^2}{2!}\phi_n''(m) + c\,\phi_{n-1}'(m) + \phi_{n-2}(m) = 0$ where $\phi_n''(m) \neq 0$, we will get two parallel asymptotes given by $y = mx + c$. The process can be similarly repeated.

7. Asymptote of the polar curve $\frac{1}{r} = f(\theta)$ is given by $r \sin(\theta - \alpha) = \frac{1}{f'(\alpha)}$, where α is a root of the equation $f(\theta) = 0$.
8. **Multiple point**—A point through which two or more than two branches of a curve pass, is called a multiple point.
9. **Double point**—A point through which only two branches of a curve pass is called a double point.
10. **Node, cusp, conjugate point**—A double point is said to be a
    (*i*) node if two tangents at the double point are real and distinct
    (*ii*) cusp if two tangents at that point are real and coincident
    (*iii*) conjugate if two tangents at that point are not real.
11. Salient features of the curve tracing (see article 1.13).

SHORT ANSWER TYPE QUESTIONS

1. Define concavity and convexity of a curve $y = f(x)$.
   [**Hint.** See art 1.2]
2. Define point of inflexion of a curve and find the point(s) of inflexion for the curve $y = x^3 + 8x^2 - 270x$. ***(P.T.U., Dec. 2005)***
   [**Hint.** S.E. 2(*i*) art. 1.4]
3. Find the points of inflexion (if any) of the following curves. ***(P.T.U., May 2005)***
   (*i*) $y = (x^3 - 6x^2 + 9x + 6)/$ [**Hint.** S.E. 6 art. 1.4]
   (*ii*) $y = \frac{x^2+1}{x^2-1}$
   [**Hint.** S.E. 5 art. 1.4]
   (*iii*) $x = 3y^4 - 4y^3 + 5$ [**Hint.** S.E. 2(*ii*) art. 1.4]
   (*iv*) $x = a(2\theta - \sin\theta)$, $y = a(2 - \cos\theta)$ [**Hint.** S.E. 3 art. 1.4]
   (*v*) $x = a - b\cos\theta$, $y = a\theta - b\sin\theta$
4. Find the values of x for which the curve $y = x^4 - 6x^3 + 12x^2 + 5x + 7$ is concave upward or downward:
   [**Hint.** S.E. 1 art. 1.4]
5. Examine the curve $y = x^3 - 9x^2 + 10x + 5$ for concavity and convexity upward.
6. Define a decreasing function and state the derivative test for decreasing function. ***(P.T.U., May 2005)***
   [**Hint.** Consult art 1.5]
7. Find the rectangular asymptotes (if any) of the following curves:
   (*i*) $x^2y^2(x^2 + y^2) = 4x^4 + 9y^4$ (*ii*) $(x + y)^2 = x^2y^2 - 7$
   (*iii*) $x^2y^2 - xy^2 - x^2y + x + y + 1 = 0$ ***(P.T.U., Dec. 2012)***
   (*iv*) $y^2x - a^2(x + a) = 0$ (*v*) $y = x(x-1)(x-2)$ ***(P.T.U., Dec. 2013)***
   (*vi*) $x^2y^3 + x^2 + 3y^2 - 9xy + 8x - 25 = 0$
   [**Hint.** S.E. 1 (*iii*) art. 1.10]

8. As x moves from left to right through the point $c = 2$, is the graph $y = x^3 - 3x + 2$ rising or is falling? Give reason for your answer.
[**Hint.** S.E. 7 art. 1.4(a)] ***(P.T.U., Dec. 2004)***

9. What do you understand by a parametric curve ? Give an example of the parametric curve involving the parameters. ***(P.T.U., Dec. 2005)***

10. State salient features of the curve tracing.
[**Hint.** Consult art 1.13]

11. Trace the curve $x^3 + y^3 = 3axy$. ***(P.T.U., May 2003)***
[**Hint.** S.E. 1 art. 1.13].

12. Draw rough sketch of the curve $y^2 = x + 5$. ***(P.T.U., May 2008)***

13. Trace the curves ***(P.T.U., Dec. 2005, 2010; May 2012)***
(i) $r = a(1 - \cos\theta)$
(ii) $r = a(1 + \cos\theta)$ ***(P.T.U., Dec. 2011, Des. 2012)***

Answers

2. $\left(-\frac{8}{3}, \frac{20464}{27}\right)$

3. (i) $\left(2, \frac{4}{3}\right)$

(ii) Nil

(iii) $(5, 0), \left(\frac{119}{27}, \frac{2}{3}\right)$

(iv) $\left\{a\left(4n\pi \pm \frac{2\pi}{3} \mp \frac{\sqrt{3}}{2}\right), \frac{3a}{2}\right\}$

(v) $x = \dfrac{a^2 - b^2}{a}, y = a\cos^{-1}\dfrac{b}{a} - \dfrac{b}{a}\sqrt{a^2 - b^2}$.

4. Concave upward in $(-\infty, 1) \cup (2, \infty)$ concave downward in $(1, 2)$

5. Concave upward for $(3, \infty)$, convex upward for $(-\infty, 3)$.

7. (i) $y = \pm 2, x = \pm 3$ (ii) $y = \pm 1, x = \pm 1$ (iii) $x = 0, x = 1, y = 0, y = 1$ (iv) $x = 0, y = \pm a$.
(v) no asymptote (vi) $x = 0, y = -1$

8. $f(x)$ is rising in $(-\infty, -1) \cup (1, \infty)$ and falling in $(-1, 1)$.

12.

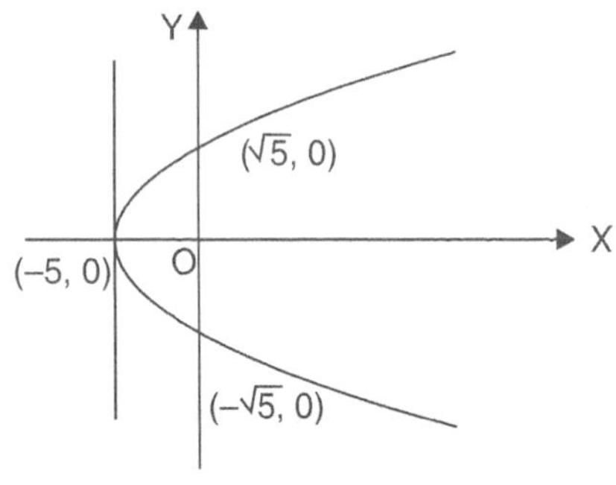

13. (i)

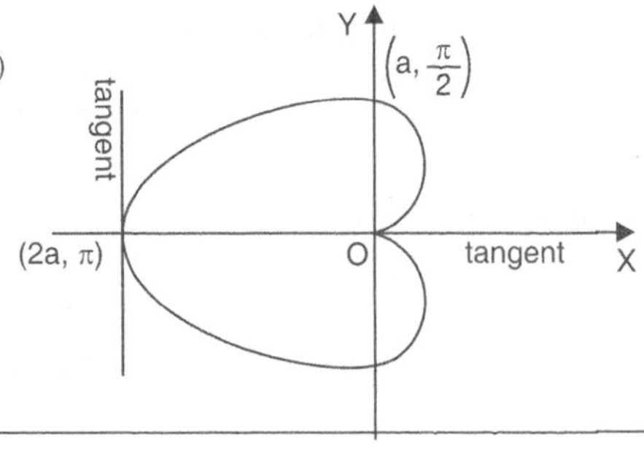

(ii)

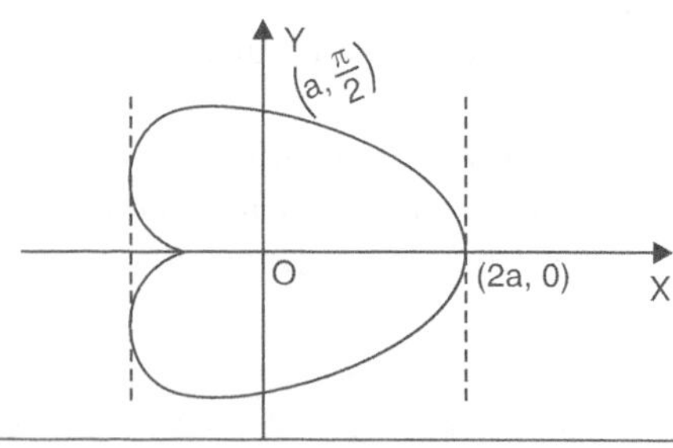

SECTION B: CURVATURE

1.17. DERIVATIVE OF AN ARC LENGTH

(1) If $y = f(x)$ is the cartesian equation of any curve and s is the length of any arc of the curve, then

$$\frac{ds}{dx} = \sqrt{1+\left(\frac{dy}{dx}\right)^2} \; ; \; \frac{ds}{dy} = \sqrt{1+\left(\frac{dx}{dy}\right)^2} .$$

(2) If ψ is the angle which the tangent at any point $P(x, y)$ of the curve $y = f(x)$ makes with the positive direction of x-axis, then

$$\frac{dx}{ds} = \cos \psi$$

$$\frac{dy}{ds} = \sin \psi$$

$$\therefore \qquad \tan \psi = \frac{dy}{dx}.$$

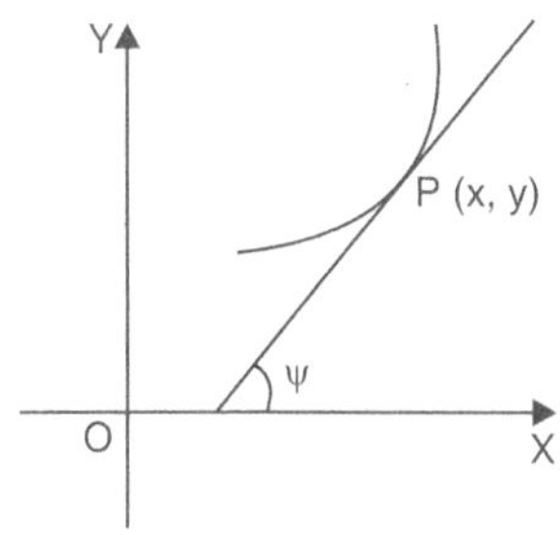

(3) From above, we have

$$\left(\frac{dx}{ds}\right)^2 + \left(\frac{dy}{ds}\right)^2 = 1$$

or $$\left(\frac{dx}{dt} \cdot \frac{dt}{ds}\right)^2 + \left(\frac{dy}{dt} \cdot \frac{dt}{ds}\right)^2 = 1$$

or $$\left(\frac{dx}{dt}\right)^2 + \left(\frac{dy}{dt}\right)^2 = \left(\frac{ds}{dt}\right)^2$$

or $$\frac{ds}{dt} = \sqrt{\left(\frac{dx}{dt}\right)^2 + \left(\frac{dy}{dt}\right)^2} .$$

(4) If radius vector at P makes an angle θ with the positive direction of x-axis; tangent at P makes an angle ψ with the positive direction of x-axis and ϕ is the angle between tangent and radius vector, then

$$\psi = \theta + \phi$$

Also, $$\sin \phi = \frac{r\,d\theta}{ds} ; \cos \phi = \frac{dr}{ds}$$

$$\therefore \qquad \tan \phi = \frac{r\,d\theta}{dr}.$$

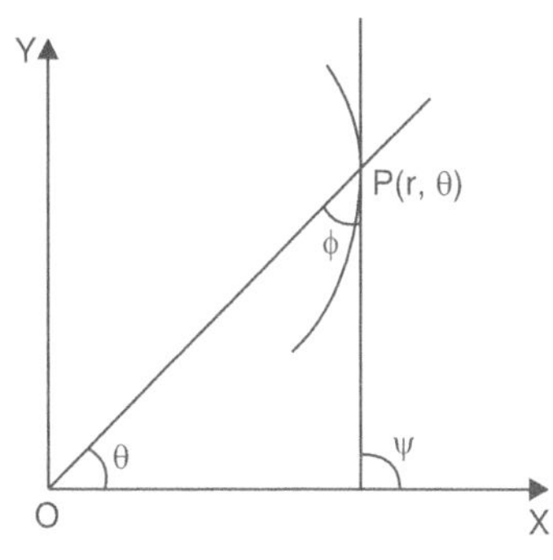

1.18. LENGTHS OF TANGENT, SUBTANGENT, NORMAL, AND SUBNORMAL

(1) Length of tangent at P

$$= \text{PT} = y \operatorname{cosec} \psi$$

$$= y\sqrt{1+\cot^2 \psi}$$

$$= y\sqrt{1+\frac{1}{\tan^2 \psi}}$$

$$= y\sqrt{1+\frac{1}{\left(\frac{dy}{dx}\right)^2}}$$

$$= \frac{y\sqrt{1+\left(\frac{dy}{dx}\right)^2}}{\frac{dy}{dx}} = \frac{y\sqrt{1+y_1^2}}{y_1},$$

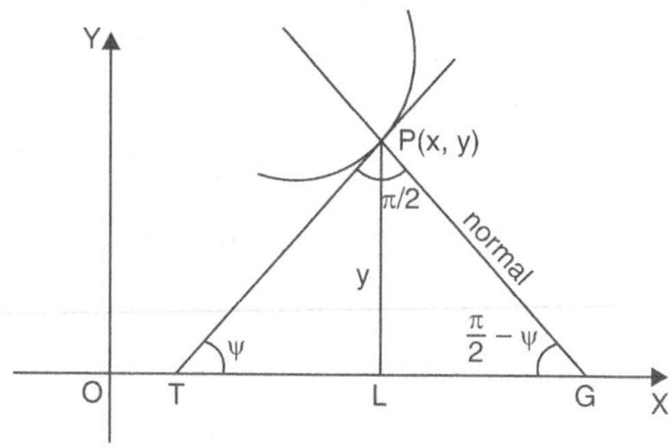

where $y_1 = \dfrac{dy}{dx}$.

(2) Length of the subtangent $= \text{TL} = y \cot \psi = \dfrac{y}{\tan \psi} = \dfrac{y}{\frac{dy}{dx}} = \dfrac{y}{y_1}$.

(3) Length of the normal at P

$$= \text{PG} = y \sec \psi$$

$$= y\sqrt{1+\tan^2 \psi} = y\sqrt{1+\left(\frac{dy}{dx}\right)^2} = y\sqrt{1+y_1^2}.$$

(4) Length of the subnormal

$$= \text{LG} = y \tan \psi$$

$$= y\,\frac{dy}{dx} = yy_1.$$

1.19. CURVATURE OF A CURVE AT A POINT

Let AB be the given curve and P and Q be any two neighbouring points on the curve such that AP = s and AQ = $s + \delta s$ $\therefore$ arc PQ = δs. Let the tangents to the curve at P and Q subtend angles ψ and $\psi + \delta\psi$ with the positive direction of x-axis. Then $\angle \text{RST} = \delta\psi$. Then the following definitions follow :

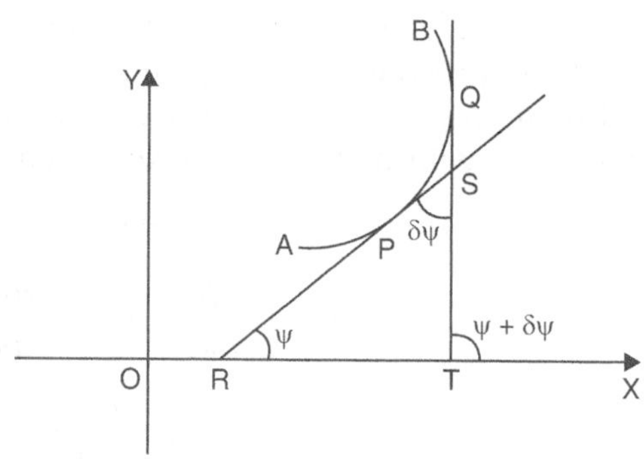

(1) $\delta\psi$, the angle measured in radians, is called the *total curvature* or *total bending* of the arc PQ or *the angle of contigence* of arc PQ.

(2) The ratio $\dfrac{\delta\psi}{\delta s}$ is called the *average curvature* of the arc PQ.

(3) $\underset{\delta s \to 0}{\text{Lt}} \dfrac{\delta\psi}{\delta s}$; if it exists ; is called the **Curvature of the curve** at P. It is denoted by K.

(4) The reciprocal of the curvature at any point P is called the **Radius of curvature** and is generally denoted by ρ (the Greek letter)

$$\therefore \qquad \rho = \frac{1}{\frac{d\psi}{ds}} = \frac{ds}{d\psi}.$$

1.20. RADIUS OF CURVATURE FOR THE CARTESIAN CURVE $y = f(x)$

Proof. The equation of the curve is $y = f(x)$. Let ψ be the angle which the tangent to the curve at P makes with x-axis,

then
$$\tan \psi = \frac{dy}{dx}$$

Differentiating w.r.t. s, we get

$$\sec^2 \psi \cdot \frac{d\psi}{ds} = \frac{d^2y}{dx^2} \cdot \frac{dx}{ds}$$

$$\sec^2 \psi \, \frac{1}{\rho} = \frac{d^2y}{dx^2} \cos \psi$$

$$\therefore \qquad \frac{1}{\rho} = \frac{d^2y}{dx^2} \cdot \frac{1}{\sec^3 \psi} = \frac{d^2y}{dx^2} \frac{1}{(1+\tan^2 \psi)^{3/2}} = \frac{d^2y}{dx^2} \cdot \frac{1}{\left[1+\left(\frac{dy}{dx}\right)^2\right]^{3/2}}$$

$$\therefore \qquad \rho = \frac{\left[1+\left(\frac{dy}{dx}\right)^2\right]^{3/2}}{\frac{d^2y}{dx^2}}$$

or
$$\rho = \frac{(1+{y_1}^2)^{3/2}}{y_2}, \quad \text{where } y_1 = \frac{dy}{dx};\ y_2 = \frac{d^2y}{dx^2}.$$

As we know that curvature of the curve at any point is dependent upon the shape of the curve and is independent of the co-ordinate system.

$\therefore$ On interchanging y with x, we get $\rho = \dfrac{(1+{x_1}^2)^{3/2}}{x_2}$, where $x_1 = \dfrac{dx}{dy}$ and $x_2 = \dfrac{d^2x}{dy^2}$.

1.21. RADIUS OF CURVATURE FOR THE PARAMETRIC CURVE

Find the radius of curvature at any point of the curve $x = f(t)$, $y = g(t)$.

Proof.
$$x = f(t) \qquad y = g(t)$$

$$\frac{dx}{dt} = f'(t) \qquad \frac{dy}{dt} = g'(t)$$

$$\therefore \qquad \frac{dy}{dx} = \frac{\frac{dy}{dt}}{\frac{dx}{dt}} = \frac{g'(t)}{f'(t)}$$

$$\frac{d^2y}{dx^2} = \frac{d}{dt}\left[\frac{g'(t)}{f'(t)}\right] \cdot \frac{dt}{dx} = \frac{f'(t)\, g''(t) - g'(t)\, f''(t)}{[f'(t)]^2} \cdot \frac{1}{f'(t)}$$

$$= \frac{f'(t)\,g''(t) - g'(t)\,f''(t)}{[f'(t)]^3} = \frac{\left\{1 + \left[\dfrac{g'(t)}{f'(t)}\right]^2\right\}^{3/2}}{\dfrac{f'(t)\,g''(t) - g'(t)\,f''(t)}{[f'(t)]^3}}$$

$$\therefore \qquad \rho = \frac{\left\{[f'(t)]^2 + [g'(t)]^2\right\}^{3/2}}{f'(t)\,g''(t) - f''(t)\,g'(t)}$$

1.22. RADIUS OF CURVATURE WHEN *x*, *y* ARE FUNCTIONS OF *s*

Proof. Let the equation of the curve be

$$x = f(s),\ y = g(s).$$

We know that

$$\frac{dx}{ds} = \cos\psi \quad \text{and} \quad \frac{dy}{ds} = \sin\psi \qquad ...(1)$$

$$\frac{d^2x}{ds^2} = -\sin\psi \cdot \frac{d\psi}{ds},\ \frac{d^2y}{ds^2} = \cos\psi \cdot \frac{d\psi}{ds}$$

$$\therefore \qquad \frac{d^2x}{ds^2} = -\sin\psi\,\frac{1}{\rho}, \qquad \frac{d^2y}{ds^2} = \cos\psi \cdot \frac{1}{\rho} \qquad ...(2)$$

Squaring and adding

$$\left(\frac{d^2x}{ds^2}\right)^2 + \left(\frac{d^2y}{ds^2}\right)^2 = \frac{1}{\rho^2}$$

$$\therefore \qquad \frac{1}{\rho^2} = \left(\frac{d^2x}{ds^2}\right)^2 + \left(\frac{d^2y}{ds^2}\right)^2.$$

Cor. From 2;
$$\rho = \frac{-\sin\psi}{\dfrac{d^2x}{ds^2}} = -\frac{\dfrac{dy}{ds}}{\dfrac{d^2x}{ds^2}}$$

Also
$$\rho = \frac{\cos\psi}{\dfrac{d^2y}{ds^2}} = \frac{\dfrac{dx}{ds}}{\dfrac{d^2y}{ds^2}}.$$

1.23. RADIUS OF CURVATURE FOR IMPLICIT EQUATION *f* (*x*, *y*) = 0

$$\rho = -\frac{(f_x^2 + f_y^2)^{3/2}}{f_{xx}f_y^2 - 2f_{xy}f_xf_y + f_{yy}f_x^2},$$

where
$$f_x = \frac{\partial f}{\partial x},\quad f_y = \frac{\partial f}{\partial y},\quad f_{xx} = \frac{\partial^2 f}{\partial x^2},\ f_{yy} = \frac{\partial^2 f}{\partial y^2}$$

$$f_{xy} = f_{yx} = \frac{\partial^2 f}{\partial y \partial x}.$$

Proof. $f(x, y) = 0$ is an implicit function of x, y

Differentiate it

$$\frac{\partial f}{\partial x} dx + \frac{\partial f}{\partial y} dy = 0 \quad \therefore \quad \frac{dy}{dx} = -\frac{f_x}{f_y}$$

Now,
$$\frac{d^2y}{dx^2} = -\frac{f_y \dfrac{d}{dx}(f_x) - f_x \dfrac{d}{dx}(f_y)}{f_y^2}$$

$$= -\frac{f_y\left[\dfrac{\partial f_x}{\partial x} + \dfrac{\partial f_x}{\partial y}\cdot\dfrac{dy}{dx}\right] - f_x\left[\dfrac{\partial}{\partial x}(f_y) + \dfrac{\partial}{\partial y}(f_y)\dfrac{dy}{dx}\right]}{f_y^2}$$

$$= -\frac{f_y f_{xx} + f_y f_{xy}\dfrac{dy}{dx} - f_x f_{xy} - f_x f_{yy}\dfrac{dy}{dx}}{f_y^2}$$

$$= -\frac{f_y f_{xx} - f_x f_{xy} + [f_y f_{xy} - f_x f_{yy}]\left(\dfrac{-f_x}{f_y}\right)}{f_y^2} \quad \text{by using the value of } \frac{dy}{dx}$$

$$= -\frac{f_y{}^2 f_{xx} - f_x f_y f_{xy} - f_x f_y f_{xy} + f_x{}^2 f_{yy}}{f_y^3} = -\frac{f_x^2 f_{yy} - 2f_x f_y f_{xy} + f_y^2 f_{xx}}{f_y^3}$$

Now,
$$\rho = \frac{\left(1 + y_1^2\right)^{3/2}}{y_2} = -\frac{\left(1 + \dfrac{f_x^2}{f_y{}^2}\right)^{3/2}}{\dfrac{f_x^2 f_{yy} - 2f_x f_y f_{xy} + f_y^2 f_{xx}}{f_y^3}}$$

$$= -\frac{(f_x^2 + f_y^2)^{3/2}}{f_{xx} f_y^2 - 2f_x f_y f_{xy} + f_{yy} f_x^2}$$

ILLUSTRATIVE EXAMPLES

Example 1. *(a) Prove that* $\dfrac{1}{\rho} = \dfrac{d}{dx}\left(\dfrac{dy}{ds}\right)$.

(b) Find radius of curvature at any point of the curve.

$$s = a \log (\sec \psi + \tan \psi) + a \sec \psi \tan \psi.$$

Sol. *(a)* We know that

$$\frac{dy}{ds} = \sin \psi$$

Differentiate both sides w.r.t.x

$$\frac{d}{dx}\left[\frac{dy}{ds}\right] = \frac{d}{dx}(\sin\psi) = \cos\psi\,\frac{d\psi}{dx} = \cos\psi\,\frac{d\psi}{ds}\cdot\frac{ds}{dx}$$

$$= \cos\psi\,.\,\frac{1}{\rho}\cdot\frac{1}{\cos\psi} \qquad \because \cos\psi = \frac{dx}{ds}$$

$$= \frac{1}{\rho}$$

$$\therefore \qquad \frac{1}{\rho} = \frac{d}{dx}\left(\frac{dy}{ds}\right)$$

(b)

$$\frac{ds}{d\psi} = \frac{a}{\sec\psi + \tan\psi}(\sec\psi\tan\psi + \sec^2\psi) + a\sec\psi(\sec^2\psi) + a(\sec\psi\tan\psi)\tan\psi$$

$$= \frac{a\sec\psi(\sec\psi + \tan\psi)}{\sec\psi + \tan\psi} + a\sec\psi\,[\sec^2\psi + \tan^2\psi]$$

$$= a\sec\psi + a\sec\psi(\sec^2\psi + \tan^2\psi)$$

$$= a\sec\psi\,[1 + \sec^2\psi + \tan^2\psi]$$

$$= a\sec\psi[2\sec^2\psi] \qquad (\because\ 1 + \tan^2\psi = \sec^2\psi)$$

$$= 2a\sec^3\psi$$

Example 2. (*i*) *Find radius of curvature at the point* $\left(\frac{3a}{2}, \frac{3a}{2}\right)$ *on the curve* $x^3 + y^3 = 3axy$.

(*ii*) *Find the curvature of the curve* $y^2 = x^3 + 8$ *at the point (1, 3).*

(P.T.U., May 2006, May 2009)

Sol. (*i*) Equation of the curve is

$$x^3 + y^3 = 3axy \qquad ...(1)$$

Differentiate w.r.t. x,

$$3x^2 + 3y^2\frac{dy}{dx} = 3ax\frac{dy}{dx} + 3ay$$

$$(y^2 - ax)\frac{dy}{dx} = ay - x^2 \quad \text{or} \quad \frac{dy}{dx} = \frac{ay - x^2}{y^2 - ax}$$

$$\left(\frac{dy}{dx}\right)_{\left(\frac{3a}{2}, \frac{3a}{2}\right)} = \frac{\frac{3a^2}{2} - \frac{9a^2}{4}}{\frac{9a^2}{4} - \frac{3a^2}{2}} = -1$$

$$\frac{d^2y}{dx^2} = \frac{(y^2 - ax)\left[a\frac{dy}{dx} - 2x\right] - (ay - x^2)\left[2y\frac{dy}{dx} - a\right]}{(y^2 - ax)^2}$$

$$\left(\frac{d^2y}{dx^2}\right)_{\left(\frac{3a}{2}, \frac{3a}{2}\right)} = \frac{\left(\frac{9a^2}{4} - \frac{3a^2}{2}\right)(-a - 3a) - \left(\frac{3a^2}{2} - \frac{9a^2}{4}\right)(-3a - a)}{\left(\frac{9a^2}{4} - \frac{3a^2}{2}\right)^2}$$

$$= \frac{2 \cdot \left(\frac{3a^2}{4}\right)(-4a)}{\left(\frac{3a^2}{4}\right)^2} = \frac{-8a}{3a^2/4} = -\frac{32}{3a}$$

$$\therefore \qquad \rho = \frac{(1+y_1^2)^{3/2}}{y_2} = \frac{(1+1)^{3/2}}{\frac{-32}{3a}} = -\frac{3a}{32} \cdot 2^{3/2} = -\frac{3a}{32} \cdot 2\sqrt{2} = -\frac{3\sqrt{2}a}{16}$$

$$\therefore \qquad \rho \text{ at } P\left(\frac{3a}{2}, \frac{3a}{2}\right) = \frac{3\sqrt{2}a}{16} \text{ (in magnitude).}$$

(*ii*) $y^2 = x^3 + 8$

Differentiate w.r.t. x; $2y\frac{dy}{dx} = 3x^2 \quad \therefore \quad \frac{dy}{dx} = \frac{3}{2}\frac{x^2}{y}$

$$\frac{d^2y}{dx^2} = \frac{3}{2} \frac{y \,.\, 2x - x^2 \,.\, \frac{dy}{dx}}{y^2} = \frac{3}{2y^2}\left[2xy - x^2 \,.\, \frac{3}{2}\frac{x^2}{y}\right]$$

$$= \frac{3}{2y^2}\left[2xy - \frac{3}{2}\frac{x^4}{y}\right]$$

At $(1, 3)$; $\frac{dy}{dx} = \frac{1}{2}$, $\frac{d^2y}{dx^2} = \frac{11}{12}$

radius of curvature at (1, 3) $= \frac{(1+y_1^2)^{3/2}}{y_2} = \frac{\left(1+\frac{1}{4}\right)^{3/2}}{11/12}$

$$= \frac{12}{11}\frac{5^{3/2}}{4^{3/2}} = \frac{12}{11}\frac{5\sqrt{5}}{8} = \frac{15\sqrt{5}}{22} \quad \therefore \quad \text{Curvature} = \frac{22}{15\sqrt{5}}.$$

Example 3. *Find the radius of curvature at any point (x, y) of the curve $x^{2/3} + y^{2/3} = a^{2/3}$.* **(P.T.U., May 2006)**

Also prove that the radius of curvature is three times the perpendicular distance from the origin to the tangent at (x, y). **(P.T.U., Dec. 2011)**

Sol. Equation of the curve is

$$x^{2/3} + y^{2/3} = a^{2/3} \qquad \ldots(1)$$

Differentiate w.r.t. x; $\frac{2}{3}x^{-1/3} + \frac{2}{3}y^{-1/3}\frac{dy}{dx} = 0$

$$\therefore \qquad \frac{dy}{dx} = -\frac{x^{-1/3}}{y^{-1/3}} = -\left(\frac{y}{x}\right)^{1/3} \quad \therefore \quad y_1 = -\left(\frac{y}{x}\right)^{1/3}$$

$$\frac{d^2y}{dx^2} = -\frac{1}{3}\left(\frac{y}{x}\right)^{-2/3} \frac{x.\frac{dy}{dx} - y.1}{x^2} = -\frac{y^{-2/3}}{3x^{4/3}}\left(-x\left(\frac{y}{x}\right)^{1/3} - y\right)$$

$$= \frac{y^{-2/3}}{3x^{4/3}}(x^{2/3}y^{1/3} + y) = \frac{y^{-2/3}}{3x^{4/3}} \cdot y^{1/3}(x^{2/3} + y^{2/3}) = \frac{y^{-1/3}}{3x^{4/3}} \cdot a^{2/3}$$

$$\therefore \qquad y_2 = \frac{1}{3}\frac{a^{2/3}}{x^{4/3}y^{1/3}}$$

$$\rho = \frac{(1+y_1^2)^{3/2}}{y_2} = \frac{\left(1+\frac{y^{2/3}}{x^{2/3}}\right)^{3/2}}{\frac{a^{2/3}}{3x^{4/3}\ y^{1/3}}} = \frac{\left(x^{2/3}+y^{2/3}\right)^{3/2}}{x \cdot \frac{a^{2/3}}{3x^{4/3}\ y^{1/3}}}$$

$$= \frac{(a^{2/3})^{3/2} \cdot 3x^{1/3}\ y^{1/3}}{a^{2/3}} = 3 \cdot a^{1/3}\ x^{1/3}\ y^{1/3}$$

$$= 3\ (axy)^{1/3}. \qquad ...(2)$$

Equation of the tangent at (x, y) to (1) is

$$Y - y = -\left(\frac{y}{x}\right)^{1/3} (X - x)$$

or $$y^{1/3}X + x^{1/3}Y - x^{1/3}y - y^{1/3}x = 0$$

or $$y^{1/3}X + x^{1/3}Y - x^{1/3}y^{1/3}(y^{2/3} + x^{2/3}) = 0$$

or $$y^{1/3}X + x^{1/3}Y - x^{1/3}y^{1/3}a^{2/3} = 0 \qquad \text{[Using (1)]}$$

Now $\perp$ distance from (0, 0) on tangent

$$= \frac{|0 + 0 - x^{1/3}y^{1/3}a^{2/3}|}{\sqrt{y^{2/3} + x^{2/3}}}$$

$$= \frac{x^{1/3}y^{1/3}a^{2/3}}{\sqrt{a^{2/3}}} = \frac{x^{1/3}y^{1/3}a^{2/3}}{a^{1/3}} = x^{1/3}y^{1/3}a^{1/3}$$

$$= (axy)^{1/3}$$

Hence from (2) we conclude that radius of curvature is 3 times the perpendicular distance from the origin on the tangent at (x, y).

Example 4. (*i*) *Find the radius of curvature at any point on the parabola* $y^2 = 4ax$.

(P.T.U., Dec. 2010)

(*ii*) *If P is any point on the parabola* $y^2 = 4ax$, *whose focus is S(a, 0), then prove that* ρ *varies as* $(SP)^{3/2}$. **(P.T.U., Dec. 2015)**

(*iii*) *If normal at any point P on the parabola meets the directrix at L, prove that* $\rho = 2PL$.

Sol. (*i*) Equation of the parabola is $y^2 = 4ax$

$$2y\frac{dy}{dx} = 4a$$

$$\therefore \quad \frac{dy}{dx} = \frac{2a}{y} = \frac{2a}{2\sqrt{ax}} = \sqrt{\frac{a}{x}}$$

$$\frac{d^2y}{dx^2} = \sqrt{a}\left(-\frac{1}{2}x^{-3/2}\right) = -\frac{\sqrt{a}}{2x^{3/2}}$$

$$\therefore \quad \rho = \frac{(1+y_1^2)^{3/2}}{y_2} = \frac{\left(1+\frac{a}{x}\right)^{3/2}}{-\frac{\sqrt{a}}{2x^{3/2}}}$$

$$\rho = \frac{-2(x+a)^{3/2}}{\sqrt{a}}$$

$\therefore \qquad \rho = \dfrac{2(x+a)^{3/2}}{\sqrt{a}}$ (in magnitude). ...(1)

(*ii*) Now SP = distance between P(x, y) and S(a, 0)

$$= \sqrt{(x-a)^2+(y-0)^2} = \sqrt{(x-a)^2+y^2} = \sqrt{(x-a)^2+4ax}$$

$$= \sqrt{(x+a)^2} = x+a$$

$\therefore$ From (1), $\qquad \rho = \dfrac{2(\text{SP})^{3/2}}{\sqrt{a}} = \dfrac{2}{\sqrt{a}}(\text{SP})^{3/2}$

$\therefore$ ρ is proportional to $(\text{SP})^{3/2}$.

(*iii*) Equation of the normal at P(x, y) is

$$\text{Y} - y = -\frac{1}{\left(\frac{dy}{dx}\right)}(\text{X} - x)$$

$$\text{Y} - y = -\frac{1}{\sqrt{a/x}}(\text{X} - x)$$

or $\qquad \text{Y} - y = -\sqrt{\dfrac{x}{a}}(\text{X} - x)$

It meets directrix X = – a, where

$$\text{Y} - y = -\sqrt{\frac{x}{a}}(-a-x) = \sqrt{\frac{x}{a}}(x+a)$$

$\therefore \qquad \text{Y} = y + \sqrt{\dfrac{x}{a}}(x+a)$

$\therefore$ Co-ordinates of L are $\left\{-a,\ y+\sqrt{\dfrac{x}{a}}(x+a)\right\}$

Now $\qquad \text{PL} = \sqrt{(x+a)^2+\left[y-y-\sqrt{\dfrac{x}{a}}(x+a)\right]^2}$

$$= \sqrt{(x+a)^2+\frac{x}{a}(x+a)^2} = \sqrt{\left(1+\frac{x}{a}\right)(x+a)^2}$$

$$= \sqrt{\frac{(x+a)^3}{a}} = \frac{(x+a)^{3/2}}{\sqrt{a}}$$

$\therefore$ From (1), ρ = 2PL.

Example 5. *Find the least value of the radius of curvature for the curve* $y = \log x$.

Sol. Given $\qquad y = \log x$

$$y_1 = \frac{1}{x};\ y_2 = -\frac{1}{x^2}$$

Radius of curvature $\qquad \rho = \dfrac{(1+y_1^2)^{3/2}}{y_2}$

$\therefore \qquad \rho = \dfrac{\left(1+\dfrac{1}{x^2}\right)^{3/2}}{-1/x^2} = -x^2\,\dfrac{(1+x^2)^{3/2}}{x^3} = -\dfrac{(1+x^2)^{3/2}}{x}$ ...(1)

ρ will be least when $\dfrac{d\rho}{dx} = 0$ and $\dfrac{d^2\rho}{dx^2} > 0$

$$\frac{d\rho}{dx} = -\frac{x \cdot \frac{3}{2}(1+x^2)^{1/2}(2x) - (1+x^2)^{3/2} \cdot 1}{x^2}$$

$$= -\frac{(1+x^2)^{1/2}(3x^2 - 1 - x^2)}{x^2} = -\frac{(1+x^2)^{1/2}(2x^2 - 1)}{x^2}$$

$\dfrac{d\rho}{dx} = 0$ gives $1 + x^2 = 0$ or $2x^2 - 1 = 0$ *i.e.*, $x^2 = \dfrac{1}{2}$ or $x = \pm\dfrac{1}{\sqrt{2}}$

$\dfrac{d^2\rho}{dx^2} > 0$ for $x = \pm\dfrac{1}{\sqrt{2}}$

$\therefore$ ρ is least for $x = \pm\dfrac{1}{\sqrt{2}}$

$\therefore$ From (1),
$$\rho = -\frac{\left(1+\frac{1}{2}\right)^{3/2}}{\pm\frac{1}{\sqrt{2}}} = \sqrt{2}\left(\frac{3}{2}\right)^{3/2} = \sqrt{2}\,\frac{3\sqrt{3}}{2\sqrt{2}} = \frac{3\sqrt{3}}{2}$$

Hence least $\rho = \dfrac{3\sqrt{3}}{2}$.

Example 6. *Prove that (i) the radius of curvature at any point P(x, y) on the ellipse* $\dfrac{x^2}{a^2} + \dfrac{y^2}{b^2} = 1$ *is* $\dfrac{a^2b^2}{p^3}$ *where p is the length of the perpendicular from the centre of the ellipse on the tangent at P.*

(ii) The radius of curvature at the end of the major axis is equal to the semi-latus rectum of the ellipse.

(iii) The radius of curvature is $\dfrac{CD^3}{ab}$, *where CD is the semiconjugate diameter to CP.*

Sol. (*i*) Equation of the ellipse is $\dfrac{x^2}{a^2} + \dfrac{y^2}{b^2} = 1$ ...(1)

Diff. w.r.t. x, $\dfrac{2x}{a^2} + \dfrac{2y}{b^2}\dfrac{dy}{dx} = 0$

$$\therefore \quad \frac{dy}{dx} = -\frac{b^2x}{a^2y}$$

$$\frac{d^2y}{dx^2} = -\frac{b^2}{a^2} \cdot \frac{y - x \cdot \frac{dy}{dx}}{y^2} = -\frac{b^2}{a^2} \cdot \frac{y + x \cdot \frac{b^2x}{a^2y}}{y^2}$$

$$= -\frac{b^2}{a^2} \cdot \frac{b^2}{y} \cdot \frac{\frac{y^2}{b^2} + \frac{x^2}{a^2}}{y^2} = -\frac{b^4}{a^2y^3} \cdot 1 \qquad \Big|\ \because \text{ of } (1)$$

$$\rho = \frac{(1+y_1^2)^{3/2}}{y_2} = \frac{\left(1+\frac{b^4x^2}{a^4y^2}\right)^{3/2}}{-\frac{b^4}{a^2y^3}}$$

$$\rho = -\frac{(a^4y^2+b^4x^2)^{3/2}}{a^6y^3\frac{b^4}{a^2y^3}} = \frac{-(a^4y^2+b^4x^2)^{3/2}}{a^4b^4}$$

$\therefore$ $$\rho = \frac{(a^4y^2+b^4x^2)^{3/2}}{a^4b^4} \quad \text{(in magnitude)} \qquad ...(2)$$

Equation of the tangent at P(x, y) to the ellipse

$$Y - y = \frac{dy}{dx}(X-x)$$

or $$Y - y = -\frac{b^2}{a^2}\frac{x}{y}(X-x)$$

or $$(b^2x)X + (a^2y)Y - (a^2y^2+b^2x^2) = 0$$

$$(b^2x)X + (a^2y)Y - a^2b^2\left(\frac{y^2}{b^2}+\frac{x^2}{a^2}\right) = 0$$

or $$(b^2x)X + (a^2y)Y - a^2b^2 = 0 \quad \because \text{ of (1)} \qquad ...(3)$$

p = length of perpendicular from (0, 0) on (3)

$$= \frac{|0+0-a^2b^2|}{\sqrt{b^4x^2+a^4y^2}} = \frac{a^2b^2}{\sqrt{b^4x^2+a^4y^2}}$$

$\therefore$ $$\sqrt{b^4x^2+a^4y^2} = \frac{a^2b^2}{p}$$

$\therefore$ From (2), $$\rho = \frac{\left(\frac{a^2b^2}{p}\right)^3}{a^4b^4} = \frac{a^2b^2}{p^3}.$$

(*ii*) Radius of curvature at the end of major axis *i.e.*, at (a, 0) or ($-a$, 0)

$$= \frac{(0+b^4a^2)^{3/2}}{a^4b^4} \quad (\text{put } x = a \text{ or } -a, y = 0 \text{ in (2)})$$

$$= \frac{b^6a^3}{a^4b^4} = \frac{b^2}{a} = \text{semi-latus rectum of the ellipse.}$$

(*iii*) We know from co-ordinate geometry that if P, any point on the ellipse, has parametric co-ordinates (a cos θ, b sin θ), then D has co-ordinates ($-a$ sin θ, b cos θ). $\because$ Conjugate diameter is rotated through an angle $\pi/2$

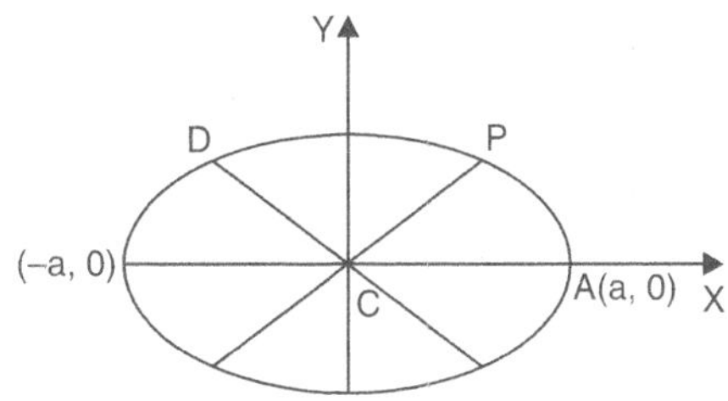

$\therefore$ $$CD^2 = a^2\sin^2\theta + b^2\cos^2\theta$$

$$= a^2\cdot\frac{y^2}{b^2} + b^2\cdot\frac{x^2}{a^2} \quad \because x = a\cos\theta,\ y = b\sin\theta$$

$$= \frac{a^4y^2 + b^4x^2}{a^2b^2}$$

$\therefore$ From (2), $\rho = \dfrac{[a^2 b^2 (CD)^2]^{3/2}}{a^4b^4} = \dfrac{a^3b^3(CD)^3}{a^4b^4} = \dfrac{CD^3}{ab}$.

Example 7. *If ρ_1 and ρ_2 be the radii of curvature at the extremities of the conjugate diameter of the ellipse, prove that $(\rho_1^{2/3} + \rho_2^{2/3})(ab)^{2/3} = a^2 + b^2$.*

(P.T.U., May 2003, Dec. 2011)

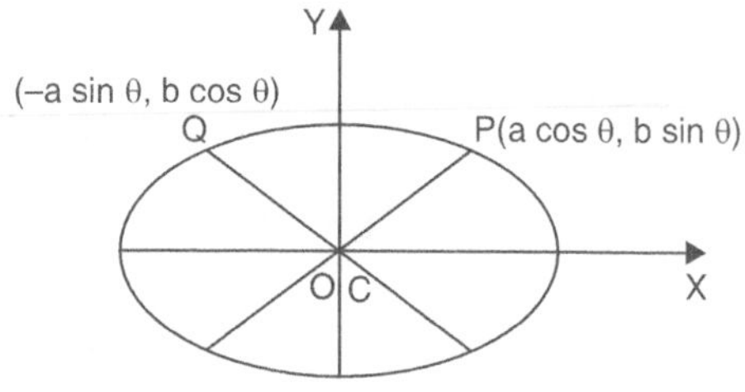

Sol. Let P(a cos θ, b sin θ) be any point on the ellipse. Then the co-ordinates of the extremity of the conjugate diameter will be Q (– a sin θ, b cos θ).

Radius of curvature of the ellipse at P(x, y) is

$$\rho = \frac{(a^4y^2 + b^4x^2)^{3/2}}{a^4b^4}$$

[proved in example 6]

for ρ_1 replace (x, y) by (a cos θ, b sin θ)

$$\rho_1 = \frac{(a^4b^2 \sin^2\theta + b^4a^2\cos^2\theta)^{3/2}}{a^4b^4}$$

for ρ_2 replace (x, y) by (– a sin θ, b cos θ)

$$\rho_2 = \frac{(a^4b^2\cos^2\theta + b^4a^2\sin^2\theta)^{3/2}}{a^4b^4}$$

Now $\rho_1^{2/3} + \rho_2^{2/3} = \dfrac{a^4b^2\sin^2\theta + b^4a^2\cos^2\theta}{(ab)^{4\cdot\frac{2}{3}}} + \dfrac{a^4b^2\cos^2\theta + b^4a^2\sin^2\theta}{(ab)^{4\cdot\frac{2}{3}}}$

$$= \frac{a^4b^2 + b^4a^2}{(ab)^{8/3}} = \frac{a^2b^2(a^2+b^2)}{(ab)^{8/3}} = \frac{a^2+b^2}{(ab)^{2/3}}$$

$\therefore$ $(\rho_1^{2/3} + \rho_2^{2/3})(ab)^{2/3} = a^2 + b^2$.

Example 8. *Find the radius of curvature at any point of the catenary $y = \dfrac{a}{2}[e^{x/a} + e^{-x/a}]$ and show that it is equal to the length of the normal at the same point and that it varies as the square of the ordinate.*

Sol. Equation of the catenary is

$$y = \frac{a}{2}(e^{x/a} + e^{-x/a}) \qquad ...(1)$$

$$\frac{dy}{dx} = \frac{a}{2}\left[\frac{1}{a}e^{x/a} - \frac{1}{a}e^{-x/a}\right] = \frac{1}{2}[e^{x/a} - e^{-x/a}]$$

$$\frac{d^2y}{dx^2} = \frac{1}{2}\cdot\left[\frac{1}{a}e^{x/a} + \frac{1}{a}e^{-x/a}\right] = \frac{1}{2a}[e^{x/a} + e^{-x/a}]$$

$$\rho = \frac{(1+y_1^2)^{3/2}}{y_2} = \frac{\left[1 + \frac{1}{4}(e^{x/a} - e^{-x/a})^2\right]^{3/2}}{\frac{1}{2a}\cdot[e^{x/a} + e^{-x/a}]}$$

$$= \frac{[4 + e^{2x/a} + e^{-2x/a} - 2]^{3/2}}{8 \cdot \frac{1}{2a} \cdot [e^{x/a} + e^{-x/a}]} = \frac{[(e^{x/a} + e^{-x/a})^2]^{3/2}}{\frac{4}{a} \cdot [e^{x/a} + e^{-x/a}]} = \frac{a\left(e^{x/a} + e^{-x/a}\right)^2}{4}$$

$$\therefore \qquad \rho = \frac{a}{4}\left(e^{\frac{x}{a}} + e^{\frac{-x}{a}}\right)^2 \qquad \text{...(2)}$$

Now let length of the normal at P$(x, y) = l$

$$\therefore \qquad l = y\sqrt{1 + \left(\frac{dy}{dx}\right)^2} = y\sqrt{1 + \frac{1}{4}(e^{x/a} - e^{-x/a})^2} \qquad \text{...(3) (art 1.18)}$$

$$= \frac{y}{2}\sqrt{(e^{x/a} + e^{-x/a})^2} = \frac{y}{2}(e^{x/a} + e^{-x/a}) = \frac{a}{4}(e^{x/a} + e^{-x/a})^2 \qquad \text{| From (1)}$$

From (2), $\rho = l$

Hence radius of curvature = Length of normal.

From (2), $\rho = \frac{a}{4}\,(e^{x/a} + e^{-x/a})^2$

$$= \frac{a}{4}\left(\frac{2y}{a}\right)^2 \qquad \text{| From (1)}$$

$$= \frac{y^2}{a}$$

$\therefore$ ρ varies as square of the ordinate.

Example 9. *If ρ_1, ρ_2 are the radii of curvature at the extremities of a focal chord of a parabola, where semilatus rectum is l. Prove that $(\rho_1)^{-2/3} + (\rho_2)^{-2/3} = (l)^{-2/3}$.*

(P.T.U., Jan. 2009, May 2012)

Sol. (Chord of the parabola which passes through the focus is called focal chord)

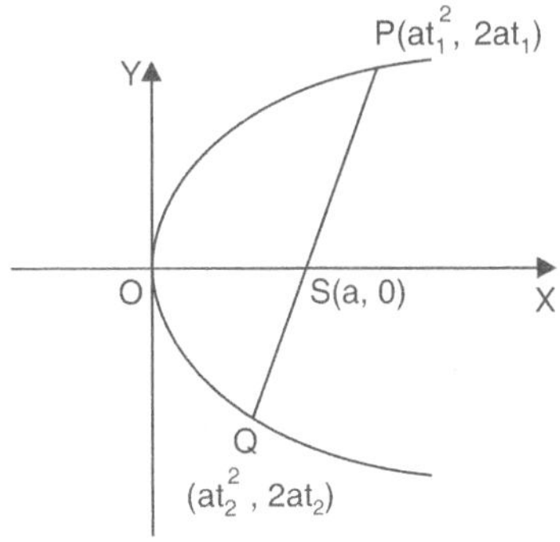

Let equation of the parabola be

$y^2 = 4ax$, where semilatus rectum = $2a = l$ (given)

$\therefore \quad a = l/2$ ...(1)

Let P$(at_1^2, 2at_1)$ be any point on the parabola.

Let Q, the other extremity of the focal chord has co-ordinates Q$(at_2^2, 2at_2)$. Equation of the chord PQ is

$$y - 2at_1 = \frac{2at_2 - 2at_1}{at_2^2 - at_1^2}\,(x - at_1^2) \qquad \left[y - y_1 = \frac{y_2 - y_1}{x_2 - x_1}(x - x_1)\right]$$

It passes through $(a, 0)$

$$-2at_1 = \frac{2a(t_2 - t_1)}{a(t_2 - t_1)(t_2 + t_1)}\,(a - at_1^2)$$

$$= \frac{2a}{t_2 + t_1}\,(1 - t_1^2)$$

$$-t_1(t_2 + t_1) = 1 - t_1^2 \quad \text{or} \quad -t_1t_2 - t_1^2 = 1 - t_1^2$$

$$\therefore \qquad t_1t_2 = -1 \quad \therefore \quad t_2 = -\frac{1}{t_1} \qquad \text{...(2)}$$

Now as proved in Example 4 (i part) the radius of curvature $= \dfrac{2(x+a)^{3/2}}{\sqrt{a}}$

for ρ_1 ; put $(x, y) = (at_1^2, 2at_1)$ in (3)

$$\rho_1 = \frac{2(at_1^2 + a)^{3/2}}{\sqrt{a}} ;$$

for ρ_2 ; put $(x, y) = (at_2^2, 2at_2)$ in (3)

$$\rho_2 = \frac{2(at_2^2 + a)^{3/2}}{\sqrt{a}}$$

$$\rho_1^{2/3} = \left(\frac{2}{\sqrt{a}}\right)^{2/3} \cdot (at_1^2 + a); \; \rho_2^{2/3} = \left(\frac{2}{\sqrt{a}}\right)^{2/3} \cdot (at_2^2 + a)$$

$$\rho_1^{-2/3} + \rho_2^{-2/3} = \left(\frac{2}{\sqrt{a}}\right)^{-2/3} \left[\frac{1}{a(t_1^2+1)} + \frac{1}{a(t_2^2+1)}\right]$$

$$= \frac{2^{-2/3}}{a^{-1/3}} \cdot \frac{1}{a} \cdot \frac{t_2^2 + 1 + t_1^2 + 1}{t_1^2 t_2^2 + t_1^2 + t_2^2 + 1}$$

$$= \frac{2^{-2/3}}{a^{2/3}} \cdot \frac{t_1^2 + t_2^2 + 2}{1 + t_1^2 + t_2^2 + 1} \qquad \because \; t_1 t_2 = -1 \qquad | \text{ from (2)}$$

$$= (2a)^{-2/3} \cdot \frac{t_1^2 + t_2^2 + 2}{t_1^2 + t_2^2 + 2} = l^{-2/3}$$

Hence $\quad \rho_1^{-2/3} + \rho_2^{-2/3} = l^{-2/3}$

Example 10. *(i) Find the point on the curve $y = e^x$ at which the curvature is maximum (ii) Show that the tangent at this point forms with axes of the co-ordinates a triangle whose sides are in the ratio $1 : \sqrt{2} : \sqrt{3}$.*

Sol. (i) Equation of the curve is

$$y = e^x \qquad \text{...(1)}$$

$$\frac{dy}{dx} = e^x = y, \quad \frac{d^2y}{dx^2} = e^x = y$$

$\therefore$ Curvature

$$\text{K} = \frac{y_2}{(1 + y_1^2)^{3/2}}$$

$$\text{K} = \frac{y}{(1 + y^2)^{3/2}}$$

K will be maximum, when $\dfrac{d\text{K}}{dy} = 0$

$$\frac{d\text{K}}{dy} = \frac{(1+y^2)^{3/2} \cdot 1 - y \cdot \frac{3}{2}(1+y^2)^{1/2} \cdot 2y}{(1+y^2)^3}$$

$$= \frac{(1+y^2)^{1/2}\,[1 + y^2 - 3y^2]}{(1+y^2)^3} = \frac{1 - 2y^2}{(1+y^2)^{5/2}}$$

$$\frac{d\text{K}}{dy} = 0 \;\Rightarrow\; 1 - 2y^2 = 0 \quad \therefore \quad y^2 = \frac{1}{2} \;\Rightarrow\; y = \pm\frac{1}{\sqrt{2}}$$

y cannot be –ve $\;\because\;$ y is exponential function e^x

$\therefore \quad y = \dfrac{1}{\sqrt{2}}$ ...(2)

From (1), $\dfrac{1}{\sqrt{2}} = e^x \quad \therefore \quad x = \log \dfrac{1}{\sqrt{2}} = -\log \sqrt{2} = -\dfrac{1}{2}\log 2$

Also $\dfrac{d^2K}{dy^2}$ is –ve at $y = \dfrac{1}{\sqrt{2}}$

$\therefore$ Point on the curve where curvature is maximum is $\left(-\dfrac{1}{2}\log 2, \dfrac{1}{\sqrt{2}}\right)$

(*ii*) Tangent at P(*x*, *y*) to (1) is

$$Y - y = \frac{dy}{dx}(X - x)$$

$$Y - y = y(X - x)$$

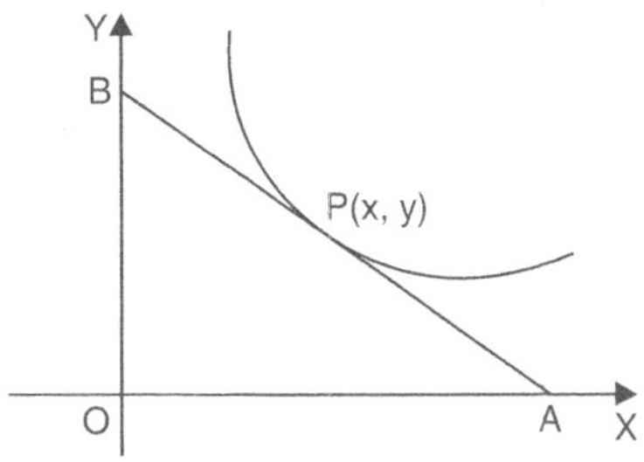

It intersects *x*-axis at A where Y = 0

$\therefore \quad -y = y\,(X - x)$

$\therefore \quad -1 = X - x$

$\therefore \quad X = x - 1$

$\therefore \quad OA = x - 1$

It intersects *y*-axis at B when X = 0

$$Y - y = y(-x)$$

$\therefore \quad Y = y(1 - x) \quad \therefore \quad OB = y(1 - x)$

Now $\quad AB = \sqrt{(x-1)^2 + y^2(1-x)^2} = |\,x - 1\,|\sqrt{1 + y^2}$

and $\quad OA : OB : AB = |\,x - 1\,| : y|\,1 - x\,| : |\,x - 1\,|\sqrt{1 + y^2}$

$$= 1 : y : \sqrt{1 + y^2}$$

Substituting $\quad y = \dfrac{1}{\sqrt{2}} \quad \because \quad$ P is $\left(-\dfrac{1}{2}\log 2, \dfrac{1}{\sqrt{2}}\right)$

$\therefore \quad OA : OB : AB = 1 : \dfrac{1}{\sqrt{2}} : \sqrt{1 + \dfrac{1}{2}}$

$$= \sqrt{2} : 1 : \sqrt{3}$$

or $\quad OB : OA : AB = 1 : \sqrt{2} : \sqrt{3}$.

Example 11. *Show that radius of curvature at each point of the curve x = a(cos t + log tan t/2), y = a sin t is inversely proportional to the length of the normal intercepted between the point on the curve and x-axis.*

Sol. Equation of the curve is $x = a(\cos t + \log \tan t/2)$; $y = a \sin t$ ...(1)

$$x' = \frac{dx}{dt} = a\left(-\sin t + \frac{\frac{1}{2}\sec^2 t/2}{\tan t/2}\right);\ y' = a\cos t$$

$$= a\left(-\sin t + \frac{1}{2}\cdot\frac{1}{\sin t/2\cos t/2}\right);\ y' = a\cos t$$

$$= a\left(-\sin t + \frac{1}{\sin t}\right);\ y' = a\cos t$$

$$x' = a\,\frac{\cos^2 t}{\sin t};\; y' = a\cos t$$

$$x'' = a\,\frac{\sin t\, 2\cos t\,(-\sin t) - \cos^2 t\cos t}{\sin^2 t};\quad y'' = -a\sin t$$

$$= a\,\frac{-2\sin^2 t\cos t - \cos^3 t}{\sin^2 t};\; y'' = -a\sin t$$

$$= \frac{-a\cos t}{\sin^2 t}(2\sin^2 t + \cos^2 t);\quad y'' = -a\sin t$$

$$x'' = \frac{-a\cos t\,(1+\sin^2 t)}{\sin^2 t};\quad y'' = -a\sin t$$

$$\rho = \frac{[(x')^2 + (y')^2]^{3/2}}{x'y'' - x''y'} = \frac{\left[a^2\,\dfrac{\cos^4 t}{\sin^2 t} + a^2\cos^2 t\right]^{3/2}}{a\,\dfrac{\cos^2 t}{\sin t}(-a\sin t) + \dfrac{a\cos t}{\sin^2 t}(1+\sin^2 t)\,a\cos t}$$

$$= \frac{\left(\dfrac{a^2\cos^2 t}{\sin^2 t}\right)^{3/2}\left(\cos^2 t + \sin^2 t\right)^{3/2}}{a^2\left[-\cos^2 t + \dfrac{\cos^2 t}{\sin^2 t}(1+\sin^2 t)\right]}$$

$$= \frac{a^3\cos^3 t\cdot 1}{\sin^3 t\cdot a^2\cdot\dfrac{\cos^2 t}{\sin^2 t}[-\sin^2 t + 1 + \sin^2 t]} = \frac{a\cos t}{\sin t}$$

$\therefore \qquad \rho = a\cot t$ ...(2)

Now
$$\frac{dy}{dx} = \frac{\dfrac{dy}{dt}}{\dfrac{dx}{dt}} = \frac{y'}{x'} = \frac{a\cos t}{\dfrac{a\cos^2 t}{\sin t}} = \frac{\sin t}{\cos t} = \tan t$$

l = length of the normal at given point

$$= y\sqrt{1+\left(\frac{dy}{dx}\right)^2} = a\sin t\sqrt{1+\tan^2 t} = a\sin t\sec t = a\tan t$$

$\therefore$ From (2), $\rho = a\cdot\dfrac{a}{l} = \dfrac{a^2}{l}$

$\therefore$ ρ is inversely proportional to the length of the normal.

Example 12. *Show that the ellipse $x = a\cos t$, $y = b\sin t$, $a > 0$, $b > 0$ has its largest curvature on its major axis and its smallest curvature on its minor axis.* **(P.T.U., Dec. 2004)**

Sol. Equation of the ellipse is

$$x = a\cos t;\qquad y = b\sin t \qquad ...(1)$$

$$x' = -a\sin t;\qquad y' = b\cos t$$

$$x'' = -a\cos t;\qquad y'' = -b\sin t$$

$$\text{Curvature K} = \frac{x'y'' - x''y'}{(x'^2 + y'^2)^{3/2}} = \frac{(-a \sin t)(-b \sin t) - (-a \cos t)(b \cos t)}{(a^2 \sin^2 t + b^2 \cos^2 t)^{3/2}}$$

$$\text{K} = \frac{ab(\sin^2 t + \cos^2 t)}{(a^2 \sin^2 t + b^2 \cos^2 t)^{3/2}}$$

$$\text{K} = \frac{ab}{\left\{a^2 \frac{(1-\cos 2t)}{2} + b^2 \frac{(1+\cos 2t)}{2}\right\}^{3/2}}$$

$$\text{K} = \frac{2\sqrt{2}ab}{\{(a^2+b^2) - (a^2-b^2)\cos 2t\}^{3/2}}$$

(*i*) K will be maximum, when $\cos 2t$ is maximum and max $\cos 2t = 1$

i.e., $\cos 2t = \cos 0, \cos 2\pi \quad \therefore \quad t = 0, \pi$

$\therefore$ From (1), $x = \pm a, y = 0$

$\therefore$ K is maximum at $(a, 0)$ and $(-a, 0)$

i.e., K is maximum on its major axis.

(*ii*) K will be minimum, when $\cos 2t$ is minimum and minimum value of $\cos 2t = -1$

$\therefore \quad 2t = \pi, -\pi \qquad \therefore \qquad t = \frac{\pi}{2}, -\frac{\pi}{2} \qquad \therefore \qquad x = 0, y = \pm b$

$\therefore$ K is minimum on minor axis.

Example 13. *Find the radius of curvature for the catenary*

$$x = c \log(s + \sqrt{c^2 + s^2}), \quad y = \sqrt{c^2 + s^2}.$$

Sol.

$$\frac{dx}{ds} = c \cdot \frac{1}{s + \sqrt{c^2+s^2}}\left(1 + \frac{2s}{2\sqrt{c^2+s^2}}\right); \frac{dy}{ds} = \frac{1}{2}\frac{2s}{\sqrt{c^2+s^2}}$$

$$\frac{dx}{ds} = c\frac{\sqrt{c^2+s^2} + s}{(s + \sqrt{c^2+s^2})\sqrt{c^2+s^2}}; \frac{dy}{ds} = \frac{s}{\sqrt{c^2+s^2}}$$

$$\frac{dx}{ds} = \frac{c}{\sqrt{c^2+s^2}}; \frac{dy}{ds} = \frac{s}{\sqrt{c^2+s^2}}$$

$$\frac{d^2x}{ds^2} = c\left(-\frac{1}{2}\right)(c^2+s^2)^{-3/2}(2s); \frac{d^2y}{ds^2} = \frac{\sqrt{c^2+s^2} - s\left(\frac{1}{2}\right)\frac{2s}{\sqrt{c^2+s^2}}}{c^2+s^2}$$

$$\frac{d^2x}{ds^2} = -\frac{cs}{(c^2+s^2)^{3/2}}; \frac{d^2y}{ds^2} = \frac{c^2+s^2-s^2}{(c^2+s^2)^{3/2}}$$

$$\frac{d^2x}{ds^2} = -\frac{cs}{(c^2+s^2)^{3/2}}; \frac{d^2y}{ds^2} = \frac{c^2}{(c^2+s^2)^{3/2}}$$

Now,

$$\frac{1}{\rho^2} = \left(\frac{d^2x}{ds^2}\right)^2 + \left(\frac{d^2y}{ds^2}\right)^2 = \frac{c^2s^2}{(c^2+s^2)^3} + \frac{c^4}{(c^2+s^2)^3}$$

$$= \frac{c^2s^2 + c^4}{(c^2+s^2)^3} = \frac{c^2(s^2+c^2)}{(c^2+s^2)^3} = \frac{c^2}{(c^2+s^2)^2}$$

$$\therefore \qquad \rho = \frac{c^2+s^2}{c}.$$

Example 14. *Find radius of curvature at (2a, 0) of the curve*

$$(x^2 + y^2)^2 - 2ax(x^2 + y^2) - a^3y = 0.$$

Sol. Equation of the curve is

$$f(x, y) = (x^2 + y^2)^2 - 2ax\,(x^2 + y^2) - a^3y = 0$$

$$f(x, y) = x^4 + y^4 + 2x^2y^2 - 2ax^3 - 2axy^2 - a^3y$$

$$f_x = 4x^3 + 4xy^2 - 6ax^2 - 2ay^2\ ;\ f_x(2a, 0) = 8a^3$$

$$f_y = 4y^3 + 4x^2y - 4axy - a^3\ ;\ f_y\,(2a, 0) = -\,a^3$$

$$f_{xx} = 12x^2 + 4y^2 - 12ax\ ;\ f_{xx}\,(2a, 0) = 24a^2$$

$$f_{yy} = 12y^2 + 4x^2 - 4ax\ ;\ f_{yy}\,(2a, 0) = 8a^2$$

$$f_{xy} = 8xy - 4ay;\ f_{xy}\,(2a, 0) = 0$$

$$\rho = -\frac{(f_x^2 + f_y^2)^{3/2}}{f_{xx}f_y^2 - 2f_{xy}\,f_xf_y + f_{yy}f_x^2} = -\frac{(64a^6 + a^6)^{3/2}}{(24a^2)(a^6) - 0 + (8a^2)(64a^6)}$$

$$= -\frac{(65)^{3/2}a}{536}$$

$$\rho = \frac{(65)^{3/2}a}{536} \text{ (in magnitude).}$$

Example 15. (*a*) *Find the curvature and radius of curvature of the curve $x = a\,(\theta - \sin\theta)$, $y = a(1 - \cos\theta)$* **(P.T.U., Jan 2010)**

(*b*) *The tangents at two points of P, Q on the cycloid $x = a(\theta - \sin\theta)$, $y = a\,(1 - \cos\theta)$ are at right angles ; show that if ρ_1, ρ_2 be the radii of curvature at these points then $\rho_1^2 + \rho_2^2 = 16a^2$.*

Sol. (*a*) Equation of the cycloids

$$x = a(\theta - \sin\theta) \qquad y = a(1 - \cos\theta)$$

$$x' = \frac{dx}{d\theta} = a\,(1 - \cos\theta), \qquad y' = a\sin\theta$$

$$x'' = \frac{d^2x}{d\theta^2} = a\sin\theta, \qquad y'' = a\cos\theta$$

$$\rho = \frac{(x'^2 + y'^2)^{3/2}}{x'y'' - x''y'} = \frac{[a^2\,(1 - \cos\theta)^2 + a^2\sin^2\theta]^{3/2}}{a(1 - \cos\theta)\,a\cos\theta - a\sin\theta\,.\,a\sin\theta}$$

$$\rho = \frac{a^3[1 - 2\cos\theta + 1]^{3/2}}{a^2[\cos\theta - \cos^2\theta - \sin^2\theta]} = a\,\frac{(2 - 2\cos\theta)^{3/2}}{\cos\theta - 1} = -\,a\,2^{3/2}\,(1 - \cos\theta)^{1/2}$$

$$\rho = -\,2\sqrt{2}\,a\,(2\sin^2\theta/2)^{1/2} = -\,4a\sin\theta/2$$

$$\rho = 4a\sin\theta/2 \text{ (in Magnitude)} \qquad ...(1)$$

$$\therefore \quad \text{Radius of curvature} = 4a\sin\frac{\theta}{2}$$

and

$$\text{curvature} = \frac{1}{4a\sin\theta/2}$$

(*b*) Let tangent at P makes an angle θ_1 with the positive direction of x-axis and the tangent then at Q makes an angle θ_2 with the +ve direction of x-axis.

Slope of tangent at P $= \left(\dfrac{dy}{dx}\right)_{\text{at } \theta_1}$,

$$= \frac{a \sin \theta_1}{a(1 - \cos \theta_1)} = \frac{2 \sin \theta_1/2 \,.\, \cos \theta_1/2}{2 \sin^2 \frac{\theta_1}{2}} = \cot \frac{\theta_1}{2}$$

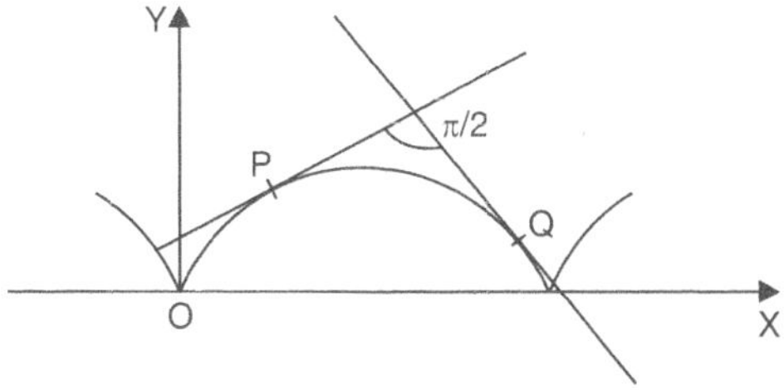

Similarly, slope of the tangent at Q $= \cot \dfrac{\theta_2}{2}$

$\because$ Tangents are perpendicular

$\therefore$ product of their slopes = –1

$$\therefore \quad \cot \frac{\theta_1}{2} \cot \frac{\theta_2}{2} = -1$$

$$\therefore \quad = -\tan \frac{\theta_1}{2} = \cot\left(\frac{\pi}{2} + \frac{\theta_1}{2}\right)$$

$$\therefore \quad \theta_2 = \pi + \theta_1$$

Put $\theta = \theta_1$ and $\theta = \pi + \theta_1$ in (1) successively, we get

$$\rho_1 = 4a \sin \frac{\theta_1}{2}$$

$$\rho_2 = 4a \sin \frac{\pi + \theta_1}{2} = 4a \cos \frac{\theta_1}{2}$$

$$\therefore \quad \rho_1^2 + \rho_2^2 = 16a^2.$$

Example 16. *Show that radius of curvature at the point* $(-2a, 2a)$ *on the curve* $x^2y = a(x^2 + y^2)$ *is* $-2a$. **(P.T.U., Dec. 2006)**

Sol. Let $f(x, y) = x^2y - a(x^2 + y^2)$

$$f_x = \frac{\partial f}{\partial x} = 2xy - 2ax = 2x(y - a);\ f_x(-2a, 2a) = -4a^2$$

$$f_y = \frac{\partial f}{\partial y} = x^2 - 2ay\ ;\ f_y(-2a, 2a) = 0$$

$$f_{xx} = 2(y - a)\ ;\ f_{xx}(-2a, 2a) = 2a^2$$

$$f_{xy} = 2x\ ;\ f_{xy}(-2a, 2a) = -4a$$

$$f_{yy} = -2a\ ;\ f_{yy}(-2a, 2a) = -2a$$

$$\rho = -\frac{(f_x^2 + f_y^2)^{3/2}}{f_{xx} f_y^2 - 2f_{xy} f_x f_y + f_{yy} f_x^2}$$

$$= -\frac{(16a^4 + 0)^{3/2}}{0 - 0 + (-2a)\,16a^4} = -\frac{64a^6}{-32a^5} = 2a$$

Hence $\rho = 2a$.

TEST YOUR KNOWLEDGE

1. Find radius of curvature at (s, ψ) of

   (*i*) $s = 8a \sin^2 \dfrac{\psi}{6}$

   (*ii*) $s = c \log \tan \left(\dfrac{\pi}{4} + \dfrac{\psi}{2}\right)$

2. Find radius of curvature of the following curves at any point (x, y) of the curve $xy = c^2$.

3. Find the radius of curvature at any point of the curve $y = c \cosh \dfrac{x}{c}$ and show that the curvature at any point varies inversely as the square of the ordinate.

4. Find the points on the parabola $y^2 = 8x$ at which the radius of curvature is $7\dfrac{13}{16}$.

5. Find the radius of curvature for $\sqrt{\dfrac{x}{a}} - \sqrt{\dfrac{y}{b}} = 1$ at the points, where it touches the co-ordinate axes.
   [**Hint.** curve touches x-axis at $(a, 0)$ and y-axis at $(0, b)$]

6. Find the radius of curvature at any point of the following curves:

   (*i*) $x = t - c \sinh \dfrac{t}{c} \cosh \dfrac{t}{c}, y = 2c \cosh \dfrac{t}{c}$

   (*ii*) $x = a (\cos t + t \sin t); y = a (\sin t - t \cos t)$

   (*iii*) $x = a \sin 2\theta (1 + \cos 2\theta); y = a \cos 2\theta (1 - \cos 2\theta)$

   (*iv*) $x = a (\theta + \sin \theta), y = a (1 - \cos \theta)$. ***(P.T.U., May 2004)***

   [Consult solved example 15]

7. Show that for the curve $s = f(x)$

$$\rho = \frac{\left(\dfrac{ds}{dx}\right)^2 \sqrt{\left(\dfrac{ds}{dx}\right)^2 - 1}}{\dfrac{d^2s}{dx^2}}$$

   , hence prove that for the curve $s = ce^{x/c}$, $c\rho = s\sqrt{s^2 - c^2}$.

8. Find ρ for the curve $s^2 = 8ay$.

9. Prove that for any curve $\dfrac{r}{\rho} = \sin \phi \left(\dfrac{d\phi}{d\theta}\right)$, where ρ is the radius of curvature and $\tan \phi = r \dfrac{d\theta}{dr}$.

Answers

1. (*i*) $4a \sin \dfrac{\psi}{3}$ (*ii*) $a \sec \psi$

2. $\dfrac{(x^4 + c^4)^{3/2}}{2c^2x^3}$

3. $c \cosh^2 \dfrac{x}{2}$

4. $\left(\dfrac{9}{8}, 3\right), \left(\dfrac{9}{8}, -3\right)$

5. $\dfrac{2a^2}{b}, \dfrac{2(b^2 + 4a^2)^{3/2}}{ab}$

6. (*i*) $2c \cosh^2 \dfrac{t}{c} \sinh \dfrac{t}{c}$ (*ii*) at (*iii*) $4a \cos 3\theta$ (*iv*) $4a \cos \theta/2$.

8. $\sqrt{8a(2a - y)}$

1.24. RADIUS OF CURVATURE AT THE ORIGIN

We are already familiar with the method of finding the radius of curvature at the origin *i.e.*, putting $x = 0, y = 0$ in the value of $\rho = \dfrac{(1+y_1^2)^{3/2}}{y_2}$ but now we will have a direct method to find the radius of curvature at the origin.

1. Method of Expansion

By Maclaurin's theorem, we have $y = f(x) = f(0) + xf'(0) + \dfrac{x^2}{2!} f''(0) + \cdots$.

Since the curve passes through the origin

$$\therefore \quad f(0) = 0$$

$$y = xf'(0) + \frac{x^2}{2!} f''(0) + \cdots$$

or
$$y = px + q\frac{x^2}{2!} + \cdots$$

where
$$p = f'(0) = \left(\frac{dy}{dx}\right)_{(0,0)}, \text{ value of } y_1 \text{ at } (0, 0);$$

$$q = f''(0) = \left(\frac{d^2y}{dx^2}\right)_{(0,0)}; \text{ value of } y_2 \text{ at } (0, 0)$$

$$\therefore \quad \rho \text{ at origin} = \frac{(1+y_1^2)^{3/2}}{y_2} = \frac{(1+p^2)^{3/2}}{q}.$$

Rules to find radius of curvature at the origin

(*a*) Put $y = px + \dfrac{qx^2}{2!} + \cdots$ in the equation of the curve.

(*b*) Equate the coefficients of like powers of x on both sides and find p and q.

(*c*) Substitute the values of p and q in $\rho = \dfrac{(1+p^2)^{3/2}}{q}$ and get radius of curvature.

2. Newton's Method to Find Radius of Curvature at the Origin

Case I. If the curve passes through the origin and the axis of x is a tangent to the curve at the origin, then we have at the origin $x = 0, y = 0, \dfrac{dy}{dx} = 0$ *i.e.*, $p = 0$.

$\therefore$ by Maclaurin's theorem (as proved above)

$$y = px + \frac{qx^2}{2!} \cdots$$

$$y = \frac{qx^2}{2!} + \cdots$$

$$\frac{2y}{x^2} = q + \text{terms containing } x.$$

Take limit when $x \to 0$ and $\therefore$ y also $\to 0$

We have $$\underset{x \to 0}{\text{Lt}} \frac{2y}{x^2} = q$$

Substituting the values of p and q in $\rho = \dfrac{(1+p^2)^{3/2}}{q} = \dfrac{(1+0)^{3/2}}{\frac{2y}{x^2}} = \dfrac{1}{\frac{2y}{x^2}}$

$$\therefore \qquad \rho = \underset{x \to 0}{\text{Lt}} \frac{x^2}{2y}.$$

Case II. Similarly, if axis of y is the tangent at (0, 0) to the curve, we have

$$\rho = \underset{x \to 0}{\text{Lt}} \frac{y^2}{2x}$$

Rules to find radius of curvature at the origin

(1) Obtain the equation of the tangent to the curve at the origin (tangent at origin is obtained by equating the lowest degree terms in the equation of the curve to zero)

(2) If x-axis is the tangent at the origin, then $\rho = \underset{x \to 0}{\text{Lt}} \dfrac{x^2}{2y}$.

(3) If y-axis is the tangent at the origin, then $\rho = \underset{x \to 0}{\text{Lt}} \dfrac{y^2}{2x}$.

(4) If neither x-axis nor y-axis is the tangent at the origin, then put $y = px + \dfrac{ax^2}{2!}$ ... and proceed as in 1.24 (1st part).

ILLUSTRATIVE EXAMPLES

Example 1. *Find the radius of curvature at the origin for the curve*

$$y^2 = x^2 \frac{a+x}{a-x}.$$

Sol. Equation of the curve is

$$y^2(a - x) = x^2 (a + x) \quad \text{or} \quad y^2 (a - x) - x^2(a + x) = 0$$

or $$(- y^2x - x^3) + a(y^2 - x^2) = 0 \quad \text{or} \quad (y^2x + x^3) + a(x^2 - y^2) = 0 \qquad ...(1)$$

Now tangent at the origin are given by $x^2 - y^2 = 0$ *i.e.*, $x = y, x = -y$ which are neither x-axis nor y-axis.

$\therefore$ put $y = px + \dfrac{qx^2}{2!} + \cdots$ in (1)

$$x\left[px + \frac{qx^2}{2!} + \cdots\right]^2 + x^3 + ax^2 - a\left[px + \frac{qx^2}{2!} + \cdots\right]^2 = 0$$

Equate coeffs. of x^2, x^3 to zero

$$a - ap^2 = 0 \qquad \therefore \qquad p^2 = 1 \qquad \therefore \qquad p = \pm 1$$

also $$p^2 + 1 - apq = 0 \qquad \therefore \qquad q = \frac{p^2 + 1}{ap}$$

$$p = 1, \quad q = \frac{2}{a}$$

$$p = -1, \; q = \frac{2}{-a}$$

$$\therefore \quad \rho \text{ at } (0, 0) = \frac{(1+p^2)^{3/2}}{q} = \frac{(1+1)^{3/2}}{\pm 2/a} = \pm \frac{a}{2} \cdot 2^{3/2} = \pm\sqrt{2}\,a.$$

Example 2. *Find radius of curvature at the origin of the curve $x = 1 - t^2$, $y = t - t^3$.*

Sol. Equation of the curve is $x = 1 - t^2$, $y = t - t^3$

Change the equation to cartesian form by eliminating t $\therefore$ $t = \sqrt{1-x}$

$$\therefore \quad y = \sqrt{1-x} - (1-x)^{3/2}$$

or $$y = \sqrt{1-x}\,\{1 - 1 + x\} = x\sqrt{1-x}$$

or $$y^2 = x^2(1-x) \quad \text{or} \quad x^3 + y^2 - x^2 = 0 \qquad \ldots(1)$$

Tangent at the origin are

$y^2 - x^2 = 0$ *i.e.,* $y = \pm x$ which are neither x-axis nor y-axis.

$\therefore$ Put $y = px + \frac{qx^2}{2!} + \cdots$ in (1)

$$x^3 + \left[px + \frac{qx^2}{2}\right]^2 - x^2 = 0$$

Equating coeffs. of x^2, x^3 to zero

$$p^2 - 1 = 0 \quad \therefore \quad p = \pm 1$$

$$1 + pq = 0 \quad \therefore \quad pq = -1 \; \therefore \quad q = -\frac{1}{p}$$

when $p = 1, q = -1; p = -1, q = 1$

For $p = 1, q = -1$

$$\therefore \quad \rho = \frac{(1+p^2)^{3/2}}{q} \quad \therefore \quad \rho = \frac{(1+1)^{3/2}}{-1} = \frac{2^{3/2}}{-1} = -2\sqrt{2}$$

for $p = -1, q = 1$

$$\rho = \frac{(1+1)^{3/2}}{1} = 2\sqrt{2}$$

$\therefore$ Radius of curvature at (0, 0) = $2\sqrt{2}$ in magnitude.

Example 3. *Find the radius of curvature at origin for the curve*

$$y^4 + x^3 + a(x^2 + y^2) - a^2y = 0. \qquad \textbf{(P.T.U., May 2015)}$$

Sol. Equation of the tangent at origin to the curve is $-a^2y = 0$ *i.e.,* $y = 0$ *i.e.,* x-axis

$\therefore$ By Newton's method $\rho = \underset{x \to 0}{\text{Lt}} \frac{x^2}{2y}$.

Divide the equation by $2y$

$$\frac{y^3}{2} + \frac{x^3}{2y} + \frac{ax^2}{2y} + \frac{ay}{2} - \frac{a^2}{2} = 0 \quad \text{or} \quad \frac{y^3}{2} + x \cdot \frac{x^2}{2y} + a \cdot \frac{x^2}{2y} + \frac{ay}{2} - \frac{a^2}{2} = 0$$

Take $y \to 0$ (then $x \to 0$) and $\frac{x^2}{2y} \to \rho$, we get

$$0 + 0 \cdot \rho + a\rho + 0 - \frac{a^2}{2} = 0 \quad \text{or} \quad a\rho - \frac{a^2}{2} = 0$$

$$\therefore \qquad \rho = \frac{a}{2}$$

TEST YOUR KNOWLEDGE

1. Find radius of curvature at origin of the following curves:
   (*i*) $a(y^2 - x^2) = x^3$ (*ii*) $y^2 - 3xy + 2x^2 - x^3 + y^4 = 0$
   (*iii*) $y^2(y + 1) = 4x^2$ (*iv*) $y^2 - 3xy - 4x^2 + x^3 + x^4y + y^5 = 0$.
2. Apply Newton's method to find the radius of curvature at the origin of the following curves:
   (*i*) $3x^2 + 4x^3 - 12y = 0$ (*ii*) $x^3 - 2x^2y + 3xy^2 - 4y^3 + 5x^2 - 6xy + 7y^2 - 8y = 0$
   (*iii*) $x^3 + y^3 = 3axy$ (*iv*) $x^4 - y^4 + 4x^2y + xy - y^2 + 2x = 0$.

Answers

1. (*i*) $2\sqrt{2}\,a$ (*ii*) $\sqrt{2}, \frac{5\sqrt{5}}{2}$ (*iii*) $\frac{5\sqrt{3}}{4}$ (*iv*) $\frac{85\sqrt{17}}{2}$
2. (*i*) 2 (*ii*) $\frac{4}{5}$ (*iii*) $\frac{3a}{2}$ (*iv*) 1.

1.25. CENTRE OF CURVATURE, CIRCLE OF CURVATURE AND CHORD OF CURVATURE

Definitions. (*a*) The *centre of curvature* of a curve at any point P is a point C on the positive direction of the normal at P and at a distance ρ from P.

(*b*) *Circle of curvature* of a curve at a point is a circle with C (the centre of curvature) as the centre and ρ (the radius of curvature) as the radius.

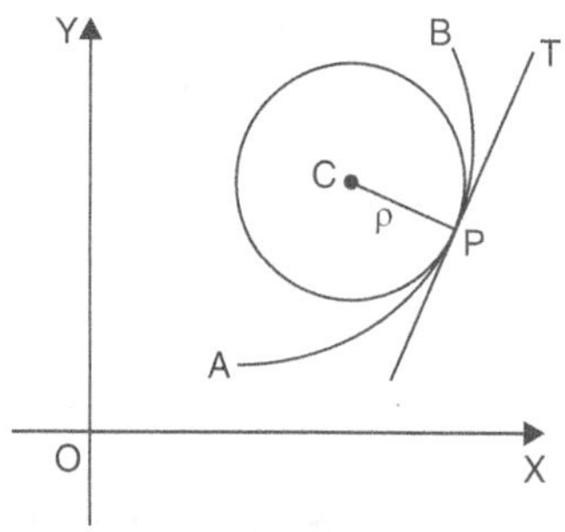

Circle of curvature touches the curve at P.

The given curve and the circle of curvature have the same curvature at that point P.

(*c*) *Chord of curvature* is any chord of the circle of curvature passing through the point P.

1.26. (a) CO-ORDINATES OF THE CENTRE OF CURVATURE FOR ANY POINT P(*x*, *y*) OF THE CURVE *y* = *f* (*x*)

Proof. Let P(x, y) be any point on the curve. Let C($\bar{x}, \bar{y}$) be the centre of curvature and ρ, the radius of curvature.

Let tangent at P makes an angle ψ with the positive direction of x-axis.

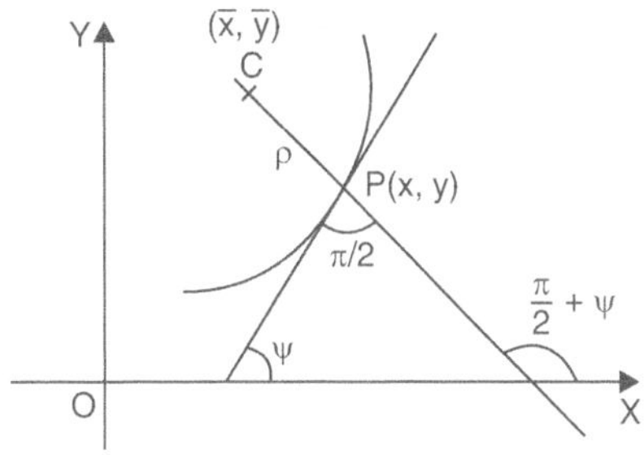

As centre C lies on the normal at P, $\therefore$ the normal at P makes an angle of $\frac{\pi}{2} + \psi$ with the positive direction of x-axis.

Equation of the normal to the curve at P(x, y) is

$$\frac{X - x}{\cos\left(\frac{\pi}{2} + \psi\right)} = \frac{Y - y}{\sin\left(\frac{\pi}{2} + \psi\right)}$$ where X, Y are current co-ordinates of any point on this curve

or $\frac{X - x}{-\sin\psi} = \frac{Y - y}{\cos\psi} = r$, where r is the distance between the variable point (X, Y) and (x, y).

Any point on the normal is $(x - r\sin\psi, y + r\cos\psi)$.

As $C(\bar{x}, \bar{y})$ is a point on the normal at a distance ρ from P

$$\therefore \qquad \bar{x} = x - \rho\sin\psi, \ \bar{y} = y + \rho\cos\psi$$

Substituting the values of ρ, $\sin\psi$ and $\cos\psi$

where
$$\rho = \frac{(1 + y_1^2)^{3/2}}{y_2},$$

$$\sin\psi = \frac{dy}{ds} = \frac{dy}{dx} \cdot \frac{dx}{ds} = y_1 \frac{1}{\frac{ds}{dx}} = \frac{y_1}{\sqrt{1 + y_1^2}}$$

$$\cos\psi = \frac{dx}{ds} = \frac{1}{\frac{ds}{dx}} = \frac{1}{\sqrt{1 + y_1^2}}$$

$$\therefore \qquad \bar{x} = x - \frac{(1 + y_1^2)^{3/2}}{y_2} \cdot \frac{y_1}{\sqrt{1 + y_1^2}}, \ \bar{y} = y + \frac{(1 + y_1^2)^{3/2}}{y_2} \cdot \frac{1}{\sqrt{1 + y_1^2}}$$

$$\therefore \qquad \bar{x} = x - \frac{y_1(1 + y_1^2)}{y_2}, \quad \bar{y} = y + \frac{1 + y_1^2}{y_2}$$

Hence centre of curvature for the point (x, y) is $\left(x - \frac{y_1(1 + y_1^2)}{y_2}, y + \frac{1 + y_1^2}{y_2}\right)$.

1.26. (b) CIRCLE OF CURVATURE

Equation of the circle of curvature is

$(x - \bar{x})^2 + (y - \bar{y})^2 = \rho^2$, where $(\bar{x}, \bar{y})$ are the co-ordinates of the centre of curvature corresponding to P(x, y), and ρ is the radius of curvature at P.

1.26. (c) LENGTHS OF THE CHORDS OF CURVATURE OF A CURVE PARALLEL TO *x*-AXIS AND *y*-AXIS

Proof. Let P(x, y) be any point on the curve, PT tangent at P to the curve and let this tangent makes, an angle of ψ with the positive direction of x-axis. Let $C(\bar{x}, \bar{y})$ be the centre of curvature corresponding to P. Draw the circle of curvature.

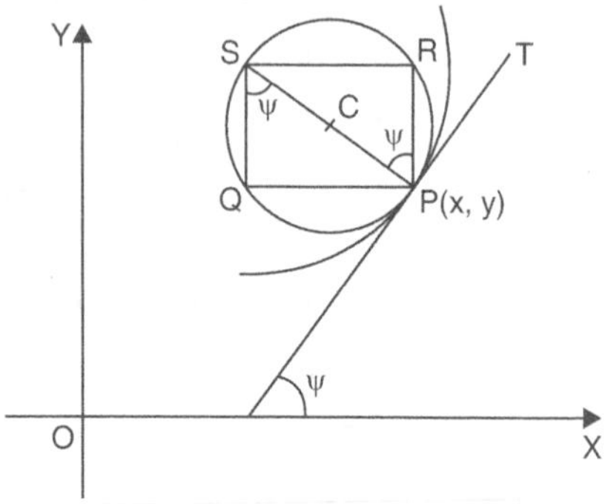

Let PQ and PR be two chords of curvature, one parallel to x-axis other parallel to y-axis.

$\angle PSQ = \angle RPS = \psi$

In ΔPQS ; $PQ = PS \sin \psi = 2\rho \sin \psi$

$$PQ = 2 \cdot \frac{(1+y_1^2)^{3/2}}{y_2} \cdot \frac{y_1}{\sqrt{1+y_1^2}} = \frac{2y_1(1+y_1^2)}{y_2}$$

$\therefore$ Chord parallel to x-axis is of length

$$= \frac{2y_1(1+y_1^2)}{y_2}.$$

Now $PR = PS \cos \psi = 2\rho \cos \psi$

$$= 2 \cdot \frac{(1+y_1^2)^{3/2}}{y_2} \cdot \frac{1}{\sqrt{1+y_1^2}} = \frac{2(1+y_1^2)}{y_2}$$

$\therefore$ Chord parallel to y-axis is of length $\dfrac{2(1+y_1^2)}{y_2}$.

1.27. EVOLUTE AND INVOLUTE

The locus of the centre of curvature of a curve is called the *Evolute* and the given curve is called the *Involute*.

ILLUSTRATIVE EXAMPLES

Example 1. *Find the co-ordinates of the centre of curvature of any point of the hyperbola* $\dfrac{x^2}{a^2} - \dfrac{y^2}{b^2} = 1$. *Hence obtain its evolute.*

Sol. Parametric equation of the hyperbola is $x = a \sec \theta$, $y = b \tan \theta$

$$\frac{dx}{d\theta} = a \sec \theta \tan \theta, \quad \frac{dy}{d\theta} = b \sec^2 \theta$$

$$y_1 = \frac{dy}{dx} = \frac{b}{a} \cdot \frac{\sec^2 \theta}{\sec \theta \tan \theta} = \frac{b}{a} \text{cosec } \theta$$

$$y_2 = -\frac{b}{a} \text{cosec } \theta \cot \theta \cdot \frac{d\theta}{dx}$$

$$= -\frac{b}{a} \text{cosec } \theta \cot \theta \cdot \frac{1}{a \sec \theta \tan \theta} = -\frac{b}{a^2} \frac{\cos \theta}{\sin^2 \theta} \cdot \frac{\cos^2 \theta}{\sin \theta}$$

$$y_2 = -\frac{b}{a^2} \cot^3 \theta$$

$$\bar{x} = x - \frac{y_1(1+y_1^2)}{y_2} = a \sec \theta - \frac{\frac{b}{a} \text{cosec } \theta \left(1 + \frac{b^2 \text{cosec}^2 \theta}{a^2}\right)}{-\frac{b}{a^2} \cot^3 \theta}$$

$$\bar{x} = a \sec \theta + a \frac{\sin^2 \theta}{\cos^3 \theta}\left(1 + \frac{b^2}{a^2 \sin^2 \theta}\right)$$

$$= \frac{a}{\cos \theta} + \frac{a \sin^2 \theta}{\cos^3 \theta} + \frac{b^2}{a \cos^3 \theta}$$

$$= \frac{a^2 \cos^2 \theta + a^2 \sin^2 \theta + b^2}{a \cos^3 \theta} = \frac{a^2 + b^2}{a \cos^3 \theta}$$

$$\bar{x} = \frac{a^2 + b^2}{a} \sec^3 \theta$$

$$\bar{y} = y + \frac{1 + y_1^2}{y_2} = b \tan \theta + \frac{1 + \frac{b^2 \operatorname{cosec}^2 \theta}{a^2}}{-\frac{b}{a^2} \cot^3 \theta}$$

$$= b \tan \theta - \frac{a^2}{b} \tan^3 \theta - b \tan \theta \sec^2 \theta$$

$$= b \tan \theta (1 - \sec^2 \theta) - \frac{a^2}{b} \tan^3 \theta$$

$$= -b \tan^3 \theta - \frac{a^2}{b} \tan^3 \theta = -\frac{a^2 + b^2}{b} \tan^3 \theta$$

Centre of curvature is $\left[\frac{a^2 + b^2}{a} \sec^3 \theta, -\frac{a^2 + b^2}{b} \tan^3 \theta\right]$.

To find Evolute we have to eliminate θ from the values of $\bar{x}$ and $\bar{y}$

where $\bar{x} = \frac{a^2 + b^2}{a} \sec^3 \theta, \; \bar{y} = -\frac{a^2 + b^2}{b} \tan^3 \theta$

$\therefore$ $\sec^3 \theta = \frac{a\bar{x}}{a^2 + b^2}$

and $\tan^3 \theta = -\frac{b\bar{y}}{a^2 + b^2}$

We know, $\sec^2 \theta - \tan^2 \theta = 1$

$$\left(\frac{a\bar{x}}{a^2 + b^2}\right)^{2/3} - \left(\frac{b\bar{y}}{a^2 + b^2}\right)^{2/3} = 1$$

Locus of $(\bar{x}, \bar{y})$ is $(ax)^{2/3} - (by)^{2/3} = (a^2 + b^2)^{2/3}$.

Example 2. *Find the circle of curvature at* $\left(\frac{a}{4}, \frac{a}{4}\right)$ *of the curve* $\sqrt{x} + \sqrt{y} = \sqrt{a}$.

Sol. Equation of the curve is

$$\sqrt{x} + \sqrt{y} = \sqrt{a} \qquad \ldots(1)$$

Differentiate w.r.t. x,

$$\frac{1}{2\sqrt{x}} + \frac{1}{2\sqrt{y}} \frac{dy}{dx} = 0 \quad \therefore \quad y_1 = \frac{dy}{dx} = -\frac{\sqrt{y}}{\sqrt{x}}$$

$$y_2 = \frac{d^2y}{dx^2} = -\frac{\sqrt{x}\cdot\frac{1}{2\sqrt{y}}\cdot y_1 - \sqrt{y}\cdot\frac{1}{2\sqrt{x}}}{x} = -\frac{-\frac{\sqrt{x}}{\sqrt{y}}\cdot\frac{\sqrt{y}}{\sqrt{x}} - \frac{\sqrt{y}}{\sqrt{x}}}{2x}$$

$$= \frac{1+\frac{\sqrt{y}}{\sqrt{x}}}{2x} = \frac{\sqrt{x}+\sqrt{y}}{2x^{3/2}} = \frac{\sqrt{a}}{2x^{3/2}}$$

At $P\left(\frac{a}{4}, \frac{a}{4}\right)$, $\quad y_1 = -1, y_2 = \frac{4}{a}$

$$\rho = \frac{(1+{y_1}^2)^{3/2}}{y_2} = \frac{(1+1)^{3/2}}{4/a} = \frac{a}{4}2\sqrt{2} = \frac{\sqrt{2}a}{2} = \frac{a}{\sqrt{2}}$$

Let $(\bar{x}, \bar{y})$ be the co-ordinates of the centre.

$$\therefore \quad \bar{x} = x - \frac{y_1(1+y_1^2)}{y_2} = \frac{a}{4} - \frac{(-1)\cdot 2}{4/a} = \frac{a}{4} + \frac{a}{2} = \frac{3a}{4}$$

$$\bar{y} = y + \frac{1+y_1^2}{y_2} = \frac{a}{4} + \frac{1+1}{4/a} = \frac{a}{4} + \frac{a}{2} = \frac{3a}{4}$$

$\therefore$ Centre of curvature is $\left(\frac{3a}{4}, \frac{3a}{4}\right)$.

$\therefore$ Equation of the circle of curvature is

$$\left(x - \frac{3a}{4}\right)^2 + \left(y - \frac{3a}{4}\right)^2 = \left(\frac{a}{\sqrt{2}}\right)^2 \quad \text{or} \quad \left(x - \frac{3a}{4}\right)^2 + \left(y - \frac{3a}{4}\right)^2 = \frac{a^2}{2}.$$

Example 3. *If C_x and C_y be the chords of curvatures parallel to axes of x and y respectively at any point of the curve $y = ae^{x/a}$, prove that* $\frac{1}{{C_x}^2} + \frac{1}{{C_y}^2} = \frac{1}{2aC_x}$.

Sol. Equation of the curve is $y = ae^{x/a}$

$$y_1 = \frac{dy}{dx} = e^{x/a} = \frac{y}{a}$$

$$y_2 = \frac{1}{a}e^{x/a} = \frac{y}{a^2}$$

$$\rho = \frac{(1+{y_1}^2)^{3/2}}{y_2} = \frac{\left(1+\frac{y^2}{a^2}\right)^{3/2}}{y/a^2} = \frac{(y^2+a^2)^{3/2}}{ay}$$

Now, $\quad \tan\psi = \frac{dy}{dx} = y_1 = \frac{y}{a}$

$$\sin\psi = \frac{y}{\sqrt{a^2+y^2}}; \cos\psi = \frac{a}{\sqrt{a^2+y^2}}$$

$$\therefore \quad C_x = 2\rho\sin\psi = \frac{2(y^2+a^2)^{3/2}}{ay}\cdot\frac{y}{\sqrt{a^2+y^2}} = \frac{2(a^2+y^2)}{a}$$

$$C_y = 2\rho\cos\psi = \frac{2(y^2+a^2)^{3/2}}{ay}\cdot\frac{a}{\sqrt{a^2+y^2}} = \frac{2(a^2+y^2)}{y}$$

$$\frac{1}{C_x^2}+\frac{1}{C_y^2}=\frac{a^2}{4(a^2+y^2)^2}+\frac{y^2}{4(a^2+y^2)^2}$$

$$=\frac{a^2+y^2}{4(a^2+y^2)^2}=\frac{1}{4(a^2+y^2)}=\frac{1}{4.\frac{aC_x}{2}}=\frac{1}{2aC_x}.$$

TEST YOUR KNOWLEDGE

1. Find centre of curvature at any point P(x, y) of the following curves and hence find their evolutes

   (*i*) $\frac{x^2}{a^2}+\frac{y^2}{b^2}=1$ (*ii*) $y^2 = 4ax$.

2. Show that centre of curvature at $\left(\frac{3a}{2},\frac{3a}{2}\right)$ of the curve $x^3 + y^3 = 3axy$ is $\left(\frac{21a}{16},\frac{21a}{16}\right)$.
3. Show that centre of curvature of the curve $x^4 + y^4 = 2a^2xy$ at P(a, a) divides OP in the ratio 6 : 1 ; 0 being the origin.
4. Show that circle of curvature at the origin for the curve $x + y = ax^3 + by^3 + cx^3$ is $(a + b)(x^2 + y^2) = 2(x + y)$.
5. If C_x, C_y be the lengths of chords of curvature parallel to x-axis, y-axis respectively at any point of the curve $y = c \cosh \frac{x}{c}$ then prove that $4c^2(C_x^2 + C_y^2) = C_y^4$.

Answer

1. (*i*) $\left(\frac{(a^2-b^2)x^3}{a^4}, -\frac{(a^2-b^2)y^3}{b^4}\right)$; $(ax)^{2/3} + (by)^{2/3} = (a^2 - b^2)^{2/3}$

   (*ii*) $(3x + 2a, -2a^{-1/2}x^{3/2})$; $27ay^2 = 4(x - 2a)^3$.

1.28. RADIUS OF CURVATURE FOR POLAR CURVES

(P.T.U., Dec. 2006, May 2007)

Find radius of curvature at any point P(r, θ) of the curve $r = f(\theta)$.

Proof. Let P(r, θ) be any point on the curve $r = f(\theta)$. Let ψ be the angle which tangent at P makes with positive direction of OX. ϕ, be the angle between tangent and radius vector.

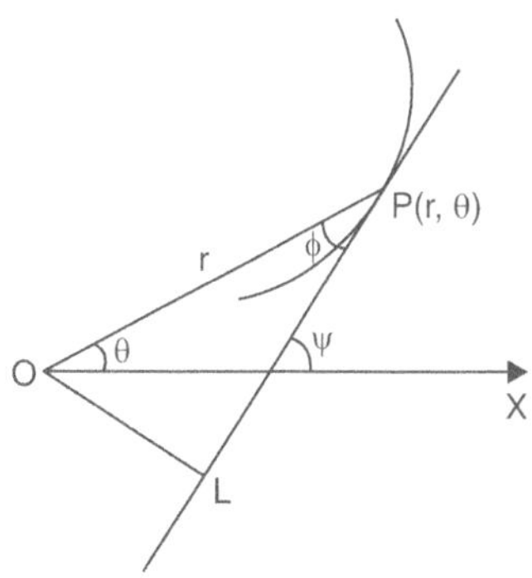

Then, $\psi = \theta + \phi$

Diff. w.r.t. s, $\frac{d\psi}{ds}=\frac{d\theta}{ds}+\frac{d\phi}{ds}$

$$=\frac{d\theta}{ds}\left[1+\frac{d\phi}{ds}\cdot\frac{ds}{d\theta}\right]$$

$$=\frac{d\theta}{ds}\left[1+\frac{d\phi}{d\theta}\right] \quad \text{...(1)}$$

We know that, $\tan\phi=\frac{r\,d\theta}{dr}$ [art. 1.17(4)]

$$= r\cdot\frac{1}{\frac{dr}{d\theta}}=\frac{r}{r_1}, \quad \text{where } r_1=\frac{dr}{d\theta}$$

$$\therefore \qquad \phi = \tan^{-1} \frac{r}{r_1}$$

Differentiate w.r.t. θ,

$$\frac{d\phi}{d\theta} = \frac{1}{1 + \frac{r^2}{r_1^2}} \cdot \frac{d}{d\theta}\left(\frac{r}{r_1}\right) = \frac{r_1^2}{r^2 + r_1^2} \cdot \frac{r_1 \, . \, r_1 - r r_2}{r_1^2} = \frac{r_1^2 - r\, r_2}{r^2 + r_1^2}$$

Also, $$\frac{ds}{d\theta} = \sqrt{r^2 + r_1^2}$$

∴ From (1), $$\frac{1}{\rho} = \frac{d\psi}{ds} = \frac{1}{\sqrt{r^2 + r_1^2}}\left[1 + \frac{r_1^2 - r r_2}{r^2 + r_1^2}\right]$$

or $$\frac{1}{\rho} = \frac{1}{\sqrt{r^2 + r_1^2}} \frac{r^2 + r_1^2 + r_1^2 - r r_2}{r^2 + r_1^2}$$

or $$\rho = \frac{(r^2 + r_1^2)^{3/2}}{r^2 + 2r_1^2 - r r_2}.$$

ILLUSTRATIVE EXAMPLES

Example 1. *Find the radius of curvature for the parabola* $\frac{2a}{r} = 1 + \cos\theta$.

Sol. Polar equation of the curve is

$$\frac{2a}{r} = 1 + \cos\theta \qquad \therefore \qquad r = \frac{2a}{1 + \cos\theta}$$

$$r_1 = \frac{dr}{d\theta} = \frac{-2a}{(1+\cos\theta)^2} \cdot (-\sin\theta) = \frac{2a \sin\theta}{(1+\cos\theta)^2}$$

$$r_2 = 2a \, \frac{(1+\cos\theta)^2 \cdot \cos\theta - \sin\theta \cdot 2(1+\cos\theta)(-\sin\theta)}{(1+\cos\theta)^4}$$

$$= 2a \, \frac{\cos\theta\,(1+\cos\theta) + 2\sin^2\theta}{(1+\cos\theta)^3}$$

$$= 2a \, \frac{\cos\theta + (\cos^2\theta + \sin^2\theta) + \sin^2\theta}{(1+\cos\theta)^3}$$

$$= 2a \, \frac{1 + \cos\theta + \sin^2\theta}{(1+\cos\theta)^3}$$

$$\therefore \qquad \rho = \frac{\left[\frac{4a^2}{(1+\cos\theta)^2} + \frac{4a^2 \sin^2\theta}{(1+\cos\theta)^4}\right]^{3/2}}{\frac{4a^2}{(1+\cos\theta)^2} + 2 \cdot \frac{4a^2\sin^2\theta}{(1+\cos\theta)^4} - \frac{2a}{1+\cos\theta} \cdot 2a \frac{1+\cos\theta+\sin^2\theta}{(1+\cos\theta)^3}}$$

$$= \frac{\left[\dfrac{4a^2}{(1+\cos\theta)^4}\right]^{3/2}\left[(1+\cos\theta)^2+\sin^2\theta\right]^{3/2}}{\left[\dfrac{4a^2}{(1+\cos\theta)^4}\right][(1+\cos\theta)^2+2\sin^2\theta-(1+\cos\theta+\sin^2\theta)]}$$

$$= \left[\frac{4a^2}{(1+\cos\theta)^4}\right]^{1/2}\cdot\frac{\left[(1+\cos\theta)^2+(1-\cos^2\theta)\right]^{3/2}}{(1+\cos\theta)^2+(1-\cos^2\theta)-(1+\cos\theta)}$$

$$= \frac{2a}{(1+\cos\theta)^2}\cdot\frac{(1+\cos\theta)^{3/2}\,[1+\cos\theta+1-\cos\theta]^{3/2}}{(1+\cos\theta)\,[1+\cos\theta+1-\cos\theta-1]}$$

$$= \frac{2a}{(1+\cos\theta)^{3/2}}\cdot\frac{2^{3/2}}{1} = 2a\left(\frac{2}{1+\cos\theta}\right)^{3/2} = \frac{2}{\sqrt{a}}\left(\frac{2a}{1+\cos\theta}\right)^{3/2} = \frac{2}{\sqrt{a}}\cdot r^{3/2}$$

$$\therefore \qquad \rho = 2\cdot\sqrt{\frac{r^3}{a}}.$$

Example 2. *If ρ_1, ρ_2 be the radii of curvature at the extremities of any chord of the cardioid $r = a(1 + \cos\theta)$ which passes through the pole, show that $\rho_1^2 + \rho_2^2 = \dfrac{16a^2}{9}$.*

(P.T.U., May 2003, 2009, 2014)

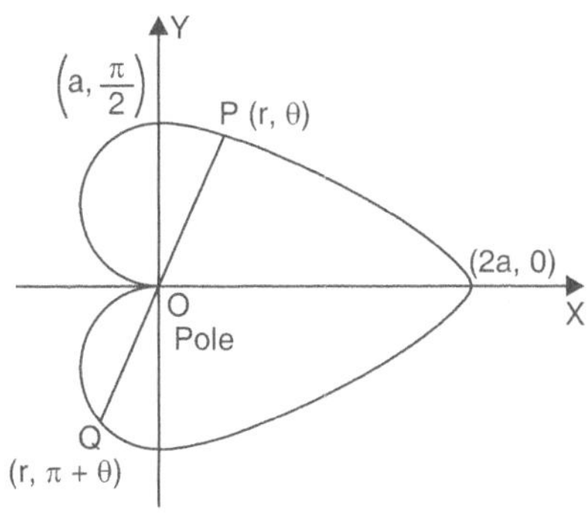

Sol. Equation of the cardioid is

$$r = a(1 + \cos\theta)$$

Differentiate w.r.t. θ twice, we get

$$r_1 = -a\sin\theta$$
$$r_2 = -a\cos\theta$$

Let PQ be any chord passing through the pole.

∴ If θ is the vectorial angle of P,

then π + θ is the vectorial angle of Q

∴ Radius of curvature at P (r, θ) is $\rho_1 = \dfrac{(r^2+r_1^2)^{3/2}}{r^2+2r_1^2-rr_2}$

$$= \frac{[a^2(1+\cos\theta)^2+a^2\sin^2\theta]^{3/2}}{a^2(1+\cos\theta)^2+2a^2\sin^2\theta-a(1+\cos\theta)(-a\cos\theta)}$$

$$= \frac{[a^2(1+\cos\theta)]^{3/2}[1+\cos\theta+(1-\cos\theta)]^{3/2}}{a^2(1+\cos\theta)\,[1+\cos\theta+2(1-\cos\theta)+\cos\theta]} \qquad (\because\ \sin^2\theta = 1-\cos^2\theta)$$

$$= \{a^2(1+\cos\theta)\}^{1/2}\,\frac{2^{3/2}}{3} = 2^{3/2}\,\frac{a\sqrt{1+\cos\theta}}{3}$$

$$\therefore \qquad \rho_1^2 = \frac{8a^2}{9}(1+\cos\theta)$$

Similarly, radius of cirvature at θ

$$\rho_2 = \frac{8a^2}{9}\,[1 + \cos(\pi + \theta)] = \frac{8a^2}{9}\,(1 - \cos\theta)$$

$$\therefore \quad \rho_1^2 + \rho_2^2 = \frac{16a^2}{9}.$$

Example 3. *Show that at the point of intersection of the curves $r = a\theta$ and $r\theta = a$ the curvatures are in the ratio 3 : 1 ($0 < \theta < 2\pi$).*

Sol. Equations of the curves are $r = a\theta$, $r\theta = a$. Their intersection is given by (eliminating r)

$$\therefore \quad a\theta^2 = a \quad i.e., \quad \theta^2 = 1 \quad i.e., \quad \theta = \pm 1$$

Now let ρ_1, ρ_2 be the radii of curvature of $r = a\theta$ and $r = \dfrac{a}{\theta}$, respectively.

For $r = a\theta$; $r_1 = a$, $r_2 = 0$

We know that $$\rho = \frac{(r^2 + r_1^2)^{3/2}}{r^2 + 2r_1^2 - rr_2}$$

$$\therefore \quad \text{For } r = a\theta; \qquad \rho_1 = \frac{(a^2\theta^2 + a^2)^{3/2}}{a^2\theta^2 + 2a^2 - 0}$$

$$\therefore \quad \text{For } r\theta = a; \qquad r_1 = -\frac{a}{\theta^2},\ r_2 = \frac{2a}{\theta^3} \qquad \therefore \quad \rho_2 = \frac{\left(\dfrac{a^2}{\theta^2} + \dfrac{a^2}{\theta^4}\right)^{3/2}}{\dfrac{a^2}{\theta^2} + 2\cdot\dfrac{a^2}{\theta^4} - \dfrac{2a^2}{\theta^4}}$$

Put $\theta = \pm 1$

$$\rho_1 = \frac{(2a^2)^{3/2}}{3a^2}; \qquad \rho_2 = \frac{(2a^2)^{3/2}}{a^2 + 2a^2 - 2a^2} = \frac{(2a^2)^{3/2}}{a^2}$$

$$\frac{\rho_1}{\rho_2} = \frac{a^2}{3a^2} = \frac{1}{3}$$

Radii of curvature are in the ratio 1 : 3

$\therefore$ Curvatures are in the ratio $\dfrac{1}{1} : \dfrac{1}{3}$ *i.e.*, 3 : 1.

Example 4. *For any curve $r = f(\theta)$, prove that $\dfrac{r}{\rho} = \sin\phi\left[1 + \dfrac{d\phi}{d\theta}\right]$.*

Sol. We know that

$$\psi = \theta + \phi.$$

$$\frac{1}{\rho} = \frac{d\psi}{ds} = \frac{d\theta}{ds} + \frac{d\phi}{ds}$$

$$= \frac{d\theta}{ds} + \frac{d\phi}{d\theta}\cdot\frac{d\theta}{ds} = \frac{d\theta}{ds}\left(1 + \frac{d\phi}{d\theta}\right)$$

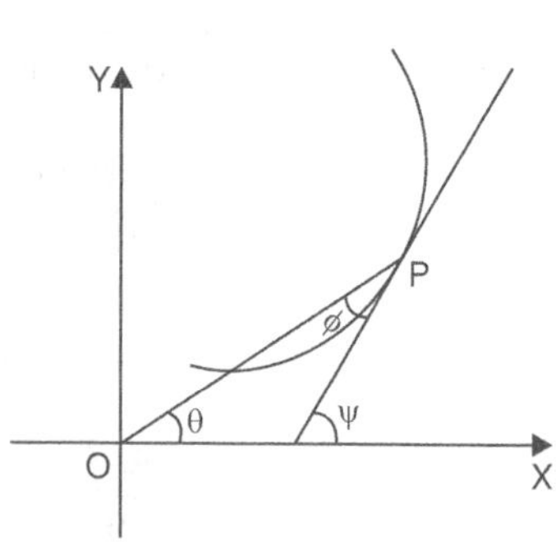

Also we know $\sin\phi = \dfrac{r\,d\theta}{ds}$ $\quad\therefore\quad \dfrac{d\theta}{ds} = \dfrac{\sin\phi}{r}$

$$\therefore \quad \frac{1}{\rho} = \frac{\sin\phi}{r}\left[1 + \frac{d\phi}{d\theta}\right] \quad \therefore \quad \frac{r}{\rho} = \sin\phi\left[1 + \frac{d\phi}{d\theta}\right].$$

Example 5. *Show that at each point of the lemniscate $r^2 = a^2 \cos 2\theta$, the radius of curvature is inversely proportional to the radius vector at that point.* **(P.T.U., May 2007)**

Sol. Equation of the curve is $r^2 = a^2 \cos 2\theta$ ...(1)

Taking logs of both sides

$$2 \log r = \log a^2 + \log \cos 2\theta$$

Differentiate w.r.t. θ

$$2.\frac{1}{r}\frac{dr}{d\theta} = \frac{1}{\cos 2\theta}(-2 \sin 2\theta)$$

$$\therefore \quad \frac{r_1}{r} = -\tan 2\theta \quad \therefore \quad r_1 = -r \tan 2\theta$$

$$r_2 = -\left\{r\, 2 \sec^2 2\theta + \tan 2\theta . \frac{dr}{d\theta}\right\}$$

$$= -\{2r \sec^2 2\theta + \tan 2\theta\,(-r \tan 2\theta)\}$$

$$= -r\,\{2 \sec^2 2\theta - \tan^2 2\theta\}$$

$$\rho = \frac{(r^2 + r_1^2)^{3/2}}{r^2 + 2r_1^2 - rr_2} = \frac{(r^2 + r^2 \tan^2 2\theta)^{3/2}}{r^2 + 2r^2 \tan^2 2\theta + r^2\,(2 \sec^2 2\theta - \tan^2 2\theta)}$$

$$= \frac{r^3 \sec^3 2\theta}{r^2\,[1 + 2\tan^2 2\theta + 2\sec^2 2\theta - \tan^2 2\theta]}$$

$$= \frac{r \sec^3 2\theta}{1 + \tan^2 2\theta + 2\sec^2 2\theta} = \frac{r \sec^3 2\theta}{3 \sec^2 2\theta} = \frac{r}{3 \cos 2\theta}$$

$$= \frac{r}{3r^2/a^2} \qquad \Big| \text{ Using (1)}$$

$$= \frac{a^2}{3r} = \frac{a^2}{3}\left(\frac{1}{r}\right).$$

$$\therefore \quad \rho \propto \frac{1}{r}$$

Hence radius of curvature is inversely proportional to the radius vector at that point.

TEST YOUR KNOWLEDGE

1. Find the radius of curvature at (r, θ) on each of the following curve,
   (*i*) $r = a(1 - \cos \theta)$, ***(P.T.U., Dec. 2005)*** (*ii*) $r^n = a^n \cos n\theta$
   (*iii*) $r = a \sin \theta$.
2. For the cardioid $r = a(1 + \cos \theta)$, show that $\frac{\rho^2}{r}$ is constant. ***(P.T.U., Dec. 2002)***
   [**Hint.** Consult solved example 2]

Answers

1. (*i*) $\frac{4a}{3} \sin \frac{\theta}{2}$ (*ii*) $\frac{a^n}{(n+1)r^{n-1}}$ (*iii*) $a/2$.

REVIEW OF THE CHAPTER

1. $\dfrac{ds}{dx} = \sqrt{1+\left(\dfrac{dy}{dx}\right)^2}$, $\dfrac{ds}{dy} = \sqrt{1+\left(\dfrac{dx}{dy}\right)^2}$.

2. $\tan\psi = \dfrac{dy}{dx}$, where ψ is the angle which the tangent at any point of the curve makes with the positive direction of x-axis.

$$\sin\psi = \frac{dy}{ds}, \quad \cos\psi = \frac{dx}{ds}$$

3. $$(ds)^2 = (dx)^2 + (dy)^2$$

$$\left(\frac{ds}{dt}\right)^2 = \left(\frac{dx}{dt}\right)^2 + \left(\frac{dy}{dt}\right)^2.$$

4. $\tan\phi = \dfrac{r\,d\theta}{dr}$, where ϕ is the angle between radius vector and tangent at any point of the curve.

$$\sin\phi = \frac{r\,d\theta}{ds}, \quad \cos\phi = \frac{dr}{ds}.$$

5. (*a*) Length of tangent $= \dfrac{y\sqrt{1+{y_1}^2}}{y_1}$; $y_1 = \dfrac{dy}{dx}$

(*b*) length of subtangent $= \dfrac{y}{y_1}$,

(*c*) length of normal $= y\sqrt{1+y_1^2}$

(*d*) length of subnormal $= yy_1$.

6. Curvature at any point of the curve $= \dfrac{d\psi}{ds}$

7. Radius of curvature $= \dfrac{ds}{d\psi}$

8. Radius of curvature

(*a*) In cartesian co-ordinates $\rho = \dfrac{(1+y_1^2)^{3/2}}{y_2}$; $y_1 = \dfrac{dy}{dx}, y_2 = \dfrac{d^2y}{dx^2}$

(*b*) In parametric co-ordinates $\rho = \dfrac{[[f'(t)]^2 + [g'(t)]^2]^{3/2}}{f'(t)\,g''(t) - f''(t)\,g'(t)}$, where $x = f(t)$, $y = g(t)$ and dashes represent derivatives.

(*c*) In polar co-ordinates $\rho = \dfrac{(r^2 + r_1^2)^{3/2}}{r^2 + 2r_1^2 - rr_2}$, where $r_1 = \dfrac{dr}{d\theta}, r_2 = \dfrac{d^2r}{d\theta^2}$.

9. Radius of curvature at the origin

(*a*) Put $y = px + \frac{qx^2}{2!}$ ... in the equation, equate like powers of x on both sides (start from lowest power of x) and find p, q then, $\rho = \frac{(1+p^2)^{3/2}}{q}$.

(*b*) By Newton's method

If tangent at origin is x-axis then

$$\rho = \underset{x \to 0}{\text{Lt}} \frac{x^2}{2y} \quad (y \text{ also tends to zero})$$

If tangent at origin is y-axis then, $\rho = \underset{x \to 0}{\text{Lt}} \frac{y^2}{2x}$ (x also tends to zero).

10. Co-ordinates of the centre of curvature

$$\left(x - \frac{y_1(1+y_1{}^2)}{y_2},\, y + \frac{1+y_1{}^2}{y_2} \right) = (\bar{x}, \bar{y}).$$

11. Circle of curvature $(x - \bar{x})^2 + (y - \bar{y})^2 = \rho^2$.

12. Length of chords:

(*a*) Length of chord parallel to x-axis $= \frac{2y_1(1+y_1^2)}{y_2}$.

(*b*) Length of chord parallel to y-axis $= \frac{2(1+y_1^2)}{y_2}$.

13. Evolute, Involute: Locus of the centre of curvature is Evolute and the curve itself is Involute.

SHORT ANSWER TYPE QUESTIONS

1. Write formula for radius of curvature for $y = f(x)$.
[**Hint.** Art 1.20]

2. Write down the formula for radius of curvature for implicit equation $f(x, y) = 0$.
[**Hint.** Art 1.23]

3. Find the radius of curvature for parametric curve. ***(P.T.U., Dec. 2006)***
[**Hint.** Art 1.21]

4. Write down the radius of curvature in polar coordinates.
[**Hint.** Art 1.28]

5. Prove that $\frac{1}{\rho} = \frac{d}{dx}\left(\frac{dy}{ds}\right)$, where s is the length of the arc measured from some fixed point and ρ, the radius of curvature.
[**Hint.** S.E. 1(*i*) art. 1.23]

6. The curvature of a circle of radius r is r^{-1}. Justify.
[**Hint.** Equation of the circle is $x^2 + y^2 = r^2$]

7. Find the curvature of the curve $y^2 = x^3 + 8$ at (1, 3). ***(P.T.U., May 2009)***
[**Hint.** S.E. 2(*ii*) art. 1.23]

8. Find point on the curve $y = e^x$ at which curvature is maximum.
[**Hint.** S.E. 10; art. 1.23]

9. Find radius of curvature at the origin for the curve $y^4 + x^3 + a(x^2 + y^2) - a^2y = 0$.
[**Hint.** S.E. 3; art. 1.24] ***(P.T.U., May 2015)***

10. Obtain the radius of curvature at any point $(at^2, 2at)$ of the parabola $y^2 = 4ax$ ***(P.T.U., Dec. 2010)***
[**Hint.** S.E. 4(1); art. 1.23]

11. Find radius of curvature at any point on the following curves:
(*i*) $r = a(1 + \cos\theta)$ ***(P.T.U., Dec. 2011)***
(*ii*) $r = a(1 - \cos\theta)$
(*iii*) $r^n = a^n \cos n\theta$
(*iv*) $r = a \sin\theta$
(*v*) $r^2 = a^2 \cos 2\theta$ ***(P.T.U., May 2007)***
[**Hint.** S.E. 5 art. 1.28]

12. Write down the coordinates of the centre of curvature for the point (x, y) of the curve $y = f(x)$.
[**Hint.** art. 1.26]

13. Find the centre of curvature at any point P(x, y) of the curve $y^2 = 4ax$.
[**Hint.** $y_1 = \sqrt{\dfrac{a}{x}}$; $y_2 = -\dfrac{\sqrt{a}}{x^{3/2}}$; $y = 2\sqrt{ax}$; substitute all these values in the centre of curvature]

Answers

7. $\dfrac{22}{15\sqrt{5}}$ **8.** $\left(-\dfrac{1}{2}\log 2, \dfrac{1}{\sqrt{2}}\right)$ **9.** $\dfrac{a}{2}$

10. $2a(1 + t^2)^{3/2}$

11. (*i*) $\dfrac{4}{3}a\cos\dfrac{\theta}{2}$ (*ii*) $\dfrac{4}{3}a\sin\dfrac{\theta}{2}$ (*iii*) $\dfrac{a^n}{r^{n-1}(n+1)}$

(*iv*) $\dfrac{a}{2}$ (*v*) $\dfrac{a^2}{3r}$

12. $\left(x - \dfrac{y_1(1+y_1^2)}{y_2}, y + \dfrac{1+y_1^2}{y_2}\right)$ **13.** $\left(2x + 2a, -\dfrac{2}{\sqrt{a}}x^{3/2}\right)$.

2

Integral Calculus

RECTIFICATION

2.1. DEFINITION

Rectification is the process of determining the lengths of the arcs of the plane curves between two given points.

2.2. ARC FORMULA FOR CARTESIAN EQUATION $y = f(x)$

Prove that the length of the arc of the curve $y = f(x)$ included between two points whose abscissae are 'a' and 'b' is

$$\int_a^b \sqrt{1+\left(\frac{dy}{dx}\right)^2}\,dx \quad or \quad \int_a^b \sqrt{1+y_1^2}\,dx \quad or \quad \int_a^b \sqrt{1+[f'(x)]^2}\,dx$$

where $y_1 = \frac{dy}{dx}$ *and* $f'(x) = \frac{dy}{dx}$.

Proof. Let A, B be two points on the curve $y = f(x)$ representing the abscissae a and b respectively. Let $P(x, y)$ be any point on the curve such that length of the arc AP = s.

Let $Q(x + \delta x, y + \delta y)$ be a neighbouring point to P such that AQ = $s + \delta s$

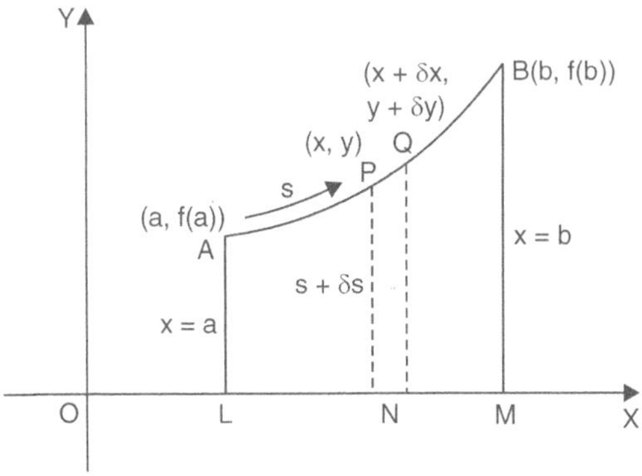

$$\therefore \quad \widehat{PQ} = s + \delta s - s = \delta s$$

from Differential Calculus we know that

$$(ds)^2 = (dx)^2 + (dy)^2 \qquad ...(1)$$

(consult art 1.17(3))

$$\therefore \quad \left(\frac{ds}{dx}\right)^2 = 1+\left(\frac{dy}{dx}\right)^2$$

$$\therefore \quad \frac{ds}{dx} = \sqrt{1+\left(\frac{dy}{dx}\right)^2}$$

Integrating both sides w.r.t. x between the limits a and b

$$\therefore \quad \int_a^b \frac{ds}{dx}\,dx = \int_a^b \sqrt{1+\left(\frac{dy}{dx}\right)^2}\,dx \qquad \text{or} \qquad s\Big|_a^b = \int_a^b \sqrt{1+\left(\frac{dy}{dx}\right)^2}\,dx$$

Value of s for $(x = b)$ – value of s for $(x = a) = \int_a^b \sqrt{1+\left(\frac{dy}{dx}\right)^2}\, dx$

or $\qquad \widehat{AB} - O = \int_a^b \sqrt{1+\left(\frac{dy}{dx}\right)^2}\, dx \qquad$ or $\qquad \widehat{AB} = \int_a^b \sqrt{1+\left(\frac{dy}{dx}\right)^2}\, dx$

Hence length of the arc AB $= \int_a^b \sqrt{1+\left(\frac{dy}{dx}\right)^2}\, dx$ which can also be put in the form

$$= \int_a^b \sqrt{1+y_1^2}\, dx \text{ where } y_1 = \frac{dy}{dx}$$

$$= \int_a^b \sqrt{1+[f'(x)]^2}\, dx \text{ where } f'(x) = \frac{dy}{dx}.$$

2.3. OTHER EXPRESSIONS FOR LENGTHS OF ARCS

(*a*) From (1), $(ds)^2 = (dx)^2 + (dy)^2$

Divide by $(dy)^2$, we get $\left(\frac{ds}{dy}\right)^2 = \left(\frac{dx}{dy}\right)^2 + 1$

$$\frac{ds}{dy} = \sqrt{1+\left(\frac{dx}{dy}\right)^2}$$

Integrate w.r.t. y

$$\int_A^B \frac{ds}{dy}\, dy = \int_{y=f(a)}^{f(b)} \sqrt{1+\left(\frac{dx}{dy}\right)^2}\, dy$$

$$\therefore \qquad \text{Arc AB} = \int_{y=f(a)}^{f(b)} \sqrt{1+\left(\frac{dx}{dy}\right)^2}\, dy \qquad ...(2)$$

(*b*) If equation of curve in parametric co-ordinates *i.e.*, x and y are function of t, then $(ds)^2 = (dx)^2 + (dy)^2$ becomes **(*P.T.U., Dec. 2006*)**

$$\left(\frac{ds}{dt}\right)^2 = \left(\frac{dx}{dt}\right)^2 + \left(\frac{dy}{dt}\right)^2$$

$$\frac{ds}{dt} = \sqrt{\left(\frac{dx}{dt}\right)^2 + \left(\frac{dy}{dt}\right)^2}$$

Integrate both sides w.r.t. t

$$\int_A^B \frac{ds}{dt}\, dt = \int_{t=\alpha}^{t=\beta} \sqrt{\left(\frac{dx}{dt}\right)^2 + \left(\frac{dy}{dt}\right)^2}\, dt$$

or

$$\text{Arc AB} = \int_{t=\alpha}^{t=\beta} \sqrt{\left(\frac{dx}{dt}\right)^2 + \left(\frac{dy}{dt}\right)^2}\, dt \qquad ...(3)$$

(*c*) If equation of curve is in polar co-ordinates *i.e.*, in terms of r, θ. **(*P.T.U., Dec. 2006*)**

Thus $\qquad x = r\cos\theta, \quad y = r\sin\theta$

$$\frac{dx}{d\theta} = -r\sin\theta + \cos\theta \cdot \frac{dr}{d\theta}; \frac{dy}{d\theta} = r\cos\theta + \sin\theta\,\frac{dr}{d\theta}$$

$\therefore \quad ds^2 = dx^2 + dy^2$

$$\left(\frac{ds}{d\theta}\right)^2 = \left(\frac{dx}{d\theta}\right)^2 + \left(\frac{dy}{d\theta}\right)^2 = (-r\sin\theta + \cos\theta\cdot r_1)^2 + (r\cos\theta + \sin\theta\, r_1)^2$$

where $r_1 = \dfrac{dr}{d\theta}$

$$\left(\frac{ds}{d\theta}\right)^2 = r^2 + r_1^2 = r^2 + \left(\frac{dr}{d\theta}\right)^2 \qquad \text{or} \qquad \frac{ds}{d\theta} = \sqrt{r^2 + \left(\frac{dr}{d\theta}\right)^2}$$

Integrate w.r.t. θ, we get

$$\int_A^B ds = \int_{\theta=\alpha}^{\theta=\beta} \sqrt{r^2 + \left(\frac{dr}{d\theta}\right)^2}\, d\theta$$

$$\therefore \quad \text{Arc AB} = \int_{\theta=\alpha}^{\beta} \sqrt{r^2 + \left(\frac{dr}{d\theta}\right)^2}\, d\theta \qquad \ldots(4)$$

Hence expression for lengths of the arcs are

(i) $s = \displaystyle\int_{x=a}^{b} \sqrt{1 + \left(\frac{dy}{dx}\right)^2}\, dx$ $\qquad$ (ii) $s = \displaystyle\int_{y=f(a)}^{f(b)} \sqrt{1 + \left(\frac{dx}{dy}\right)^2}\, dy$

(iii) $s = \displaystyle\int_{t=\alpha}^{\beta} \sqrt{\left(\frac{dx}{dt}\right)^2 + \left(\frac{dy}{dt}\right)^2}\, dt$ $\qquad$ (iv) $s = \displaystyle\int_{\theta=\alpha}^{\beta} \sqrt{r^2 + \left(\frac{dr}{d\theta}\right)^2}\, d\theta.$

ILLUSTRATIVE EXAMPLES

Example 1. *Find the length of the arc of the curve* $y = \log \dfrac{e^x - 1}{e^x + 1}$ *from* $x = 1$ *to* $x = 2$.

(P.T.U., Dec. 2004)

Sol. Equation of the curve is

$$y = \log\frac{e^x - 1}{e^x + 1} = \log(e^x - 1) - \log(e^x + 1)$$

$$\frac{dy}{dx} = \frac{1}{e^x - 1} e^x - \frac{1}{e^x + 1} e^x = e^x \left\{\frac{e^x + 1 - e^x + 1}{e^{2x} - 1}\right\}$$

$$\therefore \quad \frac{dy}{dx} = \frac{2e^x}{e^{2x} - 1}$$

$$\text{Length of the arc} = \int_1^2 \sqrt{1 + \left(\frac{dy}{dx}\right)^2}\, dx$$

$$= \int_1^2 \sqrt{1 + \frac{4e^{2x}}{(e^{2x} - 1)^2}}\, dx = \int_1^2 \sqrt{\frac{(e^{2x} - 1)^2 + 4e^{2x}}{e^{2x} - 1}}\, dx$$

$$= \int_1^2 \frac{\sqrt{(e^{2x}+1)^2}}{e^{2x}-1}\, dx = \int_1^2 \frac{e^{2x}+1}{e^{2x}-1}\, dx$$

$$= \int_1^2 \frac{e^x + e^{-x}}{e^x - e^{-x}}\, dx$$ **Note the integration**

$$= \log (e^x - e^{-x}) \Big|_1^2 \qquad \left|\ \because \int \frac{f'(x)}{f(x)}\, dx = \log f(x)\right.$$

$$= \log (e^2 - e^{-2}) - \log (e - e^{-1})$$

$$= \log \frac{e^{-2}(e^4 - 1)}{e^{-1}(e^2 - 1)} = \log \frac{1}{e} \frac{(e^2-1)(e^2+1)}{e^2 - 1} = \log \frac{e^2+1}{e}$$

$$= \log \left(e + \frac{1}{e}\right)$$

Hence length of the arc $= \log \left(e + \dfrac{1}{e}\right)$.

Example 2. *Find the length of arc of the parabola $y^2 - 4y + 2x = 0$ which lies in the first quadrant.*

Sol. Equation of the parabola is

$$y^2 - 4y + 2x = 0 \qquad ...(1)$$

or

$$(y-2)^2 = -2x + 4 = -2(x-2) \qquad ...(2)$$

$$Y^2 = -2X$$

where $Y = y - 2$, $X = x - 2$

$\therefore$ Vertex of the parabola is (2, 2) and axis parallel to x-axis

Given parabola intersects x-axis (put $y = 0$) at $x = 0$ *i.e.*, at (0, 0)

and intersects y-axis (put $x = 0$) at $y^2 - 4y = 0$ *i.e.*, at $y = 0$, $y = 4$ *i.e.*, at (0, 0), (0, 4).

$\therefore$ Portion of the curve which lies in 1st octant is OAB

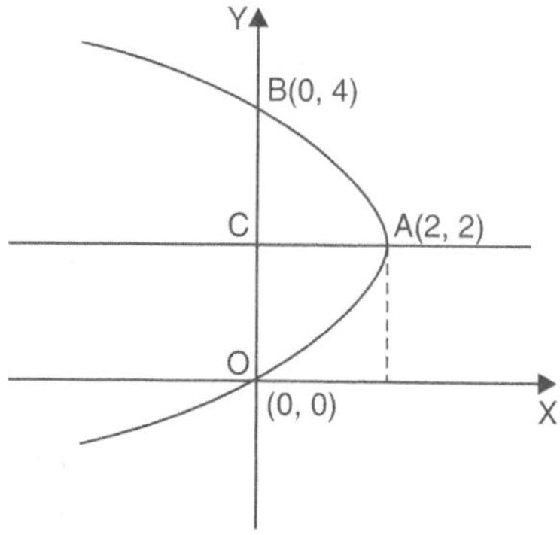

$\therefore$ Length of arc OAB $= 2\, \widehat{OA}$

$$= 2 \int_0^2 \sqrt{1 + \left(\frac{dx}{dy}\right)^2}\, dy$$

Differentiate (1) w.r.t. y, $2y - 4 + 2 \cdot \dfrac{dx}{dy} = 0$

$$\therefore \qquad \frac{dx}{dy} = 2 - y$$

$$\therefore \qquad \widehat{OAB} = 2 \int_0^2 \sqrt{1 + (2-y)^2}\, dy = 2 \int_0^2 \sqrt{1+(y-2)^2}\, dy$$

$$= 2 \left\{ \frac{(y-2)\sqrt{1+(y-2)^2}}{2} + \frac{1}{2} \sinh^{-1}(y-2) \right\}_0^2$$

$$= 2\left\{0 + 0 - \frac{-2}{2}\sqrt{1+4} - \frac{1}{2}\sinh^{-1}(-2)\right\}$$

$$= 2\left\{\sqrt{5} + \frac{1}{2}\sinh^{-1} 2\right\} = 2\sqrt{5} + \log(2+\sqrt{5}).$$

Example 3. *Show that length of the arc of the parabola $y^2 = 4ax$ cut off by the line $3y = 8x$ is $a\left[\log_e 2 + \frac{15}{16}\right]$* **(P.T.U., Dec. 2006)**

Sol. Equation of the parabola is $y^2 = 4ax$ ...(1)

Equation of the line is $3y = 8x$ ...(2)

(1) and (2) intersect at points given by

$$\left(\frac{8x}{3}\right)^2 = 4ax \quad \text{or} \quad x\left(\frac{64x}{9} - 4a\right) = 0$$

i.e., either $x = 0$ or $x = \frac{9a}{16}$

When $x = 0, y = 0$ and when $x = \frac{9a}{16}, y = \frac{3a}{2}$

$\therefore$ Points of intersection are $(0, 0), \left(\frac{9a}{16}, \frac{3a}{2}\right)$

$\therefore$ Length of the arc

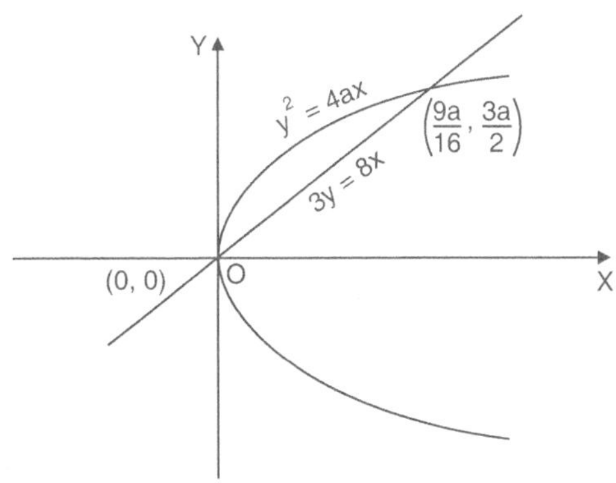

$$= \int_0^{\frac{3a}{2}} \sqrt{1 + \left(\frac{dx}{dy}\right)^2}\, dy$$

From (1), $x = \frac{y^2}{4a}$

$$\therefore \quad \frac{dx}{dy} = \frac{2y}{4a} = \frac{y}{2a}$$

$\therefore$ Length of the arc $= \int_0^{\frac{3a}{2}} \sqrt{1 + \frac{y^2}{4a^2}}\, dy$

$$= \frac{1}{2a}\int_0^{\frac{3a}{2}} \sqrt{y^2 + (2a)^2}\, dy$$

$$= \frac{1}{2a}\left\{\frac{y\sqrt{y^2+(2a)^2}}{2} + \frac{(2a)^2}{2}\log_e\left[y + \sqrt{y^2+4a^2}\right]\right|_0^{3a/2}\Bigg\}$$

$$= \frac{1}{2a}\left[\frac{1}{2}\,\frac{3a}{2}\,\frac{5a}{2} + 2a\left\{\log_e\left(\frac{3a}{2} + \frac{5a}{2}\right) - \log_e 2a\right\}\right]$$

$$= \frac{1}{2a}\left[\frac{15a^2}{8} + 2a\log_e\frac{4a}{2a}\right] = \frac{15}{16}a + \log_e 2.$$

Example 4. *Find the perimeter of the loop of the following curves*
(*i*) $9ay^2 = (x-2a)(x-5a)^2$ (*ii*) $3ay^2 = x(x-a)^2$

Sol. (*i*) Equation of the curve is $9ay^2 = (x-2a)(x-5a)^2$ ...(1)

[first of all we will have rough sketch of the curve]

1. Symmetry. Curve is symmetric about x-axis.

2. Intersection with Axes. It intersects x-axis at $(2a, 0)$; $(5a, 0)$ and y-axis at imaginary points *i.e.*, does not intersect y-axis.

3. Tangents to the Curve. Differentiate (1) w.r.t. x

$$\therefore \quad 18ay\frac{dy}{dx} = (x-2a)\,2(x-5a) + (x-5a)^2$$

or
$$\frac{dy}{dx} = \frac{(x-5a)[2x-4a+x-5a]}{18ay}$$

$$\frac{dy}{dx} = \frac{(x-5a)(3x-9a)}{\dfrac{18a\sqrt{x-2a}\,|(x-5a)|}{3\sqrt{a}}} = \pm\frac{x-3a}{\dfrac{6a\sqrt{x-2a}}{3\sqrt{a}}} = \pm\frac{x-3a}{2\sqrt{a}\sqrt{x-2a}}$$

Slope of the tangent at $(2a, 0)$ is ∞

$\therefore$ At $(2a, 0)$ tangent is parallel to y-axis

At $(5a, 0)$ slope of the tangent

$$= \pm\frac{5a-3a}{5\sqrt{a}\sqrt{5a-3a}} = \pm\frac{2\sqrt{a}}{5\sqrt{2}\sqrt{a}} = \pm\frac{\sqrt{2}}{5}$$

$\therefore$ There are two tangents at $(5a, 0)$ one making an angle of $\tan^{-1}\dfrac{\sqrt{2}}{5}$ with +ve direction of x-axis and other making an angle $\pi - \tan^{-1}\dfrac{\sqrt{2}}{5}$ with x-axis.

4. Region. y will exist when $x > 2a$

$\therefore$ curve lies on the R.H.S. of $x = 2a$.

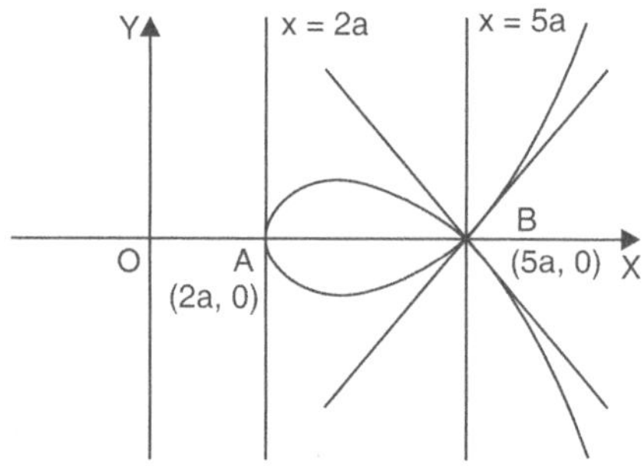

Length of the loop of the curve = 2 length of the part of the loop above x-axis

[$\because$ curve is symmetric about x-axis]

$$\therefore \quad \text{Length of the loop} = 2\int_{2a}^{5a}\sqrt{1+\left(\frac{dy}{dx}\right)^2}\,dx$$

$$= 2\int_{2a}^{5a}\sqrt{1+\frac{(x-3a)^2}{4a(x-2a)}}\,dx$$

$$= 2\int_{2a}^{5a}\frac{\sqrt{4ax-8a^2+x^2-6ax+9a^2}}{2\sqrt{a}\sqrt{x-2a}}\,dx$$

$$= 2\int_{2a}^{5a}\frac{\sqrt{x^2-2ax+a^2}}{2\sqrt{a}\sqrt{x-2a}}\,dx$$

$$= 2\int_{2a}^{5a}\frac{x-a}{2\sqrt{a}\sqrt{x-2a}}\,dx \qquad (\sqrt{x^2-2ax+a^2} = x-a \quad \because \quad x > a)$$

$$= \frac{1}{\sqrt{a}}\cdot\int_{2a}^{5a}\left[\sqrt{x-2a}+\frac{a}{\sqrt{x-2a}}\right]dx$$

$$= \frac{1}{\sqrt{a}} \cdot \left[\frac{(x-2a)^{3/2}}{3/2} + a\,\frac{(x-2a)^{1/2}}{1/2} \right]_{2a}^{5a}$$

$$= \frac{2}{\sqrt{a}} \cdot \left\{ \frac{(3a)^{3/2}}{3} + a(3a)^{1/2} \right\}$$

$$= \frac{2}{\sqrt{a}} \sqrt{3a} \cdot \left\{ \frac{3a}{3} + a \right\} = 2\sqrt{3} \cdot 2a = 4\sqrt{3}\,a \text{ units.}$$

(*ii*) Equation of the curve is $3ay^2 = x(x-a)^2$ ...(1)

1. Symmetry. The curve is symmetric about x–axis.

($\because$ on changing y to $-y$ the equation remains unchanged).

2. Origin. The curve passes through the origin and the tangent at the origin is $x = 0$ (obtained by equating to zero the lowest degree terms is (1)).

3. Intersection with Axes. Curve intersect x-axis (put $y = 0$) at (0, 0) and (a, 0) and y-axis at (0, 0).

4. Tangents to the Curves. Differentiate (1) w.r.t. x, we get

$$6ay \frac{dy}{dx} = (x-a)^2 + 2x(x-a)$$

$$\therefore \quad \frac{dy}{dx} = \frac{(x-a)(x-a+2x)}{6a\sqrt{\dfrac{x(x-a)^2}{3a}}} = \frac{(x-a)(3x-a)}{2\sqrt{3a}\,\sqrt{x}\,|x-a|}$$

$$\frac{dy}{dx} = \pm \frac{(3x-a)}{2\sqrt{3a}\,\sqrt{x}}$$

Tangent is parallel to x-axis where $\frac{dy}{dx} = 0$ *i.e.*, at $x = \frac{a}{3}$ *i.e.*, at $\left(\frac{a}{3}, \pm\frac{2a}{9}\right)$

Tangent is parallel to y-axis where $\frac{dy}{dx} = \infty$ *i.e.*, at $\sqrt{x} = 0$ *i.e.*, at (0, 0)

Slope of the tangent at (a, 0) is $\tan\theta = \pm \frac{3a-a}{2\sqrt{3a}\,.\sqrt{a}} = \pm \frac{2a}{2\sqrt{3}\,.\,a} = \pm \frac{1}{\sqrt{3}}$

$$\therefore \quad \theta = \pm \frac{\pi}{6}.$$

5. Asymptote. There is no asymptote to the curve.

6. Region. y exists only for positive values of x

$\therefore$ curve lies only on the R.H.S. of y-axis.

Sketch of the curve is in the figure

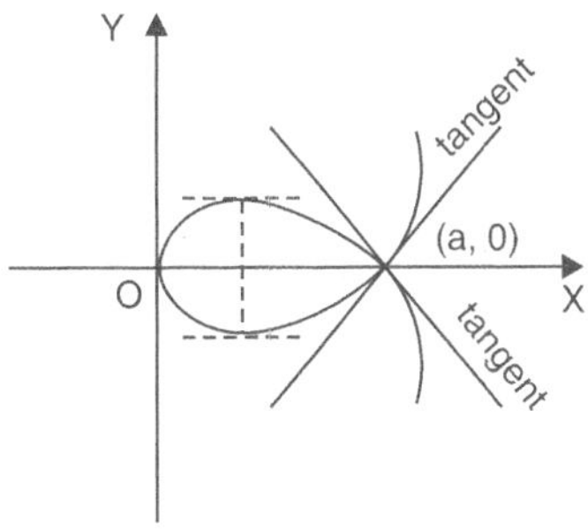

Now perimeter of the curve

$$= 2 \int_0^a \sqrt{1 + \left(\frac{dy}{dx}\right)^2}\, dx$$

$$= 2\int_0^a \sqrt{1+\frac{(3x-a)^2}{12ax}}\,dx$$

$$= 2\int_0^a \sqrt{\frac{12ax+(3x-a)^2}{2\sqrt{3ax}}}\,dx = \int_0^a \frac{\sqrt{(3x+a)^2}}{\sqrt{3ax}}\,dx$$

$$= \frac{1}{\sqrt{3a}}\int_0^a \frac{3x+a}{\sqrt{x}}\,dx = \frac{1}{\sqrt{3a}}\int_0^a \left(3\sqrt{x}+\frac{a}{\sqrt{x}}\right)dx$$

$$= \frac{1}{\sqrt{3a}}\left\{\frac{3x^{3/2}}{3/2}+a\cdot\frac{x^{1/2}}{1/2}\right\}_0^a = \frac{1}{\sqrt{3a}}\,[2a^{3/2}+2a\cdot a^{1/2}] = \frac{1}{\sqrt{3a}}\cdot 4a^{3/2}$$

$$= \frac{4}{\sqrt{3}}\,a.$$

Example 5. *Find the length of the curve $y = \log \sec x$ between $x = 0$ and $x = \frac{\pi}{3}$.*

(P.T.U., May 2007, Dec. 2011)

Sol. Equation of the curve is

$$y = \log \sec x$$

$$\frac{dy}{dx} = \frac{1}{\sec x}\sec x \tan x = \tan x$$

$$\text{Length of the arc} = \int_0^{\frac{\pi}{3}} \sqrt{1+\left(\frac{dy}{dx}\right)^2}\,dx$$

$$= \int_0^{\frac{\pi}{3}} \sqrt{1+\tan^2 x}\,dx = \int_0^{\frac{\pi}{3}} \sec x\,dx$$

$$= \log(\sec x + \tan x)\Big|_0^{\frac{\pi}{3}}$$

$$= \log\left(\sec\frac{\pi}{3}+\tan\frac{\pi}{3}\right) - \log(1+0)$$

$$= \log(2+\sqrt{3}) - 0 = \log(2+\sqrt{3}).$$

Example 6. *Rectify the curve $y = x^{2/3}$ from $x = -1$ to $x = 8$.*

Sol. Equation of the curve is $y = x^{2/3}$ or $y^3 = x^2$

(First of all we will have rough sketch of the curve)

Curve is symmetrical about y-axis. Curve passes through the origin. Curve is defined only for +ve values of y *i.e.*, curve lies above x-axis.

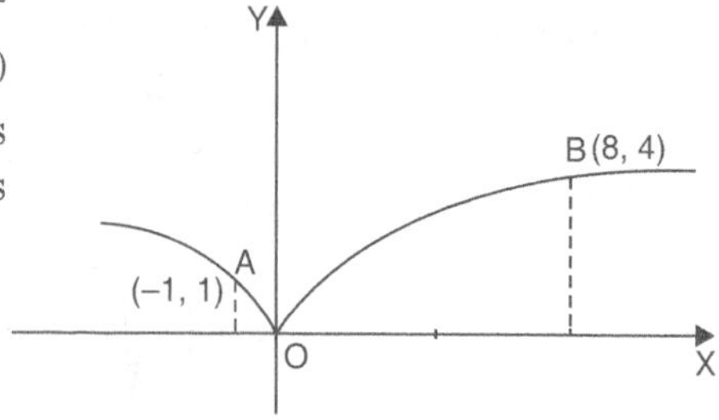

Now $\frac{dy}{dx} = \frac{2}{3}x^{-1/3} = \frac{2}{3x^{1/3}}$

at (0, 0) ; $\frac{dy}{dx} = \infty$ $\quad\therefore$ at (0, 0) tangent is y-axis.

when $x = -1, y^3 = 1 \quad \therefore \quad y = 1$

and when $x = 8, y = 4$

As length of the arc from A to B is required and we see in the figure that one portion of the arc lies in 2nd quadrant and the other in 1st quadrant.

$\therefore$ Length of the arc will be calculated in two parts

$$\therefore \text{ Length of the Arc from A to B} = \int_{\text{from A}}^{\text{to O}} \sqrt{1+\left(\frac{dy}{dx}\right)^2}\, dx + \int_{\text{from O}}^{\text{to B}} \sqrt{1+\left(\frac{dy}{dx}\right)^2}\, dx$$

$$= \int_{-1}^{0} \sqrt{1+\frac{4}{9x^{2/3}}}\, dx + \int_{0}^{8} \sqrt{1+\frac{4}{9x^{2/3}}}\, dx$$

$$= \int_{-1}^{0} \frac{\sqrt{4+9x^{2/3}}}{|3x^{1/3}| = -3x^{1/3}}\, dx + \int_{0}^{8} \frac{\sqrt{4+9x^{2/3}}}{|3x^{1/3}| = 3x^{1/3}}\, dx \qquad \textbf{Note this step}$$

$\because$ x is –ve in (–1, 0) $\qquad \because$ x is +ve in (0, 8)

Put $x^{2/3} = y \quad \therefore \quad \frac{2}{3} x^{-1/3}\, dx = dy$

i.e., $$\frac{dx}{3x^{1/3}} = \frac{1}{2} dy$$

when $x = -1, \quad y = 1$ when $x = 8, \quad y = 4$

when $x = 0, \quad y = 0$

$\therefore$ Length of the arc from A to B

$$= \int_{1}^{0} -\frac{\sqrt{4+9y}}{2}\, dy + \int_{0}^{4} \frac{\sqrt{4+9y}}{2}\, dy$$

$$= \frac{1}{2}\left\{ -\left|\frac{(4+9y)^{3/2}}{\frac{3}{2}.9}\right|_{1}^{0} \right\} + \frac{1}{2}\left|\frac{(4+9y)^{3/2}}{\frac{3}{2}.9}\right|_{0}^{4}$$

$$= \frac{1}{27}.\{-4^{3/2} + 13^{3/2}\} + \frac{1}{27}.\{(40)^{3/2} - 4^{3/2}\}$$

$$= \frac{1}{27}.\{-4^{3/2} + (13)^{3/2} + (40)^{3/2} - 4^{3/2}\}$$

$$= \frac{1}{27}.\{40\sqrt{40} + 13\sqrt{13} - 2.4\sqrt{4}\}$$

$$= \frac{1}{27}.\{80\sqrt{10} + 13\sqrt{13} - 8.2\}$$

$$= \frac{8}{27}.\left\{10\sqrt{10} + \frac{13\sqrt{13}}{8} - 2\right\}.$$

Example 7. *Rectify the ellipse $x = a \cos \theta, y = b \sin \theta$.*

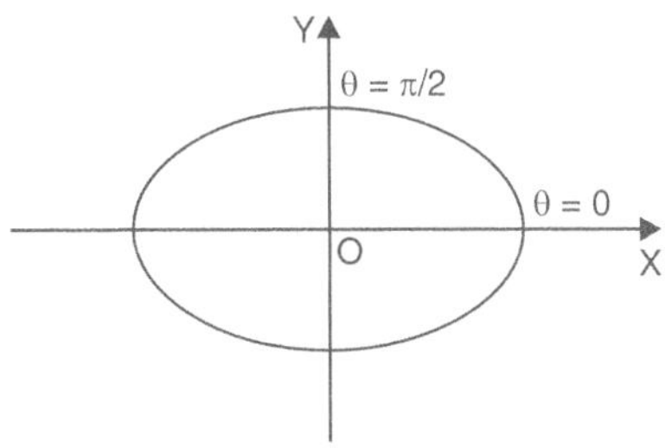

Sol. $\frac{dx}{d\theta} = -a \sin \theta, \quad \frac{dy}{d\theta} = b \cos \theta$

$\therefore$ Perimeter of the ellipse

$$= 4 \int_{0}^{\pi/2} \sqrt{\left(\frac{dx}{d\theta}\right)^2 + \left(\frac{dy}{d\theta}\right)^2}\, d\theta$$

$$= 4 \int_0^{\pi/2} \sqrt{a^2 \sin^2 \theta + b^2 \cos^2 \theta}\, d\theta.$$

It is not possible to evaluate the integral in this form $\therefore$ Put $b^2 = a^2 (1 - e^2)$

$$= 4 \int_0^{\pi/2} \sqrt{a^2 \sin^2 \theta + a^2(1-e^2) \cos^2 \theta}\, d\theta$$

$$= 4a \int_0^{\pi/2} \sqrt{1 - \cos^2 \theta + \cos^2 \theta - e^2 \cos^2 \theta}\, d\theta$$

$$= 4a \int_0^{\pi/2} \sqrt{1 - e^2 \cos^2 \theta}\, d\theta$$

$$= 4a \int_0^{\pi/2} (1 - e^2 \cos^2 \theta)^{1/2}\, d\theta \quad \text{expand by binomial theorem}$$

$$= 4a \int_0^{\pi/2} \left[1 - \frac{1}{2}(e^2 \cos^2 \theta) + \frac{\frac{1}{2}\left(-\frac{1}{2}\right)}{2!}(-e^2 \cos^2 \theta)^2 + \frac{\frac{1}{2}\left(-\frac{1}{2}\right)\left(-\frac{3}{2}\right)}{3!}(-e^2 \cos^2 \theta)^3 \ldots \right] d\theta$$

$$= 4a \int_0^{\pi/2} \left[1 - \frac{e^2}{2} \cos^2 \theta - \frac{1}{8} e^4 \cos^4 \theta - \frac{1}{16} e^6 \cos^6 \theta \ldots \right] d\theta$$

$$\left[\begin{array}{l} \text{We know that} \\ \displaystyle\int_0^{\pi/2} \cos^n \theta\, d\theta = \frac{(n-1)(n-3)\ldots}{n(n-2)\ldots} \cdot \frac{\pi}{2} \text{ if } n \text{ is even} \end{array} \right]$$

$$= 4a \left\{ \frac{\pi}{2} - \frac{e^2}{2} \cdot \frac{1}{2} \cdot \frac{\pi}{2} - \frac{e^4}{8} \frac{3 \cdot 1}{4 \cdot 2} \frac{\pi}{2} - \frac{e^6}{16} \cdot \frac{5 \cdot 3 \cdot 1}{6 \cdot 4 \cdot 2} \frac{\pi}{2} \ldots \right\}$$

$$= 4a \cdot \frac{\pi}{2} \left\{ 1 - \frac{e^2}{2^2} - \frac{1^2 \cdot 3}{2^2 \cdot 4^2} e^4 - \frac{1^2 \cdot 3^2 \cdot 5}{2^2 \cdot 4^2 \cdot 6^2} e^6 \ldots \right\}$$

$$= 2a\pi \left[1 - \frac{e^2}{2^2} - \frac{1^2 \cdot 3}{2^2 \cdot 4^2} e^4 - \frac{1^2 \cdot 3^2 \cdot 5}{2^2 \cdot 4^2 \cdot 6^2} e^6 \ldots \right].$$

Example 8. *Find the length of the boundary of the region bounded by the curve* $y = \frac{1}{2} x^2 + 1$ *and the lines* $y = x$, $x = 0$ *and* $x = 2$.

Sol. Equation of the curve is

$$y = \frac{1}{2} x^2 + 1 \quad \text{or} \quad x^2 = 2(y - 1) \qquad \ldots(1)$$

which is an upward parabola with vertex at (0, 1) and axis along to y-axis. Equations of the lines are $y = x$, $x = 0$, $x = 2$.

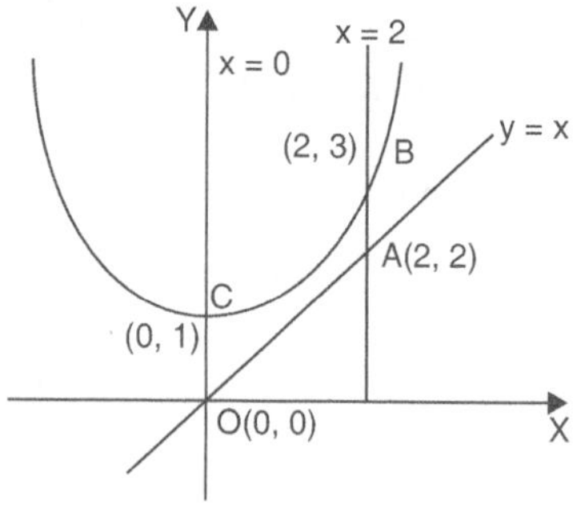

Length of the boundary of the required region

$$= \text{OA} + \text{AB} + \widehat{\text{BC}} + \text{CO} \quad \text{(see figure)}$$

Now A is the intersection of $x = 2$ and $y = x$ $\therefore$ A has co-ordinates (2, 2)

$$\therefore \qquad \text{OA} = \sqrt{2^2 + 2^2} = 2\sqrt{2}$$

B is the intersection of $x = 2$ with $y = \frac{1}{2}x^2 + 1$

$\therefore$ B has co-ordinates (2, 3) $\because$ $y = \frac{4}{2} + 1 = 3$

$$\therefore \quad \text{AB} = \sqrt{(2-2)^2 + (3-2)^2} = 1$$

$$\text{OC} = 1$$

$$\widehat{\text{BC}} = \int_0^2 \sqrt{1 + \left(\frac{dy}{dx}\right)^2}\, dx = \int_0^2 \sqrt{1 + x^2}\, dx \qquad \left(\because \text{ from (1), } \frac{dy}{dx} = x\right)$$

$$= \left.\frac{x\sqrt{1+x^2}}{2} + \frac{1}{2}\sinh^{-1} x\right|_0^2 = \frac{2\sqrt{5}}{2} + \frac{1}{2}\sinh^{-1} 2$$

$$= \sqrt{5} + \frac{1}{2}\sinh^{-1} 2$$

$\therefore$ Length of the required boundary

$$= 2\sqrt{2} + 1 + \sqrt{5} + \frac{1}{2}\sinh^{-1} 2 + 1$$

$$= 2\sqrt{2} + 2 + \sqrt{5} + \frac{1}{2}\sinh^{-1} 2.$$

Example 9. *Find the length of an arc of the cycloid* $x = a(\theta + \sin\theta), y = a(1 - \cos\theta)$ *from vertex to the point (x, y). Hence deduce the length of one arc of the cycloid.* **(P.T.U., Dec. 2013)**

Sol. Equation of the curve is

$$x = a(\theta + \sin\theta),\ y = a(1 - \cos\theta)$$

Curve is symmetric about y-axis. One arc of the cycloid is shown in the figure (for details see chapter 1 Section A of curve tracing)

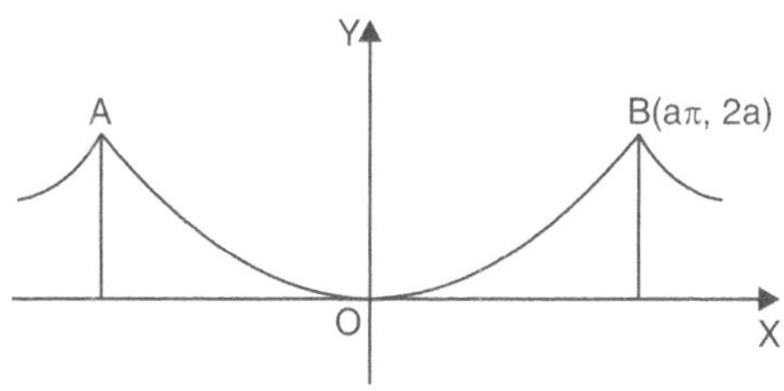

$$\frac{dx}{d\theta} = a(1 + \cos\theta) \qquad \frac{dy}{d\theta} = a\sin\theta$$

Length of the arc from (0, 0) to (x, y)

$$= \int_0^\theta \sqrt{\left(\frac{dx}{d\theta}\right)^2 + \left(\frac{dy}{d\theta}\right)^2}\, d\theta = \int_0^\theta \sqrt{a^2(1+\cos\theta)^2 + a^2\sin^2\theta}\, d\theta$$

$$= a\int_0^\theta \sqrt{1 + 2\cos\theta + \cos^2\theta + \sin^2\theta}\, d\theta$$

$$= a\int_0^\theta \sqrt{2(1+\cos\theta)}\, d\theta = a\int_0^\theta \sqrt{2 \cdot 2\cos^2\theta/2}\, d\theta$$

$$= 2a \int_0^{\theta} \cos \theta/2 \, d\theta = 2a \left. \frac{\sin \theta/2}{1/2} \right|_0^{\theta} = 4a \sin \theta/2$$

$$= 4a \sqrt{\frac{1 - \cos \theta}{2}} = 4a \sqrt{\frac{1}{2} \cdot \frac{y}{a}} = \sqrt{8ay}$$

Length of one arc of the cycloid

$$= 2 \int_0^{\pi} \sqrt{\left(\frac{dx}{d\theta}\right)^2 + \left(\frac{dy}{d\theta}\right)^2} \, d\theta$$

$$= 2 \int_0^{\pi} 2a \cos \frac{\theta}{2} \, d\theta = 4a \left. \frac{\sin \theta/2}{1/2} \right|_0^{\pi}$$

$$= 8a \sin \left. \frac{\theta}{2} \right|_0^{\pi} = 8a \left[\sin \frac{\pi}{2} - 0 \right] = 8a.$$

Example 10. *Find a curve through (1, 1) whose length integral is*

$$L = \int_1^4 \sqrt{1 + \frac{1}{4x}} \, dx$$

how many such curves are there? Give reasons for your answer. **(P.T.U., Dec. 2004)**

Sol. Let the required curve be $y = f(x)$; then length of the arc of the curve between

$$x = 1 \quad \text{and} \quad x = 4 \quad \text{is} \quad \int_1^4 \sqrt{1 + [f'(x)]^2} \, dx.$$

$$= \int_1^4 \sqrt{1 + \left(\frac{dy}{dx}\right)^2} \, dx$$

But it is given that length integral $= \int_1^4 \sqrt{1 + \frac{1}{4x}} \, dx$

$$\therefore \quad \left(\frac{dy}{dx}\right)^2 = \frac{1}{4x} \quad \therefore \quad \frac{dy}{dx} = \pm \frac{1}{2\sqrt{x}}$$

Integrate both sides w.r.t. x

$$y = \pm \frac{1}{2} \cdot \frac{\sqrt{x}}{1/2} + C \quad \text{where C is constant of integration}$$

$$\therefore \quad y = \pm \sqrt{x} + C$$

Curve passes through (1, 1)

$$\therefore \quad 1 = \pm 1 + C$$

$\therefore$ either $\quad C = 0 \quad$ or $\quad C = 2$

$$\therefore \quad y = \sqrt{x} \quad \text{or} \quad y = -\sqrt{x} + 2$$

$$\therefore \quad y^2 = x \quad \text{or} \quad (y - 2)^2 = x$$

$\therefore$ There are two curves through the point (1, 1). Both are parabolas.

Example 11. *Find the entire length of the cardioid $r = a(1 + \cos\theta)$ and show that the upper half of the curve is bisected by the line $\theta = \frac{\pi}{3}$.* **(P.T.U., May 2003, May 2012)**

Sol. Equation of the curve is $r = a(1 + \cos\theta)$. Curve is symmetric about initial line. It intersects initial line where $\theta = 0$ *i.e.*, at $r = 2a$ *i.e.*, at $(2a, 0)$ and intersects $\theta = \frac{\pi}{2}$ at $\left(a, \frac{\pi}{2}\right)$

$$\frac{dr}{d\theta} = -a\sin\theta$$

$$\tan\phi = r\frac{d\theta}{dr} = \frac{a(1+\cos\theta)}{-a\sin\theta} = -\frac{2\cos^2\theta/2}{2\sin\theta/2\cos\theta/2} = -\cot\theta/2$$

$$\therefore \quad \phi = \frac{\pi}{2} + \frac{\theta}{2}$$

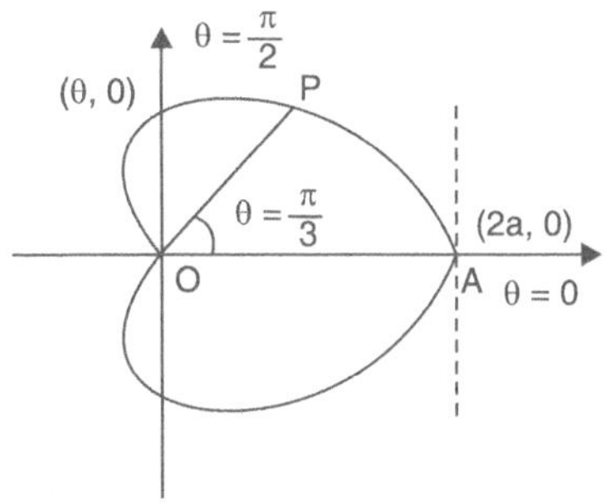

Angle between tangent and radius vector $(\theta = 0)$ is $\frac{\pi}{2}$ and angle between the tangent and $\left(\text{radius vector } \theta = \frac{\pi}{2}\right)$

is
$$\phi = \frac{\pi}{2} + \frac{\pi}{4} = \frac{3\pi}{4}$$

$$\text{Length of whole curve} = 2\int_0^{\pi}\sqrt{r^2 + \left(\frac{dr}{d\theta}\right)^2}\,d\theta$$

$$= 2\int_0^{\pi}\sqrt{a^2(1+\cos\theta)^2 + a^2\sin^2\theta}\,d\theta = 2\int_0^{\pi} a\sqrt{1+2\cos\theta+1}\,d\theta$$

$$= 2a\int_0^{\pi}\sqrt{2\cdot 2\cos^2\theta/2}\,d\theta = 4a\int_0^{\pi}\cos\theta/2\,d\theta$$

$$= 4a\cdot\frac{\sin\theta/2}{1/2}\Bigg|_0^{\pi} = 8a\left(\sin\frac{\pi}{2} - \sin 0\right) = 8a$$

$$\text{Length of the curve AP} = \int_0^{\pi/3}\sqrt{r^2 + \left(\frac{dr}{d\theta}\right)^2}\,d\theta = \int_0^{\pi/3} 2a\cos\frac{\theta}{2}\,d\theta$$

$$= 2a\cdot\frac{\sin\theta/2}{1/2}\Bigg|_0^{\frac{\pi}{3}} = 4a\cdot\sin\frac{\pi}{6} = 2a = \frac{1}{2}(4a)$$

$$= \frac{1}{2}\text{ length of upper half.}$$

Example 12. *Find the length of the arc of the parabola $\frac{2a}{r} = 1 + \cos\theta$ cut off by its latus rectum.*

Sol. Equation of the parabola is $\frac{2a}{r} = 1 + \cos\theta$. Change it to Cartesian form by taking

$$x = r\cos\theta,\ y = r\sin\theta \quad \therefore \quad r = \sqrt{x^2+y^2}$$

or
$$2a = r + r\cos\theta$$

or
$$2a = r + x \quad \text{or} \quad (2a - x)^2 = r^2 = x^2 + y^2$$

or $\quad 4a^2 - 4ax + x^2 = x^2 + y^2$

or $\quad y^2 = -4a(x - a)$, which is of the forms ...(1)

$$Y^2 = -4aX \quad \text{where} \quad X = x - a, Y = y$$

Latus rectum of a parabola is a line through the focus of the parabola ⊥ to the axis of the parabola.

Coordinates of the focus arc $X = -a, Y = 0$

i.e., $\quad x - a = -a, y = 0 \quad$ *i.e.*, $x = 0, y = 0$

Vertex of the parabola is $\quad X = 0, Y = 0 \quad \therefore \quad x = a, y = 0$

Latus rectum intersects the parabola at A(0, 2a) and A′(0, – 2a)

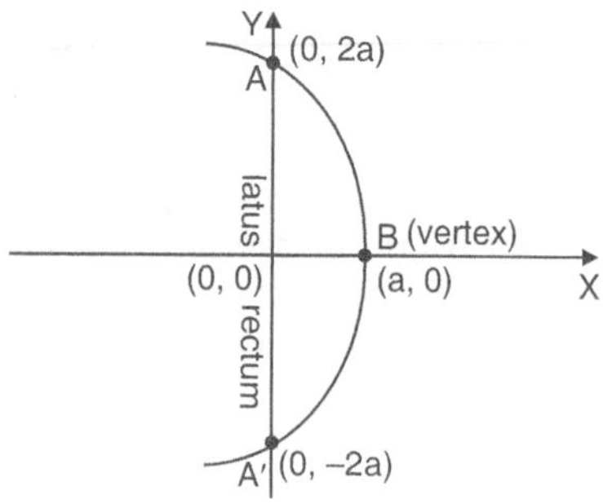

Length of the arc of the parabola

$$= \widehat{A'BA} = 2\widehat{BA}$$

$$= 2\int_0^{2a} \sqrt{1 + \left(\frac{dx}{dy}\right)^2}\, dy$$

From (1), $\quad x - a = -\dfrac{y^2}{4a}$

$$\therefore \quad \frac{dx}{dy} = -\frac{y}{2a}$$

$$\therefore \quad \text{Length of arc} = 2\int_0^{2a} \sqrt{1 + \frac{y^2}{4a^2}}\, dy = \frac{1}{a}\int_0^{2a} \sqrt{y^2 + 4a^2}\, dy$$

$$= \frac{1}{a}\left\{\frac{y\sqrt{y^2 + 4a^2}}{2} + \frac{4a^2}{2}\log(y + \sqrt{y^2 + 4a^2})\Bigg|_0^{2a}\right\}$$

$$= \frac{1}{a}\cdot\{a\, 2\sqrt{2}a + 2a^2 \log(2a + 2\sqrt{2a}) - 2a^2 \log 2a\}$$

$$= 2a\,\{\sqrt{2} + \log(1 + \sqrt{2})\}.$$

Example 13. *Show that the whole length of the lamicon $r = a + b\cos\theta$, $b < a$ is equal to that of an ellipse whose semi axes are equal in length to maximum and minimum radii vectors of the lamicon.*

Sol. Equation of the curve is $r = a + b \cos \theta$, $b < a$ for figure see S.E. 4 art. 4.16 curve tracing

$$\frac{dr}{d\theta} = -b\sin\theta$$

$$\text{Length of the lamicon} = 2\int_0^{\pi} \sqrt{r^2 + \left(\frac{dr}{d\theta}\right)^2}\, d\theta = 2\int_0^{\pi} \sqrt{(a + b\cos\theta)^2 + b^2\sin^2\theta}\, d\theta$$

$$= 2\int_0^{\pi} \sqrt{a^2 + 2ab\cos\theta + b^2}\, d\theta \qquad ...(1)$$

Maximum radius vector of lamicon = $a + b$ $\quad | \because \;$ Max $\cos\theta = 1$

Minimum radius vector of lamicon = $a - b$ $\quad | \because \;$ Min $\cos\theta = -1$

Parametric equation of the ellipse with semi major axis $= a + b$ and semi minor axis $= a - b$ is

$$x = (a + b) \cos t, y = (a - b) \sin t$$

Whole length of the ellipse

$$= 4\int_0^{\pi/2} \sqrt{\left(\frac{dx}{dt}\right)^2 + \left(\frac{dy}{dt}\right)^2}\, dt$$

$$= 4\int_0^{\frac{\pi}{2}} \sqrt{(a+b)^2 \sin^2 t + (a-b)^2 \cos^2 t}\, dt$$

$$= 4\int_0^{\pi/2} \sqrt{a^2 + b^2 + 2ab(\sin^2 t - \cos^2 t)}\, dt$$

$$= 4\int_0^{\frac{\pi}{2}} \sqrt{a^2 - 2ab \cos 2t + b^2}\, dt$$

Put $2t = \theta$ $\quad\therefore\quad$ $2dt = d\theta$

$$= 4 \int_0^{\pi} \sqrt{a^2 - 2ab \cos \theta + b^2}\, \frac{d\theta}{2}$$

$$= 2\int_0^{\pi} \sqrt{a^2 - 2ab \cos \theta + b^2}\, d\theta$$

which is same as (1). Hence the result.

TEST YOUR KNOWLEDGE

1. Find the lengths of the arcs of the following curves:
   (*i*) $y = x^{3/2}$ from (0, 0) to (4, 8). (*ii*) $y = \log(\cos x)$ from $0 \le x \le \pi/4$.
2. Find the length of the arc of the parabola $x^2 = 4ay$ extending from the vertex to one extremity of the latus rectum.
3. Find the perimeter of the loop of the curve
   $9ay^2 = x(x - 3a)^2$; $a > 0$. ***(P.T.U., May 2011)***
4. Find the length of the boundary of the region bounded by the curve
   $y = x^2 + 1$ and the lines $y = x$, $x = 0$ and $y = 2$.
5. Find the whole length of the curve
   $x^2(a^2 - x^2) = 8a^2y^2$.
6. Prove that length of the arc of the curve $x = a \sin 2\theta(1 + \cos 2\theta)$ $y = a \cos 2\theta(1 - \cos 2\theta)$ measured from (0, 0) to (x, y) is equal to $\frac{4}{3} a \sin 3\theta$.
7. Prove that loop of the curve $x = t^2, y = t - \frac{1}{3}t^3$ is of length $4\sqrt{3}$.
8. If 's' is the length of the arc of the curve $x = a(\theta + \sin\theta \cos\theta)$
   $y = a(1 + \sin\theta)^2$ measured from the point $\theta = -\frac{\pi}{2}$ to a point 'θ', show that s^4 varies as y^3.
9. Find the length of the arc of the curve $x = a(\theta - \sin\theta)$, $y = a(1 - \cos\theta)$ from 0 to 2π and show that the point $\theta = \frac{2\pi}{3}$ divides it in the ratio 1 : 3. ***(P.T.U., Jan. 2010)***
   [**Hint.** Consult S.E. 9]

10. (*i*) Rectify the curve $x^{2/3} + y^{2/3} = a^{2/3}$ **(*P.T.U., Dec. 2014*)**
[**Hint.** Parametric equation of the curve is $x = a\cos^3\theta, y = a\sin^3\theta$]
(*ii*) Using integration, find the perimeter of the curve $x^2 + y^2 = 4$ **(*P.T.U., Jan. 2009*)**

[**Hint.** Put $x = 2\cos\theta, y = 2\sin\theta$; perimeter $= 4\int_0^{\frac{\pi}{2}} \sqrt{4\cos^2\theta + 4\sin^2\theta}\, d\theta = 4\pi$]

11. Find the perimeter of the cardiod $r = a(1 - \cos\theta)$ **(*P.T.U., Dec. 2010, Dec. 2012*)**
[**Hint.** Consult S.E. 11]

12. Show that arc of the upper half of the curve $r = a(1 + \cos\theta)$ is bisected at $\theta = \frac{\pi}{3}$.

13. Find the length of the spiral $r = ae^{\theta\cot\alpha}$ between the points at which radii vectors are r_1 and r_2.

Answers

1. (*i*) $\frac{8}{27}[10\sqrt{10} - 1]$ (*ii*) $\log(\sqrt{2} + 1)$ **2.** $a[\sqrt{2} + \log(\sqrt{2} + 1)]$

3. $4\sqrt{3}\,a$ **4.** $2 + 2\sqrt{2} + \frac{1}{4}(2\sqrt{5} + \sinh^{-1} 2)$ **5.** $\pi a\sqrt{2}$

9. $8a$ **10.** (*i*) $6a$ units (*ii*) 4π **11.** $8a$

13. $\sec\alpha\,(r_2 - r_1)$.

2.4. AREAS UNDER CURVES

Prove that the area bounded by the curve $y = f(x)$, the x-axis and the ordinates $x = a$, $x = b$ is $\int_a^b y\,dx$ where $f(x)$ is a non-negative continuous functions in $[a, b]$.

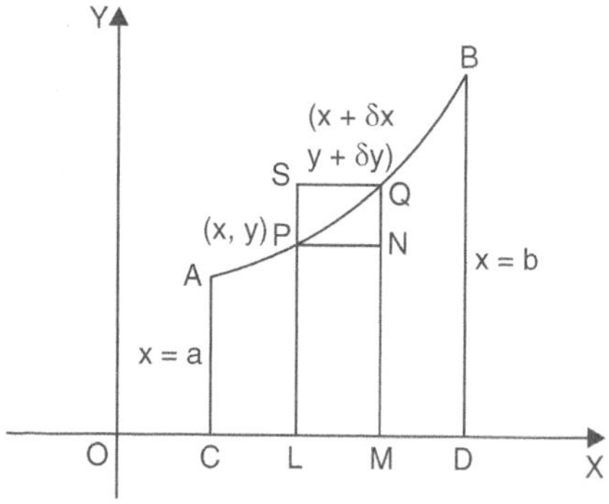

Proof. Let AB represents the curve $y = f(x)$.

Let AC, BD be the ordinates $x = a$, $x = b$ respectively.

Let P(x, y) be any point on the curve and Q$(x + \delta x, y + \delta y)$ be its neighbouring point on the curve. From P, Q draw PL, QM ⊥ on x-axis complete the rectangles PLMN and SLMQ.

Let δA be the area enclosed between PQ and the ordinates PL, QM and x-axis *i.e.*, the area PLMQ. Now area PLMQ lies between PLMN and SLMQ.

i.e., δA lies between $y\,\delta x$ and $(y + \delta y)\delta x$

or $\frac{\delta A}{\delta x}$ lies between y and $y + \delta y$

Taking limit when $\delta x \to 0$; $\underset{\delta x \to 0}{\text{Lt}} \frac{\delta A}{\delta x} = \frac{dA}{dx}$

$\therefore$ $\frac{dA}{dx}$ lies between y and a quantity $\to y$ ($\because$ when $\delta x \to 0$, δy also $\to 0$)

$$\therefore \quad \frac{dA}{dx} = y \quad \therefore \quad \int_a^b y\,dx = \int_a^b \frac{dA}{dx}\,dx = A \Big|_a^b$$

= (value of A when $x = b$) – (value of A when $x = a$)

= area ACDB – 0 = area ACDB

$$\therefore \quad A = \text{area ACDB} = \int_a^b y\,dx \quad \text{or} \quad A = \int_a^b f(x)\,dx.$$

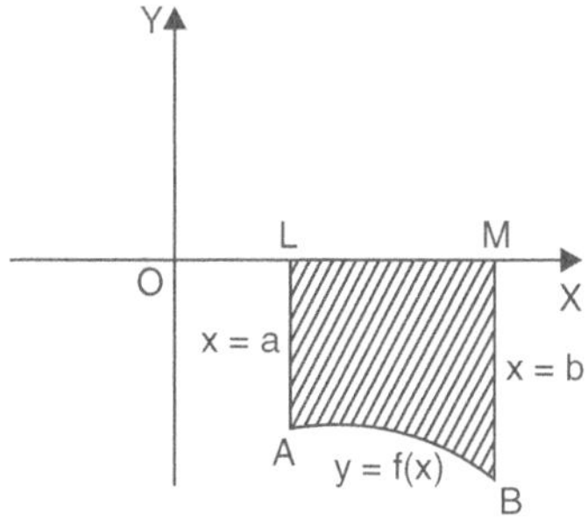

Cor. 1. *If f be a non-positive continuous function defined on [a, b], then area bounded by y = f(x) ; x-axis and x = a, x = b is* $\int_a^b |f(x)|\, dx$.

Proof. Given $f(x)$ to be a non-positive continuous function on $[a, b]$

$\therefore \quad f(x) \le 0 \ \forall x[a, b] \quad \therefore \quad -f(x) \ge 0 \ \forall x[a, b]$

$\Rightarrow -f(x)$ is non-negative continuous function defined on $[a, b]$

$\therefore$ By the above theorem, area of the region bounded by the curve $y = f(x)$, the x-axis and ordinates $x = a$, $x = b$

$$= \int_a^b \{-f(x)\}\, dx \quad \text{or} \quad \int_a^b |f(x)|\, dx \qquad \{\because \ |f(x)| = -f(x) \text{ if } f(x) \le 0\}$$

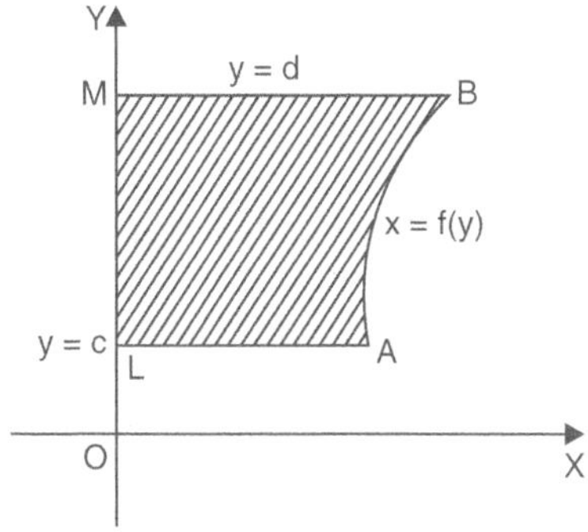

Cor. 2. If $x = f(y)$ is single valued continuous function for $c \le y \le d$ then the area of the region bounded between $x = f(y)$, y-axis and the absiccae $y = c$, $y = d$ is

$$\int_c^d |f(y)|\, dy.$$

Cor. 3. If f and g be two continuous functions defined on $[a, b]$ such that $g(x) \le f(x)$ on $[a, b]$ then the area bounded by the curve $y = f(x)$, $y = g(x)$ and the ordinates $x = a$, $x = b$ is $\int_a^b [f(x) - g(x)]\, dx$.

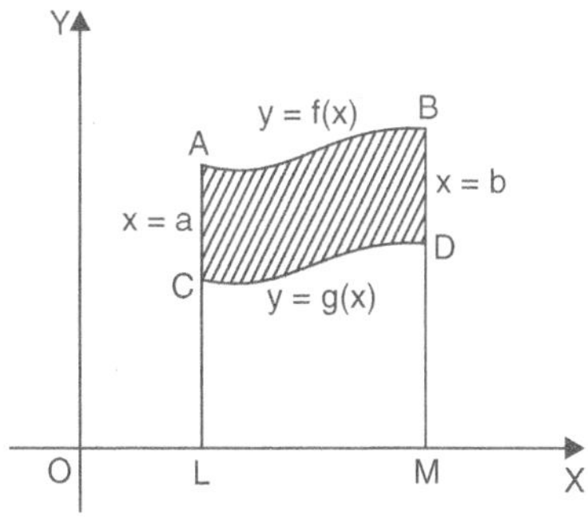

Case I. *Both f(x) and g(x) are non-negative on [a, b]*

Then required area

$$= ACDB$$
$$= \text{area ALMB} - \text{area CLMD}$$
$$= \int_a^b f(x)\, dx - \int_a^b g(x)\, dx$$
$$= \int_a^b [f(x) - g(x)]\, dx.$$

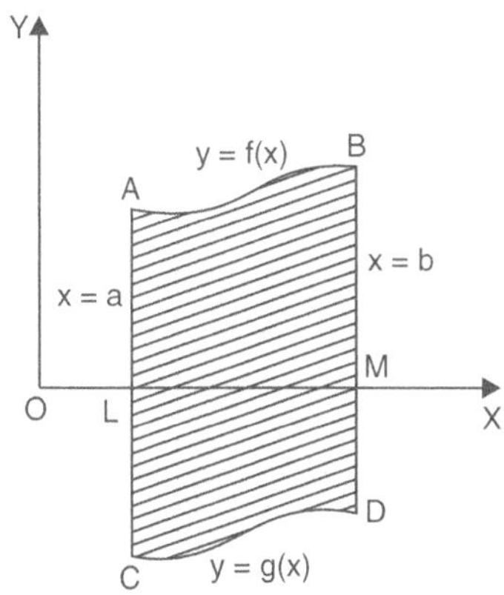

Case II. *When* $f(x) \ge 0$ *and* $g(x) \le 0 \ \forall\, x \in [a, b]$

Required area = ACDB

$$= \text{area ALMB} + \text{area LCDM}$$
$$= \int_a^b f(x)\, dx + \int_a^b |g(x)|\, dx$$
$$= \int_a^b f(x)\, dx + \int_a^b -g(x)\, dx$$
$$= \int_a^b [f(x) - g(x)]\, dx.$$

Case III. *When $f(x) \le 0$, $g(x) \le 0$ $\forall$ $x \in [a, b]$*

Then required area = ACDB

= area LCDM – area LABM

$$= \int_a^b |g(x)|\,dx - \int_a^b |f(x)|\,dx$$

$$= \int_a^b -g(x)\,dx - \int_a^b -f(x)\,dx$$

$\because$ $|g(x)| = -g(x)$ $\because$ $g(x) \le 0$

$|f(x)| = -f(x)$ $\because$ $f(x) \le 0$

$$= \int_a^b -g(x)\,dx + \int_a^b f(x)\,dx$$

$$= \int_a^b [f(x) - g(x)]\,dx.$$

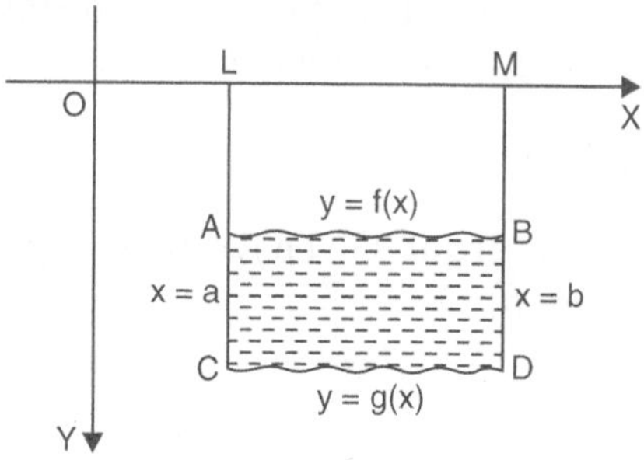

Note 1. For finding areas bounded by curves first have a rough sketch of the curve.

Note 2. Areas of the curves can also be obtained by double integrals discussed in the next chapter of Multiple Integrals article Number 6.7.

2.5. AREA FORMULA FOR PARAMETRIC EQUATION

Let $x = f(t)$, $y = g(t)$ be parametric equation of the curve then area bounded by the curve, the x-axis and the ordinates at the points where $t = a$, $t = b$ is $\int_{t=a}^{b} y\frac{dx}{dt}\,dt$. Similarly area bounded by curve, the y-axis and the absiccae at points where $t = a$, $t = b$ is $\int_a^b x\frac{dy}{dt}\,dt$.

2.6. AREA FORMULA FOR POLAR EQUATIONS

If $r = f(\theta)$ be a continuous function of θ defined on $[\theta_1, \theta_2]$, then the area bounded by the curve $r = f(\theta)$ and radii vectors $\theta = \theta_1$, $\theta = \theta_2$ is $\frac{1}{2}\int_{\theta_1}^{\theta_2} r^2\,d\theta$.

Proof. Let AB be the curve $r = f(\theta)$ and OA, OB be the radii $\theta = \theta_1$, $\theta = \theta_2$.

Let P(r, θ) be any point on the curve and Q$(r + \delta r, \theta + \delta\theta)$ be a neighbouring point to P. Now with O as centre and OP and OQ as radii draw two circular arcs to intersect OQ and OP in R and S respectively. Then sectorial area OPQ lies between the areas of sectors OPR and OSQ

i.e., δA lies between $\frac{1}{2}r^2\,\delta\theta$ and $\frac{1}{2}(r + \delta r)^2\,\delta\theta$

$\therefore$ $\dfrac{\delta A}{\delta\theta}$ lies between $\frac{1}{2}r^2$ and $\frac{1}{2}(r + \delta r)^2$

Taking limit when $\delta\theta \to 0$

$\therefore$ $\dfrac{dA}{d\theta}$ lies between $\frac{1}{2}r^2$ and a quantity $\to \dfrac{1}{2}r^2$

$$\therefore \quad \frac{dA}{d\theta} = \frac{1}{2}r^2$$

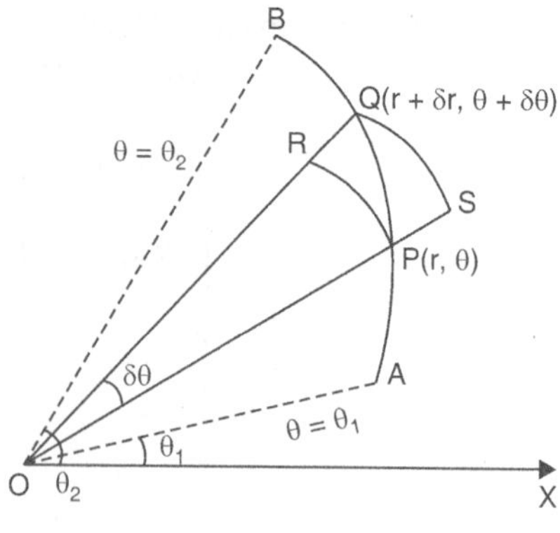

Integrating w.r.t. θ between the limits $\theta = \theta_1$ and $\theta = \theta_2$

$$\therefore \qquad \int_{\theta_1}^{\theta_2} \frac{dA}{d\theta} = \int_{\theta_1}^{\theta_2} \frac{1}{2} r^2 \, d\theta$$

$$\therefore \qquad A \Big|_{\theta_1}^{\theta_2} = \int_{\theta_1}^{\theta_2} \frac{1}{2} r^2 \, d\theta$$

$$\text{area OAB} - O = \int_{\theta_1}^{\theta_2} \frac{1}{2} r^2 \, d\theta. \quad \therefore \quad \text{area OAB} = \frac{1}{2} \int_{\theta_1}^{\theta_2} r^2 \, d\theta.$$

Cor. Area bounded by the curve $y = f(\theta)$, $r = g(\theta)$ and the radii vectors $\theta = \theta_1$, $\theta = \theta_2$ is $\frac{1}{2} \int_{\theta_1}^{\theta_2} (r_1^2 - r_2^2) \, d\theta$ where r_1 is value of 'r' of outer curve and r_2 that of inner curve.

2.7. AREA OF THE CLOSED CURVE

Prove that area of the closed curve $x = f(t)$, $y = g(t)$ *is* $\frac{1}{2} \int_{t_1}^{t_2} \left(x \frac{dy}{dt} - y \frac{dx}{dt} \right) dt$; t_1, t_2 *being the values of t such that the point (x, y) travels once completely round the curve.*

Proof. Let P(x, y) be any point on the closed curve and let its polar co-ordinates be (r, θ)

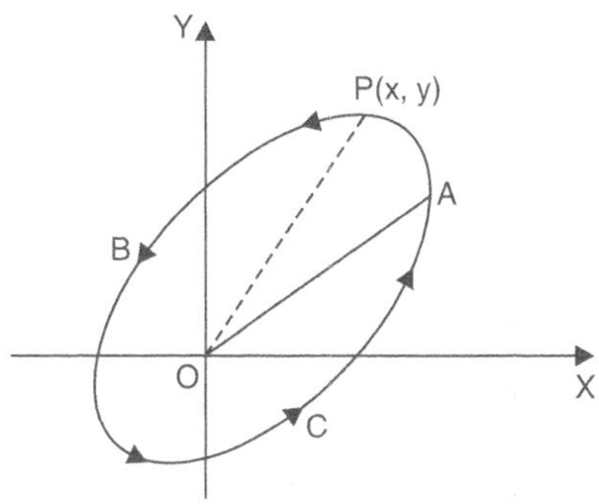

$$\therefore \qquad x = r \cos \theta, \; y = r \sin \theta$$

Area of the curve $= \frac{1}{2} \int r^2 d\theta$ and the limits of integration are the values of θ for which the point P travels completely round the curve from A to B, B to C, C to A and again.

$$\therefore \qquad \text{area of the curve} = \frac{1}{2} \int_{t=t_1}^{t_2} r^2 \frac{d\theta}{dt} \, dt$$

to express r, θ in terms of x, y we have

$$r^2 = x^2 + y^2, \quad \tan \theta = \frac{y}{x}, \quad \theta = \tan^{-1} \frac{y}{x}$$

$$\therefore \qquad \frac{d\theta}{dt} = \frac{1}{1 + \dfrac{y^2}{x^2}} \; \frac{x \dfrac{dy}{dt} - y \dfrac{dx}{dt}}{x^2} = \frac{x \dfrac{dy}{dt} - y \dfrac{dx}{dt}}{x^2 + y^2}$$

$$\therefore \qquad \text{Area} = \frac{1}{2} \int_{t=t_1}^{t_2} (x^2 + y^2) \frac{\left(x \dfrac{dy}{dt} - y \dfrac{dx}{dt} \right)}{(x^2 + y^2)} \, dt$$

$$= \frac{1}{2} \int_{t_1}^{t_2} \left(x \frac{dy}{dt} - y \frac{dx}{dt} \right) dt.$$

ILLUSTRATIVE EXAMPLES

Example 1. *Show that area bounded by the parabola $y^2 = 4ax$, the x-axis and the ordinate $x = c$ is equal to $\frac{2}{3}$ of the rectangle of which the extreme ordinate and the absiccae of the area are the adjecent sides.*

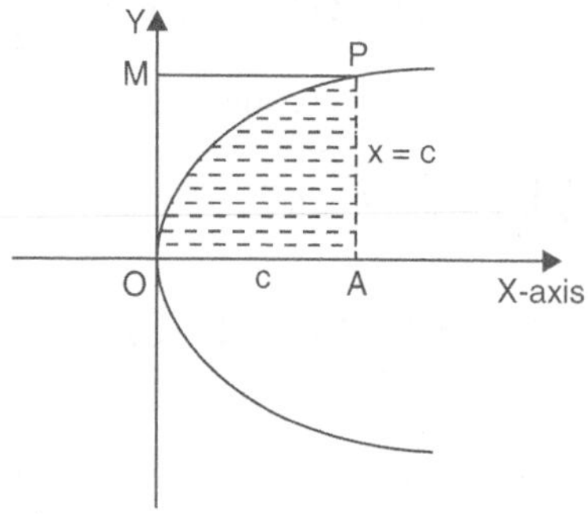

Sol. Equation of the parabola is $y^2 = 4ax$.

Let O be the vertex and PA the given ordinate $x = c$

$$\therefore \text{ Area OAP} = \int_0^c y\,dx = \int_0^c 2\sqrt{ax}\,dx$$

$$= 2\sqrt{a}\cdot\left.\frac{x^{3/2}}{3/2}\right|_0^c$$

$$= \frac{4\sqrt{a}}{3}c^{3/2} = \frac{4}{3}\sqrt{a}\,c^{3/2} = A_1(\text{say})$$

Now for the rectangle OAPM ; OA = c

(find AP) Let length of AP = d

$\therefore$ (c, d) lies on the parabola $\qquad \therefore \quad d^2 = 4ac$

$\therefore \quad d = \pm 2\sqrt{ac} \qquad \therefore \quad P\,[c, 2\sqrt{ac}\,]$

$\therefore \quad PA = 2\sqrt{ac}$

Area of the rectangle = OA . AP = $c\,2\sqrt{ac}$

$$= 2\sqrt{a}\,.\,c^{3/2} = A_2\text{ (say)}$$

Now
$$\frac{A_1}{A_2} = \frac{\frac{4}{3}\sqrt{a}\,c^{3/2}}{2\sqrt{a}\,c^{3/2}} = \frac{2}{3} \quad \therefore \quad A_1 = \frac{2}{3}A_2.$$

Example 2. *Find the area bounded by the ellipse $\frac{x^2}{a^2} + \frac{y^2}{b^2} = 1$, the ordinates $x = c$, $x = d$ and the x-axis. Deduce the area of the whole ellipse.*

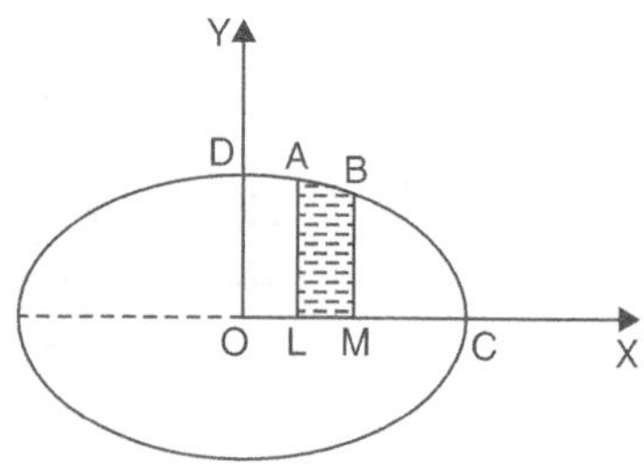

Sol. Equation of the ellipse is $\frac{x^2}{a^2} + \frac{y^2}{b^2} = 1$

Let AL represent $x = c$ and BM represent $x = d$

$$\therefore \quad \text{area ALMB} = \int_c^d y\,dx$$

$$= \int_c^d b\sqrt{1 - \frac{x^2}{a^2}}\,dx = \frac{b}{a}\int_c^d \sqrt{a^2 - x^2}\,dx$$

$$= \frac{b}{a}\cdot\left\{\frac{x\sqrt{a^2 - x^2}}{2} + \frac{a^2}{2}\sin^{-1}\frac{x}{a}\right\}_c^d$$

$$= \frac{b}{a}\left[\frac{d\sqrt{a^2 - d^2}}{2} - \frac{c\sqrt{a^2 - c^2}}{2} + \frac{a^2}{2}\left\{\sin^{-1}\frac{d}{a} - \sin^{-1}\frac{c}{a}\right\}\right]$$

Area of the whole ellipse = 4 area of the quadrant OCD for which $c = 0$ and $d = a$

$$= 4\,\frac{b}{a}\left[0 - 0 + \frac{a^2}{2}\left\{\sin^{-1}\frac{a}{a} - 0\right\}\right] = \frac{4b}{a}\cdot\left[\frac{a^2}{2}\cdot\frac{\pi}{2}\right] = \pi\,ab.$$

Example 3. *If A is the vertex, O the centre and P(x, y) a point on the hyperbola* $\frac{x^2}{a^2} - \frac{y^2}{b^2} = 1$ *prove that* $x = a \cosh \frac{2s}{ab}$, $y = b \sinh \frac{2s}{ab}$ *where s is the sectorial area OPA.*

Sol. Equation of the hyperbola is $\frac{x^2}{a^2} - \frac{y^2}{b^2} = 1$

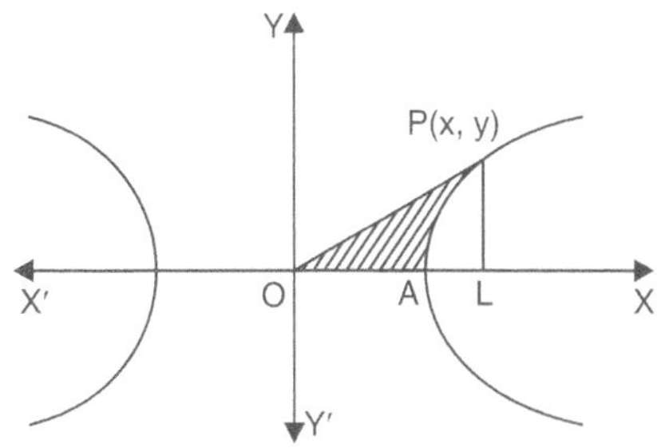

Let L be foot of ⊥ from P on x-axis

Now s = sectorial area OAP

= area of the triangle OLP – area ALP

Area of Δ OLP = $\frac{1}{2}\cdot$ OL $\cdot$ PL = $\frac{1}{2}\,xy$

Area ALP = area bounded between hyperbola, AL and ordinate PL

$$= \int_a^x y\,dx = \int_a^x b\sqrt{\frac{x^2}{a^2} - 1}\,dx$$

$$= \frac{b}{a}\int_a^x \sqrt{x^2 - a^2}\,dx = \frac{b}{a}\left[\frac{x\sqrt{x^2 - a^2}}{2} - \frac{a^2}{2}\cosh^{-1}\frac{x}{a}\right]_a^x$$

$$= \frac{b}{a}\cdot\left\{\frac{x\sqrt{x^2 - a^2}}{2} - \frac{a^2}{2}\cosh^{-1}\frac{x}{a} - 0 - 0\right\} \quad \because \quad \cosh^{-1} 1 = 0$$

$$\therefore \qquad s = \frac{1}{2}xy - \frac{b}{a}\left\{\frac{x\sqrt{x^2 - a^2}}{2} - \frac{a^2}{2}\cosh^{-1}\frac{x}{a}\right\}$$

$$= \frac{1}{2}x\,.\,b\sqrt{\frac{x^2}{a^2} - 1} - \frac{b}{a}\,\frac{x\sqrt{x^2 - a^2}}{2} + \frac{ab}{2}\cosh^{-1}\frac{x}{a}$$

$$s = \frac{ab}{2}\cosh^{-1}\frac{x}{a} \quad \therefore \quad \frac{x}{a} = \cosh\frac{2s}{ab}$$

$$\therefore \qquad x = a\cosh\frac{2s}{ab} \quad \text{Now} \quad y = b\sqrt{\frac{x^2}{a^2} - 1}$$

$$y = b\sqrt{\cosh^2\frac{2s}{ab} - 1} = b\sinh\frac{2s}{ab}$$

$$\therefore \qquad x = a\cosh\frac{2s}{ab} \quad \text{and} \quad y = b\sinh\frac{2s}{ab}.$$

Example 4. *Find the area enclosed by the curves* $x^2 = 4ay$ *and* $x^2 + 4a^2 = \dfrac{8a^3}{y}$.

Sol. Equations of the curves are

$$x^2 = 4ay \quad \text{...(1)}$$

$$x^2 + 4a^2 = \frac{8a^3}{y} \quad \text{...(2)}$$

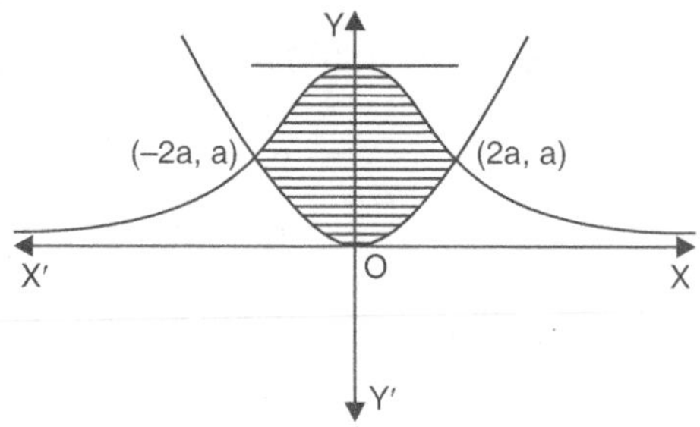

(1) is a parabola and for the rough sketch of (2) we find some points. Curve (2) is symmetric about y-axis.

It intersects y-axis (put $x = 0$) at $(0, 2a)$

Asymptote to the curve (2) parallel to x-axis is

$y = 0 \quad \because \quad yx^2 + 4a^2y = 8a^3$ is the equation of the curve.

Region of the curve is given by $0 < y \leq 2a$

$$\frac{dy}{dx} = \frac{16a^3x}{(x^2 + 4a^2)^2}$$

$\therefore$ Tangent is || to x-axis at $x = 0$ *i.e.*, at $(0, 2a)$

(1) and (2) intersect at points given by $4ay + 4a^2 = \dfrac{8a^3}{y}$

i.e.,
$$y^2 + ay = 2a^2$$
$$y^2 + 2ay - ay - 2a^2 = 0$$
$$y(y + 2a) - a(y + 2a) = 0$$

i.e.,
$$y = a, \; y = -2a$$

y is not –ve

$\therefore$ $y \neq -2a$ and $y = a$.

$\therefore$ (1) and (2) intersect at $(\pm 2a, a)$

$$\text{Required area} = \int_{-2a}^{2a} \left(\frac{8a^3}{x^2 + 4a^2} - \frac{x^2}{4a} \right) dx$$

$$= 2\int_0^{2a} \left(\frac{8a^3}{x^2 + 4a^2} - \frac{x^2}{4a} \right) dx \quad \because \text{ integrand is even and } \int_{-a}^{a} f(x)\,dx = 2\int_0^{a} f(x)\,dx$$

$$= 2\left\{ 8a^3 \cdot \frac{1}{2a} \tan^{-1} \frac{x}{2a} - \frac{x^3}{12a} \Bigg|_0^{2a} \right\}$$

$$= 2\left\{ 4a^2 \tan^{-1} 1 - \frac{8a^3}{12a} \right\} = 2\left\{ 4a^2 \frac{\pi}{4} - \frac{2a^2}{3} \right\}$$

$$= \left(2\pi - \frac{4}{3} \right) a^2.$$

Example 5. *Prove that the area of the region bounded by the curve $a^4y^2 = x^5(2a - x)$ is to that of the circle whose radius is a is 5 : 4.*

Sol. First of all we will have rough sketch of the curve $a^4y^2 = x^5(2a - x)$

(*i*) Curve is symmetric about x-axis

(*ii*) It passes through the origin

(*iii*) It intersects x-axis is $(2a, 0)$ and $(0, 0)$ and y-axis as $(0, 0)$

(*iv*) Domain of the curve is $0 \le x \le 2$; no –ve value of x:

(*v*) Slope of the tangent at any point (x, y) is $\frac{dy}{dx}$ $\therefore$ $\left(\text{find } \frac{dy}{dx}\right)$

i.e., $$a^4 . 2y \frac{dy}{dx} = 2a . 5x^4 - 6x^5$$

or $$\frac{dy}{dx} = \frac{2x^4(5a-3x)}{2a^4y} = \frac{2x^4(5a-3x)}{2a^4 . \frac{x^{5/2}}{a^2}\sqrt{2a-x}} = \frac{x^{3/2}(5a-3x)}{a^2\sqrt{2a-x}}$$

tangent is parallel to x-axis where $\frac{dy}{dx} = 0$

i.e., at $x = 0, x = \frac{5a}{3}$

i.e., at $(0, 0), \left(\frac{5a}{3}, \frac{25\sqrt{5}a}{27}\right)$

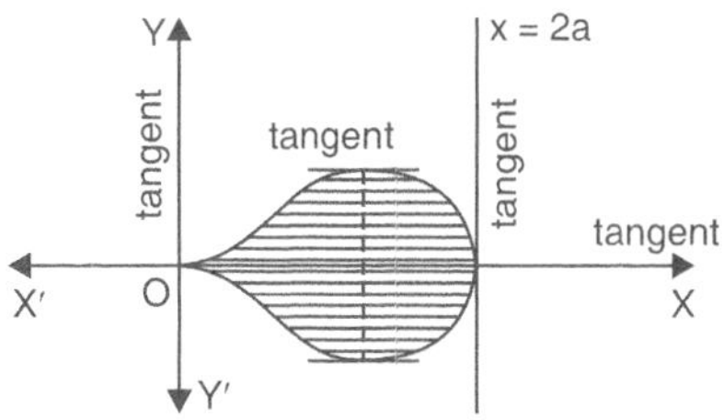

Tangent is parallel to y-axis where $\frac{dy}{dx} = \infty$ *i.e.*, at $x = 2a$ *i.e.*, at $(2a, 0)$

$\therefore$ Area of the region bounded by the curve $= 2\int_0^{2a} \frac{x^{5/2}\sqrt{2a-x}}{a^2}\, dx.$

Put $x = 2a \sin^2 \theta$

$dx = 4a \sin \theta \cos \theta \, d\theta$

$$\text{Area} = 2\int_0^{\pi/2} \frac{(2a)^{5/2} \cdot \sin^5 \theta \sqrt{2a - 2a\sin^2 \theta}}{a^2} \cdot 4a \sin \theta \cos \theta \, d\theta$$

$$= \frac{2}{a^2}\int_0^{\pi/2} (2a)^{5/2} \sqrt{2a} \cdot 4a \cdot \sin^6 \theta \cos^2 \theta \, d\theta$$

$$= \frac{2}{a^2} \cdot 8a^3 \cdot 4a \int_0^{\pi/2} \sin^6 \theta \cos^2 \theta \, d\theta$$

$$= 64a^2 \frac{5 \cdot 3 \cdot 1 \cdot 1}{8 \cdot 6 \cdot 4 \cdot 2} \frac{\pi}{2} = \frac{5}{4}\pi a^2$$

area of the circle with radius $a = \pi a^2$

$\therefore$ $\frac{\text{Area of the given curve}}{\text{area of the circle}} = \frac{5}{4}$ *i.e.*, two areas are in the ratio 5 : 4.

Example 6. *Find the area between the curve $y^2(2a - x) = x^3$ and its asymptote*

(P.T.U., Dec. 2014)

Sol. Equation of the curve is $y^2 (2a - x) = x^3$

First of all we will draw a rough sketch of the curve.

(*i*) Curve is symmetric about x-axis

(*ii*) Curve passes through the origin and the tangent at the origin is $y = 0$

(*iii*) $y^2 = \dfrac{x^3}{2a - x}$; $y = \dfrac{\pm x^{3/2}}{\sqrt{2a - x}}$ Domain of the curve is $0 < x < 2a$

(*iv*) Asymptote to the curve parallel to y-axis is $2a - x = 0$ *i.e.*, $x = 2a$.

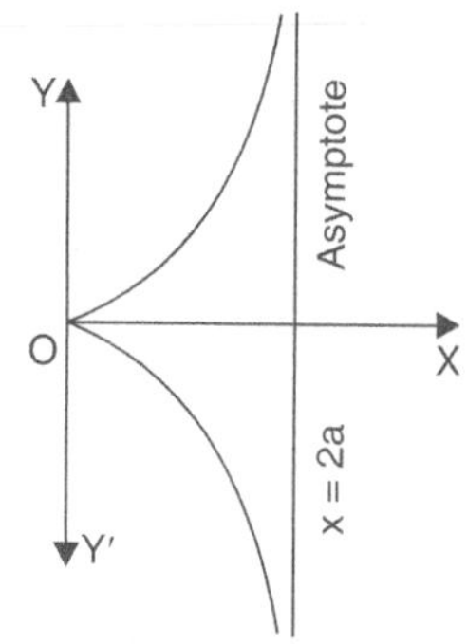

Required area $= 2\int_0^{2a} y\, dx$

$$= 2\int_0^{2a} \frac{x^{3/2}}{\sqrt{2a - x}} dx$$

Put $x = 2a \sin^2\theta$

$$dx = 4a \sin\theta \cos\theta\, d\theta$$

$$= 2\int_0^{\pi/2} \frac{(2a)^{3/2} \sin^3\theta}{(2a)^{1/2} \cos\theta} 4a \sin\theta \cos\theta\, d\theta$$

$$= 2 \cdot 2a \cdot 4a \int_0^{\pi/2} \sin^4\theta\, d\theta$$

$$= 16a^2 \frac{3 \cdot 1}{4 \cdot 2} \frac{\pi}{2} = 3\pi a^2.$$

Example 7. *Find the area between the curve $x(x^2 + y^2) = a(x^2 - y^2)$ and its asymptote. Also find the area of its loop.*

Sol. Equation of the curve is $x(x^2 + y^2) = a(x^2 - y^2)$

(*i*) It is symmetric about x-axis

(*ii*) It passes through the origin, tangent at origin is $x^2 - y^2 = 0$, *i.e.*, $y = \pm x$

(*iii*) It intersects x-axis (put $y = 0$) $x^3 = ax^2$, *i.e.*, at $(0, 0)$ and $(a, 0)$

It intersects y-axis (put $x = 0$) $y = 0$, *i.e.*, at $(0, 0)$

(*iv*) From given equation $y^2 = \dfrac{x^2(a - x)}{a + x}$

$\therefore$ Domain of the curve is $0 \le x \le a$ and $- a < x \le 0$

$\therefore$ Curve lies between $(- a, 0] \cup [0, a]$

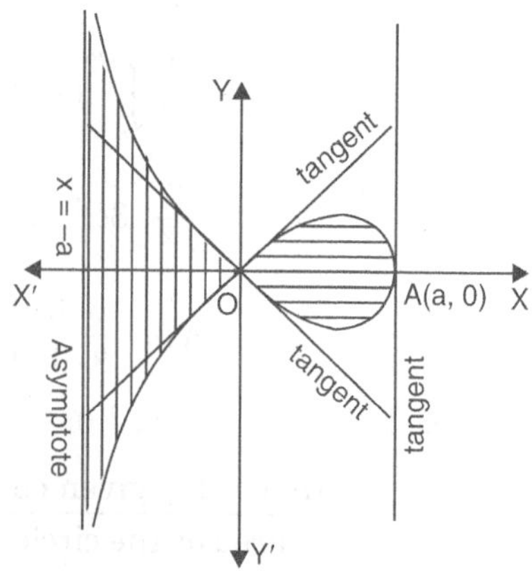

(v) Asymptote to the curve is $a + x = 0$ *i.e.*, $x = -a$

Area between the curve and its asymptote

$$= 2\int_{-a}^{0} y\, dx \text{ (double } \because \text{ curve is symmetric about } x\text{-axis)}$$

$$= 2\int_{-a}^{0} x\sqrt{\frac{a-x}{a+x}}\, dx$$

$$= 2\int_{-a}^{0} x \frac{a-x}{\sqrt{a^2-x^2}}\, dx$$

Put $x = a \sin \theta$,

$$dx = a \cos \theta d\theta$$

$$= 2\int_{-\frac{\pi}{2}}^{0} (a \sin \theta) \frac{a - a \sin \theta}{a \cos \theta} \cdot a \cos \theta\, d\theta$$

$$= 2a^2 \int_{-\frac{\pi}{2}}^{0} (\sin \theta - \sin^2 \theta)\, d\theta$$

$$= 2a^2 \int_{-\frac{\pi}{2}}^{0} \sin \theta - \frac{1 - \cos 2\theta}{2}\, d\theta$$

$$= 2a^2 \left\{ -\cos \theta - \frac{1}{2}\theta + \frac{\sin 2\theta}{4} \Bigg|_{-\frac{\pi}{2}}^{0} \right\} = 2a^2 \left[-1 - \frac{\pi}{4} \right] = -\frac{4+\pi}{2} a^2$$

$$= \frac{4+\pi}{2} a^2 \text{ (in magnitude)}$$

$$\text{Area of its loop } = 2\int_0^a y\, dx = 2\int_0^a x\sqrt{\frac{a-x}{a+x}}\, dx = 2\int_0^a \frac{x(a-x)}{\sqrt{a^2-x^2}}\, dx$$

Put $x = a \sin \theta$, $dx = a \cos \theta\, d\theta$

$$= 2\int_0^{\pi/2} \frac{a \sin \theta \cdot a(1 - \sin \theta)}{a \cos \theta} a \cos \theta\, d\theta$$

$$= 2a^2 \int_0^{\pi/2} (\sin \theta - \sin^2 \theta)\, d\theta = 2a \left[-\cos \theta \Big|_0^{\pi/2} - \frac{1}{2}\frac{\pi}{2} \right]$$

$$= 2a^2 \left[1 - \frac{\pi}{4} \right] = \frac{4-\pi}{2} a^2.$$

Example 8. *Show that the larger of the two areas into which the circle* $x^2 + y^2 = 64a^2$ *is divided by the parabola* $y^2 = 12ax$ *is* $\frac{16}{3} a^2 (8\pi - \sqrt{3})$.

Sol. Equation of the circle is $x^2 + y^2 = 64a^2$.

Equation of the parabola is $y^2 = 12ax$

Their intersection is given by $x^2 + 12ax = 64a^2$

or $$x^2 + 16ax - 4ax - 64a^2 = 0 \quad \text{or} \quad x(x + 16a) - 4a(x + 16a) = 0$$

$\therefore$ either $x = 4a$, or $x = -16a$. But x cannot be –ve

($\because$ Domain of parabola is only +ve values of x)

$\therefore$ When $x = 4a$, $y^2 = 48a^2$ *i.e.*, $y = \pm 4\sqrt{3}\,a$

area of the larger portion = area of circle – area of smaller portion

$$= \pi(8a)^2 - \text{area OBAC} \qquad ...(1)$$

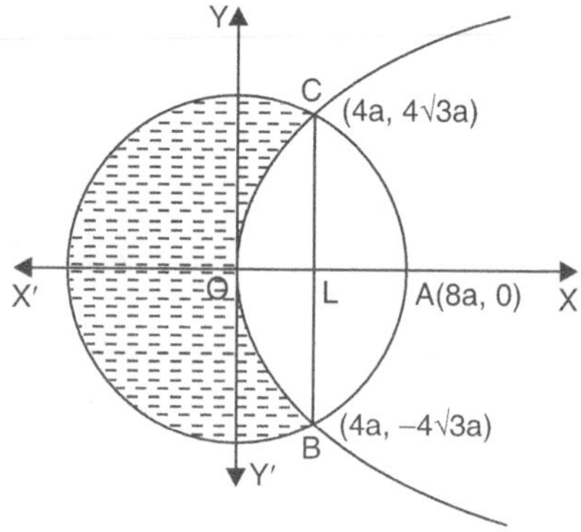

$$\text{area OBAC} = 2 \text{ area OAC}$$

$$= 2[\text{area OLC} + \text{area CLA} \qquad ...(2)$$

$$\text{area OLC} = \int_0^{4a} y\,dx = \int_0^{4a} \sqrt{12ax}\,dx$$

$$= 2\sqrt{3a} \cdot \frac{x^{3/2}}{3/2}\Bigg|_0^{4a}$$

$$= \frac{4\sqrt{3a}}{3} \cdot (4a)^{3/2} = \frac{4\sqrt{3a}}{3} \cdot 8 \cdot a^{3/2}$$

$$= \frac{32}{3}\sqrt{3}\,a^2$$

$$\text{area CLA} = \int_{4a}^{8a} \sqrt{64a^2 - x^2}\,dx = \int_{4a}^{8a} \sqrt{(8a)^2 - x^2}\,dx$$

$$= \frac{x\sqrt{(8a)^2 - x^2}}{2} + \frac{64a^2}{2}\sin^{-1}\frac{x}{8a}\Bigg|_{4a}^{8a}$$

$$= 32a^2 \cdot \sin^{-1} 1 - \frac{4a}{2} \cdot \sqrt{48a^2} - 32a^2 \sin^{-1}\frac{1}{2}$$

$$= 32a^2 \cdot \frac{\pi}{2} - 8\sqrt{3}\,a^2 - 32a^2 \cdot \frac{\pi}{6}$$

$$= 32a^2 \cdot \frac{\pi}{3} - 8\sqrt{3}\,a^2$$

$\therefore$ From (2),

$$\text{area OBAC} = 2 \cdot \left[\frac{32\sqrt{3}}{3}a^2 + \frac{32\pi}{3}a^2 - 8\sqrt{3}\,a^2\right] = 2\left[\frac{8}{3}\sqrt{3}\,a^2 + \frac{32}{3}\pi a^2\right]$$

$\therefore$ From (1),

$$\text{Area of the larger portion} = 64\pi a^2 - \frac{16}{3}\sqrt{3}\,a^2 - \frac{64}{3}\pi a^2$$

$$= \frac{128}{3}\pi a^2 - \frac{16}{3}\sqrt{3}\,a^2 = \frac{16}{3}a^2(8\pi - \sqrt{3}).$$

Example 9. (*i*) *Find the area between the curve y = cos x, the x-axis from x = 0 to x = π*

(*ii*) *Find the area contained between x-axis and one arch of the curve y = cos 3x.*

(P.T.U., Dec. 2005)

Sol. (*i*) Equation of the curve is $y = \cos x$

Required area = area OAC + area ADB

$$= \int_0^{\pi/2} \cos x \, dx + \int_{\pi/2}^{\pi} - \cos x \, dx$$

$$= \sin x \Big|_0^{\pi/2} - \sin x \Big|_{\pi/2}^{\pi}$$

$$= (1 - 0) - (0 - 1) = 2$$

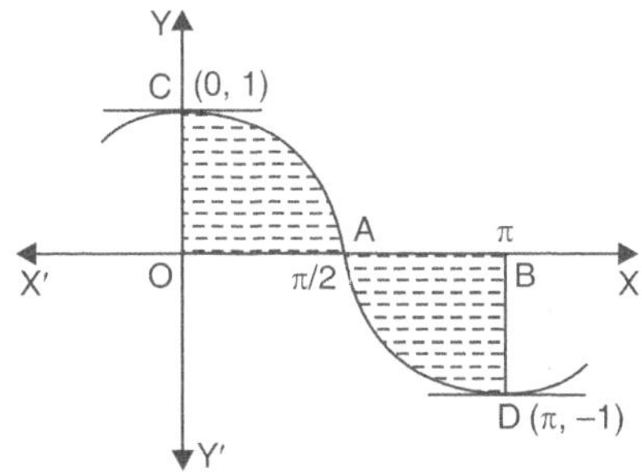

(*ii*) Equation of the curve is $y = \cos 3x$ one arch of the curve lies between 0 and $\frac{\pi}{6}$.

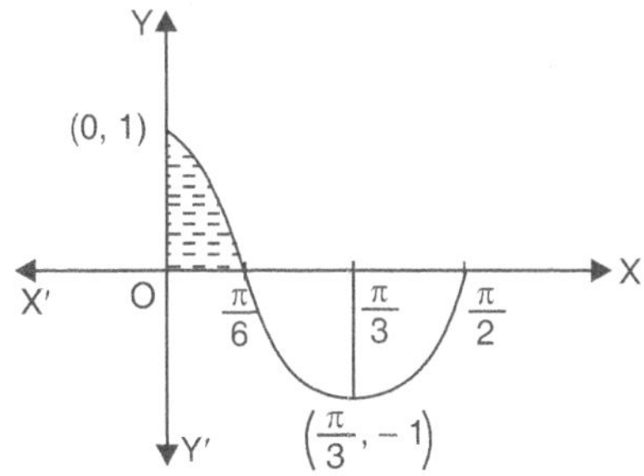

Required area

$$= \int_0^{\frac{\pi}{6}} y dx = \int_0^{\frac{\pi}{6}} \cos 3x \, dx$$

$$= \frac{\sin 3x}{3} \Big|_0^{\frac{\pi}{6}} = \frac{1}{3}\left(\sin \frac{3\pi}{6}\right)$$

$$= \frac{1}{3} \sin \frac{\pi}{2} = \frac{1}{3}.$$

Example 10. *Find the area enclosed between one arc of the cycloid x = a(θ – sin θ), y = a(1 – cos θ) and its base.*

Sol. Equation of the cycloid is $x = a(\theta - \sin \theta), y = a(1 - \cos \theta)$

Required area $= \int_{x=0}^{2a\pi} y \, dx$

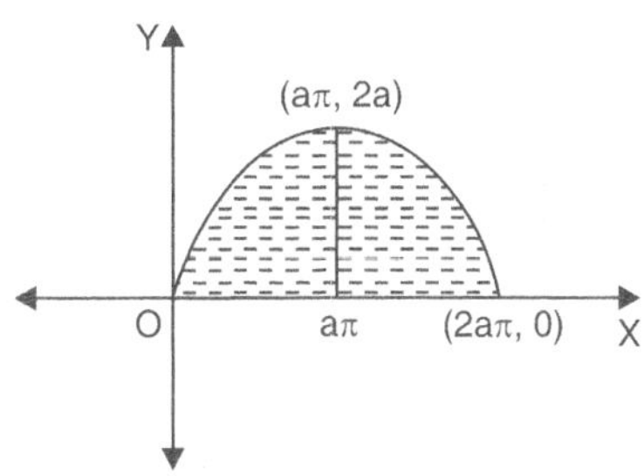

$$= \int_0^{2\pi} a(1 - \cos \theta) \, a(d\theta - \cos \theta \, d\theta)$$

$$= a^2 \int_0^{2\pi} (1 - \cos \theta)^2 \, d\theta$$

$$= a^2 \int_0^{2\pi} (1 - 2\cos \theta + \cos^2 \theta) \, d\theta$$

$$= a^2 \left[\theta - 2\sin \theta \right|_0^{2\pi} + a^2 \int_0^{2\pi} \frac{1 + \cos 2\theta}{2} d\theta$$

$$= 2\pi a^2 + \frac{a^2}{2} \left[\theta + \frac{\sin 2\theta}{2} \right|_0^{2\pi} = 2\pi a^2 + \frac{a^2}{2} \cdot 2\pi = 3\pi a^2.$$

Example 11. *Find the area of the curve $x = a\cos^3 t$, $y = b\sin^3 t$.*

Sol. Equation of the curve is $x = a \cos^3 t, y = b \sin^3 t$

As curve is symmetric about both the axes

$\therefore$ area of the curve

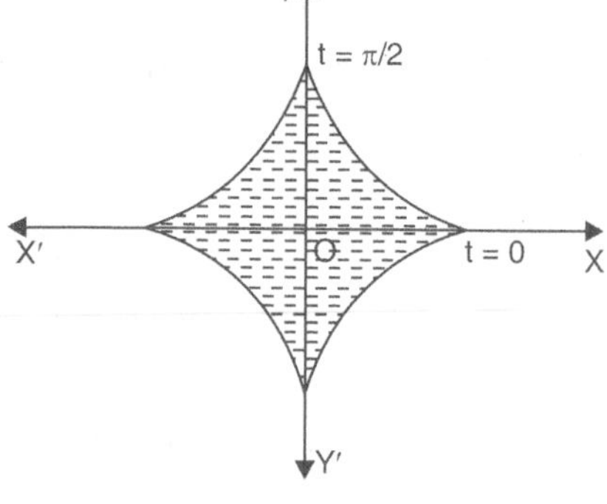

$$= 4\int_0^{\pi/2} y \frac{dx}{dt}\, dt$$

$$= 4\int_0^{\pi/2} (b \sin^3 t)(-3a \cos^2 t \sin t)\, dt$$

$$= -12ab \int_0^{\pi/2} \sin^4 t \cos^2 t\, dt$$

$$= -12ab\, \frac{3\cdot 1\cdot 1}{6\cdot 4\cdot 2}\frac{\pi}{2} = -\frac{3ab}{8}$$

$$= \frac{3}{8}\pi ab \text{ in magnitude.}$$

Example 12. *Find the area lying outside the cardioid $r = a(1 - \cos\theta)$ and inside the circle $r = a \sin\theta$.* **(P.T.U., May 2014)**

Sol. Equation of the cardioid is $r = a(1 - \cos\theta)$

Equation of the circle is $r = a \sin\theta$

For tracing of the cardioid. See example 1 art 1.16

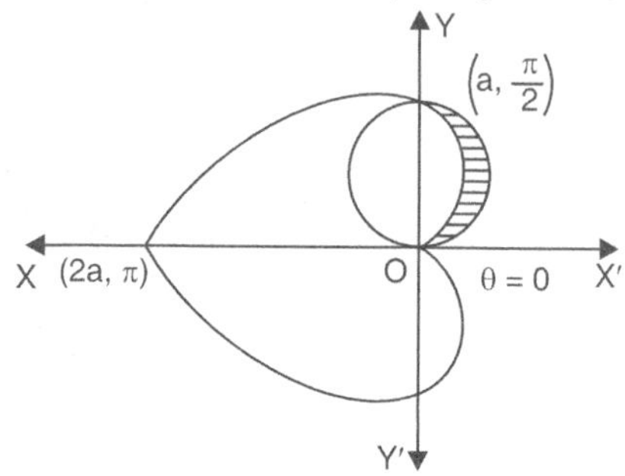

$r = a \sin\theta$ is a circle having centre at $\left(0, \frac{a}{2}\right)$ and radius $= \left(\frac{a}{2}\right)$ [as $r = a\sin\theta$ is $r^2 = ar\sin\theta$ *i.e.*, $x^2 + y^2 = ay$]

Required area = area of the circle on the RHS of $\theta = \frac{\pi}{2}$ – area of the cardioid on the RHS of $\theta = \frac{\pi}{2}$.

$$= \frac{1}{2}\int_0^{\frac{\pi}{2}} (a^2 \sin^2\theta)\, d\theta - \frac{1}{2}\int_0^{\frac{\pi}{2}} a^2(1 - \cos\theta)^2\, d\theta$$

$$= \frac{1}{2}a^2 \int_0^{\frac{\pi}{2}} [\sin^2\theta - (1 - \cos\theta)^2]\, d\theta$$

$$= \frac{1}{2}a^2 \int_0^{\frac{\pi}{2}} (\sin^2\theta - 1 + 2\cos\theta - \cos^2\theta)\, d\theta$$

$$= \frac{a^2}{2}\int_0^{\frac{\pi}{2}} [-1 + 2\cos\theta - (\cos^2\theta - \sin^2\theta)]\, d\theta$$

$$= \frac{a^2}{2}\int_0^{\frac{\pi}{2}} (-1 + 2\cos\theta - \cos 2\theta)\, d\theta$$

$$= \frac{a^2}{2}\left\{-\theta + 2\sin\theta - \frac{\sin 2\theta}{2}\right\}\Bigg|_0^{\frac{\pi}{2}}$$

$$= \frac{a^2}{2}\left[\frac{-\pi}{2}+2\right] = \frac{a^2}{2}\,\frac{4-\pi}{2} = a^2\left(1-\frac{\pi}{4}\right).$$

Note. For another method see S.E. 6(*ii*) art 6.7 chapter Multiple Integrals.

Example 13. *Prove that the area common to the circle $r = a\sqrt{2}$ and $r = 2a\cos\theta$ is $a^2(\pi - 1)$.*

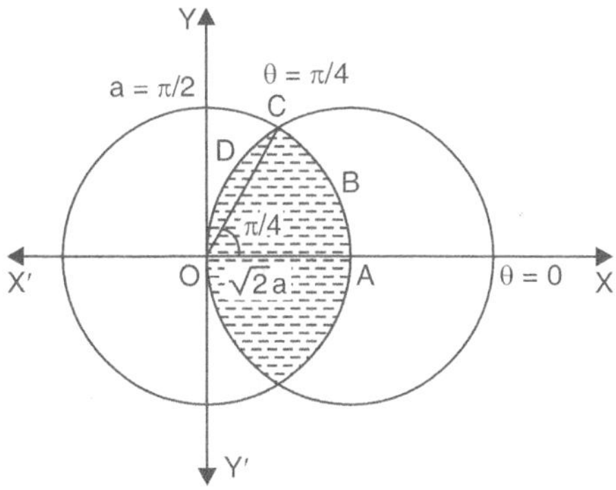

Sol. Equations of the circles are

$$r = a\sqrt{2} \qquad ...(1)$$

$$r = 2a\cos\theta \qquad ...(2)$$

(1) and (2) intersect at

$$2a\cos\theta = a\sqrt{2} \quad i.e., \cos\theta = \frac{1}{\sqrt{2}}$$

$$\theta = \pm\frac{\pi}{4}$$

$\therefore$ Vertical angle of C $= \frac{\pi}{4}$

Required area = 2[OABCDO] $\because$ both circles are symmetrical w.r.t. initial line

= 2[area OABCO + area OCDO]

= 2[area of the circle $r = \sqrt{2}a$ included between initial line and $\theta = \pi/4$

+ area of the circle $r = 2a\cos\theta$ included between $\theta = \frac{\pi}{4}$ and $\theta = \frac{\pi}{2}$

$$= \frac{2}{2}\int_0^{\pi/4}(2a^2)\,d\theta + \frac{2}{2}\int_{\pi/4}^{\pi/2} 4a^2\cos^2\theta\,d\theta$$

$$= 2a^2\,\theta\Big|_0^{\pi/4} + 4a^2\int_{\pi/4}^{\pi/2}\frac{1+\cos 2\theta}{2}\,d\theta$$

$$= 2a^2\,\frac{\pi}{4} + 2a^2\left[\theta + \frac{\sin 2\theta}{2}\right]_{\pi/4}^{\pi/2} = \frac{a^2\pi}{2} + 2a^2\left[\frac{\pi}{2}-\frac{\pi}{4}-\frac{1}{2}\right]$$

$$= a^2\pi - a^2 = a^2(\pi - 1).$$

Example 14. *Obtain the area of a loop of the curve $r = a\sin 3\theta$. Also find the total area of the curve.*

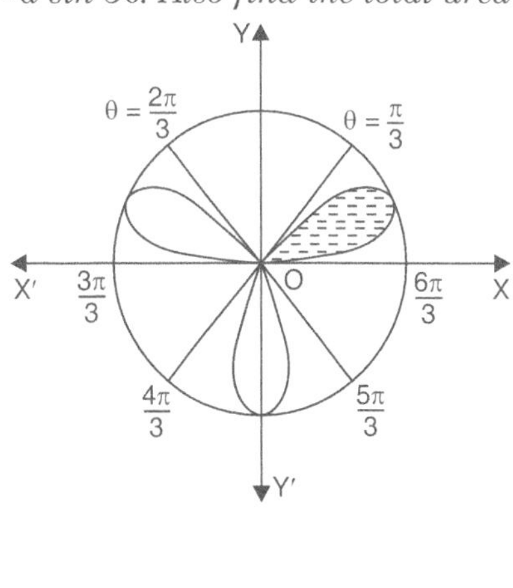

Sol. Equation of the curve is $r = a\sin 3\theta$ (For curve tracing S.E. 3 art 1.16)

$$\text{Area of one loop of the curve} = \int_0^{\pi/3}\frac{1}{2}(a\sin 3\theta)^2\,d\theta$$

$$= \frac{a^2}{2}\int_0^{\pi/3}\sin^2 3\theta\,d\theta$$

$$= \frac{a^2}{2}\int_0^{\pi/3}\frac{1-\cos 6\theta}{2}\,d\theta$$

$$= \frac{a^2}{4}\left[\theta - \frac{\sin 6\theta}{6}\right]_0^{\pi/3} = \frac{a^2}{4}\left[\frac{\pi}{3}\right] = \frac{\pi a^2}{12}$$

Total area of the curve $= 3 \cdot \frac{\pi a^2}{12} = \frac{\pi a^2}{4}$.

Example 15. *Using parametric equation of a circle show that the area of the circle of radius a is πa^2.* **(P.T.U., Dec. 2005)**

Sol. Parametric equation of a circle of radius a is $x = a \cos \theta$, $y = a \sin \theta$. As circle is symmetric about both the axes.

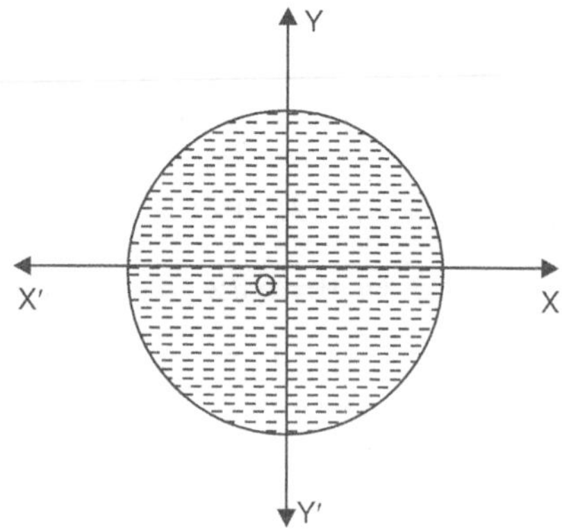

$\therefore$ Area of the circle.

$$= 4 \int_0^{\frac{\pi}{2}} \frac{1}{2}\left(x\frac{dy}{d\theta} - y\frac{dx}{d\theta}\right) d\theta$$

$$= \frac{4}{2}\int_0^{\frac{\pi}{2}} [a \cos \theta \, a \cos \theta - a \sin \theta \, (- a \sin \theta)] \, d\theta$$

$$= 2 \int_0^{\frac{\pi}{2}} a^2 \, d\theta = 2a^2 \cdot \theta \Big|_0^{\frac{\pi}{2}} = 2a^2 \frac{\pi}{2} = \pi a^2.$$

Example 16. *Calculate the area enclosed by the loop of the curve*

$$x = \frac{1-t^2}{1+t^2}, \quad y = \frac{t-t^3}{1+t^2} \; ; -1 \le t \le 1.$$

Sol. Equation of the curve is

$$x = \frac{1-t^2}{1+t^2}, \quad y = \frac{t-t^3}{1+t^2}$$

$$\frac{dx}{dt} = \frac{(1+t^2)(-2t) - (1-t^2)\,2t}{(1+t^2)^2};$$

$$= \frac{-2t[1+t^2+1-t^2]}{(1+t^2)^2};$$

$$= \frac{-4t}{(1+t^2)^2};$$

$$\frac{dy}{dt} = \frac{(1+t^2)(1-3t^2) - (t-t^3)\,2t}{(1+t^2)^2}$$

$$\frac{dy}{dt} = \frac{1 - 2t^2 - 3t^4 - 2t^2 + 2t^4}{(1+t^2)^2}$$

$$= \frac{1-4t^2-t^4}{(1+t^2)^2}$$

Required area $= \frac{1}{2}\int_{-1}^{1}\left(x\frac{dy}{dt} - y\frac{dx}{dt}\right) dt$

$$= \frac{1}{2}\int_{-1}^{1}\left\{\frac{1-t^2}{1+t^2}\,\frac{1-4t^2-t^4}{(1+t^2)^2} - \frac{t-t^3}{1+t^2}\,\frac{-4t}{(1+t^2)^2}\right\} dt$$

$$= \frac{1}{2}\int_{-1}^{1}\frac{1-t^2}{(1+t^2)^3}\,\{1 - 4t^2 - t^4 + 4t^2\}\, dt = \frac{1}{2}\int_{-1}^{1}\frac{(1-t^2)(1-t^4)}{(1+t^2)^3}\, dt$$

$$= \frac{1}{2} \cdot 2\int_0^1 \frac{(1-t^2)^2\,(1+t^2)}{(1+t^2)^3}\, dt = \int_0^1 \frac{(1-t^2)^2}{(1+t^2)^2}\, dt$$

$$= \int_0^1 \frac{(1+t^2)^2 - 4t^2}{(1+t^2)^2}\, dt = \int_0^1 \left\{1 - \frac{4t^2}{(1+t^2)^2}\right\} dt = t \Big|_0^1 - 4\int_0^1 \frac{t^2}{(1+t^2)^2}\, dt$$

$$= 1 - 4\int_0^1 \frac{t^2}{(1+t^2)^2}\, dt \quad \text{Put } t = \tan\theta,\ dt = \sec^2\theta\, d\theta.$$

$$= 1 - 4\int_0^{\pi/4} \frac{\tan^2\theta}{\sec^4\theta} \sec^2\theta\, d\theta = 1 - 4\int_0^{\pi/4} \sin^2\theta\, d\theta$$

$$= 1 - 4.\int_0^{\pi/4} \frac{1-\cos 2\theta}{2}\, d\theta = 1 - 2\left[\theta - \frac{\sin 2\theta}{2}\right]_0^{\pi/4}$$

$$= 1 - 2\left[\frac{\pi}{4} - \frac{1}{2}\right] = 1 - \frac{\pi}{2} + 1 = 2 - \frac{\pi}{2}.$$

Example 17. *Find the area of the loop of the curve $x^3 + y^3 = 3axy$.* **(P.T.U., May 2006)**

Sol. Equation of the curve is $x^3 + y^3 = 3axy$ (As it is an implicit function of x, y so we can not easily express y as a function of x) so we change it into polar form by putting $x = r\cos\theta$, $y = r\sin\theta$.

$$r^3(\cos^3\theta + \sin^3\theta) = 3ar^2 \sin\theta\cos\theta$$

or
$$r = 3a\, \frac{\sin\theta\cos\theta}{\sin^3\theta + \cos^3\theta} \quad ...(1)$$

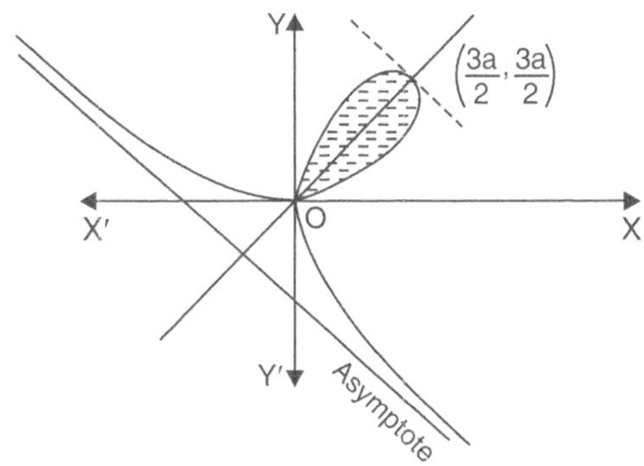

(For tracing of the figure see example 1 art 1.13)

Loop of the curve lies in the first quadrant only. To find limits of θ. Put $r = 0$ in (1) we have

$$\frac{3a\sin\theta\cos\theta}{\sin^3\theta + \cos^3\theta} = 0$$

i.e.,
$$\sin\theta\cos\theta = 0$$

$$\therefore \quad \text{either } \theta = 0 \quad \text{or} \quad \theta = \frac{\pi}{2}$$

$$\therefore \text{ Required area} = \int_0^{\frac{\pi}{2}} \frac{1}{2} r^2\, d\theta$$

$$= \int_0^{\frac{\pi}{2}} \frac{1}{2} 9a^2 \frac{\sin^2\theta\cos^2\theta}{(\sin^3\theta + \cos^3\theta)^2}\, d\theta = \frac{9a^2}{2}\int_0^{\frac{\pi}{2}} \frac{\sin^2\theta\cos^2\theta}{(\cos^6\theta\,(\tan^3\theta + 1)^2}\, d\theta$$

$$= \frac{9a^2}{2}\int_0^{\frac{\pi}{2}} \frac{\tan^2\theta\sec^2\theta}{(1+\tan^3\theta)^2}\, d\theta \ \Big|\ \text{Put} \quad \tan^3\theta = t,\ 3\tan^2\theta\sec^2\theta d\theta = dt$$

$$= \frac{9a^2}{2}\int_0^{\infty} \frac{\frac{1}{3}dt}{(1+t)^2} = \frac{3a^2}{2}\int_0^{\infty} (1+t)^{-2}\, dt$$

$$= \frac{3a^2}{2} \frac{(1+t)^{-1}}{-1}\Bigg|_0^{\infty} = -\frac{3a^2}{2}\left\{\frac{1}{1+t}\Bigg|_0^{\infty}\right\} = -\frac{3a^2}{2}\{0 - 1\} = \frac{3a^2}{2}$$

Hence area of the loop of the curve $= \dfrac{3a^2}{2}$.

TEST YOUR KNOWLEDGE

1. (*i*) Find the area of the region enclosed by the curve $y^2 = 4x$ and its latus rectum. **(*P.T.U., Dec. 2013*)**

   (*ii*) Find the area bounded by the parabolas $y^2 = 4x$, $x^2 = 4y$. **(*P.T.U., Dec. 2012*)**

2. Show that the area of the loop of the curve $3ay^2 = x(x - a)^2$ is $\dfrac{8a^2}{15\sqrt{3}}$. **(*P.T.U., May 2011*)**

3. Find the whole area included between the curve $x^2y^2 = a^2(y^2 - x^2)$ and its asymptote.

4. (*i*) Show that the area of the infinite region bounded between the curve $x^2(1 - y)y = 1$ and its asymptote is 2π.

   (*ii*) Find the area included between the curve $xy^2 = 4a^2 (2a - x)$ and its asymplote. **(*P.T.U., Dec. 2011*)**

5. Show that the area enclosed by the curves $xy^2 = a^2(a - x)$ and $(a - x)y^2 = a^2x$ is $(\pi - 2)a^2$.

6. Find the area of smaller portion enclosed by the curves $x^2 + y^2 = 9$, $y^2 = 8x$.

7. (*i*) Find the area between the curve $y = \sin x$, the x-axis from $x = 0$ to $x = 2\pi$.

   (*ii*) Find the area of the region bounded by $y = \sin x$, $y = \tan x$, $x = \dfrac{\pi}{4}$.

   $\left[\textbf{Hint. } \text{reqd. area} = \int_0^{\frac{\pi}{4}} (\tan x - \sin x)\, dx\right]$

8. Find the whole area of the curves

   (*i*) $y^2 = x^2 \dfrac{a^2 - x^2}{a^2 + x^2}$

   [**Hint.** For curve tracing see Q 2 of exercise ; art 1.13]

   (*ii*) $a^2x^2 = y^3(2a - y)$.

   [**Hint.** For curve tracing see Q 16 of exercise ; art 1.13]

9. Find the area of the loop of the curves

   (*i*) $ay^2 = x^2(a - x)$

   [**Hint.** For curve tracing see Q 4 of exercise ; art 1.13] **(*P.T.U., Dec. 2010*)**

   (*ii*) $a^3y^2 = x^4(b + x)$; $a, b > 0$

10. Prove that the area bounded by the hyperbola $\dfrac{x^2}{a^2} - \dfrac{y^2}{b^2} = 1$, the x-axis and the ordinate from P$(a \cosh \theta, b \sinh \theta)$ is $\dfrac{1}{4}\, ab(\sinh 2\theta - 2\theta)$.

11. Find the total area of the curve $\left(\dfrac{x}{a}\right)^{2/3} + \left(\dfrac{y}{b}\right)^{2/3} = 1$ where $a > 0$, $b > 0$.

    [**Hint.** S.E. 11]

12. Prove that the whole area between the four infinite branches of the tractrix

    $x = a \cos t + \dfrac{a}{2} \log\left(\tan^2 \dfrac{t}{2}\right)$, $y = a \sin t$, $0 < t < 2\pi$, $t \neq \pi$ is πa^2. $\left[\textbf{Hint. } \text{Apply formula } \int_0^{2\pi} y \dfrac{dx}{dt}\, dt\right]$

13. Prove that the area bounded by the cissoid $x = a \sin^2 t$, $y = \dfrac{a \sin^3 t}{\cos t}$; $-\dfrac{\pi}{2} < t < \dfrac{\pi}{2}$ and its asymptote is $\dfrac{3\pi a^2}{4}$.

    $$\left[\textbf{Hint. } \text{area} = \int_{-\frac{\pi}{2}}^{\frac{\pi}{2}} \left(\frac{a \sin^3 t}{\cos t}\right)(2a \sin t \cos t)\, dt = 2a^2\, 2\int_0^{\frac{\pi}{2}} \sin^4 t\, dt = 4a^2\, \frac{3 \cdot 1}{4 \cdot 2} \frac{\pi}{2}\right]$$

14. (*i*) Show that the area of the region included between the cardioids $r = a(1 + \cos\theta)$ and $r = a(1 - \cos\theta)$ is $\dfrac{a^2}{2}(3\pi - 8)$. **(P.T.U., Dec. 2013)**

(*ii*) Find the area outside the circle $r = 2a\cos\theta$ and inside the cardioid $r = a(1 + \cos\theta)$

(*iii*) Find the area of the region that lies inside the cardioid $r = a(1 + \cos\theta)$ and outside the circle $r = a$. **(P.T.U., Dec. 2004)**

15. Find the area of the loop of the following curves:

(*i*) $r^2 = a^2\cos 2\theta$ **(P.T.U., May 2012)** (*ii*) $r = a\sin 2\theta$

Find also the total area in each case.

[**Hint.** (*i*) for curve tracing see S.E. 6 art. 1.16
(*ii*) for curve tracing see Q. 5 of exercise ; art. 1.16]

16. Show that the ratio of the area of the larger to the area of the smaller loop of the curve $r = \dfrac{1}{2} + \cos 2\theta$ is $(4\pi + 3\sqrt{3})/(2\pi - 3\sqrt{3})$.

17. Prove that the area of the loop of the folium $x^3 + y^3 = 3axy$ is three times the area of one of the loops of the lemniscate $r^2 = a^2\cos 2\theta$.

18. Show that area of the loop of the curve $x = a(1 - t^2)$, $y = at(1 - t^2)$; $-1 \le t \le 1$ is $\dfrac{8a^2}{15}$.

19. Prove that area of the loop of the curve $x = t - t^3$, $y = 1 - t^4$ is $\dfrac{16}{35}$.

Answers

1. (*i*) $\dfrac{8}{3}$ (*ii*) $\dfrac{16}{3}$ **3.** $4a^2$ **4.** (*ii*) $4\pi a^2$

6. $2\left(\dfrac{\sqrt{2}}{3} + \dfrac{9\pi}{4} - \dfrac{9}{2}\sin^{-1}\dfrac{1}{3}\right)$ **7.** (*i*) 4 units (*ii*) $\dfrac{1}{2}\log 2 + \dfrac{1}{\sqrt{2}} - 1$

8. (*i*) $a^2(\pi - 2)$ (*ii*) πa^2

9. (*i*) $\dfrac{8a^2}{15}$ (*ii*) $\dfrac{32b^{7/2}}{105\,a^{3/2}}$ **11.** $\dfrac{3}{8}\pi ab$

14. (*ii*) $\dfrac{a^2\pi}{2}$ (*iii*) $a^2\left(\dfrac{\pi}{4} + 2\right)$

15. (*i*) $\dfrac{a^2}{2}, a^2$ (*ii*) $\dfrac{\pi a^2}{8}, \dfrac{\pi a^2}{4}$.

2.8. VOLUMES AND SURFACES OF REVOLUTION OF CURVES

Introduction. Let AB be a given curve and l a given line.

Draw AL, BM perpendiculars from A and B respectively on l. A solid is generated by revolving the area ALMB about the line l. *This solid is said to be obtained by the revolution of the arc AB about the line l.*

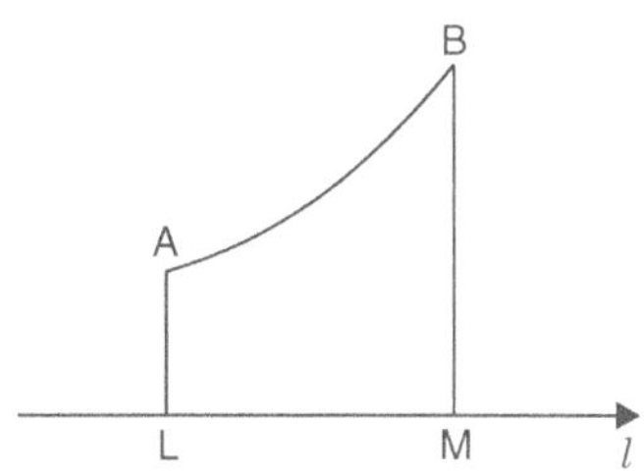

The line l about which the curve resolves is called *the axis of revolution.*

Firstly we will take x-axis or y-axis as the axis of revolution and obtain expressions for the volume and the surface of the solid. Later on we will take any line for the axis of revolution and obtain expressions for volume and surface of revolution.

2.9. VOLUME OBTAINED BY REVOLUTION OF CURVES

Show that volume obtained by revolving about x-axis the curve $y = f(x)$, intercepted between the points where abscissae are a, b is $\int_a^b \pi y^2\, dx$ or $\int_a^b \pi[f(x)]^2\, dx$; it is assumed that the curve does not cut x-axis.

Proof. Let $y = f(x)$ be the curve AB. From A, B draw AL, BM perpendiculars on x-axis such that OL = a, OM = b.

$\therefore$ Co-ordinates of A are $[a, f(a)]$ and that of B are $[b, f(b)]$.

Let P(x, y) be any variable put on the curve and Q$(x + \delta x, y + \delta y)$ be a neighbouring point to P. Complete the rectangle with ordinate PR *i.e.*, PRST and with ordinate QS *i.e.*, KRSQ.

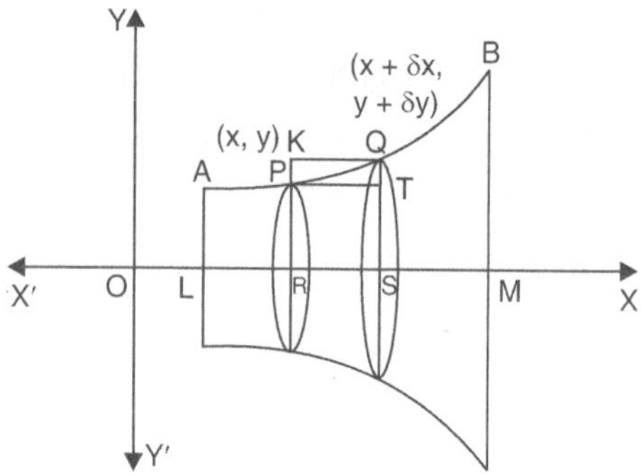

Let V be the volume of the solid obtained by revolving arc AP about x-axis and δV be the volume of the solid obtained by revolving arc PQ about x-axis.

Now we see that δV lies between the volumes of the two cylindrical bodies obtained by revolving the rectangles PRST and KRSQ about x-axis

$$\therefore \quad \pi y^2\, \delta x < \delta V < \pi(y + \delta y)^2\, \delta x$$

(volume of cylinder = $\pi r^2 h$)

or
$$\pi y^2 < \frac{\delta V}{\delta x} < \pi(y + \delta y)^2$$

Take limit when $\delta x \to 0$ (δy will also $\to 0$)

$$\therefore \quad \pi y^2 < \underset{\delta x \to 0}{\text{Lt}} \frac{\delta V}{\delta x} < \underset{\delta y \to 0}{\text{Lt}} \pi(y + \delta y)^2$$

$$\pi y^2 < \frac{dV}{dx} < \text{a quantity} \to \pi y^2$$

$$\therefore \quad \frac{dV}{dx} = \pi y^2$$

Integrating both sides w.r.t. x between the limits a and b, we get $\int_a^b \frac{dV}{dx}\, dx = \int_a^b \pi y^2\, dx$

or
$$V \Big|_a^b = \int_a^b \pi y^2\, dx$$

or
$$[\text{value of V at } x = b] - [\text{value of V at } x = a] = \int_a^b \pi y^2\, dx$$

[volume of the solid obtained by revolving arc AB about x-axis] – 0 = $\int_a^b \pi y^2\, dx$

Hence volume of the solid obtained by revolving the curve $y = f(x)$ about x-axis $= \int_a^b \pi y^2\, dx$.

Note 1. Similarly, if we revolve the curve $x = f(y)$ bounded by $y = a$ and $y = b$ about y-axis then volume of the solid of revolution about y-axis = $\int_a^b \pi x^2\, dy$; it is assumed arc does not intersects y-axis. **(P.T.U., Dec. 2012)**

Note 2. If the area bounded by the curve $y = f(x)$ and the lines $x = a$, $x = b$ and $y = p$ is revolved about the line $y = p$ (*i.e.*, a line parallel to x-axis), then the volume of the solid obtained is $\int_a^b \pi(y-p)^2\, dx$.

Note 3. If the area bounded by the curve $x = f(y)$ and the lines $y = a$, $y = b$ and $x = q$ is revolved about the line $x = q$ (*i.e.*, a line parallel to y-axis), then the volume of the solid so obtained is $\int_a^b \pi(x-q)^2\, dy$.

Note 4. If the area bounded by the curve $y = f(x)$ and the lines $x = a, y = b$ is revolved about any line (say l) then the process to find the volume of the solid of revolution is as follows:

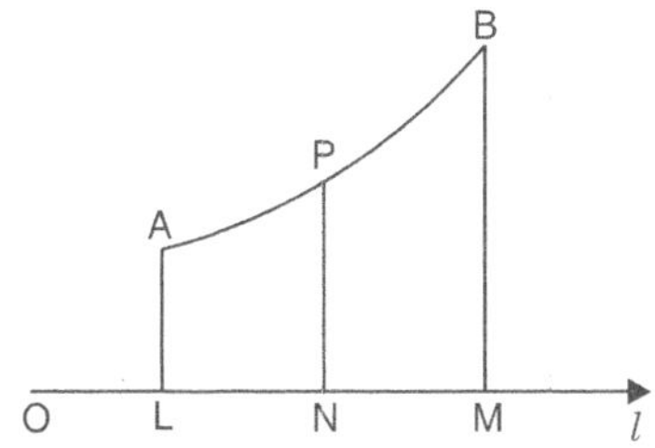

Take O, a fixed point on the axis of revolution *i.e.*, l

so that a, b are distances of the feet of $\perp$ from A, B.

Let P be any point on the curve $y = f(x)$

then from P draw PN perpendicular on l, the axis of revolution then volume of the solid of revolution about l

$$= \int_{\mathbf{OL}}^{\mathbf{OM}} \pi(\mathbf{PN})^2\, \mathbf{d(ON)}.$$

Note 5. Volumes of the solid of revolution can be obtained by double integrals discussed in the chapter of 'Multiple Integrals' 6.10.

ILLUSTRATIVE EXAMPLES

Example 1. *The curve $y^2 (a + x) = x^2 (3a - x)$ is revolved about the axis of x. Find the volume generated by the loop.* **(P.T.U., Dec 2011)**

Sol. Equation of the curve is $y^2(a + x) = x^2 (3a - x)$. First of all we will have rough sketch of the curve

(*i*) Curve is symmetric about x-axis.

(*ii*) Curve intersects x-axis (Put $y = 0$) at points given by $x^2 (3a - x) = 0$ *i.e.*, at $x = 0, x = 3a$ *i.e.*, at (0, 0) and (3*a*, 0)

It intersects y-axis (put $x = 0$), *i.e.*, at (0, 0)

(*iii*) Curve passes through the origin and the tangent at the origin is given by

$$ay^2 - 3ax^2 = 0 \qquad \text{(Equating the lowest degree terms to zero)}$$

i.e., $$y^2 = 3x^2 \quad \text{or} \quad y = \pm\sqrt{3}\,x$$

$\therefore$ At (0, 0) slopes of the tangents are $\pm\sqrt{3}$ *i.e.*, tangents make angles of 60° and – 60° with the +ve direction of x-axis.

(*iv*) $y = \pm x\sqrt{\dfrac{3a-x}{a+x}}$; $x \neq -a$

Domain of the curve is given by

$$3a - x > 0,\ a + x > 0 \ \textit{i.e.}, -a < x < 3a$$

$\therefore$ Curve lies between $x = -a$ and $x = 3a$

(v) Asymptote to the curve parallel to y-axis is given by $a + x = 0$. i.e., $x = -a$ is asymptote to the curve

$$\frac{dy}{dx} = \pm\frac{\sqrt{a+x}\left\{x\left(-\frac{1}{2\sqrt{3a-x}}+\sqrt{3a-x}\right)\right\} - x\sqrt{3a-x}\,\frac{1}{2\sqrt{a+x}}}{a+x}$$

$$= \pm\frac{6a^2-2x^2}{2(a+x)^{3/2}\sqrt{3a-x}} = \pm\frac{3a^2-x^2}{(a+x)^{3/2}\sqrt{3a-x}}$$

Tangent is parallel to x-axis where $\frac{dy}{dx} = 0$

i,e., when $3a^2 - x^2 = 0$ or $x = \pm\sqrt{3}\,a$.

$\therefore$ Tangent is parallel to x-axis at $x = \pm\sqrt{3}\,a$.

Tangent is parallel to y-axis where $\frac{dy}{dx} = \infty$

i.e., $(a + x)^{3/2}\sqrt{3a-x} = 0$ i.e., at $x = 3a$ ($x = -a$ is the asymptote)

Rough sketch of the curve is shown in the figure.

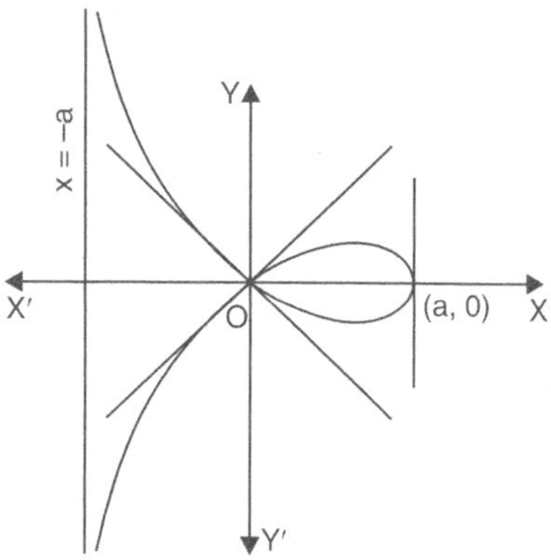

The volume generated by the loop

$$= \int_0^{3a} \pi y^2 dx$$

$$= \int_0^{3a} \pi\frac{x^2(3a-x)}{a+x}\,dx$$

$$= \pi\int_0^{3a}\left(-x^2+4ax-4a^2+\frac{4a^3}{x+a}\right)dx$$

(Obtained by dividing num. by den.)

$$= \pi\left[-\frac{x^3}{3}+2ax^2-4a^2x+4a^3\log(x+a)\right]_0^{3a}$$

$$= \pi\left[-9a^3+18a^3-12a^3+4a^3\log 4a-4a^3\log a\right]$$

$$= \pi a^3\,[4\log 4 - 3] = \pi a^3\,[8\log 2 - 3]$$

Example 2. *Trace the curves $y = \sin x$ and $y = \cos x$ as x varies from 0 to $\pi/2$ and show that the volume of the solid obtained by revolving about x-axis the region enclosed by them and the x-axis is $\pi(\pi - 2)/4$.*

Sol. Equations of the curves are $y = \sin x$; $y = \cos x$

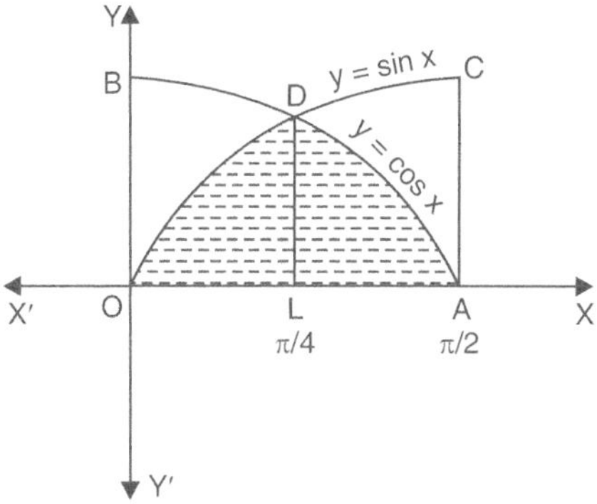

These intersect as $\sin x = \cos x$

$$\tan x = 1 \quad i.e., \quad x = \frac{\pi}{4}$$

$\therefore$ Volume of the solid obtained by revolving the region OAD about x-axis = volume of the solid obtained by revolving the region OLD + Volume of the solid obtained by revolving DLA

$$= \pi \int_0^{\pi/4} \sin^2 x \, dx + \pi \int_{\pi/4}^{\pi/2} \cos^2 x \, dx$$

$$= \pi \int_0^{\pi/4} \frac{1 - \cos 2x}{2} \, dx + \pi \int_{\pi/4}^{\pi/2} \frac{1 + \cos 2x}{2} \, dx$$

$$= \frac{\pi}{2} \left[x - \frac{\sin 2x}{2} \right]_0^{\pi/4} + \frac{\pi}{2} \left[x + \frac{\sin 2x}{2} \right]_{\pi/4}^{\pi/2}$$

$$= \frac{\pi}{2} \left[\frac{\pi}{4} - \frac{1}{2} \right] + \frac{\pi}{2} \left[\frac{\pi}{2} + 0 - \frac{\pi}{4} - \frac{1}{2} \right]$$

$$= \frac{\pi}{2} \cdot \left[\frac{\pi}{4} - \frac{1}{2} + \frac{\pi}{2} - \frac{\pi}{4} - \frac{1}{2} \right] = \frac{\pi}{2} \left[\frac{\pi}{2} - 1 \right] = \frac{\pi}{4} (\pi - 2).$$

Example 3. *Find the volume of the solid obtained by revolving the cycloid $x = a(\theta + \sin \theta)$, $y = a(1 - \cos \theta)$ about the tangent at the vertex.* **(P.T.U., May 2014)**

Sol. Equation of the cycloid is $x = a(\theta + \sin \theta)$, $y = a(1 - \cos \theta)$

For curve tracing see solved example 1, art 1.15

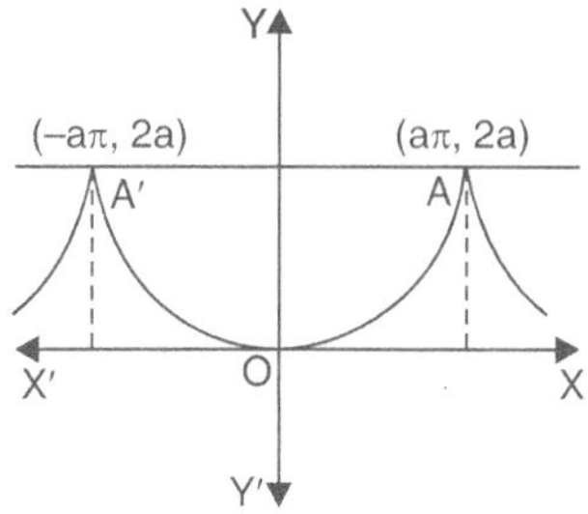

Volume of the solid obtained by revolving the arc AOA′ about the tangent at the vertex *i.e.*, x-axis

$$= 2\pi \int_0^{a\pi} y^2 \, dx$$

Substituting the values of x, y

$$y = a(1 - \cos \theta), \quad x = a(\theta + \sin \theta)$$

$$dx = a(1 + \cos \theta) \, d\theta$$

When $x = 0, \theta = 0$

When $x = a\pi, \quad \theta + \sin \theta = \pi \quad \therefore \quad \theta = \pi$

$$\therefore \qquad \text{Volume} = 2\pi \int_0^{\pi} a^2 (1 - \cos \theta)^2 \, . \, a(1 + \cos \theta) \, d\theta$$

$$= 2\pi a^3 \int_0^{\pi} (1 - \cos \theta) \sin^2 \theta \, d\theta$$

$$= 2\pi a^3 \int_0^{\pi} 2 \sin^2 \frac{\theta}{2} \, . \, 4 \sin^2 \frac{\theta}{2} \cos^2 \frac{\theta}{2} \, d\theta$$

$$= 16\pi a^3 \int_0^{\pi} \sin^4 \frac{\theta}{2} \, . \, \cos^2 \frac{\theta}{2} \, d\theta$$

Put $\frac{\theta}{2} = t \quad \therefore \quad d\theta = 2dt$

$$= 16\pi a^3 \int_0^{\pi/2} \sin^4 t \,.\, \cos^2 t \cdot 2dt$$

$$= 32\pi a^3 \frac{3\cdot 1\cdot 1}{6\cdot 4\cdot 2}\frac{\pi}{2} = \pi^2 a^3.$$

Example 4. *Find the volume of the solid obtained by the revolution of the cissoid $y^2(2a - x) = x^3$ about its asymptote.* **(P.T.U., May 2006)**

Sol. Equation of the cissoid is $y^2 (2a - x) = x^3$ ...(1)

For curve tracing consult see S.E. 5 art 1.13

Here $x = 2a$ is the asymptote.

In this case the curve is revolved about a line parallel to y-axis whose equation is $x = 2a$

Also curve is symmetric about x-axis.

$\therefore$ Volume obtained by revolving the whole curve is 2 times the volume obtained by revolving the upper half only.

$\therefore$ Volume of the solid generated by the curve

$$= 2\pi \int_0^{\infty} (x - 2a)^2 \, dy \qquad \text{(Note 3 art. 2.9)}$$

Change variable y to x.

From (1), $$y = \frac{x^{3/2}}{\sqrt{2a - x}}$$

$$\therefore \quad dy = \frac{\sqrt{2a - x}\cdot \frac{3}{2}x^{1/2} - x^{3/2}\frac{-1}{2\sqrt{2a - x}}}{2a - x}\, dx$$

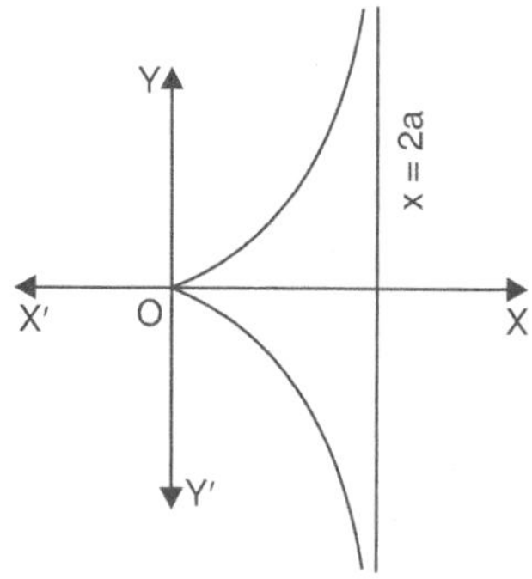

x varies from 0 to $2a$

$$\text{Volume} = 2\pi \int_0^{2a} (x - 2a)^2 \frac{\frac{3}{2}\sqrt{x}(2a - x) + \frac{x^{3/2}}{2}}{(2a - x)^{3/2}}\, dx$$

$$= 2\pi \int_0^{2a} (2a - x)^2 \frac{\sqrt{x}}{2}\cdot\frac{6a - 3x + x}{(2a - x)^{3/2}}\, dx$$

$$= 2\frac{\pi}{2} \int_0^{2a} \sqrt{x}\,\sqrt{2a - x}\,(6a - 2x)\, dx$$

$$= 2\pi \int_0^{2a} \sqrt{x}\sqrt{2a - x}\,(3a - x)\, dx$$

Put $x = 2a \sin^2 \theta \quad dx = 4a \sin\theta \cos\theta \, d\theta$

$$= 2\pi \int_0^{\pi/2} \sqrt{2a}\sin\theta \cdot \sqrt{2a}\cos\theta\,(3a - 2a\sin^2\theta)\, 4a \sin\theta\cos\theta\, d\theta$$

$$= 2\pi \int_0^{\pi/2} (2a)\sin\theta\cos\theta\, a\,(3 - 2\sin^2\theta)\,(4a)\sin\theta\cos\theta\, d\theta$$

$$= 16a^3\pi \int_0^{\pi/2} \left(3\sin^2\theta\cos^2\theta - 2\sin^4\theta\cos^2\theta\right) d\theta$$

$$= 16\pi a^3 \left\{3\frac{1\cdot 1}{4\cdot 2}\frac{\pi}{2} - 2\frac{3\cdot 1\cdot 1}{6\cdot 4\cdot 2}\frac{\pi}{2}\right\} = 16\pi a^3 \cdot \frac{\pi}{2}\left[\frac{3}{8} - \frac{1}{8}\right]$$

$$= 8\pi^2 a^3 \cdot \frac{1}{4} = 2\pi^2 a^3.$$

Example 5. *The loop of the curve $2ay^2 = x(x-a)^2$ revolves about the straight line $y = a$, find the volume of the solid generated.*

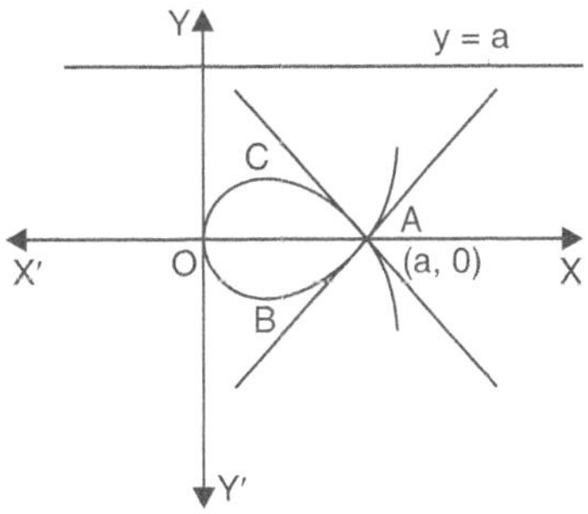

Sol. Equation of the curve is $2ay^2 = x(x-a)^2$

$$y = \pm \frac{\sqrt{x}\,(x-a)}{\sqrt{2a}}$$

Loop of the curve lies between 0 and a.

$$\therefore \qquad y = \pm \frac{\sqrt{x}\,(a-x)}{\sqrt{2a}} \qquad (\because \; a - x \text{ is +ve})$$

Hence curve consists of two parts OCA and OBA

For the curve OCA, $y = \frac{\sqrt{x}\,(a-x)}{\sqrt{2a}}$ and for the curve OBA, $y = -\frac{\sqrt{x}\,(a-x)}{\sqrt{2a}}$, curve revolves about $y = a$; a line parallel to x-axis.

$\therefore$ Volume generated by curve OCA $= \pi \int_0^a (y-a)^2\, dx$

Substitute the value of y

$$= \pi \int_0^a \left[\frac{\sqrt{x}\,(a-x)}{\sqrt{2a}} - a\right]^2 dx$$

Volume generated by the curve OBA

$$= \pi \int_0^a \left[-\frac{\sqrt{x}\,(a-x)}{\sqrt{2a}} - a\right]^2 dx = \pi \int_0^a \left[\frac{\sqrt{x}\,(a-x)}{\sqrt{2a}} + a\right]^2 dx$$

$\therefore$ Required volume = Difference of the two volumes

$$= \pi \int_0^a \left\{\left[\frac{\sqrt{x}\,(a-x)}{\sqrt{2a}} + a\right]^2 - \left[\frac{\sqrt{x}\,(a-x)}{\sqrt{2a}} - a\right]^2\right\} dx$$

$\mid (a+b)^2 - (a-b)^2 = 4ab$

$$= \pi \int_0^a 4 \cdot \frac{\sqrt{x}\,(a-x)}{\sqrt{2a}} \cdot a\, dx = 2\sqrt{2a}\pi \int_0^a (a\sqrt{x} - x^{3/2})\, dx$$

$$= 2\sqrt{2a} \cdot \pi \left[\frac{ax^{3/2}}{3/2} - \frac{x^{5/2}}{5/2}\right]_0^a = 2\sqrt{2a} \cdot \pi \left[\frac{2a}{3} a^{3/2} - \frac{2}{5} a^{5/2}\right]$$

$$= 2\sqrt{2a}\, \pi \cdot \frac{4}{15} a^{5/2} = \frac{8}{15} \sqrt{2}\, \pi a^3.$$

Example 6. *Find the volume of the solid generated by revolving the finite region bounded by the curves $y = x^2 + 1$, $y = 5$ about the line $x = 3$.*

Sol. Equation of the curve is $y = x^2 + 1$ or $x^2 = y - 1$

$$x = \pm \sqrt{y-1}.$$

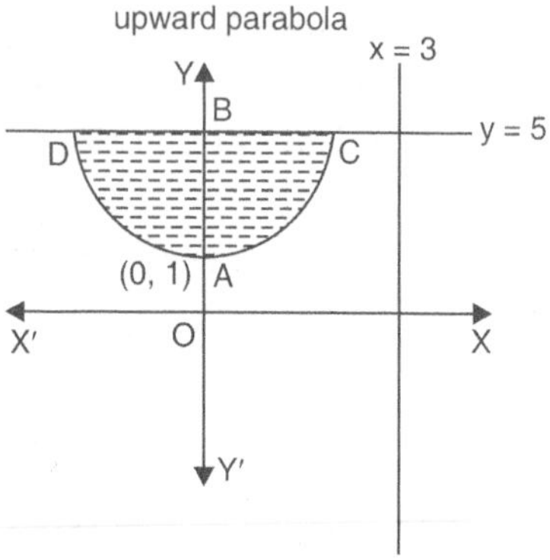

Region included between the curve and $y = 5$ consists of two portions ACB and ABD

For ACB; $x = \sqrt{y-1}$ $\because$ in 1st quad x is +ve

and for ABD; $x = -\sqrt{y-1}$ $\because$ in 2nd quad x is –ve.

$\therefore$ Volume generated by the revolution of ACB about $x = 3$ (a line parallel to y-axis) is

$$= \pi \int_1^5 (\sqrt{y-1} - 3)^2\, dy$$

$$\because \quad V = \pi \int (x-a)^2\, dy \qquad \text{(art 2.9 note 3)}$$

$$= \pi \int_{y=1}^5 (3 - \sqrt{y-1})^2\, dy \quad \because \quad x < 3 \text{ i.e., } \sqrt{y-1} < 3 \qquad \textbf{| Note this step}$$

Volume generated by the revolution of ABD about $x = 3$

$$= \pi \int_{y=1}^5 (-\sqrt{y-1} - 3)^2\, dy = \pi \int_{y=1}^5 (\sqrt{y-1} + 3)^2\, dy$$

Required volume = Difference of the two volumes

$$= \pi \int_1^5 (\sqrt{y-1} + 3)^2\, dy - \pi \int_1^5 (3 - \sqrt{y-1})^2\, dy$$

$$= \pi \int_1^5 4 \cdot 3\sqrt{y-1}\, dy = 12\pi \int_1^5 \sqrt{y-1}\, dy$$

$$= 12\pi \left. \frac{(y-1)^{3/2}}{3/2} \right|_1^5 = 8\pi (y-1)^{3/2} \Big|_1^5 = 8\pi \cdot 4^{3/2} = 64\pi \text{ cubic unit.}$$

Example 7. *The smaller segment of the ellipse* $\frac{x^2}{a^2} + \frac{y^2}{b^2} = 1$ *cut off by the chord* $\frac{x}{a} + \frac{y}{b} = 1$ *revolves completely about the chord, show that the volume generated is*

$$\frac{1}{6} \pi (10 - 3\pi)\, a^2 b^2 (a^2 + b^2)^{-1/2}.$$

Sol. Equation of the ellipse is $\frac{x^2}{a^2} + \frac{y^2}{b^2} = 1$

Equation of the chord is $\frac{x}{a} + \frac{y}{b} = 1.$

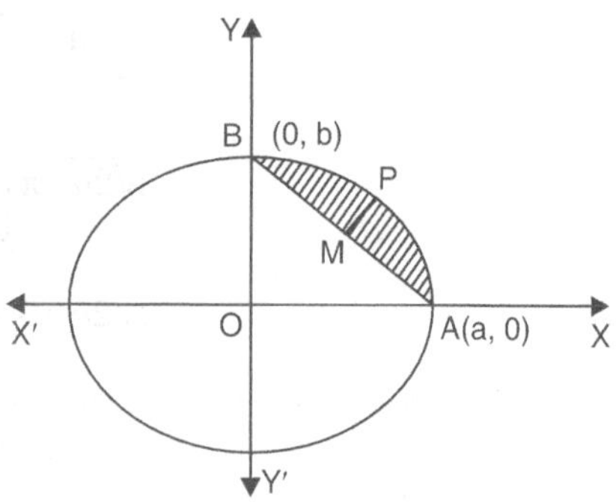

This chord passes through A(a, 0) and B (0, b) of the ellipse. The shaded region revolves about line AB (any line *i.e.*, neither || to x-axis nor || to y-axis)

$\therefore$ Take A as the fixed point on the line AB

$\therefore$ Volume of the solid generated by region

$$= \pi \int PM^2\, d(AM)$$

Let P($a \cos \theta$, $b \sin \theta$) be any point on the ellipse then PM = Length of perpendicular distance from ($a \cos \theta$, $b \sin \theta$) on $\frac{x}{a} + \frac{y}{b} = 1$

$$= \frac{\frac{a\cos\theta}{a} + \frac{b\sin\theta}{b} - 1}{\sqrt{\frac{1}{a^2} + \frac{1}{b^2}}} = \frac{\cos\theta + \sin\theta - 1}{\sqrt{a^2+b^2}}\, ab$$

$$AM^2 = AP^2 - PM^2 = (a\cos\theta - a)^2 + (b\sin\theta)^2 - \left(\frac{\cos\theta + \sin\theta - 1}{\sqrt{a^2+b^2}}\, ab\right)^2$$

$$= a^2(\cos\theta - 1)^2 + b^2\sin^2\theta - \frac{a^2b^2}{a^2+b^2}((\cos\theta - 1) + \sin\theta)^2$$

$$= a^2(\cos\theta - 1)^2 + b^2\sin^2\theta - \frac{a^2b^2}{a^2+b^2}(\cos\theta - 1)^2$$

$$- \frac{a^2b^2}{a^2+b^2}\sin^2\theta - \frac{2a^2b^2}{a^2+b^2}\sin\theta(\cos\theta - 1)$$

$$= \frac{a^4}{a^2+b^2}(\cos\theta - 1)^2 + \frac{b^4}{a^2+b^2}\sin^2\theta - \frac{2a^2b^2}{a^2+b^2}\sin\theta(\cos\theta - 1)$$

$$= \left(\frac{a^2(\cos\theta - 1) - b^2\sin\theta}{\sqrt{a^2+b^2}}\right)^2$$

$$\therefore \qquad AM = \pm\frac{a^2(\cos\theta - 1) - b^2\sin\theta}{\sqrt{a^2+b^2}} = \frac{a^2(1-\cos\theta) + b^2\sin\theta}{\sqrt{a^2+b^2}}$$

$\therefore$ Required volume

$$= \pi\int_{\theta=0}^{\pi/2} \frac{(\cos\theta + \sin\theta - 1)^2\, a^2b^2 d}{a^2+b^2}\left[\frac{a^2(1-\cos\theta) + b^2\sin\theta}{\sqrt{a^2+b^2}}\right] d\theta$$

$$= \frac{\pi a^2b^2}{(a^2+b^2)^{3/2}}\int_0^{\pi/2} (\cos\theta + \sin\theta - 1)^2 (a^2\sin\theta + b^2\cos\theta)\, d\theta$$

$$= \frac{\pi a^2b^2}{(a^2+b^2)^{3/2}}\int_0^{\pi/2} (\cos^2\theta + \sin^2\theta + 1 + 2\sin\theta\cos\theta - 2\sin\theta - 2\cos\theta)(a^2\sin\theta + b^2\cos\theta)\, d\theta$$

$$= \frac{\pi a^2b^2}{(a^2+b^2)^{3/2}}\int_0^{\pi/2} (2 + 2\sin\theta\cos\theta - 2\sin\theta - 2\cos\theta)(a^2\sin\theta + b^2\cos\theta)\, d\theta$$

$$= \frac{2\pi a^2b^2}{(a^2+b^2)^{3/2}}\int_0^{\pi/2} (a^2\sin\theta + a^2\sin^2\theta\cos\theta - a^2\sin^2\theta - a^2\sin\theta\cos\theta + b^2\cos\theta + b^2\sin\theta\cos^2\theta - b^2\sin\theta\cos\theta - b^2\cos^2\theta)\, d\theta$$

$$= \frac{2\pi a^2b^2}{(a^2+b^2)^{3/2}}\left\{a^2\cdot 1 + a^2\frac{1}{3\cdot 1} - a^2\frac{1}{2}\frac{\pi}{2} - (a^2+b^2)\frac{1}{2} + b^2 + b^2\frac{1}{3} - b^2\frac{1}{2}\frac{\pi}{2}\right\}$$

$$= \frac{2\pi a^2b^2}{(a^2+b^2)^{3/2}}\left[a^2\left(1 + \frac{1}{3} - \frac{1}{2}\right) + b^2\left(-\frac{1}{2} + 1 + \frac{1}{3}\right) - (a^2+b^2)\frac{\pi}{4}\right]$$

$$= \frac{2\pi a^2b^2}{(a^2+b^2)^{3/2}}\left[\frac{5}{6}(a^2+b^2) - \frac{\pi}{4}(a^2+b^2)\right] = \frac{\pi a^2b^2}{\sqrt{a^2+b^2}}\left[\frac{5}{3} - \frac{\pi}{2}\right].$$

Example 8. *Find the volume of the solid generated by revolving the region bounded by the curves $y = 1 + \sqrt{x}$ and $y = 1 + x$ about y-axis.*

Sol. Equation of the curve is $y = 1 + \sqrt{x}$.

Equation of the line is $y = 1 + x$

line intersects the curve $(y - 1)^2 = x$; a parabola in $y = 1 + (y - 1)^2 = 1 + y^2 - 2y + 1$ or $y^2 - 3y + 2 = 0$

$\therefore \quad (y - 1)(y - 2) = 0 \quad \therefore \quad y = 1, y = 2$

$\therefore$ when $y = 1, x = 0$ when $y = 2, x = 1$

$\therefore$ line and curve intersect at (0, 1) and (1, 2)

Shaded region is to revolve about y-axis

$\therefore$ this region consists of the line AB and the arc

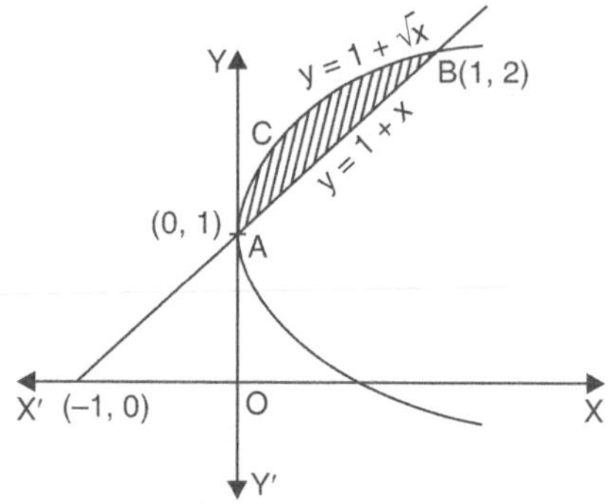

$\overparen{ACB}$. Volume of the solid generated by

$$y = 1 + x \text{ about } y\text{-axis is } = \int_1^2 \pi x^2 \, dy$$

$$= \int_1^2 \pi(y - 1)^2 \, dy = \pi \frac{(y - 1)^3}{3} \Bigg|_1^2 = \frac{\pi}{3}$$

Volume of the solid generated by $y = 1 + \sqrt{x}$ about y-axis

$$= \int_1^2 \pi x^2 \, dy = \int_1^2 \pi(y - 1)^4 \, dy = \pi \frac{(y - 1)^5}{5} \Bigg|_1^2 = \pi \cdot \frac{1}{5}$$

$\therefore$ Required volume = Difference of the above volumes $= \dfrac{\pi}{3} - \dfrac{\pi}{5} = \dfrac{2\pi}{15}$ cubic unit.

Example 9. *The region bounded by the curve $y = x^2 + 1$ and the line $y = -x + 3$ is revolved about x-axis to generate a solid. Find the volume of the solid.* **(P.T.U., May 2004)**

Sol. Equation of the curve is $y = x^2 + 1$

or $\quad x^2 = y - 1$ (upward parabola)

Equation of the line is

$y = -x + 3$

Line intersects the given curve at points:

$x^2 = -x + 3 - 1 \quad x^2 + x - 2 = 0 \quad (x - 1)(x + 2) = 0;$
$x = 1, x = -2$ when $x = 1, y = 2$ when $x = -2, y = 5$

$\therefore$ Line and curve intersect at (1, 2) and (– 2, 5)

Volume of the solid generated by the region ABC consists of the volume generated by the line AB. The volume generated by the curve ACB about x-axis.

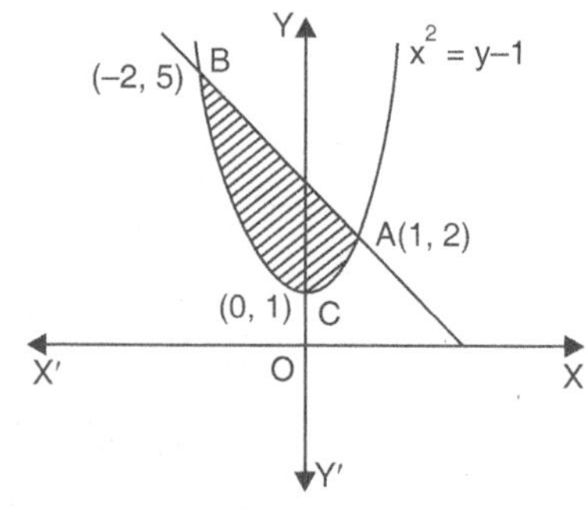

$$= \int_{-2}^{1} \pi y^2\, dx_{\text{ along line AB}} - \int_{-2}^{1} \pi y^2\, dx_{\text{ along curve ABC}} = \pi \int_{-2}^{1} (-x+3)^2\, dx - \int_{-2}^{1} \pi (x^2+1)^2\, dx$$

$$= \pi \int_{-2}^{1} (x^2 - 6x + 9 - x^4 - 2x^2 - 1)\, dx$$

$$= \pi \int_{-2}^{1} (-x^4 - x^2 - 6x + 8)\, dx = \pi \left[-\frac{x^5}{5} - \frac{x^3}{3} - 3x^2 + 8x \right]_{-2}^{1}$$

$$= \pi \left[\left(-\frac{1}{5} - \frac{1}{3} - 3 + 8 \right) - \left(\frac{32}{5} + \frac{8}{3} - 12 - 16 \right) \right]$$

$$= \pi \left[\frac{67}{15} - \frac{136}{15} + 28 \right] = \pi \left[-\frac{69}{15} + 28 \right] = \pi \left[-\frac{23}{5} + 28 \right] = \frac{117}{5}\, \pi \text{ cubic unit.}$$

Example 10. *The area cut off from the parabola $y^2 = 4ax$ by the chord joining the vertex to an end of the latus rectum is rotated through four right angles about the chord. Show that the volume of the solid so formed is $\frac{2\sqrt{5}}{75}\pi a^3$.*

Sol. Equation of the parabola is

$$y^2 = 4ax \quad \quad ...(1)$$

Let LL′ be the latus rectum of the parabola.

We know that length of the latus rectum is $4a$.

$\therefore$ Coordinates of L are $(a, 2a)$ equation of OL is

$$y - 0 = \frac{2a - 0}{a - 0}(x - 0) \quad \text{or} \quad y = 2x$$

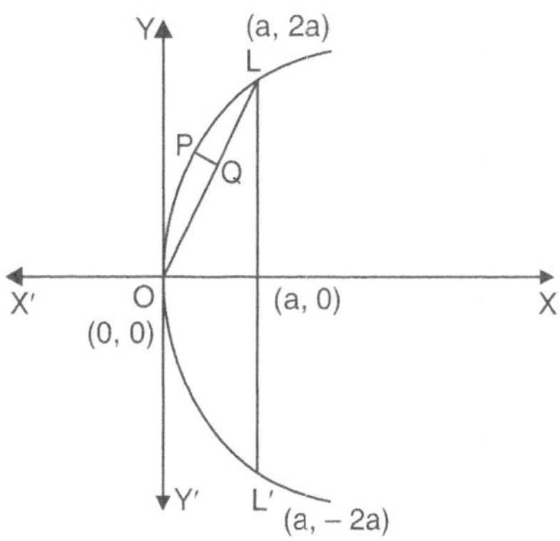

Let P(x, y) be any point on the parabola then

PQ = $\perp$ distance from P on OL

$$= \frac{|2x - y|}{\sqrt{4+1}} = \frac{|2x - y|}{\sqrt{5}}$$

$$OQ^2 = OP^2 - PQ^2 = (x^2 + y^2) - \frac{(2x-y)^2}{5} = \frac{5x^2 + 5y^2 - 4x^2 - y^2 + 4xy}{5}$$

$$\therefore \quad OQ^2 = \frac{x^2 + 4y^2 + 4xy}{5} = \frac{(x+2y)^2}{5}$$

$$\therefore \quad OQ = \frac{x + 2y}{\sqrt{5}}$$

When we rotate the curve about OL then volume of the solid

$$= \int \pi (PQ)^2\, d(OQ) = \pi \int \frac{(2x-y)^2}{5}\, d\left(\frac{x+2y}{\sqrt{5}} \right)$$

Replace y by $2\sqrt{ax}$ [using (1)]

$$= \pi \int \frac{(2x - 2\sqrt{ax})^2}{5} \, d\left[\frac{x + 4\sqrt{ax}}{\sqrt{5}}\right] = \pi \int_0^a \frac{4}{5\sqrt{5}} (x - \sqrt{ax})^2 \left(1 + \frac{4\sqrt{9}}{2\sqrt{x}}\right) dx$$

$$= \frac{4\pi}{5\sqrt{5}} \int_0^a (x^2 - 2\sqrt{a}\, x^{3/2} + ax) \left(\frac{\sqrt{x} + 2\sqrt{a}}{\sqrt{x}}\right) dx$$

$$= \frac{4\pi}{5\sqrt{5}} \int_0^a (x^{3/2} - 2\sqrt{a}x + ax^{1/2})\,(x^{1/2} + 2\sqrt{a})\, dx$$

$$= \frac{4\pi}{5\sqrt{5}} \int_0^a [x^2 - 2\sqrt{a}\, x^{3/2} + ax + 2\sqrt{a}\, x^{3/2} - 4ax + 2a\sqrt{a}\, x^{1/2}]\, dx$$

$$= \frac{4\pi}{5\sqrt{5}} \int_0^a [x^2 - 3ax + 2a\sqrt{a}\, x^{1/2}]\, dx$$

$$= \frac{4\pi}{5\sqrt{5}} \left[\frac{x^3}{3} - 3a\frac{x^2}{2} + 2a^{3/2} \cdot \frac{x^{3/2}}{3/2}\Bigg|_0^a\right] = \frac{4\pi}{5\sqrt{5}} \left[\frac{a^3}{3} - \frac{3a^3}{2} + \frac{4}{3}a^3\right]$$

$$= \frac{4\pi}{5\sqrt{5}} \, \frac{2 - 9 + 8}{6} a^3 = \frac{2\pi a^3}{15\sqrt{5}} = \frac{2\sqrt{5}}{75} \pi a^3$$

Hence volume of the solid $= \dfrac{2\sqrt{5}}{75} \pi a^3$.

Example 11. *Find the volume of the solid generated by the revolution of the curve $r = a\,(1 + \cos\theta)$ about the initial line.* **(P.T.U., Dec. 2014)**

Sol. Equation of the cardiod is $r = a\,(1 + \cos\theta)$

Symmetry: Curve is symmetrical about initial line ($\because$ if we change θ to $-\theta$ equation remains unchanged)

Pole or origin: When $r = 0$, $\theta = \pi$ (real) $\because$ curve passes through the origin and tangent at the origin is $\theta = \pi$.

Points of intersections: Curve intersects initial line $\theta = 0$ at $r = 2a$ *i.e.*, at $(2a, 0)$

Curve intersects $\theta = \dfrac{\pi}{2}$ at $r = a$ *i.e.*, at $\left(a, \dfrac{\pi}{2}\right)$

Curve intersects $\theta = \pi$ at $r = 0$ *i.e.*, at $(0, \pi)$

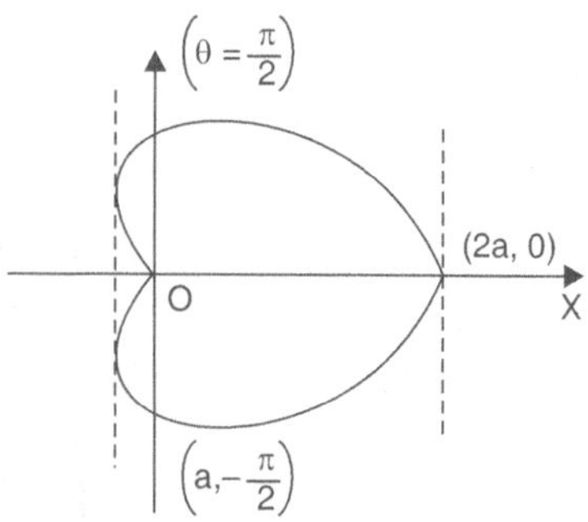

Region: We know $-1 \le \cos\theta \le 1$ $\therefore$ $0 \le 1 + \cos\theta \le 2$

i.e., $0 \le a\,(1 + \cos\theta) \le 2a$ *i.e.*, $0 \le r \le 2a$

$\therefore$ Curve lies within the circle $r = 2a$

Required volume $= \pi \int_0^{2a} y^2\, dx$ where OA $= 2a$

Change x, y to polar coordinates

$$x = r\cos\theta, \; y = r\sin\theta$$

$$x = a(1 + \cos\theta)\cos\theta$$

$$y = a(1 + \cos\theta)\sin\theta$$
$$dx = a(-\sin\theta - 2\cos\theta\sin\theta)\,d\theta$$

$$V = \pi\int_0^{\pi} a^2(1+\cos\theta)^2\sin^2\theta(-a\sin\theta)(1+2\cos\theta)\,d\theta$$

$$= -\pi a^3\int_0^{\pi}\sin^3\theta(1+\cos\theta)^2(1+2\cos\theta)\,d\theta$$

Put $\cos\theta = t$; $-\sin\theta\,d\theta = dt$

$$\theta = 0, t = 1, \theta = \pi, t = -1$$

$$= -\pi a^3\int_1^{-1}(1-t^2)(1+t)^2(1+2t)(-dt)$$

$$= \pi a^3\int_1^{-1}(1+4t+4t^2-2t^3-5t^4-2t^5)\,dt$$

$$= \pi a^3\left[t + 2t^2 + 4\frac{t^3}{3} - \frac{t^4}{2} - t^5 - \frac{t^6}{3}\right]_1^{-1}$$

$$= \pi a^3\left[-1+2-\frac{4}{3}-\frac{1}{2}+1-\frac{1}{3}-1-2-\frac{4}{3}+\frac{1}{2}+1+\frac{1}{3}\right]$$

$$= \pi a^3\left[-2-\frac{8}{3}+2\right] = -\frac{8}{3}\pi a^3 = \frac{8}{3}\pi a^3 \text{ (in magnitude)}$$

Example 12. *Find the volume of the solid obtained by revolving the lemniscate $r^2 = a^2\cos 2\theta$ about the initial line.*

Sol. Equation of the lemniscate is $r^2 = a^2\cos 2\theta$ ...(1)

1. **Symmetry:** On changing θ to $-\theta$ and r to $-r$ curve remains unchanged $\therefore$ curve is symmetric about initial line as well as $\theta = \dfrac{\pi}{2}$.

2. **Pole or origin:** Put $r = 0$, we get $\cos 2\theta = 0$ *i.e.,* $\cos 2\theta = \cos\left(2n\pi \pm \dfrac{\pi}{2}\right)$

$\therefore \quad \theta = n\pi \pm \dfrac{\pi}{4}$ for $n = 0, 1, 2, 3, ...$

$\therefore \quad \theta$ has values $-\dfrac{\pi}{4}, \dfrac{\pi}{4}, \dfrac{3\pi}{4}, \dfrac{5\pi}{4}, ...$

Tangents at poles are $\theta = \dfrac{\pi}{4}, \dfrac{3\pi}{4}, ...$

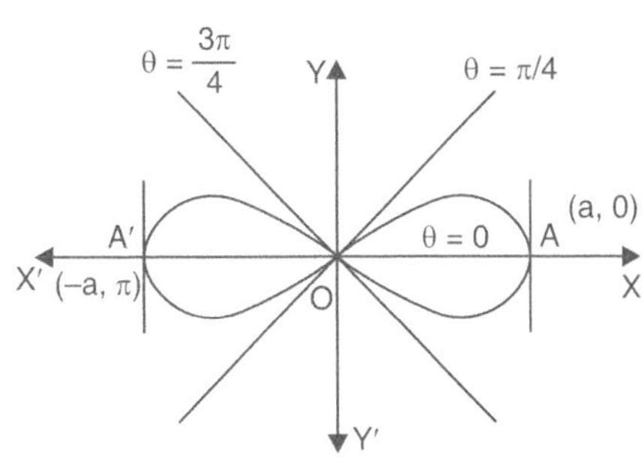

3. **Region:** $\cos 2\theta$ is negative in 2nd and 3rd quadrant

i.e., $\quad \dfrac{\pi}{2} < 2\theta < \pi$ and $\pi < 2\theta < \dfrac{3\pi}{2}$ *i.e.,* $\dfrac{\pi}{4} < \theta < \dfrac{\pi}{2}$ and $\dfrac{\pi}{2} < \theta < \dfrac{3\pi}{4}$

$\therefore$ Curve does not lie in the portion between $\left(\frac{\pi}{4}, \frac{\pi}{2}\right)$ and $\left(\frac{\pi}{2}, \frac{3\pi}{4}\right)$

4. **Points of intersection:** Curve intersects initial line *i.e.*, $\theta = 0$ at $r^2 = a^2$, *i.e.*, $r = \pm a$ i.e., at $(-a, 0)$ and $(a, 0)$

Volume of the solid obtained by revolving the whole curve about x-axis = 2 volume of the solid obtained by revolving half the curve on right hand side of O.

$\therefore$ Volume of the solid

$$= 2\int_0^a \pi y^2\, dx$$

where $x = r\cos\theta \qquad y = r\sin\theta$

Substituting the value of $r = a\sqrt{\cos 2\theta}$, we get

$$x = a\sqrt{\cos 2\theta}\cdot\cos\theta$$

$$y = a\sqrt{\cos 2\theta}\cdot\sin\theta$$

$$\therefore \quad dx = a\left\{\sqrt{\cos 2\theta}\,(-\sin\theta) + \frac{1}{2\sqrt{\cos 2\theta}}(-2\sin 2\theta)\cos\theta\right\}d\theta$$

$$= a\left\{\frac{-\sin\theta\cos 2\theta - \sin 2\theta\cos\theta}{\sqrt{\cos 2\theta}}\right\}d\theta = -a\,\frac{\sin 3\theta}{\sqrt{\cos 2\theta}}\,d\theta.$$

Also in 1st quadrant θ varies from 0 to $\frac{\pi}{4}$

$\therefore$ Required volume

$$= 2\int_0^{\pi/4} \pi a^2\cos 2\theta\sin^2\theta\cdot\frac{-a\sin 3\theta}{\sqrt{\cos 2\theta}}\,d\theta$$

$$= -2\pi a^3\int_0^{\pi/4}\sqrt{\cos 2\theta}\,\sin^2\theta\sin 3\theta\,d\theta$$

$$= -2\pi a^3\int_0^{\pi/4}\sqrt{2\cos^2\theta - 1}\,(1-\cos^2\theta)(3\sin\theta - 4\sin^3\theta)\,d\theta$$

$$= -2\pi a^3\int_0^{\pi/4}\sqrt{2\cos^2\theta - 1}\,(1-\cos^2\theta)(3 - 4 + 4\cos^2\theta)\sin\theta\,d\theta$$

Put $\cos\theta = t \quad \therefore \quad -\sin\theta\,d\theta = dt$

$$= -2\pi a^3\int_1^{1/\sqrt{2}}\sqrt{2t^2 - 1}\,(1-t^2)(-1+4t^2)(-dt)$$

$$= 2\pi a^3\int_1^{1/\sqrt{2}}\frac{(-1+t^2+4t^2-4t^4)(2t^2-1)}{\sqrt{2t^2-1}}\,dt$$

$$= 2\pi a^3\int_1^{1/\sqrt{2}}\frac{(-8t^6+14t^4-7t^2+1)}{\sqrt{2t^2-1}}\,dt \qquad ...(1)$$

Note the integration

Let $\int\frac{-8t^6+14t^4-7t^2+1}{\sqrt{2t^2-1}}dt = (at^5+bt^4+ct^3+dt^2+et+f)\sqrt{2t^2-1} + g\int\frac{1}{\sqrt{2t^2-1}}\,dt$

Differentiate both sides

$$\frac{-8t^6+14t^4-7t^2+1}{\sqrt{2t^2-1}} = (5at^4+4bt^3+3ct^2+2dt+e)\sqrt{2t^2-1}$$
$$+(at^5+bt^4+ct^3+dt^2+et+f)\frac{4t}{2\sqrt{2t^2-1}}+\frac{g}{\sqrt{2t^2-1}}$$

Multiply by $\sqrt{2t^2-1}$

$$-8t^6+14t^4-7t^2+1 = (5at^4+4bt^3+3ct^2+2dt+e)(2t^2-1)$$
$$+2(at^5+bt^4+ct^3+dt^2+et+f)t+g$$

Equating coefficients of like powers of t

Coeff. of $t^6 \to -8 = 10a+2a$ $\quad\therefore\quad a = -\frac{2}{3}$

Coeff. of $t^5 \to 0 = 8b+b$ $\quad\therefore\quad b = 0$

Coeff. of $t^4 \to 14 = -5a+6c+2c$ $\quad\therefore\quad c = \frac{4}{3}$

Coeff. of $t^3 \to 0 = -4b+4d+2d$ $\quad\therefore\quad d = 0$

Coeff. of $t^2 \to -7 = -3c+2e+2e$ $\quad\therefore\quad e = -\frac{3}{4}$

Coeff. o $t \to 0 = -2d+2f$ $\quad\therefore\quad f = 0$

Constant $1 = -e+g$ $\quad\therefore\quad g = \frac{1}{4}$

$\therefore$ From (1),

$$\text{Volume} = 2\pi a^3\left\{\left(-\frac{2}{3}t^5+\frac{4}{3}t^3-\frac{3}{4}t\right)\sqrt{2t^2-1}\,\Bigg|_1^{\frac{1}{\sqrt{2}}}+\frac{1}{4}\int_1^{\frac{1}{\sqrt{2}}}\frac{1}{\sqrt{2t^2-1}}\,dt\right\}$$

$$= 2\pi a^3\left[0+\frac{1}{12}+\frac{1}{4\sqrt{2}}\int_1^{\frac{1}{\sqrt{2}}}\frac{1}{\sqrt{t^2-\left(\frac{1}{\sqrt{2}}\right)^2}}\,dt\right]$$

$$= 2\pi a^3\left[\frac{1}{12}+\frac{1}{4\sqrt{2}}\left\{\log\left(t+\sqrt{t^2-\left(\frac{1}{\sqrt{2}}\right)^2}\right)\right\}_1^{\frac{1}{\sqrt{2}}}\right]$$

$$= 2\pi a^3\left[\frac{1}{12}+\frac{1}{4\sqrt{2}}\left\{\log\frac{1}{\sqrt{2}}-\log\left(1+\frac{1}{\sqrt{2}}\right)\right\}\right]$$

$$= 2\pi a^3\left[\left\{\frac{1}{12}+\frac{1}{4\sqrt{2}}\left\{\log\frac{1}{\sqrt{2}}-\log\frac{\sqrt{2}+1}{\sqrt{2}}\right\}\right]\right]$$

$$= 2\pi a^3\left[\frac{1}{12}-\frac{1}{4\sqrt{2}}\log\frac{\sqrt{2}+1}{\sqrt{2}\cdot\frac{1}{\sqrt{2}}}\right] = 2\pi a^3\left[\frac{1}{12}-\frac{1}{4\sqrt{2}}\log(\sqrt{2}+1)\right]$$

$$= \frac{\pi a^3}{6}\cdot\left[1-\frac{3}{\sqrt{2}}\log(\sqrt{2}+1)\right] = \frac{\pi a^3}{6\sqrt{2}}\cdot[\sqrt{2}-3\log(\sqrt{2}+1)]$$

$$= -\frac{\pi a^3}{6\sqrt{2}}\,[3\log(\sqrt{2}+1)-\sqrt{2}\,]$$

$$= \frac{\pi a^3}{6\sqrt{2}}\,[3\log(\sqrt{2}+1)-\sqrt{2}\,] \text{ in Magnitude.}$$

TEST YOUR KNOWLEDGE

1. Find the volume of the solid obtained by revolving the ellipse $\frac{x^2}{a^2}+\frac{y^2}{b^2}=1$ about axis of x.

(*P.T.U., Jan. 2010*)

2. (*i*) Find the volume of the solid formed by revolution of the curve $xy^2 = 4(2-x)$ about x-axis.

   (*ii*) Find the volume formed by revolution of the loop of the curve $y^2 = x^2\,\frac{a-x}{a+x}$ about x-axis.

3. Find the volume of the solid obtained by revolving the loop of the curve $a^2y^2 = x^2(2a-x)(x-a)$ about x-axis.

4. Prove that the volume of the solid generated by the revolution of the curve $y = \frac{a^3}{a^2+x^2}$ about its asymptote is $\frac{\pi^2a^3}{2}$.

5. Show that volume of the solid obtained by revolving the area included between the curves $y^2 = x^3$ and $x^2 = y^3$ about x-axis is $\frac{5\pi}{28}$.

   [**Hint.** for curve tracing see Q. 13 of exercise art 1.14]

6. Find the volume of the solid generated by revolving the region bounded by the curves $y = 3 - x^2$ and $y = -1$ about the line $y = -1$

   [**Hint.** Reqd. Vol. $= 2\pi\int_0^2 (3-x^2+1)^2\,dx$]

7. Show that volume of the solid generated by the revolution of the cycloid $x = a(\theta + \sin\theta)$, $y = a(1+\cos\theta)$ about x-axis is $5\pi^2a^3$. **(*P.T.U., Jan. 2009*)**

8. Find the volume of the solid of revolution of the area enclosed between $y = \sqrt{x}$, $y = 0$ for $x = 0$ to $x = 4$ about the line $y = 2$.

9. Find the volume of the solid of revolution of the region bounded by $x = 2t + 1$, $y = 4t^2 - 1$; $-\frac{1}{2} \le t \le 0$, $y = 0$ about the line $x = 1$.

10. Find the volume of the solid of revolution of the region bounded by $y = 2x - x^2$ and $y = x$ about y-axis.

11. Find the volume of the solid generated by revolving the region bounded by $y = \sqrt{x}$, $y = 0$ and $x = 9$ about y-axis

    [**Hint.** Reqd. Volume $= \int_0^3 81\,\pi dy - \int_0^3 \pi y^4 dy$]

12. Find the volume of the solid of revolution of the curve $(a - x) y^2 = a^2x$ about its asymptote.

13. Find the volume of the solid obtained by revolving the cardioid $r = a (1 - \cos \theta)$ about the initial line **(*P.T.U., Dec. 2011*)**

Answers

1. $\frac{4}{3} \pi a^2b^2$ 2. (*i*) $8\pi (\log 2 - 1)$ (*ii*) $\frac{2\pi a^3}{3} \log \frac{8}{e^2}$

3. $\frac{23\pi a^3}{60}$

6. $\frac{512}{15} \pi$ radiaus 8. $\frac{40}{3} \pi$ 9. $\frac{\pi}{2}$

10. $\frac{\pi}{6}$ 11. $\frac{972}{5} \pi$ cubic units 12. $\frac{1}{2} \pi^2 a^3$

13. $\frac{8}{3} \pi a^3$.

2.10. SURFACE OF REVOLUTION

Before finding the area of the surface of the solid obtained on revolving about x-axis the arc of a curve, we will find area of the surface of the frustum of a cone.

2.11. AREA OF THE SURFACE OF THE FRUSTUM OF A CONE

Let OCD be a right circular cone with vertex O, radius r and slant side l. We know that surface area of the cone is πlr.

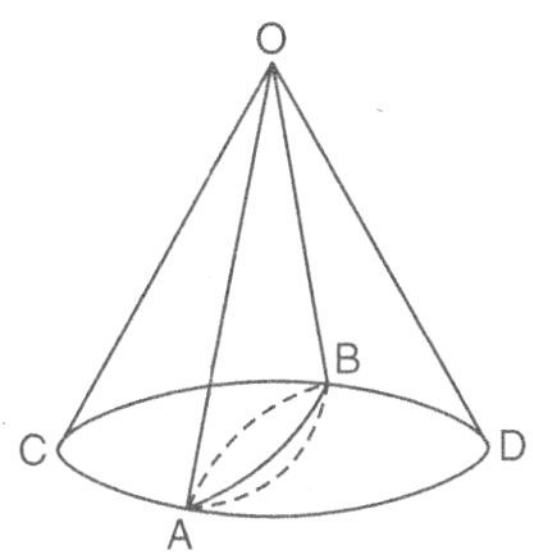

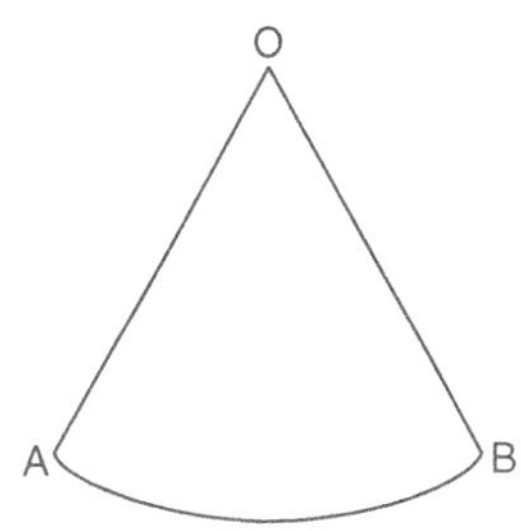

Cut off the cone along one of its generators, we get a circular sector whose radius is OA = l and arc AB = circumference of the circular base of the cone.

Area of the sector OAB = πlr which is same as the surface area of the cone.

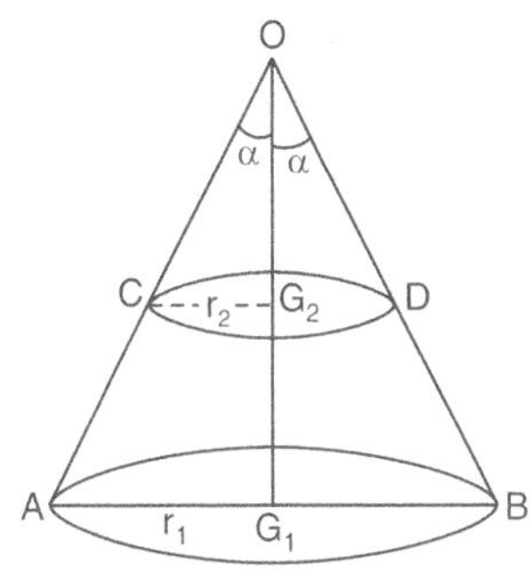

If α is the semi-vertical angle of the cone. We have $\frac{r}{l} = \sin \alpha$.

$\therefore$ Surface area of the cone $= \pi l.l \sin \alpha = \pi l^2 \sin \alpha$

Now, Let ABDC be the frustum of the cone. Let $r_1 = AG_1$ and $r_2 = CG_2$ be the radii of two circular bases.

Let slant side of frustum = l_1

Then surface area of this frustum

$$= \pi(\text{AO})^2 \sin\alpha - \pi(\text{CO})^2 \sin\alpha$$
$$= \pi \sin\alpha\,(\text{AO}^2 - \text{CO}^2)$$
$$= \pi \sin\alpha\,(\text{AO} - \text{CO})\,(\text{AO} + \text{CO})$$
$$= \pi\,(\text{AC})\,[\text{AO} \sin\alpha + \text{CO} \sin\alpha] = \pi l_1\,(r_1 + r_2)$$
$$= \pi \times \text{slant side} \times \text{sum of the radii of the two bases.}$$

2.12. AREA OF THE SURFACE OBTAINED BY REVOLVING THE CURVE ABOUT AXES

Prove that the area of the surface of the solid obtained on revolving about x-axis, the arc of the curve $y = f(x)$ intercepted between the points whose abscissae are a, b is $\int_a^b 2\pi y \frac{ds}{dx}\,dx$

$$= \int_a^b 2\pi y\,ds = \int_a^b 2\pi\, f(x)\sqrt{1 + [f'(x)]^2}\;dx.$$

Proof. Let AB represents the curve $y = f(x)$ where A, B have abscissae a, b.

$\therefore$ A has co-ordinates $(a, f(a))$

B has co-ordinates $(b, f(b))$

Let $P(x, y)$ be any point of the curve and $Q(x + \delta x, y + \delta y)$ be a neighbouring point to P. From P, Q draw PM, QN perpendiculars on x-axis. Join PQ by a straight line.

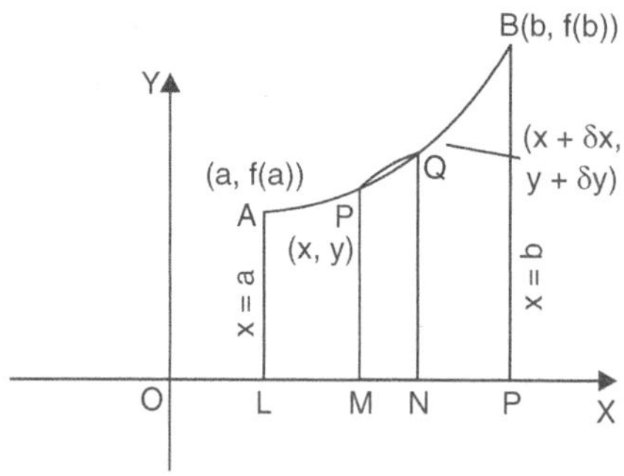

Let S be the surface area of the solid obtained by revolving the arc AP about x-axis and let δS be the surface area of the solid obtained by revolving the arc PQ about x-axis.

By revolving the chord PQ about x-axis we get a frustum of the cone whose slant side is | PQ | and the radii of whose circular ends are MP and NQ

$\therefore$ area of the frustum = π | PQ | [MP + NQ] [art. 2.11]

$$= \pi\,|\,\text{PQ}\,|\,\{y + y + \delta y\} = \delta\text{S}' \text{ (say)} \quad \ldots(1)$$

$\therefore$ δS′ is the surface area enclosed by | (PQ) | and δS is the surface area enclosed by $\overset{\frown}{\text{PQ}}$

From (1), $$\frac{\delta\text{S}'}{\delta x} = \pi\{2y + \delta y\}\,\frac{|\text{PQ}|}{\delta x} = \pi\,[2y + \delta y]\,\frac{|\text{PQ}|}{\overset{\frown}{\text{PQ}}}\cdot\frac{\overset{\frown}{\text{PQ}}}{\delta x}$$

Take limit when $\delta x \to 0$ *i.e.,* $\text{P} \to \text{Q}$

We know that $\underset{\text{P}\to\text{Q}}{\text{Lt}}\ \dfrac{\text{chord PQ}}{\text{arc PQ}} = 1$ and $\delta\text{S}' \to \delta S$

$$\therefore \quad \underset{\delta x \to 0}{\text{Lt}}\ \frac{\delta\text{S}}{\delta x} = \pi\,2y \cdot 1 \cdot \underset{\delta x \to 0}{\text{Lt}}\ \frac{\delta s}{\delta x} \quad \text{where } \overset{\frown}{\text{PQ}} = \delta s \text{ when } \delta x \to 0,\ \delta y \text{ also} \to 0$$

$$\therefore \quad \frac{dS}{dx} = 2\pi y \frac{ds}{dx}$$

Integrating both sides w.r.t. x between the limits a and b

$$\therefore \quad \int_a^b \frac{dS}{dx}\,dx = 2\pi \int_a^b y\,\frac{ds}{dx}\,dx \quad \text{or} \quad S\Big|_a^b = \int_a^b 2\pi y \frac{ds}{dx}\,dx$$

$$[\text{Value of S at } x = b] - [\text{value of S at } x = a] = \int_a^b 2\pi y \frac{ds}{dx}\,dx$$

i.e., [Surface area obtained by revolving arc AB about x-axis] $- 0 = \int_a^b 2\pi y \frac{ds}{dx}\,dx$

$$S = \int_a^b 2\pi y \frac{ds}{dx}\,dx$$

$$= \int_a^b 2\pi y \sqrt{1 + \left(\frac{dy}{dx}\right)^2}\,dx \qquad \left[\frac{ds}{dx} = \sqrt{1 + \left(\frac{dy}{dx}\right)^2}\right]$$

or

$$S = \int_a^b 2\pi f(x)\sqrt{1 + [f'(x)]^2}\,dx.$$

Note 1. If equation of the curve is $x = f(y)$ and is bounded by $y = a$, $y = b$ then surface area

$$S = \int_a^b 2\pi x \frac{ds}{dy}\,dy \quad \text{or} \quad = \int_a^b 2\pi x \sqrt{1 + \left(\frac{dx}{dy}\right)^2}\,dy.$$

Note 2. If the curve is given in parametric coordinates $x = f(t)$, $y = g(t)$ where $t_1 \le t \le t_2$

then

$$S = \int_{t_1}^{t_2} 2\pi y \frac{ds}{dt}\,dt = \int_{t_1}^{t_2} 2\pi\, g(t)\sqrt{\left(\frac{dx}{dt}\right)^2 + \left(\frac{dy}{dt}\right)^2}\,dt.$$

Note 3. If curve is given in polar co-ordinates *i.e.*, $r = f(\theta)$ $\theta_0 \le \theta \le \theta_1$

Then

$$S = \int_{\theta_0}^{\theta_1} 2\pi y \frac{ds}{d\theta}\,d\theta = \int_{\theta_0}^{\theta_1} 2\pi (r \sin\theta)\cdot\sqrt{r^2 + \left(\frac{dr}{d\theta}\right)^2}\,d\theta.$$

ILLUSTRATIVE EXAMPLES

Example 1. (*i*) *Find the surface area of the solid generated by revolving the circle* $x^2 + (y - b)^2 = a^2$, $b \ge a$ *about the x-axis.* **(P.T.U., Dec. 2004)**

(*ii*) *Find the surface area of the solid formed by revolving the astroid* $x^{2/3} + y^{2/3} = a^{2/3}$ about x-axis. **(P.T.U., May 2012)**

Sol. (*i*) Here solution will be easier if we write the equation of the circle in parametric form *i.e.*, $x = a\cos\theta$, $\quad y = b + a\sin\theta$

$$\frac{dx}{d\theta} = -a\sin\theta \qquad \frac{dy}{d\theta} = a\cos\theta$$

Surface area obtained by revolving the circle about x-axis

$$S = \int_0^{2\pi} 2\pi(b + a\sin\theta)\sqrt{\left(\frac{dx}{d\theta}\right)^2 + \left(\frac{dy}{d\theta}\right)^2}\,d\theta$$

$$= \int_0^{2\pi} 2\pi(b + a \sin \theta) \sqrt{a^2 \sin^2 \theta + a^2 \cos^2 \theta}\, d\theta$$

$$= 2\pi a \int_0^{2\pi} (b + a \sin \theta)\, d\theta = 2\pi a \left[b\theta - a \cos \theta \right]_0^{2\pi}$$

$$= 2\pi a\, (2\pi b - a + a) = 4\pi^2 ab.$$

Second Method:

By Cartesian Co-ordinates: The surface of revolution consists of two parts = surface area obtained by revolution of the upper half of circle (*i.e.*, above AB) + Surface area obtained by revolution of the lower half of the circle.

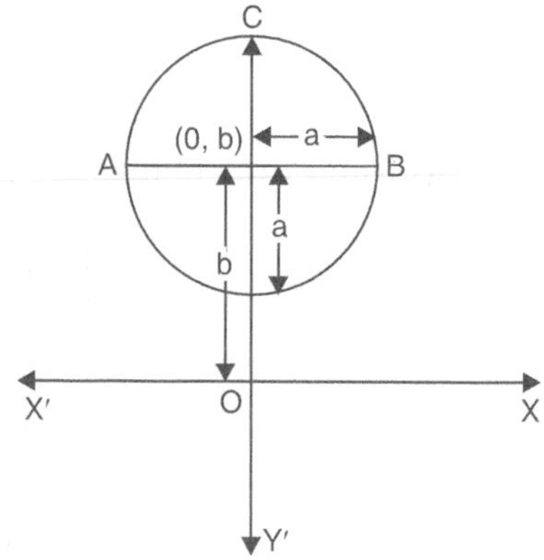

For the upper half ; $y = b + \sqrt{a^2 - x^2}$

$$\because \quad y - b = \pm \sqrt{a^2 - x^2}$$

and for the lower half $y = b - \sqrt{a^2 - x^2}$

Let S_1 be the surface area obtained by upper half

$$\therefore \qquad S_1 = \int_{-a}^{a} 2\pi y \sqrt{1 + y'^2}\, dx$$

where $$y' = \frac{1}{2\sqrt{a^2 - x^2}}(-2x) = \frac{-x}{\sqrt{a^2 - x^2}}$$

$$\therefore \qquad S_1 = \int_{-a}^{a} 2\pi \left(b + \sqrt{a^2 - x^2} \right) \sqrt{1 + \frac{x^2}{a^2 - x^2}}\, dx$$

$$= 2 \int_0^a 2\pi[b + \sqrt{a^2 - x^2}\,] \frac{a}{\sqrt{a^2 - x^2}}\, dx$$

Similarly, surface area obtained by lower half $S_2 = 2\int_0^a 2\pi \left[b - \sqrt{a^2 - x^2} \right] \dfrac{a}{\sqrt{a^2 - x^2}}\, dx$

Total surface area = $S_1 + S_2$

$$= 2\int_0^a 2\pi \cdot 2b \frac{a}{\sqrt{a^2 - x^2}}\, dx = 8\, \pi ab \sin^{-1} \frac{x}{a} \bigg|_0^a = 8\, \pi ab\, \frac{\pi}{2} = 4\pi^2 ab.$$

(*ii*) Equation of the astroid in parametric form is

$$x = a \cos^3 \theta,\ y = a \sin^3 \theta$$

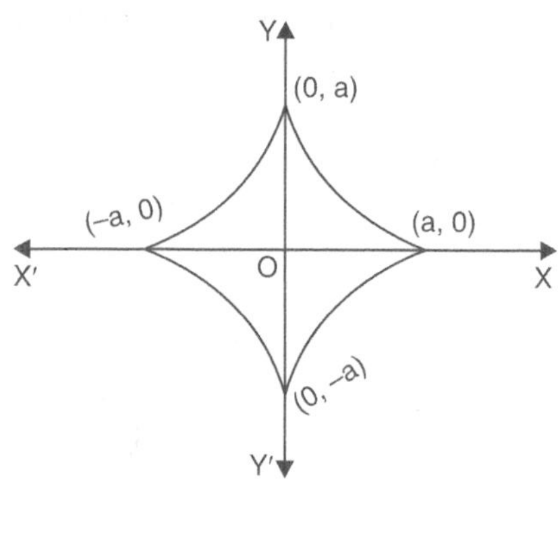

$$\frac{dx}{d\theta} = 3a \cos^2 \theta(-\sin \theta)$$

$$\frac{dy}{d\theta} = 3a \sin^2 \theta \cos \theta$$

Surface area obtained by revolving the astroid about x-axis

$$= 2\int_0^{\pi/2} 2\pi y \sqrt{\left(\frac{dx}{d\theta}\right)^2 + \left(\frac{dy}{d\theta}\right)^2}\, d\theta$$

$$= 2\int_0^{\pi/2} 2\pi \cdot a \sin^3\theta \sqrt{9a^2\cos^4\theta\sin^2\theta + 9a^2\sin^4\theta\cos^2\theta}\, d\theta$$

$$= 4\pi a \int_0^{\pi/2} \sin^3\theta \cdot 3a\sin\theta\cos\theta\, d\theta$$

$$= 12\pi a^2 \int_0^{\pi/2} \sin^4\theta\cos\theta\, d\theta = 12\pi a^2 \frac{3\cdot 1}{5\cdot 3\cdot 1} = \frac{12}{5}\pi a^2.$$

Example 2. *Prove that the surface of the protate spheroid formed by revolution of the ellipse of eccentricity about major axis is 2 area of the ellipse* $\left[\sqrt{1-e^2} + \frac{1}{e}\sin^{-1} e\right]$.

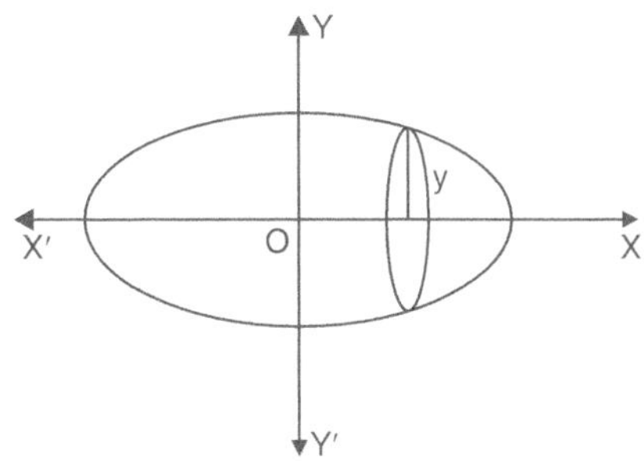

Sol. Major axis of the ellipse is x-axis.

$\therefore$ Let the parametric equation of the ellipse be

$$x = a\cos\theta,\ y = b\sin\theta$$

$\therefore$ Required surface area

$$= 2\int_0^{\frac{\pi}{2}} 2\pi y \frac{ds}{dt} . dt$$

$$= 4\pi\int_0^{\frac{\pi}{2}} b\sin\theta \sqrt{\left(\frac{dx}{d\theta}\right)^2 + \left(\frac{dy}{d\theta}\right)^2}\, d\theta$$

$$= 4\pi\int_0^{\frac{\pi}{2}} b\sin\theta\sqrt{a^2\sin^2\theta + b^2\cos^2\theta}\, d\theta$$

$$= 4\pi b\int_0^{\frac{\pi}{2}} \sin\theta\sqrt{a^2\sin^2\theta + a^2(1-e^2)\cos^2\theta\, d\theta} \qquad \left|\ \because\ b^2 = a^2(1-e^2)\right.$$

$$= 4\pi b\int_0^{\frac{\pi}{2}} \sin\theta\sqrt{a^2 - a^2e^2\cos^2\theta}\, d\theta$$

$$= 4\pi ab\int_0^{\frac{\pi}{2}} \sin\theta\sqrt{1-e^2\cos^2\theta}\, d\theta$$

Put $e\cos\theta = t \quad \therefore \quad -e\sin\theta\, d\theta = dt$

$$= 4\pi ab\int_e^0 \sqrt{1-t^2}\,\frac{dt}{-e} = -\frac{4\pi ab}{e}\int_e^0 \sqrt{1-t^2}\, dt$$

$$= -\frac{4\pi ab}{e}\left\{\frac{t\sqrt{1-t^2}}{2} + \frac{1}{2}\sin^{-1} t \Bigg|_e^0\right\}$$

$$= +\frac{4\pi ab}{e}\left\{\frac{e}{2}\sqrt{1-e^2} + \frac{1}{2}\sin^{-1} e\right\} = 2(\pi ab)\left[\sqrt{1-e^2} + \frac{1}{e}\sin^{-1} e\right]$$

$$= 2\,(\text{area of the ellipse})\left[\sqrt{1-e^2} + \frac{1}{e}\sin^{-1} e\right].$$

Example 3. *The line segment $x = \sin^2 t$, $y = \cos^2 t$, $0 \le t \le \frac{\pi}{2}$ is revolved about y-axis. Find the surface area of the solid generated.*

Sol. $x = \sin^2 t, y = \cos^2 t$

$$\frac{dx}{dt} = 2 \sin t \cos t, \quad \frac{dy}{dt} = -2 \cos t \sin t$$

Required surface area $\quad S = \int_0^{\pi/2} 2\pi x \sqrt{\left(\frac{dx}{dt}\right)^2 + \left(\frac{dy}{dt}\right)^2}\, dt$

$$= \int_0^{\pi/2} 2\pi \cdot \sin^2 t \sqrt{4 \sin^2 t \cos^2 t + 4 \cos^2 t \sin^2 t}\, dt$$

$$= 2\pi \int_0^{\pi/2} \sin^2 t \cdot 2\sqrt{2} \sin t \cos t\, dt$$

$$= 4\sqrt{2}\pi \int_0^{\pi/2} \sin^3 t \cos t\, dt = 4\sqrt{2}\pi \frac{2}{4 \cdot 2} = \sqrt{2}\,\pi \quad \text{square unit.}$$

Example 4. *Prove that the area of the surface generator by revolution of the tractrix; $x = a \cos t + \frac{a}{2} \log \tan^2 \frac{t}{2}$, $y = a \sin t$ about its asymptote is equal to the surface of sphere of radius a.*

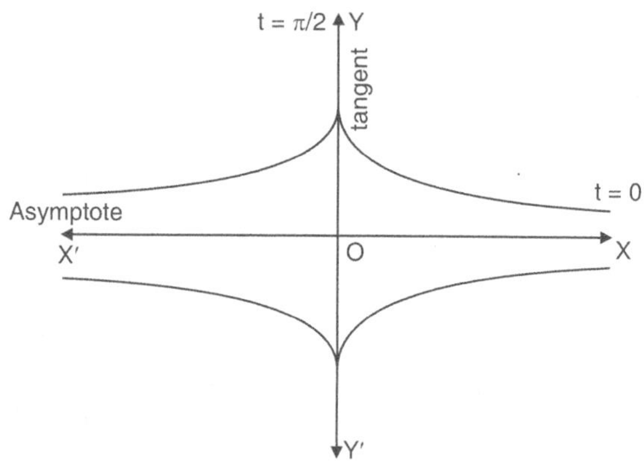

Sol. Equation of the tractrix is

$$x = a \cos t + \frac{a}{2} \log \tan^2 \frac{t}{2}; \; y = a \sin t \quad ...(1)$$

[For curve tracing see S.E. 3 art. 1.15]

Required surface area about asymptote *i.e.*, *x*-axis

$$= 2 \int_0^{\frac{\pi}{2}} 2\pi y \frac{ds}{dt}\, dt$$

$$= 4\pi \int_0^{\frac{\pi}{2}} a \sin t \sqrt{\left(\frac{dx}{dt}\right)^2 + \left(\frac{dy}{dt}\right)^2}\, dt$$

From (1), $\quad \frac{dx}{dt} = -a \sin t + \frac{a}{2} \frac{1}{\tan^2 \frac{t}{2}} \cdot 2 \tan \frac{t}{2} \sec^2 \frac{t}{2} \cdot \frac{1}{2}$

$$= -a \sin t + \frac{a}{2} \frac{\sec^2 t/2}{\tan \frac{t}{2}} = -a \sin t + \frac{a}{2} \frac{1}{\frac{\sin t/2}{\cos t/2} \cos^2 t/2}$$

$$= -a \sin t + \frac{a}{\sin t} = \frac{a}{\sin t}(1 - \sin^2 t) = \frac{a \cos^2 t}{\sin t}$$

$$\frac{dy}{dt} = a \cos t$$

$\therefore$ Required surface area $= 4\pi a \int_0^{\pi/2} \sin t \sqrt{\frac{a^2 \cos^4 t}{\sin^2 t} + a^2 \cos^2 t}\, dt$

$$= 4\pi a^2 \int_0^{\frac{\pi}{2}} \cos t \, dt = 4\pi a^2 \sin t \Big|_0^{\frac{\pi}{2}} = 4\pi a^2$$

= Surface area of the sphere of radius a.

Example 5. *The part of the lemniscate $r^2 = 2a^2 \cos 2\theta$, $0 \le \theta \le \frac{\pi}{4}$ is revolved about x-axis. Find the surface area of the solid generated. Find also the surface area if lemniscate is revolved about tangent at the pole.*

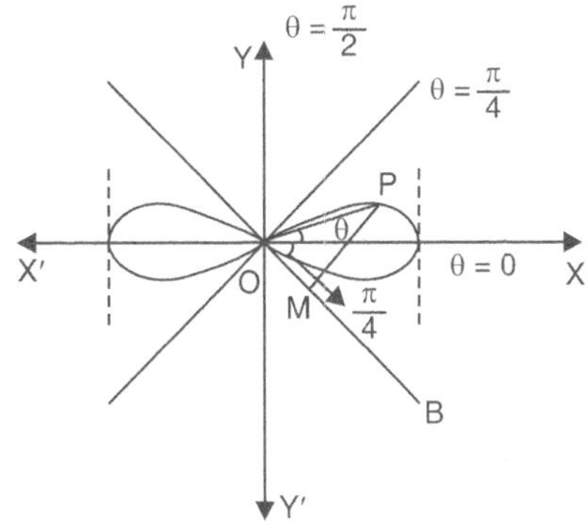

Sol. Equation of the *lemniscate* is $r^2 = 2a^2 \cos 2\theta$

(For graph of the curve see S.E. 6 art 1.16)

$$2r \frac{dr}{d\theta} = 2a^2 (-\sin 2\theta) \cdot 2$$

$$\therefore \quad \frac{dr}{d\theta} = \frac{-2a^2 \sin 2\theta}{\sqrt{2}a\sqrt{\cos 2\theta}} = -\sqrt{2}a \frac{\sin 2\theta}{\sqrt{\cos 2\theta}}$$

Surface area about x-axis

$$S = \int_0^{\pi/4} 2\pi (r \sin\theta) \sqrt{r^2 + \left(\frac{dr}{d\theta}\right)^2} \, d\theta$$

$$= \int_0^{\pi/4} 2\pi r \sin\theta \sqrt{2a^2 \cos 2\theta + \frac{2a^2 \sin^2 2\theta}{\cos 2\theta}} \, d\theta$$

$$= \int_0^{\pi/4} 2\pi \sqrt{2}a \sqrt{\cos 2\theta} \sin\theta . \frac{\sqrt{2}a}{\sqrt{\cos 2\theta}} \, d\theta$$

$$= 4\pi a^2 \int_0^{\pi/4} \sin\theta \, d\theta = 4\pi a^2 (-\cos\theta) \Big|_0^{\pi/4}$$

$$= 4\pi a^2 \left(-\frac{1}{\sqrt{2}} + 1\right) = \frac{4\pi a^2(\sqrt{2} - 1)}{\sqrt{2}} = 2\sqrt{2}(\sqrt{2} - 1)\pi a^2 = 2(2 - \sqrt{2})\pi a^2.$$

Let P be any point on the curve and OB, the tangent at the pole. Let PM be ⊥ from P on the tangent then $PM = OP \sin\left(\theta + \frac{\pi}{4}\right)$

$$= r \sin\left(\theta + \frac{\pi}{4}\right).$$

$$\text{Required surface area} = 2 \int 2\pi (PM) \, ds \text{ when revolved about OB}$$

$$= 4\pi \int_{-\frac{\pi}{4}}^{\frac{\pi}{4}} r \sin\left(\theta + \frac{\pi}{4}\right) \sqrt{r^2 + \left(\frac{dr}{d\theta}\right)^2} \, d\theta$$

$$= 4\pi \int_{-\frac{\pi}{4}}^{\frac{\pi}{4}} r \sin\left(\theta + \frac{\pi}{4}\right) \sqrt{2a^2 \cos 2\theta + \frac{2a^2 \sin^2 2\theta}{\cos 2\theta}} \, d\theta$$

$$= 4\pi \int_{-\frac{\pi}{4}}^{\frac{\pi}{4}} \sqrt{2}\, a \sqrt{\cos 2\theta} \sin\left(\theta + \frac{\pi}{4}\right) \sqrt{2} \cdot \frac{a}{\sqrt{\cos 2\theta}} \, d\theta$$

$$= 8\pi a^2 \int_{-\frac{\pi}{4}}^{\frac{\pi}{4}} \sin\left(\theta + \frac{\pi}{4}\right) d\theta$$

$$= 8\pi a^2 \left(-\cos\left(\theta + \frac{\pi}{4}\right)\Bigg|_{-\frac{\pi}{4}}^{\frac{\pi}{4}} \right) = -8\pi a^2 \left(\cos\frac{\pi}{2} - \cos 0\right) = 8\pi a^2.$$

Hence surface area when curve is rotated about the tangent at the pole = $8\pi a^2$.

2.13. THEOREMS OF PAPPUS *(P.T.U., Dec. 2004)*

(*a*) **Volume of Revolution.** If a closed plane curve revolves about a straight line in its plane, (the straight line not intersecting the curve) then the volume of the solid of revolution thus formed is obtained on multiplying the area of the region enclosed by the curve with the length of the path described by the centroid of the region.

Proof. Let $y = f(x)$ be the closed curve lying between the ordinates $x = a$, $x = b$. Let x-axis be the axis of revolution. Suppose that the curve is such that every line drawn parallel to y-axis and lying between $x = a$, $x = b$ meets the curve in two and only two points P_1 and P_2. Let $MP_1 = y_1$ and $MP_2 = y_2$ so that y_1, y_2 are functions of x.

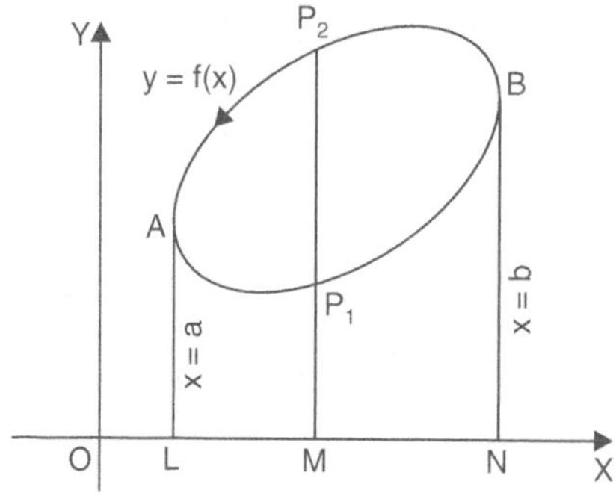

Now the volume V, of the solid of revolution is given by

$$V = \pi \int_a^b y_2^2\, dx - \pi \int_a^b y_1^2\, dx$$

$$V = \pi \int_a^b (y_2^2 - y_1^2)\, dx \qquad ...(1)$$

If $(\bar{x}, \bar{y})$ are the coordinates of the centroid of the region enclosed by the curve then we know that $\bar{y} = \dfrac{\int_a^b \frac{1}{2}(y_1 + y_2)(y_2 - y_1)\, dx}{A}$ where A is the area of the region.

$$\therefore \quad \bar{y} = \frac{\frac{1}{2}\int_a^b (y_2^2 - y_1^2)\, dx}{A}$$

$$\therefore \quad \int_a^b (y_2^2 - y_1^2)\, dx = 2A\bar{y}$$

From (1) $\qquad V = \pi \cdot 2A\bar{y} = (2\pi\bar{y})A$

$2\pi\bar{y}$ is the length of the path described by the centroid

$\therefore$ V = (area of the region) (length of the path described by centroid of the region).

(*b*) **Surface of Revolution.** If a closed plane curve revolves about a straight line in its plane (line not intersecting the curve) then the surface of the solid of revolution thus formed is obtained on multiplying the length of the curve with that of the path described by the centroid of the curve. *(P.T.U., May 2004)*

Proof. First paragraph is exactly same as that of part (*a*):

The surface S of the solid of revolution is given by $S = \int_0^l 2\pi y \, ds$ ... (1)

where l denotes the length of the arc of the curve

The centroid $\bar{y}$ of the curve is given by

$$\bar{y} = \frac{\int_0^l y \, ds}{l} \qquad \therefore \quad \int_0^l y \, ds = l\bar{y}$$

$\therefore$ From (1), $S = 2\pi \int_0^l y \, ds = 2\pi l \bar{y} = (2\pi \bar{y})\, l$ where $2\pi\bar{y}$ is the length of the path described by the centroid.

Hence S = length of the curve × length of the path described by the centroid of the curve.

Note. We can solve all the problems of obtaining volume of the solid of revolution and surface of area of the solid of revolution by directly applying Pappus theorem provided centroid of given region or curve is known.

TEST YOUR KNOWLEDGE

1. Prove that the surface of the solid generated by revolving the arc of the parabola $y^2 = 4ax$ bounded by its latus rectum about x-axis is $4\pi a^2$.
2. The arc of the curve $8a^2y^2 = x^2 (a^2 - 2x^2)$ in the first quadrant, revolves about the axes of x and y and generates the surfaces A and B show that the surface A is to the surface B as $3\sqrt{2}$ is to 2.
3. Show that the surface of the solid obtained by revolving the arc of the curve $y = \sin x$ from $x = 0$ to $x = \pi$ about x-axis is $\pi^2 \left[\sqrt{2} + \log (1 + \sqrt{2})\right]$.
4. Prove that the surface area of the solid generated by revolving the cycloid .

   $x = a(\theta - \sin \theta)$, $y = a(1 - \cos \theta)$ about the line $y = 0$ is $\frac{64\pi a^2}{3}$.
5. Show that the ratio of the areas of surface formed by the rotation of the arc of the cycloid $x = a(\theta + \sin \theta)$, $y = a(1 + \cos \theta)$ between two consecutive cusps about axis of x to the area enclosed by the cycloid and axis of x is $\frac{64}{9}$.
6. Prove that the surface of the solid formed by revolving the cardioid $r = a(1 + \cos \theta)$ about the initial line is $\frac{32}{5} \pi a^2$.
7. Prove that the surface of the solid formed by the revolution of the circle $r = 2a \cos \theta$ about initial line is $4\pi a^2$.
8. Show that surface of a sphere of radius a is $4\pi a^2$.
9. Show that the area of the sphere generated by revolution of the circle $x^2 + y^2 = a^2$ about x-axis is $4\pi a^2$.

2.14. CENTRE OF GRAVITY

It is proved in statics that if particles of masses $m_1, m_2, \cdots m_n$ be distributed at points (x_1, y_1), (x_2, y_2) ... (x_n, y_n) referred to a system of rectangular axes then C.G. of the system is given by

$$\bar{x} = \frac{\sum_{i=1}^{n} m_i x_i}{\sum_{i=1}^{n} m_i}, \quad \bar{y} = \frac{\sum_{i=1}^{n} m_i y_i}{\sum_{i=1}^{n} m_i}$$

If instead of group of isolated particles we have a continuous distribution of matter, as in the case of a rigid body the summation can be replaced by integration.

$$\therefore \qquad \bar{x} = \frac{\int x\, dm}{\int dm}\ ; \ \bar{y} = \frac{\int y\, dm}{\int dm}$$

The same method is applicable for masses in space located at (x_1, y_1, z_1) (x_2, y_2, z_2) ... . The centre of mass is

$$\bar{x} = \frac{\int x\, dm}{\int dm}, \ \bar{y} = \frac{\int y\, dm}{\int dm}, \ \bar{z} = \frac{\int z\, dm}{\int dm}$$

where limits of integration are taken to cover the whole body.

2.15. CENTRE OF GRAVITY OF A UNIFORM PLANE CURVE

Proof. Let $y = f(x)$ be a plane curve intercepted between points A and B with abscissae a and b. From A and B draw perpendiculars on x-axis and let L, M be their feet of perpendiculars then A has co-ordinates $(a, f(a))$ and B has coordinates $(b, f(b))$. Divide the arc AB into a number of very small portions and let $\overset{\frown}{PQ}$ be any such portion of the arc so that P has co-ordinates (x, y) and Q has co-ordinates $(x + \delta x, y + \delta y)$ let $\overset{\frown}{AP} = s$ and $\overset{\frown}{AQ} = s + \delta s$ $\therefore$ arc $PQ = s + \delta s - s = \delta s$. Let m be the mass per unit length then mass of $\overset{\frown}{PQ} = m.\ \delta s$ and C.G. of $\overset{\frown}{PQ}$ is (x, y). Let $(\bar{x}, \bar{y})$ be the co-ordinates of the C.G. of the arc AB then by art 2.14 $(\bar{x}, \bar{y})$ are given by

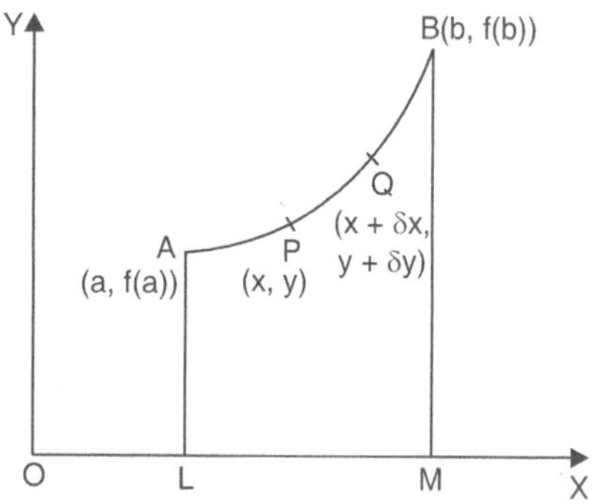

$$\bar{x} = \frac{\Sigma(m\delta s)x}{\Sigma m\delta s}, \quad \bar{y} = \frac{\Sigma(m\delta s)y}{\Sigma m\delta s}$$

As AB is a continuous curve

$\therefore$ Change summation to integration, we get

$$\bar{x} = \frac{\int_0^l mx\, ds}{\int_0^l m\, ds}; \quad \bar{y} = \frac{\int_0^l my\, ds}{\int_0^l m\, ds} \quad \text{where} \quad \overset{\frown}{AB} = l$$

$$\bar{x} = \frac{m\int_{x=a}^{b} x\frac{ds}{dx}dx}{m\cdot l}; \quad \bar{y} = \frac{m\int_a^b y\frac{ds}{dx}dx}{ml}$$

$\therefore$ C.G. of the plane curve is given by

$$\bar{x} = \frac{\int_{x=a}^{b} x \frac{ds}{dx} dx}{l}; \qquad \bar{y} = \frac{\int_{x=a}^{x=b} y \frac{ds}{dx} dx}{l}$$

where $$\frac{ds}{dx} = \sqrt{1 + \left(\frac{dy}{dx}\right)^2} = \sqrt{1 + f'^2(x)}.$$

2.16. CENTRE OF GRAVITY OF A UNIFORM PLANE AREA

Proof. Consider the plane area bounded by the curve $y = f(x)$, x-axis and the two ordinates $x = a$, $x = b$ we have to find the C.G. of the region ACDB.

Divide the region into a number of elementary strips drawn parallel to y-axis and let PLMQ be one such strip. Let the coordinates of P be (x, y) and that of Q be $(x + \delta x, y + \delta y)$. Area of the strip PLMQ is approximately same as the area of the rectangle PLMN

$\therefore$ Area of PLMN = PL · LM = $y. \delta x$

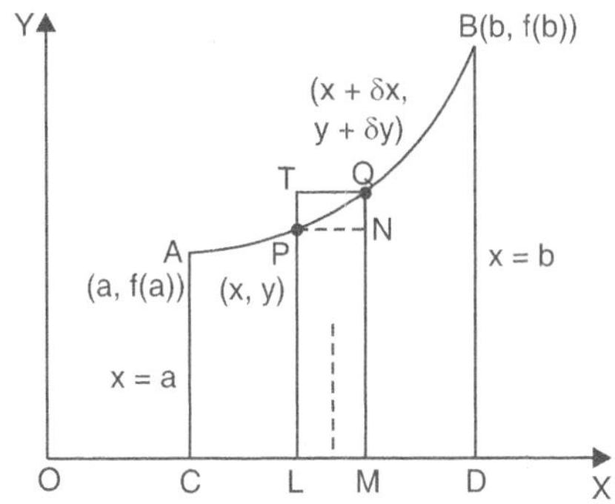

Let m be the mass per unit area

$\therefore$ mass of the rectangle PLMN = $my\ \delta x$

C.G. of this rectangle is at $\left(x, \frac{y}{2}\right)$

$\therefore$ If $(\bar{x}, \bar{y})$ be the C.G. of the whole area then

$$\bar{x} = \frac{\Sigma(my\delta x)x}{\Sigma(my\delta x)};$$

$$\bar{y} = \frac{\Sigma(my\delta x)y/2}{\Sigma my\delta x}$$

Change summation to integration

$$\bar{x} = \frac{\int_{x=a}^{b} mxydx}{\int_a^b my\, dx}; \qquad \bar{y} = \frac{\int_a^b \frac{my^2}{2} dx}{\int_a^b my\, dx}$$

$\therefore$ $$\bar{x} = \frac{\int_a^b xydx}{\int_a^b y\, dx}; \qquad \bar{y} = \frac{\int_a^b \frac{y^2}{2} dx}{\int_a^b y\, dx}$$

or $$\bar{x} = \frac{\int_a^b xy\, dx}{A}; \qquad \bar{y} = \frac{\int_a^b \frac{y^2}{2} dx}{A}$$

where A is the area of the region ACDB.

Note. If the area is bounded by the curve $x = f(y)$, y-axis and the line $y = c$, $y = d$ then C. G. is given by

$$\bar{x} = \frac{\int_c^d x^2\, dy}{\int_c^d x\, dy}; \qquad \bar{y} = \frac{\int_c^d yx\, dy}{\int_c^d x\, dy}$$

or $$\bar{x} = \frac{\int_c^d x^2\, dy}{A}; \qquad \bar{y} = \frac{\int_c^d yx\, dy}{A}$$

2.17. CENTRE OF GRAVITY OF A SECTORIAL AREA

Let sectorial area be bounded by $r = f(\theta)$ and the two radii vectors $\theta = \alpha$ and $\theta = \beta$ divide the whole area into a number of elementary strips all passing through O. Let OPQ be such one strip so that if P has coordinates (r, θ) then Q has coordinates $(r + \delta r, \theta + \delta\theta)$ then area of the elementary portion OPQ is same as area of the triangle OPQ.

$\therefore$ Elementary area

$$= \frac{1}{2} \text{OP} \cdot \text{OQ} \sin \angle\text{POQ}$$

$$= \frac{1}{2} r(r + \delta r) \sin \delta\theta = \frac{1}{2} r\,(r + \delta r)\,\delta\theta$$

$$= \frac{1}{2} r^2\,\delta\theta$$

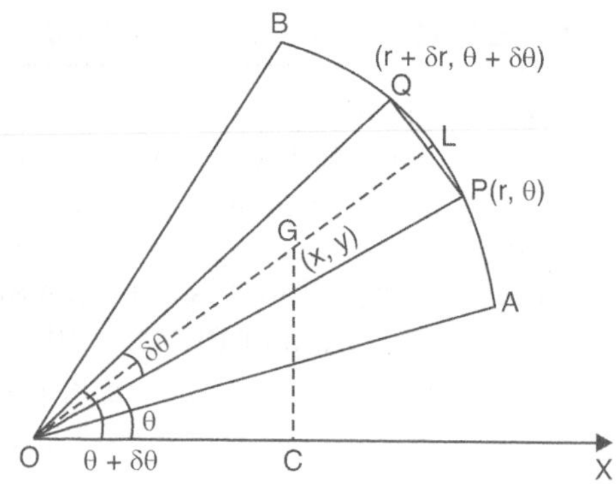

$\because$ $\delta r \delta\theta$ is very small = 0 approximately

Let m be the mass per unit area. The mass of OPQ $= m \cdot \dfrac{1}{2} r^2\,\delta\theta$. Let G$(x, y)$ be the C.G. of the strip OPQ (triangle). We know that C.G. of a triangular lamina divides each median in the ratio 2 : 1.

$$\therefore \qquad \text{OG} = \frac{2}{3}\,\text{OL} = \frac{2}{3} r$$

$\therefore$ for ΔOCG ; $\quad x = \text{OC} = \text{OG} \cos\theta = \dfrac{2}{3} r \cos\theta \quad y = \text{GC} = \left(\dfrac{2}{3} r\right) \sin\theta$

[you can take even $\theta + \delta\theta$ but $\cos(\theta + \delta\theta) = \cos\theta$ and $\sin(\theta + \delta\theta) = \sin\theta$ in (limiting) value]

$$\therefore \qquad (x, y) = \left(\frac{2}{3} r \cos\theta, \frac{2}{3} r \sin\theta\right)$$

If $(\bar{x}, \bar{y})$ be the C.G. of the whole sectorial area then

$$\bar{x} = \frac{\Sigma\left(m \frac{1}{2} r^2\,\delta\theta\right)\left(\frac{2}{3} r \cos\theta\right)}{\Sigma\left(m \frac{1}{2} r^2\,\delta\theta\right)}$$

$$\bar{y} = \frac{\Sigma\left(m \frac{1}{2} r^2\,\delta\theta\right)\left(\frac{2}{3} r \sin\theta\right)}{\Sigma\left(m \cdot \frac{1}{2} r^2\,\delta\theta\right)}$$

Change summation to integration as region is continuous

$$\therefore \qquad \bar{x} = \frac{\int_{\theta=\alpha}^{\beta} \frac{m}{3} r^3 \cos\theta\, d\theta}{\int_{\alpha}^{\beta} \frac{m}{2} r^2\, d\theta}, \quad \bar{y} = \frac{\int_{\alpha}^{\beta} \frac{m}{3} r^3 \sin\theta\, d\theta}{\int_{\alpha}^{\beta} \frac{m}{2} r^2\, d\theta}$$

or

$$\bar{x} = \frac{\int_{\alpha}^{\beta} \frac{2}{3} r^3 \cos\theta\, d\theta}{\int_{\alpha}^{\beta} r^2\, d\theta}, \quad \bar{y} = \frac{\int_{\alpha}^{\beta} \frac{2}{3} r^3 \sin\theta\, d\theta}{\int_{\alpha}^{\beta} r^2\, d\theta}.$$

2.18. CENTRE OF GRAVITY OF A VOLUME OF REVOLUTION

Let the curve $y = f(x)$ intercepted by the ordinates $x = a$, $x = b$ be revolved about x-axis. We have to find the position of the C.G. of the solid of revolution thus obtained. It is obvious that C.G. of this solid will lie on x-axis, the axis of revolution.

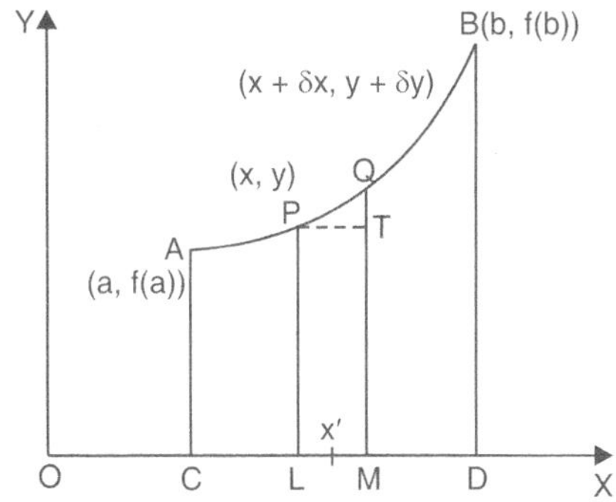

Divide the whole region into a number of elementary strips and let one such strip be PLMQ. When we revolve the curve about x-axis then the volume of the solid obtained by revolving the strip PLMQ will be same as that obtained by revolving the rectangle PLMT about x-axis.

$\therefore$ Volume of the solid of revolution of rectangle PLMT

$$= \pi\, PL^2 \cdot LM$$

[$\pi r^2 h$ $\because$ it will be a cylindric body with height LM]

$$= \pi\, y^2\, \delta x$$

Let m be the mass per unit volume

$\therefore$ Mass of elementary volume $= \pi y^2\, \delta x\, m$

Its C.G. will lie on x-axis and its abscissa is x or we can take $(x + \delta x)$

Let $\overline{x}$ is the abscissa of the C.G. of the volume of the solid of revolution then

$$\overline{x} = \frac{\Sigma\,(\pi y^2\, \delta x \cdot m)(x)}{\Sigma(\pi y^2 \delta x m)}$$

Change summation to integration as our curve is continuous

$$\therefore \qquad \overline{x} = \frac{\int_a^b \pi y^2 x m\, dx}{\int_a^b \pi y^2 m\, dx} = \frac{\int_a^b x y^2\, dx}{\int_a^b y^2\, dx}$$

Hence C.G. of the volume of the solid of revolution is $\left[\dfrac{\int_a^b x y^2\, dx}{\int_a^b y^2\, dx}, 0\right]$.

2.19. CENTRE OF GRAVITY OF A SURFACE OF REVOLUTION

Let $y = f(x)$ be the curve AB intercepted between two parts A and B with abscissae a and b respectively. Divide the arc AB into a number of elementary portions and let PQ be one such portion of the arc AB such that $\widehat{AP} = s$ and $\widehat{AQ} = s + \delta s$. Let P has coordinates (x, y) and Q has $(x + \delta x, y + \delta y)$.

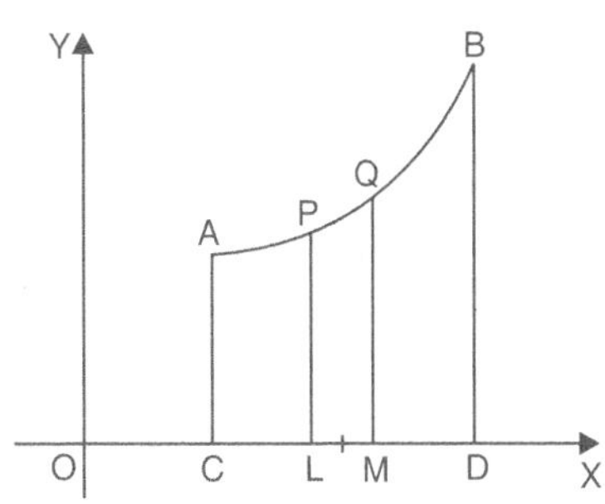

When arc AB is rotated about x-axis we get the surface of revolution and it is obvious that the C.G. of the surface of the revolution lies on x-axis.

Now surface area formed by revolution of arc PQ $= \pi(\widehat{PQ})\, y$

(Surface area of cylinder $\pi l r$ here $r = y$, $l = \widehat{PQ}$)

Let m be the mass per unit surface area

$\therefore$ Mass of the elementary surface are = $\pi y \widehat{PQ} \cdot m = \pi y \delta s m$

Its C.G. is on x-axis at a distance x from O. If $\bar{x}$ be the abscissa of the C.G. of the surface area of revolution of the arc AB about x-axis then

$$\bar{x} = \frac{\Sigma(\pi y \delta s m)x}{\Sigma \pi y \delta s m}$$

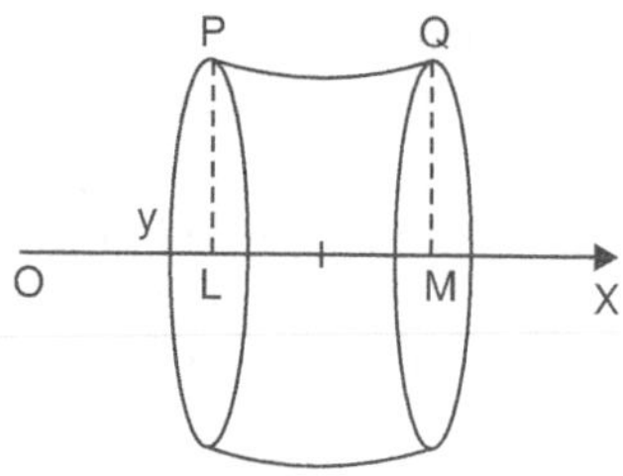

As curve is continuous

$\therefore$ Summation can be changed to integration

$$\therefore \quad \bar{x} = \frac{\int_{s=\theta}^{l} \pi y m x \, ds}{\int_{s=0}^{l} \pi y m \, ds} \quad \text{when} \quad \widehat{AB} = l$$

$$= \frac{\pi m \int_0^l xy \, ds}{\pi m \int_0^l y \, ds} = \frac{\int_{x=a}^{b} xy \frac{ds}{dx} dx}{\int_{x=a}^{b} y \frac{ds}{dx} dx}$$

Hence C.G. of the surface area of revolution about x-axis is given by

$$\bar{x} = \frac{\int_{x=a}^{b} xy \frac{ds}{dx} dx}{\int_{x=a}^{b} y \frac{ds}{dx} dx} \quad \text{where} \quad \frac{ds}{dx} = \sqrt{1 + \left(\frac{dy}{dx}\right)^2}.$$

Note. C.G. can be obtained by double integrals discussed in the chapter of 'Multiple Integrals art 6.12

ILLUSTRATIVE EXAMPLES

C.G. of Plane Curves:

Example 1. *Find the centre of gravity of the arc of the parabola $y^2 = 4ax$ bounded by the vertex and an extremity (a, 2a) of its latus rectum.*

Sol. Equation of the parabola is

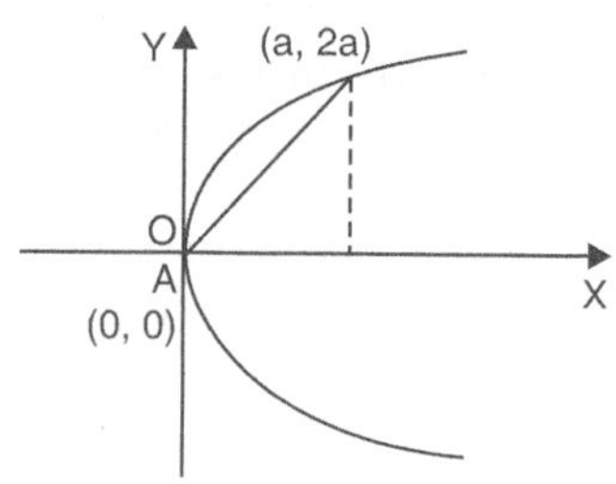

$$y^2 = 4ax$$

$$y = 2\sqrt{a}\sqrt{x}$$

$$\frac{dy}{dx} = 2\sqrt{a}\,\frac{1}{2\sqrt{x}} = \sqrt{\frac{a}{x}}$$

C.G. of the arc of a curve is given by

$$\bar{x} = \frac{\int_a^b x \frac{ds}{dx} dx}{\int_a^b \frac{ds}{dx} dx}, \quad \bar{y} = \frac{\int_a^b y \frac{ds}{dx} dx}{\int_a^b \frac{ds}{dx} dx}$$

$$\frac{ds}{dx} = \sqrt{1 + \left(\frac{dy}{ds}\right)^2} = \sqrt{1 + \frac{a}{x}} = \sqrt{\frac{x+a}{x}}$$

$$\therefore \qquad \bar{x} = \frac{\int_0^a x\sqrt{\frac{x+a}{x}}\,dx}{\int_0^a \sqrt{\frac{x+a}{x}}\,dx}; \qquad \bar{y} = \frac{\int_0^a 2\sqrt{a}\sqrt{x}\sqrt{\frac{x+a}{x}}\,dx}{\int_0^a \sqrt{\frac{x+a}{x}}\,dx}$$

$$= \frac{\int_0^a \sqrt{x^2+ax}\,dx}{\int_0^a \sqrt{\frac{x+a}{x}}\,dx} \qquad \bar{y} = \frac{2\sqrt{a}\int_0^a \sqrt{x+a}\,dx}{\int_0^a \sqrt{\frac{x+a}{x}}\,dx}$$

for integrands in denominators

Put $\sqrt{x} = \sqrt{a}t \quad \therefore \quad \frac{1}{2\sqrt{x}}\,dx = \sqrt{a}dt$

$$\bar{x} = \frac{\int_0^a \sqrt{\left(x+\frac{a}{2}\right)^2 - \left(\frac{a}{2}\right)^2}}{\int_0^1 \sqrt{a+at^2}\;2\sqrt{a}\,dt}, \quad \bar{y} = \frac{2\sqrt{a}\,\frac{2}{3}(x+a)^{3/2}\Big|_0^a}{\int_0^1 \sqrt{a+at^2}\;2\sqrt{a}\,dt}$$

$$\bar{x} = \frac{\left.\frac{\left(x+\frac{a}{2}\right)\sqrt{\left(x+\frac{a}{2}\right)^2-\left(\frac{a}{2}\right)^2}}{2} - \frac{\left(\frac{a}{2}\right)^2}{2}\log\left\{\left(x+\frac{a}{2}\right)+\sqrt{\left(x+\frac{a}{2}\right)^2-\left(\frac{a}{2}\right)^2}\right\}\right|_0^a}{2a\cdot\left\{\frac{t\sqrt{1+t^2}}{2}+\frac{1}{2}\log\left(t+\sqrt{1+t^2}\right)\right\}\Big|_0^1}$$

$$= \frac{\frac{3a}{2}\,\frac{\sqrt{\left(\frac{3a}{2}\right)^2-\left(\frac{a}{2}\right)^2}}{2} - \frac{a^2}{8}\log\left(\frac{3a}{2}+\sqrt{\left(\frac{3a}{2}\right)^2-\left(\frac{a}{2}\right)^2}\right) - \frac{a^2}{8}\log\frac{a}{2}}{2a\left\{\frac{\sqrt{2}}{2}+\frac{1}{2}\log(1+\sqrt{2})\right\}}$$

$$= \frac{\frac{3a}{2}\cdot\frac{\sqrt{2}a}{2} - \frac{a^2}{8}\log\frac{\frac{3}{2}a+\sqrt{2}a}{a/2}}{a\left[\sqrt{2}+\log\left(\sqrt{2}+1\right)\right]}$$

$$= \frac{\frac{a^2}{4}\cdot\left[3\sqrt{2}-\frac{1}{2}\log(3+2\sqrt{2})\right]}{a\left[\sqrt{2}+\log\left(\sqrt{2}+1\right)\right]} = \frac{a}{4}\,\frac{3\sqrt{2}-\frac{1}{2}\log(\sqrt{2}+1)^2}{\sqrt{2}+\log(\sqrt{2}+1)}$$

$$= \frac{a}{4}\cdot\frac{3\sqrt{2}-\log(\sqrt{2}+1)}{\sqrt{2}+\log(\sqrt{2}+1)}$$

$$\bar{y} = \frac{\frac{4}{3}\sqrt{a}\left[(2a)^{3/2}-a^{3/2}\right]}{a\left[\sqrt{2}+\log\left(\sqrt{2}+1\right)\right]} = \frac{4a^2}{3a}\cdot\frac{2\sqrt{2}-1}{\sqrt{2}+\log(\sqrt{2}+1)}$$

$$= \frac{4a}{3} \cdot \frac{2\sqrt{2} - 1}{\sqrt{2} + \log(\sqrt{2} + 1)}$$

Hence C.G. of the arc of the parabola is

$$\bar{x} = \frac{a}{4} \cdot \frac{3\sqrt{2} - \log(\sqrt{2} + 1)}{\sqrt{2} + \log(\sqrt{2} + 1)}, \qquad \bar{y} = \frac{4a}{3} \cdot \frac{2\sqrt{2} - 1}{\sqrt{2} + \log(\sqrt{2} + 1)}.$$

Example 2. *Show that the centre of gravity of the arc of the catenary $y = c \cosh \frac{x}{c}$ included between its vertex (0, c) and a point (x, y) is $\left\{x - \frac{c(y - c)}{s}, \frac{y}{2} + \frac{cx}{2s}\right\}$, where s is the length of the arc.*

Sol. Equation of the catenary is $y = c \cosh \frac{x}{c}$

$$\frac{dy}{dx} = c \sinh \frac{x}{c} \cdot \frac{1}{c} = \sinh \frac{x}{c}$$

$$\therefore \qquad \frac{ds}{dx} = \sqrt{1 + \left(\frac{dy}{dx}\right)^2} = \sqrt{1 + \sinh^2 \frac{x}{2}} = \cosh \frac{x}{c}$$

$$\therefore \qquad s = \int_0^x \cosh \frac{x}{c}\, dx = c \sinh \frac{x}{c}$$

Coordinates of the C.G. of the arc are given by

$$\bar{x} = \frac{\int_0^x x \frac{ds}{dx} dx}{\int_0^s ds}, \qquad \bar{y} = \frac{\int_0^x y \frac{ds}{dx} dx}{\int_0^s ds}$$

$$= \frac{\int_0^x x \cosh \frac{x}{c}\, dx}{s}, \qquad \bar{y} = \frac{\int_0^x c \cosh \frac{x}{c} \cosh \frac{x}{c}\, dx}{s}$$

$$= \frac{1}{s} \cdot \left\{x\, c \sinh \frac{x}{c} - c^2 \cosh \frac{x}{c}\right|_0^x ; \qquad \bar{y} = \frac{c}{s} \cdot \int_0^x \cosh^2 \frac{x}{c}\, dx$$

$$= \frac{1}{s} \cdot \left\{cx \sinh \frac{x}{c} - c^2 \cosh \frac{x}{c} + c^2\right\} \qquad = \frac{c}{s} \cdot \int_0^x \frac{1 + \cosh \frac{2x}{c}}{2} dx$$

$$= \frac{1}{s} \cdot \left\{cx \frac{s}{c} - c^2 \frac{y}{c} + c^2\right\} \qquad = \frac{c}{2s} \cdot \left\{x + \frac{c}{2} \sinh \frac{2x}{c}\right\}_0^x$$

$$= \frac{1}{s} \cdot \left\{sx - cy + c^2\right\} \qquad = \frac{c}{2s} \cdot \left\{x + \frac{c}{2} 2 \sinh \frac{x}{c} \cosh \frac{x}{c}\right\}$$

$$\bar{x} = x - \frac{c(y - c)}{s} \qquad = \frac{c}{2s}\left[x + c \frac{s}{c} \cdot \frac{y}{c}\right]$$

$$= \frac{cx}{2s} + \frac{y}{2}$$

Example 3. *Find the centre of gravity of the arc of the curve $x = a \sin^3 \theta$, $y = a \cos^3 \theta$ lying in the first quadrant.*

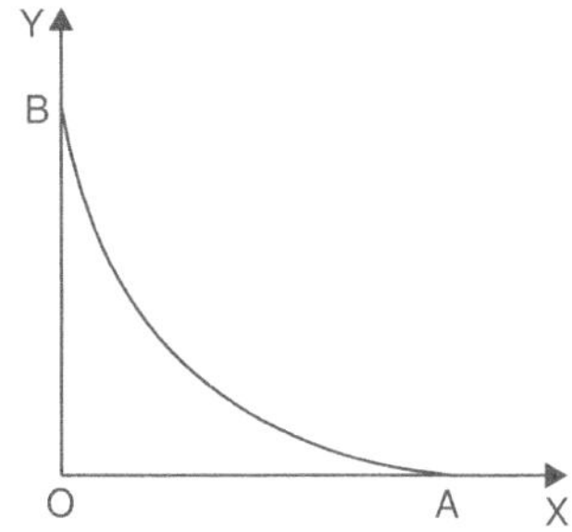

Sol. Equation of the curve is $x = a \sin^3 \theta$,

$$y = a \cos^3 \theta$$

$$\frac{dx}{d\theta} = 3a \sin^2 \theta \cos \theta, \quad \frac{dy}{d\theta} = -3a \cos^2 \theta \sin \theta$$

C.G. of the arc is given by

$$\bar{x} = \frac{\int_0^{\pi/2} x \frac{ds}{d\theta} d\theta}{\int_0^{\pi/2} \frac{ds}{d\theta} d\theta} \qquad \bar{y} = \frac{\int_0^{\pi/2} y \frac{ds}{d\theta} d\theta}{\int_0^{\pi/2} \frac{ds}{d\theta} d\theta}$$

$$\frac{ds}{d\theta} = \sqrt{\left(\frac{dx}{d\theta}\right)^2 + \left(\frac{dy}{d\theta}\right)^2} = \sqrt{9a^2 \sin^4 \theta \cos^2 \theta + 9a^2 \cos^4 \sin^2 \theta} = 3a \sin \theta \cos \theta.$$

$$\bar{x} = \frac{\int_0^{\pi/2} a \sin^3 \theta \cdot 3a \sin \theta \cos \theta \, d\theta}{\int_0^{\pi/2} 3a \sin \theta \cos \theta \, d\theta}; \qquad \bar{y} = \frac{\int_0^{\pi/2} a \cos^3 \theta \cdot 3a \sin \theta \cos \theta \, d\theta}{\int_0^{\pi/2} 3a \sin \theta \cos \theta \, d\theta}$$

$$= \frac{3a^2 \int_0^{\pi/2} \sin^4 \theta \cos \theta \, d\theta}{3a \int_0^{\pi/2} \sin \theta \cos \theta \, d\theta}; \qquad \bar{y} = \frac{3a^2 \int_0^{\pi/2} \sin \theta \cos^4 \theta \, d\theta}{3a \int_0^{\pi/2} \sin \theta \cos \theta \, d\theta}$$

$$= a \cdot \frac{\frac{3 \cdot 1}{5 \cdot 3 \cdot 1}}{\frac{1}{2}} \qquad \bar{y} = a \cdot \frac{\frac{3 \cdot 1}{5 \cdot 3}}{\frac{1}{2}}$$

$$\bar{x} = \frac{2a}{5} \qquad \bar{y} = \frac{2a}{5}$$

$\therefore$ C.G. of the arc is $\left(\frac{2a}{5}, \frac{2a}{5}\right)$.

C.G. of Areas:

Example 4. *Find the C.G. of a uniform Lamina bounded by the coordinate axes and arc of the ellipse $\frac{x^2}{a^2} + \frac{y^2}{b^2} = 1$ in first quadrant.*

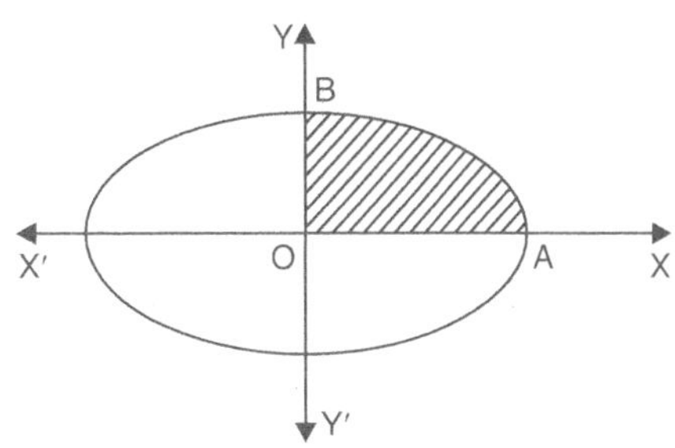

Sol. Equation of the ellipse is

$$\frac{x^2}{a^2} + \frac{y^2}{b^2} = 1$$

C.G. of the area OAB is

$$\bar{x} = \frac{\int_0^a xy\,dx}{\int_0^a y\,dx}; \qquad \bar{y} = \frac{\int_0^a \frac{y^2}{2}\,dx}{\int_0^a y\,dx}$$

$$= \frac{\int_0^a xb\sqrt{1-\frac{x^2}{a^2}}\,dx}{\int_0^a b\sqrt{1-\frac{x^2}{a^2}}\,dx}; \qquad \bar{y} = \frac{\frac{1}{2}\int_0^a b^2\left(1-\frac{x^2}{a^2}\right)dx}{\int_0^a b\sqrt{1-\frac{x^2}{a^2}}\,dx}$$

or

$$\bar{x} = \frac{\frac{b}{a}\int_0^a x\sqrt{a^2-x^2}\,dx}{\frac{b}{a}\int_0^a \sqrt{a^2-x^2}\,dx}; \qquad \bar{y} = \frac{\frac{1}{2}\frac{b^2}{a^2}\int_0^a (a^2-x^2)\,dx}{\frac{b}{a}\int_0^a \sqrt{a^2-x^2}\,dx}$$

$$= \frac{-\frac{1}{2}\frac{(a^2-x^2)^{3/2}}{3/2}\Big|_0^a}{\frac{x\sqrt{a^2-x^2}}{2}+\frac{a^2}{2}\sin^{-1}\frac{x}{a}\Big|_0^a}; \qquad \bar{y} = \frac{1}{2}\frac{b}{a}\cdot\frac{a^2x-\frac{x^3}{3}\Big|_0^a}{\frac{x\sqrt{a^2-x^2}}{2}+\frac{a^2}{2}\sin^{-1}\frac{x}{a}\Big|_0^a}$$

$$\bar{x} = \frac{0+\frac{1}{3}a^3}{\frac{a^2}{2}\frac{\pi}{2}-0}; \qquad \bar{y} = \frac{b}{2a}\cdot\frac{\frac{2}{3}a^3}{\frac{a^2}{2}\cdot\frac{\pi}{2}}$$

$$= \frac{4a}{3\pi} \qquad\qquad = \frac{4b}{3\pi}$$

$\therefore$ C.G. of given area is $\left(\frac{4a}{3\pi}, \frac{4b}{3\pi}\right)$.

Example 5. *Find the C.G. of the area between the curve $y^2(2a-x) = x^3$ and its asymptote.* **(P.T.U., May 2014)**

Sol. Equation of the curve is $y^2(2a-x) = x^3$ ...(1)

The curve is symmetric about x-axis. It passes through the origin the tangent at the origin is (equate lowest degree terms to zero) $y^2 = 0$ *i.e.*, $y = 0$ *i.e.*, x-axis.

Asymptote parallel to y-axis is (equate coeff. of highest power of y to zero) $2a - x = 0$ *i.e.*, $x = 2a$.

We have to find C.G. of the area of the curve bounded between $x = 2a$ and y-axis.

Symmetry of the figure shows that C.G. lies on x-axis.

$\therefore$ If $\bar{x}$ is the abscissae of the C.G. of the area

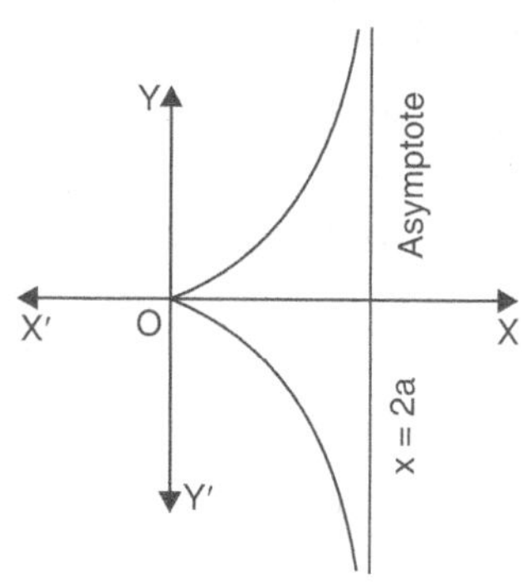

Then
$$\bar{x} = \frac{\int_0^{2a} xy\,dx}{\int_0^{2a} y\,dx}$$

$$= \frac{\int_0^{2a} x\cdot\frac{x^{3/2}}{\sqrt{2a-x}}\,dx}{\int_0^{2a}\frac{x^{3/2}}{\sqrt{2a-x}}\,dx}$$

Put $x = 2a \sin^2 \theta$
$dx = 4a \sin \theta \cos \theta \, d\theta$

$$\therefore \qquad \bar{x} = \frac{\int_0^{\pi/2} (2a \sin^2 \theta)^{5/2} \dfrac{4a \sin \theta \cos \theta \, d\theta}{\sqrt{2a} \cos \theta}}{\int_0^{\pi/2} (2a \sin^2 \theta)^{3/2} \dfrac{4a \sin \theta \cos \theta \, d\theta}{\sqrt{2a} \cos \theta}}$$

$$= \frac{\int_0^{\pi/2} 16a^3 \sin^6 \theta \, d\theta}{\int_0^{\pi/2} 8a^2 \sin^4 \theta \, d\theta} = 2a \frac{\dfrac{5 \cdot 3 \cdot 1}{6 \cdot 4 \cdot 2} \dfrac{\pi}{2}}{\dfrac{3 \cdot 1}{4 \cdot 2} \dfrac{\pi}{2}} = \frac{5}{3} a.$$

$\therefore$ Co-ordinates of C.G. of the area are $\left(\frac{5}{3}a, 0\right)$.

C.G. of Sectorial Areas:

Example 6. *Show that the centroid of the area bounded by the cardioid $r = a(1 + \cos \theta)$ is on the initial line at a distance $\frac{5a}{6}$ from the pole.*

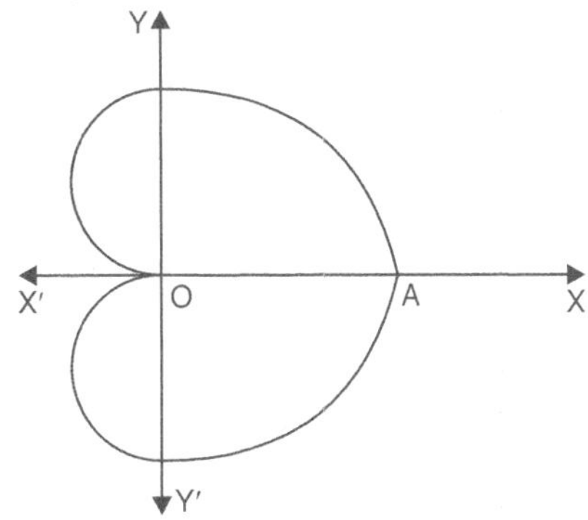

Sol. Equation of the cardioid is

$$r = a(1 + \cos \theta)$$

$$\frac{dr}{d\theta} = -a \sin \theta$$

C.G. of the area of cardioid is

$$\bar{x} = \frac{\frac{2}{3}\int_0^{\pi} r^3 \cos \theta \, d\theta}{\int_0^{\pi} r^2 \, d\theta}, \quad \bar{y} = \frac{\frac{2}{3}\int_0^{\pi} r^3 \sin \theta \, d\theta}{\int_0^{\pi} r^2 \, d\theta}$$

$$\bar{x} = \frac{2}{3} \frac{\int_0^{2\pi} a^3 (1 + \cos \theta)^3 \cos \theta \, d\theta}{\int_0^{2\pi} a^2 (1 + \cos \theta)^2}; \qquad \bar{y} = \frac{2}{3} \frac{a^3 \int_0^{2\pi} (1 + \cos \theta)^3 \sin \theta \, d\theta}{\int_0^{2\pi} a^2 (1 + \cos \theta)^2 \, d\theta}$$

$$\bar{x} = \frac{2}{3} a \frac{\int_0^{2\pi} \left(2 \cos^2 \frac{\theta}{2}\right)^3 \left(2 \cos^2 \frac{\theta}{2} - 1\right) d\theta}{\int_0^{2\pi} \left(2 \cos^2 \frac{\theta}{2}\right)^2 d\theta}; \qquad \bar{y} = \frac{2}{3} a \frac{-\dfrac{(1 + \cos \theta)^4}{4}\Bigg|_0^{2\pi}}{\int_0^{2\pi} (1 + \cos \theta)^2 \, d\theta} = 0$$

Put $\frac{\theta}{2} = t$; $\theta = 2t \; \therefore \; d\theta = 2dt$

$$= \frac{2}{3} a \frac{\int_0^{\pi} 8 \cos^6 t (2 \cot^2 t - 1) \, 2dt}{\int_0^{\pi} 4 \cos^4 t \, . \, 2dt}$$

$$= \frac{4}{3} a \frac{\int_0^{\pi} (2 \cos^8 t - \cos^6 t) \, dt}{\int_0^{\pi} \cos^4 t \, dt} = \frac{4}{3} a \frac{2 \int_0^{\pi/2} (2 \cos^8 t - \cos^6 t) \, dt}{2 \int_0^{\pi/2} \cos^4 t \, dt}$$

$$\left| \because \int_0^{2a} f(x) \, dx = 2 \int_0^{a} f(x) \, dx \text{ if } f(2a - x) = f(x) \right|$$

$$= \frac{4}{3}a \frac{2\times\left\{\frac{7\cdot5\cdot3\cdot1}{8\cdot6\cdot4\cdot2}\frac{\pi}{2} - \frac{5\cdot3\cdot1}{6\cdot4\cdot2}\frac{\pi}{2}\right\}}{\frac{3\cdot1}{4\cdot2}\frac{\pi}{2}}$$

$$= \frac{4}{3}a\cdot\frac{5\cdot3\cdot1}{6\cdot4\cdot2}\frac{8}{3}\cdot\left(\frac{7}{4}-1\right) = \frac{5}{6}a$$

$$\therefore \quad \bar{x} = \frac{5}{6}a, \qquad \bar{y} = 0$$

So C.G. lies on the initial line and is at a distance $\frac{5}{6}a$ from the pole.

Example 7. *Find the centroid of a loop of the lemniscate* $r^2 = a^2 \cos 2\theta$

Sol. The loop being symmetrical about the initial line *i.e.,* x-axis $\therefore$ its C.G. lies on OX, so that $\bar{y} = 0$ $\therefore$ C.G. of the loop of lemniscate is

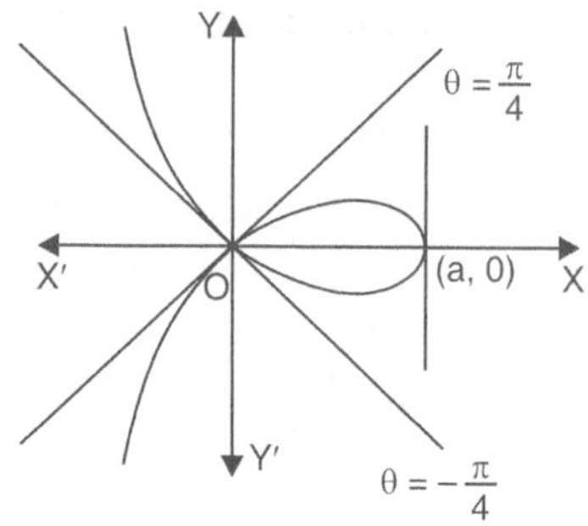

$$\bar{x} = \frac{\frac{2}{3}\int_{-\frac{\pi}{4}}^{\frac{\pi}{4}} r^3 \cos\theta\, d\theta}{\int_{-\frac{\pi}{4}}^{\frac{\pi}{4}} r^2\, d\theta}$$

$$= \frac{2}{3}\frac{\int_{-\frac{\pi}{4}}^{\frac{\pi}{4}} a^3(\cos 2\theta)^{3/2}\cos\theta\, d\theta}{\int_{-\frac{\pi}{4}}^{\frac{\pi}{4}} a^2\cos 2\theta\, d\theta}$$

$$= \frac{2}{3}a\frac{2\int_0^{\frac{\pi}{4}} (\cos 2\theta)^{3/2}\cos\theta\, d\theta}{2\int_0^{\frac{\pi}{4}} \cos 2\theta\, d\theta}$$

[$\because$ Integrand is even function

$\therefore$ Use $\int_{-a}^{a} f(x)\,dx = 2\int_0^a f(x)\,dx$]

$$= \frac{2}{3}a\frac{\int_0^{\frac{\pi}{4}} (1-2\sin^2\theta)^{3/2}\cos\theta\, d\theta}{\left.\frac{\sin 2\theta}{2}\right|_0^{\frac{\pi}{4}}}$$

$$= \frac{4a}{3}\int_0^{\frac{\pi}{4}} (1-2\sin^2\theta)^{3/2}\cos\theta\, d\theta$$

Put $\sqrt{2}\sin\theta = \sin t$

$\sqrt{2}\cos\theta\, d\theta = \cos t\, dt$

$$\therefore \quad \bar{x} = \frac{4a}{3}\int_0^{\frac{\pi}{2}} (1-\sin^2 t)^{3/2}\frac{\cos t}{\sqrt{2}}\,dt = \frac{2\sqrt{2}a}{3}\int_0^{\frac{\pi}{2}} (\cos^2 t)^{3/2}\cos t\, dt$$

$$= \frac{2\sqrt{2}a}{3}\int_0^{\frac{\pi}{2}} \cos^4 t\, dt = \frac{2\sqrt{2}a}{3}\,\frac{3\cdot 1}{4\cdot 2}\,\frac{\pi}{2} = \frac{\sqrt{2}a}{8}\pi$$

Hence C.G. of the loop of the lemniscate is $\left(\frac{\sqrt{2}a}{8}\pi, 0\right)$

Note. We can find C.G. by double integral method see S.E. 7 art 6.12.

C. G. of Volumes:

Example 8. *Find the C.G. of a uniform solid hemisphere of radius a.*

Sol. Let ACB be the solid hemisphere and OC, its axis taken as x-axis. As hemisphere is symmetric about x-axis

$\therefore \quad \bar{y} = 0$

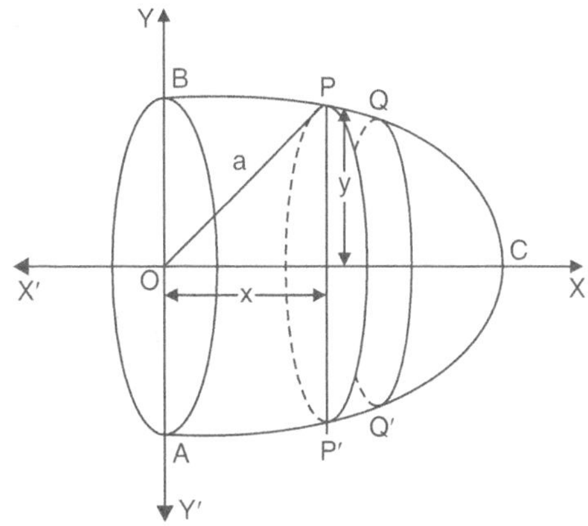

Let PP′ and QQ′ be two dises of the hemisphere (very close to one another) $\therefore$ if P is (x, y) then Q is $(x + \delta x, y + \delta y)$

Elementary volume of the hemisphere = $\pi y^2 . \delta x$ (it is a cylindrical shape surface)

$$= \pi(a^2 - x^2)\,\delta x \quad \because \quad x^2 + y^2 = a^2$$

Let m be the mass per unit volume

$$\therefore \qquad \delta m = \pi(a^2 - x^2)\, m\, \delta x$$

$$\therefore \qquad \bar{x} = \frac{\Sigma x \cdot \delta m}{\Sigma \delta m}, \; \bar{y} = 0$$

$$\bar{x} = \frac{\Sigma x\, \pi(a^2 - x^2) m\, \delta x}{\Sigma\, \pi (a^2 - x^2) m\, \delta x} = \frac{\pi m \int_0^a x(a^2 - x^2)\, dx}{\pi m \int_0^a (a^2 - x^2)\, dx}$$

$$= \frac{\left. a^2 . \frac{x^2}{2} - \frac{x^4}{4} \right|_0^a}{\left. a^2 x - \frac{x^3}{3} \right|_0^a} = \frac{\frac{a^4}{2} - \frac{a^4}{4}}{a^3 - a^3/3} = \frac{a^4}{4\cdot \frac{2a^3}{3}} = \frac{3a}{8}$$

Hence C.G. of a hemisphere lies on the axis of the hemisphere and is at a distance $\frac{3a}{8}$ from the centre of the plane face of the hemisphere.

Aliter. We can regard the hemisphere as generated by revolving through x-axis the arc of the circle $x^2 + y^2 = a^2$ lying in the first quadrant.

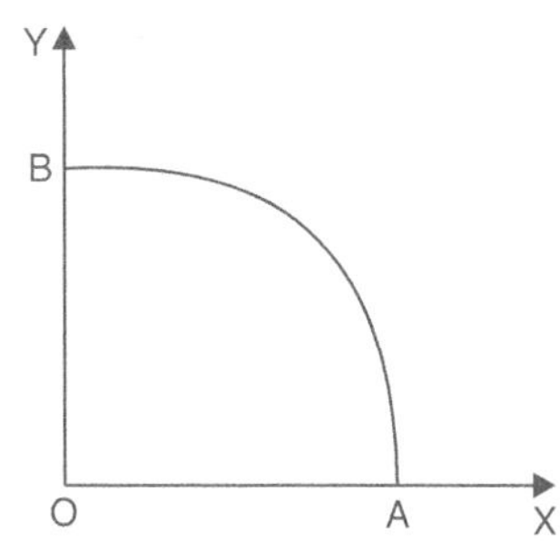

$$\therefore \qquad \bar{x} = \frac{\int_0^a xy^2\, dx}{\int_0^a y^2\, dx} = \frac{\int_0^a x[a^2 - x^2]\, dx}{\int_0^a (a^2 - x^2)\, dx} = \frac{\left. \frac{a^2x^2}{2} - \frac{x^4}{4} \right|_0^a}{\left. a^2 x - \frac{x^3}{3} \right|_0^a}$$

$$= \frac{a^4}{4} \cdot \frac{3}{2a^3} = \frac{3}{8}\, a.$$

Example 9. *Find the C.G of a solid cone.*

Sol. As cone is symmetric about its axis taken as x-axis

$\therefore \quad \bar{y} = 0$

Let PP′, QQ′ be two strips (discs) of the cone at a distance x and $x + \delta x$ from O then δm = elementary mass of the cone

$= (\pi y^2 \delta x)m$, where m is the mass per unit volume

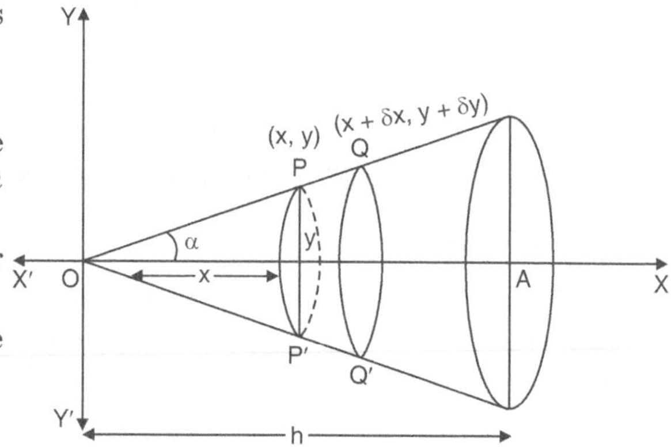

Let α be the semi vertical angle of the cone and h, be the height then $y = x \tan \alpha$.

$\therefore \quad \delta m = \pi x^2 \tan^2 \alpha\, m \,.\, \delta x$

$$\therefore \quad \bar{x} = \frac{\Sigma x \cdot \delta m}{\Sigma \delta m} = \frac{\Sigma\, \pi x^3 \tan^2 \alpha\, m\, \delta x}{\Sigma\, \pi x^2 \tan^2 \alpha\, m\, \delta x}$$

$$= \frac{\pi\, m \tan^2 \alpha \int_0^h x^3\, dx}{\pi\, m \tan^2 \alpha \int_0^h x^2\, dx} = \frac{x^4/4 \Big|_0^h}{x^3/3 \Big|_0^h} = \frac{3}{4}h$$

$\therefore$ C.G. of the cone lies on the axis of the cone and is at a distance $\frac{3}{4}h$ from the vertex of the cone or C.G. divides the axis in the ratio 3 : 1.

C.G. of Surface Areas:

Example 10. *Find the centroid of a hemispherical surface.*

Sol. Let ACB be the hemispherical surface and OC its axis is taken as x-axis. Divide the hemispherical surface into a number of strips (rings) drawn perpendicular of its axis.

Let one such strip PP′ be at a distance x for O and the other strip QQ′ be at a distance $x + \delta x$ for O. Let $\angle QOX = \theta$ and $\angle POX = \theta + \delta\theta$. Surface area of the elementary portion of the hemisphere *i.e.*, PP′ QQ′ = $2\pi(\text{PL})\, \overset{\frown}{PQ}$

Surface area of a cylinder = $2\pi rh$

$= 2\pi(a \sin \theta) \cdot \overset{\frown}{PQ}$

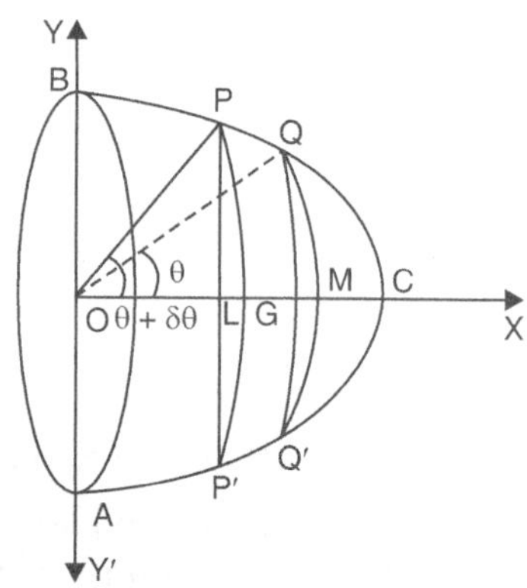

We know that $\quad \delta\theta = \dfrac{\overset{\frown}{PQ}}{a}$

$\therefore \quad \overset{\frown}{PQ} = a\delta\theta$ elementary

$\therefore \quad$ Surface area $= 2\pi(a \sin \theta) \cdot a\, \delta\theta$

Let m be the mass per unit area

$\therefore$ Mass of elementary surface area $= 2\pi a^2 \sin\theta\, \delta\theta\, m$;

Symmetry of the figure shows that C.G. will lie on x-axis, the axis of the hemisphere

$\therefore$ if $\quad \bar{x} = \text{OG}$

then $$\bar{x} = \frac{\Sigma\, (2\pi a^2 \sin \theta\, \delta\theta\, m)\, (a \cos \theta)}{\Sigma\, (2\pi a^2 \sin \theta\, \delta\theta\, m)} \qquad \Big|\ \because\ x = a \cos \theta$$

As hemisphere is a continuous body so change summation to integration, we get

$$\bar{x} = \frac{\int_0^{\pi/2} (2\pi am) \sin\theta \cos\theta \, d\theta}{\int_0^{\pi/2} (2\pi am) \sin\theta \, d\theta} = \frac{a \left. \frac{\sin^2\theta}{2} \right|_0^{\pi/2}}{\left. -\cos\theta \right|_0^{\pi/2}}$$

$$\bar{x} = \frac{\frac{a}{2}\cdot 1}{1} = \frac{a}{2}$$

∴ C.G. divides the axis of the hemispherical surface in the ratio 1 : 1.

Example 11. *Find C.G. of the curved surface of a right circular cone.*

Sol. Let OAB be the right circular cone with vertex at O and semi vertical angle α, height h. Divide the cone into a number of circular rings drawn perpendicular to the axis OL of the cone taken as x-axis. Let one such ring PP′ be at a distance x from O and another ring QQ′ be at a distance $x + \delta x$ from O.

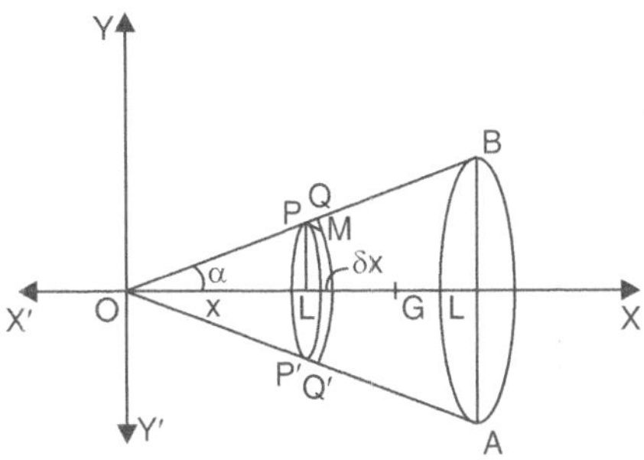

∴ Elementary surface area of the cone

PP′ Q′Q = 2π PL PQ = $2\pi . x \sin\alpha$. (PM $\sec\alpha$)

$= 2\pi x \sin\alpha \,.\, \delta x \sec\alpha$

Let m be the mass per unit area.

∴ Mass of the elementary portion of the curved surface of cone

$= (2\pi \tan\alpha \sec\alpha \, m)\, x \, \delta x$

Symmetry of the figure shows that the C.G. of the cone will lie on x-axis, its axis.

∴ If $\bar{x}$ = OG then $\bar{x} = \dfrac{\Sigma(2\pi \sin\alpha \sec\alpha \, m)\, x \, \delta x \,.\, x}{\Sigma\, 2\pi \sin\alpha \sec\alpha \, m \,.\, x \, \delta x}$

$$= \frac{2\pi \tan\alpha \sec\alpha \, m \int_0^h x^2 \, dx}{2\pi \tan\alpha \sec\alpha \,.\, m \int_0^h x \, dx} \quad \text{(as cone is a continuous body)}$$

$$= \frac{\left. x^3/3 \right|_0^h}{\left. x^2/2 \right|_0^h} = \frac{2}{3} h = \frac{2}{3}\text{OL}$$

∴ $\text{GL} = \text{OL} - \text{OG} = \text{OL} - \frac{2}{3}\,\text{OL} = \frac{1}{3}\text{OL}$

∴ $\dfrac{\text{OG}}{\text{GL}} = \dfrac{2}{1}$

∴ G divides the axis of the cone in the ratio 2 : 1.

TEST YOUR KNOWLEDGE

1. Find the C.G. of the arc of a circle of radius 'a' subtending an angle 2θ at its centre.
2. Show that the centroid of the arc of the cycloid $x = a(\theta + \sin\theta)$, $y = a(1 - \cos\theta)$ in the positive quadrant is $\left\{\left(\pi - \frac{4}{3}\right)a, \frac{2}{3}a\right\}$.
3. Show that the centroid of the area bounded by the parabola $x^2 = 4ay$, x-axis and the ordinate $x = b$ is $\left(\frac{3b}{4}, \frac{3}{40}\frac{b^2}{a}\right)$.
4. Find the C.G. of the area under one arch of the sine curve $y = \sin x$.
5. Find the C.G. of the area in the first quadrant bounded by the axes and the curve $y = 2 + x - x^2$.
6. Show that the centroid of the area bounded by the loop of the curve $y^2(a + x) = x^2(a - x)$ is $\frac{a}{3}(3\pi - 8)/4 - \pi$
7. Find the C.G. of the area between $y = 6x - x^2$ and $y = x$.
8. Find the C.G. of the area of the circle $x^2 + y^2 = a^2$ lying in the first quadrant.
9. Find the centre of gravity of the area in first quadrant lying between the curve $y^2 = x^3$ and line $y = x$

   [**Hint.** For curve see Q 13 art 1.13]
10. Find the C.G. of the area bounded by the parabola $y^2 = x$ and line $x + y = 2$.
11. A flat thin plate of uniform density is bounded by the two curves $y = x^2$, and $y = -x^2$ and the line $x = 2$. Find the co-ordinates of the C.G. of the plate.
12. Find the C.G. of the solid formed by revolution about x-axis of the parabola $y^2 = 4ax$ cut off by the ordinate $x = b$.
13. Find the C.G. of the solid formed by revolving about x-axis the area bounded by the lines $y = 0$, $x = \frac{\pi}{4}$ and the curve $y = \sin x$.
14. Determine the C.G. of the area bounded by the parabolas $y^2 = ax$, $x^2 = by$.
15. A quadrant of the ellipse $\frac{x^2}{a^2} + \frac{y^2}{b^2} = 1$ revolves about the major axis, find the C.G. of the solid thus obtained.

Answers

1. at a distance $\frac{a\sin\theta}{\theta}$ from the centre on the bisecting radius.

4. $\left(\frac{\pi}{2}, \frac{\pi}{8}\right)$ 5. $\left(\frac{4}{5}, \frac{24}{25}\right)$ 7. $\left(\frac{5}{2}, 5\right)$

8. $\left(\frac{4a}{3\pi}, \frac{4a}{3\pi}\right)$ 9. $\left(\frac{10}{21}, \frac{5}{12}\right)$ 10. $\left(\frac{8}{5}, -\frac{1}{2}\right)$

11. $\left(\frac{3}{2}, 0\right)$ 12. $\bar{x} = \frac{2}{3}b$ 13. $\bar{x} = (\pi^2 - 4\pi + 8)/8(\pi - 2)$

14. $\left(\frac{9}{20}a^{1/3}b^{5/3}, \frac{9}{20}a^{2/3}b^{1/3}\right)$ 15. $\bar{x} = \frac{3a}{8}$.

2.20. MOMENTS OF INERTIA

Def. If a particle of mass m be at a distance r from a given line, then mr^2 is called the moment of inertia of the particle about the given line. Again the moment of inertia about a given line of a finite number of particles with masses $m_1, m_2, \ldots m_n$ at distances $r_1, r_2, \ldots r_n$ from the given line is $\sum_{i=1}^{n} m_i r_i^2$.

For obtaining the moment of inertia of a body of mass M, we divide the body into a number of elementry portions and if δm represents the mass of this portion then M.I. of this portion = $\delta m \cdot r^2$ where r is distance of δm from the given line then M.I. of the whole body = $\Sigma \delta m \cdot r^2 = \int r^2 \, dm$. The process is indicated by the following examples:

Note. Moments of inertia of a body can also be obtained by double integration discussed in the next chapter 'Multiple Integrals article 6.13

ILLUSTRATIVE EXAMPLES

Example 1. *Find the moment of Inertia of a thin uniform rod of length 2a about (i) a line through its one extremity perpendicular to the rod (ii) about the line through its middle point perpendicular to the rod.* **(P.T.U., Dec. 2011)**

Sol. (*i*) Let AB the given line of length $2a$ be taken as x-axis and a line $\perp$ to it as y-axis. Divide the rod into elementry parts and let PQ be one such part so that if AP = x then

$$AQ = x + \delta x \quad \therefore \quad PQ = \delta x.$$

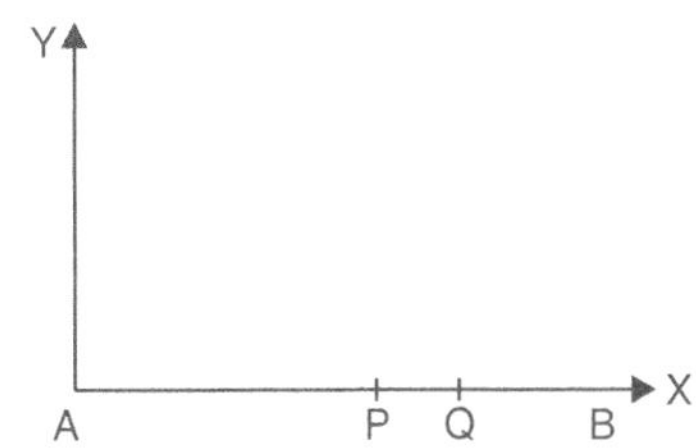

Let m be the mass per unit length

$$\therefore \qquad \delta m = m \,.\, \delta x$$

Moment of inertia of PQ about AY

$$= (m\delta x)\,(\text{AP})^2 = (m\delta x)x^2$$

Moment of inertia of the whole rod

$$= \Sigma(m\delta x)x^2 = \int_0^{2a} mx^2 \, dx$$

[$\because$ Rod is a continuous body $\therefore$ summation can be changed to integral]

$$= m \cdot \left.\frac{x^3}{3}\right|_0^{2a} = \frac{8am}{3} = \frac{4a^2}{3}(2am)$$

$$= \frac{4a^2}{3} \text{ M where M is the mass of the rod} \qquad (\because \quad \text{M} = 2am.)$$

(*ii*) Let mid point of AB be taken as origin and OY $\perp$ to AB as y-axis then M.I. of the rod about OY.

$$= \int_{-a}^{a} mx^2 \, dx = 2\int_0^{a} mx^2 \, dx$$

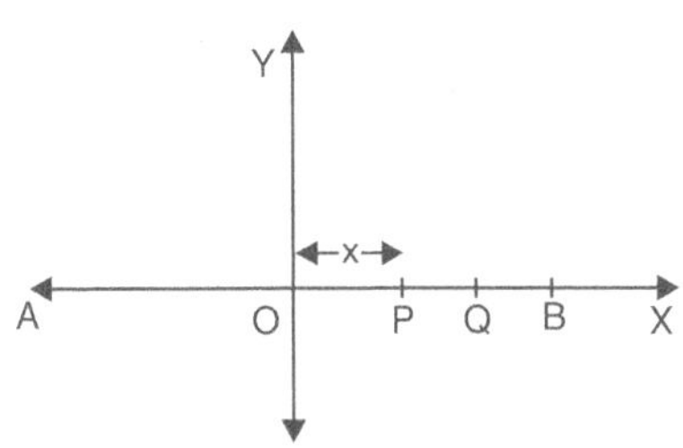

$$= 2m\frac{a^3}{3} = \frac{a^2}{3}(2ma)$$

$$= \frac{Ma^2}{3}.$$

Example 2. *Find the moment of inertia of a rectangular area A of dimensions a and b with respect to a side.* **(P.T.U., May 2010)**

Sol. Let OABC be the given rectangle with sides a and b in lengths. Take OA as x-axis and OC as y-axis.

Divide the rectangular lamina into a number of strips drawn parallel to y-axis and let PQQ′P′ be one such strip

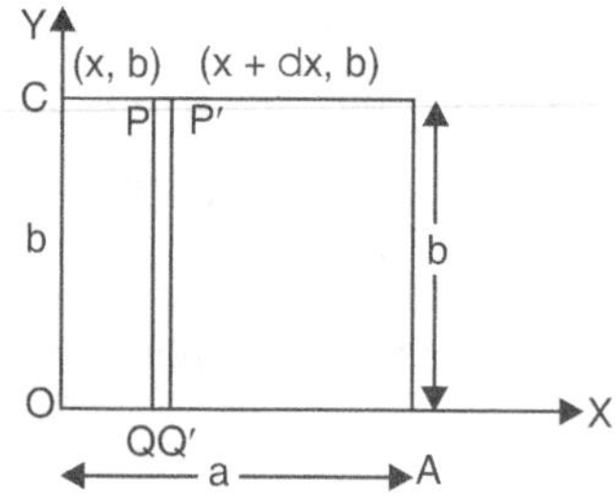

$\therefore$ Area of PQQ′P′ = $b\ \delta x$ as P is (x, b) and P′ is $(x + \delta x, b)$

Let ρ be the density

$\therefore$ Elementary mass, $\delta m = b\ \delta x \rho$

M.I. of the strip about y-axis = $(b\ \delta x\ \rho)x^2$

M.I. of the whole rectangular lamina

$$= \Sigma b\rho\, x^2\, \delta x$$

$$= \int_0^a b\rho x^2\, dx = b\rho\frac{a^3}{3} = \frac{a^2}{3}(ab)\,\rho = \frac{a^2}{3}\,\text{A}\rho$$

$$= \frac{a^2}{3}\text{M} \text{ where M is the mass of the rectangular lamina}$$

Hence M.I. = $\frac{1}{3}$ M (other side of the rectangle)^2.

Example 3. *Find the moment of inertia of a thin uniform circular disc of radius 'a' about any diameter thereof.*

Sol. Let AB be the diameter about which M.I. of the whole disc is to taken. Regard AB as y-axis and a line through O, (the mid pt. of AB) $\perp$ to AB as x-axis.

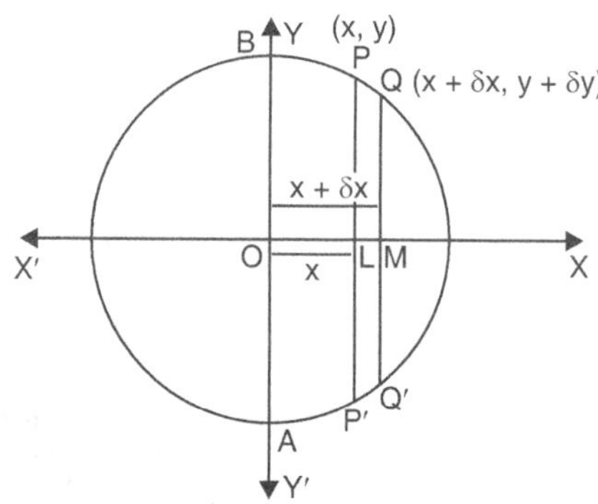

Divide the disc into a number of rods drawn parallel to AB and let PP′, QQ′ be two such rods so that if distance of PP′ from O is x then distance of QQ′ from O is $x + \delta x$, let PL = y, QM = $y + \delta y$

$\therefore$ P is (x, y) and Q is $(x + \delta x, y + \delta y)$.

Area of the elementary portion of the disc = area of the rectangle PP′Q′Q

$$= (2y)\,\delta x$$

Let m be the mass per unit area

$$\therefore \qquad \delta m = 2y\delta x \cdot m$$

$\therefore$ Moment of inertia of δm about AB

$$= (\delta m)\left(x + \frac{\delta x}{2}\right)^2$$

$$= (2y\delta x m)\cdot x^2 \text{ approximately.}$$

Moment of inertia of whole disc about AB = $\Sigma 2y\delta x m x^2$

As disc is a uniform continuous body

$\therefore$ Moment of inertia $= \int_{-a}^{a} 2x^2 ym\, dx$

$$= 2m\int_{-a}^{a} x^2\sqrt{a^2 - x^2}\, dx$$

$$= 4m\int_{0}^{a} x^2\sqrt{a^2 - x^2}\, dx$$

$\because$ Equation of circular disc is $x^2 + y^2 = a^2$

Put $x = a \sin \theta$

$dx = a \cos \theta\, d\theta$

$$= 4m\int_{0}^{\pi/2} a^2 \sin^2 \theta \cdot a \cos \theta\, a \cos \theta\, d\theta$$

$$= 4m \cdot a^4 \int_{0}^{\pi/2} \sin^2 \theta \cos^2 \theta\, d\theta = 4ma^4\, \frac{1\cdot 1}{4\cdot 2}\frac{\pi}{2}$$

$$= \frac{\pi ma^4}{4} = \frac{a^2}{4}(\pi a^2 m) = \frac{a^2}{4}\text{M}$$ where M is the mass of the circular disc.

Hence M.I. of the circular disc about any diameter $= \dfrac{\text{M}a^2}{4}$.

Example 4. *Prove that the moment of inertia of a uniform circular disc of mass M and radius 'a' about the line through its centre perpendicular to its plane is* $\dfrac{Ma^2}{4}$.

Sol. Let O be the centre of the given disc and l be any line through O $\perp$ to the plane of the disc. Divide the circular disc into a number of concentric rings and let one ring be at a distance x from O and the other ring be at a distance $x + \delta x$ from O. Then the elementary area of the circular disc $= (2\pi x) \cdot \delta x$ (circumference of the ring × thickness of the ring)

Let m be the mass per unit area

$\therefore$ Elementary mass $= \delta m = 2\pi x\, \delta x$ m

M.I. of δm about the line $l = \delta m.\, x^2 = 2\pi x\delta x\, mx^2$

M.I. of the whole disc about $l = \Sigma 2\pi x\delta x m x^2$

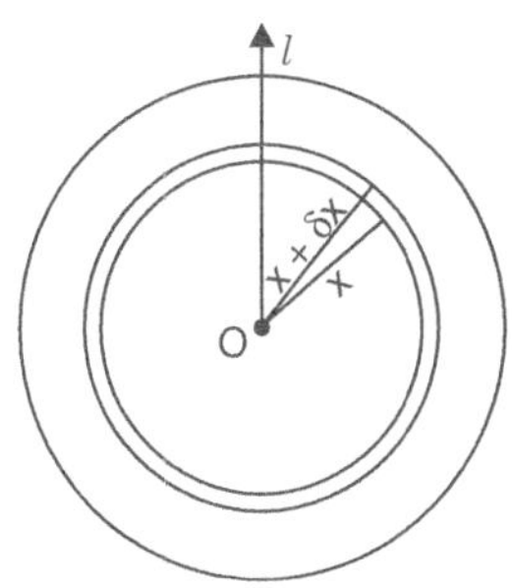

$$= \int_{0}^{a} 2\pi m\, x^3\, dx = 2\pi m \cdot \frac{x^4}{4}\Bigg|_0^a = \frac{\pi m a^4}{2}$$

$$= \frac{a^2}{2}(\pi a^2 m) = \frac{\text{M}a^2}{2}.$$

Example 5. *Find M.I. of a uniform circular ring of mass M about a line through its centre perpendicular to its plane.*

Sol. Let O be the centre of the circular ring and l be the line $\perp$ to the plane of the circular ring. Draw a line through O perpendicular to l. Divide the circular arc into a number of elementary arcs and let $\overset{\frown}{PQ}$ be one such arc so that if OP makes an angle θ with OX, then OQ makes an angle $\theta + \delta\theta$ with OX

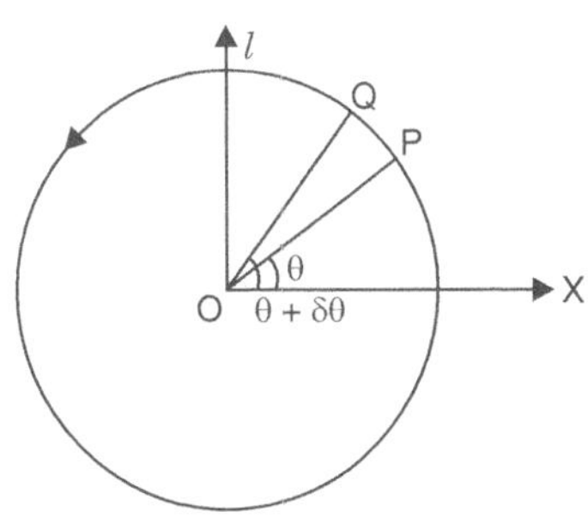

$\therefore$ $\angle POQ = \delta\theta$

$\therefore$ $\overset{\frown}{PQ} = a\delta\theta$ $(\because\ \theta = l/r)$

Let m be the mass per unit length then $\delta m = a\delta\theta\, m$

M.I. of δm about $l = (a\delta\theta\, m)a^2$

M.I. of whole circular ring about l

$$= \Sigma a^3\, m\delta\theta = \int_0^{2\pi} a^3 m\, d\theta = a^3 m \cdot \theta \Big|_0^{2\pi} = 2a^3 m\pi$$

$$= a^2\,(2\pi a m) = a^2\text{ M where M is the mass of the circular ring.}$$

Hence M.I. of circular ring about a line though the centre $\perp$ to its plane $= a^2$M.

Note. Students are advised to remember this result, it is going to be used in three dimensional hollow bodies.

Example 6. *Find the moment of inertia about y-axis of a uniform lamina of density* ρ *bounded by those parts of x-axis and the curve* $y = \cos x$ *which lie between* $x = -\dfrac{\pi}{2}$ *and* $x = \dfrac{\pi}{2}$.

Sol. Divide the region included between $y = \cos x$, $x = -\dfrac{\pi}{2}, x = \dfrac{\pi}{2}$ into a number of strips drawn parallel to y-axis Let PQQ′ P′ be one such elementary strip whose thickness is δx *i.e.*, if P has coordinates (x, y) P′ has coordinates $(x + \delta x, y + \delta y)$ thus area of this elementary strip is $y\,.\,\delta x$

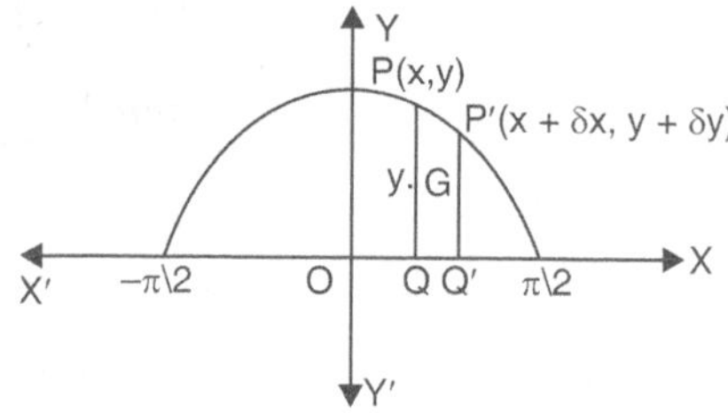

$\therefore$ Elementary mass of PQQ′ P′ $= y\delta x \cdot \rho$ and it acts at G, the C.G. of rectangle whose distance for y-axis $= x + \dfrac{\delta x}{2} = x$ app.

$\therefore$ Moment of inertia of this strip about y-axis $= (y\ \delta x\ \rho) \cdot x^2$

Moment of inertia of the whole region $= \Sigma y\delta x\rho x^2$

$$= \int_{-\pi/2}^{\pi/2} yx^2\ \rho\, dx = \rho\int_{-\pi/2}^{\pi/2} x^2 \cos x\, dx = 2\rho\int_0^{\pi/2} x^2 \cos x\, dx$$

$$\left[\because\ x^2 \cos x \text{ is even function } \therefore\ \int_{-\frac{\pi}{2}}^{\frac{\pi}{2}} x^2 \cos x\, dx = 2\int_0^{\frac{\pi}{2}} x^2 \cos x\, dx\right]$$

$$= 2\rho\left\{x^2\,(\sin x) - (2x)(-\cos x) + (2)(-\sin x)\right\}\Big|_0^{\pi/2}$$

$$= 2\rho\left[\frac{\pi^2}{4} - 2\right] = \rho\,\frac{\pi^2 - 8}{2}.$$

Now we will find M.I. of three dimensional uniform bodies.

Example 7. *Find the M.I. of a uniform solid sphere of radius 'a' about a diameter.*

Sol. Let AB be the diameter about which M.I. is to be taken. Take AB as x-axis, O its centre and OY as y-axis (shown in fig.). Divide the sphere into a number of circular discs drawn perpendicular to AB. Let PP′ be one such disc at a distance x for O and QQ′ be another disc at a distance $x + \delta x$ from O let PL $= y$, QM $= y + \delta y$. Then PP′ Q′Q is elementary volume of the sphere.

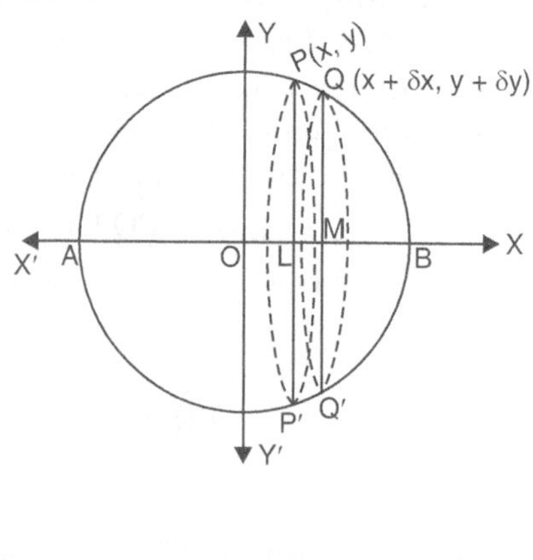

$\therefore$ Volume of PP′Q′Q = $\pi y^2\, \delta x$ (PP′ Q′Q is a cylindrical body)

Let m be the mass per unit volume.

$$\therefore \qquad \delta m = \pi y^2\, \delta x\, m$$

Now M.I. of δm about AB means M.I. of a circular disc of thickness δx about a line through its centre $\perp$ to its plane

$$= \frac{Ma^2}{2} \qquad \text{(see example 4)}$$

M.I. of δm about $\quad \text{AB} = \dfrac{\delta m \cdot y^2}{2}$

M.I. of the whole sphere about AB

$$= \frac{\Sigma \delta m y^2}{2}$$

$$= \Sigma \frac{y^2}{2} \pi y^2 \delta x m = \int_{-a}^{a} \frac{\pi m}{2} y^4\, dx = \int_{-a}^{a} \frac{\pi m}{2} (a^2 - x)^2\, dx \quad (\because \quad x^2 + y^2 = a^2)$$

$$= \frac{2\pi m}{2} \int_0^a (a^4 - 2a^2x^2 + x^4)\, dx$$

$$= \pi m \int_0^a (a^4 - 2a^2x^2 + x^4)\, dx = \pi m \left[a^4 x - \frac{2a^2x^3}{3} + \frac{x^5}{5} \right]_0^a$$

$$= \pi m\, \frac{8}{15} a^5 = \frac{2}{5} a^2 \left(\frac{4}{3} \pi a^3 m \right) = \frac{2}{5} a^2\, \text{M}$$

where M is the mass of the sphere.

Example 8. *Find the M.I. of a thin hollow sphere of radius 'a' about any diameter.*

Sol. Let AB be the diameter of the sphere about which M.I. is to be obtained. From O, the centre of the sphere, draw a line $\perp$ to AB and take it as y-axis. Divide the hollow sphere into a number of circular rings drawn $\perp$ to AB and let PP′, QQ′ be two such rings (very close to each other) then elementary surface area = area included between the rings PP′ and QQ′ = $2\pi\, \text{PL}\, \overset{\frown}{\text{PQ}}$; Let OQ makes an angle θ with OX and OP makes an angle $\theta + \delta\theta$ with OX

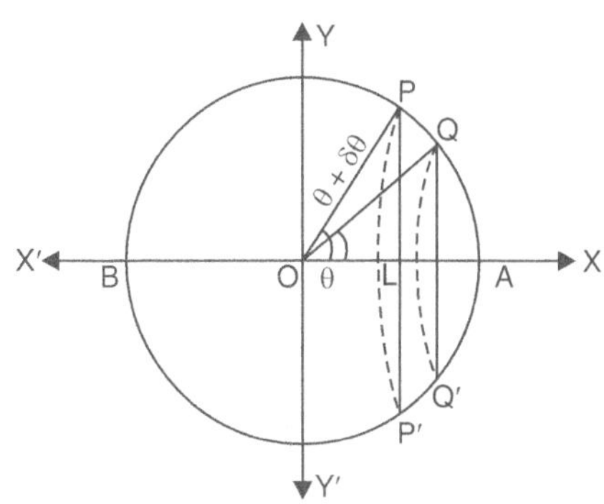

$$\therefore \qquad \delta\theta = \frac{\overset{\frown}{\text{PQ}}}{a}$$

$$\therefore \qquad \overset{\frown}{\text{PQ}} = a\, \delta\theta.$$

Let m be the mass per unit area

$$\therefore \qquad \delta m = 2\pi\,(a \sin\theta)\, a\, \delta\theta \cdot m$$

M.I. of δm about AB = M.I. of a circular ring of thickness $(a\ \delta\theta)$ about a line $\perp$ to its plane

$$= (\delta m)\, a^2 = 2\pi a^2\, m \sin\theta\, \delta\theta \cdot a^2 \quad \text{(See example 5)}$$

M.I. of the whole sphere $= \Sigma 2\pi m a^4 \sin\theta\, \delta\theta$

$$= \int_0^{\pi/2} 2\pi m a^4 \sin\theta\, d\theta$$

$$= 2\pi m a^4\, (-\cos\theta)\Big|_0^{\pi/2} = 2\pi m a^4$$

$$= \frac{a^2}{2}(4\pi a^2\, m) = \frac{a^2 \mathrm{M}}{2} \text{ where M is the mass of the sphere.}$$

TEST YOUR KNOWLEDGE

1. Find the M.I. of the area bounded by the parabola $x^2 = 4y$ and the line $y = x$ in first quadrant about y-axis.
2. Find the M.I. of the area between $y = \sin x$ from $x = 0$ to $x = \pi$ and x-axis about each axis.
3. Find the M.I. with respect to its axis of a right circular cylinder of height 'h' and radius of base 'a'.

Answers

1. $\frac{24}{5}$ M where M is the mass of the bounded region (M $= \rho$A where A $= \frac{8}{3}$ and M.I. $= \frac{64}{5}\rho$).
2. $\frac{4}{9}\rho$, $(\pi^2 - 4)\rho$ where ρ is the density.
3. $\frac{1}{2}$ Ma^2 where M is the mass of the cylinder (M $= \pi a^2 h\rho$).

REVIEW OF THE CHAPTER

1. Length of the arc of the curve is as follows:

(*i*) If the curve is $y = f(x)$; length of the arc of the curve $= \int_a^b \sqrt{1 + \left(\frac{dy}{dx}\right)^2}\, dx$

(*ii*) If the curve is $x = f(y)$; length of the arc of the curve $= \int_a^b \sqrt{1 + \left(\frac{dx}{dy}\right)^2}\, dy$

(*iii*) If the curve is in the parametric form *i.e.*, $x = f(t)$, $y = g(t)$, then length of the arc of the curve $= \int_{t=\alpha}^{t=\beta} \sqrt{\left(\frac{dx}{dt}\right)^2 + \left(\frac{dy}{dt}\right)^2}\, dt$

(*iv*) If the curve is in the polar form $r = f(\theta)$, then length of the arc of the curve

$$= \int_{\theta=\alpha}^{\beta} \sqrt{r^2 + \left(\frac{dr}{d\theta}\right)^2}\, d\theta$$

2. (*i*) Area bounded by $y = f(x)$; $x = a$, $x = b$ and x-axis is $\int_a^b y\,dx$. In case $f(x)$ is non positive continuous function then area $= \int_a^b |f(x)|\,dx$

(*ii*) If curve is $x = f(y)$; then area bounded between $y = a$, $y = b$ is $\int_{y=a}^{b} |f(y)|\,dy$

(*iii*) Area bounded between $y = f(x)$, $y = g(x)$; $x = a$, $x = b$ is $\int_a^b [f(x) - g(x)]\,dx$

(*iv*) If the curve is in the parametric form then area $= \int_{t=a}^{b} x\frac{dy}{dt}\,dt$

3. Volume obtained by revolution of

(*i*) The curve $y = f(x)$ bounded by $x = a$, $x = b$ about x-axis is $V = \int_a^b \pi y^2 dx$

(*ii*) The curve $x = f(y)$ bounded by $y = a$, $y = b$ about y-axis is then $V = \int_{y=c}^{d} \pi x^2\,dy$

(*iii*) The curve $y = f(x)$ bounded by $x = a$, $x = b$ revolves about a line $y = p$, then $V = \int_a^b \pi\,(y-p)^2\,dx$.

(*iv*) The curve $x = f(y)$ bounded by $y = a, y = b$ revolves about $x = q$ then $V = \int_a^b \pi\,(y-q)^2\,dy$

(*v*) The curve $y = f(x)$ bounded by $x = a$, $x = b$ revolves about a line l then $V = \int_{OL}^{OM} \pi\,(PN)^2\,d\,(ON)$ where O is any fixed point on the line l and L, M, N are the feet of the perpendiculars from A, B, P where A, B are extremities of the curve and P is any point on the curve.

4. (*i*) Surface area of a frustum of a right circular cone = π × slant side × sum of the radii of two bases.

(*ii*) Surface area of solid of revolution of $y = f(x)$; $x = a$, $x = b$ about x-axis is

$$S = \int_a^b 2\pi y\sqrt{1 + \left(\frac{dy}{dx}\right)^2}\,dx$$

(*iii*) If the curve is $x = f(y)$ and $y = a$, $y = b$ then $S = \int_a^b 2\pi x\sqrt{1 + \left(\frac{dx}{dy}\right)^2}\,dy$

(*iv*) If the curve is in parametric form $x = f(t)$, $y = g(t)$; $t_1 \le t \le t_2$, then

$$S = \int_{t_1}^{t_2} 2\pi g(t) \sqrt{\left(\frac{dx}{dt}\right)^2 + \left(\frac{dy}{dt}\right)^2}\, dt$$

(*v*) If the curve is in polar form $r = f(\theta)$; $\theta = \theta_1$, $\theta = \theta_2$, then

$$S = \int_{\theta_1}^{\theta_2} 2\pi (r \sin \theta) \sqrt{r^2 + \left(\frac{dr}{d\theta}\right)^2}\, d\theta$$

5. Pappus theorems:

(*i*) If a closed plane curve revolves about a line in its plane but not intersecting the curve then volume of revolution = Area of the region × length of the path described by the centroid of the region.

(*ii*) The surface of revolution = length of the curve × path described by the centroid of the curve.

6. Centre of gravity of a uniform plane curve

$$\bar{x} = \frac{\int_a^b x \frac{ds}{dx}\, dx}{\int_a^b \frac{ds}{dx}\, dx \text{ or } l}; \qquad \bar{y} = \frac{\int_a^b y \frac{ds}{dx}\, dx}{\int_a^b \frac{ds}{dx}\, dx \text{ or } l}$$

where l is the length of the curve.

7. C.G. of a uniform plane area

$$\bar{x} = \frac{\int_a^b xy\, dx}{\int_a^b y\, dx \text{ or A}}; \qquad \bar{y} = \frac{\int_a^b \frac{1}{2} y^2\, dx}{\int_a^b y\, dx \text{ or A}}$$

where A is the area of the given region.

8. C.G. of a sectorial area subtending an angle θ at the centre

$$\bar{x} = \frac{\int_{\theta=\alpha}^{\beta} \frac{2}{3} r^3 \cos\theta\, d\theta}{\int_{\theta=\alpha}^{\beta} r^2\, d\theta}; \qquad \bar{y} = \frac{\int_{\theta=\alpha}^{\beta} \frac{2}{3} r^3 \sin\theta\, d\theta}{\int_{\theta=\alpha}^{\beta} r^2\, d\theta}.$$

9. C.G. of a volume of the solid of revolution about x-axis

$$\bar{x} = \frac{\int_{x=a}^{b} xy^2\, dx}{\int_{x=a}^{b} y^2\, dx}; \qquad \bar{y} = 0.$$

10. C.G. of a surface area of the solid of revolution about x-axis

$$\bar{x} = \frac{\int_a^b xy \frac{ds}{dx}\, dx}{\int_a^b y \frac{ds}{dx}\, dx}; \qquad \bar{y} = 0.$$

11. Moment of inertia of a particle about a given line = (mass of the particle) × square of the distance of the particle from the given line.

12. (*i*) M.I. of a uniform rod of length $2a$ about a line through one extremity and ⊥ to it

$= \frac{4a^2}{3}$ where M is the mass of the rod.

(*ii*) M.I. of a uniform rod of length $2a$ about a line through its middle point and ⊥ to it

$= \frac{Ma^2}{3}$.

(*iii*) M.I. of a circular disc about its diameter $= \frac{Ma^2}{4}$ where M is the mass of the disc.

(*iv*) M.I. of a circular disc about a line passing through the centre of the disc and ⊥ to the plane of the disc $= \frac{Ma^2}{2}$.

(*v*) M.I. of a circular ring about a line through the centre ⊥ to the plane of the ring $= Ma^2$.

SHORT ANSWER TYPE QUESTIONS

1. Write formulae for the length of the arc of the curve in

   (*i*) Cartesian coordinates (*ii*) Parametric coordinates

   (*iii*) Polar coordinates [**Hint.** arts 2.2, 2.3] ***(P.T.U., Dec. 2006)***

2. Write down formulae for the area of the curve in

   (*i*) Cartesian co-ordinates (*ii*) Parametric coordinates

   (*iii*) Polar coordinates [**Hint.** arts 2.4, 2.5, 2.6]

3. Write formulae for the centre of gravity $(\overline{x}, \overline{y})$ of the following:

   (*i*) uniform plane curve

   [**Hint.** See art. 2.15]

   (*ii*) uniform plane area

   [**Hint.** See art. 2.16]

   (*iii*) sectorial area subtending an angle θ at the centre

   [**Hint.** See art. 2.17]

   (*iv*) volume of solid of revolution about x-axis

   [**Hint.** See art. 2.18]

   (*v*) surface area of the solid of revolution about x-axis.

   [**Hint.** See art. 2.19]

4. (*a*) Write down the lengths of the arcs of the following curves

(*i*) $y = \log \dfrac{e^x - 1}{e^x + 1}$ from $x = 1$ to $x = 2$. **(*P.T.U., Dec. 2004*)**

[**Hint.** S.E. 1 art. 2.3]

(*ii*) $y = \log \sec x$ between $x = 0, x = \dfrac{\pi}{3}$ **(*P.T. U., May 2007, Dec. 2011*)**

[**Hint:** S.E. 5 art. 2.3]

(*iii*) $x = a \cos^3 t, y = a \sin^3 t$ **(*P.T. U., Dec. 2014*)**

(*b*) Using integration, find the perimeter of the curve $x^2 + y^2 = 4$. **(*P.T.U., Jan. 2009*)**

5. Find a curve through (1, 1) whose length integral is

$L = \int_1^4 \sqrt{1 + \dfrac{1}{4x}}\, dx$. How many such curves are there? **(*P.T.U., Dec. 2004*)**

[**Hint.** S.E. 10 art. 2.3]

6. Show that the area bounded by the parabola $y^2 = 4ax$, and $x^2 = 4ay$ is $\dfrac{16}{3}a^2$.

(*P.T.U., Dec. 2012, May 2015*)

7. Find the area bounded by the ellipse $\dfrac{x^2}{a^2} + \dfrac{y^2}{b^2} = 1$, the ordinates $x = c$, $x = d$ and the x-axis. Deduce the area of the whole ellipse.

[**Hint.** S.E. 2 art. 2.7]

8. (*i*) Find the area of the region enclosed by the curve $y = x^2$ and the lines $x = 0, y = 0$ and $x = 2$. **(*P.T.U., May 2011*)**

(*ii*) Find the area of the region enclosed by the curve $y^2 = x$ and its latus rectum.

(*P.T.U., Dec. 2013*)

9. Using parametric equation of a circle show that area of the circle of radius a is πa^2.

[**Hint:** S.E. 15 art 2.7] **(*P.T.U., Dec 2005*)**

10. (*i*) Obtain the area of the loop of the curve $r = a \sin 3\theta$.

[**Hint:** S.E. 14 art 2.7]

(*ii*) Find the area of the cardioid $r = a\,(1 - \cos \theta)$ **(*P.T.U., May 2015*)**

$$\left[\textbf{Hint.}\ \text{Area} = 2 \cdot \frac{1}{2} \int_0^{\pi} a^2(1 - \cos\theta)^2\, d\theta = 4a^2 \int_0^{\pi} \sin^4 \frac{\theta}{2}\, d\theta \left(\because \frac{\theta}{2} = \phi \ \therefore\ d\theta = 2d\phi\right)\right.$$

$$\left. = 8a^2 \int_0^{\pi/2} \sin^4 \phi\, d\phi = 8a^2\, \frac{3.1}{4.2}\, \frac{\pi}{2} = \frac{3}{2}\pi a^2 \right]$$

(*iii*) Find the area enclosed by the leminiscate $r^2 = 4 \cos 2\theta$. **(*P.T.U., May 2012*)**

11. (*i*) Write a formula to find the volume of the solid generated by the revolution, about y-axis of the area bounded by the curve $x = f(y)$ the y-axis and the abscissae $y = a$ and $y = b$.

[**Hint.** Art 2.9 note I]

(*ii*) Find the volume of the solid obtained by revolution of the part of the parabola $y^2 = 4ax$ cut off by the latus rectum about y-axis ***(P.T.U., Dec. 2010)***

$$\left[\textbf{Hint.}\text{ Required volume } = 2\int_0^{2a} \pi \frac{y^4}{16a^2}\,dy\right]$$

12. Find the C.G. of the area bounded by the cardiod $r = a\,(1 + \cos\theta)$.

[**Hint.** S.E. 6 art. 2.19]

13. Find moment of inertia of a thin uniform rod of length $2a$ about

(*i*) a line through its one extremity prependicular to the rod. ***(P.T.U., Dec. 2011)***

(*ii*) about the line through its middle point perpendicular to the rod.

[**Hint.** S.E. 1 art. 2.20]

14. Find the M.I. of a rectangular lamina about its side ***(P.T.U., May 2010)***

[**Hint.** S.E. 2 art. 2.20]

15. Find M.I of a uniform circular ring of mass M about a line through its centre perpendicular to its plane.

[**Hint.** S.E. 4 art. 2.20]

Answers

4. (*a*) (*i*) $\log\left(e + \frac{1}{e}\right)$ (*ii*) $\log\,(2 + \sqrt{3})$ (*iii*) $6a$

(*b*) 4π

5. $y^2 = x$; $(y - 2)^2 = x$

7. $\frac{b}{a}\left[\frac{d\sqrt{a^2 - d^2}}{2} - \frac{c\sqrt{a^2 - c^2}}{2}\right] + \frac{ab}{2}\left[\sin^{-1}\frac{d}{a} - \sin^{-1}\frac{c}{a}\right]$

8. (*i*) $\frac{8}{3}$ (*ii*) $\frac{8}{3}$

10. (*i*) $\frac{\pi a^2}{12}$; (*ii*) $\frac{3}{2}\pi a^2$; (*iii*) 4

11. (*i*) $\int_a^b \pi x^2 dy$ (*ii*) $\frac{4}{5}\pi a^3$

12. $\left(\frac{5}{6}a, 0\right)$

13. (*i*) $\frac{4a^2}{3}$M (*ii*) $\frac{Ma^2}{3}$; M being the mass of the rod.

14. $\frac{1}{3}$M (other side)^2

15. a^2M.

3

Partial Differentiation

3.1. FUNCTIONS OF TWO VARIABLES

If three variables x, y, z are so related that the value of z depends upon the values of x and y, then z is called a function of two variables x and y, and this is denoted by $z = f(x, y)$.

z is called the dependent variable while x and y are called independent variables.

For example, the area of a triangle is determined when its base and altitude are known. Thus, area of a triangle is a function of two variables, base and altitude.

(In a similar way, a function of more than two variables can be defined).

Geometrically. Let $z = f(x, y)$ be a function of two independent variables x and y defined for all pairs of values of x and y which belong to an area A of the xy-plane. Then to each point (x, y) of this area corresponds a value of z given by the relation $z = f(x, y)$. Representing all these values (x, y, z) by points in space, we get a surface.

Hence the function z = f(x, y) represents a surface.

3.2. (a) LIMIT

A function $f(x, y)$ is said to tend to the limit l iff corresponding to a positive number ε, however small, there exists another positive number δ such that $| f(x, y) - l | < \varepsilon$ for every point (x, y) within the circle with its centre at (a, b) and radius δ *i.e.*, whenever
$0 < (x - a)^2 + (y - b)^2 < \delta^2$.

Alternatively. Function $f(x, y)$ is said to tend to the limit l as $x \to a$ and $y \to b$ if and only if l is independent of the path followed by the point (x, y) as $x \to a$ and $y \to b$ and it is written as $\underset{\substack{x \to a \\ y \to b}}{\text{Lt}} f(x, y) = l$.

Properties of the limits

If $\underset{\substack{x \to a \\ y \to b}}{\text{Lt}} f(x, y) = l$ and $\underset{\substack{x \to a \\ y \to b}}{\text{Lt}} g(x, y) = m$

Then (*i*) $\underset{\substack{x \to a \\ y \to b}}{\text{Lt}} \{f(x, y) \pm g(x, y)\} = l \pm m$

(*ii*) $\underset{\substack{x \to a \\ y \to b}}{\text{Lt}} \{f(x, y)\, g(x, y)\} = lm$

(*iii*) $$\underset{\substack{x \to a \\ y \to b}}{\text{Lt}} \left\{\frac{f(x, y)}{g(x, y)}\right\} = \frac{l}{m} \quad (m \neq 0).$$

3.2. (b) CONTINUITY

A function $f(x, y)$ is said to be continuous at a point (a, b) if, for any arbitrarily chosen positive number ε, however small, we can find a corresponding number δ such that $| f(x, y) - f(a, b) | < \varepsilon$ for every point (x, y) within the circle with its centre at (a, b) and radius δ *i.e.*, whenever

$$0 < (x - a)^2 + (y - b)^2 < \delta^2.$$

Alternatively, $f(x, y)$ is said to be continuous at (a, b) if $\underset{\substack{x \to a \\ y \to b}}{\text{Lt}} f(x, y) = f(a, b)$ irrespective of the path along which $x \to a$, $y \to b$.

It should not be assumed that the path along which the point (x, y) tends to (a, b) is immaterial, because $\underset{x \to a}{\text{Lt}} \{ \underset{y \to b}{\text{Lt}} f(x, y)\}$ is not always equal to $\underset{y \to b}{\text{Lt}} \{ \underset{x \to a}{\text{Lt}} f(x, y)\}$.

For example consider, $\underset{\substack{x \to 0 \\ y \to 0}}{\text{Lt}} \dfrac{x + y}{x + 2y}$

$$\underset{x \to 0}{\text{Lt}} \left\{ \underset{y \to 0}{\text{Lt}} \frac{x + y}{x + 2y} \right\} = \underset{x \to 0}{\text{Lt}} \frac{x}{x} = \underset{x \to 0}{\text{Lt}}\ 1 = 1$$

$$\underset{y \to 0}{\text{Lt}} \left\{ \underset{x \to 0}{\text{Lt}} \frac{x + y}{x + 2y} \right\} = \underset{y \to 0}{\text{Lt}} \left\{ \frac{y}{2y} \right\} = \underset{y \to 0}{\text{Lt}} \frac{1}{2} = \frac{1}{2}$$

$$\therefore \qquad \underset{x \to 0}{\text{Lt}}\ \underset{y \to 0}{\text{Lt}} \frac{x + y}{x + 2y} \neq \underset{y \to 0}{\text{Lt}}\ \underset{x \to 0}{\text{Lt}} \frac{x + y}{x + 2y}$$

Otherwise also we see that

$$\underset{\substack{x \to 0 \\ y \to 0}}{\text{Lt}} \frac{x + y}{x + 2y} \quad \text{let } x \to 0,\ y \to 0 \text{ along the line } y = mx$$

$$= \underset{x \to 0}{\text{Lt}} \frac{x + mx}{x + 2mx} = \underset{x \to 0}{\text{Lt}} \frac{(1 + m)x}{(1 + 2m)x} = \frac{1 + m}{1 + 2m}$$

Which is different for different values of m

$$\therefore \qquad \underset{x \to 0}{\text{Lt}}\ \underset{y \to 0}{\text{Lt}} \frac{x + y}{x + 2y} \neq \underset{y \to 0}{\text{Lt}}\ \underset{x \to 0}{\text{Lt}} \frac{x + y}{x + 2y}$$

However, usually, the limit is the same irrespective of the path of approach.

In what follows, we shall assume that the functions considered are continuous and their partial differential coefficients as defined in the next section, exist.

Properties of continuity of functions : If $f(x, y)$, $g(x, y)$ are continuous at (a, b) then so are the functions

(*i*) $f(x, y) \pm g(x, y)$ (*ii*) $f(x, y)\, g(x, y)$

(*iii*) $\dfrac{f(x, y)}{g(x, y)}$ $g(x, y) \neq 0.$

3.3. (a) PARTIAL DERIVATIVES OF FIRST ORDER

Let $z = f(x, y)$ be a function of two independent variables x and y. If y is kept constant and x alone is allowed to vary, then z becomes a function of x only. The derivative of z, with respect to x, treating y as constant, is called partial derivative of z w.r.t. x and is denoted by

$$\frac{\partial z}{\partial x} \quad \text{or} \quad \frac{\partial f}{\partial x} \quad \text{or} \quad f_x.$$

Thus, $$\frac{\partial z}{\partial x} = \underset{h \to 0}{\text{Lt}} \frac{f(x+h, y) - f(x, y)}{h}$$

Similarly, the derivative of z, with respect to y, treating x as constant, is called partial derivative of z w.r.t. y and is denoted by $\frac{\partial z}{\partial y}$ or $\frac{\partial f}{\partial y}$ or f_y.

Thus, $$\frac{\partial z}{\partial y} = \underset{k \to 0}{\text{Lt}} \frac{f(x, y+k) - f(x, y)}{k}$$

$\frac{\partial z}{\partial x}$ and $\frac{\partial z}{\partial y}$ are called **first order partial derivatives of z.**

[**In general,** if z is a function of two or more independent variables, then the **partial derivative of z** w.r.t. **any one** of the independent variables is the **ordinary derivative of z** w.r.t. that variable, **treating all other variables as constant.**]

3.3. (b) GEOMETRICAL MEANING OF PARTIAL DERIVATIVES OF FIRST ORDER

(P.T.U. May 2011)

Let $z = f(x, y)$ be a function of two variables x and y. Then by Art. 3.1, it represents a surface S. If $y = k$, a constant, then $y = k$ represents a plane parallel to the zx-plane.

$\therefore$ $z = f(x, y)$ and $y = k$ represent a plane curve C which is the section of S by $y = k$.

$\frac{\partial z}{\partial x}$ represents the slope of tangent to C at (x, k, z).

Thus, $\frac{\partial z}{\partial x}$ gives the slope of the tangent drawn to the curve of intersection of the surface $z = f(x, y)$ and a plane parallel to zx-plane.

Similarly, $\frac{\partial z}{\partial y}$ gives the slope of the tangent drawn to the curve of intersection of the surface $z = f(x, y)$ and a plane parallel to yz-plane.

3.4. PARTIAL DERIVATIVES OF HIGHER ORDER

Since the first order partial derivatives $\frac{\partial z}{\partial x}$ and $\frac{\partial z}{\partial y}$ are themselves functions of x and y, they can be further differentiated partially w.r.t. x as well as y. These are called second order partial derivatives of z. The usual notations for these second order partial derivatives are:

$$\frac{\partial}{\partial x}\left(\frac{\partial z}{\partial x}\right) = \frac{\partial^2 z}{\partial x^2} \text{ or } f_{xx}; \quad \frac{\partial}{\partial y}\left(\frac{\partial z}{\partial y}\right) = \frac{\partial^2 z}{\partial y^2} \text{ or } f_{yy}$$

$$\frac{\partial}{\partial x}\left(\frac{\partial z}{\partial y}\right) = \frac{\partial^2 z}{\partial x \partial y} \text{ or } f_{xy}; \quad \frac{\partial}{\partial y}\left(\frac{\partial z}{\partial x}\right) = \frac{\partial^2 z}{\partial y \partial x} \text{ or } f_{yx}$$

In general, $$\frac{\partial^2 z}{\partial x \partial y} = \frac{\partial^2 z}{\partial y \partial x} \text{ or } f_{xy} = f_{yx}.$$

Note 1. If $z = f(x)$, a function of single independent variable x, we get $\frac{dz}{dx}$.

If $z = f(x, y)$, a function of two independent variables x and y, we get $\frac{\partial z}{\partial x}$ and $\frac{\partial z}{\partial y}$.

Similarly, for a function of more than two independent variables $x_1, x_2, ..., x_n$, we get $\frac{\partial z}{\partial x_1}, \frac{\partial z}{\partial x_2}, \ldots, \frac{\partial z}{\partial x_n}$.

Note 2. (*i*) If $z = u + v$, where $u = f(x, y)$, $v = \phi(x, y)$ then z is a function of x and y.

$$\therefore \qquad \frac{\partial z}{\partial x} = \frac{\partial u}{\partial x} + \frac{\partial v}{\partial x}; \quad \frac{\partial z}{\partial y} = \frac{\partial u}{\partial y} + \frac{\partial v}{\partial y}.$$

(*ii*) If $z = uv$, where $u = f(x, y)$, $v = \phi(x, y)$ then $\frac{\partial z}{\partial x} = \frac{\partial}{\partial x}(uv) = u\frac{\partial v}{\partial x} + v\frac{\partial u}{\partial x}$

$$\frac{\partial z}{\partial y} = \frac{\partial}{\partial y}(uv) = u\frac{\partial v}{\partial y} + v\frac{\partial u}{\partial y}$$

(*iii*) If $z = \frac{u}{v}$, where $u = f(x, y)$, $v = \phi(x, y)$ then $\frac{\partial z}{\partial x} = \frac{\partial}{\partial x}\left(\frac{u}{v}\right) = \frac{v\frac{\partial u}{\partial x} - u\frac{\partial v}{\partial x}}{v^2}$

$$\frac{\partial z}{\partial y} = \frac{\partial}{\partial y}\left(\frac{u}{v}\right) = \frac{v\frac{\partial u}{\partial y} - u\frac{\partial v}{\partial y}}{v^2}$$

(*iv*) If $z = f(u)$, where $u = \phi(x, y)$ then $\frac{\partial z}{\partial x} = \frac{dz}{du}\cdot\frac{\partial u}{\partial x}; \frac{\partial z}{\partial y} = \frac{dz}{du}\cdot\frac{\partial u}{\partial y}$.

ILLUSTRATIVE EXAMPLES

Example 1. *Evaluate the following limits:*

(*i*) $\underset{\substack{x \to 1 \\ y \to 2}}{Lt} \frac{2x^2y}{x^2 + y^2 + 1}$ (*ii*) $\underset{\substack{x \to \infty \\ y \to 2}}{Lt} \frac{xy + 4}{x^2 + 2y^2}$

(*iii*) $f(x, y) = \frac{2x - y}{2x + y}$ *when* $x \to 0, y \to 0$.

Sol. (*i*) $$\underset{\substack{x \to 1 \\ y \to 2}}{\text{Lt}} \frac{2x^2y}{x^2 + y^2 + 1} = \underset{x \to 1}{\text{Lt}} \left\{\underset{y \to 2}{\text{Lt}} \frac{2x^2y}{x^2 + y^2 + 1}\right\}$$

$$= \underset{x \to 1}{\text{Lt}} \frac{4x^2}{x^2 + 5} = \frac{4}{6} = \frac{2}{3}$$

$$\underset{\substack{x \to 1 \\ y \to 2}}{\text{Lt}} \frac{2x^2y}{x^2 + y^2 + 1} = \underset{y \to 2}{\text{Lt}} \left\{\underset{x \to 1}{\text{Lt}} \frac{2x^2y}{x^2 + y^2 + 1}\right\} = \underset{y \to 2}{\text{Lt}} \frac{2y}{y^2 + 2} = \frac{4}{4 + 2} = \frac{2}{3}$$

Hence $$\underset{\substack{x \to 1 \\ y \to 2}}{\text{Lt}} \frac{2x^2y}{x^2 + y^2 + 1} = \frac{2}{3}$$

(*ii*) $$\underset{\substack{x \to \infty \\ y \to 2}}{\text{Lt}} \frac{xy + 4}{x^2 + 2y^2} = \underset{x \to \infty}{\text{Lt}} \left\{\underset{y \to 2}{\text{Lt}} \frac{xy + 4}{x^2 + 2y^2}\right\}$$

$$= \underset{x \to \infty}{\text{Lt}} \frac{2x + 4}{x^2 + 8} = \underset{x \to 0}{\text{Lt}} \frac{2 + \frac{4}{x}}{x\left(1 + \frac{8}{x^2}\right)} = \frac{2}{\infty} = 0$$

$$\underset{\substack{x \to \infty \\ y \to 2}}{\text{Lt}} \frac{xy + 4}{x^2 + 2y^2} = \underset{y \to 2}{\text{Lt}} \left\{\underset{x \to \infty}{\text{Lt}} \frac{xy + 4}{x^2 + 2y^2}\right\}$$

$$= \underset{y \to 2}{\text{Lt}} \left\{ \underset{x \to \infty}{\text{Lt}} \frac{y + \frac{4}{x}}{x\left(1 + \frac{2y^2}{x^2}\right)} \right\} = \underset{y \to 2}{\text{Lt}} \frac{y}{\infty} = \underset{y \to 2}{\text{Lt}} = \frac{2}{\infty} = 0$$

$$\therefore \quad \underset{\substack{x \to \infty \\ y \to 2}}{\text{Lt}} \frac{xy + 4}{x^2 + 2y^2} = 0.$$

(*iii*)
$$\underset{\substack{x \to 0 \\ y \to 0}}{\text{Lt}} \frac{2x - y}{2x + y} = \underset{x \to 0}{\text{Lt}} \left\{ \underset{y \to 0}{\text{Lt}} \frac{2x - y}{2x + y} \right\} = \underset{x \to 0}{\text{Lt}} \left(\frac{2x}{2x}\right) = 1$$

$$\underset{\substack{x \to 0 \\ y \to 0}}{\text{Lt}} \frac{2x - y}{2x + y} = \underset{y \to 0}{\text{Lt}} \left\{ \underset{x \to 0}{\text{Lt}} \frac{2x - y}{2x + y} \right\} = \underset{y \to 0}{\text{Lt}} \frac{-y}{y} = -1$$

$$\underset{x \to 0}{\text{Lt}} \left\{ \underset{y \to 0}{\text{Lt}} f(x, y) \right\} \neq \underset{y \to 0}{\text{Lt}} \left\{ \underset{x \to 0}{\text{Lt}} f(x, y) \right\}$$

$\therefore$ limit does not exist.

Example 2. *Show that the function f(x, y)*

$$f(x, y) = \begin{cases} 2x^2 + y; & (x, y) \neq (1, 2) \\ 0 \quad ; & (x, y) = (1, 2) \end{cases}$$

is discontinuous at (1, 2).

Sol.
$$\underset{\substack{x \to 1 \\ y \to 2}}{\text{Lt}} f(x, y) = \underset{\substack{x \to 1 \\ y \to 2}}{\text{Lt}} (2x^2 + y) = 3$$

But $f(1, 2) = 0$ (given)

$$\therefore \quad \underset{\substack{x \to 1 \\ y \to 2}}{\text{Lt}} f(x, y) \neq f(1, 2)$$

$\therefore$ function is discontinuous at (1, 2).

Example 3. *Show that the function*

$$f(x, y) = \begin{cases} \dfrac{xy}{x^2 + 2y^2}; & (x, y) \neq (0, 0) \\ 0 \quad ; & (x, y) = (0, 0) \end{cases}$$

is not continuous at (0, 0) but its partial derivatives f_x, f_y exist at (0, 0).

(P.T.U., Dec. 2004)

Sol.
$$\underset{\substack{x \to 0 \\ y \to 0}}{\text{Lt}} f(x, y) = \underset{\substack{x \to 0 \\ y \to 0}}{\text{Lt}} \frac{xy}{x^2 + 2y^2}$$

$\therefore$ Let $x \to 0, y \to 0$ along $y = mx$

$$\therefore \quad \underset{\substack{x \to 0 \\ y \to 0}}{\text{Lt}} \frac{xy}{x^2 + 2y^2} = \underset{x \to 0}{\text{Lt}} \frac{x(mx)}{x^2 + 2m^2x^2} = \frac{m}{1 + 2m^2}$$

which is different for different values of m

$\therefore$ limit of $f(x, y)$ does not exist at (0, 0)

$\therefore$ function is discontinuous at (0, 0).

For partial derivative at (0, 0)

$$f_x(0, 0) = \left(\frac{\partial f}{\partial x}\right)_{\text{at }(0,0)} = \underset{h \to 0}{\text{Lt}} \frac{f(0+h, 0) - f(0, 0)}{h}$$

$$= \underset{h \to 0}{\text{Lt}} \frac{f(h, 0) - f(0, 0)}{h} = \underset{h \to 0}{\text{Lt}} \frac{0 - 0}{h} = 0 \qquad [\because \ h \neq 0]$$

$$f_y(0, 0) = \left(\frac{\partial f}{\partial y}\right)_{\text{at }(0,0)} = \underset{k \to 0}{\text{Lt}} \frac{f(0, 0+k) - f(0, 0)}{k}$$

$$= \underset{k \to 0}{\text{Lt}} \frac{f(0, k) - f(0, 0)}{k} = \underset{k \to 0}{\text{Lt}} \frac{0 - 0}{k} = 0 \qquad [\because \ k \neq 0]$$

$\therefore$ Partial derivatives exist at (0, 0).

Example 4. *Compute $f_{xy}(0, 0)$ and $f_{yx}(0, 0)$ for the function*

$$f(x, y) = \begin{cases} \dfrac{xy^3}{x + y^2}; & (x, y) \neq (0, 0) \\ 0 \ ; & (x, y) = (0, 0) \end{cases}$$

Also discuss the continuity of f_{yx} and f_{xy} at (0, 0).

Sol.
$$f(x, y) = \begin{cases} \dfrac{xy^3}{x + y^2}; & (x, y) \neq (0, 0) \\ 0 \ ; & (x, y) = (0, 0) \end{cases}$$

$$f_x(0, 0) = \underset{h \to 0}{\text{Lt}} \frac{f(h, 0) - f(0, 0)}{h} = \underset{h \to 0}{\text{Lt}} \frac{0 - 0}{h} = 0 \qquad [\because \ h \neq 0]$$

$$f_y(0, 0) = \underset{k \to 0}{\text{Lt}} \frac{f(0, k) - f(0, 0)}{k} = \underset{k \to 0}{\text{Lt}} \frac{0 - 0}{k} = 0 \qquad [\because \ k \neq 0]$$

Now
$$\left.\begin{aligned} f_{xy}(0, 0) &= \underset{h \to 0}{\text{Lt}} \frac{f_y(h, 0) - f_y(0, 0)}{h} \\ \text{and} \quad f_{yx}(0, 0) &= \underset{k \to 0}{\text{Lt}} \frac{f_x(0, k) - f_x(0, 0)}{k} \end{aligned}\right\} \qquad ...(1)$$

To find f_{xy} and f_{yx} at (0, 0), we have to find $f_y(h, 0)$ and $f_x(0, k)$

$$f_y(h, 0) = \underset{k \to 0}{\text{Lt}} \frac{f(h, k) - f(h, 0)}{k} = \underset{k \to 0}{\text{Lt}} \frac{\dfrac{hk^3}{h + k^2} - 0}{k} = \underset{k \to 0}{\text{Lt}} \frac{hk^2}{h + k^2} = 0$$

$$f_x(0, k) = \underset{h \to 0}{\text{Lt}} \frac{f(h, k) - f(0, k)}{h} = \underset{h \to 0}{\text{Lt}} \frac{\dfrac{hk^3}{h + k^2} - 0}{h} = \underset{h \to 0}{\text{Lt}} \frac{k^3}{h + k^2} = k \ (\neq 0)$$

From (1), $f_{xy}(0, 0) = \underset{h \to 0}{\text{Lt}} \dfrac{0 - 0}{h} = 0$

$$f_{yx}(0, 0) = \underset{k \to 0}{\text{Lt}} \frac{k - 0}{k} = 1$$

$\therefore$ $f_{xy}(0, 0) = 0$ but $f_{yx}(0, 0) = 1$

$f_{xy}(0, 0) \neq f_{yx}(0, 0)$ $\therefore$ f_{xy} and f_{yx} are not continuous at (0, 0).

Example 5. (*i*) *If* $f(x, y) = x^3y - xy^3$, *find* $\left\{\dfrac{1}{\dfrac{\partial f}{\partial x}} + \dfrac{1}{\dfrac{\partial f}{\partial y}}\right\}_{\substack{x=1\\y=2}}$.

(*ii*) *If* $f(x, y, z) = \begin{vmatrix} x^2 & y^2 & z^2 \\ x & y & z \\ 1 & 1 & 1 \end{vmatrix}$, *prove that* $f_x + f_y + f_z = 0$.

Sol. (*i*)

$$f(x, y) = x^3y - xy^3$$

$$\frac{\partial f}{\partial x} = 3x^2y - y^3; \quad \frac{\partial f}{\partial y} = x^3 - 3xy^2$$

$$\therefore \quad \frac{1}{\dfrac{\partial f}{\partial x}} + \frac{1}{\dfrac{\partial f}{\partial y}} = \frac{1}{3x^2y - y^3} + \frac{1}{x^3 - 3xy^2}$$

$$\left\{\frac{1}{\dfrac{\partial f}{\partial x}} + \frac{1}{\dfrac{\partial f}{\partial y}}\right\}_{\substack{x=1\\y=2}} = \frac{1}{6-8} + \frac{1}{1-12} = -\frac{1}{2} - \frac{1}{11} = \frac{-13}{22}.$$

(*ii*)

$$f(x, y, z) = \begin{vmatrix} x^2 & y^2 & z^2 \\ x & y & z \\ 1 & 1 & 1 \end{vmatrix}$$

First of all evaluate the determinant operate $C_2 - C_1$, $C_3 - C_2$

$$f(x, y, z) = \begin{vmatrix} x^2 & y^2 - x^2 & z^2 - y^2 \\ x & y - x & z - y \\ 1 & 0 & 0 \end{vmatrix} = (y - x)(z - y)\begin{vmatrix} x^2 & y + x & z + y \\ x & 1 & 1 \\ 1 & 0 & 0 \end{vmatrix}$$

$$= (y - x)(z - y)\{y + x - z - y\}$$

$$= (x - y)(y - z)(z - x)$$

Now,

$$f_x = \frac{\partial f}{\partial x} = (y - z)\{(x - y)(-1) + 1(z - x)\} = (y - z)(y + z - 2x)$$

$$\therefore \quad f_x = (y^2 - z^2) - 2(xy - xz)$$

Similarly,

$$f_y = (z^2 - x^2) - 2(yz - yx)$$

$$f_z = (x^2 - y^2) - 2(zx - zy)$$

$$\therefore \quad f_x + f_y + f_z = 0 - 2 \cdot 0 = 0$$

Example 6. (*i*) *If* $z(x + y) = x^2 + y^2$, *show that* $\left(\dfrac{\partial z}{\partial x} - \dfrac{\partial z}{\partial y}\right)^2 = 4\left(1 - \dfrac{\partial z}{\partial x} - \dfrac{\partial z}{\partial y}\right)$.

(*ii*) *If* $z = e^{ax+by} f(ax - by)$ *find* $b\dfrac{\partial z}{\partial x} + a\dfrac{\partial z}{\partial y}$ **(P.T.U., May 2006)**

Sol. (*i*) $z = \dfrac{x^2 + y^2}{x + y}$ [z is symmetrical w.r.t. x and y]

$$\frac{\partial z}{\partial x} = \frac{(x+y)\frac{\partial}{\partial x}(x^2+y^2) - (x^2+y^2)\frac{\partial}{\partial x}(x+y)}{(x+y)^2}$$

$$= \frac{(x+y)\cdot 2x - (x^2+y^2)\cdot 1}{(x+y)^2} = \frac{x^2+2xy-y^2}{(x+y)^2}$$

Similarly, $$\frac{\partial z}{\partial y} = \frac{y^2+2xy-x^2}{(x+y)^2}$$

Now, $$\left(\frac{\partial z}{\partial x} - \frac{\partial z}{\partial y}\right)^2 = \left[\frac{2x^2-2y^2}{(x+y)^2}\right]^2 = \frac{4(x+y)^2(x-y)^2}{(x+y)^4} = \frac{4(x-y)^2}{(x+y)^2}$$

$$4\left(1 - \frac{\partial z}{\partial x} - \frac{\partial z}{\partial y}\right) = 4\left[1 - \frac{x^2+2xy-y^2}{(x+y)^2} - \frac{y^2+2xy-x^2}{(x+y)^2}\right]$$

$$= 4\left[\frac{x^2+2xy+y^2-x^2-2xy+y^2-y^2-2xy+x^2}{(x+y)^2}\right]$$

$$= \frac{4(x^2-2xy+y^2)}{(x+y)^2} = \frac{4(x-y)^2}{(x+y)^2}$$

$$\therefore \quad \left(\frac{\partial z}{\partial x} - \frac{\partial z}{\partial y}\right)^2 = 4\left(1 - \frac{\partial z}{\partial x} - \frac{\partial z}{\partial y}\right).$$

(*ii*) $$z = e^{ax+by} f(ax-by)$$

$$\frac{\partial z}{\partial x} = e^{ax+by} f'(ax-by)\, a + ae^{ax+by} f(ax-by)$$

$$\frac{\partial z}{\partial y} = e^{ax+by} f'(ax-by)\,(-b) + be^{ax+by} f(ax-by)$$

$$b\frac{\partial z}{\partial x} + a\frac{\partial z}{\partial y} = abe^{ax+by} f'(ax-by) + abe^{ax+by} f(ax-by)$$

$$- ab\, e^{ax+by} f'(ax-by) + ab\, e^{ax+by} f(ax-by)$$

$$= abe^{ax+by}[f'(ax-by) + f(ax-by) - f'(ax-by) + f(ax-by)]$$

$$= ab\, e^{ax+by}\cdot 2f(ax-by) = 2ab\, e^{ax+by} f(ax-by) = 2ab.\ z.$$

Example 7. *(i) If* $z = \log(x^2+xy+y^2)$ *prove that* $x\frac{\partial z}{\partial x} + y\frac{\partial z}{\partial y} = 2$ **(P.T.U., Dec. 2013)**

(ii) If $u = e^{x^2+y^2}$ *prove that* $x\frac{\partial u}{\partial x} + y\frac{\partial v}{\partial y} = 2u \log u$ **(P.T.U., May 2014)**

(iii) If $z = x f(x+y) + y g(x+y)$, *show that* $\frac{\partial^2 z}{\partial x^2} - 2\frac{\partial^2 z}{\partial x \partial y} + \frac{\partial^2 z}{\partial y^2} = 0.$

Sol. (*i*) $$z = \log(x^2+xy+y^2)$$

$$\frac{\partial z}{\partial x} = \frac{1}{x^2+xy+y^2}(2x+y), \quad \frac{\partial z}{\partial y} = \frac{1}{x^2+xy+y^2}(x+2y)$$

$$x\frac{\partial z}{\partial x} + y\frac{\partial z}{\partial y} = \frac{2x^2+xy}{x^2+xy+y^2} + \frac{xy+2y^2}{x^2+xy+y^2} = \frac{2(x^2+xy+y^2)}{x^2+xy+y^2}$$

$$\therefore \qquad x\frac{\partial z}{\partial x} + y\frac{\partial z}{\partial y} = 2.$$

(ii)
$$u = e^{x^2+y^2}$$
$$\frac{\partial u}{\partial x} = e^{x^2+y^2} \,.\, 2x$$
$$\frac{\partial u}{\partial y} = e^{x^2+y^2} \,.\, 2y$$
$$x\frac{\partial u}{\partial x} + y\frac{\partial u}{\partial y} = 2e^{x^2+y^2}(x^2+y^2)$$
$$= 2u \,.\, \log u \qquad \left|\begin{array}{l} \because \; e^{x^2+y^2} = u \\ \therefore \; x^2+y^2 = \log u \end{array}\right.$$

(iii)
$$z = x\,f(x+y) + y\,g(x+y)$$
$$\frac{\partial z}{\partial x} = x\,f'(x+y) + f(x+y) + y\,g'(x+y)$$
$$\frac{\partial^2 z}{\partial x^2} = x\,f''(x+y) + f'(x+y) + f'(x+y) + y\,g''(x+y)$$
$$= x\,f''(x+y) + 2f'(x+y) + y\,g''(x+y)$$
$$\frac{\partial z}{\partial y} = x\,f'(x+y) + y\,g'(x+y) + g(x+y)$$
$$\frac{\partial^2 z}{\partial y^2} = x\,f''(x+y) + y\,g''(x+y) + 1.g'(x+y) + g'(x+y)$$
$$= x\,f''(x+y) + 2g'(x+y) + y\,g''(x+y)$$
$$\frac{\partial^2 z}{\partial x \partial y} = \frac{\partial}{\partial x}[x\,f'(x+y) + y\,g'(x+y) + g(x+y)]$$
$$= x\,f''(x+y) + f'(x+y) + y\,g''(x+y) + g'(x+y)$$

Now
$$\frac{\partial^2 z}{\partial x^2} - 2\frac{\partial^2 z}{\partial x \partial y} + \frac{\partial^2 z}{\partial y^2} = x\,f''(x+y) + 2f'(x+y) + yg''(x+y)$$
$$- 2x\,f''(x+y) - 2f'(x+y) - 2yg''(x+y) - 2g'(x+y)$$
$$+ x\,f''(x+y) + 2g'(x+y) + y\,g''(x+y) = 0$$

Hence
$$\frac{\partial^2 z}{\partial x^2} - 2\frac{\partial^2 z}{\partial x \partial y} + \frac{\partial^2 z}{\partial y^2} = 0$$

Example 8. *Prove that if* $f(x, y) = \frac{1}{\sqrt{y}} \cdot e^{-\frac{(x-a)^2}{4y}}$*, then* $f_{xy} = f_{yx}$.

Sol.
$$f(x, y) = \frac{1}{\sqrt{y}} \cdot e^{-\frac{(x-a)^2}{4y}} = y^{-\frac{1}{2}}\, e^{-\frac{(x-a)^2}{4y}}$$
$$f_x = \frac{\partial f}{\partial x} = y^{-\frac{1}{2}} \cdot e^{-\frac{(x-a)^2}{4y}} \cdot \frac{\partial}{\partial x}\left[-\frac{(x-a)^2}{4y}\right]$$
$$= y^{-\frac{1}{2}} \cdot e^{-\frac{(x-a)^2}{4y}} \cdot \left[-\frac{2(x-a)}{4y}\right] = -\frac{1}{2}\, y^{-\frac{3}{2}}\,(x-a)\, e^{-\frac{(x-a)^2}{4y}}$$

$$f_y = \frac{\partial f}{\partial y} = -\frac{1}{2} y^{-\frac{3}{2}} \cdot e^{-\frac{(x-a)^2}{4y}} + y^{-\frac{1}{2}} \cdot e^{-\frac{(x-a)^2}{4y}} \cdot \frac{\partial}{\partial y}\left[-\frac{(x-a)^2}{4y}\right]$$

$$= e^{-\frac{(x-a)^2}{4y}} \left[-\frac{1}{2} y^{-\frac{3}{2}} + y^{-\frac{1}{2}} . \frac{(x-a)^2}{4y^2}\right] = \frac{1}{4} y^{-\frac{3}{2}} e^{-\frac{(x-a)^2}{4y}} \quad [-2 + y^{-1}(x-a)^2]$$

$$f_{xy} = \frac{\partial}{\partial x}\left(\frac{\partial f}{\partial y}\right)$$

$$= \frac{1}{4} y^{-\frac{3}{2}} \left\{ e^{-\frac{(x-a)^2}{4y}} \cdot \frac{\partial}{\partial x}\left[-\frac{(x-a)^2}{4y}\right] \cdot [-2 + y^{-1}(x-a)^2] + e^{-\frac{(x-a)^2}{4y}} \cdot 2y^{-1}(x-a) \right\}$$

$$= \frac{1}{4} y^{-\frac{3}{2}} \cdot e^{-\frac{(x-a)^2}{4y}} \left\{ -\frac{2(x-a)}{4y} [-2 + y^{-1}(x-a)^2] + 2y^{-1}(x-a) \right\}$$

$$= \frac{1}{4} y^{-\frac{3}{2}} \cdot e^{-\frac{(x-a)^2}{4y}} \cdot \frac{x-a}{y} \left\{ -\frac{1}{2} [-2 + y^{-1}(x-a)^2] + 2 \right\}$$

$$= \frac{1}{4} y^{-\frac{5}{2}} (x-a)\, e^{-\frac{(x-a)^2}{4y}} \left[3 - \frac{(x-a)^2}{2y}\right]$$

$$f_{yx} = \frac{\partial}{\partial y}\left(\frac{\partial f}{\partial x}\right) = -\frac{1}{2}(x-a)\left[-\frac{3}{2} y^{-\frac{5}{2}} \cdot e^{-\frac{(x-a)^2}{4y}} + y^{-\frac{3}{2}} \cdot e^{-\frac{(x-a)^2}{4y}} \cdot \frac{(x-a)^2}{4y^2}\right]$$

$$= -\frac{1}{4}(x-a)\, y^{-\frac{5}{2}} \cdot e^{-\frac{(x-a)^2}{4y}} \left[-3 + \frac{(x-a)^2}{2y}\right]$$

$$= \frac{1}{4} y^{-\frac{5}{2}} (x-a)\, e^{-\frac{(x-a)^2}{4y}} \left[3 - \frac{(x-a)^2}{2y}\right]$$

$\therefore \quad f_{xy} = f_{yx}$.

Example 9. *If* $u = x^y$, *show that* $\dfrac{\partial^3 u}{\partial x^2 \partial y} = \dfrac{\partial^3 u}{\partial x \partial y \partial x}$.

Sol. $u = x^y$

Take logs on both sides

$$\log u = y \log x \qquad ...(1)$$

Differentiate (1) partially w.r.t. y,

$$\therefore \quad \frac{1}{u}\frac{\partial u}{\partial y} = \log x \qquad \therefore \quad \frac{\partial u}{\partial y} = x^y \log x = u \log x$$

Differentiate (1) partially w.r.t. x,

$$\frac{1}{u}\frac{\partial u}{\partial x} = \frac{y}{x} \qquad \therefore \quad \frac{\partial u}{\partial x} = x^y \frac{y}{x} = yx^{y-1} \text{ or } u\frac{y}{x}$$

Now, $$\frac{\partial^2 u}{\partial x \partial y} = \frac{\partial}{\partial x}\left(\frac{\partial u}{\partial y}\right) = \frac{\partial}{\partial x}[u \log x] = u \cdot \frac{1}{x} + \log x \cdot \frac{\partial u}{\partial x} = \frac{x^y}{x} + yx^{y-1} \log x$$

$$= yx^{y-1}\log x + x^y \cdot \frac{1}{x} = x^{y-1}(y\log x + 1)$$

$$\frac{\partial^3 u}{\partial x^2 \partial y} = \frac{\partial}{\partial x}\left(\frac{\partial^2 u}{\partial x \partial y}\right) = \frac{\partial}{\partial x}[x^{y-1}(y\log x + 1)] \qquad ...(2)$$

$$\frac{\partial^2 u}{\partial y \partial x} = \frac{\partial}{\partial y}\left(\frac{\partial u}{\partial x}\right) = \frac{\partial}{\partial y}\left(\frac{uy}{x}\right) = \frac{1}{x}\left\{u \cdot 1 + y\frac{\partial u}{\partial y}\right\}$$

$$= \frac{1}{x}[x^y + yx^y \log x] = x^{y-1} + yx^{y-1}\log x = x^{y-1}(y\log x + 1)$$

$$\frac{\partial^3 u}{\partial x \partial y \partial x} = \frac{\partial}{\partial x}\left(\frac{\partial^2 u}{\partial y \partial x}\right) = \frac{\partial}{\partial x}[x^{y-1}(y\log x + 1)] \qquad ...(3)$$

From (2) and (3), $\dfrac{\partial^3 u}{\partial x^2 \partial y} = \dfrac{\partial^3 u}{\partial x \partial y \partial x}$.

Example 10. *Find p, q if $x = \sqrt{a}$ (sin u + cos v); $y = \sqrt{a}$ (cos u – sin v), z = 1 + sin (u – v) where p, q means $\frac{\partial z}{\partial x}, \frac{\partial z}{\partial y}$ respectively.*

Sol. Given

$$x = \sqrt{a}\,(\sin u + \cos v)$$

$$y = \sqrt{a}\,(\cos u - \sin v)$$

$$z = 1 + \sin(u - v)$$

$$x^2 + y^2 = a\{\sin^2 u + \cos^2 v + 2\sin u\cos v + \cos^2 u + \sin^2 v - 2\cos u\sin v\}$$

$$= a[2 + 2\sin(u - v)]$$

$$= 2a \cdot z$$

$$\therefore \quad z = \frac{1}{2a}(x^2 + y^2)$$

$$p = \frac{\partial z}{\partial x} = \frac{2x}{2a} = \frac{x}{a}$$

$$q = \frac{\partial z}{\partial y} = \frac{2y}{2a} = \frac{y}{a}.$$

Example 11. *If $\theta = t^n e^{-\frac{r^2}{4t}}$, find the value of n which will make $\frac{1}{r^2}\frac{\partial}{\partial r}\left(r^2\frac{\partial \theta}{\partial r}\right) = \frac{\partial \theta}{\partial t}$.*

(P.T.U., May 2002, Dec. 2011)

Sol.

$$\theta = t^n e^{-\frac{r^2}{4t}}$$

$$\frac{\partial \theta}{\partial r} = t^n \cdot e^{-\frac{r^2}{4t}} \cdot \left(-\frac{2r}{4t}\right) = -\frac{1}{2}rt^{n-1}e^{-\frac{r^2}{4t}}$$

$$\therefore \quad r^2\frac{\partial \theta}{\partial r} = -\frac{1}{2}r^3 \cdot t^{n-1}e^{-\frac{r^2}{4t}}$$

$$\frac{\partial}{\partial r}\left(r^2\frac{\partial \theta}{\partial r}\right) = -\frac{1}{2}t^{n-1}\left[3r^2 e^{-\frac{r^2}{4t}} + r^3 e^{-\frac{r^2}{4t}}\left(-\frac{2r}{4t}\right)\right] = -\frac{1}{2}t^{n-1}r^2 e^{-\frac{r^2}{4t}}\left[3 - \frac{r^2}{2t}\right]$$

$$\therefore \quad \frac{1}{r^2}\frac{\partial}{\partial r}\left(r^2\frac{\partial \theta}{\partial r}\right) = \frac{1}{2}t^{n-1}e^{-\frac{r^2}{4t}}\left(\frac{r^2}{2t} - 3\right)$$

Also, $$\frac{\partial\theta}{\partial t} = nt^{n-1}e^{-\frac{r^2}{4t}} + t^n\, e^{-\frac{r^2}{4t}} \cdot \left(\frac{r^2}{4t^2}\right) = t^{n-1}\, e^{-\frac{r^2}{4t}} \left(n + \frac{r^2}{4t}\right)$$

Since, $$\frac{1}{r^2}\frac{\partial}{\partial r}\left(r^2 \frac{\partial\theta}{\partial r}\right) = \frac{\partial\theta}{\partial t} \qquad \text{[given]}$$

$\therefore$ $$\frac{1}{2}t^{n-1}e^{-\frac{r^2}{4t}}\left(\frac{r^2}{2t} - 3\right) = t^{n-1}e^{-\frac{r^2}{4t}}\left(n + \frac{r^2}{4t}\right)$$

$\Rightarrow$ $$\frac{r^2}{4t} - \frac{3}{2} = n + \frac{r^2}{4t} \quad \therefore \quad n = -\frac{3}{2}.$$

Example 12. *If* $u = (1 - 2xy + y^2)^{-1/2}$, *prove that* $\frac{\partial}{\partial x}\left\{(1 - x^2)\frac{\partial u}{\partial x}\right\} + \frac{\partial}{\partial y}\left\{y^2 \frac{\partial u}{\partial y}\right\} = 0.$

Sol. $$u = (1 - 2xy + y^2)^{-1/2} = V^{-1/2}, \quad \text{where } V = 1 - 2xy + y^2$$

$$\frac{\partial u}{\partial x} = -\frac{1}{2}V^{-3/2} \cdot \frac{\partial V}{\partial x} = -\frac{1}{2}V^{-3/2}\,(-2y) = yV^{-3/2}$$

$$\frac{\partial^2 u}{\partial x^2} = y \cdot \frac{\partial}{\partial x}(V^{-3/2}) = y \cdot \left(-\frac{3}{2}\right)V^{-5/2} \cdot \frac{\partial V}{\partial x} = -\frac{3}{2}\,yV^{-5/2}\,(-2y) = 3y^2\,V^{-5/2}$$

$$\therefore \quad \frac{\partial}{\partial x}\left\{(1 - x^2)\frac{\partial u}{\partial x}\right\} = (1 - x^2) \cdot \frac{\partial^2 u}{\partial x^2} + \frac{\partial u}{\partial x} \cdot \frac{\partial}{\partial x}\,(1 - x^2)$$

$$= (1 - x^2) \cdot 3y^2V^{-5/2} + yV^{-3/2}\,(-2x) = yV^{-3/2}\,[3yV^{-1}\,(1 - x^2) - 2x] \qquad \text{...(1)}$$

Also, $$\frac{\partial u}{\partial y} = -\frac{1}{2}V^{-3/2}\frac{\partial V}{\partial y} = -\frac{1}{2}\,V^{-3/2} \cdot (-2x + 2y) = V^{-3/2} \cdot (x - y)$$

$$\frac{\partial^2 u}{\partial y^2} = V^{-3/2} \cdot \frac{\partial}{\partial y}(x - y) + (x - y) \cdot \frac{y}{\partial y}\,(V^{-3/2})$$

$$= V^{-3/2} \cdot (-1) + (x - y) \cdot \left(-\frac{3}{2}V^{-5/2}\right) \cdot \frac{\partial V}{\partial y}$$

$$= -V^{-3/2} - \frac{3}{2}\,(x - y)\,V^{-5/2} \cdot (-2x + 2y) = -V^{-3/2} + 3(x - y)^2\,V^{-5/2}$$

$$\therefore \quad \frac{\partial}{\partial y}\left\{y^2\frac{\partial u}{\partial y}\right\} = y^2 \cdot \frac{\partial^2 u}{\partial y^2} + \frac{\partial u}{\partial y} \cdot \frac{\partial}{\partial y}\,(y^2)$$

$$= y^2[-V^{-3/2} + 3(x - y)^2\,V^{-5/2}] + V^{-3/2}\,(x - y)\,.\,2y$$

$$= y\,V^{-3/2}\,[-y + 3y(x - y)^2\,V^{-1} + 2(x - y)]$$

$$= y\,V^{-3/2}\,[3y\,(x - y)^2\,V^{-1} + (2x - 3y)] \qquad \text{...(2)}$$

Adding (1) and (2), we have

$$\frac{\partial}{\partial x}\left\{(1 - x^2)\frac{\partial u}{\partial x}\right\} + \frac{\partial}{\partial y}\left\{y^2\frac{\partial u}{\partial y}\right\} = yV^{-3/2}\,[3yV^{-1}(1 - x^2) - 2x + 3y\,(x - y)^2\,V^{-1} + 2x - 3y]$$

$$= y\,V^{-3/2}[3yV^{-1}\,(1 - x^2 + x^2 - 2xy + y^2) - 3y]$$

$$= yV^{-3/2}[3yV^{-1}(1 - 2xy + y^2) - 3y]$$

$$= yV^{-3/2}\,[3y - 3y] \qquad \mid \because \quad V = 1 - 2xy + y^2$$

$$= 0.$$

Example 13. *If* $u = \log (x^3 + y^3 + z^3 - 3xyz)$, *show that*

(*i*) $\left(\frac{\partial}{\partial x} + \frac{\partial}{\partial y} + \frac{\partial}{\partial z}\right)^2 u = -\frac{9}{(x+y+z)^2}$.

(P.T.U., May 2007, Jan. 2009, May 2012, Dec. 2015)

(*ii*) $\frac{\partial^2 u}{\partial x^2} + \frac{\partial^2 u}{\partial y^2} + \frac{\partial^2 u}{\partial z^2} + 2\frac{\partial^2 u}{\partial y \partial z} + 2\frac{\partial^2 u}{\partial z \partial x} + 2\frac{\partial^2 u}{\partial x \partial y} = \frac{-9}{(x+y+z)^2}$.

Sol. (*i*) $u = \log (x^3 + y^3 + z^3 - 3xyz)$

$$\frac{\partial u}{\partial x} = \frac{3x^2 - 3yz}{x^3 + y^3 + z^3 - 3xyz}; \quad \frac{\partial u}{\partial y} = \frac{3y^2 - 3zx}{x^3 + y^3 + z^3 - 3xyz}$$

$$\frac{\partial u}{\partial z} = \frac{3z^2 - 3xy}{x^3 + y^3 + z^3 - 3xyz}$$

Adding, $\frac{\partial u}{\partial x} + \frac{\partial u}{\partial y} + \frac{\partial u}{\partial z} = \frac{3(x^2 + y^2 + z^2 - xy - yz - zx)}{x^3 + y^3 + z^3 - 3xyz} = \frac{3}{x+y+z}$

$[\because \; x^3 + y^3 + z^3 - 3xyz = (x+y+z)(x^2 + y^2 + z^2 - xy - yz - zx)]$

Now, $\left(\frac{\partial}{\partial x} + \frac{\partial}{\partial y} + \frac{\partial}{\partial z}\right)^2 u = \left(\frac{\partial}{\partial x} + \frac{\partial}{\partial y} + \frac{\partial}{\partial z}\right)\left(\frac{\partial}{\partial x} + \frac{\partial}{\partial y} + \frac{\partial}{\partial z}\right) u$

$$= \left(\frac{\partial}{\partial x} + \frac{\partial}{\partial y} + \frac{\partial}{\partial z}\right)\left(\frac{\partial u}{\partial x} + \frac{\partial u}{\partial y} + \frac{\partial u}{\partial z}\right) = \left(\frac{\partial}{\partial x} + \frac{\partial}{\partial y} + \frac{\partial}{\partial z}\right)\left(\frac{3}{x+y+z}\right)$$

$$= -\frac{3}{(x+y+z)^2} - \frac{3}{(x+y+z)^2} - \frac{3}{(x+y+z)^2} = -\frac{9}{(x+y+z)^2} \quad \ldots(1)$$

(*ii*) $\left(\frac{\partial}{\partial x} + \frac{\partial}{\partial y} + \frac{\partial}{\partial z}\right)^2 u = \left(\frac{\partial}{\partial x} + \frac{\partial}{\partial y} + \frac{\partial}{\partial z}\right)\left(\frac{\partial}{\partial x} + \frac{\partial}{\partial y} + \frac{\partial}{\partial z}\right) u = \left(\frac{\partial}{\partial x} + \frac{\partial}{\partial y} + \frac{\partial}{\partial z}\right)\left(\frac{\partial u}{\partial x} + \frac{\partial u}{\partial y} + \frac{\partial u}{\partial z}\right)$

$$= \frac{\partial}{\partial x}\left(\frac{\partial u}{\partial x} + \frac{\partial u}{\partial y} + \frac{\partial u}{\partial z}\right) + \frac{\partial}{\partial y}\left(\frac{\partial u}{\partial x} + \frac{\partial u}{\partial y} + \frac{\partial u}{\partial z}\right) + \frac{\partial}{\partial z}\left(\frac{\partial u}{\partial x} + \frac{\partial u}{\partial y} + \frac{\partial u}{\partial z}\right)$$

$$= \frac{\partial^2 u}{\partial x^2} + \frac{\partial^2 u}{\partial x \partial y} + \frac{\partial^2 u}{\partial x \partial z} + \frac{\partial^2 u}{\partial y \partial x} + \frac{\partial^2 u}{\partial y^2} + \frac{\partial^2 u}{\partial y \partial z} + \frac{\partial^2 u}{\partial z \partial x} + \frac{\partial^2 u}{\partial z \partial y} + \frac{\partial^2 u}{\partial z^2}$$

$$= \frac{\partial^2 u}{\partial x^2} + \frac{\partial^2 u}{\partial y^2} + \frac{\partial^2 u}{\partial z^2} + 2\frac{\partial^2 u}{\partial y \partial z} + 2\frac{\partial^2 u}{\partial z \partial x} + 2\frac{\partial^2 u}{\partial x \partial y}$$

$$\left[\because \frac{\partial^2 u}{\partial y \partial x} = \frac{\partial^2 u}{\partial x \partial y}, \frac{\partial^2 u}{\partial z \partial y} = \frac{\partial^2 u}{\partial y \partial z}, \frac{\partial^2 u}{\partial x \partial z} = \frac{\partial^2 u}{\partial z \partial x}\right]$$

$$\therefore \quad \frac{\partial^2 u}{\partial x^2} + \frac{\partial^2 u}{\partial y^2} + \frac{\partial^2 u}{\partial z^2} + 2\frac{\partial^2 u}{\partial y \partial z} + 2\frac{\partial^2 u}{\partial z \partial x} + 2\frac{\partial^2 u}{\partial x \partial y} = \left(\frac{\partial}{\partial x} + \frac{\partial}{\partial y} + \frac{\partial}{\partial z}\right)^2 u = \frac{-9}{(x+y+z)^2}$$

[from (1)]

Example 14. *If* $x^x y^y z^z = c$, *show that at* $x = y = z$, $\frac{\partial^2 z}{\partial x \partial y} = -(x \log ex)^{-1}$.

Sol. $x^x y^y z^z = c$ defines z as a function of x and y.

Taking logs on both sides $x \log x + y \log y + z \log z = \log c$

Differentiating partially w.r.t. y, we have

$$y \cdot \frac{1}{y} + 1 \cdot \log y + z \cdot \frac{1}{z} \cdot \frac{\partial z}{\partial y} + 1 \cdot \log z \cdot \frac{\partial z}{\partial y} = 0$$

or $$1 + \log y + (1 + \log z) \frac{\partial z}{\partial y} = 0 \qquad \ldots(1)$$

or $$\frac{\partial z}{\partial y} = -\frac{1+\log y}{1+\log z}$$

Similarly, $$\left.\begin{aligned} \frac{\partial z}{\partial y} &= -\frac{1+\log y}{1+\log z} \\ \frac{\partial z}{\partial x} &= -\frac{1+\log x}{1+\log z} \end{aligned}\right\} \qquad \ldots(2)$$

Differentiating (1) partially w.r.t. x, we have

$$\left(\frac{1}{z}\frac{\partial z}{\partial x}\right)\frac{\partial z}{\partial y} + (1+\log z)\frac{\partial^2 z}{\partial x \partial y} = 0 \quad \text{or} \quad \frac{\partial^2 z}{\partial x \partial y} = -\frac{1}{z(1+\log z)}\frac{\partial z}{\partial x} \cdot \frac{\partial z}{\partial y} \qquad \ldots(3)$$

When $x = y = z$

From (2), $$\frac{\partial z}{\partial y} = -1, \; \frac{\partial z}{\partial x} = -1$$

From (3), $$\frac{\partial^2 z}{\partial x \partial y} = -\frac{1}{x(1+\log x)}(-1)(-1) = -\frac{1}{x(\log e + \log x)} = -\frac{1}{x(\log ex)} = -(x \log ex)^{-1}.$$

Example 15. *(i) If $V = f(r)$ and $r^2 = x^2 + y^2 + z^2$ prove that $V_{xx} + V_{yy} + V_{zz} = f''(r) + \frac{2}{r} f'(r)$.*

(ii) If $V = r^m$ where $r^2 = x^2 + y^2 + z^2$, show that $V_{xx} + V_{yy} + V_{zz} = m(m+1)\, r^{m-2}$.

(P.T.U., Dec. 2006, 2010)

Sol. *(i)* $$V = f(r), \; r^2 = x^2 + y^2 + z^2$$

$\therefore$ $$2r\frac{\partial r}{\partial x} = 2x \quad \therefore \quad \frac{\partial r}{\partial x} = \frac{x}{r}. \text{ Similarly, } \frac{\partial r}{\partial y} = \frac{y}{r}, \; \frac{\partial r}{\partial z} = \frac{z}{r}$$

$$V_x = \frac{\partial V}{\partial x} = f'(r) \cdot \frac{\partial r}{\partial x} = \frac{x}{r} f'(r)$$

$$V_{xx} = x \cdot \frac{1}{r} \cdot f''(r)\frac{\partial r}{\partial x} + x f'(r)\left(\frac{-1}{r^2}\right)\frac{\partial r}{\partial x} + \frac{f'(r)}{r}$$

$$= \frac{x}{r} f''(r) \cdot \frac{x}{r} - f'(r)\frac{x}{r^2} \cdot \frac{x}{r} + \frac{f'(r)}{r}$$

$$= \frac{x^2}{r^2} f''(r) + f'(r)\left[\frac{1}{r} - \frac{x^2}{r^3}\right]$$

$$= \frac{x^2}{r^2} f''(r) + \frac{1}{r^3}(r^2 - x^2) f'(r)$$

Similarly, $$V_{yy} = \frac{y^2}{r^2} f''(r) + \frac{r^2 - y^2}{r^3} f'(r)$$

$$V_{zz} = \frac{z^2}{r^2} f''(r) + \frac{r^2 - z^2}{r^3} f'(r)$$

Adding, $$V_{xx} + V_{yy} + V_{zz} = \frac{x^2 + y^2 + z^2}{r^2} f''(r) + \frac{3r^2 - x^2 - y^2 - z^2}{r^3} f'(r)$$

$$= \frac{r^2}{r^2} f''(r) + \frac{3r^2 - r^2}{r^3} f'(r)$$

$$= f''(r) + \frac{2}{r} f'(r)$$

(*ii*) $V = r^m$, $\quad r^2 = x^2 + y^2 + z^2 \quad \therefore \quad \frac{\partial r}{\partial x} = \frac{x}{r}, \frac{\partial r}{\partial y} = \frac{y}{r}, \frac{\partial r}{\partial y} = \frac{z}{r}$

$$V_x = m\, r^{m-1} \frac{\partial r}{\partial x} = m\, r^{m-1} \cdot \frac{x}{r} = m\, r^{m-2}\, x$$

$$V_{xx} = m \left\{ r^{m-2} \cdot 1 + x \cdot (m-2)\, r^{m-3} \cdot \frac{\partial r}{\partial x} \right\}$$

$$= m\, r^{m-2} + m(m-2)\, r^{m-3} \cdot \frac{x^2}{r}$$

$$V_{xx} = m\, r^{m-2} + m(m-2)\, r^{m-4} \cdot x^2$$

Similarly, $\quad V_{yy} = m\, r^{m-2} + m(m-2)\, r^{m-4}\, y^2$

$$V_{zz} = m\, r^{m-2} + m\,(m-2)\, r^{m-4}\, z^2$$

$$V_{xx} + V_{yy} + V_{zz} = 3m\, r^{m-2} + m(m-2)\, r^{m-4}\, (x^2 + y^2 + z^2)$$

$$= 3m\, r^{m-2} + m(m-2)\, r^{m-4}\, r^2$$

$$= 3m\, r^{m-2} + m\,(m-2)\, r^{m-2}$$

$$= m\,(m+1)\, r^{m-2}$$

Hence $\quad V_{xx} + V_{yy} + V_{zz} = m\,(m+1)\, r^{m-2}$.

Example 16. *If* $x = r \cos \theta$, $y = r \sin \theta$, *prove that*

(*i*) $\frac{\partial r}{\partial x} = \frac{\partial x}{\partial r}$ $\qquad$ (*ii*) $\frac{1}{r} \cdot \frac{\partial x}{\partial \theta} = r \frac{\partial \theta}{\partial x}$ $\qquad$ (*iii*) $\frac{\partial^2 \theta}{\partial x^2} + \frac{\partial^2 \theta}{\partial y^2} = 0.$

(*iv*) $\frac{\partial^2 r}{\partial x^2} \cdot \frac{\partial^2 r}{\partial y^2} = \left(\frac{\partial^2 r}{\partial x \partial y} \right)^2$ $\qquad$ (*v*) $\frac{\partial^2 r}{\partial x^2} + \frac{\partial^2 r}{\partial y^2} = \frac{1}{r} \left[\left(\frac{\partial r}{\partial x} \right)^2 + \left(\frac{\partial r}{\partial y} \right)^2 \right].$

Sol. (*i*) $\frac{\partial r}{\partial x}$ means $\left(\frac{\partial r}{\partial x} \right)_y$ = the partial derivative of r w.r.t. x, treating y as constant.

$\therefore$ We express r in terms of x and y.

Squaring and adding the given relations, $r^2 = x^2 + y^2$

Differentiating partially w.r.t. x, we get $2r \frac{\partial r}{\partial x} = 2x$ or $\frac{\partial r}{\partial x} = \frac{x}{r}$

$\frac{\partial x}{\partial r}$ means $\left(\frac{\partial x}{\partial r} \right)_\theta$ = the partial derivative of x w.r.t. r treating θ as constant.

$\therefore$ we express x in terms of r and θ.

Thus, $\quad x = r \cos \theta$ $\qquad$ (given)

$$\frac{\partial x}{\partial r} = \cos \theta = \frac{x}{r} \qquad \left(\because \cos \theta = \frac{x}{r} \right)$$

$$\therefore \quad \frac{\partial r}{\partial x} = \frac{\partial x}{\partial r}$$

(*ii*) Expressing x in terms of r and θ, we have $x = r \cos \theta$

$$\Rightarrow \quad \frac{\partial x}{\partial \theta} = -r \sin \theta = -y \quad \Rightarrow \quad \frac{1}{r} \frac{\partial x}{\partial \theta} = -\frac{y}{r}$$

Expressing θ in terms of x and y, we have $\tan\theta = \dfrac{y}{x}$ or $\theta = \tan^{-1}\dfrac{y}{x}$

$$\frac{\partial\theta}{\partial x} = \frac{1}{1+\dfrac{y^2}{x^2}}\cdot\left(-\frac{y}{x^2}\right) = -\frac{y}{x^2+y^2} = \frac{-y}{r^2(\cos^2\theta+\sin^2\theta)} = -\frac{y}{r^2}$$

$$\Rightarrow \qquad r\frac{\partial\theta}{\partial x} = -\frac{y}{r} \qquad \therefore \quad \frac{1}{r}\cdot\frac{\partial x}{\partial\theta} = r\frac{\partial\theta}{\partial x}.$$

(*iii*) Expressing θ in terms of x and y, we have $\tan\theta = \dfrac{y}{x}$ or $\theta = \tan^{-1}\dfrac{y}{x}$

$$\frac{\partial\theta}{\partial x} = \frac{1}{1+\dfrac{y^2}{x^2}}\cdot\left(-\frac{y}{x^2}\right) = -\frac{y}{x^2+y^2} = -y(x^2+y^2)^{-1}$$

$$\frac{\partial^2\theta}{\partial x^2} = y(x^2+y^2)^{-2}\cdot 2x = \frac{2xy}{(x^2+y^2)^2}$$

$$\frac{\partial\theta}{\partial y} = \frac{1}{1+\dfrac{y^2}{x^2}}\cdot\frac{1}{x} = \frac{x}{x^2+y^2} = x(x^2+y^2)^{-1}$$

$$\frac{\partial^2\theta}{\partial y^2} = -x(x^2+y^2)^{-2}\cdot 2y = -\frac{2xy}{(x^2+y^2)^2}$$

$$\therefore \qquad \frac{\partial^2\theta}{\partial x^2} + \frac{\partial^2\theta}{\partial y^2} = 0.$$

(*iv*) from (*i*) part $r^2 = x^2 + y^2 \quad \therefore \quad r = (x^2+y^2)^{1/2}$

$$\frac{\partial r}{\partial x} = \frac{1}{2}(x^2+y^2)^{-1/2}\cdot 2x = \frac{x}{\sqrt{x^2+y^2}}$$

$$\frac{\partial r}{\partial y} = \frac{1}{2}(x^2+y^2)^{-1/2}\cdot 2y = \frac{y}{\sqrt{x^2+y^2}}$$

$$\frac{\partial^2 r}{\partial x^2} = \frac{\sqrt{x^2+y^2}\cdot 1 - x\cdot\dfrac{1}{2\sqrt{x^2+y^2}}\cdot 2x}{x^2+y^2} = \frac{x^2+y^2-x^2}{(x^2+y^2)^{3/2}} = \frac{y^2}{(x^2+y^2)^{3/2}}$$

$$\frac{\partial^2 r}{\partial y^2} = \frac{\sqrt{x^2+y^2}\cdot 1 - y\cdot\dfrac{1}{2\sqrt{x^2+y^2}}\cdot 2y}{x^2+y^2} = \frac{x^2+y^2-y^2}{(x^2+y^2)^{3/2}} = \frac{x^2}{(x^2+y^2)^{3/2}}$$

$$\frac{\partial^2 r}{\partial x\partial y} = y\cdot\frac{-1}{2}(x^2+y^2)^{-3/2}\cdot 2x = \frac{-xy}{(x^2+y^2)^{3/2}}$$

Now $$\frac{\partial^2 r}{\partial x^2}\cdot\frac{\partial^2 r}{\partial y^2} = \frac{y^2}{(x^2+y^2)^{3/2}}\cdot\frac{x^2}{(x^2+y^2)^{3/2}} = \frac{x^2y^2}{(x^2+y^2)^{3/2}} = \left(\frac{\partial^2 r}{\partial x\partial y}\right)^2$$

(*v*) And $$\frac{\partial^2 r}{\partial x^2} + \frac{\partial^2 r}{\partial y^2} = \frac{y^2}{(x^2+y^2)^{3/2}} + \frac{x^2}{(x^2+y^2)^{3/2}} = \frac{x^2+y^2}{(x^2+y^2)^{3/2}} = \frac{1}{(x^2+y^2)^{1/2}} = \frac{1}{r}$$

$$\left(\frac{\partial r}{\partial x}\right)^2 + \left(\frac{\partial r}{\partial y}\right)^2 = \frac{x^2}{x^2+y^2} + \frac{y^2}{x^2+y^2} = \frac{x^2+y^2}{x^2+y^2} = 1$$

$$\therefore \quad \frac{\partial^2 r}{\partial x^2} + \frac{\partial^2 r}{\partial y^2} = \frac{1}{r} \cdot 1 = \frac{1}{r}\left\{\left(\frac{\partial r}{\partial x}\right)^2 + \left(\frac{\partial r}{\partial y}\right)^2\right\}.$$

Example 17. *If* $x = e^{r\cos\theta}\cos(r\sin\theta)$ *and* $y = e^{r\cos\theta}\sin(r\sin\theta)$, *prove that*

$$\frac{\partial x}{\partial r} = \frac{1}{r}\cdot\frac{\partial y}{\partial \theta} \cdot \frac{\partial y}{\partial r} = -\frac{1}{r}\cdot\frac{\partial x}{\partial \theta}.$$

Hence deduce that $\dfrac{\partial^2 x}{\partial r^2} + \dfrac{1}{r}\dfrac{\partial x}{\partial r} + \dfrac{1}{r^2}\dfrac{\partial^2 x}{\partial \theta^2} = 0.$

Sol. $\quad x = e^{r\cos\theta}\cos(r\sin\theta)$

$$\therefore \quad \frac{\partial x}{\partial r} = e^{r\cos\theta}\cdot\cos\theta\cdot\cos(r\sin\theta) - e^{r\cos\theta}\sin(r\sin\theta)\cdot\sin\theta$$

$$= e^{r\cos\theta}[\cos\theta\cos(r\sin\theta) - \sin\theta\sin(r\sin\theta)]$$

$$= e^{r\cos\theta}\cos(\theta + r\sin\theta) \qquad \text{...(1)}$$

$$\frac{\partial x}{\partial \theta} = e^{r\cos\theta}\cdot(-r\sin\theta)\cdot\cos(r\sin\theta) - e^{r\cos\theta}\sin(r\sin\theta)\cdot r\cos\theta$$

$$= -re^{r\cos\theta}[\sin\theta\cos(r\sin\theta) + \cos\theta\sin(r\sin\theta)]$$

$$= -re^{r\cos\theta}\sin(\theta + r\sin\theta) \qquad \text{...(2)}$$

Also, $\quad y = e^{r\cos\theta}\sin(r\sin\theta)$

$$\therefore \quad \frac{\partial y}{\partial r} = e^{r\cos\theta}\cdot\cos\theta\cdot\sin(r\sin\theta) + e^{r\cos\theta}\cdot\cos(r\sin\theta)\sin\theta$$

$$= e^{r\cos\theta}[\sin\theta\cos(r\sin\theta) + \cos\theta\sin(r\sin\theta)]$$

$$= e^{r\cos\theta}\sin(\theta + r\sin\theta) \qquad \text{...(3)}$$

$$\frac{\partial y}{\partial \theta} = e^{r\cos\theta}(-r\sin\theta)\sin(r\sin\theta) + e^{r\cos\theta}\cos(r\sin\theta)\times r\cos\theta$$

$$= re^{r\cos\theta}[\cos\theta\cos(r\sin\theta) - \sin\theta\sin(r\sin\theta)]$$

$$= re^{r\cos\theta}\cos(\theta + r\sin\theta) \qquad \text{...(4)}$$

From (1) and (4), $\quad \dfrac{\partial x}{\partial r} = \dfrac{1}{r}\cdot\dfrac{\partial y}{\partial \theta} \qquad \text{...(5)}$

From (2) and (3), $\quad \dfrac{\partial y}{\partial r} = -\dfrac{1}{r}\cdot\dfrac{\partial x}{\partial \theta} \qquad \text{...(6)}$

From (5), $\quad \dfrac{\partial^2 x}{\partial r^2} = -\dfrac{1}{r^2}\cdot\dfrac{\partial y}{\partial \theta} + \dfrac{1}{r}\cdot\dfrac{\partial^2 y}{\partial r\partial \theta} \qquad \text{...(7)}$

From (6), $\quad \dfrac{\partial x}{\partial \theta} = -r\dfrac{\partial y}{\partial r}$

$$\therefore \quad \frac{\partial^2 x}{\partial \theta^2} = -r\frac{\partial^2 y}{\partial \theta\partial r} = -r\frac{\partial^2 y}{\partial r\partial \theta} \qquad \text{...(8)}$$

$$\therefore \quad \frac{\partial^2 x}{\partial r^2} + \frac{1}{r}\cdot\frac{\partial x}{\partial r} + \frac{1}{r^2}\cdot\frac{\partial^2 x}{\partial \theta^2} = -\frac{1}{r^2}\cdot\frac{\partial y}{\partial \theta} + \frac{1}{r}\cdot\frac{\partial^2 y}{\partial r\partial \theta} + \frac{1}{r^2}\cdot\frac{\partial y}{\partial \theta} - \frac{1}{r}\cdot\frac{\partial^2 y}{\partial r\partial \theta} = 0.$$

[Using (5), (7), (8)]

Example 18. *If* $\dfrac{x^2}{a^2+u} + \dfrac{y^2}{b^2+u} + \dfrac{z^2}{c^2+u} = 1$, *prove that*

$$\left(\frac{\partial u}{\partial x}\right)^2 + \left(\frac{\partial u}{\partial y}\right)^2 + \left(\frac{\partial u}{\partial z}\right)^2 = 2\left(x\frac{\partial u}{\partial x} + y\frac{\partial u}{\partial y} + z\frac{\partial u}{\partial z}\right).$$ **(P.T.U., Dec. 2003)**

Sol. Given $$\frac{x^2}{a^2+u} + \frac{y^2}{b^2+u} + \frac{z^2}{c^2+u} = 1 \qquad ...(1)$$

or $$x^2(a^2+u)^{-1} + y^2(b^2+u)^{-1} + z^2(c^2+u)^{-1} = 1$$

Differentiating partially w.r.t. x, we have

$$2x(a^2+u)^{-1} - x^2(a^2+u)^{-2}\cdot\frac{\partial u}{\partial x} - y^2(b^2+u)^{-2}\cdot\frac{\partial u}{\partial x} - z^2(c^2+u)^{-2}\cdot\frac{\partial u}{\partial x} = 0$$

or $$\frac{2x}{a^2+u} = \left[\frac{x^2}{(a^2+u)^2} + \frac{y^2}{(b^2+u)^2} + \frac{z^2}{(c^2+u)^2}\right]\frac{\partial u}{\partial x}$$

or $$\frac{2x}{a^2+u} = V\frac{\partial u}{\partial x} \text{ where } V = \frac{x^2}{(a^2+u)^2} + \frac{y^2}{(b^2+u)^2} + \frac{z^2}{(c^2+u)^2}$$

or $$\frac{\partial u}{\partial x} = \frac{2x}{V(a^2+u)}$$

Similarly, $$\frac{\partial u}{\partial y} = \frac{2y}{V(b^2+u)} \quad\text{and}\quad \frac{\partial u}{\partial z} = \frac{2z}{V(c^2+u)}$$

$$\therefore \quad \left(\frac{\partial u}{\partial x}\right)^2 + \left(\frac{\partial u}{\partial y}\right)^2 + \left(\frac{\partial u}{\partial z}\right)^2 = \frac{4}{V^2}\left[\frac{x^2}{(a^2+u)^2} + \frac{y^2}{(b^2+u)^2} + \frac{z^2}{(c^2+u)^2}\right]$$

$$= \frac{4}{V^2}(V) = \frac{4}{V} \qquad ...(2)$$

Now, $$2\left(x\frac{\partial u}{\partial x} + y\frac{\partial u}{\partial y} + z\frac{\partial u}{\partial z}\right) = 2\left[\frac{2x^2}{V(a^2+u)} + \frac{2y^2}{V(b^2+u)} + \frac{2z^2}{V(c^2+u)}\right]$$

$$= \frac{4}{V}\left[\frac{x^2}{a^2+u} + \frac{y^2}{b^2+u} + \frac{z^2}{c^2+u}\right]$$

$$= \frac{4}{V}(1) \qquad \text{[Using (1)]}$$

$$= \left(\frac{\partial u}{\partial x}\right)^2 + \left(\frac{\partial u}{\partial y}\right)^2 + \left(\frac{\partial u}{\partial z}\right)^2 \qquad \text{[Using (2)]}$$

Example 19. *If* $x^2 = au + bv$, $y^2 = au - bv$, *prove that* $\left(\dfrac{\partial u}{\partial x}\right)_y \left(\dfrac{\partial x}{\partial u}\right)_v = \dfrac{1}{2} = \left(\dfrac{\partial v}{\partial y}\right)_x \left(\dfrac{\partial y}{\partial v}\right)_u$.

Sol. As $$x^2 = au + bv \qquad ...(1)$$

$$y^2 = au - bv \qquad ...(2)$$

Add and subtract, we get $u = \dfrac{x^2+y^2}{2a}$, $v = \dfrac{x^2-y^2}{2b}$

$$\left(\frac{\partial u}{\partial x}\right)_y = \frac{x}{a}, \left(\frac{\partial v}{\partial y}\right)_x = -\frac{y}{b}$$

From (1), $2x\left(\frac{\partial x}{\partial u}\right)_v = a \therefore \left(\frac{\partial x}{\partial u}\right)_v = \frac{a}{2x}$

From (2), $2y\left(\frac{\partial y}{\partial v}\right)_u = -b \quad \therefore \left(\frac{\partial y}{\partial v}\right)_u = -\frac{b}{2y}$

$$\text{L.H.S.} = \left(\frac{\partial u}{\partial x}\right)_y \left(\frac{\partial x}{\partial u}\right)_v = \frac{x}{a} \cdot \frac{a}{2x} = \frac{1}{2}$$

$$\text{R.H.S.} = \left(\frac{\partial v}{\partial y}\right)_x \left(\frac{\partial y}{\partial v}\right)_u = \left(-\frac{y}{b}\right)\left(-\frac{b}{2y}\right) = \frac{1}{2}$$

Hence, $\left(\frac{\partial u}{\partial x}\right)_y \left(\frac{\partial x}{\partial u}\right)_v = \frac{1}{2} = \left(\frac{\partial v}{\partial y}\right)_x \left(\frac{\partial y}{\partial v}\right)_u$.

Example 20. *If $u = lx + my$, $v = mx - ly$, show that*

$$\left(\frac{\partial u}{\partial x}\right)_y \left(\frac{\partial x}{\partial u}\right)_v = \frac{l^2}{l^2 + m^2}, \left(\frac{\partial y}{\partial v}\right)_x \left(\frac{\partial v}{\partial y}\right)_u = \frac{l^2 + m^2}{l^2}.$$

Sol. Given $u = lx + my$ ...(1)

$v = mx - ly$ ...(2)

(*i*) $\left(\frac{\partial u}{\partial x}\right)_y$ = The partial derivative of u w.r.t. x keeping y constant.

$\therefore$ We need a relation expressing u as a function of x and y.

From (1), $\left(\frac{\partial u}{\partial x}\right)_y = l$

$\left(\frac{\partial x}{\partial u}\right)_v$ = The partial derivative of x w.r.t. u keeping v constant.

$\therefore$ We need a relation expressing x as a function of u and v.

Eliminating y between (1) and (2) by multiplying (1) by l, (2) by m and adding the products, we have

$$lu + mv = (l^2 + m^2)x \quad \text{or} \quad x = \frac{lu + mv}{l^2 + m^2}$$

$$\therefore \quad \left(\frac{\partial x}{\partial u}\right)_v = \frac{l}{l^2 + m^2}$$

Hence $\left(\frac{\partial u}{\partial x}\right)_y \left(\frac{\partial x}{\partial u}\right)_v = \frac{l^2}{l^2 + m^2}$

(*ii*) $\left(\frac{\partial y}{\partial v}\right)_x$ = the partial derivative of y w.r.t. v keeping x constant.

$\therefore$ We need a relation expressing y as a function of v and x.

From (2), $\qquad y = \dfrac{mx - v}{l} \qquad \therefore \quad \left(\dfrac{\partial y}{\partial v}\right)_x = -\dfrac{1}{l}$

Also $\left(\dfrac{\partial v}{\partial y}\right)_u$ = partial derivative of v w.r.t. y keeping u constant.

$\therefore$ We need a relation expressing v as a function of y and u.

Eliminating x between (1) and (2), we have $v = \dfrac{mu - (l^2 + m^2)y}{l}$

$$\therefore \qquad \left(\frac{\partial v}{\partial y}\right)_u = -\frac{l^2 + m^2}{l}$$

Hence $\qquad \left(\dfrac{\partial y}{\partial v}\right)_x \left(\dfrac{\partial v}{\partial y}\right)_u = \left(-\dfrac{1}{l}\right)\left(-\dfrac{l^2 + m^2}{l}\right) = \dfrac{l^2 + m^2}{l^2}.$

Example 21. *If* $u = f(ax^2 + 2hxy + by^2)$, $v = \phi\,(ax^2 + 2hxy + by^2)$, *prove that*

$$\frac{\partial}{\partial y}\left(u\,\frac{\partial v}{\partial x}\right) = \frac{\partial}{\partial x}\left(u\,\frac{\partial v}{\partial y}\right).$$

Sol. Let $\qquad z = ax^2 + 2hxy + by^2 \qquad$ ...(1)

$\therefore \qquad u = f(z),\ v = \phi(z) \qquad$ ...(2)

We have to prove $\quad \dfrac{\partial}{\partial y}\left(u\,\dfrac{\partial v}{\partial x}\right) = \dfrac{\partial}{\partial x}\left(u\,\dfrac{\partial v}{\partial y}\right)$

i.e., to prove $\quad u\,\dfrac{\partial^2 v}{\partial y\,\partial x} + \dfrac{\partial u}{\partial y}\cdot\dfrac{\partial v}{\partial x} = u\cdot\dfrac{\partial^2 v}{\partial x\,\partial y} + \dfrac{\partial u}{\partial x}\cdot\dfrac{\partial v}{\partial y}$

As $\qquad \dfrac{\partial^2 v}{\partial y\,\partial x} = \dfrac{\partial^2 v}{\partial x\,\partial y}$

$\therefore$ We have to prove only $\dfrac{\partial u}{\partial y}\cdot\dfrac{\partial v}{\partial x} = \dfrac{\partial u}{\partial x}\cdot\dfrac{\partial v}{\partial y}$

From (2), $\qquad \dfrac{\partial u}{\partial x} = f'(z)\,\dfrac{\partial z}{\partial x};\ \dfrac{\partial u}{\partial y} = f'(z)\,\dfrac{\partial z}{\partial y}$

and $\qquad \dfrac{\partial v}{\partial x} = \phi'(z)\,\dfrac{\partial z}{\partial x};\ \dfrac{\partial v}{\partial y} = \phi'(z)\,\dfrac{\partial z}{\partial y}$

From (1), $\qquad \dfrac{\partial z}{\partial x} = 2ax + 2hy$

$$\frac{\partial z}{\partial y} = 2hx + 2by$$

$$\therefore \qquad \frac{\partial u}{\partial y}\cdot\frac{\partial v}{\partial x} = f'(z)\,(2hx + 2by)\,\phi'(z)\,(2ax + 2hy)$$

$$= 4\,f'(z)\,\phi'(z)\,(hx + by)\,(ax + hy)$$

And $\qquad \dfrac{\partial u}{\partial x}\cdot\dfrac{\partial v}{\partial y} = f'(z)\,(2ax + 2hy).\ \phi'(z)\,(2hx + 2by)$

$$= 4\,f'(z)\,\phi'(z)\,(hx + by)\,(ax + hy)$$

Hence $\frac{\partial u}{\partial y}\cdot\frac{\partial v}{\partial x}=\frac{\partial u}{\partial x}\cdot\frac{\partial v}{\partial y}$

$\therefore$ $\frac{\partial}{\partial y}\left(u\frac{\partial v}{\partial x}\right)=\frac{\partial}{\partial x}\left(u\frac{\partial v}{\partial y}\right).$

TEST YOUR KNOWLEDGE

1. Show that the function $f(x, y) = \begin{cases} \frac{x^2+y^2}{|x|+|y|}; & (x,y)\neq 0 \\ 0 \quad ; & (x,y)=0 \end{cases}$ is continuous at (0, 0) but its partial derivatives do not exist at (0, 0).

2. For the function $f(x, y) = \begin{cases} \frac{xy(2x^2-3y^2)}{x^2+y^2}; & (x,y)\neq(0,0) \\ 0 \quad ; & (x,y)=(0,0) \end{cases}$

   find $f_{xy}(0, 0)$ and $f_{yx}(0, 0)$ and prove that f_{xy}, f_{yx} are discontinuous at (0, 0).

3. Find the first order partial derivatives of the following functions:

   (i) $u = y^x$ (ii) $u = \log(x^2 + y^2)$

   (iii) $u = x^2 \sin\frac{y}{x}$ (iv) $u = \frac{x}{y}\tan^{-1}\left(\frac{y}{x}\right)$.

4. If $u = x^2 + y^2 + z^2$, prove that $xu_x + yu_y + zu_z = 2u$.

5. If $z = \log(x^2 + xy + y^2)$, prove that $x\frac{\partial z}{\partial x}+y\frac{\partial z}{\partial y}=2.$ **(P.T.U., Dec. 2013)**

6. If $u = x^2y + y^2z + z^2x$, prove that $u_x + u_y + u_z = (x + y + z)^2$.

7. If $u=\frac{y}{z}+\frac{z}{x}+\frac{x}{y}$, show that $x\frac{\partial u}{\partial x}+y\frac{\partial u}{\partial y}+z\frac{\partial u}{\partial z}=0.$ **(P.T.U., Dec. 2012)**

8. If $u = \log(\tan x + \tan y)$, prove that $\sin 2x\frac{\partial u}{\partial x}+\sin 2y\frac{\partial u}{\partial y}=2.$

9. Verify that $\frac{\partial^2 u}{\partial x\partial y}=\frac{\partial^2 u}{\partial y\partial x}$ for the following functions:

   (i) $u = ax^2 + 2hxy + by^2$ (ii) $u=\tan^{-1}\left(\frac{x}{y}\right)$ (iii) $u=\log\left(\frac{x^2+y^2}{xy}\right)$

   (iv) $u = e^{ax}\sin by$ (v) $u = \log(x \sin y + y \sin x)$

   (vi) $\sin\left(\frac{y}{x}\right)$ (vii) $\log x \tan^{-1}(x^2 + y^2)$.

10. If $z = \log(e^x + e^y)$, show that $r\,t - s^2 = 0$; where $r=\frac{\partial^2 z}{\partial x^2}$, $t=\frac{\partial^2 z}{\partial y^2}$, $s=\frac{\partial^2 z}{\partial x\partial y}$.

11. If $u=\tan^{-1}\frac{xy}{\sqrt{1+x^2+y^2}}$, show that $\frac{\partial^2 u}{\partial x\partial y}=(1+x^2+y^2)^{-3/2}$.

12. If $u = e^{xyz}$, prove that $\frac{\partial^3 u}{\partial x\partial y\partial z}=(1+3xyz+x^2y^2z^2)e^{xyz}$.

13. If $u = \log (x^2 + y^2) + \tan^{-1} \dfrac{y}{x}$, prove that $\dfrac{\partial^2 u}{\partial x^2} + \dfrac{\partial^2 u}{\partial y^2} = 0$.

14. If $u = \tan^{-1} \left(\dfrac{2xy}{x^2 - y^2} \right)$, prove that $\dfrac{\partial^2 u}{\partial x^2} + \dfrac{\partial^2 u}{\partial y^2} = 0$.

15. Find the value of n so that the equation $V = r^n (3 \cos^2 \theta - 1)$ satisfies the relation

$$\frac{\partial}{\partial r}\left(r^2 \frac{\partial V}{\partial r}\right) + \frac{1}{\sin\theta} \frac{\partial}{\partial \theta}\left(\sin\theta \frac{\partial V}{\partial \theta}\right) = 0.$$

16. If $z = \tan (y + ax) - (y - ax)^{3/2}$, show that $\dfrac{\partial^2 z}{\partial x^2} = a^2 \dfrac{\partial^2 z}{\partial y^2}$.

17. If $u = \log (x^2 + y^2 + z^2)$, prove that $x \dfrac{\partial^2 u}{\partial y \partial z} = y \dfrac{\partial^2 u}{\partial z \partial x} = z \dfrac{\partial^2 u}{\partial x \partial y}$.

18. If $u = \log \sqrt{x^2 + y^2 + z^2}$, prove that $(x^2 + y^2 + z^2) \left(\dfrac{\partial^2 u}{\partial x^2} + \dfrac{\partial^2 u}{\partial y^2} + \dfrac{\partial^2 u}{\partial z^2} \right) = 1$.

19. (*a*) If $V = f(r)$, $x = r \cos \theta$, $y = r \sin \theta$, prove that $\dfrac{\partial^2 V}{\partial x^2} + \dfrac{\partial^2 V}{\partial y^2} = f''(r) + \dfrac{1}{r} f'(r)$. (*P.T.U. May 2015*)

[**Hint.** $r^2 = x^2 + y^2$; proceed as in (S.E. 15(*i*)]

(*b*) If $x^2 + y^2 + z^2 = \dfrac{1}{u^2}$, prove that $\dfrac{\partial^2 u}{\partial x^2} + \dfrac{\partial^2 u}{\partial y^2} + \dfrac{\partial^2 u}{\partial z^2} = 0$.

[**Hint.** Consult S.E. 15(*ii*) replace m by -1]

20. If $u = \sqrt{x^2 + y^2 + z^2}$, show that

(*i*) $\left(\dfrac{\partial u}{\partial x}\right)^2 + \left(\dfrac{\partial u}{\partial y}\right)^2 + \left(\dfrac{\partial u}{\partial z}\right)^2 = 1$ (*ii*) $\dfrac{\partial^2 u}{\partial x^2} + \dfrac{\partial^2 u}{\partial y^2} + \dfrac{\partial^2 u}{\partial z^2} = \dfrac{2}{u}$.

21. (*i*) If $u = e^{x-at} \cos (x - at)$, show that $\dfrac{\partial^2 u}{\partial t^2} = a^2 \dfrac{\partial^2 u}{\partial x^2}$.

(*ii*) If $v = \dfrac{1}{\sqrt{t}} e^{-\frac{x^2}{4a^2 t}}$, prove that $\dfrac{\partial v}{\partial t} = a^2 \dfrac{\partial^2 v}{\partial x^2}$.

22. (*i*) If $u = e^x(x \cos y - y \sin y)$, show that $\dfrac{\partial^2 u}{\partial x^2} + \dfrac{\partial^2 u}{\partial y^2} = 0$.

(*ii*) If $z = f(x + ay) + \phi(x - ay)$ prove that $\dfrac{\partial^2 z}{\partial y^2} = a^2 \dfrac{\partial^2 z}{\partial x^2}$.

23. If $u = \log (x^3 + y^3 - x^2 y - xy^2)$, show that $\dfrac{\partial^2 u}{\partial x^2} + 2 \dfrac{\partial^2 u}{\partial x \partial y} + \dfrac{\partial^2 u}{\partial y^2} = -4(x + y)^{-2}$.

[**Hint.** $u = \log \{x^2(x - y) - y^2(x - y)\} = \log (x - y)(x^2 - y^2) = \log (x - y)^2(x + y)$
$= 2 \log (x - y) + \log (x + y)$]

Answers

2. $f_{xy}(0, 0) = 0$, $f_{yx}(0, 0) = -3$.

3. (*i*) $y^x \log y$, xy^{x-1} ; (*ii*) $\dfrac{2x}{x^2 + y^2}, \dfrac{2y}{x^2 + y^2}$ (*iii*) $2x \sin \dfrac{y}{x} - y \cos \dfrac{y}{x}$, $x \cos \dfrac{y}{x}$

(*iv*) $\dfrac{-x}{x^2 + y^2} + \dfrac{1}{y} \tan^{-1} \dfrac{y}{x}$, $\dfrac{x^2}{y(x^2 + y^2)} - \dfrac{x}{y^2} \tan^{-1} \dfrac{y}{x}$

15. $n = -3, 2$.

3.5. HOMOGENEOUS FUNCTIONS *(P.T.U., May 2005, May 2009)*

A function $f(x, y)$ is said to be homogeneous of degree (or order) n in the variables x and y if it can be expressed in the form $x^n \phi\left(\frac{y}{x}\right)$ or $y^n \phi\left(\frac{x}{y}\right)$.

An alternative test for a function $f(x, y)$ to be homogeneous of degree (or order) n is that

$$f(tx, ty) = t^n f(x, y).$$

For example, if $\quad f(x, y) = \dfrac{x + y}{\sqrt{x} + \sqrt{y}}$, then

$$(i)\ f(x, y) = \frac{x\left(1+\dfrac{y}{x}\right)}{\sqrt{x}\left(1+\sqrt{\dfrac{y}{x}}\right)} = x^{1/2}\,\phi\left(\frac{y}{x}\right)$$

$\Rightarrow$ $f(x, y)$ is a homogeneous function of degree $\frac{1}{2}$ in x and y.

$$(ii)\ f(x, y) = \frac{y\left(\dfrac{x}{y}+1\right)}{\sqrt{y}\left(\sqrt{\dfrac{x}{y}}+1\right)} = y^{1/2}\,\phi\left(\frac{x}{y}\right)$$

$\Rightarrow$ $f(x, y)$ is a homogeneous function of degree $\frac{1}{2}$ in x and y.

$$(iii)\ f(tx, ty) = \frac{tx + ty}{\sqrt{tx} + \sqrt{ty}} = \frac{t(x + y)}{\sqrt{t}\,(\sqrt{x} + \sqrt{y})} = t^{1/2} f(x, y)$$

$\Rightarrow$ $f(x, y)$ is a homogeneous function of degree $\frac{1}{2}$ in x and y.

Similarly, a function $f(x, y, z)$ is said to be homogeneous of degree (or order) n in the variables x, y, z if

$$f(x, y, z) = x^n\,\phi\left(\frac{y}{x}, \frac{z}{x}\right) \quad \text{or} \quad y^n\,\phi\left(\frac{x}{y}, \frac{z}{y}\right) \quad \text{or} \quad z^n\,\phi\left(\frac{x}{z}, \frac{y}{z}\right).$$

Alternative test is $f(tx, ty, tz) = t^n f(x, y, z)$.

3.6. EULER'S THEOREM ON HOMOGENEOUS FUNCTIONS

(*P.T.U. May 2003, Dec. 2004, May 2005, May 2006, Jan. 2009, May 2009, Dec. 2010*)

If u is a homogeneous function of degree n in x and y, then $x\dfrac{\partial u}{\partial x} + y\dfrac{\partial u}{\partial y} = nu.$

Proof. Since u is a homogeneous function of degree n in x and y, it can be expressed as

$$u = x^n f\left(\frac{y}{x}\right)$$

$$\therefore \qquad \frac{\partial u}{\partial x} = nx^{n-1} f\left(\frac{y}{x}\right) + x^n f'\left(\frac{y}{x}\right)\cdot\left(-\frac{y}{x^2}\right)$$

$\Rightarrow$ $$x\frac{\partial u}{\partial x} = nx^n f\left(\frac{y}{x}\right) - x^{n-1} yf'\left(\frac{y}{x}\right) \qquad ...(1)$$

Also, $$\frac{\partial u}{\partial y} = x^n f'\left(\frac{y}{x}\right) . \frac{1}{x} = x^{n-1} f'\left(\frac{y}{x}\right)$$

$\Rightarrow$ $$y\frac{\partial u}{\partial y} = x^{n-1} yf'\left(\frac{y}{x}\right) \qquad ...(2)$$

Adding (1) and (2), we get $x\frac{\partial u}{\partial x} + y\frac{\partial u}{\partial y} = nx^n f\left(\frac{y}{x}\right) = nu.$

Note. Euler's theorem can be extended to a homogeneous function of any number of variables. Thus, if u is a homogeneous function of degree n in x, y and z, then $x\frac{\partial u}{\partial x} + y\frac{\partial u}{\partial y} + z\frac{\partial u}{\partial z} = nu.$

3.7. IF u IS A HOMOGENEOUS FUNCTION OF DEGREE n IN x AND y, THEN $x^2\frac{\partial^2 u}{\partial x^2} + 2xy\frac{\partial^2 u}{\partial x\,\partial y} + y^2\frac{\partial^2 u}{\partial y^2} = n(n-1)u$

Proof. Since u is a homogeneous function of degree n in x and y

$\therefore$ By Euler's Theorem, we have $$x\frac{\partial u}{\partial x} + y\frac{\partial u}{\partial y} = nu \qquad ...(1)$$

Differentiating (1) partially w.r.t. x, we have $$1\cdot\frac{\partial u}{\partial x} + x\frac{\partial^2 u}{\partial x^2} + y\cdot\frac{\partial^2 u}{\partial x\partial y} = n\frac{\partial u}{\partial x} \qquad ...(2)$$

Differentiating (1) partially, w.r.t. y, we have $$x\frac{\partial^2 u}{\partial y\partial x} + 1\cdot\frac{\partial u}{\partial y} + y\cdot\frac{\partial^2 u}{\partial y^2} = n\cdot\frac{\partial u}{\partial y}$$

But $$\frac{\partial^2 u}{\partial y\partial x} = \frac{\partial^2 u}{\partial x\partial y}$$

$\therefore$ $$x\frac{\partial^2 u}{\partial x\partial y} + \frac{\partial u}{\partial y} + y\frac{\partial^2 u}{\partial y^2} = n\frac{\partial u}{\partial y} \qquad ...(3)$$

Multiplying (2) by x, (3) by y and adding

$$x^2\frac{\partial^2 u}{\partial x^2} + 2xy\frac{\partial^2 u}{\partial x\partial y} + y^2\frac{\partial^2 u}{\partial y^2} + \left(x\frac{\partial u}{\partial x} + y\frac{\partial u}{\partial y}\right) = n\left(x\frac{\partial u}{\partial x} + y\frac{\partial u}{\partial y}\right)$$

or $$x^2\frac{\partial^2 u}{\partial x^2} + 2xy\frac{\partial^2 u}{\partial x\partial y} + y^2\frac{\partial^2 u}{\partial y^2} + nu = n\cdot nu \qquad \text{[Using (1)]}$$

$\Rightarrow$ $$x^2\frac{\partial^2 u}{\partial x^2} + 2xy\frac{\partial^2 u}{\partial x\partial y} + y^2\frac{\partial^2 u}{\partial y^2} = n^2u - nu = n(n-1)u.$$

ILLUSTRATIVE EXAMPLES

Example 1. *Verify Euler's theorem for the functions:*

(*i*) $u = (x^{1/2} + y^{1/2})(x^n + y^n)$ (*ii*) $u = \sin^{-1}\dfrac{x}{y} + \tan^{-1}\dfrac{y}{x}$

(*iii*) $f(x, y, z) = 3x^2yz + 5xy^2z + 4z^4$ **(P.T.U., Dec., 2005)**

Sol. (*i*) $u = (x^{1/2} + y^{1/2})(x^n + y^n)$ ...(1)

$$= x^{1/2}\left(1+\frac{y^{1/2}}{x^{1/2}}\right)x^n\left(1+\frac{y^n}{x^n}\right) = x^{n+1/2}\left[1+\left(\frac{y}{x}\right)^{1/2}\right]\left[1+\left(\frac{y}{x}\right)^n\right] = x^{n+1/2} f\left(\frac{y}{x}\right)$$

[or $f(tx, ty) = t^{n+1/2} f(x, y)$]

$\Rightarrow$ u is a homogeneous function of degree $\left(n+\dfrac{1}{2}\right)$ in x and y

$\therefore$ By Euler's theorem, we should have $x\dfrac{\partial u}{\partial x} + y\dfrac{\partial u}{\partial y} = \left(n+\dfrac{1}{2}\right)u$ ...(2)

From (1),
$$\frac{\partial u}{\partial x} = \frac{1}{2}x^{-1/2}(x^n + y^n) + nx^{n-1}(x^{1/2} + y^{1/2})$$

$$x\frac{\partial u}{\partial x} = \frac{1}{2}x^{1/2}(x^n + y^n) + nx^n(x^{1/2} + y^{1/2})$$

$$\frac{\partial u}{\partial y} = \frac{1}{2}y^{-1/2}(x^n + y^n) + ny^{n-1}(x^{1/2} + y^{1/2})$$

$$y\frac{\partial u}{\partial y} = \frac{1}{2}y^{1/2}(x^n + y^n) + ny^n(x^{1/2} + y^{1/2})$$

$$\therefore \quad x\frac{\partial u}{\partial x} + y\frac{\partial u}{\partial y} = \frac{1}{2}(x^{1/2} + y^{1/2})(x^n + y^n) + n(x^n + y^n)(x^{1/2} + y^{1/2})$$

$= \frac{1}{2}u + nu = (n + \frac{1}{2})u$ which is the same as (2). Hence the verification.

(*ii*)
$$u = \sin^{-1}\frac{x}{y} + \tan^{-1}\frac{y}{x} \qquad ...(1)$$

$$= \text{cosec}^{-1}\frac{y}{x} + \tan^{-1}\frac{y}{x} = x^0 f\left(\frac{y}{x}\right)$$

[OR $f(tx, ty) = f(x, y) = t^0 f(x, y)$]

$\Rightarrow$ u is a homogeneous function of degree 0 in x and y.

$\therefore$ By Euler's theorem, we should have $x\dfrac{\partial u}{\partial x} + y\dfrac{\partial u}{\partial y} = 0 \times u = 0$ ...(2)

From (1),
$$\frac{\partial u}{\partial x} = \frac{1}{\sqrt{1-\dfrac{x^2}{y^2}}}\cdot\frac{1}{y} + \frac{1}{1+\dfrac{y^2}{x^2}}\left(-\frac{y}{x^2}\right) = \frac{1}{\sqrt{y^2 - x^2}} - \frac{y}{x^2 + y^2}$$

$$x\frac{\partial u}{\partial x} = \frac{x}{\sqrt{y^2 - x^2}} - \frac{xy}{x^2 + y^2}$$

$$\frac{\partial u}{\partial y} = \frac{1}{\sqrt{1-\frac{x^2}{y^2}}}\left(-\frac{x}{y^2}\right) + \frac{1}{1+\frac{y^2}{x^2}}\cdot\frac{1}{x} = -\frac{x}{y\sqrt{y^2-x^2}} + \frac{x}{x^2+y^2}$$

$$y\frac{\partial u}{\partial y} = -\frac{x}{\sqrt{y^2-x^2}} + \frac{xy}{x^2+y^2}$$

$\therefore \quad x\frac{\partial u}{\partial x} + y\frac{\partial u}{\partial y} = 0$ which is the same as (2). Hence the verification.

(*iii*) $f(x, y, z) = 3x^2yz + 5xy^2z + 4z^4$ ...(1)

$f(x, y, z)$ is a homogeneous function of x, y, z of degree 4.

By Euler's theorem $x\frac{\partial f}{\partial x} + y\frac{\partial f}{\partial y} + z\frac{\partial f}{\partial z} = nf = 4f$

Differentiable (1) partially w.r.t. x, y, z successively, we get

$$\frac{\partial f}{\partial x} = 6xyz + 5y^2z$$

$$\frac{\partial f}{\partial y} = 3x^2z + 10xyz$$

$$\frac{\partial f}{\partial z} = 3x^2y + 5xy^2 + 16z^3$$

$$x\frac{\partial f}{\partial x} + y\frac{\partial f}{\partial y} + z\frac{\partial f}{\partial z} = 6x^2yz + 5xy^2z + 3x^2yz + 10xy^2z + 3x^2yz + 5xy^2z + 16z^4$$

$$= 12x^2yz + 20xy^2z + 16z^4 = 4(3x^2yz + 5xy^2z + 4z^4) = 4f(x, y, z).$$

Example 2. *If* $u = xf\left(\frac{y}{x}\right) + g\left(\frac{y}{x}\right)$, *prove that* $x^2\frac{\partial^2 u}{\partial x^2} + 2xy\frac{\partial^2 u}{\partial x \partial y} + y^2\frac{\partial^2 u}{\partial y^2} = 0.$

(P.T.U., May 2012)

Sol. $u = xf\left(\frac{y}{x}\right) + g\left(\frac{y}{x}\right) = u_1 + u_2$ where

$$u_1 = xf\left(\frac{y}{x}\right),\ u_2 = g\left(\frac{y}{x}\right)$$

u_1 is a homogeneous function of x, y of degree 1

$\therefore$ By Euler's theorem

$$x\frac{\partial u_1}{\partial x} + y\frac{\partial u_1}{\partial y} = 1 \cdot u_1 \quad ...(1)$$

u_2 is a homogeneous function of x, y of degree zero

$$\therefore \quad x\frac{\partial u_2}{\partial x} + y\frac{\partial u_2}{\partial y} = 0 \quad ...(2)$$

Differentiate (1) and (2) partially w.r.t. x and y

$$x\frac{\partial^2 u_1}{\partial x^2}+\frac{\partial u_1}{\partial x}+y\frac{\partial^2 u_1}{\partial x\partial y}=\frac{\partial u_1}{\partial x}$$

or $$x\frac{\partial^2 u_1}{\partial x^2}+y\frac{\partial^2 u_1}{\partial x\partial y}=0 \qquad ...(3)$$

$$x\frac{\partial^2 u_1}{\partial x\partial y}+y\frac{\partial^2 u_1}{\partial y^2}+\frac{\partial u_1}{\partial y}=\frac{\partial u_1}{\partial y}$$

or $$x\frac{\partial^2 u_1}{\partial x\partial y}+y\frac{\partial^2 u_1}{\partial y^2}=0 \qquad ...(4)$$

$$x\frac{\partial^2 u_2}{\partial x^2}+\frac{\partial u_2}{\partial x}+y\frac{\partial^2 u_2}{\partial x\partial y}=0 \qquad ...(5)$$

$$x\frac{\partial^2 u_2}{\partial x\partial y}+y\frac{\partial^2 u_2}{\partial y^2}+\frac{\partial u_2}{\partial y}=0 \qquad ...(6)$$

Multiple (3), (5) each by x and (4), (6) each by y and add

$$x^2\frac{\partial^2 u_1}{\partial x^2}+xy\frac{\partial^2 u_1}{\partial x\partial y}+xy\frac{\partial^2 u_1}{\partial x\partial y}+y^2\frac{\partial^2 u_1}{\partial y^2}+x^2\frac{\partial^2 u_2}{\partial x^2}+x\frac{\partial u_2}{\partial x}+xy\frac{\partial^2 u_2}{\partial x\partial y}$$
$$+xy\frac{\partial^2 u_2}{\partial x\partial y}+y^2\frac{\partial^2 u_2}{\partial y^2}+y\frac{\partial u_2}{\partial y}=0$$

or $$x^2\left(\frac{\partial^2 u_1}{\partial x^2}+\frac{\partial^2 u_2}{\partial x^2}\right)+2xy\left(\frac{\partial^2 u_1}{\partial x\partial y}+\frac{\partial^2 u_2}{\partial x\partial y}\right)+y^2\left(\frac{\partial^2 u_1}{\partial y^2}+\frac{\partial^2 u_2}{\partial y^2}\right)+\left(x\frac{\partial u_2}{dx}+y\frac{\partial u_2}{\partial y}\right)=0$$

or $$x^2\frac{\partial^2 u}{\partial x^2}+2xy\frac{\partial^2 u}{\partial x\partial y}+y^2\frac{\partial^2 u}{\partial y^2}+0=0 \qquad (\because \text{ of } 2)$$

or $$x^2\frac{\partial^2 u}{\partial x^2}+2xy\frac{\partial^2 u}{\partial x\partial y}+y^2\frac{\partial^2 u}{\partial y^2}=0$$

Example 3. (*i*) *If* $u=\sin^{-1}\dfrac{x^3+y^3+z^3}{ax+by+cz}$, *prove that* $xu_x+yu_y+zu_z=2\tan u$.

(P.T.U., May 2008)

(*ii*) *If* $u=\tan^{-1}\dfrac{x^3+y^3}{x-y}$, *prove that* $x^2\dfrac{\partial^2 u}{\partial x^2}+2xy\dfrac{\partial^2 u}{\partial x\partial y}+y^2\dfrac{\partial^2 u}{\partial y^2}=\sin 4u-\sin 2u=$ *2 cos 3u sin u.* **(P.T.U., Jan. 2010, Dec. 2011, Dec. 2015)**

Sol. (*i*) $$u=\sin^{-1}\frac{x^3+y^3+z^3}{ax+by+cz}$$

u is not a homogeneous function

$$\therefore \quad \sin u=\frac{x^3+y^3+z^3}{ax+by+cz}=\frac{x^3\left[1+\left(\frac{y}{x}\right)^3+\left(\frac{z}{x}\right)^3\right]}{x\left[a+b\frac{y}{x}+c\frac{z}{x}\right]}=x^2f\left(\frac{y}{x},\frac{z}{x}\right)$$

$\therefore$ $\sin u$ is a homogeneous function of order 2

$\therefore$ By Euler's theorem

$$x \frac{\partial}{\partial x}(\sin u) + y \frac{\partial}{\partial y}(\sin u) + z \frac{\partial}{\partial z}(\sin u) = 2 \sin u$$

or $$x \cos u \frac{\partial u}{\partial x} + y \cos u \frac{\partial u}{\partial y} + z \cos u \frac{\partial u}{\partial z} = 2 \sin u$$

or $$x \frac{\partial u}{\partial x} + y \frac{\partial u}{\partial y} + z \frac{\partial u}{\partial z} = 2 \tan u$$

(*ii*) $u = \tan^{-1} \dfrac{x^3 + y^3}{x - y}$

u is not a homogeneous function of x and y

$$\therefore \qquad \tan u = \frac{x^3 + y^3}{x - y} = x^2 \frac{1 + \left(\frac{y}{x}\right)^3}{1 - \left(\frac{y}{x}\right)} = x^2 f\left(\frac{y}{x}\right)$$

$\tan u$ is a homogeneous function of degree 2

$\therefore$ By Euler's theorem

$$x \frac{\partial}{\partial x}(\tan u) + y \frac{\partial}{\partial y}(\tan u) = 2 \tan u$$

or $$x \sec^2 u \frac{\partial u}{\partial x} + y \sec^2 u \frac{\partial u}{\partial y} = 2 \tan u$$

or $$x \frac{\partial u}{\partial x} + y \frac{\partial u}{\partial y} = 2 \frac{\tan u}{\sec^2 u} = 2 \sin u \cos u = \sin 2u \qquad \ldots(1)$$

Differentiate (1) partially w.r.t. x and y

$$x \frac{\partial^2 u}{\partial x^2} + \frac{\partial u}{\partial x} + y \frac{\partial^2 u}{\partial x \partial y} = \cos 2u \cdot 2 \frac{\partial u}{\partial x} \qquad \ldots(2)$$

$$x \frac{\partial^2 u}{\partial y \partial x} + y \frac{\partial^2 u}{\partial y^2} + \frac{\partial u}{\partial y} = \cos 2u \cdot 2 \frac{\partial u}{\partial y} \qquad \ldots(3)$$

Multiply (2) by x and (3) by y and add

$$x^2 \frac{\partial^2 u}{\partial x^2} + 2xy \frac{\partial^2 u}{\partial x \partial y} + \left(x \frac{\partial u}{\partial x} + y \frac{\partial u}{\partial y}\right) + y^2 \frac{\partial^2 u}{\partial y^2} = 2 \cos 2u \left(x \frac{\partial u}{\partial x} + y \frac{\partial u}{\partial y}\right)$$

or $$x^2 \frac{\partial^2 u}{\partial x^2} + 2xy \frac{\partial^2 u}{\partial x \partial y} + y^2 \frac{\partial^2 u}{\partial y^2} = (2 \cos 2u - 1)\left(x \frac{\partial u}{\partial x} + y \frac{\partial u}{\partial y}\right)$$

$$= (2 \cos 2u - 1)(\sin 2u) \qquad | \text{ Using (1)}$$

$$= \sin 4u - \sin 2u$$

$$= 2 \cos 3u \sin u.$$

Example 4. *(i) If* $u = \cos\left(\dfrac{xy + yz + zx}{x^2 + y^2 + z^2}\right)$, *prove that* $x\dfrac{\partial u}{\partial x} + y\dfrac{\partial u}{\partial y} + z\dfrac{\partial u}{\partial z} = 0.$

(ii) If $u = \log\dfrac{x^5 + y^5 + z^5}{x^2 + y^2 + z^2}$, *prove that* $x\dfrac{\partial u}{\partial x} + y\dfrac{\partial u}{\partial y} + z\dfrac{\partial u}{\partial z} = 3.$

Sol. (*i*)

$$u = \cos\frac{xy + yz + zx}{x^2 + y^2 + z^2}$$

$$= \cos\frac{x^2\left(\frac{y}{x} + \frac{y}{x}\cdot\frac{z}{x} + \frac{z}{x}\right)}{x^2\left[1 + \left(\frac{y}{x}\right)^2 + \left(\frac{z}{x}\right)^2\right]}$$

$$u = x^0 \cos\frac{\frac{y}{x} + \frac{y}{x}\cdot\frac{z}{x} + \frac{z}{x}}{1 + \left(\frac{y}{x}\right)^2 + \left(\frac{z}{x}\right)^2} = x^0 f\left(\frac{y}{x}, \frac{z}{x}\right)$$

which is a homogeneous function of x, y, z of degree 0 $\therefore$ by Euler's theorem

$\therefore$ By Euler's theorem $x\dfrac{\partial u}{\partial x} + y\dfrac{\partial u}{\partial y} + z\dfrac{\partial u}{\partial z} = 0 \cdot u = 0$

Hence $x\dfrac{\partial u}{\partial x} + y\dfrac{\partial u}{\partial y} + z\dfrac{\partial u}{\partial z} = 0.$

Aliter: Let $u = f(x, y, z)$

$$\therefore \quad f(x, y, z) = \cos\frac{xy + yz + zx}{x^2 + y^2 + z^2}$$

$$f(tx, ty, tz) = \cos\frac{t^2xy + t^2yz + t^2zx}{t^2x^2 + t^2y^2 + t^2z^2} = t^0 \cos\frac{xy + yz + zx}{x^2 + y^2 + z^2} = t^0 f(x, y, z)$$

$\therefore$ $f(x, y, z)$ is homogeneous function of x, y, z of degree 0.

$\therefore$ By Euler's theorem $x\dfrac{\partial u}{\partial x} + y\dfrac{\partial u}{\partial y} + z\dfrac{\partial u}{\partial z} = 0$

(*ii*) $u = \log\dfrac{x^5 + y^5 + z^5}{x^2 + y^2 + z^2} = \log x^3 \cdot \dfrac{1 + \left(\frac{y}{x}\right)^5 + \left(\frac{z}{x}\right)^5}{1 + \left(\frac{y}{x}\right)^2 + \left(\frac{z}{x}\right)^2}$ it is not homogeneous function of

x, y, z $\because$ it cannot be expressed as $x^n f\left(\dfrac{y}{x}, \dfrac{z}{x}\right)$

Now $e^u = x^3 \cdot \dfrac{1 + \left(\frac{y}{x}\right)^5 + \left(\frac{z}{x}\right)^5}{1 + \left(\frac{y}{x}\right)^2 + \left(\frac{z}{x}\right)^2}$, let $\phi(u) = e^u$ $\therefore$ $\phi(u) = x^3 f\left(\dfrac{y}{x}, \dfrac{z}{x}\right)$;

$\phi(u)$ is homogeneous function of x, y, z of degree 3 $\therefore$ by Euler's theorem

$$x\frac{\partial\phi}{\partial x}+y\frac{\partial\phi}{\partial y}+z\frac{\partial\phi}{\partial z}=3\phi \quad \text{or} \quad x\frac{\partial}{\partial x}(e^u)+y\frac{\partial}{\partial y}(e^u)+z\frac{\partial}{\partial z}(e^u)=3e^u$$

or $$x\,.\,e^u\frac{\partial u}{\partial x}+ye^u\frac{\partial u}{\partial y}+ze^u\frac{\partial u}{\partial z}=3e^u$$

or $$x\frac{\partial u}{\partial x}+y\frac{\partial u}{\partial y}+z\frac{\partial u}{\partial z}=3.$$

Example 5. *Given* $z=x^n f_1\left(\frac{y}{x}\right)+y^{-n}f_2\left(\frac{x}{y}\right)$, *prove that*

$$x^2\frac{\partial^2 z}{\partial x^2}+2xy\frac{\partial^2 z}{\partial x\partial y}+y^2\frac{\partial^2 z}{\partial y^2}+x\frac{\partial z}{\partial y}+y\frac{\partial z}{\partial y}=n^2 z.$$

Sol. Let $$u=x^n f_1\left(\frac{y}{x}\right),\ v=y^{-n}f_2\left(\frac{x}{y}\right)$$

$\therefore$ $$z=u+v$$

u is a homogeneous function of x, y of degree n.

$\therefore$ $$x\frac{\partial u}{\partial x}+y\frac{\partial u}{\partial y}=nu \qquad ...(1)$$

v is a homogeneous function of x, y of degree $-n$.

$\therefore$ $$x\frac{\partial v}{\partial x}+y\frac{\partial v}{\partial y}=-nv \qquad ...(2)$$

Diff. both (1) and (2) partially w.r.t. x and y

$$x\frac{\partial^2 u}{\partial x^2}+\frac{\partial u}{\partial x}+y\frac{\partial^2 u}{\partial x\partial y}=n\frac{\partial u}{\partial x} \qquad ...(3)$$

$$x\frac{\partial^2 u}{\partial y\partial x}+y\frac{\partial^2 u}{\partial y^2}+\frac{\partial u}{\partial y}=n\frac{\partial u}{\partial y} \qquad ...(4)$$

$$x\frac{\partial^2 v}{\partial x^2}+\frac{\partial v}{\partial x}+y\frac{\partial^2 v}{\partial x\partial y}=-n\frac{\partial v}{\partial x} \qquad ...(5)$$

$$x\frac{\partial^2 v}{\partial y\partial x}+y\frac{\partial^2 v}{\partial y^2}+\frac{\partial v}{\partial y}=-n\frac{\partial v}{\partial y} \qquad ...(6)$$

Multiply (3) and (5) by x and (4), (6) by y and adding,

$$x^2\left(\frac{\partial^2 u}{\partial x^2}+\frac{\partial^2 v}{\partial x^2}\right)+x\left(\frac{\partial u}{\partial x}+\frac{\partial v}{\partial x}\right)+xy\left(\frac{\partial^2 u}{\partial x\partial y}+\frac{\partial^2 v}{\partial x\partial y}\right)+xy\left(\frac{\partial^2 u}{\partial y\partial x}+\frac{\partial^2 v}{\partial y\partial x}\right)$$

$$+y^2\left(\frac{\partial^2 u}{\partial y^2}+\frac{\partial^2 v}{\partial y^2}\right)+y\left(\frac{\partial u}{\partial y}+\frac{\partial v}{\partial y}\right)$$

$$=n\left(x\frac{\partial u}{\partial x}+y\frac{\partial u}{\partial y}\right)-n\left(x\frac{\partial v}{\partial x}+y\frac{\partial v}{\partial y}\right)$$

$$x^2 \frac{\partial^2}{\partial x^2}(u+v) + y^2 \frac{\partial^2}{\partial y^2}(u+v) + 2xy \frac{\partial^2}{\partial x \partial y}(u+v)$$
$$+ x \frac{\partial}{\partial x}(u+v) + y \frac{\partial}{\partial y}(u+v) = n \,.\, nu - n\,(-nv)$$

or
$$x^2 \frac{\partial^2 z}{\partial x^2} + y^2 \frac{\partial^2 z}{\partial y^2} + 2xy \frac{\partial^2 z}{\partial x \partial y} + x \frac{\partial z}{\partial x} + y \frac{\partial z}{\partial y} = n^2 z.$$

Example 6. *If* $u = \tan^{-1} \frac{y^2}{x}$, *show that* $x^2 \frac{\partial^2 u}{\partial x^2} + 2xy \frac{\partial^2 u}{\partial x \partial y} + y^2 \frac{\partial^2 u}{\partial y^2} = -\sin 2u \sin^2 u.$

(P.T.U., May 2006, May 2011)

Sol.
$$u = \tan^{-1} \frac{y^2}{x}$$
$$\tan u = \frac{y^2}{x}$$

Let $f(x, y) = \tan u = \frac{y^2}{x} = x \frac{y^2}{x^2} = x^1 \left(\frac{y}{x}\right)^2$ which is a homogeneous function in x, y of degree 1.

$\therefore$ By Euler's theorem $x\frac{\partial f}{\partial x} + y\frac{\partial f}{\partial y} = 1 \cdot f$

or
$$x \frac{\partial}{\partial x}(\tan u) + y \frac{\partial}{\partial y}(\tan u) = \tan u$$

or
$$x \sec^2 u \cdot \frac{\partial u}{\partial x} + y \sec^2 u \frac{\partial u}{\partial y} = \tan u$$

or
$$x\frac{\partial u}{\partial x} + y\frac{\partial u}{\partial y} = \tan u \cos^2 u = \sin u \cos u = \frac{1}{2} \sin 2u \qquad ...(1)$$

Differentiating it partially w.r.t. x and y

$$x\frac{\partial^2 u}{\partial x^2} + \frac{\partial u}{\partial x} + y\frac{\partial^2 u}{\partial x \partial y} = \cos 2u \cdot \frac{\partial u}{\partial x} \qquad ...(2)$$

$$x\frac{\partial^2 u}{\partial y \partial x} + y\frac{\partial^2 u}{\partial y^2} + \frac{\partial u}{\partial y} = \cos 2u \cdot \frac{\partial u}{\partial y} \qquad ...(3)$$

Multiply (2) by x and (3) by y and add,

$$x^2 \frac{\partial^2 u}{\partial x^2} + y^2 \frac{\partial^2 u}{\partial y^2} + 2xy\frac{\partial^2 u}{\partial x \partial y} + \left(x\frac{\partial u}{\partial x} + y\frac{\partial u}{\partial y}\right) = \cos 2u \left[x\frac{\partial u}{\partial x} + y\frac{\partial u}{\partial y}\right]$$

$$x^2 \frac{\partial^2 u}{\partial x^2} + 2xy\frac{\partial^2 u}{\partial x \partial y} + y^2 \frac{\partial^2 u}{\partial y^2} = (\cos 2u - 1)\left(x\frac{\partial u}{\partial x} + y\frac{\partial u}{\partial y}\right) = -2 \sin^2 u \cdot \frac{1}{2} \sin 2u$$

[Using (1)]

Hence
$$x^2 \frac{\partial^2 u}{\partial x^2} + 2xy\frac{\partial^2 u}{\partial x \partial y} + y^2 \frac{\partial^2 u}{\partial y^2} = -\sin 2u \sin^2 u.$$

Example 7. *If* $u = \text{cosec}^{-1}\left(\dfrac{x^{1/2}+y^{1/2}}{x^{1/3}+y^{1/3}}\right)^{1/2}$*, prove that*

$$x^2\frac{\partial^2 u}{\partial x^2}+2xy\frac{\partial^2 u}{\partial x\partial y}+y^2\frac{\partial^2 u}{\partial y^2}=\frac{13+\tan^2 u}{144}.$$

Sol.
$$u = \text{cosec}^{-1}\left(\frac{x^{1/2}+y^{1/2}}{x^{1/3}+y^{1/3}}\right)^{1/2}$$

$$f(x, y) = \text{cosec}\, u = \left(\frac{x^{1/2}+y^{1/2}}{x^{1/3}+y^{1/3}}\right)^{1/2} = \frac{x^{1/4}\left[1+\left(\frac{y}{x}\right)^{1/2}\right]^{1/2}}{x^{1/6}\left[1+\left(\frac{y}{x}\right)^{1/3}\right]^{1/2}} = x^{\frac{1}{12}}\,\phi\left(\frac{y}{x}\right)$$

∴ $f(x, y)$ is a homogeneous function of degree $\dfrac{1}{12}$

∴ By Euler's theorem

$$x\frac{\partial f}{\partial x}+y\frac{\partial f}{\partial y}=\frac{1}{12}f$$

or
$$x\frac{\partial}{\partial x}(\text{cosec}\, u)+y\frac{\partial}{\partial y}(\text{cosec}\, u)=\frac{1}{12}\text{cosec}\, u$$

$$-x\,\text{cosec}\, u\cot u\frac{\partial u}{\partial x}-y\,\text{cosec}\, u\cot u\frac{\partial u}{\partial y}=\frac{1}{12}\text{cosec}\, u$$

or
$$x\frac{\partial u}{\partial x}+y\frac{\partial u}{\partial y}=\frac{-1}{12}\text{cosec}\, u\cdot\frac{1}{\text{cosec}\, u\cot u}=-\frac{1}{12}\tan u$$

∴
$$x\frac{\partial u}{\partial x}+y\frac{\partial u}{\partial y}=-\frac{1}{12}\tan u \quad \text{...(1)}$$

Differentiate (1) partially w.r.t. x and y

$$x\frac{\partial^2 u}{\partial x^2}+\frac{\partial u}{\partial y}+y\frac{\partial^2 u}{\partial x\partial y}=-\frac{1}{12}\sec^2 u\frac{\partial u}{\partial x} \quad \text{...(2)}$$

$$x\frac{\partial^2 u}{\partial y\partial x}+\frac{\partial u}{\partial y}+y\frac{\partial^2 u}{\partial y^2}=-\frac{1}{12}\sec^2 u\frac{\partial u}{\partial y} \quad \text{...(3)}$$

Multiply (2) by x and (3) by y and add

$$x^2\frac{\partial^2 u}{\partial x^2}+y^2\frac{\partial^2 u}{\partial y^2}+2xy\frac{\partial^2 u}{\partial x\partial y}+\left(x\frac{\partial u}{\partial x}+y\frac{\partial u}{\partial y}\right)=-\frac{1}{12}\sec^2 u\left(x\frac{\partial u}{\partial x}+y\frac{\partial u}{\partial y}\right)$$

or
$$x^2\frac{\partial^2 u}{\partial x^2}+y^2\frac{\partial^2 u}{\partial y^2}+2xy\frac{\partial^2 u}{\partial x\partial y}=-\left[1+\frac{1}{12}\sec^2 u\right]\left[u\frac{\partial u}{\partial x}+y\frac{\partial u}{\partial y}\right]=-\left[\frac{12+1+\tan^2 u}{12}\right]\left[-\frac{1}{12}\right]$$

$$=\frac{13+\tan^2 u}{144}$$

Hence $x^2\dfrac{\partial^2 u}{\partial x^2}+2xy\dfrac{\partial^2 u}{\partial x\partial y}+y^2\dfrac{\partial^2 u}{\partial y^2}=\dfrac{13+\tan^2 u}{144}.$

TEST YOUR KNOWLEDGE

1. Verify Euler's theorem for the functions:

(*i*) $f(x, y) = ax^2 + 2hxy + by^2$ (*ii*) $u = \dfrac{x^{1/4} + y^{1/4}}{x^{1/5} + y^{1/5}}$

[**Hint.** (*ii*) Homogeneous function of degree $\dfrac{1}{20}$]

(*iii*) $f(x, y) = \dfrac{x^2 (x^2 - y^2)^3}{(x^2 + y^2)^3}$.

[**Hint.** Homogeneous function of degree 2]

2. (*i*) If $V = \dfrac{x^3 y^3}{x^3 + y^3}$, show that $x \dfrac{\partial V}{\partial x} + y \dfrac{\partial V}{\partial y} = 3V$.

(*ii*) If $V = \log \dfrac{x^4 + y^4}{x + y}$, show that $x \dfrac{\partial V}{\partial x} + y \dfrac{\partial V}{\partial y} = 3$.

[**Hint.** $e^V = \dfrac{x^4 + y^4}{x + y}$ is a homogeneous function of degree 3]

(*iii*) If $V = f\left(\dfrac{y}{x}\right)$ show that $x \dfrac{\partial V}{\partial x} + y \dfrac{\partial V}{\partial y} = 0$.

(*iv*) If $z = \log (x^2 + xy + y^2)$ show that $x \dfrac{\partial z}{\partial x} + y \dfrac{\partial z}{\partial y} = 2$. **(*P.T.U., Dec. 2013*)**

[**Hint.** $e^z = x^2 + xy + y^2$ is a homogeneous function of degree 2]

(*v*) If $u = e^{x^2 + y^2}$ prove that $x \dfrac{\partial u}{\partial x} + y \dfrac{\partial u}{\partial y} = 2u \log u$. **(*P.T.U., Dec. 2014*)**

[**Hint.** $\log u = x^2 + y^2$ is a homogeneous function of degree 2]

3. If $u = \cos^{-1} \dfrac{x + y}{\sqrt{x} + \sqrt{y}}$, show that $x \dfrac{\partial u}{\partial x} + y \dfrac{\partial u}{\partial y} + \dfrac{1}{2} \cot u = 0$.

4. (*i*) If $u = \sin^{-1} \left(\dfrac{x^2 + y^2}{x + y}\right)$, prove that $x \dfrac{\partial u}{\partial x} + y \dfrac{\partial u}{\partial y} = \tan u$ **(*P.T.U., Dec. 2013*)**

(*ii*) If $u = \sin^{-1} \dfrac{x}{y} + \tan^{-1} \dfrac{y}{x}$, prove that $x \dfrac{\partial u}{\partial x} + y \dfrac{\partial u}{\partial y} = 0$ **(*P.T.U., Dec. 2010*)**

[**Hint.** $u = \text{cosec}^{-1} \dfrac{y}{x} + \tan^{-1} \dfrac{y}{x}$ which is homogeneous function of degree 0]

5. (*i*) If $u = \sin^{-1} \dfrac{\sqrt{x} - \sqrt{y}}{\sqrt{x} + \sqrt{y}}$, show that $\dfrac{\partial u}{\partial x} = -\dfrac{y}{x} \dfrac{\partial u}{\partial y}$.

(*ii*) If $\sin u = \dfrac{x^2 y^2}{x + y}$, show that $x \dfrac{\partial u}{\partial x} + y \dfrac{\partial u}{\partial y} = 3 \tan u$. **(*P.T.U., Dec. 2012*)**

(*iii*) If $f(x, y) = \dfrac{1}{x^2} + \dfrac{1}{xy} + \dfrac{\log x - \log y}{x^2 + y^2}$, show that $x \dfrac{\partial f}{\partial x} + y \dfrac{\partial f}{\partial y} = -2f$. **(*P.T.U., May 2004*)**

[**Hint.** $f(x, y) = x^{-2} \left[1 + \dfrac{x}{y} + \dfrac{\log \dfrac{x}{y}}{1 + \left(\dfrac{y}{x}\right)^2} \right] = x^{-2} \left[1 + \dfrac{1}{\dfrac{y}{x}} - \dfrac{\log \dfrac{y}{x}}{1 + \left(\dfrac{y}{x}\right)^2} \right]$

which is homogeneous function of degree – 2]

6. If $u = \dfrac{x^2y^2}{x+y}$, show that

(i) $x \dfrac{\partial^2 u}{\partial x^2} + y \dfrac{\partial^2 u}{\partial x \partial y} = 2 \dfrac{\partial u}{\partial x}$ (ii) $x \dfrac{\partial^2 u}{\partial x \partial y} + y \dfrac{\partial^2 u}{\partial y^2} = 2 \dfrac{\partial u}{\partial y}$.

7. (i) If $u = (x^2 + y^2)^{1/3}$, show that $x^2 \dfrac{\partial^2 u}{\partial x^2} + 2xy \dfrac{\partial^2 u}{\partial x \partial y} + y^2 \dfrac{\partial^2 u}{\partial y^2} = -\dfrac{2u}{9}$.

[**Hint.** Homogeneous function of degree 2/3]

(ii) If $u = x f\left(\dfrac{y}{x}\right)$ show that $x \dfrac{\partial u}{\partial x} + y \dfrac{\partial u}{\partial y} = u$ and $x^2 \dfrac{\partial^2 u}{\partial x^2} + 2xy \dfrac{\partial^2 u}{\partial x \partial y} + y^2 \dfrac{\partial^2 u}{\partial y^2} = 0$.

(P.T.U., May 2008)

8. If $u = \sin^{-1} \dfrac{x+y}{\sqrt{x} + \sqrt{y}}$, prove that $x^2 \dfrac{\partial^2 u}{\partial x^2} + 2xy \dfrac{\partial^2 u}{\partial x \partial y} + y^2 \dfrac{\partial^2 u}{\partial y^2} = -\dfrac{\sin u \cos 2u}{4 \cos^3 u}$.

(P.T.U., May 2008)

9. If $u = x^2 \tan^{-1}\left(\dfrac{y}{x}\right) - y^2 \tan^{-1}\left(\dfrac{x}{y}\right)$, prove that $x^2 \dfrac{\partial^2 u}{\partial x^2} + 2xy \dfrac{\partial^2 u}{\partial x \partial y} + y^2 \dfrac{\partial^2 u}{\partial y^2} = 2u$.

10. (i) If $f(x, y) = \sqrt{x^2 - y^2} \sin^{-1} \dfrac{y}{x}$, show that $x \dfrac{\partial f}{\partial x} + y \dfrac{\partial f}{\partial y} = f(x, y)$

(ii) If $f(x, y) = \sqrt{y^2 - x^2} \sin^{-1} \dfrac{x}{y} + \dfrac{x^2 - y^2}{\sqrt{x^2 + y^2}}$, show that $x \dfrac{\partial f}{\partial x} + y \dfrac{\partial f}{\partial y} = f(x, y)$.

3.8. COMPOSITE FUNCTIONS; TOTAL DERIVATIVE ***(P.T.U., May 2007)***

(i) If $u = f(x, y)$ where $x = \phi(t)$, $y = \psi(t)$

then u is called a composite function of the **single variable** t and we can find $\dfrac{du}{dt}$ which is called the **Total Derivative** of u.

(ii) If $z = f(x, y)$ where $x = \phi(u, v)$, $y = \psi(u, v)$

then z is called a composite function of **two variables** u and v so that we can find $\dfrac{\partial z}{\partial u}$ and $\dfrac{\partial z}{\partial v}$.

3.9. (a) DIFFERENTIATION OF COMPOSITE FUNCTIONS

If u is composite function of t, defined by the relations $u = f(x, y)$; $x = \phi(t)$, $y = \psi(t)$, then find the total derivative of u w.r.t. t without actually substituting the values of x, y in $f(x, y)$.

OR

To prove

$$\frac{du}{dt} = \frac{\partial u}{\partial x} \cdot \frac{dx}{dt} + \frac{\partial u}{\partial y} \cdot \frac{dy}{dt}.$$

Proof. Here $u = f(x, y)$ ...(1)

Let δt be an increment in t and δx, δy, δu the corresponding increments in x, y and u respectively. Then, we have

$$u + \delta u = f(x + \delta x, y + \delta y) \quad ...(2)$$

Subtracting (1) from (2), we get

$$\delta u = f(x + \delta x, y + \delta y) - f(x, y)$$
$$= f(x + \delta x, y + \delta y) - f(x, y + \delta y) + f(x, y + \delta y) - f(x, y)$$

$$\frac{\delta u}{\delta t} = \frac{f(x+\delta x, y+\delta y) - f(x, y+\delta y)}{\delta t} + \frac{f(x, y+\delta y) - f(x, y)}{\delta t}$$

$$= \frac{f(x+\delta x, y+\delta y) - f(x, y+\delta y)}{\delta x} \cdot \frac{\delta x}{\delta t} + \frac{f(x, y+\delta y) - f(x, y)}{\delta y} \cdot \frac{\delta y}{\delta t} \qquad \text{...(3)}$$

As $\delta t \to 0$, δx and δy both $\to 0$, so that

$$\underset{\delta t \to 0}{\text{Lt}} \frac{\delta u}{\delta t} = \frac{du}{dt}, \quad \underset{\delta t \to 0}{\text{Lt}} \frac{\delta x}{\delta t} = \frac{dx}{dt}, \quad \underset{\delta t \to 0}{\text{Lt}} \frac{\delta y}{\delta t} = \frac{dy}{dt}$$

$$\underset{\delta x \to 0}{\text{Lt}} \frac{f(x+\delta x, y+\delta y) - f(x, y+\delta y)}{\delta y} = \frac{\partial f}{\partial x} = \frac{\partial u}{\partial x}$$

$$\underset{\delta y \to 0}{\text{Lt}} \frac{f(x, y+\delta y) - f(x, y)}{\delta y} = \frac{\partial f}{\partial y} = \frac{\partial u}{\partial y}$$

$\therefore$ From (1),
$$\mathbf{\frac{du}{dt} = \frac{\partial u}{\partial x} \cdot \frac{dx}{dt} + \frac{\partial u}{\partial y} \cdot \frac{dy}{dt}}$$

Cor. 1. If $u = f(x, y, z)$ and x, y, z are function of t, then u is a composite function of t and

$$\mathbf{\frac{du}{dt} = \frac{\partial u}{\partial x} \cdot \frac{dx}{dt} + \frac{\partial u}{\partial y} \cdot \frac{dy}{dt} + \frac{\partial u}{\partial z} \cdot \frac{dz}{dt}}.$$

Cor. 2. If $z = f(x, y)$ and x, y are functions of u and v, then

$$\mathbf{\frac{\partial z}{\partial u} = \frac{\partial z}{\partial x} \cdot \frac{\partial x}{\partial u} + \frac{\partial z}{\partial y} \cdot \frac{\partial y}{\partial u}; \quad \frac{\partial z}{\partial v} = \frac{\partial z}{\partial x} \cdot \frac{\partial x}{\partial v} + \frac{\partial z}{\partial y} \cdot \frac{\partial y}{\partial v}}.$$

Cor. 3. If $u = f(x, y)$ where $y = \phi(x)$ then since $u = \psi(x)$, u is a composite function of x.

$$\therefore \quad \frac{du}{dx} = \frac{\partial u}{\partial x} \cdot \frac{dx}{dx} + \frac{\partial u}{\partial y} \cdot \frac{dy}{dx} \quad \Rightarrow \quad \mathbf{\frac{du}{dx} = \frac{\partial u}{\partial x} + \frac{\partial u}{\partial y} \cdot \frac{dy}{dx}}.$$

3.9. (b) DIFFERENTIATION OF IMPLICIT FUNCTIONS *(P.T.U., Dec. 2004)*

If we are given an implicit function $f(x, y) = c$, then $u = f(x, y)$ where $u = c$

Using Cor. 3, we have $\frac{du}{dx} = \frac{\partial u}{\partial x} + \frac{\partial u}{\partial y} \cdot \frac{dy}{dx}$

But $\frac{du}{dx} = 0 \quad \therefore \quad \frac{\partial u}{\partial x} + \frac{\partial u}{\partial y} \cdot \frac{dy}{dx} = 0$ or $\frac{\partial f}{\partial x} + \frac{\partial f}{\partial y} \cdot \frac{dy}{dx} = 0$

$$\therefore \quad \frac{dy}{dx} = -\frac{\frac{\partial f}{\partial x}}{\frac{\partial f}{\partial y}} = -\frac{f_x}{f_y}$$

Hence $\mathbf{\frac{dy}{dx} = -\frac{f_x}{f_y}}$.

which gives, the important formula for the first differential coefficient of an implicit equation.

Cor. 1. If $f(x, y) = c$ then, we have $\frac{dy}{dx} = -\frac{f_x}{f_y}$ (proved above)

Differentiating again w.r.t. x, we get

$$\frac{d^2y}{dx^2} = -\frac{f_y \frac{d}{dx}(f_x) - f_x \frac{d}{dx}(f_y)}{f_y^2} = -\frac{f_y\left[\frac{\partial f_x}{\partial x} + \frac{\partial f_x}{\partial y} \cdot \frac{dy}{dx}\right] - f_x\left[\frac{\partial f_y}{\partial x} + \frac{\partial f_y}{\partial y} \cdot \frac{dy}{dx}\right]}{f_y^2}$$

$$= -\frac{f_y\left[f_{xx} - f_{yx} \cdot \frac{f_x}{f_y}\right] - f_x\left[f_{xy} - f_{yy} \cdot \frac{f_x}{f_y}\right]}{f_y^2} = -\frac{f_{xx}f_y^2 - f_xf_yf_{xy} - f_xf_yf_{xy} - f_{yy}f_x^2}{f_y^3}$$

Hence $\mathbf{\frac{d^2y}{dx^2} = -\frac{f_{xx}f_y^2 - 2f_xf_yf_{xy} + f_{yy}f_x^2}{f_y^3}}$.

which is the important formula for the second differential of an implicit function.

3.9. (c) CHANGE OF VARIABLES

If $z = f(x, y)$ and $x = \phi(u, v)$, $y = \psi(u, v)$, it is often necessary to change the expressions involving $z, x, y, \frac{\partial z}{\partial x}, \frac{\partial z}{\partial y}$ etc to expressions involving $z, u, v, \frac{\partial z}{\partial u}, \frac{\partial z}{\partial v}$ etc.

The necessary formulae for change of variables are easily obtained. If v is regarded constant then x, y, z all will be functions of u only.

∴ From Cor. 2 (3.9 *a*)

$$\frac{\partial z}{\partial u} = \frac{\partial z}{\partial x} \cdot \frac{\partial x}{\partial u} + \frac{\partial z}{\partial y} \cdot \frac{\partial y}{\partial u} \qquad \text{...(1)}$$

Similarly if u is regarded constant then x, y, z all will be functions of v only

$$\therefore \qquad \frac{\partial z}{\partial v} = \frac{\partial z}{\partial x} \cdot \frac{\partial x}{\partial v} + \frac{\partial z}{\partial y} \cdot \frac{\partial y}{\partial v} \qquad \text{...(2)}$$

On solving (1) and (2) as simultaneous functions in $\frac{\partial z}{\partial x}$ and $\frac{\partial z}{\partial y}$, we get their values in terms of $\frac{\partial z}{\partial u}$ and $\frac{\partial z}{\partial v}$, z, u, v.

If instead of x, y being functions of u, v it is given that u, v are functions of x, y it is easier to use the formula.

$$\frac{\partial z}{\partial x} = \frac{\partial z}{\partial u} \cdot \frac{\partial u}{\partial x} + \frac{\partial z}{\partial v} \cdot \frac{\partial v}{\partial x} \qquad \text{...(3)}$$

$$\frac{\partial z}{\partial y} = \frac{\partial z}{\partial u} \cdot \frac{\partial u}{\partial y} + \frac{\partial z}{\partial v} \cdot \frac{\partial v}{\partial y} \qquad \text{...(4)}$$

The higher derivatives of u can be found by repeated application of the formula (1) and (2).

ILLUSTRATIVE EXAMPLES

Example 1. *Find* $\frac{du}{dt}$ *when* $u = x^2 + y^2$, $x = at^2$, $y = 2at$. *Also verify by direct substitution.*

Sol. $$u = x^2 + y^2, \quad x = at^2, \ y = 2at$$

u is a function of x, y which are further functions of t

$\therefore$ u is a composite function of t

$$\therefore \quad \frac{du}{dt} = \frac{\partial u}{\partial x}\frac{dx}{dt} + \frac{\partial u}{\partial y}\frac{dy}{dt}$$

$$= (2x)(2at) + (2y) \cdot 2a$$

$$= 2at^2 \cdot 2at + 2 \cdot 2at \cdot 2a$$

$$= 4a^2 t^3 + 8a^2 t = 4a^2 t (t^2 + 2) \qquad ...(1)$$

Direct substitution gives

$$u = (at^2)^2 + (2at)^2 = a^2 t^4 + 4a^2 t^2$$

$$\frac{du}{dt} = a^2 . 4t^3 + 4a^2 . 2t = 4 a^2 t^3 + 8a^2 t$$

$$= 4a^2 t + (t^2 + 2) \text{ which is same as (1)}$$

Example 2. (*i*) *If* $u = \sin^{-1}(x - y)$, $x = 3t$, $y = 4t^3$, *show that* $\frac{du}{dt} = \frac{3}{\sqrt{1-t^2}}$.

(P.T.U., Dec. 2014, May 2014)

(*ii*) *If* $u = \sin \frac{x}{y}$, $x = e^t$, $y = t^2$, *find* $\frac{du}{dt}$. **(P.T.U., May 2006)**

Sol. (*i*) The given equations define u as a composite function of t.

$$\therefore \quad \frac{du}{dt} = \frac{\partial u}{\partial x} \cdot \frac{dx}{dt} + \frac{\partial u}{\partial y} \cdot \frac{dy}{dt} \text{ where } u = \sin^{-1}(x - y),\ x = 3t,\ y = 4t^3$$

$$= \frac{1}{\sqrt{1-(x-y)^2}} \cdot 3 + \frac{1}{\sqrt{1-(x-y)^2}} (-1) \cdot 12t^2$$

$$= \frac{3(1-4t^2)}{\sqrt{1-(x-y)^2}} = \frac{3(1-4t^2)}{\sqrt{1-(3t-4t^3)^2}} = \frac{3(1-4t^2)}{\sqrt{1-9t^2+24t^4-16t^6}}$$

$$= \frac{3(1-4t^2)}{\sqrt{(1-t^2)(1-8t^2+16t^4)}} = \frac{3(1-4t^2)}{\sqrt{(1-t^2)(1-4t^2)^2}} = \frac{3}{\sqrt{1-t^2}}.$$

(*ii*) $$u = \sin \frac{x}{y}\ ; x = e^t, \ y = t^2$$

We know that $$\frac{du}{dt} = \frac{\partial u}{\partial x} \cdot \frac{dx}{dt} + \frac{\partial u}{\partial y} \cdot \frac{dy}{dt}$$

$$\therefore \quad \frac{du}{dt} = \frac{1}{y} \cos \frac{x}{y} \cdot e^t + \cos \frac{x}{y} \left(-\frac{x}{y^2} \right) 2t$$

$$= \frac{e^t}{t^2}\cos\frac{e^t}{t^2} - \frac{2t\,e^t}{t^4}\cos\frac{e^t}{t^2}$$

$$= e^t\cos\frac{e^t}{t^2}\left\{\frac{1}{t^2}-\frac{2}{t^3}\right\} = e^t\cos\frac{e^t}{t^2}\left\{\frac{t-2}{t^3}\right\}$$

Hence $$\frac{du}{dt} = e^t\,\frac{t-2}{t^3}\cos\frac{e^t}{t^2}.$$

Example 3. (*i*) *If* $w = x^2 + y^2$; $x = r - s$, $y = r + s$ *find* $\frac{\partial w}{\partial r}, \frac{\partial w}{\partial s}$ *in terms of r, s.*

(P.T.U., May 2012)

(*ii*) *If z is a function of x and y, where* $x = e^u + e^{-v}$ *and* $y = e^{-u} - e^v$, *show that*

$$\frac{\partial z}{\partial u} - \frac{\partial z}{\partial v} = x\frac{\partial z}{\partial x} - y\frac{\partial z}{\partial y}.$$

Sol. (*i*) w is a composite function of r and s.

$$\therefore \quad \frac{\partial w}{\partial r} = \frac{\partial w}{\partial x}\cdot\frac{\partial x}{\partial r} + \frac{\partial w}{\partial y}\cdot\frac{\partial y}{\partial r} \qquad ...(1)$$

and $$\frac{\partial w}{\partial s} = \frac{\partial w}{\partial x}\cdot\frac{\partial x}{\partial s} + \frac{\partial w}{\partial y}\cdot\frac{\partial y}{\partial s} \qquad ...(2)$$

We have $w = x^2 + y^2$, $x = r - s$, $y = r + s$

$$\therefore \quad \frac{\partial w}{\partial x} = 2x; \quad \frac{\partial w}{\partial y} = 2y$$

$$\frac{\partial x}{\partial r} = 1, \quad \frac{\partial x}{\partial s} = -1, \quad \frac{\partial y}{\partial r} = 1, \quad \frac{\partial y}{\partial s} = 1$$

$\therefore$ From (1) and (2)

$$\frac{\partial w}{\partial r} = 2x\cdot 1 + 2y\cdot 1 = 2(x+y) = 2(2r) = 4r$$

$$\frac{\partial w}{\partial s} = 2x\cdot(-1) + 2y\cdot 1 = -2(x-y) = -2(-2s) = 4s$$

$$\therefore \quad \frac{\partial w}{\partial r} = 4r, \quad \frac{\partial w}{\partial s} = 4s.$$

(*ii*) Given $x = e^u + e^{-v}$, $y = e^{-u} - e^v$.

Here z is a composite function of u and v.

$$\therefore \quad \frac{\partial z}{\partial u} = \frac{\partial z}{\partial x}\cdot\frac{\partial x}{\partial u} + \frac{\partial z}{\partial y}\cdot\frac{\partial y}{\partial u} = \frac{\partial z}{\partial x}\cdot e^u + \frac{\partial z}{\partial y}(-e^{-u})$$

and $$\frac{\partial z}{\partial v} = \frac{\partial z}{\partial x}\cdot\frac{\partial x}{\partial v} + \frac{\partial z}{\partial y}\cdot\frac{\partial y}{\partial v} = \frac{\partial z}{\partial x}(-e^{-v}) + \frac{\partial z}{\partial y}(-e^v)$$

Subtracting, $$\frac{\partial z}{\partial u} - \frac{\partial z}{\partial v} = (e^u + e^{-v})\frac{\partial z}{\partial x} - (e^{-u} - e^v)\frac{\partial z}{\partial y} = x\frac{\partial z}{\partial x} - y\frac{\partial z}{\partial y}.$$

Example 4. (*i*) *If* $u = f(y - z, z - x, x - y)$, *prove that* $\frac{\partial u}{\partial x} + \frac{\partial u}{\partial y} + \frac{\partial u}{\partial z} = 0$.

(P.T.U., Dec. 2003, Dec. 2011)

(*ii*) *If* $u = f\left(\frac{x}{y}, \frac{y}{z}, \frac{z}{x}\right)$, *then prove that* $x\frac{\partial u}{\partial x} + y\frac{\partial u}{\partial y} + z\frac{\partial u}{\partial z} = 0$.

(P.T.U., May 2003, 2014, Dec. 2010)

Sol. (*i*) Here $u = f(X, Y, Z)$, where $X = y - z$, $Y = z - x$, $Z = x - y$

$\therefore$ u is a composite function of x, y and z.

$$\frac{\partial u}{\partial x} = \frac{\partial u}{\partial X}\cdot\frac{\partial X}{\partial x} + \frac{\partial u}{\partial Y}\cdot\frac{\partial Y}{\partial x} + \frac{\partial u}{\partial Z}\cdot\frac{\partial Z}{\partial x} = \frac{\partial u}{\partial X}(0) + \frac{\partial u}{\partial Y}(-1) + \frac{\partial u}{\partial Z}(1)$$

$$\frac{\partial u}{\partial y} = \frac{\partial u}{\partial X}\cdot\frac{\partial X}{\partial y} + \frac{\partial u}{\partial Y}\cdot\frac{\partial Y}{\partial y} + \frac{\partial u}{\partial Z}\cdot\frac{\partial Z}{\partial y} = \frac{\partial u}{\partial X}(1) + \frac{\partial u}{\partial Y}(0) + \frac{\partial u}{\partial Z}(-1)$$

$$\frac{\partial u}{\partial z} = \frac{\partial u}{\partial X}\cdot\frac{\partial X}{\partial z} + \frac{\partial u}{\partial Y}\cdot\frac{\partial Y}{\partial z} + \frac{\partial u}{\partial Z}\cdot\frac{\partial Z}{\partial z} = \frac{\partial u}{\partial X}(-1) + \frac{\partial u}{\partial Y}(1) + \frac{\partial u}{\partial Z}(0)$$

Adding, $\frac{\partial u}{\partial x} + \frac{\partial u}{\partial y} + \frac{\partial u}{\partial z} = 0$.

(*ii*) $$u = f\left(\frac{x}{y}, \frac{y}{z}, \frac{z}{x}\right)$$

Let $X = \frac{x}{y}$, $Y = \frac{y}{z}$, $Z = \frac{z}{x}$

$\therefore$ $u = f(X, Y, Z)$

$\therefore$ u is a composite function of x, y, z

(Here u is a function of X, Y, Z and further X, Y, Z are functions of x, y, z)

$$\frac{\partial u}{\partial x} = \frac{\partial u}{\partial X}\cdot\frac{\partial X}{\partial x} + \frac{\partial u}{\partial Y}\frac{\partial Y}{\partial x} + \frac{\partial u}{\partial Z}\frac{\partial Z}{\partial x} = \frac{\partial u}{\partial X}\left(\frac{1}{y}\right) + \frac{\partial u}{\partial Y}(0) + \frac{\partial u}{\partial Z}\left(-\frac{z}{x^2}\right)$$

$$\frac{\partial u}{\partial y} = \frac{\partial u}{\partial X}\cdot\frac{\partial X}{\partial y} + \frac{\partial u}{\partial Y}\cdot\frac{\partial Y}{\partial y} + \frac{\partial u}{\partial Z}\cdot\frac{\partial Z}{\partial y} = \frac{\partial u}{\partial X}\left(-\frac{x}{y^2}\right) + \frac{\partial u}{\partial Y}\left(\frac{1}{z}\right) + \frac{\partial u}{\partial Z}(0)$$

$$\frac{\partial u}{\partial z} = \frac{\partial u}{\partial X}\cdot\frac{\partial X}{\partial z} + \frac{\partial u}{\partial Y}\frac{\partial Y}{\partial z} + \frac{\partial u}{\partial Z}\frac{\partial Z}{\partial z} = \frac{\partial u}{\partial X}(0) + \frac{\partial u}{\partial Y}\left(-\frac{y}{z^2}\right) + \frac{\partial u}{\partial Z}\left(\frac{1}{x}\right)$$

$$x\frac{\partial u}{\partial x} + y\frac{\partial u}{\partial y} + z\frac{\partial u}{\partial z} = \frac{\partial u}{\partial X}\left(\frac{x}{y} - \frac{x}{y} + 0\right) + \frac{\partial u}{\partial Y}\left(0 + \frac{y}{z} - \frac{y}{z}\right) + \frac{\partial u}{\partial Z}\left(\frac{-z}{x} + 0 + \frac{z}{x}\right)$$

$$= \frac{\partial u}{\partial X}\cdot 0 + \frac{\partial u}{\partial Y}\cdot 0 + \frac{\partial u}{\partial Z}\cdot 0 = 0$$

Aliter:

Given $$u = f\left(\frac{x}{y}, \frac{y}{z}, \frac{z}{x}\right)$$

i.e., $$u(x, y, z) = f\left(\frac{x}{y}, \frac{y}{z}, \frac{z}{x}\right)$$

$$u(tx, ty, tz) = f\left(\frac{tx}{ty}, \frac{ty}{tz}, \frac{tz}{tx}\right) = f\left(\frac{x}{y}, \frac{y}{z}, \frac{z}{x}\right) = u\,(x, y, z)$$

$\therefore$ $u(tx, ty, tz) = t^0\, u(x, y, z)$

$\because$ u is a homogeneous functions of x, y, z of degree zero

$\therefore$ By Euler's theorem, $x\frac{\partial u}{\partial x} + y\frac{\partial u}{\partial y} + z\frac{\partial u}{\partial z} = nu = 0\cdot u = 0$.

Example 5. (*i*) *If* $x = u + v + w$, $y = vw + wu + uv$, $z = uvw$ *and F is a function of* x, y, z, *show that* $u\frac{\partial F}{\partial u} + v\frac{\partial F}{\partial v} + w\frac{\partial F}{\partial w} = x\frac{\partial F}{\partial x} + 2y\frac{\partial F}{\partial y} + 3z\frac{\partial F}{\partial z}$.

(*ii*) *If* $u = f(x^2 + 2yz,\ y^2 + 2zx)$, *prove that:* $(y^2 - zx)\frac{\partial u}{\partial x} + (x^2 - yz)\frac{\partial u}{\partial y} + (z^2 - xy)\frac{\partial u}{\partial z} = 0$.

Sol. (*a*) F is a function of x, y, z and x, y, z are further functions of u, v, w

$\therefore$ F is a composite function of u, v, w.

$$x = u + v + w, \quad y = vw + wu + uv,\ z = uvw$$

$$\therefore \quad \frac{\partial F}{\partial u} = \frac{\partial F}{\partial x}\cdot\frac{\partial x}{\partial u} + \frac{\partial F}{\partial y}\cdot\frac{\partial y}{\partial u} + \frac{\partial F}{\partial z}\cdot\frac{\partial z}{\partial u} = \frac{\partial F}{\partial x}\cdot 1 + \frac{\partial F}{\partial y}(w + v) + \frac{\partial F}{\partial z}\cdot vw \quad ...(1)$$

$$\frac{\partial F}{\partial v} = \frac{\partial F}{\partial x}\cdot\frac{\partial x}{\partial v} + \frac{\partial F}{\partial y}\cdot\frac{\partial y}{\partial v} + \frac{\partial F}{\partial z}\cdot\frac{\partial z}{\partial v} = \frac{\partial F}{\partial x}\cdot 1 + \frac{\partial F}{\partial y}\cdot(w + u) + \frac{\partial F}{\partial z}\cdot uw \quad ...(2)$$

$$\frac{\partial F}{\partial w} = \frac{\partial F}{\partial x}\cdot\frac{\partial x}{\partial w} + \frac{\partial F}{\partial y}\cdot\frac{\partial y}{\partial w} + \frac{\partial F}{\partial z}\cdot\frac{\partial z}{\partial w} = \frac{\partial F}{\partial x}\cdot 1 + \frac{\partial F}{\partial y}\cdot(v + u) + \frac{\partial F}{\partial z}\cdot uv \quad ...(3)$$

Multiply (1) by u, (2) by v, (3) by w and add

$$u\frac{\partial F}{\partial u} + v\frac{\partial F}{\partial v} + w\frac{\partial F}{\partial w} = (u + v + w)\frac{\partial F}{\partial x} + (uw + uv + vw + vu + wv + wu)\frac{\partial F}{\partial y} + (uvw + vuw + wuv)\frac{\partial F}{\partial z}$$

$$= x\frac{\partial F}{\partial x} + 2y\frac{\partial F}{\partial y} + 3z\frac{\partial F}{\partial z}$$

Hence $\quad u\frac{\partial F}{\partial u} + v\frac{\partial F}{\partial v} + w\frac{\partial F}{\partial w} = x\frac{\partial F}{\partial x} + 2y\frac{\partial F}{\partial y} + 3z\frac{\partial F}{\partial z}$.

(*ii*) Let $\quad X = x^2 + 2yz,\ Y = y^2 + 2zx \quad \therefore \quad u = f(X, Y)$

u is a function of X, Y ; X, Y are functions of x, y, z

$\therefore$ u is a composite function of x, y, z

$$\therefore \quad \frac{\partial u}{\partial x} = \frac{\partial u}{\partial X}\cdot\frac{\partial X}{\partial x} + \frac{\partial u}{\partial Y}\cdot\frac{\partial Y}{\partial x} = \frac{\partial u}{\partial X}\cdot 2x + \frac{\partial u}{\partial Y}\cdot 2z \quad ...(1)$$

$$\frac{\partial u}{\partial y} = \frac{\partial u}{\partial X}\cdot\frac{\partial X}{\partial y} + \frac{\partial u}{\partial Y}\cdot\frac{\partial Y}{\partial y} = \frac{\partial u}{\partial X}.\,2z + \frac{\partial u}{\partial Y}.\,2y \quad ...(2)$$

$$\frac{\partial u}{\partial z} = \frac{\partial u}{\partial X}\cdot\frac{\partial X}{\partial z} + \frac{\partial u}{\partial Y}\cdot\frac{\partial Y}{\partial z} = \frac{\partial u}{\partial X}.\,2y + \frac{\partial u}{\partial Y}.\,2x \quad ...(3)$$

Multiply (1) by $y^2 - zx$; (2) by $(x^2 - yz)$, (3) by $z^2 - xy$ and add

$$(y^2 - zx)\frac{\partial u}{\partial x} + (x^2 - yz)\frac{\partial u}{\partial y} + (z^2 - xy)\frac{\partial u}{\partial z}$$

$$= \frac{\partial u}{\partial X}(2xy^2 - 2zx^2 + 2x^2z - 2yz^2 + 2yz^2 - 2xy^2)$$

$$+ \frac{\partial u}{\partial Y}[2y^2z - 2z^2x + 2x^2y - 2y^2z + 2z^2x - 2x^2y]$$

$$= \frac{\partial u}{\partial X}\cdot 0 + \frac{\partial u}{\partial Y}\cdot 0 = 0.$$

Example 6. *If $w = f(x, y)$, $x = r\cos\theta$, $y = r\sin\theta$ show that*

$$\left(\frac{\partial w}{\partial r}\right)^2 + \frac{1}{r^2}\left(\frac{\partial w}{\partial \theta}\right)^2 = \left(\frac{\partial f}{\partial x}\right)^2 + \left(\frac{\partial f}{\partial y}\right)^2.$$ **(P.T.U., Dec. 2007)**

Sol. The given equations define w as a composite function of r and θ.

$$\frac{\partial w}{\partial r} = \frac{\partial w}{\partial x}\cdot\frac{\partial x}{\partial r} + \frac{\partial w}{\partial y}\cdot\frac{\partial y}{\partial r} = \frac{\partial w}{\partial x}\cdot\cos\theta + \frac{\partial w}{\partial y}\cdot\sin\theta$$

or $$\frac{\partial w}{\partial r} = \frac{\partial f}{\partial x}\cos\theta + \frac{\partial f}{\partial y}\sin\theta \quad \text{...(1)} \quad [\because \; w = f(x, y)]$$

Also $$\frac{\partial w}{\partial \theta} = \frac{\partial w}{\partial x}\cdot\frac{\partial x}{\partial \theta} + \frac{\partial w}{\partial y}\cdot\frac{\partial y}{\partial \theta} = \frac{\partial w}{\partial x}(-r\sin\theta) + \frac{\partial w}{\partial y}(r\cos\theta)$$

or $$\frac{1}{r}\frac{\partial w}{\partial \theta} = -\frac{\partial f}{\partial x}\sin\theta + \frac{\partial f}{\partial y}\cos\theta \quad \text{...(2)}$$

Squaring and adding (1) and (2), we get

$$\left(\frac{\partial w}{\partial r}\right)^2 + \frac{1}{r^2}\left(\frac{\partial w}{\partial \theta}\right)^2 = \left(\frac{\partial f}{\partial x}\right)^2 + \left(\frac{\partial f}{\partial y}\right)^2.$$

Example 7. *If $u = xe^y z$, where $y = \sqrt{a^2 - x^2}$, $z = \sin^2 x$, find $\frac{du}{dx}$.*

Sol. Here u is a function of x, y and z while y and z are functions of x.

$\therefore$ u is a composite function of x.

$\therefore$ $$\frac{du}{dx} = \frac{\partial u}{\partial x}\cdot\frac{dx}{dx} + \frac{\partial u}{\partial y}\cdot\frac{dy}{dx} + \frac{\partial u}{\partial z}\cdot\frac{dz}{dx}$$

$$= e^y z\cdot 1 + xe^y z\cdot\tfrac{1}{2}(a^2 - x^2)^{-1/2}(-2x) + xe^y\cdot 2\sin x\cos x$$

$$= e^y\left[z - \frac{x^2 z}{\sqrt{a^2 - x^2}} + x\sin 2x\right].$$

Example 8. *Find $\frac{du}{dx}$ if $u = \sin(x^2 + y^2)$, where $a^2x^2 + b^2y^2 = c^2$.*

Sol. The given equations are the form $u = f(x, y)$ and $\phi(x, y) = k$

$\therefore$ u is a composite function of x.

$$\frac{du}{dx} = \frac{\partial u}{\partial x}\cdot\frac{dx}{dx} + \frac{\partial u}{\partial y}\cdot\frac{dy}{dx} = \frac{\partial u}{\partial x} + \frac{\partial u}{\partial y}\cdot\frac{dy}{dx} \quad \text{...(1)}$$

Now, $$\frac{\partial u}{\partial x} = 2x\cos(x^2 + y^2), \quad \frac{\partial u}{\partial y} = 2y\cos(x^2 + y^2)$$

Also, differentiating $a^2x^2 + b^2y^2 = c$ w.r.t. x, we have

$$2a^2x + 2b^2y \cdot \frac{dy}{dx} = 0 \quad \text{or} \quad \frac{dy}{dx} = -\frac{a^2x}{b^2y}$$

$\therefore$ From (1),

$$\frac{du}{dx} = 2x \cos (x^2 + y^2) + 2y \cos (x^2 + y^2) \cdot \left[-\frac{a^2x}{b^2y}\right]$$

$$= 2\left[x - \frac{a^2x}{b^2}\right] \cos (x^2 + y^2) = \frac{2(b^2 - a^2)x}{b^2} \cdot \cos (x^2 + y^2).$$

Example 9. *Find* $\dfrac{dy}{dx}$, *when*

(*i*) $x^y + y^x = c$ (*ii*) $(\cos x)^y = (\sin y)^x$.

Sol. (*i*) Let $f(x, y) = x^y + y^x$, then $f(x, y) = c$

[Using art. 3.9*b*] $\dfrac{dy}{dx} = -\dfrac{f_x}{f_y}$

(Find f_x and f_y)

$$f_x = \frac{\partial f}{\partial x} = \frac{\partial}{\partial x}(x^y) + \frac{\partial}{\partial x}(y^x) = yx^{y-1} + y^x \log y$$

$$f_y = \frac{\partial}{\partial y}(x^y) + \frac{\partial}{\partial y}(y^x) = x^y \log x + xy^{x-1}$$

$$\Rightarrow \quad \frac{dy}{dx} = -\frac{yx^{y-1} + y^x \log y}{x^y \log x + xy^{x-1}}.$$

(*ii*) Let $f(x, y) = (\cos x)^y - (\sin y)^x = 0$

$$\frac{dy}{dx} = -\frac{f_x}{f_y} = -\frac{y(\cos x)^{y-1} \cdot (-\sin x) - (\sin y)^x \log (\sin y)}{(\cos x)^y \log (\cos x) - x (\sin y)^{x-1} \cdot \cos y}$$

$$= \frac{y(\cos x)^{y-1} \sin x + (\cos x)^y \log (\sin y)}{(\cos x)^y \log (\cos x) - x (\cos x)^y (\sin y)^{-1} \cos y}$$

[$\because$ $(\sin y)^x = (\cos x)^y$]

$$= \frac{(\cos x)^y \left[y \cdot \dfrac{\sin x}{\cos x} + \log \sin y\right]}{(\cos x)^y [\log \cos x - x \cot y]} = \frac{y \tan x + \log \sin y}{\log \cos x - x \cot y}.$$

Example 10. *If* $z = xy\, f\left(\dfrac{y}{x}\right)$ *and* z *is constant show that* $\dfrac{f'\left(\dfrac{y}{x}\right)}{f\left(\dfrac{y}{x}\right)} = \dfrac{x\left[y + x\dfrac{dy}{dx}\right]}{y\left[y - x\dfrac{dy}{dx}\right]}$.

Sol. $z = xy\, f\left(\dfrac{y}{x}\right)$; z is constant

$\therefore$ Let $\phi(x, y) = xy\, f\left(\dfrac{y}{x}\right)$, $\phi(x, y) = c$

[Using art. 3.9(*b*)] $\dfrac{dy}{dx} = -\dfrac{\phi_x}{\phi_y}$ ...(1)

$$\phi_x = \frac{\partial \phi}{\partial x} = \frac{\partial}{\partial x}\left[xy\, f\left(\frac{y}{x}\right)\right]$$

$$= y\left\{x\, f'\left(\frac{y}{x}\right)\left(\frac{-y}{x^2}\right) + f\left(\frac{y}{x}\right)\right\}$$

$$= y\left[-\frac{y}{x} f'\left(\frac{y}{x}\right) + f\left(\frac{y}{x}\right)\right]$$

$$\phi_y = \frac{\partial \phi}{\partial y} = \frac{\partial}{\partial y}\left[xy\, f\left(\frac{y}{x}\right)\right] = x\left[y\, f'\left(\frac{y}{x}\right)\cdot\frac{1}{x} + f\left(\frac{y}{x}\right)\right]$$

$$= x\left[\frac{y}{x} f'\left(\frac{y}{x}\right) + f\left(\frac{y}{x}\right)\right]$$

From (1),
$$\frac{dy}{dx} = -\frac{y\left[-\frac{y}{x} f'\left(\frac{y}{x}\right) + f\left(\frac{y}{x}\right)\right]}{x\left[\frac{y}{x} f'\left(\frac{y}{x}\right) + f\left(\frac{y}{x}\right)\right]}$$

or
$$y\, f'\left(\frac{y}{x}\right)\frac{dy}{dx} + xf\left(\frac{y}{x}\right)\frac{dy}{dx} = \frac{y^2}{x} f'\left(\frac{y}{x}\right) - yf\left(\frac{y}{x}\right)$$

or
$$f'\left(\frac{y}{x}\right)\left\{y\frac{dy}{dx} - \frac{y^2}{x}\right\} = f\left(\frac{y}{x}\right)\left[-x\frac{dy}{dx} - y\right]$$

or
$$\frac{f'\left(\frac{y}{x}\right)}{f\left(\frac{y}{x}\right)} = \frac{-\left(y + x\frac{dy}{dx}\right)}{-\frac{y}{x}\left(y - x\frac{dy}{dx}\right)} = \frac{x}{y}\cdot\frac{y + x\frac{dy}{dx}}{y - x\frac{dy}{dx}}.$$

Example 11. *If* $\phi(x, y, z) = 0$, *show that* $\left(\frac{\partial y}{\partial z}\right)_x \left(\frac{\partial z}{\partial x}\right)_y \left(\frac{\partial x}{\partial y}\right)_z = -1.$

(P.T.U., Dec. 2011)

Sol. The given relation defines y as a function of x and z. Treating x as constant

$$\left(\frac{\partial y}{\partial z}\right)_x = -\frac{\frac{\partial \phi}{\partial z}}{\frac{\partial \phi}{\partial y}}$$

The given relation defines z as a function of x and y. Treating y as constant

$$\left(\frac{\partial z}{\partial x}\right)_y = -\frac{\frac{\partial \phi}{\partial x}}{\frac{\partial \phi}{\partial z}}$$

Similarly,
$$\left(\frac{\partial x}{\partial y}\right)_z = -\frac{\frac{\partial \phi}{\partial y}}{\frac{\partial \phi}{\partial x}}$$

Multiplying, we get

$$\left(\frac{\partial y}{\partial z}\right)_x \left(\frac{\partial z}{\partial x}\right)_y \left(\frac{\partial x}{\partial y}\right)_z = -1.$$

Example 12. *If z is a function of x and y and u, v be two other variables such that $u = lx + my$, $v = ly - mx$, show that* $\dfrac{\partial^2 z}{\partial x^2} + \dfrac{\partial^2 z}{\partial y^2} = (l^2 + m^2)\left(\dfrac{\partial^2 z}{\partial u^2} + \dfrac{\partial^2 z}{\partial v^2}\right)$. **(P.T.U., May 2004)**

Sol. z is a composite function of x, y

$$\therefore \qquad \frac{\partial z}{\partial x} = \frac{\partial z}{\partial u}\cdot\frac{\partial u}{\partial x} + \frac{\partial z}{\partial v}\cdot\frac{\partial v}{\partial x} = l\frac{\partial z}{\partial u} - m\frac{\partial z}{\partial v}$$

$$\Rightarrow \qquad \frac{\partial}{\partial x} = l\frac{\partial}{\partial u} - m\frac{\partial}{\partial v}$$

Now

$$\frac{\partial^2 z}{\partial x^2} = \frac{\partial}{\partial x}\left(\frac{\partial z}{\partial x}\right) = \left(l\frac{\partial}{\partial u} - m\frac{\partial}{\partial v}\right)\left(l\frac{\partial z}{\partial u} - m\frac{\partial z}{\partial v}\right)$$

$$= l^2\frac{\partial^2 z}{\partial u^2} - lm\frac{\partial^2 z}{\partial u \partial v} - lm\frac{\partial^2 z}{\partial v \partial u} + m^2\frac{\partial^2 z}{\partial v^2}$$

$$\therefore \qquad \frac{\partial^2 z}{\partial x^2} = l^2\frac{\partial^2 z}{\partial u^2} - 2lm\frac{\partial^2 z}{\partial u \partial v} + m^2\frac{\partial^2 z}{\partial v^2} \qquad ...(1)$$

Now

$$\frac{\partial z}{\partial y} = \frac{\partial z}{\partial u}\cdot\frac{\partial u}{\partial y} + \frac{\partial z}{\partial v}\cdot\frac{\partial v}{\partial y} = m\frac{\partial z}{\partial u} + l\frac{\partial z}{\partial v}$$

$$\therefore \qquad \frac{\partial}{\partial y} = m\frac{\partial}{\partial u} + l\frac{\partial}{\partial v}$$

$$\frac{\partial^2 z}{\partial y^2} = \frac{\partial}{\partial y}\left(\frac{\partial z}{\partial y}\right) = \left(m\frac{\partial}{\partial u} + l\frac{\partial}{\partial v}\right)\left(m\frac{\partial z}{\partial u} + l\frac{\partial z}{\partial v}\right)$$

$$= m^2\frac{\partial^2 z}{\partial u^2} + 2lm\frac{\partial^2 u}{\partial u \partial v} + l^2\frac{\partial^2 z}{\partial v^2} \qquad ...(2)$$

Adding (1) and (2),

$$\frac{\partial^2 z}{\partial x^2} + \frac{\partial^2 z}{\partial y^2} = (l^2 + m^2)\left(\frac{\partial^2 z}{\partial u^2} + \frac{\partial^2 z}{\partial v^2}\right).$$

Example 13. *Prove that* $\dfrac{\partial^2 z}{\partial x^2} + \dfrac{\partial^2 z}{\partial y^2} = \dfrac{\partial^2 z}{\partial u^2} + \dfrac{\partial^2 z}{\partial y^2}$

where $x = u \cos \alpha - v \sin \alpha$, $y = u \sin \alpha + v \cos \alpha$ **(P.T.U., Dec. 2011)**

Or

By changing the independent variables u and v to x and y by means of the relations $x = u \cos \alpha - v \sin \alpha$, $y = u \sin \alpha + v \cos \alpha$, show that $\dfrac{\partial^2 z}{\partial u^2} + \dfrac{\partial^2 z}{\partial v^2}$ *transforms into* $\dfrac{\partial^2 z}{\partial x^2} + \dfrac{\partial^2 z}{\partial y^2}$.

Sol. Here z is a composite function of u and v

$$\frac{\partial z}{\partial u} = \frac{\partial z}{\partial x}\cdot\frac{\partial x}{\partial u} + \frac{\partial z}{\partial y}\cdot\frac{\partial y}{\partial u} = \cos\alpha\,\frac{\partial z}{\partial x} + \sin\alpha\,\frac{\partial z}{\partial y}$$

or $$\frac{\partial}{\partial u}(z) = \left(\cos\alpha\frac{\partial}{\partial x} + \sin\alpha\frac{\partial}{\partial y}\right)z \Rightarrow \frac{\partial}{\partial u} \equiv \cos\alpha\frac{\partial}{\partial x} + \sin\alpha\frac{\partial}{\partial y} \quad ...(1)$$

Also $$\frac{\partial z}{\partial v} = \frac{\partial z}{\partial x}\cdot\frac{\partial x}{\partial v} + \frac{\partial z}{\partial y}\cdot\frac{\partial y}{\partial v} = -\sin\alpha\frac{\partial z}{\partial x} + \cos\alpha\frac{\partial z}{\partial y}$$

or $$\frac{\partial}{\partial v}(z) = \left(-\sin\alpha\frac{\partial}{\partial x} + \cos\alpha\frac{\partial}{\partial y}\right)z$$

$\Rightarrow$ $$\frac{\partial}{\partial v} \equiv -\sin\alpha\frac{\partial}{\partial x} + \cos\alpha\frac{\partial}{\partial y} \quad ...(2)$$

Now we shall make use of the equivalence of operators as given by (1) and (2).

$$\frac{\partial^2 z}{\partial u^2} = \frac{\partial}{\partial u}\left(\frac{\partial z}{\partial u}\right) = \left(\cos\alpha\frac{\partial}{\partial x} + \sin\alpha\frac{\partial}{\partial y}\right)\left(\cos\alpha\frac{\partial z}{\partial x} + \sin\alpha\frac{\partial z}{\partial y}\right)$$

$$= \cos^2\alpha\frac{\partial^2 z}{\partial x^2} + \cos\alpha\sin\alpha\frac{\partial^2 z}{\partial x\partial y} + \sin\alpha\cos\alpha\frac{\partial^2 z}{\partial y\partial x} + \sin^2\alpha\frac{\partial^2 z}{\partial y^2}$$

$$= \cos^2\alpha\frac{\partial^2 z}{\partial x^2} + 2\cos\alpha\sin\alpha\frac{\partial^2 z}{\partial x\partial y} + \sin^2\alpha\frac{\partial^2 z}{\partial y^2} \quad ...(3)$$

$$\frac{\partial^2 z}{\partial v^2} = \frac{\partial}{\partial v}\left(\frac{\partial z}{\partial v}\right) = \left(-\sin\alpha\frac{\partial}{\partial x} + \cos\alpha\frac{\partial}{\partial y}\right)\left(-\sin\alpha\frac{\partial z}{\partial x} + \cos\alpha\frac{\partial z}{\partial y}\right)$$

$$= \sin^2\alpha\frac{\partial^2 z}{\partial x^2} - \sin\alpha\cos\alpha\frac{\partial^2 z}{\partial x\partial y} - \cos\alpha\sin\alpha\frac{\partial^2 z}{\partial y\partial x} + \cos^2\alpha\frac{\partial^2 z}{\partial y^2}$$

$$= \sin^2\alpha\frac{\partial^2 z}{\partial x^2} - 2\cos\alpha\sin\alpha\frac{\partial^2 z}{\partial x\partial y} + \cos^2\alpha\frac{\partial^2 z}{\partial y^2} \quad ...(4)$$

Adding (3) and (4), $\dfrac{\partial^2 z}{\partial u^2} + \dfrac{\partial^2 z}{\partial v^2} = \dfrac{\partial^2 z}{\partial x^2} + \dfrac{\partial^2 z}{\partial y^2}$.

Example 14. *Transform the equation* $\dfrac{\partial^2 u}{\partial x^2} + \dfrac{\partial^2 u}{\partial y^2} = 0$ *into polar co-ordinates.*

Sol. The relations connecting cartesian co-ordinates, (x, y) with polar co-ordinates (r, θ) are

$$x = r\cos\theta,\ y = r\sin\theta$$

Squaring and adding, $r^2 = x^2 + y^2$

Dividing, $$\tan\theta = \frac{y}{x}$$

$\therefore$ $$r = \sqrt{x^2 + y^2},\ \theta = \tan^{-1}\left(\frac{y}{x}\right)$$

$$\frac{\partial r}{\partial x} = \frac{x}{\sqrt{x^2 + y^2}} = \frac{r\cos\theta}{r} = \cos\theta$$

$$\frac{\partial r}{\partial y} = \frac{y}{\sqrt{x^2 + y^2}} = \frac{r\sin\theta}{r} = \sin\theta$$

$$\frac{\partial\theta}{\partial x} = \frac{1}{1 + \frac{y^2}{x^2}}\cdot\left(-\frac{y}{x^2}\right) = -\frac{y}{x^2 + y^2} = -\frac{r\sin\theta}{r^2} = -\frac{\sin\theta}{r}$$

$$\frac{\partial \theta}{\partial y} = \frac{1}{1+\frac{y^2}{x^2}} \cdot \frac{1}{x} = \frac{x}{x^2+y^2} = \frac{r\cos\theta}{r^2} = \frac{\cos\theta}{r}$$

Here u is a composite function of x and y

$$\frac{\partial u}{\partial x} = \frac{\partial u}{\partial r} \cdot \frac{\partial r}{\partial x} + \frac{\partial u}{\partial \theta} \cdot \frac{\partial \theta}{\partial x} = \cos\theta \frac{\partial u}{\partial r} - \frac{\sin\theta}{r} \frac{\partial u}{\partial \theta}$$

or $$\frac{\partial}{\partial x}(u) = \left(\cos\theta \frac{\partial}{\partial r} - \frac{\sin\theta}{r}\frac{\partial}{\partial \theta}\right) u$$

$\Rightarrow$ $$\frac{\partial}{\partial x} \equiv \cos\theta \frac{\partial}{\partial r} - \frac{\sin\theta}{r}\frac{\partial}{\partial \theta} \qquad \ldots(1)$$

Also $$\frac{\partial u}{\partial y} = \frac{\partial u}{\partial r} \cdot \frac{\partial r}{\partial y} + \frac{\partial u}{\partial \theta} \cdot \frac{\partial \theta}{\partial y} = \sin\theta \frac{\partial u}{\partial r} + \frac{\cos\theta}{r}\frac{\partial u}{\partial \theta}$$

or $$\frac{\partial}{\partial y}(u) = \left(\sin\theta \frac{\partial}{\partial r} + \frac{\cos\theta}{r}\frac{\partial}{\partial \theta}\right) u \quad \Rightarrow \quad \frac{\partial}{\partial y} \equiv \sin\theta \frac{\partial}{\partial r} + \frac{\cos\theta}{r}\frac{\partial}{\partial \theta} \qquad \ldots(2)$$

Now we shall make use of the equivalence of cartesian and polar operators as given by (1) and (2),

$$\frac{\partial^2 u}{\partial x^2} = \frac{\partial}{\partial x}\left(\frac{\partial u}{\partial x}\right) = \left(\cos\theta \frac{\partial}{\partial r} - \frac{\sin\theta}{r}\frac{\partial}{\partial \theta}\right)\left(\cos\theta \frac{\partial u}{\partial r} - \frac{\sin\theta}{r}\frac{\partial u}{\partial \theta}\right)$$

$$= \cos\theta \frac{\partial}{\partial r}\left(\cos\theta \frac{\partial u}{\partial r} - \frac{\sin\theta}{r}\frac{\partial u}{\partial \theta}\right) - \frac{\sin\theta}{r}\frac{\partial}{\partial \theta}\left(\cos\theta \frac{\partial u}{\partial r} - \frac{\sin\theta}{r}\frac{\partial u}{\partial \theta}\right)$$

$$= \cos\theta \left[\cos\theta \frac{\partial^2 u}{\partial r^2} - \sin\theta \frac{\partial u}{\partial \theta}\left(-\frac{1}{r^2}\right) - \frac{\sin\theta}{r} \cdot \frac{\partial^2 u}{\partial r \partial \theta}\right]$$

$$- \frac{\sin\theta}{r}\left[-\sin\theta \frac{\partial u}{\partial r} + \cos\theta \frac{\partial^2 u}{\partial \theta \partial r} - \frac{\cos\theta}{r}\frac{\partial u}{\partial \theta} - \frac{\sin\theta}{r}\frac{\partial^2 u}{\partial \theta^2}\right]$$

$$= \cos^2\theta \frac{\partial^2 u}{\partial r^2} + \frac{2\cos\theta\sin\theta}{r^2}\frac{\partial u}{\partial \theta} + \frac{\sin^2\theta}{r}\frac{\partial u}{\partial r} - \frac{2\cos\theta\sin\theta}{r^2}\frac{\partial^2 u}{\partial r \partial \theta} + \frac{\sin^2\theta}{r^2}\frac{\partial^2 u}{\partial \theta^2} \qquad \ldots(3)$$

$$\frac{\partial^2 u}{\partial y^2} = \frac{\partial}{\partial y}\left(\frac{\partial u}{\partial y}\right) = \left(\sin\theta \frac{\partial}{\partial r} + \frac{\cos\theta}{r}\frac{\partial}{\partial \theta}\right)\left(\sin\theta \frac{\partial u}{\partial r} + \frac{\cos\theta}{r}\frac{\partial u}{\partial \theta}\right)$$

$$= \sin\theta \frac{\partial}{\partial r}\left(\sin\theta \frac{\partial u}{\partial r} + \frac{\cos\theta}{r}\frac{\partial u}{\partial \theta}\right) + \frac{\cos\theta}{r}\frac{\partial}{\partial \theta}\left(\sin\theta \frac{\partial u}{\partial r} + \frac{\cos\theta}{r}\frac{\partial u}{\partial \theta}\right)$$

$$= \sin\theta \left[\sin\theta \frac{\partial^2 u}{\partial r^2} - \frac{\cos\theta}{r^2}\frac{\partial u}{\partial \theta} + \frac{\cos\theta}{r}\frac{\partial^2 u}{\partial r \partial \theta}\right]$$

$$+ \frac{\cos\theta}{r}\left[\cos\theta \frac{\partial u}{\partial r} + \sin\theta \frac{\partial^2 u}{\partial \theta \partial r} - \frac{\sin\theta}{r}\frac{\partial u}{\partial \theta} + \frac{\cos\theta}{r}\frac{\partial^2 u}{\partial \theta^2}\right]$$

$$= \sin^2\theta \frac{\partial^2 u}{\partial r^2} - \frac{2\cos\theta\sin\theta}{r^2}\frac{\partial u}{\partial \theta} + \frac{\cos^2\theta}{r}\frac{\partial u}{\partial r} + \frac{2\cos\theta\sin\theta}{r}\frac{\partial^2 u}{\partial r \partial \theta} + \frac{\cos^2\theta}{r^2}\frac{\partial^2 u}{\partial \theta^2} \qquad \ldots(4)$$

Adding (3) and (4), $\dfrac{\partial^2 u}{\partial x^2}+\dfrac{\partial^2 u}{\partial y^2}=\dfrac{\partial^2 u}{\partial r^2}+\dfrac{1}{r}\dfrac{\partial u}{\partial r}+\dfrac{1}{r^2}\dfrac{\partial^2 u}{\partial \theta^2}$

$\therefore\quad \dfrac{\partial^2 u}{\partial x^2}+\dfrac{\partial^2 u}{\partial y^2}=0$ transforms into $\dfrac{\partial^2 u}{\partial r^2}+\dfrac{1}{r}\dfrac{\partial u}{\partial r}+\dfrac{1}{r^2}\dfrac{\partial^2 u}{\partial \theta^2}=0.$

Example 15. *If* $x=e^r\cos\theta,\ y=e^r\sin\theta$, *show that* $\dfrac{\partial^2 u}{\partial x^2}+\dfrac{\partial^2 u}{\partial y^2}=e^{-2r}\left[\dfrac{\partial^2 u}{\partial r^2}+\dfrac{\partial^2 u}{\partial \theta^2}\right]$ *where* $u=f(x,y)$.

Sol. $u=f(x,y)\ ;\ x=e^r\cos\theta,\ y=e^r\sin\theta$

u is a function of x, y; x, y are functions of r, θ

$\therefore$ u is a composite function of r, θ

$$\therefore\quad \frac{\partial u}{\partial r}=\frac{\partial u}{\partial x}\cdot\frac{\partial x}{\partial r}+\frac{\partial u}{\partial y}\cdot\frac{\partial y}{\partial r}=(e^r\cos\theta)\frac{\partial u}{\partial x}+(e^r\sin\theta)\frac{\partial u}{\partial y}=x\frac{\partial u}{\partial x}+y\frac{\partial u}{\partial y}$$

and

$$\frac{\partial u}{\partial \theta}=\frac{\partial u}{\partial x}\cdot\frac{\partial x}{\partial \theta}+\frac{\partial u}{\partial y}\cdot\frac{\partial y}{\partial \theta}=(-e^r\sin\theta)\frac{\partial u}{\partial x}+(e^r\cos\theta)\frac{\partial u}{\partial y}$$

$$=-y\frac{\partial u}{\partial x}+x\frac{\partial u}{\partial y}$$

$$\therefore\quad \frac{\partial u}{\partial r}=\left(x\frac{\partial}{\partial x}+y\frac{\partial}{\partial y}\right)u \quad\text{or}\quad \frac{\partial}{\partial r}=x\frac{\partial}{\partial x}+y\frac{\partial}{\partial y}$$

$$\frac{\partial u}{\partial \theta}=\left(-y\frac{\partial}{\partial x}+x\frac{\partial}{\partial y}\right)u \quad\text{or}\quad \frac{\partial}{\partial \theta}=-y\frac{\partial}{\partial x}+x\frac{\partial}{\partial y}$$

Now

$$\frac{\partial^2 u}{\partial r^2}=\frac{\partial}{\partial r}\left(\frac{\partial u}{\partial r}\right)=\left(x\frac{\partial}{\partial x}+y\frac{\partial}{\partial y}\right)\left(x\frac{\partial u}{\partial x}+y\frac{\partial u}{\partial y}\right)$$

$$=x^2\frac{\partial^2 u}{\partial x^2}+2xy\frac{\partial^2 u}{\partial x\partial y}+y^2\frac{\partial^2 u}{\partial y^2}\qquad\ldots(1)$$

$$\frac{\partial^2 u}{\partial \theta^2}=\frac{\partial}{\partial \theta}\left(\frac{\partial u}{\partial \theta}\right)=\left(-y\frac{\partial}{\partial x}+x\frac{\partial}{\partial y}\right)\left(-y\frac{\partial u}{\partial x}+x\frac{\partial u}{\partial y}\right)$$

$$=y^2\frac{\partial^2 u}{\partial x^2}-2xy\frac{\partial^2 u}{\partial x\partial y}+x^2\frac{\partial^2 u}{\partial y^2}\qquad\ldots(2)$$

Adding (1) and (2),

$$\frac{\partial^2 u}{\partial r^2}+\frac{\partial^2 u}{\partial \theta^2}=(x^2+y^2)\left[\frac{\partial^2 u}{\partial x^2}+\frac{\partial^2 u}{\partial y^2}\right]$$

$$=[e^{2r}\cos^2\theta+e^{2r}\sin^2\theta]\left[\frac{\partial^2 u}{\partial x^2}+\frac{\partial^2 u}{\partial y^2}\right]=e^{2r}\left[\frac{\partial^2 u}{\partial x^2}+\frac{\partial^2 u}{\partial y^2}\right]$$

Hence $\dfrac{\partial^2 u}{\partial x^2}+\dfrac{\partial^2 u}{\partial y^2}=e^{-2r}\left[\dfrac{\partial^2 u}{\partial r^2}+\dfrac{\partial^2 u}{\partial \theta^2}\right].$

Example 16. *If $u = f(x, y, z)$ and $x = r \sin\theta \cos\phi$, $y = r \sin\theta \sin\phi$, $z = r\cos\theta$, then show that* $\left(\frac{\partial f}{\partial x}\right)^2 + \left(\frac{\partial f}{\partial y}\right)^2 + \left(\frac{\partial f}{\partial z}\right)^2 = \left(\frac{\partial f}{\partial r}\right)^2 + \frac{1}{r^2}\left(\frac{\partial f}{\partial \theta}\right)^2 + \frac{1}{r^2 \sin^2\theta}\left(\frac{\partial f}{\partial \phi}\right)^2$. **(P.T.U., Dec. 2004)**

Sol. $u = f(x, y, z)$

$x = r \sin\theta \cos\phi,\ y = r \sin\theta \sin\phi,\ z = r\cos\theta$

$$\frac{\partial f}{\partial r} = \frac{\partial f}{\partial x}\cdot\frac{\partial x}{\partial r} + \frac{\partial f}{\partial y}\cdot\frac{\partial y}{\partial r} + \frac{\partial f}{\partial z}\cdot\frac{\partial z}{\partial r}$$

$$\frac{\partial f}{\partial \theta} = \frac{\partial f}{\partial x}\cdot\frac{\partial x}{\partial \theta} + \frac{\partial f}{\partial y}\cdot\frac{\partial y}{\partial \theta} + \frac{\partial f}{\partial z}\cdot\frac{\partial z}{\partial \theta}$$

$$\frac{\partial f}{\partial \phi} = \frac{\partial f}{\partial x}\cdot\frac{\partial x}{\partial \phi} + \frac{\partial f}{\partial y}\cdot\frac{\partial y}{\partial \phi} + \frac{\partial f}{\partial z}\cdot\frac{\partial z}{\partial \phi}$$

or
$$\frac{\partial f}{\partial r} = \frac{\partial f}{\partial x}(\sin\theta\cos\phi) + \frac{\partial f}{\partial y}(\sin\theta\sin\phi) + \frac{\partial f}{\partial z}(\cos\theta)$$

$$\frac{\partial f}{\partial \theta} = \frac{\partial f}{\partial x}(r\cos\theta\cos\phi) + \frac{\partial f}{\partial y}(r\cos\theta\sin\phi) + \frac{\partial f}{\partial z}(-r\sin\theta)$$

$$\frac{\partial f}{\partial \phi} = \frac{\partial f}{\partial x}(-r\sin\theta\sin\phi) + \frac{\partial f}{\partial y}(r\sin\theta\cos\phi) + 0$$

or
$$\frac{\partial f}{\partial r} = \frac{\partial f}{\partial x}(\sin\theta\cos\phi) + \frac{\partial f}{\partial y}(\sin\theta\sin\phi) + \frac{\partial f}{\partial z}(\cos\theta)$$

$$\frac{1}{r}\frac{\partial f}{\partial \theta} = \frac{\partial f}{\partial x}(\cos\theta\cos\phi) + \frac{\partial f}{\partial y}(\cos\theta\sin\phi) + \frac{\partial f}{\partial z}(-\sin\theta)$$

$$\frac{1}{r\sin\theta}\frac{\partial f}{\partial \phi} = \frac{\partial f}{\partial x}(-\sin\phi) + \frac{\partial f}{\partial y}(\cos\phi)$$

Squaring and adding all the above three functions, we get

$$\left(\frac{\partial f}{\partial r}\right)^2 + \left(\frac{1}{r}\frac{\partial f}{\partial \theta}\right)^2 + \frac{1}{r^2\sin^2\theta}\left(\frac{\partial f}{\partial \phi}\right)^2 = \left(\frac{\partial f}{\partial x}\right)^2 + \left(\frac{\partial f}{\partial y}\right)^2 + \left(\frac{\partial f}{\partial z}\right)^2$$

all other terms cancel with each other proved.

Example 17. *Show that* $a\frac{\partial w}{\partial y} = b\frac{\partial w}{\partial x}$ *if a and b are constants and $w = f(ax + by)$ is a differentiable function of $u = ax + by$.* **(P.T.U., May 2005)**

Sol. Given $w = f(ax + by)$

and $u = ax + by$ ...(1)

$\therefore$ $w = f(u)$ where u is a function of x and y

Now
$$\frac{\partial w}{\partial x} = \frac{\partial}{\partial x}f(u) = \frac{\partial f}{\partial u}\times\frac{\partial u}{\partial x} = \frac{\partial f}{\partial u}\cdot a \qquad \left[\text{from (1)},\ \frac{\partial u}{\partial x} = a\right]$$

$$\frac{\partial w}{\partial y} = \frac{\partial}{\partial y}f(u) = \frac{\partial f}{\partial u}\times\frac{\partial u}{\partial y} = \frac{\partial f}{\partial u}\times b \qquad \left[\text{from (1)},\ \frac{\partial u}{\partial y} = b\right]$$

$$\therefore \qquad a\frac{\partial w}{\partial y} = ab\frac{\partial f}{\partial u};$$

$$b\frac{\partial w}{\partial x} = ab\frac{\partial f}{\partial u}$$

Hence
$$a\frac{\partial w}{\partial y} = b\frac{\partial w}{\partial x}.$$

TEST YOUR KNOWLEDGE

1. (*i*) If $u = x^2 + y^2 + z^2$ and $x = e^{2t}$, $y = e^{2t}\cos 3t$, $z = e^{2t}\sin 3t$; find $\frac{du}{dt}$ as a total derivative and verify the result by direct substitution.

   (*ii*) If $u = x^3 + y^3$ where $x = a\cos t$, $y = b\sin t$, find $\frac{du}{dt}$ and verify result by direct substitution.

2. (*i*) If $z = u^2 + v^2$, $u = r\cos\theta$, $v = r\sin\theta$, find $\frac{\partial z}{\partial r}$ and $\frac{\partial z}{\partial \theta}$.

   (*ii*) If $z = \log(u^2 + v)$; $u = e^{x^2+y^2}$, $v = x^2 + y$, find $\frac{\partial z}{\partial x}$ and $\frac{\partial z}{\partial y}$.

3. If $u = f(r, s)$, $r = x + y$, $s = x - y$, show that $\frac{\partial u}{\partial x} + \frac{\partial u}{\partial y} = 2\frac{\partial u}{\partial r}$.

4. If $x = u + v$, $y = uv$ and z is a function of x, y ; show that $u\frac{\partial z}{\partial u} + v\frac{\partial z}{\partial v} = x\frac{\partial z}{\partial x} + 2y\frac{\partial z}{\partial y}$.

5. If $u = f(r, s)$, $r = x + at$, $s = y + bt$, show that $\frac{\partial u}{\partial t} = a\frac{\partial u}{\partial x} + b\frac{\partial u}{\partial y}$.

6. If $u = x\log(xy)$, where $x^3 + y^3 + 3xy = 1$, find $\frac{du}{dx}$. ***(P.T.U., May 2015)***

7. (*i*) If $u = f(r, s, t)$ and $r = \frac{x}{y}$, $s = \frac{y}{z}$, $t = \frac{z}{x}$, prove that $x\frac{\partial u}{\partial x} + y\frac{\partial u}{\partial y} + z\frac{\partial u}{\partial z} = 0$.

   [**Hint.** S.E. 4(*ii*)] ***(P.T.U., May 2003)***

   (*ii*) If $u = f(2x - 3y, 3y - 4z, 4z - 2x)$, prove that $\frac{1}{2}\frac{\partial u}{\partial x} + \frac{1}{3}\frac{\partial u}{\partial y} + \frac{1}{4}\frac{\partial u}{\partial z} = 0$. ***(P.T.U., Dec. 2011)***

8. (*i*) If $z = x^2y$ and $x^2 + xy + y^2 = 1$, show that $\frac{dz}{dx} = 2xy - \frac{x^2(2x+y)}{x+2y}$.

   (*ii*) If $z = \sqrt{x^2 + y^2}$ and $x^3 + y^3 + 3axy = 5a^2$, find the value of $\frac{dz}{dx}$ when $x = y = a$.

   (P.T.U., May 2009)

9. If $x^3 + 3x^2y + 6xy^2 + y^3 = 1$, find $\frac{dy}{dx}$.

10. If $x^y = y^x$, show that $\frac{dy}{dx} = \frac{y(y - x\log y)}{x(x - y\log x)}$ using partial derivative method.

    [**Hint.** Consult S.E. 9(*a*)]

11. If $x^3 + y^3 - 3axy = 0$, find $\frac{dy}{dx}$ and $\frac{d^2y}{dx^2}$ using partial derivative method.

12. Prove that if $y^3 - 3ax^2 + x^3 = 0$, then $\frac{d^2y}{dx^2} + \frac{2a^2x^2}{y^5} = 0$.

13. If $f(x, y) = 0$, $\phi(y, z) = 0$, show that $\frac{\partial f}{\partial y} \cdot \frac{\partial \phi}{\partial z} \cdot \frac{dz}{\partial x} = \frac{\partial f}{\partial x} \cdot \frac{\partial \phi}{\partial y}$

14. If $x = \rho \cos \phi$, $y = \rho \sin \phi$, and $v = f(x, y)$ show that $\left(\frac{\partial v}{\partial x}\right)^2 + \left(\frac{\partial v}{\partial y}\right)^2 = \left(\frac{\partial v}{\partial \rho}\right)^2 + \frac{1}{\rho^2}\left(\frac{\partial v}{\partial \phi}\right)^2$

[**Hint.** See S.E. 6]

15. By changing the independent variables x and y to u and v by means of the relations $u = x - ay$, $v = x + ay$, show that $a^2 \frac{\partial^2 z}{\partial x^2} - \frac{\partial^2 z}{\partial y^2}$ transforms into $4a^2 \frac{\partial^2 z}{\partial u \partial v}$.

[**Hint.** Consult S.E. 13]

16. Prove that $\frac{\partial^2 u}{\partial x^2} + \frac{\partial^2 u}{\partial y^2} = \frac{\partial^2 u}{\partial s^2} + \frac{\partial^2 u}{\partial t^2}$, where $x = s \cos \alpha - t \sin \alpha$ and $y = s \sin \alpha + t \cos \alpha$.

[**Hint.** Consult S.E. 13]

17. If by substitution $u = x^2 - y^2$, $v = 2xy$, $f = \phi(u, v)$, show that $\frac{\partial^2 f}{\partial x^2} + \frac{\partial^2 f}{\partial y^2} = u(x^2 + y^2)\left[\frac{\partial^2 \phi}{\partial u^2} + \frac{\partial^2 \phi}{\partial v^2}\right]$.

Answers

1. (*i*) $8e^{4t}$ (*ii*) $3 \sin t \cos t\,(b^3 \sin t - a^3 \cos t)$

2. (*i*) $2r$, 0 (*ii*) $\frac{2x}{u^2 + v}(2u^2 + 1), \frac{1}{u^2 + v}(4yu^2 + 1)$

6. $1 + \log(xy) - \frac{x(x^2 + y)}{y(y^2 + x)}$

8. (*ii*) = 0

9. $-\frac{x^2 + 2xy + 2y^2}{x^2 + 4xy + y^2}$

11. $\frac{x^2 - ay}{ax - y^2}, \frac{2a^3xy}{(ax - y^2)^3}$.

3.10. JACOBIANS

If u and v are functions of two independent variables x and y, then the determinant

$$\begin{vmatrix} \frac{\partial u}{\partial x} & \frac{\partial u}{\partial y} \\ \frac{\partial v}{\partial x} & \frac{\partial v}{\partial y} \end{vmatrix}$$

is called Jacobian of u, v with respect to x, y and is denoted by the symbol $J\left(\frac{u, v}{x, y}\right)$ or $\frac{\partial(u, v)}{\partial(x, y)}$.

Similarly, if u, v, w be functions of x, y, z, then the Jacobian of u, v, w with respect to x, y, z is

$$J\left(\frac{u, v, w}{x, y, z}\right) \quad \text{or} \quad \frac{\partial(u, v, w)}{\partial(x, y, z)} = \begin{vmatrix} \dfrac{\partial u}{\partial x} & \dfrac{\partial u}{\partial y} & \dfrac{\partial u}{\partial z} \\ \dfrac{\partial v}{\partial x} & \dfrac{\partial v}{\partial y} & \dfrac{\partial v}{\partial z} \\ \dfrac{\partial w}{\partial x} & \dfrac{\partial w}{\partial y} & \dfrac{\partial w}{\partial z} \end{vmatrix}.$$

3.11. PROPERTIES OF JACOBIANS

I. *If u, v are functions of r, s where r, s are functions of x, y, then*

$$J\left(\frac{u, v}{x, y}\right) = \frac{\partial(u, v)}{\partial(x, y)} = \frac{\partial(u, v)}{\partial(r, s)} \times \frac{\partial(r, s)}{\partial(x, y)}.$$

Proof. Since u, v are composite functions of x, y

$$\therefore \quad \left.\begin{aligned} \frac{\partial u}{\partial x} &= \frac{\partial u}{\partial r}\cdot\frac{\partial r}{\partial x} + \frac{\partial u}{\partial s}\cdot\frac{\partial s}{\partial x} = u_r r_x + u_s s_x \\ \frac{\partial u}{\partial y} &= \frac{\partial u}{\partial r}\cdot\frac{\partial r}{\partial y} + \frac{\partial u}{\partial s}\cdot\frac{\partial s}{\partial y} = u_r r_y + u_s s_y \\ \frac{\partial v}{\partial x} &= \frac{\partial v}{\partial r}\cdot\frac{\partial r}{\partial x} + \frac{\partial v}{\partial s}\cdot\frac{\partial s}{\partial x} = v_r r_x + v_s s_x \\ \frac{\partial v}{\partial y} &= \frac{\partial v}{\partial r}\cdot\frac{\partial r}{\partial y} + \frac{\partial v}{\partial s}\cdot\frac{\partial s}{\partial y} = v_r r_y + v_s s_y \end{aligned}\right] \quad \text{...(A)}$$

Now $$\frac{\partial(u,v)}{\partial(r,s)} \times \frac{\partial(r,s)}{\partial(x,y)} = \begin{vmatrix} u_r & u_s \\ v_r & v_s \end{vmatrix} \cdot \begin{vmatrix} r_x & r_y \\ s_x & s_y \end{vmatrix}$$

Interchanging rows and columns in the second determinant

$$= \begin{vmatrix} u_r & u_s \\ v_r & v_s \end{vmatrix} \cdot \begin{vmatrix} r_x & s_x \\ r_y & s_y \end{vmatrix} = \begin{vmatrix} u_r r_x + u_s s_x & u_r r_y + u_s s_y \\ v_r r_x + v_s s_x & v_r r_y + v_s s_y \end{vmatrix} = \begin{vmatrix} \dfrac{\partial u}{\partial x} & \dfrac{\partial u}{\partial y} \\ \dfrac{\partial v}{\partial x} & \dfrac{\partial v}{\partial y} \end{vmatrix} \quad \text{[Using (A)]}$$

$$= \frac{\partial(u, v)}{\partial(x, y)}.$$

II. *If J_1 is the Jacobian of u, v, with respect to x, y and J_2 is the Jacobian of x, y, with respect to u, v, then $J_1 J_2 = 1$ i.e., $\dfrac{\partial(u,v)}{\partial(x,y)} \cdot \dfrac{\partial(x,y)}{\partial(u,v)} = 1$.* **(*P.T.U., Dec. 2012*)**

Proof. Let $u = u(x, y)$ and $v = v(x, y)$, so that u and v are functions of x, y.

Differentiating partially w.r.t. u and v, we get

$$\left.\begin{aligned} 1 &= \frac{\partial u}{\partial x}\cdot\frac{\partial x}{\partial u} + \frac{\partial u}{\partial y}\cdot\frac{\partial y}{\partial u} = u_x x_u + u_y y_u \\ 0 &= \frac{\partial u}{\partial x}\cdot\frac{\partial x}{\partial v} + \frac{\partial u}{\partial y}\cdot\frac{\partial y}{\partial v} = u_x x_v + u_y y_v \\ 0 &= \frac{\partial v}{\partial x}\cdot\frac{\partial x}{\partial u} + \frac{\partial v}{\partial y}\cdot\frac{\partial y}{\partial u} = v_x x_u + v_y y_u \\ 1 &= \frac{\partial v}{\partial x}\cdot\frac{\partial x}{\partial v} + \frac{\partial v}{\partial y}\cdot\frac{\partial y}{\partial v} = v_x x_v + v_y y_v \end{aligned}\right] \quad \text{...(B)}$$

Now, $$\frac{\partial(u,v)}{\partial(x,y)}\cdot\frac{\partial(x,y)}{\partial(u,v)}=\begin{vmatrix} u_x & u_y \\ v_x & v_y \end{vmatrix}\begin{vmatrix} x_u & x_v \\ y_u & y_v \end{vmatrix}$$

Interchanging rows and columns in the second determinant

$$=\begin{vmatrix} u_x & u_y \\ v_x & v_y \end{vmatrix}\begin{vmatrix} x_u & y_u \\ x_v & y_v \end{vmatrix}=\begin{vmatrix} u_x x_u + u_y y_u & u_x x_v + u_y y_v \\ v_x x_u + v_y y_u & v_x x_v + v_y y_v \end{vmatrix}$$

$$=\begin{vmatrix} 1 & 0 \\ 0 & 1 \end{vmatrix}=1.$$ [Using (B)]

ILLUSTRATIVE EXAMPLES

Example 1. *(i) If $x = r\cos\theta$, $y = r\sin\theta$, verify* $\frac{\partial(x,y)}{\partial(r,\theta)}\cdot\frac{\partial(r,\theta)}{\partial(x,y)}=1$.

(P.T.U., May 2005, Dec. 2011)

(ii) If $x = r\cos\theta$, $y = r\sin\theta$, $z = z$ evaluate $\frac{\partial(x,y,z)}{\partial(r,\theta,z)}$. **(P.T.U., Dec. 2013)**

(iii) If $x = u(1+v)$, $y = v(1+u)$, find the value of $\frac{\partial(x,y)}{\partial(u,v)}$. **(P.T.U., May 2014)**

Sol. (*i*) $x = r\cos\theta,\ y = r\sin\theta$

Squaring and adding $r^2 = x^2 + y^2 \quad \therefore \quad r=\sqrt{x^2+y^2}$

Dividing $\tan\theta=\frac{y}{x} \quad \therefore \quad \theta=\tan^{-1}\frac{y}{x}$

Now, $$\frac{\partial(x,y)}{\partial(r,\theta)}=\begin{vmatrix} \frac{\partial x}{\partial r} & \frac{\partial x}{\partial\theta} \\ \frac{\partial y}{\partial r} & \frac{\partial y}{\partial\theta} \end{vmatrix}=\begin{vmatrix} \cos\theta & -r\sin\theta \\ \sin\theta & r\cos\theta \end{vmatrix}=r$$

$$\frac{\partial r}{\partial x}=\frac{1}{2\sqrt{x^2+y^2}}\cdot 2x=\frac{x}{\sqrt{x^2+y^2}}=\frac{x}{r}$$

$$\frac{\partial r}{\partial y}=\frac{1}{2\sqrt{x^2+y^2}}\cdot 2y=\frac{y}{\sqrt{x^2+y^2}}=\frac{y}{r}$$

$$\frac{\partial\theta}{\partial x}=\frac{1}{1+\frac{y^2}{x^2}}(-y/x^2)=\frac{y}{x^2+y^2}=-\frac{y}{r^2}$$

$$\frac{\partial\theta}{\partial y}=\frac{1}{1+\frac{y^2}{x^2}}\left(\frac{1}{x}\right)=\frac{x}{x^2+y^2}=\frac{x}{r^2}$$

$$\therefore \quad \frac{\partial(r,\theta)}{\partial(x,y)}=\begin{vmatrix} \frac{\partial r}{\partial x} & \frac{\partial r}{\partial y} \\ \frac{\partial\theta}{\partial x} & \frac{\partial\theta}{\partial y} \end{vmatrix}=\begin{vmatrix} \frac{x}{r} & \frac{y}{r} \\ -\frac{y}{r^2} & \frac{x}{r^2} \end{vmatrix}=\frac{x^2+y^2}{r^3}=\frac{r^2}{r^3}=\frac{1}{r}$$

$$\therefore \qquad \frac{\partial(x,y)}{\partial(r,\theta)} \cdot \frac{\partial(r,\theta)}{\partial(x,y)} = r \cdot \frac{1}{r} = 1.$$

(ii) $$x = r\cos\theta,\ y = r\sin\theta,\ z = z$$

$$\frac{\partial(x,y,z)}{\partial(r,\theta,z)} = \begin{vmatrix} \frac{\partial x}{\partial r} & \frac{\partial x}{\partial \theta} & \frac{\partial x}{\partial z} \\ \frac{\partial y}{\partial r} & \frac{\partial y}{\partial \theta} & \frac{\partial y}{\partial z} \\ \frac{\partial z}{\partial r} & \frac{\partial z}{\partial \theta} & \frac{\partial z}{\partial z} \end{vmatrix} = \begin{vmatrix} \cos\theta & -r\sin\theta & 0 \\ \sin\theta & r\cos\theta & 0 \\ 0 & 0 & 1 \end{vmatrix} = r\cos^2\theta + r\sin^2\theta = r.$$

(iii) $$x = u(1+v) \qquad y = v(1+u)$$

$$\frac{\partial x}{\partial u} = 1 + v \qquad \frac{\partial y}{\partial u} = v$$

$$\frac{\partial x}{\partial v} = u \qquad \frac{\partial y}{\partial v} = 1 + u$$

$$\frac{\partial(x,y)}{\partial(u,v)} = \begin{vmatrix} \frac{\partial x}{\partial u} & \frac{\partial x}{\partial v} \\ \frac{\partial y}{\partial u} & \frac{\partial y}{\partial v} \end{vmatrix} = \begin{vmatrix} 1+v & u \\ v & 1+u \end{vmatrix}$$

$$= (1+v)(1+u) - uv$$

$$= 1 + v + u.$$

Example 2. *If* $u = \dfrac{x+y}{1-xy}$, $v = \tan^{-1} x + \tan^{-1} y$, *find* $\dfrac{\partial(u,v)}{\partial(x,y)}$. **(P.T.U., May 2012)**

Sol. $$u = \frac{x+y}{1-xy} \qquad \therefore \qquad \frac{\partial u}{\partial x} = \frac{(1-xy)\cdot 1 - (x+y)(-y)}{(1-xy)^2} = \frac{1+y^2}{(1-xy)^2}$$

$$\frac{\partial u}{\partial y} = \frac{(1-xy)\cdot 1 - (x+y)(-x)}{(1-xy)^2} = \frac{1+x^2}{(1-xy)^2}$$

$$v = \tan^{-1} x + \tan^{-1} y;\ \frac{\partial v}{\partial x} = \frac{1}{1+x^2},\ \frac{\partial v}{\partial y} = \frac{1}{1+y^2}$$

$$\frac{\partial(u,v)}{\partial(x,y)} = \begin{vmatrix} \frac{\partial u}{\partial x} & \frac{\partial u}{\partial y} \\ \frac{\partial v}{\partial x} & \frac{\partial v}{\partial y} \end{vmatrix} = \begin{vmatrix} \frac{1+y^2}{(1-xy)^2} & \frac{1+x^2}{(1-xy)^2} \\ \frac{1}{1+x^2} & \frac{1}{1+y^2} \end{vmatrix} = 0.$$

Example 3. *If* $x = r\sin\theta\cos\phi$, $y = r\sin\theta\sin\phi$, $z = r\cos\theta$, *show that* $\dfrac{\partial(x,y,z)}{\partial(r,\theta,\phi)} = r^2\sin\theta$.

Sol. $$\frac{\partial(x,y,z)}{\partial(r,\theta,\phi)} = \begin{vmatrix} \frac{\partial x}{\partial r} & \frac{\partial x}{\partial \theta} & \frac{\partial x}{\partial \phi} \\ \frac{\partial y}{\partial r} & \frac{\partial y}{\partial \theta} & \frac{\partial y}{\partial \phi} \\ \frac{\partial z}{\partial r} & \frac{\partial z}{\partial \theta} & \frac{\partial z}{\partial \phi} \end{vmatrix} = \begin{vmatrix} \sin\theta\cos\phi & r\cos\theta\cos\phi & -r\sin\theta\sin\phi \\ \sin\theta\sin\phi & r\cos\theta\cos\phi & r\sin\theta\sin\phi \\ \cos\theta & -r\sin\theta & 0 \end{vmatrix}$$

Taking out common factors (r from second column and $r \sin\theta$ from third column)

$$= r^2 \sin\theta \begin{vmatrix} \sin\theta\cos\phi & \cos\theta\cos\phi & -\sin\phi \\ \sin\theta\sin\phi & \cos\theta\sin\phi & \cos\phi \\ \cos\theta & -\sin\theta & 0 \end{vmatrix}$$

Expanding by third row

$$= r^2 \sin\theta \left\{ \cos\theta \begin{vmatrix} \cos\theta\cos\phi & -\sin\phi \\ \cos\theta\sin\phi & \cos\phi \end{vmatrix} + \sin\theta \begin{vmatrix} \sin\theta\cos\phi & -\sin\phi \\ \sin\theta\sin\phi & \cos\phi \end{vmatrix} \right\}$$

$$= r^2 \sin\theta\, [\cos\theta\, (\cos\theta\cos^2\phi + \cos\theta\sin^2\phi) + \sin\theta\, (\sin\theta\cos^2\phi + \sin\theta\sin^2\phi)]$$

$$= r^2 \sin\theta\, (\cos^2\theta + \sin^2\theta) = r^2 \sin\theta.$$

Example 4. *If* $u = x \sin y$, $v = y \sin x$ *find* $\dfrac{\partial(u,v)}{\partial(x,y)}$ **(P.T.U., May 2006)**

Sol. $u = x \sin y \qquad v = y \sin x$

$$\therefore \quad \frac{\partial u}{\partial x} = \sin y \qquad \frac{\partial v}{\partial x} = y \cos x$$

$$\frac{\partial u}{\partial y} = x \cos y \qquad \frac{\partial v}{\partial y} = \sin x$$

$$\frac{\partial(u,v)}{\partial(x,y)} = \begin{vmatrix} \dfrac{\partial u}{\partial x} & \dfrac{\partial u}{\partial y} \\ \dfrac{\partial v}{\partial x} & \dfrac{\partial v}{\partial y} \end{vmatrix} = \begin{vmatrix} \sin y & x\cos y \\ y\cos x & \sin x \end{vmatrix}$$

$$= \sin x \sin y - xy \cos x \cos y.$$

Example 5. *If* $u = \dfrac{x}{y-z}$, $v = \dfrac{y}{z-x}$, $w = \dfrac{z}{x-y}$, *show that* $\dfrac{\partial(u, v, w)}{\partial(x, y, z)} = 0$.

Sol.

$$u = \frac{x}{y-z};\ \frac{\partial u}{\partial x} = \frac{1}{y-z},\ \frac{\partial u}{\partial y} = \frac{-x}{(y-z)^2},\ \frac{\partial u}{\partial z} = \frac{x}{(y-z)^2}$$

$$v = \frac{y}{z-x};\ \frac{\partial v}{\partial x} = \frac{y}{(z-x)^2},\ \frac{\partial v}{\partial y} = \frac{1}{z-x},\ \frac{\partial v}{\partial z} = \frac{-y}{(z-x)^2}$$

$$w = \frac{z}{x-y};\ \frac{\partial w}{\partial x} = \frac{-z}{(x-y)^2},\ \frac{\partial w}{\partial y} = \frac{z}{(x-y)^2},\ \frac{\partial w}{\partial z} = \frac{1}{x-y}$$

$$\frac{\partial(u,v,w)}{\partial(x,y,z)} = \begin{vmatrix} \dfrac{1}{y-z} & \dfrac{-x}{(y-z)^2} & \dfrac{x}{(y-z)^2} \\ \dfrac{y}{(z-x)^2} & \dfrac{1}{z-x} & \dfrac{-y}{(z-x)^2} \\ \dfrac{-z}{(x-y)^2} & \dfrac{z}{(x-y)^2} & \dfrac{1}{x-y} \end{vmatrix}$$

$$= \frac{1}{(x-y)^2(y-z)^2(z-x)^2} \begin{vmatrix} y-z & -x & x \\ y & z-x & -y \\ -z & z & x-y \end{vmatrix}$$

Operate $R_1 \to R_1 - R_2 - R_3$

$$= \frac{1}{(x-y)^2(y-z)^2(z-x)^2} \begin{vmatrix} 0 & -2z & 2y \\ y & z-x & -y \\ -z & z & x-y \end{vmatrix}$$

Expand w.r.t. R_1

$$= \frac{1}{(x-y)^2(y-z)^2(z-x)^2}\begin{bmatrix} 2z(xy-y^2-yz) \\ +2y(yz+z^2-zx) \end{bmatrix}$$

$$= \frac{2}{(x-y)^2(y-z)^2(z-x)^2}\begin{bmatrix} xyz-y^2z-yz^2 \\ +y^2z+yz^2-xyz \end{bmatrix}$$

$$= \frac{2}{(x-y)^2(y-z)^2(z-x)^2}\cdot 0$$

$$= 0$$

Hence $\dfrac{\partial(u, v, w)}{\partial(x, y, z)} = 0.$

Example 6. *If $u = x + y + z$, $uv = y + z$, $uvw = z$, show that* $\dfrac{\partial(x, y, z)}{\partial(u, v, w)} = u^2 v.$

(P.T.U., Jan. 2010, Dec. 2011)

Sol. $u = x + y + z,\ uv = y + z \cdot uvw = z$

$\therefore$ $z = uvw,\ y = uv - uvw,\ x = u - (y + z) = u - uv$

$\therefore$ $x = u - uv \quad y = uv - uvw,\ z = uvw$

$$\frac{\partial x}{\partial u} = 1 - v,\ \frac{\partial x}{\partial v} = -u,\ \frac{\partial x}{\partial w} = 0$$

$$\frac{\partial y}{\partial u} = v - vw,\ \frac{\partial y}{\partial v} = u - uw,\ \frac{\partial y}{\partial w} = -uv$$

$$\frac{\partial z}{\partial u} = vw,\ \frac{\partial z}{\partial v} = uw,\ \frac{\partial z}{\partial w} = uv$$

$$\frac{\partial(x, y, z)}{\partial(u, v, w)} = \begin{vmatrix} \frac{\partial x}{\partial u} & \frac{\partial x}{\partial v} & \frac{\partial x}{\partial w} \\ \frac{\partial y}{\partial u} & \frac{\partial y}{\partial v} & \frac{\partial y}{\partial w} \\ \frac{\partial z}{\partial u} & \frac{\partial z}{\partial v} & \frac{\partial z}{\partial w} \end{vmatrix} = \begin{vmatrix} 1-v & -u & 0 \\ v(1-w) & u(1-w) & -uv \\ vw & uw & uv \end{vmatrix}$$

Operate $R_2 \to R_2 + R_3$

$$= \begin{vmatrix} 1-v & -u & 0 \\ v & u & 0 \\ vw & uw & uv \end{vmatrix} = u\begin{vmatrix} 1-v & -1 & 0 \\ v & 1 & 0 \\ vw & w & uv \end{vmatrix}$$

Expand with $C_3 = u \cdot uv\,(1 - v + v) = u \cdot uv \cdot 1 = u^2 v$.

Example 7. *If $u = xyz$, $v = xy + yz + zx$, $w = x + y + z$, show that*

$$\frac{\partial(x, y, z)}{\partial(u, v, w)} = \frac{1}{(x-y)(y-z)(z-x)}.$$

Sol. $u = xyz,\ v = xy + yz + zx,\ w = x + y + z$

As x, y, z cannot be easily expressed in terms of u, v, w $\therefore$ we apply the Jacobian property

that $$\frac{\partial(x, y, z)}{\partial(u, v, w)} = \frac{1}{\dfrac{\partial(u, v, w)}{\partial(x, y, z)}}$$

$\therefore$ $$\frac{\partial u}{\partial x} = yz,\ \frac{\partial u}{\partial y} = zx,\ \frac{\partial u}{\partial z} = xy$$

$$\frac{\partial v}{\partial x} = y + z, \ \frac{\partial v}{\partial y} = z + x, \ \frac{\partial v}{\partial z} = x + y$$

$$\frac{\partial w}{\partial x} = 1, \ \frac{\partial w}{\partial y} = 1, \ \frac{\partial w}{\partial z} = 1$$

$$\frac{\partial(u, v, w)}{\partial(x, y, z)} = \begin{vmatrix} \frac{\partial u}{\partial x} & \frac{\partial u}{\partial y} & \frac{\partial u}{\partial z} \\ \frac{\partial v}{\partial x} & \frac{\partial v}{\partial y} & \frac{\partial v}{\partial z} \\ \frac{\partial w}{\partial x} & \frac{\partial w}{\partial y} & \frac{\partial w}{\partial z} \end{vmatrix} = \begin{vmatrix} yz & zx & xy \\ y+z & z+x & x+y \\ 1 & 1 & 1 \end{vmatrix}$$

$$= \begin{vmatrix} yz & z(x-y) & x(y-z) \\ y+z & x-y & y-z \\ 1 & 0 & 0 \end{vmatrix}, \text{ operate } C_2 - C_1, \ C_3 - C_2$$

$$= (x-y)(y-z) \begin{vmatrix} yz & z & x \\ y+z & 1 & 1 \\ 1 & 0 & 0 \end{vmatrix} = (x-y)(y-z)(z-x)$$

$$\therefore \qquad \frac{\partial(x, y, z)}{\partial(u, v, w)} = \frac{1}{(x-y)(y-z)(z-x)}.$$

TEST YOUR KNOWLEDGE

1. If $u = x(1 - y)$, $v = xy$, prove that $JJ' = 1$.
2. If $u = \dfrac{yz}{x}$, $v = \dfrac{zx}{y}$ and $w = \dfrac{xy}{z}$, show that $\dfrac{\partial(u, v, w)}{\partial(x, y, z)} = 4$.
3. (*i*) If $u = \dfrac{y^2}{2x}$, $v = \dfrac{x^2 + y^2}{2x}$, find $\dfrac{\partial(u, v)}{\partial(x, y)}$.

   (*ii*) If $u = x^3 + xy$, $v = xy$, find $\dfrac{\partial(u, v)}{\partial(x, y)}$. ***(P.T.U., May 2009)***
4. If $x = a \cos \xi \cosh \eta$, $y = a \sinh \xi \sin \eta$, show that $\dfrac{\partial(x, y)}{\partial(\xi, \eta)} = \dfrac{a^2}{2}(\cosh 2\xi - \cos 2\eta)$.

   $\left[\textbf{Hint. } \cosh^2 x = \dfrac{\cosh 2x + 1}{2}, \sinh^2 x = \dfrac{\cosh 2x - 1}{2}\right]$
5. If $F = xu + v - y$, $G = u^2 + vy + w$, $H = zu - v + vw$, compute $\dfrac{\partial(F, G, H)}{\partial(u, w, v)}$.
6. If $u = x^2 - 2y$, $v = x + y + z$, $w = x - 2y + 3z$, find $\dfrac{\partial(u, v, w)}{\partial(x, y, z)}$.

Answers

3. (*i*) $-\dfrac{y}{2x}$ (*ii*) $3x^3$ **5.** $xw - x - xyv + 2uv - z$ **6.** $10x + 4$.

3.12. JACOBIAN OF IMPLICIT FUNCTIONS

If x, y, u, v are connected by implicit functions $f_1(x, y, u, v) = 0$, $f_2(x, y, u, v) = 0$ where u, v are implicit functions of x and y then

$$\frac{\partial(u, v)}{\partial(x, y)} = (-1)^2 \frac{\dfrac{\partial(f_1, f_2)}{\partial(x, y)}}{\dfrac{\partial(f_1, f_2)}{\partial(u, v)}}.$$

Proof. Given

$$f_1 (x, y, u, v) = 0 \quad \text{...(1)}$$

$$f_2 (x, y, u, v) = 0 \quad \text{...(2)}$$

Differentiate (1), (2) w.r.t. x, y

$$\frac{\partial f_1}{\partial x} + \frac{\partial f_1}{\partial u} \cdot \frac{\partial u}{\partial x} + \frac{\partial f_1}{\partial v} \cdot \frac{\partial v}{\partial x} = 0 \quad \text{...(3)}$$

$$\frac{\partial f_1}{\partial y} + \frac{\partial f_1}{\partial u} \cdot \frac{\partial u}{\partial y} + \frac{\partial f_1}{\partial v} \cdot \frac{\partial v}{\partial y} = 0 \quad \text{...(4)}$$

$$\frac{\partial f_2}{\partial x} + \frac{\partial f_2}{\partial u} \cdot \frac{\partial u}{\partial x} + \frac{\partial f_2}{\partial v} \cdot \frac{\partial v}{\partial x} = 0 \quad \text{...(5)}$$

$$\frac{\partial f_2}{\partial y} + \frac{\partial f_2}{\partial u} \cdot \frac{\partial u}{\partial y} + \frac{\partial f_2}{\partial v} \cdot \frac{\partial v}{\partial y} = 0 \quad \text{...(6)}$$

Now $$\frac{\partial(f_1, f_2)}{\partial(u, v)} \times \frac{\partial(u, v)}{\partial(x, y)} = \begin{vmatrix} \dfrac{\partial f_1}{\partial u} & \dfrac{\partial f_1}{\partial v} \\ \dfrac{\partial f_2}{\partial u} & \dfrac{\partial f_2}{\partial v} \end{vmatrix} \times \begin{vmatrix} \dfrac{\partial u}{\partial x} & \dfrac{\partial u}{\partial y} \\ \dfrac{\partial v}{\partial x} & \dfrac{\partial v}{\partial y} \end{vmatrix}$$

$$= \begin{vmatrix} \dfrac{\partial f_1}{\partial u} \cdot \dfrac{\partial u}{\partial x} + \dfrac{\partial f_1}{\partial v} \cdot \dfrac{\partial v}{\partial x} & \dfrac{\partial f_1}{\partial u} \cdot \dfrac{\partial u}{\partial y} + \dfrac{\partial f_1}{\partial v} \cdot \dfrac{\partial v}{\partial y} \\ \dfrac{\partial f_2}{\partial u} \cdot \dfrac{\partial u}{\partial x} + \dfrac{\partial f_2}{\partial v} \cdot \dfrac{\partial v}{\partial x} & \dfrac{\partial f_2}{\partial u} \cdot \dfrac{\partial u}{\partial y} + \dfrac{\partial f_2}{\partial v} \cdot \dfrac{\partial v}{\partial y} \end{vmatrix}$$

$$= \begin{vmatrix} -\dfrac{\partial f_1}{\partial x} & -\dfrac{\partial f_1}{\partial y} \\ -\dfrac{\partial f_2}{\partial x} & -\dfrac{\partial f_2}{\partial y} \end{vmatrix} = (-1)^2 \frac{\partial(f_1, f_2)}{\partial(x, y)} \quad \text{[using (3), (4), (5), (6)]}$$

$$\therefore \quad \frac{\partial(f_1, f_2)}{\partial(u, v)} \times \frac{\partial(u, v)}{\partial(x, y)} = (-1)^2 \frac{\partial(f_1, f_2)}{\partial(x, y)}$$

$$\therefore \quad \frac{\partial(u, v)}{\partial(x, y)} = (-1)^2 \frac{\partial(f_1, f_2)}{\partial(x, y)} \Big/ \frac{\partial(f_1, f_2)}{\partial(u, v)}$$

In general the variables $x_1, x_2, \cdots, x_n$ are connected with $u_1, u_2, \cdots, u_n$ implicity as

$$f_1 (x_1, x_2, \cdots, x_n, u_1, u_2, \cdots, u_n) = 0$$

$$f_2 (x_1, x_2, \cdots, x_n, u_1, u_2, \cdots, u_n) = 0$$

....

....

$$f_n(x_1, x_2, \cdots, x_n, u_1, u_2, \cdots, u_n) = 0$$

then we have $\dfrac{\partial(u_1,u_2,\cdots,u_n)}{\partial(x_1,x_2,\cdots,x_n)} = (-1)^n \dfrac{\partial(f_1,f_2,\cdots,f_n)}{\partial(x_1,x_2,\cdots,x_n)} \Bigg/ \dfrac{\partial(f_1,f_2,\cdots,f_n)}{\partial(u_1,u_2,\cdots,u_n)}$.

ILLUSTRATIVE EXAMPLES

Example 1. *If $u^3 + v + w = x + y^2 + z^2$; $u + v^3 + w = x^2 + y + z^2$, $u + v + w^3 = x^2 + y^2 + z$. Prove that*

$$\frac{\partial(u, v, w)}{\partial(x, y, z)} = \frac{1 - 4(xy + yz + zx) + 16xyz}{2 - 3(u^2 + v^2 + w^2) + 27u^2v^2w^2}.$$

Sol. Let
$$f_1 = u^3 + v + w - x - y^2 - z^2$$
$$f_2 = u + v^3 + w - x^2 - y - z^2$$
$$f_3 = u + v + w^3 - x^2 - y^2 - z$$

$$\frac{\partial(f_1,f_2,f_3)}{\partial(x,y,z)} = \begin{vmatrix} -1 & -2y & -2z \\ -2x & -1 & -2z \\ -2x & -2y & -1 \end{vmatrix} = -1 + 4(yz + zx + xy) - 16xyz$$

and
$$\frac{\partial(f_1, f_2, f_3)}{\partial(u, v, w)} = \begin{vmatrix} 3u^2 & 1 & 1 \\ 1 & 3v^2 & 1 \\ 1 & 1 & 3w^2 \end{vmatrix} = 2 - 3(u^2 + v^2 + w^2) + 27u^2v^2w^2$$

Now
$$\frac{\partial(u, v, w)}{\partial(x, y, z)} = (-1)^3 \frac{\dfrac{\partial(f_1, f_2, f_3)}{\partial(x, y, z)}}{\dfrac{\partial(f_1, f_2, f_3)}{\partial(u, v, w)}} = (-1)^3 \frac{-1 + 4(yz + zx + xy) - 16xyz}{2 - 3(u^2 + v^2 + w^2) + 27u^2v^2w^2}$$

$$= \frac{1 - 4(yz + zx + xy) + 16xyz}{2 - 3(u^2 + v^2 + w^2) + 27u^2v^2w^2}.$$

Example 2. *If u, v, w are the roots of the equation $(\lambda - x)^3 + (\lambda - y)^3 + (\lambda - z)^3 = 0$. Find $\dfrac{\partial(u, v, w)}{\partial(x, y, z)}$.*

Sol. $(\lambda - x)^3 + (\lambda - y)^3 + (\lambda - z)^3 = 0$

or
$$3\lambda^3 - 3(x + y + z)\lambda^2 + 3(x^2 + y^2 + z^2)\lambda - (x^3 + y^3 + z^3) = 0$$

Sum of the roots $u + v + w = x + y + z$

Sum of the roots taken two at a time, $uv + vw + wu = x^2 + y^2 + z^2$

Product of the roots $uvw = \dfrac{1}{3}(x^3 + y^3 + z^3)$

Let
$$f_1 = u + v + w - x - y - z$$
$$f_2 = uv + vw + wu - x^2 - y^2 - z^2$$
$$f_3 = uvw - \frac{1}{3}(x^3 + y^3 + z^3)$$

$$\frac{\partial(f_1,f_2,f_3)}{\partial(x,y,z)}=\begin{vmatrix}-1 & -1 & -1\\ -2x & -2y & -2z\\ -x^2 & -y^2 & -z^2\end{vmatrix}=-2(x-y)(y-z)(z-x)$$

$$\frac{\partial(f_1,f_2,f_3)}{\partial(u,v,w)}=\begin{vmatrix}1 & 1 & 1\\ v+w & w+u & u+v\\ vw & wu & uv\end{vmatrix}=-(u-v)(v-w)(w-u)$$

$$\frac{\partial(u,v,w)}{\partial(x,y,z)}=(-1)^3\frac{\dfrac{\partial(f_1,f_2,f_3)}{\partial(x,y,z)}}{\dfrac{\partial(f_1,f_2,f_3)}{\partial(u,v,w)}}=(-1)^3\frac{-2(x-y)(y-z)(z-x)}{-(u-v)(v-w)(w-u)}$$

$$=-2\frac{(x-y)(y-z)(z-x)}{(u-v)(v-w)(w-u)}.$$

Example 3. *If* $u=\dfrac{x}{\sqrt{1-r^2}}$, $v=\dfrac{y}{\sqrt{1-r^2}}$, $w=\dfrac{z}{\sqrt{1-r^2}}$, *show that* $\dfrac{\partial(u,v,w)}{\partial(x,y,z)}=\dfrac{1}{(1-r^2)^{5/2}}$ *where* $r^2=x^2+y^2+z^2$.

Sol. $u=\dfrac{x}{\sqrt{1-r^2}}$, $\dfrac{\partial u}{\partial x}=\dfrac{\sqrt{1-r^2}\cdot 1-x\cdot\dfrac{1}{2\sqrt{1-r^2}}(-2r)\dfrac{\partial r}{\partial x}}{1-r^2}=\dfrac{1-r^2+xr\left(\dfrac{x}{r}\right)}{(1-r^2)^{3/2}}$

$$=\frac{1-r^2+x^2}{(1-r^2)^{3/2}}=\frac{1+x^2-x^2-y^2-z^2}{(1-r^2)^{3/2}}=\frac{1-y^2-z^2}{(1-r^2)^{3/2}}$$

$$\therefore\quad \frac{\partial u}{\partial x}=\frac{1-y^2-z^2}{(1-r^2)^{3/2}};$$

$$\frac{\partial u}{\partial y}=x\left(-\frac{1}{2}\right)(1-r^2)^{-3/2}(-2r)\frac{\partial r}{\partial y}=\frac{xr}{(1-r^2)^{3/2}}\cdot\frac{y}{r}=\frac{xy}{(1-r^2)^{3/2}}$$

$$\frac{\partial u}{\partial z}=x\left(-\frac{1}{2}\right)(1-r^2)^{-3/2}(-2r)\frac{\partial r}{\partial z}=\frac{xz}{(1-r^2)^{3/2}}$$

$$\therefore\quad \frac{\partial u}{\partial x}=\frac{1-y^2-z^2}{(1-r^2)^{3/2}},\ \frac{\partial u}{\partial y}=\frac{xy}{(1-r^2)^{3/2}},\ \frac{\partial u}{\partial z}=\frac{xz}{(1-r^2)^{3/2}}$$

Similarly, $\dfrac{\partial v}{\partial x}=\dfrac{xy}{(1-r^2)^{3/2}};\ \dfrac{\partial v}{\partial y}=\dfrac{1-z^2-x^2}{(1-r^2)^{3/2}};\ \dfrac{\partial v}{\partial z}=\dfrac{yz}{(1-r^2)^{3/2}}$

$$\frac{\partial w}{\partial x}=\frac{xz}{(1-r^2)^{3/2}};\ \frac{\partial w}{\partial y}=\frac{yz}{(1-r^2)^{3/2}};\ \frac{\partial w}{\partial z}=\frac{1-x^2-y^2}{(1-r^2)^{3/2}}$$

$$\frac{\partial(u,v,w)}{\partial(x,y,z)}=\begin{vmatrix}\dfrac{\partial u}{\partial x} & \dfrac{\partial u}{\partial y} & \dfrac{\partial u}{\partial z}\\ \dfrac{\partial v}{\partial x} & \dfrac{\partial v}{\partial y} & \dfrac{\partial v}{\partial z}\\ \dfrac{\partial w}{\partial x} & \dfrac{\partial w}{\partial y} & \dfrac{\partial w}{\partial z}\end{vmatrix}=\begin{vmatrix}\dfrac{1-y^2-z^2}{(1-r^2)^{3/2}} & \dfrac{xy}{(1-r^2)^{3/2}} & \dfrac{xz}{(1-r^2)^{3/2}}\\ \dfrac{xy}{(1-r^2)^{3/2}} & \dfrac{1-z^2-x^2}{(1-r^2)^{3/2}} & \dfrac{yz}{(1-r^2)^{3/2}}\\ \dfrac{xz}{(1-r^2)^{3/2}} & \dfrac{yz}{(1-r^2)^{3/2}} & \dfrac{1-x^2-y^2}{(1-r^2)^{3/2}}\end{vmatrix}$$

$$= \frac{1}{(1-r^2)^{9/2}} \begin{vmatrix} 1-y^2-z^2 & xy & xz \\ xy & 1-z^2-x^2 & yz \\ xz & yz & 1-x^2-y^2 \end{vmatrix}$$

Operate $R_1(x)$, $R_2(y)$, $R_3(z)$

$$= \frac{1}{(1-r^2)^{9/2}} \frac{1}{xyz} \begin{vmatrix} x(1-y^2-z^2) & x^2y & x^2z \\ xy^2 & y(1-z^2-x^2) & y^2z \\ xz^2 & yz^2 & z(1-x^2-y^2) \end{vmatrix}$$

Operate $C_1\left(\frac{1}{x}\right)$, $C_2\left(\frac{1}{y}\right)$, $C_3\left(\frac{1}{z}\right)$

$$= \frac{1}{(1-r^2)^{9/2}} \frac{xyz}{xyz} \begin{vmatrix} 1-y^2-z^2 & x^2 & x^2 \\ y^2 & 1-z^2-x^2 & y^2 \\ z^2 & z^2 & 1-x^2-y^2 \end{vmatrix}$$

Operate $R_1 + R_2 + R_3$

$$= \frac{1}{(1-r^2)^{9/2}} \begin{vmatrix} 1 & 1 & 1 \\ y^2 & 1-z^2-x^2 & y^2 \\ z^2 & z^2 & 1-x^2-y^2 \end{vmatrix}$$

Operate $C_2 - C_1$, $C_3 - C_1$

$$= \frac{1}{(1-r^2)^{9/2}} \begin{vmatrix} 1 & 0 & 0 \\ y^2 & 1-x^2-y^2-z^2 & 0 \\ z^2 & 0 & 1-x^2-y^2-z^2 \end{vmatrix}$$

$$= \frac{1}{(1-r^2)^{9/2}} \begin{vmatrix} 1 & 0 & 0 \\ y^2 & 1-r^2 & 0 \\ z^2 & 0 & 1-r^2 \end{vmatrix}$$

Expand with first row

$$= \frac{1}{(1-r^2)^{9/2}} (1-r^2)^2 = \frac{1}{(1-r^2)^{5/2}}$$

Hence $$\frac{\partial(u, v, w)}{\partial(x, y, z)} = \frac{1}{(1-r^2)^{5/2}}.$$

3.13. PARTIAL DERIVATIVE OF IMPLICIT FUNCTIONS BY JACOBIAN

Statement. *Given* $f_1(x, y, u, v) = 0$, $f_2(x, y, u, v) = 0$

to prove $$\frac{\partial u}{\partial x} = -\frac{\partial(f_1, f_2)}{\partial(x, v)} \bigg/ \frac{\partial(f_1, f_2)}{\partial(u, v)}$$

$$\frac{\partial v}{\partial x} = -\frac{\partial(f_1, f_2)}{\partial(u, x)} \bigg/ \frac{\partial(f_1, f_2)}{\partial(v, u)}.$$

Proof. $f_1(x, y, u, v) = 0$; $f_2(x, y, u, v) = 0$ ...(1)

$$\therefore \qquad \frac{\partial f_1}{\partial u}\cdot\frac{\partial u}{\partial x}+\frac{\partial f_1}{\partial v}\cdot\frac{\partial v}{\partial x}+\frac{\partial f_1}{\partial x}\cdot 1=0$$

$$\frac{\partial f_2}{\partial u}\cdot\frac{\partial u}{\partial x}+\frac{\partial f_2}{\partial v}\cdot\frac{\partial v}{\partial x}+\frac{\partial f_2}{\partial x}.1=0$$

Solve for $\frac{\partial u}{\partial x}, \frac{\partial v}{\partial x}$

$$\frac{\frac{\partial u}{\partial x}}{\begin{vmatrix}\frac{\partial f_1}{\partial v} & \frac{\partial f_1}{\partial x}\\ \frac{\partial f_2}{\partial v} & \frac{\partial f_2}{\partial x}\end{vmatrix}}=\frac{\frac{\partial v}{\partial x}}{-\begin{vmatrix}\frac{\partial f_1}{\partial u} & \frac{\partial f_1}{\partial x}\\ \frac{\partial f_2}{\partial u} & \frac{\partial f_2}{\partial x}\end{vmatrix}}=\frac{1}{\begin{vmatrix}\frac{\partial f_1}{\partial u} & \frac{\partial f_1}{\partial v}\\ \frac{\partial f_2}{\partial u} & \frac{\partial f_2}{\partial v}\end{vmatrix}}$$

$$\frac{\frac{\partial u}{\partial x}}{\frac{\partial(f_1,f_2)}{\partial(v,x)}}=-\frac{\frac{\partial v}{\partial x}}{\frac{\partial(f_1,f_2)}{\partial(u,x)}}=\frac{1}{\frac{\partial(f_1,f_2)}{\partial(u,v)}}$$

$$\therefore \qquad \frac{\partial u}{\partial x}=\frac{\frac{\partial(f_1,f_2)}{\partial(v,x)}}{\frac{\partial(f_1,f_2)}{\partial(u,v)}}=-\frac{\frac{\partial(f_1,f_2)}{\partial(x,v)}}{\frac{\partial(f_1,f_2)}{\partial(u,v)}}$$

$$\frac{\partial v}{\partial x}=-\frac{\partial(f_1,f_2)}{\partial(u,x)}\bigg/\frac{\partial(f_1,f_2)}{\partial(u,v)}$$

Note: To remember this result note that numerator of $\frac{\partial u}{\partial x}$ is obtained from the denominator by replacing u by x in the Jacobian $\frac{\partial(f_1,f_2)}{\partial(u,v)}$. And the numerator of $\frac{\partial v}{\partial x}$ is obtained from the denominator by replacing v by x in the Jacobian.

Similarly on differentiating (1) partially w.r.t. y, we get

$$\frac{\partial u}{\partial y}=-\frac{\frac{\partial(f_1,f_2)}{\partial(y,v)}}{\frac{\partial(f_1,f_2)}{\partial(u,v)}};\quad \frac{\partial v}{\partial y}=-\frac{\frac{\partial(f_1,f_2)}{\partial(u,y)}}{\frac{\partial(f_1,f_2)}{\partial(u,v)}}$$

To get $\frac{\partial x}{\partial u}, \frac{\partial x}{\partial v}$ and $\frac{\partial y}{\partial u}, \frac{\partial y}{\partial v}$ differentiate partially w.r.t. u and v respectively, we get

$$\frac{\partial x}{\partial u}=-\frac{\frac{\partial(f_1,f_2)}{\partial(u,y)}}{\frac{\partial(f_1,f_2)}{\partial(x,y)}};\quad \frac{\partial y}{\partial u}=-\frac{\frac{\partial(f_1,f_2)}{\partial(x,u)}}{\frac{\partial(f_1,f_2)}{\partial(x,y)}}$$

and

$$\frac{\partial x}{\partial v}=-\frac{\frac{\partial(f_1,f_2)}{\partial(v,y)}}{\frac{\partial(f_1,f_2)}{\partial(x,y)}};\quad \frac{\partial y}{\partial v}=-\frac{\frac{\partial(f_1,f_2)}{\partial(x,v)}}{\frac{\partial(f_1,f_2)}{\partial(x,y)}}$$

Example 4. *If* $x = u^2 - v^2$, $y = 2uv$, *find* $\frac{\partial u}{\partial x}, \frac{\partial u}{\partial y}, \frac{\partial v}{\partial x}, \frac{\partial v}{\partial y}, \frac{\partial(u, v)}{\partial(x, y)}$.

Sol. Let
$$f_1 = x - u^2 + v^2$$
$$f_2 = y - 2uv$$

$$\frac{\partial u}{\partial x} = -\frac{\frac{\partial(f_1,f_2)}{\partial(x,v)}}{\frac{\partial(f_1,f_2)}{\partial(u,v)}} = -\frac{\begin{vmatrix} \frac{\partial f_1}{\partial x} & \frac{\partial f_1}{\partial v} \\ \frac{\partial f_2}{\partial x} & \frac{\partial f_2}{\partial v} \end{vmatrix}}{\begin{vmatrix} \frac{\partial f_1}{\partial u} & \frac{\partial f_1}{\partial v} \\ \frac{\partial f_2}{\partial u} & \frac{\partial f_2}{\partial v} \end{vmatrix}} = -\frac{\begin{vmatrix} 1 & 2v \\ 0 & -2u \end{vmatrix}}{\begin{vmatrix} -2u & 2v \\ -2v & -2u \end{vmatrix}}$$

$$= \frac{2u}{4(u^2+v^2)} = \frac{u}{2(u^2+v^2)}$$

$$\frac{\partial u}{\partial y} = -\frac{\frac{\partial(f_1,f_2)}{\partial(y,v)}}{\frac{\partial(f_1,f_2)}{\partial(u,v)}} = -\frac{\begin{vmatrix} \frac{\partial f_1}{\partial y} & \frac{\partial f_1}{\partial v} \\ \frac{\partial f_2}{\partial y} & \frac{\partial f_2}{\partial v} \end{vmatrix}}{\begin{vmatrix} \frac{\partial f_1}{\partial u} & \frac{\partial f_1}{\partial v} \\ \frac{\partial f_2}{\partial u} & \frac{\partial f_2}{\partial v} \end{vmatrix}} = -\frac{\begin{vmatrix} 0 & 2v \\ 1 & -2u \end{vmatrix}}{4(u^2+v^2)}$$

$$\therefore \quad \frac{\partial u}{\partial y} = \frac{2v}{4(u^2+v^2)} = \frac{v}{2(u^2+v^2)}$$

$$\frac{\partial v}{\partial x} = -\frac{\frac{\partial(f_1,f_2)}{\partial(u,x)}}{\frac{\partial(f_1,f_2)}{\partial(u,v)}} = -\frac{\begin{vmatrix} \frac{\partial f_1}{\partial u} & \frac{\partial f_1}{\partial x} \\ \frac{\partial f_2}{\partial u} & \frac{\partial f_2}{\partial x} \end{vmatrix}}{\begin{vmatrix} \frac{\partial f_1}{\partial u} & \frac{\partial f_1}{\partial v} \\ \frac{\partial f_2}{\partial u} & \frac{\partial f_2}{\partial v} \end{vmatrix}} = -\frac{\begin{vmatrix} -2u & 1 \\ -2v & 0 \end{vmatrix}}{4(u^2+v^2)}$$

$$\therefore \quad \frac{\partial v}{\partial x} = \frac{-v}{2(u^2+v^2)}$$

$$\frac{\partial v}{\partial y} = -\frac{\frac{\partial(f_1,f_2)}{\partial(u,y)}}{\frac{\partial(f_1,f_2)}{\partial(u,v)}} = -\frac{\begin{vmatrix} \frac{\partial f_1}{\partial u} & \frac{\partial f_1}{\partial y} \\ \frac{\partial f_2}{\partial u} & \frac{\partial f_2}{\partial y} \end{vmatrix}}{\begin{vmatrix} \frac{\partial f_1}{\partial u} & \frac{\partial f_1}{\partial v} \\ \frac{\partial f_2}{\partial u} & \frac{\partial f_2}{\partial v} \end{vmatrix}} = -\frac{\begin{vmatrix} -2u & 0 \\ -2x & 1 \end{vmatrix}}{4(u^2+v^2)}$$

$$\frac{\partial v}{\partial y} = \frac{2u}{4(u^2+v^2)} = \frac{u}{2(u^2+v^2)}$$

$$\frac{\partial(u, v)}{\partial(x, y)} = \begin{vmatrix} \frac{\partial u}{\partial x} & \frac{\partial u}{\partial y} \\ \frac{\partial v}{\partial x} & \frac{\partial v}{\partial y} \end{vmatrix} \text{ (As } u, v \text{ cannot be expressed in terms of } x, y \text{ easily)}$$

$\therefore$ We can write

$$\frac{\partial(u, v)}{\partial(x, y)} = \frac{1}{\frac{\partial(x, y)}{\partial(u, v)}} = \frac{1}{\begin{vmatrix} \frac{\partial x}{\partial u} & \frac{\partial x}{\partial v} \\ \frac{\partial y}{\partial u} & \frac{\partial y}{\partial v} \end{vmatrix}}$$

$$= \frac{1}{\begin{vmatrix} 2u & -2v \\ 2v & 2u \end{vmatrix}} = \frac{1}{4(u^2+v^2)}.$$

Example 5. *Use Jacobian to find* $\frac{\partial u}{\partial x}$ *if* $u^2 + xv^2 - xy = 0$ *and* $u^2 + xyv + v^2 = 0$.

Sol. Let $f_1 = u^2 + xv^2 - xy$

$f_2 = u^2 + xyv + v^2$

$$\frac{\partial u}{\partial x} = -\frac{\frac{\partial(f_1, f_2)}{\partial(x, v)}}{\frac{\partial(f_1, f_2)}{\partial(u, v)}} = -\frac{\begin{vmatrix} \frac{\partial f_1}{\partial x} & \frac{\partial f_1}{\partial v} \\ \frac{\partial f_2}{\partial x} & \frac{\partial f_2}{\partial v} \end{vmatrix}}{\begin{vmatrix} \frac{\partial f_1}{\partial u} & \frac{\partial f_1}{\partial v} \\ \frac{\partial f_2}{\partial u} & \frac{\partial f_2}{\partial v} \end{vmatrix}} = -\frac{\begin{vmatrix} v^2 - y & 2xv \\ yv & xy + 2v \end{vmatrix}}{\begin{vmatrix} 2u & 2xv \\ 2u & xy + 2v \end{vmatrix}}$$

$$= -\frac{(v^2 - y)(xy + 2v) - 2xyv^2}{2\{xyu + 2uv - 2xuv\}} = -\frac{xyv^2 - xy^2 + 2v^3 - 2yv - 2xyv^2}{2xyu + 4uv - 4xuv}$$

$$= \frac{xyv^2 + xy^2 - 2v^3 + 2yv}{2xyu + 4uv - 4xuv}.$$

Example 6. *If* $x = u + v + w$, $y = u^2 + v^2 + w^2$, $z = u^3 + v^3 + w^3$, *show that* $\frac{\partial u}{\partial x} = \frac{vw}{(u-v)(u-w)}$.

Sol. Let $f_1 = u + v + w - x$

$f_2 = u^2 + v^2 + w^2 - y$

$f_3 = u^3 + v^3 + w^3 - z$

$$\frac{\partial(f_1, f_2, f_3)}{\partial(u, v, w)} = \begin{vmatrix} 1 & 1 & 1 \\ 2u & 2v & 2w \\ 3u^2 & 3v^2 & 3w^2 \end{vmatrix} = 6(u - v)(v - w)(w - u)$$

Replace u by x

$$\frac{\partial(f_1,f_2,f_3)}{\partial(x,v,w)} = \begin{vmatrix} -1 & 1 & 1 \\ 0 & 2v & 2w \\ 0 & 3v^2 & 3w^2 \end{vmatrix} = 6vw\,(v-w)$$

Now
$$\frac{\partial u}{\partial x} = -\frac{\dfrac{\partial(f_1,f_2,f_3)}{\partial(x,v,w)}}{\dfrac{\partial(f_1,f_2,f_3)}{\partial(u,v,w)}} = -\frac{6vw(v-w)}{6(u-v)(v-w)(w-u)}$$

$$= \frac{-6vw}{(u-v)(w-u)} = \frac{6vw}{(u-v)(u-w)}.$$

Example 7. *If* $u = x + y^2$, $v = y + z^2$, $w = z + x^2$, *prove that* $\dfrac{\partial x}{\partial u} = \dfrac{1}{1+8xyz}$. *Also find* $\dfrac{\partial^2 x}{\partial u^2}$.

Sol. Let
$$f_1 = u - x - y^2$$
$$f_2 = v - y - z^2$$
$$f_3 = w - z - x^2$$

$$\frac{\partial(f_1,f_2,f_3)}{\partial(x,y,z)} = \begin{vmatrix} -1 & -2y & 0 \\ 0 & -1 & -2z \\ -2x & 0 & -1 \end{vmatrix} = -1 - 8\,xyz$$

Replace x by u.

$$\frac{\partial(f_1,f_2,f_3)}{\partial(u,y,z)} = \begin{vmatrix} 1 & -2y & 0 \\ 0 & -1 & -2z \\ 0 & 0 & -1 \end{vmatrix} = 1$$

$\therefore$
$$\frac{\partial x}{\partial u} = -\frac{\dfrac{\partial(f_1,f_2,f_3)}{\partial(u,y,z)}}{\dfrac{\partial(f_1,f_2,f_3)}{\partial(x,y,z)}} = -\frac{1}{-1-8xyz} = \frac{1}{1+8xyz}.$$

Also
$$\frac{\partial^2 x}{\partial u^2} = \frac{\partial}{\partial u}\left[\frac{1}{1+8xyz}\right] = -\frac{1}{(1+8xyz)^2}\left[\frac{\partial}{\partial u}(1+8xyz)\right]$$

$$= -\frac{1}{(1+8xyz)^2}\,8\left\{xy\frac{\partial z}{\partial u} + yz\frac{\partial x}{\partial u} + zx\frac{\partial y}{\partial u}\right\} \quad \ldots(1)$$

Find $\dfrac{\partial y}{\partial u}, \dfrac{\partial z}{\partial u}$

$$\frac{\partial y}{\partial u} = -\frac{\dfrac{\partial(f_1,f_2,f_3)}{\partial(x,u,z)}}{\dfrac{\partial(f_1,f_2,f_3)}{\partial(x,y,z)}}, \quad \frac{\partial z}{\partial u} = -\frac{\dfrac{\partial(f_1,f_2,f_3)}{\partial(x,y,u)}}{\dfrac{\partial(f_1,f_2,f_3)}{\partial(x,y,z)}}$$

$$\frac{\partial(f_1,f_2,f_3)}{\partial(x,u,z)} = \begin{vmatrix} -1 & 1 & 0 \\ 0 & 0 & -2z \\ -2x & 0 & -1 \end{vmatrix} = 4xz$$

$$\frac{\partial(f_1, f_2, f_3)}{\partial(x, y, u)} = \begin{vmatrix} -1 & -2y & 1 \\ 0 & -1 & 0 \\ -2x & 0 & 0 \end{vmatrix} = -2x$$

$$\therefore \quad \frac{\partial y}{\partial u} = -\frac{4xz}{-1-8xyz} = \frac{4xz}{1+8xyz}$$

$$\frac{\partial z}{\partial u} = -\frac{-2x}{-1-8xyz} = -\frac{2x}{1+8xyz}$$

$\therefore$ From (1),

$$\frac{\partial^2 x}{\partial u^2} = \frac{-8}{(1+8xyz)^2}\left[xy\frac{-2x}{1+8xyz} + yz\frac{1}{1+8xyz} + zx \cdot \frac{4xz}{1+8xyz}\right]$$

$$= \frac{-8}{(1+8xyz)^3}[-2x^2y + yz + 4x^2z^2].$$

TEST YOUR KNOWLEDGE

1. If $x^2 + y^2 + u^2 - v^2 = 0$ and $uv + xy = 0$, prove that $\dfrac{\partial(u, v)}{\partial(x, y)} = \dfrac{x^2 - y^2}{u^2 + v^2}$.
2. If $x + y + z = u$, $y + z = uv$, $z = uvw$, show that $\dfrac{\partial(x, y, z)}{\partial(u, v, w)} = u^2 v$.
3. If u, v, w are the roots of the equation in λ and $\dfrac{x}{a+\lambda} + \dfrac{y}{b+\lambda} + \dfrac{z}{c+\lambda} = 1$, find $\dfrac{\partial(x, y, z)}{\partial(u, v, w)}$.

   [**Hint.** Consult Ex. 2]

Answer

3. $\dfrac{(u-v)(v-w)(w-u)}{(a-b)(b-c)(c-a)}$.

REVIEW OF THE CHAPTER

1. **Limit of a Function of Two Variables *i.e.*, *f*(*x*, *y*).** A function $f(x, y)$ is said to tend to the limit l iff corresponding to a positive number ε, $\exists$s another positive number δ s.t. $|f(x, y) - l| < \varepsilon \ \forall \ (x, y)$ within the circle with its centre at (a, b) and radius δ, *i.e.* $0 < (x-a)^2 + (y-b)^2 < \delta^2$

   or $f(x, y)$ is said to tend limit l as $x \to a$, $y \to b$ iff l is independent of the path followed by the point (x, y) as $x \to a$ and $y \to b$ and is written as $\underset{\substack{x \to a \\ y \to b}}{\text{Lt}} f(x, y) = l$

   Properties: If $\underset{\substack{x \to a \\ y \to b}}{\text{Lt}} f(x, y) = l$ and $\underset{\substack{x \to a \\ y \to b}}{\text{Lt}} g(x, y) = m$

   then (*i*) $\underset{\substack{x \to a \\ y \to b}}{\text{Lt}} [f(x, y) \pm g(x, y)] = l \pm m$

   (*ii*) $\underset{\substack{x \to a \\ y \to b}}{\text{Lt}} f(x, y), g(x, y) = lm.$

   (*iii*) $\underset{\substack{x \to a \\ y \to b}}{\text{Lt}} \dfrac{f(x, y)}{g(x, y)} = \dfrac{l}{m} \quad (m \neq 0).$

2. **Continuity of f(x, y).** A function is said to be continuous at a point (a, b) iff corresponding to a positive number ε, however small, $\exists s$ a +ve number δ s.t.

$$|f(x, y) - f(a, b)| < \varepsilon \ \forall\ (x, y)$$

within the circle with its centre at (a, b) and radius δ *i.e.*,

$$\underset{\substack{x \to a \\ y \to b}}{\text{Lt}}\ f(x,y) = f(a,b).$$

Properties. If $f(x, y)$ and $g(x, y)$ are continuous functions of x, y at (a, b) then $f(x) \pm g(x)$; $f(x)\, g(x)$; $f(x, y)/g(x, y)$ $[g(x, y) \neq 0]$ are also continuous functions of x, y.

3. **Partial Derivatives.** If $z = f(x, y)$ then

$$\frac{\partial z}{\partial x} = \underset{h \to 0}{\text{Lt}}\ \frac{f(x+h, y) - f(x, y)}{h} \text{ and } \frac{\partial z}{\partial y} = \underset{k \to 0}{\text{Lt}}\ \frac{f(x, y+k) - f(x, y)}{k}$$

$\dfrac{\partial z}{\partial x}$ is represented by f_x and $\dfrac{\partial z}{\partial y}$ is represented by f_y

$\dfrac{\partial^2 z}{\partial x^2} = \dfrac{\partial}{\partial x}\left(\dfrac{\partial z}{\partial x}\right)$ and is represented by f_{xx}

$\dfrac{\partial^2 z}{\partial y^2}$ is represented by f_{yy} and $\dfrac{\partial^2 z}{\partial x \partial y} = \dfrac{\partial^2 z}{\partial y \partial x}$ is represented by f_{xy} or f_{yx}.

4. **Homogeneous Function.** A function $f(x, y)$ is said to be homogeneous function of degree n if it can be expressed as $x^n\, \phi\left(\dfrac{y}{x}\right)$ or $y^n\, \phi\left(\dfrac{x}{y}\right)$ or alternatively if

$$f(tx, ty) = t^n f(x, y).$$

5. **Euler's Theorem on Homogeneous Functions.** If u is a homogeneous function of degree n in x and y then

$$x\frac{\partial u}{\partial x} + y\frac{\partial u}{\partial y} = nu \text{ and } x^2 \frac{\partial^2 u}{\partial x^2} + 2xy \frac{\partial^2 u}{\partial x \partial y} + y^2 \frac{\partial^2 u}{\partial y^2} = n(n-1)\, u.$$

6. **Composite Functions.** If $u = f(x, y)$ and $x = \phi(t)$, $y = \psi(t)$, then u is called composite function of single variable t.

7. **Total Derivative.** If $u = f(x, y)$, $x = \phi(t)$, $y = \psi(t)$ *i.e.*, u is a composite funciton of t then $\dfrac{du}{dt}$ is called total derivative and can be obtained without actually substituting the values of x, y in $f(x, y)$. Total derivative is $\dfrac{du}{dt} = \dfrac{\partial u}{\partial x} \cdot \dfrac{dx}{dt} + \dfrac{\partial u}{\partial y} \cdot \dfrac{dy}{dt}$.

8. **Differentiation of Implicit Functions.** If given implicit function $f(x, y) = c$; we can write $u = f(x, y)$ where $u = c$

then $$\frac{dy}{dx} = -\frac{\frac{\partial f}{\partial x}}{\frac{\partial f}{\partial y}} = -\frac{f_x}{f_y}$$

and $$\frac{d^2y}{dx^2} = -\frac{f_{xx}f_y^2 - 2f_xf_yf_{xy} + f_{yy}f_x^2}{f_y^3}$$

9. **Change of Variables.** If $z = f(x, y)$ and $x = \phi(u, v)$, $y = \psi(u, v)$, then

$$\frac{\partial z}{\partial u} = \frac{\partial z}{\partial x} \cdot \frac{\partial x}{\partial u} + \frac{\partial z}{\partial y} \cdot \frac{\partial y}{\partial u}$$

$$\frac{\partial z}{\partial v} = \frac{\partial z}{\partial x} \cdot \frac{\partial x}{\partial v} + \frac{\partial z}{\partial y} \cdot \frac{\partial y}{\partial v}$$

or we can express

$$\frac{\partial z}{\partial x} = \frac{\partial z}{\partial u} \cdot \frac{\partial u}{\partial x} + \frac{\partial z}{\partial v} \cdot \frac{\partial v}{\partial x}$$

$$\frac{\partial z}{\partial y} = \frac{\partial z}{\partial u} \cdot \frac{\partial u}{\partial y} + \frac{\partial z}{\partial v} \cdot \frac{\partial v}{\partial y}.$$

10. **Jacobians.** If u, v are functions of two variables x and y then Jacobian of u, v w.r.t. x, y

is defined as $$J\left(\frac{u, v}{x, y}\right) = \begin{vmatrix} \frac{\partial u}{\partial x} & \frac{\partial u}{\partial y} \\ \frac{\partial v}{\partial x} & \frac{\partial v}{\partial y} \end{vmatrix} = \frac{\partial(u, v)}{\partial(x, y)}$$

and $$J\left(\frac{u, v, w}{x, y, z}\right) = \begin{vmatrix} \frac{\partial u}{\partial x} & \frac{\partial u}{\partial y} & \frac{\partial u}{\partial z} \\ \frac{\partial v}{\partial x} & \frac{\partial v}{\partial y} & \frac{\partial v}{\partial z} \\ \frac{\partial w}{\partial x} & \frac{\partial w}{\partial y} & \frac{\partial w}{\partial z} \end{vmatrix} = \frac{\partial(u, v, w)}{\partial(x, y, z)}.$$

Properties of Jacobians:

(*i*) If u, v are functions of r, s where r, s are functions of x, y then

$$J\left(\frac{u, v}{x, y}\right) = \frac{\partial(u, v)}{\partial(x, y)} = \frac{\partial(u, v)}{\partial(r, s)} \times \frac{\partial(r, s)}{\partial(x, y)}.$$

(*ii*) If J_1 is Jacobian of u, v w.r.t. x, y and J_2 is the Jacobian of x, y w.r.t. u, v then $J_1 J_2 = 1$

i.e, $$\frac{\partial(u, v)}{\partial(x, y)} \times \frac{\partial(x, y)}{\partial(u, v)} = 1.$$

11. Jacobian of Implicit Functions. If x, y, u, v are connected by implicit functions $f_1(x, y, u, v) = 0$ and $f_2(x, y, u, v) = 0$ where u, v are implicit functions of x, y, then

$$\frac{\partial(u, v)}{\partial(x, y)} = (-1)^2 \frac{\dfrac{\partial(f_1, f_2)}{\partial(x, y)}}{\dfrac{\partial(f_1, f_2)}{\partial(u, v)}}$$

Generalised form is $\dfrac{\partial(u_1, u_2, \cdots, u_n)}{\partial(x_1, x_2, \cdots, x_n)} = (-1)^n \dfrac{\partial(f_1, f_2, \cdots f_n)}{\partial(x_1, x_2, \cdots, x_n)} \Big/ \dfrac{\partial(f_1, f_2, \cdots, f_n)}{\partial(u_1, u_2, \cdots, u_n)}$.

12. Partial Derivatives of Implicit Functions by Jacobian. If $f_1(x, y, u, v) = 0$ and $f_2(x, y, u, v) = 0$.

Then $\dfrac{\partial u}{\partial x} = -\dfrac{\partial(f_1, f_2)}{\partial(x, v)} \Big/ \dfrac{\partial(f_1, f_2)}{\partial(u, v)}$

(Numerator is obtained by replacing u by x in denominator)

$$\frac{\partial v}{\partial x} = -\frac{\partial(f_1, f_2)}{\partial(u, x)} \Big/ \frac{\partial(f_1, f_2)}{\partial(u, v)}$$

$$\frac{\partial u}{\partial y} = -\frac{\dfrac{\partial(f_1, f_2)}{\partial(y, v)}}{\dfrac{\partial(f_1, f_2)}{\partial(u, v)}}, \quad \frac{\partial v}{\partial y} = -\frac{\dfrac{\partial(f_1, f_2)}{\partial(u, y)}}{\dfrac{\partial(f_1, f_2)}{\partial(u, v)}}.$$

SHORT ANSWER TYPE QUESTIONS

1. If $z = f(x, y)$ be a surface then what is the geometrical meaning of $\dfrac{\partial z}{\partial x}$? **(*P.T.U., May 2011*)**

   [**Hint.** Art. 3.3(*i*)]

2. (*i*) If $z = e^{ax + by} f(ax - by)$, find $b\dfrac{\partial z}{\partial x} + a\dfrac{\partial z}{\partial y}$. **(*P.T.U., May 2006*)**

   [**Hint.** S.E. 6(*ii*) art. 3.4]

   (*ii*) If $u = (1 - 2x + y^2)^{-1/2}$, prove that $x\dfrac{\partial u}{\partial x} - y\dfrac{\partial u}{\partial y} = y^2 u^3$.

   [**Hint.** S.E. 12(*i*) art. 3.4]

3. If $x = r\cos\theta$, $y = r\sin\theta$, prove that

   (*i*) $\dfrac{\partial r}{\partial x} = \dfrac{\partial x}{\partial r}$ (*ii*) $\dfrac{1}{r}\dfrac{\partial x}{\partial \theta} = r\dfrac{\partial \theta}{\partial x}$.

   [**Hint.** S.E. 16 art. 3.4]

4. Find the first order derivatives of the following functions:

   (*i*) $u = x^2 \sin\dfrac{y}{x}$ (*ii*) $u = \displaystyle\sum_{n=0}^{\infty} (xy)^n$ (*iii*) $u = y^x$.

   (*P.T.U., Dec. 2014*)

   [**Hint.** $u = 1 + xy + (xy)^2 + \ldots \infty$, $u = (1 - xy)^{-1} = \dfrac{1}{1 - xy}$]

5. (*i*) If $u = x^2y + y^2 z + z^2 x$ prove that $u_x + u_y + u_z = (x + y + z)^2$

(*ii*) If $u = \frac{y}{z} + \frac{z}{x} + \frac{x}{y}$ prove that $xu_x + yu_y + zu_z = 0$.

6. If $u = \log (\tan x + \tan y)$; prove that $2x \frac{\partial u}{\partial x} + \sin 2y \frac{\partial u}{\partial y} = 2$.

7. Verify $\frac{\partial^2 u}{\partial x \partial y} = \frac{\partial^2 u}{\partial y \partial x}$ for the following functions:

(*i*) $u = \tan^{-1} \frac{x}{y}$

(*ii*) $u = ax^2 + 2hxy + by^2$

(*iii*) $u = e^{ax} \sin by$

(*iv*) $u = \log (x \sin y + y \sin x)$

(*v*) $u = \sin^{-1} \frac{x}{y}$

(*vi*) $u = \log \frac{x^2 + y^2}{xy}$.

$\left[\textbf{Hint. } \log \frac{x^2 + y^2}{xy} = \log (x^2 + y^2) - \log x - \log y\right]$

8. If $u = x^y$, find $\frac{\partial^2 u}{\partial x \partial y}$ at (1, 2) **(*P.T.U., May 2011*)**

[**Hint:** S.E. 9 art 3.4]

9. If $z = \log (e^x + e^y)$ show that $rt - s^2 = 0$, where $r = \frac{\partial^2 z}{\partial x^2}$, $s = \frac{\partial^2 z}{\partial x \partial y}$, $t = \frac{\partial^2 z}{\partial y^2}$.

10. If $u = \log (x^2 + y^2) + \tan^{-1} \frac{y}{x}$ prove that $\frac{\partial^2 u}{\partial x^2} + \frac{\partial^2 u}{\partial y^2} = 0$.

11. If $u = (x^2 + y^2 + z^2)^{-1/2}$ prove that $u_{xx} + u_{yy} + u_{zz} = 0$.

12. (*i*) If $z (x + y) = x^2 + y^2$ show that $\left(\frac{\partial z}{\partial x} - \frac{\partial z}{\partial y}\right)^2 = 4\left(1 - \frac{\partial z}{\partial x} - \frac{\partial z}{\partial y}\right)$. [**Hint.** S.E.6(*i*) art. 3.4]

(*ii*) If $z = \log (x^2 + xy + y^2)$ prove that $x\frac{\partial z}{\partial y} + y\frac{\partial z}{\partial y} = 2$. **(*P.T.U., Dec. 2013*)**

[**Hint.** S.E.7(*ii*) art. 3.4]

13. Define homogeneous function with the help of an example. **(*P.T.U., 2005, May 2009, May 2010*)**
[**Hint.** Art. 3.5]

14. (*i*) State and prove, Euler's theorem on homogeneous functions
[**Hint.** Art. 3.6] **(*P.T.U., 2006, Jan. 2009, May 2009*)**

(*ii*) State Euler's theorem and use it to prove $x \frac{\partial u}{\partial x} + y \frac{\partial u}{\partial y} = 0$ where $u = \sin^{-1} \frac{x}{y} + \tan^{-1} \frac{y}{x}$.

(*P.T.U., Dec. 2010*)

[**Hint.** S.E. I (*ii*) art. 3.7]

15. Verify Euler's theorem for $f(x, y, z) = 3x^2yz + 5xy^2z + 4z^4$ **(*P.T.U., Dec. 2005*)**
[**Hint.** S.E. I (*iii*) art 3.7]

16. (*i*) If $u = f\left(\frac{y}{x}\right)$ show that $u\frac{\partial u}{\partial x} + y\frac{\partial u}{\partial y} = 0$

[**Hint.** S.E. 2 art. 3.7]

(*ii*) If $u = xf\left(\frac{y}{x}\right)$, show that $x\frac{\partial u}{\partial x} + y\frac{\partial u}{\partial y} = u$ and $x^2\frac{\partial^2 u}{\partial x^2} + 2xy\frac{\partial^2 u}{\partial x \partial y} + y^2\frac{\partial^2 u}{\partial y^2} = 0$

(P.T.U., May 2008)

17. Verify Euler's theorem for the function $z = ax^2 + 2hxy + by^2$.

18. State Euler's theorem and use it to prove $x\frac{\partial u}{\partial x} + y\frac{\partial u}{\partial y} = 2u \log u$, where $u = e^{x^2+y^2}$.

(P.T.U., May 2014)

19. If $f(x, y) = \sqrt{x^2 - y^2}\ \sin^{-1}\left(\frac{y}{x}\right)$ prove that $x\frac{\partial f}{\partial x} + y\frac{\partial f}{\partial y} = f(x, y)$.

20. (*i*) If $u = \cos^{-1}\frac{x+y}{\sqrt{x}+\sqrt{y}}$, show that $x\frac{\partial u}{\partial x} + y\frac{\partial u}{\partial y} + \frac{1}{2}\cot u = 0$.

(*ii*) If $u = \sin^{-1}\frac{x+y}{\sqrt{x}+\sqrt{y}}$, show that $x\frac{\partial u}{\partial x} + y\frac{\partial u}{\partial y} = \frac{1}{2}\tan u$. **(P.T.U., May 2008)**

[**Hint.** In each case u is not homogeneous function but cos u and sin u both are homogeneous function of degree $\frac{1}{2}$ apply Euler's theorem to each]

(*iii*) If $u = \sin^{-1}\frac{x^2+y^2}{x+y}$, prove that $x\frac{\partial u}{\partial x} + y\frac{\partial u}{\partial y} = \tan u$. **(P.T.U., Dec. 2013)**

21. If $u = \log\frac{x^4+y^4}{x+y}$, show that $x\frac{\partial u}{\partial x} + y\frac{\partial u}{\partial y} = 3$.

22. Define composite function of single and double variables. **(P.T.U., May 2007)**
[**Hint.** Art. 3.8]

23. State Euler's theorem and establish the same for any suitable function.
[**Hint.** Art. 3.6] **(P.T.U., May 2005)**

24. If given $f(x, y) = c$, an implicit function prove that $\frac{dy}{dx} = -\frac{f_x}{f_y}$.

[**Hint.** Art. 3.9(*b*)]

25. (*i*) If $u = f(y-z, z-x, x-y)$, prove that $x\frac{\partial u}{\partial x} + y\frac{\partial u}{\partial y} + z\frac{\partial u}{\partial z} = 0$ **(P.T.U., Dec. 2011)**

[**Hint.** S.E.4 (*a*) art 3.9 (*c*)]

(*ii*) If $u = f\left(\frac{x}{y}, \frac{y}{z}, \frac{z}{x}\right)$ prove that $x\frac{\partial u}{\partial x} + y\frac{\partial u}{\partial y} + z\frac{\partial u}{\partial z} = 0$. **(P.T.U., Dec. 2010, May 2014)**

[**Hint.** S.E. 4(*ii*) art. 3.9 (*c*)]

26. Find $\frac{du}{dt}$ when

(*i*) $u = x^2 + y^2$, $x = at^2$, $y = 2at$.

[**Hint.** S.E. 1 art. 3.9(*c*)]

(*ii*) $u = x^2 + y^2 + z^2$; $x = e^{2t}$, $y = e^{2t} \cos 3t$, $z = e^{2t} \sin 3t$.

(*iii*) $u = \sin \frac{x}{y}$; $x = e^t$, $y = t^2$. **(P.T.U., May 2006)**

[**Hint.** S.E. 2 art. 3.9(*c*)]

(*iv*) $u = x^3 + y^3$, $x = a \cos t$, $y = b \sin t$.

27. If $w = x^2 + y^2$, $x = r - s$, $y = r + s$ find $\frac{\partial w}{\partial r}, \frac{\partial w}{\partial s}$ in terms of r, s. **(P.T.U., May 2012)**

[**Hint.** S.E. 3(*i*) art. 3.9]

28. If $x = u + v$, $y = uv$ and z is a function of x, y show that $u\frac{\partial z}{\partial u} + v\frac{\partial z}{\partial v} = x\frac{\partial z}{\partial x} + 2y\frac{\partial z}{\partial y}$.

29. If $u = x \log (xy)$ where $x^3 + y^3 + 3xy = 1$, find $\frac{du}{dx}$. **(P.T.U., May 2015)**

30. If $z = \sqrt{x^2 + y^2}$ and $x^3 + y^3 + z\,axy = 5a^2$, find the value of $\frac{dz}{dx}$ at $x = y = a$. **(P.T.U., May 2009)**

31. When $x^y + y^x = \text{C}$, show that $\frac{dy}{dx} = \frac{yx^{y-1} + y^x \log y}{x^y \log x + xy^{x-1}}$.

[**Hint.** S.E. 9(*i*) art. 3.9(*c*)]

32. When $(\cos x)^y = (\sin y)^x$, prove that $\frac{dy}{dx} = \frac{x \tan x + \log \sin y}{\log \cos x - x \cot y}$.

[**Hint.** S.E. 9(*ii*) art. 3.9(*c*)]

33. If $\text{F}(x, y, z) = 0$, prove that $\left(\frac{\partial y}{\partial x}\right)_z \left(\frac{\partial z}{\partial y}\right)_x \left(\frac{\partial x}{\partial z}\right)_y = -1$. **(P.T.U., Dec. 2011)**

[**Hint.** S.E. 11 art 3.9 (*i*)]

34. Find total derivative of $z = \tan^{-1}\left(\frac{x}{y}\right)$ when $(x, y) \neq (0, 0)$. **(P.T.U., Dec. 2007)**

$\left[\textbf{Hint. } dz = \frac{\partial z}{\partial x} dx + \frac{\partial z}{\partial y} dy = \frac{y\,dx - x\,dy}{x^2 + y^2}\right]$

35. If $x = r \cos \theta$, $y = r \sin \theta$ verify $\frac{\partial(x, y)}{\partial(r, \theta)} \times \frac{\partial(r, \theta)}{\partial(x, y)} = 1$. **(P.T.U., Dec. 2005, May 2005, Dec. 2011)**

[**Hint.** S.E. 1(*i*) art. 3.11]

36. If $r = \sqrt{x^2 + y^2}$, $\theta = \tan^{-1} \frac{y}{x}$ evaluate $\frac{\partial(r, \theta)}{\partial(x, y)}$.

37. $x = r \cos \theta$, $y = r \sin \theta$, $z = z$ prove that $\frac{\partial(x, y, z)}{\partial(r, \theta, z)} = r$. **(P.T.U., Dec. 2013)**

[**Hint.** S.E. 1 (*ii*) art. 3.11]

38. (*i*) If $u = x - y$, $v = xy$, prove that $\text{JJ}' = 1$.

(*ii*) If $x = u(1 + v)$, $y = v(1 + u)$, find the value of $\frac{\partial(x, y)}{\partial(u, v)}$. **(P.T.U., May 2014)**

[**Hint.** S.E. 1 (*iii*) art. 3.11]

39. (*i*) If $u = x^2 - 2y$, $v = x + y$, prove that $\dfrac{\partial(u, v)}{\partial(x, y)} = 2x + 2.$ **(*P.T.U., May 2006*)**

(*ii*) If $u = \dfrac{y^2}{2x}$, $v = \dfrac{x^2 + y^2}{2x}$, prove that $\dfrac{\partial(u, v)}{\partial(x, y)} = -\dfrac{y}{2x}$.

(*iii*) If $u = \dfrac{yz}{x}$, $v = \dfrac{zx}{y}$, $w = \dfrac{xy}{z}$, prove that $\dfrac{\partial(u, v, w)}{\partial(x, y, z)} = 4.$

(*iv*) If $u = x \sin y$, $v = y \sin x$, prove that $\dfrac{\partial(u, v)}{\partial(x, y)} = \sin x \sin y - xy \cos x \cos y$. **(*P.T.U., May 2006*)**

[**Hint.** S.E. 4 art. 3.11]

(*v*) If $u = x^3 + xy$, $v = xy$, prove that $\dfrac{\partial(u, v)}{\partial(x, y)} = 3x^3$. **(*P.T.U., May 2009*)**

(*vi*) If $u = \dfrac{x + y}{1 - xy}$; $v = \tan^{-1} x + \tan^{-1} y$, prove that $\dfrac{\partial(u,v)}{\partial(x, y)} = 0.$ **(*P.T.U., May 2012*)**

Answers

2. (*i*) $2abz$

4. (*i*) $2x \sin \dfrac{y}{x} - y \cos \dfrac{y}{x}, x \cos \dfrac{y}{x}$ (*ii*) $\dfrac{y}{(1 - xy)^2}, \dfrac{x}{(1 - xy)^2}$ (*iii*) $y^x \log y, xy^{x-1}$

8. 1

26. (*i*) $4a^2 t (t^2 + 2)$ (*ii*) $8e^{4t}$

(*iii*) $\cos \dfrac{e^t}{t^2}$ (*iv*) $-3a^3 \cos^2 t \sin t$.

27. $4r, 4s$ **29.** $1 + \log xy - \dfrac{x(x^2 + y)}{y(x + y^2)}$. **30.** 0

34. $\dfrac{ydx - xdy}{x^2 + y^2}$ **36.** $\dfrac{1}{r}$ **38.** (*ii*) $1 + u + v$.

4

Application of Partial Differentiation

4.1. TANGENT AND NORMAL TO A SURFACE

Let P(x, y, z) and Q($x + \delta x, y + \delta y, z + \delta z$) be two neighbouring points on the surface

$$F(x, y, z) = 0 \qquad ...(1)$$

Consider any curve on the surface joining the points P and Q. As Q → P along this curve, the chord PQ → the tangent line PT to the surface at P. Since there is no end to the number of curves joining P and Q, we get a number of tangent lines to the surface at P. All these tangent lines at P lie in a plane called the tangent plane to the surface at P.

Let δs denote the length of arc PQ and δc denote the length of chord PQ. For plane curves

$$\underset{Q \to P}{\text{Lt}} \left(\frac{\delta s}{\delta c}\right) = 1.$$

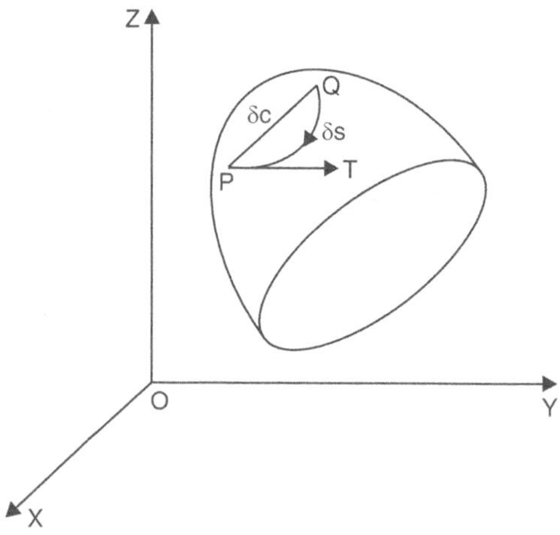

The direction ratios of the chord PQ are $\delta x, \delta y, \delta z$. (using $x_2 - x_1,\ y_2 - y_1,\ z_2 - z_1$)

Dividing by δc, the direction cosines of the chord PQ are

$$\frac{\delta x}{\delta c}, \frac{\delta y}{\delta c}, \frac{\delta z}{\delta c} \quad i.e., \quad \frac{\delta x}{\delta s} \cdot \frac{\delta s}{\delta c}, \frac{\delta y}{\delta s} \cdot \frac{\delta s}{\delta c}, \frac{\delta z}{\delta s} \cdot \frac{\delta s}{\delta c}$$

When $\delta s \to 0$, Q → P, $\frac{\delta s}{\delta c} \to 1$ and the chord PQ tends to a tangent line PT. Also the co-ordinates x, y, z of a moving point on the curve can be expressed as functions of a single parameter s

$$\therefore \quad \text{The direction cosines of PT are } \left\langle \frac{dx}{ds}, \frac{dy}{ds}, \frac{dz}{ds} \right\rangle \qquad ...(2)$$

Differentiating (1) w.r.t s, we have $\frac{\partial F}{\partial x}\frac{dx}{ds} + \frac{\partial F}{\partial y} \cdot \frac{dy}{ds} + \frac{\partial F}{\partial z} \cdot \frac{dz}{ds} = 0$

But this is the condition that the tangent line whose direction cosines are given by (2) is perpendicular to the line having direction ratios

$$\left\langle \frac{dF}{dx}, \frac{dF}{dy}, \frac{dF}{dz} \right\rangle \qquad ...(3)$$

Since the curve PQ on the surface (1) was taken in an arbitrary direction, it follows that the line having direction ratios (3) is perpendicular to all the tangent lines at P. Therefore all the tangent lines at P lie in a plane through P at right angles to the line (3). Thus the direction ratios of a normal to the tangent plane are

$$\left\langle \frac{\partial F}{\partial x}, \frac{\partial F}{\partial y}, \frac{\partial F}{\partial z} \right\rangle.$$

Hence the equation of the tangent plane to the surface (1) at the point P is

$$\frac{\partial F}{\partial x}(X - x) + \frac{\partial F}{\partial y}(Y - y) + \frac{\partial F}{\partial z}(Z - z) = 0$$

where (X, Y, Z) are the current coordinates of any point on this tangent plane.

Also the equation of the normal to the surface (1) at the point P (*i.e.*, line through P, perpendicular to the tangent plane at P) is $\dfrac{X - x}{\dfrac{\partial F}{\partial x}} = \dfrac{Y - y}{\dfrac{\partial F}{\partial y}} = \dfrac{Z - z}{\dfrac{\partial F}{\partial z}}$.

ILLUSTRATIVE EXAMPLES

Example 1. *Find the equation of the tangent plane and normal line to the surface* $xyz = a^3$ *at* (x_1, y_1, z_1). **(P.T.U., May 2005, Dec. 2005)**

Sol. $f(x, y, z) = xyz - a^3 = 0$

$$\frac{\partial f}{\partial x} = yz, \quad \frac{\partial f}{\partial y} = zx, \quad \frac{\partial f}{\partial z} = xy$$

At (x_1, y_1, z_1) ; $\quad \dfrac{\partial f}{\partial x} = y_1 z_1; \ \dfrac{\partial f}{\partial y} = z_1 x_1; \ \dfrac{\partial f}{\partial z} = x_1 y_1$

Equation of the tangent plane at (x_1, y_1, z_1) is

$$(x - x_1)\, y_1 z_1 + (y - y_1) z_1 x_1 + (z - z_1) x_1 y_1 = 0$$

or

$$\frac{x - x_1}{x_1} + \frac{y - y_1}{y_1} + \frac{z - z_1}{z_1} = 0$$

or

$$\frac{x}{x_1} + \frac{y}{y_1} + \frac{z}{z_1} = 3$$

Normal line at (x_1, y_1, z_1) is

$$\frac{x - x_1}{\left(\frac{\partial f}{\partial x}\right)_{(x_1, y_1, z_1)}} = \frac{y - y_1}{\left(\frac{\partial f}{\partial y}\right)_{(x_1, y_1, z_1)}} = \frac{z - z_1}{\left(\frac{\partial f}{\partial z}\right)_{(x_1, y_1, z_1)}}$$

or $$\frac{x - x_1}{y_1 z_1} = \frac{y - y_1}{z_1 x_1} = \frac{z - z_1}{x_1 y_1}.$$

Example 2. *Show that the surface $x^2 - 2yz + y^3 = 4$ is perpendicular to any member of the family of surfaces $x^2 + 1 = (2 - 4a)y^2 + az^2$ at the point of intersection (1, –1, 2).*

Sol. The given surfaces are $F(x, y, z) \equiv x^2 - 2yz + y^3 - 4 = 0$ ...(1)

and $$f(x, y, z) \equiv x^2 + 1 + (4a - 2)y^2 - az^2 = 0 \quad ...(2)$$

$$\frac{\partial F}{\partial x} = 2x, \quad \frac{\partial F}{\partial y} = -2z + 3y^2, \quad \frac{\partial F}{\partial z} = -2y$$

At the point (1, –1, 2), $\frac{\partial F}{\partial x} = 2, \quad \frac{\partial F}{\partial y} = -2(2) + 3(-1)^2 = -1, \quad \frac{\partial F}{\partial z} = -2(-1) = 2$

$\Rightarrow$ d.r.'s of the normal line to (1) at (1, –1, 2) are $<2, -1, 2>$

From (2) $$\frac{\partial f}{\partial x} = 2x, \quad \frac{\partial f}{\partial y} = 2(4a - 2)y, \quad \frac{\partial f}{\partial z} = -2az$$

At the point (1, –1, 2), $\frac{\partial f}{\partial x} = 2, \frac{\partial f}{\partial y} = -8a + 4, \frac{\partial f}{\partial z} = -4a$

$\Rightarrow$ d.r.'s of the normal line to (2) at (1, –1, 2) are $< 2, 4 - 8a, -4a >$.

The surface (1) will be perpendicular to any member of the family of surfaces (2) if the normal line to (1) at (1, –1, 2) is perpendicular to the normal line to (2) at (1, –1, 2) for all values a, the parameter.

i.e., if $(2)(2) + (-1)(4 - 8a) + (2)(-4a) = 0$ $\quad | a_1a_2 + b_1b_2 + c_1c_2 = 0$

or if $4 - 4 + 8a - 8a = 0$

which is true for all values of a. Hence the result.

Example 3. *Find the equation of the tangent plane for the surface $x^3 + y^3 + 3xyz = 3$ at (1, 2, –1).* **(P.T.U., May 2006, 2015)**

Sol. Equation of the surface is $F(x, y, z) = x^3 + y^3 + 3xyz - 3 = 0$ ...(1)

Differentiate partially w.r.t. x, y, z

$$\frac{\partial F}{\partial x} = 3x^2 + 3yz \; ; \; \frac{\partial F}{\partial y} = 3y^2 + 3zx \; ; \; \frac{\partial F}{\partial z} = 3z^2 + 3xy$$

At the point (1, 2, –1)

$$\frac{\partial F}{\partial x} = -3, \frac{\partial F}{\partial y} = 18, \frac{\partial F}{\partial z} = 6$$

Equation of the tangent plane at (1, 2, –1) is $(x - 1)(-3) + (y - 2)(18) + (z + 1)6 = 0$

or $$-3x + 18y + 6z - 27 = 0$$

or $$x - 6y - 2z + 9 = 0.$$

Example 4. *Show that the plane $ax + by + cz + d = 0$ touches the surface*

$$px^2 + qy^2 + 2z = 0 \text{ if } \frac{a^2}{p} + \frac{b^2}{q} + 2cd = 0.$$

Sol. Let the plane $ax + by + cz + d = 0$ ...(1)

touches the surface $F(x, y, z) = px^2 + qy^2 + 2z = 0$ at (x_1, y_1, z_1) ...(2)

From (2), $$\frac{\partial F}{\partial x} = 2px, \frac{\partial F}{\partial y} = 2qy, \frac{\partial F}{\partial z} = 2$$

At (x_1, y_1, z_1); $$\frac{\partial F}{\partial x} = 2px_1, \frac{\partial F}{\partial y} = 2qy_1, \frac{\partial F}{\partial z} = 2$$

Equation of the tangent plane at (x_1, y_1, z_1) is

$$(x - x_1)(2px_1) + (y - y_1)(2qy_1) + (z - z_1)\,2 = 0$$

cancel 2 throughout

or $$(px_1)x + (qy_1)y + z = px_1^2 + qy_1^2 + z_1$$

As (x_1, y_1, z_1) lies on (2) $\therefore$ $px_1^2 + qy_1^2 + 2z_1 = 0$

$\therefore$ $$px_1^2 + qy_1^2 + z_1 = -z_1$$

$\therefore$ Equation of the tangent plane is

$$(px_1)x + (qy_1)y + z = z_1$$

or $(px_1)x + (qy_1)y + z + z_1 = 0$ which is same as (1)

$\therefore$ Comparing coefficients $\dfrac{px_1}{a} = \dfrac{qy_1}{b} = \dfrac{1}{c} = \dfrac{+z_1}{d}$

$\therefore$ $$x_1 = \frac{a}{cp}, y_1 = \frac{b}{cq}, z_1 = \frac{d}{c}$$

$\therefore$ Point of contact is $\left(\dfrac{a}{cp}, \dfrac{b}{cq}, \dfrac{d}{c}\right)$

It satisfies (2) $\therefore$ $p \cdot \dfrac{a^2}{c^2p^2} + q\dfrac{b^2}{c^2q^2} + \dfrac{2d}{c} = 0$ or $\dfrac{a^2}{p} + \dfrac{b^2}{q} + 2cd = 0.$

TEST YOUR KNOWLEDGE

Find the equations of the tangent plane and the normal line to each of the following surfaces at the given points: [Q. 1 to Q. 7]

1. $xyz = 6$ at (1, 2, 3).

2. $2xz^2 - 3xy - 4x = 7$ at (1, –1, 2)

3. $\dfrac{x^2}{4} + \dfrac{y^2}{9} - \dfrac{z^2}{25} = 1$ at (2, 3, 5)

4. $z^2 = 4(1 + x^2 + y^2)$ at (2, 2, 6). ***(P.T.U., Jan. 2009, Dec. 2012)***

5. $\dfrac{x^2}{2} - \dfrac{y^2}{3} = z$ at (2, 3, –1). ***(P.T.U., Dec. 2013)***

6. $2x^2 + y^2 + 2z = 3$ at (2, 1, –3)

7. $x^2 + y^2 + z^2 = a^2$ at (x_1, y_1, z_1) **(*P.T.U., May 2010*)**

8. Show that the plane $3x + 12y - 6z - 17 = 0$ touches the conicoid $3x^2 - 6y^2 + 9z^2 + 17 = 0$. Find also the point of contact.

[**Hint.** Consult S.E. 4]

Answers

1. $6x + 3y + 2z = 18; \frac{x-1}{6} = \frac{y-2}{3} = \frac{z-3}{2}$

2. $7x - 3y + 8z = 26; \frac{x-1}{7} = \frac{y+1}{-3} = \frac{z-2}{8}$

3. $15x + 10y - 6z = 30; \frac{x-2}{15} = \frac{y-3}{10} = \frac{z-5}{-6}$

4. $4x + 4y - 3z + 2 = 0; \frac{x-2}{4} = \frac{y-2}{4} = \frac{z-6}{-3}$

5. $2x - 2y - z + 1 = 0; \frac{x-2}{2} = \frac{y-3}{-2} = \frac{z+1}{-1}$

6. $4x + y + z = 6; \frac{x-2}{4} = \frac{y-1}{1} = \frac{z+3}{1}$

7. $xx_1 + yy_1 + zz_1 = a^2; \frac{x-x_1}{x_1} = \frac{y-y_1}{y_1} = \frac{z-z_1}{z_1}$

8. $\left(-1, 2, \frac{2}{3}\right)$.

4.2. TAYLOR'S AND MACLAURIN'S THEOREMS FOR A SINGLE VARIABLE FUNCTION

Satement. *If $f(x)$ defined in the closed interval $[a, b]$ is such that*

(i) $f, f', f'' \ldots f^{n-1}$ are continuous in $[a, b]$

(ii) $f^n(x)$ exists in the open interval (a, b) then there exists at least one real number $c \in (a, b)$ such that

$$f(b) = f(a) + (b-a) f'(a) + \frac{(b-a)^2}{2!} f''(a) + \cdots + \frac{(b-a)^{n-1}}{(n-1)!} f^{n-1}(a) + \frac{(b-a)^n}{n!} f^n(c)$$

Proof. Consider a function $\phi(x)$ defined by

$$\phi(x) = f(x) + (b-x) f'(x) + \frac{(b-x)^2}{2!} f''(x) + \cdots + \frac{(b-x)^{n-1}}{(n-1)!} f^{n-1}(x) + \frac{(b-x)^n}{n!} \text{A} \quad \ldots(1)$$

where A is constant to be so chosen that $\phi(b) = \phi(a)$

i.e.,
$$f(b) = f(a) + (b-a) f'(a) + \frac{(b-a)^2}{2!} f''(a) + \cdots + \frac{(b-a)^{n-1}}{(n-1)!} f^{n-1}(a) + \frac{(b-a)^n}{n!} \text{A} \quad \ldots(2)$$

$\because$ $f, f', f'', \cdots, f^{n-1}(x)$ are all continuous functions in $[a, b]$ and derivable in (a, b) so is $\phi(x)$ as $\phi(x)$ is a linear combinations of all these functions.

(*i*) $\phi(x)$ is continuous in $[a, b]$

(*ii*) $\phi(x)$ is derivable in (a, b)

(*iii*) $\phi(a) = \phi(b)$

$\therefore$ $\phi(x)$ satisfies all the conditions of Rolle's theorem in $[a, b]$

$\therefore$ $\exists$s at least one $c \in (a, b)$ such that $\phi'(c) = 0$

From (1),

$$\phi'(x) = f'(x) + [-f'(x) + (b-x)f''(x)] + \left[-(b-x)f''(x) + (b-x)f''(x) + \frac{(b-x)^2}{2!}f'''(x)\right]$$

$$+ \cdots + \left[-\frac{(b-x)^{n-2n-1}}{(n-2)!}f^{n-1}(x) + \frac{(b-x)^{n-1}}{(n-1)!}f^n(x)\right] + \left[-\frac{(b-x)^{n-1}}{(n-1)!}A\right]$$

$$\therefore \quad \phi'(x) = \frac{(b-x)^{n-1}}{(n-1)!}f^n(x) - \frac{(b-x)^{n-1}}{(n-1)!}A$$

[all other terms cancel in pairs]

$$\therefore \quad \phi'(c) = 0 \Rightarrow \frac{(b-c)^{n-1}}{(n-1)!}f^n(c) - \frac{(b-c)^{n-1}}{(n-1)!}A = 0$$

$$\therefore \quad A = f^n(c) \qquad \because \quad b - c \neq 0 \quad i.e., c \neq b$$

Substituting the value of A in (2)

$$f(b) = f(a) + (b-a)f'(a) + \frac{(b-a)^2}{2!}f''(a) + \cdots + \frac{(b-a)^{n-1}}{(n-1)!}f^{n-1}(a) + \frac{(b-a)^n}{n!}f^n(c) \qquad \ldots(3)$$

which proves Taylor's theorem with **Lagrange's form of remainder**

$\frac{(b-a)^n}{n!}f^n(c)$ is called Lagrange's Remainder after n terms and is generally denoted by R_n.

Cor. 1. Taylor's Formula

If in (3) we keep 'a' as fixed and replace 'b' by independent variable 'x' t. Then Taylor's theorem becomes

$$f(x) = f(a) + (x-a)f'(a) + \frac{(x-a)^2}{2!}f''(a) + \cdots + \frac{(x-a)^{n-1}}{(n-1)!}f^{n-1}(a) + R_n(x) \qquad \ldots(4)$$

where $R_n(x) = \frac{(x-a)^n}{n!}f^n(c)$; $c \in (a, x)$ is the remainder term of the expansion or error term of the expansion. Equation (4) is called Taylor's formula.

Cor. 2. If in (3) we keep 'a' as fixed and replace 'b' by '$a + h$' then Taylor's theorem becomes

$$f(a+h) = f(a) + hf'(a) + \frac{h^2}{2!}f''(a) + \cdots + \frac{h^{n-1}}{(n-1)!}f^{n-1}(a) + R_n$$

where $$R_n = \frac{h^n}{n!}f^n(a + \theta h) ; 0 < \theta < 1$$

This is called another form of Taylor's theorem (with Lagrange's form of remainder) in the interval $[a, a + h]$.

Cor. 3. Taylor's Polynomial. If we neglect error term in (4), we obtain

$$f(x) = f(a) + (x-a)f'(a) + \frac{(x-a)^2}{2!}f''(a) + \cdots + \frac{(x-a)^{n-1}}{(n-1)!}f^{n-1}(a)$$

$$= \sum_{m=0}^{n-1} \frac{(x-a)^m}{m!} f^m(a),$$

which is called $(n - 1)$th degree Taylor's polynomial approximation of $f(x)$.

Cor. 4. Maclaurin's Theorem. If in (3) we put $a = 0$ and $b = x$, we get

$$f(x) = f(0) + x f'(0) + \frac{x^2}{2!} f''(0) + \cdots + \frac{x^{n-1}}{(h-1)!} f^{n-1}(0) + R_n$$

where $$R_n = \frac{x^n}{n!} f^n(\theta x); \ 0 < \theta < 1$$

This is called Maclaurin's theorem with Lagrange's form of remainder.

Cor. 5. Taylor's Series and Maclaurin's Series

If (*i*) $f(x)$ is a function possessing derivatives of all orders in $[a, x]$ and (*ii*) $R_n = \frac{(x-a)^n}{n!}$ $f^n(c) \to 0$ as $n \to \infty$ then $f(x) = f(a) + (x-a) f'(a) + \frac{(x-a)^2}{2!} f''(a) + \cdots + \frac{(x-a)^{n-1}}{(n-1)!} f^{n-1}(a) + \cdots \infty$ is called Taylor's infinite series.

It can also be written in the form

$$f(a+h) = f(a) + hf'(a) + \frac{h^2}{2!} f''(a) + \cdots + \frac{n^{n-1}}{(n-1)!} f^{n-1}(a) + \cdots \infty$$

If we put $a = 0$ in Taylor's series then $f(x) = f(0) + x f'(a) + \frac{x^2}{2!} f''(a) + \frac{x^{n-1}}{(n-1)!} f^{n-1}(a) + \cdots \infty$ is called Maclaurin's infinite series.

Cor. 6. (*a*) **Failure of Taylor's Expansion.** Expansion of $f(a + h)$ in ascending powers of h is not possible if

(*i*) any of $f(x), f'(x), f''(x)$ ... fails to exist finitely for any value of x is $[a, a + h]$

or (*ii*) $R^n = \frac{h^n}{n!} f^n(a + \theta h)$ does not tend to zero as $n \to \infty$.

(*b*) **Failure of Maclaurin's Expansion.** Expansion of $f(x)$ in ascending powers of x is not possible if

(*i*) any of $f(x), f'(x), f''(x)$ .... fails to exist finitely for any value of x in $[0, x]$

(*ii*) $R^n = \frac{x^n}{n!} f^n(\theta x)$ does not tend to zero as $n \to \infty$.

ILLUSTRATIVE EXAMPLES

Example 1. *Use Taylor's theorem with Lagrange's form of remainder to the function* $f(x) = \log x$ *in* $[1, x]$ *and hence show that* $\log x = (x-1) - \frac{(x-1)^2}{2} + \frac{(x-1)^3}{3} - \frac{(x-1)^4}{4} + \cdots$

$$+ \frac{(-1)^{n-2}(x-1)^{n-1}}{n-1} + \frac{(-1)^{n-1}(x-1)^n}{n[1+\theta(n-1)]^n}$$

Sol. Here

$f(x) = \log x$	$f(1) = 0$
$f'(x) = \dfrac{1}{x}$	$f'(1) = 1$
$f''(x) = -\dfrac{1}{x^2}$	$f''(1) = -1 = -1\,!$
$f'''(x) = \dfrac{2}{x^3}$	$f'''(1) = 2 = 2\,!$
$f^{iv}(x) = \dfrac{-2\cdot 3}{x^4}$	$f^{iv}(1) = \dfrac{-2\cdot 3}{1} = -3\,!$
....	....
$f^{n-1}(x) = \dfrac{(-1)^{n-2}\, 2\cdot 3\cdot 4 \cdots (n-2)}{x^{n-1}}$	$f^{n-1}(1) = (-1)^{n-2}\,(n-2)\,!$
$f^{n}(x) = \dfrac{(-1)^{n-1}\, 2\cdot 3\cdot 4 \cdots (n-1)}{x^{n}}$	$f^{n}\,[1+\theta(x-1)] = \dfrac{(-1)^{n-1}(n-1)\,!}{[1+\theta(x-1)]^n}$

Hence by Taylor's theorem with Lagrange's form of remainder

$$f(x) = f(1) + (x-1)\,f'(1) + \frac{(x-1)^2}{2!}\,f''(1) + \frac{(x-1)^3}{3!}\,f'''(1) + \cdots + \frac{(x-1)^{n-1}}{(n-1)!}\,f^{n-1}(1)$$

$$+ \frac{(x-1)^n}{n!}\,f^n\,[1+\theta(x-1)] \qquad \text{(by putting } a = 1,\ a+h = x \text{ in cor. 2)}$$

$$\therefore\quad \log(x) = 0 + (x-1)\,.\,1 + \frac{(x-1)^2}{2!}\,(-1) + \frac{(x-1)^3}{3!}\,2\,! + \frac{(x-1)^4}{4!}\,(-3\,!)$$

$$+ \cdots + \frac{(x-1)^{n-1}}{(n-1)!}\,(-1)^{n-2}\,(n-2)\,! + \frac{(x-1)^n}{n!}\,(-1)^{n-1}\,\frac{(x-1)!}{[1+\theta\,(x-1)]^n}$$

$$= (x-1) - \frac{(x-1)^2}{2} + \frac{(x-1)^3}{3} - \frac{(x-1)^4}{4} + \cdots + \frac{(-1)^{n-2}(x-1)^{n-1}}{n-1} + \frac{(-1)^{n-1}(x-1)^n}{n[1+\theta(x-1)]^n}$$

Example 2. *Expand the following in powers of x by Maclaurin's series*

(i) sin x *(ii)* $tan^{-1}\,x$

Sol. *(i)*

$f(x) = \sin x$	$f(0) = 0$
$f'(x) = \cos x$	$f'(0) = 1$
$f''(x) = -\sin x$	$f''(0) = 0$
$f'''(x) = -\cos x$	$f'''(0) = -1$
$f^{iv}(x) = \sin x$	$f^{iv}(0) = 0$
$f^{v}(x) = \cos x$	$f^{v}(0) = 1$

By Maclaurin's Series

$$f(x) = f(0) + xf'(0) + \frac{x^2}{2!}\,f''(0) + \frac{x^3}{3!}\,f'''(0) + \frac{x^4}{4!}\,f^{iv}(0) + \frac{x^5}{5!}\,f^{v}(0) + \cdots \infty$$

$$= 0 + x\cdot 1 + \frac{x^2}{2!}\cdot 0 + \frac{x^3}{3!}\,(-1) + \frac{x^4}{4!}\cdot 0 + \frac{x^5}{5!}\cdot 1$$

$$= x - \frac{x^3}{3!} + \frac{x^5}{5!} - \cdots \infty.$$

(*ii*) $f(x) = \tan^{-1} x$

$f(x) = \tan^{-1} x$	$f(0) = 0$
$f'(x) = \dfrac{1}{1+x^2}$	$f'(0) = 1$
$f''(x) = \dfrac{-2x}{(1+x)^2}$	$f''(0) = 0$
$f'''(x) = -2\dfrac{1-3x^2}{(1+x^2)^3}$	$f'''(0) = -2 = -2\,!$
$f^{iv}(x) = \dfrac{12x\,(2-2x^2)}{(1+x^2)^4}$	$f^{iv}(0) = 0$
$f^{v}(x) = \dfrac{24\,(1-10x^2+5x^4)}{(1+x^2)^5}$	$f^{v}(0) = 24 = 4\,!$

By Maclaurin's Series

$$f(x) = f(0) + x f'(0) + \frac{x^2}{2!} f''(0) + \frac{x^3}{3!} f'''(0) + \frac{x^4}{4!} f^{iv}(0) + \frac{x^5}{5!} f^{v}(0) + \cdots \infty$$

$$= 0 + x \cdot 1 + \frac{x^2}{2!} \cdot 0 + \frac{x^3}{3!}(-2\,!) + \frac{x^4}{4!} \cdot 0 + \frac{x^5}{5!}\, 4\,! + \cdots \infty$$

$$= x - \frac{x^3}{3} + \frac{x^5}{5} - \cdots \infty.$$

Example 3. *Write nth degree Taylor Polynomial of $f(x) = \cos x$ at $x = \dfrac{\pi}{2}$.*

Sol.

$f(x) = \cos x$	$f\left(\dfrac{\pi}{2}\right) = 0$
$f'(x) = -\sin x = \cos\left(\dfrac{\pi}{2} + x\right)$	$f'\left(\dfrac{\pi}{2}\right) = -1$
$f''(x) = -\cos x = \cos\left(2\dfrac{\pi}{2} + x\right)$	$f''\left(\dfrac{\pi}{2}\right) = 0$
$f'''(x) = \sin x = \cos\left(3\dfrac{\pi}{2} + x\right)$	$f'''\left(\dfrac{\pi}{2}\right) = 1$
$f^{iv}(x) = \cos x = \cos\left(4\dfrac{\pi}{2} + x\right)$	$f^{iv}\left(\dfrac{\pi}{2}\right) = 0$
…………………	…………………
$f^{n}(x) = \cos\left(\dfrac{n\pi}{2} + x\right)$	$f^{n}\left(\dfrac{\pi}{2}\right) = \cos\left(\dfrac{n\pi}{2} + \dfrac{\pi}{2}\right) = \cos\,(n+1)\,\dfrac{\pi}{2}$

$$f(x) = \cos x = f\left(\frac{\pi}{2}\right) + \left(x - \frac{\pi}{2}\right) f'\left(\frac{\pi}{2}\right) + \frac{\left(x - \frac{\pi}{2}\right)^2}{2!} f''\left(\frac{\pi}{2}\right)$$

$$+ \frac{\left(x - \frac{\pi}{2}\right)^3}{3!} f'''\left(\frac{\pi}{2}\right) + \cdots + \frac{\left(x - \frac{\pi}{2}\right)^n}{n!} f^{n}\left(\frac{\pi}{2}\right)$$

$$\cos x = 0 + \left(x - \frac{\pi}{2}\right)(-1) + \frac{\left(x - \frac{\pi}{2}\right)^2}{2!} \cdot 0 + \frac{\left(x - \frac{\pi}{2}\right)^3}{3!} \cdot 1 + \frac{\left(x - \frac{\pi}{2}\right)^4}{4!} \cdot 0$$

$$+ \frac{\left(x - \frac{\pi}{2}\right)^2}{5!} \cdot (-1) + \cdots + \frac{\left(x - \frac{\pi}{2}\right)^n}{n!} \cos (n + 1)\frac{\pi}{2}$$

$$= -\left(x - \frac{\pi}{2}\right) + \frac{\left(x - \frac{\pi}{2}\right)^3}{3!} - \frac{\left(x - \frac{\pi}{2}\right)^5}{5!} + \cdots \frac{\left(x - \frac{\pi}{2}\right)^n}{n!} \cos (n + 1)\frac{\pi}{2}.$$

Example 4. *Obtain the Taylor's polynomial expansion of the function f(x) = sin x about $x = \frac{\pi}{4}$ show that the error term tends to zero as $n \to \infty$ for all real x. Hence write Taylor's series expansion of f(x).*

Sol.

$$f(x) = \sin x \qquad f\left(\frac{\pi}{4}\right) = \frac{1}{\sqrt{2}}$$

$$f'(x) = \cos x \qquad f'\left(\frac{\pi}{4}\right) = \frac{1}{\sqrt{2}}$$

$$f''(x) = -\sin x \qquad f''\left(\frac{\pi}{4}\right) = -\frac{1}{\sqrt{2}}$$

$$f'''(x) = -\cos x \qquad f'''\left(\frac{\pi}{4}\right) = -\frac{1}{\sqrt{2}}$$

$$f^{iv}(x) = \sin x \qquad f^{iv}\left(\frac{\pi}{4}\right) = \frac{1}{\sqrt{2}}$$

....

$$f^n(x) = (-1)^{n/2} \sin x \quad \text{when } n \text{ is even}$$

$$= (-1)^{\frac{n-1}{2}} \cos x \quad \text{when } n \text{ is odd.}$$

Taylor's polynomial expansion about $x = \frac{\pi}{4}$ is

$$f(x) = f\left(\frac{\pi}{4}\right) + \left(x - \frac{\pi}{4}\right) f'\left(\frac{\pi}{4}\right) + \frac{\left(x - \frac{\pi}{4}\right)^2}{2!} f''\left(\frac{\pi}{4}\right) + \frac{\left(x - \frac{\pi}{4}\right)^3}{3!} f'''\left(\frac{\pi}{4}\right) +$$

$$\cdots + \frac{\left(x - \frac{\pi}{4}\right)^{n-1}}{(n-1)!} f^{n-1}\left(\frac{\pi}{4}\right) + R_n \qquad \text{...(1) (art. 4.2)}$$

where $R_n = \frac{\left(x - \frac{\pi}{4}\right)^n}{n!} f^n(c); \ \frac{\pi}{4} < c < x$

Now $$| R_n | = \left| \frac{\left(x - \frac{\pi}{4}\right)^n}{n!} (-1)^{n/2} \sin c \right| \text{ if } n \text{ is even}$$

and $$\left| \frac{\left(x - \frac{\pi}{4}\right)^n}{n!} (-1)^{\frac{n-1}{2}} \cos c \right| \text{ if } n \text{ is odd.}$$

In each case $$| R_n | \le \left| \frac{\left(x - \frac{\pi}{4}\right)^n}{n!} \right| \quad \because \ | (-1)^{n/2} \sin c | \le 1$$

and $$\left| (-1)^{\frac{n-1}{2}} \cos c \right| \le 1 \text{ and } n! > 1 \quad \therefore \ \frac{1}{n!} < 1$$

$$\therefore \quad | R_n | \le \frac{\left| \left(x - \frac{\pi}{4}\right)^n \right|}{n!} \to 0 \text{ as } n \to \infty \quad \because \ x - \frac{\pi}{4} < 1 \quad \therefore \ \left(x - \frac{\pi}{4}\right)^n \to 0 \text{ as } n \to \infty$$

$\therefore$ From (1), the Taylor's series expansion is

$$f(x) = \sin x = \frac{1}{\sqrt{2}} + \left(x - \frac{\pi}{4}\right) \cdot \frac{1}{\sqrt{2}} + \frac{\left(x - \frac{\pi}{4}\right)^2}{2!} \cdot \left(-\frac{1}{\sqrt{2}}\right) + \frac{\left(x - \frac{\pi}{4}\right)^3}{3!} \cdot \left(-\frac{1}{\sqrt{2}}\right) + \cdots \infty$$

$$= \frac{1}{\sqrt{2}} \left\{ 1 + \left(x - \frac{\pi}{4}\right) - \frac{\left(x - \frac{\pi}{4}\right)^2}{2!} - \frac{\left(x - \frac{\pi}{4}\right)^3}{3!} + \ldots \infty \right\}$$

Example 5. *For the Taylor's polynomial approximation of degree $\le n$ about the point $x = 0$ for the function $f(x) = e^x$. Determine the value of n such that the error satisfies $| R_n | \le 0.005$ when $-1 \le x \le 1$.*

Sol.

$$f(x) = e^x \qquad f(0) = 1$$
$$f'(x) = e^x \qquad f'(0) = 1$$
$$f''(x) = e^x \qquad f''(0) = 1$$
$$\cdots\cdots\cdots\cdots\cdots\cdots\cdots\cdots$$
$$f^n(x) = e^x \qquad f^n(0) = 1$$

Taylor's polynomial approximation of e^x of degree $\le n$ is

$$f(x) = e^x = f(0) + x f'(0) + \frac{x^2}{3!} f''(0) + \cdots + \frac{x^{n-1}}{(n-1)!} f^{n-1}(0) + R_n$$

$$= 1 + x + \frac{x^2}{2!} + \cdots + \frac{x^{n-1}}{(n-1)!} + R_n$$

where $\quad R_n = \dfrac{x^n}{n!} f^n(c) \quad$ where $\quad -1 < c < 1$

$$= \frac{x^n}{n!} e^c$$

$$| R_n | = \left| \frac{x^n}{n!} e^c \right|$$

Now $\quad -1 < c < 1 \quad \therefore \quad e^{-1} < e^c < e^1 \quad \therefore \quad | e^c | < e$

Also $\quad -1 < c < 1 \quad \Rightarrow \quad -1 < x < 1 \quad \Rightarrow \quad | x | < 1 \quad \Rightarrow \quad | x^n | < 1$

$\therefore \quad | R_n | \le \dfrac{e}{n!}$ in $[-1, 1]$

given $\quad | R_n | \le 0.005 \quad \therefore \quad \dfrac{e}{n!} \le 0.005 = \dfrac{5}{1000} = \dfrac{1}{200}$

$\therefore \quad n! \ge 200e$

We know that $e > 2 \quad \therefore \quad n! \ge 400$

As $\quad 5! < 400 < 6! \quad \therefore \quad n! > 5! \quad \therefore \quad n > 5$

$\therefore$ We will require at least six terms in Taylor's polynomial approximation to achieve the given accuracy.

Example 6. *Expand $4x^2 + 5x + 3$ in powers of $x - 1$ by using Taylor's series.*

Sol. Here $\quad f(x) = 4x^2 + 5x + 3$

$$f(x) = 4x^2 + 5x + 3 \qquad f(1) = 12$$
$$f'(x) = 8x + 5 \qquad f'(1) = 13$$
$$f''(x) = 8 \qquad f''(1) = 8$$
$$f'''(x) = 0 \qquad f'''(1) = 0$$

All higher derivatives of $f(x)$ other than second vanish

$\therefore$ By Taylor's series expansion

$$f(x) = f(1) + (x-1) f'(1) + \frac{(x-1)^2}{2!} f''(1)$$
$$= 12 + (x-1)\,13 + \frac{(x-1)^2}{2!} \cdot 8$$
$$= 12 + 13(x-1) + 4(x-1)^2.$$

Example 7. *For what value of x can we replace $\sin x$ by $x - \dfrac{x^3}{3!}$ with an error of magnitude no greater than 3×10^{-4}. Explain the steps.* **(P.T.U., Dec. 2004)**

Sol.

$$f(x) = \sin x \qquad f(0) = 0$$
$$f'(x) = \cos x \qquad f'(0) = 1$$

$$f''(x) = -\sin x \qquad f''(0) = 0$$
$$f'''(x) = -\cos x \qquad f'''(0) = -1$$
$$f^{iv}(x) = \sin x \qquad f^{iv}(0) = 0$$

Taylor's expansion about $x = 0$ upto x^3 is

$$f(x) = f(0) + xf'(0) + \frac{x^2}{2!}f''(0) + \frac{x^3}{3!}f'''(0) + R_4$$

$$= 0 + x + \frac{x^2}{2!}\cdot 0 + \frac{x^3}{3!}(-1) + R_4 = x - \frac{x^3}{3!} + R_4$$

where $$R_4 = \frac{x^4}{4!}f^{iv}(c); \quad 0 < c < x$$

$$= \frac{x^4}{4!}\sin c$$

$$|R_4| = \left|\frac{x^4}{4!}\sin c\right| \le \frac{|x^4|}{24} \qquad \because |\sin c| \le 1$$

given $$\frac{|x^4|}{24} < 3 \times 10^{-4} \quad \therefore \quad |x^4| < \frac{72}{10^4}$$

i.e., $$|x| < \frac{(72)^{1/4}}{10} = \frac{2.913}{10} = 0.2913 \quad \textit{i.e.,} \quad -0.2913 < x < 0.2913$$

$\therefore$ When $x \in (-0.2913, 0.2913)$ error remains $< 3 \times 10^{-4}$.

Example 8. *Compute the values of the following correct to four decimal places by using Taylor's series*

(*i*) *cos 32°* (*ii*) $\sqrt{10}$

Sol. (*i*) Let $f(x) = \cos x$

$$a = 30° = \frac{\pi}{6} \quad \text{and} \quad h = 2° = \frac{\pi}{90} = 0.035$$

Taylor's series is

$$f(a+h) = f(a) + hf'(a) + \frac{h^2}{2!}f''(a) + \frac{h^3}{3!}f'''(a) + \cdots \infty \qquad ...(1)$$

$$f(x) = \cos x \qquad f\left(\frac{\pi}{6}\right) = \cos\frac{\pi}{6} = \frac{\sqrt{3}}{2}$$

$$f'(x) = -\sin x \qquad f'\left(\frac{\pi}{6}\right) = -\sin\frac{\pi}{6} = -\frac{1}{2}$$

$$f''(x) = -\cos x \qquad f''\left(\frac{\pi}{6}\right) = -\cos\frac{\pi}{6} = -\frac{\sqrt{3}}{2}$$

$$f'''(x) = \sin x \qquad f'''\left(\frac{\pi}{6}\right) = \sin\frac{\pi}{6} = \frac{1}{2}$$

Substituting all these values in (1),

$$\cos 32° = \frac{\sqrt{3}}{2} + \frac{\pi}{90}\left(-\frac{1}{2}\right) + \frac{1}{2!}\left(\frac{\pi}{90}\right)^2\left(-\frac{\sqrt{3}}{2}\right) + \frac{1}{3!}\left(\frac{\pi}{90}\right)^3\frac{1}{2} + \cdots$$

$$= 0.866025 + (0.035)(-0.5) + \frac{1}{2}(0.035)^2(-0.866025) + \cdots$$

$$= 0.866025 - 0.0175 - 0.0005304 + \cdots$$

$$= 0.848048 = 0.8481 \text{ (app.)}$$

(*ii*) Let $f(x) = \sqrt{x}$

$a = 9,\ h = 1$

$$f(a+h) = f(a) + hf'(a) + \frac{h^2}{2!}f''(a) + \cdots \infty \qquad \ldots(1)$$

$$f(x) = \sqrt{x} \qquad f(a) = \sqrt{a} = 3$$

$$f'(x) = \frac{1}{2\sqrt{x}} \qquad f'(a) = \frac{1}{2\sqrt{a}} = \frac{1}{6}$$

$$f''(x) = -\frac{1}{4x^{3/2}} \qquad f''(a) = -\frac{1}{4.9.3} = -\frac{1}{108}$$

$$f'''(x) = \frac{1}{4}\cdot\frac{3}{2}\cdot\frac{1}{x^{5/2}} \qquad f'''(a) = \frac{3}{8.81.3} = \frac{1}{648}$$

Substituting all these values in (1),

$$\sqrt{10} = 3 + 1\cdot\frac{1}{6} + \frac{1}{2}\left(\frac{-1}{108}\right) + \frac{1}{6}\left(\frac{1}{648}\right) + \cdots$$

$$= 3 + 0.16666 - 0.00463 + 0.00025 \ldots$$

$$= 3.1023 \text{ (app.)}.$$

TEST YOUR KNOWLEDGE

1. Apply Taylor's theorem with Lagrange's form of remainder to $f(x) = e^x$ in $[0, a]$ and show that $0 < R_{n+1} < \dfrac{e^a\, a^{n+1}}{(n+1)!}$.

[**Hint.** $e^a = 1 + a + \dfrac{a^2}{2!} + \dfrac{a^3}{3!} + \cdots + \dfrac{a^n}{n!} + R_{n+1}$ where $R_{n+1} = \dfrac{a^{n+1}}{(n+1)!}e^{\theta a}$; $0 < \theta < 1$]

2. Use Maclaurin's theorem with Lagrange's form of the remainder to $f(x) = \dfrac{1}{(1+x)^2}$ and show that

$$\frac{1}{(1+x)^2} = 1 - 2x + 3x^2 - 4x^3 + \cdots + (-1)^{n-1} nx^{n-1} + \frac{(-1)^n (n+1)!}{(1+\theta x)^{n+2}} x^n.$$

3. The function $f(x) = \sin x$ is approximate by Taylor's polynomial of degree three about the point $x = 0$. Find c such that the error satisfies $|R_4(x)| \le 0.001$ for all x in the interval $(0, c)$.

4. Obtain a fourth degree Taylor's polynomial approximation to $f(x) = e^{2x}$ about $x = 0$. Find the maximum error when $0 \le x \le 0.5$.

5. Calculate e with an error of less than 10^{-6}.

6. Expand the following in powers of x by Maclaurin's theorem

(*i*) e^{mx} (*ii*) a^{mx} (*iii*) $\log(1 + x)$

(*iv*) $\sin^{-1} x$ (*v*) $\tan x$ **(*P.T.U., May 2009*)**

7. Using Taylor's expansion expand the following upto first four terms

(*i*) $\tan x$ in powers of $x - \dfrac{\pi}{4}$ (*ii*) $\log \sin x$ in powers of $x - \dfrac{\pi}{4}$

(*iii*) $\log x$ in powers of $x - 1$; $x \in (0, 2)$ (*iv*) $\log \cos(x + h)$ in powers of h

(*v*) $\sin^{-1}(x + h)$ in powers of h.

8. Use Taylor's series to express $2x^3 + 7x^2 + x + 6$ in powers of $x - 2$.

9. Calculate the approximate value of the following, to four decimal places using Taylor's expansion

(*i*) $\sin 31°$ (*ii*) $\tan 46° 30'$ (*iii*) $\sqrt[3]{.029}$.

10. Expand $\tan x$ by Maclaurin's series and hence evaluate $\tan 16° 30'$.

$$\left[\textbf{Hint. } 16°30' = \frac{33}{2} \times \frac{\pi}{180} \text{ radius} = 0.288 \text{ rad.}\right]$$

Answers

3. $c = 0.3936$.

4. $1 + 2x + 2x^2 + \dfrac{4}{3}x^3 + \dfrac{2}{3}x^4$; Maximum Error $= \dfrac{e}{120}$.

5. 2.718282.

6. (*i*) $1 + mx + \dfrac{m^2x^2}{2!} + \dfrac{m^3x^3}{3!} + \cdots + \dfrac{m^nx^n}{n!} + \cdots \infty$

(*ii*) $1 + (m\log a)x + \dfrac{(m\log a)^2}{2!}x^2 + \dfrac{(m\log a)^3}{3!}x^3 + \cdots + \dfrac{(m\log a)^n}{n!}x^n + \cdots \infty$

(*iii*) $x - \dfrac{x^2}{2} + \dfrac{x^3}{3} - \dfrac{x^4}{4} + \cdots + \dfrac{(-1)^n x^n}{n} + \cdots \infty$

(*iv*) $x + \dfrac{1}{2}\dfrac{x^3}{3} + \dfrac{1}{2}\cdot\dfrac{3}{4}\cdot\dfrac{x^5}{5} + \dfrac{1}{2}\cdot\dfrac{3}{4}\cdot\dfrac{5}{8}\cdot\dfrac{x^7}{7} + \cdots \infty$ valid only for $|x| < 1$

(*v*) $x - \dfrac{x^3}{3} + \dfrac{2x^5}{15} - \cdots \infty$

7. (*i*) $1 + 2\left(x - \frac{\pi}{4}\right) + 2\left(x - \frac{\pi}{4}\right)^2 + \frac{8}{3}\left(x - \frac{\pi}{4}\right)^3.$

(*ii*) $-\frac{1}{2}\log 2 + \left(x - \frac{\pi}{4}\right) - \left(x - \frac{\pi}{4}\right)^2 + \frac{2}{3}\left(x - \frac{\pi}{4}\right)^3$

(*iii*) $(x-1) - \frac{(x-1)^2}{2} + \frac{(x-1)^3}{3} - \frac{(x-1)^4}{4} + \cdots$

(*iv*) $\log\cos x - h\tan x - \frac{h^2}{2}\sec^2 x - \frac{h^3}{3}\sec^2 x\, dx$

(*v*) $\sin^{-1} x + h \,.\, \frac{1}{\sqrt{1-x^2}} + \frac{h^2}{2!}\cdot\frac{x}{(1-x^2)^{3/2}} + \frac{h^3}{3!}\frac{2x^2+1}{(1-x^2)^{5/2}} + \cdots \infty$

8. $52 + 53(x-2) + 19(x-2)^2 + 2(x-2)^3.$

9. (*i*) .515 (*ii*) 1.0538 (*iii*) 0.3074

10. $\tan x = x - \frac{x^3}{3} + \frac{2x^5}{15} \cdots \infty$; $\tan 16° 30' = 0.296$ app.

4.3. TAYLOR'S AND MACLAURIN'S THEOREMS FOR A FUNCTION OF TWO VARIABLES (*P.T.U., Dec. 2006*)

We know that from art 4.2 Taylor's theorem for a function $f(x)$ of single variable x is

$$f(x+h) = f(x) + hf'(x) + \frac{h^2}{2!}f''(x) + \frac{h^3}{3!}f'''(x) + \cdots$$

Now let $f(x, y)$ be a function of two independent variables x and y. If y is kept constant, then by Taylor's theorem for a function of a single variable x, we have

$$f(x+h, y+k) = f(x, y+k) + h\frac{\partial}{\partial x}f(x, y+k) + \frac{h^2}{2!}\frac{\partial^2}{\partial x^2}f(x, y+k) + \frac{h^3}{3!}\frac{\partial^3}{\partial x^3}f(x, y+k) + \cdots \quad \ldots(1)$$

Now keeping x constant and applying Taylor's theorem for a function of a single variable y, we have

$$f(x, y+k) = f(x, y) + k\frac{\partial}{\partial y}f(x, y) + \frac{k^2}{2!}\frac{\partial^2}{\partial y^2}f(x, y) + \frac{k^3}{3!}\frac{\partial^3}{\partial y^3}f(x, y) + \cdots \quad \ldots(2)$$

Using (2), we can write (1) as

$$f(x+h, y+k) = \left[f(x,y) + k\frac{\partial}{\partial y}f(x,y) + \frac{k^2}{2!}\frac{\partial^2}{\partial y^2}f(x,y) + \frac{k^3}{3!}\frac{\partial^3}{\partial y^3}f(x,y) + \cdots\right]$$
$$+ h\frac{\partial}{\partial x}\left[f(x,y) + k\frac{\partial}{\partial y}f(x,y) + \frac{k^2}{2!}\frac{\partial^2}{\partial y^2}f(x,y) + \cdots\right]$$
$$+ \frac{h^2}{2!}\frac{\partial^2}{\partial x^2}\left[f(x,y) + k\frac{\partial}{\partial y}f(x,y) + \cdots\right] + \frac{h^3}{3!}\frac{\partial^3}{\partial x^3}[f(x,y) + \cdots] + \cdots$$

$$= \left[f(x,y) + k\frac{\partial f}{\partial y} + \frac{k^2}{2!}\frac{\partial^2 f}{\partial y^2} + \frac{k^3}{3!}\frac{\partial^3 f}{\partial y^3} + \cdots \right] + \left[h\frac{\partial f}{\partial x} + hk\frac{\partial^2 f}{\partial x\,\partial y} + \frac{hk^2}{2!}\frac{\partial^3 f}{\partial x\,\partial y^2} + \cdots \right]$$

$$+ \left[\frac{h^2}{2!}\frac{\partial^2 f}{\partial x^2} + \frac{h^2 k}{2!}\frac{\partial^3 f}{\partial x^2\,\partial y} + \cdots \right] + \frac{h^3}{3!}\frac{\partial^3 f}{\partial x^3} + \cdots$$

$$= f(x, y) + \left(h\frac{\partial f}{\partial x} + k\frac{\partial f}{\partial y} \right) + \left(\frac{h^2}{2!}\frac{\partial^2 f}{\partial x^2} + hk\frac{\partial^2 f}{\partial x\,\partial y} + \frac{k^2}{2!}\frac{\partial^2 f}{\partial y^2} \right)$$

$$+ \left(\frac{h^3}{3!}\frac{\partial^3 f}{\partial x^3} + \frac{h^2 k}{2!}\frac{\partial^3 f}{\partial x^2 \partial y} + \frac{hk^2}{2!}\frac{\partial^3 f}{\partial x \partial y^2} + \frac{k^3}{3!}\frac{\partial^3 f}{\partial y^3} \right) + \cdots$$

$$= f(x, y) + \left(h\frac{\partial f}{\partial x} + k\frac{\partial f}{\partial y} \right) + \frac{1}{2!}\left(h^2\frac{\partial^2 f}{\partial x^2} + 2hk\frac{\partial^2 f}{\partial x\,\partial y} + k^2\frac{\partial^2 f}{\partial y^2} \right)$$

$$+ \frac{1}{3!}\left(h^3\frac{\partial^3 f}{\partial x^3} + 3h^2 k\frac{\partial^3 f}{\partial x^2\,\partial y} + 3hk^2\frac{\partial^3 f}{\partial x\,\partial y^2} + k^3\frac{\partial^3 f}{\partial y^3} \right) + \cdots$$

$$= f(x, y) + \left(h\frac{\partial}{\partial x} + k\frac{\partial}{\partial y} \right) f + \frac{1}{2!}\left(h\frac{\partial}{\partial x} + k\frac{\partial}{\partial y} \right)^2 f + \frac{1}{3!}\left(h\frac{\partial}{\partial x} + k\frac{\partial}{\partial y} \right)^3 f + \cdots$$

Cor. 1. Putting $x = a$ and $y = b$, we have

$$f(a + h, b + k) = f(a, b) + [hf_x(a, b) + kf_y(a, b)] + \frac{1}{2!}[h^2 f_{xx}(a, b) + 2hkf_{xy}(a, b) + k^2 f_{yy}(a, b)]$$

$$+ \frac{1}{3!}[h^3 f_{xxx}(a, b) + 3h^2 kf_{xxy}(a, b) + 3hk^2 f_{xyy}(a, b) + k^3 f_{yyy}(a, b)] + \cdots$$

Cor. 2. In Cor. 1, putting $a + h = x$ and $b + k = y$ so that $h = x - a$ and $k = y - b$, we have

$$f(x, y) = f(a, b) + [(x - a) f_x (a, b) + (y - b) f_y(a, b)] + \frac{1}{2!}[(x - a)^2 f_{xx}(a, b)$$

$$+ 2(x - a)(y - b) f_{xy}(a, b) + (y - b)^2 f_{yy}(a, b)] + \cdots$$

Cor. 3. Putting $a = 0, b = 0$ in Cor. 2, we have

$$f(x, y) = f(0, 0) + [xf_x(0, 0) + yf_y(0, 0)] + \frac{1}{2!}[x^2 f_{xx}(0, 0) + 2xy\, f_{xy}(0, 0) + y^2 f_{yy}(0, 0)] + \cdots$$

This is called **Maclaurin's theorem** for two variables.

Note 1. Cor. 3 is used to expand $f(x, y)$ in powers of x and y [or to expand $f(x, y)$ in the neighbourhood of origin [0, 0].

Cor. 2 is used to expand $f(x, y)$ in the neighbourhood of (a, b).

Note 2. Taylor's and Maclaurin's series can be easily extended to functions of any number of variables say $f(x_1, x_2, x_3, \ldots, x_n)$.

e.g., If f is a function of three variables then Taylor's series in the neighbourhood of (h, k, l) becomes

$$f(x+h, y+k, z+l) = f(x, y, z) + \left(h\frac{\partial}{\partial x} + k\frac{\partial}{\partial y} + l\frac{\partial}{\partial z}\right) f + \frac{1}{2!}\left[h\frac{\partial}{\partial x} + k\frac{\partial}{\partial y} + l\frac{\partial}{\partial z}\right]^2 f$$

$$+ \frac{1}{3!}\left[h\frac{\partial}{\partial x} + k\frac{\partial}{\partial y} + l\frac{\partial}{\partial z}\right]^3 f + \cdots$$

$$= f + [h f_x + k f_y + l f_z] + \frac{1}{2!}[h^2 f_{xx} + k^2 f_{yy} + l^2 f_{zz} + 2hk f_{xy} + 2kl f_{yz} + 2lh f_{zx}]$$

$$+ \frac{1}{3!}[h^3 f_{xxx} + k^3 f_{yyy} + l^3 f_{zzz} + 3h^2k f_{xxy} + 3h^2l f_{xxz} + 3k^2h f_{yyx}$$

$$+ 3k^2l f_{yyz} + 3l^2h f_{zzx} + 3l^2k f_{zzy} +] \cdots$$

Cor. 1. Putting $x = a, y = b, z = c$

$$f(a+h, b+k, c+l) = f(a, b, c) + [hf_x(a, b, c) + kf_y(a, b, c) + l f_z(a, b, c)]$$

$$+ \frac{1}{2!}[h f_x(a, b, c) + k f_y(a, b, c) + l f_z(a, b, c)]^2 + \cdots$$

Cor. 2. Putting $a + h = x, b + k = y, c + l = z$ we have from cor 1.

$$f(x, y, z) = f(a, b, c) + [(x-a) f_x + (y-b) f_y + (z-c) f_z]$$

$$+ \frac{1}{2!}[(x-a) f_x + (y-b) f_y + (z-c) f_z]^2 + \cdots$$

[All partial derivatives are to be evaluated at (a, b, c)]

Cor. 3. Put $a = 0, b = 0, c = 0$ in Cor 2, we get Maclaurin's theorem as

$$f(x, y, z) = f(0, 0, 0) + [x f_x + y f_y] + \frac{1}{2!}[x f_y + y f_y]^2 + \cdots$$

[all partial derivatives to be evaluated at $(0, 0, 0)$].

ILLUSTRATIVE EXAMPLES

Example 1. *Expand $e^x \cos y$ at $\left(1, \frac{\pi}{4}\right)$ using Taylor's series up to second degree terms.*

(P.T.U., Dec 2011)

Sol. Here $f(x, y) = e^x \cos y$; $\quad f\left(1, \frac{\pi}{4}\right) = \frac{e}{\sqrt{2}}$

$f_x(x, y) = e^x \cos y$; $\quad f_x\left(1, \frac{\pi}{4}\right) = \frac{e}{\sqrt{2}}$

$f_y(x, y) = -e^x \sin y$; $\quad f_y\left(1, \frac{\pi}{4}\right) = -\frac{e}{\sqrt{2}}$

$$f_{xx}(x, y) = e^x \cos y; \qquad f_{xx}\left(1, \frac{\pi}{4}\right) = \frac{e}{\sqrt{2}}$$

$$f_{xy}(x, y) = -e^x \sin y; \qquad f_{xy}\left(1, \frac{\pi}{4}\right) = -\frac{e}{\sqrt{2}}$$

$$f_{yy}(x, y) = -e^x \cos y; \qquad f_{yy}\left(1, \frac{\pi}{4}\right) = -\frac{e}{\sqrt{2}}$$

$\therefore$ By Taylor's series

$$e^x \cos y = f(x, y) = f\left(1, \frac{\pi}{4}\right) + \left[(x-1)f_x\left(1, \frac{\pi}{4}\right) + \left(y - \frac{\pi}{4}\right)f_y\left(1, \frac{\pi}{4}\right)\right]$$

$$+ \frac{1}{2!}\left[(x-1)^2 f_{xx}\left(1, \frac{\pi}{4}\right) + 2(x-1)\left(y - \frac{\pi}{4}\right)f_{xy}\left(1, \frac{\pi}{4}\right) + \left(y - \frac{\pi}{4}\right)^2 f_{yy}\left(1, \frac{\pi}{4}\right)\right] + \cdots \infty$$

$$= \frac{e}{\sqrt{2}} + \left[(x-1)\frac{e}{\sqrt{2}} + \left(y - \frac{\pi}{4}\right)\left(-\frac{e}{\sqrt{2}}\right)\right]$$

$$+ \frac{1}{2!}\left[(x-1)^2 \frac{e}{\sqrt{2}} + 2(x-1)\left(y - \frac{\pi}{4}\right)\left(-\frac{e}{\sqrt{2}}\right) + \left(y - \frac{\pi}{4}\right)^2\left(-\frac{e}{\sqrt{2}}\right)\right] + \cdots \infty$$

$$= \frac{e}{\sqrt{2}}\left[1 + (x-1) - \left(y - \frac{\pi}{4}\right) + \frac{(x-1)^2}{2} - (x-1)\left(y - \frac{\pi}{4}\right) - \frac{1}{2}\left(y - \frac{\pi}{4}\right)^2 + \cdots \infty\right]$$

Example 2. *Expand $\tan^{-1} xy$ in the neighbourhood of (1, 1).*

Or

Expand $\tan^{-1} xy$ in powers of $(x - 1)$ and $(y - 1)$ using Taylor's theorem

(P.T.U., Dec. 2014)

Sol. Here, $f(x, y) = \tan^{-1} xy$, $f(1, 1) = \tan^{-1} 1 = \frac{\pi}{4}$

$$f_x(x, y) = \frac{1}{1 + x^2y^2}\, y; \qquad f_x(1, 1) = \frac{1}{2}$$

$$f_y(x, y) = \frac{x}{1 + x^2y^2}; \qquad f_y(1, 1) = \frac{1}{2}$$

$$f_{xx}(x, y) = y\frac{-1}{(1 + x^2y^2)^2}\, 2xy^2 = \frac{-2xy^3}{(1 + x^2y^2)^2}; \; f_{xx}(1, 1) = \frac{-2}{4} = -\frac{1}{2}$$

$$f_{yy}(x, y) = x\,\frac{-1}{(1 + x^2y^2)^2}\, 2x^2y = -\frac{2x^3y}{(1 + x^2y^2)^2}; f_{yy}(1,1) = -\frac{1}{2}$$

$$f_{xy}(x, y) = \frac{(1+x^2y^2)\cdot 1 - x(2xy^2)}{(1+x^2y^2)^2} = \frac{1-x^2y^2}{(1+x^2y^2)^2}\,;\; f_{xy}(1,1) = 0$$

$$\therefore \quad \tan^{-1} xy = f(x, y) = f(1, 1) + [(x-1) f_x(1, 1) + (y-1) f_y(1, 1)]$$

$$+ \frac{1}{2!}[(x-1)^2 f_{xx}(1, 1) + 2(x-1)(y-1) f_{xy}(1, 1)$$

$$+ (y-1)^2 f_{yy}(1, 1)] + \cdots$$

$$= \frac{\pi}{4} + \left[(x-1)\frac{1}{2} + (y-1)\frac{1}{2}\right]$$

$$+ \frac{1}{2!}\left[(x-1)^2\left(-\frac{1}{2}\right) + 2(x-1)(y-1)\cdot 0 + (y-1)^2\left(-\frac{1}{2}\right)\right] + \cdots$$

$$= \frac{\pi}{4} + \frac{1}{2}(x-1) + \frac{1}{2}(y-1) - \frac{1}{4}(x-1)^2 - \frac{1}{4}(y-1)^2 + \cdots$$

Example 3. *Expand e^{xy} at (1, 1).*

Sol.

$$f(x, y) = e^{xy} \qquad f(1, 1) = e$$

$$f_x(x, y) = y\, e^{xy} \qquad f_x(1, 1) = e$$

$$f_y(x, y) = x\, e^{xy} \qquad f_y(1, 1) = e$$

$$f_{xx}(x, y) = y^2 e^{xy} \qquad f_{xx}(1, 1) = e$$

$$f_{xy}(x, y) = y(x\, e^{xy}) + e^{xy} = e^{xy}(xy + 1); \qquad f_{xy}(1, 1) = 2e$$

$$f_{yy}(x, y) = x^2 e^{xy} \qquad f_{yy}(1, 1) = e$$

By Taylor's Theorem

$$e^{xy} = f(x, y) = f(1, 1) + [(x-1) f_x(1, 1) + (y-1) f_y(1, 1)]$$

$$+ \frac{1}{2!}[(x-1)^2 f_{xx}(1, 1) + 2(x-1)(y-1) f_{xy}(1, 1) + (y-1)^2 f_{yy}(1, 1)] + \cdots]$$

$$= e + [(x-1)e + (y-1)e] + \frac{1}{2!}[(x-1)^2 e + 2(x-1)(y-1)\,2e + (y-1)^2 e] + \cdots$$

$$= e\left[1 + (x-1) + (y-1) + \frac{1}{2}\{(x-1)^2 + 4(x-1)(y-1) + (y-1)^2\} + \cdots\right]$$

Example 4. *Expand y^x up to second degree terms at (1, 1).* **(P.T.U., Dec. 2004)**

Sol.

$$f(x, y) = y^x; \qquad f(1, 1) = 1$$

$$f_x(x, y) = y^x \log y; \qquad f_x(1, 1) = 0$$

$$f_y(x, y) = xy^{x-1}; \qquad f_y(1, 1) = 1$$

$$f_{xx}(x, y) = y^x (\log y)^2; \qquad f_{xx}(1, 1) = 0$$

$$f_{xy}(x, y) = y^x \cdot \frac{1}{y} + \log y \cdot xy^{x-1} = y^{x-1} + xy^{x-1}\log y;\; f_{xy}(1, 1) = 1$$

$$f_{yy}(x, y) = x(x-1)\,y^{x-2};\; f_{yy}(1, 1) = 0$$

Now, $$y^x = f(x, y) = f_{xy}(1, 1) + [(x-1) f_x(1, 1) + (y-1) f_y(1, 1)]$$
$$+ \frac{1}{2!}[(x-1)^2 f_{xx}(1, 1) + 2(x-1)(y-1) f_{xy}(1, 1) + (y-1)^2 f_{yy}(1, 1)] + \cdots$$
$$= 1 + [(x-1)0 + (y-1)\cdot 1] + 1 + \frac{1}{2!}[(x-1)^2 \cdot 0 + 2(x-1)(y-1) + (y-1)^2 \cdot 0]$$
$$= 1 + (y-1) + \frac{1}{2}\cdot 2(x-1)(y-1) + \cdots$$
$$= 1 + (y-1) + (x-1)(y-1) + \cdots$$

Example 5. *Expand* $(1+x+y^2)^{1/2}$ *at (1, 0).*

Sol. $f(x, y) = (1+x+y^2)^{1/2}$ $\qquad f(1, 0) = \sqrt{2}$

$$f_x = \frac{1}{2\sqrt{1+x+y^2}} \qquad f_x(1, 0) = \frac{1}{2\sqrt{2}}$$

$$f_y = \frac{1}{2\sqrt{1+x+y^2}} \qquad f_y(1, 0) = 0$$

$$f_{xx} = \frac{-1}{4(1+x+y^2)^{3/2}} \qquad f_{xx}(1, 0) = -\frac{1}{4\cdot 2^{3/2}}$$

$$f_{xy} = \frac{-2y}{4(1+x+y^2)^{3/2}} \qquad f_{xy}(1, 0) = 0$$

$$f_{yy} = \frac{\sqrt{1+x+y^2}\cdot 1 - y\cdot\dfrac{1}{2\sqrt{1+x+y^2}}\cdot 2y}{1+x+y^2} \qquad f_{yy}(1, 0)\ \frac{2}{2^{3/2}} = \frac{1}{\sqrt{2}}$$

$$= \frac{1+x+y^2-y^2}{(1+x+y^2)^{3/2}} = \frac{1+x}{(1+x+y^2)^{3/2}}$$

$$f_{xxx} = \frac{3}{8(1+x+y^2)^{5/2}} \qquad f_{xxx}(1, 0) = \frac{3}{8(2)^{5/2}} = \frac{3}{32\sqrt{2}}$$

$$f_{xxy} = \frac{3\cdot 2y}{8(1+x+y^2)^{5/2}} \qquad f_{xxy}(1, 0) = 0$$

$$f_{yyx} = \frac{(1+x+y^2)^{3/2}\, 1 - (1+x)\cdot\dfrac{3}{2}(1+x+y^2)^{1/2}}{(1+x+y^2)^3} \qquad f_{yyx}(1, 0) = \frac{-1}{2^{5/2}} = \frac{-1}{4\sqrt{2}}$$

$$= \frac{2(1+x+y^2) - 3(1+x)}{2(1+x+y^2)^{5/2}} = \frac{2y^2 - x - 1}{2(1+x+y^2)^{5/2}}$$

$$f_{yyy} = \frac{-3(1+x)\cdot 2y}{2(1+x+y^2)^{5/2}} \qquad f_{yyy}(1, 0) = 0.$$

By Taylor's Theorem

$$(1 + x + y^2)^{1/2} = f(x, y) = f(1, 0) + [(x-1) f_x(1, 0) + (y-0) f_y(1, 0)]$$

$$+ \frac{1}{2!} \cdot [(x-1)^2 f_{xx}(1, 0) + 2(x-1)(y-0) f_{xy}(1, 0) + (y-0)^2 f_{yy}(1, 0)]$$

$$+ \frac{1}{3!} [(x-1)^3 f_{xxx}(1, 0) + 3(x-1)^2 (y-0) f_{xxy}(1, 0) + 3(x-1)(y-0)^2 f_{xyy}(1, 0)$$

$$+ (y-0)^3 f_{yyy}(1, 0)] + \cdots$$

$$= \sqrt{2} + \left[(x-1)\frac{1}{2\sqrt{2}} + 0\right] + \frac{1}{2!}\left[(x-1)^2\left(-\frac{1}{8\sqrt{2}}\right) + 0 + y^2 \frac{1}{\sqrt{2}}\right]$$

$$+ \frac{1}{3!}\left[(x-1)^3 \frac{3}{32\sqrt{2}} + 0 + 3(x-1)y^2\left(\frac{-1}{4\sqrt{2}}\right) + 0\right] + \cdots \infty$$

$$= \sqrt{2}\left[1 + \frac{x-1}{4} - \frac{(x-1)^2}{32} + \frac{y^2}{4} + \ldots \infty\right].$$

Example 6. *Find the first six terms of the expansions of the function $e^x \log(1 + y)$ in a Taylor series in the neighbourhood of the point (0, 0).* **(P.T.U., Jan. 2010, May 2014)**

Sol. Here

$$f(x, y) = e^x \log(1 + y), \qquad f(0, 0) = 0$$

$$f_x(x, y) = e^x \log(1 + y), \qquad f_x(0, 0) = 0$$

$$f_y(x, y) = \frac{e^x}{1+y}, \qquad f_y(0, 0) = 1$$

$$f_{xx}(x, y) = e^x \log(1 + y), \qquad f_{xx}(0, 0) = 0$$

$$f_{xy}(x, y) = \frac{e^x}{1+y}, \qquad f_{xy}(0, 0) = 1$$

$$f_{yy}(x, y) = -\frac{e^x}{(1+y)^2}, \qquad f_{yy}(0, 0) = -1$$

$$f_{xxx}(x, y) = e^x \log(1 + y), \qquad f_{xxx}(0, 0) = 0$$

$$f_{xxy}(x, y) = \frac{e^x}{1+y}, \qquad f_{xxy}(0, 0) = 1$$

$$f_{xyy}(x, y) = -\frac{e^x}{(1+y)^2}, \qquad f_{xyy}(0, 0) = -1$$

$$f_{yyy}(x, y) = \frac{2e^x}{(1+y)^3}, \qquad f_{yyy}(0, 0) = 2$$

....

$\therefore \quad e^x \log(1+y) = f(x, y)$

$$= f(0,0) + [xf_x(0,0) + yf_y(0,0)] + \frac{1}{2!}[x^2 f_{xx}(0,0) + 2xy\, f_{xy}(0,0)$$

$$+ y^2 f_{yy}(0,0)] + \frac{1}{3!}[x^3 f_{xxx}(0,0) + 3x^2y\, f_{xxy}(0,0) + 3xy^2 f_{xyy}(0,0)$$

$$+ y^3 f_{yyy}(0,0)] + \cdots$$

$$= 0 + [x \cdot 0 + y \cdot 1] + \frac{1}{2}[x^2 \cdot 0 + 2xy \cdot 1 + y^2(-1)] + \frac{1}{6}[x^3 \cdot 0 + 3x^2y \cdot 1$$

$$+ 3xy^2 \cdot (-1) + y^3 \cdot 2] + \cdots$$

$$= y + xy - \frac{1}{2}y^2 + \frac{1}{2}x^2y - \frac{1}{2}xy^2 + \frac{1}{3}y^2 \ldots$$

Example 7. *Expand* $\dfrac{(x+h)(y+k)}{x+h+y+k}$ *in powers of h, k upto and inclusive of the second degree terms.*

Sol. Here $f(x+h, y+k) = \dfrac{(x+h)(y+k)}{x+h+y+k}$

Putting $h = k = 0$, we have $f(x, y) = \dfrac{xy}{x+y}$

$$f_x = \frac{(x+y)\cdot y - xy \cdot 1}{(x+y)^2} = \frac{y^2}{(x+y)^2}; \qquad f_y = \frac{x^2}{(x+y)^2}, \text{ by symmetry}$$

$$f_{xx} = -\frac{2y^2}{(x+y)^3},\ f_{yy} = -\frac{2x^2}{(x+y)^3}$$

$$f_{xy} = \frac{(x+y)^2 \cdot 2x - x^2 \cdot 2(x+y)}{(x+y)^4} = \frac{2x(x+y) - 2x^2}{(x+y)^3} = \frac{2xy}{(x+y)^3}$$

$$\therefore \quad \frac{(x+h)(y+k)}{x+h+y+k} = f(x+h, y+k) = f(x,y) + [hf_x + kf_y] + \frac{1}{2!}[h^2 f_{xx} + 2hk\, f_{xy} + k^2 f_{yy}] + \cdots$$

$$= \frac{xy}{x+y} + \left[h \cdot \frac{y^2}{(x+y)^2} + k \cdot \frac{x^2}{(x+y)^2}\right]$$

$$+ \frac{1}{2}\left[h^2 \cdot \frac{-2y^2}{(x+y)^3} + 2hk \frac{2xy}{(x+y)^3} + k^2 \cdot \frac{-2x^2}{(x+y)^3}\right] + \cdots$$

$$= \frac{xy}{x+y} + \frac{y^2}{(x+y)^2} \cdot h + \frac{x^2}{(x+y)^2} \cdot k - \frac{y^2}{(x+y)^3} \cdot h^2 + \frac{2xy}{(x+y)^3} \cdot hk - \frac{x^2}{(x+y)^3} \cdot k^2 + \cdots$$

Example 8. *Expand $f(x, y) = \sin xy$ in powers of $(x - 1)$ and $\left(y - \frac{\pi}{2}\right)$ upto second degree terms.* **(P.T.U., Dec. 2010)**

Sol. $f(x, y) = \sin xy$; $\quad f\left(1, \frac{\pi}{2}\right) = 1$

$f_x(x, y) = y \cos xy$; $\quad f_x\left(1, \frac{\pi}{2}\right) = 0$

$f_y(x, y) = x \cos xy$; $\quad f_y\left(1, \frac{\pi}{2}\right) = 0$

$f_{xx}(x, y) = -y^2 \sin xy$; $\quad f_{xx}\left(1, \frac{\pi}{2}\right) = -\frac{\pi^2}{4}$

$f_{xy}(x, y) = y(-x \sin xy) + \cos xy$

$= \cos xy - xy \sin xy$; $\quad f_{xy}\left(1, \frac{\pi}{2}\right) = -\frac{\pi}{2}$

$f_{yy}(x, y) = -x^2 \sin xy$; $\quad f_{yy}\left(1, \frac{\pi}{2}\right) = -1$

By Taylor's theorem $\left(a = 1, b = \frac{\pi}{2}\right)$

$$\sin xy = f(x, y) = f\left(1, \frac{\pi}{2}\right) + \left[(x-1) f_x\left(1, \frac{\pi}{2}\right) + \left(y - \frac{\pi}{2}\right) f_y\left(1, \frac{\pi}{2}\right)\right]$$

$$+ \frac{1}{2!}\left\{(x-1)^2 f_{xx}\left(1, \frac{\pi}{2}\right) + 2(x-1)\left(y - \frac{\pi}{2}\right) f_{xy}\left(1, \frac{\pi}{2}\right) + \left(y - \frac{\pi}{2}\right)^2 f_{yy}\left(1, \frac{\pi}{2}\right)\right\}$$

$$+ \cdots\cdots$$

$$= 1 + \left[(x-1)\,0 + \left(y - \frac{\pi}{2}\right).\,0\right] + \frac{1}{2}\left[(x-1)^2\left(-\frac{\pi^2}{4}\right) + 2(x-1)\left(y - \frac{\pi}{2}\right)\left(-\frac{\pi}{2}\right) + \left(y - \frac{\pi}{2}\right)^2 (-1)\right] + \cdots$$

Hence, $\sin xy = 1 - \frac{\pi^2}{8}(x-1)^2 - \frac{\pi}{2}(x-1)\left(y - \frac{\pi}{2}\right) - \frac{1}{2}\left(y - \frac{\pi}{2}\right)^2 + \cdots$

Example 9. *Expand $x^2y + 3y - 2$ in powers of $(x - 1)$ and $(y + 2)$ using Taylor's Theorem.*
(P.T.U., May 2006; May 2009, Dec. 2011, May 2012)

Sol. Expansion of $f(x, y)$ in powers of $(x - a)$ and $(y - b)$ is given by

$$f(x, y) = f(a, b) + [(x - a) f_x(a, b) + (y - b) f_y(a, b)] + \frac{1}{2!} [(x - a)^2 f_{xx}(a, b)$$
$$+ 2(x - a)(y - b) f_{xy}(a, b) + (y - b)^2 f_{yy}(a, b)] + \frac{1}{3!} [(x - a)^3 f_{xxx}(a, b)$$
$$+ 3(x - a)^2 (y - b) f_{xxy}(a, b) + 3(x - a)(y - b)^2 f_{xyy}(a, b)$$
$$+ (y - b)^3 + f_{yyy}(a, b)] + \cdots \quad ...(1)$$

Here $f(x, y) = x^2y + 3y - 2, \quad a = 1, b = -2$

$f(1, -2) = 1^2 \times (-2) + 3(-2) - 2 = -10$

$f_x = 2xy, \quad f_x(1, -2) = 2(1)(-2) = -4;$ $\quad f_y = x^2 + 3, \quad f_y(1, -2) = 1^2 + 3 = 4$

$f_{xx} = 2y, \quad f_{xx}(1, -2) = 2(-2) = -4;$ $\quad f_{xy} = 2x, \quad f_{xy}(1, -2) = 2(1) = 2$

$f_{yy} = 0, \quad f_{yy}(1, -2) = 0;$ $\quad f_{xxx} = 0, \quad f_{xxx}(1, -2) = 0$

$f_{xxy} = 2, \quad f_{xyy}(1, -2) = 2;$ $\quad f_{xyy} = 0, \quad f_{xyy}(1, -2) = 0$

$f_{yyy} = 0, \quad f_{yyy}(1, -2) = 0$

All higher order partial derivatives vanish.

$\therefore$ From (1), we have

$x^2y + 3y - 2 = f(x, y)$

$$= -10 + [(x - 1)(-4) + (y + 2)(4)] + \frac{1}{2} [(x - 1)^2 (-4) + 2(x - 1)(y + 2)(2) + (y + 2)^2 (0)]$$
$$+ \frac{1}{6} [(x - 1)^3 (0) + 3(x - 1)^2 (y + 2)(2) + 3(x - 1)(y + 2)^2(0) + (y + 2)^3(0)]$$
$$= -10 - 4(x - 1) + 4(y + 2) - 2(x - 1)^2 + 2(x - 1)(y + 2) + (x - 1)^2 (y + 2).$$

Example 10. *Expand $f(x, y) = 21 + x - 20y - 4x^2 + xy + 6y^2$ in Taylor's series of maximum order about the point $(-1, 2)$.* **(P.T.U., Dec. 2007)**

Sol. $f(x, y) = 21 + x - 20y - 4x^2 + xy + 6y^2;$ $\quad f(-1, 2) = -2$

$f_x = 1 - 8x + y;$ $\quad f_x(-1, 2) = 11$

$f_y = -20 + x + 12y;$ $\quad f_y(-1, 2) = 3$

$f_{xx} = -8;$ $\quad f_{xx}(-1, 2) = -8$

$f_{xy} = 1;$ $\quad f_{xy}(-1, 2) = 1$

$f_{yy} = 12;$ $\quad f_{yy}(-1, 2) = 12$

Next derivatives all vanish

By Taylor's Series

$$f(x, y) = f(-1, 2) + [(x + 1) f_x(-1, 2) + (y - 2) f_y(-1, 2)]$$
$$+ \frac{1}{2!} [(x + 1)^2 f_{xx}(-1, 2) + 2(x + 1)(y - 2) f_{xy}(-1, 2) + (y - 2)^2 f_{yy}(-1, 2)]$$
$$= -2 + (x + 1) . 11 + (y - 2) . 3 + \frac{1}{2} [(x + 1)^2 (-8) + 2(x + 1)(y - 2) . 1 + (y - 2)^2 12]$$
$$= -2 + 11(x + 1) + 3(y - 2) -4 (x + 1)^2 + (x + 1)(y - 2) + 6(y - 2)^2.$$

Example 11. *Find approximate value of* $\sqrt{(298)^2 + (401)^2}$ *by using Taylor series.*

(P.T.U., Dec. 2007)

Sol. Let $$f(x, y) = \sqrt{x^2 + y^2} \quad \ldots(1)$$

$$f(x + h, y + k) = \sqrt{(x + h)^2 + (y + k)^2}$$

where $$x + h = 298, y + k = 401$$

Take $$x = 300, y = 400$$

$\therefore$ $$h = -2, k = 1$$

By Taylor's series

$$f(x + h, y + k) = f(x, y) + \left(h \frac{\partial f}{\partial x} + k \frac{\partial f}{\partial y}\right) + \cdots \infty$$

From (1), $$\frac{\partial f}{\partial x} = \frac{1}{2\sqrt{x^2 + y^2}} 2x = \frac{x}{\sqrt{x^2 + y^2}}; \frac{\partial f}{\partial y} = \frac{y}{\sqrt{x^2 + y^2}}$$

$\therefore$ $$f(x + h, y + k) = f(x, y) + \left(h \cdot \frac{x}{\sqrt{x^2 + y^2}} + k \cdot \frac{y}{\sqrt{x^2 + y^2}}\right) + \cdots$$

$$= f(x, y) + \frac{hx + ky}{\sqrt{x^2 + y^2}}$$

At (300, 400),

$$f(298, 401) = f(300, 400) + \frac{(-2)300 + 1(400)}{\sqrt{90000 + 160000}}$$

$$\sqrt{(298)^2 + (401)^2} = \sqrt{90000 + 160000} - \frac{200}{500}$$

$$= 500 - \frac{2}{5} = \frac{2498}{5} = 499.6$$

Hence $\sqrt{(298)^2 + (401)^2} = 499.6$.

Note. Above example can be done by the method of differentiation. See S.E. 12 art. 4.6.

TEST YOUR KNOWLEDGE

1. Show that $e^y \log (1 + x) = x + xy - \frac{x^2}{2}$ approximately.

   [**Hint.** Find the expansion at (0, 0) consult S.E. 6]
2. Expand $e^x \cos y$ in powers of x and y as far as the terms of third degree.
   [**Hint.** Consult S.E. 1]
3. Expand $e^{ax} \sin by$ in powers of x and y as far as the terms of third degree.
4. Expand $e^x \sin y$ is powers of x and y as far as terms of the third degree.

5. Find Taylor's series expansion of maximum order for the function $f(x, y) = x^2 + 3y^2 - 9x - 9y + 26$ about the point (2, 2).
[**Hint.** consult S.E 10]

6. Expand $\sin (x + h)(y + k)$ by Taylor's theorem.

7. Expand $\tan^{-1} \dfrac{y}{x}$ in the neighbourhood of (1, 1) ***(P.T.U. May 2008)***

Answers

2. $1 + x + \dfrac{1}{2!}(x^2 - y^2) + \dfrac{1}{3!}(x^3 - 3xy^2) + \cdots$ **3.** $by + abxy + \dfrac{1}{3!}(3a^2bx^2y - b^3y^3) + \cdots$

4. $y + xy + \dfrac{1}{2}x^2y - \dfrac{1}{6}y^3 + \cdots$

5. $f(x, y) = 6 - 5(x - 2) + 3(y - 2) + (x - 2)^2 + 3(y - 2)^2$

6. $\sin xy + (hy + kx) \cos xy + hk \cos xy - \dfrac{1}{2}(hy + kx)^2 \sin xy + \cdots \infty$

7. $\dfrac{\pi}{4} - \dfrac{1}{2}(x - 1) + \dfrac{1}{2}(y - 1) + \dfrac{1}{4}(x - 1)^2 - \dfrac{1}{4}(y - 1)^2 + \ldots \infty$

4.4. ERROR TERMS IN TAYLOR'S SERIES

Taylor's series in two variables with remainder R_n in the neighbourhood of (a, b) is

$$f(a + h, b + k) = f(a, b) + \left(h\frac{\partial}{\partial x} + k\frac{\partial}{\partial y}\right) f(a, b) + \frac{1}{2!}\left(h\frac{\partial}{\partial x} + k\frac{\partial}{\partial y}\right)^2 f(a, b) + \cdots$$
$$+ \frac{1}{n!}\left(h\frac{\partial}{\partial x} + k\frac{\partial}{\partial y}\right)^n f(a,b) + R_n$$

where $$R_n = \frac{1}{(n+1)!}\left(h\frac{\partial}{\partial x} + k\frac{\partial}{\partial y}\right)^{n+1} f(a + \theta h, b + \theta k), 0 < \theta < 1$$

R_n is the **error term.**

Cor. 1. If we write

$$a + h = x, b + k = y \quad i.e., h = x - a, k = y - b$$

Taylor series becomes

$$f(x, y) = f(a, b) + \left[(x - a)\frac{\partial}{\partial x} + (y - b)\frac{\partial}{\partial y}\right] f(a, b)$$
$$+ \frac{1}{2!}\left[(x - a)\frac{\partial}{\partial x} + (y - b)\frac{\partial}{\partial y}\right]^2 f(a, b) + \cdots$$
$$+ \frac{1}{n!}\left[(x - a)\frac{\partial}{\partial x} + (y - b)\frac{\partial}{\partial y}\right]^n f(a, b) + R_n \quad \ldots(1)$$

where $$R_n = \frac{1}{(n+1)!}\left[(x - a)\frac{\partial}{\partial x} + (y - b)\frac{\partial}{\partial y}\right]^{n+1} f[a + \theta(x - a), b + \theta(y - b)]$$

or $$R_n = \frac{1}{(n+1)!}\left[(x - a)\frac{\partial}{\partial x} + (y - b)\frac{\partial}{\partial y}\right]^{n+1} f\{\xi, \eta\} \quad \ldots(2)$$

where $\xi = a + \theta(x - a), \qquad \eta = b + \theta(y - b)$

i.e., $\xi = (1 - \theta)\, a + \theta x; \qquad \eta = (1 - \theta)b + \theta y, \quad 0 < \theta < 1$

Cor. 2. Linear Polynomial Approximation to $f(x, y)$

Linear polynomial approximation to $f(x, y)$ is obtained by putting $n = 1$ in (1)

$$f(x, y) = f(a, b) + \left[(x-a)\frac{\partial}{\partial x} + (y-b)\frac{\partial}{\partial y}\right] f(a, b)$$

$$= f(a, b) + [(x - a) f_x + (y - b) f_y];\ f_x, f_y \text{ to be evaluated at } (a, b)$$

Quadratic Polynomial Approximation to $f(x, y)$ quadratic polynomial approximation to $f(x, y)$ can be obtained by putting $n = 2$ in (1)

$$f(x, y) = f(a, b) + [(x - a) f_x + (y - b)f_y] + \frac{1}{2!}[(x - a) f_x + (y - b)f_y]^2$$

$[(f_x, f_y, f_{xx}, f_{yy}$ and f_{xy} are to be evaluated at $(a, b)]$

$$= f(a, b) + [(x - a) f_x + (y - b)f_y]$$

$$+ \frac{1}{2!}[(x - a)^2 f_{xx} + 2(x - a)(y - b) f_{xy} + (y - b)^2 f_{yy}]$$

Cor. 3. If we put $(a, b) = (0, 0)$, we obtain Maclaurin's theorem

$$f(x, y) = f(0, 0) + \left(x\frac{\partial}{\partial x} + y\frac{\partial}{\partial y}\right) f(0,0)$$

$$+ \frac{1}{2!}\left[x\frac{\partial}{\partial x} + y\frac{\partial}{\partial y}\right]^2 f(0, 0) \cdots \frac{1}{n!}\left[x\frac{\partial}{\partial x} + y\frac{\partial}{\partial y}\right]^n f(0, 0) + R_n$$

where $$R_n = \frac{1}{(n+1)!}\left[x\frac{\partial}{\partial x} + y\frac{\partial}{\partial y}\right]^{n+1} f(\theta x, \theta y);\ 0 < \theta < 1.$$

4.5. ERROR ESTIMATE

Error term in Taylor's series of the function of two variables in the neighbourhood of (a, b) is

$$R_n = \frac{1}{(n+1)!}\left[(x-a)\frac{\partial}{\partial x} + (y-b)\frac{\partial}{\partial y}\right]^{n+1} f(\xi, \eta) \qquad \ldots(1)$$

where $\xi = (1 - \theta)a + \theta x,\ \eta = (1 - \theta)b + \theta y;\quad 0 < \theta < 1$

Since the point (ξ, η) or the value of θ in (1) is not known, we cannot evaluate the error term exatly. But it is always possible to find a bound of the error term in a given rectangular region R : $|\, x - a \,| < \delta_1$, $|\, y - b \,| < \delta_2$. We assume that all the partial derivatives of $f(x, y)$ of the required order are continuous throughout this region.

For $n = 1$ (linear approximation) the error term is given by

$$R_1 = \frac{1}{2!}\left[(x-a)\frac{\partial}{\partial x} + (y-b)\frac{\partial}{\partial y}\right]^2 f$$

where all partial derivatives are to be evaluated at (ξ, η)

$$\therefore \qquad R_1 = \frac{1}{2!}\{(x - a)^2 f_{xx} + 2(x - a)(y - b) f_{xy} + (y - b)^2 f_{yy}]$$

$$|R_1| = \frac{1}{2}\,|(x - a)^2 f_{xx} + 2(x - a)(y - b) f_{xy} + (y - b)^2 f_{yy}|$$

$$\le \frac{1}{2}\{|\, x - a\,|^2 |f_{xx}| + 2|x - a|\ |y - b|\ |f_{xy}| + |(y - b)^2|\ |f_{yy}|\}$$

Assume $\quad B = \text{Max.}\ \{|f_{xx}|, |f_{xy}|, |f_{yy}|\}$

Then $\quad |R_1| \leq \frac{1}{2} B\{|x-a|^2 + 2\,|x-a|\,|y-b| + |y-b|^2\}$

$$= \frac{1}{2} B\{|x-a|+|y-b|\}^2 \leq \frac{B}{2}(\delta_1+\delta_2)^2$$

This value of $|R_1|$ is called **Maxmimum Absolute Error in the linear approximation** of $f(x, y)$ about the point (a, b).

For n = 2 (Quadratic Approximation) the error term is given by

$$R_2 = \frac{1}{3!}\left\{(x-a)\frac{\partial}{\partial x} + (y-b)\frac{\partial}{\partial y}\right\}^3 f$$

$$= \frac{1}{3!}\{(x-a)^3 f_{xxx} + 3(x-a)^2(y-b) f_{xxy} + 3(x-a)(y-b)^2 f_{xyy} + (y-b)^3 f_{yyy}\}$$

where partial derivatives are evaulated at (ξ, η)

Now $\quad |R_2| = \frac{1}{3!}\,|(x-a)^3 f_{xxx} + 3(x-a)^2(y-b) f_{xxy}$

$$+ 3\,(x-a)\,(y-b)^2 f_{xyy} + (y-b)^3 f_{yyy}|$$

$$\leq \frac{1}{6}\{|x-a|^3\,|f_{xxx}| + 3\,|x-a|^2\,|y-b|\,|f_{xxy}|$$

$$+ 3|x-a|\,|y-b|^2\,|f_{xyy}| + |y-b|^3\,|f_{yyy}|\}$$

If $\quad B = \text{Max.}\ \{|f_{xxx}|, |f_{xxy}|, |f_{xyy}|, |f_{yyy}|\}$

Then $\quad |R_2| < \frac{B}{6}\{|x-a|^3 + 3|x-a|^2\,|y-b| + 3|x-a|\,|y-b|^2 + |y-b|^3\}$

$$= \frac{B}{6}\{|x-a| + |y-b|\}^3$$

$$\leq \frac{B}{6}(\delta_1+\delta_2)^3$$

This value of $|R_2|$ is **called the Maximum Absolute Error in the Quadratic Approximation.**

Cor. If f is a function of more than two variables, the error estimates can also be obtained in the similar manner as done in the case of two variables.

e.g., If f is a function of three variable x, y, z and is to approximated by a linear polynomial about the point (a, b, c) then

$$|R_1| \leq \frac{B}{2}\{|x-a| + |y-b| + |z-c|\}^2 \leq \frac{B}{2}\{\delta_1+\delta_2+\delta_3\}^2$$

where $\quad |x-a| \leq \delta_1,\ |y-b| \leq \delta_2;\ |z-c| \leq \delta_3$

ILLUSTRATIVE EXAMPLES

Example 1. *Find the linear and the quadratic Taylor Series polynomial approximation to the function $f(x, y) = 2x^3 + 3y^3 - 4x^2y$ about the point (1, 2). Obtain the maximum absolute errors in the region $|x - 1| < 0.01$ and $|y - 2| < 0.1$.*

Sol.

$$f(x, y) = 2x^3 + 3y^3 - 4x^2y \qquad f(1, 2) = 18$$
$$f_x = 6x^2 - 8xy \qquad f_x(1, 2) = -10$$
$$f_y = 9y^2 - 4x^2 \qquad f_y(1, 2) = 32$$
$$f_{xy} = -8x \qquad f_{xy}(1, 2) = -8$$
$$f_{xx} = 12x - 8y \qquad f_{xx}(1, 2) = -4$$
$$f_{yy} = 18y \qquad f_{yy}(1, 2) = 36$$
$$f_{xxx} = 12 \qquad f_{xxy} = -8$$
$$f_{yyy} = 18 \qquad f_{xyy} = 0$$

First approximation is given by

$$f(x, y) = f(1, 2) + [(x - 1)f_x + (y - 2)f_y]$$
$$= 18 + (x - 1)(-10) + (y - 2)(32)$$
$$= 18 - 10(x - 1) + 32(y - 2)$$

Maximum absolute error in first approximation is given by

$$|R_1| \le \frac{B}{2}[|x - 1| + |y - 2|]^2$$
$$\le \frac{B}{2}\{(0.01) + (0.1)]^2$$
$$\le \frac{B}{2}\{0.11\}^2 = 0.00605B$$

[Find B] B is the maximum of $|f_{xx}|$, $|f_{yy}|$, $|f_{xy}|$ in the given region $|x - 1| < 0.01$ and $|y - 2| < 0.1$

$$\text{Max. } |f_{xx}| = \text{Max. } |12x - 8y| = \text{Max. } |12(x - 1) - 8(y - 2) - 4|$$
$$\le \text{Max. } \{12|x - 1| + 8|y - 2| + 4\}$$
$$< 12(0.01) + 8(0.1) + 4 = 4.92$$
$$\text{Max. } |f_{yy}| = \text{Max. } |18y| = \text{Max. } |18(y - 2) + 36|$$
$$\le \text{Max. } \{18|y - 2| + 36\}$$
$$\le 18(0.1) + 36 = 37.8$$
$$\text{Max. } |f_{xy}| = \text{Max. } |-8x| = \text{Max. } |-8(x - 1) - 8|$$
$$\le \text{Max. } \{|8(x - 1)| + 8\} \le 8(0.01) + 8 = 8.08$$
$$B = \text{Max. } \{4.92, 37.8, 8.08\} = 37.8$$

$$\therefore \quad |R_1| \le (0.00605)(37.8) = 0.23.$$

Quadratic Approximation is given by

$$f(x, y) = f(1, 2) + \{(x - 1)f_x + (y - 2)f_y\} + \frac{1}{2!}\{(x - 1)^2 f_{xx} + 2(x - 1)(y - 2)f_{xy} + (y - 2)^2 f_{yy}\} \text{ all derivates at } (1, 2)$$

$$= 18 + \{(x-1)(-10) + (y-2)32\}$$
$$+ \frac{1}{2}\{(x-1)^2(-4) + 2(x-1)(y-2)(-8) + (y-2)^2\, 36\}$$
$$= 18 - 10(x-1) + 32(y-2) - 2(x-1)^2 - 8(x-1)(y-2) + 18(y-2)^2$$

Maximum absolute error in quadratic approximation is given by

$$|R_2| \le \frac{B}{6}\{|x-1| + |y-2|\}^3 < \frac{B}{6}\{0.01 + 0.1\}^3 = \frac{B}{6}\{0.001331\}$$
$$B = \text{Max.}\{|f_{xxx}|, |f_{yyy}|, |f_{xxy}|, |f_{xyy}|\}$$

all these third order partial derivatives are constant

$$\therefore \quad B = \text{Max.}\{12, 18, 8, 0\} = 18$$

$$\therefore \quad |R_2| \le \frac{18}{6}(0.001331) = 0.003993 = 0.004.$$

Example 2. *The function $x^2 - xy + y^2$ is approximated by a first degree Taylor's polynomial about the point (2, 3) Find a square $|x-2| < \delta$; $|y-3| < \delta$ with centre at (2, 3) such that error of approximation is ≤ 0.1 in magnitude for all points within this square.* **(P.T.U., Dec. 2004)**

Sol. $f(x, y) = x^2 - xy + y^2$

Its first degree (Linear) approximation is given by

$$f(x, y) = f(2, 3) + [(x-2)f_x(2, 3) + (y-3)f_y(2, 3)] \text{ where } f(2, 3) = 7$$

$f_x = 2x - y$ $\qquad f_x(2, 3) = 1$

$f_y = -x + 2y$ $\qquad f_y(2, 3) = 4$

$$\therefore \quad f(x, y) = 7 + (x-2)\,.\,1 + (y-3)\,.\,4$$

Now Maximum absolute error in linear approximation is given by

$$|R_1| \le \frac{B}{2}\{|x-2| + |y-3|\}^2$$
$$\le \frac{B}{2}[\delta + \delta]^2 = 2B\delta^2 \text{ where B is the maximum of } |f_{xx}|, |f_{yy}|, |f_{xy}|$$

Now $\quad f_{xx} = 2, f_{xy} = -1, f_{yy} = 2$

$|f_{xx}| = 2, \quad |f_{xy}| = 1, \quad |f_{yy}| = 2$

all are constant $\therefore$ B = Max . {2, 1, 2} = 2

$\therefore \quad |R_1| \le 2.2\delta^2$ or $|R_1| \le 4\delta^2$

But it is given $|R_1| \le 0.1$

$\therefore \quad 4\delta^2 < 0.1$ or $\delta^2 < 0.025$

$\therefore \quad \delta < \sqrt{0.025} = 0.1581$. Hence the square is $|x-2| < 0.1581$

and $\quad |y-3| < 0.1581$

Example 3. *Expand $f(x, y, z) = e^z \sin(x + y)$ in Taylor's series upto second order terms about the point (0, 0, 0). Obtain the maximum error in the region $|x| < 0.1$, $|y| < 0.1$; $|z| < 0.1$.* **(P.T.U., Dec. 2004)**

Sol. $f(x, y, z) = e^z \sin(x + y)$

$f(0, 0, 0) = 0$

$f_x = e^z \cos(x + y)$ $\qquad f_x(0, 0, 0) = 1$

$f_y = e^z \cos(x + y)$ $\qquad f_y(0, 0, 0) = 1$

$$f_z = e^z \sin (x + y); \qquad f_z (0, 0, 0) = 0$$
$$f_{xx} = -e^z \sin (x + y); \qquad f_{xx} (0, 0, 0) = 0$$
$$f_{yy} = -e^z \sin (x + y); \qquad f_{yy} (0, 0, 0) = 0$$
$$f_{zz} = e^z \sin (x + y); \qquad f_{zz} (0, 0, 0) = 0$$
$$f_{xy} = -e^z \sin (x + y); \qquad f_{xy} (0, 0, 0) = 0$$
$$f_{yz} = e^z \cos (x + y); \qquad f_{yz} (0, 0, 0) = 1$$
$$f_{zx} = e^z \cos (x + y); \qquad f_{zx} (0, 0, 0) = 1$$
$$f_{xxx} = -e^z \cos (x + y), \qquad f_{xyx} = -e^z \cos (x + y),$$
$$f_{yyy} = -e^z \cos (x + y), \qquad f_{zzz} = e^z \sin (x + y)$$

and similar other third order partial derivatives

i.e., all third order derivatives will either contain $\pm e^z \cos (x + y)$ or $\pm e^z \sin (x + y)$. Second order Approximation is

$$f(x, y, z) = f(0, 0, 0) + [xf_x(0, 0, 0) + yf_y(0, 0, 0) + z f_z(0, 0, 0)]$$
$$+ \frac{1}{2!} \cdot \{x^2 f_{xx}(0, 0, 0) + y^2 f_{yy}(0, 0, 0) + z^2 f_{zz}(0, 0, 0)$$
$$+ 2yz f_{yz}(0, 0, 0) + 2zx f_{zx}(0, 0, 0) + 2xy f_{xy}(0, 0, 0)]$$

$$f(x, y, z) = 0 + \{x \cdot 1 + y \cdot 1 + z \cdot 0] + \frac{1}{2!} \{x^2 \cdot 0 + y^2 \cdot 0 + z^2 \cdot 0$$
$$+ 2yz \cdot 1 + 2zx \cdot 1 + 2xy \cdot 0\}$$

$$= (x + y) + \frac{1}{2}(2yz + 2zx) = (x + y) + (yz + zx)$$

Maximum Absolute error in quadratic Approximation

$$| R_2 | < \frac{B}{6} \{|x| + |y| + |z|\}^3 \qquad ...(1)$$

where B is maximum of $\{|f_{xxx}|, |f_{yyy}|, |f_{zzz}|, |f_{xxy}| \; ..... \}$ *i.e.*, modulus of all third order derivatives.

We know that modulus of third order derivatives is either $|e^z \sin (x + y)|$ or $|e^z \cos (x + y)|$

$$|e^z \sin (x + y)| \le |e^z| \cdot 1 \quad \because \quad | \sin (x + y)| \le 1$$
$$\le e^{|z|} < e^{0.1} = 1.11$$

Similarly $|e^z \cos (x + y)| < 1.11$

$$\therefore \qquad B = 1.11$$

$$\therefore \text{ from (1),} \qquad |R_2| < \frac{1.11}{6} (0.1 + 0.1 + 0.1)^3$$
$$< \frac{1.11}{6} (0.027) = 0.004995 = 0.005$$

Hence Maximum absolute error = 0.005.

Example 4. *If $f(x, y) = \tan^{-1} (xy)$, find an approximate value of $f(1.1, 0.8)$ using Taylor's series quadratic approximation.*

Sol. For the application of Taylor's series first of all we have to change the point (1.1, 0.8) into $(a + h, b + k)$ $\therefore$ if we choose (a, b) to be (1, 1) then $h = 0.1$, $k = -0.2$

$\therefore$ Quadratic approximation of $f(x, y)$ at (1, 1) is

$$f(1.1, 0.8) = f\{1 + 0.1, 1 + (-0.2)\}$$

$$= f(1, 1) + [(0.1) f_x + (-0.2) f_y] + \frac{1}{2!} [(0.1)^2 f_{xx} + 2(0.1)(-0.2) f_{xy} + (-0.2)^2 f_{yy}(1, 1)]$$

Now $\quad f(x, y) = \tan^{-1} xy; \qquad f(1, 1) = \frac{\pi}{4}$ ...(1)

$$f_x = \frac{1}{1+x^2y^2} \cdot y \qquad f_x(1, 1) = \frac{1}{2}$$

$$f_y = \frac{x}{1+x^2y^2} \qquad f_y(1, 1) = \frac{1}{2}$$

$$f_{xx} = y \frac{-1(2x)}{(1+x^2y^2)^2} \qquad f_{xx}(1, 1) = -\frac{2}{4} = -\frac{1}{2}$$

$$f_{yy} = x \cdot \frac{-2y}{(1+x^2y^2)^2} \qquad f_{yy}(1, 1) = -\frac{1}{2}$$

$$f_{xy} = \frac{(1+x^2y^2) . 1 - y(2x^2y)}{(1+x^2y^2)^2} = \frac{1-x^2y^2}{(1+x^2y^2)^2} \; ; f_{xy}(1, 1) = 0$$

$\therefore$ From (1),

$$f(1.1, 0.8) = \frac{\pi}{4} + \left[(0.1) \cdot \frac{1}{2} - (0.2)\frac{1}{2}\right] + \frac{1}{2}\left[(0.01)\left(-\frac{1}{2}\right) + 0 + (0.04)\left(-\frac{1}{2}\right)\right]$$

$$= \frac{\pi}{4} + [0.05 - 0.1] + \frac{1}{2} [0.005 - 0.02]$$

$$= 0.7854 - 0.05 + 0.0125 = 0.7229.$$

TEST YOUR KNOWLEDGE

1. Obtain the first degree Taylor's series approximation to the function $f(x, y) = 2x^2 - xy + y^2 + 3x - 4y + 1$ about the point (–1, 1). Find the maximum error in the region $|x + 1| < 0.1$; $|y - 1| < 0.1$.
2. Obtain the linear approximation of Taylor's series to the function $f(x, y) = e^y \log (x + y)$ about the point (1, 0). Estimate the maximum absolute error over the rectangle $|x - 1| < 0.1$ and $|y| < 0.1$.
3. Expand $f(x, y, z) = e^x \sin (y + z)$ in Taylor Series upto second terms about the point $\left(0, 1, \frac{\pi}{2}\right)$ obtain the maximum error in the region $|x| < 0.1$, $|y - 1| < 0.1$, $\left|z - \frac{\pi}{2}\right| < 0.1$.

   [**Hint.** Consult S.E. 3]
4. Obtain the second order Taylor's series approximation of the function $f(x, y) = xy^2 + y \cos (x - y)$ about the point (1, 1). Find the maximum absolute error in the region $|x - 1| < 0.05$; $|y - 1| < 0.1$.

Answers

1. $f(x, y) = -2 - 2(x + 1) - (y - 1)$; B = 4; $|R_1| \leq 0.08$
2. $f(x, y) = (x - 1) + y$; B = 4.6912; $|R_1| \leq 0.0938$
3. $f(x, y, z) = 1 + x + \frac{1}{2}\left[x^2 - \frac{\pi^2}{4}(y-1)^2 - \left(z - \frac{\pi}{2}\right)^2 - \pi(y-1)\left(z - \frac{\pi}{2}\right)\right]$

   B = 7.0817, $|R_2| \leq 0.0319$
4. $f(x, y) = 2 + (x - 1) + 3(y - 1) + \frac{1}{2}[-(x-1)^2 + 6(x-1)(y-1) + (y-1)^2] + \cdots \infty$

   $B = 5.1; |R_2| \leq 0.0029$

4.6. ERRORS AND APPROXIMATIONS

Let $$z = f(x, y) \quad ...(1)$$

If δx, δy are small increments is x and y respectively and δz, the corresponding increment is z, then

$$z + \delta z = f(x + \delta x, y + \delta y) \quad ...(2)$$

Subtracting (1) from (2), we get

$$\delta z = f(x + \delta x, y + \delta y) - f(x, y) = f(x, y) + \delta x \frac{\partial f}{\partial y} + \delta y \frac{\partial f}{\partial y} + \cdots - f(x, y)$$

$$= \delta x \frac{\partial f}{\partial x} + \delta y \frac{\partial f}{\partial y} \quad \text{(Approximately) [neglecting higher powers of } \delta x, \delta y]$$

$$\therefore \quad \delta z = \frac{\partial f}{\partial x}.\delta x + \frac{\partial f}{\partial y}.\delta y \text{ (Approximately)}$$

$\Rightarrow$ If δx and δy are small changes (or errors) in x and y respectively, then an approximate change (or errror) in z is δz.

Replacing δx, δy, δz by dx, dy, dz respectively, we have $\mathbf{dz = \frac{\partial f}{\partial x} dx + \frac{\partial f}{\partial y} dy}$

Note 1. If δx is the error in x, then **relative error** $= \frac{\delta x}{x}$

percentage error $= \frac{\delta x}{x} \times 100.$

Note 2. $f'(x)\, dx$ is called the **differential of** $f(x)$.

ILLUSTRATIVE EXAMPLES

Example 1. (*i*) *Find the percentage error in the area of an ellipse when an error of + 1 per cent is made in measuring the major and minor axes.* **(P.T.U., Dec. 2012)**

(*ii*) *If an error committed in measuring the sides of squares is 2%. Find the error in calculating the area.* **(P.T.U., May 2007)**

(*iii*) *What error in the common logarithm of a number will be produced by an error of 1% in the number.* **(P.T.U., May 2004)**

Sol. (*i*) If x and y are semi-major and semi-minor axes of the ellipse, then its area A is given by $A = \pi xy$.

Taking logs, $\log A = \log \pi + \log x + \log y$

Taking differentials $\dfrac{dA}{A} = 0 + \dfrac{dx}{x} + \dfrac{dy}{y}$

or $$\frac{dA}{A} \times 100 = \frac{dx}{x} \times 100 + \frac{dy}{y} \times 100 = 1 + 1 = 2.$$

$\therefore$ Error in the area = 2%.

(*ii*) Let the side of the square be x

$$\text{Area} = A = x^2$$

$$dA = 2x dx$$

Given $$dx = 2\% \text{ of } x = \frac{2}{100} x = \frac{x}{50}$$

$\therefore$ $$dA = 2x \cdot \frac{x}{50} = \frac{1}{25} x^2 = \frac{1}{25} A$$

$\therefore$ $$\frac{dA}{A} = \frac{1}{25} = 0.04$$

$\therefore$ error in the area = 0.04

(*iii*) Let x be the number and δx be its error then $dx = 1\%$ of $x = \dfrac{x}{100}$

Common logarithm of $x = \log_{10} x = A$ (say)

Then $$A = \log_{10} x = \frac{\log_e x}{\log_e 10}$$

$$dA = \frac{1}{\log_e 10} \left(\frac{1}{x} dx \right)$$

$$= \frac{1}{\log_e 10} \cdot \frac{1}{100} = \frac{1}{100} \times \frac{1}{2.303} = 0.004343$$

Example 2. *If* $P = \dfrac{E^2}{R}$ *then show that % error in P = 2% error in E – 1 % error in R.*

Hence find the approximate percentage change in P when E is increased by 3% and R is decreased by 2%. **(P.T.U., Dec., 2003; Dec., 2004)**

Sol. $$P = \frac{E^2}{R}$$

Taking logs of both sides,

$$\log P = 2 \log E - \log R$$

Differentiate both sides,

$$\frac{1}{P} dP = 2 \cdot \frac{1}{E} dE - \frac{1}{R} dR$$

or $$\frac{dP}{P} \times 100 = 2 \cdot \frac{dE}{E} \times 100 - \frac{dR}{R} \times 100$$

$\therefore$ % error in P = 2% error in E –% error in R.

Now given $$dE = \frac{3}{100}E \quad \text{and} \quad dR = -\frac{2}{100}R$$

$$\therefore \quad \frac{dP}{P} \times 100 = 2(3) - (-2) = 6 + 2 = 8$$

$$\therefore \quad dP = \frac{8}{100}P$$

$\therefore$ Change in P = 8%.

Example 3. *The work that must be done to propel a ship of displacement D for a distance 'S' in time 't' is proportional to* $\dfrac{S^2D^{2/3}}{t^2}$.

Find approximately the increase of work necessary when the displacement is increased by 1%, the time diminished by 1% and the distance diminished by 2%.

Sol. Let the work done be W, then $W \propto \dfrac{S^2D^{2/3}}{t^2}$

$$\Rightarrow \quad W = k \cdot \frac{S^2D^{2/3}}{t^2}, \text{ where } k \text{ is a constant.}$$

Taking logs, $$\log W = \log k + 2\log S + \frac{2}{3}\log D - 2\log t$$

Taking differentials,

$$\frac{dW}{W} = 0 + 2\frac{dS}{S} + \frac{2}{3}\frac{dD}{D} - 2\frac{dt}{t}$$

or $$\frac{dW}{W} \times 100 = 2\frac{dS}{S} \times 100 + \frac{2}{3}\frac{dD}{D} \times 100 - 2\frac{dt}{t} \times 100$$

$$= 2(-2) + \frac{2}{3}(1) - 2(-1) = -\frac{4}{3}$$

$\therefore$ Approximate increase of work $= -\dfrac{4}{3}\%$.

Example 4. *The radius of a sphere is found to be 10 cm with a possible error of 0.02 cm. What is the relative error in computing the volume?* **(P.T.U. Dec. 2013)**

Sol. Let r be the radius of the sphere

Volume of the sphere

$$V = \frac{4}{3}\pi r^3$$

Differentiating both sides,

$$dV = \frac{4}{3}\pi \cdot 3r^2\, dr$$

$$\frac{dV}{V} = \frac{3}{r}dr$$

Given $r = 10$ cm and $dr = 0.02$ cm

$$\therefore \quad \frac{dV}{V} = \frac{3}{10}(0.02) = 0.006$$

Hence relative error in computing the volume = 0.006 cc.

Example 5. *The period T of a simple pendulum of length l is given by* $T = 2T_1\sqrt{l/g}$. *Find (i) the error and (ii) percent error (relative error) made in computing T by using* $l = 2$ *ft and* $g = 32$ *ft/sec*^{2} *if the true values are* $l = 1.95$ *and* $g = 32.2$ *ft/sec*^{2}.

Sol. (*i*) $$T = 2\pi\sqrt{\frac{l}{g}}$$

Taking logs of both sides

$$\log T = \log 2\pi + \frac{1}{2}\log l - \frac{1}{2}\log g$$

Differentiate both sides

$$\frac{1}{T}dT = 0 + \frac{1}{2}\cdot\frac{1}{l}\,dl - \frac{1}{2}\frac{1}{g}\,dg$$

$$\frac{dT}{T} = \frac{1}{2}\cdot\left[\frac{dl}{l} - \frac{dg}{g}\right]$$

given $l = 1.95,\ l + dl = 2 \quad \therefore \quad dl = 2 - 1.95 = 0.05$

$g = 32.2,\ g + dg = 32 \quad \therefore \quad dg = 32 - 32.2 = -0.2$

$$\therefore \quad \frac{dT}{T} = \frac{1}{2}\cdot\left[\frac{0.05}{1.95} + \frac{0.2}{32.2}\right] = \frac{1}{2}[0.025 + 0.0063]$$

$$dT = \frac{T}{2}[0.031] = \frac{1}{2}\cdot 2\pi\sqrt{\frac{1.95}{32.2}}(0.031)$$

$$dT = (3.14)\sqrt{\frac{1.95}{32.2}}(0.031) = 0.0245$$

(*ii*) Percentage error $= \frac{dT}{T}100 = (0.0155)\,100 = 1.55$

$$\text{Relative error} = \frac{dT}{T} = 0.0155$$

Example 6. *The diameter and altitude of a can in the shape of a right circular cylinder are measured as 4 cm and 6 cm respectively. The possible error in each measurement is 0.1 cm. Find approximately the maximum possible error in the value computed for the volume and lateral surface.*

Sol. Let the diameter and altitude of the can be denoted by D and H respectively.

Then radius $= \frac{D}{2}$

(*i*) The volume V of the can is given by

$$V = \pi r^2 h = \frac{\pi}{4} D^2 H = f(D, H)$$

$$\therefore \quad dV = \frac{\partial V}{\partial D} dD + \frac{\partial V}{\partial H} dH = \frac{\pi}{4}[2DH dD + D^2 dH]$$

$$= \frac{\pi}{4}[2 \times 4 \times 6 \times 0.1 + 4^2 \times 0.1] = \frac{\pi}{4}(6.4) = 1.6\,\pi \text{ cu cm.}$$

(*ii*) The lateral surface S of the can is given by

$$S = 2\pi rh = \pi DH = f(D, H)$$

$$\therefore \quad dS = \frac{\partial S}{\partial D} dD + \frac{\partial S}{\partial H} dH$$

$$= \pi[H dD + D dH] = \pi[6 \times 0.1 + 4 \times 0.1] = \pi \text{ sq. cm.}$$

Example 7. *In determining the specific gravity by the formula* $S = \dfrac{A}{A - W}$ *where A is the weight in air and W is the weight in water A can be read within 0.01 lb and W within 0.02 lb. Find approximately the maximum error in S, if the readings are A = 1.1 lb, W = 0.6 lb. Find also the maximum relative error.*

Sol. $$S = \frac{A}{A - W}$$

Differentiate partially w.r.t. A and W

$$\frac{\partial S}{\partial A} = \frac{(A - W)\cdot 1 - A \cdot 1}{(A - W)^2} = \frac{-W}{(A - W)^2}$$

$$\frac{\partial S}{\partial W} = \frac{A}{(A - W)^2}$$

$$dS = \frac{\partial S}{\partial A} \cdot dA + \frac{\partial S}{\partial W} dW = \frac{1}{(A - W)^2}[-W\,dA + A dW]$$

Now error in S *i.e.*, dS is maximum when dA is –ve

$$\therefore \quad \text{Max. } dS = \frac{1}{(A - W)^2}[W dA + A dW]$$

given $$A = 1.1,\ W = 0.6,\ dA = 0.01,\ dW = 0.02$$

$$\therefore \quad \text{Max. } dS = \frac{1}{(1.1 - 0.6)^2}[(0.6)(0.01) + (1.1)(0.02)] = 0.112$$

$$\text{Maximum relative error} = \frac{\text{max. } dS}{S} = \frac{0.112}{A/A - W}$$

$$= (0.112)\,\frac{1.1 - 0.6}{1.1} = 0.051.$$

Example 8. *The height of a tower is determined by observing the elevation θ and φ of its summit from two points in a direct line with the foot of the tower and at a distance 'a' apart. Show that the error in the calculated height due to small errors dθ and dφ is approximately* $a(\sin^2\theta\, d\phi - \sin^2\phi\, d\theta)\,\text{cosec}^2(\theta - \phi)$.

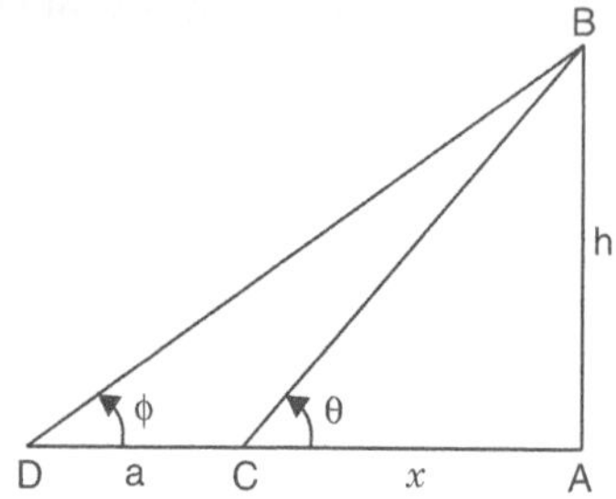

Sol. Let h be the height of the tower AB and C and D, the two points of observation so that

$$CD = a, \angle ACB = \theta,$$
$$\angle ADB = \phi. \text{ Let } AC = x$$

From rt. $\angle d$. ΔBAC, $\quad x = h \cot \theta \quad$ ...(1)

From rt. $\angle d$. ΔBAD, $\quad x + a = h \cot \phi \quad$ ...(2)

Subtracting (1) from (2),

$$a = h (\cot \phi - \cot \theta) = h \left(\frac{\cos \phi}{\sin \phi} - \frac{\cos \theta}{\sin \theta} \right)$$

$$= h \left[\frac{\sin \theta \cos \phi - \cos \theta \sin \phi}{\sin \theta \sin \phi} \right] = \frac{h \sin (\theta - \phi)}{\sin \theta \sin \phi}$$

$$\Rightarrow \qquad h = \frac{a \sin \theta \sin \phi}{\sin (\theta - \phi)} = f(\theta, \phi) \qquad \text{...(1)}$$

Taking logs, $\quad \log h = \log a + \log \sin \theta + \log \sin \phi - \log \sin (\theta - \phi)$

Taking differentials

$$\frac{dh}{h} = 0 + \frac{\cos \theta}{\sin \theta} d\theta + \frac{\cos \phi}{\sin \phi} d\phi - \frac{\cos (\theta - \phi)}{\sin (\theta - \phi)} (d\theta - d\phi)$$

or
$$\frac{dh}{h} = \left[\frac{\cos \theta}{\sin \theta} - \frac{\cos (\theta - \phi)}{\sin (\theta - \phi)} \right] d\theta + \left[\frac{\cos \phi}{\sin \phi} + \frac{\cos (\theta - \phi)}{\sin (\theta - \phi)} \right] d\phi$$

$$= \frac{\sin (\theta - \phi) \cos \theta - \cos (\theta - \phi) \sin \theta}{\sin \theta \sin (\theta - \phi)} d\theta + \frac{\sin (\theta - \phi) \cos \phi + \cos (\theta - \phi) \sin \phi}{\sin \phi \sin (\theta - \phi)} d\phi$$

$$= \frac{\sin [(\theta - \phi) - \theta]}{\sin \theta \sin (\theta - \phi)} d\phi + \frac{\sin [(\theta - \phi) + \phi]}{\sin \phi \sin (\theta - \phi)} d\phi$$

$$= \frac{\sin (-\phi)}{\sin \theta \sin (\theta - \phi)} d\theta + \frac{\sin \theta}{\sin \phi \sin (\theta - \phi)} d\phi$$

$$= \frac{\sin^2 \theta \, d\phi - \sin^2 \phi \, d\theta}{\sin \theta \sin \phi \sin (\theta - \phi)} \qquad | \because \sin (-\phi) = -\sin \phi$$

$$\therefore \qquad dh = h \cdot \frac{\sin^2 \theta \, d\phi - \sin^2 \phi \, d\theta}{\sin \theta \sin \phi \sin (\theta - \phi)} = \frac{a \sin \theta \sin \phi}{\sin (\theta - \phi)} \cdot \frac{\sin^2 \theta \, d\phi - \sin^2 \phi \, d\theta}{\sin \theta \sin \phi \sin (\theta - \phi)}$$

| Using (1)

$$= a(\sin^2 \theta \, d\phi - \sin^2 \phi \, d\theta) \operatorname{cosec}^2 (\theta - \phi).$$

Example 9. *The height h and semi-vertical angle α of a cone are measured, and from them A, the total area of the cone, including the base is calculated. If A and α are in error by small quantities δA and $\delta\alpha$ respectively, find the corresponding error in the area. Show further that, if $\alpha = \frac{\pi}{6}$, an error of + 1% in h will be approximately compensated by an error of –0.33° in α.*

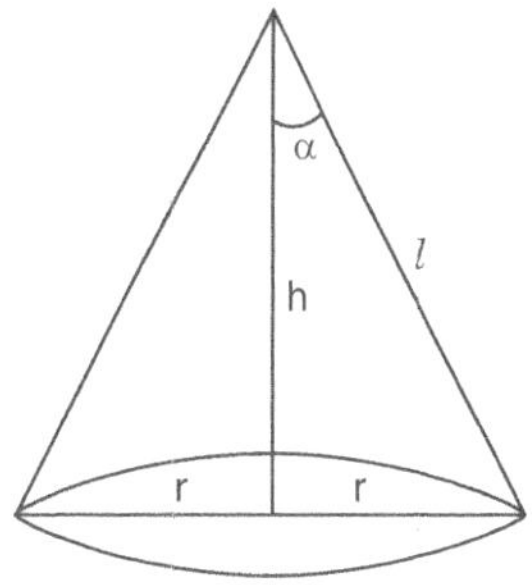

Sol. Base radius $r = h \tan \alpha$

Slant height $l = h \sec \alpha$

Area of base $= \pi r^2$

Area of curved surface $= \pi r l$

$\therefore$ Total area $A = \pi r^2 + \pi r l = \pi r(r + l)$

$$= \pi h \tan \alpha(h \tan \alpha + h \sec \alpha)$$

$$= \pi h^2 (\tan^2 \alpha + \tan \alpha \sec \alpha) = f(h, \alpha)$$

$$\delta A = \frac{\partial A}{\partial h} \delta h + \frac{\partial A}{\partial \alpha} \delta \alpha$$

$$= 2\pi h (\tan^2 \alpha + \tan \alpha \sec \alpha)\, \delta h + \pi h^2 (2 \tan \alpha \sec^2 \alpha + \sec^3 \alpha + \tan \alpha \sec \alpha \tan \alpha)\, \delta \alpha$$

$$= 2\pi h \tan \alpha(\tan \alpha + \sec \alpha)\, \delta h + \pi h^2 \sec \alpha(2 \tan \alpha \sec \alpha + \sec^2 \alpha + \tan^2 \alpha)\, \delta \alpha \quad ...(1)$$

which gives the error in A.

Putting $\alpha = \frac{\pi}{6}$ and $\delta h = 1\%$ of $h = \frac{h}{100}$ in (1), we have

$$\delta A = 2\pi h \cdot \frac{1}{\sqrt{3}} \left(\frac{1}{\sqrt{3}} + \frac{2}{\sqrt{3}} \right) \cdot \frac{h}{100} + \pi h^2 \cdot \frac{2}{\sqrt{3}} \left(\frac{2}{\sqrt{3}} \cdot \frac{2}{\sqrt{3}} + \frac{4}{3} + \frac{1}{3} \right) \delta \alpha$$

$$= \frac{2\pi h^2}{100} + 2\sqrt{3}\pi h^2 \delta \alpha$$

Since the error in h is to be compensated by the error in α, $\delta A = 0$

$$\Rightarrow \quad \frac{1}{100} + \sqrt{3}\, \delta \alpha = 0 \quad \text{or} \quad \delta \alpha = -\frac{1}{100\sqrt{3}} \text{ radians}$$

or $$\delta \alpha = -\frac{0.01}{1.732} \times 57.3 \text{ degrees} \qquad [\because \ 1 \text{ radian} = 57.3° \text{ nearly}]$$

$$= -0.33 \text{ degrees.}$$

Example 10. *If the sides and angles of a triangle ABC vary in such a way that its circum-radius remains constant, prove that* $\frac{da}{\cos A} + \frac{db}{\cos B} + \frac{dc}{\cos C} = 0.$ **(P.T.U., Dec. 2011)**

Sol. The circum-radius R of a ΔABC is given by

$$R = \frac{a}{2 \sin A} = \frac{b}{2 \sin B} = \frac{c}{2 \sin C}$$

Now $a = 2R \sin A$ (R is constant)

Taking differentials, $da = 2R \cos A\, dA$ or $\frac{da}{\cos A} = 2R\, dA$

Similarly, $\frac{db}{\cos B} = 2R\, dB$; $\frac{dc}{\cos C} = 2R\, dC$

Adding, $$\frac{da}{\cos A} + \frac{db}{\cos B} + \frac{dc}{\cos C} = 2R\,(dA + dB + dC) \quad ...(1)$$

Also $\qquad A + B + C = \pi$

Taking differentials, $\qquad dA + dB + dC = 0$

$\therefore$ From (1), $\qquad \dfrac{da}{\cos A} + \dfrac{db}{\cos B} + \dfrac{dc}{\cos C} = 0.$

Example 11. *In a plane triangle, if the sides a, b be constant, prove that the variations of its angles are given by the relations* $\dfrac{dA}{\sqrt{a^2 - b^2 \sin^2 A}} = \dfrac{dB}{\sqrt{b^2 - a^2 \sin^2 B}} = -\dfrac{dC}{c}$, *the letters having their usual significance.*

Sol. By sine formula, $\qquad \dfrac{a}{\sin A} = \dfrac{b}{\sin B}$

$$a \sin B = b \sin A \qquad ...(1)$$

Taking differentials ; $a \cos B \,.\, dB = b \cos A \,.\, dA$

or
$$\frac{dA}{a \cos B} = \frac{dB}{b \cos A} = \frac{dA + dB}{a \cos B + b \cos A} \qquad ...(2) \qquad \left[\because \text{ If } \frac{a}{b} = \frac{c}{d} \text{ then each } \frac{a+c}{b+d}\right]$$

Now $\qquad a \cos B = a\sqrt{1 - \sin^2 B} = \sqrt{a^2 - a^2 \sin^2 B} = \sqrt{a^2 - b^2 \sin^2 A}$ $\qquad$ [Using (1)]

$\qquad b \cos A = b\sqrt{1 - \sin^2 A} = \sqrt{b^2 - b^2 \sin^2 A} = \sqrt{b^2 - a^2 \sin^2 B}$ $\qquad$ [Using (1)]

$a \cos B + b \cos A = c$ $\qquad$ [By projection formula]

Also $\quad A + B + C = \pi \;\Rightarrow\; A + B = \pi - C \;$ so that $dA + dB = -dC$

$\therefore$ From (2), $\qquad \dfrac{dA}{\sqrt{a^2 - b^2 \sin^2 A}} = \dfrac{dB}{\sqrt{b^2 - a^2 \sin^2 B}} = \dfrac{-dC}{c}.$

Example 12. *Using differentials, find the approximate value of* $\sqrt{(298)^2 + (401)^2}$.

(P.T.U., Dec. 2007)

Sol. Let $\qquad f(x, y) = \sqrt{x^2 + y^2}$ then $f(x + \delta x, y + \delta y) = \sqrt{(x + \delta x)^2 + (y + \delta y)^2}$

where $\qquad x + \delta x = 298, \quad y + \delta y = 401$

Take $x = 300$, $\qquad y = 400$

$\therefore \qquad \delta x = -2, \delta y = 1$

$$df = \frac{\partial f}{\partial x} dx + \frac{\partial f}{\partial y} dy = \frac{1}{2\sqrt{x^2 + y^2}} 2x dx + \frac{1}{2\sqrt{x^2 + y^2}} 2y dy$$

$\therefore$
$$df = \frac{x\,dx + y\,dy}{\sqrt{x^2 + y^2}}$$

at $x = 300, y = 400$; $\qquad df = \dfrac{300\,dx + 400\,dy}{\sqrt{90000 + 160000}} = \dfrac{300\,dx + 400\,dy}{500} = \dfrac{3dx + 4dy}{5}$

By Taylor's series

$$f(x + \delta x, y + \delta y) = f(x, y) + \left(\delta x \frac{\partial f}{\partial x} + \delta y \frac{\partial f}{\partial y}\right) + \cdots \infty = f(x, y) + \delta f$$

$$\therefore \quad f(298, 401) = f(300, 400) + \frac{3(-2) + 4(1)}{5}$$

$$= \sqrt{(300)^2 + (400)^2} + \frac{-2}{5} = 500 - \frac{2}{5} = \frac{2498}{5}$$

$$\therefore \quad \sqrt{(298)^2 + (401)^2} = 499.6.$$

TEST YOUR KNOWLEDGE

1. Find the percentage error in the area of a rectangle when an error of + 1 per cent is made in measuring its length and breadth. ***(P.T.U., Dec. 2012)***
2. The radius of a circle is found to be 100 cm. Find the relative error in the area of the circle due to an error of 1 mm in measuring the radius. ***(P.T.U., Dec. 2013)***
3. Find the possible percentage error in computing the resistance r from the formula $\frac{1}{r} = \frac{1}{r_1} + \frac{1}{r_2}$, r_1 and r_2 are both in error by 2%. ***(P.T.U., May 2002, Dec. 2014)***
4. The density of the atmosphere at height z metres is given by $\rho = \rho_0 e^{\frac{-z}{800}}$, where ρ_0 is the densi at sea level. Find the approximate increase in height corresponding to a fall of 1% in the density.
5. If the kinetic energy $k = \frac{wv^2}{2g}$, find approximately the change in the kinetic energy as w changes from 49 to 49.5 and v changes from 1600 to 1590.
6. If the H.P. required to propel a steamer varies as the cube of the velocity and the square of length, prove that a 3% increase in velocity and 4% increase in length will require an increase about 17% in H.P.
   [**Hint.** If H denotes the horse power, v the velocity and l, the length, $\Rightarrow$ $H = kv^3l^2$, where k is a constant.]
7. The indicated horse-power I of an engine is calculated from the formula $I = \frac{PLAN}{33000}$, where $A = \frac{\pi}{4} d^2$. Assuming that errors of r per cent may have been made in measuring P, L, N and d, find the greatest possible error in I.

   $$\left[\textbf{Hint. } \log I = \log P + \log L + \log A + \log N - \log 33000 \text{ Diff. and use } dA = \frac{\pi}{4} 2d; \text{ and error in each } = r\%\right]$$
8. The period T of a simple pendulum is $T = 2\pi \sqrt{\frac{1}{g}}$. Find the maximum error in T due to possible errors upto 1% in l and 2.5% in g.
9. The quantity Q of water flowing over a notch is given by $Q = \frac{8}{15} \times 0.64 \times \sqrt{2} g H^{5/2}$, where H is the head of the notch. What is the percentage error in Q caused by measuring H as 0.198 instead of 0.2? ***(P.T.U., May 2003)***
   [**Hint.** Taking log and differentiate given $dH = -0.002$]

10. The voltage V across a resistor is measured with an error h and the resistance R is measured with an error k. Show that the error in calculating the power $W = \frac{V^2}{R}$ generated in the resistor is

$\frac{V}{R^2}(2Rh - Vk)$.

[**Hint.** Here $dV = h$, $dR = k$, $dW = ?$]

11. The torsional rigidity of a length of a wire is obtained from the formula $N = \frac{8\pi I l}{t^2 r^4}$. If l is decreased by 2%, r increased by 2%, t increased by 1.5%, show that the value of N is diminished by 13% approximately.

12. If a triangle ABC be slightly varied but so as to remain inscribed in the same circle, prove that

$$\frac{\delta a}{\cos A} + \frac{\delta b}{\cos B} + \frac{\delta c}{\cos C} = 0.$$

13. The range R of a projectile which starts with a velocity v at an elevation α is given by $R = \frac{v^2 \sin 2\alpha}{g}$. Find the percentage error in R due to an error of 1% in v and an error of $\frac{1}{2}$% in α.

14. If $ax^2 + by^2 + cz^2 = 1$ and $lx + my + nz = 0$, prove that $\frac{dx}{bny - cmz} = \frac{dy}{clz - anx} = \frac{dz}{amx - bly}$.

15. Two sides a, b of a triangle and included angle C are measured. Show that the error δc in the computed length of third side c due to a small error in the angle C is given by $\delta c = a \sin B \cdot \delta C$.

16. If $z = xy(x + y)$, find d^2z, when x and y are the independent variables.

17. At a distance of 50 metres from the foot of the tower the elevation of its top is 30°. If the possible errors in measuring the distance and elevation are 2 cm and 0.05 degrees, find the approximate error in calculating the height.

18. If $f(x, y) = \tan^{-1} xy$, compute $f(0.9, -1.2)$ approximately by the use of differentiation.
[**Hint.** Consult S.E. 14, take $x + \delta x = 0.9$, and $y + \delta y = -1.2$; $x = 1, y = -1$ $\therefore$ $\delta x = -0.1, \delta y = -0.2$

$$\delta f = \frac{y\,\delta x + x\delta y}{1 + x^2y^2} = -0.05 \text{ at } (1, -1)\Bigg]$$

and $f(1, -1) = \tan^{-1}(-1) = \frac{3\pi}{4} = 2.355$

Apply Taylor's theorem.

19. The deflection at the centre of a rod of length l and diameter d supported at its ends and loaded at the centre with a weight w varies as wl^3d^{-4}, what is the increase in the deflection corresponding to p% increase in w, q% decrease in l and r% increase in d.

20. In estimating the cost of pile of bricks measured as 6 m × 50 m × 4 m, the tape is streched 1% beyond the standard length. If the count is 12 bricks to 1 m^3 and bricks cost by ₹ 100 per 1,000, find the approximate error in the cost.

Answers

1. 2% **2.** 0.002 sq. cm. **3.** 0.006 cc **4.** 8 m
5. – 4500 units **7.** $5r$% **8.** 1.75% **9.** – 2.5%
13. $\cot 2\alpha + \frac{2}{v}$ **16.** $2y(dx)^2 + 2x(dy)^2 + 4(x + y)\,dxdy$ **17.** 7 cm
18. 2.305 **19.** $p - 3q - 4r$% **20.** 43.2 rupees.

4.7. MAXIMA AND MINIMA OF FUNCTIONS OF TWO VARIABLES

A function $f(x, y)$ is said to have a **maximum value at $x = a, y = b$ if $f(a, b) > f(a + h, b + k)$**, for small and independent values of h and k, positive or negative.

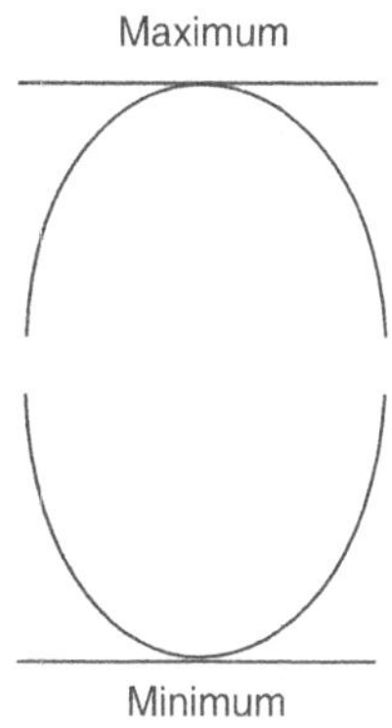

A function $f(x, y)$ is said to have a **minimum value at $x = a, y = b$** if $f(a, b) < f(a + h, b + k)$, for small and independent values of h and k, positive or negative.

Thus $f(x, y)$ has a maximum or minimum value at a point (a, b) according as

$$\Delta f = f(a + h, b + k) - f(a, b) < \text{ or } > 0.$$

Geometrically, the surface $z = f(x, y)$ has a maximum at the point (a, b) if $f(a, b)$ is the greatest in the small neighbourhood of the point (a, b). As such, the surface descends in all directions at this point. The point of maxima may well be compared with the highest point of a dome.

The surface $z = f(x, y)$ has a minimum at the point (a, b) if $f(a, b)$ is the smallest in the small neighbourhood of the point (a, b). As such, the surface ascends in all directions at this point. The point of minima may well be compared with the lowest point of a bowl.

Saddle Point: ***(P.T.U., Dec. 2005)***

A point (a, b) is said to be saddle point of a surface if the tangent plane at (a, b) is horizontal and surface descends in some direction and ascends in other direction.

Extreme Value:

A maximum or a minimum value of a function is called its *extreme value.*

4.8. NECESSARY CONDITION FOR A FUNCTION $f(x, y)$ TO HAVE AN EXTREMUM AT (a, b) ***(P.T.U., May 2004)***

By Taylor's theorem, we have

$$\Delta f = f(a + h, b + k) - f(a, b)$$

$$= [hf_x(a, b) + kf_y(a, b)] + \frac{1}{2!}[h^2f_{xx}(a, b) + 2hkf_{xy}(a, b) + k^2f_{yy}(a, b)] + \cdots \quad ...(1)$$

For small values of h and k, the second and higher order terms are still smaller and may be neglected.

Thus, sign of Δf = sign of $[hf_x(a, b) + kf_y(a, b)]$

Taking $h = 0$, the sign of Δf changes with the sign of k. Similarly taking $k = 0$, the sign of Δf changes with the sign of h. Since Δf changes sign with h and k, $f(x, y)$ cannot have a maximum or a minimum value at (a, b) unless $f_x(a, b) = 0 = f_y(a, b)$.

Hence the necessary conditions for $f(x, y)$ to have an extreme value at (a, b) are

$$f_x(a, b) = 0, f_y(a, b) = 0$$

4.9. CONDITION FOR $f(x, y)$ TO BE MAXIMUM OR MINIMUM ***(P.T.U., 2004)***

Firstly prove art 4.8 and then if the conditions for an extreme value at (a, b) *i.e.*, $f_x(a, b) = 0$, $f_y(a, b) = 0$ are satisfied, then for small values of h and k, we have, from (1),

$$\Delta f = \frac{1}{2!}\,[h^2 f_{xx}(a, b) + 2hkf_{xy}(a, b) + k^2 f_{yy}(a, b)] + \cdots$$

$$= \frac{1}{2!}\,[h^2 r + 2hks + k^2 t] + \cdots \quad \text{where } r = f_{xx}(a, b),\ s = f_{xy}(a, b),\ t = f_{yy}(a, b)$$

or $$\Delta f = \frac{1}{2r}\,[h^2 r^2 + 2hkrs + k^2 rt] + \cdots = \frac{1}{2r}\,[(h^2 r^2 + 2hkrs + k^2 s^2) + k^2 rt - k^2 s^2)] + \cdots$$

$$= \frac{1}{2r}\,[(hr + ks)^2 + k^2(rt - s^2)] + \cdots \qquad \ldots(2)$$

Now $(hr + ks)^2$ is always positive and $k^2(rt - s^2)$ will be positive if $rt - s^2 > 0$. In this case, Δf will have the same sign as that of r for all values of h and k.

Hence if $rt - s^2 > 0$, then $f(x, y)$ has a maximum or a minimum at (a, b) according as $r < 0$ or $r > 0$.

If $rt - s^2 < 0$, then Δf changes sign with h and k. Hence there is neither a maximum nor a minimum value at (a, b). The point (a, b) is a saddle point in this case.

If $rt - s^2 = 0$, no conclusion can be drawn about a maximum or a minimum value at (a, b) and hence further investigation is required.

Note. The point (a, b) is called a ***stationary point*** if $f_x(a, b) = 0$, $f_y(a, b) = 0$. The value $f(a, b)$ is called a ***stationary value***. Thus every extreme value is a stationary value but the converse may not be true.

4.10. WORKING RULE TO FIND THE EXTREME VALUES OF A FUNCTION $z = f(x, y)$

(P.T.U., May 2008, Jan. 2009)

(*i*) Find $\dfrac{\partial z}{\partial x}$ and $\dfrac{\partial z}{\partial y}$.

(*ii*) Solve $\dfrac{\partial z}{\partial x} = 0$ and $\dfrac{\partial z}{\partial y} = 0$ simultaneously.

Let (a, b) ; (c, d) ; ... be the solutions of these equations.

(*iii*) For each solution in step (*ii*), find $r = \dfrac{\partial^2 f}{\partial x^2}$, $s = \dfrac{\partial^2 f}{\partial x \partial y}$, $t = \dfrac{\partial^2 f}{\partial y^2}$.

(*iv*) (*a*) If $rt - s^2 > 0$ and $r < 0$ for a particular solution (a, b) of step (*ii*), then z has a maximum value at (a, b).

(*b*) If $rt - s^2 > 0$ and $r > 0$ for a particular solution (a, b) of step (*ii*), then z has a minimum value at (a, b).

(*c*) If $rt - s^2 < 0$ for a particular solution (a, b) of step (*ii*), then z has no extreme value at (a, b).

(*d*) If $rt - s^2 = 0$, the case is doubtful and requires further investigation.

ILLUSTRATIVE EXAMPLES

Example 1. (*i*) *Examine the function* $x^3 + y^3 - 3axy$ *for maxima and minima.*

(P.T.U., May 2006)

(*ii*) *Examine for extreme values the function* $f(x, y) = x^2 + 2xy$. **(P.T.U., Dec. 2014)**

Sol. (*i*) Here $f(x, y) = x^3 + y^3 - 3axy$

$$f_x = 3x^2 - 3ay,\ f_y = 3y^2 - 3ax,\ r = f_{xx} = 6x,\ s = f_{xy} = -3a,\ t = f_{yy} = 6y$$

Now for extreme values $f_x = 0$ and $f_y = 0$

$$\Rightarrow \quad x^2 - ay = 0 \quad ...(1) \qquad y^2 - ax = 0 \quad ...(2)$$

From (1), $y = \dfrac{x^2}{a}$

$\therefore$ From (2), $\dfrac{x^4}{a^2} - ax = 0$ or $x(x^3 - a^3) = 0$ or $x = 0, a$

when $x = 0,\ y = 0$; when $x = a,\ y = a$

$\therefore$ There are two stationary points (0, 0) and (a, a).

Now $rt - s^2 = 36xy - 9a^2$

At (0, 0), $rt - s^2 = -9a^2 < 0$

$\Rightarrow$ There is no extreme value at (0, 0).

At (*a*, *a*), $rt - s^2 = 36a^2 - 9a^2 = 27a^2 > 0$

$\Rightarrow$ $f(x, y)$ has extreme value at (a, a)

Now $r = 6a$

If $a > 0$, $r > 0$ so that $f(x, y)$ has a minimum value at (a, a).

Minimum value $= f(a, a) = a^3 + a^3 - 3a^3 = -a^3$

If $a < 0$, $r < 0$ so that $f(x, y)$ has a maximum value at (a, a).

Maximum value $= f(a, a) = a^3 + a^3 - 3a^3 = -a^3$.

(*ii*) Here, $f(x, y) = x^2 + 2xy$

$$f_x = 2x + 2y$$
$$f_y = 2x$$
$$r = f_{xx} = 2$$
$$s = f_{xy} = 2$$
$$t = f_{yy} = 0$$

For extreme values $f_x = 0, f_y = 0$

$$\therefore \quad 2x + 2y = 0,\ 2x = 0 \ \Rightarrow\ x = 0, y = 0$$

$\therefore$ Stationary point is (0, 0)

At (0, 0), $rt - s^2 = 0 - 4 = -4 < 0$

$\Rightarrow$ There is no extreme value at (0, 0)

Example 2. *Discuss maxima and minima of $x^3y^2\ (1 - x - y)$.* **(P.T.U., May 2009)**

Sol. Here $f(x, y) = x^3y^2(1 - x - y) = x^3y^2 - x^4y^2 - x^3y^3$

$$f_x = 3x^2y^2 - 4x^3y^2 - 3x^2y^3, \quad f_y = 2x^3y - 2x^4y - 3x^3y^2$$

$$r = f_{xx} = 6xy^2 - 12x^2y^2 - 6xy^3;\ s = f_{xy} = 6x^2y - 8x^3y - 9x^2y^2;$$

$$t = f_{yy} = 2x^3 - 2x^4 - 6x^3y$$

Now for extreme values $f_x = 0$ and $f_y = 0$

$\Rightarrow \quad x^2y^2(3 - 4x - 3y) = 0 \quad ...(1) \qquad x^3y(2 - 2x - 3y) = 0 \quad ...(2)$

Solving (1) and (2),

From (1), $x = 0$ or $y = 0$ or $3 - 4x - 3y = 0$

From (2), $x = 0$ or $y = 0$ or $2 - 2x - 3y = 0$

Solving, $4x + 3y - 3 = 0$ and $2x + 3y - 2 = 0$, we get

$$x = 1/2, \quad y = 1/3$$

$\therefore$ The stationary points are (0, 0) and $\left(\frac{1}{2}, \frac{1}{3}\right)$.

Now $\quad rt - s^2 = 6xy^2(1 - 2x - y)\ 2x^3\ (1 - x - 3y) - [x^2y\ (6 - 8x - 9y)]^2$

$$= 12x^4y^2\ (1 - 2x - y)\ (1 - x - 3y) - x^4y^2(6 - 8x - 9y)^2$$

At (0, 0), $rt - s^2 = 0$

$\therefore$ At (0, 0) further investigation is needed.

Consider $f(a + h, b + k) - f(a, b) = f(h, k) - f(0, 0) = h^3k^2(1 - h - k)$

$= h^3k^2$ (neglecting $h^4k^2 + h^3k^3$ which is very small as compared to h^3k^2)

which is > 0 if $h > 0$ and < 0 if $h < 0$

$\therefore$ $f(a + h, b + k) - f(a, b)$ does not keep same sign for all small values of h and k (positive or negative).

$\therefore$ There is no extreme value at (0, 0).

At $\left(\frac{1}{2}, \frac{1}{3}\right)$, $\quad rt - s^2 = 12 \cdot \frac{1}{16} \cdot \frac{1}{9}\left(-\frac{1}{3}\right)\left(-\frac{1}{2}\right) - \frac{1}{16} \cdot \frac{1}{9}(1) = \frac{1}{144}(2 - 1) = \frac{1}{14} > 0.$

Also $\quad r = 6 \cdot \frac{1}{2} \cdot \frac{1}{9} - 12 \cdot \frac{1}{4} \cdot \frac{1}{9} - 6 \cdot \frac{1}{2} \cdot \frac{1}{8} = \frac{1}{3} - \frac{1}{3} - \frac{3}{8} = -\frac{3}{8} < 0.$

$\therefore$ $f(x, y)$ has a maximum value at $\left(\frac{1}{2}, \frac{1}{3}\right)$

Maximum value $= f\left(\frac{1}{2}, \frac{1}{3}\right) = \frac{1}{8} \cdot \frac{1}{9}\left(1 - \frac{1}{2} - \frac{1}{3}\right) = \frac{1}{432}.$

Example 3. *Locate the stationary points of $x^4 + y^4 - 2x^2 + 4xy - 2y^2$ and determine their nature.* **(P.T.U., May 2005, Dec. 2010)**

Sol. Here $f(x, y) = x^4 + y^4 - 2x^2 + 4xy - 2y^2$

$$f_x = 4x^3 - 4x + 4y, \quad f_y = 4y^3 + 4x - 4y$$

$$r = f_{xx} = 12x^2 - 4, \quad s = f_{xy} = 4, \quad t = f_{yy} = 12y^2 - 4$$

Now for extreme values $f_x = 0$ and $f_y = 0$

$\Rightarrow \quad x^3 - x + y = 0 \quad ...(1) \qquad y^3 + x - y = 0 \quad ...(2)$

Adding (1) and (2), $x^3 + y^3 = 0 \Rightarrow y = -x.$

Putting $y = -x$ in (1), we get $x^3 - 2x = 0$ or $x(x^2 - 2) = 0$ $\therefore$ $x = 0, \pm\sqrt{2}$.

Since $y = -x$, the stationary points are $(0, 0)$, $(\sqrt{2}, -\sqrt{2})$ and $(-\sqrt{2}, \sqrt{2})$.

Now $\quad rt - s^2 = (12x^2 - 4)(12y^2 - 4) - 16$

At (0, 0), $\quad rt - s^2 = 16 - 16 = 0.$

$\therefore$ At (0, 0), further investigation is needed.

Now $\quad f(0, 0) = 0$

$$f(x, y) = x^4 + y^4 - 2(x^2 - 2xy + y^2) = x^4 + y^4 - 2(x - y)^2$$

When *h* and *k* are small, $f(h, k) = 2h^4 > 0$ for $h = k$

and $\quad f(h, k) = -2(h - k)^2 < 0$ for $h \neq k$ (neglecting $h^4 + k^4$ which is very small)

$\Rightarrow \quad f(0, 0) < f(h, k) \quad$ for $h = k$

$\quad f(0, 0) > f(h, k) \quad$ for $h \neq k$

$\Rightarrow$ There is no extreme value at (0, 0).

At $(\sqrt{2}, -\sqrt{2})$, $\quad rt - s^2 = (12 \times 2 - 4)^2 - 16 = 384 > 0$

$$r = 12 \times 2 - 4 = 20 > 0$$

$\therefore$ $f(x, y)$ has a minimum value at $(\sqrt{2}, -\sqrt{2})$.

Minimum Value $= (\sqrt{2})^4 + (-\sqrt{2})^4 - 2(\sqrt{2})^2 + 4\sqrt{2}\,(-\sqrt{2}) - 2(-\sqrt{2})^2$

$$= 4 + 4 - 4 - 8 - 4 = -8.$$

Hence, $f(x, y)$ has a minimum value -8 at $(-\sqrt{2}, \sqrt{2})$.

Example 4. *Examine for minimum and maximum values: sin x + sin y + sin (x + y).*

Sol. Here $f(x, y) = \sin x + \sin y + \sin (x + y)$

$$f_x = \cos x + \cos (x + y), f_y = \cos y + \cos (x + y)$$

$$r = f_{xx} = -\sin x - \sin (x + y), s = f_{xy} = -\sin (x + y),$$

$$t = f_{yy} = -\sin y - \sin (x + y)$$

Now $\quad f_x = 0$ and $f_y = 0$

$\Rightarrow \quad \cos x + \cos (x + y) = 0 \quad$ ...(1) $\quad \cos y + \cos (x + y) = 0 \quad$ ...(2)

Subtracting (2) from (1), $\cos x - \cos y = 0$ or $\cos x = \cos y$ $\therefore$ $x = y$

From (1), $\cos x + \cos 2x = 0$ or $\cos 2x = -\cos x = \cos (\pi - x)$ or $2x = \pi - x$

$\therefore \quad x = \dfrac{\pi}{3}$

$\therefore \quad x = y = \dfrac{\pi}{3}$ is a stationary point.

At $\left(\dfrac{\pi}{3}, \dfrac{\pi}{3}\right)$, $\quad r = -\dfrac{\sqrt{3}}{2} - \dfrac{\sqrt{3}}{2} = -\sqrt{3} < 0$, $\quad s = \dfrac{\sqrt{3}}{2}$, $\quad t = -\dfrac{\sqrt{3}}{2} - \dfrac{\sqrt{3}}{2} = -\sqrt{3}$

or $\quad rt - s^2 = 3 - \dfrac{3}{4} = \dfrac{9}{4} > 0$

Also $\quad r < 0$

$\therefore$ $f(x, y)$ has a maximum value at $\left(\frac{\pi}{3}, \frac{\pi}{3}\right)$.

Maximum value $= f\left(\frac{\pi}{3}, \frac{\pi}{3}\right) = \sin \frac{\pi}{3} + \sin \frac{\pi}{3} + \sin \frac{2\pi}{3} = \frac{\sqrt{3}}{2} + \frac{\sqrt{3}}{2} + \frac{\sqrt{3}}{2} = \frac{3\sqrt{3}}{2}$.

Example 5. *The temperature T at any point (x, y, z) in space is T = 400 xyz². Find the highest temperature on the surface of the unit sphere $x^2 + y^2 + z^2 = 1$.* **(P.T.U., Dec. 2011)**

Sol. Given $T = 400\, xyz^2$

and $x^2 + y^2 + z^2 = 1 \quad \therefore \quad z^2 = 1 - x^2 - y^2$

$$T = 400\, xy\,(1 - x^2 - y^2)$$

For extreme values $T_x = 0, T_y = 0$

$\therefore$ $400y\,(1 - 3x^2 - y^2) = 0$ and $400x\,\{1 - x^2 - 3y^2\} = 0$

$400y\,(1 - 3x^2 - y^2) = 0$ gives either $y = 0$ or $1 - 3x^2 - y^2 = 0$

$400x\,(1 - x^2 - 3y^2) = 0$ gives either $x = 0$ or $1 - x^2 - 3y^2 = 0$

$\therefore$ Stationary points are $(0, 0)$; $\left(\pm\frac{1}{2}, \pm\frac{1}{2}\right)$; $(\pm 1, 0)$; $(0, \pm 1)$

$$r = T_{xx} = 400y(-6x) = -2400\, xy$$
$$s = T_{xy} = 400[1 - 3x^2 - 3y^2]$$
$$t = T_{yy} = 400x\,[-6y] = -2400xy$$

At (0, 0); $r = 0, s = 400, t = 0; rt - s^2 = 0 - 400 < 0 \quad \therefore$ (0, 0) is not an extreme point

At (± 1, 0); $r = 0, s = -800, t = 0; rt - s^2 < 0 \quad \therefore$ $(\pm 1, 0)$ is not an extreme point

At (0, ± 1); $r = 0, s = -800, t = 0 \quad \therefore \quad rt - s^2 < 0 \quad \therefore$ $(0, \pm 1)$ is not an extreme point

At $\left(\frac{1}{2}, \frac{1}{2}\right)$; $r = -600, s = -200, t = -600 \quad \therefore \quad rt - s^2 = 32000 > 0$ and $r < 0$

$\therefore$ T is maximum at $\left(\frac{1}{2}, \frac{1}{2}\right)$

At $\left(\frac{1}{2}, -\frac{1}{2}\right)$; $r = 600, s = -200, t = 600$; $rt - s^2 = 32000 > 0$ but $r > 0$

$\therefore$ T is minimum at $\left(\frac{1}{2}, -\frac{1}{2}\right)$

At $\left(-\frac{1}{2}, \frac{1}{2}\right)$; $r = 600, s = -200, t = 600$

$$rt - s^2 = 32000 > 0 \text{ and } r > 0$$

$\therefore$ T is minimum at $\left(-\frac{1}{2}, \frac{1}{2}\right)$

At $\left(-\frac{1}{2}, -\frac{1}{2}\right)$; $r = -600, s = -200, t = -600$

$$rt - s^2 = 32000 > 0 \text{ but } r < 0$$

$\therefore$ T is maximum at $\left(-\frac{1}{2}, -\frac{1}{2}\right)$

$\therefore$ T has minimum value at $\left(\frac{1}{2}, \frac{1}{2}\right)$ and $\left(-\frac{1}{2}, \frac{1}{2}\right)$

and temperature is highest at $\left(\frac{1}{2}, \frac{1}{2}\right)$ and at $\left(\frac{-1}{2}, \frac{-1}{2}\right)$ and the highest value of temperature at each point is:

$$400 \cdot \frac{1}{4}\left(1 - \frac{1}{4} - \frac{1}{4}\right) = 100\left(\frac{1}{2}\right) = 50.$$

Example 6. *A rectangular box, open at the top, is to have a volme of 32 cubic feet. Find the dimensions of the box requiring least material for its construction.*

(P.T.U., May 2008, May 2012)

Sol. Let x, y and z be the length, breadth and height respectively. Let V be the given capacity and S, the surface.

Given V = 32 cubic feet

$$V = xyz \text{ or } z = \frac{V}{xy} = \frac{32}{xy} \qquad \text{...(1)}$$

$$S = xy + 2xz + 2yz = xy + \frac{64}{y} + \frac{64}{x} \qquad \text{(by using (1))}$$

$$= f(x, y)$$

$$fx = \frac{\partial S}{\partial x} = y - \frac{64}{x^2}$$

$$fy = \frac{\partial S}{\partial y} = x - \frac{64}{y^2}$$

$$r = fxx = \frac{128}{x^3},\ s = fxy = 1,\ t = f_{yy} = \frac{128}{y^3}$$

For extreme values $\quad f_x = 0, f_y = 0$

$$\therefore \quad y - \frac{64}{x^2} = 0 \qquad \text{...(2)}$$

and

$$x - \frac{64}{y^2} = 0 \qquad \text{...(3)}$$

Solving (2) and (3), we get $x = 4, y = 4$

$\therefore$ (4, 4) is a stationary point

At (4, 4); $\quad r = 2, s = 1, t = 2$

$$rt - s^2 = 4 - 1 = 3 > 0 \text{ and } r > 0$$

$\therefore$ S is minimum at $x = 4, y = 4$

$\therefore$ Dimensions of the box for least surface area are

$$x = 4, y = 4 \quad \text{and} \quad z = \frac{32}{16} = 2 \text{ i.e., } x = 4, y = 4, z = 2.$$

Example 7. *Prove that if the perimeter of a triangle is constant, its area is maximum when the triangle is equilateral.*

Sol. Let a, b, c be the sides of a triangle whose perimeter $2s$ is constant.

Then $2s = a + b + c$ or $c = 2s - a - b$ ...(1)

$$\Delta = \sqrt{s(s-a)(s-b)(s-c)} = \sqrt{s(s-a)(s-b)(a+b-s)} \quad | \text{Using (1)}$$

Let $z = \Delta^2 = s(s-a)(s-b)(a+b-s) = f(a, b)$

$$f_a = s(s-b)\frac{\partial}{\partial a}[(s-a)(a+b-s)]$$

$$= s(s-b)[-(a+b-s)+(s-a)] = s(s-b)(2s-2a-b)$$

$$f_b = s(s-a)\frac{\partial}{\partial b}[(s-b)(a+b-s)]$$

$$= s(s-a)[-(a+b-s)+(s-b)] = s(s-a)(2s-a-2b)$$

$$r = f_{aa} = -2s(s-b),$$

$$s = f_{ab} = s[-(2s-a-2b)-(s-a)] = s(2a+2b-3s)$$

$$t = f_{bb} = -2s(s-a)$$

Now $f_a = 0$ and $f_b = 0$

$\Rightarrow$ $s(s-b)(2s-2a-b) = 0$ and $s(s-a)(2s-a-2b) = 0$

$(s-b)(2s-2a-b) = 0$ ...(1)

$(s-a)(2s-a-2b) = 0$ ...(2) $|\because s \neq 0$

From (1), $s = b$ or $2s = 2a + b$

When $s = b$, from (2), $(b-a)(-a) = 0$ or $b = a$ $|\because a \neq 0$

when $2s = 2a + b$, from (2), $\frac{b}{2}(a-b) = 0$ or $a = b$ $|\because b \neq 0$

If we express z as a function of b and c, we similarly get $b = c$

$$\therefore \quad a = b = c = \frac{2s}{3}$$

$$r = -2s\left(\frac{s}{3}\right) = -\frac{2s^2}{3} < 0$$

$$\text{`}s\text{'} = s\left(\frac{4s}{3} + \frac{4s}{3} - 3s\right) = s\left(-\frac{s}{3}\right) = -\frac{s^2}{3}$$

$$t = -2s\left(\frac{s}{3}\right) = -\frac{2s^2}{3}$$

$$\text{`}rt - s^2\text{'} = \frac{4s^4}{9} - \frac{s^4}{9} = \frac{s^4}{3} > 0.$$

Also $r < 0$

$\therefore$ z and hence Δ is maximum when $a = b = c = \frac{2s}{3}$ *i.e.*, when the triangle is equilateral.

Example 8. *Prove that all rectangular parallelopiped with same volume, the cube has the least surface.*

Sol. Let x, y, z be the length, breadth and height of the rectangular parallelopiped then

$$V = xyz \quad \text{(constant)} \qquad ...(1)$$

$$S = xy + yz + zx = xy + (y + x)\frac{V}{xy} = xy + V\left(\frac{1}{x} + \frac{1}{y}\right)$$

For extreme values $\frac{\partial S}{\partial x} = 0, \frac{\partial S}{\partial y} = 0$

$$\frac{\partial S}{\partial x} = y - \frac{V}{x^2};\ \frac{\partial S}{\partial y} = x - \frac{V}{y^2}$$

$$\frac{\partial S}{\partial x} = 0, \frac{\partial S}{\partial y} = 0 \Rightarrow y - \frac{V}{x^2} = 0, x - \frac{V}{y^2} = 0$$

or
$$\left.\begin{matrix} x^2y = V \\ xy^2 = V \end{matrix}\right\} \Rightarrow x\frac{V^2}{x^4} = V \text{ or } x^3 = V \text{ or } x = V^{1/3}$$

and
$$y = V^{1/3}$$

$\therefore$ Stationary point is $(V^{1/3}, V^{1/3})$

$$r = \frac{\partial^2 S}{\partial x^2} = \frac{2V}{x^3}, s = \frac{\partial^2 S}{\partial x \partial y} = 1, t = \frac{\partial^2 S}{\partial y^2} = \frac{2V}{y^3}$$

$$rt - s^2 = \frac{4V^2}{x^3y^3} - 1$$

At $(V^{1/3}, V^{1/3})$; $\quad rt - s^2 = \frac{4V^2}{V \cdot V} - 1 = 4 - 1 = 3 > 0$ and $r > 0$

$\therefore$ S is minimum at $x = V^{1/3}, y = V^{1/3}$

From (1), $z = \frac{V}{xy} = \frac{V}{V^{1/3} \cdot V^{1/3}} = V^{1/3}$

$\therefore$ S is minimum when $x = y = z = V^{1/3}$

$\therefore$ Rectangular parallelopiped is a cube.

Example 9. *Show that the rectangular solid of maximum volume that can be inscribed in a given sphere is a cube.* **(P.T.U., May 2003)**

Sol. Let x, y, z be the length, breadth and height of the rectangular solid. If V is the volume of the solid, then

$$V = xyz \qquad ...(1)$$

Since each diagonal of the solid passes through the centre of the sphere

$\therefore$ each diagonal = diameter of sphere = d (say)

$$\Rightarrow \sqrt{x^2 + y^2 + z^2} = d \text{ or } x^2 + y^2 + z^2 = d^2 \text{ or } z = \sqrt{d^2 - x^2 - y^2}$$

$\therefore$ From (1), $\quad V = xy\sqrt{d^2 - x^2 - y^2}$

or
$$V^2 = x^2y^2(d^2 - x^2 - y^2) = d^2x^2y^2 - x^4y^2 - x^2y^4 = f(x, y)$$

$$f_x = 2d^2xy^2 - 4x^3y^2 - 2xy^4 = 2xy^2(d^2 - 2x^2 - y^2)$$

$$f_y = 2d^2x^2y - 2x^4y - 4x^2y^3 = 2x^2y\,(d^2 - x^2 - 2y^2)$$
$$r = f_{xx} = 2d^2y^2 - 12x^2y^2 - 2y^4,\ s = f_{xy} = 4d^2xy - 8x^3y - 8xy^3$$
$$t = f_{yy} = 2d^2x^2 - 2x^4 - 12x^2y^2$$

Now $\quad f_x = 0$ and $f_y = 0$

$\Rightarrow \quad d^2 - 2x^2 - y^2 = 0 \quad$ ...(1) [$\because\ x \neq 0, y \neq 0$]

$d^2 - x^2 - 2y^2 = 0 \quad$ ...(2)

Subtracting (2) from (1),

$$-x^2 + y^2 = 0 \Rightarrow y = x$$

$\therefore$ From (1), $\quad x = y = \dfrac{d}{\sqrt{3}}$

$$z = \sqrt{d^2 - \frac{d^2}{3} - \frac{d^2}{3}} = \frac{d}{\sqrt{3}}$$

At $x = y = \dfrac{d}{\sqrt{3}}$,

$$r = 2d^2 \cdot \frac{d^2}{3} - 12 \cdot \frac{d^2}{3} \cdot \frac{d^2}{3} - 2 \cdot \frac{d^4}{9} = -\frac{8d^4}{9} < 0$$

$$s = 4d^2 \cdot \frac{d^2}{3} - 8 \cdot \frac{d^3}{3\sqrt{3}} \cdot \frac{d}{\sqrt{3}} - 8 \cdot \frac{d}{\sqrt{3}} \cdot \frac{d^3}{3\sqrt{3}} = -\frac{4d^4}{9}$$

$$t = 2d^2 \cdot \frac{d^2}{3} - 2 \cdot \frac{d^4}{9} - 12 \cdot \frac{d^2}{3} \cdot \frac{d^2}{3} = -\frac{8d^4}{9}$$

$$rt - s^2 = \frac{64d^8}{81} - \frac{16d^8}{81} = \frac{16d^8}{27} > 0, \text{ Also } r < 0.$$

$\therefore$ V^2 and hence, V is maximum when $x = y = z$ *i.e.*, when the rectangular solid is a cube.

Example 10. *Find the volume of the largest rectangular parallelopiped that can be inscribed in the ellipsoid,* $\dfrac{x^2}{a^2} + \dfrac{y^2}{b^2} + \dfrac{z^2}{c^2} = 1.$ **(P.T.U., May 2007, 2014)**

Sol. Let (x, y, z) be a vertex of the parallelopiped,

then it lies on the ellipsoid $\dfrac{x^2}{a^2} + \dfrac{y^2}{b^2} + \dfrac{z^2}{c^2} = 1 \quad$ ...(1)

As the rectangular parallelopiped inscribed in the ellipsoid is the largest

$\therefore$ (0, 0, 0) is also the centre of the parallelopiped.

If (x, y, z) are the coordinates of one of the vertices of the parallelopiped. Then coordinates of its opposite vertex is $(-x, -y, -z)$

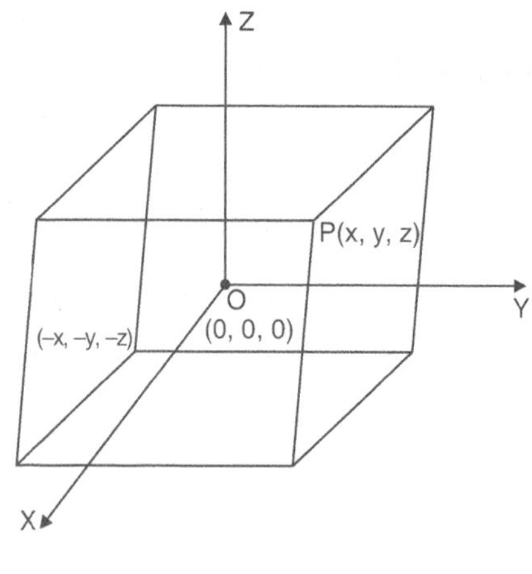

$\therefore$ Sides of the parallelopiped are of dimensions $2x$, $2y$, $2z$

$\therefore$ Volume $\quad V = 2x \cdot 2y \cdot 2z = 8xyz.$

$$\Rightarrow \quad V^2 = 64x^2y^2z^2 = 64x^2y^2c^2\left(1 - \frac{x^2}{a^2} - \frac{y^2}{b^2}\right) = 64c^2\left(x^2y^2 - \frac{x^4y^2}{a^2} - \frac{x^2y^4}{b^2}\right) = f(x, y)$$

$$f_x = 64c^2\left(2xy^2 - \frac{4x^3y^2}{a^2} - \frac{2xy^4}{b^2}\right); \qquad f_y = 64c^2\left(2x^2y - \frac{2x^4y}{a^2} - \frac{4x^2y^3}{b^2}\right)$$

$$r = f_{xx} = 64c^2\left(2y^2 - \frac{12x^2y^2}{a^2} - \frac{2y^4}{b^2}\right); \qquad s = f_{xy} = 64c^2\left(4xy - \frac{8x^3y}{a^2} - \frac{8xy^3}{b^2}\right)$$

$$t = f_{yy} = 64c^2\left(2x^2 - \frac{2x^4}{a^2} - \frac{12x^2y^2}{b^2}\right)$$

Now $\quad f_x = 0$ and $f_y = 0$

$$\Rightarrow \quad 128c^2xy^2\left(1 - \frac{2x^2}{a^2} - \frac{y^2}{b^2}\right) = 0$$

and

$$128c^2x^2y\left(1 - \frac{x^2}{a^2} - \frac{2y^2}{b^2}\right) = 0$$

$$\Rightarrow \quad 1 - \frac{2x^2}{a^2} - \frac{y^2}{b^2} = 0 \qquad \ldots(1) \quad [\because \; x \neq 0, y \neq 0]$$

$$1 - \frac{x^2}{a^2} - \frac{2y^2}{b^2} = 0 \qquad \ldots(2)$$

Subtracting (2) from (1), $\quad -\frac{x^2}{a^2} + \frac{y^2}{b^2} = 0$ or $y = \frac{bx}{a}$

$\therefore$ From (1), $\quad 1 - \frac{2x^2}{a^2} - \frac{x^2}{a^2} = 0$ or $x^2 = \frac{a^2}{3}$ or $x = \frac{a}{\sqrt{3}}$

$$\therefore \quad y = \frac{b}{a} \cdot \frac{a}{\sqrt{3}} = \frac{b}{\sqrt{3}}$$

and

$$z^2 = c^2\left(1 - \frac{x^2}{a^2} - \frac{y^2}{b^2}\right) = c^2\left(1 - \frac{1}{3} - \frac{1}{3}\right) = \frac{c^2}{3}$$

$$\therefore \quad z = \frac{c}{\sqrt{3}}$$

Thus $x = \frac{a}{\sqrt{3}}$, $y = \frac{b}{\sqrt{3}}$ is a stationary point.

At this point $\left(\frac{a}{\sqrt{3}}, \frac{b}{\sqrt{3}}\right)$, $r = 64c^2\left[\frac{2b^2}{3} - \frac{12}{a^2}\cdot\frac{a^2}{3}\cdot\frac{b^2}{3} - \frac{2}{b^2}\cdot\frac{b^4}{9}\right] = -\frac{512}{9}b^2c^2 < 0$

$$s = 64c^2\left[4\cdot\frac{a}{\sqrt{3}}\cdot\frac{b}{\sqrt{3}} - \frac{8}{a^2}\cdot\frac{a^3}{3\sqrt{3}}\cdot\frac{b}{\sqrt{3}} - \frac{8}{b^2}\cdot\frac{a}{\sqrt{3}}\cdot\frac{b^3}{3\sqrt{3}}\right] = -\frac{256}{9}abc^2$$

$$t = 64c^2\left[2\cdot\frac{a^2}{3} - \frac{2}{a^2}\cdot\frac{a^4}{9} - \frac{12}{b^2}\cdot\frac{a^2}{3}\cdot\frac{b^2}{3}\right] = -\frac{512}{9}a^2c^2$$

$$rt - s^2 = \left(\frac{512}{9}\right)^2 a^2b^2c^4 - \left(\frac{256}{9}\right)^2 a^2b^2c^4 = \left(\frac{256}{9}\right)^2 a^2b^2c^4(4-1) > 0.$$

Also $\quad r < 0$

$\therefore\ V^2$ and hence V is maximum when $x = \frac{a}{\sqrt{3}}, y = \frac{b}{\sqrt{3}}, z = \frac{c}{\sqrt{3}}$

Maximum volume $= 8\cdot\frac{a}{\sqrt{3}}\cdot\frac{b}{\sqrt{3}}\cdot\frac{c}{\sqrt{3}} = \frac{8abc}{3\sqrt{3}}$.

Example 11. *A tent on a square base of side x has its sides vertical of height y and the top is regular pyramid of height h. Find x and y in terms of h, if the canvas required for its construction is to be minimum for the tent to have a given capacity.*

Sol. Volume of the tent-Volume of the rectangular parallopiped + Volume of the top

$$V = x^2y + \frac{1}{3}x^2z \qquad \text{...(1)}$$

where z is the height of the pyramid (given $z = h$)

Surface area of the tent = Surface area of the rectangular parallopiped + area of the top *i.e.* area of four triangles.

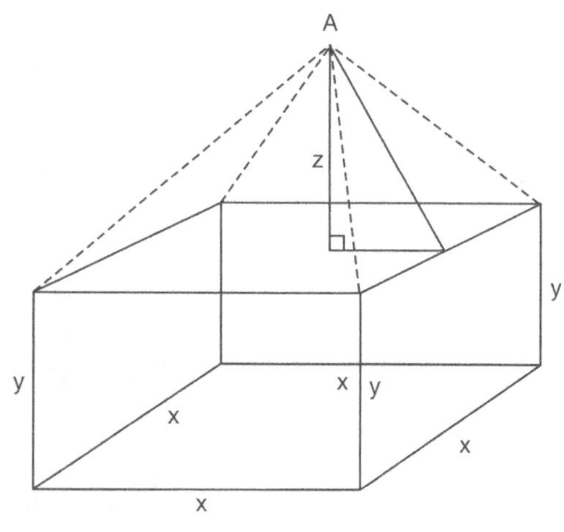

$$S = 4xy + 4\left[\frac{1}{2}xh'\right] \text{ where } h'^2 = z^2 + \left(\frac{x}{2}\right)^2$$

$$\therefore \quad S = 4xy + 2x\sqrt{z^2 + \frac{x^2}{4}}$$

$$= 4x\left[\frac{V - \frac{x^2z}{3}}{x^2}\right] + x\sqrt{z^2 + x^2}$$

$$\because \text{ from (1), } y = \frac{V - \frac{1}{3}x^2z}{x^2}$$

$$= \frac{4V}{x} - \frac{4}{3}xz + x\sqrt{z^2 + x^2}$$

For extreme values $\frac{\partial S}{\partial x} = 0,\ \frac{\partial S}{\partial z} = 0,$

$$\frac{\partial S}{\partial x} = -\frac{4V}{x^2} - \frac{4}{3}z + x \cdot \frac{1}{2\sqrt{4z^2 + x^2}}(2x) + \sqrt{4z^2 + x^2}$$

$$= -\frac{4V}{x^2} - \frac{4}{3}z + \frac{2x^2 + 4z^2}{\sqrt{4z^2 + x^2}}$$

$$\therefore \quad \frac{\partial S}{\partial x} = 0 \text{ given } -\frac{4V}{x^2} - \frac{4}{3}z + \frac{2x^2 + 4z^2}{\sqrt{4z^2 + x^2}} = 0 \qquad ...(2)$$

$$\frac{\partial S}{\partial z} = -\frac{4}{3}x + x \cdot \frac{1}{2\sqrt{4z^2 + x^2}} 8z = -\frac{4x}{3} + \frac{4xz}{\sqrt{4z^2 + x^2}}$$

$$\therefore \quad \frac{\partial S}{\partial z} = 0 \text{ given } -\frac{4x}{3} + \frac{4xz}{\sqrt{4z^2 + x^2}} = 0 \quad \text{or} \quad 4z^2 + x^2 = 9z^2 \quad \text{or} \quad 5z^2 = x^2$$

$$\therefore \quad x = \sqrt{5}\, z$$

Substituting the value of x in (2),

$$\frac{4V}{5z^2} + \frac{4}{3}z = \frac{10z^2 + 4z^2}{\sqrt{4z^2 + 5z^2}}$$

or
$$\frac{4V}{5z^2} + \frac{4}{3}z = \frac{14z^2}{3z} = \frac{14}{3}z$$

or
$$\frac{4V}{5z^2} = \frac{14}{3}z - \frac{4}{3}z = \frac{10}{3}z \qquad \therefore \quad z^3 = \frac{6}{25}V \qquad \text{or} \quad V = \frac{25}{6}z^3$$

Substituting the values of V and x^2 in (1),

$$\frac{25}{6}z^3 = 5z^2\left[y + \frac{z}{3}\right]$$

or
$$\frac{5}{6}z = y + \frac{z}{3} \quad \text{or} \quad y = \left(\frac{5}{6} - \frac{1}{3}\right)z = \frac{3}{6}z = \frac{1}{2}z$$

$$\therefore \quad x = \sqrt{5}\, z \quad \text{and} \quad y = \frac{1}{2}z \qquad ...(3)$$

$$r = \frac{\partial^2 S}{\partial x^2} = \frac{\partial}{\partial x}\left[-\frac{4V}{x^2} - \frac{4}{3}z + \frac{2x^2 + 4z^2}{\sqrt{4z^2 + x^2}}\right]$$

$$= \frac{8V}{x^3} + \frac{\sqrt{4z^2 + x^2} \cdot 4x - (2x^2 + 4z^2)\dfrac{1}{2\sqrt{4z^2 + x^2}}(2x)}{4z^2 + x^2}$$

$$r = \frac{8V}{x^3} + \frac{4x(4z^2 + x^2) - x(2x^2 + 4z^2)}{(4z^2 + x^2)^{3/2}}$$

$$r = \frac{8V}{x^3} + x \cdot \frac{12z^2 + 2x^2}{(4z^2 + x^2)^{3/2}}$$

At $x = \sqrt{5z}$, $\quad V = \frac{25}{6}z^3$

$$r = \frac{8}{5\sqrt{5}z^3}\,\frac{25}{6}z^3 + \sqrt{5}z\,\frac{12z^2 + 10z^2}{(4z^2 + 5z^2)^{3/2}}$$

$$= \frac{20}{3\sqrt{5}} + \frac{\sqrt{5}\,(22)z^3}{27z^3} = \frac{20}{3\sqrt{5}} + \frac{22\sqrt{5}}{27} = \frac{180 + 110}{27\sqrt{5}}$$

$$= \frac{290}{27\sqrt{5}} = \frac{58\sqrt{5}}{27}$$

$$s = \frac{\partial^2 S}{\partial x \partial z} = \frac{\partial}{\partial x}\left\{-\frac{4x}{3} + \frac{4xz}{\sqrt{4z^2 + x^2}}\right\}$$

$$= -\frac{4}{3} + 4z\,\frac{\sqrt{4z^2 + x^2}\cdot 1 - x\,\dfrac{1}{2\sqrt{4z^2 + x^2}}\cdot 2x}{4z^2 + x^2}$$

$$= -\frac{4}{3} + 4z\,\frac{4z^2 + x^2 - x^2}{\left(4z^2 + x^2\right)^{3/2}}$$

$$= -\frac{4}{3} + \frac{16z^3}{\left(4z^2 + x^2\right)^{3/2}}$$

Substitute $x^2 = 5z^2$

$$s = -\frac{4}{3} + \frac{16z^3}{27z^3} = -\frac{20}{27}$$

$$t = \frac{\partial^2 s}{\partial z^2} = \frac{\partial}{\partial z}\left[-\frac{4x}{3} + \frac{4xz}{\sqrt{4z^2 + x^2}}\right]$$

$$= 4x\,\frac{\sqrt{4z^2 + x^2}\cdot 1 - z\,\dfrac{1}{2\sqrt{4z^2 + x^2}}(8z)}{4z^2 + x^2}$$

$$= 4x\,\frac{4z^2 + x^2 - 4z^2}{(4z^2 + x^2)^{3/2}} = \frac{4x^3}{(4z^2 + x^2)^{3/2}}$$

Put $x^2 = 5z^2$, $\quad t = \dfrac{4.5\sqrt{5}\,z^3}{27z} = \dfrac{20\sqrt{5}}{27}$

$$rt - s^2 = \frac{58\sqrt{5}}{27}\cdot\frac{20\sqrt{5}}{27} - \left(\frac{20}{27}\right)^2 > 0$$

r is also > 0

$\therefore$ S is minimum and the values of x and y are

$$x = \sqrt{5}z,\ y = \frac{1}{2}z$$

Given $z = h$ $\therefore$ $\quad x = \sqrt{5}\,h,\ y = \dfrac{1}{2}h.$

Example 12. *In a plane triangle ABC, find the maximum value of cos A cos B cos C.*

Sol. $\cos A \cos B \cos C = \cos A \cos B \cos [\pi - (A + B)]$ $\quad | \because \; A + B + C = \pi$

$$= -\cos A \cos B \cos (A + B) = f(A, B)$$

$$\frac{\partial f}{\partial A} = -\cos B\,[-\sin A \cos (A + B) - \cos A \sin (A + B)]$$

$$= \cos B \sin [A + (A + B)] = \cos B \sin (2A + B)$$

$$\frac{\partial f}{\partial B} = -\cos A[-\sin B \cos (A + B) - \cos B \sin (A + B)]$$

$$= \cos A \sin [B + (A + B)] = \cos A \sin (A + 2B)$$

$$r = 2 \cos B \cos (2A + B)$$

$$s = -\sin A \sin (A + 2B) + \cos A \cos (A + 2B)$$

$$= \cos [A + (A + 2B)] = \cos (2A + 2B)$$

$$t = 2 \cos A \cos (A + 2B)$$

Now $\quad \dfrac{\partial f}{\partial A} = 0 \quad \text{and} \quad \dfrac{\partial f}{\partial B} = 0$

$\Rightarrow \quad \cos B \sin (2A + B) = 0$ ...(1)

$\quad\; \cos A \sin (A + 2B) = 0$ ...(2)

If $\quad \cos B = 0, \quad$ then $\quad B = \dfrac{\pi}{2}$

From (2), $\quad \cos A \sin (A + \pi) = 0 \quad$ or $\quad \cos A(-\sin A) = 0$

$\Rightarrow$ either $\cos A = 0$ *i.e.*, $A = \dfrac{\pi}{2}$ which is not possible

$(\because \; A + B + C = \pi \;\Rightarrow\; C = 0)$

or $\quad \sin A = 0$ *i.e.*, $A = 0$ or π which is not possible.

$\therefore \quad \cos B \neq 0.$

Similarly, $\cos A \neq 0$

$\therefore$ From (1), $\quad \sin (2A + B) = 0 \quad$ or $\quad 2A + B = \pi$

From (2), $\quad \sin (A + 2B) = 0 \quad$ or $\quad A + 2B = \pi$

Solving these equations,

$$A = B = \frac{\pi}{3}$$

At $A = B = \dfrac{\pi}{3}$,

$$r = 2 \cos \frac{\pi}{3} \cos \pi = -1,$$

$$s = \cos \frac{4\pi}{3} = \cos \left(\pi + \frac{\pi}{3}\right) = -\cos \pi = -\frac{1}{2}$$

$$t = 2 \cos \frac{\pi}{3} \cos \pi = -1$$

$$rt - s^2 = 1 - \frac{1}{4} = \frac{3}{4} > 0$$

Also $\qquad r = -1 < 0$

$\Rightarrow$ $f(A, B)$ is maximum at $A = B = \dfrac{\pi}{3}$

$$\text{Maximum value} = f\left(\frac{\pi}{3}, \frac{\pi}{3}\right) = -\cos\frac{\pi}{3}\cos\frac{\pi}{3}\cos\frac{2\pi}{3} = -\left(\frac{1}{2}\right)\left(\frac{1}{2}\right)\left(-\frac{1}{2}\right) = \frac{1}{8}.$$

Example 13. *The sum of three positive numbers is constant. Prove that their product is maximum when they are equal.* **(P.T.U., Dec. 2006)**

Sol. Let the three positive numbers be x, y, z

Given $x + y + z = k$ (constant)

Their product $\qquad P = xyz = xy\,(k - x - y)$

Let $\qquad f(x, y) = P = xy(k - x - y)$

$$f_x = y\{x(-1) + (k - x - y)\cdot 1\} = y[k - 2x - y]$$

$$f_y = x\{y(-1) + (k - x - y)\cdot 1\} = x[k - x - 2y]$$

For extreme values $\qquad f_x = 0, f_y = 0$

$\therefore \qquad y(k - 2x - y) = 0 \;\Rightarrow\;$ either $y = 0$ or $k - 2x - y = 0$

$\qquad\qquad x(k - x - 2y) = 0 \;\Rightarrow\;$ either $x = 0$ or $k - x - 2y = 0$

i.e. $\quad$ either $x = 0,\; y = 0$ or $2x + y - k = 0, x + 2y - k = 0$ which gives $x = \dfrac{k}{3}, y = \dfrac{k}{3}$

$\therefore$ Stationary points are $(0, 0)$ and $\left(\dfrac{k}{3}, \dfrac{k}{3}\right)$

Now
$$r = f_{xx} = -2y,$$
$$s = f_{xy} = y(-1) + (k - 2x - y)\,1 = k - 2x - 2y$$
$$t = f_{yy} = -2x$$
$$rt - s^2 = 4xy - (k - 2x - 2y)^2$$

At $(0, 0)$; $\qquad rt - s^2 = -k^2 < 0 \quad \therefore \quad f(x, y)$ has no extreme value at $(0, 0)$

At $\left(\dfrac{k}{3}, \dfrac{k}{3}\right)$; $\qquad rt - s^2 = 4 - \dfrac{k^2}{9} - \left(k - \dfrac{2k}{3} - \dfrac{2k}{3}\right)^2$

$$= \frac{4k^2}{9} - \frac{k^2}{9} = \frac{3k^2}{9} = \frac{k^2}{3} > 0 \quad \text{and} \quad r = -\frac{2k}{3} < 0$$

$\therefore \quad f(x, y)$ has maximum value at $\left(\dfrac{k}{3}, \dfrac{k}{3}\right)$

$\therefore \quad$ Product P has maximum value at

$$x = \frac{k}{3}, y = \frac{k}{3} \quad \text{and} \quad z = k - x - y$$

$$= k - \frac{k}{3} - \frac{k}{3} = \frac{k}{3}$$

i.e. $$x = y = z = \frac{k}{3} \text{ (each)}$$

Hence product is maximum when the numbers are equal.

4.11. LAGRANGE'S METHOD OF UNDETERMINED MULTIPLIERS

(P.T.U., Dec. 2007)

To find the maximum or minimum values of a function of three (or more) variables, when the variables are not independent but are connected by some given relation, we try to convert the given function to the one, having least number of independent variables with the help of the given relation. [as is done in previous examples]

When this procedure is not practicable,we use Lagrange's method.

Let $f(x, y, z)$ be a function of x, y, z which is to be examined for maximum or minimum value.

Let the variables x, y, z be connected by the relation $\phi(x, y, z) = 0$ ...(1)

For $f(x, y, z)$ to have a maximum or minimum value, the necessary condition is

$$\frac{\partial f}{\partial x} = 0, \frac{\partial f}{\partial y} = 0, \frac{\partial f}{\partial z} = 0$$

$$\therefore \quad \frac{\partial f}{\partial x} dx + \frac{\partial f}{\partial y} dy + \frac{\partial f}{\partial z} dz = 0 \quad \text{...(2)}$$

Also, from (1), taking differentials, we get

$$\frac{\partial \phi}{\partial x} dx + \frac{\partial \phi}{\partial y} dy + \frac{\partial \phi}{\partial z} dz = 0 \quad \text{...(3)}$$

Multiplying (3) by a parameter λ and adding to (2), we get

$$\left(\frac{\partial f}{\partial x} + \lambda \frac{\partial \phi}{\partial x}\right) dx + \left(\frac{\partial f}{\partial y} + \lambda \frac{\partial \phi}{\partial y}\right) dy + \left(\frac{\partial f}{\partial z} + \lambda \frac{\partial \phi}{\partial z}\right) dz = 0$$

This equation will hold good if

$$\frac{\partial f}{\partial x} + \lambda \frac{\partial \phi}{\partial x} = 0, \frac{\partial f}{\partial y} + \lambda \frac{\partial \phi}{\partial y} = 0, \frac{\partial f}{\partial z} + \lambda \frac{\partial \phi}{\partial z} = 0 \quad \text{...(4)}$$

These equations together with equation (1), give the values of x, y, z and λ for a maximum or minimum.

Note. The above equations can be easily obtained by considering Lagrange's function

$$F(x, y, z) = f(x, y, z) + \lambda\phi(x, y, z) \qquad \text{(Called Auxiliary equation)}$$

and considering the stationary values of $F(x, y, z)$. For stationary values of $F(x, y, z)$, $dF = 0$

$$\Rightarrow \quad \left(\frac{\partial f}{\partial x} + \lambda \frac{\partial \phi}{\partial x}\right) dx + \left(\frac{\partial f}{\partial y} + \lambda \frac{\partial \phi}{\partial y}\right) dy + \left(\frac{\partial f}{\partial z} + \lambda \frac{\partial \phi}{\partial z}\right) dz = 0$$

$$\Rightarrow \quad \frac{\partial f}{\partial x} + \lambda \frac{\partial \phi}{\partial x} = 0, \frac{\partial f}{\partial y} + \lambda \frac{\partial \phi}{\partial y} = 0, \frac{\partial f}{\partial z} + \lambda \frac{\partial \phi}{\partial z} = 0.$$

4.12. ADVANTAGES AND DISADVANTAGES OF LAGRANGE'S METHOD OF UNDETERMINED MULTIPLIERS

(P.T.U., May 2010)

Advantages

(1) The stationary values of $f(x, y, z)$ can be determined from $\phi(x, y, z) = 0$ and $\frac{\partial f}{\partial x} + \lambda \frac{\partial \phi}{\partial x} = 0$, $\frac{\partial f}{\partial y} + \lambda \frac{\partial \phi}{\partial y} = 0$ and $\frac{\partial f}{\partial z} + \lambda \frac{\partial \phi}{\partial z} = 0$, even without determining x, y, z explicitly.

(2) This method can be extended to a function of several variables (say) $x_1, x_2, \cdots, x_n$ and subject to any number of constraints say (m)

Disadvantage

Lagrange's method does not enable us to find whether stationary point is maximum or minimum. To determine this fact further investigation are needed. Mainly; either physical consideration of the problem tells us about this fact or we have to find d^2F. For maximum value $d^2F < 0$ and for minimum value $d^2F > 0$. Of course this is a lengthy process.

Example 14. *Find the shortest and longest distances from the point (1, 2, –1) to the sphere* $x^2 + y^2 + z^2 = 24$. **(P.T.U. Dec. 2011, Dec. 2013)**

Sol. Let $P(x, y, z)$ be any point on the sphere $x^2 + y^2 + z^2 = 24$.

Distance of the point A(1, 2, –1) from $P(x, y, z)$

$$AP = \sqrt{(x-1)^2 + (y-2)^2 + (z+1)^2}$$

If the distance is maximum or minimum so will be the square of the distance

Let $f(x, y, z) = (x - 1)^2 + (y - 2)^2 + (z + 1)^2$ subject to the condition that

$$\phi(x, y, z) = x^2 + y^2 + z^2 - 24 = 0 \qquad ...(1)$$

Consider Lagrange's function

$$F(x, y, z) = f(x, y, z) + \lambda\phi(x, y, z)$$

$$= (x - 1)^2 + (y - 2)^2 + (z + 1)^2 + \lambda(x^2 + y^2 + z^2 - 24)$$

For stationary points $dF = 0$

i.e., $$\frac{\partial F}{\partial x} dx + \frac{\partial F}{\partial y} dy + \frac{\partial F}{\partial z} dz = 0$$

$$\Rightarrow \qquad \frac{\partial F}{\partial x} = 0, \frac{\partial F}{\partial y} = 0, \frac{\partial F}{\partial z} = 0$$

i.e., $$2(x - 1) + 2\lambda x = 0 \qquad ...(2)$$

$$2(y - 2) + 2\lambda y = 0 \qquad ...(3)$$

$$2(z + 1) + 2\lambda z = 0 \qquad ...(4)$$

Multiply (2) by x, (3) by y (4) by z and add

$$2(x^2 + y^2 + z^2 - x - 2y + z) + 2\lambda(x^2 + y^2 + z^2) = 0$$

or $$2(24 - x - 2y + z) + 2\lambda \cdot 24 = 0 \qquad \text{[Using (1)]}$$

or $$x + 2y - z = 24(1 + \lambda) \qquad ...(5)$$

From (2), (3), (4), we get

$$x = \frac{1}{1+\lambda}, y = \frac{2}{1+\lambda}, z = -\frac{1}{1+\lambda}$$

Substituting the values of x, y, z in (5), we get

$$\frac{1}{1+\lambda} + 2 \cdot \frac{2}{1+\lambda} + \frac{1}{1+\lambda} = 24(1 + \lambda)$$

$$6 = 24(1 + \lambda)^2$$

or $$(1 + \lambda)^2 = \frac{1}{4} \quad \text{or} \quad 1 + \lambda = \pm\frac{1}{2}$$

$$\therefore \qquad \lambda = -\frac{1}{2}, \lambda = -\frac{3}{2}$$

When $\lambda = -\frac{1}{2}$; $x = 2, \quad y = 4, \quad z = -2$

When $\lambda = -\frac{3}{2}$; $x = -2, \quad y = -4, z = 2$

We get two points P(2, 4, –2) and Q(–2, –4, 2) on the sphere

Now distance

$$AP = \sqrt{(2-1)^2 + (4-2)^2 + (-2+1)^2} = \sqrt{1+4+1} = \sqrt{6}$$

$$AQ = \sqrt{(-2-1)^2 + (-4-2)^2 + (2+1)^2} = \sqrt{9+36+9} = 3\sqrt{6}$$

$\therefore$ P(2, 4, –2) is at a minimum distance from A and minimum distance $= \sqrt{6}$.

Q(–2, –4, 2) is at a maximum distance from A and maximum distance $= 3\sqrt{6}$.

Example 15. *Find the minimum value of* $x^2 + y^2 + z^2$ *given that* $xyz = a^3$

(P.T.U., Jan. 2010)

Sol. Let $\quad u = x^2 + y^2 + z^2 \quad$ ...(1)

Subject to the condition

$$\phi(x, y, z) = xyz - a^3 = 0 \quad ...(2)$$

Consider Lagrange's function,

$$F(x, y, z) = u + \lambda\phi = (x^2 + y^2 + z^2) + \lambda(xyz - a^3)$$

For stationary points, $dF = 0$

i.e., $\frac{\partial F}{\partial x} dx + \frac{\partial F}{\partial y} dy + \frac{\partial F}{\partial z} dz = 0$

$$\frac{\partial F}{\partial x} = 2x + \lambda yz\,; \frac{\partial F}{\partial y} = 2y + \lambda zx\,; \frac{\partial F}{\partial z} = 2z + \lambda xy \quad ...(3)$$

$\therefore \quad (2x + \lambda yz)\, dx + (2y + \lambda zx)\, dy + (2z + \lambda xy)\, dz = 0$

$\Rightarrow \quad 2x + \lambda yz = 0 \quad$ ...(4)

$2y + \lambda zx = 0 \quad$ ...(5)

$2z + \lambda xy = 0 \quad$ ...(6)

Multiply (4) by x, (5) by y, (6) by z and add

$$2(x^2 + y^2 + z^2) + 3\lambda xyz = 0$$

or $\quad 2u + 3\lambda a^3 = 0$

or $\quad \lambda = -\frac{2u}{3a^3} \quad$ ...(7)

From (4), (5) and (6), we have

$$x = \frac{u}{3a^3} yz, y = \frac{u}{3a^3} zx, z = \frac{u}{3a^3} xy$$

From (2), $\frac{u}{3a^3} yz \cdot \frac{u}{3a^3} zx \cdot \frac{u}{3a^3} xy = a^3$

or $\quad \frac{u^3 (xyz)^2}{27a^9} = a^3$

or $\quad \frac{u^3 a^6}{27a^9} = a^3$

or $\quad u^3 = 27a^6$

or $\quad u = 3a^2 \quad$ ...(8)

This is the extreme value of u.

For maximum or minimum value we have to find d^2F

$$d^2F = F_{xx}(dx)^2 + F_{yy}(dy)^2 + F_{zz}(dz)^2 + 2F_{xy}\, dx\, dy + 2F_{yz}\, dy\, dz + 2F_{zx}\, dz\, dx$$
$$= 2(dx)^2 + 2(dy)^2 + 2(dz)^2 + 2\lambda z dx\, dy + 2\lambda x dy\, dz + 2\lambda y\, dz\, dx$$
$$= 2[(dx)^2 + (dy)^2 + (dz)^2] + 2\lambda[z dx\, dy + x dy\, dz + y dz\, dx]$$

From (3), $F_{xx} = F_{yy} = F_{zz} = 2$

$F_{xy} = \lambda z,\ F_{yz} = \lambda x\ F_{zx} = \lambda y$

Given $\phi(x, y, z) = xyz - a^3 = 0$

$$d^2\phi = 0$$

$\therefore\quad \phi_{xx}(dx)^2 + \phi_{yy}(dy)^2 + \phi_{zz}(dz)^2 + 2\phi_{xy}\, dx\, dy + 2\phi_{yz}\, dy\, dz + 2\phi_{zx}\, dz\, dx = 0$

or $\quad 0(dx)^2 + 0(dy)^2 + 0 \cdot (dz)^2 + 2[z dx dy + x dy dz + y dz dx = 0$

$[\phi_x = yz,\ \phi_y = zx,\ \phi_z = xy,\ \phi_{xx} = \phi_{yy} = \phi_{zz} = 0$

$\phi_{xy} = z;\ \phi_{yz} = x,\ \phi_{zx} = y]$

$\therefore\quad d^2F = 2[(dx)^2 + (dy)^2 + (dz)^2] + 0$

$\therefore\quad d^2F > 0$

Hence u is minimum and minimum value of $u = 3a^2$.

Example 16. *Find the shortest distance between the line $y = 10 - 2x$ and the ellipse $\frac{x^2}{4} + \frac{y^2}{9} = 1$.* **(P.T.U., Dec. 2004, 2005)**

Sol. Let $P(x, y)$ be any point on the line $y = 10 - 2x$ ...(1)

and $Q(u, v)$ be any point on the ellipse $\frac{x^2}{4} + \frac{y^2}{9} = 1$ ...(2)

Distance between P and Q $= \sqrt{(x-u)^2 + (y-v)^2}$

Distance PQ will be shortest if PQ^2 is shortest.

$\therefore$ Let $\quad f(x, y, u, v) = (x - u)^2 + (y - v)^2$

Subject to the conditions

$$\varphi_1(x, y) = 2x + y - 10 = 0 \qquad ...(3)$$

$$\varphi_2(u, v) = \frac{u^2}{4} + \frac{v^2}{9} - 1 = 0 \qquad ...(4)$$

Consider Lagrange's function

$$F(x, y, u, v, \varphi_1, \varphi_2) = (x-u)^2 + (y-v)^2 + \lambda_1(2x + y - 10) + \lambda_2\left(\frac{u^2}{4} + \frac{v^2}{9} - 1\right)$$

For extreme values, we have $dF = 0$

i.e., $$\frac{\partial F}{\partial x}dx + \frac{\partial F}{\partial y}dy + \frac{\partial F}{\partial u}du + \frac{\partial F}{\partial v}dv = 0$$

i.e., $$\frac{\partial F}{\partial x} = 0,\quad \frac{\partial F}{\partial y} = 0,\quad \frac{\partial F}{\partial u} = 0,\quad \frac{\partial F}{\partial v} = 0$$

i.e., $$2(x-u) + 2\lambda_1 = 0 \qquad ...(5)$$

$$2(y-v) + \lambda_1 = 0 \qquad ...(6)$$

$$-2(x-u) + \frac{u}{2}\lambda_2 = 0 \qquad ...(7)$$

$$-2(y-v)+\frac{2v}{9}\lambda_2 = 0 \qquad ...(8)$$

Eliminate λ_1, λ_2 from these equation

from (5) and (6) $\quad 2(x-u)-4(y-v)=0 \quad$ or $\quad x-u=2(y-v) \qquad ...(9)$

from (7) and (8)

$$-\frac{4v}{9}(x-u)+2\cdot\frac{u}{2}(y-v)=0 \quad \text{or} \quad 4v(x-u)=9u(y-v) \qquad ...(10)$$

Dividing (10) by (9), we get

$$4v = \frac{9u}{2}$$

$$\therefore \quad v = \frac{9}{8}u$$

from (4), $$\frac{u^2}{4}+\frac{81u^2}{64}\cdot\frac{1}{9}=1$$

$$\therefore \quad u^2\left[\frac{1}{4}+\frac{9}{64}\right]=1 \quad \therefore \quad u^2=\frac{64}{25} \qquad \therefore \quad u=\pm\frac{8}{5}$$

$$v=\frac{9}{8}\left(\pm\frac{8}{5}\right) \quad v=\pm\frac{9}{5} \qquad \therefore \quad u, v \text{ are obtained}$$

from (9), $\quad x-2y=u-2v$
from (3), $\quad 2x+y=10$ $\quad\Big]$ solve for x, y

$$y=\frac{-2u+4v+10}{5}, \quad x=\frac{u-2v+20}{5} \qquad ...(11)$$

For $$u=\frac{8}{5}, v=\frac{9}{5}; \quad y=\frac{14}{5}, \quad x=\frac{18}{5}$$

$\therefore$ when $$(u, v)=\left(\frac{8}{5},\frac{9}{5}\right); \quad (x, y)=\left(\frac{18}{5},\frac{14}{5}\right)$$

The distance between PQ $= \sqrt{\left(\frac{18}{5}-\frac{8}{5}\right)^2+\left(\frac{14}{5}-\frac{9}{5}\right)^2} = \sqrt{4+1}=\sqrt{5}$

Corresponding to $$u=-\frac{8}{5}, v=-\frac{9}{5}$$

We have $$y=\frac{6}{5}, \quad x=\frac{22}{5}$$

$$\therefore \quad (u, v)=\left(-\frac{8}{5},-\frac{9}{5}\right); \quad (x, y)=\left(\frac{22}{5},\frac{6}{5}\right)$$

$$\therefore \quad PQ=\sqrt{\left(\frac{22}{5}+\frac{8}{5}\right)^2+\left(\frac{6}{5}+\frac{9}{5}\right)^2} = \sqrt{36+9} = \sqrt{45}=3\sqrt{5}$$

$\sqrt{5}$ is less than $3\sqrt{5}$

$\therefore$ Shortest distance between the line and the ellipse is $\sqrt{5}$.

Example 17. *Find the point on the surface of $z = x^2 + y^2 + 10$ nearest to the plane $x + 2y - z = 0$.* **(P.T.U., May 2003)**

Sol. Let $P(x, y, z)$ be any point on the surface $z = x^2 + y^2 + 10$

and $Q(u, v, w)$ be any point on the plane $x + 2y - z = 0$

Then distance between these two points $PQ = \sqrt{(x-u)^2 + (y-v)^2 + (z-w)^2}$

P will be nearest to the plane if PQ is least *i.e.*, PQ^2 is least.

Let $f(x, y, z) = (x - u)^2 + (y - v)^2 + (z - w)^2$ subject to the conditions

$$\phi_1(x, y, z) = z - x^2 - y^2 - 10 = 0$$

and
$$\phi_2(u, v, w) = u + 2v - w = 0$$

Consider Lagrange's function

$$F(x, y, z, u, v, w) = f(x, y, z) + \lambda\phi_1 + \mu\phi_2 = (x - u)^2 + (y - v)^2 + (z - w)^2 + \lambda(z - x^2 - y^2 - 10) + \mu(u + 2v - w)$$

For extreme values $dF = 0$

i.e.,
$$\frac{\partial F}{\partial x} dx + \frac{\partial F}{\partial y} dy + \frac{\partial F}{\partial z} dz + \frac{\partial F}{\partial u} du + \frac{\partial F}{\partial v} dv + \frac{\partial F}{\partial w} dw = 0$$

i.e.,
$$\frac{\partial F}{\partial x} = 0, \frac{\partial F}{\partial y} = 0, \frac{\partial F}{\partial z} = 0, \frac{\partial F}{\partial u} = 0, \frac{\partial F}{\partial v} = 0, \frac{\partial F}{\partial w} = 0$$

$$\frac{\partial F}{\partial x} = 2(x - u) - 2\lambda x = 0 \quad ...(1)$$

$$\frac{\partial F}{\partial y} = 2(y - v) - 2\lambda y = 0 \quad ...(2)$$

$$\frac{\partial F}{\partial z} = 2(z - w) + \lambda = 0 \quad ...(3)$$

$$\frac{\partial F}{\partial u} = -2(x - u) + \mu = 0 \quad ...(4)$$

$$\frac{\partial F}{\partial v} = -2(y - v) + 2\mu = 0 \quad ...(5)$$

$$\frac{\partial F}{\partial w} = -2(z - w) - \mu = 0 \quad ...(6)$$

From (4), (5), (6) $\quad 2(x - u) = \mu$

$$2(y - v) = 2\mu$$

and
$$2(z - w) = -\mu$$

Substituting the values of $2(x - u)$, $2(y - v)$, $2(z - w)$ in (1), (2) and (3) respectively, we get

$$\mu - 2\lambda x = 0 \quad i.e., \quad x = \frac{\mu}{2\lambda}$$

$$2\mu - 2\lambda y = 0 \quad i.e., \quad y = \frac{\mu}{\lambda}$$

$$-\mu + \lambda = 0 \quad i.e., \quad \lambda = \mu \quad \therefore \quad x = \frac{1}{2}, y = 1$$

Substituting the values of x, y in the equation of the surface, we get $z = \frac{1}{4} + 1 + 10 = \frac{45}{4}$

$\therefore$ Point on the surface nearest to the given plane has coordinates $\left(\frac{1}{2}, 1, \frac{45}{4}\right)$.

To prove this is nearest to Q, we have to prove $d^2F > 0$.

We know that

$$d^2F = F_{xx}\,dx^2 + F_{yy}\,dy^2 + F_{zz}\,dz^2 + 2F_{xy}\,dxdy + 2F_{yz}\,dydz + 2F_{zx}\,dzdx + F_{uu}\,du^2 + F_{vv}\,dv^2 + F_{ww}\,dw^2 + 2F_{uv}\,dudv + 2F_{vw}\,dvdw + 2F_{wu}\,dwdu$$

Now $F_{xx} = \dfrac{\partial^2 F}{\partial x^2} = 2(1-\lambda)$; $F_{yy} = \dfrac{\partial^2 F}{\partial y^2} = 2(1-\lambda)$; $F_{zz} = \dfrac{\partial^2 F}{\partial z^2} = 2$

$F_{xy} = F_{yz} = F_{zx} = 0$

$F_{uu} = 2,\quad F_{vv} = 2,\quad F_{ww} = 2$

$F_{uv} = F_{vw} = F_{wu} = 0$

$\therefore\quad d^2F = 2(1-\lambda)dx^2 + 2(1-\lambda)dy^2 + 2dz^2 + 2du^2 + 2dv^2 + 2dw^2$ ...(7)

Now to find λ Substitute the values of u, v, w from (1), (2), (3) in $u + 2v - w = 0$

$$(1-\lambda)x + 2(1-\lambda)y - z - \frac{\lambda}{2} = 0$$

At $\left(\dfrac{1}{2}, 1, \dfrac{45}{4}\right)$, $\quad (1-\lambda)\left(\dfrac{1}{2} + 2\right) - \dfrac{45}{4} - \dfrac{\lambda}{2} = 0$

$$\frac{5}{2} - \frac{5}{2}\lambda - \frac{45}{4} - \frac{\lambda}{2} = 0$$

or $$-3\lambda = \frac{45}{4} - \frac{5}{2} = \frac{35}{4}$$

$\therefore$ $$\lambda = -\frac{35}{4}$$

λ is –ve $\therefore$ from (7) d^2F is positive *i.e.*, $d^2F > 0$

Hence point $\left(\dfrac{1}{2}, 1, \dfrac{45}{4}\right)$ is nearest to the plane $x + 2y - z = 0$.

Example 18. *Prove that the stationary values of* $u = \dfrac{x^2}{a^4} + \dfrac{y^2}{b^4} + \dfrac{z^2}{c^4}$ *where* $lx + my + nz = 0$ *and* $\dfrac{x^2}{a^2} + \dfrac{y^2}{b^2} + \dfrac{z^2}{c^2} = 1$ *are the roots of the equation* $\dfrac{l^2a^4}{1-a^2u} + \dfrac{m^2b^4}{1-b^2u} + \dfrac{n^2c^4}{1-c^2u} = 0.$

Sol. Consider Lagrange's function,

$$F(x, y, z) = \left(\frac{x^2}{a^4} + \frac{y^2}{b^4} + \frac{z^2}{c^4}\right) + \lambda(lx + my + nz) + \mu\left(\frac{x^2}{a^2} + \frac{y^2}{b^2} + \frac{z^2}{c^2} - 1\right)$$

For stationary values, $dF = 0$

$$\Rightarrow\quad \left(\frac{2x}{a^4} + \lambda l + \frac{2\mu x}{a^2}\right)dx + \left(\frac{2y}{b^4} + \lambda m + \frac{2\mu y}{b^2}\right)dy + \left(\frac{2z}{c^4} + \lambda n + \frac{2\mu z}{c^2}\right)dz = 0$$

$$\Rightarrow\quad \frac{2x}{a^4} + \lambda l + \frac{2\mu x}{a^2} = 0 \qquad ...(1)$$

$$\Rightarrow\quad \frac{2y}{b^4} + \lambda m + \frac{2\mu y}{b^2} = 0 \qquad ...(2)$$

$$\frac{2z}{c^4} + \lambda n + \frac{2\mu z}{c^2} = 0 \qquad ...(3)$$

Multiplying (1), (2), (3) by x, y, z, respectively and adding, we get

$$2\left(\frac{x^2}{a^4}+\frac{y^2}{b^4}+\frac{z^2}{c^4}\right)+\lambda\,(lx+my+nz)+2\mu\left(\frac{x^2}{a^2}+\frac{y^2}{b^2}+\frac{z^2}{c^2}\right)=0$$

$\Rightarrow$ $$2u+\lambda(0)+2\mu(1)=0 \qquad |\text{from given relations}$$

$\Rightarrow$ $$\mu=-u$$

$\therefore$ Equation (1) becomes

$$\frac{2x}{a^4}+\lambda l-\frac{2ux}{a^2}=0 \quad\text{or}\quad \frac{2x}{a^4}(1-a^2u)=-\lambda l \quad\text{or}\quad x=-\frac{\lambda la^4}{2(1-a^2u)}$$

Similarly, $$y=-\frac{\lambda mb^4}{2(1-b^2u)},\quad z=-\frac{\lambda nc^4}{2(1-c^2u)}$$

To eliminate λ between them, multiply these values of x, y, z by l, m, n, respectively and add. Then

$$lx+my+nz=-\frac{\lambda}{2}\left[\frac{l^2a^4}{1-a^2u}+\frac{m^2b^4}{1-b^2u}+\frac{n^2c^4}{1-c^2u}\right]$$

Since $lx+my+nz=0$, we have $$\frac{l^2a^4}{1-a^2u}+\frac{m^2b^4}{1-b^2u}+\frac{n^2c^4}{1-c^2u}=0$$

which is a quadratic in u and gives two stationary values of u.

Example 19. *Find the maximum value of $u = x^p y^q z^r$ when the variables x, y, z are subject to the condition $ax + by + cz = p + q + r$.* **(P.T.U., Dec. 2004)**

Sol. Consider Lagrange's function

$$F(x, y, z) = x^p y^q z^r + \lambda(ax + by + cz - p - q - r)$$

For maximum or minimum values $dF = 0$

i.e., $$\frac{\partial F}{\partial x}dx+\frac{\partial F}{\partial y}dy+\frac{\partial F}{\partial z}dz=0$$

where $\frac{\partial F}{\partial x}=px^{p-1}y^qz^r,\ \frac{\partial F}{\partial y}=x^p\,qy^{q-1}z^r,\ \frac{\partial F}{\partial z}=x^py^q\,rz^{r-1}$

i.e., $$[px^{p-1}y^qz^r+\lambda a]\,dx+[x^pqy^{q-1}z^r+\lambda b]\,dy+[x^py^q\,rz^{r-1}+\lambda c]\,dz=0$$

$\Rightarrow$ $$p\,x^{p-1}y^qz^r+\lambda a=0 \qquad ...(1)$$

$$q\,x^py^{q-1}z^r+\lambda b=0 \qquad ...(2)$$

$$r\,x^py^qz^{r-1}+\lambda c=0 \qquad ...(3)$$

Multiply (1) by x, (2) by y, (3) by z and add,

$$(p+q+r)\,x^py^qz^r+\lambda(ax+by+cz)=0$$

or $$(p+q+r)\cdot u+\lambda(p+q+r)=0 \text{ where } u=x^py^qz^r \text{ and } ax+by+cz=p+q+r$$

$\therefore$ $$\lambda=-u$$

From (1), $p\,x^{p-1}y^qz^r-ua=0$

Multiply by x; $p\,x^py^qz^r-uax=0$

or $$pu-uax=0 \quad\text{or}\quad p=ax$$

or $$x=p/a$$

Similarly, $$y=q/b,\ z=r/c \qquad ...(4)$$

Substituting the values of x, y, z in $u = x^p y^q z^r$, we get stationary value of

$$u=\left(\frac{p}{a}\right)^p\left(\frac{q}{b}\right)^q\left(\frac{r}{c}\right)^r.$$

Now u will be maximum if $d^2F < 0$

$$d^2F = F_{xx}(dx)^2 + F_{yy}(dy)^2 + F_{zz}(dz)^2 + 2F_{xy}\,dx\,dy + 2F_{yz}\,dy\,dz + 2F_{zx}\,dz\,dx$$

$$F_{xx} = p(p-1)\,x^{p-2}y^qz^r = \frac{p(p-1)}{x^2}x^py^qz^r$$

$$= \frac{p(p-1)}{x^2}u = \frac{p(p-1)}{p^2/a^2}u = \left(a^2 - \frac{a^2}{p}\right)u \qquad \text{[Using (4)]}$$

$$\therefore \qquad F_{xx} = a^2\left(1-\frac{1}{p}\right)u.$$

Similarly, $$Fyy = b^2\left(1-\frac{1}{q}\right)u;\ F_{zz} = c^2\left(1-\frac{1}{r}\right)u$$

$$F_{xy} = px^{p-1}\,qy^{q-1}\,z^r$$

$$= \frac{pq}{xy}x^py^az^r = \frac{pq}{xy}u = \frac{pq}{pq/ab}u = abu \qquad \text{[Using (4)]}$$

Similarly, $$Fyz = bcu\ ;\ Fzx = cau$$

$$\therefore \qquad d^2F = a^2\left(1-\frac{1}{p}\right)u\,(dx)^2 + b^2\left(1-\frac{1}{q}\right)u\,(dy)^2 + c^2\left(1-\frac{1}{r}\right)u\,(dz)^2$$

$$+ 2\,abu\,dx\,dy + 2\,bcu\,dy\,dz + 2cau\,dz\,dx$$

$$= u[a^2(dx)^2 + b^2(dy)^2 + c^2(dz)^2 + 2ab\,dx\,dy + 2bc\,dy\,dz$$

$$+ 2ca\,dz\,dx - u\left[\frac{a^2}{p}(dx)^2 + \frac{b^2}{q}(dy)^2 + \frac{c^2}{r}(dz)^2\right]$$

$$= u(adx + bdy + cdz)^2 - u\left(\frac{a^2}{p}(dx)^2 + \frac{b^2}{q}(dy)^2 + \frac{c^2}{r}(dz)^2\right)$$

given $ax + by + cz = p + q + r$

Differentiate; $adx + bdy + cdz = 0$

$$\therefore \qquad d^2F = 0 - u\left[\frac{a^2}{p}(dx)^2 + \frac{b^2}{q}(dy)^2 + \frac{c^2}{r}(dz)^2\right] < 0$$

$\therefore$ u is maximum and the maximum value of $u = \left(\frac{p}{a}\right)^p\left(\frac{q}{b}\right)^q\left(\frac{r}{c}\right)^r$

Example 20. *Find the points on the surface $z^2 = xy + 1$ rearrest to the origin.* **(P.T.U. Dec. 2015)**

Sol. Let $P(x, y, z)$ be any point on the surface $z^2 = xy + 1$

Let $\phi(x, y, z) = z^2 - xy - 1$

Distance of P from the origin O(0, 0)

$$OP = \sqrt{x^2 + y^2 + z^2}$$

P will be nearest to the origin O if OP is least *i.e.*, OP^2 is least

Let $f(x, y, z) = (x - 0)^2 + (y - 0)^2 + (z - 0)^2 = x^2 + y^2 + z^2$

Consider Lagrange's function

$F(x, y, z) = f(x, y, z) + \lambda\phi(x, y, z) = x^2 + y^2 + z^2 + \lambda(z^2 - xy - 1)$

For stationary points $dF = 0$

i.e., $$\frac{\partial F}{\partial x}dx + \frac{\partial F}{\partial y}dy + \frac{\partial F}{\partial z}dz = 0$$

$$\Rightarrow \qquad \frac{\partial F}{\partial x} = 0, \frac{\partial F}{\partial y} = 0, \frac{\partial F}{\partial z} = 0$$

i.e.,

$$2x - \lambda y = 0 \qquad \text{....(1)}$$
$$2y - \lambda x = 0 \qquad \text{....(2)}$$
$$2z + 2\lambda z = 0 \qquad \text{....(3)}$$

From (3) either $z = 0$ or $\lambda = -1$

If $z = 0$ then from (1) and (2) $\frac{2x}{y} = \lambda = \frac{2y}{x}$

or $$2x^2 = 2y^2 \text{ i.e., } x^2 = y^2$$

The $\lambda = \frac{2x}{\lambda} = \pm\frac{2y}{y} = \pm 2$ which is not so $\qquad (\because \ \lambda = -1)$

$\therefore \quad \lambda = -1$

Form (1) and (2), $\left.\begin{matrix} 2x + y = 0 \\ 2y + x = 0 \end{matrix}\right] \Rightarrow \begin{matrix} x = 0 \\ y = 0 \end{matrix}$

$$z^2 = xy + 1 \quad \Rightarrow \quad z^2 = 0 + 1 = 1$$

$\therefore$ $$z = \pm 1$$

$\therefore$ Points on the given surface are $(0, 0, \pm 1)$.

Example 21. *If* $u = \frac{a^3}{x^2} + \frac{b^3}{y^2} + \frac{c^3}{z^2}$ *where* $x + y + z = 1$*, prove that the stationary value of* u *is given by* $x = \frac{a}{a+b+c}, y = \frac{b}{a+b+c}, z = \frac{c}{a+b+c}$. **(P.T.U., May 2004)**

Sol. Consider $F(x, y, z) = \frac{a^3}{x^2} + \frac{b^3}{y^2} + \frac{c^3}{z^2} + \lambda(x + y + z - 1)$

For stationary values $dF = 0$

i.e., $$\frac{\partial F}{\partial x}dx + \frac{\partial F}{\partial y}dy + \frac{\partial F}{\partial z}dz = 0$$

i.e., $$\left(-\frac{2a^3}{x^3} + \lambda\right)dx + \left(-\frac{2b^3}{y^3} + \lambda\right)dy + \left(-\frac{2c^3}{z^3} + \lambda\right)dz = 0$$

$$\Rightarrow \qquad -\frac{2a^3}{x^3} + \lambda = 0, -\frac{2b^3}{y^3} + \lambda = 0, -\frac{2c^3}{z^3} + \lambda = 0$$

$$\Rightarrow \qquad \lambda = \frac{2a^3}{x^3}, \lambda = \frac{2b^3}{y^3}, \lambda = \frac{2c^3}{z^3}$$

$$\Rightarrow \qquad \frac{2a^3}{x^3} = \frac{2b^3}{y^3} = \frac{2c^3}{z^3}$$

$$\Rightarrow \qquad \frac{a}{x} = \frac{b}{y} = \frac{c}{z} = \frac{a+b+c}{x+y+z} = \frac{a+b+c}{1}$$

$$\Rightarrow \qquad x = \frac{a}{a+b+c}; y = \frac{b}{a+b+c}; z = \frac{c}{a+b+c}.$$

Example 22. *Use Lagrange's method to find the minimum value of $x^2 + y^2 + z^2$ subject to the conditions $x + y + z = 1$ and $xyz + 1 = 0$.* **(P.T.U., Dec. 2005)**

Sol. Consider Lagrange's function

$$F(x, y, z) = x^2 + y^2 + z^2 + \lambda(x + y + z - 1) + \mu(xyz + 1) \qquad ...(1)$$

For maximum or minimum value $dF = 0$

i.e., $$\frac{\partial F}{\partial x} dx + \frac{\partial F}{\partial y} dy + \frac{\partial F}{\partial z} dz = 0$$

i.e., $$(2x + \lambda + \mu yz)\, dx + (2y + \lambda + \mu zx)\, dy + (2z + \lambda + \mu xy) = 0$$

i.e., $$2x + \lambda + \mu yz = 0 \qquad ...(2)$$

$$2y + \lambda + \mu zx = 0 \qquad ...(3)$$

$$2z + \lambda + \mu xy = 0 \qquad ...(4)$$

(2) – (3) gives $\quad 2(x - y) + \mu z(y - x) = 0$

or $$(x - y)(2 - \mu z) = 0$$

either $\quad x - y = 0 \quad$ or $\quad 2 - \mu z = 0$

i.e., either $\quad x = y \quad$ or $\quad \mu = \dfrac{2}{z}$

Similarly, from (3) and (4) either $y = z$ or $\mu = \dfrac{2}{x}$

and from (4) and (2) either $z = x$ or $\mu = \dfrac{2}{y}$

Combining, we get

either $\quad x = y = z \quad$ or $\quad \mu = \dfrac{2}{x} = \dfrac{2}{y} = \dfrac{2}{z}$

given $$x + y + z = 1 \quad \therefore \quad 3x = 1 \quad \therefore \quad x = \frac{1}{3}, y = \frac{1}{3}, z = \frac{1}{3}$$

and $$\mu = \frac{2}{1/3} = 6$$

$\therefore \quad \left(\dfrac{1}{3}, \dfrac{1}{3}, \dfrac{1}{3}\right)$ is the stationary point.

Also $$d^2F = F_{xx}\, dx^2 + F_{yy}\, dy^2 + F_{zz}\, dx^2 + 2F_{xy}\, dxdy + 2F_{yz}\, dy\, dz + 2F_{zx}\, dz\, dx.$$

$$= 2\, dx^2 + 2\, dy^2 + 2\, dz^2 + 2\, \mu z\, dxdy + 2\mu\, x\, dydz + 2\, \mu\, y\, dz\, dx$$

$$= 2dx^2 + 2dy^2 + 2dz^2 + 2 \cdot 6 \cdot \frac{1}{3}\, dx\, dy + 2 \cdot 6 \cdot \frac{1}{3}\, dy\, dz + 2 \cdot 6 \cdot \frac{1}{3}\, dz\, dx$$

$$= 2(dx + dy + dz)^2 > 0$$

$\therefore \quad F(x, y, z)$ is minimum at $\left(\dfrac{1}{3}, \dfrac{1}{3}, \dfrac{1}{3}\right)$ and the minimum value of $x^2 + y^2 + z^2$ is

$$\frac{1}{9} + \frac{1}{9} + \frac{1}{9} = \frac{1}{3}$$

TEST YOUR KNOWLEDGE

1. Examine for extreme values:

   (*i*) $x^2 + y^2 + 6x + 12$ (*ii*) $x^3 + y^3 + 3xy$ (*iii*) $3x^2 - y^2 + x^3$ **(P.T.U., Dec., 2003)**

   (*iv*) $x^2y^2 - 5x^2 - 8xy - 5y^2$ (*v*) $x^3 + 3xy^2 - 15x^2 - 15y^2 + 72x$

   (*vi*) $x^2 - xy + y^2 + 3x - 2y + 1$ (*vii*) $x^3 + y^3 - 63(x + y) + 12xy$ (*viii*) $xy + \frac{a^3}{x} + \frac{a^3}{y}$

   (*ix*) $\sin x \sin y \sin (x + y)$ (*x*) $\cos x + \cos y + \cos (x + y)$.

2. Find the stationary points of the function $z = x^3y^2 (12 - x - y)$ satisfying the condition $x > 0, y > 0$ and examine their nature.

3. Find the minimum values of $x^2 + y^2 + z^2$ when

   (*i*) $x + y + z = 3a$ (*ii*) $ax + by + cz = p$.

4. Find the minimum value of the function $x + y + z$ subject to the condition $\frac{a}{x} + \frac{b}{y} + \frac{c}{z} = 1$.

5. Find the minimum value of $ax + by + cz$ subject to the condition $\frac{a^2}{x} + \frac{b^2}{y} + \frac{c^2}{z} = 1$.

6. Find the minimum value of the function $x^2 + y^2 + z^2$ subject to the condition

   (*i*) $ax + by + cz = a + b + c$ (*ii*) $xy + yz + zx = 3a^2$.

   (P.T.U., May 2011)

7. Find the greatest and the least values that the function $f(x, y) = xy$ takes on the ellipse $\frac{x^2}{8} + \frac{y^2}{2} = 1$.

8. Given $f(x, y, z) = \frac{5xyz}{x + 2y + 4z}$, find the values of x, y, z for which $f(x, y, z)$ is a maximum, subject to the condition $xyz = 8$.

9. A rectangular box, open at the top, is to have a given capacity. Find the dimensions of the box requiring least material for its construction.

   [**Hint.** Consult S.E. 6]

10. Find the dimensions of the rectangular box, open at the top, of maximum capacity whose surface is 432 sq. cm.

11. Find the shortest and the longest distances from the point (3, 4, 12) to the sphere $x^2 + y^2 + z^2 = 1$.

    [**Hint.** Consult S.E. 14] **(P.T.U., Dec. 2004, May 2015)**

12. Find the maximum and minimum distances from the origin to the curve $5x^2 + 6xy + 5y^2 - 8 = 0$.

    [**Hint.** Consult S.E. 14]

13. Prove that of all the rectangulars parallelopipeds of given surface, cube has the maximum volume.

14. Divide 24 into three parts such that the continued product of the first, square of the second and the cube of the third may be maximum.

    [**Hint.** Given $x + y + z = 24$, $P = xy^2z^3 = (24 - y - z) y^2z^3 = f(y, z)$ for extreme values, $f_y = 0, f_z = 0$, we get $y = 8$, $z = 12$ $\therefore$ $x = 4$]

15. (*a*) Find the stationary values of $x^2 + y^2 + z^2$ subjected to $ax^2 + by^2 + cz^2 = 1$ and $lx + my + nz = 0$.

    (*b*) If $u = ax^2 + by^2 + cz^2$ where $x^2 + y^2 + z^2 = 1$ and $lx + my + nz = 0$, prove that the stationary value of u satisfies the equation $\frac{l^2}{a - u} + \frac{m^2}{b - u} + \frac{n^2}{c - u} = 0$. **(P.T.U., Dec. 2003)**

    [**Hint.** Consult S.E. 18]

16. Find the maximum value of $x^my^nz^p$, when $x + y + z = a$.

    [**Hint.** Consult S.E. 19]

17. If $u = a^3x^2 + b^3y^2 + c^3z^2$ where $x^{-1} + y^{-1} + z^{-1} = 1$, show that the stationary value of u is given by

$$x = \frac{\Sigma a}{a}, \; y = \frac{\Sigma a}{b}, \; z = \frac{\Sigma a}{c}.$$

[**Hint.** Consult S.E. 21] ***(P.T.U., Jan. 2009)***

Answers

1. (*i*) Min. value = 3 at (–3, 0) (*ii*) Max. value = 1 at (–1, –1) (*iii*) Max. value = 4 at(–2, 0)
(*iv*)Max. value = 0 at (0, 0) (*v*) Max. value = 112 at (4, 0); Min. value = 108 at (6, 0)

(*vi*) Min. value = $-\frac{4}{3}$ at $\left(-\frac{4}{3}, \frac{1}{3}\right)$

(*vii*) Max. value = 784 at (–7, –7); Min. value = –216 at (3, 3)

(*viii*) Min. value = $3a^2$ at (a, a) (*ix*) Max. value = $\frac{3\sqrt{3}}{8}$ at $\left(\frac{\pi}{3}, \frac{\pi}{3}\right)$; Min. value = $-\frac{3\sqrt{3}}{8}$ at $\left(\frac{2\pi}{3}, \frac{2\pi}{3}\right)$

(*x*) Min. value = $-\frac{3}{2}$ at $\left(\frac{2\pi}{3}, \frac{2\pi}{3}\right)$

2. Maximum at (6, 4) **3.** (*i*) $3a^2$ (*ii*) $\frac{p^2}{a^2 + b^2 + c^2}$

4. $(\sqrt{a} + \sqrt{b} + \sqrt{c})^2$ at $(\sqrt{a}\,(\sqrt{a} + \sqrt{b} + \sqrt{c}),\ \sqrt{b}\,(\sqrt{a} + \sqrt{b} + \sqrt{c}),\ \sqrt{c}\,(\sqrt{a} + \sqrt{b} + \sqrt{c})$

5. $(a^{3/2} + b^{3/2} + c^{3/2})^2$ **6.** (*i*) $\frac{(a+b+c)^2}{a^2 + b^2 + c^2}$ (*ii*) $3a^2$ at (a, a, a) and $(-a, -a, -a)$

7. Greatest value = 2 and least value = – 2 **8.** (4, 2, 1)

9. $(2v)^{1/3}, (2v)^{1/3}, \frac{1}{2}(2v)^{1/3}$ **10.** Length = Breadth = 6 cm, Height = 3 cm

11. 12, 14 **12.** 4, 1 **14.** 4, 8, 12

15. (*a*) Two stationary values of u are roots of $\frac{l^2}{au-1} + \frac{m^2}{bu-1} + \frac{n^2}{cu-1} = 0$

16. $m^m n^n p^p a^{m+n+p} / (m + n + p)^{m+n+p}$.

REVIEW OF THE CHAPTER

1. Tangent and Normal to a Surface. The equation of the tangent plane at (x_1, y_1, z_1) to the surface $F(x, y, z) = 0$ is

$$(x - x_1)\left(\frac{\partial F}{\partial x}\right)_{\text{at}(x_1, y_1, z_1)} + (y - y_1)\left(\frac{\partial F}{\partial y}\right)_{(x_1, y_1, z_1)} + (z - z_1)\left(\frac{\partial F}{\partial z}\right)_{(x_1, y_1, z_1)} = 0$$

2. Equation of the Normal at (x_1, y_1, z_1) is

$$\frac{x - x_1}{\left(\frac{\partial F}{\partial x}\right)_{(x_1, y_1, z_1)}} = \frac{y - y_1}{\left(\frac{\partial F}{\partial y}\right)_{(x_1, y_1, z_1)}} = \frac{z - z_1}{\left(\frac{\partial F}{\partial z}\right)_{(x_1, y_1, z_1)}}$$

3. Taylor's Theorem for a Single Variable. If $f(x)$ defined in the closed interval $[a, b]$ is such that

(*i*) $f, f', f'', ..., f^{n-1}$ are continuous in $[a, b]$

(*ii*) $f^n(x)$ exists in the open interval (a, b) then ∃, at least one real number $c \in (a, b)$ such that

$$f(b) = f(a) + (b - a) f'(a) + \frac{(b-a)^2}{2!} f''(a) + \cdots + \frac{(b-a)^{n-1}}{(n-1)!} f^{n-1}(a) + \frac{(b-a)^n}{n!} f^n(c)$$

4. **Taylor's Formula**

$$f(x) = f(a) + (x - a) f'(a) + \frac{(x-a)^2}{2!} f''(a) + \cdots + \frac{(x-a)^{n-1}}{(n-1)!} f^{n-1}(a) + R_n(n) \text{ where}$$

$$R_n(x) = \frac{(x-a)^n}{n!} f^n(c); \; c \in (a, x)$$

R_n is known as remainder term or error term.

5. **Maclaurin's Theorem.** (same conditions for $f(x)$ as in Taylor's theorem) then Maclaurin's theorem is

$$f(x) = f(0) + x f'(0) + \frac{x^2}{2!} f''(0) + \cdots + \frac{x^{n-1}}{(n-1)!} f^{n-1}(0) + R_n \text{ where } R_n = \frac{x^n}{n!} f^n(\theta x); \; 0 < \theta < 1.$$

6. Taylor's and Maclaurins theorems for functions of two variables

$$f(x+h, y+k) = f(x, y) + \left(h\frac{\partial f}{\partial x} + k\frac{\partial f}{\partial y}\right) + \frac{1}{2!}\left(h^2\frac{\partial^2 f}{\partial x^2} + 2hk\frac{\partial^2 f}{\partial x \partial y} + k^2\frac{\partial^2 f}{\partial y^2}\right)$$

$$+ \frac{1}{3!}\left(h^3 + \frac{\partial^3 f}{\partial x^3} + 3h^2 k\frac{\partial^3 f}{\partial x^2 \partial y} + 3hk^2\frac{\partial^3 f}{\partial x \partial y^2} + k^3\frac{\partial^3 f}{\partial y^3}\right) + \cdots \text{ is Taylor's theorem}$$

$$f(x, y) = f(0, 0) + [xf_x(0, 0) + yf_y(0, 0)] + \frac{1}{2!}[x^2 f_{xx}(0, 0) + 2xy\, f_{xy}(0, 0) + y^2 f_{yy}(0, 0)]$$

$$+ \frac{1}{3!}[x^3 f_{xxx}(0, 0) + 3x^2 y\, f_{xxy}(0, 0) + 3xy^2 f_{xyy}(0, 0) + y^3 f_{yyy}(0, 0)] + \cdots \infty$$

is Maclaurin's theorem.

Note. These can be extended to any number of variables.

7. **Error terms in Taylor's Series**

(*i*) $| R_1 | \le \frac{1}{2} B\{ | x - a | + | y - b | \}^2$ where $B = \text{Max}\{ | f_{xx} |, | f_{xy} |, | f_{yy} | \}$ is called Maximum absolute error in the linear approximation.

(*ii*) $| R_2 | \le \frac{B}{6}\{ | x - a | + | y - b | \}^3$ where $B = \text{Max.}\{| f_{xxx} |, | f_{xxy} |, | f_{xyy} |, | f_{yyy} |\}$ is called maximum absolute error in the quadratic approximation.

8. **Errors and Approximations.** If $z = f(x, y)$ then $\delta z = \frac{\partial f}{\partial x}\delta x + \frac{\partial f}{\partial y}\delta y$;

error in x is δx and error in y is δy.

Relative error in $x = \frac{\delta x}{x}$

Percentage error is $x = \frac{\delta x}{x} \times 100$

Similar result for y.

9. **Maxima and Minima.** A function $f(x, y)$ is said to have a maximum value at $x = a$, $y = b$ if $f(a, b) > f(a + h, h + k)$ for small and independent values of h and k, +ve or –ve. A function $f(x, y)$ is said to have a minimum value at $x = a, y = b$ if $f(a, b) < f(a + h, b + k)$ for small and independent values of h and k, +ve or –ve.

10. **Saddle Point.** If at any point (a, b) on the surface $z = f(x, y)$ tangent plane is horizontal and the surface descends in certain directions and ascends in other directions then (a, b) is called saddle point.

11. Conditions for Extremum. Necessary condition for a function $f(x, y)$ to have extremum (extreme values) at (a, b) is both $\frac{\partial f}{\partial x} = 0, \frac{\partial f}{\partial y} = 0$ at (a, b).

12. Condition for Maxima and Minima. A function $f(x, y)$ is said to have maximum and minimum values at (a, b) if $\frac{\partial f}{\partial x} = 0, \quad \frac{\partial f}{\partial y} = 0$ at (a, b)

and (*a*) if $rt - s^2 > 0$ and $r < 0$ for (a, b) then z has a maximum value at (a, b).

(*b*) if $rt - s^2 > 0$ and $r > 0$ for (a, b) then z has a minimum value at (a, b)

(*c*) if $rt - s^2 < 0$ for (a, b) then z has no extreme value at (a, b).

(*d*) If $rt - s^2 = 0$, the case is doubtful and requires further investigations.

13. Lagrange's Method of Undetermined Multiples. If the function to be examined for maxima, minima is $f(x, y, z) = 0$ where x, y, z are connected by $\phi(x, y, z) = 0$ then let $F = f + \lambda\phi$; λ is known as Lagrange's multiplier, and for extremum $F_x = 0, F_y = 0, F_z = 0$.

i.e. $\frac{\partial f}{\partial x} + \lambda\frac{\partial \phi}{\partial x} = 0, \frac{\partial f}{\partial y} + \lambda\frac{\partial \phi}{\partial y} = 0, \frac{\partial f}{\partial z} + \lambda\frac{\partial \phi}{\partial z} = 0$

These equations together with $\phi(x, y, z) = 0$ give the values of x, y, z, λ for extreme value of $f(x, y, z)$.

14. Advantages of Lagrange's Method

(1) The stationary values of $f(x, y, z)$ can be deterimed directly even without determining x, y, z explicity

(2) This method can be extended to a function of several variables and subject to any number of constraints

15. Disadvantages of Lagrange's Method

(1) This method does not enable us to find whether stationary point in maximum or minimum. Further investigations are needed

(2) $f_x + \lambda\phi_x = 0, f_y + \lambda\phi_y = 0, f_z + \lambda\phi_z = 0$ are only necessary conditions but not sufficient

SHORT ANSWER TYPE QUESTIONS

1. Find the equation of the tangent planes and normal lines to the following surfaces:

(*i*) $xyz = a^3$ at $(x_1 y_1 z_1)$ **(P.T.U., May 2005, Dec. 2005)**

[**Hint.** S.E. 1(*a*) art. 4.2)

(*ii*) $z^2 = 4(1 + x^2 + y^2)$ at (2, 2, 6) **(P.T.U., Dec. 2012)**

(*iii*) $x^3 + y^3 + 3xyz = 3$ at (1, 2, –1) **(P.T.U., May 2006, 2015)**

[**Hint.** S.E, 3 art. 4.2]

(*iv*) $x^2 + y^2 + z^2 = a^2$ at (x_1, y_1, z_1) **(P.T.U., May 2010)**

(*v*) $2xz^2 - 3xy - 4x = 7$ at (1, –1, 2) (*vi*) $\frac{x^2}{2} - \frac{y^2}{3} = z$ at (2, 3, –1) **(P.T.U., Dec. 2013)**

2. Show that the plane $3x + 12y - 6z = 17$ touches the conicoid $3x^2 - 6y^2 + 9z^2 + 17 = 0$.

3. Expand the following in power of x

(*i*) $\sin x$ **(P.T.U., Dec. 2003)**

(*ii*) $\tan^{-1} x$ [**Hint.** S.E. 2 art. 4.2] **(P.T.U., Dec. 2003)**

(*iii*) $\tan x$ **(P.T.U., May 2009)**

4. For what value of x can we replace $\sin x$ by $x - \frac{x^3}{3!}$ with an error of magnitude not greater than 3×10^{-4}. Explain the steps. [**Hint.** S.E. 7 art. 4.2] **(P.T.U., Dec. 2004)**

5. Expand $4x^2 + 5x + 3$ in powers of $x - 1$ by using Taylor's series.
[**Hint.** S.E. 6 art. 4.2]

6. Compute the value of the following to four decimal places by using Taylor series.
(*i*) $\sqrt{10}$ (*ii*) $\cos 32°$
[**Hint.** S.E. 8 art. 4.2]

7. Write Taylor's theorem for a function of two variables. ***(P.T.U., Dec. 2006)***
[**Hint.** Consult art. 4.3]

8. Write Maclaurine's theorem for a function of two variables.
[**Hint.** Consult art. 4.3]

9. Expand $e^x \cos y$ about $\left(1, \frac{\pi}{4}\right)$ upto second depsee terms. ***(P.T.U., Dec. 2011)***
[**Hint.** Consult S.E. art. 4.3]

10. (*i*) Find approximate value of $\sqrt{(298)^2 + (401)^2}$ by using Taylor's series. ***(P.T.U., Dec. 2007)***
[**Hint.** Consult S.E. 11 art. 4.3]
(*ii*) Using differentials find the approximate value of $\sqrt{(298)^2 + (401)^2}$ ***(P.T.U., Dec. 2007)***
[**Hint.** Consult S.E. 12 art 4.6]

11. If $f(x, y) = \tan^{-1}(xy)$ find an approximate value of $f(1.1, 0.8)$ by using Taylor's series.
[**Hint.** Consult S.E. 4 art 4.5]

12. Find the percentage error in the area of a rectangle when an error of +1 percent is made in measuring the length and breadth. ***(P.T.U., Dec. 2012)***

13. If an error committed in measuring the sides of a square is 2%. Find the error in calculating the area. ***(P.T.U., May 2007)***
[**Hint.** Consult S.E. 1(*ii*) art. 4.6]

14. What error in the common logarithm of a number will be produced by an error of 1% in the number. ***(P.T.U., May 2004)***
[**Hint.** Consult S.E. 1(*iii*) art. 4.6]

15. (*i*) If $P = \frac{E^2}{R}$, show that % error in P = 2% error in E – 1% error in R.
[**Hint.** Consult S.E. 2, art. 4.6]
(*ii*) The power dissipated in a resistor is given by $P = \frac{E^2}{R}$ find approximate percentage change in P when E is increased by 3% and R is decreased by 2%. ***(P.T.U., Dec. 2003, Dec. 2004)***
[**Hint.** Consult S.E. 2, art. 4.6]

16. (*i*) The radius of a circle is found to be 100 cm. Find the relative error in the area of the circle due to an error of 1 mm in measuring the radius. [**Hint.** 1 mm = 1/10 cm] ***(P.T.U. Dec. 2013)***
(*ii*) The radius of a sphere is found to be 10 cm with possible error of 0.02 cm. What is the % error in computing the volume of the sphere. [**Hint.** Consult S.E.4, art. 4.6] ***(P.T.U. Dec. 2013)***

17. Find the possible percentage error in computing the resistance r from the formula $\frac{1}{r} = \frac{1}{r_1} + \frac{1}{r_2}$ if r_1 and r_2 are both in error by 2%. ***(P.T.U., May 2002, Dec. 2014)***

18. If $x^2 + y^2 + z^2 - 2xyz = 1$, show that $\frac{dx}{\sqrt{1-x^2}} + \frac{dy}{\sqrt{1-y^2}} + \frac{dz}{\sqrt{1-z^2}} = 0$

19. If the sides and angles of a trangle ABC vary in such a way that its circumradius remains constant, prove that $\frac{da}{\cos A} + \frac{db}{\cos B} + \frac{dc}{\cos C} = 0$ ***(P.T.U., Dec. 2011)***
[**Hint.** Consult S.E. 10 art 4.6]

20. (*i*) If $V = \frac{\pi}{4} x^2 y$, prove that % in V = 2 (% error in x) + (% error in y).

(*ii*) If $T = 2\pi\sqrt{\frac{l}{g}}$, prove that % in $T = \frac{1}{2}$(% error in l – % error in g).

[**Hint.** Take logs of both sides differentiate and multiply by 100]

21. Define maxima and minima of function of two variables.

[**Hint.** Art. 4.7]

22. Define Saddle point of a function of two variables with example. ***(P.T.U., Dec. 2005)***

[**Hint.** Art. 4.7]

23. (*i*) What is the necessary condition for a function $f(x, y)$ to have an extremum at (a, b).

[**Hint.** Consult art 4.8] ***(P.T.U., 2004)***

(*ii*) State the method to find maximum and minimum values of $z = f(x, y)$ using partial derivatives.

[**Hint.** Consult art 4.10] ***(P.T.U., May 2008, Jan. 2009)***

24. Examine the following functions for maxima and minima

(*i*) $f(x, y) = x^3 + y^3 - 3axy$ ***(P.T.U., May 2006)***

[**Hint.** S.E. I(*i*) art. 4.10]

(*ii*) $f(x, y) = x^2 + y^2 + 6x + 12$. ***(P.T.U., Dec. 2003)***

(*iii*) $f(x, y) = x^4 + y^4 - 2x^2 + 4xy - 2y^2$. ***(P.T.U., May 2005)***

[**Hint.** S.E. 3 art. 4.10]

(*iv*) $f(x, y) = x^2 + 2xy$. ***(P.T.U., Dec. 2014)***

[**Hint.** S.E. (*ii*) art. 4.10]

(*v*) $f(x, y) = \sin x + \sin y + \sin (x + y)$

[**Hint.** S.E.4 art. 4.10]

(*vi*) $f(x, y) = \sin x \sin y \sin (x + y)$

25. Explain Lagrange's method of multipliers for finding maxima and minima of a function $f(x, y)$. What are its draw backs? ***(P.T.U., Dec. 2007)***

[**Hint.** Consult art. 4.12]

26. Why is Lagrange's method for undetermined multiplier convenient to use?

[**Hint.** Consult art. 4.12] ***(P.T.U., May 2010)***

Answers

1.

Equations of the tangent plane;	Equations of the normal line
(*i*) $\frac{x}{x_1} + \frac{y}{y_1} + \frac{z}{z_1} = 3$;	$\frac{x - x_1}{y_1 z_1} = \frac{y - y_1}{z_1 x_1} = \frac{z - z_1}{x_1 y_1}$
(*ii*) $4x + y + z = 6$;	$\frac{x - 2}{4} = \frac{y - 1}{1} = \frac{z - 3}{1}$
(*iii*) $x - 6y - 2z + 9 = 0$;	$\frac{x - 1}{1} = \frac{y - 2}{-6} = \frac{z + 1}{-2}$
(*iv*) $xx_1 + yy_1 + zz_1 = a^2$;	$\frac{x - x_1}{x_1} = \frac{y - y_1}{y_1} = \frac{z - z_1}{z_1}$

(*v*) $7x - 3y + 8z = 26$; $\frac{x-1}{7} = \frac{y+1}{-3} = \frac{z-2}{8}$

(*vi*) $2x - 2y - z + 1 = 0$; $\frac{x-2}{2} = \frac{y-3}{-2} = \frac{z+1}{-1}$

3. (*i*) $x - \frac{x^3}{3!} + \frac{x^5}{5!} \cdots \infty$ (*ii*) $x - \frac{x^3}{3} + \frac{x^5}{5} \cdots \infty$ (*iii*) $x - \frac{x^3}{3} + \frac{2x^5}{15} \cdots \infty$

4. $x \in (-.2913, .2913)$ **5.** $12 + 13(x-1) + 4(x-1)^2$

6. (*i*) 3.1023 (*ii*) 0.8481

9. $e[1 + (x-1) + (y-1) + \frac{1}{2}\{(x-1)^2 + 4(x-1)(y-1) + (y-1)^2\} + \ldots]$

10. (*i*)499.6 (*ii*) 499.6

11. 0.7229 **12.** 2% **13.** 0.04

14. 0.004343 **15.** (*ii*) 8% **16.** (*i*) 0.002 sq cm (*ii*) 0.006 c.c

17. 2%

24. (*i*) Max. value $= -a^3$ at (a, a) if $a > 0$; Min value $= -a^3$ at (a, a) if $a < 0$

(*ii*) Min. value = 3 at $(-3, 0)$

(*iii*) Min. value $= -8$ at $(-\sqrt{2}, \sqrt{2})$

(*iv*) No. max. value, no min. value

(*v*) Max. value $= \frac{3\sqrt{3}}{2}$ at $\left(\frac{\pi}{3}, \frac{\pi}{3}\right)$

(*vi*) Max. value $= \frac{3\sqrt{3}}{8}$ at $\left(\frac{\pi}{3}, \frac{\pi}{3}\right)$; Min. value $= -\frac{3\sqrt{3}}{8}$ at $\left(\frac{2\pi}{3}, \frac{2\pi}{3}\right)$.

PART B

- A Brief Introduction of Three Dimensional Geometry
- Multiple Integrals
- Vector Calculus
- Application of Vector Calculus

5

A Brief Introduction of Three Dimensional Geometry

5.1. INTRODUCTION

Students have already studied, in details, some topics of three dimensional geometry viz. plane, straight line, sphere etc. in their lower classes. In this chapter we shall introduce, in brief, the topics of cone, cylinder and conicoid which play a significant role in multiple integrals.

5.2. DEFINITION OF THE CONE

A cone is the surface generated by a straight line which passes through a fixed point and is subjected to one more condition; for example it may intersect a given curve, touch a given surface or make a given angle with a line through the fixed point.

The fixed point is called *the vertex* of the cone, the moving line in any position is called *the generator* of the cone and if the generator cuts a fixed curve, then the curve is called *the Guiding Curve*.

5.3. HOMOGENEOUS EQUATION

An equation $f(x, y, z) = 0$ is said to be homogeneous in x, y, z if $f(rx, ry, rz) = 0$ for all values of r or in other words we may define it as follows.

An equation in x, y, z is said to be homogeneous if the sum of powers of x, y, z in each term is the same *i.e.,* each term is of the same degree.

5.4. EQUATION OF THE CONE WITH VERTEX AT THE ORIGIN

Show that equation of a cone, whose vertex is the origin, is homogeneous in x, y, z and conversely *i.e.*, every homogeneous equation in x, y, z represents a cone whose vertex is the origin.

Proof. Let equation of the cone with vertex as origin be $f(x, y, z) = 0$ ...(1)

Let $P(x_1, y_1, z_1)$ be any point on the cone, then $f(x_1, y_1, z_1) = 0$ ...(2)

Equation of the generator OP is $\dfrac{x}{x_1} = \dfrac{y}{y_1} = \dfrac{z}{z_1}$ ...(3)

Any point on OP is $Q(rx_1, ry_1, rz_1)$. Since the generator completely lies on the cone, the point Q lies on the cone for all values of r.

Therefore $f(rx_1, ry_1, rz_1) = 0$ for all values of r ...(4)

From (2) and (4), it follows that $f(x_1, y_1, z_1) = 0$ is homogeneous in x_1, y_1, z_1.

$\therefore$ $f(x, y, z) = 0$ is homogeneous in x, y, z.

Conversely. *Every homogeneous equation in x, y, z represents a cone whose vertex is the origin.*

Let the homogeneous equation be $f(x, y, z) = 0$ ...(5)

If $P(x_1, y_1, z_1)$ is a point on the above surface, then $f(x_1, y_1, z_1) = 0$ ...(6)

Also since the equation (5) is homogeneous, we have $f(rx_1, ry_1, rz_1) = 0$ ...(7)
for all values of r.

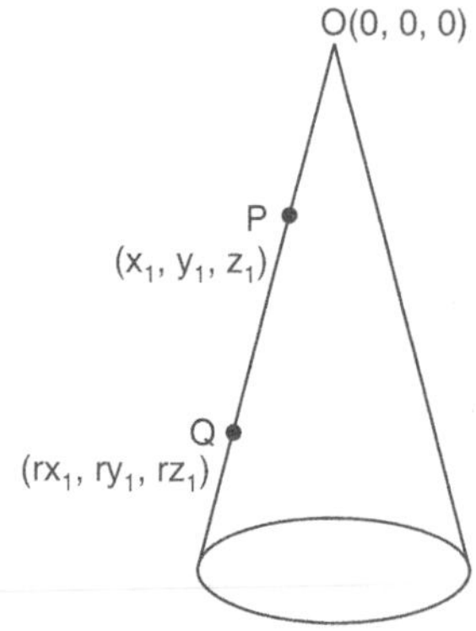

But the point $Q(rx_1, ry_1, rz_1)$ is any point on the line OP. Thus every point of the line OP lies on the surface (5). Hence the surface is being generated by the lines through O, and therefore represents a cone with vertex at the origin. Hence the result.

Quadric Cone. *The second degree homogeneous equation* $ax^2 + by^2 + cz^2 + 2fyz + 2gzx + 2hxy = 0$ *represents a cone (called a quadric cone) with vertex at the origin.*

ILLUSTRATIVE EXAMPLES

Example 1. *Find the equations to the cones with vertex at the origin and which pass through the curve*

(*i*) $ax^2 + by^2 + cz^2 = 1,\ lx + my + nz = p$ **(P.T.U., May 2006)**

(*ii*) $x^2 + y^2 + z^2 - x - 1 = 0,\ x^2 + y^2 + z^2 + y - 2 = 0.$

Sol. [*Make one equation homogeneous with the help of the other equation.*]

(*i*) The equations are $ax^2 + by^2 + cz^2 = 1$ ...(1)

and $lx + my + nz = p$ ...(2)

From (2), we have $\dfrac{lx + my + nz}{p} = 1$ ...(3)

[*Take first degree terms in x, y, z on LHS and take constant terms on RHS and make RHS = 1*] **Note this step.**

Now (1) can be written as $ax^2 + by^2 + cz^2 = (1)^2$

or $$ax^2 + by^2 + cz^2 = \left(\frac{lx + my + nz}{p}\right)^2$$ | *Using* (3) to make (1) homogeneous in x, y, z

or $$p^2(ax^2 + by^2 + cz^2) = (lx + my + nz)^2,$$
which is the required equation of the cone.

(*ii*) The equations of the guiding curve are

$$x^2 + y^2 + z^2 - x - 1 = 0 \quad ...(1)$$

and $$x^2 + y^2 + z^2 + y - 2 = 0 \quad ...(2)$$

The guiding curve is a circle whose plane is $x + y = 1$ ...(3) obtained by subtracting (1) from (2)]

[Equations (1) and (2) or equations (1) and (3) or equations (2) and (3) represent the same curve.

Making (1) homogeneous with the help of (3), we have

$$x^2 + y^2 + z^2 - x(x + y) - (x + y)^2 = 0$$

or $z^2 - 3xy - x^2 = 0$ or $x^2 + 3xy - z^2 = 0$ which is the required equation of the cone.

Example 2. *Planes through OX and OY include an angle α show that their line of intersection lies on the cone $z^2 (x^2 + y^2 + z^2) = x^2y^2 \tan^2 \alpha$.*

Sol. Equations of the plane through OX *i.e.*, $y = 0, z = 0$ is

$$y + \lambda z = 0 \qquad ...(1)$$

and equation of the plane through OY *i.e.*, $x = 0, z = 0$ is

$$z + \mu x = 0 \qquad ...(2)$$

d.r'. s of (1) are $<0, 1, \lambda>$

d.r'. s of (2) are $<\mu, 0, 1>$

$$\tan \alpha = \frac{\sqrt{(1-0)^2 + (\lambda\mu - 0)^2 + (0-\mu)^2}}{0\cdot\mu + 1\cdot 0 + \lambda\cdot 1} \quad i.e., \tan^2 \alpha = \frac{1 + \lambda^2\mu^2 + \mu^2}{\lambda^2} \qquad ...(3)$$

Line of intersection of (1) and (2) is $y = -\lambda z, z = -\mu x$

or

$$\frac{x}{-\frac{1}{\mu}} = \frac{y}{-\lambda} = \frac{z}{1} \qquad ...(4)$$

To find locus of the line (4), eliminate λ, μ from (3) and (4)

∴ from (4), $\lambda = -\frac{y}{z}, \mu = -\frac{z}{x}$

$$\text{Substitute in (3)}, \tan^2 \alpha = \frac{1 + \frac{y^2}{z^2} \times \frac{z^2}{x^2} + \frac{z^2}{x^2}}{\frac{y^2}{z^2}}$$

or

$$\tan^2 \alpha = \frac{x^2z^2 + y^2z^2 + z^4}{x^2y^2} \quad \text{or} \quad z^2(x^2 + y^2 + z^2) = x^2y^2 \tan^2 \alpha.$$

Example 3. *A is a point on OX and B is a point on OY so that angle OAB is constant α. On AB as diameter, a circle is described whose plane is parallel to OZ. Prove that as AB varies, the circle generates the cone $2xy - z^2 \sin 2\alpha = 0$.*

Sol. A is any point on OX, let A $(a, 0, 0)$

and B is any point on OY, let B $(0, b, 0)$

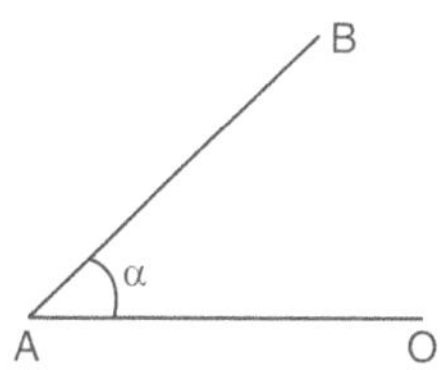

d.r'.s of OA are $<a, 0, 0>$ and that of AB are $<-a, b, 0>$

$$\alpha = \angle OAB \quad \therefore \quad \tan \alpha = \frac{\sqrt{0 + 0 + (ab)^2}}{-a^2 + 0 + 0}$$

$$\therefore \quad \tan^2 \alpha = \frac{a^2b^2}{a^4} \quad \therefore \quad b^2 = a^2 \tan^2 \alpha \quad \text{or} \quad \tan \alpha = b/a \qquad ...(1)$$

Equation of the sphere with AB as diameter is

$$(x - a)\, x + y\, (y - b) + z - z = 0 \quad \text{or} \quad x^2 + y^2 + z^2 - ax - by = 0 \qquad ...(2)$$

Equation of the plane through $(a, 0, 0)$ and $(0, b, 0)$ is

$$A\,(x - a) + B\,(y - 0) + C\,(z - 0) = 0 \quad \text{or} \quad A\,(x - a) + By + Cz = 0 \qquad ...(3)$$

where $A(-a) + Bb = 0$ *i.e.*, $Bb = Aa$ ...(4)

Plane (3) is parallel to OZ; $\frac{x}{0}=\frac{y}{0}=\frac{z}{1}$

$\therefore$ Normal to the plane with d.r'.s <A, B, C> is $\perp$ to the line OZ

$\therefore$ $\quad A \cdot 0 + B \cdot 0 + C \cdot 1 = 0$ *i.e.*, $C = 0$

($\because$ from (3) $A(x-a) + By = 0$)

Using (4) $\quad A(x-a) + \frac{Aa}{b}y = 0$ or $x - a + \frac{a}{b}y = 0$ or $\frac{x}{a}+\frac{y}{b} = 1$ ...(5)

To find equation of the cone, homogenize (2) and (5)

from (2) $x^2+y^2+z^2-(ax+by)\left(\frac{x}{a}+\frac{y}{b}\right)=0$

or $$(x^2+y^2+z^2) = x^2 + \left(\frac{b}{a}+\frac{a}{b}\right)xy + y^2$$

or $$z^2 = \left(\tan\alpha + \frac{1}{\tan\alpha}\right)xy = \frac{1+\tan^2\alpha}{\tan\alpha}xy$$

or $$z^2 = 2 \cdot \frac{1}{\sin 2\alpha}xy \text{ or } 2xy = z^2\sin 2\alpha.$$

5.5. THE DIRECTION COSINES (or Direction Ratios) OF A GENERATOR OF A CONE SATISFY THE EQUATION OF THE CONE WHOSE VERTEX IS THE ORIGIN

Let the generator be $\quad \frac{x}{l}=\frac{y}{m}=\frac{z}{n}$ ...(1)

and the cone be $\quad f(x, y, z) = 0$ ...(2)

Since the cone has vertex at the origin, $\therefore$ equation (2) is homogeneous.

Any point on (1) is (lr, mr, nr). Since the generator completely lies on the surface (2) $\therefore$ the point (lr, mr, nr) lies on (2) for all values of r and hence $f(lr, mr, nr) = 0$ ...(3) for all values of r.

But (2) is homogeneous

$\therefore$ When $r = 1$

$f(l, m, n) = 0$. This proves the result.

Conversely. *If the d.r.'s of a st. line which passes through the origin satisfy a homogeneous equation, then the line is a generator of the cone whose vertex is the origin.*

Let the d.r.'s $<l, m, n>$ of a line satisfy the homogeneous equation $f(l, m, n) = 0$ ...(1)

Then equation of any line through the origin is $\frac{x}{l}=\frac{y}{m}=\frac{z}{n}=\frac{1}{r}$ (say) ...(2)

$\Rightarrow \quad l = rx, m = ry, n = rz$

Putting these values of l, m, n in (1), we have $f(rx, ry, rz) = 0$ for the values of r.

i.e., $\quad f(x, y, z) = 0$ when $r = 1$

which being a homogeneous equation in x, y, z represents a cone whose vertex is the origin.

5.6. QUADRIC CONE THROUGH THE AXES

Show that the general equation of a cone of second degree which passes through the axes is

$$fyz + gzx + hxy = 0 \qquad \textbf{(P.T.U., Dec. 2006)}$$

The cone which passes through the axes will have origin as its vertex. The general equation of a cone of second degree having vertex as origin is

$$ax^2 + by^2 + cz^2 + 2fyz + 2gzx + 2hxy = 0 \qquad \text{...(1)}$$

Since it passes through x-axis.

$\therefore$ the d.c.'s <1, 0, 0> of x-axis must satisfy (1). (Art. 5.4)

$\therefore \qquad a(1)^2 + b(0) + c(0) + 2f(0) + 2g(0) + h(0) = 0 \quad \text{or} \quad a = 0.$

Similarly, as the cone passes through axes of y and z, we have $b = 0$, $c = 0$

$\therefore$ (1) becomes, $\quad 2fyz + 2gzx + 2hxy = 0 \quad \text{or} \quad fzy + gzx + hxy = 0$

which is the required equation.

Example 4. *Find the equation to the quadric cone which passes through the three co-ordinates axes and the three mutually perpendicular lines*

$$\frac{x}{1} = \frac{y}{-2} = \frac{z}{3}, \frac{x}{1} = \frac{y}{-1} = \frac{z}{-1}, \frac{x}{5} = \frac{y}{4} = \frac{z}{1}.$$

Sol. Any cone through the co-ordinates axes is $fyz + gzx + hxy = 0$ ...(1)

Since it passes through the line $\dfrac{x}{1} = \dfrac{y}{-2} = \dfrac{z}{3}$

$\therefore \qquad f(-2)(3) + g(3)(1) + h(1)(-2) = 0 \qquad \Big|\ \because$ *direction ratios of the generator satisfy the equation of cone. Art. 5.5*

or $$6f - 3g + 2h = 0 \qquad \text{...(2)}$$

Similarly, as (1) passes through $\quad \dfrac{x}{1} = \dfrac{y}{-1} = \dfrac{z}{-1}$

$\therefore \qquad f(-1)(-1) + g(-1)(1) + h(1)(-1) = 0 \quad \text{or} \quad f - g - h = 0$ ...(3)

Solving (2) and (3) by cross-multiplication, we have

$$\frac{f}{3+2} = \frac{g}{2+6} = \frac{h}{-6+3} \quad \text{or} \quad \frac{f}{5} = \frac{g}{8} = \frac{h}{-3} \qquad \text{...(4)}$$

Eliminating f, g, h from (1) by using (4), we have $5yz + 8zx - 3xy = 0$ which is the required equation of the cone.

d.r.'s of the third generator $\dfrac{x}{5} = \dfrac{y}{4} = \dfrac{z}{1}$ also satisfy the equation of this cone.

5.7. RIGHT CIRCULAR CONE

A right circular cone *is the surface generated by a straight line which passes through a fixed point and makes a constant angle with a fixed line through the fixed point.*

(*i*) *The fixed point is called the* **vertex.**

(*ii*) *The constant angle is called the* **semi-vertical angle.**

(*iii*) The fixed line through the fixed point *i.e., vertex is called the* **axis of the cone.**

Remark. *The section of a right circular cone by any plane* ⊥ *to its axis is a circle.*

Equations of Right Circular Cone

(a) Standard Form

To show that the equation of the right circular cone whose vertex is at the origin, semi-vertical angle is α and axis along Z-axis is $x^2 + y^2 = z^2 \tan^2 \alpha$.

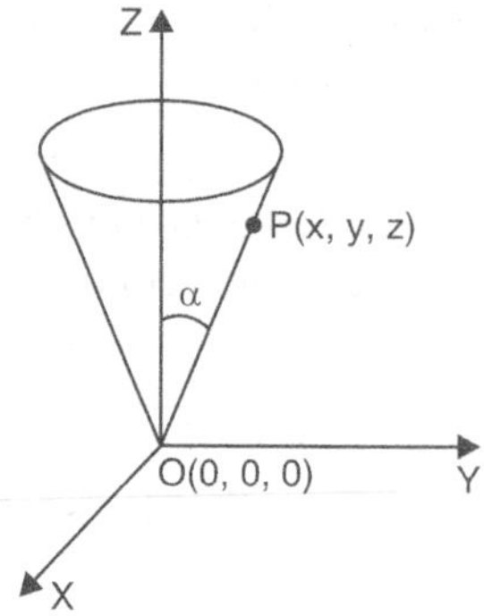

Let P(x, y, z) be any point on the cone.

d.r.'s of OP are $<x, y, z>$

d.c.'s of OZ are $<0, 0, 1>$

$$\therefore \qquad \cos \alpha = \frac{x(0) + y(0) + z(1)}{\sqrt{x^2 + y^2 + z^2}} = \frac{z}{\sqrt{x^2 + y^2 + z^2}}$$

Squaring and cross-multiplying, we have

$$(x^2 + y^2 + z^2) \cos^2 \alpha = z^2$$

or $$x^2 + y^2 + z^2 = z^2 \sec^2 \alpha$$

or $$x^2 + y^2 + z^2 = z^2(1 + \tan^2 \alpha)$$

or $$x^2 + y^2 = z^2 \tan^2 \alpha$$

which is the required equation.

(b) General Form

To find the equation of the right circular cone whose vertex is at (α, β, γ) semi-vertical angle θ, and axis the line $\frac{x-\alpha}{l} = \frac{y-\beta}{m} = \frac{z-\gamma}{n}$.

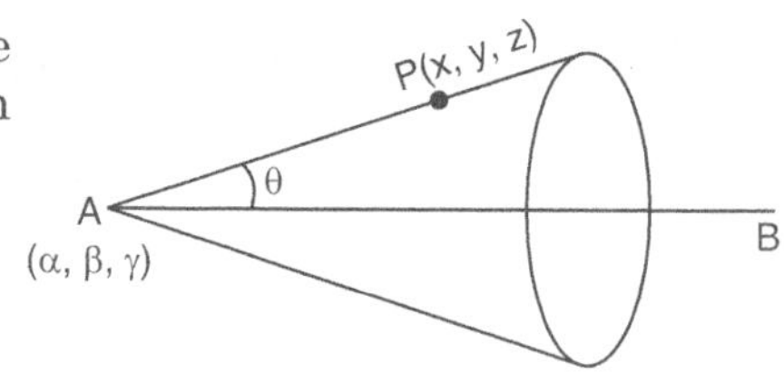

Let P(x, y, z) be any point on the cone and AB, the axis of cone whose d.r.'s are $<l, m, n>$ and passes through the vertex A (α, β, γ).

$\angle$PAB = θ, the semi-vertical angle.

d.r.'s of AP are $<x - \alpha, y - \beta, z - \gamma>$

d.r.'s of AB are $<l, m, n>$

$$\therefore \qquad \cos \theta = \frac{l(x-\alpha) + m(y-\beta) + n(z-\gamma)}{\sqrt{l^2 + m^2 + n^2} \cdot \sqrt{(x-\alpha)^2 + (y-\beta)^2 + (z-\gamma)^2}}$$

or $$(l^2 + m^2 + n^2) [(x - \alpha)^2 + (y - \beta)^2 + (z - \gamma)^2] \cos^2 \theta = [l(x - \alpha) + m(y - \beta) + n(z - \gamma)]^2$$

which is the required equation of the cone.

5.8. EQUATION OF THE RIGHT CIRCULAR CONES WHICH CONTAIN THE THREE COORDINATE AXES AS GENERATORS

As right circular cone contains three coordinate axes as generators $\therefore$ the common point of intersection of these generators is the vertex of the cone $\therefore$ (0, 0, 0) is the vertex of the cone.

Let $\frac{x}{l} = \frac{y}{m} = \frac{z}{n}$ be the axis of the cone and α be the semi-vertical angle of the cone.

Equation of the cone is

$$(l^2 + m^2 + n^2) [x^2 + y^2 + z^2] \cos^2 \alpha = [lx + my + nz]^2 \qquad ...(1)$$

Now α is angle between axis and each of the three rectangular axes whose d.c'.s are $<1, 0, 0>$; $<0, 1, 0>$; $<0, 0, 1>$

$$\therefore \quad \cos\alpha = \frac{l \cdot 1}{\sqrt{l^2+m^2+n^2}} = \frac{m}{\sqrt{l^2+m^2+n^2}} = \frac{n}{\sqrt{l^2+m^2+n^2}}$$

$\therefore \quad \cos^2\alpha\,(l^2 + m^2 + n^2) = l^2 = m^2 = n^2 = k$ (say)

$$\therefore \quad l = m = n = \pm\sqrt{k}$$

Substituting the values of l, m, n in (1)

$$k\,(x^2 + y^2 + z^2) = [\pm\sqrt{k}x \pm \sqrt{k}y \pm \sqrt{k}z\,]^2$$

or $\quad x^2 + y^2 + z^2 = x^2 + y^2 + z^2 \pm 2xy \pm 2yz \pm 2zx$

or $\quad xy \pm yz \pm zx = 0$ is the required equation of the cone.

Example 5. *Lines are drawn from the origin with direction cosines proportional to <1, 2, 2>; <2, 3, 6>; <3, 4, 12>; find the direction cosines of the axis of right circular cone through them, and prove that the semi-vertical angle of the cone is* $\cos^{-1}\frac{1}{\sqrt{3}}$. *Also find the equation of the right circular cone.*

Sol. Let <l, m, n> be the d.c.'s of the axis of the right circular cone, through three lines say OA, OB, OC whose d.c.'s are proportional to <1, 2, 2>; <2, 3, 6> and <3, 4, 12> respectively.

Since the axis makes the same angle (say α) with the generators OA, OB, OC.

$$\therefore \quad \frac{l(1)+m(2)+n(2)}{\sqrt{1+4+4}} = \frac{l(2)+m(3)+n(6)}{\sqrt{4+9+36}} = \frac{l(3)+m(4)+n(12)}{\sqrt{9+16+144}} \text{ (each} = \cos\alpha)$$

or $$\frac{l+2m+2n}{3} = \frac{2l+3m+6n}{7} = \frac{3l+4m+12n}{13} = \cos\alpha \qquad ...(1)$$

From first and second members,

$$7l + 14m + 14n = 6l + 9m + 18n \quad \text{or} \quad l + 5m - 4n = 0 \qquad ...(2)$$

From the second and third members of (1),

$$26l + 39m + 78n = 21l + 28m + 84n \quad \text{or} \quad 5l + 11m - 6n = 0 \qquad ...(3)$$

Solving (2) and (3), we have $\frac{l}{-1} = \frac{m}{1} = \frac{n}{1}$.

Hence the d.c.'s of the axis are proportional to <–1, 1, 1>.

Dividing each by $\sqrt{1+1+1} = \sqrt{3}$, the d.c.'s are $\left\langle \frac{-1}{\sqrt{3}}, \frac{1}{\sqrt{3}}, \frac{1}{\sqrt{3}} \right\rangle$.

Putting these values of l, m, n in (1), we have

$$\cos\alpha = \frac{l+2m+2n}{3} = \frac{\frac{-1}{\sqrt{3}} + \frac{2}{\sqrt{3}} + \frac{2}{\sqrt{3}}}{3} = \frac{\frac{3}{\sqrt{3}}}{3} = \frac{1}{\sqrt{3}}$$

$$\therefore \quad \alpha = \cos^{-1}\frac{1}{\sqrt{3}}$$

Let P(x, y, z) be any point on the cone.

d.r.'s of OP are < x, y, z >.

d.r.'s of axis of cone are <–1, 1, 1>

Since OP makes an angle α with the axis

$$\therefore \quad \cos\alpha = \frac{x(-1)+y(1)+z(1)}{\sqrt{x^2+y^2+z^2}\cdot\sqrt{1+1+1}} \quad \text{or} \quad \frac{1}{\sqrt{3}} = \frac{-x+y+z}{\sqrt{x^2+y^2+z^2}\cdot\sqrt{3}}$$

Squaring both sides

$$x^2+y^2+z^2 = x^2+y^2+z^2-2xy+2yz-2zx \quad \text{or} \quad xy-yz+zx=0$$

which is the required equation of the cone.

Example 6. *Find the equation of the right circular cone which passes through the point (1, 1, 2) and has vertex at the origin and axis the line* $\frac{x}{2}=\frac{y}{-4}=\frac{z}{3}$.

Sol. Vertex of the cone is (0, 0, 0)

Axis of the cone is

$$\frac{x}{2}=\frac{y}{-4}=\frac{z}{3}.$$

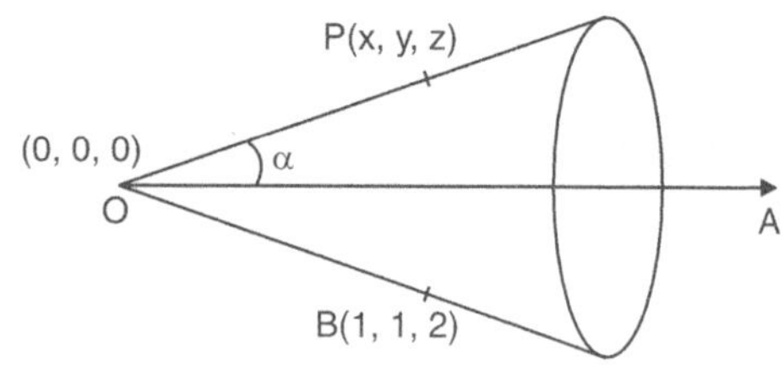

Its d.r'.s are <2, –4, 3>

Let α be the semi vertical angle of the cone. As B (1, 1, 2) is any point on the cone.

$\therefore$ OB is one of the generators of the cone

$\therefore$ d.r'.s of OB are <1 – 0, 1 – 0, 2 – 0> = <1, 1, 2>

$$\therefore \quad \cos\alpha = \frac{1.2+1(-4)+2(3)}{\sqrt{4+16+9}\sqrt{1+1+4}} = \frac{4}{\sqrt{29}\sqrt{6}}$$

Let P(x, y, z) be any point on the cone.

$\therefore$ d.r'.s of OP are $<x, y, z>$

$\therefore$ angle between OP and axis is also α

$$\therefore \quad \cos\alpha = \frac{x\cdot 2+y(-4)+z\cdot 3}{\sqrt{4+16+9}\sqrt{x^2+y^2+z^2}}$$

$$\therefore \quad \frac{4}{\sqrt{29}\sqrt{6}} = \frac{2x-4y+3z}{\sqrt{26}\sqrt{x^2+y^2+z^2}}$$

or $\quad 16(x^2+y^2+z^2) = 6(2x-4y+3z)^2$

or $\quad 16x^2+16y^2+16z^2 = 24x^2+96y^2+54z^2-96xy-144yz+72zx$

$\quad 8x^2+80y^2+38z^2-96xy-144yz+72zx=0$

or $\quad 4x^2+40y^2+19z^2-48xy-72yz+36zx=0.$

Example 7. *Find the equation of the right circular cone whose vertex is at the origin, whose axis is the line* $\frac{x}{1}=\frac{y}{2}=\frac{z}{3}$, *and which has semi-vertical angle of 30°.*

(P.T.U., Dec. 2004)

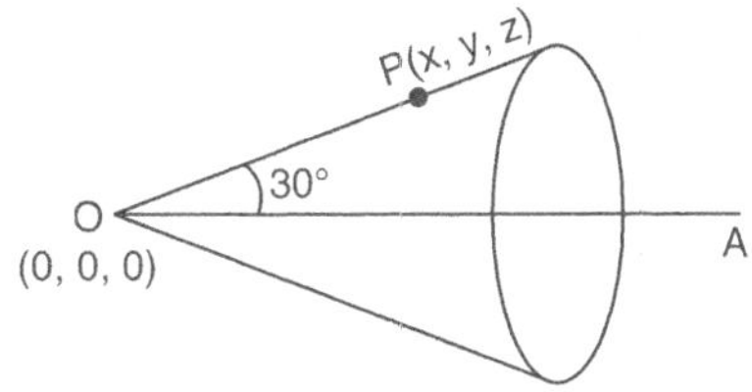

Sol. The axis, say OA, of the cone is $\frac{x}{1} = \frac{y}{2} = \frac{z}{3}$.

The d.c.'s of the axis OA are proportional to $<1, 2, 3>$.

Let P(x, y, z) be any point on the cone.

d.r.'s of OP are $<x, y, z>$

Since $\angle POA = \frac{1}{2}(\text{Vertical angle}) = 30°$

$$\therefore \quad \cos 30° = \frac{x \cdot 1 + y \cdot 2 + z \cdot 3}{\sqrt{x^2 + y^2 + z^2} \cdot \sqrt{1 + 4 + 9}}$$

or
$$\frac{\sqrt{3}}{2} = \frac{x + 2y + 3z}{\sqrt{x^2 + y^2 + z^2} \cdot \sqrt{14}}$$

or $21(x^2 + y^2 + z^2) = 2(x + 2y + 3z)^2$ or $19x^2 + 13y^2 + 3z^2 - 8xy - 24yz - 12zx = 0$

which is the required equation of the cone.

Example 8. *Prove that $x^2 - y^2 + z^2 - 2x + 4y + 6z + 6 = 0$ represents a right circular cone whose vertex is (1, 2, –3), where axis is parallel to OY and whose semi vertical angle is 45°.*

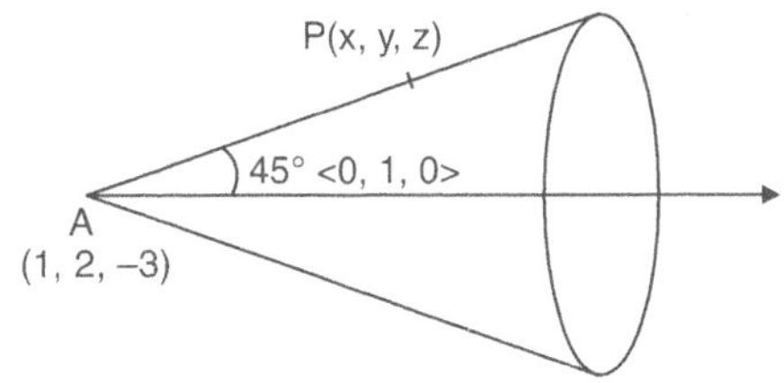

Sol. Vertex of the cone is (1, 2, –3)

Axis is || to OY *i.e.*, $\frac{x}{0} = \frac{y}{1} = \frac{z}{0}$

$\therefore$ d.r'.s of axis are $<0, 1, 0>$

Let P(x, y, z) be any point on the cone

The d.r'. s of AP are $<x - 1, y - 2, z + 3>$

$$\therefore \quad \cos 45° = \frac{(x-1).0 + (y-2).1 + (z+3).0}{\sqrt{(x-1)^2 + (y-2)^2 + (z+3)^2}}$$

or
$$\frac{1}{\sqrt{2}} = \frac{(y-2)}{\sqrt{(x-1)^2 + (y-2)^2 + (z+3)^2}}$$

Squaring and simplify,

$$x^2 - 2x + 1 + y^2 - 4y + 4 + z^2 + 6z + 9 = 2(y^2 - 4y + 4)$$

or
$$x^2 - y^2 + z^2 - 2x + 4y + 6z + 6 = 0.$$

Example 9. *The vertex of a right circular cone is (1, –1, 1), the axis is parallel to $x = -\frac{y}{2} = -z$ and one of its generators has direction cosines proportional to (2, 2, 1). Find its equation.*

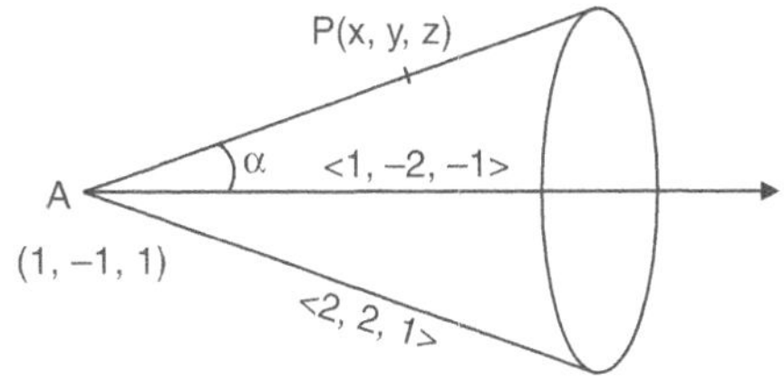

Sol. Let α be the semi vertical angle of the cone. Vertex is (1, –1, 1) and d.r'.s of one of the generators are $<2, 2, 1>$

d.r'.s of the axis of the cone are $<1, -2, -1>$

$$\therefore \quad \cos \alpha = \frac{2(1) + 2(-2) + (1)(-1)}{\sqrt{4+4+1}\sqrt{1+4+1}} = \frac{-3}{3\sqrt{6}} = -\frac{1}{\sqrt{6}}$$

Let P(x, y, z) be any point on the cone

$\therefore$ d.r'.s of OP are $< x - 1, y + 1, z - 1 >$

$$\cos \alpha = \frac{(x-1)\cdot 1 + (y+1)(-2) + (z-1)(-1)}{\sqrt{6}\sqrt{(x-1)^2 + (y+1)^2 + (z-1)^2}}$$

$$-\frac{1}{\sqrt{6}} = \frac{(x-1) - 2(y+1) - (z-1)}{\sqrt{6}\sqrt{(x-1)^2 + (y+1)^2 + (z-1)^2}}$$

Squaring both sides and simplifies

$$x^2 - 2x + 1 + y^2 + 2y + 1 + z^2 - 2z + 1$$
$$= x^2 - 2x + 1 + 4(y^2 + 2y + 1) + (z^2 - 2z + 1) - 4(x-1)(y+1) + 4(y+1)(z-1) - 2(x-1)(z-1)$$

or $$3y^2 + 6y + 3 - 4(xy + x - y - 1) + 4(yz + z - y - 1) - 2(xz - x - z + 1) = 0$$

or $$3y^2 + 4yz - 4xy - 2xz - 2x + 6y + 6z + 1 = 0.$$

TEST YOUR KNOWLEDGE

1. Find the equations of the cones with vertex at the origin and which pass through the curves given by

   (i) $x^2 + y^2 = 4, z = 2.$ (ii) $x^2 + y^2 + z^2 = 9, x + y + z = 1.$ ***(P.T.U., May 2008)***

   [**Hint.** Consult S.E. 1(*i*)]

   (iii) $\frac{x^2}{4} + \frac{y^2}{9} + \frac{z^2}{1} = 1, x + y + z = 1.$

   (iv) $x^2 + y^2 + z^2 + 2x - y + 3z - 1 = 0$ and $x^2 + y^2 + z^2 + x + 2z - 5 = 0.$

2. (i) Find the equation of the cone where vertex is the origin and base the circle $x = a, y^2 + z^2 = b^2$ and show that the section of the cone by a plane parallel to the plane XOY is a hyperbola.

   (ii) The plane $x + y + z = 1$ meets the co-ordinate axes in A, B and C. Prove that the equation of the cone generated by lines through O, the origin , to meet the circle ABC is $yz + zx + xy = 0.$

3. Show that the line $\frac{x}{l} = \frac{y}{m} = \frac{z}{n}$ where $2l^2 + 3m^2 - 4n^2 = 0$ is a generator of the cone

   $$2x^2 + 3y^2 - 4z^2 = 0.$$

4. Show that the lines drawn through a point (α, β, γ) whose d.c.'s satisfy $al^2 + bm^2 + cn^2 = 0$ generate the cone $a(x - \alpha)^2 + b(y - \beta)^2 + c(z - \gamma)^2 = 0.$

5. (i) Find the equation of the cone which passes through the three co-ordinate axes and the three mutually perpendicular lines

   $$\frac{x}{2} = \frac{y}{1} = \frac{z}{-1}, \frac{x}{1} = \frac{y}{3} = \frac{z}{5}, \frac{x}{8} = \frac{y}{-11} = \frac{z}{5}.$$

   [**Hint.** Consult solved example 4]

   (ii) Find the equation of the cone which passes through the three co-ordinate axes and the lines

   $$\frac{x}{1} = \frac{y}{-1} = \frac{z}{1}\ ;\ x = 2y = 3z.$$

6. Find the equation of the cone of revolution with vertex at the origin, axis the y-axis and semi-vertical angle 30°.

7. The axis of a right circular cone whose vertex is at origin O, makes equal angles with the co-ordinate axes and the cone passes through the line drawn from O with direction cosines proportional to <1, –2, 2>. Find the equation of the cone.

   $$\left[\textbf{Hint.}\ \text{The d.c.'s of the line which makes equal angles with co-ordinate axes are } \left\langle \frac{1}{\sqrt{3}}, \frac{1}{\sqrt{3}}, \frac{1}{\sqrt{3}} \right\rangle\right].$$

8. The vertex of a right circular cone is (0, 0, 0). Its generators intersect the circle through the points (1, 2, 2), (1, –2, 2), (2, –1, –2). Find the equation of the cone.

9. Find the equation of a right circular cone whose vertex is (1, –1, 2), axis the line $\frac{x-1}{2} = \frac{y+1}{1} = \frac{z-2}{-2}$ and semi-vertical angle 45°.

[**Hint.** Consult solved example 8]

Answers

1. (*i*) $x^2 + y^2 = z^2$ (*ii*) $4(x^2 + y^2 + z^2) + 9(xy + yz + zx) = 0$

(*iii*) $27x^2 + 32y^2 + 72(yz + zx + xy) = 0$ (*iv*) $3x^2 + 11y^2 + 3z^2 + 14xy + 18yz - 30zx = 0$

2. (*i*) $a^2(y^2 + z^2) = b^2x^2$

5. (*i*) $16yz - 33zx - 25xy = 0$ (*ii*) $yz + 4zx + 3xy = 0$

6. $3(x^2 + z^2) = y^2$ **7.** $4(x^2 + y^2 + z^2) + 9(xy + yz + zx) = 0$

8. $12x^2 - 4y^2 - 3z^2 + 8xz = 0$ **9.** $x^2 + y^2 + z^2 - 8xy + 8yz + 16zx - 4x + 6y - 12z + 36 = 0$.

5.9. EQUATION OF THE CONE WHOSE VERTEX IS THE POINT (α, β, γ) AND BASE THE CONIC $f(x, y) = ax^2 + by^2 + 2hxy + 2gx + 2fy + c = 0, z = 0$

Proof. Equation of the conic is $ax^2 + by^2 + 2hxy + 2gx + 2fy + c = 0, z = 0$ ...(1)

The equations of *any line* through (α, β, γ) are

$$\frac{x-\alpha}{l} = \frac{y-\beta}{m} = \frac{z-\gamma}{n} \qquad ...(2)$$

It meets the plane $z = 0$, where $\frac{x-\alpha}{l} = \frac{y-\beta}{m} = \frac{0-\gamma}{n}$

or $$x - \alpha = \frac{-l}{n}\gamma,\ y - \beta = \frac{-m}{n}\gamma \quad \text{or} \quad x = \alpha - \frac{l\gamma}{n},\ y = \beta - \frac{m}{n}\gamma.$$

If this point lies on the given conic, then

$$a\left(\alpha - \frac{l\gamma}{n}\right)^2 + b\left(\beta - \frac{m}{n}\gamma\right)^2 + 2h\left(\alpha - \frac{l}{n}\gamma\right)\left(\beta - \frac{m}{n}\gamma\right) + 2f\left(\beta - \frac{m}{n}\gamma\right) + 2g\left(\alpha - \frac{l}{n}\gamma\right) + c = 0 \qquad ...(3)$$

This is the condition for the line (2) to intersect the conic (1).

To find the locus of lines (2), we have to eliminate l, m, n from (2) and (3).

Now putting the values of l, m, n, from (2) in (3), we have

$$a\left(\alpha - \frac{x-\alpha}{z-\gamma}\cdot\gamma\right)^2 + b\left(\beta - \frac{y-\beta}{z-\gamma}\cdot\gamma\right)^2 + 2h\left(\alpha - \frac{x-\alpha}{z-\gamma}\cdot\gamma\right)\left(\beta - \frac{y-\beta}{z-\gamma}\cdot\gamma\right)$$
$$+ 2f\left(\beta - \frac{y-\beta}{z-\gamma}\cdot\gamma\right) + 2g\left(\alpha - \frac{x-\alpha}{z-\gamma}\cdot\gamma\right) + c = 0$$

or $$a\frac{(\alpha z - \gamma x)^2}{(z-\gamma)^2} + b\frac{(\beta z - \gamma y)^2}{(z-\gamma)^2} + 2h\frac{(\alpha z - \gamma x)(\beta z - \gamma y)}{(z-\gamma)^2} + 2f\frac{(\beta z - \gamma y)}{z-\gamma} + 2g\frac{(\alpha z - \gamma x)}{z-\gamma} + c = 0$$

or $$a(\alpha z - \gamma x)^2 + b(\beta z - \gamma y)^2 + 2h(\alpha z - \gamma x)(\beta z - \gamma y) + 2f(\beta z - \gamma y)(z - \gamma) + 2g(\alpha z - \gamma x)(z - \gamma) + c(z - \gamma)^2 = 0$$

which is the required equation of the cone.

ILLUSTRATIVE EXAMPLES

Example 1. (*i*) *Find the equation of the cone whose vertex is at the point (1, 1, 1) and base the circle* $x^2 + y^2 = 4$, $z = 2$. **(P.T.U., May 2005)**

(*ii*) *Find the equation of the cone whose vertex is* (α, β, γ) *and guiding curve*

$$x^2 + y^2 + 2gx + 2fy + c = 0,\ z = 0.$$ **(P.T.U., Dec. 2005)**

Sol. (*i*) The given base conic is the circle

$$x^2 + y^2 = 4,\ z = 2 \quad ...(1)$$

Any line through the vertex (1, 1, 1) is $\dfrac{x-1}{l} = \dfrac{y-1}{m} = \dfrac{z-1}{n}$ ...(2)

It meets the plane $z = 2$ where $\dfrac{x-1}{l} = \dfrac{y-1}{m} = \dfrac{2-1}{n} = \dfrac{1}{n}$

$$\therefore \quad x - 1 = \frac{l}{n},\ y - 1 = \frac{m}{n} \quad \therefore\ x = 1 + \frac{l}{n},\ y = 1 + \frac{m}{n}$$

This point lies on the circle (1)

$$\therefore \quad \left(1 + \frac{l}{n}\right)^2 + \left(1 + \frac{m}{n}\right)^2 = 4 \quad ...(3)$$

To find locus of the line (2) eliminate l, m, n

From (2); $\dfrac{l}{n} = \dfrac{x-1}{z-1},\ \dfrac{m}{n} = \dfrac{y-1}{z-1}$

Substituting in (3)

$$\left(1 + \frac{x-1}{z-1}\right)^2 + \left(1 + \frac{y-1}{z-1}\right)^2 = 4$$

or $$(x + z - 2)^2 + (y + z - 2)^2 = 4\ (z - 1)^2$$

or $x^2 + y^2 - 2z^2 + 2yz + 2zx - 4x - 4y + 4 = 0$ which is the required equation of the cone.

(*ii*) Equation of the guiding curve is

$$x^2 + y^2 + 2gx + 2fy + c = 0,\ z = 0 \quad ...(1)$$

Equation of any line through the vertex (α, β, γ) is

$$\frac{x-\alpha}{l} = \frac{y-\beta}{m} = \frac{z-\gamma}{n} \quad ...(2)$$

It meets (1) $\therefore$ put $z = 0$

$$\frac{x-\alpha}{l} = \frac{y-\beta}{m} = \frac{-\gamma}{n}$$

$$\therefore \quad x = \alpha - \frac{l}{n}\gamma,\ y = \beta - \frac{m}{n}\gamma$$

This point lies on (1)

$$\therefore \qquad \left(\alpha - \frac{l}{n}\gamma\right)^2 + \left(\beta - \frac{m}{n}\gamma\right)^2 + 2g\left(\alpha - \frac{l}{n}\gamma\right) + 2f\left(\beta - \frac{m}{n}\gamma\right) + C = 0 \qquad ...(3)$$

To find locus of (2) eliminate l, m, n from (2) and (3)

$$\left(\alpha - \frac{x-\alpha}{z-\gamma}\gamma\right)^2 + \left(\beta - \frac{y-\beta}{z-\gamma}\gamma\right)^2 + 2g\left(\alpha - \frac{x-\alpha}{z-\gamma}\gamma\right) + 2f\left(\beta - \frac{y-\beta}{z-\gamma}\gamma\right) + C = 0$$

or $$(\alpha z - \gamma x)^2 + (\beta z - \gamma y)^2 + 2g\,(z - \gamma)\,(\alpha z - \gamma x) + 2f\,(z - \gamma)\,(\beta\gamma - \gamma y) + C(z - \gamma)^2 = 0$$

which is the required equation of the cone.

Example 2. (*i*) *Find the equation of the cone formed by rotating the line $2y + 3z = 6$, $x = 0$ about z-axis.* **(P.T.U., May 2003)**

(*ii*) *Find the equation of the right circular cone generated by rotating the line $\frac{x}{1} = \frac{y}{2} = \frac{z}{3}$ about the line $\frac{x}{-1} = \frac{y}{1} = \frac{z}{2}$.* **(P.T.U., May 2004)**

Sol. (*i*) The given line is $2y + 3z = 6$, $x = 0$ or $2y = 6 - 3z$, $x = 0$

or $$2y = -3(z - 2),\ x = 0 \text{ or } \frac{y}{3} = \frac{z-2}{-2} = \frac{x}{0} \text{ or } \frac{x}{0} = \frac{y}{3} = \frac{z-2}{-2} \qquad ...(1)$$

Equation of z-axis is $\frac{x}{0} = \frac{y}{0} = \frac{z}{1}$ ...(2)

Point of intersection of (1) and (2) is the vertex of the cone and the vertex of the cone is obtained by taking any point on (2) *i.e.* $(0, 0, t)$

Substitute in (1), $\frac{0}{0} = \frac{0}{3} = \frac{t-2}{-2}$, we get $t = 2$

$\therefore$ Vertex is (0, 0, 2)

Let θ be the angle which line (1) makes with line (2)

$$\therefore \qquad \cos\theta = \frac{0\cdot 0 + 3\cdot 0 + (-2)\cdot 1}{\sqrt{0+9+4}\,\sqrt{0+0+1}} = \frac{-2}{\sqrt{13}}$$

Let the equation of any other generator through (0, 0, 2) be $\frac{x}{l} = \frac{y}{m} = \frac{z-2}{n}$ ...(3)

Now generator given by (3) also makes an angle θ with the axis of the cone.

$$\therefore \qquad \cos\theta = \frac{l.0 + m\cdot 0 + n\cdot 1}{\sqrt{l^2+m^2+n^2}} = \frac{n}{\sqrt{l^2+m^2+n^2}}$$

Substituting the value of $\cos\theta$, we get

$$\frac{-2}{\sqrt{13}} = \frac{n}{\sqrt{l^2+m^2+n^2}} \quad \text{or} \quad 4(l^2 + m^2 + n^2) = 13n^2$$

or $$4l^2 + 4m^2 - 9n^2 = 0 \qquad ...(4)$$

As cone is the locus of generator (*ii*) eliminate l, m, n from (3) and (4)

$4(x)^2 + 4(y)^2 - 9(z - 2)^2 = 0$ or $4x^2 + 4y^2 - 9z^2 + 36z - 36 = 0$ is the equation of the cone.

(*ii*) Equations of the lines are

$$\frac{x}{1} = \frac{y}{2} = \frac{z}{3}$$

$$\frac{x}{-1} = \frac{y}{1} = \frac{z}{2}$$

Both lines pass through (0, 0, 0)

$\therefore$ their point of intersection is (0, 0, 0)

$\therefore$ Vertex of the cone is (0, 0, 0)

Axis of the cone is $\frac{x}{-1} = \frac{y}{1} = \frac{z}{2}$

Let θ be the semi vertical angle of the cone.

$\therefore$ θ is the angle between (1) and (2)

d.r'.s of (1) are <1, 2, 3>

d.r'.s of (2) are <–1, 1, 2>

$$\therefore \qquad \cos\theta = \frac{1(-1) + 2 \cdot 1 + 3 \cdot 2}{\sqrt{1+4+9}\,\sqrt{1+1+4}} = \frac{7}{\sqrt{14}\,\sqrt{6}} = \frac{7}{\sqrt{84}}$$

Equation of any generator of the cone is

$$\frac{x-0}{l} = \frac{y-0}{m} = \frac{z-0}{n} \text{ i.e., } \frac{x}{l} = \frac{y}{m} = \frac{z}{n} \qquad \text{...(1)}$$

$$\therefore \qquad \cos\theta = \frac{l(-1) + m(1) + n(2)}{\sqrt{l^2 + m^2 + n^2}\,\sqrt{6}}$$

or

$$\frac{7}{\sqrt{84}} = \frac{-l + m + 2n}{\sqrt{6}\,\sqrt{l^2 + m^2 + n^2}}$$

Squaring and simplifying

$$\frac{49}{14} = \frac{(-l + m + 2n)^2}{l^2 + m^2 + n^2}$$

or $$7(l^2 + m^2 + n^2) = 2(l^2 + m^2 + 4n^2 - 2lm + 4mn - 4nl)$$

or $$5l^2 + 5m^2 - n^2 + 4lm - 8mn + 8nl = 0$$

As cone is the locus of the generator (*i*) eliminate l, m, n with the help of (1)

$$\therefore \qquad 5x^2 + 5y^2 - z^2 + 4xy - 8yz + 8xz = 0.$$

Example 3. *Find the equation of the cone whose vertex is (1, 2, 3) and which passes through the circle* $x^2 + y^2 + z^2 = 4$, $x + y + z = 1$. **(P.T.U., Dec. 2010, Jan. 2010)**

Sol. Let any line through (1, 2, 3) be

$$\frac{x-1}{l} = \frac{y-2}{m} = \frac{z-3}{n} \qquad \text{...(1)}$$

Any point on (1) is P$(lr + 1, mr + 2, nr + 3)$

It will intersect the circle

$$x^2 + y^2 + z^2 = 4, \quad x + y + z = 1$$

if P$(lr + 1, mr + 2, nr + 3)$ satisfies this equation

i.e., $$(lr + 1)^2 + (mr + 2)^2 + (nr + 3)^2 = 4 \quad ...(2)$$

$$lr + 1 + mr + 2 + nr + 3 = 1 \quad ...(3)$$

Eliminate 'r' from (2) and (3)

From (3), $$r = \frac{-5}{l + m + n}$$

Substituting in (2), $$r^2(l^2 + m^2 + n^2) + 2r(l + 2m + 3n) + 10 = 0$$

i.e., $$(l^2 + m^2 + n^2)\frac{25}{(l + m + n)^2} - \frac{10(l + 2m + 3n)}{l + m + n} + 10 = 0$$

or $$25(l^2 + m^2 + n^2) - 10(l + 2m + 3n)(l + m + n) + 10(l + m + n)^2 = 0$$

or $$5l^2 + 3m^2 + n^2 - 2lm - 6mn - 4nl = 0 \quad ...(4)$$

Eliminate l, m, n from (4) with the help of (1)

$$5(x - 1)^2 + 3(y - 2)^2 + (z - 3)^2 - 2(x - 1)(y - 2) - 6(y - 2)(z - 3) - 4(z - 3)(x - 1) = 0$$

or $$5x^2 + 3y^2 + z^2 - 2xy - 6yz - 4zx + 6x + 8y + 10z - 26 = 0$$

which is the required equation of the cone.

Example 4. *Find the semi-vertical angle and the equation of the right circular cone having its vertex at the origin and passing through the circle $x^2 + z^2 = 25$, $y = 4$.*

Sol. Axis of the cone is a line through the vertex ⊥ to the plane of the circle *i.e.*, $y = 4$.

∴ axis of the cone is y-axis *i.e.*, $\frac{x}{0} = \frac{y}{1} = \frac{z}{0}$.

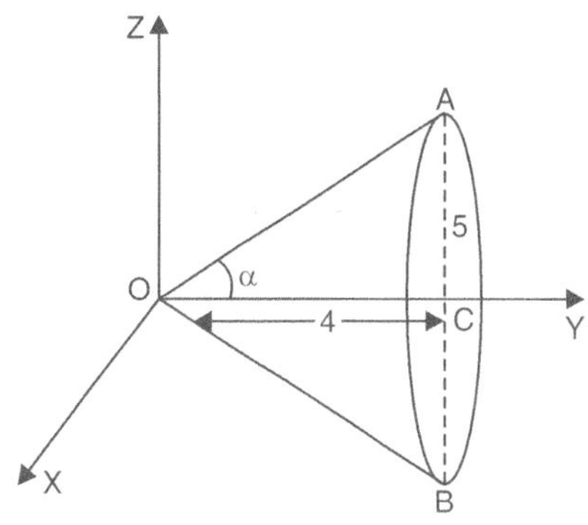

If α is the semi-vertical angle of the cone then

$$\tan \alpha = \frac{AC}{OC} = \frac{5}{4}$$

To find equation of the cone homogenise $x^2 + z^2 = 25$ with the help of $y = 4$. *i.e.*, $\frac{y}{4} = 1$

∴ $$x^2 + z^2 = 25(1)^2 = 25\frac{y^2}{16}$$

or $16(x^2 + z^2) = 25y^2$ is the required equation of the cone.

5.10. ENVELOPING CONE

(*a*) Def. *The locus of the tangent lines from a given point to a sphere is a cone called the* **enveloping cone** *from the point to the sphere.*

(*b*) Equation of the enveloping cone.

To find the equation of the enveloping cone from the point (x_1, y_1, z_1) to the sphere

$$x^2 + y^2 + z^2 = a^2$$

The equation of sphere is $x^2 + y^2 + z^2 = a^2$ ...(1)

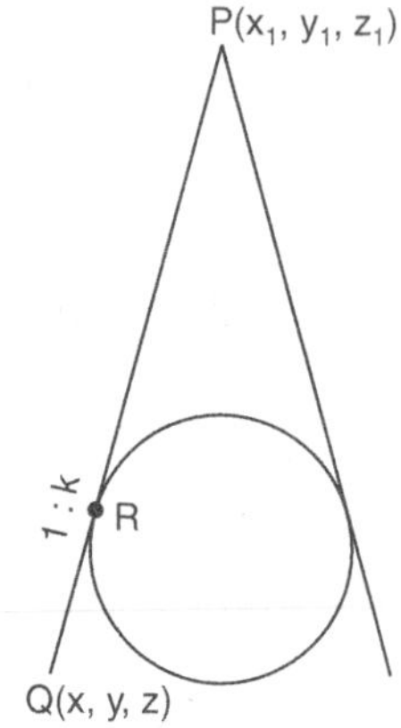

Let $P(x_1, y_1, z_1)$ be the given point.

Let $Q(x, y, z)$ be *any* point on a tangent from P to the sphere.

Let PQ be divided by the point of contact R in the ratio $k : 1$.

Then the co-ordinates of R are $\left(\frac{kx + x_1}{k+1}, \frac{ky + y_1}{k+1}, \frac{kz + z_1}{k+1}\right)$

Since R lies on the sphere (1)

$$\therefore \quad \left(\frac{kx + x_1}{k+1}\right)^2 + \left(\frac{ky + y_1}{k+1}\right)^2 + \left(\frac{kz + z_1}{k+1}\right)^2 = a^2$$

or $$(kx + x_1)^2 + (ky + y_1)^2 + (kz + z_1)^2 = a^2(k + 1)^2$$

or $$k^2x^2 + 2kxx_1 + x_1^2 + k^2y^2 + 2kyy_1 + y_1^2 + k^2z^2 + 2kzz_1 + z_1^2 = a^2(k^2 + 2k + 1)$$

or $$k^2(x^2 + y^2 + z^2 - a^2) + 2k(xx_1 + yy_1 + zz_1 - a^2) + (x_1^2 + y_1^2 + z_1^2 - a^2) = 0 \quad ...(2)$$

which is a quadratic in k.

Since the line PQ touches the sphere, the two values of k must be equal,

$\therefore$ Discriminant of (2) = 0

or $$4(xx_1 + yy_1 + zz_1 - a^2)^2 - 4(x^2 + y^2 + z^2 - a^2)(x_1^2 + y_1^2 + z_1^2 - a^2) = 0$$

| *Using* $b^2 - 4ac = 0$

or $$(x^2 + y^2 + z^2 - a^2)(x_1^2 + y_1^2 + z_1^2 - a^2) = (xx_1 + yy_1 + zz_1 - a^2)^2$$

which is the required equation of the enveloping cone.

Aid to Memory. *If* $S \equiv x^2 + y^2 + z^2 - a^2$, *so that* $S = 0$ *is the equation of sphere, then* $S_1 = x_1^2 + y_1^2 + z_1^2 - a^2$, *i.e.,* S_1 *is the result obtained by substituting the point* (x_1, y_1, z_1) *in S: and* $T \equiv xx_1 + yy_1 + zz_1 - a^2$; *the expression of tangent plane at* (x_1, y_1, z_1) *to the sphere, then the enveloping cone is* $\mathbf{SS_1 = T^2}$.

Note. Check; The equation of the enveloping cone is satisfied by the coordinates of the given point.

Example 5. *Show that the plane* $z = 0$ *cuts the enveloping cone of the sphere* $x^2 + y^2 + z^2 = 11$ *which has its vertex at (2, 4, 1) in a rectangular hyperbola.*

Sol. The equation of the sphere is $x^2 + y^2 + z^2 - 11 = 0$ and the vertex is (2, 4, 1).

Here $$S = x^2 + y^2 + z^2 - 11, x_1 = 2, y_1 = 4, z_1 = 1.$$

$$S_1 = x_1^2 + y_1^2 + z_1^2 - 11 = 4 + 16 + 1 - 11 = 10$$

and $$T = xx_1 + yy_1 + zz_1 - 11 = 2x + 4y + z - 11.$$

$\therefore$ Equation of the enveloping cone is

$$(x^2 + y^2 + z^2 - 11)(10) = (2x + 4y + z - 11)^2$$ | *Using* $SS_1 = T^2$

or $$10(x^2 + y^2 + z^2 - 11) - (2x + 4y + z - 11)^2 = 0$$

This meets the plane $z = 0$ in the curve

$$10(x^2 + y^2 - 11) - (2x + 4y - 11)^2 = 0$$

or $$10(x^2 + y^2) - (4x^2 + 16y^2 + 11) = 0$$

This represents a rectangular hyperbola in the XY-plane

if co-eff. of x^2 + co-eff. of $y^2 = 0$ or if $(10 - 4) + (10 - 16) = 0$ which is true.

Hence the result.

5.11. INTERSECTION OF A LINE AND A CONE

To find the intersection of the line $\frac{x - x_1}{l} = \frac{y - y_1}{m} = \frac{z - z_1}{n}$ and the cone

$$ax^2 + by^2 + cz^2 + 2fyz + 2gzx + 2hxy = 0.$$

Proof. Equation of the line is $\frac{x - x_1}{l} = \frac{y - y_1}{m} = \frac{z - z_1}{n}$ ...(1)

Equation of the cone is $ax^2 + by^2 + cz^2 + 2fyz + 2gzx + 2hxy = 0$ ...(2)

Any point on the line is $(x_1 + lr, y_1 + mr, z_1 + nr)$

It lies on the cone (2)

$$\therefore \quad a(x_1 + lr)^2 + b(y_1 + mr)^2 + c(z_1 + nr)^2 + 2f(y_1 + mr)(z_1 + nr) + 2g(z_1 + nr)(x_1 + lr) + 2h(x_1 + lr)(y_1 + mr) = 0$$

or
$$r^2(al^2 + bm^2 + cn^2 + 2fmn + 2gnl + 2hlm) + 2r[(ax_1 + hy_1 + gz_1)l + (hx_1 + by_1 + fz_1)m + (gx_1 + fy_1 + cz_1)n] + [ax_1^2 + by_1^2 + cz_1^2 + 2fy_1z_1 + 2gz_1x_1 + 2hx_1y_1] = 0 \quad ...(3)$$

which is quadratic in r, which shows that the line (1) intersects the cone in two points.

If the two values of r are r_1 and r_2 then the co-ordinates of the points of intersection are $P(x_1 + lr_1, y_1 + mr_1, z_1 + nr_1)$ and $Q(x_1 + lr_2, y_1 + mr_2, z_1 + nr_2)$ where $AP = r_1$ and $AQ = r_2$.

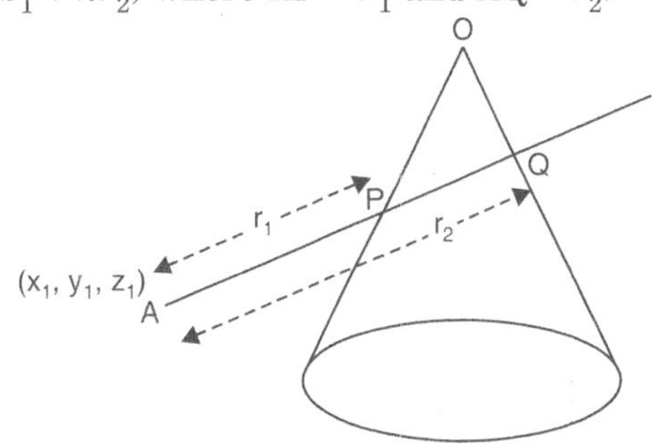

Cor. If the point $A(x_1, y_1, z_1)$ lies on the cone

i.e., $ax_1^2 + by_1^2 + cz_1^2 + 2fy_1z_1 + 2gz_1x_1 + 2hx_1y_1 = 0$

then from (3)

$$r^2(al^2 + bm^2 + cn^2 + 2fmn + 2gnl + 2hlm) + 2r[(ax_1 + hy_1 + gz_1)l + (hx_1 + by_1 + fz_1)m + (gx_1 + fy_1 + cz_1)n] = 0$$

$\Rightarrow$ one of the values of $r = 0$

In case $(ax_1 + hy_1 + gz_1)l + (hx_1 + by_1 + fz_1)m + (gx_1 + fy_1 + cz_1)n = 0$

then the other value of r is also zero *i.e.,* the two points of intersections P and Q coincide with A.

Then the line becomes tangent line to the cone at (x_1, y_1, z_1).

Conclusion line $\frac{x - x_1}{l} = \frac{y - y_1}{m} = \frac{z - z_1}{n}$ is a tangent line to the cone

$$ax^2 + by^2 + cz^2 + 2fyz + 2gzx + 2hxy = 0$$

iff
$$(ax_1 + hy_1 + gz_1)l + (hx_1 + by_1 + fz_1)m + (gx_1 + fy_1 + cz_1)n = 0$$

and
$$ax_1^2 + by_1^2 + cz_1^2 + 2fy_1z_1 + 2gz_1x_1 + 2hx_1y_1 = 0$$

which can be written as

$$(ax_1 + hy_1 + gz_1)x_1 + (hx_1 + by_1 + fz_1)y_1 + (gx_1 + fy_1 + cz_1)z_1 = 0.$$

5.12. EQUATION OF THE TANGENT PLANE AT $P(x_1, y_1, z_1)$ OF THE CONE $ax^2 + by^2 + cz^2 + 2fyz + 2gzx + 2hxy = 0$

Proof. We know that tangent plane at any point is the locus of the tangent line at that point

Let $$\frac{x-x_1}{l} = \frac{y-y_1}{m} = \frac{z-z_1}{n} \qquad ...(1)$$

be any tangent line to the cone

$$ax^2 + by^2 + cz^2 + 2fyz + 2gzx + 2hxy = 0 \text{ at } (x_1, y_1, z_1) \qquad ...(2)$$

$\therefore$ Any point on the line is $(x_1 + lr, y_1 + mr, z_1 + nr)$ it lies on (2)

$$\therefore \quad a(x_1 + lr)^2 + b(y_1 + mr)^2 + c(z_1 + nr)^2 + 2f(y_1 + mr)(z_1 + nr) + 2g(z_1 + nr)(x_1 + lr) + 2h(x_1 + lr)(y_1 + mr) = 0$$

or $$r^2(al^2 + bm^2 + cn^2 + 2fmn + 2gnl + 2hlm) + 2r\,[(ax_1 + hy_1 + gz_1)l + (hx_1 + by_1 + fz_1)m + (gx_1 + fy_1 + cz_1)n] + (ax_1^2 + by_1^2 + cz_1^2 + 2fy_1z_1 + 2gz_1x_1 + 2hx_1y_1) = 0 \qquad ...(3)$$

$\because$ (x_1, y_1, z_1) lies on the cone

$$\therefore \quad ax_1^2 + byx_1^2 + cz_1^2 + 2fy_1z_1 + 2gz_1x_1 + 2hx_1y_1 = 0 \qquad ...(4)$$

Substituting (4) in (3), we get

$$r^2(al^2 + bm^2 + cn^2 + 2fmn + 2gnl + 2hlm) + 2r\,[(ax_1 + hy_1 + gz_1)l + (hx_1 + by_1 + fz_1)m + (gx_1 + fy_1 + cz_1)n] = 0 \qquad ...(5)$$

Now one of the values of r is zero and $\because$ line is a tangent line

$\therefore$ Both the values of r are equal

$\therefore$ Other value of r is also zero

$\therefore$ $(ax_1 + hy_1 + gz_1)l + (hx_1 + by_1 + fz_1)m + (gx_1 + fy_1 + cz_1)n = 0$

To find locus of line (1) eliminate l, m, n with help of (1), we get

$$(ax_1 + hy_1 + gz_1)(x - x_1) + (hx_1 + by_1 + fz_1)(y - y_1) + (gx_1 + fy_1 + cz_1)(z - z_1) = 0$$

or $$(ax_1 + hy_1 + gz_1)x + (hx_1 + by_1 + fz_1)y + (gx_1 + fy_1 + cz_1)z = ax_1^2 + by_1^2 + cz_1^2 + 2hx_1y_1 + 2gz_1x_1 + 2fy_1z = 0 \qquad | \because \text{ of } (4)$$

$\therefore$ Equation of the tangent plane at (x_1, y_1, z_1) to the cone (2) is

$$(ax_1 + hy_1 + gz_1)x + (hx_1 + by_1 + fz_1)y + (gx_1 + fy_1 + cz_1)z = 0 \qquad ...(6)$$

Cor. 1. The tangent plane at any point of a cone passes through its vertex.

[$\because$ (0, 0, 0) Satisfies (6)]

Cor. 2. The tangent plane at any point of a cone touches the cone along the generator through P.

For verification we see that generator through P to the above cone with vertex (0, 0, 0) is $\frac{x}{x_1} = \frac{y}{y_1} = \frac{z}{z_1}$

Any point on this generator is $Q(rx_1, ry_1, rz_1)$.

Tangent plane at Q is

$$(arx_1 + hry_1 + grz_1)x + (hrx_1 + bry_1 + frz_1)y + (grx_1 + fry_1 + crz_1)z = 0$$

$$r \neq 0 \quad \therefore \quad (ax_1 + hy_1 + gz_1)x + (hx_1 + by_1 + fz_1)y + (gx_1 + fy_1 + cz_1)z = 0$$

which is same as tangent plane at (x_1, y_1, z_1) hence the result.

Working rule to write down the equation of the tangent plane at $P(x_1, y_1, z_1)$ to the cone with vertex (0, 0, 0)

Step 1. Represent equation of the cone by

$$F(x, y, z) = ax^2 + by^2 + cz^2 + 2fyz + 2gzx + 2hxy = 0.$$

Step 2. Find $\frac{\partial F}{\partial x}, \frac{\partial F}{\partial y}, \frac{\partial F}{\partial z}$.

Step 3. Find the values of $\frac{\partial F}{\partial x}, \frac{\partial F}{\partial y}, \frac{\partial F}{\partial z}$ at (x_1, y_1, z_1).

Step 4. Equation of the tangent plane at (x_1, y_1, z_1) is

$$\left(\frac{\partial F}{\partial x}\right)_{\text{at }(x_1, y_1, z_1)} x + \left(\frac{\partial F}{\partial y}\right)_{\text{at }(x_1, y_1, z_1)} y + \left(\frac{\partial F}{\partial z}\right)_{\text{at }(x_1, y_1, z_1)} z = 0$$

Working rule to write down the equation of the tangent plane at (x_1, y_1, z_1) to general form of the cone $ax^2 + by^2 + cz^2 + 2fyz + 2gzx + 2hxy + 2ux + 2vy + 2wz + d = 0$

Change x^2 to xx_1, y^2 to yy_1, z^2 to zz_1, $2yz$ to $yz_1 + y_1z$, $2zx$ to $zx_1 + z_1x$, $2xy$ to $xy_1 + x_1y$, $2x$ to $x + x_1$, $2y$ to $y + y_1$, $2z$ to $z + z_1$.

Example 6. *Find the equation of the tangent plane at the point (–3, 0, –1) of the cone*

$$4x^2 - y^2 + 2z^2 + 2xy - 3yz + 12x - 11y + 6z + 4 = 0.$$

Sol. Equation of the cone is

$$4x^2 - y^2 + 2z^2 + 2xy - 3yz + 12x - 11y + 6z + 4 = 0$$

Equation of the tangent plane at (–3, 0, –1) is

$$4(-3x) - (0 \cdot y) + 2 \cdot [z(-1)] + [x \cdot 0 + y(-3)]$$
$$-\frac{3}{2} \cdot [y(-1) + z \cdot 0] + 6 \cdot (x - 3) - \frac{11}{2} \cdot (y + 0) + 3 \cdot (z - 1) + 4 = 0$$

(by applying working rule)

$$-12x - 2z - 3y + \frac{3y}{2} + 6x - 18 - \frac{11}{2}y + 3z - 3 + 4 = 0$$

or $$-6x - 7y + z - 17 = 0$$

or $$6x + 7y - z + 17 = 0.$$

Example 7. *Find the plane which touches the cone $x^2 + 2y^2 + 4z^2 - 6yz + 4zx - 5xy = 0$, along the generator whose d.r.'s are <1, 1, 1>.*

Sol. Equation of the generator of the cone is $\frac{x-0}{1} = \frac{y-0}{1} = \frac{z-0}{1}$

[Vertex of the cone is (0, 0, 0)]

or $$x = y = z$$

Any point on generator (r, r, r)

Equation of the tangent plane at (r, r, r) to the cone

$$F(x, y, z) = x^2 + 2y^2 + 4z^2 - 6yz + 4zx - 5xy = 0 \text{ is}$$

$$\left(\frac{\partial F}{\partial x}\right)_{\text{at }(r, r, r)} x + \left(\frac{\partial F}{\partial y}\right)_{\text{at }(r, r, r)} y + \left(\frac{\partial F}{\partial z}\right)_{\text{at }(r, r, r)} z = 0$$

$$(2x + 4z - 5y)_{\text{at }(r, r, r)}\, x + (4y - 6z - 5x)_{\text{at }(r, r, r)}\, y + (8z + 6y - 4x)_{\text{at }(r, r, r)}\, z = 0$$

or $$rx - 7ry + 6rz = 0$$

or $$x - 7y + 6z = 0 \quad \text{is the required tangent plane.}$$

Some Important Results from Determinants

1. $$\Delta = \begin{vmatrix} a & h & g \\ h & b & f \\ g & f & c \end{vmatrix} = abc + 2fgh - af^2 - bg^2 - ch^2$$

2. A, B, C, F, G, H are cofactors of a, b, c, f, g, h

$$A = bc - f^2 \qquad F = gh - af$$

$$B = ca - g^2 \qquad G = hf - bg$$

$$C = ab - h^2 \qquad H = fg - ch.$$

3. $$BC - F^2 = a\Delta, \quad CA - G^2 = b\Delta, \quad AB - H^2 = c\Delta$$

4. $$\begin{vmatrix} a & h & g & u \\ h & b & f & v \\ g & f & c & w \\ u & v & w & 0 \end{vmatrix} = -\,[Au^2 + Bv^2 + Cw^2 + 2Fvw + 2hwu + 2Huv].$$

5.13. CONDITION OF TANGENCY OF A PLANE AND A CONE

Find the condition that the plane $lx + my + nz = 0$ may touch the cone

$$ax^2 + by^2 + cz^2 + 2fyz + 2gzx + 2hxy = 0.$$

Proof. Equation of the cone is

$$ax^2 + by^2 + cz^2 + 2fyz + 2gzx + 2hxy = 0 \qquad ...(1)$$

Equation of the tangent plane is

$$lx + my + nz = 0 \qquad ...(2)$$

Let (2) touches (1) at $P(x_1, y_1, z_1)$ equation of the tangent plane at P is

$$(ax_1 + hy_1 + gz_1)\,x + (hx_1 + by_1 + fz_1)\,y + (gx_1 + fy_1 + cz_1)z = 0 \qquad ...(3)$$

(3) is same as (2)

$$\therefore \quad \frac{ax_1 + hy_1 + gz_1}{l} = \frac{hx_1 + by_1 + fz_1}{m} = \frac{gx_1 + fy_1 + cz_1}{n} = -k \text{ (say)}$$

$$\therefore \quad ax_1 + hy_1 + gz_1 + kl = 0 \qquad ...(4)$$

$$hx_1 + by_1 + fz_1 + km = 0 \qquad ...(5)$$

$$gx_1 + fy_1 + cz_1 + kn = 0 \qquad ...(6)$$

$P(x_1, y_1, z_1)$ also lies on (2)

$$\therefore \quad lx_1 + my_1 + nz_1 + k \cdot 0 = 0 \qquad ...(7)$$

Eliminate x_1, y_1, z_1, from (4), (5), (6), (7), we get

$$\begin{vmatrix} a & h & g & l \\ h & b & f & m \\ g & f & c & n \\ l & m & n & 0 \end{vmatrix} = 0$$

or $$-(Al^2 + Bm^2 + Cn^2 + 2Fmn + 2Gnl + 2Hlm) = 0$$

or $$Al^2 + Bm^2 + Cn^2 + 2Fmn + 2Gnl + 2Hlm = 0$$

where A, B, C, F, G, H have their usual meanings hence the conditions obtained.

Example 8. *Show that the plane $3x - 6y + 4z = 0$ touches the cone $2x^2 - 4y^2 + 5z^2 + 4zx - 2xy = 0$*

Sol. Let $3x - 6y + 4z = 0$ ...(1)

touches the cone

$$2x^2 - 4y^2 + 5z^2 + 4zx - 2xy = 0 \text{ at } (x_1, y_1, z_1)$$

$\therefore$ Equation of the tangent plane at (x_1, y_1, z_1) to the cone in which $a = 2$, $b = -4$, $c = 5$, $f = 0$, $g = 2$, $h = -1$ is

$$x(2x_1 - y_1 + 2z_1) + y(-x_1 - 4y_1 + 0.z_1) + z(2x_1 + 0.y_1 + 5z_1) = 0$$

which is same as (1)

$$\therefore \quad \frac{2x_1 - y_1 + 2z_1}{3} = \frac{-x_1 - 4y_1}{-6} = \frac{2x_1 + 5z_1}{4} = \lambda \text{ (say)}$$

$$\therefore \quad 2x_1 - y_1 + 2z_1 - 3\lambda = 0 \quad ...(2)$$

$$-x_1 - 4y_1 + 6\lambda = 0 \quad ...(3)$$

$$2x_1 + 5z_1 - 4\lambda = 0 \quad ...(4)$$

(x_1, y_1, z_1) also lies on (1)

$$\therefore \quad 3x_1 - 6y_1 + 4z_1 = 0 \quad ...(5)$$

Eliminate x_1, y_1, z_1, λ from (2), (3), (4), and (5)

Operate $R_1 + 2R_2$, $R_3 + 2R_2$, $R_4 + 3R_2$

$$\begin{vmatrix} 2 & -1 & 2 & -3 \\ -1 & -4 & 0 & 6 \\ 2 & 0 & 5 & -4 \\ 3 & -6 & 4 & 0 \end{vmatrix} = 0 \quad \text{or} \quad \begin{vmatrix} 0 & -9 & 2 & 9 \\ -1 & -4 & 0 & 6 \\ 0 & -8 & 5 & 8 \\ 0 & -18 & 4 & 18 \end{vmatrix} = 0$$

R_1 and R_4 are identical

$\therefore$ Value of the determinant = 0

Hence 0 = 0 ; plane touches the cone.

5.14. ANGLE BETWEEN TWO LINES IN WHICH A PLANE THROUGH THE VERTEX (Origin) CUTS A CONE

Method. To find the angle between two lines in which a plane through the vertex *i.e.*, origin cuts a cone, we have to find the d.r.'s of the two lines passing through the origin. These d.r's satisfy the equation of the plane as well as equation of the cone. Solve the two equations for d.r.'s and hence angle can be obtained.

The method is explained by the following solved examples:

Example 9. *Find the angle between the lines of section of the plane*

$3x + y + 5z = 0$ and the cone $6yz - 2zx + 5xy = 0$.

Sol. The given plane is $3x + y + 5z = 0$ ...(1)

and the cone is $6yz - 2zx + 5xy = 0$ ...(2)

Let the equation of a line of section be $\frac{x}{l} = \frac{y}{m} = \frac{z}{n}$ ...(3)

Then $\because$ it lies on the plane (1),

$$\therefore \quad 3l + m + 5n = 0 \quad ...(4)$$

Again, $\because$ (3) lies on cone (2), $\therefore$ its d.c.'s satisfy the equation of cone. Thus,

$$6mn - 2nl + 5lm = 0 \quad ...(5)$$

To eliminate m from (4) and (5)

From (4), $m = -(3l + 5n)$. Putting the value in (5)

$$-6n(3l + 5n) - 2nl - 5l(3l + 5n) = 0 \quad \text{or} \quad 15l^2 + 45ln + 30n^2 = 0$$

or $l^2 + 3nl + 2n^2 = 0$ or $(l + n)(l + 2n) = 0$

∴ Either $l + n = 0$ | or $l + 2n = 0$

or $l + 0.\,m + n = 0$ | or $l + 0.m + 2n = 0$

Also from (4), $3l + m + 5n = 0$ | and $3l + m + 5n = 0$ | *From* (4)

∴ $\dfrac{l}{0-1} = \dfrac{m}{3-5} = \dfrac{n}{1-0}$ | ∴ $\dfrac{l}{0-2} = \dfrac{m}{6-5} = \dfrac{n}{1}$

or $\dfrac{l}{1} = \dfrac{m}{2} = \dfrac{n}{-1}$ | or $\dfrac{l}{-2} = \dfrac{m}{1} = \dfrac{n}{1}$

Thus, the d.r.'s of the two lines are <1, 2, –1> and <–2, 1, 1>.

∴ If θ is the angle between the two lines of section, then

$$\cos\theta = \frac{1(-2) + 2(1) + (-1)(1)}{\sqrt{1+4+1}\,\sqrt{4+1+1}} = \frac{-1}{6} = \frac{1}{6} \text{ (numerically)}$$

∴ $$\theta = \cos^{-1}\left(\frac{1}{6}\right).$$

Example 10. *Prove that the plane $ax + by + cz = 0$ cuts the cone $yz + zx + xy = 0$ in perpendicular lines if $\frac{1}{a} + \frac{1}{b} + \frac{1}{c} = 0$.* **(P.T.U., Dec. 2004)**

Sol. The given plane is $ax + by + cz = 0$ ...(1)

and the cone is $yz + zx + xy = 0$ ...(2)

Let one of the lines of section be $\dfrac{x}{l} = \dfrac{y}{m} = \dfrac{z}{n}$.

Then this line lies on (1),

∴ $$al + bm + cn = 0 \quad ...(3)$$

Also the line lies on the cone, ∴ its d.c.'s satisfy the equation of the cone, and thus

$$mn + nl + lm = 0 \quad ...(4)$$

From (3), $$n = -\frac{al + bm}{c}$$

Putting this value in (4), we get

$$-m\left(\frac{al + bm}{c}\right) - l\left(\frac{al + bm}{c}\right) + lm = 0$$

or $$m(al + bm) + l(al + bm) - clm = 0 \quad \text{or} \quad al^2 + lm(a + b - c) + bm^2 = 0$$

Dividing throughout by m^2, we get $a \cdot \dfrac{l^2}{m^2} + \dfrac{l}{m}(a + b - c) + b = 0$ which is a quadratic in $\dfrac{l}{m}$.

Let $\dfrac{l_1}{m_1} \cdot \dfrac{l_2}{m_2}$ be its two roots, then

Product of the roots $= \dfrac{l_1}{m_1} \cdot \dfrac{l_2}{m_2} = \dfrac{b}{a}$ or $\dfrac{l_1 l_2}{m_1 m_2} = \dfrac{1}{a} \Big/ \dfrac{1}{b}$ or $\dfrac{l_1 l_2}{\frac{1}{a}} = \dfrac{m_1 m_2}{\frac{1}{b}} = \dfrac{n_1 n_2}{\frac{1}{c}}$

| *By symmetry*

Now the lines will be at right angles if $l_1l_2 + m_1m_2 + n_1n_2 = 0$ or if $\frac{1}{a} + \frac{1}{b} + \frac{1}{c} = 0$ hence the result.

Example 11. *Find the equations of the lines in which the plane $3x + 4y + z = 0$ cuts the cone $15x^2 - 32y^2 - 7z^2 = 0$.*

Sol. Let the equation of line in which the plane $3x + 4y + z = 0$ ...(1)

cuts the cone $15x^2 - 32y^2 - 7z^2 = 0$ ...(2)

be
$$\frac{x}{l} = \frac{y}{m} = \frac{z}{n} \quad ...(3)$$

(3) lies on (1) and (2)

$$\therefore \quad 3l + 4m + n = 0 \quad ...(4)$$
$$15l^2 - 32m^2 - 7n^2 = 0 \quad ...(5)$$

Eliminate n from (4) and (5)

$$n = -(3l + 4m)$$
$$\therefore \quad 15l^2 - 32m^2 - 7(3l + 4m)^2 = 0$$
or
$$48l^2 + 168\, lm + 144\, m^2 = 0$$
$$6l^2 + 21lm + 18m^2 = 0$$
or
$$2l^2 + 7lm + 6m^2 = 0$$
or
$$2l^2 + 4lm + 3lm + 6m^2 = 0$$
or
$$2l(l + 2m) + 3m(l + 2m) = 0$$
or
$$2l + 3m = 0 \quad \text{or} \quad l + 2m = 0$$

Combine these with (4)

or $2l + 3m + 0 \,.\, n = 0$	$l + 2m + 0 \,.\, n = 0$
and $3l + 4m + n = 0$	$3l + 4m + n = 0$
$\frac{l}{3} = \frac{m}{-2} = \frac{n}{-1}$	$\frac{l}{2} = \frac{m}{-1} = \frac{n}{-2}$

$\therefore$ Equations of the two lines are

$$\frac{x}{3} = \frac{y}{-2} = \frac{z}{-1} \quad \text{and} \quad \frac{x}{2} = \frac{y}{-1} = \frac{z}{-2}.$$

TEST YOUR KNOWLEDGE

1. Find the equation of the cone whose vertex is at the point (3, 1, 2) and whose guiding curve is the ellipse $2x^2 + 3y^2 = 1$, $z = 0$.
2. Find the equation of the cone whose vertex is (α, β, γ) and guiding curve given by

   (*i*) $y^2 = 4ax$, $z = 0$ (*ii*) $\frac{x^2}{a^2} + \frac{y^2}{b^2} = 1$, $z = 0$
3. Prove that a line which passes through point (α, β, γ) and meets the parabola $z^2 = 4ax$, $y = 0$ lies on the cone $(\beta z - \gamma y)^2 = 4a(\beta - \gamma)(\beta x - \alpha y)$.
4. Find the equation of a right circular cone generated when the straight line $2x + 3y = 6$, $z = 0$ revolves about y-axis.

   [**Hint.** Consult solved example 2]

5. Find the equation of the cone with vertex at (1, 1, 1) and passing through the curve of intersection of $x^2 + y^2 + z^2 = 1$ and $x + y + z = 1$.

   [**Hint.** Consult solved example 3]

6. Find the equation of the surface generated by revolving the line $y - z = 0, x = 0$ about the axis $z = 2, x = 0$.

   [**Hint.** Consult solved example 2)

7. Find the enveloping cone of the sphere $x^2 + y^2 + z^2 - 2x + 4z - 1 = 0$ with its vertex at (1, 1, 1).

8. Find the angle between the lines of section of the plane $6x - y - 2z = 0$ and the cone $108\,x^2 - 7y^2 - 20z^2 = 0$.

9. Find the condition that the plane $ux + vy + wz = 0$ may cut the cone $ax^2 + by^2 + cz^2 = 0$ in perpendicular lines.

   [**Hint.** Consult S.E. 10]

10. Find the equation of the plane which touches the cone $x^2 + 2y^2 - 3z^2 - 2yz + 5zx + 3xy = 0$ along the generator whose d.r.'s are <1, 1, –1>

    [**Hint.** Consult S.E. 7]

Answers

1. $2x^2 + 3y^2 + 5z^2 - 3yz - 6zx + z - 1 = 0$
2. (*i*) $(\beta z - \gamma y)^2 = 4a\,(z - \gamma)\,(\alpha z - \gamma x)$ (*ii*) $\dfrac{1}{a^2}(\alpha z - \gamma x) + \dfrac{1}{b^2}(\beta z - \gamma y)^2 = (z - \gamma)^2$
4. $4x^2 - 9y^2 + 4z^2 + 36y - 36 = 0$
5. $x^2 + y^2 + z^2 - 2(xy + yz + zx) + 2(x + y + z) - 3 = 0$
6. $x^2 - y^2 + z^2 + 4y - 4z = 0$ 7. $4x^2 + 3y^2 - 5z^2 - 6yz - 8x + 16z - 4 = 0$
8. $\cos^{-1}\left(\dfrac{20}{21}\right)$ 9. $(b + c)u^2 + (c + a)\,v^2 + (a + b)\,w^2 = 0$
10. $y + z = 0$.

THE CYLINDER

5.15. THE CYLINDER

(*a*) **Definitions.** A cylinder is the surface generated by a straight line which is parallel to a fixed straight line and satisfies one more condition; for instance, it may intersect a fixed curve or touch a given surface.

The moving line in any direction is called the *generator* of the cylinder and if the generator intersects a fixed curve, then the curve is called the *Guiding Curve.*

(*b*) **Quadratic cylinder.** A cylinder whose equation is second degree in x, y, z is called a quadratic cylinder.

(*c*) **A right circular cylinder.** A right circular cylinder is the surface generated by a straight line which is parallel to a fixed line and is at a constant distance from it. The fixed line is called *the Axis* and constant distance is called the radius of the cylinder. The section of a right circular cylinder by any plane perpendicular to its axis is a circle.

5.16. EQUATION TO THE CYLINDER WHOSE GENERATORS ARE PARALLEL TO THE LINE $\frac{x}{l} = \frac{y}{m} = \frac{z}{n}$ AND INTERSECT THE CURVE $ax^2 + by^2 + 2hxy + 2gx + 2fy + c = 0, \quad z = 0$

The given line is $\frac{x}{l} = \frac{y}{m} = \frac{z}{n}$ ...(1)

and the base conic is $ax^2 + by^2 + 2hxy + 2gx + 2fy + c = 0, z = 0$ ...(2)

Let (x_1, y_1, z_1) be any point on the cylinder.

The equations of the generator through (x_1, y_1, z_1) are $\frac{x - x_1}{l} = \frac{y - y_1}{m} = \frac{z - z_1}{n}$ ...(3)

It meets the plane $z = 0$, where putting $z = 0$ in (3), we have

$$\frac{x - x_1}{l} = \frac{y - y_1}{m} = \frac{z - z_1}{n} \quad \text{or} \quad x - x_1 = -\frac{1}{n}z_1, y - y_1 = -\frac{m}{n}z_1$$

or
$$x = x_1 - \frac{1}{n}z_1, y = y_1 - \frac{m}{n}z_1.$$

If this point lies on the conic (2), then

$$a\left(x_1 - \frac{l}{n}z_1\right)^2 + b\left(y_1 - \frac{m}{n}z_1\right)^2 + 2h\left(x_1 - \frac{l}{n}z_1\right)\left(y_1 - \frac{m}{n}z_1\right) + 2g\left(x_1 - \frac{l}{n}z_1\right) + 2f\left(y_1 - \frac{m}{n}z_1\right) + c = 0.$$

$\therefore$ Locus of (x_1, y_1, z_1) is [changing (x_1, y_1, z_1) to (x, y, z)],

$$a\left(x - \frac{l}{n}z\right)^2 + b\left(y - \frac{m}{n}z\right)^2 + 2h\left(x - \frac{l}{n}z\right)\left(y - \frac{m}{n}z\right) + 2g\left(x - \frac{l}{n}z\right) + 2f\left(y - \frac{m}{n}z\right) + c = 0.$$

or $a(nx - lz)^2 + b(ny - mz)^2 + 2h(nx - lz)(ny - mz) + 2ng(nx - lz) + 2nf(ny - mz) + cn^2 = 0$

which is the required equation of the cylinder.

Cor. If the generators are || to z-axis, so that $l = 0$, $m = 0$ and $n = 1$, the above equation of the cylinder becomes $ax^2 + by^2 + 2hxy + 2gx + 2fy + c = 0$ which is *free from z* and is already known to the reader.

Thus, if we are required to find the equation of the cylinder whose generators are || to z-axis, and intersect a given conic, *then eliminate z from the equations of the conic,* which will give the equation of the cylinder.

If the generators are || to x-axis then eliminate x and if the generators are || to y-axis, eliminate y from the equations of the conic to get the equation of cylinder.

ILLUSTRATIVE EXAMPLES

Example 1. (*i*) *Find the equation of the quadric cylinder whose generators intersect the curve $ax^2 + by^2 = 2z$, $lx + my + nz = p$ and are parallel to z-axis.*

(*ii*) *Find the equation of the quadric cylinder with generators parallel to x-axis and passing through the curve $ax^2 + by^2 + cz^2 = l$, $lx + my + nz = p$.*

Sol. (*i*) The given base conic is $ax^2 + by^2 = 2z$, $lx + my + nz = p$ ...(1)

Since the generators of the cylinder are || to z-axis, the equation of the cylinder will be free from z. *Thus we eliminate z from the equations (1) to get the equation of cylinder.*

Now from the first equation of (1), we have $z = \dfrac{ax^2 + by^2}{2}$

Putting in the second equation, $lx + my + n\left(\dfrac{ax^2 + by^2}{2}\right) = p$

or $\quad 2lx + 2my + n(ax^2 + by^2) = 2p \quad$ or $\quad n(ax^2 + by^2) + 2lx + 2my - 2p = 0$

which is the required equation of cylinder.

(*ii*) The given conic is $ax^2 + by^2 + cz^2 = 1,\ lx + my + nz = p$

Since the generators of the cylinder are || to x-axis, the equation of the cylinder should be free from x. *Thus, we eliminate x from the given equations.*

Now from the second equation $x = \dfrac{p - my - nz}{l}$

Putting it in the first equation, we have

$$a\left(\frac{p - my - nz}{l}\right)^2 + by^2 + cz^2 = 1 \quad \text{or} \quad a(p - my - nz)^2 + bl^2y^2 + cl^2z^2 = l^2$$

or $\quad a(p^2 + m^2y^2 + n^2z^2 - 2mpy - 2npz + 2mnyz) + bl^2y^2 + cl^2z^2 - l^2 = 0$

or $\quad (bl^2 + am^2)y^2 + (cl^2 + an^2)z^2 + 2amnyz - 2ampy - 2anpz + (ap^2 - l^2) = 0$

which is the required equation of the cylinder.

Example 2. (*i*) *Find the equation of the cylinder whose generators are parallel to the line $x = -\dfrac{y}{2} = \dfrac{z}{3}$ and whose guiding curve is the ellipse $x^2 + 2y^2 = 1$, $z = 3$.*

(*ii*) *Find the equation of the cylinder whose generating lines have the direction cosines, l, m, n and which passes through the circumference of the fixed circle $x^2 + z^2 = a^2$, in the ZOX plane.* **(P.T.U., May 2008)**

Sol. (*i*) Let (x_1, y_1, z_1) be *any* point on the cylinder. Then equations of the generator through $(x_1\ y_1, z_1)$ is $\dfrac{x - x_1}{1} = \dfrac{y - y_1}{-2} = \dfrac{z - z_1}{3}$ ...(1)

$\left[\because \text{ generators are parallel to the line } \dfrac{x}{1} = \dfrac{y}{-2} = \dfrac{z}{3}\right]$

The equation of the guiding curve is $x^2 + 2y^2 = 1$, $z = 3$

Put $z = 3$ in (1) $\quad \dfrac{x - x_1}{1} = \dfrac{y - y_1}{-2} = \dfrac{3 - z_1}{3}$

or
$$x = x_1 + \frac{1}{3}(3 - z_1) = x_1 - \frac{1}{3}z_1 + 1$$
$$y = y_1 - \frac{2}{3}(3 - z_1) = y_1 + \frac{2}{3}z_1 - 2$$

Substituting the values of x, y in $x^2 + 2y^2 = 1$

$$\left[x_1 - \frac{1}{3}z_1 + 1\right]^2 + 2\left[y_1 + \frac{2}{3}z_1 - 2\right]^2 = 1$$

Locus of (x_1, y_1, z_1) is $\dfrac{(3x - z + 3)^2}{9} + 2\dfrac{(3y + 2z - 6)^2}{9} = 1$

or $$9x^2 + 18y^2 + 9z^2 - 6xz + 24yz + 18x - 72y - 54z + 72 = 0$$

or $$3x^2 + 6y^2 + 3z^2 - 2xz + 8yz + 6x - 24y - 18z + 24 = 0.$$

(*ii*) Let (x_1, y_1, z_1) be *any* point on the cylinder. The equations of the generator through (x_1, y_1, z_1) are

$$\frac{x - x_1}{l} = \frac{y - y_1}{m} = \frac{z - z_1}{n} \qquad ...(1)$$

The given circle is $x^2 + z^2 = a^2$ in the ZOX plane, *i.e.*, $y = 0$ plane.

Thus the given circle is $x^2 + z^2 = a^2, y = 0$ ...(2)

The line (1) meets the plane $y = 0$, where

$$\frac{x - x_1}{l} = \frac{y - y_1}{m} = \frac{z - z_1}{n} \quad \text{or} \quad x = x_1 - \frac{l}{m}y_1, z = z_1 - \frac{n}{m}y_1$$

If this point lies on (2), then

$$\left(x_1 - \frac{l}{m}y_1\right)^2 + \left(z_1 - \frac{n}{m}y_1\right)^2 = a^2 \quad \text{or} \quad (mx_1 - ly_1)^2 + (mz_1 - ny_1)^2 = m^2a^2$$

$\therefore$ Locus of (x_1, y_1, z_1) is $(mx - ly)^2 + (mz - ny)^2 = m^2a^2$

which is the equation of the cylinder.

Example 3. *A cylinder cuts the plane $z = 0$ in the curve $x^2 + \frac{y^2}{4} = \frac{1}{4}$ and has its axis parallel to $3x = -6y = 2z$. Find its equation.*

Sol. Let $(x_1, y_1 z_1)$ be any point on the cylinder.

Equation of the line to which axis is parallel to

$$\frac{x}{1/3} = \frac{y}{-1/6} = \frac{z}{1/2} \quad \text{or} \quad \frac{x}{2} = \frac{y}{-1} = \frac{z}{3}$$

Equation of the generator through (x_1, y_1, z_1) is

$$\frac{x - x_1}{2} = \frac{y - y_1}{-1} = \frac{z - z_1}{3} \qquad ...(1)$$

Cylinder cuts the plane $z = 0$ and the curve $x^2 + \frac{y^2}{4} = \frac{1}{4}$ ...(2)

$\therefore$ $z = 0$ satisfies (1)

$$\frac{x - x_1}{2} = \frac{y - y_1}{-1} = \frac{-z_1}{3} \quad \therefore \quad x = x_1 - \frac{2}{3}z_1, y = y_1 + \frac{1}{3}z_1$$

this point lies on (2)

$$\therefore \quad \left(x_1 - \frac{2}{3}z_1\right)^2 + \frac{1}{4}\left(y_1 + \frac{1}{3}z_1\right)^2 = \frac{1}{4}$$

or $$x_1^2 + \frac{y_1^2}{4} + \frac{17z_1^2}{36} - \frac{4}{3}x_1z_1 + \frac{1}{6}y_1z_1 = \frac{1}{4}$$

or $$36x_1^2 + 9y_1^2 + 17z_1^2 - 48x_1z_1 + 6y_1z_1 - 9 = 0$$

Locus of (x_1, y_1, z_1) is

$$36x^2 + 9y^2 + 17z^2 - 48xz + 6yz - 9 = 0.$$

5.17. EQUATION OF RIGHT CIRCULAR CYLINDER

(*a*) Standard Form. To show that the equation of the right circular cylinder whose axis is the *z*-axis and radius *a*, is $x^2 + y^2 = a^2$.

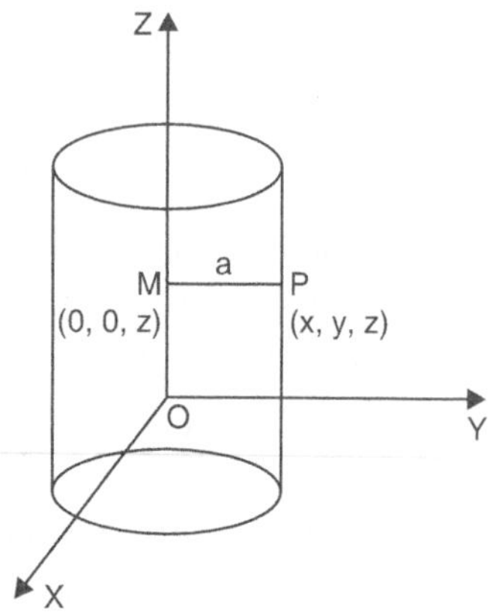

Let P(x, y, z) be *any* point on the cylinder. Draw PM ⊥ Z-axis, so that OM = z. Then the co-ordinates of M are (0, 0, z).

Also MP = radius of cylinder = a (given)

But $$MP = \sqrt{(x-0)^2 + (y-0)^2 + (z-z)^2}$$

$$= \sqrt{x^2 + y^2} \qquad | \textit{ distance formula}$$

$$\therefore \quad \sqrt{x^2 + y^2} = a \quad \text{or} \quad x^2 + y^2 = a^2$$

which is the required equation.

(*b*) General Form

To find the equation to the right circular cylinder whose radius is *r* and axis is the line $\dfrac{x-\alpha}{l} = \dfrac{y-\beta}{m} = \dfrac{z-\gamma}{n}$.

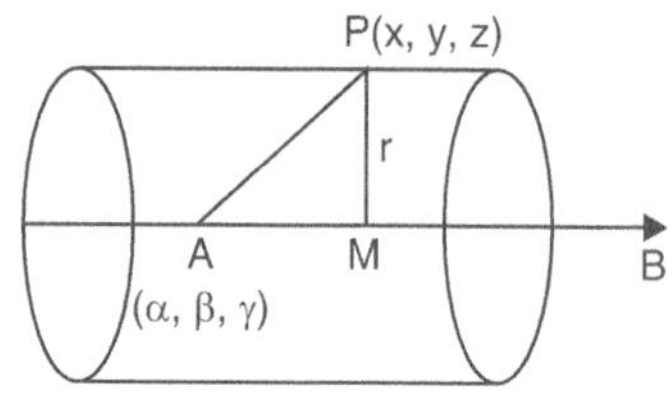

Let AB be the axis of the cylinder whose equations are

$$\frac{x-\alpha}{l} = \frac{y-\beta}{m} = \frac{z-\gamma}{n}. \qquad ...(1)$$

where A(α, β, γ) is a point on it.

The d.r.'s of the axis are $<l, m, n>$. Dividing each by

$\sqrt{l^2 + m^2 + n^2} = \sqrt{\Sigma l^2}$, its actual d.c.'s are $\left\langle \dfrac{l}{\sqrt{\Sigma l^2}}, \dfrac{m}{\sqrt{\Sigma l^2}}, \dfrac{n}{\sqrt{\Sigma l^2}} \right\rangle$.

Let P(x, y, z) be any point on the cylinder. Draw PM ⊥ on the axis AB and join PA.

Then PM = radius of cylinder = r *(given)*

AM = projection of AP on the axis

$$= \frac{l(x-\alpha) + m(y-\beta) + n(z-\gamma)}{\sqrt{l^2 + m^2 + n^2}}$$

| *Using* $l(x_2 - x_1) + m(y_2 - y_1) + n(z_2 - z_1)$

$$AP^2 = (x-\alpha)^2 + (y-\beta)^2 + (z-\gamma)^2$$

Now $$AP^2 = AM^2 + PM^2$$

$$\therefore \quad (x-\alpha)^2 + (y-\beta)^2 + (z-\gamma)^2 = \frac{[l(x-\alpha) + m(y-\beta) + n(z-\gamma)]^2}{l^2 + m^2 + n^2} + r^2$$

which is the required equation of the cylinder.

Example 4. *Find the equation of the right circular cylinder of radius 2 and whose axis passes through the point (1, 2, 3) and has direction cosines proportional to (2, 1, 2).*

(P.T.U., Dec. 2003)

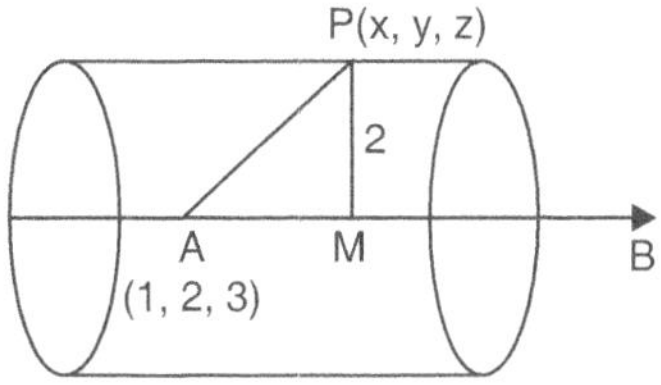

Sol. Let AB be the axis of the cylinder whose equation is

$$\frac{x-1}{2}=\frac{y-2}{1}=\frac{z-3}{2} \quad ...(1)$$

Its d.r.'s are <2, 1, 2>

Dividing each by $\sqrt{4+1+4}$ = 3, its actual d.c.'s are $\left\langle \frac{2}{3}, \frac{1}{3}, \frac{2}{3} \right\rangle$.

Let P(x, y, z) be any point on the cylinder. Join AP and draw PM ⊥ on AB so that

MP = radius of cylinder = 2 (given)

Now in rt. ∠d ΔAMP,

$$AP^2 = AM^2 + MP^2 = AM^2 + 4 \quad ...(2) \quad | \because \ MP = 2$$

Also $AP^2 = (x-1)^2 + (y-2)^2 + (z-3)^2$ | *distance formula*

and AM = projection of AP on the axis AB

$$= \frac{2}{3}(x-1) + \frac{1}{3}(y-2) + \frac{2}{3}(z-3) = \frac{2x+y+2z-10}{3}$$

$\therefore$ From (2), $(x-1)^2 + (y-2)^2 + (z-3)^2 = \dfrac{(2x+y+2z-10)^2}{9} + 4$

or $9[(x-1)^2 + (y-2)^2 + (z-3)^2] = (\overline{2x+y+2z} - 10)^2 + 36$

$$= (2x+y+2z)^2 + 100 - 20(2x+y+2z) + 36$$

or $5x^2 + 8y^2 + 5z^2 - 4yz - 8zx - 4xy + 22x - 16y - 14z - 10 = 0$

which is the required equation of the cylinder.

Example 5. *The radius of a normal section of (a right circular) cylinder is 2, the axis lies along the line* $\frac{x-1}{2}=\frac{y+3}{-1}=\frac{z-2}{5}$; *Find its equation.*

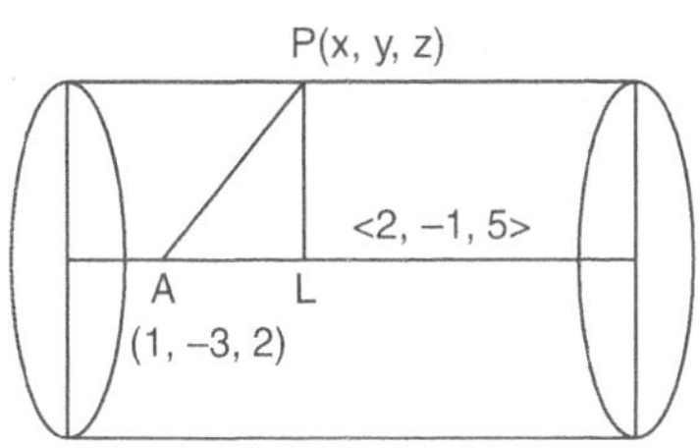

Sol. Equation of the axis of the cylinder is $\frac{x-1}{2}=\frac{y+3}{-1}=\frac{z-2}{5}$. It passes through the fixed point A (1, –3, 2). Let P (x, y, z) be any point on the cylinder and PL, the perpendicular from P on axis = Radius of the cylinder = 2

$$AP^2 = AL^2 + PL^2 \quad ...(1)$$

$AP^2 = (x-1)^2 + (y+3)^2 + (z-2)^2$ by distance formula

$$AL = \text{Projection of AP on axis} = \frac{(x-1).2 + (y+3)(-1) + (z-2).5}{\sqrt{4+1+25}}$$

$\therefore$ from (1), $(x-1)^2 + (y+3)^2 + (z-2)^2 = \dfrac{1}{30}\{2(x-1) - (y+3) + 5(z-2)\}^2 + 4$

or $30(x-1)^2 + 30(y+3)^2 + 30(z-2)^2 = 4(x-1)^2 + (y+3)^2 + 25(z-2)^2$

$$- 4(x-1)(y+3) - 10(y+3)(z-2) + 20(x-1)(z-2) + 120$$

or $\quad 26(x-1)^2 + 29(y+3)^2 + 5(z-2)^2 + 4(x-1)(y+3) + 10(y+3)(z-2)$

$$- 20(x-1)(z-2) - 120 = 0$$

or $$26x^2 + 29y^2 + 5z^2 + 4xy + 10yz - 20zx + 150y + 30z + 75 = 0$$

Example 6. *Find the equation to the right circular cylinder whose guiding circle is*

$$x^2 + y^2 + z^2 = 9,\ x - y + z = 3.$$

Sol. [*The axis of the cylinder is the line through the centre of sphere and* ⊥ *to the plane of the circle and radius of the cylinder is equal to radius of circle.*]

The sphere is $\quad x^2 + y^2 + z^2 = 9 \quad$ ...(1)

and plane is $\quad x - y + z = 3 \quad$ ...(2)

The centre of sphere is O(0, 0, 0) and its radius is OB = 3.

$$\text{OA} = \perp \text{ distance of O(0, 0, 0) from plane (2)}$$

$$= \frac{0-0+0-3}{\sqrt{1+1+1}}$$

$$= -\sqrt{3} = \sqrt{3} \quad [(\textit{numerically})]$$

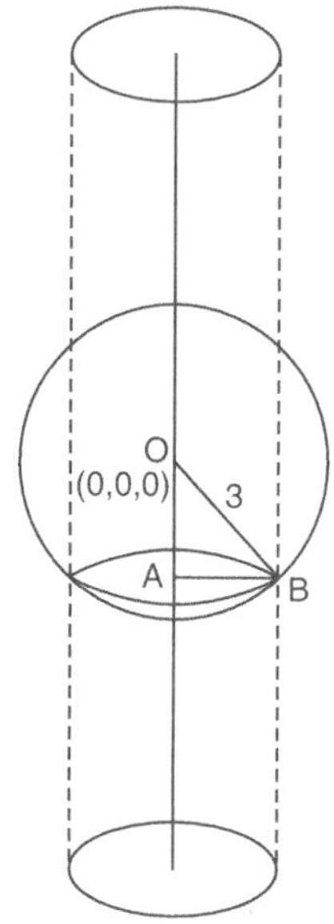

∴ $\quad \text{AB} = \text{radius of circle} = \sqrt{\text{OB}^2 - \text{OA}^2} \quad$ | *From* ΔOAB

$$= \sqrt{9-3} = \sqrt{6}$$

Again, equations of the line through the centre O(0, 0, 0) of the sphere and ⊥ to plane (2) are $\frac{x}{1} = \frac{y}{-1} = \frac{z}{1}$ which is the axis of the cylinder.

Its d.c.'s are proportional to <1, –1, 1>. Dividing each by $\sqrt{1+1+1} = \sqrt{3}$, its actual d.c.'s are $\left\langle \frac{1}{\sqrt{3}}, \frac{-1}{\sqrt{3}}, \frac{1}{\sqrt{3}} \right\rangle$.

Let P(*x*, *y*, *z*) be any point on the cylinder. Join OP and draw MP ⊥ to the axis OA so that

MP = radius of cylinder or circle = $\sqrt{6}$ $\quad$ [*proved above*]

Then in the rt. ∠*d* Δ OMP,

$$\text{OP}^2 = \text{MP}^2 + \text{OM}^2 = 6 + \text{ON}^2 \quad | \because \quad \text{MP} = \sqrt{6} \quad ...(3)$$

Now $\quad \text{OP}^2 = x^2 + y^2 + z^2$

and $\quad \text{OM} = \text{projection of OP on the line OA} = \frac{x-y+z}{\sqrt{3}}$

| *Using* $l(x_2 - x_1) + m(y_2 - y_1) + n(z_2 - z_1)$

∴ From (3), $\quad x^2 + y^2 + z^2 = \left(\frac{x-y+z}{\sqrt{3}}\right)^2 + 6$

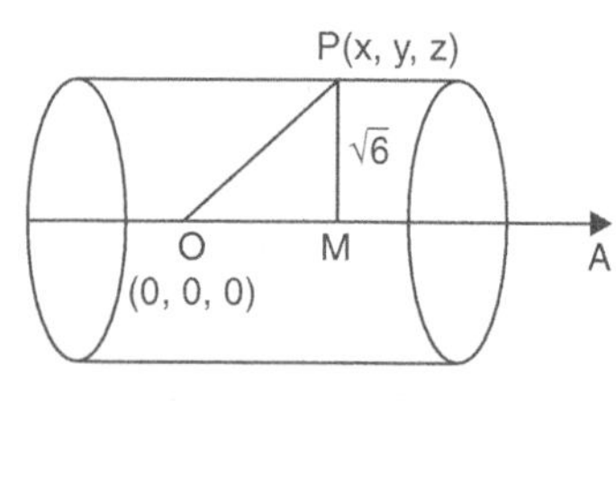

or $$x^2 + y^2 + z^2 + xy + yz - zx - 9 = 0$$

which is the required equation of the cylinder.

Example 7. *A right circular cylinder is such that its section by the plane $x + y = z$ is a circle of radius 3 and centre (1, 2, 3). Find its equation.* **(P.T.U., May 2003)**

Sol. Axis of the cylinder is a line passing through the centre (1, 2, 3) of the circle and perpendicular to the plane of the circle $x + y - z = 0$ *d.r.*'s of the normal to the plane are <1, 1, –1>

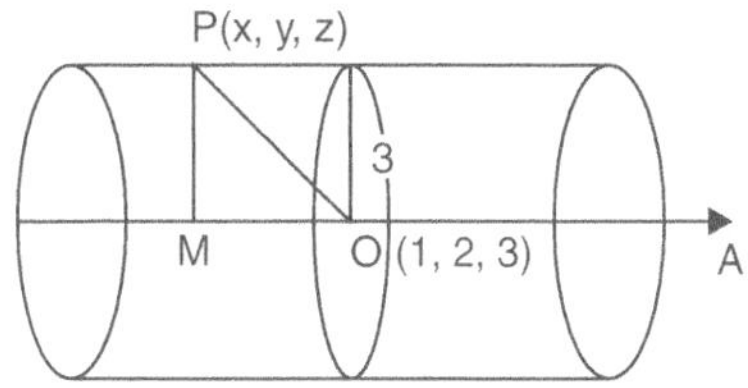

∴ Equation of the axis of the cylinder is

$$\frac{x-1}{1} = \frac{y-2}{1} = \frac{z-3}{-1}$$

i.e.,

$$\frac{x-1}{\frac{1}{\sqrt{3}}} = \frac{y-2}{\frac{1}{\sqrt{3}}} = \frac{z-3}{\frac{-1}{\sqrt{3}}}$$

(by changing *d.r.*'s to *d.c.*'s)

Let P(x, y, z) be any point on the surface of the cylinder ∴ $PO^2 = PM^2 + MO^2$

or $$(x-1)^2 + (y-2)^2 + (z-3)^2 = 9 + \left\{(x-1)\frac{1}{\sqrt{3}} + (y-2)\frac{1}{\sqrt{3}} + (z-3)\left(\frac{-1}{\sqrt{3}}\right)\right\}^2$$

$$(x-1)^2 + (y-2)^2 + (z-3)^2 = 9 + \frac{1}{3}(x+y-z)^2$$

$$3(x^2 - 2x + 1) + 3(y^2 - 4y + 4) + 3(z^2 - 6z + 9) = 27 + x^2 + y^2 + z^2 + 2xy - 2yz - 2zx$$

or $2x^2 + 2y^2 + 2z^2 - 2xy + 2yz + 2zx - 6x - 12y - 18z + 15 = 0$

which is the equation of the cylinder.

5.18. ENVELOPING CYLINDER

Def. *The locus of the tangent lines to sphere which are parallel to a given line is a cylinder known as the enveloping cylinder of the sphere.*

Equation of the enveloping cylinder.

To find the equation of the enveloping cylinder of the sphere $x^2 + y^2 + z^2 = a^2$, whose generators are parallel to the line $\frac{x}{l} = \frac{y}{m} = \frac{z}{n}$.

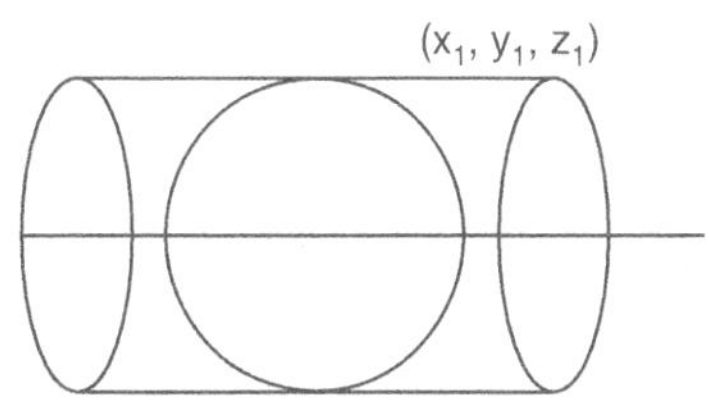

The given sphere is $$x^2 + y^2 + z^2 = a^2 \quad ...(1)$$

and the line is $$\frac{x}{l} = \frac{y}{m} = \frac{z}{n} \quad ...(2)$$

Let P(x_1, y_1, z_1) be any point on a tangent || to the line (2).

Then the equations of this tangent are

$$\frac{x-x_1}{l} = \frac{y-y_1}{m} = \frac{z-z_1}{n} \quad ...(3)$$

Any point on this line is $(lr + x_1, mr + y_1, nr + z_1)$.

If it lies on the sphere (1), then

$$(lr + x_1)^2 + (mr + y_1)^2 + (nr + z_1)^2 = a^2$$

or $$r^2(l^2 + m^2 + n^2) + 2r(lx_1 + my_1 + nz_1) + (x_1^2 + y_1^2 + z_1^2 - a^2) = 0 \quad ...(4)$$

which is a quadratic in r.

Since the line (3) touches the sphere (1), the two values of r must be equal, *i.e.*, the quadratic (4) must have equal roots for which discriminant = 0 or $$4(lx_1 + my_1 + nz_1)^2 - 4(l^2 + m^2 + n^2)(x_1^2 + y_1^2 + z_1^2 - a^2) = 0 \qquad | \; b^2 - 4ac = 0$$

or $$(lx_1 + my_1 + nz_1)^2 = (l^2 + m^2 + n^2)(x_1^2 + y_1^2 + z_1^2 - a^2)$$

$\therefore$ Locus of (x_1, y_1, z_1) is [changing (x_1, y_1, z_1) to (x, y, z)],

$$(lx + my + nz)^2 = (l^2 + m^2 + n^2)(x^2 + y^2 + z^2 - a^2),$$

which is the required equation of the enveloping cylinder.

Example 8. *Find the equation of the enveloping cylinder of the sphere $x^2 + y^2 + z^2 - 2x + 4y = 1$, having its generators parallel to the line $x = y = z$.*

Sol. The given sphere is $x^2 + y^2 + z^2 - 2x + 4y - 1 = 0$ ...(1)

and the line is $$\frac{x}{1} = \frac{y}{1} = \frac{z}{1}. \qquad \text{...(2)}$$

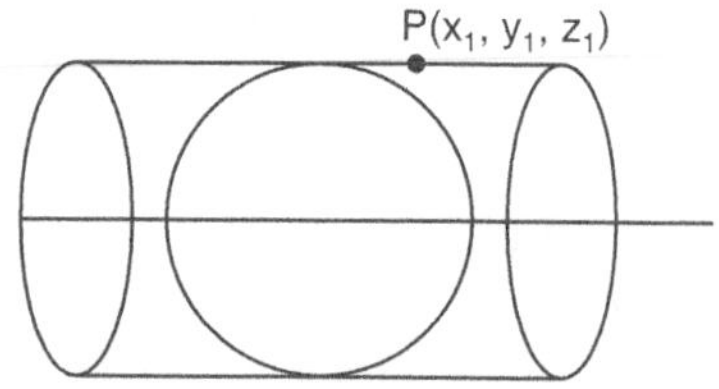

Let $P(x_1, y_1, z_1)$ be *any* point on a tangent which is || to line (2).

$\therefore$ Equations of the tangent line is

$$\frac{x - x_1}{1} = \frac{y - y_1}{1} = \frac{z - z_1}{1} \qquad \text{...(3)}$$

Any point on this line is $(r + x_1, r + y_1, r + z_1)$.

If it lies on the sphere (1), then

$$(r + x_1)^2 + (r + y_1)^2 + (r + z_1)^2 - 2(r + x_1) + 4(r + y_1) - 1 = 0$$

or $$3r^2 + 2r(x_1 + y_1 + z_1 - 1) + (x_1^2 + y_1^2 + z_1^2 - 2x_1 + 4y_1 - 1) = 0 \qquad \text{...(4)}$$

which is a quadratic in r.

Since the line (3) touches the sphere (1), so it meets the sphere only in one point, *i.e.*, the two values of r in (4) must be equal for which the discriminant = 0

$$4(x_1 + y_1 + z_1 - 1)^2 - 12(x_1^2 + y_1^2 + z_1^2 - 2x_1 + 4y_1 - 1) = 0 \qquad | \; b^2 - 4ac = 0$$

or $$3(x_1^2 + y_1^2 + z_1^2 - 2x_1 + 4y_1 - 1) = (x_1 + y_1 + z_1 - 1)^2$$

$\therefore$ Locus of (x_1, y_1, z_1) is [change (x_1, y_1, z_1) to (x, y, z)]

$$3(x^2 + y^2 + z^2 - 2x + 4y - 1) = (\overline{x + y + z} - 1)^2 = (x + y + z)^2 + 1 - 2(x + y + z)$$

or $$x^2 + y^2 + z^2 - xy - yz - zx - 2x + 7y + z - 2 = 0$$

which is the required equation.

TEST YOUR KNOWLEDGE

1. Find the equation of the cylinder whose generators are parallel to the line $\frac{x}{1} = \frac{y}{2} = \frac{z}{3}$ and pass through the curve $x^2 + y^2 = 16, z = 0$.

   [**Hint.** Consult solved example 2(*a*)]

2. Find the equation of the cylinder whose generators are parallel to z-axis and which pass through the curve of intersection of $x^2 + y^2 + z^2 = 1$ and $x + y + z = 1$.

   [**Hint.** Consult solved example 2(*a*)]

3. Find the equation of the right circular cylinder of radius 2 whose axis passes through (1, 2, 3) and has direction cosines proportional to 2, – 3, 6.

4. (*i*) Find the equation of the right circular cylinder of radius 3 and axis $\frac{x-1}{2} = \frac{y-3}{2} = \frac{z-5}{-1}$.

   [**Hint.** Consult solved example 5]

   (*ii*) Find the equation of the right circular cylinder whose axis is the line $x = 2y = -z$ and radius 4.

5. Find the equation of the right circular cylinder of radius a, whose axis passes through the origin and makes equal angles with the co-ordinate axis.

$\left[\textbf{Hint.} \text{ Equation of axis } \frac{x}{1} = \frac{y}{1} = \frac{z}{1}\right]$

6. Find the equation of the enveloping cylinder of the sphere $x^2 + y^2 + z^2 - 2y - 4z - 11 = 0$ having its gnerators parallel to the line $x = -2y - 2z$.

[**Hint.** Consult solved example 8]

Answers

1. $9x^2 + 9y^2 + 5z^2 - 6zx - 12yz - 144 = 0$ **2.** $x^2 + y^2 + xy - x - y = 0$

3. $45x^2 + 40y^2 + 13z^2 + 12xy + 36yz - 24zx - 42x - 280y - 126z + 294 = 0$

4. (*i*) $5x^2 + 5y^2 + 8z^2 + 4yz + 4zx - 8xy - 6x - 42y - 96z + 225 = 0$

(*ii*) $5x^2 + 8y^2 + 5z^2 - 4xy + 4yz + 8zx - 144 = 0$

5. $2(x^2 + y^2 + z^2 - xy - yz - zx) = 3a^2$

6. $5(x^2 + y^2 + z^2) + 2(xy + yz - zx + x - 7y - 11z) - 67 = 0.$

5.19. QUADRIC SURFACE

Def. The locus of the general equation of the second degree in x, y, z is called a quadric surface or a quadratic surface.

$\therefore$ The equation of the quadric surface is $f(x, y, z) = ax^2 + by^2 + cz^2 + 2fyz + 2gzx + 2hxy + 2ux + 2vy + 2wz + d = 0$ which contains ten arbitrary constants (at least one of then is non-zero).

The quadric surfaces are of two types

(1) Degenerate quadrics

(2) Non-degenerate quadrics

Degenerate quadrics. A quadric is said to be degenerate if it represents : (*i*) empty set (*ii*) or a single point (*iii*) or a single line (*iv*) or a single plane (*v*) or a pair of intersecting planes (*vi*) or a pair of parallel planes.

Non-degenerate quadrics. If a quadric is not a degenerate quadric, it is called non degenerate quadric.

5.20. CONICOID

A non degenerate quadric surface is called a **Conicoid**.

The conicoid can be classified into the following types:

Name of the surface	Standard form	
(1) Ellipsoid	$\frac{x^2}{a^2} + \frac{y^2}{b^2} + \frac{z^2}{c^2} = 1$	
(2) Hyperboloid of one sheet	$\frac{x^2}{a^2} + \frac{y^2}{b^2} - \frac{z^2}{c^2} = 1$	where $a > b > c > 0$
(3) Hyperboloid of two sheets	$\frac{x^2}{a^2} - \frac{y^2}{b^2} - \frac{z^2}{c^2} = 1$	where $a > b > c > 0$

or $$\frac{x^2}{a^2}+\frac{y^2}{b^2}-\frac{z^2}{c}=-1$$

(4) Elliptic cone $$\frac{x^2}{a^2}+\frac{y^2}{b^2}-\frac{z^2}{c^2}=0$$

(5) Elliptic paraboloid $$\frac{x^2}{a^2}+\frac{y^2}{b^2}=\frac{2z}{c}$$

(6) Hyperbolic paraboloid $$\frac{x^2}{a^2}-\frac{y^2}{b^2}=\frac{2z}{c}$$

(7) Parabolic cylinder $$y^2=4ax$$

(8) Elliptic cylinder $$\frac{x^2}{a^2}+\frac{y^2}{b^2}=1$$

ILLUSTRATIVE EXAMPLES

Example 1. *What type of surface is represented by $z^2 = 4(1 + x^2 + y^2)$?*

Sol. Equation of the surface is

$$z^2 = 4\,(1 + x^2 + y^2)$$

$$\frac{z^2}{4} = 1 + x^2 + y^2$$

or $$x^2 + y^2 - \frac{z^2}{4} = -1 \quad \text{or} \quad \frac{x^2}{1}+\frac{y^2}{1}-\frac{z^2}{4}=-1$$

Compare it with $\dfrac{x^2}{a^2}+\dfrac{y^2}{b^2}-\dfrac{z^2}{c^2}=-1$

$$a = 1,\ b = 1,\ c = 2$$

which is a hyperboloid of two sheets.

Example 2. *Identify the following surfaces:*

(i) $x^2 + 4y^2 + 3z^2 + 2x - 8y + 9z - 10 = 0$ *(ii)* $x^2 + 2y^2 - z^2 + 2x + 4y - 6z - 18 = 0$

(iii) $3x^2 - y^2 + z^2 + 18x - 16y - 8z - 3 = 0$ *(iv)* $3x^2 + 4y^2 - 6x + 16y - 4z + 27 = 0$

(v) $4x^2 + y^2 - 4x - 4y + 1 = 0$

Sol. (*i*) $$x^2 + 4y^2 + 3z^2 + 2x - 8y + 9z - 10 = 0$$

or $$(x^2 + 2x) + 4(y^2 - 2y) + 3(z^2 + 3z) = 10$$

or $$(x + 1)^2 + 4(y - 1)^2 + 3\left(z+\frac{3}{2}\right)^2 = 10 + 1 + 4 + \frac{27}{4}$$

or $$(x + 1)^2 + 4(y - 1)^2 + 3\left(z+\frac{3}{2}\right)^2 = \frac{87}{4}$$

Shifting the origin to $\left(-1, 1, \frac{-3}{2}\right)$, we get

$$X^2 + 4Y^2 + 3Z^2 = \frac{87}{4}$$

or $$\frac{X^2}{\frac{87}{4}}+\frac{Y^2}{\frac{87}{16}}+\frac{Z^2}{\frac{87}{12}} = 1$$ which is an ellipsoid (*ii*) $x^2 + 2y^2 - z^2 + 2x + 4y - 6z - 18 = 0$

or $(x^2 + 2x) + 2(y^2 + 2y) - (z^2 + 6z) = 18$

or $(x + 1)^2 + 2(y + 1)^2 - (z + 3)^2 = 18 + 1 + 2 - 9$

or $(x + 1)^2 + 2(y + 1)^2 - (z + 3)^2 = 12$

Shifting the origin to (–1, –1, –3), we get

$$X^2 + 2Y^2 - Z^2 = 12$$

or $\frac{X^2}{12} + \frac{Y^2}{6} - \frac{Z^2}{12} = 1$, which is a hyperboloid of one sheet.

(*iii*) $3x^2 - y^2 + z^2 + 18x - 16y - 8z - 3 = 0$

or $3(x^2 + 6x) - (y^2 + 16y) + (z^2 - 8z) = 3$

or $3(x + 3)^2 - (y + 8)^2 + (z - 4)^2 = 3 + 27 - 64 + 16$

or $3(x + 3)^2 - (y + 8)^2 + (z - 4)^2 = -18$

Shift the origin to (–3, –8, 4)

$$3X^2 - Y^2 + Z^2 = -18$$

or $-\frac{X^2}{6} + \frac{Y^2}{18} - \frac{Z^2}{18} = 1$, which is hyperboloid of two sheets.

(*iv*) $3x^2 + 4y^2 - 6x + 16y - 4z + 27 = 0$

or $3(x^2 - 2x) + 4(y^2 + 4y) = 4z - 27$

or $3(x - 1)^2 + 4(y + 2)^2 = 4z - 27 + 3 + 16$

or $3(x - 1)^2 + 4(y + 2)^2 = 4z - 8$

or $3(x - 1)^2 + 4(y + 2)^2 = 4(z - 2)$

Shift the origin to (1, –2, 2), we get

$$3X^2 + 4Y^2 = 4Z$$

Divide by 12; $\frac{X^2}{4} + \frac{Y^2}{3} = \frac{Z}{3}$, which is elliptic paraboloid.

(*v*) $4x^2 + y^2 - 4x - 4y + 1 = 0$

or $4(x^2 - x) + (y^2 - 4y) = -1$

or $4\left(x - \frac{1}{2}\right)^2 + (y - 2)^2 = -1 + 1 + 4$

or $4\left(x - \frac{1}{2}\right)^2 + (y - 2)^2 = 4$

Shift the origin to $\left(\frac{1}{2}, 2, 0\right)$, we get

$$4X^2 + Y^2 = 4$$

or $\frac{X^2}{1} + \frac{Y^2}{4} = 1$, which is an elliptic cylinder.

TEST YOUR KNOWLEDGE

What type of surfaces are represented by the following equations:

1. $\frac{y^2}{2}+\frac{z^2}{3}-\frac{x^2}{2}=1$ **(P.T.U., Dec. 2011)**
2. $36x^2 + 9y^2 + 4z^2 = 36$ **(P.T.U., May 2014)**
3. $4x^2 + y^2 - 9z^2 = 16.$
4. $z^2 = 4(1 + x^2 + y^2)$
5. $x^2 + y^2 = 9z^2.$
6. $\frac{x^2}{36}+\frac{y^2}{9}-\frac{z^2}{4}=0.$ **(P.T.U., Dec. 2011)**
7. $\frac{x^2}{16}+\frac{y^2}{4}=2z$
8. $9(x^2 + y^2) = 16z$ **(P.T.U., Dec. 2011)**
9. $\frac{x^2}{2}-\frac{y^2}{3}=z$ **(P.T.U., Dec. 2011)**
10. $x^2 + y^2 = 2.$ **(P.T.U., May 2004)**
11. $4x^2 + 9y^2 = 36$
12. $y^2 = 4x$
13. $3x^2 + 2y^2 + 6z^2 - 6x - 4y - 12z + 10 = 0$
14. $9x^2 - 4y^2 + 36x + 8y - 4 = 0$
15. $6x^2 + 4y^2 - 3z^2 - 12x + 16y + 18z - 17 = 0$
16. $4x^2 - 3y^2 + 2z^2 - 24x + 30y + 8z - 19 = 0.$

Answers

1. Hyperboloid of one sheet.
2. Ellipsoid.
3. Hyperboloid of one sheet.
4. Hyperboloid of two sheets.
5. Ellipticone.
6. Ellipticone.
7. Elliptic paraboloid.
8. Elliptic paraboloid.
9. Hyperbolic paraboloid.
10. Right circular cylinder.
11. Elliptic cylinder.
12. Parabolic cylinder.
13. Ellipsoid
14. Hyperbolic cylinder
15. Hyperboloid of one sheet
16. Hyperboloid of one sheet

REVIEW OF THE CHAPTER

1. Equation of a cone, whose vertex is the origin is homogeneous in x, y, z and conversly every homogeneous equation in x, y, z represents a cone whose vertex is the origin.
2. The direction ratios of a generator of a cone satisfy the equation of the cone whose vertex is the origin.
3. General equation of a cone of second degree which passes through axes is
$$fyz + gzx + hxy = 0.$$
4. Equation of a right circular cone which passes through axes is $yz \pm zx \pm xy = 0$.
5. A right circular cone is the surface generated by a straight line which passes through a fixed point (vertex) and makes a constant angle with a fixed line (axis) through the fixed point.
6. The equation of the right circular cone whose vertex is (α, β, γ), semivertical angle θ and axis the line $\frac{x-\alpha}{l}=\frac{y-\beta}{m}=\frac{z-\gamma}{n}$ is:
$$(l^2 + m^2 + n^2)\{(x-\alpha)^2 + (y-\beta)^2 + (z-\gamma)^2\}\cos^2\theta = \{l(x-\alpha) + m(y-\beta) + n(z-\gamma)\}^2.$$
7. Equation of the cone whose vertex is the point (α, β, γ) and base the conic $f(x, y) = ax^2 + by^2 + 2hxy + 2gx + 2fy + c = 0, z = 0$ is $a(\alpha z - \gamma x)^2 + b(\beta z - \gamma y)^2 + 2h(\alpha z - \gamma x)(\beta z - \gamma y) + 2f(\beta z - \gamma y)(z - \gamma) + 2g(\alpha z - \gamma x)(z - \gamma) + c(z - \gamma)^2 = 0$.

8. **Enveloping Cone.** The locus of the tangent lines from a given point to a sphere is a cone called enveloping cone from the point to the sphere. Equation of the enveloping cone from the point (x_1, y_1, z_1) to the sphere $x^2 + y^2 + z^2 = a^2$ is $SS_1 = T^2$ where $S = x^2 + y^2 + z^2 - a^2$; $S_1 = x_1^2 + y_1^2 + z_1^2 - a^2$ and $T = xx_1 + yy_1 + zz_1 - a^2$.

9. A line $\frac{x - x_1}{l} = \frac{y - y_1}{m} = \frac{z - z_1}{n}$ is a tangent line to the cone $ax^2 + by^2 + cz^2 + 2fyz + 2gzx + 2hxy = 0$ if $(ax_1 + hy_1 + gz_1)x_1 + (hx_1 + by_1 + fz_1)y_1 + (gx_1 + fy_1 + cz_1)z_1 = 0$.

10. Equation of the tangent plane at (x_1, y_1, z_1) to the cone $F(x, y, z) = ax^2 + by^2 + cz^2 + 2fyz + 2gzx + 2hxy = 0$ is $(ax_1 + hy_1 + gz_1)\, x + (hx_1 + by_1 + fz_1)\, y + (gx_1 + fy_1 + cz_1)\, z = 0$

or $$\left(\frac{\partial F}{\partial x}\right)_{\text{at}(x_1, y_1, z_1)} x + \left(\frac{\partial F}{\partial y}\right)_{(x_1, y_1, z_1)} y + \left(\frac{\partial F}{\partial z}\right)_{(x_1, y_1, z_1)} z = 0$$

11. Equation of the cylinder whose generators are, parallel to $\frac{x}{l} = \frac{y}{m} = \frac{z}{n}$ and intersect the curve $ax^2 + by^2 + cz^2 + 2hxy + 2gx + 2fy + c = 0$, $z = 0$ is $a(nx - lz)^2 + b\,(ny - mz)^2 + 2h\,(nx - lz)\,(ny - mz) + 2ng\,(nx - lz) + 2nf\,(ny - mz) + cn^2 = 0$.

12. **Right Circular Cylinder:** Equation of the right circular cylinder whose axis is the z-axis and radius a is $x^2 + y^2 = a^2$.

13. Equation of the right circular cylinder whose axis is $\frac{x - \alpha}{l} = \frac{y - \beta}{m} = \frac{z - \gamma}{n}$ and radius r

is $(x - \alpha)^2 + (y - \beta)^2 + (z - \gamma)^2 = \dfrac{[l(x - \alpha) + m(y - \beta) + n(z - \gamma)]^2}{l^2 + m^2 + n^2} + r^2$.

14. **Enveloping Cylinder:** Equation of the enveloping cylinder of the sphere $x^2 + y^2 + z^2 = a^2$ whose generators are parallel to the line $\frac{x}{l} = \frac{y}{m} = \frac{z}{n}$ is $(lx + my + nz)^2 = (l^2 + m^2 + n^2)(x^2 + y^2 + z^2 - a^2)$.

15. **Conicoid:** A non degenerate quadric surface is conicoid.

16. Types of conicoid

(*i*) Ellipsoid $\frac{x^2}{a^2} + \frac{y^2}{b^2} + \frac{z^2}{c^2} = 1$

(*ii*) Hyperboloid of one sheet $\frac{x^2}{a^2} + \frac{y^2}{b^2} - \frac{z^2}{c^2} = 1$

(*iii*) Hyperboloid of two sheets $\frac{x^2}{a^2} - \frac{y^2}{b^2} - \frac{z^2}{c^2} = 1$

(*iv*) Elliptic cone $\frac{x^2}{a^2} + \frac{y^2}{b^2} - \frac{z^2}{c^2} = 0$

(*v*) Elliptic paraboloid $\frac{x^2}{a^2} + \frac{y^2}{b^2} = \frac{2z}{c}$

(*vi*) Hyperbolic paraboloid $\frac{x^2}{a^2} - \frac{y^2}{b^2} = \frac{2z}{c}$

(*vii*) Parabolic cylinder $y^2 = 4ax$

(*viii*) Elliptic cylinder $\frac{x^2}{a^2} + \frac{y^2}{b^2} = 1$

17. Types of Cylinder

(*i*) Parabolic cylinder (generated by a straight line parallel to z-axis) $y^2 = 4ax$

(*ii*) Elliptic cylinder (generated by a straight line parallel to z-axis) $\dfrac{x^2}{a^2} + \dfrac{y^2}{b^2} = 1$

(*iii*) Hyperbolic cylinder (generated by a straight line parallel to z-axis) $\dfrac{x^2}{a^2} - \dfrac{y^2}{b^2} = 1.$

SHORT ANSWER TYPE QUESTIONS

1. Show that the equation of the cone whose vertex is the origin is homogeneous in x, y, z.
[**Hint.** art. 5.4]
2. Find the equation of the cone with vertex at the origin and which passes through the curve $ax^2 + by^2 + cz^2 = 1, lx + my + nz = p$. ***(P.T.U., May 2006)***
[**Hint.** S.E1(*i*) art. 5.4]
3. Find the equations of the cones with vertex at the origin and which pass through the following curves
(*i*) $x^2 + y^2 + z^2 - x - 1 = 0, x^2 + y^2 + z^2 + y - 2 = 0.$
[**Hint.** S.E 1(*ii*) art. 5.4]
(*ii*) $x^2 + y^2 = 4, z = 2$ (*iii*) $x^2 + y^2 + z^2 = 9, x + y + z = 1$
(*iv*) $x = a, y^2 + z^2 = b^2.$ ***(P.T.U., May 2008)***
[**Hint.** S.E. 2 art. 5.4]
4. Show that the general equation of a cone of second degree which passes through the axes is $fyz + gzx + hxy = 0$ [art. 5.6] ***(P.T.U., Dec. 2006)***
5. Show that the cone which touches three coordinate plane is $\sqrt{fx} \pm \sqrt{gy} \pm \sqrt{hz} = 0.$
6. Find the equation to the quadric cone which passes through the three coordinate axes and three mutually perpendicular lines $\dfrac{x}{1} = \dfrac{y}{-2} = \dfrac{z}{3}; \dfrac{x}{1} = \dfrac{y}{-1} = \dfrac{z}{-1}; \dfrac{x}{5} = \dfrac{y}{4} = \dfrac{z}{1}.$
[**Hint.** S.E. 4 art. 5.6]
7. Define a right circular cone.
[**Hint.** art. 5.7]
8. Show that equation of the right circular cone whose vertex is at the origin, semivertical angle α and axis along z-axis is $x^2 + y^2 = z^2 \tan^2 \alpha$.
[**Hint.** art. 5.7]
9. Find the equation of the right circular cone whose vertex is (α, β, γ), semivertical angle θ and axis the line $\dfrac{x-\alpha}{l} = \dfrac{y-\beta}{m} = \dfrac{z-\gamma}{n}$.
[**Hint.** art. 5.7(*b*)]
10. Find the equation of the right circular cone whose vertex is at the origin, axis is the line $\dfrac{x}{1} = \dfrac{y}{2} = \dfrac{z}{3}$ and which has semivertical angle of 30°. ***(P.T.U., 2004)***
[**Hint.** S.E. 7(*i*) art. 5.8]

11. Find the equation of a right circular cone whose vertex is (1, –1, 2), axis is the line $\frac{x-1}{2}=\frac{y+1}{1}=\frac{z-2}{-2}$ and semivertical angle 45°.

12. Find equation of the right circular cone whose vertex is at the point (1, 1, 1) and base the circle $x^2+y^2=4, z=2.$ ***(P.T.U., May 2005)***
[**Hint.** S.EI(*i*) art. 5.9]

13. Find the semivertical angle and equation of the right circular cone having its vertex at the origin and passing through the circle $x^2+z^2=25, y=4.$
[**Hint.** S.E. 4 art. 5.9]

14. Define an enveloping cone and write its equation. [art. 5.10]

15. Find the enveloping cone of the sphere $x^2+y^2+z^2+2x-2y=2$ with its vertex at (1, 1, 1).

16. Find the equation of the following cones:
(*i*) whose vertex is (0, 0, 5) and guiding curve $x^2+2xy+3y^2=1, z=3$
(*ii*) whose vertex is (α, β, γ) and guiding curve $y^2=4ax, z=0$
(*iii*) whose vertex is (3, 1, 2) and guiding curve $2x^2+3y^2=1, z=0.$

17. The generators of a cone pass through the point (1, 1, 1) and their direction cosines l, m, n satisfy the equation $l^2+m^2=3n^2$; obtain the equation of the cone.

[**Hint.** Equation of the generator is $\frac{x-1}{l}=\frac{y-1}{m}=\frac{z-1}{n}$ eliminate l, m, n from $l^2+m^2=3n^2$ *i.e,* $(x-1)^2+(y-1)^2=3(z-1)^2$].

18. Write the equation of the tangent plane at $P(x_1, y_1, z_1)$ of the cone
$ax^2+by^2+cz^2+2fyz+2gzx+2hxy=0.$
[**Hint.** art. 5.12]

19. Find the equation of the tangent plane at the point (– 3, 0, – 1) of the cone
$4x^2-y^2+2z^2+2xy-3yz+12x-11y+6z+4=0.$
[**Hint.** S.E 6 art. 5.12]

20. Find the plane which touches the cone $x^2+2y^2+4z^2-6yz+4zx-5xy=0$ along the generator whose direction ratios are <1, 1, 1>.
[**Hint.** S.E. 7 art. 5.12]

21. Find the equation of cylinder whose generators are parallel to x-axis and pass through the curve $x^2+y^2+2z^2=1, x-3y+z=4.$
[**Hint.** Consult S.E. 1 (*b*) art. 5.16) eliminate x from the two equation.]

22. Find the equation of the cylinder whose generators are parallel to the line with d.c.'s $< l, m, n>$ and intersecting the conic $ax^2+2hxy+by^2=1, z=0.$
[**Hint.** Consult S.E. 2 art. 5.16; equation of the generator through (x_1, y_1, z_1) is

$\frac{x-x_1}{l}=\frac{y-y_1}{m}=\frac{z-z_1}{n}$ put $z=0; x=x_1-\frac{l}{n}z_1 y=y_1-\frac{m}{n}z_1$ substitute x, y in $ax^2+2hxy+by^2=1$ and find locus of (x_1, y_1, z_1)].

23. Find the equation of the cylinder whose generators are parallel to the line $\frac{x}{3}=\frac{y-4}{5}=\frac{z-1}{-4}$ and which has for its guiding curve, the hyperbola $3x^2-4y^2=5, z=2$
[**Hint.** Consult S.E. 2 art. 5.16].

24. A right circular cylinder is such that its section by the plane $x+y=z$ is a circle of radius 3 and centre (1, 2, 3). Find its equation.
[**Hint.** S.E. 7 art. 5.17] ***(P.T.U., May 2003)***

25. Define an enveloping cylinder.

26. Find the equation of the enveloping cylinder of $x^2 + y^2 + z^2 = a^2$ whose generators are parallel to $\frac{x}{l} = \frac{y}{m} = \frac{z}{n}$.

27. Write standard equations of the following:

(*i*) ellipsoid

(*ii*) Hyperboloid of one sheet **(*P.T.U., Dec. 2007*)**

(*iii*) Hyperboloid of two sheets

(*iv*) elliptic cone

(*v*) elliptic paraboloid

(*vi*) hyperbolic paraboloid

(*vii*) parabolic cylinder

(*viii*) elliptic cylinder

(*ix*) hyperbolic cylinder

28. What type of surfaces are represented by the following equations:

(*i*) $z^2 = 4(1 + x^2 + y^2)$ [**Hint.** S.E. 1 art 5.20]

(*ii*) $36x^2 + 9y^2 + 4z^2 = 36$ [**Hint.** Divide by 36]

(*iii*) $4x^2 + y^2 - 9z^2 = 16$ $\left[\textbf{Hint.}\text{ Divide by 16 and write } \frac{x^2}{4} + \frac{y^2}{16} - \frac{z^2}{16/9} = 1\right]$

(*iv*) $\frac{x^2}{36} + \frac{y^2}{9} - \frac{z^2}{4} = 0$

(*v*) $9(x^2 + y^2) = 16z$ **(*P.T.U., Dec. 2011*)**

(*vi*) $\frac{y^2}{2} + \frac{z^2}{3} - \frac{x^2}{2} = 1$ **(*P.T.U., Dec. 2011*)**

(*vii*) $\frac{x^2}{a^2} - \frac{y^2}{b^2} = \frac{2z}{c}$ **(*P.T.U., May 2010*)**

(*viii*) $x^2 + y^2 = a^2$ **(*P.T.U., May 2004*)**

(*ix*) $x^2 + 2y^2 - z^2 + 2x + 4y - 6z - 18 = 0$

(*x*) $3x^2 + 2y^2 + 6z^2 - 6x - 4y - 12z + 10 = 0$

Answers

2. $p^2(ax^2 + by^2 + cz^2) = (lx + my + nz)^2$

3. (*i*) $x^2 + 3xy - z^2 = 0$ (*ii*) $x^2 + y^2 = z^2$

(*iii*) $4(x^2 + y^2 + z^2) + 9(xy + yz + zx) = 0$ (*iv*) $a^2(y^2 + z^2) = b^2x^2$

6. $5yz + 8zx - 3xy = 0$

9. $(l^2 + m^2 + n^2)\,[(x - \alpha)^2 + (y - \beta)^2 + (z - \gamma)^2] \cos^2\theta = [l(x - \alpha) + m(y - \beta) + n(z - \gamma)^2]$

10. $19x^2 + 13y^2 + 3z^2 - 8xy - 24yz - 12zx = 0$

11. $x^2 + y^2 + z^2 - 8xy + 8yz + 16zx - 4x + 6y - 12z + 36 = 0$

12. $x^2 + y^2 - 2z^2 + 2yz + 2zx - 4x - 4y + 4 = 0$

13. $\tan^{-1}\frac{5}{4}$, $16(x^2 + z^2) = 25y^2$

15. $3x^2 - y^2 + 4zx - 10x + 2y - 4z + 6 = 0$

16. (*i*) $4x^2 + 8xy + 12y^2 - z^2 + 10z - 25 = 0$ (*ii*) $(\beta z - \gamma y)^2 = 4a(z - \gamma)(\alpha z - \gamma x)$

(*iii*) $2x^2 + 3y^2 + 5z^2 - 3yz - 6zx + z - 1 = 0$

17. $x^2 + y^2 - 3z^2 - 2x - 2y + 6z - 1 = 0$ **19.** $6x + 7y - z + 17 = 0$

20. $x - 7y + 6z = 0.$

21. $10y^2 - 6yz + 3z^2 + 24y - 8z + 15 = 0$

22. $a(nx - lz)^2 + 2h(nx - lz)(ny - mz) + b(xy - mz)^2 = n^2$

23. $48x^2 - 64y^2 - 73z^2 + 72xz - 160yz - 144x + 320y + 292z - 372 = 0$

24. $2x^2 + 2y^2 + 2z^2 - 2xy + 2yz + 2zx - 6x - 12y - 18z + 15 = 0$

26. $(lx + my + nz)^2 = (l^2 + m^2 + n^2)(x^2 + y^2 + z^2 - a^2)$

28. (*i*) Hyperboloid of two sheets

(*ii*) ellipsoid

(*iii*) hyperboloid of one sheet

(*iv*) elliptic cone

(*v*) elliptic paraboloid

(*vi*) Hyperboloid of one sheet

(*vii*) Hyperbolic paraboloid

(*viii*) Right circular cylinder

(*ix*) Hyperboloid of one sheet

(*x*) ellipsoid

6

Multiple Integrals

6.1. DOUBLE INTEGRALS

The definite integral $\int_a^b f(x)\,dx$ is defined as the limits of the sum $f(x_1)\delta x_1 + f(x_2)\delta x_2 + \cdots + f(x_n)\delta x_n$ when $n \to \infty$ and each of the lengths $\delta x_1, \delta x_2, \cdots, \delta x_n$ tends to zero. Here $\delta x_1, \delta x_2, \cdots, \delta x_n$ are n sub-intervals into which the range $b - a$ has been divided and $x_1, x_2, \cdots, x_n$ are values of x lying respectively in the first, second, $\cdots$, nth sub-interval.

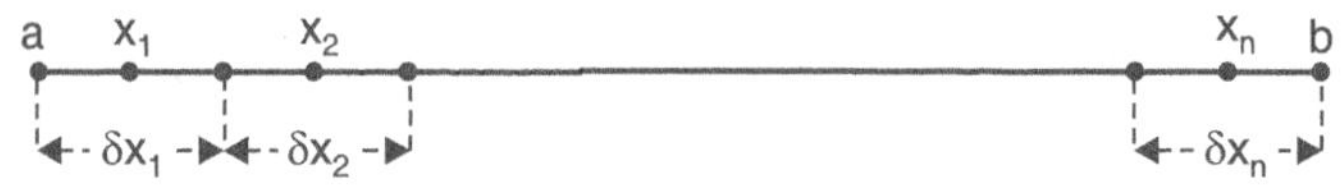

A double integral is its counterpart in two dimensions. Let a single-valued and bounded function $f(x, y)$ of two independent variables x, y be defined in a closed region R of the xy-plane. Divide the region R into sub-regions by drawing lines parallel to co-ordinate axes. Number the rectangles which lie entirely inside the region R, from 1 to n. Let (x_r, y_r) be any point inside the rth rectangle whose area is δA_r.

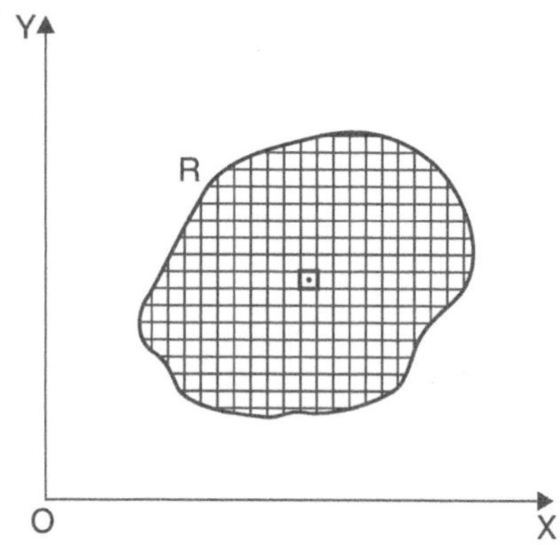

Consider the sum

$$f(x_1, y_1)\delta A_1 + f(x_2, y_2)\delta A_2 + \cdots + f(x_n, y_n)\delta A_n$$

$$= \sum_{r=1}^{n} f(x_r, y_r)\delta A_r \qquad \text{...(1)}$$

Let the number of these sub-regions increase indefinitely, such that the largest linear dimension (*i.e.*, diagonal) of δA_r approaches zero. The limit of the sum (1), if it exists, irrespective of the mode of sub-division, is called the *double integral* of $f(x, y)$ over the region R and is denoted by $\iint_R f(x, y)\,dA$

In other words,
$$\underset{\substack{n \to \infty \\ \delta A_r \to 0}}{\text{Lt}} \sum_{r=1}^{n} f(x_r, y_r)\,\delta A_r = \iint_R f(x, y)\,dA$$

which is also expressed as
$$\iint_R f(x, y)\,dxdy \quad \text{or} \quad \iint_R f(x, y)\,dy\,dx$$

6.2. EVALUATION OF DOUBLE INTEGRALS

The methods of evaluating the double integrals depend upon the nature of the curves bounding the region R. Let the region R be bounded by the curves $x = x_1$, $x = x_2$ and $y = y_1$, $y = y_2$.

(*i*) **When x_1, x_2 are functions of y and y_1, y_2 are constants.** Let AB and CD be the curves ($x_1 = \phi_1(y)$) and $x_2 = \phi_2(y)$.

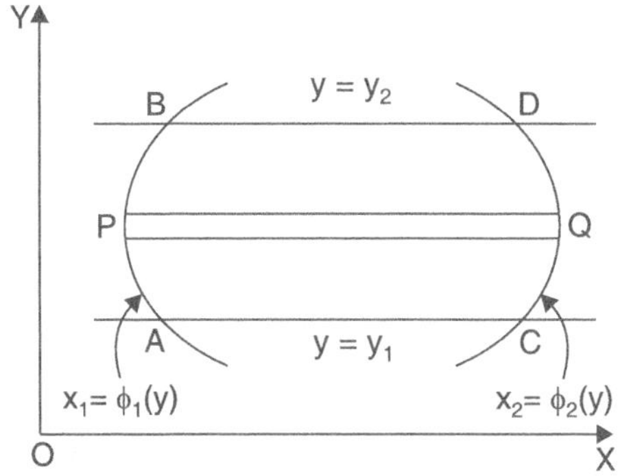

Take a horizontal strip PQ of width δy. Here the double integral is evaluated first w.r.t. x (treating y as a constant). The resulting expression which is a function of y is integrated w.r.t. y between the limits $y = y_1$ and $y = y_2$. Thus

$$\iint_R f(x, y)\, dx\, dy = \boxed{\int_{y_1}^{y_2} \boxed{\int_{x_1 = \phi_1(y)}^{x_2 = \phi_2(y)} f(x,y)\, dx}\; dy}$$

the integration being carried from the inner to the outer rectangle. Geometrically, the integral in the inner rectangle indicates that the integration is performed along the horizontal strip PQ (keeping y constant) while the outer rectangle corresponds to the sliding of the strip PQ from AC to BD thus covering the entire region ABDC of integration.

(*ii*) **When y_1, y_2 are functions of x and x_1, x_2 are constants.** Let AB and CD be the curves ($y_1 = \phi_1(x)$) and $y_2 = \phi_2(x)$. Take a vertical strip PQ of width δx. Here the double integral is evaluated first w.r.t. y (treating x as constant). The resulting expression which is a function of x is integrated w.r.t. x between the limits $x = x_1$ and $x = x_2$. Thus.

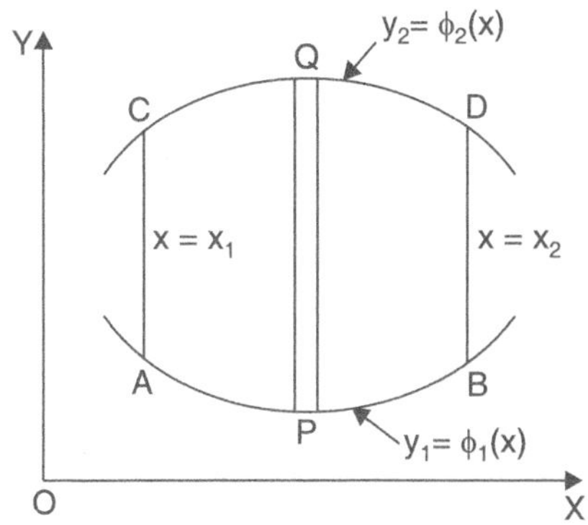

$$\iint_R f(x, y)\, dx\, dy = \boxed{\int_{x_1}^{x_2} \boxed{\int_{y_1 = \phi_1(x)}^{y_2 = \phi_2(x)} f(x, y)\, dy}\; dx}$$

the integration being carried from the inner to the outer rectangle. Geometrically, the integral in the inner rectangle indicates that the integration is performed along the vertical strip PQ (keeping x constant) while the outer rectangle corresponds to the sliding of the strip PQ from AC to BD thus covering the entire region ABDC of integration.

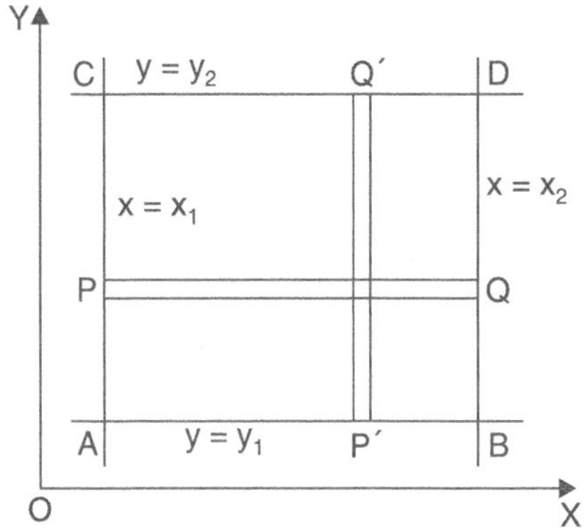

(*iii*) **When x_1, x_2, y_1, y_2 are constants.** Here the region of integration R is the rectangle ABDC. It is immaterial whether we integrate first along the horizontal strip PQ and then slide it from AB to CD; or we integrate first along the vertical strip P′Q′ and then slide it from AC to BD. Thus the order of integration is immaterial, provided the limits of integration are changed accordingly.

$$\iint_R f(x, y)\, dx\, dy = \boxed{\int_{y_1}^{y_2} \boxed{\int_{x_1}^{x_2} f(x, y)\, dx}\, dy} = \boxed{\int_{x_1}^{x_2} \boxed{\int_{y_1}^{y_2} f(x, y)\, dy}\, dx}$$

Note 1. From cases (*i*) and (*ii*) above, we observe that *integration is to be performed w.r.t. that variable having variable limits first and then w.r.t. the variable with constant limits.*

Note 2. If $f(x, y)$ has discontinuities within or on the boundary of the region of integration, then the change of the order of integration does not result into the same integrals.

ILLUSTRATIVE EXAMPLES

Example 1. *Prove that* $\int_1^2 \int_3^4 (xy + e^y)\, dy\, dx = \int_3^4 \int_1^2 (xy + e^y)\, dx\, dy.$

Sol. $$\int_1^2 \int_3^4 (xy + e^y)\, dy\, dx = \int_1^2 \left[\int_3^4 (xy + e^y)\, dy\right] dx$$

$$= \int_1^2 \left[\frac{xy^2}{2} + e^y\right]_3^4 dx = \int_1^2 \left(8x + e^4 - \frac{9}{2}x - e^3\right) dx$$

$$= \int_1^2 \left(\frac{7}{2}x + e^4 - e^3\right) dx = \left[\frac{7x^2}{4} + (e^4 - e^3)x\right]_1^2$$

$$= 7 + 2(e^4 - e^3) - \frac{7}{4} - (e^4 - e^3) = \frac{21}{4} + e^4 - e^3$$

$$\int_3^4 \int_1^2 (xy + e^y)\, dx\, dy = \int_3^4 \left[\int_1^2 (xy + e^y)\, dx\right] dy = \int_3^4 \left[\frac{yx^2}{2} + xe^y\right]_1^2 dy$$

$$= \int_3^4 \left(2y + 2e^y - \frac{y}{2} - e^y\right) dy = \int_3^4 \left(\frac{3y}{2} + e^y\right) dy$$

$$= \left[\frac{3y^2}{4} + e^y\right]_3^4 = 12 + e^4 - \frac{27}{4} - e^3 = \frac{21}{4} + e^4 - e^3$$

Hence the result.

Example 2. *Evaluate the following:*

(*i*) $\int_0^1 \int_0^1 \frac{dx\, dy}{\sqrt{(1 - x^2)(1 - y^2)}}$ (*ii*) $\int_0^1 \int_x^{\sqrt{x}} (x^2 + y^2)\, dy\, dx.$ **(P.T.U., Dec. 2013)**

Sol. (*i*) $$\int_0^1 \int_0^1 \frac{dx\, dy}{\sqrt{(1-x^2)(1-y^2)}} = \int_0^1 \left[\int_0^1 \frac{dx}{\sqrt{(1-x^2)(1-y^2)}}\right] dy = \int_0^1 \frac{1}{\sqrt{1-y^2}} \left[\sin^{-1} x\right]_0^1 dy$$

$$= \int_0^1 \frac{1}{\sqrt{1-y^2}} \cdot \frac{\pi}{2}\, dy = \frac{\pi}{2} \left[\sin^{-1} y\right]_0^1 = \frac{\pi}{2} \cdot \frac{\pi}{2} = \frac{\pi^2}{4}.$$

(*ii*) $$\int_0^1 \int_x^{\sqrt{x}} (x^2 + y^2)\, dy\, dx = \int_0^1 \left[x^2 y + \frac{y^3}{3}\right]_x^{\sqrt{x}} dx$$

$$= \int_0^1 \left[x^{5/2} + \frac{x^{3/2}}{3} - \left(x^3 + \frac{x^3}{3}\right)\right] dx$$

$$= \left[\frac{2}{7}x^{7/2} + \frac{1}{3}\cdot\frac{2}{5}x^{5/2} - \frac{x^4}{4} - \frac{x^4}{12}\right]_0^1 = \frac{2}{7} + \frac{2}{15} - \frac{1}{4} - \frac{1}{12} = \frac{3}{35}.$$

Example 3. *Show that* $\int_0^1 dx \int_0^1 \frac{x-y}{(x+y)^3}\, dy \neq \int_0^1 dy \int_0^1 \frac{x-y}{(x+y)^3}\, dx.$

Sol. $$\text{LHS} = \int_0^1 dx \int_0^1 \frac{2x - (x+y)}{(x+y)^3}\, dy = \int_0^1 dx \int_0^1 \left[\frac{2x}{(x+y)^3} - \frac{1}{(x+y)^2}\right] dy$$

$$= \int_0^1 \left[2x\,.\frac{(x+y)^{-2}}{-2} - \frac{(x+y)^{-1}}{-1}\right]_0^1 dx = \int_0^1 \left[\frac{-x}{(x+y)^2} + \frac{1}{x+y}\right]_0^1 dx$$

$$= \int_0^1 \left[\frac{-x}{(x+1)^2} + \frac{1}{x+1} + \frac{1}{x} - \frac{1}{x}\right] dx = \int_0^1 \frac{-x+x+1}{(x+1)^2}\, dx = \int_0^1 \frac{1}{(x+1)^2}\, dx$$

$$= \left[-\frac{1}{x+1}\right]_0^1 = -\frac{1}{2} + 1 = \frac{1}{2}$$

$$\text{RHS} = \int_0^1 dy \int_0^1 \frac{(x+y) - 2y}{(x+y)^3}\, dx = \int_0^1 dy \int_0^1 \left[\frac{1}{(x+y)^2} - \frac{2y}{(x+y)^3}\right] dx$$

$$= \int_0^1 \left[\frac{(x+y)^{-1}}{-1} - 2y\cdot\frac{(x+y)^{-2}}{-2}\right]_0^1 dy = \int_0^1 \left[-\frac{1}{x+y} + \frac{y}{(x+y)^2}\right]_0^1 dy$$

$$= \int_0^1 \left[-\frac{1}{1+y} + \frac{y}{(1+y)^2} + \frac{1}{y} - \frac{1}{y}\right] dy = \int_0^1 \frac{-1-y+y}{(1+y)^2}\, dy = -\int_0^1 \frac{1}{(1+y)^2}\, dy$$

$$= \left[\frac{1}{1+y}\right]_0^1 = \frac{1}{2} - 1 = -\frac{1}{2}$$

$\therefore$ The two integrals are not equal.

Example 4. *Evaluate* $\int_0^1 \int_0^{\sqrt{1+x^2}} \frac{dy\, dx}{1 + x^2 + y^2}$. **(P.T.U., May 2006)**

Sol. $$\text{I} = \int_0^1 \left[\int_0^{\sqrt{1+x^2}} \frac{1}{(1+x^2) + y^2}\, dy\right] dx$$

$$= \int_0^1 \frac{1}{\sqrt{1+x^2}} \left[\tan^{-1} \frac{y}{\sqrt{1+x^2}}\right]_0^{\sqrt{1+x^2}} dx$$

$$= \int_0^1 \frac{1}{\sqrt{1+x^2}} [\tan^{-1} 1 - \tan^{-1} 0]\, dx = \frac{\pi}{4} \int_0^1 \frac{dx}{\sqrt{1+x^2}}$$

$$= \frac{\pi}{4} \left[\log (x + \sqrt{1+x^2})\right]_0^1 = \frac{\pi}{4} [\log (1 + \sqrt{2}) - \log 1] = \frac{\pi}{4} \log (\sqrt{2} + 1).$$

Example 5. *Evaluate* $\int_0^1 \int_{x^2}^{2-x} xy\, dy\, dx$ **(P.T.U., Dec. 2012)**

Sol. $\int_0^1 \int_{x^2}^{2-x} xy\, dy\, dx = \int_0^1 x \frac{y^2}{2}\Big|_{x^2}^{2-x} dx = \int_0^1 \frac{x}{2}\{(2-x)^2 - x^4\}\, dx$

$$= \int_0^1 \frac{x}{2}\{4 - 4x + x^2 - x^4\}\, dx = \frac{1}{2}\int_0^1 (4x - 4x^2 + x^3 - x^5)\, dx$$

$$= \frac{1}{2}\left\{2x^2 - \frac{4x^3}{3} + \frac{x^4}{4} - \frac{x^6}{6}\right\}\Bigg|_0^1 = \frac{1}{2}\left(2 - \frac{4}{3} + \frac{1}{4} - \frac{1}{6}\right)$$

$$= \frac{1}{2} \cdot \frac{9}{12} = \frac{9}{24}$$

Example 6. (*i*) *Evaluate* $\iint e^{2x+3y}\, dx\, dy$ *over the triangle bounded by* $x = 0$, $y = 0$ *and* $x + y = 1$.

(*ii*) *Evaluate* $\int\int_S \sqrt{xy - y^2}\, dxdy$ *where S is the triangle with vertices (0, 0), (10, 1) and (1, 1).*

Sol. (*i*) The region R of integration is the triangle OAB. Here x varies from 0 to 1 and y varies from x-axis upto the line $x + y = 1$ *i.e.*, from 0 to $1 - x$.

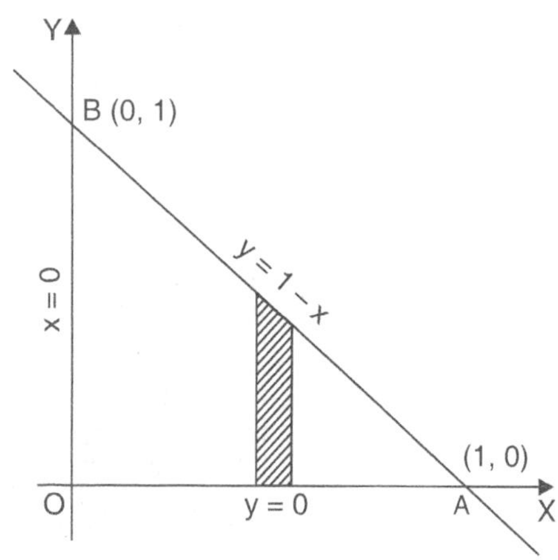

$\therefore$ The region R can be expressed as

$$0 \le x \le 1,\ 0 \le y \le 1 - x$$

$$\therefore \quad \iint_R e^{2x+3y}\, dx\, dy = \int_0^1 \int_0^{1-x} e^{2x+3y}\, dy\, dx$$

$$= \int_0^1 \left[\frac{1}{3} e^{2x+3y}\right]_0^{1-x} dx$$

$$= \frac{1}{3}\int_0^1 (e^{3-x} - e^{2x})\, dx$$

$$= \frac{1}{3}\left[- e^{3-x} - \frac{1}{2} e^{2x}\right]_0^1 = -\frac{1}{3}\left[\left(e^2 + \frac{1}{2}e^2\right) - \left(e^3 + \frac{1}{2}\right)\right]$$

$$= -\frac{1}{3}\left[- e^2(e-1) + \frac{1}{2}(e^2 - 1)\right]$$

$$= \frac{1}{6}(e-1)[2e^2-(e+1)] = \frac{1}{6}(e-1)(2e^2-e-1)$$

$$= \frac{1}{6}(e-1)(e-1)(2e+1) = \frac{1}{6}(e-1)^2(2e+1).$$

(*ii*) Region R of integration is the region of the triangle OAB (shown in the figure). Here equation of the line OA is

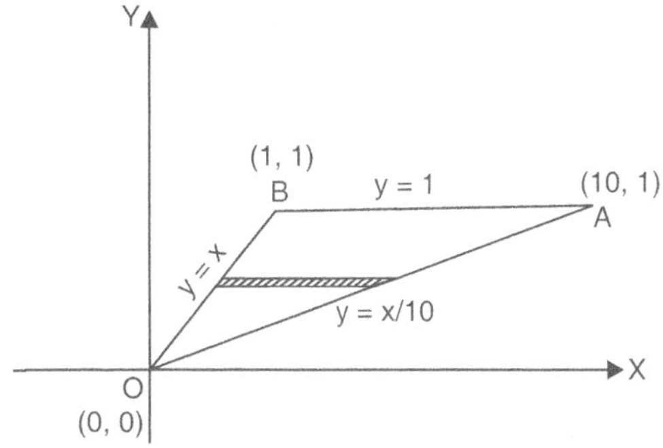

$$y - 0 = \frac{1-0}{10-0}(x-0)$$

or $\qquad y = \frac{1}{10}x$

equation of OB is $\qquad y = x$

$\therefore \qquad R = \{(x, y) ; y \le x \le 10y ; 0 \le y \le 1\}$

$$\therefore \quad \int\int_S \sqrt{xy-y^2}\,dxdy = \int_0^1 \left[\int_y^{10y} \sqrt{xy-y^2}\,dx\right] dy$$

$$= \int_0^1 \left.\frac{(xy-y^2)^{3/2}}{\frac{3}{2}y}\right|_y^{10y} dy = \frac{2}{3}\int_0^1 \left[\frac{(9y^2)^{3/2}}{y} - 0\right] dy$$

$$= \frac{2}{3}\int_0^1 27y^2\,dy = \frac{2}{3}\left[27\left.\frac{y^3}{3}\right|_0^1\right] = \frac{2}{3}\cdot 9 = 6.$$

Example 7. *Evaluate* $\iint_R y\,dx\,dy$, *where R is the region bounded by the parabolas* $y^2 = 4x$ *and* $x^2 = 4y$.

Sol. Solving $y^2 = 4x$ and $x^2 = 4y$, we have

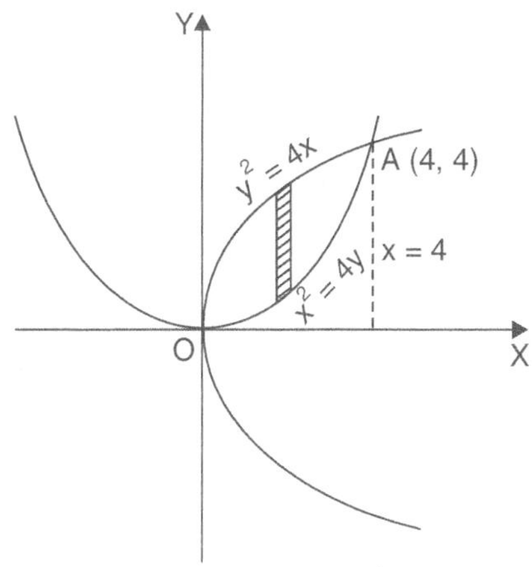

$$\left(\frac{x^2}{4}\right)^2 = 4x \quad \text{or} \quad x(x^3-64) = 0$$

$\therefore \qquad x = 0, 4$

When $\qquad x = 4,\ y = 4$

$\therefore$ Co-ordinates of A are (4, 4)

The region R can be expressed as

$$0 \le x \le 4,\ \frac{x^2}{4} \le y \le 2\sqrt{x}$$

$$\therefore \quad \iint_R y\,dx\,dy = \int_0^4 \int_{x^2/4}^{2\sqrt{x}} y\,dy\,dx$$

$$= \int_0^4 \frac{1}{2}\Big[y^2\Big]_{x^2/4}^{2\sqrt{x}} dx = \frac{1}{2}\int_0^4 \left(4x - \frac{x^4}{16}\right) dx$$

$$= \frac{1}{2}\left[2x^2 - \frac{x^5}{80}\right]_0^4 = \frac{1}{2}\left[32 - \frac{1024}{80}\right] = \frac{48}{5}.$$

Example 8. *Evaluate* $\iint xy(x+y)\,dxdy$ *over the area between* $y = x^2$ *and* $y = x$.

Sol. Region of integration is

$$R = \left\{(x, y); y \le x \le \sqrt{y}; 0 \le y \le 1\right\}$$

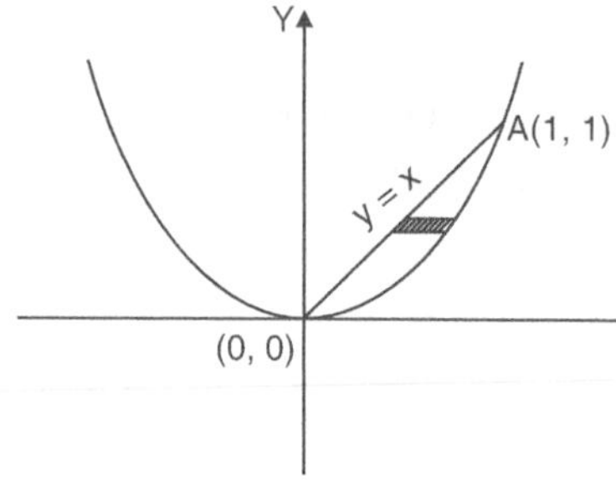

$$\therefore \quad \iint_R xy\,(x+y)\,dxdy$$

$$= \int_0^1 \left[\int_y^{\sqrt{y}} xy\,(x+y)\,dx\right] dy$$

$$= \int_0^1 y\,\frac{x^3}{3} + y^2\,\frac{x^2}{2}\Bigg|_y^{\sqrt{y}}\, dy$$

$$= \int_0^1 \left(\frac{1}{3}y^{5/2} + \frac{1}{2}y^3 - \frac{y^4}{3} - \frac{y^4}{2}\right) dy$$

$$= \int_0^1 \left(\frac{1}{3}y^{5/2} + \frac{1}{2}y^3 - \frac{5}{6}y^4\right) dy = \frac{1}{3}\left(\frac{2}{7}y^{7/2}\right) + \frac{1}{2}\cdot\frac{y^4}{4} - \frac{5}{6}\,\frac{y^5}{5}\Bigg|_0^1$$

$$= \frac{2}{21} + \frac{1}{8} - \frac{1}{6} = \frac{16 + 21 - 28}{168} = \frac{9}{168} = \frac{3}{56}.$$

Example 9. *Evaluate* $\iint (x+y)^2\ dx\,dy$ *over the area bounded by the ellipse*

$$\frac{x^2}{a^2} + \frac{y^2}{b^2} = 1.$$

Sol. For the ellipse

$$\frac{y}{b} = \pm\sqrt{1 - \frac{x^2}{a^2}} \quad \text{or} \quad y = \pm\,\frac{b}{a}\sqrt{a^2 - x^2}$$

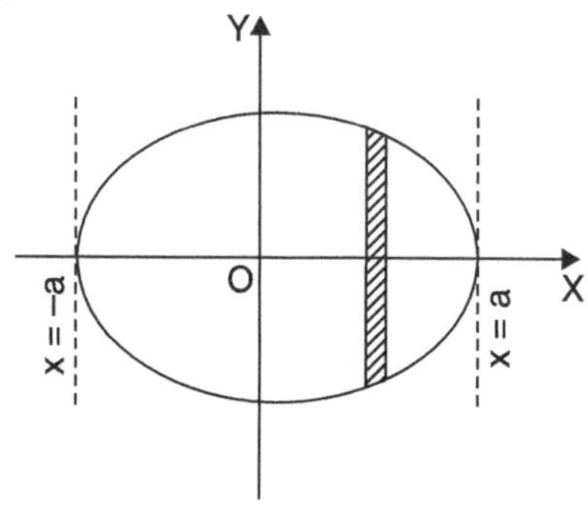

$\therefore$ The region of integration R can be expressed as

$$-a \le x \le a, -\frac{b}{a}\sqrt{a^2 - x^2} \le y \le \frac{b}{a}\sqrt{a^2 - x^2}$$

$$\therefore \quad \iint_R (x+y)^2\,dx\,dy = \iint_R (x^2 + y^2 + 2xy)\,dx\,dy$$

$$= \int_{-a}^{a}\int_{-b/a\sqrt{a^2-x^2}}^{b/a\sqrt{a^2-x^2}} (x^2 + y^2 + 2xy)\,dy\,dx$$

$$= \int_{-a}^{a}\int_{-b/a\sqrt{a^2-x^2}}^{b/a\sqrt{a^2-x^2}} (x^2 + y^2)\,dy\,dx + \int_{-a}^{a}\int_{-b/a\sqrt{a^2-x^2}}^{b/a\sqrt{a^2-x^2}} 2xy\,dy\,dx$$

$$= \int_{-a}^{a} \int_{0}^{b/a\sqrt{a^2-x^2}} 2(x^2+y^2)\,dy\,dx + 0$$

[since $(x^2 + y^2)$ is an even function of y and $2xy$ is an odd function of y]

$$= \int_{-a}^{a} \left[2\left(x^2 y + \frac{y^3}{3} \right) \right]_0^{b/a\sqrt{a^2-x^2}} dx$$

$$= 2\int_{-a}^{a} \left[x^2 \cdot \frac{b}{a}\sqrt{a^2-x^2} + \frac{1}{3}\cdot\frac{b^3}{a^3}(a^2-x^2)^{3/2} \right] dx$$

$$= 4\int_{0}^{a} \left[\frac{b}{a}x^2\sqrt{a^2-x^2} + \frac{b^3}{3a^3}(a^2-x^2)^{3/2} \right] dx$$

put $x = a\sin\theta$, $\therefore\ dx = a\cos\theta\,d\theta$

$$= 4\int_{0}^{\pi/2} \left(\frac{b}{a}\cdot a^2\sin^2\theta\cdot a\cos\theta + \frac{b^3}{3a^3}\cdot a^3\cos^3\theta \right) \times a\cos\theta\,d\theta$$

$$= 4\int_{0}^{\pi/2} \left(a^3 b\sin^2\theta\cos^2\theta + \frac{ab^3}{3}\cos^4\theta \right) d\theta$$

We know from integral calculus that

$$\int_0^{\frac{\pi}{2}} \sin^p x\cos^q x\,dx = \frac{\{(p-1)(p-3)\cdots\}\{(q-1)(q-3)\cdots\}}{(p+q)(p+q-2)\cdots}\,\frac{\pi}{2}$$

if both p, q are even otherwise no $\frac{\pi}{2}$

and $$\int_0^{\frac{\pi}{2}} \cos^p x\,dx = \int_0^{\frac{\pi}{2}} \sin^p x\,dx = \frac{(p-1)(p-3)\cdots}{p(p-2)-\cdots}\,\frac{\pi}{2}$$ if p is even otherwise no $\frac{\pi}{2}$

$$\therefore \quad \text{Given integral} = 4\left[a^3 b\cdot\frac{1\cdot 1}{4\cdot 2}\cdot\frac{\pi}{2} + \frac{ab^3}{3}\cdot\frac{3\cdot 1}{4\cdot 2}\cdot\frac{\pi}{2} \right]$$

$$= \frac{\pi}{4}(a^3 b + ab^3) = \frac{\pi}{4}ab(a^2+b^2).$$

Example 10. *Sketch the region of integration and evaluate the following integrals*

(*i*) $\displaystyle\int_1^{\log 8} \int_0^{\log y} e^{x+y}\,dxdy$

(*ii*) *Evaluate* $\displaystyle\int_0^2 \int_0^{y^2} e^{x/y}\,dx\,dy$ **(P.T.U., May 2004)**

Sol. (*i*) The given integral is

$$\int_1^{\log 8}\int_0^{\log y} e^{x+y}\,dxdy$$

Here x varies from 0 to $\log y$

i.e., $x = 0$ and $x = \log y$ or $y = e^x$

and y varies from 1 to log 8

i.e., $y = 1$ and $y = \log 8$

Region R is shown in the adjoining figure (shaded portion is the region)

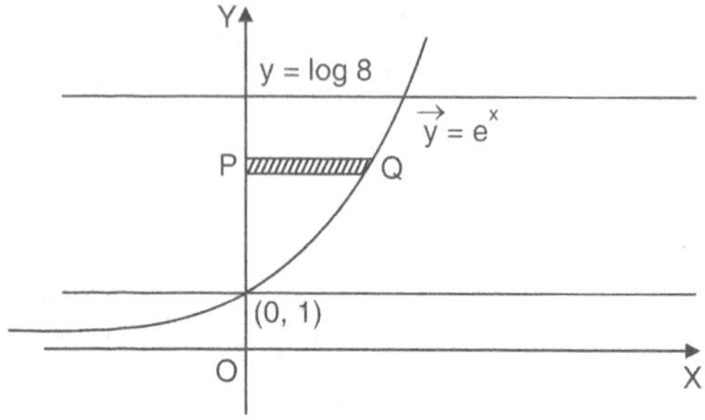

$$\int_1^{\log 8}\int_0^{\log y} e^{x+y}\,dx\,dy = \int_1^{\log 8}\int_0^{\log y} e^y\cdot e^x\,dx\,dy$$

$$= \int_1^{\log 8}\left[\int_0^{\log y} e^y\cdot e^x\,dx\right]dy = \int_1^{\log 8} e^y\cdot e^x\Big|_0^{\log y}\,dy$$

$$= \int_1^{\log 8} e^y\,[e^{\log y} - e^0]\,dy = \int_1^{\log 8} \underset{\substack{(2nd)\\(function)}}{e^y}\;\underset{\substack{Ist\\function}}{(y-1)}\,dy$$

Integrate by parts

$$= (y-1)\,e^y\Big|_1^{\log 8} - \int_1^{\log 8} 1\cdot e^y\,dy = (\log 8 - 1)\,e^{\log 8} - 0 - e^y\Big|_1^{\log 8}$$

$$= (\log 8 - 1)\cdot 8 - [e^{\log 8} - e^1]$$

$$= 8\log 8 - 8 - 8 + e$$

$$= 8\log 8 - 16 + e = 8(\log 8 - 2) + e.$$

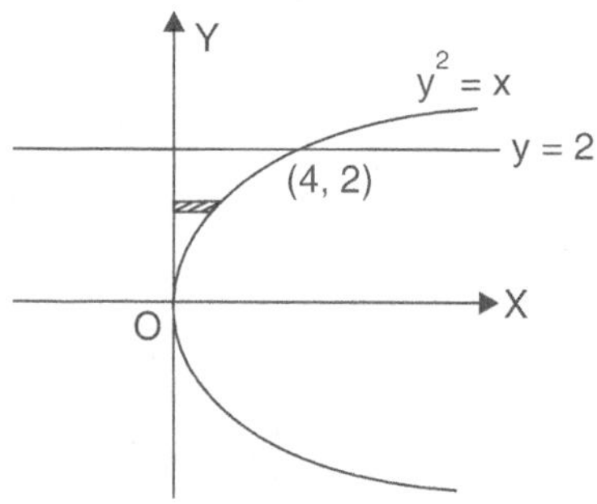

(*ii*) $\displaystyle\int_0^2\int_0^{y^2} e^{x/y}\,dx\,dy$

$$= \int_0^2\left[\frac{e^{x/y}}{1/y}\Bigg|_0^{y^2}\right]dy$$

$$= \int_0^2 (ye^y - y)\,dy = \int_0^2 ye^y - \frac{y^2}{2}\Bigg|_0^2$$

Integrating by parts,

$$= ye^y\Big|_0^2 - \int_0^2 1.e^y\,dy - 2 = 2e^2 - 0 - e^y\Big|_0^2 - 2$$

$$= 2e^2 - e^2 + 1 - 2 = e^2 - 1.$$

Example 11. (*i*) *Sketch the region of integration and evaluate* $\iint_R (y - 2x^2)\,dx\,dy$ *where R is the region inside the square* $|x| + |y| = 1$. **(P.T.U., Dec. 2004)**

(*ii*) *Write the limits of integration in* $\iint_R xy\,dx\,dy$, *where R is the region inside the square* $|x| + |y| = 1$ *and hence evaluate.* **(P.T.U., May 2003)**

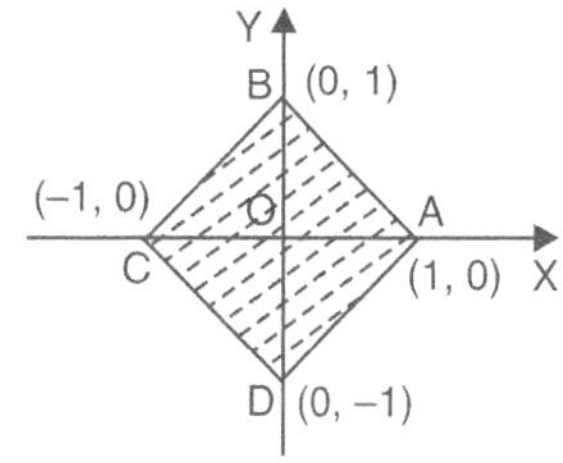

Sol. (*i*) The given region R is $|x| + |y| = 1$. It intersects x-axis (put $y = 0$) at $|x| = 1$ *i.e.*, $x = \pm 1$ *i.e.*, at $(-1, 0)$ and $(1, 0)$

Similarly it intersects y-axis (put $x = 0$) at $(0, -1)$ and $(0, 1)$

Region R is shown in the adjoining figure. (Shaded portion). It is a square with vertices at $(1, 0)$, $(0, 1)$, $(-1, 0)$ and $(0, -1)$

Limits of integration for the region OAB are

$$0 \le x \le 1 - y \quad \text{and} \quad 0 \le y \le 1$$

$$\therefore \quad \iint_R (y - 2x^2)\, dxdy = 4 \int_0^1 \int_0^{1-y} (y - 2x^2)\, dxdy$$

$$= 4 \int_0^1 \left(yx - \frac{2x^3}{3} \right) \Bigg|_0^{1-y} dy = 4 \int_0^1 \left[y(1-y) - \frac{2}{3}(1-y)^3 \right] dy$$

$$= 4 \left\{ \frac{y^2}{2} - \frac{y^3}{3} - \frac{2}{3} \frac{(1-y)^4}{4(-1)} \Bigg|_0^1 \right\}$$

$$= 4 \left\{ \frac{1}{2} - \frac{1}{3} + \frac{1}{6}(-1) \right\} = 4\, \frac{3 - 2 - 1}{6} = 0$$

(*ii*) R is the region inside the square ABCD

$\therefore$ For the total region R the limits of integration are $y - 1 \le x \le 1 - y$ and $-1 \le y \le 1$. But for the region OAB the limits of integration are $0 \le x \le 1 - y$ and $0 \le y \le 1$.

$$\iint_R xy\, dxdy = 4 \int_0^1 \int_0^{1-y} xy\, dxdy$$

$$= 4 \int_0^1 \left[y \frac{x^2}{2} \Bigg|_0^{1-y} \right] dy$$

$$= \frac{4}{2} \int_0^1 y(1 - y)^2 dy$$

$$= 2 \int_0^1 y(1 - 2y + y^2)\, dy = 2 \int_0^1 (y - 2y^2 + y^3)\, dy$$

$$= 2 \left\{ \frac{y^2}{2} - 2\frac{y^3}{3} + \frac{y^4}{4} \Bigg|_0^1 \right\} = 2 \left(\frac{1}{2} - \frac{2}{3} + \frac{7}{4} \right)$$

$$= 2\, \frac{6 - 8 + 3}{12} = \frac{2}{12} = \frac{1}{6}.$$

TEST YOUR KNOWLEDGE

Evaluate the following integrals (1–10):

1. (*i*) $\int_0^3 \int_0^1 (x^2 + 3y^2)\, dy\, dx$. (*ii*) $\int_0^1 \int_0^1 (x + 2)\, dy\, dx$ **(P.T.U., Dec. 2005)**

2. (*i*) $\int_0^3 \int_1^2 xy\,(1 + x + y)\, dy\, dx$. (*ii*) $\int_0^1 \int_0^3 (x + 3)\, dy\, dx$. **(P.T.U., May 2005)**

3. $\int_1^a \int_1^b \frac{dy\, dx}{xy}$.

4. $\int_1^2 \int_0^x \frac{dy\,dx}{x^2 + y^2}$. **(P.T.U., May 2011)**

5. $\int_0^1 dx \int_0^x e^{y/x}\, dy$.

6. $\int_0^1 \int_{x^2}^x (x^2 + 3y + 2)\, dy\, dx$.

7. (*i*) $\int_0^1 \int_0^{x^2} e^{y/x}\, dy\, dx$ (*ii*) $\int_0^2 \int_0^{y^2} e^{x/y}\, dx\, dy$.

8. (*i*) $\int_0^1 \int_y^{y^2+1} x^2 y\, dx\, dy$ (*ii*) $\int_0^2 \int_0^{\sqrt{2x}} xy\, dy\, dx$. **(P.T.U., Dec., 2003)**

9. $\int_0^{2a} \int_0^{\sqrt{2ax - x^2}} (x^2 + y^2)\, dy\, dx$.

10. $\int_0^a \int_{\sqrt{a^2 - x^2 - y^2}}^{\sqrt{a^2 - y^2}} dx\, dy$.

11. Evaluate $\iint (x^2 + y^2)\, dx\, dy$ over the region in the positive quadrant for which $x + y \le 1$.
[**Hint.** $0 \le x \le 1 - y;\ 0 \le y \le 1$]

12. Evaluate $\iint x^2 y^2\, dx\, dy$ over the circle $x^2 + y^2 = 1$.

[**Hint.** $\sqrt{1-x^2} \le y \le \sqrt{1-x^2}; -1 \le x \le 1$]

13. Evaluate $\iint xy\, dx\, dy$ over the positive quadrant of the circle $x^2 + y^2 = a^2$.

[**Hint.** $0 \le x \le \sqrt{a^2 - y^2}; 0 \le y \le a$]

14. Compute the value of $\iint_R y\, dx\, dy$, where R is the region in the first quadrant bounded by the ellipse

$$\frac{x^2}{a^2} + \frac{y^2}{b^2} = 1.$$

[**Hint.** $0 \le x \le a\sqrt{1 - \frac{y^2}{b^2}}; 0 \le y \le b$]

15. Evaluate $\iint_A xy\, dx\, dy$, where A is the domain bounded by x-axis, ordinate $x = 2a$ and the curve $x^2 = 4ay$.

[**Hint.** $0 \le y \le \frac{x^2}{4a}; 0 \le x \le 2a$]

Answers

1. (*i*) 12 (*ii*) $\frac{5}{2}$ 2. (*i*) $30\frac{3}{4}$ (*ii*) $10\frac{1}{2}$

3. $\log a \log b$ 4. $\frac{\pi}{4} \log 2$ 5. $\frac{1}{2}(e - 1)$ 6. $\frac{7}{12}$

7. (*i*) $\frac{1}{2}$ (*ii*) $e^2 - 1$ 8. (*i*) $\frac{67}{120}$ (*ii*) $\frac{8}{3}$

9. $\frac{3\pi a^4}{4}$ 10. $\frac{\pi a^3}{6}$ 11. $\frac{1}{6}$ 12. $\frac{\pi}{24}$

13. $\frac{a^4}{8}$ 14. $\frac{ab^2}{3}$ 15. $\frac{a^4}{3}$

6.3. EVALUATION OF DOUBLE INTEGRALS IN POLAR CO-ORDINATES

To evaluate $\int_{\theta_1}^{\theta_2}\int_{r_1}^{r_2} f(r,\theta)\,dr\,d\theta$ over the region bounded by the straight lines $\theta = \theta_1$, $\theta = \theta_2$ and the curves $r = r_1$, $r = r_2$, we first integrate w.r.t. r between the limits $r = r_1$ and $r = r_2$ (treating θ as a constant). The resulting expression is then integrated w.r.t. θ between the limits $\theta = \theta_1$ and $\theta = \theta_2$.

Geometrically, AB and CD are the curves $r = f_1(\theta)$ and $r = f_2(\theta)$ bounded by the lines $\theta = \theta_1$ and $\theta = \theta_2$ so that ACDB is the region of integration. PQ is a wedge of angular thickness $\delta\theta$. Then $\int_{r=r_1}^{r=r_2} f(r,\theta)\,dr$ indicates that the integration is performed along PQ (*i.e.*, r varies, θ is constant) and the integration w.r.t. θ means rotation of this strip PQ from AC to BD.

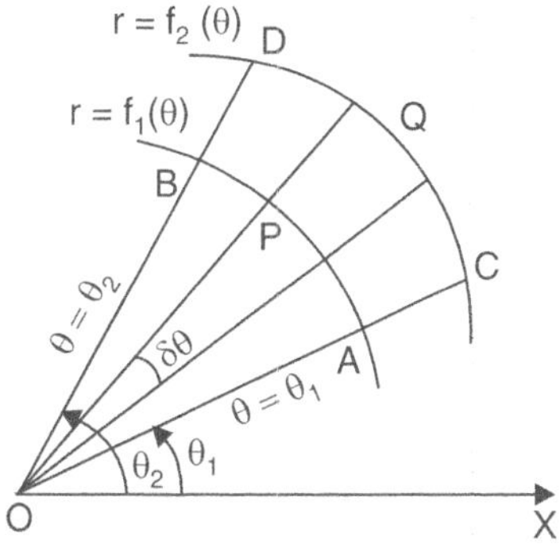

ILLUSTRATIVE EXAMPLES

Example 1. *Evaluate* $\int_0^{\pi/2}\left[\int_0^{a\cos\theta} r\sqrt{a^2 - r^2}\,dr\right]d\theta.$

Sol.

$$I = \int_0^{\pi/2}\left[\int_0^{a\cos\theta} -\frac{1}{2}(a^2 - r^2)^{1/2}(-2r)\,dr\right]d\theta.$$

$$= \int_0^{\pi/2}\left[-\frac{1}{2}\cdot\frac{(a^2 - r^2)^{3/2}}{3/2}\right]_0^{a\cos\theta} d\theta$$

$$= -\frac{1}{3}\int_0^{\pi/2}(a^3\sin^3\theta - a^3)\,d\theta = -\frac{a^3}{3}\left[\frac{2}{3} - \frac{\pi}{2}\right] = \frac{a^3}{18}(3\pi - 4).$$

Example 2. *Evaluate* $\iint r^3\,dr\,d\theta$, *over the area bounded between the circles* $r = 2\cos\theta$ *and* $r = 4\cos\theta$.

Sol. The region of integration R is shown shaded. Here r varies from $2\cos\theta$ to $4\cos\theta$ while θ varies from $-\frac{\pi}{2}$ to $\frac{\pi}{2}$.

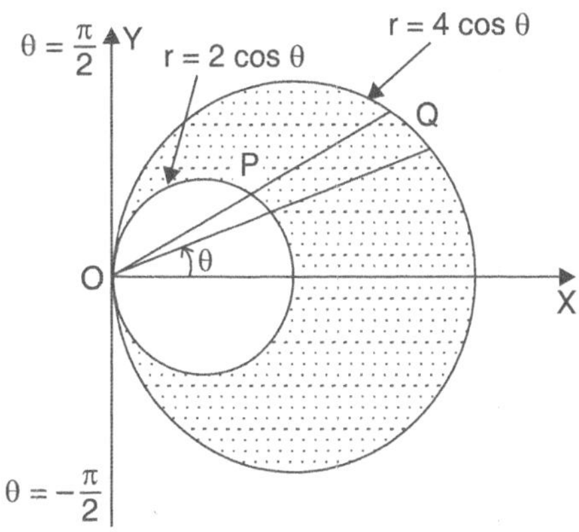

$$\therefore \quad \iint_R r^3\,dr\,d\theta = \int_{-\pi/2}^{\pi/2}\int_{2\cos\theta}^{4\cos\theta} r^3\,dr\,d\theta$$

$$= \int_{-\pi/2}^{\pi/2}\left[\frac{r^4}{4}\right]_{2\cos\theta}^{4\cos\theta} d\theta$$

$$= \int_{-\pi/2}^{\pi/2}\frac{1}{4}(256\cos^4\theta - 16\cos^4\theta)\,d\theta$$

$$= 60 \int_{-\pi/2}^{\pi/2} \cos^4 \theta \, d\theta$$

$$= 120 \int_0^{\pi/2} \cos^4 \theta \, d\theta$$

$$= 120 \times \frac{3 \times 1}{4 \times 2} \cdot \frac{\pi}{2} = \frac{45}{2} \pi.$$

Example 3. *Evaluate* $\iint r \sin \theta \, dr \, d\theta$ *over the area of the cardioid* $r = a(1 + \cos \theta)$ *above the initial line.*

Sol. The region of integration R is covered by radial strips whose ends are $r = 0$ and $r = a(1 + \cos \theta)$, the strips starting from $\theta = 0$ and ending at $\theta = \pi$.

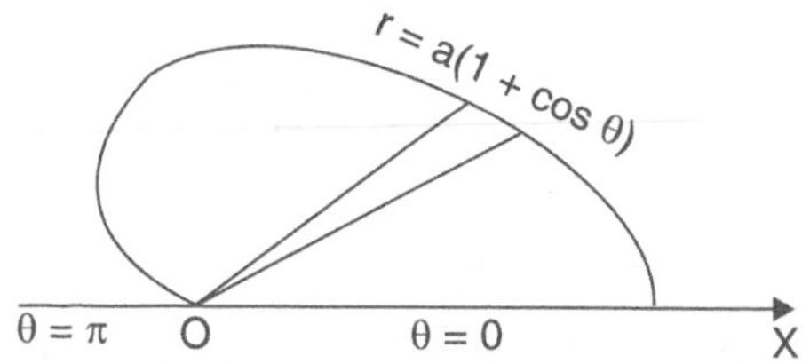

$$\therefore \quad \iint_R r \sin \theta \, dr \, d\theta = \int_0^{\pi} \int_0^{a(1+\cos\theta)} r \sin \theta \, dr \, d\theta.$$

$$= \int_0^{\pi} \sin \theta \left[\frac{r^2}{2} \right]_0^{a(1+\cos\theta)} d\theta = \frac{1}{2} \int_0^{\pi} \sin \theta \cdot a^2 (1 + \cos \theta)^2 \, d\theta$$

$$= \frac{a^2}{2} \int_0^{\pi} 2 \sin \frac{\theta}{2} \cos \frac{\theta}{2} \cdot \left(2 \cos^2 \frac{\theta}{2} \right)^2 d\theta = 4a^2 \int_0^{\pi} \sin \frac{\theta}{2} \cos^5 \frac{\theta}{2} \, d\theta$$

$$= 4a^2 \int_0^{\pi/2} 2 \sin \phi \cos^5 \phi \, d\phi \qquad \left[\text{Putting } \frac{\theta}{2} = \phi \text{ and } d\theta = 2d\phi \right]$$

$$= -8a^2 \int_0^{\pi/2} \cos^5 \phi (-\sin \phi) \, d\phi = -8a^2 \cdot \left[\frac{\cos^6 \phi}{6} \right]_0^{\pi/2}$$

$$\because \int [f(x)]^n f'(x) \, dx = \frac{[f(x)]^{n+1}}{n+1}, n \neq -1$$

$$= -\frac{4a^2}{3}(0 - 1) = \frac{4a^2}{3}.$$

Example 4. *Evaluate* $\iint \frac{r \, dr \, d\theta}{\sqrt{a^2 + r^2}}$ *over one loop of the lemniscate* $r^2 = a^2 \cos 2\theta$.

Sol. The region of integration R is covered by radial strips whose ends are $r = 0$ and $r = a\sqrt{\cos 2\theta}$, the strips starting from $\theta = -\frac{\pi}{4}$ and ending at $\theta = \frac{\pi}{4}$.

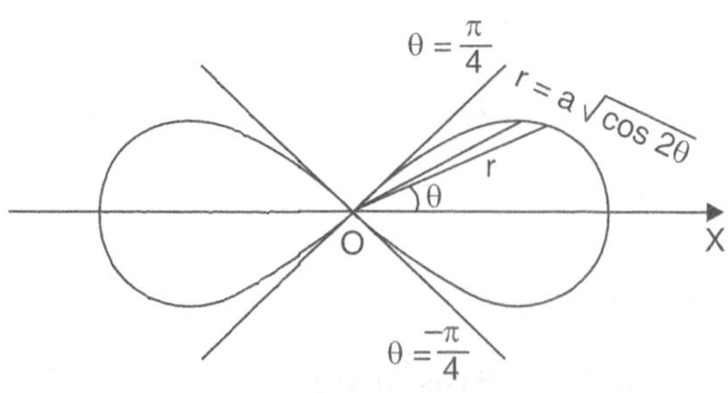

$$\therefore \quad \iint_R \frac{r\,dr\,d\theta}{\sqrt{a^2+r^2}} = \int_{-\pi/4}^{\pi/4}\int_0^{a\sqrt{\cos 2\theta}} \tfrac{1}{2}(a^2+r^2)^{-1/2}\cdot 2r\,dr\,d\theta$$

$$= \int_{-\pi/4}^{\pi/4}\left[\frac{1}{2}\cdot\frac{(a^2+r^2)^{1/2}}{1/2}\right]_0^{a\sqrt{\cos 2\theta}} d\theta = \int_{-\pi/4}^{\pi/4}[(a^2+a^2\cos 2\theta)^{1/2}-a]\,d\theta$$

$$= a\int_{-\pi/4}^{\pi/4}[(1+\cos 2\theta)^{1/2}-1]\,d\theta = a\int_{-\pi/4}^{\pi/4}[(2\cos^2\theta)^{1/2}-1]\,d\theta$$

$$= a\int_{-\pi/4}^{\pi/4}(\sqrt{2}\cos\theta-1)\,d\theta = 2a\int_0^{\pi/4}(\sqrt{2}\cos\theta-1)\,d\theta$$

| $\because$ integrand is an even function

$$= 2a\left[\sqrt{2}\sin\theta-\theta\right]_0^{\pi/4} = 2a\left[\sqrt{2}\cdot\frac{1}{\sqrt{2}}-\frac{\pi}{4}\right] = 2a\left(1-\frac{\pi}{4}\right).$$

TEST YOUR KNOWLEDGE

Evaluate the following integrals (1–4):

1. $\displaystyle\int_0^{\pi}\int_0^{a\sin\theta} r\,dr\,d\theta$

2. $\displaystyle\int_0^{\pi/2}\int_0^{a\cos\theta} r\sin\theta\,dr\,d\theta$

3. $\displaystyle\int_0^{\pi/2}\int_{a(1+\cos\theta)}^{a} r\,dr\,d\theta$

4. $\displaystyle\int_0^{\pi}\int_0^{a(1+\cos\theta)} r^2\cos\theta\,dr\,d\theta$

5. Show that $\displaystyle\iint_R r^2\sin\theta\,dr\,d\theta = \frac{2a^3}{3}$, where R is the region bounded by the semi-circle $r = 2a\cos\theta$, above the initial line.

6. Evaluate $\displaystyle\iint r^3\,dr\,d\theta$ over the area included between the circles $r = 2\sin\theta$ and $r = 4\sin\theta$.

Answers

1. $\dfrac{\pi a^2}{4}$ 2. $\dfrac{a^2}{6}$ 3. $-a^2\left(1+\dfrac{\pi}{8}\right)$

4. $\dfrac{5}{8}\pi a^3$ 6. $\dfrac{45\pi}{2}$.

6.4. CHANGE OF ORDER OF INTEGRATION

In a double integral, if the limits of integration are constant, then the order of integration is immaterial, provided the limits of integration are changed accordingly. Thus

$$\int_c^d\int_a^b f(x,y)\,dx\,dy = \int_a^b\int_c^d f(x,y)\,dy\,dx.$$

But if the limits of integration are variable, a change in the order of integration necessitates change in the limits of integration. A rough sketch of the region of integration helps in fixing the new limits of integration.

ILLUSTRATIVE EXAMPLES

Example 1. *Change the order of integration in* $\int_0^a \int_y^a \frac{x\,dx\,dy}{x^2 + y^2}$ *and hence evaluate the same.*

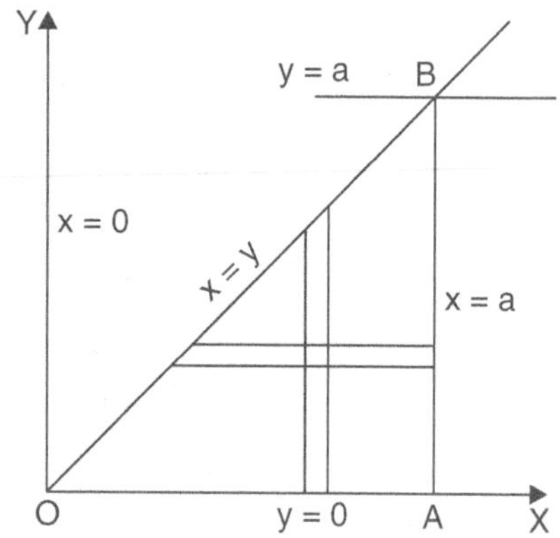

Sol. From the limits of integration, it is clear that the region of integration is bounded by $x = y$, $x = a$, $y = 0$ and $y = a$. Thus the region of integration is the ΔOAB and is divided into horizontal strips. For changing the order of integration, we divide the region of integration into vertical strips. The new limits of integration become: y varies from 0 to x and x varies from 0 to a.

$$\int_0^a \int_y^a \frac{x\,dx\,dy}{x^2 + y^2} = \int_0^a \int_0^x \frac{x\,dy\,dx}{x^2 + y^2}$$

$$= \int_0^a x \cdot \left[\frac{1}{x}\tan^{-1}\frac{y}{x}\right]_0^x dx$$

$$= \int_0^a \frac{\pi}{4}\,dx = \frac{\pi}{4}\cdot\left[x\right]_0^a = \frac{\pi a}{4}.$$

Example 2. *Change the order of integration in* $I = \int_0^1 \int_{x^2}^{2-x} xy\,dy\,dx$ *and hence evaluate the same.* **(P.T.U., May 2006, May 2009, Dec. 2011, 2015)**

Sol. From the limits of integration, it is clear that we have to integrate first with respect to y which varies from $y = x^2$ to $y = 2 - x$ and then with respect to x which varies from $x = 0$ to $x = 1$. The region of integration OAB is divided into vertical strips. For changing the order of integration, we divide the region of integration into horizontal strips.

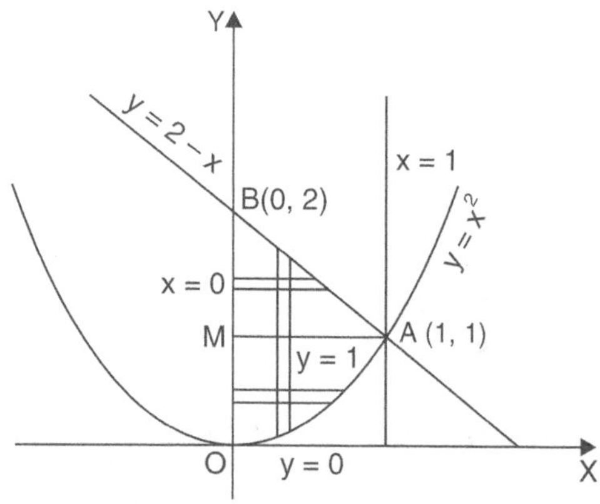

Solving $y = x^2$ and $y = 2 - x$, the co-ordinates of A are (1, 1). Draw AM ⊥ OY. The region of integration is divided into two parts, OAM and MAB.

For the region OAM, x varies from 0 to $\sqrt{y}$ and y varies from 0 to 1. For the region MAB, x varies from 0 to $2 - y$ and y varies from 1 to 2.

$$\therefore \quad \int_0^1 \int_{x^2}^{2-x} xy\,dx\,dy = \int_0^1 \int_0^{\sqrt{y}} xy\,dx\,dy + \int_1^2 \int_0^{2-y} xy\,dx\,dy$$

$$= \int_0^1 y \cdot \left[\frac{x^2}{2}\right]_0^{\sqrt{y}} dy + \int_1^2 y \cdot \left[\frac{x^2}{2}\right]_0^{2-y} dy = \frac{1}{2}\int_0^1 y^2 dy + \frac{1}{2}\int_1^2 y(2-y)^2\, dy$$

$$= \frac{1}{2}\cdot\left[\frac{y^3}{3}\right]_0^1 + \frac{1}{2}\int_1^2 (4y - 4y^2 + y^3)\, dy = \frac{1}{6} + \frac{1}{2}\left[2y^2 - \frac{4y^3}{3} + \frac{y^4}{4}\right]_1^2$$

$$= \frac{1}{6} + \frac{1}{2}\left[\left(8 - \frac{32}{3} + 4\right) - \left(2 - \frac{4}{3} + \frac{1}{4}\right)\right] = \frac{3}{8}.$$

Example 3. *Evaluate by changing the order of integration of* $\int_0^1 \int_x^{\sqrt{2-x^2}} \frac{x\,dx\,dy}{\sqrt{x^2+y^2}}$.

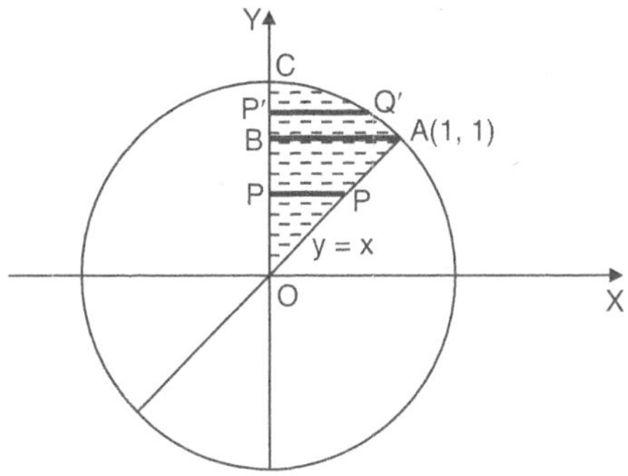

Sol. From limits of integration it is clear that we have to first integrate w.r.t. y and then w.r.t. x. y varies from $y = x$ to $y = \sqrt{2 - x^2}$ *i.e.*, $y^2 = 2 - x^2$ or $x^2 + y^2 = 2$ *i.e.*, a circle with centre at (0, 0) and radius = $\sqrt{2}$ and x varies 0 to 1. This integration is firstly performed along vertical strips and then along horizontal strips. For change of the order of integration we have to first perform integration along horizontal strips and then along vertical strips. The region of integration is shown shaded in the figure.

For horizontal strips the whole region is divided into two portion OAB and ACB. Let the horizontal strip in the portion OAB be PQ and that of in ACB be P′Q′.

The curve and the line intersect at A(1, 1).

For OAB ; x varies from 0 to x and y varies for 0 to 1.

For ACB ; x varies from 0 to $\sqrt{2-y^2}$ and y varies for 1 to $\sqrt{2}$

$$\therefore \quad \int_0^1 \int_x^{\sqrt{2-x^2}} \frac{x\,dx\,dy}{\sqrt{x^2+y^2}} = \int_0^1 \int_0^y \frac{x}{\sqrt{x^2+y^2}}\,dx\,dy + \int_0^{\sqrt{2}} \int_0^{\sqrt{2-y^2}} \frac{x\,dx\,dy}{\sqrt{x^2+y^2}}$$

$$= \frac{1}{2}\int_0^1 \int_0^y (x^2+y^2)^{-1/2}(2x)\,dx\,dy + \frac{1}{2}\int_1^{\sqrt{2}} \int_0^{\sqrt{2-y^2}} (2x)(x^2+y^2)^{-1/2}\,dx\,dy$$

$$= \frac{1}{2}\int_0^1 \left.\frac{(x^2+y^2)^{1/2}}{1/2}\right|_0^y dy + \frac{1}{2}\int_1^{\sqrt{2}} \frac{(x^2+y^2)^{1/2}}{1/2}\cdot\Bigg|_0^{\sqrt{2-y^2}} dy$$

| By using $\int [f(x)]\, f'(x)dx = \frac{[f(x)]^{n+1}}{n+1}$, $n \neq -1$

$$= \int_0^1 \left(\sqrt{2}y - y\right)dy + \int_1^{\sqrt{2}} \left(\sqrt{2} - y\right)dy$$

$$= \left(\frac{\sqrt{2}y^2}{2} - \frac{y^2}{2}\right)\Bigg|_0^1 + \left(\sqrt{2}y - \frac{y^2}{2}\right)\Bigg|_1^{\sqrt{2}} = \frac{\sqrt{2}-1}{2} + 2 - 1 - \sqrt{2} + \frac{1}{2}$$

$$= \frac{1}{2}\left[\sqrt{2} - 1 + 4 - 2 - 2\sqrt{2} + 1\right] = \frac{1}{2}\left[2 - \sqrt{2}\right] = 1 - \frac{1}{\sqrt{2}}.$$

Example 4. *Change the order of integration of*

$$\int_0^{2a}\int_{\sqrt{2ax-x^2}}^{\sqrt{2ax}} V\, dy\, dx. \qquad \textbf{(P.T.U., Dec. 2003)}$$

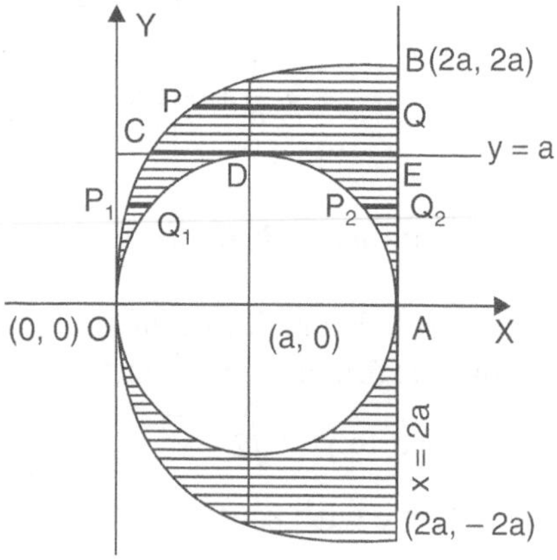

Sol. From the limits of integration it is clear that we have to first integrate w.r.t. y which varies from $\sqrt{2ax-x^2}$ to $\sqrt{2ax}$ *i.e.*, $\sqrt{2ax-x^2} \le y \le \sqrt{2ax}$ *i.e.*, $2ax - x^2 \le y^2 \le 2ax$ and then w.r.t. x which varies from 0 to $2a$ *i.e.*, $0 \le x \le 2a$. This shows that integration is firstly performed along veritcal strips. For change of the order of integration we have to perform integration firstly along horizontal strips and then along vertical. We draw rough sketch of the region $y^2 = 2ax - x^2$ or $x^2 + y^2 - 2ax = 0$ which is a circle with centre at $(a, 0)$ and radius a, $y = \sqrt{2ax}$ or $y^2 = 2ax$ is a right handed parabola. The region of integration is shown shaded in the figure.

$x = 0$ represents y-axis and $x = 2a$ represents a line parallel to y-axis.

The two curves intersect at (0, 0). For horizontal strip PQ we see that only the region BCE is covered ∴ we divide the whole region into three portions namely: BCE, ODC and AED

for BCE;

x varies from $\frac{y^2}{2a}$ to $2a$

y varies from a to $2a$

$\because\ y^2 = 2ax - x^2$ or $x^2 - 2ax + y^2 = 0$

$$x = \frac{2a \pm \sqrt{4a^2 - 4y^2}}{2} = a \pm \sqrt{a^2 - y^2}$$

for ODC ; x varies from $\frac{y^2}{2a}$ to $a + \sqrt{a^2 - y^2}$

y varies from 0 to a

for AED x varies from $a + \sqrt{a^2 - y^2}$ to $2a$

y varies from 0 to a

$$\therefore \quad \int_0^{2a}\int_{\sqrt{2ax-x^2}}^{\sqrt{2ax}} V\, dy\, dx = \int_a^{2a}\int_{y^2/2a}^{2a} V\, dx\, dy + \int_0^{a}\int_{y^2/2a}^{a+\sqrt{a^2-y^2}} V\, dx\, dy$$

$$+ \int_0^{a}\int_{a+\sqrt{a^2-y^2}}^{2a} V\, dx\, dy$$

Example 5. *Change the order of integration in the following integral and evaluate:*

(*i*) $\displaystyle\int_0^a\int_{a-\sqrt{a^2-y^2}}^{a+\sqrt{a^2-y^2}} dx\, dy$ (*ii*) $\displaystyle\int_0^{4a}\int_{x^2/4a}^{2\sqrt{ax}} dy\, dx.$ **(P.T.U., Jan. 2010)**

Sol. (*i*) $\displaystyle\int_0^a\int_{a-\sqrt{a^2-y^2}}^{a+\sqrt{a^2-y^2}} dx\, dy$

From the limits of integration it is clear that we have to first integrate w.r.t. x and then w.r.t. y

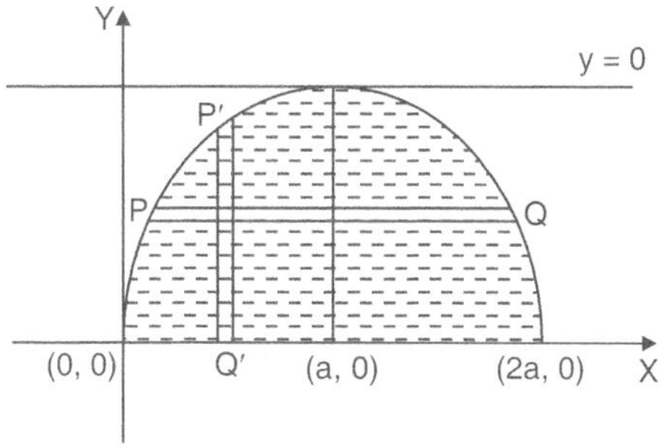

x varies from $a - \sqrt{a^2 - y^2}$ to $a + \sqrt{a^2 - y^2}$

i.e., $x = a \pm \sqrt{a^2 - y^2}$; $(x - a)^2 = a^2 - y^2$

or $\qquad x^2 + y^2 - 2ax = 0$

i.e., inside the circle with centre at $(a, 0)$ and radius a and y varies from 0 to a.

This integration is first perfomed along horizontal strips PQ. For changing the order of integration divide the region into vertical strips P′Q′ where y varies from 0 to $\sqrt{2ax - x^2}$ and x varies from 0 to $2a$. The region of integration is shown shaded in the figure.

$$\therefore \quad \int_0^a \int_{a-\sqrt{a^2-y^2}}^{a+\sqrt{a^2-y^2}} dxdy = \int_0^{2a} \int_0^{\sqrt{2ax-x^2}} dy\,dx = \int_0^{2a} y \Big|_0^{\sqrt{2ax-x^2}} dx$$

$$= \int_0^{2a} \sqrt{2ax - x^2}\,dx = \int_0^{2a} \sqrt{a^2 - (x-a)^2}\,dx$$

$$= \frac{(x-a)\sqrt{a^2 - (x-a)^2}}{2} + \frac{a^2}{2} \sin^{-1} \frac{x-a}{a} \Big|_0^{2a} = \frac{a^2}{2}\frac{\pi}{2} - \frac{a^2}{2}\left(-\frac{\pi}{2}\right) = \frac{a^2\pi}{2}.$$

(*ii*) From the limits of integration, it is clear that we have to integrate first w.r.t. y which varies from $y = \dfrac{x^2}{4a}$ to $y = 2\sqrt{ax}$ and then w.r.t. x which varies from $x = 0$ to $x = 4a$. Thus integration is first performed along the vertical strip PQ which extends from a point P on the parabola $y = \dfrac{x^2}{4a}$ (*i.e.*, $x^2 = 4ay$) to the point Q on the parabola $y = 2\sqrt{ax}$ (*i.e.*, $y^2 = 4ax$). Then the strip slides from O to A $(4a, 4a)$, the point of intersection of the two parabolas.

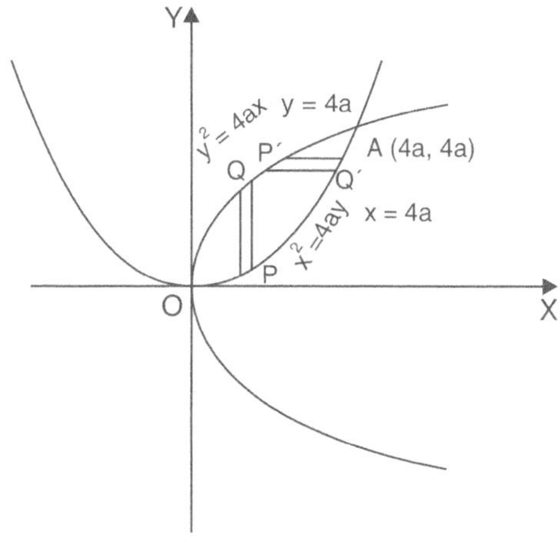

For changing the order of integration, we divide the region of integration OPAQO into horizontal strips P′Q′ which extend from P′ on the parabola $y^2 = 4ax$ *i.e.*, $x = \dfrac{y^2}{4a}$ to Q′ on the parabola $x^2 = 4ay$ *i.e.*, $x = 2\sqrt{ay}$. Then this strip slides from O to A$(4a, 4a)$, *i.e.*, y varies from 0 to $4a$.

$$\therefore \quad \int_0^{4a} \int_{x^2/4a}^{2\sqrt{ax}} dy\,dx = \int_0^{4a} \int_{y^2/4a}^{2\sqrt{ay}} dx\,dy$$

$$= \int_0^{4a} \Big[x \Big]_{y^2/4a}^{2\sqrt{ay}} dy = \int_0^{4a} \left(2\sqrt{ay} - \frac{y^2}{4a} \right) dy$$

$$= \left[2\sqrt{a} \cdot \frac{y^{3/2}}{3/2} - \frac{y^3}{12a} \right]_0^{4a} = \frac{4}{3}\sqrt{a} \cdot (4a)^{3/2} - \frac{64a^3}{12a}$$

$$= \frac{4}{3}\sqrt{a} \cdot 8a^{3/2} - \frac{16a^2}{3} = \frac{32a^2}{3} - \frac{16a^2}{3} = \frac{16a^2}{3}.$$

Example 6. *Change the order of integration in* $\int_0^a \int_{mx}^{lx} f(x, y)\, dy\, dx$.

(P.T.U. Jan. 2009)

Sol. From the limits of integration, it is clear that we have to integrate first w.r.t. y which varies from $y = mx$ to $y = lx$ and then w.r.t. x which varies from $x = 0$ to $x = a$.

Thus integration is first performed along the vertical strip PQ which extends from a point P on the line $y = mx$ to the point Q on the line $y = lx$.

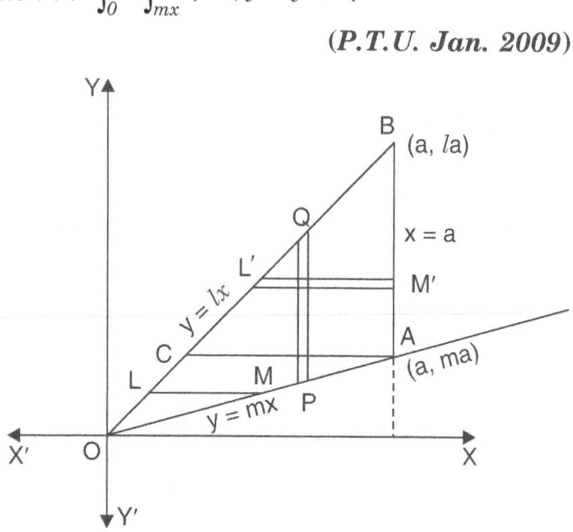

For change of order of integration we have to first perform integration along horizontal strips and then along vertical strips.

For horizontal strips the whole region is divided into two portions OAC and CAB.

Let the horizontal strip in the portion OAC be LM and that of in CAB be L′M′. For the region OAC, x varies from $\frac{y}{l}$ to $\frac{y}{m}$ and y varies from 0 to ma.

For the region CAB; x varies from $\frac{y}{l}$ to a and y varies from ma to la

$$\therefore \quad \int_0^a \int_{mx}^{lx} f(x, y)\, dy\, dx = \int_{y=0}^{ma} \int_{x=\frac{y}{l}}^{y/m} f(x, y)\, dx\, dy + \int_{y=ma}^{la} \int_{x=\frac{y}{l}}^{a} f(x, y)\, dx\, dy.$$

Example 7. *Evaluate the following*

(i) $\int_0^a \int_{y^2/a}^{y} \frac{y\, dx\, dy}{(a-x)\sqrt{ax - y^2}}$ $\qquad$ (ii) $\int_0^\infty \int_x^\infty \frac{e^{-y}}{y}\, dy\, dx$. **(P.T.U., May 2004, Dec., 2011)**

Sol. (i) $\int_0^a \int_{y^2/a}^{y} \frac{y\, dx\, dy}{(a-x)\sqrt{ax - y^2}}$

Here limits of integration are given by $0 \le y \le a$ and $\frac{y^2}{a} \le x \le y$. The limits of x are depending on y where as that of y are constant. $\therefore$ We have to integrate first w.r.t. x and then w.r.t. y but we cannot easily integrate in this order, integral of $\frac{y}{(a-x)\sqrt{ax - y^2}}$ w.r.t. y is simpler.

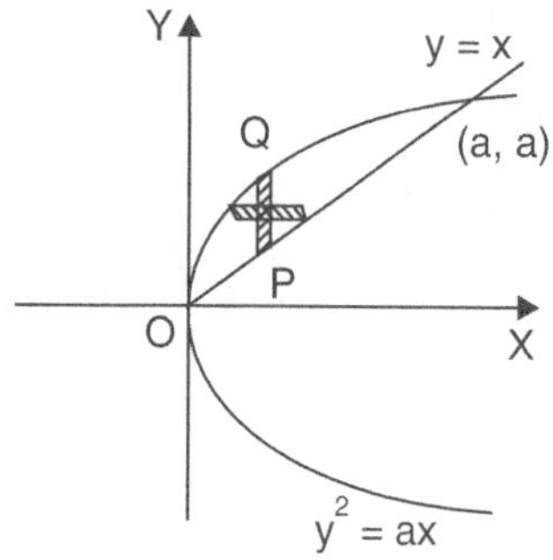

$\therefore$ We change the order of integration for this we first sketch the region given by $x = \frac{y^2}{a}$ or $y^2 = ax$ (a right handed parabola) and $x = y$ is a straight line and these two intersect at points given by $x^2 = ax$ or $x = 0, x = a$ *i.e.*, at (0, 0) and (a, a).

To change the order of integration we need limits of y as variable and that of x as constant.

$\therefore$ We divide the region into vertical strips PQ

$\therefore$ y varies from x to $\sqrt{ax}$

and x varies from 0 to a

$$\therefore \int_0^a \int_{y^2/a}^{y} \frac{y\,dx\,dy}{(a-x)\sqrt{ax-y^2}} = \int_0^a \left[\int_x^{\sqrt{ax}} \frac{y}{(a-x)\sqrt{ax-y^2}}\,dy\right] dx$$

$$= \int_0^a \left[\frac{1}{a-x}\int_x^{\sqrt{ax}} \left(-\frac{1}{2}\right)(-2y)(ax-y^2)^{-1/2}\,dy\right] dx$$

$$= \int_0^a \frac{1}{a-x}\left(-\frac{1}{2}\right)\left[\frac{(ax-y^2)^{1/2}}{1/2}\Bigg|_x^{\sqrt{ax}}\right] dx = \int_0^a \frac{-1}{2(a-x)}[(0-2(ax-x^2)^{1/2}]\,dx$$

$$= \int_0^a \frac{1}{a-x}\sqrt{x}\,\sqrt{a-x}\,dx = \int_0^a \frac{\sqrt{x}}{\sqrt{a-x}}\,dx$$

Put $x = a\sin^2\theta$; $dx = 2a\sin\theta\cos\theta\,d\theta$

$$= \int_0^{\pi/2} \frac{\sqrt{a}\sin\theta}{\sqrt{a}\cos\theta}\,2a\sin\theta\cos\theta\,d\theta = 2a\int_0^{\pi/2}\sin^2\theta\,d\theta$$

$$= 2a\,\frac{1}{2}\frac{\pi}{2} = \frac{a\pi}{2}.$$

(ii) $\displaystyle\int_0^\infty \int_x^\infty \frac{e^{-y}}{y}\,dy\,dx$

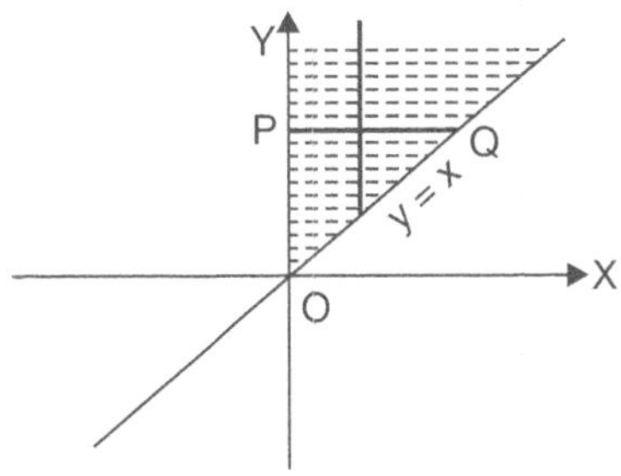

Here the limits of integrations of y are from x to ∞ and the limits of integrations of x are from 0 to ∞. As $\dfrac{e^{-y}}{y}$ cannot be integrated w.r.t. y, $\therefore$ we change the order of integration. So we first sketch the region and then divide it into horizontal strips (Region is the shaded portion).

For horizontal strip x varies from 0 to y and y varies from 0 to ∞

$$\therefore \int_0^\infty \int_x^\infty \frac{e^{-y}}{y}\,dydx = \int_0^\infty \int_0^y \frac{e^{-y}}{y}\,dx\,dy = \int_0^\infty \left[\int_0^y \frac{e^{-y}}{y}\,dx\right] dy$$

$$= \int_0^\infty \frac{e^{-y}}{y}\,x\Bigg|_0^y\,dy = \int_0^\infty \frac{e^{-y}}{y}\cdot y\,dy = \int_0^\infty e^{-y}\,dy = -e^{-y}\Bigg|_0^\infty = 0 + 1 = 1.$$

Example 8. *Evaluate the following integrals by changing the order of integration:*

(*i*) $\int_0^{\infty} \int_0^{x} xe^{-x^2/y}\, dydx$ **(P.T.U., Dec. 2010, May 2014, 2015)**

(*ii*) $\int_0^{\frac{a}{\sqrt{2}}} \int_y^{\sqrt{a^2-y^2}} \log(x^2 + y^2)\, dx\, dy;\ a > 0.$

Sol. (*i*) $\int_0^{\infty} \int_0^{x} xe^{-x^2/y}\, dy\, dx$

limits of integration show that the region is bounded by $y = 0, y = x$ and $x = 0, x = \infty$. Region is firstly integrated w.r.t. y and then w.r.t. x. So the region is divided into vertical strips. For change of integration divide the region into horizontal strips as shown in the figure.

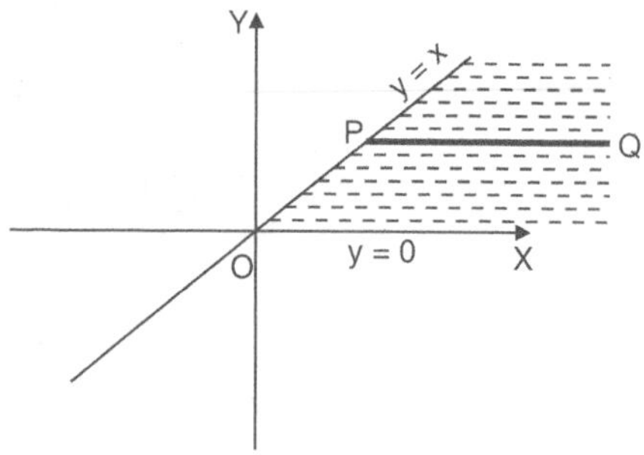

For horizontal strip x varies from y to ∞ and y varies from 0 to ∞

$$\therefore \quad \int_0^{\infty}\int_0^{x} xe^{-x^2/y} dy\, dx = \int_0^{\infty}\int_y^{\infty} xe^{-x^2/y}\, dx\, dy$$

To integrate $xe^{-x^2/y}$ w.r.t. x

Put $\frac{x^2}{y} = t \quad \therefore \quad 2x\, dx = -y\, dt$ and t varies from y to ∞

$$= \int_0^{\infty}\left[\int_y^{\infty} e^{-t}\, y \frac{dt}{2}\right] dy = \int_0^{\infty} \frac{y}{2} \cdot \frac{e^{-t}}{-1}\Bigg|_y^{\infty} dy$$

$$= \int_0^{\infty} -\frac{y}{2}\,(0 - e^{-y})\, dy$$

$$= \frac{1}{2}\int_0^{\infty} ye^{-y}\, dy$$

$$= \frac{1}{2}\left\{y\frac{e^{-y}}{-1} - \int_0^{\infty} \cdot 1 \frac{e^{-y}}{-1} dy\right\} = \frac{1}{2}\left\{-ye^{-y} + \frac{e^{-y}}{-1}\Bigg|_0^{\infty}\right\}$$

$$= \frac{1}{2}\{-0 + 0 + 0 + 1\} = \frac{1}{2}.$$

(*ii*) $\int_0^{\frac{a}{\sqrt{2}}} \int_y^{\sqrt{a^2-y^2}} \log(x^2 + y^2)\, dx\, dy$

From the limits of integration it is clear that x varies from y to $\sqrt{a^2 - y^2}$ *i.e.*, $x = y$ and $x = \sqrt{a^2 - y^2}$ or $x^2 + y^2 = a^2$ *i.e.*, a circle with centre at (0, 0) and radius a. Region is firstly integrated w.r.t. x means the integration is firstly performed along horizontal strips and then along vertical. For change of integration we have to perform integration w.r.t. y *i.e.*, vertical strips first and then horizontal strips. Region is shown in the figure.

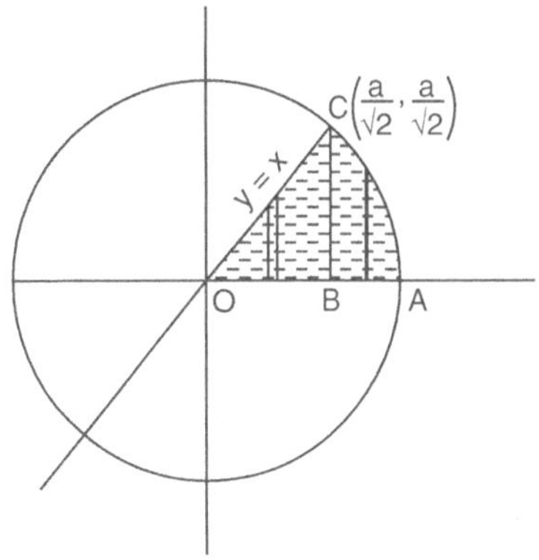

The whole region of integration is divided into two parts OBC and ACB. For OBC; y varies from 0 to x and x varies from 0 to $\frac{a}{\sqrt{2}}$.

For ACB; y varies from 0 to $\sqrt{a^2 - x^2}$ and x varies from $\frac{a}{\sqrt{2}}$ to a.

$$\therefore \quad \int_0^{\frac{a}{\sqrt{2}}} \int_y^{\sqrt{a^2-y^2}} \log (x^2 + y^2)\, dx\, dy$$

$$= \int_0^{\frac{a}{\sqrt{2}}} \int_0^x \log (x^2 + y^2)\, dy\, dx + \int_{\frac{a}{\sqrt{2}}}^a \int_0^{\sqrt{a^2-x^2}} \log (x^2 + y^2)\, dy\, dx$$

$$= I_1 + I_2 \qquad \text{...(1)}$$

$$I_1 = \int_0^{\frac{a}{\sqrt{2}}} \int_0^x 1 \cdot \log (x^2 + y^2)\, dy\, dx$$

$$= \int_0^{\frac{a}{\sqrt{2}}} \left\{ \log(x^2 + y^2) \cdot y \Big|_0^x - \int_0^x \frac{2y}{x^2 + y^2} \cdot y\, dy \right\} dx$$

$$= \int_0^{\frac{a}{\sqrt{2}}} x \log 2x^2 - 2 \int_0^x \frac{y^2 + x^2 - x^2}{y^2 + x^2}\, dy \Bigg]\, dx$$

$$= \int_0^{\frac{a}{\sqrt{2}}} \left\{ x \log 2x^2 - 2\left(y - x^2 \frac{1}{x} \tan^{-1} \frac{y}{x} \right)\Bigg|_0^x \right\} dx$$

$$= \int_0^{\frac{a}{\sqrt{2}}} (x \log 2x^2 - 2x + 2x \tan^{-1} 1)\, dx$$

$$= \int_0^{\frac{a}{\sqrt{2}}} \left\{ x \log 2x^2 - 2x + 2x \frac{\pi}{4} \right\} dx \text{ integrate by parts}$$

$$= (\log 2x^2) \frac{x^2}{2} \Bigg|_0^{\frac{a}{\sqrt{2}}} - \int_0^{\frac{a}{\sqrt{2}}} \frac{1}{2x^2} \cdot 4x \cdot \frac{x^2}{2}\, dx - x^2 + \frac{\pi x^2}{4} \Bigg|_0^{\frac{a}{\sqrt{2}}}$$

$$= \frac{a^2}{4} \log a^2 - \int_0^{\frac{a}{\sqrt{2}}} x\, dx - \frac{a^2}{2} + \frac{\pi}{4} \cdot \frac{a^2}{2}$$

$$= \frac{a^2}{4} \log a^2 - \frac{x^2}{2} \Bigg|_0^{\frac{a}{\sqrt{2}}} - \frac{a^2}{2} + \frac{\pi a^2}{8}$$

$$= \frac{a^2}{2} \log a - \frac{a^2}{4} - \frac{a^2}{2} + \frac{\pi a^2}{8} = \frac{a^2 \log a}{2} - \frac{3a^2}{4} + \frac{\pi a^2}{8}$$

$$I_2 = \int_{\frac{a}{\sqrt{2}}}^{a} \left\{\int_0^{\sqrt{a^2-x^2}} 1 \cdot \log(x^2+y^2)\,dy\right\}dx$$

$$= \int_{\frac{a}{\sqrt{2}}}^{a} \left[\log(x^2+y^2)\cdot y \Big|_0^{\sqrt{a^2-x^2}} - \int_0^{\sqrt{a^2-x^2}} \frac{1}{x^2+y^2}\cdot 2y\cdot y\,dy\right]dx$$

$$= \int_{\frac{a}{\sqrt{2}}}^{a} \left[\sqrt{a^2-x^2}\,\log a^2 - 2\int_0^{\sqrt{a^2-x^2}} \frac{y^2+x^2-x^2}{y^2+x^2}\,dy\right]dx$$

$$= \int_{\frac{a}{\sqrt{2}}}^{a} \left[(\log a^2)\sqrt{a^2-x^2} - 2\int_0^{\sqrt{a^2-x^2}} \left[1 - x^2\cdot\frac{1}{x^2+y^2}\right]dy\right]dx$$

$$= \int_{\frac{a}{\sqrt{2}}}^{a} \left[(\log a^2)\sqrt{a^2-x^2} - 2\left(y - \frac{x^2}{x}\tan^{-1}\frac{y}{x}\Big|_0^{\sqrt{a^2-x^2}}\right)\right]dx$$

$$= \int_{\frac{a}{\sqrt{2}}}^{a} \left[(\log a^2)\sqrt{a^2-x^2} - 2\sqrt{a^2-x^2} + 2x\tan^{-1}\frac{\sqrt{a^2-x^2}}{x}\right]dx$$

$$= \int_{\frac{a}{\sqrt{2}}}^{a} (2\log a - 2)\sqrt{a^2-x^2}\,dx + 2\int_{\frac{a}{\sqrt{2}}}^{a} x\tan^{-1}\frac{\sqrt{a^2-x^2}}{x}\,dx$$

(integrate by parts)

$$= (2\log a - 2)\int_{\frac{a}{\sqrt{2}}}^{a} \sqrt{a^2-x^2}\,dx + 2\left\{\left(\tan^{-1}\frac{\sqrt{a^2-x^2}}{x}\right)\cdot\frac{x^2}{2}\Big|_{\frac{a}{\sqrt{2}}}^{a}\right.$$

$$\left. - \int_{\frac{a}{\sqrt{2}}}^{a} \frac{1}{1+\frac{a^2-x^2}{x^2}}\,\frac{x\frac{-2x}{2\sqrt{a^2-x^2}} - \sqrt{a^2-x^2}}{x^2} \times \frac{x^2}{2}\,dx\right.$$

$$= (2\log a - 2)\int_{\frac{a}{\sqrt{2}}}^{a} \sqrt{a^2-x^2}\,dx + 2\left(-\frac{a^2}{4}\cdot\frac{\pi}{4}\right) - 2\int_{\frac{a}{\sqrt{2}}}^{a} \frac{x^2}{a^2}\cdot\frac{-x^2-a^2+x^2}{x^2\sqrt{a^2-x^2}}\cdot\frac{x^2}{2}\,dx$$

$$= (2\log a - 2)\int_{\frac{a}{\sqrt{2}}}^{a} \sqrt{a^2-x^2}\,dx - \frac{\pi a^2}{8} + \int_{\frac{a}{\sqrt{2}}}^{a} \frac{x^2}{\sqrt{a^2-x^2}}\,dx$$

$$= (2\log a - 2)\int_{\frac{a}{\sqrt{2}}}^{a} \sqrt{a^2-x^2}\,dx - \frac{\pi a^2}{8} - \int_{\frac{a}{\sqrt{2}}}^{a} \frac{a^2-x^2-a^2}{\sqrt{a^2-x^2}}\,dx$$

$$= (2\log a - 2)\int_{\frac{a}{\sqrt{2}}}^{a} \sqrt{a^2-x^2}\,dx - \frac{\pi a^2}{8} - \int_{\frac{a}{\sqrt{2}}}^{a} \sqrt{a^2-x^2}\,dx + a^2\int_{\frac{a}{\sqrt{2}}}^{a} \frac{dx}{\sqrt{a^2-x^2}}$$

$$= (2\log a - 3)\int_{\frac{a}{\sqrt{2}}}^{a} \sqrt{a^2-x^2}\,dx - \frac{\pi a^2}{8} + a^2\int_{\frac{a}{\sqrt{2}}}^{a} \frac{dx}{\sqrt{a^2-x^2}}$$

$$= (2 \log a - 3) \left\{ \frac{x\sqrt{a^2 - x^2}}{2} + \frac{a^2}{2} \sin^{-1} \frac{x}{a} \right\} + a^2 \sin^{-1} \frac{x}{a} \Bigg|_{\frac{a}{\sqrt{2}}}^{a} - \frac{\pi a^2}{8}$$

$$= (2 \log a - 3) \left(\frac{a^2}{2} \cdot \frac{\pi}{2} - \frac{a}{2\sqrt{2}} \frac{a}{\sqrt{2}} - \frac{a^2}{2} \sin^{-1} \frac{1}{\sqrt{2}} \right) + a^2 \left(\frac{\pi}{2} - \frac{\pi}{4} \right) - \frac{\pi a^2}{8}$$

$$= (2 \log a - 3)\, a^2 \left(\frac{\pi}{4} - \frac{1}{4} - \frac{\pi}{8} \right) + a^2 \frac{\pi}{4} - \frac{\pi a^2}{8}$$

$$= a^2 (2 \log a - 3) \left(\frac{\pi}{8} - \frac{1}{4} \right) + \frac{\pi a^2}{8}$$

Substituting the values of I_1 and I_2 in (1)

$$\int_0^{\frac{a}{\sqrt{2}}} \int_y^{\sqrt{a^2 - y^2}} \log (x^2 + y^2)\, dx\, dy = \frac{a^2 \log a}{2} - \frac{3a^2}{4} + \frac{\pi a^2}{8} + 2a^2 \log a \left(\frac{\pi}{8} - \frac{1}{4} \right)$$

$$- \frac{3a^2 \pi}{8} + \frac{3}{4} a^2 + \frac{\pi a^2}{8}$$

$$= a^2 \log a \left[\frac{1}{2} + \frac{\pi}{4} - \frac{1}{2} \right] - \frac{\pi a^2}{8} = \frac{\pi a^2}{4} \left[\log a - \frac{1}{2} \right]$$

Example 9. *Evaluate the following integrals by changing the order of integration*

(*i*) $\int_0^b \int_0^{\frac{a}{b}\sqrt{b^2 - y^2}} xy\, dx\, dy$ **(P.T.U., Dec. 2013)** (*ii*) $\int_0^1 \int_{4y}^4 e^{x^2}\, dx\, dy$ **(P.T.U., Dec. 2005)**

Sol. (*i*) From the limits of integration, it is clear that region of integration is bounded by $x = 0, y = 0, y = b$ and the ellipse $x = \frac{a}{b}\sqrt{b^2 - y^2}$ *i.e.*, $\frac{x^2}{a^2} + \frac{y^2}{b^2} = 1$.

Thus, the region of integration is OAB. As in the given integral we have to first integrate w.r.t. x and then w.r.t. y. So, the region is firstly divided along horizontal strips and then along vertical strips. For change of integration we have to first perform integration w.r.t. y and then w.r.t. x. So the region is firstly divided along vertical strips and then along horizontal strips.

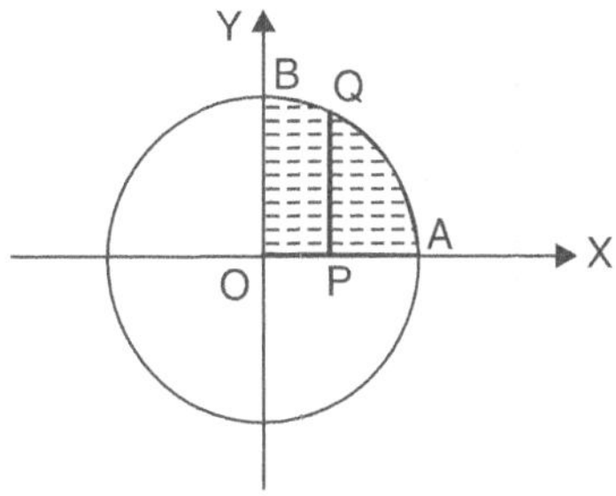

Solving, $x = \frac{a}{b}\sqrt{b^2 - y^2}$ gives $y = \frac{b}{a}\sqrt{a^2 - x^2}$

For vertical strips PQ, y varies from 0 to $\frac{b}{a}\sqrt{a^2 - x^2}$ and x varies from 0 to a

$$\therefore \quad \int_0^b \int_0^{\frac{a}{b}\sqrt{b^2 - y^2}} xy\, dx\, dy = \int_0^a \int_0^{\frac{b}{a}\sqrt{a^2 - x^2}} xy\, dy\, dx$$

$$= \int_0^a x \frac{y^2}{2} \Bigg|_0^{\frac{b}{a}\sqrt{a^2 - x^2}} dx = \int_0^a \frac{x}{2} \cdot \frac{b^2}{a^2} (a^2 - x^2)\, dx$$

$$= \frac{b^2}{2a^2} \left[a^2 \frac{x^2}{2} - \frac{x^4}{4} \right]_0^a = \frac{b^2}{2a^2} \left(\frac{a^4}{2} - \frac{a^4}{4} \right) = \frac{b^2}{2a^2} \cdot \frac{a^4}{4} = \frac{a^2 b^2}{8}.$$

(ii) $\int_0^1 \int_{4y}^4 e^{x^2}\, dx\, dy$

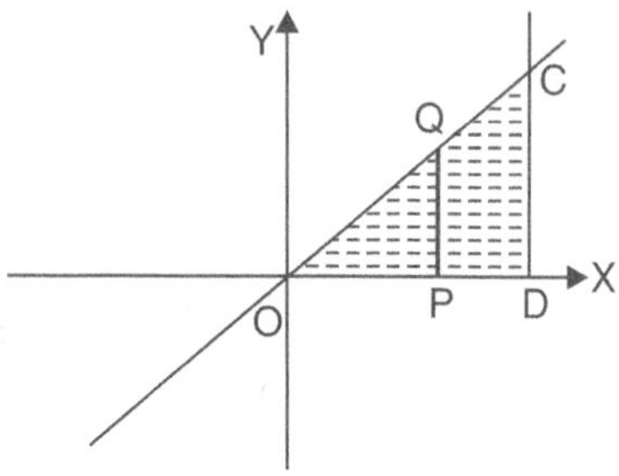

From the limits of integration, it is clear that the region of integration is $x = 4y, x = 4, y = 0, y = 1$. As in the given integral we have to first integrate w.r.t. x and then w.r.t. y, so region is divided into horizontal strips. To change order of integration we have to divide the region ODC (shown in figure) into vertical strips.

For the region ODC y varies from 0 to $\frac{x}{4}$ and x varies for 0 to 4.

$$\therefore \quad \int_0^1 \int_{4y}^4 e^{x^2}\, dx\, dy = \int_0^4 \int_0^{x/4} e^{x^2}\, dy\, dx = \int_0^4 e^{x^2}\, y \Big|_0^{x/4} dx = \int_0^4 \frac{x}{4} e^{x^2}\, dx$$

$$= \frac{1}{4}\int_0^4 x e^{x^2}\, dx \quad \text{Put } x^2 = t;\ 2\,xdx = dt$$

$$= \frac{1}{4}\int_0^{16} e^t \frac{dt}{2} = \frac{1}{8} e^t \Big|_0^{16} = \frac{e^{16} - 1}{8}.$$

TEST YOUR KNOWLEDGE

Evaluate the following integrals by changing the order of integration:

1. $\int_0^4 \int_y^4 \frac{x\, dx\, dy}{x^2 + y^2}$

2. $\int_0^{a/\sqrt{2}} \int_y^{\sqrt{a^2 - y^2}} x\, dx\, dy$

3. $\int_0^a \int_{x^2/a}^{2a - x} xy\, dy\, dx$

4. $\int_0^1 \int_0^{\sqrt{1 - x^2}} y^2\, dy\, dx$ **(P.T.U., Dec. 2004)**

5. $\int_0^a \int_{\sqrt{ax}}^a \frac{y^2\, dy\, dx}{\sqrt{y^4 - a^2 x^2}}$

6. $\int_0^3 \int_1^{\sqrt{4 - y}} (x + y)\, dx\, dy$

7. $\int_0^a \int_{x/a}^{\sqrt{x/a}} (x^2 + y^2)\, dy\, dx$

8. $\int_0^1 \int_x^1 \sin y^2\, dy\, dx$ **(P.T.U., May 2012)**

   [**Hint:** given integral = $\int_0^1 \int_0^y \sin y^2\, dx\, dy$]

9. Change the order of integration in $\int_{-a}^a \int_0^{\sqrt{a^2 - y^2}} f(x, y)\, dx\, dy$ **(P.T.U., Jan. 2009)**

Answers

1. π 2. $\frac{\sqrt{2}a^3}{6}$ 3. $\frac{3a^4}{8}$ 4. $\frac{\pi}{16}$

5. $\frac{\pi a^2}{6}$ 6. $\frac{241}{60}$ 7. $\frac{a^3}{28} + \frac{a}{20}$ 8. $\frac{1}{2}(1 - \cos 1)$

9. $\int_0^a \int_{-\sqrt{a^2 - x^2}}^{\sqrt{a^2 - x^2}} f(x, y)\, dy\, dx$

6.5. TRIPLE INTEGRALS

Consider a function $f(x, y, z)$ which is continuous at every point of a finite region V of three dimensional space. Divide the region V into n sub-regions of respective volumes $\delta V_1, \delta V_2, \ldots\ldots, \delta V_n$. Let (x_r, y_r, z_r) be an arbitrary point in the rth sub-region. Consider the sum $\sum_{r=1}^{n} f(x_r, y_r, z_r)\,\delta V_r$.

The limit of this sum as $n \to \infty$ and $\delta V_r \to 0$, if it exists, is called the *triple integral* of $f(x, y, z)$ over the region V and is denoted by $\iiint_V f(x, y, z)\, dV$.

For purposes of evaluation, it can be expressed as the repeated integral

$$\int_{z_1}^{z_2}\int_{y_1}^{y_2}\int_{x_1}^{x_2} f(x,y,z)\,dx\,dy\,dz, \qquad \ldots(1)$$

the order of integration depending upon the limits.

Let x_1, x_2 be function of y, z ; y_1, y_2 be function of z and z_1, z_2 be constants, *i.e.*,

Let $x_1 = f_1(y, z), x_2 = f_2(y, z), y_1 = \phi_1(z), y_2 = \phi_2(z)$ and $z_1 = a, z_2 = b$.

Then the integral (1) is evaluated as follows:

$$\int_{z_1=a}^{z_2=b}\left[\int_{y_1=\phi_1(z)}^{y_2=\phi_2(z)}\left[\int_{x_1=f_1(y,z)}^{x_2=f_2(y,z)} f(x,y,z)\,dx\right]dy\right]dz$$

First $f(x, y, z)$ is integrated w.r.t. x (keeping y and z constant) between the limits x_1 and x_2. The resulting expression, which is a function of y and z is then integrated w.r.t. y (keeping z constant) between the limits y_1 and y_2. The resulting expression, which is a function of z only is then integrated w.r.t. z between the limits z_1 and z_2. The order of integration is from the innermost rectangle to the outermost rectangle.

Limits involving two variables are kept innermost, then the limits involving one variable and finally the constant limits.

If x_1, x_2 ; y_1, y_2 and z_1, z_2 are all constants, then the order of integration is immaterial, provided the limits are changed accordingly. Thus

$$\int_{z_1}^{z_2}\int_{y_1}^{y_2}\int_{x_1}^{x_2} f(x,y,z)\,dx\,dy\,dz = \int_{x_1}^{x_2}\int_{z_1}^{z_2}\int_{y_1}^{y_2} f(x,y,z)\,dy\,dz\,dx.$$

$$= \int_{x_1}^{x_2}\int_{y_1}^{y_2}\int_{z_1}^{z_2} f(x,y,z)\,dz\,dy\,dx, \text{ etc.}$$

ILLUSTRATIVE EXAMPLES

Example 1. *Evaluate* $\int_0^1 \int_0^{\sqrt{1-x^2}} \int_0^{\sqrt{1-x^2-y^2}} \frac{1}{\sqrt{1-x^2-y^2-z^2}}\, dz\, dy\, dx.$

(P.T.U., May 2004)

Sol.
$$I = \int_0^1 \int_0^{\sqrt{1-x^2}} \int_0^{\sqrt{1-x^2-y^2}} \frac{1}{\sqrt{(1-x^2-y^2)-z^2}}\, dz\, dy\, dx$$

$$= \int_0^1 \int_0^{\sqrt{1-x^2}} \left[\sin^{-1} \frac{z}{\sqrt{1-x^2-y^2}}\right]_0^{\sqrt{1-x^2-y^2}} dy\, dx$$

$$= \int_0^1 \int_0^{\sqrt{1-x^2}} \frac{\pi}{2}\, dy\, dx = \int_0^1 \frac{\pi}{2}\Big[y\Big]_0^{\sqrt{1-x^2}} dx$$

$$= \frac{\pi}{2}\int_0^1 \sqrt{1-x^2}\, dx = \frac{\pi}{2}\left[\frac{x\sqrt{1-x^2}}{2} + \frac{1}{2}\sin^{-1} x\right]_0^1 = \frac{\pi}{4}[\sin^{-1} 1] = \frac{\pi}{4}\cdot\frac{\pi}{2} = \frac{\pi^2}{8}.$$

Example 2. *Evaluate* $\int_0^2 \int_1^2 \int_0^{yz} xyz\, dx\, dy\, dz.$ **(P.T.U., May 2006, May 2009)**

Sol. $\int_0^2 \int_1^2 \left[\int_0^{yz} xyz\, dx\right] dy\, dz$

$$= \int_0^2 \int_1^2 yz \left.\frac{x^2}{2}\right|_0^{yz} dy\, dz = \int_0^2 \left[\int_1^2 \frac{y^3 z^3}{2}\, dy\right] dz$$

$$= \int_0^2 \frac{z^3}{2} \cdot \left.\frac{y^4}{4}\right|_1^2 dz = \int_0^2 \frac{z^3}{8}(16-1)\, dz = \frac{15}{8}\cdot \left.\frac{z^4}{4}\right|_0^2$$

$$= \frac{15}{8}\cdot\frac{16}{4} = \frac{15}{2}.$$

Example 3. *Evaluate* $\int_1^e \int_1^{\log y} \int_1^{e^x} \log z\, dz\, dx\, dy.$

Sol.
$$I = \int_1^e \int_1^{\log y} \left[\int_1^{e^x} \log z\, dz\right] dx\, dy$$

Since $\int_1^{e^x} \log z\, dz = \int_1^{e^x} \log z \cdot 1\, dz$

Integrating by parts $= \Big[\log z \cdot z\Big]_1^{e^x} - \int_1^{e^x} \frac{1}{z}\cdot z\, dz$

$$= e^x \log e^x - 0 - \Big[z\Big]_1^{e^x} = xe^x - e^x + 1 = (x-1)\, e^x + 1$$

$$\therefore \qquad I = \int_1^e \int_1^{\log y} [(x-1)e^x + 1]\, dx\, dy$$

Now $\displaystyle\int_1^{\log y} [(x-1)e^x + 1]\, dx = \int_1^{\log y} (x-1)e^x\, dx + \Big[x\Big]_1^{\log y}$

$$= \Big[(x-1)e^x\Big]_1^{\log y} - \int_1^{\log y} 1 \cdot e^x\, dx + \log y - 1 \qquad \text{(Integration by parts)}$$

$$= (\log y - 1)e^{\log y} - \Big[e^x\Big]_1^{\log y} + \log y - 1$$

$$= y(\log y - 1) - (e^{\log y} - e) + \log y - 1 \qquad [\because\ e^{\log y} = y]$$

$$= y(\log y - 1) - y + e + \log y - 1 = (y+1)\log y - 2y + e - 1$$

$$\therefore \qquad I = \int_1^e [\log y \cdot (y+1) - 2y + e - 1]\, dy$$

$$= \left[\log y \,.\left(\frac{y^2}{2} + y\right)\right]_1^e - \int_1^e \frac{1}{y}\left(\frac{y^2}{2} + y\right) dy - \Big[y^2\Big]_1^e + (e-1)\Big[y\Big]_1^e$$

$$= \frac{e^2}{2} + e - \int_1^e \left(\frac{y}{2} + 1\right) dy - (e^2 - 1) + (e-1)^2$$

$$= \frac{e^2}{2} + e - \left[\frac{y^2}{4} + y\right]_1^e - 2e + 2 = \frac{e^2}{2} + e - \left[\left(\frac{e^2}{4} + e\right) - \left(\frac{1}{4} + 1\right)\right] - 2e + 2$$

$$= \frac{e^2}{4} - 2e + \frac{13}{4} = \frac{1}{4}(e^2 - 8e + 13).$$

Example 4. *Evaluate* $\displaystyle\int_0^{\log 2} \int_0^x \int_0^{x+\log y} e^{x+y+z}\, dz\, dy\, dx.$

Sol. $\displaystyle\int_0^{\log 2} \int_0^x \int_0^{x+\log y} e^{x+y} \cdot e^z\, dz\, dy\, dx = \int_0^{\log 2} \int_0^x e^{x+y} \cdot e^z \Big|_0^{x+\log y} dy\, dx$

$$= \int_0^{\log 2} \int_0^x e^{x+y} (e^{x+\log y} - 1)\, dy\, dx$$

$$= \int_0^{\log 2} \int_0^x e^{x+y} (e^x \cdot e^{\log y} - 1) dy\, dx = \int_0^{\log 2} \int_0^x e^{x+y} (ye^x - 1)\, dy\, dx$$

$$= \int_0^{\log 2} \int_0^x [e^{2x}(ye^y) - e^x \,.\, e^y]\, dy\, dx$$

$$= \int_0^{\log 2} \left\{e^{2x}(y-1)e^y - e^x e^y \Big|_0^x\right\} dx$$

$$\Big|\ \because \int y e^y\, dy = y\, e^{2y} - \int 1 \cdot e^y\, dy = (y-1)e^y$$

$$= \int_0^{\log 2} \{e^{2x}(x-1)e^x - e^{2x} + e^{2x} + e^x\}\, dx$$

$$= \int_0^{\log 2} \{(x-1)e^{3x} + e^x\}\, dx$$

$$= (x-1)\,\frac{e^{3x}}{3}\Bigg|_0^{\log 2} - \int_0^{\log 2} 1\cdot\frac{e^{3x}}{3}\,dx + e^x\Big|_0^{\log 2}$$

$$= (\log 2 - 1)\cdot\frac{1}{3}\,e^{3\log 2} + \frac{1}{3} - \frac{e^{3x}}{9}\Bigg|_0^{\log 2} + (e^{\log 2} - 1)$$

$$= \frac{8}{3}(\log 2 - 1) + \frac{1}{3} - \frac{e^{3\log 2}}{9} + \frac{1}{9} + 2 - 1 \qquad \Big|\ \because\ e^{3\log 2} = e^{\log 2^3} = 8$$

$$= \frac{8}{3}\log 2 - \frac{8}{3} + \frac{1}{3} - \frac{8}{9} + \frac{1}{9} + 1 = \frac{8}{3}\log 2 - \frac{19}{9}.$$

Example 5. *Evaluate* $\int_0^{\frac{\pi}{2}}\int_0^{a\cos\theta}\int_0^{\sqrt{a^2-r^2}} r\,dz\,dr\,d\theta.$

Sol. $\int_0^{\frac{\pi}{2}}\int_0^{a\cos\theta} rz\Big|_0^{\sqrt{a^2-r^2}}\,dr\,d\theta$

$$= \int_0^{\frac{\pi}{2}}\left[\int_0^{a\cos\theta} r\sqrt{a^2-r^2}\,dr\right]d\theta$$

$$= \int_0^{\frac{\pi}{2}}\left(-\frac{1}{2}\right)\frac{(a^2-r^2)^{3/2}}{3/2}\Bigg|_0^{a\cos\theta} d\theta \qquad \text{using } \int [f(x)]^n f'(x)\,dx = \frac{[f(x)]^{n+1}}{n+1},\ n \neq -1$$

$$= -\frac{1}{3}\int_0^{\frac{\pi}{2}} (a^3\sin^3\theta - a^3)\,d\theta$$

$$= -\frac{a^3}{3}\left\{\int_0^{\frac{\pi}{2}} \sin^3\theta\,d\theta - \int_0^{\frac{\pi}{2}} 1\cdot d\theta\right\}$$

$$= -\frac{a^3}{3}\left\{\frac{2}{3.1} - \frac{\pi}{2}\right\} \quad \text{using } \int_0^{\frac{\pi}{2}} \sin^n\theta\,d\theta = \frac{(n-1)(n-3)\cdots}{n(n-1)\cdots} \text{ when } n \text{ is odd}$$

$$= \frac{a^3}{3}\left\{\frac{\pi}{2} - \frac{2}{3}\right\}.$$

Example 6. *Evaluate* $\iiint \frac{dx\,dy\,dz}{(x+y+z+1)^3}$ *over the tetrahedron bounded by the coordinate planes and the plane* $x + y + z = 1.$ **(P.T.U., Dec. 2003)**

Sol. $\iiint \frac{dx\,dy\,dz}{(x+y+z+1)^3}$

$$= \int_0^1\int_0^{1-x}\int_0^{1-x-y} (x+y+z+1)^{-3}\,dz\,dy\,dx = \int_0^1\int_0^{1-x} \frac{(x+y+z+1)^{-2}}{-2}\Bigg|_0^{1-x-y} dy\,dx$$

$$= -\frac{1}{2}\int_0^1\int_0^{1-x} (x+y+1+1-x-y)^{-2} - (x+y+1)^{-2}]\,dy\,dx$$

$$= -\frac{1}{2}\int_0^1\int_0^{1-x} \frac{1}{2^2} - (x+y+1)^{-2}]\,dy\,dx$$

$$= -\frac{1}{2}\int_0^1 \frac{1}{4}y - \frac{(x+y+1)^{-1}}{-1}\Bigg|_0^{1-x} dx$$

$$= -\frac{1}{2}\int_0^1 \left[\frac{1-x}{4} + 2^{-1} - (x+1)^{-1}\right] dx$$

$$= -\frac{1}{2}\int_0^1 \left[\frac{3}{4} - \frac{x}{4} - \frac{1}{x+1}\right] dx$$

$$= -\frac{1}{2}\left[\frac{3}{4}x - \frac{x^2}{8} - \log(x+1)\Bigg|_0^1\right]$$

$$= -\frac{1}{2}\cdot\left(\frac{3}{4} - \frac{1}{8} - \log 2\right) = -\frac{1}{2}\left(\frac{5}{8} - \log 2\right) = \frac{1}{2}\log 2 - \frac{5}{16}.$$

TEST YOUR KNOWLEDGE

Evaluate the following integrals (1–10):

1. $\int_{-c}^{c}\int_{-b}^{b}\int_{-a}^{a} (x^2 + y^2 + z^2) dx\, dy\, dz$ **(P.T.U., May 2015)**

2. $\int_0^a\int_0^a\int_0^a (yz + zx + xy)\, dx\, dy\, dz$

3. $\int_{-1}^{1}\int_0^{z}\int_{x-z}^{x+z} (x + y + z)\, dy\, dx\, dz.$ **(P.T.U., May 2014)**

4. $\int_1^3\int_{1/x}^{1}\int_0^{\sqrt{xy}} xyz\, dz\, dy\, dx$

5. $\int_0^1\int_0^{1-x}\int_0^{1-x-y} xyz\, dz\, dy\, dx$ **(P.T.U., Dec. 2013)**

6. $\int_0^1\int_{y^2}^{1-x}\int_0^{1-x} x\, dz\, dx\, dy$

7. $\int_0^1\int_0^{\sqrt{1-x^2}}\int_0^{\sqrt{1-x^2-y^2}} xyz\, dz\, dy\, dx$ **(P.T.U., May 2012)**

8. $\int_0^{\pi/2} d\theta \int_0^{a\sin\theta} dr \int_0^{(a^2-r^2)} r dz$

9. $\int_0^a\int_0^x\int_0^{x+y} e^{x+y+z}\, dz\, dy\, dx$

10. $\int_0^{\pi/2}\int_0^{a\sin\theta}\int_0^{\frac{a^2-r^2}{a}} r\, dz\, dr\, d\theta$

11. Evaluate $\iiint (x + y + z)\, dx\, dy\, dz$ over the tetrahedron bounded by $x = 0, y = 0, z = 0$ and $x + y + z = 1$
[**Hint.** See S.E. 6]

Answers

1. $\frac{8}{3}abc(a^2 + b^2 + c^2)$ **2.** $\frac{3}{4}a^5$ **3.** 0 **4.** $\frac{13}{9} - \frac{1}{6}\log 3$

5. $\frac{1}{720}$ **6.** $\frac{4}{35}$ **7.** $\frac{1}{48}$ **8.** $\frac{5\pi a^3}{64}$

9. $\frac{1}{8}e^{4a} - \frac{3}{4}e^{2a} + e^a - \frac{3}{8}$ **10.** $\frac{5\pi a^3}{64}$ **11.** $\frac{1}{8}.$

6.6. CHANGE OF VARIABLES

Quite often, the evaluation of a double or triple integral is greatly simplified by a suitable change of variables.

Let the variables x, y in the double integral $\iint_R f(x, y)\, dx\, dy$ be changed to u, v by means of the relations $x = \phi(u, v)$, $y = \psi(u, v)$, then the double integral is transformed into

$$\iint_{R'} f\{\phi(u, v), \psi(u, v)\}\ |\ J\ |\ du\, dv \quad \text{where } J = \frac{\partial(x, y)}{\partial(u, v)} = \begin{vmatrix} \frac{\partial x}{\partial u} & \frac{\partial x}{\partial v} \\ \frac{\partial y}{\partial u} & \frac{\partial y}{\partial v} \end{vmatrix}$$

is the Jacobian of transformation from (x, y) to (u, v) co-ordinates and R′ is the region in the uv-plane which corresponds to the region R in the xy-plane.

(i) To change cartesian co-ordinates (x, y) to polar co-ordinates (r, θ).

Here we have $x = r\cos\theta$, $y = r\sin\theta$ so that $x^2 + y^2 = r^2$

$$J = \begin{vmatrix} \frac{\partial x}{\partial r} & \frac{\partial x}{\partial \theta} \\ \frac{\partial y}{\partial r} & \frac{\partial y}{\partial \theta} \end{vmatrix} = \begin{vmatrix} \cos\theta & -r\sin\theta \\ \sin\theta & r\cos\theta \end{vmatrix} = r(\cos^2\theta + \sin^2\theta) = r$$

$$\therefore \iint_R f(x, y)\, dx\, dy = \iint_{R'} f(r\cos\theta, r\sin\theta)\, r\, dr\, d\theta$$

i.e., replace x by $r\cos\theta$, y by $r\sin\theta$ and $dxdy$ by $rdrd\theta$.

(ii) To change cartesian co-ordinates (x, y, z) to spherical polar co-ordinates (r, θ, φ).

Here, we have $x = r\sin\theta\cos\phi$

$y = r\sin\theta\sin\phi$

$z = r\cos\theta$

so that $x^2 + y^2 + z^2 = r^2$

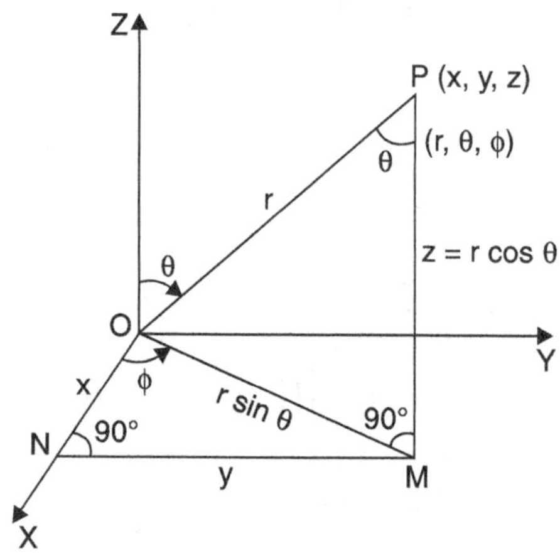

$$J = \begin{vmatrix} \frac{\partial x}{\partial r} & \frac{\partial x}{\partial \theta} & \frac{\partial x}{\partial \phi} \\ \frac{\partial y}{\partial r} & \frac{\partial y}{\partial \theta} & \frac{\partial y}{\partial \phi} \\ \frac{\partial z}{\partial r} & \frac{\partial z}{\partial \theta} & \frac{\partial z}{\partial \phi} \end{vmatrix}$$

$$= \begin{vmatrix} \sin\theta\cos\phi & r\cos\theta\cos\phi & -r\sin\theta\sin\phi \\ \sin\theta\sin\phi & r\cos\theta\sin\phi & r\sin\theta\cos\phi \\ \cos\theta & -r\sin\theta & 0 \end{vmatrix}$$

$$= r^2\sin\theta$$

$$\therefore \quad \iiint_V f(x,y,z)\,dx\,dy\,dz = \iiint_{V'} f(r\sin\theta\cos\phi, r\sin\theta\sin\phi, r\cos\theta)\, r^2 \sin\theta\, dr\, d\theta\, d\phi$$

i.e., to replace x by $r \sin \theta \cos \phi$; y by $r \sin \theta \sin \phi$; z by $r \cos \theta$ and $dx\, dy\, dz$ by $r^2 \sin \theta\, dr\, d\theta\, d\phi$

(iii) To change cartesian co-ordinates (x, y, z) to cylindrical co-ordinates (r, φ, z).

Here we have $x = r \cos \phi$

$y = r \sin \phi$

$z = z$

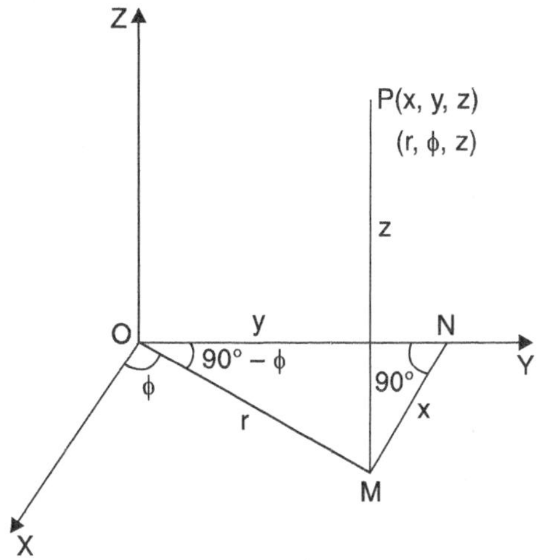

$$J = \begin{vmatrix} \frac{\partial x}{\partial r} & \frac{\partial x}{\partial \phi} & \frac{\partial x}{\partial z} \\ \frac{\partial y}{\partial r} & \frac{\partial y}{\partial \phi} & \frac{\partial y}{\partial z} \\ \frac{\partial z}{\partial r} & \frac{\partial z}{\partial \phi} & \frac{\partial z}{\partial z} \end{vmatrix}$$

$$= \begin{vmatrix} \cos\phi & -r\sin\phi & 0 \\ \sin\phi & r\cos\phi & 0 \\ 0 & 0 & 1 \end{vmatrix}$$

$$= r(\cos^2\phi + \sin^2\phi) = r$$

$$\therefore \quad \iiint_V f(x, y, z)\, dx\, dy\, dz$$

$$= \iiint_{V'} f(r\cos\phi, r\sin\phi, z)\, r\, dr\, d\phi\, dz.$$

i.e., replace x by $r \cos \phi$, y by $r \sin \phi$, z remains same replace $dx\, dy$ by $r dr\, d\phi$; dz remain same.

ILLUSTRATIVE EXAMPLES

Example 1. *Evaluate* $\iint \sqrt{\dfrac{1 - x^2 - y^2}{1 + x^2 + y^2}}\, dx\, dy$ *over the positive quadrant of the circle* $x^2 + y^2 = 1$.

Sol. Changing to polar coordinates by putting $x = r \cos \theta, y = r \sin \theta$; $x^2 + y^2 = 1$ transforms into $r = 1$. For the region of integration R, r varies from 0 to 1 and θ varies from 0 to $\dfrac{\pi}{2}$.

$$I = \iint_R \sqrt{\frac{1 - x^2 - y^2}{1 + x^2 + y^2}}\, dx\, dy$$

$$= \int_0^{\pi/2} \int_0^1 \sqrt{\frac{1 - r^2}{1 + r^2}}\, r\, dr\, d\theta \quad | \, dx\, dy \text{ is replaced by } r\, dr\, d\theta$$

$$= \int_0^{\pi/2} \int_0^1 \frac{r(1 - r^2)}{\sqrt{1 - r^4}}\, dr\, d\theta$$

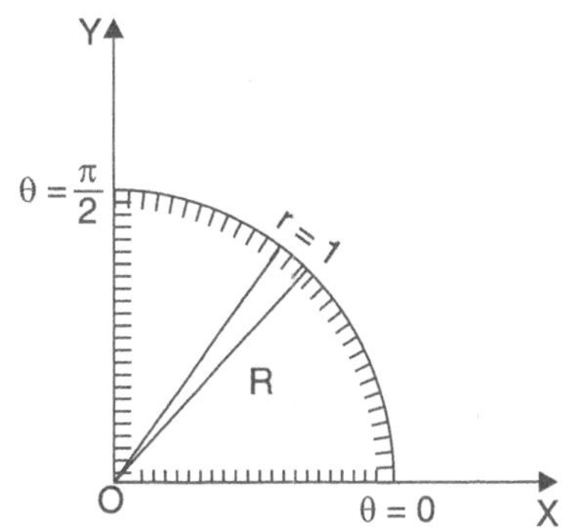

Now $$\int_0^1 \frac{r(1 - r^2)}{\sqrt{1 - r^4}}\, dr = \int_0^1 \left(\frac{r}{\sqrt{1 - r^4}} - \frac{r^3}{\sqrt{1 - r^4}} \right) dr$$

$$= \frac{1}{2}\int_0^1 \frac{2r}{\sqrt{1-r^4}}\,dr + \frac{1}{4}\int_0^1 -4r^3(1-r^4)^{-1/2}dr$$

$$= \frac{1}{2}\int_0^1 \frac{dt}{\sqrt{1-t^2}} + \frac{1}{4}\cdot\left[\frac{(1-r^4)^{1/2}}{1/2}\right]_0^1, \text{ where } t = r^2$$

$$= \frac{1}{2}\left[\sin^{-1} t\right]_0^1 + \frac{1}{2}(0-1) = \frac{1}{2}\left(\frac{\pi}{2}\right) - \frac{1}{2} = \frac{\pi}{4} - \frac{1}{2}$$

$$\therefore \quad I = \int_0^{\pi/2}\left(\frac{\pi}{4}-\frac{1}{2}\right)d\theta = \left(\frac{\pi}{4}-\frac{1}{2}\right)\left[\theta\right]_0^{\pi/2} = \left(\frac{\pi}{4}-\frac{1}{2}\right)\frac{\pi}{2} = \frac{\pi^2}{8} - \frac{\pi}{4}.$$

Example 2. *Evaluate* $\iint\sqrt{a^2-x^2-y^2}\,dx\,dy$ *over the semi-circle* $x^2 + y^2 = ax$ *in the positive quadrant.*

Sol. Changing to polar co-ordinates, $x^2 + y^2 = ax$ transforms into $r = a \cos \theta$. For the region of integration R, r varies from 0 to $a \cos \theta$ and θ varies from 0 to $\frac{\pi}{2}$.

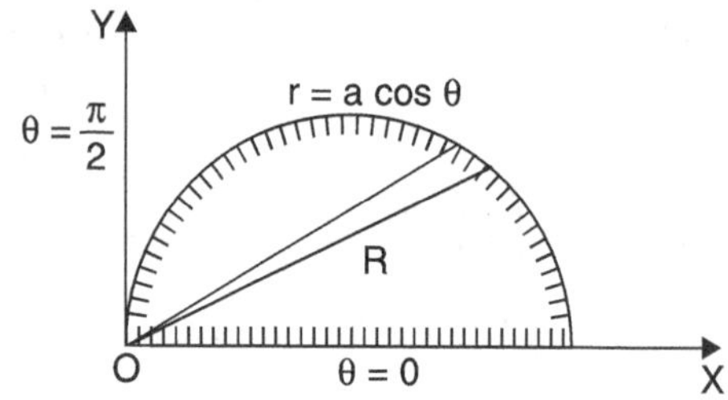

$$\therefore \quad \iint_R \sqrt{a^2-x^2-y^2}\,dx\,dy$$

$$= \int_0^{\pi/2}\int_0^{a\cos\theta}\sqrt{a^2-r^2}\,.\,r\,dr\,d\theta$$

$$= \int_0^{\pi/2}\int_0^{a\cos\theta} -\frac{1}{2}(a^2-r^2)^{1/2}(-2r)\,dr\,d\theta$$

$$= \int_0^{\pi/2} -\frac{1}{2}\cdot\left[\frac{(a^2-r^2)^{3/2}}{3/2}\right]_0^{a\cos\theta} d\theta$$

$$= -\frac{1}{3}\int_0^{\pi/2}(a^3\sin^3\theta - a^3)\,d\theta = -\frac{a^3}{3}\int_0^{\pi/2}(\sin^3\theta - 1)\,d\theta$$

$$= -\frac{a^3}{3}\left[\frac{2}{3}-\frac{\pi}{2}\right] = \frac{a^3}{3}\left(\frac{\pi}{2}-\frac{2}{3}\right).$$

Example 3. *Change into polar co-ordinates and evaluate*

$$\int_0^\infty\int_0^\infty e^{-(x^2+y^2)}\,dy\,dx.$$

(P.T.U., Dec. 2003, May 2011, Dec. 2011)

Sol. For the region of integration in cartesian co-ordinates, y varies from 0 to ∞ and x also varies from 0 to ∞. Thus the region of integration is the plane XOY. Changing to polar co-ordinates by putting $x = r \cos \theta$, $y = r \sin \theta$ so that $x^2 + y^2 = r^2$; for the region of integration r varies from 0 to ∞ and θ varies from 0 to $\frac{\pi}{2}$.

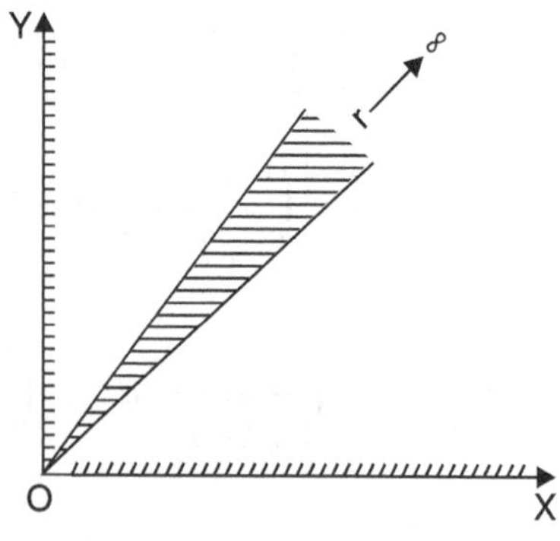

$$\therefore \int_0^\infty \int_0^\infty e^{-(x^2+y^2)}\, dy\, dx = \int_0^{\pi/2}\int_0^\infty e^{-r^2} \cdot r\, dr\, d\theta$$

$$= \int_0^{\pi/2}\int_0^\infty \frac{1}{2} \cdot e^{-r^2} \cdot 2r\, dr\, d\theta$$

$$= \int_0^{\pi/2}\int_0^\infty \frac{1}{2} e^{-t} dt\, d\theta, \text{where } t = r^2$$

$$= \int_0^{\pi/2} \left[-\frac{1}{2}e^{-t}\right]_0^\infty d\theta = -\frac{1}{2}\int_0^{\pi/2}(0-1)\,d\theta = \frac{1}{2}\Big[\theta\Big]_0^{\pi/2} = \frac{\pi}{4}.$$

Example 4. *Evaluate the following by changing to polar coordinates*

(i) $\displaystyle\int_0^2 \int_0^{\sqrt{2x-x^2}} \frac{x\, dy\, dx}{\sqrt{x^2+y^2}}$ **(P.T.U., May 2012)**

(ii) $\displaystyle\int_{-1}^{1} \int_{-\sqrt{1-y^2}}^{\sqrt{1-y^2}} \log_e (x^2+y^2+1)\, dx\, dy$ **(P.T.U., May 2003)**

Sol. (*i*) In the given integral, y varies from 0 to $\sqrt{2x-x^2}$ and x varies from 0 to 2.

$$y = \sqrt{2x-x^2} \quad \Rightarrow \quad y^2 = 2x - x^2 \quad \Rightarrow \quad x^2 + y^2 = 2x.$$

In polar co-ordinates, we have $r^2 = 2r \cos \theta$ or $r = 2 \cos \theta$.

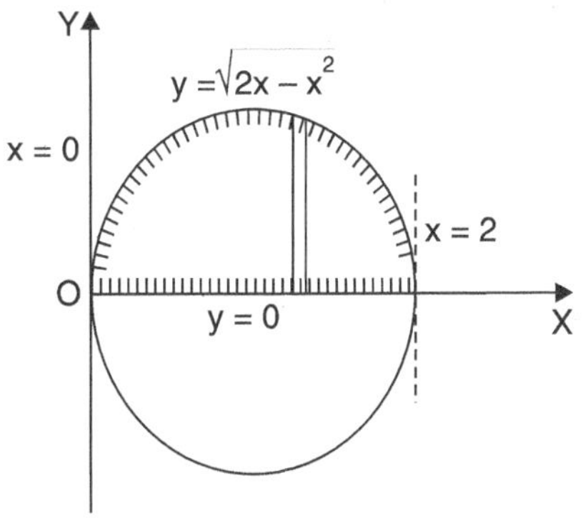

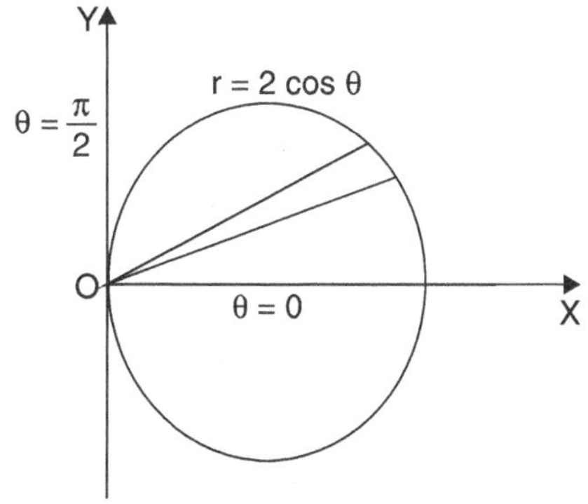

$\therefore$ For the region of integration, r varies from 0 to $2 \cos \theta$ and θ varies from 0 to $\dfrac{\pi}{2}$.

In the given integral, replacing x by $r \cos \theta$, y by $r \sin \theta$, $dy\, dx$ by $r\, dr\, d\theta$, we have

$$I = \int_0^{\pi/2}\int_0^{2\cos\theta} \frac{r\cos\theta \cdot r\, dr\, d\theta}{r} = \int_0^{\pi/2}\int_0^{2\cos\theta} r\cos\theta\, dr\, d\theta$$

$$= \int_0^{\pi/2} \cos\theta \left[\frac{r^2}{2}\right]_0^{2\cos\theta} d\theta = \int_0^{\pi/2} 2\cos^3\theta\, d\theta = 2 \cdot \frac{2}{3} = \frac{4}{3}.$$

(*ii*) $\displaystyle\int_{-1}^{1} \int_{-\sqrt{1-y^2}}^{\sqrt{1-y^2}} \log_e (x^2+y^2+1)\, dx\, dy$

In polar coordinates $x = r \cos \theta$, $y = r \sin \theta$ $dxdy = r\, dr\, d\theta$

Here x varies from $-\sqrt{1-y^2}$ to $\sqrt{1-y^2}$

i.e., $x = \sqrt{1-y^2}$ or $x^2 + y^2 = 1$ or $r^2 = 1$ or $r = 1$ and y varies from -1 to 1

i.e., $r \sin\theta = -1$ and $r \sin\theta = 1$ put $r = 1$, we get $\sin\theta = -1$ and 1 $\therefore$ $\theta = -\frac{\pi}{2}$ and $\frac{\pi}{2}$

$\therefore$ In polar coordinates r varies from 0 to 1 and θ varies from $-\frac{\pi}{2}$ to $\frac{\pi}{2}$.

$$\therefore \quad \int_{-1}^{1}\int_{-\sqrt{1-y^2}}^{\sqrt{1-y^2}} \log_e (x^2+y^2+1)\,dx\,dy = \int_{\theta=-\frac{\pi}{2}}^{\frac{\pi}{2}}\int_{r=0}^{1} \log_e (r^2+1) r\,dr\,d\theta$$

Put $r^2 + 1 = t$ $\therefore$ $2r\,dr = dt$ in the above integral

$$= \int_{\theta=-\frac{\pi}{2}}^{\frac{\pi}{2}}\int_{t=1}^{2} \log_e t\left(\frac{1}{2}dt\right)d\theta = \frac{1}{2}\int_{\theta=-\frac{\pi}{2}}^{\frac{\pi}{2}}\int_{t=1}^{2} \log_e t\,dt\,d\theta$$

Integrate by parts

$$= \frac{1}{2}\int_{-\frac{\pi}{2}}^{\frac{\pi}{2}}\left[(\log t)\cdot t\Big|_1^2 - \int_1^2 \frac{1}{t}t\,dt\right]d\theta = \frac{1}{2}\int_{-\frac{\pi}{2}}^{\frac{\pi}{2}} 2\log 2 - t\Big|_1^2 \,d\theta$$

$$= \frac{1}{2}\int_{-\frac{\pi}{2}}^{\frac{\pi}{2}} (2\log 2 - 1)\,d\theta = \frac{1}{2}(2\log 2 - 1)\theta\Big|_{-\frac{\pi}{2}}^{\frac{\pi}{2}}$$

$$= \frac{1}{2}(2\log 2 - 1)\pi = \frac{\pi}{2}(2\log 2 - 1).$$

Example 5. *Evaluate* $\iiint z(x^2+y^2+z^2)\,dx\,dy\,dz$ *through the volume of the cylinder* $x^2 + y^2 = a^2$ *intercepted by the planes* $z = 0$ *and* $z = h$.

Sol. Changing to cylindrical co-ordinates by changing x to $r\cos\phi$, y to $r\sin\phi$ and replacing $dx\,dy\,dz$ by $r\,dr\,d\phi\,dz$

$$\text{I} = \int_0^h\int_0^{2\pi}\int_0^a z(r^2+z^2)\,r\,dr\,d\phi\,dz = \int_0^h\int_0^{2\pi}\int_0^a (zr^3 + z^3 r)\,dr\,d\phi\,dz$$

$$= \int_0^h\int_0^{2\pi}\left[z\cdot\frac{r^4}{4} + z^3\cdot\frac{r^2}{2}\right]_0^a d\phi\,dz = \int_0^h\int_0^{2\pi}\left(\frac{a^4}{4}z + \frac{a^2}{2}z^3\right)d\phi\,dz$$

$$= \int_0^h\left(\frac{a^4}{4}z + \frac{a^2}{2}z^3\right)\Big[\phi\Big]_0^{2\pi} dz = \int_0^h 2\pi\left(\frac{a^4}{4}z + \frac{a^2}{2}z^3\right)dz$$

$$= 2\pi\left[\frac{a^4z^2}{8} + \frac{a^2z^4}{8}\right]_0^h = \frac{\pi}{4}(a^4h^2 + a^2h^4) = \frac{\pi}{4}a^2h^2(a^2+h^2).$$

Example 6. *Prove that* $\iiint_V \sqrt{1-\frac{x^2}{a^2}-\frac{y^2}{b^2}-\frac{z^2}{c^2}}\, dx\, dy\, dz = \frac{\pi^2}{4} abc,$

where $$V = \left\{(x, y, z);\ \frac{x^2}{a^2}+\frac{y^2}{b^2}+\frac{z^2}{c^2} \le 1\right\}.$$

Sol. Here $V = \left\{(x,y,z);\ \frac{x^2}{a^2}+\frac{y^2}{b^2}+\frac{z^2}{c^2} \le 1\right\}$ where V the volume of the ellipsoid

$$\frac{x^2}{a^2}+\frac{y^2}{b^2}+\frac{z^2}{c^2} = 1$$

Put $\frac{x}{a} = u,\ \frac{y}{b} = v,\ \frac{z}{c} = w \quad \therefore \quad x = au, y = bv, z = cw$ and $|\ J\ | = abc$

$$\therefore \quad I = \iiint_V \sqrt{1-\frac{x^2}{a^2}-\frac{y^2}{b^2}-\frac{z^2}{c^2}}\, dx\, dy\, dz = \iiint_{V'} \sqrt{1-u^2-v^2-w^2}\ abc\, du\, dv\, dw$$

where V′ is the volume of the sphere $u^2 + v^2 + w^2 = 1$.

Change it to spherical polar coordinates

$$u = r \sin\theta \cos\phi,\ v = r \sin\theta \sin\phi,\ z = r\cos\theta$$

and $$du\, dv\, dw = r^2 \sin\theta\, dr\, d\theta\, d\phi$$

$$\therefore \quad I = \int_{\phi=0}^{2\pi}\int_{\theta=0}^{\pi}\int_{r=0}^{1} \sqrt{1-r^2}\ abc\ r^2 \sin\theta\, dr\, d\theta\, d\phi$$

$$= abc \int_0^{2\pi}\int_0^{\pi}\left[\int_0^1 r^2\sqrt{1-r^2}\, dr\right] \sin\theta\, d\theta\, d\phi \qquad \text{...(1)}$$

To integrate $\int_0^1 r^2 \sqrt{1-r^2}\, dr$; put $r = \sin t \quad \therefore \quad dr = \cos t\, dt$

$$\therefore \quad \int_0^1 r^2\sqrt{1-r^2}\, dr = \int_0^{\frac{\pi}{2}} \sin^2 t \cos^2 t\, dt = \frac{1.1}{4.2}\,\frac{\pi}{2} = \frac{\pi}{16}$$

From (1) $$I = abc \int_0^{2\pi}\int_0^{\pi} \frac{\pi}{16} \sin\theta\, d\theta\, d\phi = \frac{abc\pi}{16}\int_0^{2\pi} -\cos\theta\Big|_0^{\pi}\, d\phi$$

$$= \frac{abc\pi}{16}\cdot 2\int_0^{2\pi} d\phi = \frac{abc\pi}{8}\cdot \phi\Big|_0^{2\pi} = \frac{abc\pi}{8}\, 2\pi = \frac{abc\,\pi^2}{4}.$$

Example 7. *Evaluate* $\iiint \frac{dxdydz}{\sqrt{1-x^2-y^2-z^2}}$, *the integral being extended to the positive octant of the sphere* $x^2 + y^2 + z^2 = 1$. **(P.T.U., May 2004, May 2008)**

Sol. In the positive octant of the sphere

$$0 \le z \le \sqrt{1-x^2-y^2}\ ;\ 0 \le y \le \sqrt{1-x^2}\ ;\ 0 \le x \le 1$$

$\therefore \quad \int_0^1\int_0^{\sqrt{1-y^2}}\int_0^{\sqrt{1-x^2-y^2}} \frac{dx\, dy\, dz}{\sqrt{1-x^2-y^2-z^2}}$ represents the volume of the sphere in positive octant.

Changing to spherical polar co-ordinates by putting $x = r\sin\theta\cos\phi$, $y = r\sin\theta\sin\phi$, $z = r\cos\theta$ so that $x^2 + y^2 + z^2 = r^2$.

For the volume of sphere $x^2 + y^2 + z^2 = 1$ in the positive octant, r varies from 0 to 1, θ varies from 0 to $\frac{\pi}{2}$ and ϕ varies from 0 to $\frac{\pi}{2}$.

Replacing $dz\, dy\, dx$ by $r^2 \sin\theta\, dr\, d\theta\, d\phi$, we have

$$I = \int_0^{\pi/2}\int_0^{\pi/2}\int_0^1 \frac{r^2 \sin\theta\, dr\, d\theta\, d\phi}{\sqrt{1-r^2}} = \int_0^{\pi/2}\int_0^{\pi/2}\int_0^1 \frac{1-(1-r^2)}{\sqrt{1-r^2}} \sin\theta\, dr\, d\theta\, d\phi$$

$$= \int_0^{\pi/2}\int_0^{\pi/2}\int_0^1 \left(\frac{1}{\sqrt{1-r^2}} - \sqrt{1-r^2}\right) \sin\theta\, dr\, d\theta\, d\phi$$

$$= \int_0^{\pi/2}\int_0^{\pi/2} \sin\theta \left[\sin^{-1} r - \left(\frac{r\sqrt{1-r^2}}{2} + \frac{1}{2}\sin^{-1} r\right)\right]_0^1 d\theta\, d\phi$$

$$= \int_0^{\pi/2}\int_0^{\pi/2} \sin\theta \left[\frac{\pi}{2} - \frac{1}{2}\cdot\frac{\pi}{2}\right] d\theta\, d\phi = \int_0^{\pi/2} \frac{\pi}{4}\Big[-\cos\theta\Big]_0^{\pi/2} d\phi = \int_0^{\pi/2} \frac{\pi}{4}\, d\phi$$

$$= \frac{\pi}{4}\cdot\frac{\pi}{2} = \frac{\pi^2}{8}.$$

Note. For the whole volume of the sphere $x^2 + y^2 + z^2 = a^2$.

$$0 \le r \le a,\ 0 \le \theta \le \pi,\ 0 \le \phi \le 2\pi.$$

Example 8. *Evaluate* $\iint_R (x + y)^2\, dx\, dy$, *where R is the parallelogram in the xy-plane with vertices (1, 0), (3, 1), (2, 2), (0, 1), using the transformation* $u = x + y$ *and* $v = x - 2y$.

Sol. The vertices A(1, 0), B(3, 1), C(2, 2), D(0, 1) of the parallelogram ABCD in the xy-plane become A′(1, 1), B′(4, 1), C′(4, –2), D′(1, –2) in the uv-plane under the given transformation.

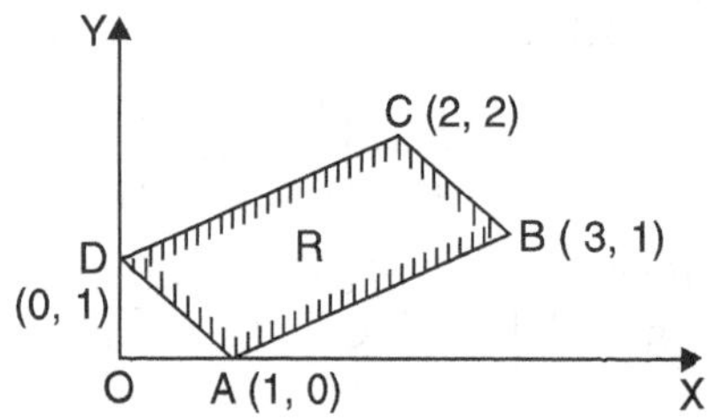

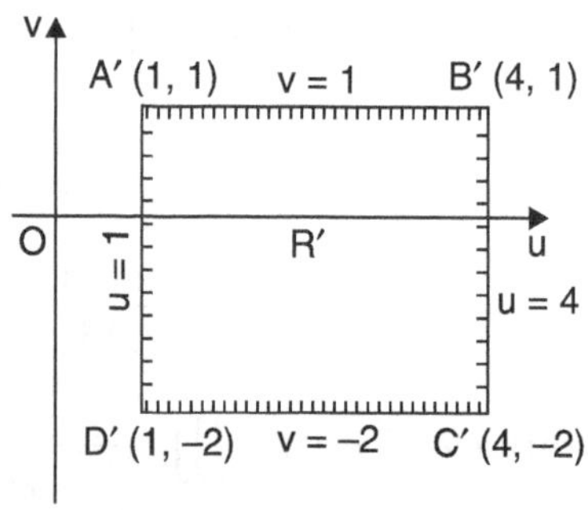

The region R in the xy-plane becomes the region R′ in the uv-plane which is a rectangle bounded by the line $u = 1$, $u = 4$ and $v = -2$, $v = 1$. Solving the given equations for x and y, we have $x = \frac{1}{3}(2u + v)$, $y = \frac{1}{3}(u - v)$

$$\therefore \qquad J = \frac{\partial(x, y)}{\partial(u, v)} = \begin{vmatrix} \frac{\partial x}{\partial u} & \frac{\partial x}{\partial v} \\ \frac{\partial y}{\partial u} & \frac{\partial y}{\partial v} \end{vmatrix} = \begin{vmatrix} \frac{2}{3} & \frac{1}{3} \\ \frac{1}{3} & -\frac{1}{3} \end{vmatrix} = -\frac{1}{3}$$

$$\therefore \quad \iint_R (x+y)^2\, dx\, dy = \iint_{R'} u^2\, |J|\, du\, dv = \int_{-2}^{1} \int_{1}^{4} u^2 \cdot \frac{1}{3} du\, dv$$

$$= \int_{-2}^{1} \frac{1}{3}\left[\frac{u^3}{3}\right]_1^4 dv = \int_{-2}^{1} 7\, dv = 7\Big[v\Big]_2^1 = 7 \times 3 = 21.$$

Example 9. (*i*) *Using the transformation $x + y = u$, $y = uv$ show that*

$\iint \sqrt{xy(1-x-y)}\, dx\, dy = \frac{2\pi}{105}$; *integration being taken over the area of the triangle bounded by the lines $x = 0$, $y = 0$, $x + y = 1$.* **(P.T.U., Dec. 2004)**

(*ii*) *Evaluate* $\int_0^1 \int_0^{1-x} e^{\frac{y}{x+y}}\, dy\, dx$ *by using the transformation $x + y = u$, $y = uv$.*

Sol. (*i*) Given relation is $x + y = u$, $y = uv$

transforms x, y to u, v we get

$$x + y = u,\ y = uv$$

$$\Rightarrow \quad x = u - uv = u(1-v)$$

i.e., $\left.\begin{aligned} x &= u(1-v) \\ y &= uv \end{aligned}\right\}$

Now $x = 0 \Rightarrow u = 0$ or $v = 1$

$y = 0 \Rightarrow u = 0$ or $v = 0$

$x + y = 1 \Rightarrow u - uv + uv = 1 \Rightarrow u = 1$

$\therefore$ Triangle transforms into a square as shown in the figure.

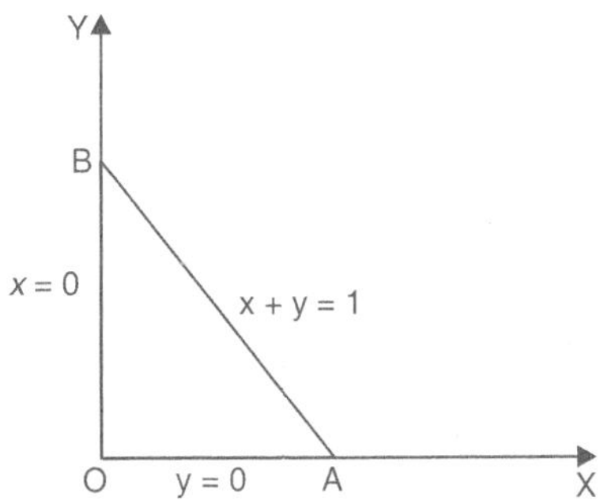

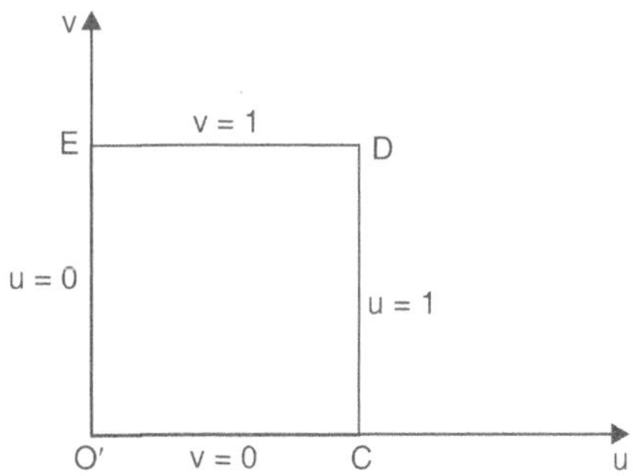

We know that

$$\iint_R f(x, y)\, dxdy = \iint_{R'} f[\phi(u, v), \psi(u, v)]\, |J|\, dudv$$

$$\therefore \quad |J| = \begin{vmatrix} \frac{\partial x}{\partial u} & \frac{\partial x}{\partial v} \\ \frac{\partial y}{\partial u} & \frac{\partial y}{\partial v} \end{vmatrix} = \begin{vmatrix} 1-v & -u \\ v & u \end{vmatrix} = u - uv + uv = u$$

$$\therefore \quad \iint [xy(1-x-y)]^{1/2}\, dxdy = \int_{u=0}^{1} \int_{v=0}^{1} [u(1-v)\, uv(1-u)]^{1/2}\, u\, du\, dv$$

$$= \int_0^1 (u^2\sqrt{1-u})\, du \int_0^1 \sqrt{v - v^2}\, dv \qquad \ldots(1)$$

Put $u = \sin^2 \theta$

$du = 2 \sin \theta \cos \theta \, d\theta$

(1) becomes

$$= \int_0^{\pi/2} \sin^4 \theta \cdot \cos \theta \, 2 \sin \theta \cos \theta \, d\theta \int_0^1 \sqrt{\left(\frac{1}{2}\right)^2 - \left(v - \frac{1}{2}\right)^2} \, dv$$

$$= \int_0^{\pi/2} 2 \sin^5 \theta \cdot \cos^2 \theta \, d\theta \left[\frac{\left(v - \frac{1}{2}\right)\sqrt{\left(\frac{1}{2}\right)^2 - \left(v - \frac{1}{2}\right)^2}}{2} + \frac{\frac{1}{4}}{2} \sin^{-1} \frac{v - \frac{1}{2}}{1/2} \right]_0^1$$

$$= 2 \frac{4 \cdot 2 \cdot 1}{7 \cdot 5 \cdot 3 \cdot 1} \left[0 + \frac{1}{8}\frac{\pi}{2} - \frac{1}{8}\left(-\frac{\pi}{2}\right) \right]$$

$$= \frac{16}{105} \times \frac{\pi}{8} = \frac{2\pi}{105}.$$

(*ii*) $\int_0^1 \int_0^{1-x} e^{\frac{y}{x+y}} \, dy \, dx$

Limits of integration of y are from 0 to $1 - x$, *i.e.*, $(y = 0)$ $y = 1 - x$ or $x + y = 1$ and limits of integration of x are from 0 to 1. In this case also the region of integration is same as in part (*i*)

$\therefore$ Under the transformation $x + y = u$, $y = uv$ the triangle OAB transforms to the square O′CDE

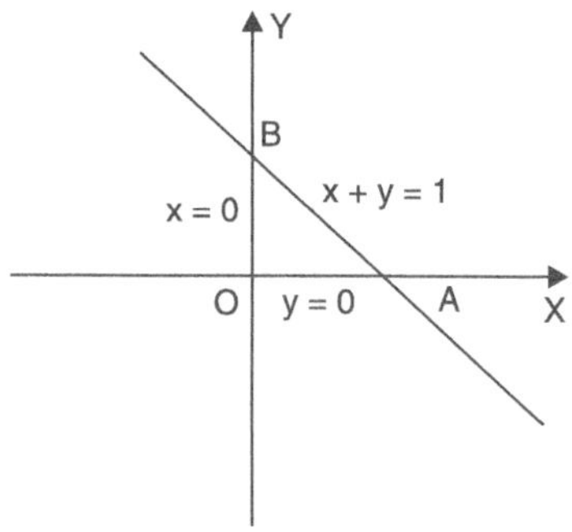

$$\therefore \int_0^1 \int_0^{1-x} e^{\frac{y}{x+y}} \, dy \, dx = \int_0^1 \int_0^1 e^{\frac{uv}{u}} \, |J| \, du \, dv$$

$| J | = u$ proved in (*i*) part

$$= \int_0^1 \int_0^1 e^v \, u \, du \, dv$$

$$= \int_0^1 \int e^v \left[\frac{u^2}{2} \Bigg|_0^1 \right] dv$$

$$= \int_0^1 e^v \cdot \frac{1}{2} dv = \frac{1}{2} e^v \Bigg|_0^1 = \frac{e^1 - e^0}{2} = \frac{e - 1}{2}.$$

Example 10. *Transform* $\int_0^{\pi} \int_0^a r^3 \sin \theta \cos \theta \, dr \, d\theta$ *to cartesian form and evaluate.*

Sol. In the given integral r varies from 0 to a and θ varies from 0 to π ; $r = a$ represents a circle

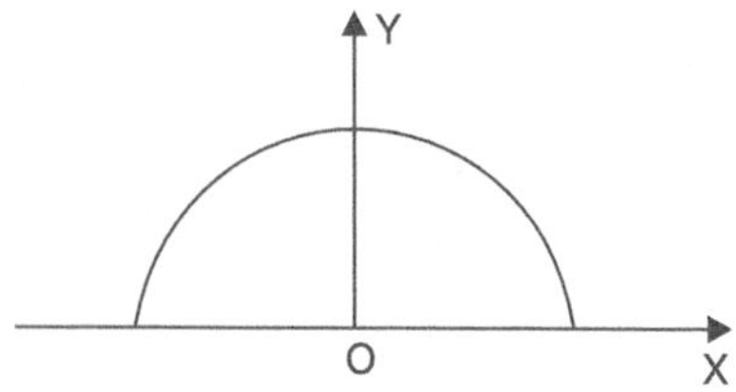

$\therefore$ given region of integration is semicircle in the upper half plane the relation between polar and cartesian system is

$$x = r \cos \theta, y = r \sin \theta$$

$$\therefore \qquad r^2 = x^2 + y^2 \quad \text{and} \quad \tan\theta = \frac{y}{x}$$

$$r = a \;\Rightarrow\; x^2 + y^2 = a^2$$

Also $\qquad dr d\theta = |\,J\,|\, dx\, dy$ where

$$|\,J\,| = \begin{vmatrix} \dfrac{\partial r}{\partial x} & \dfrac{\partial r}{\partial y} \\ \dfrac{\partial \theta}{\partial x} & \dfrac{\partial \theta}{\partial y} \end{vmatrix}$$

$$r^2 = x^2 + y^2$$

$$dr\, \frac{\partial r}{\partial x} = 2x \qquad \therefore \qquad \frac{\partial r}{\partial x} = \frac{x}{r}, \frac{\partial r}{\partial y} = \frac{y}{r}$$

$$\theta = \tan^{-1}\frac{y}{x}$$

$$\frac{\partial \theta}{\partial x} = \frac{1}{1+\dfrac{y^2}{x^2}}\left(\frac{-y}{x^2}\right) = -\frac{y}{x^2+y^2} = -\frac{y}{r^2}$$

$$\frac{\partial \theta}{\partial y} = \frac{1}{1+\dfrac{y^2}{x^2}} \cdot \frac{1}{x} = \frac{x}{x^2+y^2} = \frac{x}{r^2}$$

$$|\,J\,| = \begin{vmatrix} \dfrac{x}{r} & \dfrac{y}{r} \\ -\dfrac{y}{r^2} & \dfrac{x}{r^2} \end{vmatrix} = \frac{x^2}{r^3} + \frac{y^2}{r^3} = \frac{r^2}{r^3} = \frac{1}{r}$$

Now $\displaystyle\int_0^{\pi}\int_0^{a} r^3 \sin\theta\cos\theta\, dr\, d\theta$

$$= \int_0^{\pi}\int_0^{a} (r\sin\theta)(r\cos\theta)(r\,dr\,d\theta) = \int_{-a}^{a}\int_{y=0}^{\sqrt{a^2-x^2}} yx\,dy\,dx$$

which is required Cartesian form

$$= \int_{-a}^{a} x\left.\frac{y^2}{2}\right|_0^{\sqrt{a^2-x^2}} dx = \int_{-a}^{a} \frac{x}{2}(a^2-x^2)\,dx = \frac{1}{2}\int_{-a}^{a} (a^2x - x^3)\,dx$$

$$= \frac{1}{2}\left[\frac{a^2x^2}{2} - \frac{x^4}{4}\Big|_{-a}^{a}\right] = 0$$

TEST YOUR KNOWLEDGE

1. Evaluate $\iint \sin \pi (x^2 + y^2)\, dx\, dy$ over the region bounded by the circle $x^2 + y^2 = 1$ by changing to polar co-ordinates.

2. Evaluate $\iint e^{x^2 + y^2} dx\, dy$ where R is the semi-circular region bounded by the x-axis and the curve $y = \sqrt{1 - x^2}$ by changing to polar coordinates **(*P.T.U. Dec. 2014*)**

$$\left[\textbf{Hint. } \int_{\theta = 0}^{\pi} \int_{r = 0}^{1} e^{r^2}\, r\, dr\, d\theta \right]$$

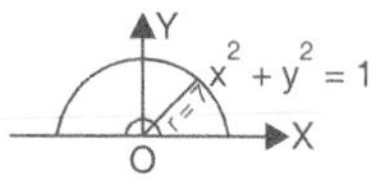

3. Evaluate $\iint xy\, (x^2 + y^2)^{3/2}\, dx\, dy$ over the positive quadrant of the circle $x^2 + y^2 = 1$.

4. Evaluate the following by changing into polar co-ordinates:

(*i*) $\displaystyle\int_0^a \int_y^a \frac{x\, dx\, dy}{x^2 + y^2}$ (*ii*) $\displaystyle\int_{-a}^{a} \int_{-\sqrt{a^2 - x^2}}^{\sqrt{a^2 - x^2}} dy\, dx$ (*iii*) $\displaystyle\int_0^{4a} \int_{y^2/4a}^{y} \frac{x^2 - y^2}{x^2 + y^2}\, dx\, dy$

$$\left[\textbf{Hint. } (iii) \int_{\theta = \frac{\pi}{4}}^{\frac{\pi}{2}} \int_{r = 0}^{\frac{4a \cos \theta}{\sin^2 \theta}} \frac{\cos^2 \theta - \sin^2 \theta}{r^2}\, r\, dr\, d\theta \right]$$

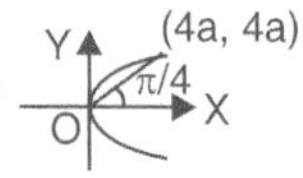

(*iv*) $\displaystyle\int_0^a \int_0^{\sqrt{a^2 - y^2}} (x^2 + y^2)\, dx\, dy$ (*v*) $\displaystyle\int_0^a \int_0^{\sqrt{a^2 - x^2}} y^2 \sqrt{x^2 + y^2}\, dy\, dx$ (*vi*) $\displaystyle\int_0^a \int_y^a \frac{x^2\, dx\, dy}{\sqrt{x^2 + y^2}}$.

$$\left[\textbf{Hint. } (iv) \int_0^{\frac{\pi}{2}} \int_0^a r^3\, dr d\theta \right] \qquad \left[\textbf{Hint. } (vi) \int_{\theta = 0}^{\frac{\pi}{4}} \int_0^{a \sec \theta} \frac{r \cos^2 \theta}{r}\, r dr d\theta \right]$$

5. Evaluate $\iint_D e^{-(x^2 + y^2)}\, dy\, dx$ where D is the region bounded by $x^2 + y^2 = a^2$.

6. Evaluate $\iint xy\, (x^2 + y^2)^{n/2}\, dx\, dy$ over the positive quadrant of $x^2 + y^2 = 4$, supposing $n + 3 > 0$.

$$\left[\textbf{Hint. } \int_0^{\frac{\pi}{2}} \int_0^2 (r \cos \theta)\, (r \sin \theta)\, r^n\, r dr d\theta \right]$$

7. Evaluate $\displaystyle\iint \sqrt{\frac{1 - \frac{x^2}{a^2} - \frac{y^2}{b^2}}{1 + \frac{x^2}{a^2} + \frac{y^2}{b^2}}}\, dx\, dy$ over the positive quadrant of the ellipse $\dfrac{x^2}{a^2} + \dfrac{y^2}{b^2} = 1$.

[**Hint.** Put $x = au,\ y = bv$ ellipse transform to circle; $\displaystyle\iint \sqrt{\frac{1 - (u^2 + v^2)}{1 + u^2 + v^2}}\, ab\, du\, dv$. Put $u = r \sin \theta$, $v = r \sin \theta$ then $du\, dv = r\, dr\, d\theta$. See S.E. 6]

8. Evaluate $\displaystyle\int_0^1 \int_0^{\sqrt{1 - y^2}} \int_0^{\sqrt{1 - x^2 - y^2}} \frac{dz\, dy\, dx}{\sqrt{1 - x^2 - y^2 - z^2}}$ by changing to spherical polar coordinates

[**Hint.** consult S.E. 7]

9. Evaluate $\iiint z(x^2 + y^2)\, dx\, dy\, dz$ over the volume of the cylinder $x^2 + y^2 = 1$ intercepted by the planes $z = 2$ and $z = 3$.

[**Hint.** See S.E. 5]

10. Evaluate the following integrals through the volume of the sphere $x^2 + y^2 + z^2 = 1$, by changing into spherical polar co-ordinates:

(*i*) $\iiint z^2\, dx\, dy\, dz$ (*ii*) $\iiint (x^2 + y^2 + z^2)^m\, dx\, dy\, dz$. $(m > 0)$

[**Hint.** See S.E. 7]

Answers

1. 2 **2.** $(e - 1)\frac{\pi}{2}$ **3.** $\frac{1}{14}$

4. (*i*) $\frac{\pi a}{4}$ (*ii*) πa^2 (*iii*) $8\left(\frac{\pi}{2} - \frac{5}{3}\right)a^2$ (*iv*) $\frac{\pi a^4}{8}$

(*v*) $\frac{\pi a^5}{20}$ (*vi*) $\frac{a^3}{3} \log(\sqrt{2} + 1)$ **5.** $\pi(1 - e^{-a^2})$ **6.** $\frac{2^{n+3}}{n+4}$

7. $\frac{\pi ab}{8}(\pi - 2)$ **8.** $\frac{\pi^2}{8}$ **9.** $\frac{5\pi}{4}$

10. (*i*) $\frac{4\pi}{15}$ (*ii*) $\frac{4\pi}{2m+3}$.

6.7. AREA BY DOUBLE INTEGRATION

(*a*) *Cartesian Co-ordinates.* The area A of the region bounded by the curves $y = f_1(x)$, $y = f_2(x)$ and the lines $x = a$, $x = b$ is given by $A = \int_a^b \int_{f_1(x)}^{f_2(x)} dy\, dx$

The area A of the region bounded by the curves $x = f_1(y)$, $x = f_2(y)$ and the lines $y = c$, $y = d$ is given by

$$A = \int_c^d \int_{f_1(y)}^{f_2(y)} dx\, dy.$$

(*b*) *Polar Co-ordinates.* The area A of the region bounded by the curves $r = f_1(\theta)$, $r = f_2(\theta)$ and the lines $\theta = \alpha$, $\theta = \beta$ is given by $A = \int_\alpha^\beta \int_{f_1(\theta)}^{f_2(\theta)} r\, dr\, d\theta$

6.8. VOLUME BY DOUBLE INTEGRATION

(*a*) *Cartesian Co-ordinates.* Consider a surface $z = f(x, y)$ ...(1)

Let the orthogonal projection of its region R′ on the xy-plane be the region R given by $\phi(x, y) = 0$. ...(2)

Now (2) represents a cylinder with generators parallel to z-axis and the guiding curve given by (2). Let V be the volume of this cylinder between R and R′.

Divide R into elementary rectangles of area $\delta x\, \delta y$ by drawing lines parallel to the axis of x and y. On each of these rectangles, erect prisms of lengths parallel to z-axis. Volume of this prism between R and R′ is $z\, \delta x\, \delta y$. The volume V is composed of such prisms.

$$\therefore \qquad V = \underset{\substack{\delta x \to 0 \\ \delta y \to 0}}{\text{Lt}} \sum z\, \delta x\, \delta y = \iint_R z\, dx\, dy.$$

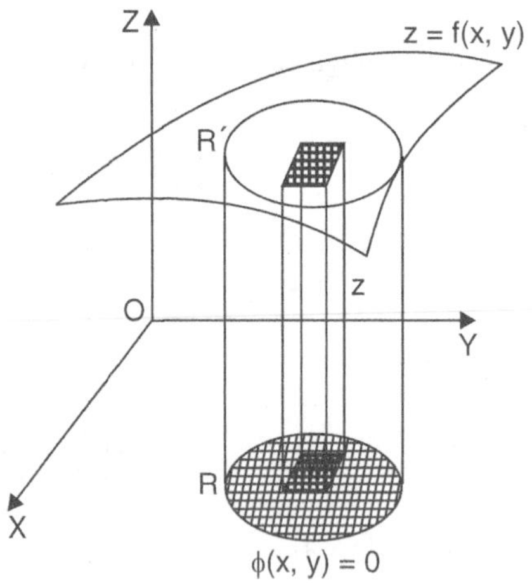

(*b*) *Cylindrical Co-ordinates.* Let the equation of the surface be $z = f(r, \phi)$. Replacing $dx\, dy$ by $r\, dr\, d\phi$, we get $V = \iint_R zr\, dr\, d\phi$.

ILLUSTRATIVE EXAMPLES

Example 1. (*a*) *Find the area lying between the parabola $y = 4x - x^2$ and the line $y = x$.* **(P.T.U., May 2003)**

(*b*) *Find by double integration the area bounded by $x = 2y - y^2$ and $x = y^2$.* **(P.T.U., Dec., 2011)**

(*c*) *Find the area between the parabolas $y^2 = 4ax$, $x^2 = 4ay$.* **(P.T.U., Dec. 2003, Dec. 2012, May 2015)**

Sol. (*a*) The two curves intersect at points whose abscissae are given by $4x - x^2 = x$

or $\qquad x^2 - 3x = 0 \quad$ *i.e.,* $\quad x = 0, 3$.

Using vertical strips, the required area lies between $x = 0$, $x = 3$ and $y = x$, $y = 4x - x^2$.

$$\therefore \text{ Required area} = \int_0^3 \int_x^{4x - x^2} dy\, dx$$

$$= \int_0^3 \Big[y \Big]_x^{4x - x^2} dx$$

$$= \int_0^3 (4x - x^2 - x)\, dx = \int_0^3 (3x - x^2)\, dx$$

$$= \left[\frac{3x^2}{2} - \frac{x^3}{3} \right]_0^3 = \frac{27}{2} - 9 = 4.5.$$

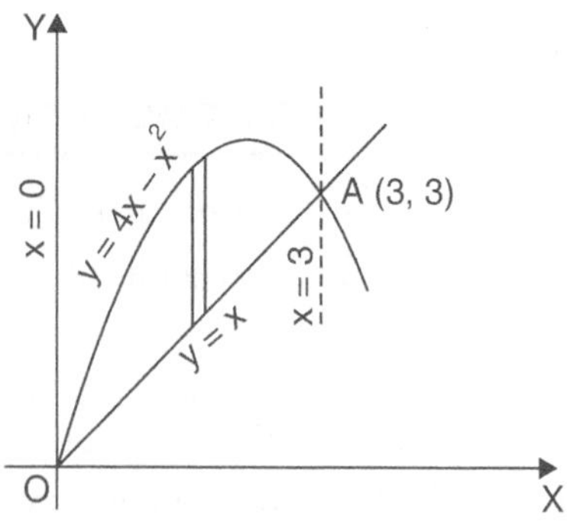

(*b*) Equations of the curves are $x = 2y - y^2$ and $x = y^2$. Both are parabolas.

They intersect at points given by

$y^2 = 2y - y^2$ or $2y^2 - 2y = 0$ or $2y(y-1) = 0$

or $y = 0, y = 1$ or at (0, 0) and (1, 1)

Equation of the parabola $x = 2y - y^2$ can be put in the form $x = -(y^2 - 2y)$

$$x - 1 = -(y^2 - 2y + 1) = -(y-1)^2$$

or $(y-1)^2 = -(x-1)$, which is a parabola having vertex at (1, 1)

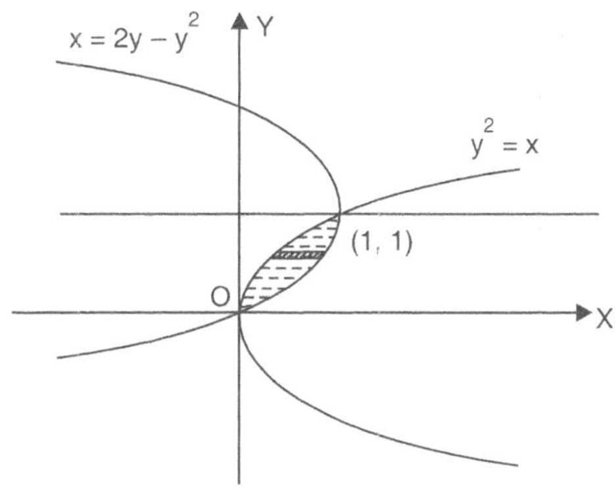

Using horizontal strip

$$\text{Required area} = \int_0^1 \int_{y^2}^{2y-y^2} dx\, dy$$

$$= \int_0^1 x \Big|_{y^2}^{2y-y^2} dy = \int_0^1 (2y - y^2 - y^2)\, dy = \int_0^1 (2y - 2y^2)\, dy$$

$$= 2\left[\frac{y^2}{2} - \frac{y^3}{3}\Big|_0^1\right] = 2\cdot\left[\frac{1}{2} - \frac{1}{3}\right] = 2\cdot\frac{1}{6} = \frac{1}{3}.$$

(*c*) The two curves $y^2 = 4ax$

and $x^2 = 4ay$

intersect at $\dfrac{x^4}{16a^2} = 4ax$

or $x^4 - 64a^3x = 0$

or $x(x^3 - 64a^3) = 0$

i.e., at $x = 0$ and $x = 4a$

i.e., at (0, 0) and $(4a, 4a)$

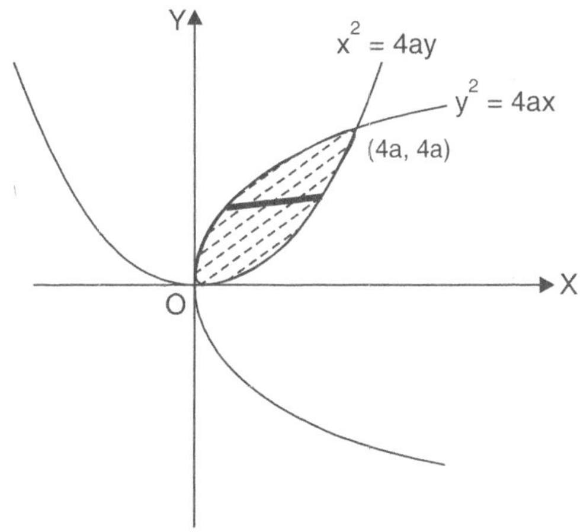

Using horizontal strips; x varies from $\dfrac{y^2}{4a}$ to $2\sqrt{ay}$ and y varies from 0 to $4a$.

$$\therefore \text{ Required area} = \int_0^{4a} \int_{\frac{y^2}{4a}}^{2\sqrt{ay}} dx\, dy = \int_0^{4a} x \Big|_{\frac{y^2}{4a}}^{2\sqrt{ay}} dy$$

$$= \int_0^{4a} \left(2\sqrt{ay} - \frac{y^2}{4a}\right) dy$$

$$= 2\sqrt{a}\,\frac{y^{3/2}}{3/2} - \frac{y^3}{12a}\Big|_0^{4a}$$

$$= \frac{4\sqrt{a}}{3}(4a)^{3/2} - \frac{(4a)^3}{12a} = \frac{4}{3}\sqrt{a}\cdot 8a^{3/2} - \frac{64a^3}{12a}$$

$$= \frac{32}{3}a^2 - \frac{16}{3}a^2 = \frac{16}{3}a^2.$$

Example 2. *Find the area of the region bounded by the lines $x = -2$, $x = 2$ and the circle $x^2 + y^2 = 9$.* **(P.T.U., May 2007)**

Sol. The equation on of the circles $x^2 + y^2 = 9$

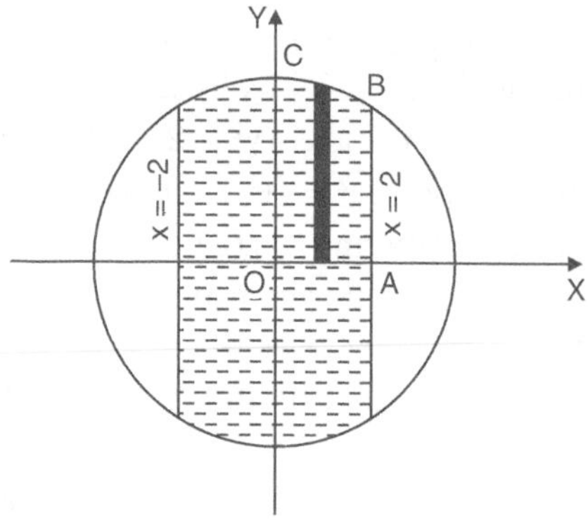

Required area = 4 OABC

$$= 4\int_0^2 \int_0^{\sqrt{9-x^2}} dy\, dx$$

$$= 4\int_0^2 y\Big|_0^{\sqrt{9-x^2}} dx$$

$$= 4\int_0^2 \sqrt{9-x^2}\, dx$$

$$= 4 . \left\{\frac{x\sqrt{9-x^2}}{2} + \frac{9}{2}\sin^{-1}\frac{x}{3}\Big|_0^2\right\}$$

$$= 4\left\{\frac{2\sqrt{5}}{2} + \frac{9}{2}\sin^{-1}\frac{2}{3}\right\}$$

$$= 4\sqrt{5} + 18\sin^{-1} 2/3.$$

Example 3. *Find the smaller of the areas bounded by the ellipse $4x^2 + 9y^2 = 36$ and the straight line $2x + 3y = 6$.*

Sol. The equation of the ellipse is

$$\frac{x^2}{9} + \frac{y^2}{4} = 1 \qquad ...(1)$$

and the line is
$$\frac{x}{3} + \frac{y}{2} = 1 \qquad ...(2)$$

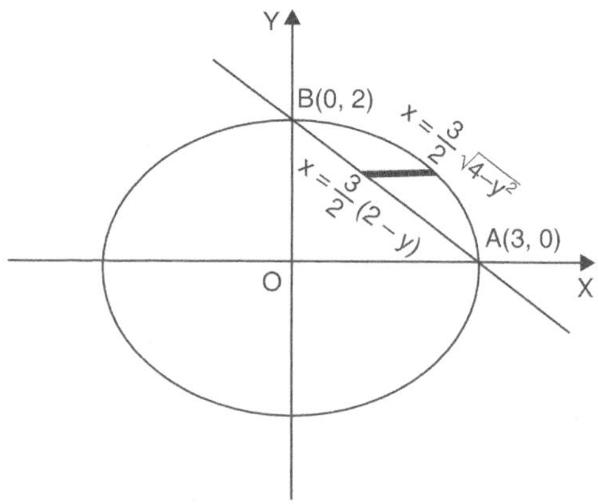

Both meet x-axis at A(3, 0) and y-axis at B(0, 2).

Using horizontal strips, the required area lies between

$$x = \frac{3}{2}(2-y),\ x = \frac{3}{2}\sqrt{4-y^2} \text{ and } y = 0, y = 2.$$

$$\therefore \quad \text{Required area} = \int_0^2 \int_{\frac{3}{2}(2-y)}^{\frac{3}{2}\sqrt{4-y^2}} dx\, dy$$

$$= \int_0^2 \Big[x\Big]_{\frac{3}{2}(2-y)}^{\frac{3}{2}\sqrt{4-y^2}} dy$$

$$= \int_0^2 \frac{3}{2}[\sqrt{4-y^2} - (2-y)]\, dy$$

$$= \frac{3}{2}\left[\frac{y\sqrt{4-y^2}}{2} + \frac{4}{2}\sin^{-1}\frac{y}{2} - 2y + \frac{y^2}{2}\right]_0^2$$

$$= \frac{3}{2}[2\sin^{-1} 1 - 4 + 2] = \frac{3}{2}\left(2\cdot\frac{\pi}{2} - 2\right) = \frac{3}{2}(\pi - 2).$$

Example 4. *Using double integration, find the area enclosed by the curves $y^2 = x^3$ and $y = x$.*

(P.T.U., Dec. 2005, May 2009)

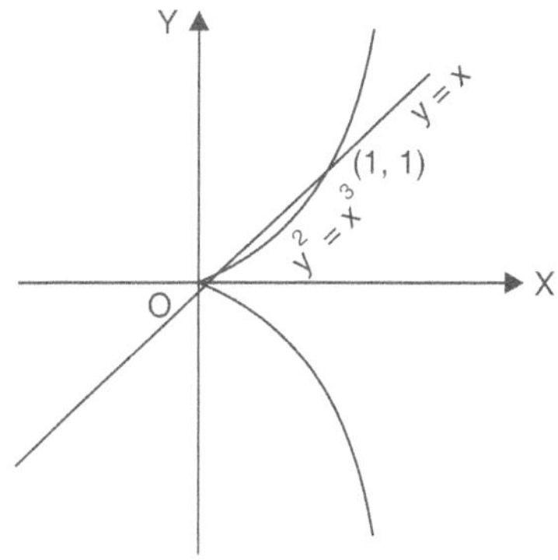

Sol. The two curves intersect at points given by $x^2 = x^3$ or $x^2(x - 1) = 0$ or $x = 1$ *i.e.,* (0, 0) and (1, 1)

Using vertical strips the required area lies between $x = 0, x = 1, y = x^{3/2}$ and $y = x$

$$\therefore \text{ Required area} = \int_0^1 \int_{x^{3/2}}^{x} dy\,dx = \int_0^1 y\Big|_{x^{3/2}}^{x} dx$$

$$= \int_0^1 (x - x^{3/2})dx = \left.\frac{x^2}{2} - \frac{x^{5/2}}{5/2}\right|_0^1$$

$$= \frac{1}{2} - \frac{2}{5} = \frac{1}{10}.$$

Example 5. *Find the area enclosed by the curves $y = \dfrac{3x}{x^2 + 2}$ and $4y = x^2$.*

Sol. Equations of the curves are $y = \dfrac{3x}{x^2+2}$ ...(1)

and $4y = x^2$ ...(2)

For rough sketch of $y = \dfrac{3x}{x^2+2}$ we see that the curve passes through the origin and the tangent at the origin is obtained by equating lowest degree terms in $yx^2 + 2y - 3x = 0$ to zero

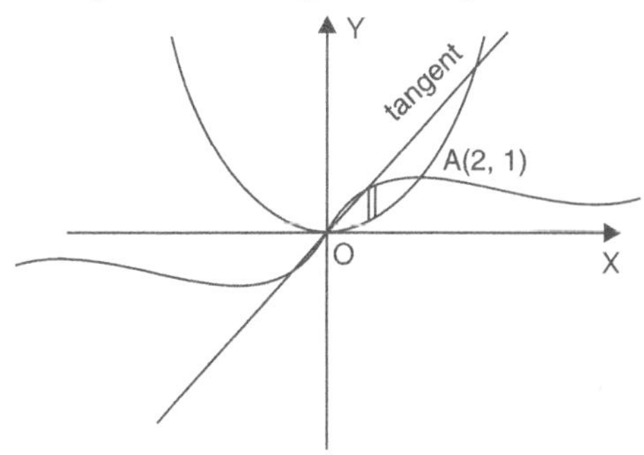

i.e., $2y - 3x = 0$ or $y = \dfrac{3}{2}x$

Asymptote to the curve is $y = 0$

When x is +ve, y is +ve and when x is –ve, y is –ve. (0, 0) is the point of inflexion.

(1) and (2) intersect at points given by

$$\frac{x^2}{4} = \frac{3x}{x^2+2} \quad i.e., x = 0 \quad \text{or} \quad x(x^2 + 2) = 12$$

or $\quad x^3 + 2x - 12 = 0 \quad$ or $\quad (x - 2)(x^2 + 2x + 6) = 0$

i.e., at (0, 0) and (2, 1)

Required region is shown in the figure

$$\text{Area OAO} = \int_{x=0}^{2} \int_{x^4/4}^{3x/x^2+2} dy\,dx \qquad \text{(vertical strip is taken)}$$

$$= \int_0^2 y \Big|_{\frac{x^4}{4}}^{\frac{3x}{x^2+4}} dx = \int_0^2 \left(\frac{3x}{x^2+4} - \frac{x^2}{4}\right) dx$$

$$= \left.\frac{3}{2}\log(x^2+2) - \frac{x^3}{12}\right|_0^2 = \frac{3}{2}\log 6 - \frac{3}{2}\log 2 - \frac{8}{12} = \frac{3}{2}\log 3 - \frac{2}{3}.$$

Example 6. (*i*) *Find, by double integration, the area lying inside the circle $r = a \sin \theta$ and outside the cardioid $r = a(1 - \cos \theta)$.* **(P.T.U., May 2014)**

(*ii*) *Find the area outside the circle $r = a$ and inside the cardioid $r = a(1 + \cos \theta)$.*

(P.T.U., May 2004, Dec. 2011)

Sol. To find intersection of the curves

(*i*) Equations of the curves are $r = a \sin \theta$; $r = a(1 - \cos \theta)$ eliminating r between the equations of two curves, $\sin \theta = 1 - \cos \theta$ or $\sin \theta + \cos \theta = 1$

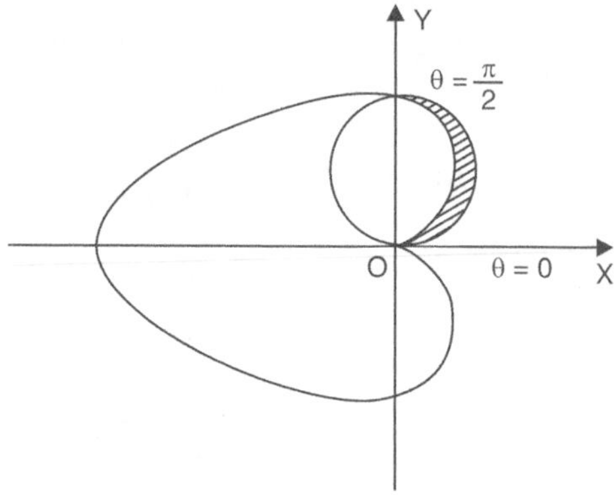

Squaring $1 + \sin 2\theta = 1$ or $\sin 2\theta = 0$

$\therefore \quad 2\theta = 0$ or π *i.e.*, $\theta = 0$ or $\dfrac{\pi}{2}$

For the required area, r varies from $a(1 - \cos \theta)$ to $a \sin \theta$ and θ varies from 0 to $\dfrac{\pi}{2}$.

$\therefore$ Required area

$$= \int_0^{\pi/2} \int_{a(1-\cos\theta)}^{a \sin \theta} r \, dr \, d\theta$$

$$= \int_0^{\pi/2} \left[\frac{r^2}{2} \right]_{a(1-\cos\theta)}^{a \sin \theta} d\theta = \frac{1}{2} \int_0^{\pi/2} a^2 \left[\sin^2 \theta - (1 - \cos \theta)^2 \right] d\theta$$

$$= \frac{a^2}{2} \int_0^{\pi/2} (\sin^2 \theta - 1 - \cos^2 \theta + 2 \cos \theta)] \, d\theta$$

$$= \frac{a^2}{2} \int_0^{\pi/2} (-2 \cos^2 \theta + 2 \cos \theta) \, d\theta = a^2 \left[-\frac{1}{2} \cdot \frac{\pi}{2} + 1 \right] = a^2 \left(1 - \frac{\pi}{4} \right).$$

(*ii*) Intersection of $r = a$ and $r = a(1 + \cos \theta)$ is given by $a = a(1 + \cos \theta)$ *i.e.*, $1 + \cos \theta = 1$ *i.e.*, $\cos \theta = 0$.

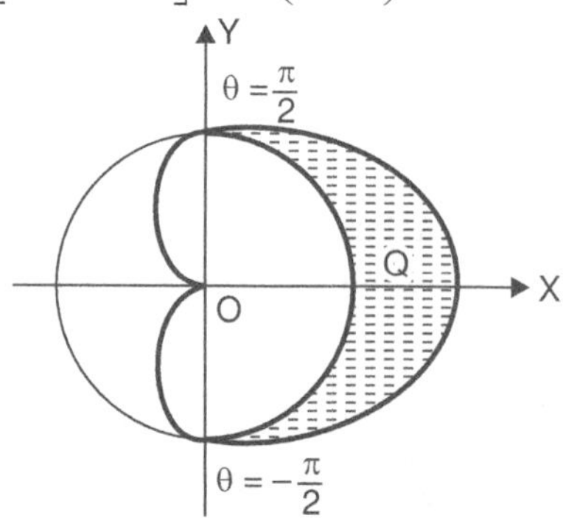

$\therefore \quad \theta = \dfrac{\pi}{2}, -\dfrac{\pi}{2}$

For the required area r varies from a to $a(1 + \cos \theta)$ and θ varies from $-\dfrac{\pi}{2}$ to $\dfrac{\pi}{2}$

$\therefore$ Required area

$$= \int_{-\frac{\pi}{2}}^{\pi/2} \int_a^{a(1+\cos\theta)} r \, dr \, d\theta$$

$$= \int_{-\frac{\pi}{2}}^{\pi/2} \frac{r^2}{2} \Bigg|_a^{a(1+\cos\theta)} d\theta = \frac{1}{2} \int_{-\frac{\pi}{2}}^{\pi/2} \left[a^2 (1 + \cos \theta)^2 - a^2 \right] d\theta$$

$$= \frac{1}{2} \cdot 2a^2 \int_0^{\pi/2} \left[(1 + \cos \theta)^2 - 1 \right] d\theta = a^2 \int_0^{\pi/2} (2 \cos \theta + \cos^2 \theta) \, d\theta$$

$$= 2a^2 \Big[\sin \theta \Big]_0^{\pi/2} + a^2 \, \frac{1}{2} \frac{\pi}{2}$$

$$= 2a^2 \cdot 1 + \frac{a^2 \pi}{4} = \frac{a^2}{4} (8 + \pi).$$

Example 7. *Find by double integration, the area of one loop of the laminscate $r^2 = a^2 \cos 2\theta$.*

(P.T.U., May 2012)

Sol. For details of figure see S.E. 6 art. 4.16

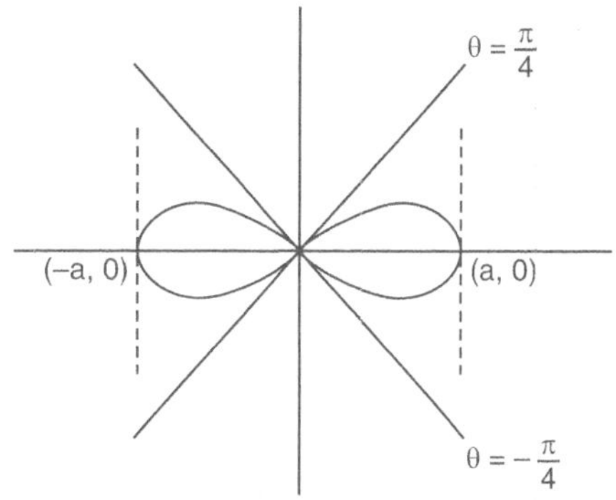

Area of one loop

$$= \int_{\theta=-\frac{\pi}{4}}^{\frac{\pi}{4}} \int_{r=0}^{a\sqrt{\cos 2\theta}} r\,dr\,d\theta$$

$$= \int_{-\frac{\pi}{4}}^{\frac{\pi}{4}} \left.\frac{r^2}{2}\right|_0^{a\sqrt{\cos 2\theta}} d\theta$$

$$= \frac{1}{2}\int_{-\frac{\pi}{4}}^{\frac{\pi}{4}} a^2 \cos 2\theta\, d\theta = \frac{a^2}{2}\cdot\left.\frac{\sin 2\theta}{2}\right|_{-\frac{\pi}{4}}^{\frac{\pi}{4}}$$

$$= \frac{a^2}{4}\left[\sin\frac{\pi}{2} - \sin\left(-\frac{\pi}{2}\right)\right] = \frac{a^2}{4}\cdot 2 = \frac{a^2}{2}.$$

Example 8. *Find the volume of the ellipsoid $\frac{x^2}{a^2}+\frac{y^2}{b^2}+\frac{z^2}{c^2} = 1$.*

(P.T.U., May 2003, Dec. 2007, Dec. 2012)

Sol. On account of symmetry, the required volume is 8 times the volume of the ellipsoid in the positive octant. The volume OABC in the positive octant lies between the ellipsoid $z = c\sqrt{1-\frac{x^2}{a^2}-\frac{y^2}{b^2}}$ and the plane XOY, and is bounded on the sides by the planes $x = 0$, $y = 0$.

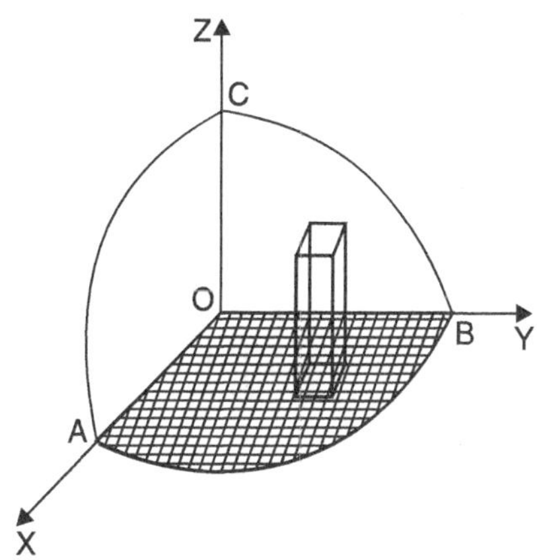

The ellipsoid cuts the plane XOY in the ellipse

$$\frac{x^2}{a^2}+\frac{y^2}{b^2} = 1, z = 0$$

∴ The region OAB above which the volume OABC lies is bounded by

$$x = 0, x = a \text{ and } y = 0, y = b\sqrt{1-\frac{x^2}{a^2}}$$

Hence the required volume of the ellipsoid

$$= 8\int_0^a \int_0^{b\sqrt{1-x^2/a^2}} z\,dy\,dx = 8\int_0^a \int_0^{b\sqrt{1-x^2/a^2}} c\sqrt{1-\frac{x^2}{a^2}-\frac{y^2}{b^2}}\,dy\,dx$$

$$= 8\int_0^a \int_0^t c\sqrt{\frac{t^2}{b^2}-\frac{y^2}{b^2}}\,dy\,dx \quad \text{where } \sqrt{1-\frac{x^2}{a^2}} = \frac{t}{b}.$$

$$= 8\int_0^a\int_0^t \frac{c}{b}\sqrt{t^2-y^2}\,dy\,dx = 8\int_0^a \frac{c}{b}\left[\frac{y\sqrt{t^2-y^2}}{2}+\frac{t^2}{2}\sin^{-1}\frac{y}{t}\right]_0^t dx$$

$$= \frac{4c}{b}\int_0^a t^2\sin^{-1}1\,dx = \frac{2\pi c}{b}\int_0^a t^2\,dx$$

$$= \frac{2\pi c}{b}\int_0^a b^2\left(1-\frac{x^2}{a^2}\right)dx = 2\pi bc\left[x-\frac{x^3}{3a^2}\right]_0^a = 2\pi bc\left(a-\frac{a}{3}\right)=\frac{4}{3}\pi abc.$$

Example 9. *Find the volume of the tetrahedron bounded by the co-ordinate planes and the plane* $\frac{x}{a}+\frac{y}{b}+\frac{z}{c}=1.$ **(P.T.U., Dec. 2003, May 2014, Dec. 2015)**

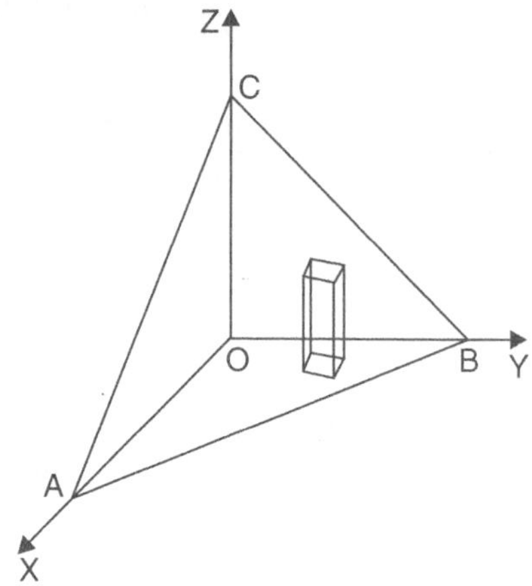

Sol. The required volume OABC lies between the plane $z=c\left(1-\frac{x}{a}-\frac{y}{b}\right)$ and the plane XOY, and is bounded on the sides by the planes $x=0, y=0$.

The plane $\frac{x}{a}+\frac{y}{b}+\frac{z}{c}=1$ cuts the plane XOY in the line $\frac{x}{a}+\frac{y}{b}=1, z=0$

$\therefore$ The region OAB above which the volume OABC lies is bounded by $x=0$, $x=a$ and $y=0$,

$$y=b\left(1-\frac{x}{a}\right)$$

Hence the required volume of the tetrahedron

$$= \int_0^a\int_0^{b(1-x/a)} z\,dy\,dx = \int_0^a\int_0^{b(1-x/a)} c\left(1-\frac{x}{a}-\frac{y}{b}\right)dy\,dx$$

$$= \int_0^a c\left[\left(1-\frac{x}{a}\right)y-\frac{y^2}{2b}\right]_0^{b(1-x/a)} dx = c\int_0^a\left[b\left(1-\frac{x}{a}\right)^2-\frac{b}{2}\left(1-\frac{x}{a}\right)^2\right]dx$$

$$= \frac{bc}{2}\int_0^a\left(1-\frac{x}{a}\right)^2dx = \frac{bc}{2}\cdot\left[\frac{\left(1-\frac{x}{a}\right)^3}{-\frac{3}{a}}\right]_0^a = -\frac{abc}{6}(0-1)=\frac{1}{6}abc.$$

Example 10. *Find the volume bounded by the paraboloid* $x^2+y^2=az$, *the cylinder* $x^2+y^2=2ay$ *and the plane* $z=0$.

Sol. Required volume $=\iint z\,dx\,dy$

$$= 2\iint \frac{x^2+y^2}{a}\, dx\, dy$$

(twice ∵ volume in both parts of the circle is same) (Change it to polar coordinates same)

$$x = r\cos\theta,\ y = r\sin\theta,\ dx\, dy = r\, dr d\theta$$

When $x^2 + y^2 = 2ay$, we have in polar coordinates

$$r^2 = 2ar\sin\theta \quad \therefore \quad r = \theta \quad \text{or} \quad r = 2a\sin\theta$$

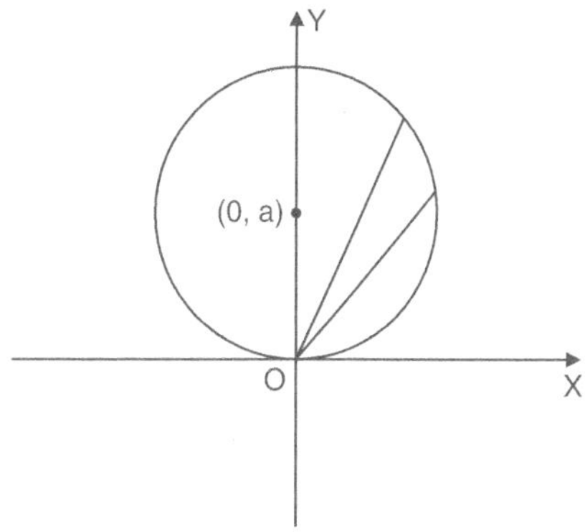

∴ r varies from 0 to $2a\sin\theta$ and θ veries from 0 to $\frac{\pi}{2}$

Required volume

$$= 2\int_{\theta=0}^{\frac{\pi}{2}}\int_{r=0}^{2a\sin\theta} \frac{r^2}{a}\, r\, dr\, d\theta$$

$$= \frac{2}{a}\int_0^{\frac{\pi}{2}} \frac{r^4}{4}\Bigg|_0^{2a\sin\theta} d\theta = \frac{1}{2a}\int_0^{\frac{\pi}{2}} 16a^4\sin^4\theta\, d\theta$$

$$= 8a^3\, \frac{3\cdot 1}{4\cdot 2}\,\frac{\pi}{2} = \frac{3a^3\pi}{2}.$$

Example 11. (*i*) *Find the volume bounded by the cylinder $x^2 + y^2 = 4$ and the planes $y + z = 4$ and $z = 0$.* **(P.T.U., Dec. 2011)**

(*ii*) *Find the volume bounded by the xy plane, the cylinder $x^2 + y^2 = 1$ and the plane $x + y + z = 3$.* **(P.T.U., Dec. 2005)**

Sol. (*i*) From the figure, it is clear that $z = 4 - y$ is to be integrated over the circle $x^2 + y^2 = 4$ in the xy-plane. To cover the shaded half of this circle, x varies from 0 to $\sqrt{4-y^2}$ while y varies from -2 to 2.

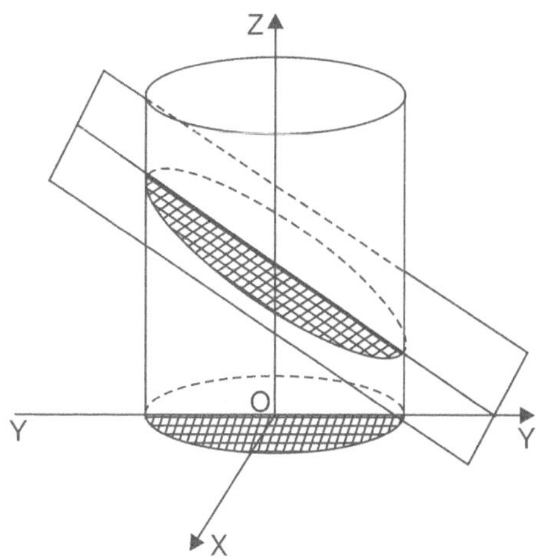

∴ Required volume

$$= 2\int_{-2}^{2}\int_0^{\sqrt{4-y^2}} z\, dx\, dy$$

$$= 2\int_{-2}^{2}\int_0^{\sqrt{4-y^2}} (4-y)\, dx\, dy$$

$$= 2\int_{-2}^{2} (4-y)\Big[x\Big]_0^{\sqrt{4-y^2}} dy$$

$$= 2\int_{-2}^{2} (4-y)\sqrt{4-y^2}\, dy$$

$$= 2\int_{-2}^{2} 4\sqrt{4-y^2}\, dy - 2\int_{-2}^{2} y\sqrt{4-y^2}\, dy$$

$$= 8\int_{-2}^{2} \sqrt{4-y^2}\, dy$$

[the second integral is zero since $y\sqrt{4-y^2}$ is an odd function of y]

$$= 16 \int_0^2 \sqrt{4 - y^2}\, dy \qquad [\because \sqrt{4 - y^2} \text{ is an even function of } y]$$

$$= 16 \left[\frac{y\sqrt{4 - y^2}}{2} + \frac{4}{2} \sin^{-1} \frac{y}{2} \right]_0^2 = 16\, [2 \sin^{-1} 1] = 32 \times \frac{\pi}{2} = 16\pi.$$

(*ii*) From the figure it is clear that $z = 3 - x - y$ is to be integrated over the circle $x^2 + y^2 = 1$ over xy plane for which x varies from 0 to $\sqrt{1 - y^2}$ and y varies from –1 to 1.

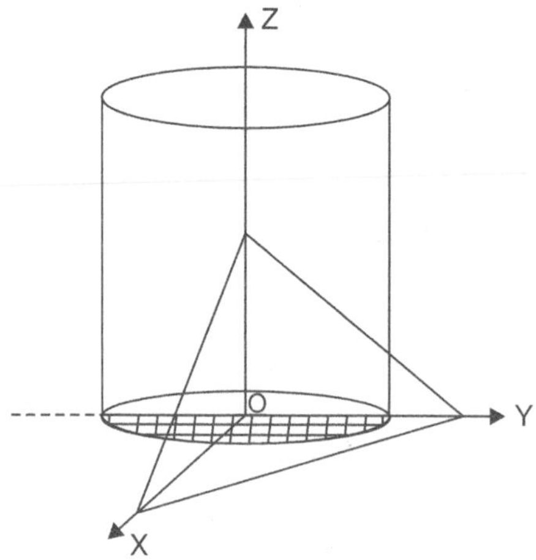

$\therefore$ Required volume

$$= 2 \int_{-1}^{1} \int_0^{\sqrt{1 - y^2}} (3 - x - y)\, dx\, dy$$

$$= 2 \int_{-1}^{1} \left[3x - \frac{x^2}{2} - yx \right]_0^{\sqrt{1 - y^2}} dy$$

$$= 2 \int_{-1}^{1} \left[3\sqrt{1 - y^2} - \frac{1 - y^2}{2} - y\sqrt{1 - y^2} \right] dy$$

$$= 6 \int_{-1}^{1} \sqrt{1 - y^2}\, dy - \frac{2}{2} \left(y - \frac{y^3}{3} \Bigg|_{-1}^{1} \right) - 2 \int_{-1}^{1} y\sqrt{1 - y^2}\, dy$$

$$= 6 \left\{ \frac{y\sqrt{1 - y^2}}{2} + \frac{1}{2} \sin^{-1} y \right\}_{-1}^{1} - \left(\frac{4}{3} \right) - 0 \quad \because\ y\sqrt{1 - y^2} \text{ is odd function}$$

$$= 6 \left\{ \frac{1}{2}\, 2\, \frac{\pi}{2} \right\} - \frac{4}{3} = 2 \left(\frac{3\pi}{2} - \frac{2}{3} \right) = \left(3\pi - \frac{4}{3} \right).$$

Example 12. *Prove that the volume enclosed between the cylinders $x^2 + y^2 = 2ax$ and $z^2 = 2ax$ is $\dfrac{128\, a^3}{15}$.*

Sol. The section of the cylinder is a circle $x^2 + y^2 - 2ax = 0$ with centre $(a, 0)$ in the xy plane $\therefore$ $z = \sqrt{2ax}$ is to be integrated over the circle $x^2 + y^2 - 2ax = 0$.

$\therefore$ In 1st quadrant y varies from 0 to $\sqrt{2ax - x^2}$ and x varies from 0 to $2a$.

$$\therefore \quad \text{Required volume} = 4 \int_0^{2a} \int_0^{\sqrt{2ax - x^2}} z\, dy dx$$

$$= 4 \int_0^{2a} \int_0^{\sqrt{2ax - x^2}} \sqrt{2ax}\, dy dx$$

$$= 4 \int_0^{2a} \sqrt{2ax}\, y \Bigg|_0^{\sqrt{2ax - x^2}} dx = 4 \int_0^{2a} \sqrt{2ax}\, \sqrt{2ax - x^2}\, dx$$

$$= 4\sqrt{2a}\int_0^{2a} x\sqrt{2a-x}\, dx$$

Put $2a - x = t^2$ $\quad\therefore\quad dx = -2t\,dt$

$$= 4\sqrt{2a}\int_{\sqrt{2a}}^{0} (2a - t^2)\cdot t(-2t)\,dt$$

when $x = 0$, $t^2 = 2a$

$t = \sqrt{2a}$, when $x = 2a$, $t = 0$

$$= -8\sqrt{2a}\int_{\sqrt{2a}}^{0} (2at^2 - t^4)\,dt$$

$$= -8\sqrt{2a}\left[\frac{2at^3}{3} - \frac{t^5}{5}\right]_{\sqrt{2a}}^{0}$$

$$= -8\sqrt{2a}\left[-\frac{2a}{3}(2a)^{3/2} + \frac{(2a)^{5/2}}{5}\right] = -8\sqrt{2a}\left(\frac{1}{3}-\frac{1}{5}\right)(2a)^{5/2}$$

$$= 8(2a)^3\,\frac{2}{15} = \frac{16}{15}\cdot 8a^3 = \frac{128}{15}a^3.$$

Example 13. *Find the volume common to the cylinders $x^2 + y^2 = a^2$ and $x^2 + z^2 = a^2$.*

(P.T.U., May 2004)

Sol. The section of the cylinder $x^2 + y^2 = a^2$ is the circle $x^2 + y^2 = a^2$ in the xy-plane. In the figure, only one-eighth (in positive octant) of the required volume is shown.

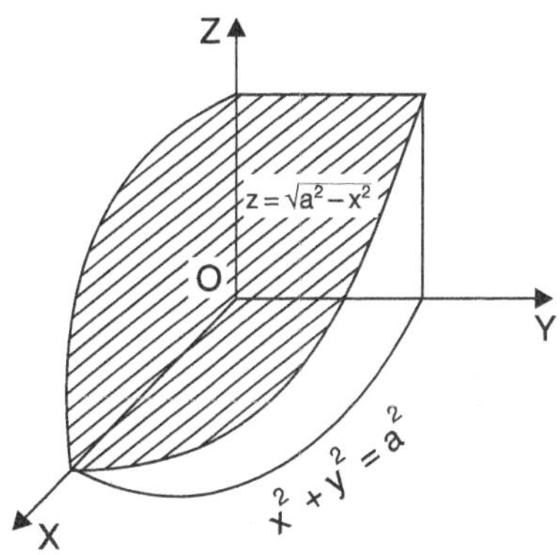

From the figure, it is evident that $z = \sqrt{a^2 - x^2}$ is to be evaluated over the quadrant of the circle $x^2 + y^2 = a^2$ in the first quadrant for which x varies from 0 to a and y varies from 0 to $\sqrt{a^2 - x^2}$.

$$\therefore\quad \text{Required volume} = 8\int_0^a \int_0^{\sqrt{a^2-x^2}} z\,dy\,dx$$

$$= 8\int_0^a \int_0^{\sqrt{a^2-x^2}} \sqrt{a^2-x^2}\,dy\,dx = 8\int_0^a \sqrt{a^2-x^2}\Big[y\Big]_0^{\sqrt{a^2-x^2}} dx$$

$$= 8\int_0^a (a^2 - x^2)\,dx = 8\left[a^2x - \frac{x^3}{3}\right]_0^a = 8\left(a^3 - \frac{a^3}{3}\right) = \frac{16a^3}{3}.$$

Example 14. *Find the volume common to the sphere $x^2 + y^2 + z^2 = a^2$ and the cylinder $x^2 + y^2 = ay$.*

Sol. The required volume is the part of the sphere $x^2 + y^2 + z^2 = a^2$ lying within the cylinder. On account of the symmetry of the sphere, half of it lies above the plane XOY and half below it.

$\therefore$ Required volume $= 2 \iint z \, dy \, dx$

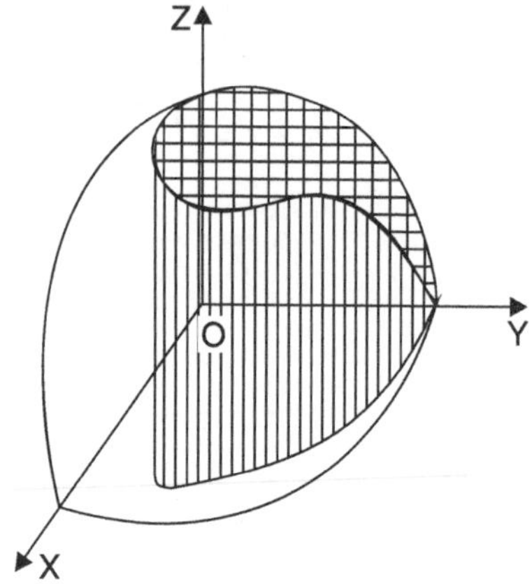

where $z = \sqrt{a^2 - x^2 - y^2}$, and the region of integration is the area inside the circle $x^2 + y^2 = ay$ in the xy-plane. ...(1)

On account of symmetry, the volumes above the two parts of circle (1) in the first and the second quadrants are equal. (The figure shows only the part in the first quadrant).

$$\therefore \text{ Required volume} = 2 \times 2 \iint_R \sqrt{a^2 - x^2 - y^2} \, dy \, dx$$

where R is the half of the circle (1) lying in the first quadrant.

Changing to polar co-ordinates by putting $x = r\cos\theta$, $y = r\sin\theta$

so that $x^2 + y^2 = r^2$, equation (1) becomes

$r^2 = ar\sin\theta$ or $r = a\sin\theta$.

The region of integration is bounded by

$$r = 0,\ r = a\sin\theta \text{ and } \theta = 0,\ \theta = \frac{\pi}{2}.$$

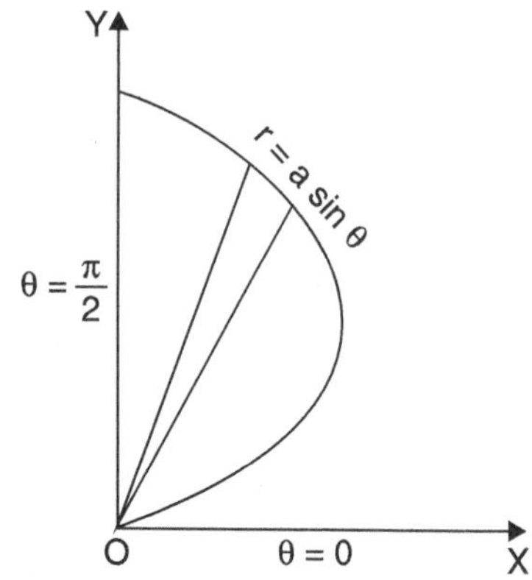

$\therefore$ Required volume

$$= 4\int_0^{\pi/2}\int_0^{a\sin\theta} \sqrt{a^2 - r^2} \cdot r \, dr \, d\theta$$

$$= 4\int_0^{\pi/2}\int_0^{a\sin\theta} -\frac{1}{2}(a^2 - r^2)^{1/2}(-2r) \, dr \, d\theta$$

$$= 4\int_0^{\pi/2} -\frac{1}{2}\cdot\left[\frac{(a^2 - r^2)^{3/2}}{3/2}\right]_0^{a\sin\theta} d\theta$$

$$= -\frac{4}{3}\int_0^{\pi/2}(a^3\cos^3\theta - a^3) = -\frac{4a^3}{3}\left[\frac{2}{3} - \frac{\pi}{2}\right] = \frac{2}{9}a^3(3\pi - 4).$$

Example 15. *A triangular prism is formed by the planes whose equations are $ay = bx$, $y = 0$ and $x = a$, prove that the volume of this prism between the plane $z = 0$ and the surface $z = c + xy$ is $\frac{ab}{8}(4c + ab)$.*

Sol. Prism is bounded between $z = 0$ and $z = c + xy$ and is formed by the planes $ay = bx$, $y = 0$ and $x = a$.

$\therefore$ Required volume $= \iint z \, dydx$ where $z = c + xy$ and y varies from 0 to $\frac{bx}{a}$ and x varies from 0 to a.

$$\therefore \qquad \text{Volume} = \int_{x=0}^{a}\int_{y=0}^{\frac{bx}{a}} (c + xy)\, dydx$$

$$= \int_0^a \left. cy + \frac{xy^2}{2} \right|_0^{\frac{bx}{a}} dx = \int_0^a \left(\frac{bc}{a}x + \frac{x}{2}\cdot\frac{b^2x^2}{a^2}\right)dx$$

$$= \int_0^a \left(\frac{bc}{a} x + \frac{b^2}{2a^2} x^3 \right) dx = \frac{bc}{a} \cdot \frac{x^2}{2} + \frac{b^2}{2a^2} \cdot \frac{x^4}{4} \Bigg|_0^a$$

$$= \frac{bc}{2a} \cdot a^2 + \frac{b^2}{8a^2} \cdot a^4 = \frac{abc}{2} + \frac{a^2b^2}{8} = \frac{ab}{8}(4c + ab).$$

TEST YOUR KNOWLEDGE

1. Find, by double integration, the area enclosed by the ellipse $\frac{x^2}{a^2} + \frac{y^2}{b^2} = 1$.
2. Find, by double integration, the smaller of the areas bounded by the circle $x^2 + y^2 = 9$ and the line $x + y = 3$.
3. Show, by double integration, that the area between the parabolas $y^2 = 4ax$ and $x^2 = 4ay$ is $\frac{16}{3} a^2$.
4. Find the area bounded by the parabola $y = x^2$ and the line $y = 2x + 3$.
5. Find the area bounded by the parabolas $y^2 = 4 - x$ and $y^2 = 4 - 4x$. **(P.T.U., Dec. 2013)**
6. Find the area bounded by the circles $r = 2 \sin \theta$ and $r = 4 \sin \theta$.
   [**Hint.** Circles are $x^2 + y^2 - 2y = 0$ and $x^2 + y^2 - 4y = 0$]
7. Find, by double integration, the volume of the sphere $x^2 + y^2 + z^2 = 9$.
8. Find the volume bounded by the xy-plane, the paraboloid $2z = x^2 + y^2$ and the cylinder $x^2 + y^2 = 4$.
   [**Hint.** See S.E. 10]
9. Find the volume of the region bounded by $z = x^2 + y^2$, $z = 0$, $x = -a$, $x = a$ and $y = -a$, $y = a$.
10. Find the volume bounded by the plane $z = 0$, surface $z = x^2 + y^2 + 2$ and the cylinder $x^2 + y^2 = 4$.

    [**Hint.** Req. volume $= 4 \int_0^2 \int_{y=0}^{\sqrt{4-x^2}} (x^2 + y^2 + 2)\, dy\, dx$]
11. Find the volume under the plane $z = x + y$ and above the area cut from the first quadrant by the ellipse $4x^2 + 9y^2 = 36$.

    [**Hint.** Req. volume $= \int_{x=0}^{3} \int_{y=0}^{\frac{2}{3}\sqrt{9-x^2}} (x + y)\, dy\, dx$]
12. Find the volume of the cylinder $x^2 + y^2 - 2ax = 0$ intercepted between the paraboloid $x^2 + y^2 = 2az$ and the xy-plane.
    [**Hint.** See S.E. 10]
13. Find the volume bounded by the cylinder $x^2 + y^2 = 4$ and the hyperboloid $x^2 + y^2 - z^2 = 1$.

Answers

1. πab	**2.** $\frac{9}{4}(\pi - 2)$	**4.** $10\frac{2}{3}$	**5.** 8
6. 3π	**7.** 36π	**8.** 4π	**9.** $\frac{8a^4}{3}$
10. 16π	**11.** 10	**12.** $\frac{3}{4}\pi a^3$	**13.** $4\sqrt{3}\pi$.

6.9. VOLUME BY TRIPLE INTEGRATION

The volume V of a three dimensional region is given by $V = \iiint_V dx\,dy\,dz$.

If the region is bounded by $x = f_1(y, z), x = f_2(y, z)$; $y = \phi_1(z), y = \phi_2(z)$ and $z = a, z = b$, then

$$V = \int_a^b \int_{\phi_1(z)}^{\phi_2(z)} \int_{f_1(y,z)}^{f_2(y,z)} dx\,dy\,dz$$

The order of integration may be changed with a suitable change in the limits of integration.

In cylindrical co-ordinates, we have $V = \iiint_V r\,dr\,d\phi\,dz$

In spherical polar co-ordinates, we have $V = \iiint_V r^2 \sin\theta\,dr\,d\theta\,d\phi$.

ILLUSTRATIVE EXAMPLES

Example 1. *Find by triple integration, the volume of the paraboloid of revolution $x^2 + y^2 = 4z$ cut off by the plane $z = 4$.*

Sol. By symmetry, the required volume is 4 times the volume in the positive octant. The volume in the positive octant is bounded on the sides by the zx and yz-planes ; from above by the plane $z = 4$ and below by the curved surface $x^2 + y^2 = 4z$.

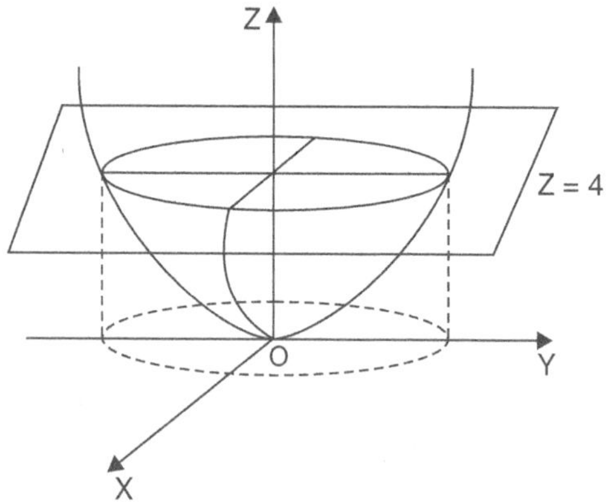

The section of the paraboloid by the plane $z = 4$ is the circle $x^2 + y^2 = 16$, $z = 4$ and its projection on the xy-plane is the circle $x^2 + y^2 = 16$, $z = 0$.

The volume in the positive octant is bounded by $z = \frac{x^2 + y^2}{4}, z = 4, y = 0, y = \sqrt{16 - x^2}$ and $x = 0, x = 4$.

∴ Required volume

$$= 4 \int_0^4 \int_0^{\sqrt{16-x^2}} \int_{\frac{(x^2+y^2)}{4}}^{4} dz\,dy\,dx = 4 \int_0^4 \int_0^{\sqrt{16-x^2}} \Big[z \Big]_{\frac{x^2+y^2}{4}}^{4} dy\,dx$$

$$= 4 \int_0^4 \int_0^{\sqrt{16-x^2}} \left(4 - \frac{x^2 + y^2}{4}\right) dy\,dx = 4 \int_0^4 \left[\left(4 - \frac{x^2}{4}\right) y - \frac{1}{4} \cdot \frac{y^3}{3} \right]_0^{\sqrt{16-x^2}} dx$$

$$= 4 \int_0^4 \left[\left(4 - \frac{x^2}{4}\right) \sqrt{16 - x^2} - \frac{1}{12} (16 - x^2)^{3/2} \right] dx$$

$$= 4\int_0^4 \left[\frac{1}{4}(16 - x^2)\sqrt{16 - x^2} - \frac{1}{12}(16 - x^2)^{3/2}\right] dx$$

$$= 4\int_0^4 \frac{1}{6}(16 - x^2)^{3/2}\, dx = \frac{2}{3}\int_0^4 (16 - x^2)^{3/2}\, dx$$

$$= \frac{2}{3}\int_0^{\pi/2} (16)^{3/2} \cdot \cos^3\theta \cdot 4\cos\theta\, d\theta, \text{where } x = 4\sin\theta$$

$$= \frac{512}{3}\int_0^{\pi/2} \cos^4\theta\, d\theta = \frac{512}{3}\cdot\frac{3\cdot 1}{4\cdot 2}\cdot\frac{\pi}{2} = 32\pi.$$

Example 2. *Find, by triple integration, the volume of the region bounded by the paraboloid* $az = x^2 + y^2$ *and the cylinder* $x^2 + y^2 = R^2$.

Sol. Changing to cylindrical co-ordinates, by putting $x = r\cos\theta$, $y = r\sin\theta$ the equation of the paraboloid becomes $az = r^2$ or $z = \dfrac{r^2}{a}$ and the equation of the cylinder becomes $r^2 = R^2$ or $r = R$. On account of symmetry, the required volume is four times the volume in the positive octant. Thus, in the common region, z varies from 0 to $\dfrac{r^2}{a}$, r varies from 0 to R and θ varies from 0 to $\dfrac{\pi}{2}$.

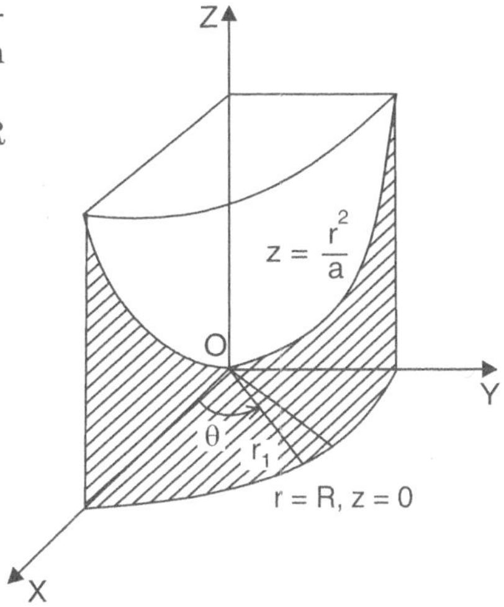

$$\therefore \quad \text{Required volume} = 4\int_0^{\pi/2}\int_0^{R}\int_0^{r^2/a} r\, dz\, dr\, d\theta$$

$$= 4\int_0^{\pi/2}\int_0^{R} r\Big[z\Big]_0^{r^2/a} dr\, d\theta$$

$$= 4\int_0^{\pi/2}\int_0^{R} r\cdot\frac{r^2}{a}\, dr\, d\theta = 4\int_0^{\pi/2}\left[\frac{r^4}{4a}\right]_0^{R} d\theta$$

$$= \frac{1}{a}\int_0^{\pi/2} R^4\, d\theta = \frac{R^4}{a}\cdot\frac{\pi}{2} = \frac{\pi R^4}{2a}.$$

Example 3. *Find, by triple integration, the volume of a sphere of radius a.* **(P.T.U., May 2003, May 2015)**

Sol. Changing to spherical polar co-ordinates by putting $x = r\sin\theta\cos\phi$, $y = r\sin\theta\sin\phi$, $z = r\cos\theta$ so that $x^2 + y^2 + z^2 = r^2$.

The equation of a sphere of radius a in cartesian co-ordinates is $x^2 + y^2 + z^2 = a^2$

The same equation in spherical polar co-ordinates is $r^2 = a^2$ or $r = a$.

On account of symmetry, the required volume is 8 times the volume of the sphere in the positive octant for which r varies from 0 to a, θ varies from 0 to $\dfrac{\pi}{2}$ and ϕ varies from 0 to $\dfrac{\pi}{2}$.

$$\therefore \quad \text{Required volume} = 8\int_0^{\pi/2}\int_0^{\pi/2}\int_0^{a} r^2\sin\theta\, dr\, d\theta\, d\phi$$

$$= 8\int_0^{\pi/2}\int_0^{\pi/2} \sin\theta \cdot \left[\frac{r^3}{3}\right]_0^a d\theta\, d\phi = 8\int_0^{\pi/2}\int_0^{\pi/2} \frac{a^3}{3}\sin\theta\, d\theta\, d\phi$$

$$= 8\int_0^{\pi/2} -\frac{a^3}{3}\left[\cos\theta\right]_0^{\pi/2} d\phi = \frac{8}{3}a^3\int_0^{\pi/2} d\phi = \frac{8}{3}a^3 \cdot \frac{\pi}{2} = \frac{4}{3}\pi a^3.$$

Example 4. *Find the volume of the region enclosed by the surfaces $z = x^2 + 3y^2$, $z = 8 - x^2 - y^2$.* **(P.T.U., May 2003)**

Sol. Equations of the surfaces are

$$z = x^2 + 3y^2,\ z = 8 - x^2 - y^2$$

Their intersections are given by

$$x^2 + 3y^2 = 8 - x^2 - y^2 \quad \text{or} \quad 2x^2 + 4y^2 = 8 \quad \text{or} \quad x^2 + 2y^2 = 4$$

$\therefore$ z varies from $x^2 + 3y^2$ to $8 - x^2 - y^2$

y varies from $-\sqrt{\dfrac{4 - x^2}{2}}$ to $\sqrt{\dfrac{4 - x^2}{2}}$

and x varies from -2 to 2

$$\text{Volume of the Region} = \int_{-2}^{2}\int_{-\sqrt{\frac{4+x^2}{2}}}^{\sqrt{\frac{4-x^2}{2}}}\int_{x^2+3y^2}^{8-x^2-y^2} dz\, dy\, dx$$

$$= \int_{-2}^{2}\int_{-\sqrt{\frac{4-x^2}{2}}}^{\sqrt{\frac{4-x^2}{2}}} (8 - x^2 - y^2 - x^2 - 3y^2)\, dy\, dx$$

$$= \int_{-2}^{2}\int_{-\sqrt{\frac{4-x^2}{2}}}^{\sqrt{\frac{4-x^2}{2}}} (8 - 2x^2 - 4y^2)\, dy\, dx$$

$$= \int_{-2}^{2} 8y - 2x^2y - \frac{4y^3}{3}\Bigg|_{-\sqrt{\frac{4-x^2}{2}}}^{\sqrt{\frac{4-x^2}{2}}} dx = \int_{-2}^{2} (8 - 2x^2)\, 2\sqrt{\frac{4 - x^2}{2}} - \frac{4}{3}\cdot 2\left(\frac{4 - x^2}{2}\right)^{3/2} dx$$

$$= \int_{-2}^{2} \frac{8}{3\sqrt{2}}(4 - x^2)^{3/2}\, dx \quad \text{Put } x = 2\sin\theta;\ dx = 2\cos\theta\, d\theta$$

$$= \int_{-\frac{\pi}{2}}^{\frac{\pi}{2}} \frac{8}{3\sqrt{2}}(4 - 4\sin^2\theta)^{3/2} \cdot 2\cos\theta\, d\theta$$

$$= \frac{128}{3\sqrt{2}}\int_{-\pi/2}^{\pi/2} \cos^4\theta\, d\theta = \frac{128}{3\sqrt{2}}\, 2\int_0^{\frac{\pi}{2}} \cos^4\theta\, d\theta = \frac{256}{3\sqrt{2}}\,\frac{3\cdot 1}{4\cdot 2}\,\frac{\pi}{2} = \frac{16}{\sqrt{2}}\pi = 8\sqrt{2}\,\pi.$$

6.10. (a) VOLUMES OF SOLIDS OF REVOLUTION

Consider an elementary area $\delta x \delta y$ at the point P(x, y) of a plane area R. If this elementary area is revolved about the x-axis, it will generate a ring of radius y. The elementary volume δV of this ring is given by

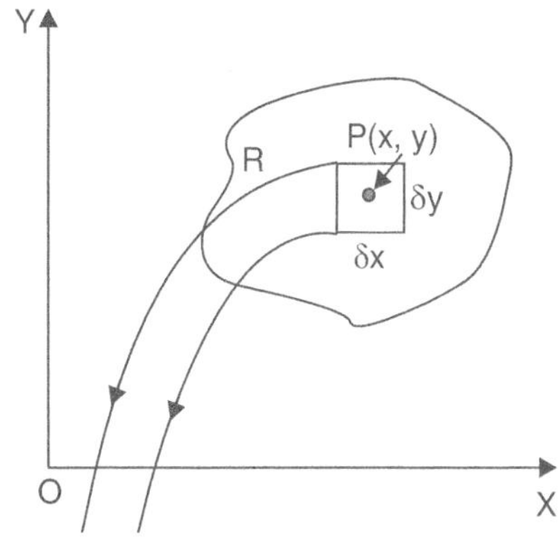

$$\delta V = 2\pi y \,.\, \delta y \, \delta x$$

$$\therefore \qquad V = 2\pi \iint_R y \, dy \, dx \qquad ...(1)$$

which is the total volume of the solid formed by the revolution of the area R about x-axis.

Similarly, if the area R revolves about y-axis, then the required volume V is given by

$$V = 2\pi \iint_R x \, dx \, dy \qquad ...(2)$$

Changing to polar co-ordinates by putting

$x = r \cos \theta$, $y = r \sin \theta$ and $dy\, dx = r\, dr\, d\theta$

Formula (1) becomes

$$V = 2\pi \iint_R r \sin \theta \cdot r \, dr \, d\theta = 2\pi \iint_R r^2 \sin \theta \, dr \, d\theta$$

Formula (2) becomes $V = 2\pi \iint_R r^2 \cos \theta \, dr \, d\theta$.

6.10. (b) SURFACE AREA BY DOUBLE INTEGRATION

If $z = f(x, y)$ be the surface, then

$$\text{Surface area} = \iint \sqrt{1 + \left(\frac{\partial z}{\partial x}\right)^2 + \left(\frac{\partial z}{\partial y}\right)^2} \, dx \, dy \, .$$

Example 5. *Find the volume generated by revolving the area enclosed by loop of the curve $y^4 = x(4 - x)$ about x-axis.* **(P.T.U., May 2008)**

Sol. First of all we will have a rough sketch of the curve

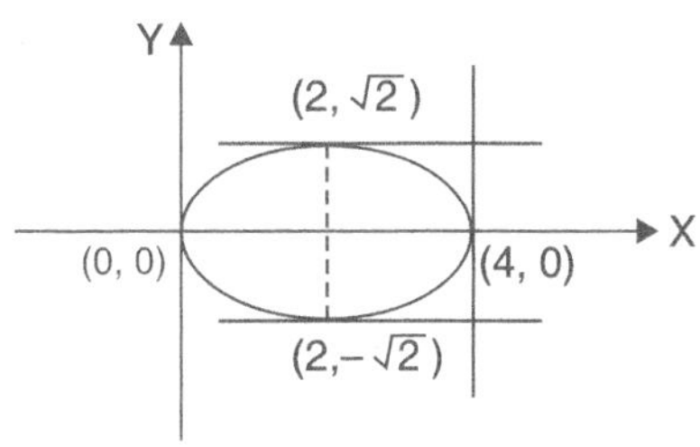

(*i*) curve is symmetric about x-axis

(*ii*) curve passes through the origin and tangent at the origin is $x = 0$ *i.e.*, y-axis

(*iii*) curve intersect x-axis at (0, 0) and (4, 0)

(*iv*) $4y^3 \dfrac{dy}{dx} = 4 - 2x$ or $\dfrac{dy}{dx} = \dfrac{2 - x}{2y^3}$

Tangent is parallel to x-axis where $\frac{dy}{dx} = 0$ *i.e.*, at $x = 2$ *i.e.*, at $(2, \pm\sqrt{2})$ and tangent is parallel to y-axis at $y = 0$ *i.e.*, at $x = 0$ and $x = 4$ *i.e.*, at $(0, 0)$ and $(4, 0)$

$\therefore$ Required volume

$$= 2\pi \int_0^4 \int_0^{[x(4-x)]^{\frac{1}{4}}} y\, dy\, dx = 2\pi \int_0^4 \frac{y^2}{2}\Big|_0^{[x(4-x)]^{\frac{1}{4}}} dx$$

$$= \pi \int_0^4 \sqrt{x(4-x)}\, dx$$

Put $x = 4\sin^2\theta \quad \therefore \quad dx = 8 \sin\theta \cos\theta\, d\theta$

$$V = \pi \int_0^{\frac{\pi}{2}} \sqrt{4\sin^2\theta \cdot 4\cos^2\theta} \cdot 8 \sin\theta \cos\theta\, d\theta$$

$$= 32\pi \int_0^{\frac{\pi}{2}} \sin^2\theta \cos^2\theta\, d\theta$$

$$= 32\pi \frac{1.1}{4.2} \frac{\pi}{2} = 2\pi^2. \quad \text{Using } \int_0^{\frac{\pi}{2}} \sin^p\theta \cos^q\theta\, d\theta$$

$$= \left[\frac{[(p-1)(p-3)\ldots]\cdot[(q-1)(q-3)\ldots]}{(p+q)(p+q-2)\ldots} \frac{\pi}{2} \; (p, q \text{ both even}) \right]$$

Example 6. *Calculate, by double integration, the volume generated by the revolution of the cardioid $r = a(1 - \cos\theta)$ about its axis.* **(P.T.U., May 2008, Dec. 2011)**

Sol. Since the upper and the lower halves of the cardioid generate the same volume, we consider the revolution of the upper half only for which r varies from 0 to $a(1 - \cos\theta)$ and θ varies from 0 to π.

$\therefore$ Required volume of revolution

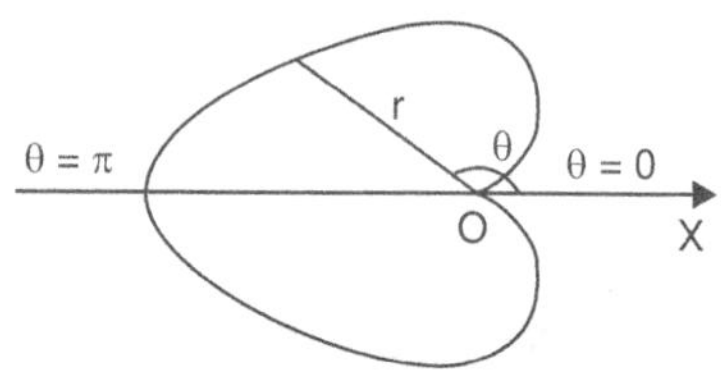

$$= 2\pi \int_0^{\pi} \int_0^{a(1-\cos\theta)} r^2 \sin\theta\, dr\, d\theta$$

$$= 2\pi \int_0^{\pi} \sin\theta \left[\frac{r^3}{3} \right]_0^{a(1-\cos\theta)} d\theta$$

$$= \frac{2\pi}{3} \int_0^{\pi} \sin\theta \cdot a^3 (1 - \cos\theta)^3\, d\theta$$

$$= \frac{2\pi a^3}{3} \int_0^{\pi} (1 - \cos\theta)^3 \sin\theta\, d\theta$$

$$= \frac{2\pi a^3}{3} \left[\frac{(1-\cos\theta)^4}{4} \right]_0^{\pi} = \frac{\pi a^3}{6} [(1 - \cos\pi)^4]$$

$$= \frac{\pi a^3}{6} \cdot 2^4 = \frac{8}{3} \pi a^3.$$

Example 7. *Find the area of the portion of the cylinder $x^2 + y^2 = 4y$ lying inside the sphere $x^2 + y^2 + z^2 = 16$.*

Sol. Equation of the sphere is $x^2 + y^2 + z^2 = 16$

Differentiate partially w.r.t. x and y

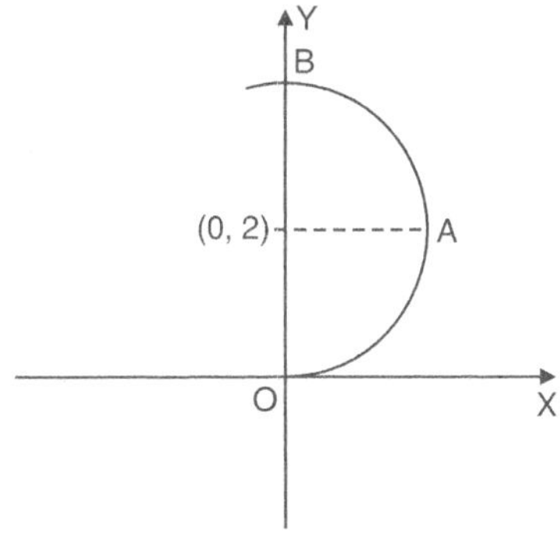

$$2x + 2z\frac{\partial z}{\partial x} = 0$$

$$\therefore \quad \frac{\partial z}{\partial x} = -\frac{x}{z}. \text{ Similarly, } \frac{\partial z}{\partial y} = -\frac{y}{z}$$

$\therefore$ Required surface area = 4 times the surface area in semi circle OAB

$$\therefore \quad \text{Surface area} = 4\iint \sqrt{1+\left(\frac{\partial z}{\partial x}\right)^2+\left(\frac{\partial z}{\partial y}\right)^2}\, dx\, dy$$

$$= 4\iint \sqrt{1+\frac{x^2}{z^2}+\frac{y^2}{z^2}}\, dx\, dy$$

$$= 4\iint \sqrt{\frac{x^2+y^2+z^2}{z^2}}\, dx\, dy = 4\iint \frac{4}{\sqrt{16-x^2-y^2}}\, dxdy$$

Change it into polar coordinates $x = r\cos\theta$, $y = r\sin\theta$, $dxdy = r\,dr\,d\theta$

$$x^2 + y^2 = 4y \Rightarrow r^2 = 4r\sin\theta \Rightarrow r = 0, r = 4\sin\theta$$

$\therefore$ r varies from 0 to $4\sin\theta$ and θ varies from 0 to $\frac{\pi}{2}$

$$\therefore \quad \text{Surface area} = 16\int_0^{\frac{\pi}{2}}\int_0^{4\sin\theta} \frac{1}{\sqrt{16-r^2}}\, r\, dr\, d\theta = -8\int_0^{\frac{\pi}{2}}\int_0^{4\sin\theta} \frac{-2r}{\sqrt{16-r^2}}\, dr\, d\theta$$

$$= -8\int_0^{\frac{\pi}{2}} \left.\frac{(16-r^2)^{1/2}}{1/2}\right|_0^{4\sin\theta} d\theta \text{ , using } \int [f(x)]^n f'(x)\, dx = \frac{[f(x)]^{n+1}}{n+1}, n \neq -1$$

$$= -16\int_0^{\frac{\pi}{2}} [[16-16\sin^2\theta]^{1/2} - 4]d\theta = -16\int_0^{\frac{\pi}{2}} [4\cos\theta - 4]\, d\theta$$

$$= -16\left[4\sin\theta - 4\theta \Big|_0^{\frac{\pi}{2}}\right] = -16\,[4 - 2\pi] = 32\pi - 64.$$

Example 8. *Show that surface area of the sphere $x^2 + y^2 + z^2 = a^2$ is $4\pi a^2$.*

Sol. Equation of the sphere is $x^2 + y^2 + z^2 = a^2$

Differentiate partially w.r.t. x and y

$$2x + 2z\frac{\partial z}{\partial x} = 0 \quad \therefore \quad \frac{\partial z}{\partial x} = -x/z \; ; \; \frac{\partial z}{\partial y} = -\frac{y}{z}$$

Required surface area is 8 times the surface area in 1st quadrant

$$\therefore \qquad \text{Surface area} = 8 \iint \sqrt{1 + \left(\frac{\partial z}{\partial x}\right)^2 + \left(\frac{\partial z}{\partial y}\right)^2}\, dxdy$$

$$= 8 \iint \sqrt{1 + \frac{x^2}{z^2} + \frac{y^2}{z^2}}\, dx\, dy = 8 \iint \sqrt{\frac{x^2 + y^2 + z^2}{z}}\, dx\, dy$$

$$= 8 \iint \frac{a}{\sqrt{a^2 - x^2 - y^2}}\, dx\, dy$$

Change it to polar co-ordinates $x = r \cos \theta, y = r \sin \theta; dx\, dy = r\, dr\, d\theta$

$$= 8a \int_{\theta=0}^{\frac{\pi}{2}} \int_{r=0}^{a} \frac{1}{\sqrt{a^2 - r^2}}\, r\, dr\, d\theta$$

$$= 8a \int_0^{\frac{\pi}{2}} \int_0^{a} \left(-\frac{1}{2}\right) (a^2 - r^2)^{-1/2} (-2r)\, dr\, d\theta \;\Bigg|\; \text{using} \int [f(x)]^n f'(x)\, dx = \frac{[f(x)]^{n+1}}{n+1}, n \neq -1$$

$$= -4a \int_0^{\frac{\pi}{2}} \left. \frac{(a^2 - r^2)^{1/2}}{1/2} \right|_0^a d\theta = -8a \int_0^{\frac{\pi}{2}} (0 - a)\, d\theta$$

$$= 8a^2 \cdot \theta \Big|_0^{\frac{\pi}{2}} = 8a^2 \frac{\pi}{2} = 4\pi\, a^2.$$

TEST YOUR KNOWLEDGE

1. (*a*) Find, by triple integration, the volume in the positive octant bounded by the co-ordinate planes and the plane $x + 2y + 3z = 4$.

$$\left[\textbf{Hint. } \text{Reqd. vol} = \int_{x=0}^{4} \int_{y=0}^{\frac{4-x}{2}} \int_{z=0}^{\frac{4-x-2y}{3}} dzdydx\right]$$

(*b*) Using triple integration find the volume of the tetrahedron bounded by the co-ordinate plane and the plane $\frac{x}{a} + \frac{y}{b} + \frac{z}{c} = 1$. **(*P.T.U. Dec. 2015*)**

$$\left[\textbf{Hint. } \text{Reqd. volume} = \int_{x=0}^{a} \int_{y=0}^{b\left(1-\frac{x}{a}\right)} \int_{z=0}^{\left(1-\frac{x}{a}-\frac{y}{b}\right)} dz\, dy\, dx\right]$$

2. Find the volume of the portion of the sphere $x^2 + y^2 + z^2 = a^2$ lying inside the cylinder $x^2 + y^2 = ax$.
3. Find the volume bounded above by the sphere $x^2 + y^2 + z^2 = 2a^2$ and below by the paraboloid $az = x^2 + y^2$.

[**Hint.** Intersection of sphere and paraboloid is $x^2 + y^2 + \left(\frac{x^2 + y^2}{a}\right)^2 = 2a^2$

or $(x^2 + y^2 - a^2)(x^2 + y^2 + 2a^2) = 0$ *i.e.*, $x^2 + y^2 = a^2$; change to cylindrical coordinates

$$V = \int_{\theta=0}^{2\pi} \int_{r=0}^{\sqrt{2}a} \int_{z=\frac{r^2}{a}}^{\sqrt{2a^2 - r^2}} dz\, (r\, dr\, d\theta)]$$

4. Find, by double integration, the volume of the solid generated by revolving the ellipse $\frac{x^2}{a^2} + \frac{y^2}{b^2} = 1$ about the x-axis.

[**Hint.** $V = \int_{-a}^{a} \int_{0}^{\frac{b}{a}\sqrt{a^2 - x^2}} y\, dy\, dx$] **(*P.T.U., Jan. 2010*)**

5. Find, by double integration, the volume generated by revolving the cardioid $r = a(1 + \cos\theta)$ about the initial line. **(*P.T.U., Dec. 2015*)**

[**Hint.** See S.E. 2]

6. Find the area of the portion of the cylinder $x^2 + z^2 = 4$ lying inside the cylinder $x^2 + y^2 = 4$.

[**Hint.** Reqd. surface area $= 8\int_0^2 \int_0^{\sqrt{4-x^2}} \frac{2}{\sqrt{4-x^2}}\, dx\, dy$]

Answers

1. (*a*) $\frac{16}{9}$ (*b*) $\frac{1}{6}abc$ **2.** $\frac{2a^3}{9}(3\pi - 4)$ **3.** $\left(\frac{4}{3}\sqrt{2} - \frac{7}{6}\right)\pi a^3$

4. $\frac{4}{3}\pi ab^2$ **5.** $\frac{8}{3}\pi a^3$ **6.** 32.

6.11. CALCULATION OF MASS

(*a*) *For a plane lamina* of area A, if the density at a point P(x, y) be $\rho = f(x, y)$, then its total mass M is given by

$$M = \iint_A \rho\, dx\, dy = \iint_A f(x, y)\, dx\, dy$$

In polar co-ordinates, if $\rho = f(r, \theta)$ at a point P(r, θ), then

$$M = \iint_A \rho\, r\, dr\, d\theta = \iint_A f(r, \theta)\, r\, dr\, d\theta.$$

(*b*) *For a solid* of volume V, if the density at a point P(x, y, z) be $\rho = f(x, y, z)$, then its total mass M is given by

$$M = \iiint_V \rho\, dx\, dy\, dz = \iiint_V f(x, y, z)\, dx\, dy\, dz.$$

ILLUSTRATIVE EXAMPLES

Example 1. *Find the mass of a plate in the shape of the curve* $\left(\frac{x}{a}\right)^{2/3} + \left(\frac{y}{b}\right)^{2/3} = 1$, *the density being given by* $\rho = \mu\, xy$.

Sol. Let the required mass be M which is four times the mass in the first quadrant.

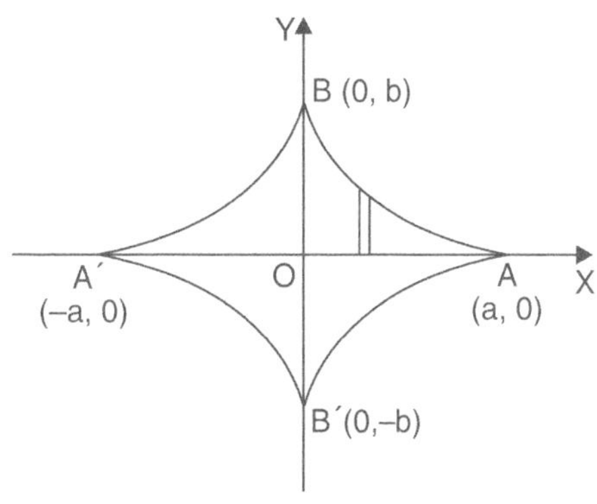

From the equation of the curve

$$\left(\frac{y}{b}\right)^{2/3} = 1 - \left(\frac{x}{a}\right)^{2/3}$$

or $$y = b\left[1-\left(\frac{x}{a}\right)^{2/3}\right]^{3/2} = y_1 \quad \text{(say)}$$

For the region OAB, x varies from 0 to a and y varies from 0 to y_1.

$$\therefore \quad M = 4\int_0^a\int_0^{y_1} \rho\, dy\, dx = 4\int_0^a\int_0^{y_1} \mu\, xy\, dy\, dx$$

$$= 4\int_0^a \mu x\cdot\left[\frac{y^2}{2}\right]_0^{y_1} dx = 2\mu\int_0^a xy_1^2\, dx$$

$$= 2\mu\int_0^a xb^2\left[1-\left(\frac{x}{a}\right)^{2/3}\right]^3 dx$$

Put $x = a\sin^3\theta$, then $dx = 3a\sin^2\theta\cos\theta\, d\theta$

When $x = 0, \theta = 0$; when $x = a, \theta = \frac{\pi}{2}$

$$\therefore \quad M = 2\mu b^2\int_0^{\pi/2} a\sin^3\theta\,(1-\sin^2\theta)^3\cdot 3a\sin^2\theta\cos\theta\, d\theta$$

$$= 6\mu a^2b^2\int_0^{\pi/2}\sin^5\theta\cos^7\theta\, d\theta = 6\mu a^2b^2\cdot\frac{4\cdot2\cdot6\cdot4\cdot2}{12\cdot10\cdot8\cdot6\cdot4\cdot2} = \frac{\mu a^2b^2}{20}.$$

Example 2. *Find the mass of a lamina in the form of the cardioid r = a(1 + cos θ) whose density at any point varies as the square of its distance from the initial line.*

Sol. Let the required mass be M which is twice the mass above the initial line.

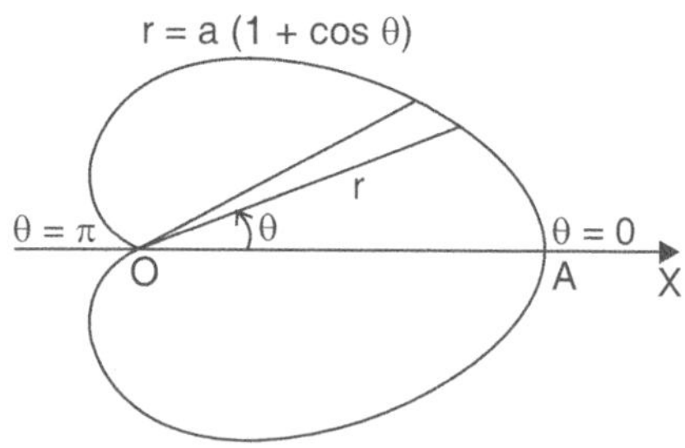

Since the distance of any point (r, θ) from the initial line is $r\sin\theta$, the density at (r, θ) is given by

$$\rho = \mu\,(r\sin\theta)^2 = \mu\, r^2\sin^2\theta.$$

For the region above the initial line, θ varies from 0 to π and r varies from 0 to $a\,(1+\cos\theta)$.

$$\therefore \quad M = 2\int_0^{\pi}\int_0^{a(1+\cos\theta)} \rho\, r\, dr\, d\theta$$

$$= 2\int_0^{\pi}\int_0^{a(1+\cos\theta)} \mu\, r^3\sin^2\theta\, dr\, d\theta$$

$$= 2\int_0^{\pi}\mu\sin^2\theta\cdot\left[\frac{r^4}{4}\right]_0^{a(1+\cos\theta)} d\theta = \frac{\mu}{2}\int_0^{\pi}\sin^2\theta\cdot a^4\,(1+\cos\theta)^4\, d\theta$$

$$= \frac{4a^4}{2}\int_0^{\pi}\left(2\sin\frac{\theta}{2}\cos\frac{\theta}{2}\right)^2\cdot\left(2\cos^2\frac{\theta}{2}\right)^4 d\theta = 32\mu a^4\int_0^{\pi}\sin^2\frac{\theta}{2}\cos^{10}\frac{\theta}{2}\, d\theta$$

Put $\frac{\theta}{2} = t$, then $d\theta = 2dt$

When $\theta = 0, t = 0$; when $\theta = \pi, t = \frac{\pi}{2}$

$$\therefore \quad M = 32\mu a^4\int_0^{\pi/2} 2\sin^2 t\cos^{10} t\, dt = 64\mu a^4\cdot\frac{1\cdot9\cdot7\cdot5\cdot3\cdot1}{12\cdot10\cdot8\cdot6\cdot4\cdot2}\cdot\frac{\pi}{2} = \frac{21}{32}\mu\pi a^4.$$

Example 3. *Find the mass of the tetrahedron bounded by the co-ordinate planes and the plane* $\frac{x}{a}+\frac{y}{b}+\frac{z}{c}=1$, *the variable density* $\rho = kxyz$.

Sol. The tetrahedron OABC is bounded by the planes $x = 0, y = 0, z = 0$ and $\frac{x}{a}+\frac{y}{b}+\frac{z}{c}=1$.

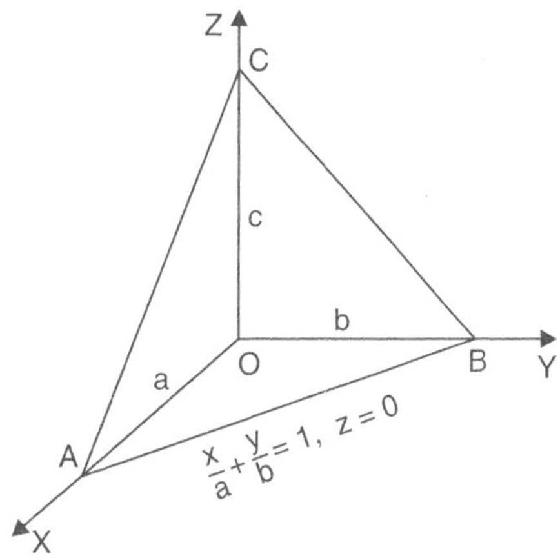

For the whole volume of this tetrahedron, x varies from 0 to $a\left(1-\frac{y}{b}-\frac{z}{c}\right)$, y varies from 0 to $b\left(1-\frac{z}{c}\right)$ and z varies from 0 to c.

$\therefore$ Required mass

$$M = \int_0^c\int_0^{b(1-z/c)}\int_0^{a(1-y/b-z/c)} \rho\, dx\, dy\, dz$$

$$= \int_0^c\int_0^{b(1-z/c)}\int_0^{a(1-y/b-z/c)} k\, xyz\, dx\, dy\, dz$$

$$= \int_0^c\int_0^{b(1-z/c)} kyz\cdot\left[\frac{x^2}{2}\right]_0^{a(1-y/b-z/c)} dy\, dz$$

$$= \int_0^c\int_0^{b(1-z/c)} \frac{kyz}{2}\cdot a^2\left(1-\frac{y}{b}-\frac{z}{c}\right)^2 dy\, dz$$

$$= \int_0^c\int_0^{b(1-z/c)} \frac{ka^2z}{2}\cdot y\left[\left(1-\frac{z}{c}\right)^2-2\left(1-\frac{z}{c}\right)\frac{y}{b}+\frac{y^2}{b^2}\right]dy\, dz$$

$$= \int_0^c\int_0^{b(1-z/c)} \frac{ka^2z}{2}\left[\left(1-\frac{z}{c}\right)^2 y-\frac{2}{b}\left(1-\frac{z}{c}\right)y^2+\frac{y^3}{b^2}\right]dy\, dz$$

$$= \int_0^c \frac{ka^2z}{2}\left[\left(1-\frac{z}{c}\right)^2\cdot\frac{y^2}{2}-\frac{2}{b}\left(1-\frac{z}{c}\right)\cdot\frac{y^3}{3}+\frac{y^4}{4b^2}\right]_0^{b(1-z/c)} dz$$

$$= \frac{ka^2}{2}\int_0^c z\left[\frac{b^2}{2}\left(1-\frac{z}{c}\right)^4-\frac{2b^2}{3}\left(1-\frac{z}{c}\right)^4+\frac{b^2}{4}\left(1-\frac{z}{c}\right)^4\right]dz = \frac{ka^2b^2}{24}\int_0^c z\left(1-\frac{z}{c}\right)^4 dz.$$

Put $\quad z = c\sin^2\theta$, then $dz = 2c\sin\theta\cos\theta\, d\theta$

When $\;z = 0, \theta = 0$; when $z = c$, $\theta = \frac{\pi}{2}$

$$\therefore \quad M = \frac{ka^2b^2}{24}\int_0^{\pi/2} c\sin^2\theta\,(1-\sin^2\theta)^4\cdot 2c\sin\theta\cos\theta\, d\theta = \frac{k\,a^2b^2c^2}{12}\int_0^{\pi/2}\sin^3\theta\cos^9\theta\, d\theta$$

$$= \frac{k\,a^2b^2c^2}{12}\cdot\frac{2\cdot 8\cdot 6\cdot 4\cdot 2}{12\cdot 10\cdot 8\cdot 6\cdot 4\cdot 2} = \frac{k\,a^2b^2c^2}{720}.$$

6.12. CENTRE OF GRAVITY (C.G.)

(*a*) *The C.G.* $(\bar{x}, \bar{y})$ *of a plane lamina* occupying an area A in the *xy* plane and having density $\rho = f(x, y)$ is given by

$$\bar{x} = \frac{\iint_A x\rho\, dx\, dy}{\iint_A \rho\, dx\, dy}, \quad \bar{y} = \frac{\iint_A y\rho\, dx\, dy}{\iint_A \rho\, dx\, dy}$$

In polar co-ordinates, if $\rho = f(r, \theta)$ at a point $P(r, \theta)$, then the C.G. $(\bar{x}, \bar{y})$ is given by

$$\bar{x} = \frac{\iint_A r\cos\theta \,.\, \rho r\, dr\, d\theta}{\iint_A \rho r\, dr\, d\theta} = \frac{\iint_A r^2 \cos\theta\, \rho d\, r\, d\theta}{\iint_A \rho\, r\, dr\, d\theta}$$

$$\bar{y} = \frac{\iint_A r\sin\theta\, \rho r\, dr\, d\theta}{\iint_A \rho r\, dr\, d\theta} = \frac{\iint_A r^2 \sin\theta\, \rho\, dr\, d\theta}{\iint_A \rho r\, dr\, d\theta}$$

(*b*) *The C.G.*$(\bar{x}, \bar{y}, \bar{z})$ *of a solid* occupying a volume V and having density $\rho = f(x, y, z)$ is given by **(P.T.U., May 2004)**

$$\bar{x} = \frac{\iiint_V x\rho\, dx\, dy\, dz}{\iint_V \rho\, dx\, dy\, dz}, \quad \bar{y} = \frac{\iint_V y\rho\, dx\, dy\, dz}{\iint_V \rho\, dx\, dy\, dz}, \quad \bar{z} = \frac{\iiint_V z\rho\, dx\, dy\, dz}{\iiint_V \rho\, dx\, dy\, dz}.$$

Example 4. *Find the C.G. of the area between* $y = 6x - x^2$ *and* $y = x$.

Sol. $y = 6x - x^2$ is a downward parabola $(x-3)^2 = -(y-9)$ having vertex at (3, 9)

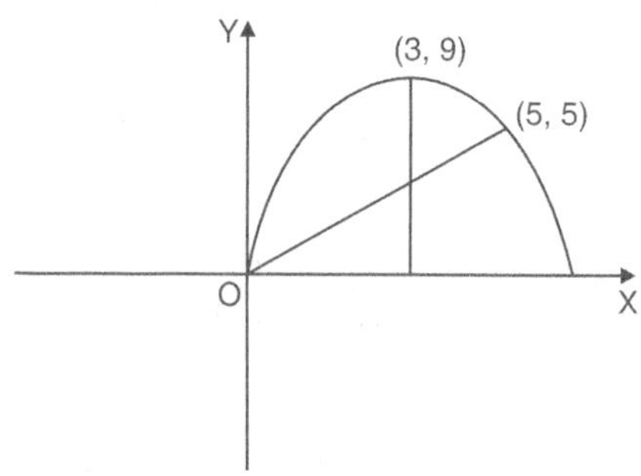

$y = 6x - x^2$ and $y = x$ intersect at $x = 6x - x^2$

i.e., at $x^2 - 5x = 0$

i.e., at $x = 0, x = 5$ *i.e.,* at (0, 0) and (5, 5)

Let the density be ρ (constant) then C.G. of the area is

$$\bar{x} = \frac{\int_0^5 \int_x^{6x-x^2} x\rho\, dy\, dx}{\int_0^5 \int_x^{6x-x^2} \rho\, dy\, dx}$$

$$= \frac{\int_0^5 x\, y\Big|_x^{6x-x^2} dx}{\int_0^5 y\Big|_x^{6x-x^2} dx} = \frac{\int_0^5 x(6x - x^2 - x)\, dx}{\int_0^5 (6x - x^2 - x)\, dx} = \frac{\int_0^5 (5x^2 - x^3)\, dx}{\int_0^5 (5x - x^2)\, dx}$$

$$= \frac{5\dfrac{x^3}{3} - \dfrac{x^4}{4}\Big|_0^5}{5\dfrac{x^2}{2} - \dfrac{x^3}{3}\Big|_0^5} = \frac{\dfrac{625}{3} - \dfrac{625}{4}}{\dfrac{125}{2} - \dfrac{125}{3}} = \frac{625}{12} \times \frac{6}{125} = \frac{5}{2}$$

$$\bar{y} = \frac{\int_0^5 \int_x^{6x-x^2} y\,\rho\,dy\,dx}{\int_0^5 \int_x^{6x-x^2} \rho\,dy\,dx} = \frac{\int_0^5 \left.\frac{y^2}{2}\right|_x^{6x-x^2} dx}{\int_0^5 y\,\Big|_x^{6x-x^2} dx} = \frac{\frac{1}{2}\int_0^5 [(6x-x^2)^2 - x^2]\,dx}{\int_0^5 (6x-x^2-x)\,dx}$$

$$= \frac{\frac{1}{2}\int_0^5 (35x^2 - 12x^3 + x^4)\,dx}{\int_0^5 (5x - x^2)\,dx} = \frac{\frac{1}{2}\left\{\frac{35x^3}{3} - 12\frac{x^4}{4} + \frac{x^5}{5}\Big|_0^5\right\}}{\left(\frac{5x^2}{2} - \frac{x^3}{3}\right)\Big|_0^5}$$

$$= \frac{1}{2}\,\frac{(125)\left[\frac{35}{3} - 15 + 5\right]}{25\left(\frac{5}{2} - \frac{5}{3}\right)} = \frac{5}{2}\,\frac{5/3}{5\cdot\frac{1}{6}} = 5 \quad \therefore \quad (\bar{x}, \bar{y}) = \left(\frac{5}{2}, 5\right).$$

Example 5. *Find by the double integration the C.G. of the area of the cardioid $r = a\,(1 + \cos\theta)$.*

(P.T.U., Dec. 2007)

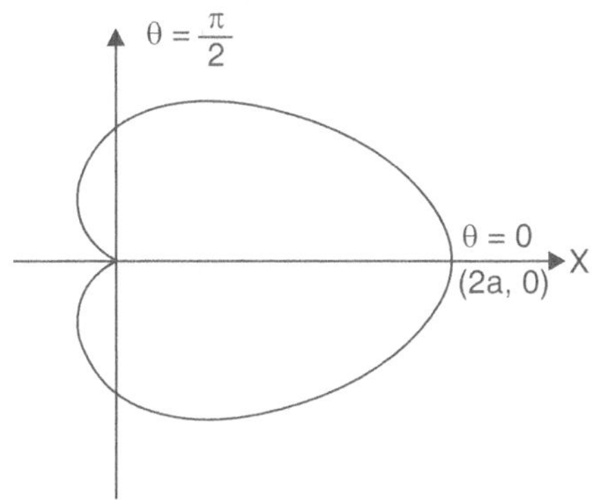

Sol. Equation of the cardioid is $r = a\,(1 + \cos\theta)$.

It is symmetric about initial line *i.e.*, x-axis $\therefore\ \bar{y} = 0$.

Let ρ be the density. For the cardioid r varies from 0 to $a\,(1 + \cos\theta)$ and θ varies from $-\pi$ to π

$$\bar{x} = \frac{\int_{-\pi}^{\pi} \int_0^{a(1+\cos\theta)} (r\cos\theta)\rho\,r\,dr\,d\theta}{\int_{-\pi}^{\pi} \int_0^{a(1+\cos\theta)} \rho\,r\,dr\,d\theta} = \frac{\int_{-\pi}^{\pi} \cos\theta \left.\frac{r^3}{3}\right|_0^{a(1+\cos\theta)} d\theta}{\int_{-\pi}^{\pi} \left.\frac{r^2}{2}\right|_0^{a(1+\cos\theta)} d\theta}$$

$$= \frac{\frac{1}{3}\int_{-\pi}^{\pi} \cos\theta \cdot a^3 (1+\cos\theta)^3\,d\theta}{\frac{1}{2}\int_{-\pi}^{\pi} a^2\,(1+\cos\theta)^2\,d\theta} = \frac{2}{3}a\,\frac{2\int_0^{\pi} \cos\theta\,(1 + 3\cos\theta + 3\cos^2\theta + \cos 3\theta)\,d\theta}{2\int_0^{\pi} (1 + 2\cos\theta + \cos^2\theta)\,d\theta}$$

$$= \frac{2}{3}a \cdot \frac{\int_0^{\pi} (\cos\theta + 3\cos^2\theta + 3\cos^3\theta + \cos^4\theta)\,d\theta}{\int_0^{\pi} (1 + 2\cos\theta + \cos^2\theta)\,d\theta}$$

We know that $\int_0^{2a} f(x)\,dx = 2\int_0^a f(x)dx$ if $f(2a - x) = f(x)$

$= 0$ if $f(2a - x) = -f(x)$

$$\therefore \qquad \int_0^{\pi} \cos\theta \, d\theta = \int_0^{\pi} \cos^3\theta \, d\theta = 0$$

$\because$ $\cos(\pi - \theta)$ and $\cos^3(\pi - \theta) = -\cos\theta$ and $-\cos^3\theta$, respectively

$$\bar{x} = \frac{2}{3}a \frac{\int_0^{\pi} (3\cos^2\theta + \cos^4\theta)\, d\theta}{\int_0^{\pi} (1 + \cos^2\theta)\, d\theta} = \frac{2}{3}a \frac{2\int_0^{\frac{\pi}{2}} (3\cos^2\theta + \cos^4\theta)\, d\theta}{2\int_0^{\frac{\pi}{2}} (1 + \cos^2\theta)\, d\theta}$$

$$= \frac{2}{3}a \frac{3\frac{1}{2}\frac{\pi}{2} + \frac{3\cdot 1}{4\cdot 2}\frac{\pi}{2}}{\frac{\pi}{2} + \frac{1}{2}\frac{\pi}{2}} = \frac{5a}{6}$$

Hence C.G. of the area of the cardioid is $\left(\frac{5a}{6}, 0\right)$.

Example 6. *Find the centroid of a loop of the lemniscate $r^2 = a^2 \cos 2\theta$.*

Sol. The loop being symmetrical about the initial line, its C.G. lies on OX, so that $\bar{y} = 0$.

Since the density is not given, let it be k, a constant.

For the loop, θ varies from $-\frac{\pi}{4}$ to $\frac{\pi}{4}$ and r varies from 0 to $a\sqrt{\cos 2\theta}$.

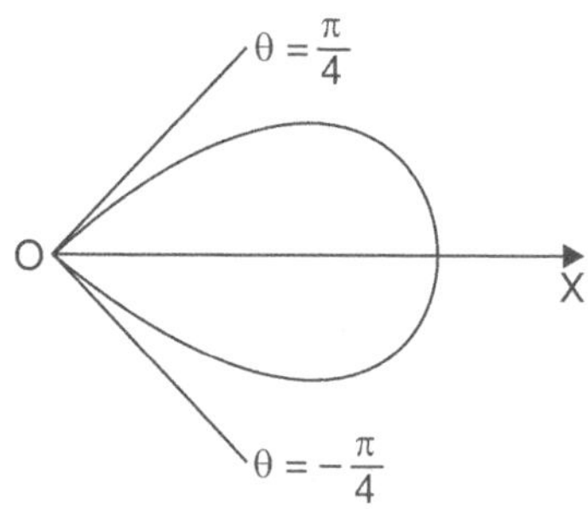

$$\therefore \qquad \bar{x} = \frac{\int_{-\pi/4}^{\pi/4} \int_0^{a\sqrt{\cos 2\theta}} r\cos\theta \cdot \rho r\, dr\, d\theta}{\int_{-\pi/4}^{\pi/4} \int_0^{a\sqrt{\cos 2\theta}} \rho r\, dr\, d\theta}$$

$$= \frac{\int_{-\pi/4}^{\pi/4} \int_0^{a\sqrt{\cos 2\theta}} k r^2 \cos\theta\, dr\, d\theta}{\int_{-\pi/4}^{\pi/4} \int_0^{a\sqrt{\cos 2\theta}} kr\, dr\, d\theta}$$

$$= \frac{\int_{-\pi/4}^{\pi/4} \cos\theta \left[\frac{r^3}{3}\right]_0^{a\sqrt{\cos 2\theta}} d\theta}{\int_{-\pi/4}^{\pi/4} \left[\frac{r^2}{2}\right]_0^{a\sqrt{\cos 2\theta}} d\theta} = \frac{\int_{-\pi/4}^{\pi/4} \frac{a^3}{3} \cos\theta\, (\cos 2\theta)^{3/2}\, d\theta}{\int_{-\pi/4}^{\pi/4} \frac{a^2}{2} \cos 2\theta\, d\theta}$$

$$= \frac{2\int_0^{\pi/4} \frac{a}{3} \cos\theta\, (\cos 2\theta)^{3/2} d\theta}{2\int_0^{\pi/4} \frac{1}{2} \cos 2\theta\, d\theta} = \frac{\frac{2a}{3}\int_0^{\pi/4} \cos\theta\, (1 - 2\sin^2\theta)^{3/2}\, d\theta}{\left[\frac{\sin 2\theta}{2}\right]_0^{\pi/4}}$$

$$= \frac{4a}{3}\int_0^{\pi/4} \cos\theta\, (1 - 2\sin^2\theta)^{3/2}\, d\theta$$

Put $\sqrt{2}\sin\theta = \sin t$, then $\sqrt{2}\cos\theta\, d\theta = \cos t\, dt$

when $\qquad \theta = 0, t = 0$; when $\theta = \frac{\pi}{4}, t = \frac{\pi}{2}$

$$\therefore \qquad \bar{x} = \frac{4a}{3}\int_0^{\pi/2}(1-\sin^2 t)^{3/2}\cdot\frac{1}{\sqrt{2}}\cos t\,dt$$

$$= \frac{2\sqrt{2}a}{3}\int_0^{\pi/2}\cos^4 t\,dt = \frac{2\sqrt{2}a}{3}\cdot\frac{3\cdot 1}{4\cdot 2}\cdot\frac{\pi}{2} = \frac{\pi a\sqrt{2}}{8}$$

$$\left|\ \text{Using } \int_0^{\frac{\pi}{2}}\cos^n t\,dt = \frac{(n-1)(n-3)\cdots}{n(n-2)\cdots}\frac{\pi}{2} \text{ where } n \text{ is even}\right.$$

Hence $\quad G \leftrightarrow \left(\dfrac{\pi a\sqrt{2}}{8}, 0\right)$.

TEST YOUR KNOWLEDGE

1. Find the mass of an elliptic plate $\dfrac{x^2}{a^2}+\dfrac{y^2}{b^2}=1$, if the density at any point (x, y) on it is $\mu\, xy$.
2. Find the mass of the area bounded by the curves $y^2 = x$ and $y = x^3$, if $\rho = \mu(x^2 + y^2)$.
3. Find the mass of a plate in the form of one loop of the lemniscate $r^2 = a^2 \cos 2\theta$, if the density at a point varies as the square of its distance from the pole.
4. Find the mass of a rectangular parallelopiped bounded by the planes $x = 0, x = a; y = 0, y = b$ and $z = 0, z = c$, if the density ρ at a point $P(x, y, z)$ is given by $\rho = x + y + z$.
5. Find the mass of the solid bounded by the planes $y = 0, z = 0$ and $z = h$ $(h > 0)$ and the cylinder $x^2 + y^2 = a^2$ $(x > 0)$, if $\rho = \mu yz$.
6. Find the centre of gravity of the area of the circle $x^2 + y^2 = a^2$ lying in the first quadrant.
7. The density at any point (x, y) of a lamina is $\dfrac{\sigma}{a}(x+y)$, where σ and a are constants. The lamina is bounded by the lines $x = 0, y = 0, x = a, y = b$. Find the position of its centre of gravity.
8. Find the centre of gravity of the area in the first quadrant lying between the curves $y^2 = x^3$ and $y = x$.
9. Find the centre of gravity of the area bounded by the parabola $y^2 = x$ and the line $x + y = 2$.

Answers

1. $\dfrac{\mu^2 a^2 b^2}{2}$ 2. $\dfrac{23\mu}{105}$ 3. $\dfrac{\mu\pi a^4}{16}$

4. $\dfrac{abc}{2}(a+b+c)$ 5. $\dfrac{1}{3}\mu a^3 h^2$ 6. $\left(\dfrac{4a}{3\pi}, \dfrac{4a}{3\pi}\right)$

7. $\left[\dfrac{a(4a+3b)}{6(a+b)}, \dfrac{b(3a+4b)}{6(a+b)}\right]$ 8. $\left(\dfrac{10}{21}, \dfrac{5}{12}\right)$ 9. $\left(\dfrac{8}{5}, -\dfrac{1}{2}\right)$

6.13. MOMENT OF INERTIA

If a particle of mass m be at a distance r from a given line, then mr^2 is called the *moment of inertia of the particle about the given line.* The sum of similar expressions for all the particles of a body *i.e.*, Σmr^2 is called the *moment of inertia of the body about the given line.*

If the moment of inertia of a body of mass M about a line is denoted by MK^2, then K is called the *radius of gyration of the body about the given line.*

Moment of inertia of a plane lamina. Let A be the area of a plane lamina and ρ its density. Consider an elementary mass $\rho\, dx\, dy$ at the point $P(x, y)$ of the lamina so that its M.I. about x-axis

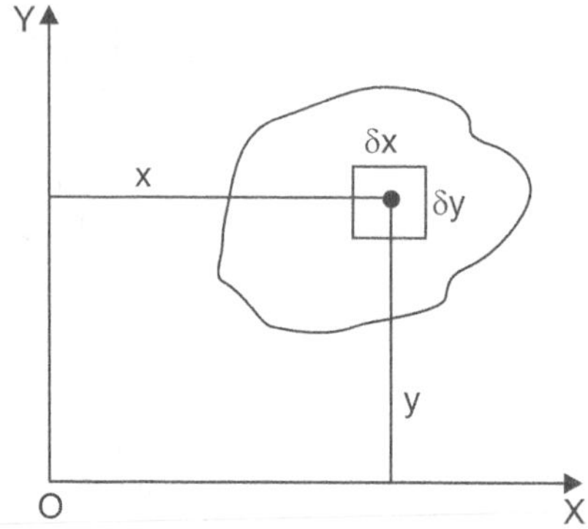

$$= \rho\, \delta x\, \delta y \cdot y^2$$

∴ M.I. of the lamina about x-axis is given by

$$I_x = \iint_A \rho\, y^2\, dx\, dy$$

Similarly, the M.I. of the lamina about y-axis is given by

$$I_y = \iint_A \rho\, x^2\, dx\, dy.$$

(*i*) **Theorem of Perpendicular Axis.** If I_x and I_y be the moments of inertia of a plane lamina about two perpendicular lines OX and OY respectively, then its moment of inertia I_z about OZ, perpendicular to OX and OY, is given by $I_z = I_x + I_y$. **(P.T.U., May 2008)**

(*ii*) **Theorem of Parallel Axis.** If I is the moment of inertia of a body of mass M about a given line through its centre of gravity, then its moment of inertia I_d about a line parallel to the given line and at a distance d from it is given by $I_d = I + Md^2$.

ILLUSTRATIVE EXAMPLES

Example 1. *Find the moment of inertia of a circular plate about a tangent line.*

Sol. Equation of the circular plate is $x^2 + y^2 = a^2$

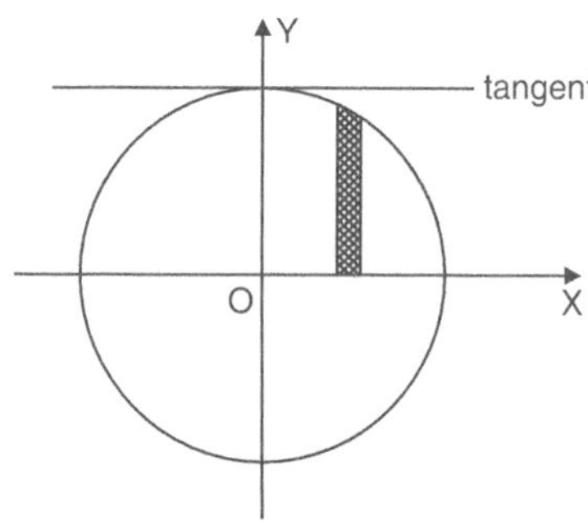

M = mass of the circular plate

$$= 4\int_0^a \int_{y=0}^{\sqrt{a^2-x^2}} \rho\, dy\, dx$$

$$= 4\rho \int_0^a \sqrt{a^2 - x^2}\, dx$$

$$= 4\rho \left\{ \frac{x\sqrt{a^2 - x^2}}{2} + \frac{a^2}{2}\sin^{-1}\frac{x}{a} \right|_0^a \Bigg\}$$

$$= 4\rho \left\{ \frac{a^2}{2} \cdot \frac{\pi}{2} \right\} = \pi a^2 \rho \quad \therefore \quad \rho = \frac{M}{\pi a^2} \qquad \text{...(1)}$$

I_x = Moment of inertia about x-axis

$$= 4\int_0^a \int_0^{\sqrt{a^2-x^2}} \rho\, y^2\, dy\, dx = 4\rho \int_0^a \left. \frac{y^3}{3} \right|_0^{\sqrt{a^2-x^2}} dx$$

$$= \frac{4\rho}{3}\int_0^a (a^2 - x^2)^{3/2}\, dx. \quad \text{Put } x = a\sin\theta \quad dx = a\cos\theta\, d\theta$$

$$= \frac{4\rho}{3}\int_0^{\frac{\pi}{2}} a^3 \cos^3 \theta \,.\, a \cos \theta \, d\theta = \frac{4\rho\, a^4}{3} \frac{3\cdot 1}{4\cdot 2} \frac{\pi}{2}$$

$$I_x = \frac{\pi a^4}{4} \cdot \rho = \frac{\pi a^4}{4} \cdot \frac{M}{\pi a^2} = \frac{Ma^2}{4} \text{ using (1)}$$

By theorem of parallel axis

M.I. about tangent $= I_x + Md^2 = I_x + Ma^2 \quad \because \quad d = a$

$$= \frac{Ma^2}{4} + Ma^2 = \frac{5}{4} Ma^2.$$

Example 2. *Find M.I. about x-axis of the arc of the parabola $y = \sqrt{x}$ lying between (0, 0) and (4, 2).* **(P.T.U., May 2005, May 2009)**

Sol. Equation of the parabola is $y = \sqrt{x}$ or $y^2 = x$

Let M be the mass of the arc OP then

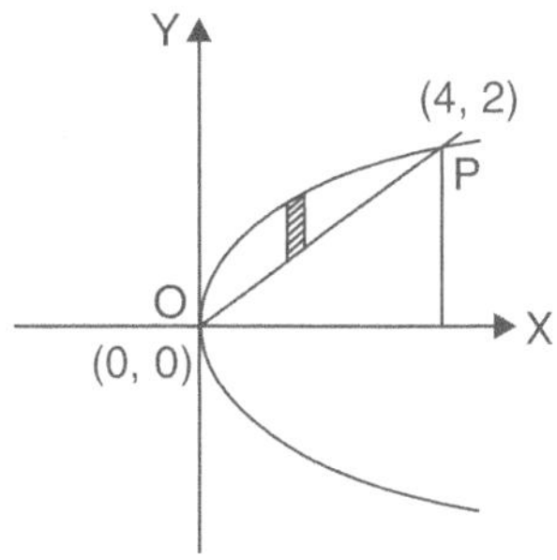

$$M = \int_{x=0}^{4} \int_0^{\sqrt{x}} \rho \, dx \, dy$$

$$= \int_0^4 \rho \, y \Big|_0^{\sqrt{x}} dx = \int_0^4 \rho\sqrt{x} \, dx$$

$$= \rho \left. \frac{x^{3/2}}{3/2} \right|_0^4 = \frac{2\rho}{3} \cdot 4^{3/2} = \frac{16\rho}{3}$$

$$\therefore \qquad \rho = \frac{3M}{16}$$

Now, $\quad I_x =$ M.I. about x-axis of $\widehat{OP}$

$$= \int_0^4 \int_0^{\sqrt{x}} \rho \, y^2 \, dx \, dy$$

$$= \rho \int_0^4 \left[\int_0^{\sqrt{x}} y^2 \, dy \right] dx = \rho \int_0^4 \left. \frac{y^3}{3} \right|_0^{\sqrt{x}} dx$$

$$= \frac{\rho}{3} \int_0^4 x^{3/2} \, dx = \frac{\rho}{3} \cdot \left. \frac{x^{5/2}}{5/2} \right|_0^4 = \frac{2\rho}{15} 4^{5/2}$$

$$= \frac{2\rho}{15} 32 = \frac{64}{15} \cdot \frac{3M}{16} = \frac{4M}{5}.$$

Example 3. *A thin plate of uniform thickness and density is in the form of a lamina bounded by parabola $x^2 = y$ and the line $y = x + 2$. Find the moment of inertia of the lamina about the coordinate axes.* **(P.T.U., Dec. 2005)**

Sol. Equation of the parabola is

$$x^2 = y$$

Equation of the line is $y = x + 2$

These intersect at points given by

$$x^2 = x + 2 \quad \text{or} \quad x^2 - x - 2 = 0$$

or $\quad (x + 1)(x - 2) = 0$

$\therefore \quad x = -1, x = 2 \quad \therefore \quad y = 1, y = 4$

$\therefore$ Points of intersection are (–1, 1) and (2, 4)

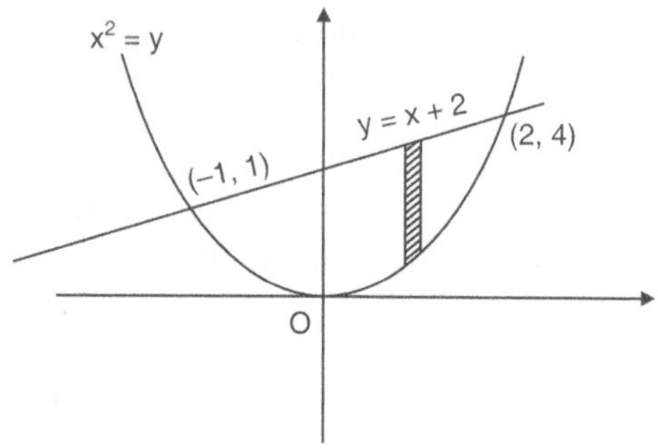

$$M = \int_{-1}^{2} \int_{x^2}^{x+2} \rho \, dy \, dx$$

$$= \rho \int_{-1}^{2} (x + 2 - x^2) \, dx$$

$$= \rho \left[2x + \frac{x^2}{2} - \frac{x^3}{3} \Big|_{-1}^{2} \right] = \rho \left[4 + 2 - \frac{8}{3} - \left(-2 + \frac{1}{2} + \frac{1}{3} \right) \right]$$

$$\therefore \quad M = \frac{9}{2} \rho$$

$$I_x = \int_{-1}^{2} \int_{x^2}^{x+2} \rho \, y^2 \, dy \, dx = \rho \int_{-1}^{2} \frac{y^3}{3} \Big|_{x^2}^{x+2} dx$$

$$= \frac{\rho}{3} \int_{-1}^{2} [(x + 2)^3 - x^6] \, dx = \frac{\rho}{3} \left[\frac{(x+2)^4}{4} - \frac{x^7}{7} \Big|_{-1}^{2} \right]$$

$$= \frac{\rho}{3} \left\{ 64 - \frac{128}{7} - \frac{1}{4} - \frac{1}{7} \right\} = \frac{\rho}{3} \left\{ \frac{255}{4} - \frac{129}{7} \right\} = \frac{1269}{84} \rho = \frac{1269}{84} \times \frac{2M}{9} = \frac{47}{14} M$$

$$I_y = \int_{-1}^{2} \int_{x^2}^{x+2} \rho \, x^2 \, dy \, dx = \int_{-1}^{2} \rho \, x^2 (x + 2 - x^2) \, dx = \rho \int_{-1}^{2} (x^3 + 2x^2 - x^4) \, dx$$

$$= \rho \left[\frac{x^4}{4} + \frac{2x^3}{3} - \frac{x^5}{5} \Big|_{-1}^{2} \right] = \rho \left\{ 4 + \frac{16}{3} - \frac{32}{5} - \frac{1}{4} + \frac{2}{3} - \frac{1}{5} \right\}$$

$$= \rho \frac{63}{20} = \frac{63}{20} \times \frac{2M}{9} = \frac{7}{10} M$$

Hence $I_x = \frac{47}{14} M, \quad I_y = \frac{7}{10} M.$

Example 4. *Find the moment of inertia of a quadrant of the ellipse* $\frac{x^2}{a^2} + \frac{y^2}{b^2} = 1$ *of mass M about the z-axis, if the density at a point is proportional to xy.*

Sol. Here the region R of integration is the quadrant OAB which is bounded by

$$x = 0, x = \frac{a}{b}\sqrt{b^2 - y^2} \text{ and } y = 0, y = b.$$

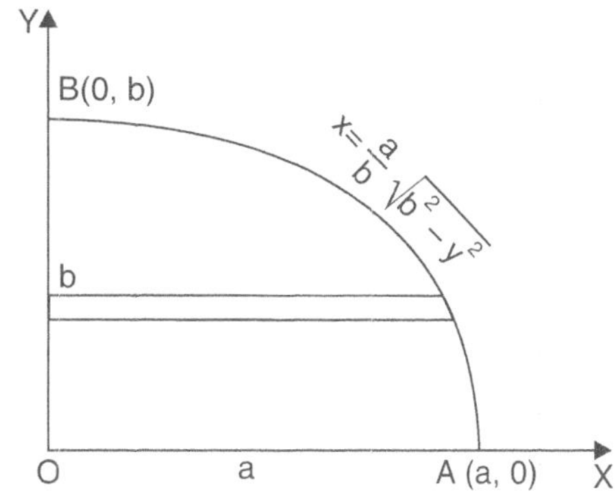

Let $\rho = \mu xy$, then the mass M of the quadrant is given by

$$M = \iint_R \rho \, dx \, dy$$

$$= \int_a^b \int_0^{a/b\sqrt{b^2 - y^2}} \mu \, xy \, dx \, dy$$

$$= \int_0^b \mu y \cdot \left[\frac{x^2}{2} \right]_0^{a/b\sqrt{b^2 - y^2}} dy$$

$$= \frac{\mu}{2} \int_0^b y \cdot \frac{a^2}{b^2} (b^2 - y^2) \, dy = \frac{\mu a^2}{2b^2} \int_0^b (b^2 y - y^3) \, dy$$

$$= \frac{\mu a^2}{2b^2} \left[b^2 \cdot \frac{y^2}{2} - \frac{y^4}{4} \right]_0^b = \frac{\mu a^2}{2b^2} \left(\frac{b^4}{2} - \frac{b^4}{4} \right) = \frac{\mu a^2 b^2}{8}.$$

$$\therefore \qquad \mu = \frac{8M}{a^2 b^2}$$

M.I. about x-axis is given by $I_x = \iint_A \rho \, y^2 \, dx \, dy = \int_0^b \int_0^{a/b\sqrt{b^2 - y^2}} \mu \, xy^3 \, dx \, dy$

$$= \int_0^b \mu y^3 \cdot \left[\frac{x^2}{2} \right]_0^{a/b\sqrt{b^2 - y^2}} dy$$

$$= \frac{\mu}{2} \int_0^b y^3 \cdot \frac{a^2}{b^2} (b^2 - y^2) \, dy = \frac{\mu a^2}{2b^2} \int_0^b (b^2 y^3 - y^5) \, dy$$

$$= \frac{\mu a^2}{2b^2} \left[b^2 \cdot \frac{y^4}{4} - \frac{y^6}{6} \right]_0^b = \frac{\mu a^2}{2b^2} \cdot \frac{b^6}{12} = \mu \cdot \frac{a^2 b^4}{24} = \frac{8M}{a^2 b^2} \cdot \frac{a^2 b^4}{24} = \frac{1}{3} M b^2$$

M.I. about y-axis is given by $I_y = \iint_A \rho \, x^2 \, dx \, dy = \int_0^b \int_0^{a/b\sqrt{b^2 - y^2}} \mu \, x^3 y \, dx \, dy$

$$= \int_0^b \mu y \cdot \left[\frac{x^4}{4} \right]_0^{a/b\sqrt{b^2 - y^2}} dy$$

$$= \frac{\mu}{4} \int_0^b y \cdot \frac{a^4}{b^4} (b^2 - y^2)^2 \, dy = \frac{\mu a^4}{4b^4} \int_0^b -\tfrac{1}{2} (b^2 - y^2)^2 (-2y) \, dy$$

$$= -\frac{\mu a^4}{8b^4} \left[\frac{(b^2 - y^2)^3}{3} \right]_0^b = \frac{\mu a^4 b^2}{24} = \frac{8M}{a^2 b^2} \cdot \frac{a^4 b^2}{24} = \frac{1}{3} M a^2.$$

$\therefore$ By the theorem of perpendicular axis, the M.I. about z-axis is given by

$$I_z = I_x + I_y = \tfrac{1}{3} M (a^2 + b^2).$$

Example 5. *Using double integration find the M.I. about x-axis of the area enclosed by the lines $x = 0$, $y = 0$, $\frac{x}{a}+\frac{y}{b}=1$.*

Sol. Here region of integration is bounded by $x = 0$, $y = 0$, $\frac{x}{a}+\frac{y}{b}=1$

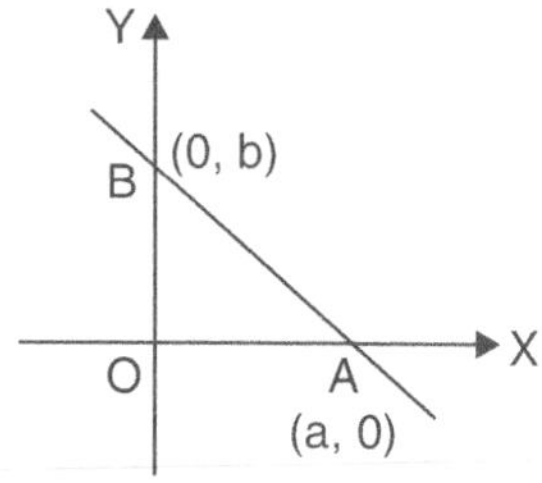

Let M be the mass of the region

$$M = \int_0^b \int_0^{a\left(1-\frac{y}{b}\right)} \rho \, dx \, dy$$

$$= \rho \int_0^b x \Big|_0^{a\left(1-\frac{y}{b}\right)} dy = \rho \int_0^b a\left(1-\frac{y}{b}\right) dy = \rho a \left(y - \frac{y^2}{2b} \right)\Bigg|_0^b$$

$$= \rho a \left(b - \frac{b^2}{2b} \right) = \frac{\rho a b}{2} \quad \therefore \quad \rho = \frac{2M}{ab}$$

Now M.I. about x-axis

$$= \int_0^b \int_0^{a\left(1-\frac{y}{b}\right)} \rho \, y^2 \, dx \, dy = \rho \int_0^b \left[\int_0^{a\left(1-\frac{y}{b}\right)} y^2 \, dx \right] dy$$

$$= \rho \int_0^b y^2 x \Big|_0^{a\left(1-\frac{y}{b}\right)} dy = \rho \int_0^b y^2 a \left(1-\frac{y}{b}\right) dy$$

$$= a\rho \int_0^b \left(y^2 - \frac{y^3}{b} \right) dy = a\rho \left[\frac{y^3}{3} - \frac{y^4}{4b} \Big|_0^b \right]$$

$$= a\rho \left(\frac{b^3}{3} - \frac{b^3}{4} \right) = a\rho \frac{b^3}{12} = \frac{ab^3}{12} \cdot \frac{2M}{ab} = \frac{Mb^2}{6}.$$

Example 6. *Find the moment of inertia of one loop of the lemniscate $r^2 = a^2 \cos 2\theta$ about*

(i) the initial line **(P.T.U., May 2008, 2015)**

(ii) an axis perpendicular to its plane at the pole.

Sol. (*i*) Let M be the mass and ρ, the density of one loop of the lemniscate $r^2 = a^2 \cos 2\theta$.

For the loop, θ varies from $-\frac{\pi}{4}$ to $\frac{\pi}{4}$ and r varies from 0 to $a^2\sqrt{\cos 2\theta}$.

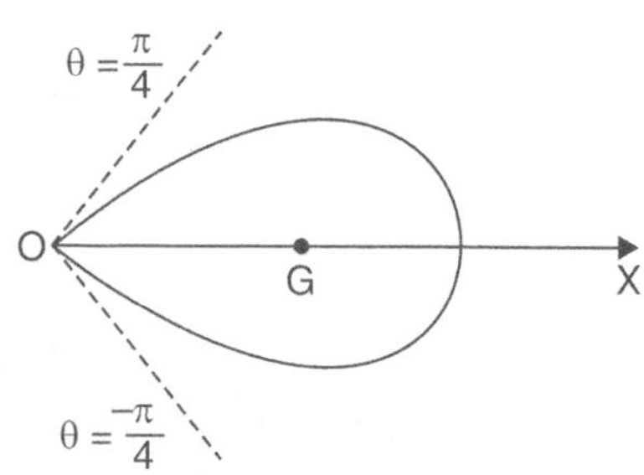

$$= \int_{-\pi/4}^{\pi/4} \int_0^{a\sqrt{\cos 2\theta}} \rho r \, dr \, d\theta$$

$$= \int_{-\pi/4}^{\pi/4} \rho \cdot \left[\frac{r^2}{2} \right]_0^{a\sqrt{\cos 2\theta}} d\theta$$

$$= \frac{\rho}{2}\int_{-\pi/4}^{\pi/4} a^2 \cos 2\theta \, d\theta = \frac{\rho a^2}{2} \cdot 2\int_0^{\pi/4} \cos 2\theta \, d\theta = \rho a^2 \left[\frac{\sin 2\theta}{2}\right]_0^{\pi/4} = \frac{\rho a^2}{2}$$

M.I. about the initial line is given by $I_x = \iint_R \rho y^2 \, dx \, dy = \iint_R \rho r^2 \sin^2\theta \cdot r \, dr \, d\theta$

$$= \int_{-\pi/4}^{\pi/4}\int_0^{a\sqrt{\cos 2\theta}} \rho \sin^2\theta \cdot r^3 \, dr \, d\theta = \int_{-\pi/4}^{\pi/4} \rho \sin^2\theta \left[\frac{r^4}{4}\right]_0^{a\sqrt{\cos 2\theta}} d\theta$$

$$= \frac{\rho a^4}{4}\int_{-\pi/4}^{\pi/4} \sin^2\theta \cos^2 2\theta \, d\theta = \frac{\rho a^4}{4} \cdot 2\int_0^{\pi/4} \sin^2\theta \cos^2 2\theta \, d\theta$$

($\because$ $\sin^2\theta \cos^2 2\theta$ is an even function of θ)

$$= \frac{\rho a^4}{2}\int_0^{\pi/2} \sin^2\frac{\phi}{2} \cos^2\phi \cdot \frac{1}{2} d\phi, \quad \text{where} \quad \theta = \frac{\phi}{2}$$

$$= \frac{\rho a^4}{4}\int_0^{\pi/2} \left(\frac{1-\cos\phi}{2}\right) \cos^2\phi \, d\phi = \frac{\rho a^4}{4}\int_0^{\pi/2} (\cos^2\phi - \cos^3\phi) \, d\phi$$

$$= \frac{\rho a^4}{8}\left[\frac{1}{2} \cdot \frac{\pi}{2} - \frac{2}{3}\right] = \frac{\pi a^4}{96} = (3\pi - 8) = \frac{2M}{a^2} \cdot \frac{a^4}{96}(3\pi - 8) \qquad \because \; M = \frac{\rho a^2}{2}$$

$$= \frac{Ma^2}{48}(3\pi - 8).$$

(*ii*) M.I. about the y-axis

$$I_y = \iint_R \rho x^2 \, dx \, dy = \rho \int_{-\frac{\pi}{4}}^{\frac{\pi}{4}} \int_0^{a\sqrt{\cos 2\theta}} r^2 \cos^2\theta \, r \, dr \, d\theta$$

$$= \rho \int_{-\frac{\pi}{4}}^{\frac{\pi}{4}} \cos^2\theta \cdot \left.\frac{r^4}{4}\right|_0^{a\sqrt{\cos 2\theta}} d\theta = \frac{\rho}{4}\int_{-\frac{\pi}{4}}^{\frac{\pi}{4}} \cos^2\theta \cdot a^4 \cos^2 2\theta \, d\theta$$

$$= \frac{\rho a^4}{4} \, 2 \int_0^{\frac{\pi}{4}} \cos^2\theta \cos^2 2\theta \, d\theta \quad \therefore \quad \text{integrand is even}$$

Put $2\theta = \phi \quad \therefore \quad 2d\theta = d\phi$

$$= \frac{\rho a^4}{2}\int_0^{\frac{\pi}{2}} \cos^2\frac{\phi}{2} \cos^2\phi \cdot \frac{d\phi}{2} = \frac{\rho a^4}{4}\int_0^{\frac{\pi}{2}} \frac{1+\cos\phi}{2} \cos^2\phi \, d\phi$$

$$= \frac{\rho a^4}{8}\int_0^{\frac{\pi}{2}} (\cos^2\phi + \cos^3\phi) \, d\phi = \frac{\rho a^4}{8}\left\{\frac{1}{2}\frac{\pi}{2} + \frac{2}{3.1}\right\}$$

$$= \frac{\rho a^4}{8}\left\{\frac{\pi}{4} + \frac{2}{3}\right\} = \frac{\rho a^4}{96}(3\pi + 8) = \frac{a^4}{96} \times \frac{2M}{a^2}(3\pi + 8) \qquad \because \; \rho = \frac{2M}{a^2}$$

$$= \frac{Ma^2}{48}(3\pi + 8)$$

M.I. about an axis ⊥ to the plane of lemniscate is $I_z = I_x + I_y$

$$= \frac{Ma^2}{48}(3\pi - 8 + 3\pi + 8) = \frac{Ma^2}{48} \cdot 6\pi = \frac{Ma^2\pi}{8} = \frac{\pi a^2}{8} \text{ M.}$$

TEST YOUR KNOWLEDGE

1. Find the moment of inertia of an elliptic disc of mass M and axes of lengths $2a$, $2b$ about the major axis.
2. Find the moment of inertia of the area bounded by the circle $r = 2a \cos \theta$ with respect to a line perpendicular to its plane through the pole.

   [**Hint.** Replace x, y by $r \cos \theta$, $r \sin \theta$ and $dxdy$ by $rdr\, d\theta$ find I_x, I_y then find $I_z = I_x + I_y$, r varies from 0 to $2a \cos \theta$ and θ varies from $-\frac{\pi}{2}$ to $\frac{\pi}{2}$]
3. Find the moment of inertia about the initial line of the cardioid $r = a(1 + \cos \theta)$.

Answers

1. $\frac{1}{4} Mb^2$ **2.** $\frac{3}{2} \rho\pi a^4$ **3.** $\frac{21}{32} \rho\pi a^4$

REVIEW OF THE CHAPTER

1. **Double Integration**

   (*i*) In cartesian co-ordinate $\iint_R f(x, y)\, dx\, dy = \int_{y_1}^{y_2} \left[\int_{x_1}^{x_2} f(x, y)\, dx\right] dx$ when x_1, x_2 are functions of y and y_1, y_2 are constant

   $\iint_R f(x, y)\, dxdy = \int_{x_1}^{x_2} \left[\int_{y_1}^{y_2} f(x, y)\, dy\right] dx$ when y_1, y_2 are functions of x and x_1, x_2 are constants.

   (*ii*) In polar co-ordinates $\int_{\theta=\theta_1}^{\theta_2} \int_{r=r_1}^{r_2} f(r, \theta)\, dr\, d\theta = \int_{\theta_1}^{\theta_2} \left[\int_{r_1}^{r_2} f(r, \theta)\, dr\right] d\theta$.

2. **Change of order of Integration.** In a double integral if limits of integration are constant then order of integration is immaterial and if limits of integration are variable then change in order of integration brings change in the limits of integration.

3. $\iiint_V f(x, y, z)\, dx\, dy\, dz = \int_{z=a}^{b} \left[\int_{y=\phi_1(z)}^{y=\phi_2(z)} \left\{\int_{x_1=f_1(y,z)}^{x_2=f_2(yz)} f(x, y, z)\, dx\right\} dy\right] dz$.

4. (*i*) **Change of Variables.** If variables x, y in $\iint_R f(x, y)\, dx\, dy$ be changed to u, v by the relation $x = \phi(u, v)$, $y = \psi(u, v)$, then $\iint_R f(x, y)\, dx\, dy$ transforms to

   $\iint f(\phi, \psi)\, |J|\, du\, dv$

   When $J = \frac{\partial(x, y)}{\partial(u, v)} = \begin{vmatrix} \frac{\partial x}{\partial u} & \frac{\partial x}{\partial v} \\ \frac{\partial y}{\partial u} & \frac{\partial y}{\partial v} \end{vmatrix}$.

(*ii*) Change (x, y) to polar (r, θ) here $J = r$; $\iint_R f(x, y)\, dx\, dy = \iint_{R'} f(r\cos\theta, r\sin\theta)\, r\, dr\, d\theta.$

5. In Triple Integration

(*i*) Change of cartesian (x, y, z) to polar spherical co-ordinates (r, θ, ϕ)

$x = r\cos\theta\cos\phi,\ y = r\sin\theta\sin\phi,\ z = r\cos\theta,\ J = r^2\sin\theta$

$$\iiint_V f(x, y, z)\, dx\, dy\, dz = \iiint f(r\sin\theta\cos\phi, r\sin\theta\sin\phi, r\cos\theta)\, r^2\sin\theta\, dr\, d\theta\, d\phi$$

(*ii*) Change of cartesian (x, y, z) to cylinderical coordinates (r, ϕ, z)

$$x = r\cos\phi,\ y = r\sin\phi,\ z = z,\ J = r$$

$$\therefore \quad \iiint_V f(x, y, z)\, dx\, dy\, dz = \iiint_{V'} f(r\cos\theta, r\sin\phi, z)\, r\, dr\, d\phi\, dz$$

6. Area by Double Integration

(*i*) in cartesian co-ordinates $= \int_a^b \int_{f_1(x)}^{f_2(x)} dy\, dx$

(*ii*) in polar form $= \int_\alpha^\beta \int_{r = f_1(\theta)}^{f_2(\theta)} r\, dr\, d\theta.$

7. Volume as a Double Integral

(*i*) In cartesian co-ordinates $V = \iint_R z\, dx\, dy$

(*ii*) In polar coordinates if $z = f(x, y)$ then $V = \int_{\theta = \alpha}^{\beta} \int_{r = f_1(\theta)}^{f_2(\theta)} f(r\cos\theta, r\sin\theta)\, r\, dr\, d\theta$

(*iii*) In cylindrical co-ordinates $V = \iint_R z\, dr\, d\phi.$

8. Volume as a Triple Integral

(*i*) in cartesian co-ordinates $\iiint_V dx\, dy\, dz$

(*ii*) in spherical polar co-ordinates $V = \iiint r^2\sin\theta\, dr\, d\theta\, d\phi$

(*iii*) in cylindrical co-ordinates $V = \iiint r\, dr\, d\phi\, dz.$

9. Volume of Solid of Revolution

(*i*) in cartesian co-ordinates $V = 2\pi \iint_R y\, dy dx$

(*ii*) in polar co-ordinates $V = 2\pi \iint r^2\cos\theta\, dr d\theta.$

10. C.G. of a Plane Lamina of Area A

(*i*) $\bar{x} = \dfrac{\iint x\rho\, dx\, dy}{\iint \rho\, dx\, dy}$ $\quad \bar{y} = \dfrac{\iint y\rho\, dx\, dy}{\iint \rho\, dx\, dy}$ where density $\rho = f(x, y)$ in polar coordinates.

(*ii*) $\bar{x} = \dfrac{\iint_A r^2\cos\theta\, \rho\, dr\, d\theta}{\iint_A \rho r\, dr\, d\theta}$, $\quad \bar{y} = \dfrac{\iint r^2\sin\theta\, \rho\, dr\, d\theta}{\iint \rho r\, dr\, d\theta}$ in polar coordinates.

11. C.G. of a Solid Occupying Volume V having Density $\rho = f(x, y, z)$

$$\bar{x} = \frac{\iiint_V x\,\rho\,dx\,dy\,dz}{\iiint_V \rho\,dx\,dy\,dz}, \quad \bar{y} = \frac{\iiint_V y\,\rho\,dx\,dy\,dz}{\iiint_V \rho\,dx\,dy\,dz}, \quad \bar{z} = \frac{\iiint_V z\,\rho\,dx\,dy\,dz}{\iiint_V \rho\,dx\,dy\,dz}.$$

12. Moment of Inertia of a Plane Lamina

(*i*) About x-axis $= \iint_A \rho y^2\,dx\,dy$ (*ii*) About y-axis $= \iint_A \rho x^2\,dx\,dy$.

13. (*i*) **Theorem of Perpendicular Axis.** If Ix and Iy be the moments of inertia of a plane lamina about two perpendicular lines OX and OY respectively, then its moment of inertia Iz about OZ, perpendicular to OX and OY, is given by $I_z = I_x + I_y$.
(P.T.U. May 2008)

(*ii*) **Theorem of Parallel Axis.** If I is the moment of inertia of a body of mass M about a given line through its centre of gravcity, then its moment of inertia I_d about a line parallel to the given line and at a distance d from it is given by $I_d = I + Md^2$.

SHORT ANSWER TYPE QUESTIONS

1. Evaluate the following:

(*i*) $\int_0^1\int_0^1 (x+2)\,dy\,dx$ ***(P.T.U., Dec. 2005)***
(*ii*) $\int_0^1\int_0^3 (x+3)\,dy\,dx$ ***(P.T.U., May 2005)***
[**Hint.** S.E. 2 art. 6.2]

(*iii*) $\int_0^1\int_0^{\sqrt{1+x^2}} \frac{dy\,dx}{1+x^2+y^2}$ ***(P.T.U., May 2006)***
[**Hint.** S.E. 4 art. 6.2]
(*iv*) $\int_1^2\int_0^x \frac{dy\,dx}{x^2+y^2}$ ***(P.T.U., May 2011)***

(*v*) $\int_0^2\int_0^{\sqrt{2x}} xy\,dy\,dx$ ***(P.T.U., Dec. 2003)***
[**Hint.** S.E. 5 art. 6.2]
(*vi*) $\int_0^1\int_x^{\sqrt{x}} (x^2+y^2)\,dy\,dx$ ***(P.T.U., Dec. 2013)***
[**Hint.** S.E. 2(*ii*) art. 6.2]

(*vii*) $\int_0^3\int_1^{\sqrt{4-x}} (x+y)\,dx\,dy$ ***(P.T.U., Dec. 2012)***
(*viii*) $\int_0^1\int_{x^2}^{2-x} xy\,dy\,dx$ ***(P.T.U. Dec. 2012)***

(*ix*) $\int_1^{\log 8}\int_0^{\log y} e^{x+y}\,dx\,dy$
[**Hint.** S.E. 10(*i*) art. 6.2]
(*x*) $\int_0^1\int_{x^2}^{x} (x^2+3y+2)\,dy\,dx$

(*xi*) $\int_0^2\int_0^{y^2} e^{x/y}\,dx\,dy$ ***(P.T.U., May 2004)***
[**Hint.** S.E. 10(*ii*) art. 6.2]

2. Prove that $\int_1^2\int_3^4 (xy+e^y)\,dy\,dx = \int_3^4\int_1^2 (xy+e^y)\,dx\,dy$
[**Hint.** S.E. 1 art. 6.2]

3. Evaluate the following:

(*i*) $\int_0^{\frac{\pi}{2}}\left[\int_0^{a\cos\theta} r\sqrt{a^2-r^2}\,dr\right]d\theta$; (*ii*) $\int_0^{\pi}\int_0^{a\sin\theta} r\,dr\,d\theta$; (*iii*) $\int_0^{\pi}\int_0^{a(1+\cos\theta)} r^2\cos\theta\,dr\,d\theta$
[**Hint.** S.E.I art. 6.3]

4. Write the limits of integration in $\iint_R xy\,dxdy$ where R is the region inside the square $|x| + |y| = 1$. **(*P.T.U., May 2003*)**

[**Hint.** S.E. 11(*ii*) art. 6.2]

5. Sketch the region of integration and determine the order of integration of the integral $\iint_R (y - 2x^2)\,dx\,dy$ where R is the region inside the square $|x| + |y| = 1$. **(*P.T.U., Dec. 2004*)**

[**Hint.** S.E. 11(*i*) art. 6.2]

6. (*i*) Calculate $\iint_R \frac{\sin x}{x}\,dx\,dy$ where R is the triangle in the *xy* plane bounded by the *x*-axis, the line $y = x$ and the line $x = 1$ **(*P.T.U., Dec. 2010*)**

$$\left[\textbf{Hint.} \int_{y=0}^{1}\int_{x=y}^{1} \frac{\sin x}{x}\,dx\,dy = \int_{x=0}^{1}\int_{y=0}^{x} \frac{\sin x}{x}\,dy\,dx \text{ (by change of order of integration)} \right.$$

$$\left. = \int_0^1 \sin x\,dx = -\cos x \Big|_0^1 = 1 - \cos 1\right].$$

(*ii*) Evaluate $\int_0^{\pi}\int_x^{\pi} \frac{\sin y}{y}\,dy\,dx$ **(*P.T.U., Dec. 2014*)**

[**Hint.** Same as 6(*i*)]

7. Evaluate the following by changing the order of integration:

(*i*) $\int_0^1\int_{x^2}^{2-x} xy\,dy\,dx$

[**Hint.** S.E. 2 art. 6.4]

(*ii*) $\int_0^{2a}\int_{\sqrt{2ax-x^2}}^{\sqrt{2ax}} dy\,dx$ **(*P.T.U., Dec. 2003*)**

[See S.E. 4 art. 6.4]

(*iii*) $\int_0^a\int_{y^2/a}^{y} \frac{y\,dx\,dy}{(a-x)\sqrt{ax-y^2}}$

[**Hint.** S.E. 7(*i*) art. 6.4]

(*iv*) $\int_0^{\infty}\int_x^{\infty} \frac{e^{-y}}{y}\,dy\,dx$ **(*P.T.U., May 2004, Dec. 2011*)**

[**Hint.** S.E. 7(*ii*) art. 6.4]

(*v*) $\int_0^1\int_0^{\sqrt{1-x^2}} y^2\,dy\,dx$ **(*P.T.U., Dec. 2004*)**

[**Hint.** S.E. 9(*i*) art. 6.4]

(*vi*) $\int_0^a\int_{a-\sqrt{a^2-y^2}}^{a+\sqrt{a^2-y^2}} dy\,dx$

[**Hint.** S.E. 5(*i*) art. 6.4]

(*vii*) $\int_0^b\int_0^{\frac{a}{b}\sqrt{b^2-y^2}} xy\,dx\,dy$ **(*P.T.U., Dec. 2013*)**

[**Hint.** S.E. 9(*i*) art 6.4]

8. Evaluate the following triple integrals.

(*i*) $\int_0^2\int_1^2\int_0^{yz} xyz\,dx\,dy\,dx$ **(*P.T.U., May 2006, 2009*)**

[**Hint.** S.E. 2 art. 6.5]

(*ii*) $\int_{-1}^1\int_0^z\int_{x-z}^{x+z} (x + y + z)\,dy\,dx\,dz$ **(*P.T.U., May 2014*)**

(*iii*) $\int_0^1\int_0^{1-x}\int_0^{1-x-y} xyz\,dz\,dy\,dx.$ **(*P.T.U., Dec. 2013*)**

(*iv*) $\int_0^1\int_0^{1-x^2}\int_0^{1-x^2-y^2} xyz\,dz\,dy\,dx.$ **(*P.T.U., May 2012*)** (*v*) $\int_{-c}^{c}\int_{-b}^{b}\int_{-a}^{a}(x^2+y^2+z^2)\,dx\,dy\,dz$ **(P.T.U., May 2015)**

[**Hint.** $\int_{-c}^{c}\int_{-b}^{b}\left.\frac{x^3}{3}+(y^2+z^2)\,x\right|_{-a}^{a}dy\,dz = 2\int_{-c}^{c}\int_{-b}^{b}\left[\frac{a^3}{3}+(y^2+z^2)\,a\right]dy\,dz$

$$= 2\int_{-c}^{c}\left.\left(\frac{a^3}{3}+az^2\right)y+\frac{y^3}{3}\right|_{-b}^{b}dz = 4\int_{-c}^{c}\left[\left(\frac{a^3}{3}+az^2\right)b+\frac{b^3}{3}\right]dz$$

$$= 4\left[\left(\frac{a^3b}{3}+\frac{b^3}{3}\right)z+\frac{abz^3}{3}\right]_{-c}^{c} = 8\left[\frac{(a^3b+b^3)c}{3}+\frac{abc^3}{3}\right]$$

$$= \frac{8}{3}abc\,(a^2+b^2+c^2)$$

9. Evaluate the following by changing into polar coordinates:

(*i*) $\int_0^\infty\int_0^\infty e^{-(x^2+y^2)}\,dxdy$ **(P.T.U., Dec. 2003, 2011)** [**Hint.** S.E. 3 art. 6.6]

(*ii*) $\int_0^2\int_0^{\sqrt{2x-x^2}}\frac{x\,dy\,dx}{\sqrt{x^2+y^2}}$ **(P.T.U., May 2012)** [S.E. 4(*i*) art. 6.6]

(*iii*) $\iint\sqrt{a^2-x^2-y^2}\,dx\,dy$ over the semi circle $x^2+y^2=ax$ in positive quadrant.
[**Hint.** S.E. 2 art. 6.6]

(*iv*) $\int_0^a\int_0^{\sqrt{a^2-y^2}}(x^2+y^2)\,dx\,dy$

(*v*) $\int_{-a}^{a}\int_{-\sqrt{a^2-x^2}}^{\sqrt{a^2-x^2}}dy\,dx$

(*vi*) $\int_{-1}^{1}\int_{-\sqrt{1-y^2}}^{\sqrt{1-y^2}}\log(x^2+y^2+1)\,dx\,dy$ **(P.T.U., May 2003)**

[**Hint.** Given integral changes to $\int_{-\frac{\pi}{2}}^{\frac{\pi}{2}}\int_0^1\log(r^2+1)\,r\,dr\,d\theta$; Integrals by parts]

10. Change the cartesian integral $\int_0^1\int_x^{\sqrt{2x-x^2}}f(x,y)dydx$ to an equivalent polar integral.
[**Hint.** S.E. 4 art 6.6] **(P.T.U., Dec. 2011)**

11. (*i*) Using double integral find the area bounded by parabola $x^2 = 4ay$ and $y^2 = 4ax$
[**Hint.** S.E. 1(*c*) art. 6.8] **(P.T.U., May 2015)**
(*ii*) Write the area bounded by the parabolas $y^2 = 4 - x$, $y^2 = 4 - 4x$ in terms of bouble integral. **(P.T.U., Dec. 2013)**

12. Find the area of the segment cut-off from the parabola $y = 4x - x^2$ by the straight line $y = x$.
[**Hint.** S.E. I(*a*) art. 6.8] **(P.T.U., May 2003)**

13. Find the area enclosed by the curves $y^2 = x^3$ and $y = x$. **(P.T.U., Dec. 2005, May 2009)**
[**Hint.** S.E. 4 art. 6.8]

14. Find the area of the region bounded by the lines $x = 2$, $x = -2$ and the circle $x^2 + y^2 = 9$.
[**Hint.** S.E. 2 art. 6.8] **(P.T.U., May 2007)**

15. (*i*) Find the area of the cardioid $r = a\,(1 - \cos\theta)$ **(P.T.U., May 2015)**
(*ii*) Find the area of one loop of the lemniscate $r^2 = a^2\cos 2\theta$. **(P.T.U., May 2012)**
[**Hint.** S.E. 7 art. 6.8]

16. (*i*) Find by triple integral the volume of the sphere $x^2 + y^2 + z^2 = b^2$. **(P.T.U., May 2003)**
[**Hint.** S.E. 3 art. 6.9]

(*ii*) Calculate the volume of the solid bounded by the planes $x = 0, y = 0, x + y + z = 1, z = 0$.
(*P.T.U., May 2015*)

17. Find by double integration the volume of the solid generated by revolving the ellipse $\frac{x^2}{a^2} + \frac{y^2}{b^2} = 1$ about y-axis.
[**Hint.** S.E. 5 art. 6.10]

18. Find by double integration the C.G. of the area of the cardiod $r = a(1 + \cos\theta)$.
[**Hint.** S.E. 6 art. 6.12]

19. Define theorem of perpendicular axis on M.I. of a body of mass M. **(*P.T.U., May 2008*)**
[**Hint.** Consult art 6.14]

20. Define theorem of parallel axis on the M.I. of a body of mass M.

21. Find by double integration the M.I. of a thin uniform circular disc of radius a about any diameter.
[**Hint.** See S.E.1, art. 6.16]

Answers

1. (*i*) $\frac{5}{2}$ (*ii*) $\frac{21}{2}$ (*iii*) $\frac{\pi}{4}\log(\sqrt{2}+1)$ (*iv*) $\frac{\pi}{4}\log 2$

(*v*) $\frac{8}{3}$ (*vi*) $\frac{3}{35}$ (*vii*) $\frac{241}{60}$ (*viii*) $\frac{9}{24}$

(*ix*) $8(\log 8 - 2) + e$ (*x*) $\frac{7}{12}$ (*xi*) $e^2 - 1$

3. (*i*) $\frac{a^3}{18}(3\pi - 4)$ (*ii*) $\frac{\pi a^2}{4}$ (*iii*) $\frac{5}{8}\pi a^3$

4. $-1 \le x \le 1, -1 \le y \le 1$ **5.** 0 **6.** (*i*) $1 - \cos 1$ (*ii*) 2

7. (*i*) $\frac{3}{8}$ (*ii*) $\frac{8}{3}a^2$ (*iii*) $\frac{a\pi}{2}$ (*iv*) 1

(*v*) $\frac{\pi}{16}$ (*vi*) $\frac{a^2\pi}{2}$ (*vii*) $\frac{1}{8}a^2b^2$

8. (*i*) $\frac{15}{2}$ (*ii*) 0 (*iii*) $\frac{1}{720}$ (*iv*) $\frac{1}{48}$

(*v*) $\frac{8}{3}abc(a^2 + b^2 + c^2)$ **9.** (*i*) $\frac{\pi}{4}$ (*ii*) $\frac{4}{3}$

(*iii*) $\frac{a^3}{3}\left(\frac{\pi}{2} - \frac{2}{3}\right)$ (*iv*) $\frac{\pi a^4}{8}$ (*v*) πa^2 (*vi*) $\frac{\pi a^4}{8}$

10. $\int_{\frac{\pi}{4}}^{\frac{\pi}{2}}\int_0^{2\cos\theta} f(r\cos\theta, r\cos\theta)\, r\,dr\,d\theta$ **11.** (*i*) $\frac{1}{3}$ (*ii*) 8 **12.** 4.5

13. $\frac{1}{10}$ **14.** $4\sqrt{5} + 18\sin^{-1}\frac{2}{3}$ **15.** (*i*) $\frac{3}{2}a^2$ (*ii*) $\frac{a^2}{2}$ **16.** (*i*) $\frac{4}{3}\pi b^3$ (*ii*) $\frac{1}{6}$

17. $\frac{4}{3}\pi a^2 b$ **18.** $\left(\frac{5a}{6}, 0\right)$ **21.** $\frac{Ma^2}{4}$.

7

Vector Calculus

7.1. RECAPITULATION OF VECTOR ALGEBRA

Before starting with vector calculus we would first like to revise the basic principles of vector algebra which students have done in their lower classes. These are of frequent use in vector calculus.

(*i*) **Position vector of a point.** Let O be any fixed point (called origin) in plane or space and P be any point then $\overrightarrow{OP}$ is called position vector (P.V.) of P w.r.t. O, *i.e.*, if $\overrightarrow{OP} = \vec{r}$ then we say that P.V. of P is $\vec{r}$ and is written as P $(\vec{r})$.

If P.V. of A is $\vec{a}$ and, that of B is $\vec{b}$ then, $\overrightarrow{AB} = \vec{b} - \vec{a}$ = Position vector of B – Position vector of A

(*ii*) **Addition of two vectors.** Let $\vec{a}, \vec{b}$ be represented by two adjacent sides of a parallelogram, then their sum $\vec{a} + \vec{b}$ is represented by the diagonal of the parallelogram which is co-initial with given vectors.

Addition of vectors is cummutative as well as associative *i.e.*, $\vec{a} + \vec{b} = \vec{b} + \vec{a}$ and

$$(\vec{a} + \vec{b}) + \vec{c} = \vec{a} + (\vec{b} + \vec{c})$$

Also, $$|\vec{a} + \vec{b}| \le |\vec{a}| + |\vec{b}| \text{ and } |\vec{a} - \vec{b}| \ge \big||\vec{a}| - |\vec{b}|\big|$$

(*iii*) If vectors are multiplied by scalars

Then, $$m(n\vec{a}) = (mn)\vec{a} = n(m\vec{a})$$

$$(m+n)\vec{a} = m\vec{a} + n\vec{a}$$

$$m(\vec{a} + \vec{b}) = m\vec{a} + m\vec{b}$$

(*iv*) **Section Formula.** If A $(\vec{a})$, B $(\vec{b})$ and C divides AB in the ratio $m : n$

Then position vector of C *i.e.*,

$$\vec{c} = \frac{m\vec{b} + n\vec{a}}{m+n}$$

From here we deduce that

$$m\vec{b} + n\vec{a} = (m+n)\vec{c}$$

(*v*) A vector $\vec{r}$ can be expressed in rectangular components as $\vec{r} = x\hat{i} + y\hat{j} + z\hat{k}$ where $\hat{i}, \hat{j}, \hat{k}$ are the unit vectors representing X-axis, Y-axis and Z-axis and x, y, z are scalar components of $\vec{r}$ along X-axis, Y-axis and Z-axis.

(*vi*) **Scalar dot product** of two vectors is defined as $\vec{a} \bullet \vec{b} = |\vec{a}||\vec{b}|\cos\theta$ where θ is the angle between $\vec{a}$ and $\vec{b}$. Dot product is cummutative.

i.e., $$\vec{a} \bullet \vec{b} = \vec{b} \bullet \vec{a}$$

$$\vec{a} \bullet \vec{a} = a^2$$

If $\vec{a} \perp \vec{b}$, then $\vec{a} \bullet \vec{b} = 0$

Dot product is distributive with addition *i.e.*, $\vec{a} \bullet (\vec{b} + \vec{c}) = \vec{a} \bullet \vec{b} + \vec{a} \bullet \vec{c}$

In rectangular components if $\vec{a} = a_1\hat{i} + a_2\hat{j} + a_3\hat{k}$; $\vec{b} = b_1\hat{i} + b_2\hat{j} + b_3\hat{k}$

Then $$\vec{a} \bullet \vec{b} = a_1 b_1 + a_2 b_2 + a_3 b_3$$

If $\vec{a} \perp \vec{b}$, then $a_1 b_1 + a_2 b_2 + a_3 b_3 = 0$

$$\cos\theta = \frac{a \quad b}{|a||b|} = \frac{a_1 b_1 + a_2 b_2 + a_3 b_3}{\sqrt{a_1^2 + a_2^2 + a_3^2}\sqrt{b_1^2 + b_2^2 + b_3^2}}$$

(*vii*) **Vector cross product** of two vectors $\vec{a}$ and $\vec{b}$ is defined as $\vec{a} \times \vec{b} = |\vec{a}||\vec{b}|\sin\theta\,\hat{n}$ where $\hat{n}$ is a unit vector perpendicular to both $\vec{a}$ and $\vec{b}$ and $\vec{a}, \vec{b}, \hat{n}$ form a right handed screw in anti-clockwise direction.

Cross product is not commutative *i.e.*, $\vec{a} \times \vec{b} \neq \vec{b} \times \vec{a}$;

$\vec{a} \times \vec{a} = \vec{0}$. Also if $\vec{a}$ is parallel to $\vec{b}$, then $\vec{a} \times \vec{b} = \vec{0}$

In rectangular components

$$\vec{a} \times \vec{b} = \begin{vmatrix} \hat{i} & \hat{j} & \hat{k} \\ a_1 & a_2 & a_3 \\ b_1 & b_2 & b_3 \end{vmatrix} = \sum \hat{i}\,(a_2 b_3 - a_3 b_2)$$

$$\sin\theta = \frac{|\vec{a} \times \vec{b}|}{|\vec{a}||\vec{b}|} = \frac{\sqrt{\sum (a_2 b_3 - a_3 b_2)^2}}{\sqrt{\sum a_1^2}\sqrt{\sum b_1^2}}$$

If $\vec{a} \parallel \vec{b}$, then $$\frac{a_1}{b_1} = \frac{a_2}{b_2} = \frac{a_3}{b_3}$$

(*viii*) **Scalar triple product** of three vectors is defined as $\left[\vec{a}\quad\vec{b}\quad\vec{c}\right] = \vec{a}\bullet\vec{b}\times\vec{c}$

It is independent of the position of dot and cross. Also $\vec{a}, \vec{b}, \vec{c}$ can be changed into cyclic order without affecting the products. In rectangular components

$$\left[\vec{a}\quad\vec{b}\quad\vec{c}\right] = \begin{vmatrix} a_1 & a_2 & a_3 \\ b_1 & b_2 & b_3 \\ c_1 & c_2 & c_3 \end{vmatrix}$$

Scalar triple product is zero when (*i*) all the three vectors are coplanar and (*ii*) if any two of them are parallel or collinear.

(*ix*) **Vector triple product** of three vectors is defined as $\vec{a}\times\left(\vec{b}\times\vec{c}\right) = (\vec{a}\bullet\vec{c})\vec{b} - \left(\vec{a}\bullet\vec{b}\right)\vec{c}$

and $\left(\vec{a}\times\vec{b}\right)\times\vec{c} = (\vec{a}\bullet\vec{c})\vec{b} - \left(\vec{b}\bullet\vec{c}\right)\vec{a}$

(*x*) Moments of a force $\vec{F}$ about a line represented by $\overrightarrow{AB}$ is $\overrightarrow{AB}\times\vec{F}$

(*xi*) Work done by a force $\vec{F}$ along a line $\overrightarrow{AB}$ is $\vec{F}\bullet\overrightarrow{AB}$.

7.2. (a) DEFINITION

Vector Function: If a vector $\vec{r}$ varies continuously with the variation of any scalar variable *t*, then $\vec{r}$ is said to be a vector function of *t* and is written as $\vec{r}$ or $r\left(\vec{t}\right)$ or $\vec{r} = f\left(\vec{t}\right)$. For example, the position vector $\vec{r}$ of a particle moving along a curved path is a vector function of time *t*, a scalar.

7.2. (b) DIFFERENTIATION OF A VECTOR FUNCTION WITH RESPECT TO A SCALAR

Differentiation of a vector function w.r.t. a scalar is defined in the same manner as in differential calculus.

Let $\vec{r} = f\left(\vec{t}\right)$ be a vector function of scalar variable *t*

Give increments to both sides

$$\vec{r} + \delta\vec{r} = f\left(\vec{t} + \delta\vec{t}\right)$$

Subtract $$\delta\vec{r} = f\left(\vec{t} + \delta\vec{t}\right) - f\left(\vec{t}\right)$$

Divide by $$\delta t \quad \frac{\delta\vec{r}}{\delta t} = \frac{f\left(\vec{t} + \delta\vec{t}\right) - f\left(\vec{t}\right)}{\delta t}$$

Taking limit when $\delta t \to 0$

$$\underset{\delta t\to 0}{\text{Lt}} \frac{\delta \vec{r}}{\delta t} = \underset{\delta t\to 0}{\text{Lt}} \frac{f(\vec{t}+\delta\vec{t})-f(\vec{t})}{\delta t}$$

If limit on the right hand side exists then this limiting value is denoted by $\frac{d\vec{r}}{dt}$

i.e.,
$$\frac{d\vec{r}}{dt} = \underset{\delta t\to 0}{\text{Lt}} \frac{f(\vec{t}+\delta\vec{t})-f(\vec{t})}{\delta t}$$

Then $\frac{d\vec{r}}{dt}$ is called derivative or differential of $\vec{r}$ w.r.t. t. Also $\frac{d\vec{r}}{dt}$ can be written as $\frac{d}{dt}f(\vec{t}) = f'(\vec{t})$.

Since $\frac{d\vec{r}}{dt}$ is itself a vector function of t, its derivative is denoted by $\frac{d^2\vec{r}}{dt^2}$ and is called second derivative of $\vec{r}$ w.r.t. t. Similarly we can define higher order derivatives of $\vec{r}$.

7.3. GENERAL RULES FOR DIFFERENTIATION

If $\vec{a},\vec{b},\vec{c}$ are the vector functions of a scalar t and ϕ is any scalar function of t, then

(*i*) $\frac{d}{dt}\left(\vec{a}\pm\vec{b}\right)=\frac{d\vec{a}}{dt}\pm\frac{d\vec{b}}{dt}$

(*ii*) $\frac{d}{dt}\left(\vec{a}\bullet\vec{b}\right)=\vec{a}\bullet\frac{d\vec{b}}{dt}+\frac{d\vec{a}}{dt}\bullet\vec{b}$

(*iii*) $\frac{d}{dt}\left(\vec{a}\times\vec{b}\right)=\vec{a}\times\frac{d\vec{b}}{dt}+\frac{d\vec{a}}{dt}\times\vec{b}$

(*iv*) $\frac{d}{dt}(\phi\vec{a})=\phi\frac{d\vec{a}}{dt}+\frac{d\phi}{dt}\vec{a}$

(*v*) $\frac{d}{dt}\left[\vec{a}\ \vec{b}\ \vec{c}\right]=\left[\frac{d\vec{a}}{dt}\vec{b}\ \vec{c}\right]+\left[\vec{a}\ \frac{d\vec{b}}{dt}\vec{c}\right]+\left[\vec{a}\ \vec{b}\ \frac{d\vec{c}}{dt}\right]$

(*vi*) $\frac{d}{dt}\left[\vec{a}\times\left(\vec{b}\times\vec{c}\right)\right]=\frac{d\vec{a}}{dt}\times\left(\vec{b}\times\vec{c}\right)+\vec{a}\times\left(\frac{d\vec{b}}{dt}\times\vec{c}\right)+\vec{a}\times\left(\vec{b}\times\frac{d\vec{c}}{dt}\right)$

Proof. (*i*)
$$\frac{d}{dt}\left(\vec{a}+\vec{b}\right) = \underset{\delta t\to 0}{\text{Lt}} \frac{\left(\vec{a}+\delta\vec{a}+\vec{b}+\delta\vec{b}\right)-\left(\vec{a}+\vec{b}\right)}{\delta t}$$

$$= \underset{\delta t\to 0}{\text{Lt}} \frac{\delta\vec{a}+\delta\vec{b}}{\delta t} = \underset{\delta t\to 0}{\text{Lt}}\left(\frac{\delta\vec{a}}{\delta t}+\frac{\delta\vec{b}}{\delta t}\right)=\frac{d\vec{a}}{dt}+\frac{d\vec{b}}{dt}$$

Similarly, $\frac{d}{dt}\left(\vec{a}-\vec{b}\right) = \frac{d\vec{a}}{dt}-\frac{d\vec{b}}{dt}$

(*ii*) $\dfrac{d}{dt}\left(\vec{a} \bullet \vec{b}\right) = \underset{\delta t \to 0}{\text{Lt}} \dfrac{\left(\vec{a} + \delta\vec{a}\right) \bullet \left(\vec{b} + \delta\vec{b}\right) - \vec{a} \bullet \vec{b}}{\delta t} = \underset{dt \to 0}{\text{Lt}} \dfrac{\vec{a} \bullet \delta\vec{b} + \delta\vec{a} \bullet \vec{b} + \delta\vec{a} \bullet \delta\vec{b}}{\delta t}$

$$= \underset{\delta t \to 0}{\text{Lt}} \left[\vec{a} \bullet \frac{\delta\vec{b}}{\delta t} + \frac{\delta\vec{a}}{\delta t} \bullet \vec{b} + \frac{\delta\vec{a}}{\delta t} \bullet \delta\vec{b}\right]$$

$$= \vec{a} \bullet \frac{d\vec{b}}{dt} + \frac{d\vec{a}}{dt} \bullet \vec{b} + \frac{d\vec{a}}{dt} \bullet \vec{0} \qquad (\because \text{ When } \delta t \to 0, \text{ also } \delta\vec{b} \to 0)$$

$$= \vec{a}\, \frac{d\vec{b}}{dt} + \frac{d\vec{a}}{dt} \bullet \vec{b}$$

(*iii*) $\dfrac{d}{dt}\left(\vec{a} \times \vec{b}\right) = \underset{\delta t \to 0}{\text{Lt}} \dfrac{\left(\vec{a} + \delta\vec{a}\right) \times \left(\vec{b} + \delta\vec{b}\right) - \vec{a} \times \vec{b}}{\delta t} = \underset{\delta t \to 0}{\text{Lt}} \dfrac{\vec{a} \times \delta\vec{b} + \delta\vec{a} \times \vec{b} + \delta\vec{a} \times \delta\vec{b}}{\delta t}$

$$= \underset{\delta t \to 0}{\text{Lt}} \left[\vec{a} \times \frac{\delta\vec{b}}{\delta t} + \frac{\delta\vec{a}}{\delta t} \times \vec{b} + \frac{\delta\vec{a}}{\delta t} \times \delta\vec{b}\right]$$

$$= \vec{a} \times \frac{d\vec{b}}{dt} + \frac{d\vec{a}}{dt} \times \vec{b} + \frac{d\vec{a}}{dt} \times \vec{0} \qquad (\because \text{ When } \delta t \to 0, \ d\vec{b} \to 0)$$

$$\therefore \quad \frac{d}{dt}\left(\vec{a} \times \vec{b}\right) = \vec{a} \times \frac{d\vec{b}}{dt} + \frac{d\vec{a}}{dt} \times \vec{b}$$

Note. Since $\vec{a} \times \vec{b} \neq \vec{b} \times \vec{a}$ $\therefore$ While evaluating $\dfrac{d}{dt}\left(\vec{a} \times \vec{b}\right)$ the order of the factors $\vec{a}$ and $\vec{b}$ must be maintained.

(*iv*) $\dfrac{d}{dt}(\phi\vec{a}) = \underset{\delta t \to 0}{\text{Lt}} \dfrac{(\phi + \delta\phi)\left(\vec{a} + \delta\vec{a}\right) - \phi\vec{a}}{\delta t}$

$$= \underset{\delta t \to 0}{\text{Lt}} \frac{\phi\delta\vec{a} + \delta\phi\vec{a} + \delta\phi.\delta\vec{a}}{\delta t} = \underset{\delta t \to 0}{\text{Lt}} \left[\phi\, \frac{\delta\vec{a}}{\delta t} + \frac{\delta\phi}{\delta t}\, \vec{a} + \frac{\delta\phi}{\delta t}\, \delta\vec{a}\right]$$

$$= \phi\, \frac{d\vec{a}}{dt} + \frac{d\phi}{dt}\, \vec{a} + \frac{d\phi}{dt}\, \vec{0} \qquad (\because \text{ When } \delta t \to 0, \text{ also } \delta\vec{a} \to 0)$$

$$= \phi\, \frac{d\vec{a}}{dt} + \frac{d\phi}{dt}\, \vec{a}$$

Note. $\phi\vec{a}$ is the product of a scalar with a vector, so generally scalar is written in first position and vector in second position.

(*v*) $\dfrac{d}{dt}\left[\vec{a}\ \vec{b}\ \vec{c}\right] = \dfrac{d}{dt}\left\{\vec{a} \bullet \left(\vec{b} \times \vec{c}\right)\right\}$

$$= \vec{a} \bullet \frac{d}{dt}\left(\vec{b} \times \vec{c}\right) + \frac{d\vec{a}}{dt} \bullet \left(\vec{b} \times \vec{c}\right)$$

$$= \vec{a} \bullet \left(\vec{b} \times \frac{d\vec{c}}{dt} + \frac{d\vec{b}}{dt} \times \vec{c}\right) + \frac{d\vec{a}}{dt} \bullet \left(\vec{b} \times \vec{c}\right)$$

$$= \vec{a} \bullet \left(\vec{b} \times \frac{d\vec{c}}{dt}\right) + \vec{a} \bullet \left(\frac{d\vec{b}}{dt} \times \vec{c}\right) + \frac{d\vec{a}}{dt} \bullet \left(\vec{b} \times \vec{c}\right)$$

Rearranging the terms, we have $\frac{d}{dt}[\vec{a}\ \vec{b}\ \vec{c}] = \left[\frac{d\vec{a}}{dt}\ \vec{b}\ \vec{c}\right] + \left[\vec{a}\ \frac{d\vec{b}}{dt}\ \vec{c}\right] + \left[\vec{a}\ \vec{b}\ \frac{d\vec{c}}{dt}\right]$

Note. $\left[\vec{a}\ \vec{b}\ \vec{c}\right]$ is the scalar product of three vectors $\vec{a}, \vec{b}, \vec{c}$ while evaluating $\frac{d}{dt}\left[\vec{a}\ \vec{b}\ \vec{c}\right]$ cyclic order of $\vec{a}, \vec{b}, \vec{c}$ should be maintained.

(*vi*) $$\frac{d}{dt}\left\{\vec{a} \times \left(\vec{b} \times \vec{c}\right)\right\} = \vec{a} \times \frac{d}{dt}\left(\vec{b} \times \vec{c}\right) + \frac{d\vec{a}}{dt} \times \left(\vec{b} \times \vec{c}\right)$$

$$= \vec{a} \times \left[\vec{b} \times \frac{d\vec{c}}{dt} + \frac{d\vec{b}}{dt} \times \vec{c}\right] + \frac{d\vec{a}}{dt} \times \left(\vec{b} \times \vec{c}\right)$$

$$= \vec{a} \times \left(\vec{b} \times \frac{d\vec{c}}{dt}\right) + \vec{a} \times \left(\frac{d\vec{b}}{dt} \times \vec{c}\right) + \frac{d\vec{a}}{dt} \times \left(\vec{b} \times \vec{c}\right)$$

Rearranging the terms, we have $= \frac{d\vec{a}}{dt} \times \left(\vec{b} \times \vec{c}\right) + \vec{a} \times \left(\frac{d\vec{b}}{dt} \times \vec{c}\right) + \vec{a} \times \left(\vec{b} \times \frac{d\vec{c}}{dt}\right)$.

7.4. DERIVATIVE OF A CONSTANT VECTOR

A vector is said to be constant if both its magnitude and direction are fixed. If either of these changes, the vector is not constant.

Let $\vec{r}$ be a constant vector function of the scalar variable t.

$\therefore \qquad \vec{r} = f\left(\vec{t}\right)$

Give increments to both the sides

$\vec{r} = f\left(\vec{t} + \delta\vec{t}\right) \quad \vec{r}$ remains the same $\quad \because$ It is constant

Subtracting $\qquad \vec{0} = f\left(\vec{t} + \delta\vec{t}\right) - f\left(\vec{t}\right)$

$$\therefore \qquad \frac{d\vec{r}}{dt} = \underset{\delta t \to 0}{\text{Lt}} \frac{f\left(\vec{t} + \delta\vec{t}\right) - f\left(\vec{t}\right)}{\delta t} = \underset{\delta t \to 0}{\text{Lt}} \frac{\vec{0}}{\delta t} = \vec{0}$$

Thus derivative of a constant vector is always a zero (null) vector.

7.5. IF $\vec{r}$ HAS CONSTANT MAGNITUDE BUT VARIABLE DIRECTION THEN $\vec{r} \cdot \frac{d\vec{r}}{dt} = 0$

(P.T.U., May 2004, 2012, Dec. 2011)

Proof. Let $\vec{r} = f(\vec{t})$

$\because$ $\vec{r}$ is constant in magnitude $\quad\therefore\quad |\vec{r}| = \text{constant}$

Now, $$\vec{r} \bullet \vec{r} = |\vec{r}|^2 = \text{constant}$$

Differentiate both sides w.r.t. t

$$\frac{d}{dt}(\vec{r} \bullet \vec{r}) = \frac{d}{dt}(\text{constant}) = 0$$

$$\therefore \quad \vec{r} \bullet \frac{d\vec{r}}{dt} + \frac{d\vec{r}}{dt} \bullet \vec{r} = 0$$

$$\Rightarrow \quad 2\vec{r} \bullet \frac{d\vec{r}}{dt} = 0 \quad \Rightarrow \quad \vec{r} \bullet \frac{d\vec{r}}{dt} = 0$$

Cor. If $\vec{r}$ is constant in magnitude then $\vec{r}$ is perpendicular to $\frac{d\vec{r}}{dt}$

$\because$ $\vec{r} \bullet \frac{d\vec{r}}{dt} = 0 \Rightarrow \vec{r}$ is $\perp$ $\frac{d\vec{r}}{dt}$ $\because$ When dot product of two vectors is zero, the vectors are perpendicular to each other.

7.6. IF $\vec{r}$ HAS CONSTANT DIRECTION BUT VARIABLE MAGNITUDE THEN $\vec{r} \times \frac{d\vec{r}}{dt} = \vec{0}$

(P.T.U., May 2014)

Proof. Let $\vec{r} = f(\vec{t})$

Let $$|\vec{r}| = |f(\vec{t})| = \phi(t)$$

Let $\hat{g}(t)$ be a unit vector in the direction of $\vec{r}$ or $f(\vec{t})$

$$\therefore \quad \vec{r} = \phi(t)\, g(\hat{t})$$

Differentiate both sides w.r.t. t

$$\frac{d\vec{r}}{dt} = \phi(t)\frac{d}{dt}g(\hat{t}) + \frac{d\phi}{dt}g(\hat{t})$$

$$= \phi(t)\,\vec{0} + \frac{d\phi}{dt} g(\hat{t}) \qquad \left(\because \; g(\hat{t}) \text{ is a constant vector. Its mag} = 1 \text{ and direction is same as that of constant vector } \vec{r}\,\right)$$

$$\frac{d\vec{r}}{dt} = \frac{d\phi}{dt} g(\hat{t})$$

Now,
$$\vec{r} \times \frac{d\vec{r}}{dt} = \{\phi(t) g(\hat{t})\} \times \frac{d\phi}{dt}\left(g(\hat{t})\right) = \phi(t)\frac{d\phi}{dt}\{g(\hat{t}) \times g(\hat{t})\} = \phi(t)\frac{d\phi}{dt}\,\vec{0} = \vec{0}.$$

7.7. GEOMETRICAL INTERPRETATIONS OF $\vec{r}$ AND $\frac{d\vec{r}}{dt}$

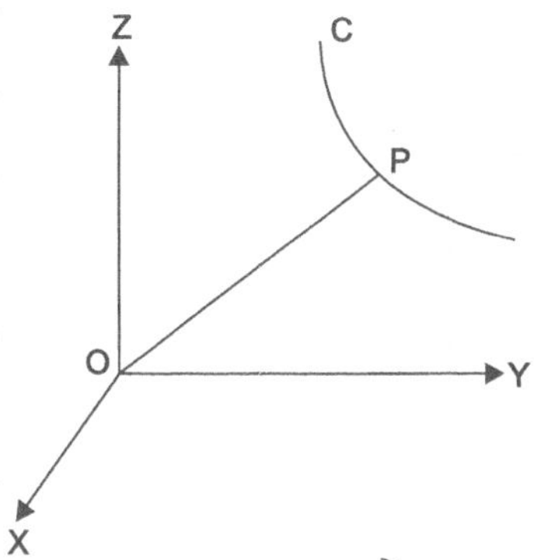

Let O be the origin of reference. Let the position vector of a point P be given by $\vec{r} = f(\vec{t})$. As t varies continuously, P traces out a curve C (see figure). Thus a vector function $\vec{r} = f(\vec{t})$ represents a curve in space *e.g.*, the vector equation $\vec{r} = at^2\hat{i} + 2at\hat{j}$ represents the parabola $y^2 = 4ax$ in the xy-plane because its parametric equations are $x = at^2$, $y = 2at$.

Now, let $\vec{r} = f(\vec{t})$ be the vector equation of a curve C in space. Let $\vec{r}$ and $\vec{r} + \delta\vec{r}$ be the position vectors of two neighbouring points P and Q on this curve then

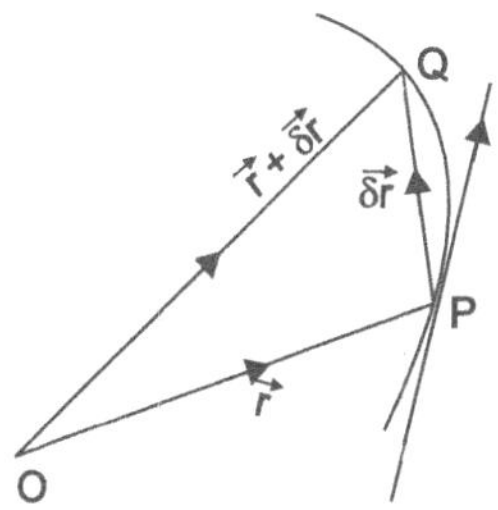

$$\overrightarrow{PQ} = \overrightarrow{OQ} - \overrightarrow{OP} = \vec{r} + \delta\vec{r} - \vec{r} = \delta\vec{r}$$

$\therefore \quad \frac{\delta\vec{r}}{\delta t}$ is directed along the chord PQ.

As $\delta t \to 0$; Q $\to$ P and chord PQ $\to$ tangent to curve at P

$\therefore \quad \underset{\delta t \to 0}{\text{Lt}} \frac{\delta\vec{r}}{\delta t} = \frac{d\vec{r}}{dt}$ is a vector along the tangent to the curve at P.

Suppose the scalar parameter is replaced by s, where s denotes the arc length from any convenient point A on the curve upto P. Thus arc AP $= s + \delta s$ so that arc PQ $= \delta s$. In this case $\frac{d\vec{r}}{ds}$ will be a vector along tangent at P. Also $\left|\frac{d\vec{r}}{ds}\right| = \underset{\delta s \to 0}{\text{Lt}} \left|\frac{\delta\vec{r}}{\delta s}\right| = \underset{Q \to P}{\text{Lt}} \frac{\text{Chord PQ}}{\text{arc PQ}} = 1$.

$\therefore \quad \frac{d\vec{r}}{ds}$ is a unit vector along tangent at P and is represented by $\hat{T}$.

7.8. VELOCITY AND ACCELERATION

If $\vec{r}$ is the position vector of a moving particle and t, a scalar variable denotes the time, then $\delta\vec{r}$ is the displacement of the particle in δt time. The vector $\dfrac{\delta\vec{r}}{\delta t}$ represents the average velocity of the particle during the interval δt and its limiting value when $\delta t \to 0$ represents the variable velocity of the particle at time t. If $\vec{v}$ represents the velocity of the particle at time t then $\vec{v} = \underset{\delta t\to 0}{\text{Lt}} \dfrac{\delta\vec{r}}{\delta t} = \dfrac{d\vec{r}}{dt}$ and its direction is along the tangent at that point. If $\delta\vec{v}$ be the change in velocity $\vec{v}$ in time δt then $\dfrac{\delta\vec{v}}{\delta t}$ is the average acceleration of the particle during the interval δt. If $\vec{a}$ represents the acceleration at time t then $\vec{a} = \underset{\delta t\to 0}{\text{Lt}} \dfrac{\delta\vec{v}}{\delta t} = \dfrac{d\vec{v}}{dt}$ or $\dfrac{d}{dt}\left(\dfrac{d\vec{r}}{dt}\right) = \dfrac{d^2\vec{r}}{dt^2}$.

7.9. TANGENTIAL AND NORMAL ACCELERATION OF A PARTICLE MOVING IN A PLANE CURVE

Proof. Let $\vec{r}$ be the position vector of a point P at any time t. Its resultant velocity be $\vec{V}$ then

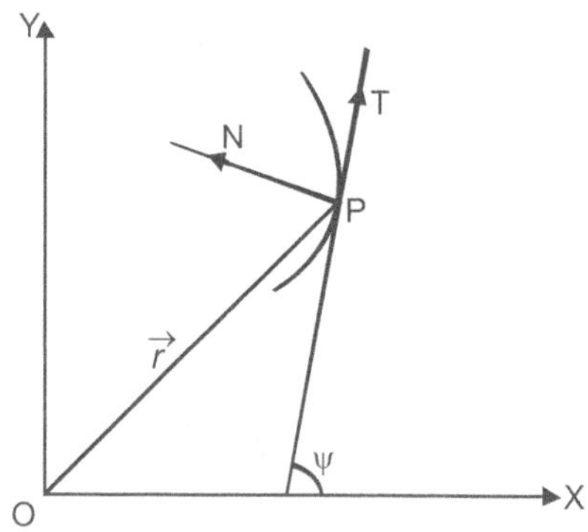

$$\vec{V} = \frac{d\vec{r}}{dt} = \frac{d\vec{r}}{ds}\frac{ds}{dt}$$

Since $\dfrac{d\vec{r}}{ds}$ is a unit vector along tangent at P

$\therefore$ Let $$\frac{d\vec{r}}{ds} = \hat{T}$$

$\therefore$ $$\vec{V} = \frac{ds}{dt}\hat{T}$$

Acceleration is given by $$\frac{d\vec{V}}{dt} = \frac{d^2s}{dt^2}\hat{T} + \frac{ds}{dt}\frac{d\hat{T}}{dt} \qquad \ldots(1)$$

Now, $$\frac{d\hat{T}}{dt} = \frac{d\hat{T}}{d\psi}\frac{d\psi}{dt}$$

As $\hat{T}$ has constant magnitude

$\therefore$ (from art 7.5) $\hat{T} \bullet \dfrac{d\hat{T}}{d\psi} = 0$ *i.e.,* $\dfrac{d\hat{T}}{d\psi}$ is $\perp$ $\hat{T}$.

$\therefore$ $\dfrac{d\hat{T}}{d\psi}$ is a unit vector along normal at P, let $\dfrac{d\hat{T}}{d\psi} = \hat{N}$

$$\therefore \quad \frac{d\hat{T}}{dt} = \frac{d\psi}{dt}\hat{N} = \frac{d\psi}{ds}\frac{ds}{dt}\hat{N} = \frac{1}{\rho}\frac{ds}{dt}\hat{N}, \text{ where } \rho \text{ is the radius of curvature at P.}$$

$$\therefore \quad \text{From (1),} \quad \frac{d\vec{V}}{dt} = \frac{d^2s}{dt^2}\hat{T} + \frac{ds}{dt}\frac{1}{\rho}\frac{ds}{dt}\hat{N} = \frac{d^2s}{dt^2}\hat{T} + \frac{1}{\rho}\left(\frac{ds}{dt}\right)^2\hat{N}$$

$$= \frac{dv}{dt}\hat{T} + \frac{1}{\rho}v^2\hat{N}, \quad \text{where} \quad \frac{ds}{dt} = u$$

Hence tangential and normal accelerations are $\frac{dv}{dt}$ and $\frac{v^2}{\rho}$

or $\frac{dv}{dt}$ can also be expressed in the form $\frac{d^2s}{dt^2}$ or $\frac{dv}{ds}$.

Note 1. The magnitude of acceleration is not always the rate of change of $|\vec{v}|$ $\because$ $\vec{A}$ (acceleration vector) is not always tangential to the path.

Note 2. The acceleration is the rate of change of $|\vec{v}|$ iff normal acceleration is zero.

7.10. RADIAL AND TRANSVERSE ACCELERATION OF A PARTICLE MOVING IN A PLANE CURVE

Let $\vec{r}$ be the position vector of any point P(r, θ) at time t.

Let $\hat{R}$ and $\hat{T}$ be the unit vectors in radial and transverse direction respectively.

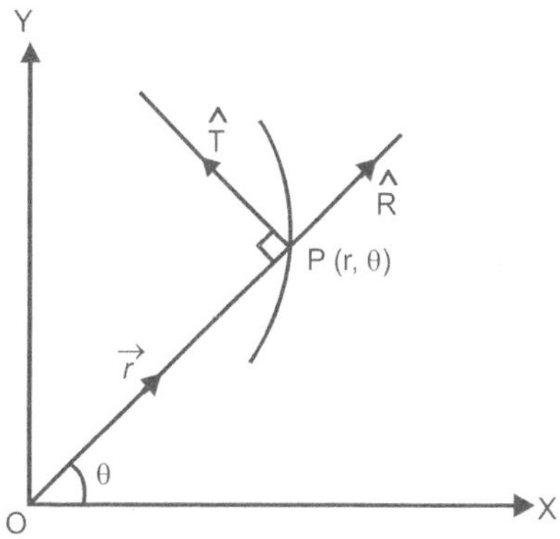

Now, $\vec{r} = r\hat{R}$

Velocity $\vec{V} = \frac{d\vec{r}}{dt} = \frac{d}{dt}\left(r\hat{R}\right)$

$$= \frac{dr}{dt}\hat{R} + r\frac{d}{dt}\hat{R} = \frac{dr}{dt}\hat{R} + r\frac{d\hat{R}}{d\theta}\frac{d\theta}{dt}$$

$$= \frac{dr}{dt}\hat{R} + r\frac{d\theta}{dt}\hat{T}$$

$\left[\begin{array}{l} \because \ \hat{R} \bullet \frac{d\hat{R}}{d\theta} = 0 \\ \therefore \ \frac{d\hat{R}}{d\theta} \text{ is } \perp \hat{R} \text{ and} \\ \because \ \hat{T} \text{ is } \perp \hat{R} \quad \therefore \ \frac{d\hat{R}}{d\theta} = \hat{T} \end{array}\right]$

$\therefore$ Components of velocity in radial and transverse direction are $\frac{dr}{dt}$ and $r\frac{d\theta}{dt}$

Also acceleration $\vec{A} = \dfrac{d\vec{V}}{dt} = \dfrac{d}{dt}\left[\dfrac{dr}{dt}\hat{R} + r\dfrac{d\theta}{dt}\hat{T}\right]$

$$= \frac{d^2r}{dt^2}\hat{R} + \frac{dr}{dt}\frac{d\hat{R}}{dt} + \frac{dr}{dt}\frac{d\theta}{dt}\hat{T} + r\frac{d^2\theta}{dt^2}\hat{T} + r\frac{d\theta}{dt}\frac{d}{dt}\hat{T}$$

$$= \frac{d^2r}{dt^2}\hat{R} + \frac{dr}{dt}\frac{d\hat{R}}{d\theta}\frac{d\theta}{dt} + \left(\frac{dr}{dt}\frac{d\theta}{dt} + r\frac{d^2\theta}{dt^2}\right)\hat{T} + r\frac{d\theta}{dt}\frac{d\hat{T}}{d\theta}\times\frac{d\theta}{dt}$$

Since $\dfrac{d\hat{R}}{d\theta} = \hat{T}$ and $\dfrac{d\hat{T}}{d\theta} = -\hat{R}$

$$\therefore \quad \vec{A} = \frac{d^2r}{dt^2}\hat{R} + \frac{dr}{dt}\frac{d\theta}{dt}\hat{T} + \left(\frac{dr}{dt}\frac{d\theta}{dt} + r\frac{d^2\theta}{dt^2}\right)\hat{T} - r\left(\frac{d\theta}{dt}\right)^2\hat{R}$$

$$= \left[\frac{d^2r}{dt^2} - r\left(\frac{d\theta}{dt}\right)^2\right]\hat{R} + \left[2\frac{dr}{dt}\frac{d\theta}{dt} + r\frac{d^2\theta}{dt^2}\right]\hat{T}$$

Hence the radial and transverse components of acceleration are

$$\frac{d^2r}{dt^2} - r\left(\frac{d\theta}{dt}\right)^2 \text{ and } 2\frac{dr}{dt}\frac{d\theta}{dt} + r\frac{d^2\theta}{dt^2}.$$

7.11. TO PROVE

(*i*) $$\frac{d^2}{dt^2}\left[\vec{r}\ \frac{d\vec{r}}{dt}\ \frac{d^2\vec{r}}{dt^2}\right] = \left[\vec{r}\ \frac{d^2\vec{r}}{dt^2}\ \frac{d^3\vec{r}}{dt^3}\right] + \left[\vec{r}\ \frac{d\vec{r}}{dt}\ \frac{d^4\vec{r}}{dt^4}\right]$$

(*ii*) $$\frac{d}{dt}\left[\vec{r}\times\left(\frac{d\vec{r}}{dt}\times\frac{d^2\vec{r}}{dt^2}\right)\right] = \frac{d\vec{r}}{dt}\times\left(\frac{d\vec{r}}{dt}\times\frac{d^2\vec{r}}{dt^2}\right) + \vec{r}\times\left(\frac{d\vec{r}}{dt}\times\frac{d^3\vec{r}}{dt^3}\right)$$

Proof. (*i*) Let $R = \left[\vec{r}\ \dfrac{d\vec{r}}{dt}\ \dfrac{d^2\vec{r}}{dt^2}\right]$ $\therefore$ R is a scalar triple product of $\vec{r}$, $\dfrac{d\vec{r}}{dt}$, $\dfrac{d^2\vec{r}}{dt^2}$

$$\therefore \quad \frac{dR}{dt} = \left[\frac{d\vec{r}}{dt}\ \frac{d\vec{r}}{dt}\ \frac{d^2\vec{r}}{dt^2}\right] + \left[\vec{r}\ \frac{d^2\vec{r}}{dt^2}\ \frac{d^2\vec{r}}{dt^2}\right] + \left[\vec{r}\ \frac{d\vec{r}}{dt}\ \frac{d^3\vec{r}}{dt^3}\right]$$

$$= \left[\vec{r}\ \frac{d\vec{r}}{dt}\ \frac{d^3\vec{r}}{dt^3}\right]$$

[Since scalar triple product having two equal vectors vanishes]

Differentiate again

$$\frac{d^2R}{dt^2} = \left[\frac{d\vec{r}}{dt}\ \frac{d\vec{r}}{dt}\ \frac{d^3\vec{r}}{dt^3}\right] + \left[\vec{r}\ \frac{d^2\vec{r}}{dt^2}\ \frac{d^3\vec{r}}{dt^3}\right] + \left[\vec{r}\ \frac{d\vec{r}}{dt}\ \frac{d^4\vec{r}}{dt^4}\right]$$

$$= \left[\vec{r}\ \frac{d^2\vec{r}}{dt^2}\ \frac{d^3\vec{r}}{dt^3}\right] + \left[\vec{r}\ \frac{d\vec{r}}{dt}\ \frac{d^4\vec{r}}{dt^4}\right]$$

(*ii*) Let $\vec{R} = \vec{r} \times \left(\frac{d\vec{r}}{dt} \times \frac{d^2\vec{r}}{dt^2}\right)$

$\therefore$ $\vec{R}$ is a vector triple product of three vectors

$$\therefore \quad \frac{d\vec{R}}{dt} = \frac{d\vec{r}}{dt} \times \left(\frac{d\vec{r}}{dt} \times \frac{d^2\vec{r}}{dt^2}\right) + \vec{r} \times \left(\frac{d^2\vec{r}}{dt^2} \times \frac{d^2\vec{r}}{dt^2}\right) + \vec{r} \times \left(\frac{d\vec{r}}{dt} \times \frac{d^3\vec{r}}{dt^3}\right)$$

$$\therefore \quad \frac{d\vec{R}}{dt} = \frac{d\vec{r}}{dt} \times \left(\frac{d\vec{r}}{dt} \times \frac{d^2\vec{r}}{dt}\right) + \vec{r} \times \left(\frac{d\vec{r}}{dt} \times \frac{d^3\vec{r}}{dt^3}\right) \quad \because \quad \frac{d^2\vec{r}}{dt^2} \times \frac{d^2\vec{r}}{dt^2} = 0$$

7.12. DERIVATIVE OF A VECTOR IN TERMS OF ITS COMPONENTS

Let $\vec{r}$ be a vector function of scalar variable t.

Let $\vec{r} = x\hat{i} + y\hat{j} + z\hat{k}$, where $\hat{i}, \hat{j}, \hat{k}$ are unit vectors in directions of x-axis, y-axis, z-axis respectively and x, y, z are components of $\vec{r}$ along these directions respectively. As $\vec{r}$ is a function of t.

$\therefore$ x, y, z are also functions of t.

Differentiate w.r.t. t, we have $\frac{d\vec{r}}{dt} = \frac{d}{dt}(x\hat{i}) + \frac{d}{dt}(y\hat{j}) + \frac{d}{dt}(z\hat{k})$

$$= x\frac{d\hat{i}}{dt} + \frac{dx}{dt}\hat{i} + y\frac{d}{dt}\hat{j} + \frac{dy}{dt}\hat{j} + z\frac{d\hat{k}}{dt} + \frac{dz}{dt}\hat{k}$$

$$= \frac{dx}{dt}\hat{i} + \frac{dy}{dt}\hat{j} + \frac{dz}{dt}\hat{k} \qquad \left[\text{Since } \frac{d\hat{i}}{dt} = \frac{d\hat{j}}{dt} = \frac{d\hat{k}}{dt} = \vec{0}\right]$$

If $x = f_1(t), y = f_2(t), z = f_3(t)$ then $\frac{d\vec{r}}{dt} = f_1'(t)\hat{i} + f_2'(t)\hat{j} + f_3'(t)\hat{k}$

$\therefore$ To differentiate a vector, differentiate its components only.

ILLUSTRATIVE EXAMPLES

Example 1. *Given* $\vec{r} = t^m \vec{a} + t^n \vec{b}$ *where* $\vec{a}$ *and* $\vec{b}$ *are constant vectors show that if* $\vec{r}$ *and* $\frac{d^2\vec{r}}{dt^2}$ *are parallel then* $m + n = 1$ *and* $m = n$.

Sol.
$$\vec{r} = t^m \vec{a} + t^n \vec{b}$$

$$\frac{d\vec{r}}{dt} = mt^{m-1}\vec{a} + nt^{n-1}\vec{b}$$

$$\left[\because \quad \vec{a}, \vec{b} \text{ are constant vectors } \therefore \quad \frac{d\vec{a}}{dt} = \vec{0}, \frac{d\vec{b}}{dt} = \vec{0}\right]$$

$$\frac{d^2\vec{r}}{dt^2} = m(m-1)t^{m-2}\vec{a} + n(n-1)t^{n-2}\vec{b}$$

$\because$ $\vec{r}$ and $\frac{d^2\vec{r}}{dt^2}$ are parallel.

$\therefore$
$$\vec{r} = k\frac{d^2\vec{r}}{dt^2}, \text{ where } k \text{ is any constant}$$

i.e.,
$$t^m\vec{a} + t^n\vec{b} = k\left[m(m-1)t^{m-2}\vec{a} + n(n-1)t^{n-2}\vec{b}\right]$$

Since $\vec{a}, \vec{b}$ are independent

$\therefore$ Comparing co-efficients of $\vec{a}$ and $\vec{b}$, we get

$$t^m = k\,[m\,(m-1)\,t^{m-2}] \quad \Rightarrow \quad t^2 = k\,m\,(m-1)$$
$$t^n = k\,[n\,(n-1)\,t^{n-2}] \quad \Rightarrow \quad t^2 = k\,n\,(n-1)$$

Eliminate k, we get $m\,(m-1) = n\,(n-1)$

or $\quad m^2 - m - n^2 + n = 0 \quad$ or $\quad (m^2 - n^2) - (m - n) = 0$

or $\quad (m - n)\,(m + n - 1) = 0 \quad$ or $\quad$ either $m = n$ or $m + n = 1$

Example 2. *Show that if* $\vec{r} = \vec{a}\sin\omega t + \vec{b}\cos\omega t$, *where* $\vec{a}, \vec{b}, \omega$ *are constants, then* $\frac{d^2\vec{r}}{dt^2} = -\omega^2\vec{r}$ *and* $\vec{r} \times \frac{d\vec{r}}{dt} = -w\vec{a} \times \vec{b}$.

Sol.
$$\vec{r} = \vec{a}\sin\omega t + \vec{b}\cos\omega t \qquad ...(1)$$

Differentiate again w.r.t. t

$$\frac{d\vec{r}}{dt} = \vec{a}(\cos\omega t)\,\omega + \vec{b}\,(-\sin\omega t)\,\omega \qquad \left[\begin{array}{l}\because \ \vec{a}, \vec{b} \text{ are constants vectors} \\ \therefore \ \frac{d\vec{a}}{dt} = \frac{d\vec{b}}{dt} = \vec{0}\end{array}\right]$$

$$\therefore \qquad \frac{d\vec{r}}{dt} = \omega\left(\vec{a}\,\cos\omega t - \vec{b}\,\sin\omega t\right)$$

Differentiate again w.r.t. t

$$\frac{d^2\vec{r}}{dt^2} = \omega\left[\vec{a}\,(-\sin\omega t)\omega - \vec{b}\,(\cos\omega t)\omega\right] = -\omega^2\left[\vec{a}\,\sin\omega t + \vec{b}\,\cos\omega t\right]$$

$$= -\omega^2\vec{r} \qquad [\because \text{ of } (1)]$$

$$\text{Hence} \qquad \frac{d^2\vec{r}}{dt^2} = -\omega^2\vec{r}$$

$$\text{Now,} \qquad \vec{r}\times\frac{d\vec{r}}{dt} = \left(\vec{a}\,\sin\omega t + \vec{b}\,\cos\omega t\right)\times\omega\left(\vec{a}\,\cos\omega t - \vec{b}\,\sin\omega t\right)$$

$$= \omega\left[-\vec{a}\times\vec{b}\,\sin^2\omega t + \vec{b}\times\vec{a}\,\cos^2\omega t\right] \qquad [\because \vec{a}\times\vec{a} = \vec{b}\times\vec{b} = \vec{0}\,]$$

$$= -\omega\left(\vec{a}\times\vec{b}\right)\,[\sin^2\omega t + \cos^2\omega t]$$

$$\vec{r}\times\frac{d\vec{r}}{dt} = -\omega\left(\vec{a}\times\vec{b}\right)$$

Example 3. *If* $\hat{R}$ *is a unit vector in the direction of* $\vec{r}$, *prove that* $\hat{R}\times\frac{d\hat{R}}{dt} = \frac{\vec{r}}{r^2}\times\frac{d\vec{r}}{dt}$ *where* $r = |\vec{r}\,|$.

Sol. $\vec{r} = r\hat{R} \quad \therefore \quad \hat{R} = \frac{\vec{r}}{r} = \left(\frac{1}{r}\right)\vec{r}$

Differentiate both sides w.r.t. t

$$\frac{d\hat{R}}{dt} = \frac{d}{dt}\left(\frac{1}{r}\,\vec{r}\right) = \frac{1}{r}\frac{d\vec{r}}{dt} - \frac{1}{r^2}\frac{dr}{dt}\,\vec{r}$$

$$\text{Now,} \qquad \hat{R}\times\frac{d\hat{R}}{dt} = \left(\frac{1}{r}\,\vec{r}\right)\times\left(\frac{1}{r}\frac{d\vec{r}}{dt} - \frac{1}{r^2}\frac{dr}{dt}\,\vec{r}\right) = \frac{1}{r^2}\,\vec{r}\times\frac{d\vec{r}}{dt} - \frac{\vec{r}}{r^3}\frac{d\vec{r}}{dt}\,\vec{r}\times\vec{r}$$

$$\text{Since,} \qquad \vec{r}\times\vec{r} = \vec{0}$$

$$\therefore \qquad \hat{R}\times\frac{d\hat{R}}{dt} = \frac{1}{r^2}\,\vec{r}\times\frac{d\vec{r}}{dt} = \frac{\vec{r}}{r^2}\times\frac{d\vec{r}}{dt}$$

Example 4. *If* $\vec{r}$ *is a vector function of a scalar t and* $\vec{a}$ *is a constant vector, differentiate the following w.r.t. t.*

(*i*) $\dfrac{\vec{r}\times\vec{a}}{\vec{r}\bullet\vec{a}}$ (*ii*) $\dfrac{\vec{r}+\vec{a}}{\vec{r}^{\,2}+\vec{a}^{\,2}}$.

Sol. (*i*) Let $\vec{R} = \dfrac{\vec{r} \times \vec{a}}{\vec{r} \bullet \vec{a}} = \left(\dfrac{1}{\vec{r} \bullet \vec{a}}\right)(\vec{r} \times \vec{a})$

$\vec{r} \bullet \vec{a}$ is a scalar function of t, $\vec{a}$ is a constant vector $\therefore \dfrac{d\vec{a}}{dt} = 0$

Differentiate both sides w.r.t. t

$$\frac{d\vec{R}}{dt} = \left(\frac{1}{\vec{r} \bullet \vec{a}}\right)\frac{d}{dt}(\vec{r} \times \vec{a}) + \frac{d}{dt}\left(\frac{1}{\vec{r} \bullet \vec{a}}\right)(\vec{r} \times \vec{a})$$

$$= \frac{1}{\vec{r} \bullet \vec{a}}\left\{\frac{d\vec{r}}{dt} \times \vec{a} + \vec{r} \times \frac{d\vec{a}}{dt}\right\} - \frac{1}{(\vec{r} \bullet \vec{a})^2}\frac{d}{dt}(\vec{r} \bullet \vec{a})(\vec{r} \times \vec{a})$$

$$\left(\because \frac{d}{dt}(\vec{r} \bullet \vec{a})^{-1} = -1(\vec{r} \bullet \vec{a})^{-2}\frac{d}{dt}(\vec{r} \bullet \vec{a})\right)$$

$$= \frac{1}{\vec{r} \bullet \vec{a}}\left\{\frac{d\vec{r}}{dt} \times \vec{a}\right\} - \frac{1}{(\vec{r} \bullet \vec{a})^2}\left\{\frac{d\vec{r}}{dt} \bullet \vec{a} + \vec{r} \bullet \frac{d\vec{a}}{dt}\right\}(\vec{r} \times \vec{a}) \qquad \left(\because \frac{d\vec{a}}{dt} = 0\right)$$

$$= \frac{1}{\vec{r} \bullet \vec{a}}\left\{\frac{d\vec{r}}{dt} \times \vec{a}\right\} - \frac{1}{(\vec{r} \bullet \vec{a})^2}\left(\frac{d\vec{r}}{dt} \bullet \vec{a}\right)(\vec{r} \times \vec{a})$$

$$= \frac{\dfrac{d\vec{r}}{dt} \times \vec{a}}{\vec{r} \bullet \vec{a}} - \frac{\left(\dfrac{d\vec{r}}{dt} \bullet \vec{a}\right)(\vec{r} \times \vec{a})}{(\vec{r} \bullet \vec{a})^2}.$$

(*ii*) Let $\vec{R} = \dfrac{\vec{r} + \vec{a}}{\vec{r}^2 + \vec{a}^2}$

$\vec{r}^2 = |\vec{r}|^2$ is a scalar function of t.

$\vec{a}^2 = |\vec{a}|^2$ is a constant, independent of t (given)

$$\therefore \quad \frac{d\vec{R}}{dt} = \frac{d}{dt}\left\{\left(\frac{1}{\vec{r}^2 + \vec{a}^2}\right)(\vec{r} + \vec{a})\right\}$$

$$= \frac{1}{\vec{r}^2 + \vec{a}^2}\frac{d}{dt}(\vec{r} + \vec{a}) + \frac{d}{dt}\left\{\frac{1}{\vec{r}^2 + \vec{a}^2}\right\}(\vec{r} + \vec{a})$$

$$= \frac{1}{\vec{r}^2 + \vec{a}^2}\left(\frac{d\vec{r}}{dt} + \frac{d\vec{a}}{dt}\right) + (-1)\frac{1}{(\vec{r}^2 + \vec{a}^2)^2}\left(2\vec{r} \bullet \frac{d\vec{r}}{dt} + 2\vec{a} \bullet \frac{d\vec{a}}{dt}\right)(\vec{r} + \vec{a})$$

$$\left[\text{Since } \frac{d}{dt}(\vec{r}^2) = \frac{d}{dt}(\vec{r} \bullet \vec{r}) = \vec{r} \bullet \frac{d\vec{r}}{dt} + \frac{d\vec{r}}{dt} \bullet \vec{r} = 2\vec{r} \bullet \frac{d\vec{r}}{dt}\right]$$

$$= \frac{1}{\vec{r}^2 + \vec{a}^2}\frac{d\vec{r}}{dt} - \frac{1}{\left(\vec{r}^2 + \vec{a}^2\right)^2}\left(2\vec{r} \bullet \frac{d\vec{r}}{dt}\right)\left(\vec{r} + \vec{a}\right) \qquad \left(\text{Put} \frac{d\vec{a}}{dt} = 0\right)$$

$$= \frac{\frac{d\vec{r}}{dt}}{\vec{r}^2 + \vec{a}^2} - \frac{2\vec{r} \bullet \frac{d\vec{r}}{dt}}{\left(\vec{r}^2 + \vec{a}^2\right)^2}\left(\vec{r} + \vec{a}\right).$$

Example 5. *If* $\frac{d\vec{u}}{dt} = \vec{\omega} \times \vec{u}$ *and* $\frac{d\vec{v}}{dt} = \vec{\omega} \times \vec{v}$, *prove that* $\frac{d}{dt}\left(\vec{u} \times \vec{v}\right) = \vec{\omega} \times \left(\vec{u} \times \vec{v}\right)$

Sol. $\frac{d}{dt}(\vec{u} \times \vec{v}) = \vec{u} \times \frac{d\vec{v}}{dt} + \frac{d\vec{u}}{dt} \times \vec{v}$

Substitute the values of $\frac{d\vec{u}}{dt}$ and $\frac{d\vec{v}}{dt}$, we get $\frac{d}{dt}(\vec{u} \times \vec{v}) = \vec{u} \times (\vec{\omega} \times \vec{v}) + (\vec{\omega} \times \vec{u}) \times \vec{v}$

$$= (\vec{u} \bullet \vec{v})\,\vec{\omega} - (\vec{u} \bullet \vec{\omega})\,\vec{v} + (\vec{v} \bullet \vec{\omega})\,\vec{u} - (\vec{v} \bullet \vec{u})\,\vec{\omega}$$

$$= (\vec{v} \bullet \vec{\omega})\vec{u} - (\vec{u} \bullet \vec{\omega})\vec{v}$$

($\because$ $\vec{u} \bullet \vec{v} = \vec{v} \bullet \vec{u}$; 1st and 4th terms cancel each other)

$$= (\vec{\omega} \bullet \vec{v})\vec{u} - (\vec{\omega} \bullet \vec{u})\vec{v} = \vec{\omega} \times (\vec{u} \times \vec{v})$$

Example 6. *If* $\vec{r} = a\cos t\,\hat{i} + a\sin t\,\hat{j} + at\tan\alpha\hat{k}$, *find* $\left|\frac{d\vec{r}}{dt} \times \frac{d^2\vec{r}}{dt^2}\right|$ *and* $\left[\frac{d\vec{r}}{dt}\ \frac{d^2\vec{r}}{dt^2}\ \frac{d^3\vec{r}}{dt^3}\right]$.

(P.T.U., Jan. 2010)

Sol.

$$\vec{r} = a\cos t\,\hat{i} + a\sin t\,\hat{j} + at\,\tan\alpha\hat{k}$$

$$\frac{d\vec{r}}{dt} = -a\sin t\hat{i} + a\cos t\hat{j} + a\tan\alpha\hat{k}$$

$$\frac{d^2\vec{r}}{dt^2} = -a\cos t\hat{i} - a\sin t\hat{j}$$

$$\frac{d^3\vec{r}}{dt^3} = a\sin t\,\hat{i} - a\cos t\hat{j}$$

$$\frac{d\vec{r}}{dt} \times \frac{d^2\vec{r}}{dt^2} = \begin{vmatrix} \hat{i} & \hat{j} & \hat{k} \\ -a\sin t & a\cos t & a\tan\alpha \\ -a\cos t & -a\sin t & 0 \end{vmatrix}$$

$$= a^2\sin t\,\tan\alpha\hat{i} - a^2\cos t\,\tan\alpha\,\hat{j} + a^2\,\hat{k}$$

$$\left|\frac{d\vec{r}}{dt} \times \frac{d^2\vec{r}}{dt^2}\right| = \sqrt{a^4\sin^2 t\tan^2\alpha + a^4\cos^2 t\,\tan^2\alpha + a^4}$$

$$= a^2\sqrt{\tan^2\alpha+1} = a^2 \sec\alpha.$$

Also, $$\left[\frac{d\vec{r}}{dt}\ \frac{d^2\vec{r}}{dt^2}\ \frac{d^3\vec{r}}{dt^3}\right] = \begin{vmatrix} -a\sin t & a\cos t & a\tan\alpha \\ -a\cos t & -a\sin t & 0 \\ a\sin t & -a\cos t & 0 \end{vmatrix}$$

$$= a \tan a\,(a^2\cos^2 t + a^2\sin^2 t) = a^3 \tan a.$$

Example 7. *At any point of the curve $x = 3\cos t$, $y = 3\sin t$, $z = 4t$ find*

(i) Tangent vector *(ii) Unit tangent vector*

(iii) Normal vector *(iv) Unit normal vector*

Sol. Let $\vec{r}$ be the position vector of the point

$$\therefore \quad \vec{r} = x\hat{i} + y\hat{j} + z\hat{k}$$

$$\therefore \quad \vec{r} = (3\cos t)\,\hat{i} + (3\sin t)\hat{j} + (4t)\,\hat{k}$$

$$(i) \quad \frac{d\vec{r}}{dt} = (-3\sin t)\,\hat{i} + (3\cos t)\hat{j} + (4)\hat{k} \qquad \ldots(1)$$

which is the required tangent vector.

$$(ii) \text{ Unit tangent vector} = \frac{\text{Tangent vector}}{\text{Magnitude of tangent vector}}$$

Now, magnitude of tangent vector $= \left|\frac{d\vec{r}}{dt}\right| = \sqrt{(-3\sin t)^2 + (3\cos t)^2 + 4^2} = \sqrt{9+16} = 5$

$$\therefore \text{ Unit tangent vector} = \frac{1}{5}\frac{d\vec{r}}{dt} = \frac{1}{5}\left\{(-3\sin t)\hat{i} + (3\cos t)\hat{j} + 4\hat{k}\right\}$$

(iii) Normal vector $= \dfrac{d\hat{T}}{ds}$, where $\hat{T}$ represent the unit tangent vector at the point.

$$= \frac{d\hat{T}}{dt}\frac{dt}{ds}$$

Now, $$\left|\frac{ds}{dt}\right| = \left|\frac{d\vec{r}}{dt}\right| = 5$$

$$\therefore \quad \text{Normal vector} = \frac{1}{5}\frac{d\hat{T}}{dt} = \frac{1}{5}\frac{d}{dt}\frac{1}{5}\left\{(-3\sin t)\,\hat{i} + (3\cos t)\hat{j} + 4\hat{k}\right\}$$

$$= \frac{1}{25}\left\{-3\cos t\,\hat{i} - 3\sin t\,\hat{j}\right\}$$

$$\therefore \quad \text{Normal vector} = -\frac{3}{25}\left\{\cos t\,\hat{i} + \sin t\,\hat{j}\right\}$$

(*iv*) Unit normal vector $= \dfrac{-\dfrac{3}{25}\left(\cos t\,\hat{i} + \sin t\,\hat{j}\right)}{\left|-\dfrac{3}{25}\left(\cos t\,\hat{i} + \sin t\,\hat{j}\right)\right|} = \dfrac{-\dfrac{3}{25}\left(\cos t\,\hat{i} + \sin t\,\hat{j}\right)}{\left|-\dfrac{3}{25}\right|\left|\left(\cos t\,\hat{i} + \sin t\,\hat{j}\right)\right|}$

$$= \frac{-\dfrac{3}{25}\left(\cos t\,\hat{i} + \sin t\,\hat{j}\right)}{\dfrac{3}{25}\sqrt{\cos^2 t + \sin^2 t}} = -\left(\cos t\,\hat{i} + \sin t\,\hat{j}\right).$$

Example 8. *Find the angle between the tangents to the curves* $\vec{r} = t^2\,\hat{i} + 2t\,\hat{j} - t^3\,\hat{k}$ *at points* $t = \pm 1$.

Sol.

$$\vec{r} = t^2\hat{i} + 2t\,\hat{j} - t^3\,\hat{k}$$

$$\frac{d\vec{r}}{dt} = 2t\hat{i} + 2\hat{j} - 3t^2\hat{k}$$

Let $\vec{T}_1$ and $\vec{T}_2$ be the vectors along tangents at $t = 1$ and $t = -1$ respectively, then

$$\vec{T}_1 = 2\hat{i} + 2\hat{j} - 3\hat{k}$$

$$\vec{T}_2 = -2\hat{i} + 2\hat{j} - 3\hat{k}$$

If θ is the angle between $\vec{T}_1$ and $\vec{T}_2$, then

$$\cos\theta = \frac{\vec{T}_1 \bullet \vec{T}_2}{\left|\vec{T}_1\right|\left|\vec{T}_2\right|} = \frac{2(-2) + 2(2) + (-3)(-3)}{\sqrt{4+4+9}\,\sqrt{4+4+9}} = \frac{-4+4+9}{17} = \frac{9}{17}$$

$$\therefore \qquad \theta = \cos^{-1}\frac{9}{17}.$$

Example 9. *A particle moves along the curve* $x = t^3 + 1$, $y = t^2$, $z = 2t + 5$ *where t is the time. Find the components of its velocity and acceleration at time* $t = 1$ *in the direction* $2\hat{i} + 3\hat{j} + 6\hat{k}$. **(P.T.U., Jan. 2009, May 2015)**

Sol. Let $\vec{r}$ be the position vector of the point at time t.

Then $\qquad \vec{r} = x\hat{i} + y\hat{j} + z\hat{k}$

$$\therefore \qquad \vec{r} = \left(t^3 + 1\right)\hat{i} + t^2\,\hat{j} + (2t+5)\hat{k}$$

Velocity at time $t = \dfrac{d\vec{r}}{dt} = \left(3t^2\right)\hat{i} + (2t)\hat{j} + (2)\hat{k}$

When $t = 1$, $\qquad$ velocity $= 3\hat{i} + 2\hat{j} + 2\hat{k}$

Unit vector along the direction $2\hat{i} + 3\hat{j} + 6\hat{k}$.

$$= \frac{2\hat{i}+3\hat{j}+6\hat{k}}{\sqrt{4+9+36}} = \frac{1}{7}\left(2\hat{i}+3\hat{j}+6\hat{k}\right)$$

$\therefore$ Component of velocity in the direction $2\hat{i}+3\hat{j}+6\hat{k}$ at $t = 1$.

$$= \left(3\hat{i}+2\hat{j}+2\hat{k}\right)\bullet\frac{1}{7}\left(2\hat{i}+3\hat{j}+6\hat{k}\right)$$

$$= \frac{1}{7}(3\cdot 2+2\cdot 3+2\cdot 6) = \frac{1}{7}(6+6+12) = \frac{24}{7}.$$

Now acceleration $= \dfrac{d^2\vec{r}}{dt^2} = (6t)\hat{i}+(2)\hat{j} = 6t\hat{i}+2\hat{j}$

Component of acceleration in the direction $2\hat{i}+3\hat{j}+6\hat{k}$

$$= \left(6t\hat{i}+2\hat{j}\right)\cdot\frac{1}{7}\left(2\hat{i}+3\hat{j}+6\hat{k}\right) = \frac{1}{7}(6t\,(2)+2\cdot 3+0\cdot 6)$$

$$= \frac{1}{7}\;(12t+6)$$

At $t = 1$,

Component of acceleration $= \dfrac{1}{7}\;(12\times 1+6) = \dfrac{18}{7}$.

Example 10. *The position vector of a particle at time t is* $\vec{r} = \cos(t-1)\hat{i} + \sinh(t-1)\hat{j} + \alpha t^3\hat{k}$. *Find the condition imposed on* α *by requiring that at* $t = 1$, *the acceleration is normal to the position vector.*

Sol. $\vec{r} = \cos(t-1)\hat{i}+\sinh(t-1)\hat{j}+\alpha t^3\hat{k}$

$$\frac{d\vec{r}}{dt} = -\sin(t-1)\hat{i}+\cosh(t-1)\hat{j}+3\alpha t^2\hat{k}$$

Acceleration $\dfrac{d^2\vec{r}}{dt^2} = -\cos(t-1)\hat{i}+\sinh(t-1)\hat{j}+6\alpha t\hat{k}$

At $t = 1$, acceleration $= \dfrac{d^2r}{dt^2} = -\hat{i}+6\alpha\hat{k}$ and $\hat{r} = \hat{i}+\alpha\hat{k}$ ...(1)

Position vector of the point at $t = 1$ is $\vec{r} = \hat{i}+\alpha\hat{k}$

$\because$ Acceleration is normal to position vector at $t = 1$ $\therefore$ $\dfrac{d^2\vec{r}}{dt^2}\cdot\vec{r} = 0$ at $t = 1$.

i.e., $\left(-\hat{i}+6\alpha\hat{k}\right)\cdot\left(\hat{i}+\alpha\hat{k}\right) = 0$ (From 1)

i.e., $-1+6\alpha^2 = 0$ or $\alpha^2 = \dfrac{1}{6}$ *i.e.,* $\alpha = \pm\dfrac{1}{\sqrt{6}}$.

Example 11. *A particle moves along the curve* $\vec{r} = \left(t^3 - 4t\right)\hat{i} + \left(t^2 + 4t\right)\hat{j} + \left(8t^2 - 3t^3\right)\hat{k}$ *where t is time. Find the magnitude of acceleration along tangent and normal at time t = 2.*

(P.T.U., May 2007, May 2011)

Sol. $$\vec{r} = \left(t^3 - 4t\right)\hat{i} + \left(t^2 + 4t\right)\hat{j} + \left(8t^2 - 3t^3\right)\hat{k}$$

$$\vec{v} = \frac{d\vec{r}}{dt} = \left(3t^2 - 4\right)\hat{i} + (2t + 4)\hat{j} + \left(16t - 9t^2\right)\hat{k}$$

$$\vec{a} = \frac{d^2\vec{r}}{dt^2} = 6t\,\hat{i} + 2\hat{j} + (16 - 18t)\hat{k}$$

Tangential component of acceleration $= \dfrac{dv}{dt} = \dfrac{d^2s}{dt^2}$

Now $$v = \frac{ds}{dt} = \left|\frac{d\vec{r}}{dt}\right| = \sqrt{\left(3t^2 - 4\right)^2 + (2t + 4)^2 + \left(16t - 9t^2\right)^2}$$

$$\frac{d^2s}{dt^2} = \frac{2\left(3t^2 - 4\right)6t + 2\,(2t + 4)\,2 + 2\left(16t - 9t^2\right)(16 - 18t)}{2\sqrt{\left(3t^2 - 4\right)^2 + (2t + 4)^2 + \left(16t - 9t^2\right)^2}}$$

At $t = 2$; $$v = \frac{ds}{dt} = \sqrt{64 + 64 + 16} = 12$$

$$\frac{d^2s}{dt^2} = \frac{1}{2.12}\{192 + 32 + 160\} = \frac{384}{24} = 16$$

Component of acceleration along tangent at $t = 2$ is $\left(\dfrac{d^2s}{dt^2} \text{ at } t = 2\right) = 16$

Component of acceleration along inward normal $= \dfrac{v^2}{\rho}$; where ρ is the radius of curvature

$$\frac{1}{\rho} = \left|\frac{d}{ds}\hat{T}\right| \quad \text{where} \quad \hat{T} = \frac{d\vec{r}}{ds} = \frac{d\vec{r}}{dt}\frac{dt}{ds} = \frac{1}{v}\frac{d\vec{r}}{dt}$$

$$\therefore \quad \hat{T} = \frac{1}{12}\left\{\left(3t^2 - 4\right)\hat{i} + (2t + 4)\hat{j} + \left(16t - 9t^2\right)\hat{k}\right\}$$

$$\frac{d\hat{T}}{ds} = \frac{d\hat{T}}{dt}\frac{dt}{ds} = \frac{1}{12}\frac{d\hat{T}}{dt} = \frac{1}{12}\left\{\frac{1}{12}\left[6t\,\hat{i} + 2\hat{j} + (16 - 18t)\hat{k}\right]\right\}$$

$$\left|\frac{d\hat{T}}{ds}\right| = \frac{1}{144}\sqrt{36t^2+4+(16-18t)^2}$$

At $t = 2$;

$$\left|\frac{d\hat{T}}{ds}\right| = \frac{1}{144}\sqrt{144+4+400} = \frac{\sqrt{548}}{144}$$

$\therefore$ Component of acceleration along normal $= \dfrac{\sqrt{548}}{144}\times 144 = \sqrt{548} = 2\sqrt{137}.$

TEST YOUR KNOWLEDGE

1. Show that $\vec{r} = \vec{a}\,e^{mt} + \vec{b}\,e^{nt}z$ is the solution of the differential equation

$$\frac{d^2\vec{r}}{dt^2} - (m+n)\frac{d\vec{r}}{dt} + mn\vec{r} = \vec{0}.$$

2. If $\vec{r} = (\cos nt)\hat{i} + (\sin nt)\hat{j}$, where n is constant and t varies, show that $\vec{r}\times\dfrac{d\vec{r}}{dt} = n\hat{k}$.

3. Verify the formula $\dfrac{d}{dt}\left(\vec{a}\cdot\vec{b}\right) = \vec{a}\cdot\dfrac{d\vec{b}}{dt} + \dfrac{d\vec{a}}{dt}\cdot\vec{b}$ for

$$\vec{a} = 5t^2\hat{i} + t\hat{j} - t^3\hat{k},\ \vec{b} = \sin t\hat{i} - \cos t\hat{j}.$$ **(*P.T.U., May 2005*)**

4. If $\vec{a} = x^2yz\hat{i} - 2xz^3\hat{j} + xz^2\hat{k}$ and $\vec{b} = 2z\hat{i} + y\hat{j} - x^2\hat{k}$, find $\dfrac{\partial^2}{\partial x\,\partial y}\left(\vec{a}\times\vec{b}\right)$ at (1, 0, –2).

5. If $\vec{a}$ is a constant vector and $\vec{r}$ is a vector function of a scalar t, differentiate w.r.t. t.

(*i*) $\vec{r}\cdot\vec{a}$ (*ii*) $\vec{r}\times\vec{a}$

(*iii*) $\vec{r}\times\dfrac{d\vec{r}}{dt}$ (*iv*) $\vec{r}\cdot\dfrac{d\vec{r}}{dt}$.

6. Prove the following:

(*i*) $$\frac{d}{dt}\left[\vec{a}\cdot\frac{d\vec{b}}{dt} - \frac{d\vec{a}}{dt}\cdot\vec{b}\right] = \vec{a}\cdot\frac{d^2\vec{b}}{dt^2} - \frac{d^2\vec{a}}{dt^2}\cdot\frac{d\vec{b}}{dt}$$

(*ii*) $$\frac{d}{dt}\left[\vec{a}\times\frac{d\vec{b}}{dt} - \frac{d\vec{a}}{dt}\times\vec{b}\right] = \vec{a}\times\frac{d^2\vec{b}}{dt^2} - \frac{d^2\vec{a}}{dt^2}\times\vec{b}.$$

7. (*i*) Find the unit tangent vector at any point on the curve $x = t^2 + 2$, $y = 4t - 5$, $z = 2t^2 - 6t$, where t is any variable. Also determine the unit tangent vector at the point $t = 2$.
   (*ii*) Find the unit tangent vector to any point on the curve $x = a\cos t$, $y = a\sin t$, $z = bt$, **(*P.T.U., Dec. 2006*)**

   [**Hint.** See solved example 7 (*ii*)]

8. Find the unit tangent and unit normal vector at $t = 2$ on the curve.
   $x = t^2 - 1$, $y = 4t - 3$, $z = 2t^2 - 6t$, where t is any variable.

9. The position vector of a moving particle at a time t is $\vec{R}(t) = t^2\hat{i} + -t^3\hat{j} + t^4\hat{k}$. Find the tangential and normal components of its acceleration at time $t = 1$. **(P.T.U., May 2010)**
[**Hint.** See S.E. 11]

10. A particle moves along the curve $x = e^{-t}$, $y = 2\cos 3t$, $z = 2\sin 3t$, where t is the time. Determine its velocity and acceleration vectors and also the magnitudes of velocity and acceleration at $t = 0$.
[**Hint.** See solved example 9] **(P.T.U., Dec. 2002)**

11. A particle moves along the curve $x = 2t^2$, $y = t^2 - 4t$, $z = 3t - 5$ where t is the time. Find the components of its velocity and acceleration at time $t = 1$, in the direction $\hat{i} - 3\hat{j} + 2\hat{k}$.

12. A particle moves so that its position vector is given by $\vec{r} = \cos\omega t\,\hat{i} + \sin\omega t\,\hat{j}$. Show that the velocity $\vec{v}$ of the particle is perpendicular to $\vec{r}$ and $\vec{r} \times \vec{v}$ is a constant vector.

13. The position vector of a point at any time t is given by $\vec{r} = e^t\left(\cos t\,\hat{i} + \sin t\,\hat{j}\right)$. Show that $\vec{a} = 2(\vec{v} - \vec{r})$ where $\vec{a}$ and $\vec{v}$ are acceleration and velocity of a particle.

14. A particle having position vector $\vec{r}$ is moving in a circle with constant angular velocity ω. Show by vector methods, that the acceleration is equal to $-\omega^2\vec{r}$.

Answers

4. $-4\left(\hat{i} + 2\hat{j}\right)$

5. (*i*) $\dfrac{d\vec{r}}{dt} \bullet \vec{a}$ (*ii*) $\dfrac{d\vec{r}}{dt} \times \vec{a}$

(*iii*) $\vec{r} \times \dfrac{d^2\vec{r}}{dt^2}$ (*iv*) $\left(\dfrac{d\vec{r}}{dt}\right)^2 + \vec{r} \bullet \dfrac{d^2\vec{r}}{dt^2}$

7. (*i*) $\dfrac{t\hat{i} + 2\hat{j} + (2t - 3)\hat{k}}{\sqrt{5t^2 - 12t + 13}}$, $\dfrac{1}{3}\left(2\hat{i} + 2\hat{j} + \hat{k}\right)$ (*ii*) $\dfrac{(-a\sin t)\hat{i} + (a\cos t)\hat{j} + b\hat{k}}{\sqrt{a^2 + b^2}}$

8. $\dfrac{1}{3}\left(2\hat{i} + 2\hat{j} + \hat{k}\right)$, $\dfrac{1}{3\sqrt{5}}\left(2\hat{i} + 2\hat{k}\right)$

9. $\dfrac{70}{\sqrt{29}}$, $2\sqrt{46}$

10. $\sqrt{37}$, $5\sqrt{13}$

11. $\dfrac{8\sqrt{14}}{7}$, $\dfrac{-\sqrt{14}}{7}$

7.13. SCALAR AND VECTOR POINT FUNCTIONS

Point Function. A variable quantity whose value at any point in a region of space depends upon the position of the point, is called a point function.

Point functions are of two types:

(*i*) Scalar Point Function (*ii*) Vector Point Function.

(*i*) **Scalar Point Function.** A function ϕ (x, y, z) is called a scalar point function if it associates a scalar with every point in region R of a space. Region R is called scalar field. The temperature distribution in a heated body, density of a body and potential due to gravity are examples of scalar point functions.

(*ii*) **Vector Point Function.** If a function $\vec{V}$ (x, y, z) defines a vector at every point of the region R of a space then $\vec{V}$ (x, y, z) is called a vector point function and R is called a vector field. Every vector $\vec{v}$ of the field is regarded as a localized vector attached to the corresponding point (x, y, z).

The velocity of a moving fluid at any instant, gravitational forces are examples of vector point function.

7.14. GRADIENT OF A SCALAR FIELD *(P.T.U., Jan. 2009)*

Let ϕ (x, y, z) be a function defining a scalar field, then the vector $\hat{i}\frac{\partial \phi}{\partial x}+\hat{j}\frac{\partial \phi}{\partial y}+\hat{k}\frac{\partial \phi}{\partial z}$ is called the gradient of the scalar field ϕ and is denoted by grad ϕ.

Thus $$\text{grad } \phi = \hat{i}\frac{\partial \phi}{\partial x}+\hat{j}\frac{\partial \phi}{\partial y}+\hat{k}\frac{\partial \phi}{\partial z}$$

Grad ϕ is a vector quantity.

The gradient of scalar field ϕ is obtained by operating on ϕ the vector operator.

$$\hat{i}\frac{\partial}{\partial x}+\hat{j}\frac{\partial}{\partial y}+\hat{k}\frac{\partial}{\partial z}$$

This operator is denoted by symbol ∇ (read as del) (also called nable)

Thus $$\text{grad } \phi = \nabla \phi.$$

7.15. GEOMETRICAL INTERPRETATION OF GRADIENT *(P.T.U., May 2010, May 2012)*

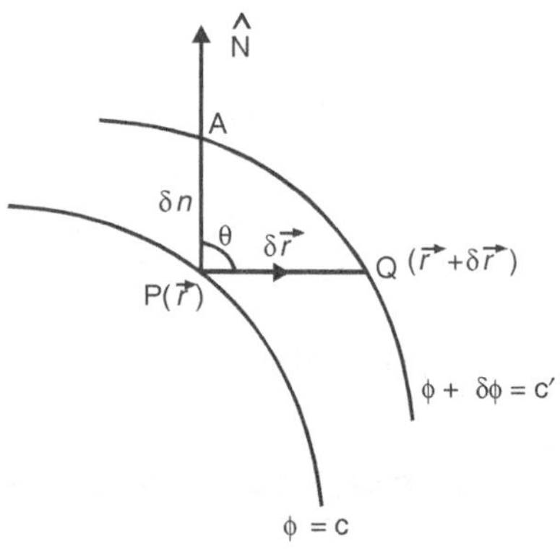

If a surface ϕ $(x, y, z) = c$ is drawn through any point P such that at each point on the surface, the function has the same value as at P, then such a surface is called a **level surface** through P.

Through any point passes one and only one level surface. Also no two level surfaces can intersect.

Consider the level surface through the point P at which the function has value ϕ and let $\phi + \delta\phi$ be another level surface through the neighbouring point Q.

Let $\vec{r}$ and $\vec{r}+\delta\vec{r}$ be the position vectors of P and Q respectively then $\overrightarrow{\mathbf{PQ}} = \delta\vec{r}$

Now, $$\nabla\phi \cdot \delta\vec{r} = \left(\hat{i}\frac{\partial \phi}{\partial x}+\hat{j}\frac{\partial \phi}{\partial y}+\hat{k}\frac{\partial \phi}{\partial z}\right)\cdot\left(\hat{i}\,\delta x+\hat{j}\,\delta y+\hat{k}\,\delta z\right)$$

$$= \frac{\partial \phi}{\partial x}\delta x+\frac{\partial \phi}{\partial y}\delta y+\frac{\partial \phi}{\partial z}\delta z = \delta\phi \quad \ldots(1)$$

If Q lies on the same surface as P, then $\delta\,\phi = 0$

$\therefore$ (1) reduces to $\nabla\phi \bullet \delta\vec{r} = 0$

$\therefore$ $\nabla\phi$ is perpendicular to $\delta\vec{r}$ which is true for all values of r.

Hence $\nabla\phi$ is normal to the surface $\phi(x, y, z) = c$.

Let $\nabla\phi = |\nabla\phi|\,\hat{N}$, where $\hat{N}$ is a unit normal to $\phi = c$ at P. Let PA $= \delta n$ be the perpendicular distance between the two level surfaces $\phi = c$ and $\phi + \delta\phi = c'$. Then rate of change of ϕ in the direction of normal to the surface through P is $\dfrac{\partial\phi}{\partial n} = \underset{\delta n \to 0}{\text{Lt}} \dfrac{\delta\phi}{\delta n} = \underset{\delta n \to 0}{\text{Lt}} \dfrac{\nabla\phi \cdot \delta\vec{r}}{\delta n}$ [$\because$ of (1)]

$$= \underset{\delta n \to 0}{\text{Lt}} \frac{|\nabla\phi|\,\hat{N} \cdot \delta\vec{r}}{\delta n}$$

Since,
$$\hat{N} \cdot \delta\vec{r} = \left|\hat{N}\right| \left|\delta\vec{r}\right| \cos\theta = 1 \cdot \text{PQ} \cos\theta = \delta n$$

$$\therefore \quad \frac{\partial\phi}{\partial n} = \underset{\delta n \to 0}{\text{Lt}} \frac{|\nabla\phi|\,\delta n}{\delta n} = |\nabla\phi|$$

$$\therefore \quad |\nabla\phi| = \frac{\partial\phi}{\partial n}.$$

Hence the gradient of a scalar field ϕ is a vector normal to the surface $\phi = c$ and has a magnitude equal to the rate of change of ϕ along the normal.

Cor 1. Equation of the tangent plane to a surface $\phi(x, y, z) = c$ at a point A (x_1, y_1, z_1) can be derived from the gradient vector at that point.

Since gradient vector at a point A (x_1, y_1, z_1) on the surface $\phi(x, y, z) = c$ represents normal to the surface at that point $\therefore$ if we take P(x, y, z) be any point on the tangent plane at (x_1, y_1, z_1) then the vector $(x - x_1)\hat{i} + (y - y_1)\hat{j} + (z - z_1)\hat{k}$ will be perpendicular to the normal vector at (x_1, y_1, z_1)

$$\therefore \quad (\nabla\phi)_{(x_1, y_1, z_1)} \bullet \left[(x - x_1)\hat{i} + (y - y_1)\hat{j} + (z - z_1)\hat{k}\right] = 0$$

i.e.,
$$\left(\frac{\partial f}{\partial x}\right)_{(x_1, y_1, z_1)} (x - x_1) + \left(\frac{\partial\phi}{\partial z}\right)_{(x_1, y_1, z_1)} (y - y_1) + \left(\frac{\partial\phi}{\partial z}\right)_{(x_1, y_1, z_1)} (z - z_1) = 0$$

which is the equation of the tangent plane at (x_1, y_1, z_1) to $\phi(x, y, z) = c$.

Cor 2. Equation of the normal at A (x_1, y_1, z_1) to the surface $\phi(x, y, z) = c$.

Let P (x, y, z) be any variable point on the normal to the surface $\phi(x, y, z) = c$. Then $\overrightarrow{\text{AP}}$ is parallel to normal vector $\nabla\phi$ at (x_1, y_1, z_1)

$$\therefore \quad \overrightarrow{\text{AP}} \times (\nabla\phi)_{(x_1, y_1, z_1)} = \vec{0}$$

i.e.,
$$\left[(x - x_1)\hat{i} + (y - y_1)\hat{j} + (z - z_1)\hat{k}\right] \times \left[\left(\frac{\partial\phi}{\partial x}\right)_{(x_1, y_1, z_1)} \hat{i} + \left(\frac{\partial\phi}{\partial y}\right)_{(x_1, y_1, z_1)} \hat{j} + \left(\frac{\partial\phi}{\partial z}\right)_{(x_1, y_1, z_1)} \hat{k}\right] = \vec{0}$$

$$\therefore \quad \begin{vmatrix} \hat{i} & \hat{j} & \hat{k} \\ x - x_1 & y - y_1 & z - z_1 \\ \left(\frac{\partial \phi}{\partial x}\right)_{(x_1, y_1, z_1)} & \left(\frac{\partial \phi}{\partial y}\right)_{(x_1, y_1, z_1)} & \left(\frac{\partial \phi}{\partial z}\right)_{(x_1, y_1, z_1)} \end{vmatrix} = \vec{0}$$

$$\therefore \quad \sum \left[(y - y_1)\left(\frac{\partial \phi}{\partial z}\right)_{(x_1, y_1, z_1)} - (z - z_1)\left(\frac{\partial \phi}{\partial y}\right)_{(x_1, y_1, z_1)} \right] \hat{i} = \vec{0}$$

$$\therefore \quad (y - y_1)\left(\frac{\partial \phi}{\partial z}\right)_{(x_1, y_1, z_1)} - (z - z_1)\left(\frac{\partial \phi}{\partial y}\right)_{(x_1, y_1, z_1)} = 0; \quad \therefore \quad \frac{y - y_1}{\left(\frac{\partial \phi}{\partial y}\right)_{(x_1, y_1, z_1)}} = \frac{z - z_1}{\left(\frac{\partial \phi}{\partial z}\right)_{(x_1, y_1, z_1)}}$$

Similarly, by equating components of $\hat{j}$ and $\hat{k}$ to zero, we get

$$\frac{z - z_1}{\left(\frac{\partial \phi}{\partial z}\right)_{(x_1, y_1, z_1)}} = \frac{x - x_1}{\left(\frac{\partial \phi}{\partial x}\right)_{(x_1, y_1, z_1)}} \text{ and } \frac{x - x_1}{\left(\frac{\partial \phi}{\partial x}\right)_{(x_1, y_1, z_1)}} = \frac{y - y_1}{\left(\frac{\partial \phi}{\partial y}\right)_{(x_1, y_1, z_1)}}$$

Combining all three, the equation of the normal at A (x_1, y_1, z_1) is

$$\frac{x - x_1}{\left(\frac{\partial \phi}{\partial x}\right)_{(x_1, y_1, z_1)}} = \frac{y - y_1}{\left(\frac{\partial \phi}{\partial y}\right)_{(x_1, y_1, z_1)}} = \frac{z - z_1}{\left(\frac{\partial \phi}{\partial z}\right)_{(x_1, y_1, z_1)}}.$$

7.16. DIRECTIONAL DERIVATIVE *(P.T.U., Jan. 2010)*

Let PQ = δr then $\underset{\delta r \to 0}{\text{Lt}} \frac{\delta \phi}{\delta r} = \frac{\partial \phi}{\partial r}$ is called directional derivative of ϕ at P in the direction of PQ.

Let $\hat{N}'$ be a unit vector in the direction of PQ then $\hat{N} \cdot \hat{N}' = \cos \theta$

$$\delta r = \frac{\delta n}{\cos \theta} = \frac{\delta n}{\hat{N} \cdot \hat{N}'}$$

$$\therefore \quad \frac{\partial \phi}{\partial r} = \underset{\delta r \to 0}{\text{Lt}} \frac{\delta \phi}{\delta n} \hat{N} \cdot \hat{N}' = \hat{N} \cdot \hat{N}' \frac{\partial \phi}{\partial n}$$

$$\therefore \quad \frac{\partial \phi}{\partial r} = \hat{N} \cdot \hat{N}' |\nabla \phi| \qquad \therefore \quad |\nabla \phi| = \frac{\partial \phi}{\partial n} \text{ from art. 7·15}$$

$$= \hat{N}' \cdot |\nabla \phi| \hat{N} = \hat{N}' \cdot \nabla \phi \qquad \because \quad \hat{N} |\nabla \phi| = \nabla \phi.$$

Thus the directional derivative $\frac{\partial \phi}{\partial r}$ *is the resolved part of* $\nabla\phi$ *in the direction of* $\hat{N}'$ *i.e.,* $\overrightarrow{PQ}$

Since $$\frac{\partial \theta}{\partial r} = \hat{N}' \bullet \nabla\phi = |\nabla\phi| \cos\theta \leq |\nabla\phi|$$

$\therefore$ **$\nabla\phi$ gives the maximum rate of change of ϕ and the magnitude of this maximum rate of change is $|\nabla\phi|$.**

7.17. PROPERTIES OF GRADIENT

(1) If ϕ is a constant scalar point function, then $\nabla\phi = \vec{0}$.

(2) If ϕ_1 and ϕ_2 are two scalar point functions, then

(*a*) $\nabla(\phi_1 \pm \phi_2) = \nabla\phi_1 \pm \nabla\phi_2$

(*b*) $\nabla(c_1\phi_1 + c_2\phi_2) = c_1\nabla\phi_1 + c_2\nabla\phi_2$, *where* c_1 *and* c_2 *are constants.*

(*c*) $\nabla(\phi_1\phi_2) = \phi_1\nabla\phi_2 + \phi_2\nabla\phi_1$

(*d*) $\nabla\left(\frac{f_1}{f_2}\right) = \frac{f_2\nabla f_1 - \phi_1\nabla f_2}{f_2^2}$, $\phi_2 \neq 0$.

Proof. (1) $$\nabla\phi = \left(\hat{i}\frac{\partial}{\partial x} + \hat{j}\frac{\partial}{\partial y} + \hat{k}\frac{\partial}{\partial z}\right)\phi = \hat{i}\frac{\partial\phi}{\partial x} + \hat{j}\frac{\partial\phi}{\partial y} + \hat{k}\frac{\partial\phi}{\partial z} = 0$$

$\left[\because \phi \text{ is constant } \therefore \frac{\partial\phi}{\partial x} = \frac{\partial\phi}{\partial y} = \frac{\partial\phi}{\partial z} = 0\right]$

(2) (*a*) $$\nabla(\phi_1 \pm \phi_2) = \left(\hat{i}\frac{\partial}{\partial x} + \hat{j}\frac{\partial}{\partial y} + \hat{k}\frac{\partial}{\partial z}\right)(\phi_1 \pm \phi_2)$$

$$= \hat{i}\frac{\partial}{\partial x}(\phi_1 \pm \phi_2) + \hat{j}\frac{\partial}{\partial y}(\phi_1 \pm \phi_2) + \hat{k}\frac{\partial}{\partial z}(\phi_1 \pm \phi_2)$$

$$= \hat{i}\left[\frac{\partial\phi_1}{\partial x} \pm \frac{\partial\phi_2}{\partial x}\right] + \hat{j}\left[\frac{\partial\phi_1}{\partial y} \pm \frac{\partial\phi_2}{\partial y}\right] + \hat{k}\left[\frac{\partial\phi_1}{\partial z} \pm \frac{\partial\phi_2}{\partial z}\right]$$

$$\nabla(\phi_1 \pm \phi_2) = \left(\hat{i}\frac{\partial\phi_1}{\partial x} + \hat{j}\frac{\partial\phi_1}{\partial y} + \hat{k}\frac{\partial\phi_1}{\partial z}\right) \pm \left(\hat{i}\frac{\partial\phi_2}{\partial x} + \hat{j}\frac{\partial\phi_2}{\partial y} + \hat{k}\frac{\partial\phi_2}{\partial z}\right)$$

$$= \nabla\phi_1 \pm \nabla\phi_2$$

(*b*) Students can easily prove it.

(*c*) $$\nabla(\phi_1\phi_2) = \left(\hat{i}\frac{\partial}{\partial x} + \hat{j}\frac{\partial}{\partial y} + \hat{k}\frac{\partial}{\partial z}\right)\phi_1\phi_2$$

$$= \hat{i}\frac{\partial}{\partial x}(\phi_1\phi_2) + \hat{j}\frac{\partial}{\partial y}(\phi_1\phi_2) + \hat{k}\frac{\partial}{\partial z}(\phi_1\phi_2)$$

$$= \hat{i}\left[\phi_1 \frac{\partial \phi_2}{\partial x} + \phi_2 \frac{\partial \phi_1}{\partial x}\right] + \hat{j}\left[\phi_1 \frac{\partial \phi_2}{\partial y} + \phi_2 \frac{\partial \phi_1}{\partial y}\right] + \hat{k}\left[\phi_1 \frac{\partial \phi_2}{\partial z} + \phi_2 \frac{\partial \phi_1}{\partial z}\right]$$

$$= \left(\hat{i}\,\phi_1 \frac{\partial \phi_2}{\partial x} + \hat{j}\,\phi_1 \frac{\partial \phi_2}{\partial y} + \hat{k}\,\phi_1 \frac{\partial \phi_2}{\partial z}\right) + \left(\hat{i}\,\phi_2 \frac{\partial \phi_1}{\partial x} + \hat{j}\,\phi_2 \frac{\partial \phi_1}{\partial y} + \hat{k}\,\phi_2 \frac{\partial \phi_1}{\partial z}\right)$$

$$= \phi_1 \nabla \phi_2 + \phi_2 \nabla \phi_1$$

(*d*) $$\nabla\left(\frac{\phi_1}{\phi_2}\right) = \left(\hat{i}\frac{\partial}{\partial x} + \hat{j}\frac{\partial}{\partial y} + \hat{k}\frac{\partial}{\partial z}\right)\left(\frac{\phi_1}{\phi_2}\right) = \hat{i}\frac{\partial}{\partial x}\left(\frac{\phi_1}{\phi_2}\right) + \hat{j}\frac{\partial}{\partial y}\left(\frac{\phi_1}{\phi_2}\right) + \hat{k}\frac{\partial}{\partial z}\left(\frac{\phi_1}{\phi_2}\right)$$

$$= \hat{i}\,\frac{\phi_2 \frac{\partial}{\partial x}\phi_1 - \phi_1 \frac{\partial}{\partial x}\phi_2}{\phi_2^2} + \hat{j}\,\frac{\phi_2 \frac{\partial}{\partial y}\phi_1 - \phi_1 \frac{\partial}{\partial y}\phi_2}{\phi_2^2} + \hat{k}\,\frac{\phi_2 \frac{\partial}{\partial z}\phi_1 - \phi_1 \frac{\partial}{\partial z}\phi_2}{\phi_2^2}$$

$$= \frac{1}{\phi_2^2}\left[\hat{i}\,\phi_2 \frac{\partial \phi_1}{\partial x} + \hat{j}\,\phi_2 \frac{\partial \phi_1}{\partial y} + \hat{k}\,\phi_2 \frac{\partial \phi_1}{\partial z}\right] - \left(\hat{i}\,\phi_1 \frac{\partial \phi_2}{\partial x} + \hat{j}\,\phi_1 \frac{\partial \phi_2}{\partial y} + \hat{k}\,\phi_1 \frac{\partial \phi_2}{\partial z}\right)$$

$$= \frac{1}{\phi_2^2}\left[\phi_2 \nabla \phi_1 - \phi_1 \nabla \phi_2\right]$$

$$\therefore \quad \nabla\left(\frac{\phi_1}{\phi_2}\right) = \frac{\phi_2 \nabla \phi_1 - \phi_1 \nabla \phi_2}{\phi_2^2}; \quad \phi_2 \neq 0.$$

ILLUSTRATIVE EXAMPLES

Example 1. *Find gradient of the following functions*

(i) $\phi = y^2 - 4xy$ *at (1, 2)* **(P.T.U., May 2008)**

(ii) $\phi = x^3 + y^3 + 3xyz$ *at (1, –2, –1).*

Sol. (*i*) $\phi = y^2 - 4xy$

$$\text{grad. } \phi = \nabla \phi = \left(\hat{i}\frac{\partial}{\partial x} + \hat{j}\frac{\partial}{\partial y} + \hat{k}\frac{\partial}{\partial z}\right)\left(y^2 - 4xy\right)$$

$$= \hat{i}\,(-4y) + \hat{j}\,(2y - 4x)$$

At (1, 2); $\quad$ grad. $\phi = -8\hat{i} + 0\hat{j} = -8\hat{i}$

(*ii*) $\quad \phi = x^3 + y^3 + 3xyz$

$$\text{grad. } \phi = \nabla \phi = \left(\hat{i}\frac{\partial}{\partial x} + \hat{j}\frac{\partial}{\partial y} + \hat{k}\frac{\partial}{\partial z}\right)\left(x^3 + y^3 + 3xyz\right)$$

$$= \hat{i}\left(3x^2 + 3yz\right) + \hat{j}\left(3y^2 + 3xz\right) + \hat{k}\left(3xy\right)$$

At (1, –2, –1) ; grad. $\phi = 9\hat{i} + 9\hat{j} - 6\hat{k} = 3\left(3\hat{i} + 3\hat{j} - 2\hat{k}\right)$.

Example 2. *Find a unit vector normal to the surface $x^3 + y^3 + 3xyz = 3$ at the point (1, 2, –1).* **(P.T.U., Dec. 2012)**

Sol. Let $\phi(x, y, z) = x^3 + y^3 + 3xyz - 3$

We know that $\nabla\phi$ is a vector normal to the surface $\phi = c$

$$\therefore \qquad \nabla\phi = \text{grad } \phi = \left(\hat{i}\frac{\partial}{\partial x} + \hat{j}\frac{\partial}{\partial y} + \hat{k}\frac{\partial}{\partial z}\right)\left(x^3 + y^3 + 3xyz - 3\right)$$

$$= \hat{i}(3x^2 + 3yz) + \hat{j}(3y^2 + 3xz) + \hat{k}(3xy).$$

Normal vector at (1, 2, –1) is $-3\hat{i} + 9\hat{j} + 6\hat{k}$

Unit normal vector at (1, 2, –1)

$$= \frac{3(-\hat{i} + 3\hat{j} + 2\hat{k}}{3\sqrt{1 + 9 + 4}} = \frac{-\hat{i} + 3\hat{j} + 2\hat{k}}{\sqrt{14}}$$

Example 3. *Find the normal vector and the equation of the tangent plane to the surface $z = \sqrt{x^2 + y^2}$ at the point $(1, 1, \sqrt{2})$.* **(P.T.U., Jan. 2008)**

Sol. Let $\phi(x, y, z) = \sqrt{x^2 + y^2} - z$

$$\frac{\partial\phi}{\partial x} = \frac{1}{2\sqrt{x^2 + y^2}} \,.\, 2x = \frac{x}{\sqrt{x^2 + y^2}}$$

$$\frac{\partial\phi}{\partial y} = \frac{1}{2\sqrt{x^2 + y^2}} \,.\, 2y = \frac{y}{\sqrt{x^2 + y^2}}$$

$$\frac{\partial\phi}{\partial z} = -1$$

We know that $\nabla\phi$ is a vector normal to the surface $\phi = C$

$$\therefore \qquad \nabla\phi = \text{grad } \phi = \left(\hat{i}\frac{\partial}{\partial x} + \hat{j}\frac{\partial}{\partial y} + \hat{k}\frac{\partial}{\partial z}\right)\phi$$

$$= \hat{i}\frac{x}{\sqrt{x^2 + y^2}} + \hat{j}\frac{y}{\sqrt{x^2 + y^2}} + \hat{k}(-1)$$

$$= \frac{x\hat{i} + y\hat{j} - \hat{k}\sqrt{x^2 + y^2}}{\sqrt{x^2 + y^2}}$$

Normal vector at $(1, 1, \sqrt{2}) = \dfrac{\hat{i} + \hat{j} - \sqrt{2}\,\hat{k}}{\sqrt{2}}$

Unit normal vector at $(1, 1, \sqrt{2}) = \dfrac{\hat{i} + \hat{j} - \sqrt{2}\hat{k}}{\sqrt{2}\sqrt{\frac{1}{2} + \frac{1}{2} + 1}} = \dfrac{\hat{i} + \hat{j} - \sqrt{2}\hat{k}}{2}$

Now the equation of the tangent plane at $(1, 1, \sqrt{2})$ is

$$(x - 1)\left(\frac{\partial\phi}{\partial x}\right)_{(1,1,\sqrt{2})} + (y - 1)\left(\frac{\partial\phi}{\partial y}\right)_{(1,1,\sqrt{2})} + (z - \sqrt{2})\left(\frac{\partial\phi}{\partial z}\right)_{(1,1,\sqrt{2})} = 0$$

i.e., $$(x-1)\frac{1}{\sqrt{2}}+(y-1)\left(\frac{1}{\sqrt{2}}\right)+(z-\sqrt{2})(-1)=0$$

or $$x-1+y-1-\sqrt{2}\,(z-\sqrt{2})=0 \text{ or } x+y-\sqrt{2}z=0.$$

Example 4. *If* $\vec{r}=x\hat{i}+y\hat{j}+z\hat{k}$, *show that*

(i) $\text{grad } r=\dfrac{\vec{r}}{r}$ *(ii)* $\text{grad}\left(\dfrac{1}{r}\right)=-\dfrac{\vec{r}}{r^3}$ **(P.T.U., Dec. 2003)**

(iii) $\nabla r^n = n\,r^{n-2}\vec{r}$

(iv) $\nabla(\vec{a}\cdot\vec{r})=\vec{a}$, *where* $\vec{a}$ *is a constant vector.* **(P.T.U., Dec. 2002)**

Sol. *(i)* $r=|\vec{r}|=\sqrt{x^2+y^2+z^2}$ or $r^2=x^2+y^2+z^2$

Differentiate partially w.r.t. x, y and z respectively, we get $2r\dfrac{\partial r}{\partial x}=2x$, $\dfrac{\partial r}{\partial x}=\dfrac{x}{r}$.

Similarly, $$\frac{\partial r}{\partial y}=\frac{y}{r},\ \frac{\partial r}{\partial z}=\frac{z}{r}$$

Now, $$\text{grad } r=\nabla r=\left(\hat{i}\frac{\partial}{\partial x}+\hat{j}\frac{\partial}{\partial y}+\hat{k}\frac{\partial}{\partial z}\right)(r)$$

$$=\hat{i}\frac{\partial r}{\partial x}+\hat{j}\frac{\partial r}{\partial y}+\hat{k}\frac{\partial r}{\partial z}=\hat{i}\frac{x}{r}+\hat{j}\frac{y}{r}+\hat{k}\frac{z}{r}=\frac{1}{r}\left(x\hat{i}+y\hat{j}+z\hat{k}\right)=\frac{\vec{r}}{r}$$

Hence $$\text{grad } r=\frac{\vec{r}}{r}$$

(ii) $$\text{grad }\frac{1}{r}=\nabla\left(\frac{1}{r}\right)=\left(\hat{i}\frac{\partial}{\partial x}+\hat{j}\frac{\partial}{\partial y}+\hat{k}\frac{\partial}{\partial z}\right)\left(\frac{1}{r}\right)$$

$$=\hat{i}\left(-\frac{1}{r^2}\right)\frac{\partial r}{\partial x}+\hat{j}\left(-\frac{1}{r^2}\right)\frac{\partial r}{\partial y}+\hat{k}\left(-\frac{1}{r^2}\right)\frac{\partial r}{\partial z}$$

$$=-\frac{1}{r^2}\left(\hat{i}\frac{x}{r}+\hat{j}\frac{y}{r}+\hat{k}\frac{z}{r}\right)=-\frac{1}{r^3}\left(x\hat{i}+y\hat{j}+z\hat{k}\right)=-\frac{\vec{r}}{r^3}$$

(iii) $$\nabla r^n=\left(\hat{i}\frac{\partial}{\partial x}+\hat{j}\frac{\partial}{\partial y}+\hat{k}\frac{\partial}{\partial z}\right)r^n$$

$$=\hat{i}\,nr^{n-1}\frac{\partial r}{\partial x}+\hat{j}\,nr^{n-1}\frac{\partial r}{\partial y}+\hat{k}\,nr^{n-1}\frac{\partial r}{\partial z}$$

$$=nr^{n-1}\left[\hat{i}\frac{x}{r}+\hat{j}\frac{y}{r}+\hat{k}\frac{z}{r}\right]=nr^{n-2}\left(x\hat{i}+y\hat{j}+z\hat{k}\right)=nr^{n-2}\vec{r}$$

(iv) $\nabla(\vec{a}\cdot\vec{r})$

Let $\vec{a}=a_1\hat{i}+a_2\hat{j}+a_3\hat{k}$, where a_1, a_2, a_3 are constants.

$$\vec{a} \cdot \vec{r} = a_1x + a_2y + a_3z \qquad \text{Since } \vec{r} = x\hat{i} + y\hat{j} + z\hat{k}$$

$$\nabla(\vec{a} \cdot \vec{r}) = \left(\hat{i}\frac{\partial}{\partial x} + \hat{j}\frac{\partial}{\partial y} + \hat{k}\frac{\partial}{\partial z}\right)(a_1x + a_2y + a_3z)$$

$$= \hat{i}\frac{\partial}{\partial x}(a_1x + a_2y + a_3z) + \hat{j}\frac{\partial}{\partial y}(a_1x + a_2y + a_3z) + \hat{k}\frac{\partial}{\partial z}(a_1x + a_2y + a_3z)$$

$$= \hat{i}\,a_1 + \hat{j}\,a_2 + \hat{k}\,a_3 = a_1\hat{i} + a_2\hat{j} + a_3\hat{k} = \vec{a}$$

Hence $\nabla(\vec{a} \cdot \vec{r}) = \vec{a}$

Example 5. *What is the directional derivative of the function $xy^2 + yz^3$ at the point $(2, -1, 1)$ (i) In the direction of the vector $\hat{i} + 2\hat{j} + 2\hat{k}$?* **(P.T.U., Dec. 2011, May 2015)**

(ii) In the direction of the normal to the surface $x \log z - y^2 = -4$ at $(-1, 2, 1)$. **(P.T.U., May 2014)**

Sol. (*i*) Let $\phi(x, y, z) = xy^2 + yz^3$

$$\text{Gradient of } \phi = \nabla\phi = \hat{i}\frac{\partial\phi}{\partial x} + \hat{j}\frac{\partial\phi}{\partial y} + \hat{k}\frac{\partial\phi}{\partial z}$$

$$= \hat{i}\,y^2 + \hat{j}(2xy + z^3) + \hat{k}(3yz^2)$$

$$\nabla\phi \text{ at } (2, -1, 1) = \hat{i} - 3\hat{j} - 3\hat{k}$$

If $\hat{n}$ is a unit vector in the direction of $\hat{i} + 2\hat{j} + 2\hat{k}$, then $\hat{n} = \dfrac{\hat{i} + 2\hat{j} + 2\hat{k}}{\sqrt{1 + 4 + 4}} = \dfrac{1}{3}(\hat{i} + 2\hat{j} + 2\hat{k})$

$\therefore$ Directional derivative of the given function ϕ at $(2, -1, 1)$ in the direction of

$$\hat{i} + 2\hat{j} + 2\hat{k} = [\nabla\phi \text{ at } (2, -1, 1)] \cdot \hat{n}$$

$$= (\hat{i} - 3\hat{j} - 3\hat{k}) \cdot \frac{1}{3}(\hat{i} + 2\hat{j} + 2\hat{k})$$

$$= \frac{1 - 6 - 6}{3} = -\frac{11}{3}$$

(*ii*) Normal to the surface $f(x, y, z) = x \log z - y^2 + 4 = \nabla f$

$$= \left(\hat{i}\frac{\partial}{\partial x} + \hat{j}\frac{\partial}{\partial y} + \hat{k}\frac{\partial}{\partial z}\right)(x \log z - y^2 + 4)$$

$$= \hat{i}(\log z) + \hat{j}(-2y) + \hat{k}\left(\frac{x}{z}\right)$$

Normal vector at (– 1, 2, 1)

$$= \hat{i}\,(0) - 2\hat{j}\,(2) + \hat{k}\,(-1)$$

$$= -\,4\hat{j} - \hat{k}$$

$$\hat{n} = \text{unit normal vector at } (-1, 2, 1)$$

$$= \frac{-4\hat{j} - \hat{k}}{\sqrt{17}}$$

Directional derivative of (x, y, z) at $(2, -1, 1)$ in the direction of normal at

$$(-1, 2, 1) = (\hat{i} - 3\hat{j} - 3\hat{k}) \cdot \frac{-4\hat{j} - \hat{k}}{\sqrt{17}}$$

$$= \frac{12 + 3}{\sqrt{17}} = \frac{15}{\sqrt{17}}.$$

Example 6. *Find all the directional derivative of the function $f = x^2 - y^2 + 2z^2$ at the point P (1, 2, 3) in the direction of the line PQ, where Q is the point (5, 0, 4).*

In what direction it will be maximum ? Find also the magnitude of this maximum.

Sol. Gradient of $f = \nabla f$

$$= \left(\hat{i}\,\frac{\partial}{\partial x} + \hat{j}\,\frac{\partial}{\partial y} + \hat{k}\,\frac{\partial}{\partial z}\right)\left(x^2 - y^2 + 2z^2\right) = \hat{i}\,(2x) + \hat{j}\left((-2y) + \hat{k}\,(4z)\right)$$

$$\nabla f \text{ at } (1, 2, 3) = 2\hat{i} - 4\hat{j} + 12\hat{k}.$$

Now $\overrightarrow{PQ}$ = P.V of Q – P.V. of P $= 5\hat{i} + 4\hat{k} - \left(\hat{i} + 2\hat{j} + 3\hat{k}\right) = 4\hat{i} - 2\hat{j} + \hat{k}$

If $\hat{n}$ is a unit vector in the direction $\overrightarrow{PQ}$, then $\hat{n} = \dfrac{4\hat{i} - 2\hat{j} + \hat{k}}{\sqrt{16 + 4 + 1}} = \dfrac{1}{\sqrt{21}}\left(4\hat{i} - 2\hat{j} + \hat{k}\right)$

Direction derivative of f at (1, 2, 3) in the direction of $\overrightarrow{PQ} = \left[\nabla f \text{ at} (1, 2, 3)\right] \cdot \hat{n}$

$$= \left(2\hat{i} + 4\hat{j} + 12\hat{k}\right) \cdot \frac{1}{\sqrt{21}}\left(4\hat{i} + 2\hat{j} + \hat{k}\right) = \frac{1}{\sqrt{21}}(8 + 8 + 12) = \frac{28}{\sqrt{21}} = \frac{28}{21}\sqrt{21} = \frac{4}{3}\sqrt{21}$$

The directional derivative of f is maximum in the direction of the normal to the given surface *i.e.*, in the direction of $\nabla f = \left(2\hat{i} - 4\hat{j} + 12\hat{k}\right)$.

The maximum value of this directional derivative $= |\,\nabla f\,| = \sqrt{4 + 16 + 144}$

$$= \sqrt{164} = 2\sqrt{41}.$$

Example 7. *Find the directional derivative of $\phi = e^{2x} \cos yz$ at the origin in the direction of the tangent to the curve $x = a \sin t$, $y = a \cos t$, $z = at$ at $t = \dfrac{\pi}{4}$.*

Sol. Gradient of $\phi = \nabla \phi = \left(\hat{i}\,\frac{\partial}{\partial x} + \hat{j}\,\frac{\partial}{\partial y} + \hat{k}\,\frac{\partial}{\partial z}\right)\left(e^{2x}\cos yz\right)$

$$= \hat{i}\left(2e^{2x}\cos yz\right) + \hat{j}\left(-e^{2x}z\sin yz\right) + \hat{k}\left\{e^{2x}(-\sin yz)\,y\right\}$$

At the origin *i.e.*, when $x = 0, y = 0, z = 0.$

$$\nabla\phi = \hat{i}\,(2) = 2\hat{i}$$

Equation of the curve is $x = a\sin t,\ y = a\cos t,\ z = at$

Any point on the curve is $\vec{r} = \hat{i}\,(a\sin t) + \hat{j}\,(a\cos t) + \hat{k}\,(at)$

Direction of the tangent is given by $= \frac{d\vec{r}}{dt} = (a\cos t)\,\hat{i} - (a\sin t)\hat{j} + a\,\hat{k}$

At $t = \frac{\pi}{4}$, direction of tangent $= \frac{a}{\sqrt{2}}\,\hat{i} - \frac{a}{\sqrt{2}}\hat{j} + a\,\hat{k}$

$\hat{n}$ = unit direction of the tangent

$$= \frac{\frac{a}{\sqrt{2}}\hat{i} - \frac{a}{\sqrt{2}}\,\hat{j} + a\,\hat{k}}{\sqrt{\frac{a^2}{2} + \frac{a^2}{2} + a^2}} = \frac{\frac{a}{\sqrt{2}}\left(\hat{i} - \hat{j} + \sqrt{2}\,\hat{k}\right)}{\sqrt{2}\,a}$$

$$= \frac{1}{2}\left(\hat{i} - \hat{j} + \sqrt{2}\,\hat{k}\right)$$

Directional derivative of ϕ at (0, 0, 0) in the direction of tangent at $t = \frac{\pi}{4}$ is $= \nabla\phi \cdot \hat{n}$ at (0, 0, 0).

$$= 2\hat{i}\cdot\frac{1}{2}\left(\hat{i} - \hat{j} + \sqrt{2}\,\hat{k}\right) = 1.$$

Example 8. *For the function* $\phi\,(x, y) = \frac{x}{x^2 + y^2}$, *find the magnitude of the directional derivative along a line making an angle 30° with the positive axis at (0, 2).*

Sol. Gradient of $\phi = \nabla\phi$

$$= \left(\hat{i}\,\frac{\partial}{\partial x} + \hat{j}\,\frac{\partial}{\partial y} + \hat{k}\,\frac{\partial}{\partial z}\right)\left(\frac{x}{x^2 + y^2}\right) = \hat{i}\left[\frac{\left(x^2 + y^2\right)\cdot 1 - x\cdot 2x}{\left(x^2 + y^2\right)^2}\right] + \hat{j}\left(\frac{-x\cdot 2y}{\left(x^2 + y^2\right)^2}\right)$$

$$= \frac{y^2 - x^2}{\left(x^2 + y^2\right)^2}\,\hat{i} - \frac{2xy}{\left(x^2 + y^2\right)^2}\,\hat{j}$$

Gradient of ϕ at (0, 2) $= \frac{4}{16}\hat{i} - 0 = \frac{\hat{i}}{4}$

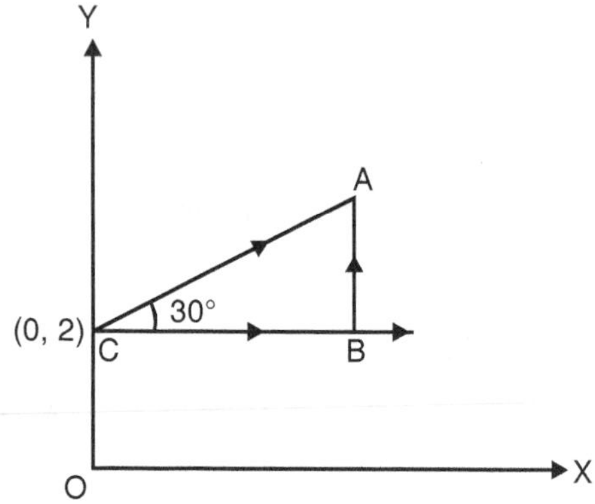

Now, $\overrightarrow{CA} = \overrightarrow{CB} + \overrightarrow{BA}$

$\overrightarrow{CB} = CA\cos 30^\circ\,\hat{i} \quad \because$ CB is || to X-axis

$\overrightarrow{BA} = CA\sin 30^\circ\,\hat{j} \quad \because$ BA is || to Y-axis

$\therefore \quad \overrightarrow{CA} = CA\left\{\cos 30^\circ\,\hat{i} + \sin 30^\circ\,\hat{j}\right\}$

$$\frac{\overrightarrow{CA}}{CA} = \frac{\sqrt{3}}{2}\hat{i} + \frac{1}{2}\hat{j} = \frac{\sqrt{3}\hat{i} + \hat{j}}{2}$$

$$\widehat{CA} = \frac{1}{2}\left(\sqrt{3}\,\hat{i} + \hat{j}\right)$$

Directional derivative of ϕ at (0, 2) in the direction of $\overrightarrow{CA} = \frac{\hat{i}}{4}\cdot\widehat{CA} = \frac{\hat{i}}{4}\cdot\frac{1}{2}\left(\sqrt{3}\hat{i} + \hat{j}\right) = \frac{\sqrt{3}}{8}$.

Example 9. *The temperature at any point in space is given by $T = xy + yz + zx$. Determine the directional derivative of T in the direction of the vector $3\hat{i} - 4\hat{k}$ at the point (1, 1, 1).*

Sol. $T = xy + yz + zx$

$$\text{Gradient of } T = \nabla T = \left(\hat{i}\frac{\partial}{\partial x} + \hat{j}\frac{\partial}{\partial y} + \hat{k}\frac{\partial}{\partial z}\right)(xy + yz + zx) = \hat{i}\,(y + z) + \hat{j}\,(z + x) + \hat{k}\,(x + y)$$

Gradient of T at (1, 1, 1) $= 2\hat{i} + 2\hat{j} + 2\hat{k}$

Directional derivative of T at (1, 1, 1) in the direction of $\left(3\hat{i} - 4\hat{k}\right)$

$$= \left(2\hat{i} + 2\hat{j} + 2\hat{k}\right)\cdot\left(\frac{3\hat{i} - 4\hat{k}}{\sqrt{9 + 16}}\right) = \frac{1}{5}(2\cdot3 - 2\cdot4) = \frac{-2}{5}.$$

Example 10. *(i) In what direction from (3, 1, – 2) is the directional derivative of $f = x^2 y^2 z^4$ maximum? Find also the magnitude of this maximum.*

(ii) Find the maximum value of directional derivative of $f = x^2 - 2y^2 + 4z^2$ at the point (1, 1, –1). **(P.T.U., May 2009)**

Sol. (*i*) Gradient of $\phi = \nabla\phi = \left(\hat{i}\frac{\partial}{\partial x} + \hat{j}\frac{\partial}{\partial y} + \hat{k}\frac{\partial}{\partial z}\right)\left(x^2 y^2 z^4\right)$

$$= \hat{i}\left(2x y^2 z^4\right) + \hat{j}\left(2x^2 y z^4\right) + \hat{k}\left(4x^2 y^2 z^3\right)$$

Gradient of ϕ at (3, 1, – 2) = $96\hat{i} + 288\hat{j} - 288\hat{k}$

Directional derivative is maximum in the direction given by $\Delta\phi$ at (3, 1, –2)
$= 96\hat{i} + 288\hat{j} - 288\hat{k}$

In any other direction the magnitude of the directional derivative will be less than its maximum value which is

$$= \sqrt{(96)^2 + (288)^2 + (288)^2} = 96\sqrt{1+9+9} = 96\sqrt{19}.$$

(*ii*) We know that the maximum value of directional derivative of $f = x^2 - 2y^2 + 4z^2$ at (1, 1, –1) is obtained from the value of gradient of f

$$\therefore \quad \text{grad } f = \nabla f$$

$$= \left(\hat{i}\frac{\partial}{\partial x} + \hat{j}\frac{\partial}{\partial y} + \hat{k}\frac{\partial}{\partial z}\right)\left(x^2 - 2y^2 + 4z^2\right)$$

$$= 2x\hat{i} - 4y\hat{j} + 8z\hat{k}$$

Value of ∇f at (1, 1, –1)

$$= 2\hat{i} - 4\hat{j} - 8\hat{k}$$

$\therefore$ Maximum value of directional derivative of f at (1, 1, –1)

$$= \sqrt{4 + 16 + 64} = \sqrt{84} = 2\sqrt{21}$$

Example 11. *Let $f(x, y, z)$ and $\phi(x, y, z)$ be two scalar functions. Find an expression for $\nabla^2 (f g)$ in terms of $\nabla^2 f$, $\nabla^2 g$, ∇f and ∇g.*

Sol.

$$\nabla(f g) = f\,\nabla(g) + g\,(\nabla f)$$

$$\nabla^2 (f g) = \nabla[\nabla(f g)] = \nabla\{f(\nabla g)\} + \nabla\{g(\nabla f)\}$$

$$= f(\nabla^2 g) + (\nabla f) \bullet (\nabla g) + (\nabla g) \bullet (\nabla f) + g(\nabla^2 f)$$

$$= f(\nabla^2 g) + 2(\nabla f) \bullet (\nabla g) + g(\nabla^2 f).$$

Example 12. *Find the angle between the surfaces $x^2 + y^2 + z^2 = 9$ and $z = x^2 + y^2 - 3$ at the point (2, –1, 2).*

Sol. Let $\phi_1 = x^2 + y^2 + z^2 - 9$

and $\phi_2 = x^2 + y^2 - z - 3$

$$\text{grad } \phi_1 = \left(\hat{i}\frac{\partial}{\partial x} + \hat{j}\frac{\partial}{\partial y} + \hat{k}\frac{\partial}{\partial z}\right)\left(x^2 + y^2 + z^2 - 9\right) = 2x\,\hat{i} + 2y\,\hat{j} + 2z\,\hat{k}$$

Similarly, $\text{grad } \phi_2 = 2x\,\hat{i} + 2y\,\hat{j} - \hat{k}$

Now, angle between the two surfaces at (2, –1, 2) is the angle between their normals at the point.

$\therefore$ Let $\vec{n}_1$ and $\vec{n}_2$ be the normal vectors to ϕ_1 and ϕ_2 respectively at (2, –1, 2).

Now,

$$\vec{n}_1 = \text{grad } \phi_1 = \nabla\phi_1 \quad \text{at} \quad (2, -1, 2) = 4\hat{i} - 2\hat{j} + 4\hat{k}.$$

$$\vec{n}_2 = \text{grad } \phi_2 = \nabla\phi_2 \quad \text{at} \quad (2, -1, 2) = 4\hat{i} - 2\hat{j} - \hat{k}$$

If θ be the angle between $\vec{n}_1$ and $\vec{n}_2$ then

$$\cos\theta = \frac{\vec{n}_1 \cdot \vec{n}_2}{|\vec{n}_1||\vec{n}_2|} = \frac{4(4) - 2(-2) + 4(-1)}{\sqrt{16+4+16}\,\sqrt{16+4+1}}$$

$$\cos\theta = \frac{16+4-4}{6\sqrt{21}} = \frac{16}{6\sqrt{21}} = \frac{8}{3\sqrt{21}}$$

$$\therefore \qquad \theta = \cos^{-1}\frac{8}{3\sqrt{21}}.$$

TEST YOUR KNOWLEDGE

1. Find grad ϕ at (1, –2, –1) if

(*i*) $\phi = 3x^2y - y^3z^2$ (*ii*) $\phi = x^2 + yz$

(*iii*) $\phi = \log(x^2 + y^2 + z^2)$

2. Find a unit vector normal to the following surfaces

(*i*) $z^2 = x^2 + y^2$ at the point (1, 0, –1)

(*ii*) $x y^3 z^2 = 4$ at the point (–1, –1, 2)

(*iii*) $x^2 y + 2xz = 4$ at the point (2, –2, 3)

(*iv*) $z = x^3 + y^2$ at (1, –2, 5) ***(P.T.U., May 2002)***

(*v*) $x^2 + 3y^2 + 2z^2 = 6$ at (1, 0, 1)

(*vi*) $z = \sqrt{x^2 + y^2}$ at (3, 4, 5) ***(P.T.U., Jan. 2008)***

[**Hint.** Consult S.E. 3]

3. If $r = |\vec{r}|$ where $\vec{r} = x\hat{i} + y\hat{j} + z\hat{k}$, prove that

(*i*) $\nabla f(r) = f'(r)\nabla(r)$ (*ii*) $\nabla(\log r) = \dfrac{\vec{r}}{r^2}$

[**Hint.** S.E. 10 art 7.23]

(*iii*) $\nabla\left(e^{r^2}\right) = 2e^{r^2}\vec{r}$ (*iv*) grad $|\vec{r}|^2 = 2\vec{r}$

4. (*i*) Find the directional derivative of $\phi = x^2 + y^2 + z^2$ at the point (2, 2, 1) in the direction of $2\hat{i} + 2\hat{j} + \hat{k}$.

(*ii*) What is the directional derivative of $2xy + z^2$ at the point (1, –1, 3) in the direction of the vector $\hat{i} + 2\hat{j} + 2\hat{k}$? ***(P.T.U., May 2004)***

5. Find the directional derivative of $\phi = 4xz^3 - 3x^2yz^2$ at the point (2, –1, 2) along Z-axis.

6. Find the directional derivative of $f = 3e^{2x-y+z}$ at the point A (1, 1, –1) in the direction $\overrightarrow{AB}$, where B is the point (–3, 5, 6).

7. (*i*) Calculate the directional derivative of the function $\phi(x, y, z) = xy^2 + yz^3$ at the point (1, –1, 1) in the direction of (3, 1, –1).

(*ii*) Find the directional derivative of $f(x, y, z) = x^2y^2z^2$ at the point (1, 1, – 1) in the direction of tangent to the curve $x = e^t$, $y = 2\sin t + 1$, $z = t - \cos t$ at $t = 0$.

[**Hint.** See solved example 7]

8. Find the direction in which the directional derivative of $f(x, y) = (x^2 - y^2)xy$ at (1, 1) is zero.

9. Find the directional derivative of $\vec{\nabla}^2$, where $\vec{\nabla} = xy^2\hat{i} + zy^2\hat{j} + xz^2\hat{k}$ at the point (2, 0, 3) in the direction of outward normal to the sphere $x^2 + y^2 + z^2 = 14$ at the point (3, 2, 1).
10. Find the directional derivative of the scalar function $f(x, y, z) = xyz$ in the direction of the outer normal to the surface $z = xy$ at the point (3, 1, 3).
11. What is the greatest rate of increase of $u = x^2 + y z^2$ at the point (1, –1, 3) ?
12. If θ is the acute angle between the surfaces $xy^2z = 3x + z^2$ and $3x^2 - y^2 + 2z = 1$ at the point (1, –2, 1), show that $\cos q = \frac{3}{7\sqrt{6}}$.
13. Calculate the angle between the normals to the surface $xy = z^2$ at the point (4, 1, 2) and (3, 3, –3).
14. Find the angle between tangent planes to the surfaces $x \log z = y^2 - 1$ and $x^2 y = 2 - z$ at the point (1, 1, 1).
15. Find the values of constants a and b so that the surfaces $ax^2 - byz = (a + 2)x$ and $4x^2 y + z^3 = 4$ may intersect orthogonally at the point (1, –1, 2).
[**Hint.** The point P (1, –1, 2) lies on both the surfaces and ($\nabla\phi_1$ at P) • ($\nabla\phi_2$ at P) = 0]
16. The temperature at any point (x, y, z) in space is given by $T(x, y, z) = x^2 + y^2 - z$. A mosquito located at (1, 1, 2) desires to fly in such a direction that it will get warm as soon as possible. In which direction should it fly ?
17. If the directional derivative of $f(x, y, z) = axy + byz + czx$ at (1, 1, 1) has the maximum magnitude 4 in a direction parallel to x-axis then find the values of a, b, c.
18. Find the equation of the tangent plane to the surface $2xz^2 - 3xy - 4x = 7$ at the point (1, –1, 2). Also find the equation of the normal at (1, –1, 2).

Answers

1. (*i*) $-12\hat{i} - 9\hat{j} - 16\hat{k}$ (*ii*) $2x\hat{i} + 2\hat{j} + y\hat{k}$ (*iii*) $\frac{1}{3}(\hat{i} - 2\hat{j} - \hat{k})$

2. (*i*) $\frac{1}{\sqrt{2}}(\hat{i} + \hat{k})$ (*ii*) $-\frac{1}{\sqrt{11}}(\hat{i} + 3\hat{j} - \hat{k})$ (*iii*) $\frac{1}{3}(-\hat{i} + 2\hat{j} + 2\hat{k})$

(*iv*) $\frac{2\hat{i} - 4\hat{j} - \hat{k}}{\sqrt{21}}$ (*v*) $\frac{\hat{i} + 2\hat{k}}{\sqrt{5}}$ (*vi*) $\frac{3\hat{i} + 4\hat{j} - 5\hat{k}}{5\sqrt{2}}$

4. (*i*) 6 (*ii*) $\frac{14}{3}$ 5. 144

6. $-\frac{5}{3}$ 7. (*i*) $\frac{5}{\sqrt{11}}$ (*ii*) $\frac{2\sqrt{6}}{3}$

8. $\frac{\hat{i} + \hat{j}}{\sqrt{2}}$ 9. $\frac{1404}{\sqrt{14}}$ 10. $\frac{27}{\sqrt{11}}$

11. $2\hat{i} + 9\hat{j} - 6\hat{k}$ 13. $\cos^{-1}\sqrt{\frac{3}{62}}$ 14. $\cos^{-1}\frac{1}{\sqrt{30}}$

15. $a = 2.5, b = 1$ 16. $\frac{1}{3}(2\hat{i} + 2\hat{j} - \hat{k})$ 17. $a = 2, b = -2, c = 2$

18. $7x - 3y + 8z - 26 = 0; \frac{x-1}{7} = \frac{y+1}{-3} = \frac{z-2}{8}$.

7.18. DIVERGENCE OF A VECTOR FUNCTION

The divergence of a continuously differentiable vector point function $\vec{V}$ is denoted by div $\vec{V}$ and is defined as

$$\text{div } \vec{V} = \nabla \cdot \vec{V} = \left(\hat{i}\frac{\partial}{\partial x} + \hat{j}\frac{\partial}{\partial y} + \hat{k}\frac{\partial}{\partial z}\right)\cdot \vec{V} = \hat{i}\cdot\frac{\partial \vec{V}}{\partial x} + \hat{j}\cdot\frac{\partial \vec{V}}{\partial y} + \hat{k}\cdot\frac{\partial \vec{V}}{\partial z}$$

Clearly, divergence of a vector point function is a scalar point function.

If $\quad \vec{V} = V_1\hat{i} + V_2\hat{j} + V_3\hat{k}$

then $\quad \text{div } \vec{V} = \nabla \cdot \vec{V} = \left(\hat{i}\frac{\partial}{\partial x} + \hat{j}\frac{\partial}{\partial y} + \hat{k}\frac{\partial}{\partial z}\right)\cdot\left(V_1\hat{i} + V_2\hat{j} + V_3\hat{k}\right) = \frac{\partial V_1}{\partial x} + \frac{\partial V_2}{\partial y} + \frac{\partial V_3}{\partial z}.$

For example, if $\vec{r} = x\hat{i} + y\hat{j} + z\hat{k}$, then $\text{div } \vec{r} = \frac{\partial}{\partial x}(x) + \frac{\partial}{\partial y}(y) + \frac{\partial}{\partial z}(z) = 1 + 1 + 1 = 3.$

7.19. PHYSICAL INTERPRETATION OF DIVERGENCE

Let us consider the case of a fluid flow. Consider a small rectangular parallelopiped of dimensions *dx, dy, dz* parallel to X-axis, Y-axis and Z-axis respectively.

Let $\vec{V} = V_x\hat{i} + V_y\hat{j} + V_z\hat{k}$ be the velocity of fluid at P (x, y, z) where (V_x, V_y, V_z) are components of $\vec{V}$ parallel to X-axis, Y-axis, Z-axis respectively.

Mass of the fluid flowing in through the face ABCD per unit time = Velocity × area of the face = $V_x\,(dy\,dz)$

$\therefore$ Mass of the fluid flowing out across the face PQRS per unit time = $V_{x+\delta x}\,(dy\,dz)$.

$$V_{x+\delta x} = V_x + \delta x\,\frac{\partial V_x}{\partial x} + \cdots \text{ by Taylor's theorem.}$$

$$\therefore \quad V_{x+\delta x}\,(dydz) = \left(V_x + \frac{\partial V_x}{\partial x}\,dx\right)(dy\,dz)$$

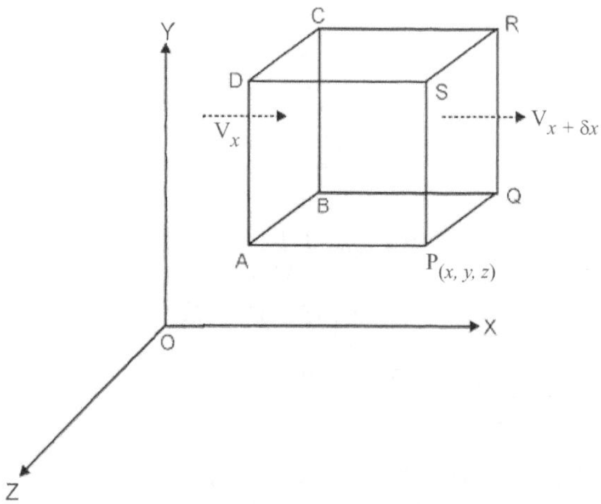

Decrease in mass of fluid in the parallelopiped corresponding to the flow along X-axis per unit time

$$= V_x\, dy\, dz - \left(V_x + \frac{\partial V_x}{\partial x}\, dx\right) dy\, dz = -\frac{\partial V_x}{\partial x}\, dx\, dy\, dz \qquad \text{(–ve sign shows decrease)}$$

Similarly, the decrease in mass of fluid to the flow along Y-axis $= \frac{\partial V_y}{\partial y}\, dx\, dy\, dz$ and decrease in mass along Z-axis $= \frac{\partial V_z}{\partial z}\, dx\, dy\, dz$.

Total decrease in mass of fluid per unit time $= \left(\frac{\partial V_x}{\partial x} + \frac{\partial V_y}{\partial y} + \frac{\partial V_z}{\partial z}\right) dx\, dy\, dz$

The rate of loss of fluid per unit volume $= \frac{\partial V_x}{\partial x} + \frac{\partial V_y}{\partial y} + \frac{\partial V_z}{\partial z}$

$$= \left(\hat{i}\frac{\partial}{\partial x} + \hat{j}\frac{\partial}{\partial y} + \hat{k}\frac{\partial}{\partial z}\right) \bullet \left(\hat{i}\, V_x + \hat{j}\, V_y + \hat{k}\, V_z\right) = \nabla \bullet \vec{V} = \text{div}\, \vec{V} \qquad \ldots(1)$$

$\therefore$ div $\vec{V}$ is the rate at which the fluid is flowing at a point per unit volume. If the flux entering any element of the space is the same as that leaving it *i.e.*, div $\vec{V} = 0$ everywhere then such a point function is called a **Solenoidal Vector Function.**

Equation (1) is also called the equation of continuity or conservation of mass.

Note. For details of the solenoidal vector function consult chapter 8 art 8.2.

7.20. CURL OF A VECTOR POINT FUNCTION *(P.T.U., May 2007)*

The curl of a continuously differentiable vector point function $\vec{V}$ is defined by the equation

$$\text{Curl}\, \vec{V} = \nabla \times \vec{V} = \left(\hat{i}\frac{\partial}{\partial x} + \hat{j}\frac{\partial}{\partial y} + \hat{k}\frac{\partial}{\partial z}\right) \times \left(\vec{V}\right)$$

$$\text{Curl}\, \vec{V} = \hat{i} \times \frac{\partial \vec{V}}{\partial x} + \hat{j} \times \frac{\partial \vec{V}}{\partial y} + \hat{k} \times \frac{\partial \vec{V}}{\partial z}$$

Clearly the curl of a vector point function is a vector point function.

If $\qquad \vec{V} = V_1\hat{i} + V_2\hat{j} + V_3\hat{k}$

Then $\qquad \text{curl}\, \vec{V} = \left(\hat{i}\frac{\partial}{\partial x} + \hat{j}\frac{\partial}{\partial y} + \hat{k}\frac{\partial}{\partial z}\right) \times \left(V_1\hat{i} + V_2\hat{j} + V_3\hat{k}\right)$

$$= \begin{vmatrix} \hat{i} & \hat{j} & \hat{k} \\ \frac{\partial}{\partial x} & \frac{\partial}{\partial y} & \frac{\partial}{\partial z} \\ V_1 & V_2 & V_3 \end{vmatrix} = \hat{i}\left(\frac{\partial V_3}{\partial y} - \frac{\partial V_2}{\partial z}\right) + \hat{j}\left(\frac{\partial V_1}{\partial z} - \frac{\partial V_3}{\partial x}\right) + \hat{k}\left(\frac{\partial V_2}{\partial x} - \frac{\partial V_1}{\partial y}\right)$$

Note. Curl of a vector point function is also called rotation of a vector point function.

7.21. PHYSICAL INTERPRETATION OF CURL OF A VECTOR POINT FUNCTION

(P.T.U., Dec. 2003, May 2010, Dec. 2011, May 2014)

Consider a rigid body rotating about a fixed axis through the point O with uniform angular velocity ω. If $\vec{V}$ be the Linear Velocity and $\vec{r}$ be the position vector of any point on the rotating body.

then $\quad \vec{V} = \vec{\omega} \times \vec{r}$

then $\quad \text{curl } \vec{V} = \nabla \times \vec{V} = \nabla \times (\vec{\omega} \times \vec{r})$

Let $\quad \vec{\omega} = \omega_1 \hat{i} + \omega_2 \hat{j} + \omega_3 \hat{k}$ and $\vec{r} = x\hat{i} + y\hat{j} + z\hat{k}$

then
$$\text{curl } \vec{V} = \nabla \times \left\{ \left(\omega_1 \hat{i} + \omega_2 \hat{j} + \omega_3 \hat{k} \right) \times \left(x\hat{i} + y\hat{j} + z\hat{k} \right) \right\}$$

$$= \nabla \times \begin{vmatrix} \hat{i} & \hat{j} & \hat{k} \\ \omega_1 & \omega_2 & \omega_3 \\ x & y & z \end{vmatrix} = \nabla \times \left\{ \hat{i}\ (\omega_2 z - \omega_3 y) + \hat{j}\ (\omega_3 x - \omega_1 z) + \hat{k}\ (\omega_1 y - \omega_2 x) \right\}$$

$$= \left(\hat{i} \frac{\partial}{\partial x} + \hat{j} \frac{\partial}{\partial y} + \hat{k} \frac{\partial}{\partial z} \right) \times \left\{ (\omega_2 z - \omega_3 y)\, \hat{i} + (\omega_3 x - \omega_1 z)\, \hat{j} + (\omega_1 y - \omega_2 x)\, \hat{k} \right\}$$

$$= \begin{vmatrix} \hat{i} & \hat{j} & \hat{k} \\ \dfrac{\partial}{\partial x} & \dfrac{\partial}{\partial y} & \dfrac{\partial}{\partial z} \\ \omega_2 z - \omega_3 y & \omega_3 x - \omega_1 z & \omega_1 y - \omega_2 x \end{vmatrix}$$

$$= \hat{i} \left\{ \frac{\partial}{\partial y} (\omega_1 y - \omega_2 x) - \frac{\partial}{\partial z} (\omega_3 x - \omega_1 z) \right\} + \hat{j} \left\{ \frac{\partial}{\partial z} (\omega_2 z - \omega_3 y) - \frac{\partial}{\partial x} (\omega_1 y - \omega_2 x) \right\}$$
$$+ \hat{k} \left\{ \frac{\partial}{\partial x} (\omega_3 x - \omega_1 z) - \frac{\partial}{\partial y} (\omega_2 z - \omega_3 y) \right\}$$

$$= \hat{i}\,(\omega_1 + \omega_1) + \hat{j}\,(\omega_2 + \omega_2) + \hat{k}\,(\omega_3 + \omega_3) = \hat{i}\, 2\omega_1 + \hat{j}\, 2\omega_2 + \hat{k}\, 2\omega_3$$

$$= 2\left(\omega_1 \hat{i} + \omega_2 \hat{j} + \omega_3 \hat{k} \right) = 2\vec{\omega}$$

Curl $\vec{V} = 2\vec{\omega}$ shows that curl of a vector field is connected with rotational properties of the vector field and justifies the name *rotation* used for *curl*.

Hence the angular velocity at any point is equal to half the curl of linear velocity at that point of the body.

Cor. ***If curl $\vec{V} = \vec{0}$, then $\vec{V}$ is called irrotational vector and the field V is termed irrotational.***

Note. For details of the irrotational vectors consult chapter 8 arts 8.3 and 8.4.

7.22. PROPERTIES OF DIVERGENCE AND CURL

1. **For a constant vector $\vec{a}$, div $\vec{a} = 0$, curl $\vec{a} = \vec{0}$**

2. **div** $(\vec{A} + \vec{B}) = \text{div}\, \vec{A} + \text{div}\, \vec{B}$ **or** $\nabla \cdot (\vec{A} + \vec{B}) = \nabla \cdot \vec{A} + \nabla \cdot \vec{B}$

Proof.

$$\text{div}(\vec{A} + \vec{B}) = \nabla \cdot (\vec{A} + \vec{B}) = \left(\hat{i}\frac{\partial}{\partial x} + \hat{j}\frac{\partial}{\partial y} + \hat{k}\frac{\partial}{\partial z}\right) \cdot (\vec{A} + \vec{B})$$

$$= \hat{i} \cdot \frac{\partial}{\partial x}(\vec{A} + \vec{B}) + \hat{j} \cdot \frac{\partial}{\partial y}(\vec{A} + \vec{B}) + \hat{k} \cdot \frac{\partial}{\partial z}(\vec{A} + \vec{B})$$

$$= \hat{i} \cdot \left(\frac{\partial \vec{A}}{\partial x} + \frac{\partial \vec{B}}{\partial x}\right) + \hat{j} \cdot \left(\frac{\partial \vec{A}}{\partial y} + \frac{\partial \vec{B}}{\partial y}\right) + \hat{k} \cdot \left(\frac{\partial \vec{A}}{\partial z} + \frac{\partial \vec{B}}{\partial z}\right)$$

$$= \left(i \cdot \frac{\partial \vec{A}}{\partial x} + \hat{j} \cdot \frac{\partial \vec{A}}{\partial y} + \hat{k} \cdot \frac{\partial \vec{A}}{\partial z}\right) + \left(\hat{i} \cdot \frac{\partial \vec{B}}{\partial x} + \hat{j} \cdot \frac{\partial \vec{B}}{\partial y} + \hat{k} \cdot \frac{\partial \vec{B}}{\partial z}\right)$$

$$= \left(\hat{i}\frac{\partial}{\partial x} + \hat{j}\frac{\partial}{\partial y} + \hat{k}\frac{\partial}{\partial z}\right) \cdot \vec{A} + \left(\hat{i}\frac{\partial}{\partial x} + \hat{j}\frac{\partial}{\partial y} + \hat{k}\frac{\partial}{\partial z}\right) \cdot \vec{B}$$

$$= \nabla \cdot \vec{A} + \nabla \cdot \vec{B}$$

$$= \text{div}\, \vec{A} + \text{div}\, \vec{B}$$

3. **Curl** $(\vec{A} + \vec{B}) = \text{curl}\, \vec{A} + \text{curl}\, \vec{B}$ **or** $\nabla \times (\vec{A} + \vec{B}) = \nabla \times \vec{A} + \nabla \times \vec{B}$

Proof. Curl $(\vec{A} + \vec{B}) = \nabla \times (\vec{A} + \vec{B})$

$$= \left(\hat{i}\frac{\partial}{\partial x} + \hat{j}\frac{\partial}{\partial y} + \hat{k}\frac{\partial}{\partial z}\right) \times (\vec{A} + \vec{B})$$

$$= \hat{i} \times \frac{\partial}{\partial x}(\vec{A} + \vec{B}) + \hat{j} \times \frac{\partial}{\partial y}(\vec{A} + \vec{B}) + \hat{k} \times \frac{\partial}{\partial z}(\vec{A} + \vec{B})$$

$$= \hat{i} \times \left(\frac{\partial \vec{A}}{\partial x} + \frac{\partial \vec{B}}{\partial x}\right) + \hat{j} \times \left(\frac{\partial \vec{A}}{\partial y} + \frac{\partial \vec{B}}{\partial y}\right) + \hat{k} \times \left(\frac{\partial \vec{A}}{\partial z} + \frac{\partial \vec{B}}{\partial z}\right)$$

$$= \left(\hat{i} \times \frac{\partial \vec{A}}{\partial x} + \hat{j}\frac{\partial \vec{A}}{\partial y} + \hat{k}\frac{\partial \vec{A}}{\partial z}\right) + \left(\hat{i} \times \frac{\partial \vec{B}}{\partial x} + \hat{j}\frac{\partial \vec{B}}{\partial y} + \hat{k}\frac{\partial \vec{B}}{\partial z}\right)$$

$$\text{Curl}\,(\vec{A} + \vec{B}) = \left(\hat{i}\frac{\partial}{\partial x} + \hat{j}\frac{\partial}{\partial y} + \hat{k}\frac{\partial}{\partial z}\right) \times \vec{A} + \left(\hat{i}\frac{\partial}{\partial x} + \hat{j}\frac{\partial}{\partial y} + \hat{k}\frac{\partial}{\partial z}\right) \times \vec{B}$$

$$= \nabla \times \vec{A} + \nabla \times \vec{B} = \text{curl}\, \vec{A} + \text{curl}\, \vec{B}$$

Hence curl $(\vec{A} + \vec{B}) = \text{curl}\, \vec{A} + \text{curl}\, \vec{B}$

4. If $\vec{A}$ is a vector function and ϕ is a scalar function, then

$$\textbf{div}\left(\phi\vec{\mathbf{A}}\right) = \phi\left(\textbf{div}\,\vec{\mathbf{A}}\right) + (\textbf{grad}\,\phi)\cdot\vec{\mathbf{A}}$$

or
$$\nabla\cdot\left(\phi\vec{\mathbf{A}}\right) = \phi\left(\nabla\cdot\vec{\mathbf{A}}\right) + (\nabla\phi)\cdot\vec{\mathbf{A}}$$

Proof. $\textbf{div}\left(\phi\vec{A}\right) = \nabla\cdot\left(\phi\vec{A}\right) = \left(\hat{i}\frac{\partial}{\partial x} + \hat{j}\frac{\partial}{\partial y} + \hat{k}\frac{\partial}{\partial z}\right)\cdot\left(\phi\vec{A}\right)$

$$= \hat{i}\cdot\frac{\partial}{\partial x}\left(\phi\vec{A}\right) + \hat{j}\cdot\frac{\partial}{\partial y}\left(\phi\vec{A}\right) + \hat{k}\cdot\frac{\partial}{\partial z}\left(\phi\vec{A}\right) = \Sigma\hat{i}\cdot\left(\phi\frac{\partial}{\partial x}\vec{A} + \frac{\partial\phi}{\partial x}\vec{A}\right)$$

$$= \phi\Sigma\hat{i}\cdot\frac{\partial}{\partial x}\vec{A} + \Sigma\left(\hat{i}\frac{\partial\phi}{\partial x}\right)\cdot\vec{A}$$

$$= \phi\left\{\hat{i}\cdot\frac{\partial}{\partial x}\vec{A} + \hat{j}\cdot\frac{\partial}{\partial y}\vec{A} + \hat{k}\cdot\frac{\partial}{\partial z}\vec{A}\right\} + \left\{\left(\hat{i}\frac{\partial\phi}{\partial x}\right)\cdot\vec{A} + \left(\hat{j}\frac{\partial\phi}{\partial y}\right)\cdot\vec{A} + \left(\hat{k}\frac{\partial\phi}{\partial z}\right)\cdot\vec{A}\right\}$$

$$= \phi\left\{\hat{i}\frac{\partial}{\partial x} + \hat{j}\frac{\partial}{\partial y} + \hat{k}\frac{\partial}{\partial z}\right\}\cdot\vec{A} + \left\{\hat{i}\frac{\partial\phi}{\partial x} + \hat{j}\frac{\partial\phi}{\partial y} + \hat{k}\frac{\partial\phi}{\partial z}\right\}\cdot\vec{A}$$

$$= \phi\left(\nabla\cdot\vec{A}\right) + (\nabla\phi)\cdot\vec{A}$$

$$\therefore\ \text{div}\left(\phi\vec{A}\right) = \phi\left(\text{div}\,\vec{A}\right) + (\text{grad}\,\phi)\cdot\vec{A}$$

5. If $\vec{A}$ is a vector function and ϕ is a scalar function then *(P.T.U., Jan. 2010)*

$$\textbf{Curl}\left(\phi\vec{\mathbf{A}}\right) = (\textbf{grad}\,\phi)\times\vec{\mathbf{A}} + \phi\,\textbf{curl}\,\vec{\mathbf{A}}$$

or
$$\nabla\times\left(\phi\vec{\mathbf{A}}\right) = (\nabla\phi)\times\vec{\mathbf{A}} + \phi\left(\nabla\times\vec{\mathbf{A}}\right).$$

Proof. $\text{Curl}\left(\phi\vec{A}\right) = \nabla\times\left(\phi\vec{A}\right) = \left(\hat{i}\frac{\partial}{\partial x} + \hat{j}\frac{\partial}{\partial y} + \hat{k}\frac{\partial}{\partial z}\right)\times\left(\phi\vec{A}\right)$

$$= \hat{i}\times\frac{\partial}{\partial x}\left(\phi\vec{A}\right) + \hat{j}\times\frac{\partial}{\partial y}\left(\phi\vec{A}\right) + \hat{k}\times\frac{\partial}{\partial z}\left(\phi\vec{A}\right)$$

$$= \hat{i}\times\left\{\frac{\partial\phi}{\partial x}\vec{A} + \phi\frac{\partial\vec{A}}{\partial x}\right\} + \hat{j}\times\left\{\frac{\partial\phi}{\partial y}\vec{A} + \phi\frac{\partial\vec{A}}{\partial y}\right\} + \hat{k}\times\left\{\frac{\partial\phi}{\partial z}\vec{A} + \phi\frac{\partial\vec{A}}{\partial z}\right\}$$

$$= \left\{\hat{i}\times\left(\frac{\partial\phi}{\partial x}\vec{A}\right) + \hat{j}\times\left(\frac{\partial\phi}{\partial y}\vec{A}\right) + \hat{k}\times\left(\frac{\partial\phi}{\partial z}\vec{A}\right)\right\} + \phi\left\{\hat{i}\times\frac{\partial\vec{A}}{\partial x} + \hat{j}\times\frac{\partial\vec{A}}{\partial y} + \hat{k}\times\frac{\partial\vec{A}}{\partial z}\right\}$$

$$= \left\{\frac{\partial\phi}{\partial x}\hat{i}\times\vec{A} + \frac{\partial\phi}{\partial y}\hat{j}\times\vec{A} + \frac{\partial\phi}{\partial z}\hat{k}\times\vec{A}\right\} + \phi\left\{\hat{i}\frac{\partial}{\partial x} + \hat{j}\frac{\partial}{\partial y} + \hat{k}\frac{\partial}{\partial z}\right\}\times\vec{A}$$

$$= \left\{\left(\hat{i}\frac{\partial}{\partial x}+\hat{j}\frac{\partial}{\partial y}+\hat{k}\frac{\partial}{\partial z}\right)\phi\right\}\times\vec{A}+\phi\left\{\hat{i}\frac{\partial}{\partial x}+\hat{j}\frac{\partial}{\partial y}+\hat{k}\frac{\partial}{\partial z}\right\}\times\vec{A}$$

$$= (\nabla\phi)\times\vec{A}+\phi\left(\nabla\times\vec{A}\right) = (\text{grad }\phi)\times\vec{A}+\phi\left(\text{curl }\vec{A}\right)$$

6. $\nabla\left(\vec{\mathbf{A}}\cdot\vec{\mathbf{B}}\right)=\left(\vec{A}\cdot\nabla\right)\vec{B}+\left(\vec{B}\cdot\nabla\right)\vec{A}+\vec{A}\times\left(\nabla\times\vec{B}\right)+\vec{B}\times\left(\nabla\times\vec{A}\right)$

Proof. $\nabla\left(\vec{A}\cdot\vec{B}\right)=\Sigma\,\hat{i}\,\frac{\partial}{\partial x}\left(\vec{A}\cdot\vec{B}\right)=\Sigma\,\hat{i}\left\{\vec{A}\cdot\frac{\partial\vec{B}}{\partial x}+\frac{\partial\vec{A}}{\partial x}\cdot\vec{B}\right\}=\Sigma\left(\vec{A}\cdot\frac{\partial\vec{B}}{\partial x}\right)\hat{i}+\Sigma\left(\vec{B}\cdot\frac{\partial\vec{A}}{\partial x}\right)\hat{i}$

Now, we know that $\vec{a}\times\left(\vec{b}\times\vec{c}\right)=(\vec{a}\cdot\vec{c})\vec{b}-\left(\vec{a}\cdot\vec{b}\right)\vec{c}$

$$\therefore \quad \left(\vec{a}\cdot\vec{b}\right)\vec{c}=(\vec{a}\cdot\vec{c})\vec{b}-\vec{a}\times\left(\vec{b}\times\vec{c}\right)$$

$$\therefore \quad \left(\vec{A}\cdot\frac{\partial\vec{B}}{\partial x}\right)\hat{i}=\left(\vec{A}\cdot\hat{i}\right)\frac{\partial\vec{B}}{\partial x}-\vec{A}\times\left(\frac{\partial\vec{B}}{\partial x}\times\hat{i}\right)=\left(\vec{A}\cdot\hat{i}\right)\frac{\partial\vec{B}}{\partial x}+\vec{A}\times\left(\hat{i}\times\frac{\partial\vec{B}}{\partial x}\right)$$

$$\therefore \quad \Sigma\left(\vec{A}\cdot\frac{\partial\vec{B}}{\partial x}\right)\hat{i}=\left(\vec{A}\cdot\Sigma\,\hat{i}\,\frac{\partial}{\partial x}\right)\vec{B}+\vec{A}\times\Sigma\left(\hat{i}\times\frac{\partial\vec{B}}{\partial x}\right)=\left(\vec{A}\cdot\nabla\right)\vec{B}+\vec{A}\times\left(\nabla\times\vec{B}\right)$$

Similarly, $\quad \Sigma\left(\vec{B}\cdot\frac{\partial\vec{A}}{\partial x}\right)\hat{i}=\left(\vec{B}\cdot\nabla\right)\vec{A}+\vec{B}\times\left(\nabla\times\vec{A}\right)$ (By interchanging A and B)

$$\therefore \quad \nabla\left(\vec{A}\cdot\vec{B}\right)=\left(\vec{A}\cdot\nabla\right)\vec{B}+\vec{A}\times\left(\nabla\times\vec{B}\right)+\left(\vec{B}\cdot\nabla\right)\vec{A}+\vec{B}\times\left(\nabla\times\vec{A}\right)$$

$$=\left(\vec{A}\cdot\nabla\right)\vec{B}+\left(\vec{B}\cdot\nabla\right)\vec{A}+\vec{A}\times\left(\nabla\times\vec{B}\right)+\vec{B}\times\left(\nabla\times\vec{A}\right)$$

7. $\nabla\cdot\left(\vec{\mathbf{A}}\times\vec{\mathbf{B}}\right)=\vec{B}\cdot\left(\nabla\times\vec{A}\right)-\vec{A}\cdot\left(\nabla\times\vec{B}\right)$ ***(P.T.U., Jan. 2010)***

Proof. $\nabla\cdot\left(\vec{A}\times\vec{B}\right)=\Sigma\,\hat{i}\cdot\frac{\partial}{\partial x}\left(\vec{A}\times\vec{B}\right)=\Sigma\,\hat{i}\cdot\left\{\vec{A}\times\frac{\partial\vec{B}}{\partial x}+\frac{\partial\vec{A}}{\partial x}\times\vec{B}\right\}$

$$=\Sigma\left\{\hat{i}\cdot\left(\vec{A}\times\frac{\partial\vec{B}}{\partial x}\right)\right\}+\Sigma\left\{\hat{i}\cdot\left(\frac{\partial\vec{A}}{\partial x}\times\vec{B}\right)\right\}=\Sigma\hat{i}\cdot\left(\frac{\partial\vec{A}}{\partial x}\times\vec{B}\right)-\Sigma\hat{i}\cdot\left(\frac{\partial\vec{B}}{\partial x}\times\vec{A}\right)$$

$$=\Sigma\left(\hat{i}\times\frac{\partial\vec{A}}{\partial x}\right)\cdot\vec{B}-\Sigma\left(\hat{i}\times\frac{\partial\vec{B}}{\partial x}\right)\cdot\vec{A}=\left(\nabla\times\vec{A}\right)\cdot\vec{B}-\left(\nabla\times\vec{B}\right)\cdot\vec{A}$$

$$=\vec{B}\cdot\left(\nabla\times\vec{A}\right)-\vec{A}\cdot\left(\nabla\times\vec{B}\right)$$

8. $\nabla\times(\vec{A}\times\vec{B})=(\nabla\bullet\vec{B})\vec{A}-(\nabla\bullet\vec{A})\vec{B}+(\vec{B}\bullet\nabla)\vec{A}-(\vec{A}\bullet\nabla)\vec{B}$ ***(P.T.U., May 2010, May 2012)***

Proof.
$$\nabla\times(\vec{A}\times\vec{B}) = \Sigma\hat{i}\times\frac{\partial}{\partial x}(\vec{A}\times\vec{B})=\Sigma\hat{i}\times\left\{\frac{\partial\vec{A}}{\partial x}\times\vec{B}+\vec{A}\times\frac{\partial\vec{B}}{\partial x}\right\}$$
$$= \Sigma\hat{i}\times\left(\frac{\partial\vec{A}}{\partial x}\times\vec{B}\right)+\Sigma\hat{i}\times\left(\vec{A}\times\frac{\partial\vec{B}}{\partial x}\right)$$
$$= \Sigma\left\{(\hat{i}\cdot\vec{B})\frac{\partial\vec{A}}{\partial x}-\left(\hat{i}\cdot\frac{\partial\vec{A}}{\partial x}\right)\vec{B}\right\}+\Sigma\left\{\left(\hat{i}\cdot\frac{\partial\vec{B}}{\partial x}\right)\vec{A}-(\hat{i}\cdot\vec{A})\frac{\partial\vec{B}}{\partial x}\right\}$$
$$= \Sigma(\vec{B}\cdot\hat{i})\frac{\partial\vec{A}}{\partial x}-\left\{\Sigma\hat{i}\cdot\frac{\partial\vec{A}}{\partial x}\right\}\vec{B}+\left\{\Sigma\hat{i}\cdot\frac{\partial\vec{B}}{\partial x}\right\}\vec{A}-\Sigma(\vec{A}\cdot\hat{i})\frac{\partial\vec{B}}{\partial x}$$
$$= \left(\vec{B}\cdot\Sigma\hat{i}\frac{\partial}{\partial x}\right)\vec{A}-(\nabla\cdot\vec{A})\vec{B}+(\nabla\cdot\vec{B})\vec{A}-\left(\vec{A}\cdot\Sigma\hat{i}\frac{\partial}{\partial x}\right)\vec{B}$$
$$= (\vec{B}\cdot\nabla)\vec{A}-(\nabla\cdot\vec{A})\vec{B}+(\nabla\cdot\vec{B})\vec{A}-(\vec{A}\cdot\nabla)\vec{B}$$
$$= (\nabla\cdot\vec{B})\vec{A}-(\nabla\cdot\vec{A})\vec{B}+(\vec{B}\cdot\nabla)\vec{A}-(\vec{A}\cdot\nabla)\vec{B}$$

7.23. REPEATED OPERATIONS BY ∇

Before starting with the repeated operations by ∇, students are advised to note the following:

If ϕ (x, y, z) and $\vec{V}$ (x, y, z) be scalar and vector point functions respectively, then

(*i*) Since ϕ is scalar we can take its gradient only.

(*ii*) Since grad ϕ and $\vec{V}$ are both vector functions we can take their divergence as well as curl.

(*iii*) Since div $\vec{V}$ is a scalar function we can take its gradient only.

1. Div (grad ϕ) = $\nabla^2\phi$ where $\nabla^2=\frac{\partial^2}{\partial x^2}+\frac{\partial^2}{\partial y^2}+\frac{\partial^2}{\partial z^2}$

Proof. Div (grad ϕ) = $\nabla\cdot(\nabla\phi)$
$$=\left(\hat{i}\frac{\partial}{\partial x}+\hat{j}\frac{\partial}{\partial y}+\hat{k}\frac{\partial}{\partial z}\right)\cdot\left(\hat{i}\frac{\partial\phi}{\partial x}+\hat{j}\frac{\partial\phi}{\partial y}+\hat{k}\frac{\partial\phi}{\partial z}\right)$$
$$=\frac{\partial}{\partial x}\left(\frac{\partial\phi}{\partial x}\right)+\frac{\partial}{\partial y}\left(\frac{\partial\phi}{\partial y}\right)+\frac{\partial}{\partial z}\left(\frac{\partial\phi}{\partial z}\right)=\frac{\partial^2\phi}{\partial x^2}+\frac{\partial^2\phi}{\partial y^2}+\frac{\partial^2\phi}{\partial z^2}$$
$$=\left(\frac{\partial^2}{\partial x^2}+\frac{\partial^2}{\partial y^2}+\frac{\partial^2}{\partial z^2}\right)\phi=\nabla^2\phi$$

Note 1. ∇^2 is called Laplacian Operator and $\nabla^2\phi=0$ is called **Laplace Equation.**

Note 2. A function satisfying Laplace Equation is called **Harmonic Function** *i.e.*, ϕ is Harmonic Function.

2. Curl (grad φ) = ∇ × (∇φ) = $\vec{0}$

Proof. Curl (grad φ) = ∇ × (∇ φ)

$$= \left(\hat{i}\frac{\partial}{\partial x} + \hat{j}\frac{\partial}{\partial y} + \hat{k}\frac{\partial}{\partial z}\right) \times \left(\hat{i}\frac{\partial \phi}{\partial x} + \hat{j}\frac{\partial \phi}{\partial y} + \hat{k}\frac{\partial \phi}{\partial z}\right)$$

$$= \begin{vmatrix} \hat{i} & \hat{j} & \hat{k} \\ \frac{\partial}{\partial x} & \frac{\partial}{\partial y} & \frac{\partial}{\partial z} \\ \frac{\partial \phi}{\partial x} & \frac{\partial \phi}{\partial y} & \frac{\partial \phi}{\partial z} \end{vmatrix} = \Sigma \hat{i} \left\{\frac{\partial}{\partial y}\left(\frac{\partial \phi}{\partial z}\right) - \frac{\partial}{\partial z}\left(\frac{\partial \phi}{\partial y}\right)\right\} = \Sigma \hat{i} \left\{\frac{\partial^2 \phi}{\partial y \partial z} - \frac{\partial^2 \phi}{\partial z \partial y}\right\} = \vec{0} \qquad \left(\because \frac{\partial^2 \phi}{\partial y \partial z} = \frac{\partial^2 \phi}{\partial z \partial y}\right)$$

Hence curl (grad φ) = $\vec{0}$

Note. Curl (grad φ) = $\vec{0}$ implies that gradient field describes an irrotational motion.

3. Div $\left(\textbf{Curl } \vec{V}\right) = \nabla \bullet \left(\nabla \times \vec{V}\right) = 0$

Proof. Div $\left(\textbf{Curl } \vec{V}\right) = \nabla \bullet \left(\nabla \times \vec{V}\right)$

Let $\vec{V} = V_1 \hat{i} + V_2 \hat{j} + V_3 \hat{k}$

$$\therefore \quad \nabla \times \vec{V} = \begin{vmatrix} \hat{i} & \hat{j} & \hat{k} \\ \frac{\partial}{\partial x} & \frac{\partial}{\partial y} & \frac{\partial}{\partial z} \\ V_1 & V_2 & V_3 \end{vmatrix} = \Sigma \hat{i} \left(\frac{\partial V_3}{\partial y} - \frac{\partial V_2}{\partial z}\right)$$

$$\therefore \quad \text{Div}\left(\text{Curl } \vec{V}\right) = \left(\hat{i}\frac{\partial}{\partial x} + \hat{j}\frac{\partial}{\partial y} + \hat{k}\frac{\partial}{\partial z}\right) \bullet \left(\Sigma \hat{i}\left(\frac{\partial V_3}{\partial y} - \frac{\partial V_2}{\partial z}\right)\right)$$

$$= \left(\hat{i}\frac{\partial}{\partial x} + \hat{j}\frac{\partial}{\partial y} + \hat{k}\frac{\partial}{\partial z}\right) \bullet \left\{\hat{i}\left(\frac{\partial V_3}{\partial y} - \frac{\partial V_2}{\partial z}\right) + \hat{j}\left(\frac{\partial V_1}{\partial z} - \frac{\partial V_3}{\partial x}\right) + \hat{k}\left(\frac{\partial V_2}{\partial x} - \frac{\partial V_1}{\partial y}\right)\right\}$$

$$= \frac{\partial}{\partial x}\left\{\frac{\partial V_3}{\partial y} - \frac{\partial V_2}{\partial z}\right\} + \frac{\partial}{\partial y}\left\{\frac{\partial V_1}{\partial z} - \frac{\partial V_3}{\partial x}\right\} + \frac{\partial}{\partial z}\left\{\frac{\partial V_2}{\partial x} - \frac{\partial V_1}{\partial y}\right\}$$

$$= \frac{\partial^2 V_3}{\partial x \partial y} - \frac{\partial^2 V_2}{\partial x \partial z} + \frac{\partial^2 V_1}{\partial y \partial z} - \frac{\partial^2 V_3}{\partial y \partial x} + \frac{\partial^2 V_2}{\partial z \partial x} - \frac{\partial^2 V_1}{\partial z \partial y} = 0$$

Hence $\text{Div} \bullet \left(\text{Curl } \vec{V}\right) = 0$

Note. Div $\left(\text{curl } \vec{V}\right) = 0$ implies that curl $\vec{V}$ is a solenoidal vector point function.

4. Curl $\left(\textbf{Curl } \vec{V}\right)$ = grad div $\vec{V} - \nabla^2 \vec{V}$ ***(P.T.U., June 2003, Dec. 2005, 2015)***

or **Curl $\left(\textbf{Curl } \vec{V}\right) = \nabla\left(\nabla \bullet \vec{V}\right) - (\nabla \bullet \nabla)\vec{V}$**

Proof. Let $\vec{V} = V_1 \hat{i} + V_2 \hat{j} + V_3 \hat{k}$

$$\text{Curl } \vec{V} = \hat{i}\left(\frac{\partial V_3}{\partial y} - \frac{\partial V_2}{\partial z}\right) + \hat{j}\left(\frac{\partial V_1}{\partial z} - \frac{\partial V_3}{\partial x}\right) + \hat{k}\left(\frac{\partial V_2}{\partial x} - \frac{\partial V_1}{\partial y}\right)$$

$$\text{Curl Curl } \vec{V} = \begin{vmatrix} \hat{i} & \hat{j} & \hat{k} \\ \dfrac{\partial}{\partial x} & \dfrac{\partial}{\partial y} & \dfrac{\partial}{\partial z} \\ \dfrac{\partial V_3}{\partial y} - \dfrac{\partial V_2}{\partial z} & \dfrac{\partial V_1}{\partial z} - \dfrac{\partial V_3}{\partial x} & \dfrac{\partial V_2}{\partial x} - \dfrac{\partial V_1}{\partial y} \end{vmatrix}$$

$$= \Sigma\hat{i}\left\{\frac{\partial}{\partial y}\left(\frac{\partial V_2}{\partial x} - \frac{\partial V_1}{\partial y}\right) - \frac{\partial}{\partial z}\left(\frac{\partial V_1}{\partial z} - \frac{\partial V_3}{\partial x}\right)\right\} = \Sigma\hat{i}\left\{\frac{\partial^2 V_2}{\partial y\,\partial x} - \frac{\partial^2 V_1}{\partial y^2} - \frac{\partial^2 V_1}{\partial z^2} + \frac{\partial^2 V_3}{\partial z\,\partial x}\right\}$$

$$= \Sigma\hat{i}\left\{\frac{\partial^2 V_2}{\partial y\,\partial x} + \frac{\partial^2 V_3}{\partial z\,\partial x} - \frac{\partial^2 V_1}{\partial y^2} - \frac{\partial^2 V_1}{\partial z^2}\right\}$$

Add and subtract $\dfrac{\partial^2 V_1}{\partial x^2}$

$$= \Sigma\hat{i}\left\{\left(\frac{\partial^2 V_1}{\partial x^2} + \frac{\partial^2 V_2}{\partial x\,\partial y} + \frac{\partial^2 V_3}{\partial x\,\partial z}\right) - \left(\frac{\partial^2 V_1}{\partial x^2} + \frac{\partial^2 V_1}{\partial y^2} + \frac{\partial^2 V_1}{\partial z^2}\right)\right\}$$

$$= \Sigma\hat{i}\left[\frac{\partial}{\partial x}\left(\frac{\partial V_1}{\partial x} + \frac{\partial V_2}{\partial y} + \frac{\partial V_3}{\partial z}\right) - \left(\frac{\partial^2 V_1}{\partial x^2} + \frac{\partial^2 V_1}{\partial y^2} + \frac{\partial^2 V_1}{\partial z^2}\right)\right]$$

$$= \Sigma\hat{i}\left\{\frac{\partial}{\partial x}\left(\nabla\cdot\vec{V}\right) - \nabla^2 V_1\right\} = \Sigma\hat{i}\frac{\partial}{\partial x}\left(\nabla\cdot\vec{V}\right) - \nabla^2\,\Sigma\hat{i}\,V_1$$

$$= \nabla\left(\nabla\cdot\vec{V}\right) - \nabla^2\vec{V} = \text{grad}\,(\text{div }\vec{V}) - \nabla^2\vec{V}$$

Cor. From above result we can also deduce

$$\textbf{grad}\left(\textbf{div }\vec{V}\right) = \text{Curl}\left(\text{Curl }\vec{V}\right) + \nabla^2\vec{V}$$ ***(P.T.U. Dec. 2011)***

or

$$\nabla\left(\nabla\cdot\vec{V}\right) = \nabla\times\left(\nabla\times\vec{V}\right) + \nabla^2\vec{V}$$

Note. For application in questions, the results of repeated application of ∇ can easily be written down (treating ∇ as a vector)

(*i*) $\nabla\cdot\nabla\,\phi = \nabla^2\phi$ $\quad\because\ \vec{a}\cdot\vec{a} = a^2$

(*ii*) $\nabla\times\nabla\,\phi = \vec{0}$ $\quad\because\ \vec{a}\times\vec{a} = \vec{0}$

(*iii*) $\nabla\cdot\left(\nabla\times\vec{V}\right) = 0$ $\quad\because$ in scalar triple product $\vec{a}\cdot\left(\vec{a}\times\vec{b}\right) = 0$

(*iv*) $\nabla\times(\nabla\times\vec{V}) = (\nabla\cdot\vec{V})\nabla - \nabla^2 V$ $\quad\because\ \vec{a}\times(\vec{b}\times\vec{c}) = (\vec{a}\cdot\vec{c})\vec{b} - (\vec{a}\cdot\vec{b})\vec{c}.$

ILLUSTRATIVE EXAMPLES

Example 1. *Evaluate the following:*

(i) $Div\left[(xy\sin z)\hat{i} + (y^2 \sin x)\hat{j} + (z^2 \sin xy)\hat{k}\right]$ *at the point* $\left(0, \frac{\pi}{2}, \frac{\pi}{2}\right)$.

(ii) *Curl Curl of* $\vec{V} = (2xz^2)\hat{i} - yz\hat{j} + (3xz^3)\hat{k}$ *at (1, 1, 1)* **(P.T.U., May 2006)**

Sol. *(i)* $\text{Div}\left[(xy\sin z)\hat{i} + (y^2 \sin x)\hat{j} + (z^2 \sin xy)\hat{k}\right]$

$$= \left(\hat{i}\frac{\partial}{\partial x} + \hat{j}\frac{\partial}{\partial y} + \hat{k}\frac{\partial}{\partial z}\right) \cdot \left[(xy\sin z)\hat{i} + (y^2 \sin x)\hat{j} + (z^2 \sin xy)\hat{k}\right]$$

$$= \frac{\partial}{\partial x}(xy\sin z) + \frac{\partial}{\partial y}(y^2 \sin x) + \frac{\partial}{\partial z}(z^2 \sin xy)$$

$$= y \sin z + 2y \sin x + 2z \sin xy$$

At the point $\left(0, \frac{\pi}{2}, \frac{\pi}{2}\right)$

$$\text{div}\left\{(xy\sin z)\hat{i} + (y^2 \sin x)\hat{j} + (z^2 \sin xy)\hat{k}\right\} = \frac{\pi}{2} + 0 + 0 = \frac{\pi}{2}$$

(ii)
$$\vec{V} = (2xz^2)\hat{i} - yz\hat{j} + (3xz^3)\hat{k}$$

$$\text{Curl } \vec{V} = \begin{vmatrix} \hat{i} & \hat{j} & \hat{k} \\ \frac{\partial}{\partial x} & \frac{\partial}{\partial y} & \frac{\partial}{\partial z} \\ 2xz^2 & -yz & 3xz^3 \end{vmatrix}$$

$$= \hat{i}\left\{\frac{\partial}{\partial y}(3xz^3) - \frac{\partial}{\partial z}(-yz)\right\} - \hat{j}\left\{\frac{\partial}{\partial x}(3xz^3) - \frac{\partial}{\partial z}(2xz^2)\right\} + \hat{k}\left\{\frac{\partial}{\partial x}(-yz) - \frac{\partial}{\partial y}(2xz^2)\right\}$$

$$= \hat{i}\{y\} - \hat{j}\{3z^3 - 4xz\} + \hat{k}\{0\} = y\hat{i} + (-3z^3 + 4xz)\hat{j}$$

$$\text{Curl Curl } \vec{V} = \begin{vmatrix} \hat{i} & \hat{j} & \hat{k} \\ \frac{\partial}{\partial x} & \frac{\partial}{\partial y} & \frac{\partial}{\partial z} \\ y & -3z^3 + 4xz & 0 \end{vmatrix}$$

$$= \hat{i}\left\{\frac{\partial}{\partial y}(0) - \frac{\partial}{\partial z}(-3z^3 + 4xz)\right\} - \hat{j}\left\{\frac{\partial}{\partial x}(0) - \frac{\partial}{\partial z}(y)\right\} + \hat{k}\left\{\frac{\partial}{\partial x}(-3z^3 + 4xz) - \frac{\partial}{\partial y}(y)\right\}$$

$$= -\hat{i}(-9z^2 + 4x) + \hat{j}(0) + \hat{k}\{4z - 1\} = (9z^2 - 4x)\hat{i} + (4z - 1)\hat{k}$$

At (1, 1, 1) curl curl $\vec{V} = 5\hat{i} + 3\hat{k}$.

Example 2. *(i) Find div $\vec{F}$ and curl $\vec{F}$, where $\vec{F}$ = grad $(x^3 + y^3 + z^3 - 3xyz)$.*

(P.T.U., Dec. 2014, May 2015)

(ii) If $\phi(x, y, z) = x - y + 2z^2$ compute the value of $\nabla \times (\nabla\phi)$. **(P.T.U., Dec. 2013)**

Sol. *(i)* $\vec{\mathbf{F}}$ = grad $(x^3 + y^3 + z^3 - 3xyz) = \left(\hat{i}\frac{\partial}{\partial x} + \hat{j}\frac{\partial}{\partial y} + \hat{k}\frac{\partial}{\partial z}\right)\left(x^3 + y^3 + z^3 - 3xyz\right)$

$$= \hat{i}\left(3x^2 - 3yz\right) + \hat{j}\left(3y^2 - 3zx\right) + \hat{k}\left(3z^2 - 3xy\right)$$

Now, $\text{div}\,\vec{\mathbf{F}} = \left(\hat{i}\frac{\partial}{\partial x} + \hat{j}\frac{\partial}{\partial y} + \hat{k}\frac{\partial}{\partial z}\right)\bullet\left\{3\left(x^2 - yz\right)\hat{i} + 3\left(y^2 - zx\right)\hat{j} + 3\left(z^2 - xy\right)\hat{k}\right\}$

$$= 3\left\{\frac{\partial}{\partial x}\left(x^2 - yz\right) + \frac{\partial}{\partial y}\left(y^2 - zx\right) + \frac{\partial}{\partial z}\left(z^2 - xy\right)\right\}$$

$$= 3\,\{2x + 2y + 2z\} = 6\,(x + y + z)$$

$$\text{Curl}\ \vec{\mathbf{F}} = \left(\hat{i}\frac{\partial}{\partial x} + \hat{j}\frac{\partial}{\partial y} + \hat{k}\frac{\partial}{\partial z}\right)\times\left\{3\left(x^2 - yz\right)\hat{i} + 3\left(y^2 - zx\right)\hat{j} + 3\left(z^2 - xy\right)\hat{k}\right\}$$

$$= \begin{vmatrix} \hat{i} & \hat{j} & \hat{k} \\ \frac{\partial}{\partial x} & \frac{\partial}{\partial y} & \frac{\partial}{\partial z} \\ x^2 - yz & y^2 - zx & z^2 - xy \end{vmatrix} = 3\Sigma\hat{i}\left\{\frac{\partial}{\partial y}\left(z^2 - xy\right) - \frac{\partial}{\partial z}\left(y^2 - zx\right)\right\} = 3\Sigma\hat{i}\,\{-x + x\}$$

$$= 3\,\Sigma\,\hat{i}\,.\,0 = \vec{0}\,.$$

(ii) $\phi(x, y, z) = x - y + 2z^2$

$$\nabla\phi = \left(\hat{i}\frac{\partial}{\partial x} + \hat{j}\frac{\partial}{\partial y} + \hat{k}\frac{\partial}{\partial z}\right)(x - y + 2z^2)$$

$$= \hat{i}\,(1) + \hat{j}(-1) + \hat{k}(4z) = \hat{i} - \hat{j} + 4z\hat{k}$$

$$\nabla \times \nabla\phi = \text{curl}\ (\hat{i} - \hat{j} + 4z\hat{k}) = \begin{vmatrix} \hat{i} & \hat{j} & \hat{k} \\ \frac{\partial}{\partial x} & \frac{\partial}{\partial y} & \frac{\partial}{\partial z} \\ 1 & -1 & 4z \end{vmatrix}$$

$$= \hat{i}\,(0 - 0) - \hat{j}\,(0 - 0) + \hat{k}\,(0 - 0) = \vec{0}$$

$$\therefore \qquad \nabla \times \nabla\phi = \vec{0}\,.$$

Example 3. *If $u = x^2 + y^2 + z^2$, $\vec{r} = x\hat{i} + y\hat{j} + z\hat{k}$, then find div $\left(u\vec{r}\right)$ in terms of u.*

(P.T.U., Dec. 2005)

Sol. $\text{div}\,\left(u\vec{r}\right) = \left(\hat{i}\frac{\partial}{\partial x} + \hat{j}\frac{\partial}{\partial y} + \hat{k}\frac{\partial}{\partial z}\right)\bullet\left(u\vec{r}\right)$

Now, $\left(u\vec{r}\right) = (x^2 + y^2 + z^2)\left(x\hat{i} + y\hat{j} + z\hat{k}\right)$

$$= x\left(x^2 + y^2 + z^2\right)\hat{i} + y\left(x^2 + y^2 + z^2\right)\hat{j} + z\left(x^2 + y^2 + z^2\right)\hat{k}$$

$$\frac{\partial}{\partial x}\left(u\vec{r}\right) = \left(3x^2 + y^2 + z^2\right)\hat{i} + (2xy)\,\hat{j} + (2xz)\,\hat{k}$$

$$\frac{\partial}{\partial y}(u\vec{r}) = (2xy)\hat{i} + \left(x^2 + z^2 + 3y^2\right)\hat{j} + (2yz)\hat{k}$$

$$\frac{\partial}{\partial z}(u\vec{r}) = (2xz)\hat{i} + (2yz)\hat{j} + \left(x^2 + y^2 + 3z^2\right)\hat{k}$$

$$\therefore \quad \text{div}\,(u\vec{r}) = \hat{i}\bullet\frac{\partial}{\partial x}(u\vec{r}) + \hat{j}\bullet\frac{\partial}{\partial y}(u\vec{r}) + \hat{k}\bullet\frac{\partial}{\partial z}(u\vec{r})$$

$$= (3x^2 + y^2 + z^2) + (x^2 + z^2 + 3y^2) + (x^2 + y^2 + 3z^2)$$

$$= 5x^2 + 5y^2 + 5z^2 = 5(x^2 + y^2 + z^2) = 5u.$$

Example 4. *Let* $\vec{r} = x\hat{i} + y\hat{j} + z\hat{k}$ *and* $\vec{a}$ *be a constant vector, find the value of*

(i) div $\dfrac{\vec{a}\times\vec{r}}{r^n}$ *(ii) curl* $\dfrac{\vec{a}\times\vec{r}}{r^n}$. **(P.T.U., May 2014)**

Sol. (*i*) Let $\vec{a} = a_1\hat{i} + a_2\hat{j} + a_3\hat{k}$

$$\vec{a}\times\vec{r} = \begin{vmatrix} \hat{i} & \hat{j} & \hat{k} \\ a_1 & a_2 & a_3 \\ x & y & z \end{vmatrix} = \hat{i}\,(a_2z - a_3y) + \hat{j}\,(a_3x - a_1z) + \hat{k}\,(a_1y - a_2x)$$

$\text{div}\left(\dfrac{\vec{a}\times\vec{r}}{r^n}\right) = \text{div}\left(\dfrac{1}{r^n}\vec{a}\times\vec{r}\right)$ is of the type div $(\phi\vec{A})$, where $\phi = \dfrac{1}{r^n}$ and $\vec{A} = \vec{a}\times\vec{r}$

We know that div $(\phi\vec{A}) = \phi$ div $\vec{A}$ + (grad ϕ) . $\vec{A}$

$$\text{div}\left(\frac{1}{r^n}\vec{a}\times\vec{r}\right) = \frac{1}{r^n}\,\textbf{div}\,(\vec{a}\times\vec{r}) + \left(\textbf{grad}\,\frac{1}{r^n}\right).\,\vec{a}\times\vec{r} \qquad \ldots(1)$$

Now,

$$\text{div}\,(\vec{a}\times\vec{r}) = \left(\hat{i}\frac{\partial}{\partial x} + \hat{j}\frac{\partial}{\partial y} + \hat{k}\frac{\partial}{\partial z}\right)[\hat{i}(a_2z - a_3y) + \hat{j}(a_3x - a_1z) + \hat{k}(a_1y - a_2x)]$$

$$= \hat{i}\,.\,0 + \hat{j}\,.\,0 + \hat{k}\,.\,0 = 0$$

$$\text{grad}\left(\frac{1}{r^n}\right) = \left(\hat{i}\,\frac{\partial}{\partial x} + \hat{j}\,\frac{\partial}{\partial y} + \hat{k}\frac{\partial}{\partial z}\right)(x^2 + y^2 + z^2)^{-n/2}$$

$$= \sum\hat{i}\left(-\frac{n}{2}\right)(x^2 + y^2 + z^2)^{-\frac{n}{2}-1}\,.\,2x$$

$$= \sum\hat{i}\,(-nx)\,r^{2\left(\frac{-n+2}{2}\right)} = -n\sum\hat{i}xr^{-(n+2)}$$

$$= -\frac{n}{r^{n+2}}(x\hat{i} + y\hat{j} + z\hat{k}) = -\frac{n\vec{r}}{r^{n+2}}$$

From (1)

$$\text{div}\,\frac{\vec{a}\times\vec{r}}{r^n} = \frac{1}{r^n}\,.\,0 - \frac{n}{r^{n+2}}(\vec{r}\,.\,\vec{a}\times\vec{r}) = \frac{1}{r^n}\,.\,0 - \frac{n}{r^{n+2}}\,.\,0 = 0$$

$$\therefore \qquad \text{div} \ \frac{\vec{a} \times \vec{r}}{r^n} = 0$$

(ii) $\text{curl} \ \dfrac{\vec{a} \times \vec{r}}{r^n} = \text{curl} \left[\dfrac{1}{r^n} (\vec{a} \times \vec{r}) \right]$ which is of the type curl $(\phi \vec{A})$

We know that $\quad \text{curl} \ (\phi \vec{A}) = (\text{grad} \ \phi) \times \vec{A} + \phi \ \text{curl} \ \vec{A}$

$$\text{Curl} \ \frac{(\vec{a} \times \vec{r})}{r^n} = \left(\text{grad} \ \frac{1}{r^n} \right) \times (\vec{a} \times \vec{r}) + \frac{1}{r^n} \text{curl} \ (\vec{a} \times \vec{r})$$

$$= -\frac{n}{r^{n+2}} [\vec{r} \times (\vec{a} \times \vec{r})] + \frac{1}{r^n} \text{curl} \ (\vec{a} \times \vec{r})$$

$$\text{Curl} \ (\vec{a} \times \vec{r}) = \begin{vmatrix} \hat{i} & \hat{j} & \hat{k} \\ \dfrac{\partial}{\partial x} & \dfrac{\partial}{\partial y} & \dfrac{\partial}{\partial z} \\ a_2 z - a_3 y & a_3 x - a_1 z & a_1 y - a_2 x \end{vmatrix}$$

$$= \hat{i} 2a_1 + \hat{j} 2a_2 + \hat{k} 2a_3 = 2\vec{a}$$

$$\text{Curl} \ \frac{\vec{a} \times \vec{r}}{r^n} = -\frac{n}{r^{n+2}} \left\{ (\vec{r} \,.\, \vec{r}) \vec{a} - (\vec{r} \,.\, \vec{a}) \vec{r} \right\} + \frac{1}{r^n} 2\vec{a}$$

$$= -\frac{n}{r^{n+2}} [r^2 \vec{a} - (\vec{r} \cdot \vec{a}) \vec{r}] + \frac{2\vec{a}}{r^n} = \frac{-n\vec{a}}{r^n} + \frac{n}{r^{n+2}} (\vec{r} \,.\, \vec{a}) \vec{r} + \frac{2\vec{a}}{r^n}$$

$$= \frac{2-n}{r^n} \vec{a} + \frac{n}{r^{n+2}} (\vec{a} \,.\, \vec{r}) \vec{r}$$

$$\therefore \qquad \text{Curl} \ \frac{\vec{a} \times \vec{r}}{r^n} = \frac{2-n}{r^n} \vec{a} + \frac{n}{r^{n+2}} (\vec{a} \,.\, \vec{r}) \vec{r}$$

Example 5. *Given the vector field* $\vec{V} = \left(x^2 - y^2 + 2xz\right)\hat{i} + (xz - xy + yz)\hat{j} + \left(z^2 + x^2\right)\hat{k}$ *find Curl* $\vec{V}$. *Show that the vectors given by Curl* $\vec{V}$ *at P (1, 2, – 3) and Q (2, 3, 12) are orthogonal.*

Sol. $\vec{V} = \left(x^2 - y^2 + 2xz\right)\hat{i} + (xz - xy + yz)\hat{j} + \left(z^2 + x^2\right)\hat{k}$

$\text{Curl} \ \vec{V} = \nabla \times \vec{V}$.

$$= \begin{vmatrix} \hat{i} & \hat{j} & \hat{k} \\ \dfrac{\partial}{\partial x} & \dfrac{\partial}{\partial y} & \dfrac{\partial}{\partial z} \\ x^2 - y^2 + 2xz & xz - xy + yz & z^2 + x^2 \end{vmatrix}$$

$$= \hat{i}\left\{\frac{\partial}{\partial y}\left(z^2+x^2\right)-\frac{\partial}{\partial z}(xz-xy+yz)\right\}+\hat{j}\left\{\frac{\partial}{\partial z}\left(x^2-y^2+2xz\right)-\frac{\partial}{\partial x}\left(z^2+x^2\right)\right\}$$

$$+\hat{k}\left\{\frac{\partial}{\partial x}(xz-xy+yz)-\frac{\partial}{\partial y}\left(x^2-y^2+2xz\right)\right\}$$

$$= \hat{i}\{-(x+y)\}+\hat{j}\{2x-2x\}+\hat{k}\{z-y+2y\} = -(x+y)\hat{i}+(y+z)\hat{k}$$

Curl $\vec{V}$ at P (1, 2, – 3) = $-3\hat{i}-\hat{k}$

Curl $\vec{V}$ at Q (2, 3, 12) = $-5\hat{i}+15\hat{k}$.

Curl $\vec{V}$ at P, Q will be orthogonal if their dot product is zero.

i.e., $$\left(-3\hat{i}-\hat{k}\right)\bullet\left(-5\hat{i}+15\hat{k}\right) = 15-15 = 0.$$

Hence curl vectors at P and Q are orthogonal.

Example 6. *If* $u\vec{F} = \nabla v$, *where* u, v *are scalars fields and* $\vec{F}$ *is a vector field, show that* $\vec{F}\bullet \text{curl}\,\vec{F} = 0$.

Sol. $$\text{Curl}\,\vec{F} = \text{Curl}\left(\frac{1}{u}\nabla v\right) = \nabla\times\left(\frac{1}{u}\nabla v\right)$$

We know that $\nabla\times\left(\phi\vec{A}\right) = \nabla\phi\times\vec{A}+\phi\nabla\times\vec{A}$

$$\therefore \quad \text{Curl}\,\vec{F} = \left(\nabla\frac{1}{u}\right)\times(\nabla v)+\frac{1}{u}\nabla\times(\nabla v) = \left(\nabla\frac{1}{u}\right)\times(\nabla v) \qquad \left(\because\ \nabla\times(\nabla v)=\vec{0}\right)$$

$$\vec{F}\bullet\text{Curl}\,\vec{F} = \left(\frac{1}{u}\nabla v\right)\bullet\left(\nabla\frac{1}{u}\right)\times(\nabla v) = \frac{1}{u}\left\{\nabla v\bullet\nabla\frac{1}{u}\times\nabla v\right\}$$

$$= \frac{1}{u}\left[\nabla v, \nabla\frac{1}{u}, \nabla v\right] = \frac{1}{u}\bullet 0 = 0$$

[$\because$ In scalar triple product two vectors are equal]

Hence $\vec{F}\bullet\text{Curl}\,\vec{F} = 0$

Example 7. *If* $\vec{A} = \nabla\times\left(\phi\hat{i}\right)$ *where* $\nabla^2\phi = 0$, *show that* $\vec{A}\bullet\nabla\times\vec{A} = \frac{\partial\phi}{\partial z}\bullet\frac{\partial^2\phi}{\partial y\,\partial x}-\frac{\partial\phi}{\partial y}\bullet\frac{\partial^2\phi}{\partial z\,\partial x}$.

Sol. $$\vec{A} = \nabla\times\left(\phi\hat{i}\right) = \nabla\phi\times\hat{i}$$

$$\nabla\times\vec{A} = \nabla\times\left(\nabla\phi\times\hat{i}\right) = \left(\nabla\bullet\hat{i}\right)\nabla\phi-(\nabla\bullet\nabla\phi)\hat{i} \text{ using } \vec{a}\times\left(\vec{b}\times\vec{c}\right) = \left(\vec{a}\bullet\vec{b}\right)\vec{b}-\left(\vec{a}\bullet\vec{b}\right)\vec{c}$$

$$\nabla\bullet\hat{i} = \left(\hat{i}\frac{\partial}{\partial x}+\hat{j}\frac{\partial}{\partial y}+\hat{k}\frac{\partial}{\partial z}\right)\bullet\hat{i} = \frac{\partial}{\partial x}$$

$$\therefore\ \nabla\times\vec{A} = \frac{\partial}{\partial x}(\nabla\phi)-\left(\nabla^2\phi\right)\hat{i} = \frac{\partial}{\partial x}\left[\hat{i}\frac{\partial\phi}{\partial x}+\hat{j}\frac{\partial\phi}{\partial y}+\hat{k}\frac{\partial\phi}{\partial z}\right]-\left(\frac{\partial^2\phi}{\partial x^2}+\frac{\partial^2\phi}{\partial y^2}+\frac{\partial^2\phi}{\partial z^2}\right)\hat{i}$$

$$= \hat{i}\frac{\partial^2\phi}{\partial x^2} + \hat{j}\frac{\partial^2\phi}{\partial x\,\partial y} + \hat{k}\frac{\partial^2\phi}{\partial x\,\partial z} - \frac{\partial^2\phi}{\partial x^2}\hat{i} - \frac{\partial^2\phi}{\partial y^2}\hat{i} - \frac{\partial^2\phi}{\partial z^2}\hat{i}$$

$$= -\left(\frac{\partial^2\phi}{\partial y^2} + \frac{\partial^2\phi}{\partial z^2}\right)\hat{i} + \frac{\partial^2\phi}{\partial x\,\partial y}\hat{j} + \frac{\partial^2\phi}{\partial x\,\partial z}\hat{k}$$

$$\vec{A}\bullet\nabla\times\vec{A} = \left(\nabla\times\phi\hat{i}\right)\bullet\left[-\left(\frac{\partial^2\phi}{\partial y^2} + \frac{\partial^2\phi}{\partial z^2}\right)\hat{i} + \frac{\partial^2\phi}{\partial x\,\partial y}\hat{j} + \frac{\partial^2\phi}{\partial x\,\partial z}\hat{k}\right]$$

$$= \left(\nabla\phi\times\hat{i}\right)\bullet\left[-\left(\frac{\partial^2\phi}{\partial y^2} + \frac{\partial^2\phi}{\partial z^2}\right)\hat{i} + \frac{\partial^2\phi}{\partial x\,\partial y}\hat{j} + \frac{\partial^2\phi}{\partial x\,\partial z}\hat{k}\right]$$

$$= \left[\left(\hat{i}\frac{\partial\phi}{\partial x} + \hat{j}\frac{\partial\phi}{\partial y} + \hat{k}\frac{\partial\phi}{\partial z}\right)\times\hat{i}\right]\bullet\left[-\left(\frac{\partial^2\phi}{\partial y^2} + \frac{\partial^2\phi}{\partial z^2}\right)\hat{i} + \frac{\partial^2\phi}{\partial x\,\partial y}\hat{j} + \frac{\partial^2\phi}{\partial x\,\partial z}\hat{k}\right]$$

$$= \left[-\frac{\partial\phi}{\partial y}\hat{k} + \frac{\partial\phi}{\partial z}\hat{j}\right]\bullet\left[-\left(\frac{\partial^2\phi}{\partial y^2} + \frac{\partial^2\phi}{\partial z^2}\right)\hat{i} + \frac{\partial^2\phi}{\partial x\,\partial y}\hat{j} + \frac{\partial^2\phi}{\partial x\,\partial z}\hat{k}\right]$$

$$\left(\because \hat{i}\times\hat{i} = \vec{0},\ \hat{j}\times\hat{i} = -\hat{k},\ \hat{k}\times\hat{i} = \hat{j}\right)$$

$$= -\frac{\partial\phi}{\partial y}\bullet\frac{\partial^2\phi}{\partial x\,\partial z} + \frac{\partial\phi}{\partial z}\bullet\frac{\partial^2\phi}{\partial x\,\partial y} = \frac{\partial\phi}{\partial z}\bullet\frac{\partial^2\phi}{\partial y\,\partial x} - \frac{\partial\phi}{\partial y}\frac{\partial^2\phi}{\partial z\,\partial x}$$

Example 8. *By taking $\vec{F} = u\,\nabla\,v$, where u and v are scalars, prove that*

$\nabla\bullet\vec{F} = u\,\nabla^2 v + \nabla u\bullet\nabla v.$

Sol.

$$\vec{F} = u\,\nabla\,v = u\left(\hat{i}\frac{\partial v}{\partial x} + \hat{j}\frac{\partial v}{\partial y} + \hat{k}\frac{\partial v}{\partial z}\right)$$

$$\nabla\bullet\vec{F} = \left(\hat{i}\frac{\partial}{\partial x} + \hat{j}\frac{\partial}{\partial y} + \hat{k}\frac{\partial}{\partial z}\right)\bullet u\left(\hat{i}\frac{\partial v}{\partial x} + \hat{j}\frac{\partial v}{\partial y} + \hat{k}\frac{\partial v}{\partial z}\right)$$

Differentiate by product rule

$$= \left(\hat{i}\frac{\partial u}{\partial x} + \hat{j}\frac{\partial u}{\partial y} + \hat{k}\frac{\partial u}{\partial z}\right)\bullet\left(\hat{i}\frac{\partial v}{\partial x} + \hat{j}\frac{\partial v}{\partial y} + \hat{k}\frac{\partial v}{\partial z}\right)$$

$$+ u\left(\hat{i}\frac{\partial}{\partial x} + \hat{j}\frac{\partial}{\partial y} + \hat{k}\frac{\partial}{\partial z}\right)\bullet\left(\hat{i}\frac{\partial v}{\partial x} + \hat{j}\frac{\partial v}{\partial y} + \hat{k}\frac{\partial v}{\partial z}\right)$$

$$= \nabla u\bullet\nabla v + u\left[\frac{\partial}{\partial x}\left(\frac{\partial v}{\partial x}\right) + \frac{\partial}{\partial y}\left(\frac{\partial v}{\partial y}\right) + \frac{\partial}{\partial z}\left(\frac{\partial v}{\partial z}\right)\right]$$

$$= \nabla u \bullet \nabla v + u\left[\frac{\partial^2 v}{\partial x^2} + \frac{\partial^2 v}{\partial y^2} + \frac{\partial^2 v}{\partial z^2}\right] = \nabla u \bullet \nabla v + u\nabla^2 v$$

$$= u\,\nabla^2 v + \nabla u \bullet \nabla v$$

Example 9. *If r is the distance of a point (x, y, z) from the origin, prove that* $curl\left(\hat{k} \times grad\,\frac{1}{r}\right) + grad\left(\hat{k}\cdot grad\,\frac{1}{r}\right) = 0$, *where* $\hat{k}$ *is a unit vector in the direction of z.*

Sol.

$$r = \sqrt{x^2 + y^2 + z^2}$$

$$\therefore \quad \frac{1}{r} = \left(x^2 + y^2 + z^2\right)^{-\frac{1}{2}}$$

$$\text{grad}\,\frac{1}{r} = \nabla\left(x^2 + y^2 + z^2\right)^{-\frac{1}{2}} = \left(\hat{i}\,\frac{\partial}{\partial x} + \hat{j}\,\frac{\partial}{\partial y} + \hat{k}\,\frac{\partial}{\partial z}\right)\left(x^2 + y^2 + z^2\right)^{-\frac{1}{2}}$$

$$= -\frac{1}{2}\left(x^2 + y^2 + z^2\right)^{-\frac{3}{2}}\left[2x\hat{i} + 2y\hat{j} + 2z\hat{k}\right]$$

$$= \frac{-1}{\left(x^2 + y^2 + z^2\right)^{\frac{3}{2}}}\left(x\hat{i} + y\hat{j} + z\hat{k}\right)$$

Now $$\hat{k} \times \text{grad}\,\frac{1}{r} = \hat{k} \times \left\{\frac{-1}{\left(x^2 + y^2 + z^2\right)^{\frac{3}{2}}}\right\}\left\{x\hat{i} + y\hat{j} + z\hat{k}\right\} = \frac{-1}{\left(x^2 + y^2 + z^2\right)^{\frac{3}{2}}}\left\{x\hat{j} - y\hat{i}\right\}$$

$$\text{Curl}\left(\hat{k} \times \text{grad}\frac{1}{r}\right) = \nabla \times \left[\frac{-1}{\left(x^2 + y^2 + z^2\right)^{\frac{3}{2}}}\right]\left[x\hat{j} - y\hat{i}\right]$$

$$= \nabla \times \left\{\frac{-x}{\left(x^2 + y^2 + z^2\right)^{\frac{3}{2}}}\,\hat{j} + \frac{y}{\left(x^2 + y^2 + z^2\right)^{\frac{3}{2}}}\,\hat{i}\right\}$$

$$= \begin{vmatrix} \hat{i} & \hat{j} & \hat{k} \\ \dfrac{\partial}{\partial x} & \dfrac{\partial}{\partial y} & \dfrac{\partial}{\partial z} \\ \dfrac{y}{\left(x^2 + y^2 + z^2\right)^{\frac{3}{2}}} & \dfrac{-x}{\left(x^2 + y^2 + z^2\right)^{\frac{3}{2}}} & 0 \end{vmatrix}$$

$$= \hat{i}\left\{0 - \frac{\partial}{\partial z}\,\frac{-x}{\left(x^2 + y^2 + z^2\right)^{\frac{3}{2}}}\right\} + \hat{j}\left\{\frac{\partial}{\partial z}\left(\frac{y}{\left(x^2 + y^2 + z^2\right)^{\frac{3}{2}}}\right) - 0\right\}$$

$$+ \hat{k}\left\{-\frac{\partial}{\partial x}\,\frac{x}{\left(x^2 + y^2 + z^2\right)^{\frac{3}{2}}} - \frac{\partial}{\partial y}\,\frac{y}{\left(x^2 + y^2 + z^2\right)^{\frac{3}{2}}}\right\}$$

$$= \hat{i}\left\{\frac{-3xz}{(x^2+y^2+z^2)^{\frac{5}{2}}}\right\} + \hat{j}\left\{\frac{-3yz}{(x^2+y^2+z^2)^{\frac{5}{2}}}\right\}$$

$$+ \hat{k}\left\{\frac{3x^2 - x^2 - y^2 - z^2 + 3y^2 - x^2 - y^2 - z^2}{(x^2+y^2+z^2)^{\frac{5}{2}}}\right\}$$

$$\therefore \quad \text{Curl}\left(\hat{k} \times \text{grad}\frac{1}{r}\right) = \frac{\hat{i}(-3xz) + \hat{j}(-3yz) + \hat{k}(x^2+y^2-2z^2)}{(x^2+y^2+z^2)^{\frac{5}{2}}} \qquad \ldots(1)$$

Now, $\quad \hat{k}\bullet\text{grad}\dfrac{1}{r} = \hat{k}\bullet\dfrac{-1}{(x^2+y^2+z^2)^{\frac{3}{2}}}(x\hat{i} + y\hat{j} + z\hat{k}) = -\dfrac{1}{(x^2+y^2+z^2)^{\frac{3}{2}}}z$

$$\text{grad}\left(\hat{k}\bullet\text{grad}\frac{1}{r}\right) = \nabla\left\{-\frac{z}{(x^2+y^2+z^2)^{\frac{3}{2}}}\right\} = \left\{\hat{i}\frac{\partial}{\partial x} + \hat{j}\frac{\partial}{\partial y} + \hat{k}\frac{\partial}{\partial z}\right\}\left(\frac{-z}{(x^2+y^2+z^2)^{\frac{3}{2}}}\right)$$

$$= \left[\frac{\hat{i}(-z)\left(-\frac{3}{2}\right)(2x)}{(x^2+y^2+z^2)^{\frac{5}{2}}} + \hat{j}\frac{(-z)\left(-\frac{3}{2}\right)(2y)}{(x^2+y^2+z^2)^{\frac{5}{2}}} + \hat{k}\left\{\frac{(-z)\left(-\frac{3}{2}\right)(2z)}{(x^2+y^2+z^2)^{\frac{5}{2}}} - \frac{1}{(x^2+y^2+z^2)^{\frac{3}{2}}}\right\}\right]$$

$$= \frac{\hat{i}(3xz) + \hat{j}(3yz) + \hat{k}(3z^2 - x^2 - y^2 - z^2)}{(x^2+y^2+z^2)^{\frac{5}{2}}}$$

$$= \frac{\hat{i}(3xz) + \hat{j}(3yz) + \hat{k}(-x^2 - y^2 + 2z^2)}{(x^2+y^2+z^2)^{\frac{5}{2}}} \qquad \ldots(2)$$

Adding (1) and (2), we get $\textit{Curl}\left(\hat{k} \times \textit{grad}\frac{1}{r}\right) + \textit{grad}\left(\hat{k}\bullet\textit{grad}\frac{1}{r}\right) = 0.$

Example 10. *Prove that (i)* $\nabla^2 f(r) = f''(r) + \frac{2}{r}f'(r).$

(ii) $\nabla^2(r^n) = n(n+1)r^{n-2}$ **(P.T.U., May 2007, 2008)**

Sol. (*i*) $\nabla^2 f(r) = \nabla \bullet \{\nabla f(r)\} = \text{div}\{\text{grad} f(r)\}$

$$\text{grad} f(r) = \left(\hat{i}\frac{\partial}{\partial x} + \hat{j}\frac{\partial}{\partial y} + \hat{k}\frac{\partial}{\partial z}\right)f(r) = \hat{i}\frac{\partial}{\partial x}f(r) + \hat{j}\frac{\partial}{\partial y}f(r) + \hat{k}\frac{\partial}{\partial z}f(r)$$

$$= \hat{i} f'(r) \frac{\partial r}{\partial x} + \hat{j} f'(r) \frac{\partial r}{\partial y} + \hat{k} f'(r) \frac{\partial r}{\partial z} = f'(r) \left[\hat{i} \frac{\partial r}{\partial x} + \hat{j} \frac{\partial r}{\partial y} + \hat{k} \frac{\partial r}{\partial z} \right]$$

$$= f'(r) \text{ grad } r \qquad \ldots(1)$$

$\because \qquad \nabla^2 f(r) = \text{div } [f'(r) \text{ grad } r]$

We know that grad $r = \dfrac{\vec{r}}{r}$ (See S.E. 4 art 7.17)

$$= \text{div} \left[f'(r) \frac{\vec{r}}{r} \right]$$

$$= \text{div} \left[\frac{f'(r)}{r} \vec{r} \right]$$

which is of the type $\text{div}(\phi \vec{a})$, where $\phi = \dfrac{f'(r)}{r}$ and $\vec{a} = \vec{r}$

$$\therefore \quad \text{div} \left[\frac{f'(r)}{r} \vec{r} \right] = \frac{f'(r)}{r} \text{div} \vec{r} + \vec{r} \bullet \text{grad} \frac{f'(r)}{r}$$

Since div $\vec{r} = 3$ and from (1) grad $f(r) = f'(r)$ grad r; Replace $f(r)$ by $\dfrac{f'(r)}{r}$, we get

$$\text{grad} \frac{f'(r)}{r} = \frac{d}{dr} \left[\frac{f'(r)}{r} \right] \text{grad } r$$

$$= \frac{f'(r)}{r} 3 + \vec{r} \bullet \left[\frac{d}{dr} \left(\frac{f'(r)}{r} \right) \text{grad} r \right]$$

$$= \frac{3f'(r)}{r} + \vec{r} \bullet \left\{ \frac{r f''(r) - f'(r)}{r^2} \right\} \bullet \frac{\vec{r}}{r}$$

$$= \frac{3f'(r)}{r} + \left\{ \frac{f''(r)}{r^2} - \frac{f'(r)}{r^3} \right\} \vec{r} \bullet \vec{r} = \frac{3f'(r)}{r} + \left\{ f''(r) - \frac{f'(r)}{r} \right\} \frac{r^2}{r^2}$$

$$= \frac{3f'(r)}{r} + f''(r) - \frac{f'(r)}{r} = f''(r) + \frac{2}{r} f'(r)$$

(*ii*) $\nabla^2 r^n = \nabla \bullet \nabla r^n = \text{div grad } r^n$

$$\text{grad } r^n = \hat{i} \frac{\partial}{\partial x} r^n + \hat{j} \frac{\partial}{\partial y} r^n + \hat{k} \frac{\partial}{\partial z} r^n$$

$$= \hat{i} \bullet n r^{n-1} \frac{\partial r}{\partial x} + \hat{j}\, n r^{n-1} \frac{\partial r}{\partial y} + \hat{k}\, n r^{n-1} \frac{\partial r}{\partial z}$$

$$= n r^{n-1} \left\{ \hat{i} \frac{x}{r} + \hat{j} \frac{y}{r} + \hat{k} \frac{z}{r} \right\} = n r^{n-1} \frac{\vec{r}}{r} = n r^{n-2} \vec{r}$$

Now, $\qquad \nabla^2 r^n = \text{div} \left[\left(n r^{n-2} \right) \vec{r} \right]$ which is of the type $\text{div}(\phi \vec{a})$

$$= \left(nr^{n-2}\right) \text{div } \vec{r} + \vec{r} \bullet \text{grad}\left(nr^{n-2}\right)$$

$$= \left(nr^{n-2}\right) 3 + \vec{r} \cdot n \text{ grad } r^{n-2}$$

$$= 3nr^{n-2} + n\vec{r} \bullet (n-2) r^{n-4} \vec{r}$$

$\because$ grad $r^n = nr^{n-2}\vec{r}$; change n to $n-2$

grad $r^{n-2} = (n-2) r^{n-4}\vec{r}$

$$= 3n\, r^{n-2} + n\,(n-2) r^{n-4} (\vec{r} \bullet \vec{r})$$
$$= 3n\, r^{n-2} + n\,(n-2)\, r^{n-4} \bullet r^2$$
$$= 3n\, r^{n-2} + n\,(n-2)\, r^{n-2}$$
$$= (3n + n^2 - 2n)\, r^{n-2}$$
$$= (n + n^2)\, r^{n-2} = n\,(n+1)\, r^{n-2}.$$

Example 11. *Find directional derivative of div $(\vec{u})$ at the point (1, 2, 2) in the direction of the outer normal of the sphere $x^2 + y^2 + z^2 = 9$ for $\vec{u} = x^4\hat{i} + y^4\hat{j} + z^4\hat{k}$.*

Sol. div $(\vec{u}) = \nabla \bullet \vec{u}$

$$= \left(\hat{i}\frac{\partial}{\partial x} + \hat{j}\frac{\partial}{\partial y} + \hat{k}\frac{\partial}{\partial z}\right) \bullet \left(x^4\hat{i} + y^4\hat{j} + z^4\hat{k}\right) = 4\,(x^3 + y^3 + z^3)$$

Outer normal to the sphere $= \nabla(x^2 + y^2 + z^2 - 9)$

$$= \left(\hat{i}\frac{\partial}{\partial x} + \hat{j}\frac{\partial}{\partial y} + \hat{k}\frac{\partial}{\partial z}\right) \bullet \left(x^2 + y^2 + z^2 - 9\right)$$

$$= \hat{i}(2x) + \hat{j}(2y) + \hat{k}\,(2z) = 2\left(x\hat{i} + y\hat{j} + z\hat{k}\right)$$

Outer normal at the point (1, 2, 2) $= 2\left(\hat{i} + 2\hat{j} + 2\hat{k}\right)$

Gradient of div $\vec{u} = \nabla\,(4\,x^3 + 4\,y^3 + 4\,z^3)$

$$= \left(\hat{i}\frac{\partial}{\partial x} + \hat{j}\frac{\partial}{\partial y} + \hat{k}\frac{\partial}{\partial z}\right)\left(4x^3 + 4y^3 + 4z^3\right) = 12\left(x^2\hat{i} + y^2\hat{j} + z^2\hat{k}\right)$$

Gradient of div $\vec{u}$ at (1, 2, 2) $= 12\left(\hat{i} + 4\hat{j} + 4\hat{k}\right)$

Directional derivative of div $\vec{u}$ in the direction of outer normal

$$= 12\left(\hat{i} + 4\hat{j} + 4\hat{k}\right) \bullet \frac{\left(2\hat{i} + 4\hat{j} + 4\hat{k}\right)}{\sqrt{4 + 16 + 16}}$$

$$= \frac{12}{6}\,(1.2 + 4.4 + 4.4) = 2\,(2 + 16 + 16) = 68.$$

TEST YOUR KNOWLEDGE

1. If $\vec{r} = x\hat{i} + y\hat{j} + z\hat{k}$, show that div $\vec{r} = 3$, curl $\vec{r} = \vec{0}$. **(*P.T.U., Dec. 2006*)**

2. (*i*) Find divergence and Curl of the vector $\vec{V} = (xyz)\hat{i} + (3x^2y)\hat{j} + (xz^2 - y^2z)\hat{k}$ at the point (2, –1, 1).

   (*ii*) Show that for the vector field $\vec{F} = 2xy\hat{i} + xe^y\hat{j} + 2z\hat{k}$, $\nabla \cdot (\nabla \times \vec{F}) = 0$. **(*P.T.U., Dec. 2013*)**

3. (*i*) If $\vec{F} = (x + y + 1)\hat{i} + \hat{j} - (x + y)\hat{k}$, show that $\vec{F} \bullet \text{curl}\, \vec{F} = 0$ **(*P.T.U., Dec. 2012*)**

   (*ii*) Find $\nabla \times (\nabla\phi)$ where $\phi(x, y, z) = -2x^3yz^2$. **(*P.T.U., Dec. 2013*)**

4. If $\vec{A} = (3xz^2)\hat{i} - (yz)\hat{j} + (x + 2z)\hat{k}$, find Curl (Curl $\vec{A}$).

5. If $\vec{V} = \dfrac{x\hat{i} + y\hat{j} + z\hat{k}}{\sqrt{x^2 + y^2 + z^2}}$, show that $\nabla \bullet \vec{V} = \dfrac{2}{\sqrt{x^2 + y^2 + z^2}}$ and $\nabla \times \vec{V} = \vec{0}$ or If $\vec{V} = \dfrac{\vec{r}}{r}$, show that divergence of $\vec{V} = \dfrac{2}{r}$ and Curl $\vec{V} = \vec{0}$.

6. If $\vec{V}_1$ and $\vec{V}_2$ be the vectors joining the fixed points (x_1, y_1, z_1) and (x_2, y_2, z_2) to a variable point (x, y, z) prove that

   (*i*) $\text{div}\left(\vec{V}_1 \times \vec{V}_2\right) = 0$

   (*ii*) $\text{grad}\left(\vec{V}_1 \bullet \vec{V}_2\right) = \vec{V}_1 + \vec{V}_2$

   (*iii*) $\text{Curl}\left(\vec{V}_1 \times \vec{V}_2\right) = 2\left(\vec{V}_1 - \vec{V}_2\right)$.

7. If $\vec{a}$ is a constant vector and $\vec{r} = x\hat{i} + y\hat{j} + z\hat{k}$ prove that

   (*i*) $\text{div}\,(\vec{a} \times \vec{r}) = 0$

   (*ii*) $\text{Curl}\left[(\vec{a} \bullet \vec{r})\vec{r}\right] = \vec{a} \times \vec{r}$

   (*iii*) $\nabla(\vec{a} \bullet \vec{a}) = 2(\vec{a} \bullet \nabla)\vec{a} + 2\vec{a} \times (\nabla \times \vec{a})$.

   (*iv*) $\text{Curl}\,(\vec{a} \times \vec{r}) = 2\vec{a}$ **(*P.T.U., May 2010*)**

   (*v*) $\text{grad}\,(\vec{a} \bullet \vec{r}) = \vec{a}$

8. If $\vec{r} = x\hat{i} + y\hat{j} + z\hat{k}$, prove that

   (*i*) $\nabla^2\left(\dfrac{1}{r}\right) = 0$ **(*P.T.U., June 2003*)**

   (*ii*) $\nabla^2(r^n)\vec{r} = n(n + 1)r^{n-2}\vec{r}$ **(*P.T.U., Dec. 2003*)**

   (*iii*) $\nabla \bullet \left\{r\nabla\left(\dfrac{1}{r^3}\right)\right\} = \dfrac{3}{r^4}$

   (*iv*) $\nabla^2\left\{\nabla \bullet \left(\dfrac{\vec{r}}{r^2}\right)\right\} = 2r^{-4}$.

9. Find the directional derivative of $\nabla \bullet (\nabla\phi)$ at the point (1, – 2, 1) in the direction of outer normal to the surface $xy^2z = 3x + z^2$ where, $\phi = 2x^3y^2z^4$.

Answers

2. (*i*) $14, 2\hat{i} - 3\hat{j} - 14\hat{k}$ (*ii*) -3 3. (*ii*) $\vec{0}$ 4. $-6x\,\hat{i} + (6z - 1)\,\hat{k}$

9. $\dfrac{1724}{\sqrt{21}}$.

7.24. INTEGRATION OF VECTORS

Definition. Let $\overrightarrow{f(t)}$ and $\overrightarrow{g(t)}$ be two vectors functions of a scalar variable t such that $\dfrac{d}{dt}\overrightarrow{g(t)} = \overrightarrow{f(t)}$ then $\overrightarrow{g(t)}$ is called an integral of $\overrightarrow{f(t)}$ with respect to t and we write

$$\int \overrightarrow{f(t)}dt = \overrightarrow{g(t)}$$

If $\vec{c}$ is any arbitrary constant vector then we know that $\dfrac{d}{dt}\left[\overrightarrow{g(t)} + \vec{c}\right] = \overrightarrow{f(t)}$

$$\therefore \qquad \int \overrightarrow{f(t)}\,dt = \overrightarrow{g(t)} + \vec{c}$$

$\overrightarrow{g(t)}$ is called the indefinite integral of $\overrightarrow{f(t)}$. The constant vector $\vec{c}$ is called constant of integration and can be determined if some initial conditions are given.

The definite integral of $\overrightarrow{f(t)}$ between the limits $t = a$ and $t = b$ is written as

$$\int_a^b \overrightarrow{f(t)}\,dt = \overrightarrow{g(t)}\Big|_a^b = g(b) - g(a)$$

Note 1. If $\overrightarrow{f(t)} = f_1(t)\hat{i} + f_2(t)\,\hat{j} + f_3(t)\,\hat{k}$

then $$\int \overrightarrow{f(t)}dt = \hat{i}\int f_1(t)dt + \hat{j}\int f_2(t)dt + \hat{k}\int f_3(t)dt$$

Thus in order to integrate a vector function integrate its components.

Note 2. We can obtain some standard results for integration of vector functions by considering the derivatives of suitable vector functions. For example:

(*i*) $\dfrac{d}{dt}(\vec{r} \bullet \vec{s}) = \dfrac{d\vec{r}}{dt} \bullet \vec{s} + \vec{r} \bullet \dfrac{d\vec{s}}{dt}$

$\Rightarrow \displaystyle\int \left(\frac{d\vec{r}}{dt} \bullet \vec{s} + \vec{r} \bullet \frac{d\vec{s}}{dt}\right)dt = \vec{r} \bullet \vec{s} + c$, where c is a scalar quantity since integrand is scalar.

(*ii*) $\dfrac{d\vec{r}}{dt}\left(\vec{r}^{\,2}\right) = 2\vec{r} \bullet \dfrac{d\vec{r}}{dt}$

$$\Rightarrow \int \left(2\vec{r} \cdot \frac{d\vec{r}}{dt}\right) dt = \vec{r}^{\,2} + c$$, where c is scalar quantity since integrand is scalar.

(*iii*) $$\frac{d}{dt}\left(\vec{r} \times \frac{d\vec{r}}{dt}\right) = \frac{d\vec{r}}{dt} \times \frac{d\vec{r}}{dt} + \vec{r} \times \frac{d^2\vec{r}}{dt^2} = \vec{r} \times \frac{d^2\vec{r}}{dt^2} \quad \because \quad \frac{d\vec{r}}{dt} \times \frac{d\vec{r}}{dt} = \vec{0}$$

$$\Rightarrow \int \left(\vec{r} \times \frac{d^2\vec{r}}{dt^2}\right) dt = \vec{r} \times \frac{d\vec{r}}{dt} + \vec{c}$$, where $\vec{c}$ is a vector quantity since integrand is vector.

(*iv*) If $\vec{a}$ is a constant vector then, $\dfrac{d}{dt}(\vec{a} \times \vec{r}) = \vec{a} \times \dfrac{d\vec{r}}{dt}$

$$\Rightarrow \int \left(\vec{a} \times \frac{d\vec{r}}{dt}\right) dt = \vec{a} \times \vec{r} + \vec{c}$$, where $\vec{c}$ is a vector quantity since integrand is vector.

ILLUSTRATIVE EXAMPLES

Example 1. *If* $\overrightarrow{f(t)} = (5t^2 - 3t)\hat{i} + 6t^3\hat{j} - 7t\hat{k}$ *evaluate* $\displaystyle\int_{t=2}^{t=4} f(\vec{t})\, dt$.

Sol. $$\int_{t=2}^{t=4} f(\vec{t})dt = \int_2^4 (5t^2 - 3t)\hat{i} + (6t^3)\hat{j} - (7t)\hat{k}\, dt$$

$$= \left(\frac{5t^3}{3} - \frac{3t^2}{2}\right)\hat{i} + \frac{6t^4}{4}\hat{j} - \frac{7t^2}{2}\hat{k}\Bigg|_2^4$$

$$= \left(\frac{320}{3} - 24\right)\hat{i} + 384\hat{j} - 56\hat{k} - \left(\frac{40}{3} - 6\right)\hat{i} - 24\hat{j} + 14\hat{k}$$

$$= \frac{248}{3}\hat{i} + 384\hat{j} - 56\hat{k} - \frac{22}{3}\hat{i} - 24\hat{j} + 14\hat{k} = \frac{226}{3}\hat{i} + 360\hat{j} - 42\hat{k}.$$

Example 2. *The acceleration of a particle at any time t is given by* $\vec{a} = 18 \cos 3t\hat{i} - 8 \sin 2t\hat{j} + 6t\hat{k}$. *If the velocity* $\vec{v}$ *and displacement* $\vec{r}$ *be zero at* $t = 0$. *Find* $\vec{v}$ *and* $\vec{r}$ *at any time t.*

Sol. $$\vec{a} = \frac{d^2\vec{r}}{dt^2} = 18\cos 3t\hat{i} - 8\sin 2t\hat{j} + 6t\hat{k}$$

Integrating w.r.t. t, we have

$$\vec{v} = \frac{d\vec{r}}{dt} = \int\left\{(18 \cos 3t)\hat{i} - (8 \sin 2t)\hat{j} + (6t)\hat{k}\right\}dt + \vec{c}$$

$$= 6\sin 3t\,\hat{i} + 4\cos 2t\,\hat{j} + 3t^2\,\hat{k} + \vec{c}$$

At $t = 0$, $\quad \vec{v} = \vec{0}$

$\therefore \quad \vec{0} = 4\hat{j} + \vec{c} \quad \therefore \quad \vec{c} = -4\hat{j}$

$$\therefore \quad \vec{v} = \frac{d\vec{r}}{dt} = 6\sin 3t\,\hat{i} + 4(\cos 2t - 1)\hat{j} + 3t^2\,\hat{k}.$$

Integrating again w.r.t. t

$$\vec{r} = -2\cos 3t\,\hat{i} + 4\left(\frac{\sin 2t}{2} - t\right)\hat{j} + t^3\,\hat{k} + \vec{c}_1$$

At $t = 0$, $\quad \vec{r} = \vec{0} \quad \therefore \quad \vec{0} = -2\hat{i} + \vec{c}_1 \quad \therefore \quad \vec{c}_1 = 2\hat{i}$

$$\therefore \quad \vec{r} = -(2\cos 3t)\hat{i} + (2\sin 2t - 4t)\hat{j} + t^3\,\hat{k} + 2\hat{i}$$

$$\vec{r} = 2(1 - \cos 3t)\hat{i} + 2(\sin 2t - 2t)\hat{j} + t^3\,\hat{k}.$$

Example 3. *If* $\overrightarrow{r(t)} = 5t^2\,\hat{i} + t\,\hat{j} - t^3\,\hat{k}$, *prove that* $\int_1^2 \vec{r} \times \frac{d^2\vec{r}}{dt^2}\,dt = -14\hat{i} + 75\hat{j} - 15\hat{k}$.

Sol. $\frac{d}{dt}\left(\vec{r} \times \frac{d\vec{r}}{dt}\right) = \frac{d\vec{r}}{dt} \times \frac{d\vec{r}}{dt} + \vec{r} \times \frac{d^2\vec{r}}{dt^2} = \vec{r} \times \frac{d^2\vec{r}}{dt^2} \quad \because \quad \frac{d\vec{r}}{dt} \times \frac{d\vec{r}}{dt} = \vec{0}$

Integrating both sides w.r.t. t between the limits $t = 1$ and $t = 2$, we get

$$\int_1^2 \left(\vec{r} \times \frac{d^2\vec{r}}{dt^2}\right) dt = \vec{r} \times \frac{d\vec{r}}{dt}\Bigg|_1^2 \qquad \ldots(1)$$

[Now find $\vec{r} \times \frac{d\vec{r}}{dt}$]

$$\vec{r} \times \frac{d\vec{r}}{dt} = \left\{(5t^2)\hat{i} + (t)\hat{j} - (t^3)\hat{k}\right\} \times \left\{(10t)\hat{i} + \hat{j} - 3t^2\,\hat{k}\right\} = \begin{vmatrix} \hat{i} & \hat{j} & \hat{k} \\ 5t^2 & t & -t^3 \\ 10t & 1 & -3t^2 \end{vmatrix}$$

$$= -2t^3\,\hat{i} + 5t^4\,\hat{j} - 5t^2\,\hat{k}$$

$$\therefore \quad \int_1^2 \vec{r} \times \frac{d^2\vec{r}}{dt^2}\,dt = -2t^3\hat{i} + 5t^4\hat{j} - 5t^2\hat{k}\Big|_1^2 = -16\hat{i} + 80\hat{j} - 20\hat{k} + 2\hat{i} - 5\hat{j} + 5\hat{k}$$

$$= -14\hat{i} + 75\hat{j} - 15\hat{k}.$$

Example 4. *Given* $\vec{r}(t) = 2\hat{i} - \hat{j} + 2\hat{k}$ *when* $t = 2$ *and* $\vec{r}(t) = 4\hat{i} - 2\hat{j} + 3\hat{k}$ *when* $t = 3$.

Show that $\int_2^3 \left(\vec{r} \cdot \frac{d\vec{r}}{dt}\right) dt = 10$.

Sol. $\frac{d}{dt}\vec{r}^{\,2} = 2\vec{r} \cdot \frac{d\vec{r}}{dt}$

Integrate both sides w.r.t. t, we have $\int_2^3 2\vec{r} \bullet \frac{d\vec{r}}{dt}\, dt = \vec{r}^{\,2} \Big|_2^3$

$$\therefore \qquad \int_2^3 \vec{r} \bullet \frac{d\vec{r}}{dt}\, dt = \frac{1}{2} \vec{r}^{\,2} \Big|_2^3 \qquad \ldots(1)$$

Now, At $t = 3$, $\vec{r} = 4\hat{i} - 2\hat{j} + 3\hat{k}$

$\therefore$ $\vec{r}^{\,2} = \vec{r} \bullet \vec{r} = 16 + 4 + 9 = 29$

At $t = 2$, $\vec{r} = 2\hat{i} - \hat{j} + 2\hat{k}$

$\vec{r}^{\,2} = 4 + 1 + 4 = 9$

$$\therefore \text{ From (1), } \int_2^3 \vec{r} \bullet \frac{d\vec{r}}{dt}\, dt = \frac{1}{2} \bullet [29 - 9] = 10.$$

7.25. LINE INTEGRAL

Definition. Any integral which is evaluated along a curve is called a line integral. Consider a continuous vector point function $\overrightarrow{F(P)}$ which is defined at each point of the curve C in space.

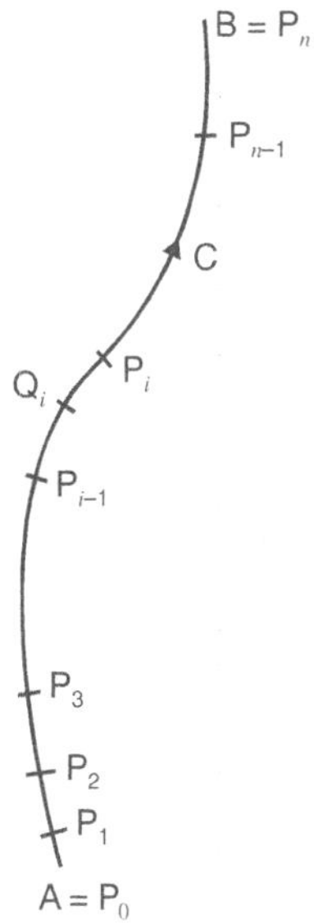

Divide the curve C into n parts by the points.

$A = P_0, P_1, P_2 \cdots P_{i-1}, P_i \cdots P_{n-1}, P_n = B$ and let $\vec{r}_0, \vec{r}_1, \vec{r}_2, \cdots, \vec{r}_{i-1}, \vec{r}_i, \cdots, \vec{r}_{n-1}, \vec{r}_n$ be the position vectors of these points.

Let Q_i be any point on the arc $P_{i-1} P_i$. Then the limit of the sum

$$\sum_{i=1}^{n} F(\overrightarrow{Q_i}) \bullet \delta\vec{r}_i \text{ , where } \delta\vec{r}_i = \vec{r}_i - \vec{r}_{i-1} \qquad \ldots(1)$$

as $n \to \infty$ and every $|\delta r_i| \to 0$ (if it exists) is called a line integral of $\vec{F}$ along C and is denoted by

$$\int_C \vec{F} \bullet \overrightarrow{dR} \quad \text{or} \quad \int_C \vec{F} \bullet \frac{\overrightarrow{dR}}{dt}\, dt$$

Clearly, it is a scalar quantity.

It is called **tangential line integral of $\vec{F}$ along C.**

Note 1. If the scalar products in (1) are replaced by vector products then the corresponding line integral is $\int_C \vec{F} \times d\vec{r}$ which is a vector.

Note 2. If the vector function $\vec{F}$ is replaced by a scalar function ϕ, then the corresponding line integral is defined as $\int_C \phi \, d\vec{r}$ which is a vector.

Note 3. If $\vec{F}(x, y, z) = f_1\hat{i} + f_2\hat{j} + f_3\hat{k}$ and $\vec{r} = x\hat{i} + y\hat{j} + z\hat{k}$ then

$$\int_C \vec{F} \cdot d\vec{r} = \int_{t_1}^{t_2} \left(f_1 \frac{dx}{dt} + f_2 \frac{dy}{dt} + f_3 \frac{dz}{dt} \right) dt$$

Note 4. If C is a closed curve, then the integral sign $\int_C$ is replaced by $\oint_C$

Example 5. *Evaluate* $\int_C x\,dx - yz\,dy + e^z dz$ *where C is the curve defined by* $x = t^3$, $y = -t$, $z = t^2$ *for t lying in the interval* $0 \le t \le 2$. **(P.T.U., Dec. 2013)**

Sol. Given $x = t^3$, $y = -t$, $z = t^2$ and $0 \le t \le 2$.

$$dx = 3t^2\,dt,\ dy = -\,dt,\ dz = 2t\,dt$$

$$\int_C x\,dx - yz\,dy + e^z dz = \int_0^2 t^3(3t^2\,dt) - (-t)(t^2)(-dt) + e^{t^2}(2t\,dt)$$

$$= \int_0^2 (3t^5 - t^3 + 2t\,e^{t^2})\,dt$$

$$= \left. \frac{3t^6}{6} - \frac{t^4}{4} + e^{t^2} \right|_0^2 \qquad \left[\because \int e^{t^2} \cdot 2t\,dt = e^{t^2} \right]$$

$$= (32 - 4 + e^4) - (1) = 27 + e^4.$$

7.26. CIRCULATION

If $\vec{V}$ represents the velocity of a fluid particle and C is a closed curve, then the line integral $\int_C \vec{V} \cdot d\vec{r}$ is called circulation of $\vec{V}$ round the curve C.

Irrotational. If the circulation of $\vec{V}$ round every closed curve in a region D vanishes, then $\vec{V}$ is said to be irrotational in D.

7.27. WORK DONE BY A FORCE

If $\vec{F}$ represents a force acting on a particle moving along an arc AB of a curve C, then the work done during a small displacement $\delta\vec{r}$ is $\vec{F} \cdot \delta\vec{r}$

$\therefore$ Total work done by $\vec{F}$ during the displacement from A to B is given by $\int_A^B \vec{F} \cdot \overrightarrow{dr}$.

If the force is conservative, then there exists a scalar function ϕ such that

$$\vec{F} = \nabla\phi = \hat{i}\frac{\partial\phi}{\partial x} + \hat{j}\frac{\partial\phi}{\partial y} + \hat{k}\frac{\partial\phi}{\partial z}$$

$\therefore$ The work done by $\vec{F}$ during displacement from A to B $= \int_A^B \vec{F} \bullet \overrightarrow{dr}$

$$= \int_A^B \left(\hat{i}\frac{\partial \phi}{\partial x} + \hat{j}\frac{\partial \phi}{\partial y} + \hat{k}\frac{\partial \phi}{\partial z} \right) \bullet \left(\hat{i}\,dx + \hat{j}\,dy + \hat{k}\,dz \right)$$

$$= \int_A^B \left(\frac{\partial \phi}{\partial x}dx + \frac{\partial \phi}{\partial y}dy + \frac{\partial \phi}{\partial z}dz \right) = \int_A^B d\phi = \phi \Big|_A^B = \phi_B - \phi_A .$$

Thus in a conservative field work done depends upon the value of ϕ at the end points A and B and **not on the path joining A and B.**

Note. If $\int_A^B \vec{F} \bullet \overrightarrow{dr}$ is to be proved to be independent of path, then prove $\vec{F} = \nabla \phi$

Here $\vec{F}$ is called conservative (irrotational) vector field and ϕ is called scalar potential.

And $\nabla \times \vec{F} = \nabla \times \nabla \phi = \vec{0}$.

Example 6. *A vector field is given by* $\vec{F} = (\sin y)\hat{i} + x(1 + \cos y)\hat{j}$. *Evaluate the line integral over a circular path given by* $x^2 + y^2 = a^2$, $z = 0$.

(P.T.U., May 2012, 2014, 2015)

Sol. Line integral of $\vec{F}$ is $\int_C \vec{F} \bullet \overrightarrow{dr}$

$$= \oint_C \left[(\sin y)\hat{i} + x(1 + \cos y)\hat{j} \right] \bullet \left[dx\,\hat{i} + dy\,\hat{j} \right]$$

$$= \oint_C \sin y\,dx + (x + x\cos y)\,dy$$

$$= \oint_C (\sin y\,dx + x\cos y\,dy) + x\,dy$$

$$= \oint_C d\,(x \sin y) + \oint_C x\,dy$$

The parametric equation of C is $x = a\cos\theta$, $y = a\sin\theta$, where $0 \le \theta \le 2\pi$.

$$\therefore \quad \text{Line integral} = \int_0^{2\pi} d\left[a\cos\theta \sin(a\sin\theta) \right] + \int_0^{2\pi} (a\cos\theta)(a\cos\theta\,d\theta)$$

$$= a\cos\theta \sin(a\sin\theta) \Big|_0^{2\pi} + a^2 \int_0^{2\pi} \cos^2\theta\,d\theta$$

$$= 0 + a^2\, 4 \int_0^{\frac{\pi}{2}} \cos^2\theta\,d\theta = 4a^2 \frac{1}{2}\frac{\pi}{2} = \pi a^2$$

Example 7. *Compute* $\int_c \vec{F} \cdot d\vec{r}$, *where* $\vec{F} = \dfrac{\hat{i}\, y - \hat{j}\, x}{x^2 + y^2}$ *and c is the circle* $x^2 + y^2 = 1$ *transversed counter clockwise.* **(P.T.U., Dec. 2004)**

Sol. $$\int_c \vec{F} \cdot d\vec{r} = \int_c \left(\frac{\hat{i}\, y - \hat{j}\, x}{x^2 + y^2} \right) \cdot \left(dx\,\hat{i} + dy\,\hat{j} + dz\,\hat{k} \right)$$

$$= \int_c \frac{y}{x^2 + y^2}\, dx - \frac{x\,dy}{x^2 + y^2} = \int_c \frac{y\,dx - x\,dy}{1} \qquad (\because x^2 + y^2 = 1)$$

Parametric equation of the circle (with radius 1) is $x = 1 . \cos\theta$, $y = 1 . \sin\theta$

$\therefore \qquad dx = -\sin\theta\, d\theta,\ dy = \cos\theta\, d\theta$

$$\therefore \quad \int_c \vec{F} \cdot d\vec{r} = \int_0^{2\pi} \sin\theta\,(-\sin\theta)\, d\theta - \cos\theta\,(\cos\theta\, d\theta)$$

$$= -\int_0^{2\pi} \left(\sin^2\theta + \cos^2\theta \right) d\theta = -\int_0^{2\pi} d\theta = -\theta \Big|_0^{2\pi} = -2\pi$$

Hence $\int_c \vec{F} \cdot d\vec{r} = -2\pi.$

Example 8. *If* $\vec{F} = 3xy\hat{i} - y^2\hat{j}$. *Evaluate* $\int_c \vec{F} \cdot d\vec{r}$, *where c is the curve in xy-plane* $y = 2x^2$ *from (0, 0) to (1, 2).* **(P.T.U., May 2004)**

Sol. $$\int_c \vec{F} \cdot d\vec{r} = \int_c \left(3xy\hat{i} - y^2\hat{j} \right) \cdot \left(dx\,\hat{i} + dy\,\hat{j} \right) = \int_c 3xy\, dx - y^2 dy$$

$$= \int_c 3xy\, dx - \int_c y^2\, dy$$

To integrate; change 1st integrand to function of x only with the help of $y = 2x^2$.

$$\therefore \quad \int_c \vec{F} \cdot d\vec{r} = \int_0^1 3x \,.\, 2x^2\, dx - \int_0^2 y^2\, dy \quad (x \text{ varies from 0 to 1 and } y \text{ varies from 0 to 2}).$$

$$= 6\,\frac{x^4}{4}\Big|_0^1 - \frac{y^3}{3}\Big|_0^2 = \frac{6}{4} - \frac{8}{3} = \frac{3}{2} - \frac{8}{3} = \frac{-7}{6}.$$

Second Method. Take parametric equation of the parabola $y = 2x^2$ which is $x = t$, $y = 2t^2$

$\therefore$ when $x = t$; $dx = dt$, when $y = 2t^2$; $dy = 4t\, dt$

At the point (0, 0), $x = 0 \quad \therefore \quad t = 0$

At the point (1, 2); $x = 1 \quad \therefore \quad t = 1$

$$\therefore \quad \int_c \vec{F} \cdot d\vec{r} = \int_c 3xy\, dx - y^2\, dy = \int_0^1 3 \,.\, t \,.\, 2t^2 - 4t^4 \,.\, 4t\, dt = \int_0^1 \left(6t^3 - 16t^5 \right) dt$$

$$= 6\,\frac{t^4}{4} - 16\,\frac{t^6}{6}\Big|_0^1 = \frac{6}{4} - \frac{8}{3} = -\frac{7}{6}.$$

Example 9. *Find the work done in moving a particle in the force field* $\vec{F} = 3x^2\hat{i} + (2xz - y)\hat{j} + z\hat{k}$ *along*

(i) the straight line from (0, 0, 0) to (2, 1, 3). **(P.T.U., May 2005, May 2010)**

(ii) the curve defined by $x^2 = 4y$, $3x^3 = 8z$ *from* $x = 0$ *to* $x = 2$.

Sol. Work done $= \int \vec{F} \cdot d\vec{r} = \int_c \left(3x^2\hat{i} + (2xz - y)\hat{j} + z\hat{k}\right) \cdot \left(\hat{i}\,dx + \hat{j}\,dy + \hat{k}\,dz\right)$

$$= \int_c 3x^2\,dx + (2xz - y)\,dy + z\,dz$$

(*i*) Equation of the line through (0, 0, 0) to (2, 1, 3) is

$$\frac{x-0}{2-0} = \frac{y-0}{1-0} = \frac{z-0}{3-0} \quad \text{or} \quad \frac{x}{2} = \frac{y}{1} = \frac{z}{3} = t \text{ (say)}$$

$\therefore\ x = 2t, \quad y = t, \quad z = 3t$

$dx = 2dt, dy = dt, \ dz = 3\,dt$

When $x = 0, \quad t = 0$

When $x = 2, \quad t = 1$

$\therefore$ t varies from 0 to 1.

$$\text{Work done} = \int_{t=0}^{1} 3\left(4t^2\right)(2dt) + \left(12t^2 - t\right)dt + 7t\,dt$$

$$= \int_0^1 \left(24t^2 + 12t^2 - t + 9t\right)dt = \int_0^1 \left(36t^2 + 8t\right)dt = 36\frac{t^3}{3} + 8\frac{t^2}{2}\Bigg|_0^1 = 12 + 4 = 16.$$

(*ii*) Along the curve $x^2 = 4y$, $3x^3 = 8z$

For parametric equation of the curve, we take $x = t \quad \therefore \quad y = \frac{t^2}{4}, z = \frac{3t^3}{8}$

$$\therefore \quad dx = dt, dy = \frac{t\,dt}{2}, \quad dz = \frac{9t^2}{8}\,dt$$

When $x = 0, \ t = 0$

When $x = 2, \ t = 2$ $\quad \therefore$ t varies from 0 to 2.

$$\therefore \quad \text{Work done} = \int_0^2 \left(3t^2\right)dt + \left(\frac{3}{4}t^4 - \frac{t^2}{4}\right)\frac{t}{2}\,dt + \left(\frac{3}{8}t^3\right)\left(\frac{9}{8}t^2\right)dt$$

$$= \int_0^2 \left(3t^2 + \frac{3}{8}t^5 - \frac{t^3}{8} + \frac{27}{64}t^5\right)dt = \int_0^2 \left(3t^2 - \frac{t^3}{8} + \frac{51}{64}t^5\right)dt$$

$$= t^3 - \frac{t^4}{32} + \frac{51}{64}\cdot\frac{t^6}{6}\Bigg|_0^2 = 8 - \frac{1}{2} + \frac{51}{6} = 8 - \frac{1}{2} + \frac{17}{2} = 8 + \frac{16}{2} = 16.$$

Example 10. *Find the work done by the force* $\vec{F} = x\hat{i} - z\hat{j} + 2y\hat{k}$ *in the displacement along the closed path c consisting of the segments* c_1, c_2, c_3 *where*

On c_1 $0 \le x \le 1$ $y = x, \quad z = 0$

On c_2 $0 \le z \le 1$ $x = 1, \quad y = 1$

On c_3 $1 \ge x \ge 0$ $y = z = x$

Sol. Total work done $= \int_c \vec{F} \bullet d\vec{r} = \int_c \left(x\hat{i} - z\hat{j} + 2y\hat{k}\right) \bullet \left(\hat{i}\,dx + \hat{j}\,dy + \hat{k}\,dz\right)$

$$= \int_c x\,dx - z\,dy + 2y\,dz$$

$$= \int_{c_1} x\,dx - z\,dy + 2y\,dz + \int_{c_2} x\,dx - z\,dy + 2y\,dz + \int_{c_3} x\,dx - z\,dy + 2y\,dz$$

$$= W_1 + W_2 + W_3$$

$$W_1 = \int_{c_1} x\,dx - z\,dy + 2y\,dz$$

On c_1 $y = x, \quad z = 0$ and x varies from 0 to 1

$\therefore$ $dy = dx, \; dz = 0$

$$W_1 = \int_0^1 x\,dx - 0 + 0 = \left.\frac{x^2}{2}\right|_0^1 = \frac{1}{2}$$

$$W_2 = \int_{c_2} x\,dx - z\,dy + 2y\,dz$$

On c_2 $x = 1, \quad y = 1$ and z varies from 0 to 1

$dx = 0, \quad dy = 0$

$$W_2 = \int_0^1 0 - 0 + 2 \cdot 1\,dz = 2z\,\Big|_0^1 = 2 \text{ and}$$

$$W_3 = \int_{c_3} x\,dx - z\,dy + 2y\,dz$$

On c_3 $y = z = x$ $\quad \therefore \left.\begin{matrix} y = x \\ z = x \end{matrix}\right\}$ and x varies from 1 to 0

$dy = dx, \quad dz = dx$

$$W_3 = \int_1^0 x\,dx - x\,dx + 2x\,dx = x^2\,\Big|_1^0 = -1$$

$\therefore$ Total work done $= \dfrac{1}{2} + 2 - 1 = \dfrac{1}{2} + 1 = \dfrac{3}{2}.$

Example 11. *Show that the integral* $\int_{(1,2)}^{(3,4)} \left(xy^2 + y^3\right)dx + \left(x^2 y + 3xy^2\right)dy$ *is independent of the path joining the points (1, 2) and (3, 4). Hence evaluate the integral.*

Sol. $\int_{(1,2)}^{(3,4)} \left(xy^2 + y^3\right)dx + \left(x^2 y + 3xy^2\right)dy$

$$= \int_{(1,2)}^{(3,4)} \left\{\left(xy^2 + y^3\right)\hat{i} + \left(x^2y + 3xy^2\right)\hat{j}\right\} \bullet \left\{dx\,\hat{i} + dy\,\hat{j}\right\}$$

$$= \int_{(1,2)}^{(3,4)} \vec{F} \bullet d\vec{r} \quad \therefore \quad \vec{F} = \left(xy^2 + y^3\right)\hat{i} + \left(x^2y + 3xy^2\right)\hat{j}$$

$$\nabla \times \vec{F} = \begin{vmatrix} \hat{i} & \hat{j} & \hat{k} \\ \dfrac{\partial}{\partial x} & \dfrac{\partial}{\partial y} & \dfrac{\partial}{\partial z} \\ xy^2 + y^3 & x^2 + 3xy^2 & 0 \end{vmatrix}$$

$$= \hat{i}\left[-\frac{\partial}{\partial z}\left(x^2y + 3xy^2\right)\right] + \hat{j}\left[\frac{\partial}{\partial z}\left(xy^2 + y^3\right) - \frac{\partial}{\partial x}(0)\right]$$

$$+ \hat{k}\left[\frac{\partial}{\partial x}\left(x^2y + 3xy^2\right) - \frac{\partial}{\partial y}\left(xy^2 + y^3\right)\right]$$

$$= \hat{i} \cdot 0 + \hat{j} \cdot 0 + \hat{k}\left[2xy + 3y^2 - 2xy - 3y^2\right] = \vec{0}$$

$\therefore \quad \nabla \times \vec{F} = \vec{0} \quad \therefore$ Integral is independent of the path of integration.

Also, if $\nabla \times \vec{F} = \vec{0}$, then $\vec{F} = \nabla\phi$

$$\left(xy^2 + y^3\right)\hat{i} + \left(x^2y + 3xy^2\right)\hat{j} = \hat{i}\,\frac{\partial\phi}{\partial x} + \hat{j}\,\frac{\partial\phi}{\partial y} + \hat{k}\,\frac{\partial\phi}{\partial z}$$

$$\therefore \quad \frac{\partial\phi}{\partial x} = xy^2 + y^3; \quad \frac{\partial\phi}{\partial y} = x^2y + 3xy^2, \quad \frac{\partial\phi}{\partial z} = 0$$

$$d\phi = \frac{\partial\phi}{\partial x}dx + \frac{\partial\phi}{\partial y}dy + \frac{\partial\phi}{\partial z}dz = (x\,y^2 + y^3)\,dx + (x^2\,y + 3x\,y^2)\,dy = \vec{F} \bullet d\vec{r}$$

$$= (xy^2\,dx + x^2\,y\,dy) + (\,y^3\,dx + 3x\,y^2\,dy)$$

$$= \frac{1}{2}\,d\left(x^2\,y^2\right) + d\left(xy^3\right) = d\left[\frac{1}{2}\,x^2\,y^2 + xy^3\right]$$

Integrate

$$\therefore \quad \int_{(1,2)}^{(3,4)} d\phi = \int_{(1,2)}^{(3,4)} d\left[\frac{1}{2}x^2y^2 + xy^3\right] = \frac{1}{2}x^2y^2 + xy^3 \Big]_{(1,2)}^{(3,4)}$$

$$= \left(\frac{1}{2}9 \cdot 16 + 3 \cdot 64\right) - (2+8) = 254$$

Hence $\displaystyle\int_{(1,2)}^{(3,4)} \left(xy^2 + y^3\right)dx + \left(x^2y + 3xy^2\right)dy = 254.$

Example 12. *Find circulation of $\vec{F}$ round the curve C, where $\vec{F} = e^x \sin y \hat{i} + e^x \cos y \hat{j}$ and C is the rectangle whose vertices are (0, 0), (1, 0), $\left(1, \frac{\pi}{2}\right)$, $\left(0, \frac{\pi}{2}\right)$.*

Sol. Circulation of $\vec{F}$ round the curve C

$$= \oint_C \vec{F} \cdot d\vec{r} = \oint_C \left\{\left(e^x \sin y\right)\hat{i} + e^x \cos y \hat{j}\right\} \cdot \left(dx\,\hat{i} + \hat{j}\,dy\right)$$

$$= \oint_C e^x \sin y\, dx + e^x \cos y\, dy$$

$$= \int_{C_1} e^x \sin y\, dx + e^x \cos y\, dy + \int_{C_2} e^x \sin y\, dx + e^x \cos y\, dy$$

$$+ \int_{C_3} e^x \sin y\, dx + e^x \cos y\, dy + \int_{C_4} e^x \sin y\, dx + e^x \cos y\, dy$$

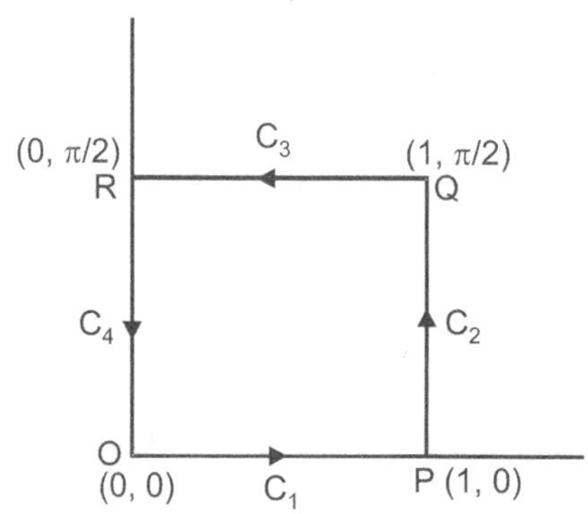

Along C_1, $y = 0$; x varies from 0 to 1

$\therefore \quad dy = 0$

$$\int_{C_1} e^x \sin y\, dx + e^x \cos y\, dy = \int_0^1 0 + e^x \cdot 0 = 0$$

$\therefore$ Integral along $C_1 = 0$

Along C_2, $x = 1$; y varies from 0 to $\frac{\pi}{2}$

$dx = 0$

$$\int_{C_2} e^x \sin y dx + e^x \cos y dy = \int_0^{\pi/2} e \sin y \cdot 0 + e \cos y dy = e \sin y \Big|_0^{\pi/2} = e$$

$\therefore$ Integral along $C_2 = e$

Along C_3; $y = \frac{\pi}{2}$; x varies from 1 to 0

$$dy = 0$$

$$\int_{C_3} e \sin y \, dx + e^x \cos y \, dy = \int_1^0 e^x \cdot 1 \, dx + e^x \cdot 0 = e^x \Big|_1^0 = 1 - e$$

$\therefore$ Integral along $\quad C_3 = 1 - e$

Along C_4; $x = 0$; y varies from $\frac{\pi}{2}$ to 0

$\therefore \quad dx = 0$

$$\int_{C_4} e^x \sin y \, dx + e^x \cos y \, dy = \int_{\frac{\pi}{2}}^{0} \sin y \cdot 0 + 1 \cdot \cos y \, dy = \sin y \Big|_{\frac{\pi}{2}}^{0} = 0 - 1 = -1$$

$\therefore$ Integral along $\quad C_4 = -1$

$\therefore$ Circulation of $\quad \vec{F} = 0 + e + 1 - e - 1 = 0.$

Example 13. *Show that* $\int x^2 y^2 ds = \frac{\pi a^5}{4}$ *around the circle* $x^2 + y^2 = a^2$. **(P.T.U., 2004)**

Sol. Parametric equation of the circle is $x = a \cos \theta$, $y = a \sin \theta$.

$$\therefore \quad \frac{ds}{d\theta} = \sqrt{\left(\frac{dx}{d\theta}\right)^2 + \left(\frac{dy}{d\theta}\right)^2} = \sqrt{(-a \sin\theta)^2 + (a \cos\theta)^2} = a$$

$$\therefore \quad \int x^2 y^2 ds = 4\int_0^{\frac{\pi}{2}} \left(a^2 \cos^2 \theta\right)\left(a^2 \sin^2 \theta\right) \bullet a d\theta$$

$$= 4a^5 \int_0^{\frac{\pi}{2}} \sin^2 \theta \cos^2 \theta \, d\theta = 4a^5 \int_0^{\frac{\pi}{2}} \frac{1}{4} (\sin 2\theta)^2 \, d\theta$$

$$= a^5 \int_0^{\frac{\pi}{2}} \frac{1 - \cos 4\theta}{2} \, d\theta = \frac{a^5}{2}\left[\theta - \frac{\sin 4\theta}{8}\right]_0^{\frac{\pi}{2}} = \frac{a^5}{2} \frac{\pi}{2} = \frac{\pi a^5}{4}$$

TEST YOUR KNOWLEDGE

1. Evaluate $\int_C (x^2 + yz)\, dz$ where C is the curve defined by $x = t$, $y = t^2$, $z = 3t$ for t lying in the interval $1 \le t \le 2$. ***(P.T.U. May 2008)***
[**Hint:** Consult S.E. 5]

2. If $\frac{d^2 \vec{p}}{dt^2} = 6t\hat{i} - 12t^2 \hat{j} + 4 \cos t \hat{k}$, find $\vec{p}$ given that $\frac{d\vec{p}}{dt} = -\hat{i} - 3\hat{k}$ and $\vec{p} = 2\hat{i} + \hat{j}$ at $t = 0$.

3. The acceleration of a particle at any point is given by $\vec{a} = 12 \cos 2t\,\hat{i} - 8 \sin 2t\,\hat{j} + 16\,t\,\hat{k}$. If the velocity $\vec{v}$ and displacement $\vec{r}$ are zero at $t = 0$. Find $\vec{v}$ and $\vec{r}$ at any time t.

4. If $\vec{r} = t\hat{i} - t^2\hat{j} + (t-1)\hat{k}$ and $\vec{s} = 2t^2\hat{i} + 6t\,\hat{k}$, evaluate

(i) $\int_0^2 \vec{r} \cdot \vec{s}\, dt$ (ii) $\int_0^2 (\vec{r} \times \vec{s})\, dt$

5. Find the value of $\vec{r}$ satisfying the equation $\dfrac{d^2 \vec{r}}{dt^2} = 6t\,\hat{i} - 24t^2\,\hat{j} + 4 \sin t\,\hat{k}$ given that $\vec{r} = 2\hat{i} + \hat{j}$ and $\dfrac{d\vec{r}}{dt} = -\hat{i} - 3\hat{k}$ at $t = 0$.

6. If $\phi = 2xyz^2$, $\vec{F} = xy\,\hat{i} - z\,\hat{j} + x^2\,\hat{k}$ and c is the curve $x = t^2$, $y = 2t$, $z = t^3$ from $t = 0$ to $t = 1$. Evaluate the

(i) $\int_c \phi\, d\vec{r}$ (ii) $\int_c \vec{F} \times d\vec{r}$

7. Evaluate $\int_c \left(x^2 + xy\right) dx + \left(x^2 + y^2\right) dy$, where c is the square formed by the lines $y = \pm 1$, $x = \pm 1$.

8. If $\vec{F} = \left(5xy - 6x^2\right)\hat{i} + (2y - 4x)\hat{j}$; evaluate $\int_c \vec{F} \cdot d\vec{r}$ along the curve c in the xy-plane $y = x^3$ from the point (1, 1) to (2, 8).

9. If $\vec{F} = \left(3x^2 + 6y\right)\hat{i} - 14\,yz\,\hat{j} + 20\,xz^2\,\hat{k}$ evaluate the integral $\int_c \vec{F} \cdot d\vec{r}$ from (0, 0, 0) to (1, 1, 1) along the following paths c :

(i) $x = t$, $y = t^2$, $z = t^3$

(ii) the straight line joining (0, 0, 0) to (1, 1, 1)

(iii) the straight line from (0, 0, 0) to (1, 0, 0) then to (1, 1, 0) and then to (1, 1, 1).

10. Compute the line integral $\int_c \left(y^2\, dx - x^2\, dy\right)$ along the triangle whose vertices are (1, 0); (0, 1) and (–1, 0). **(P.T.U Dec. 2012)**

11. Evaluate $\int_c (y + 3z)\,dx + (2z + x)\,dy + (3x + 2y)\,dz$ where c is the arc of helix $x = a \cos\theta$, $y = a \sin\theta$, $z = \dfrac{2a\theta}{\pi}$ between the points $(a, 0, 0)$ and $(0, a, a)$.

12. Evaluate $\int_c \left\{\left(2x^2y + y + z^2\right)\hat{i} + 2\left(1 + yz^3\right)\hat{j} + \left(2z + 3y^2 z^2\right)\hat{k}\right\} \cdot d\vec{r}$ along the curve c; $y^2 + z^2 = a^2$; $x = 0$.

13. If a force $\vec{F} = 2x^2y\hat{i} + 3xy\hat{j}$ displaces a particle in xy-plane from (0, 0) to (1, 4) along the curve $y = 4x^2$. Find the work done.

14. (*i*) Find the work done by a force $y\hat{i} + x\hat{j}$ which displaces a particle from origin to a point $(\hat{i} + \hat{j})$.

(*ii*) If $\vec{F} = \dfrac{-y\hat{i} + x\hat{j}}{x^2 + y^2}$, find work done by $\vec{F}$ along the upper half of the circle passing through the points (–1, 0) and (1, 0).

15. Find the work done when a force $\vec{F} = (x^2 - y^2 + x)\hat{i} - (2xy + y)\hat{j}$ moves a particle in xy-plane from (0, 0) to (1, 1) along the parabola $y^2 = x$. Is the work done different when the path is the straight line $y = x$?

16. Compute the work done by the force $\vec{F} = (2y + 3)\hat{i} + xz\hat{j} + (yz - x)\hat{k}$ when it moves a particle from the point (0, 0, 0) to the point (2, 1, 1) along the curve $x = 2t^2$, $y = t$, $z = t^3$.

17. (*i*) Find the work done in moving a particle in the force field $\vec{F} = 3x^2\hat{i} + (2xz - y)\hat{j} + z\hat{k}$ along the curve defined by $x^2 = 4y$, $3x^3 = 8z$, from $x = 0$ to $x = 2$.

(*ii*) Find the work done by the force $\vec{F} = -xy\hat{i} + y^2\hat{j} + z\hat{k}$ in moving a particle over the circular path $x^2 + y^2 = 4$, $z = 0$ from (2, 0, 0) at (0, 2, 0). ***(P.T.U., Jan. 2008, May 2010)***

18. Show that $\vec{V} = (2xy + z^3)\hat{i} + x^2\hat{j} + 3xz^2\hat{k}$ is a conservative field. Find its scalar potential ϕ such that $\vec{V}$ = grad ϕ. Find the work done by the force $\vec{V}$ in moving a particle from (1, –2, 1) to (3, 1, 4).

19. Show that the integral $\int_c (2xy + 3)\,dx + (x^2 - 4z)\,dy - 4y\,dz$, where c is any path joining (0, 0, 0) to (1, –1, 3) does not depend on the path c and evaluate the integral.

20. Find the circulation of $\vec{F}$ round the curve c where $\vec{F} = y\hat{i} + z\hat{j} + x\hat{k}$ and c is the circle $x^2 + y^2 = 1$, $z = 0$. ***(P.T.U., Dec. 2003)***

21. Evaluate (*i*) $\int_C x^2 y\, dS$, where C is curve defined by $x = 3\cos t$, $y = 3\sin t$; $0 \le t \le \frac{\pi}{2}$.

(*ii*) $\int_C (x^2 + yz)\,dS$, where C is the curve defined by $x = 4y$, $z = 3$ from $\left(2, \frac{1}{2}, 3\right)$ to (4, 1, 3).

[Hint. Let $x = t$ $\therefore$ $y = \frac{t}{4}$, $z = 3$ and $2 \le t \le 4$; $\dfrac{dS}{dt} = \sqrt{\left(\dfrac{dx}{dt}\right)^2 + \left(\dfrac{dy}{dt}\right)^2}$ **]**.

Answers

1. $\dfrac{163}{4}$

2. $(t^3 - t + 2)\hat{i} + (1 - t^4)\hat{j} + (4 - 4\cos t - 3t)\hat{k}$

3. $\vec{v} = 6\sin 2t\,\hat{i} + 4(\cos 2t - 1)\hat{j} + 8t^2\hat{k}$; $\vec{r} = 3(1 - \cos 2t)\hat{i} + 2(\sin 2t - 2t)\hat{j} + \frac{8}{3}t^3\hat{k}$

4. (*i*) 12 (*ii*) $-24\hat{i} - \frac{40}{3}\hat{j} + \frac{64}{5}\hat{k}$

5. $\vec{r} = (t^3 - t + 2)\hat{i} + (1 - 2t^4)\hat{j} + (t - 4\sin t)\hat{k}$

6. (*i*) $\frac{8}{11}\hat{i} + \frac{4}{5}\hat{j} + \hat{k}$ (*ii*) $-\frac{9}{10}\hat{i} - \frac{2}{3}\hat{j} + \frac{7}{5}\hat{k}$

7. 0

8. 35

9. (*i*) 5 (*ii*) $\frac{13}{3}$ (*iii*) $\frac{23}{3}$

10. $-\frac{2}{3}$

11. $2a^2$

12. 0

13. $\frac{104}{5}$

14. (*i*) 1 (*ii*) π

15. $-\frac{2}{3}$; No it is same

16. $8\frac{8}{35}$

17. (*i*) 16 (*ii*) $\frac{16}{3}$

18. $x^2 y + x z^3$; 202

19. 14

20. $-\pi$

21. (*i*) 27 (*ii*) $\frac{139\sqrt{17}}{24}$.

7.28. SURFACE INTEGRAL

Definition. Any integral which is to be evaluated over a surface is called a surface integral. Let $\vec{F}$ be a continuous vector point function and S be a surface. Divide S into a finite number of subsurfaces $\delta S_1, \delta S_2 \cdots \delta S_k$. Let P_i be any point in δS_i and $\hat{n}_i$ be a unit vector at P_i in the direction of outward normal to the surface at P_i.

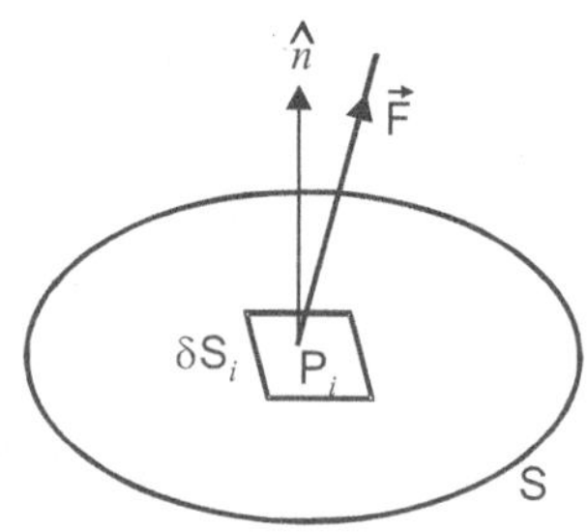

Then the limiting value of the sum $\sum_{i=1}^{k} \vec{F}(P_i) \cdot \vec{n}_i \, \delta S_i$ as $k \to \infty$ and $\delta S_i \to 0 \ \forall i$ is called normal surface integral of $\vec{F}(P)$ over S and is denoted by $\iint_S \vec{F} \cdot \vec{n} \, dS$.

The surface element $\overrightarrow{\delta S}$ surrounding any point P can be regarded as a vector whose magnitude = area of δS and the direction that of outward unit normal $\hat{n}$ *i.e.*, $\overrightarrow{\delta S} = \hat{n}\,\delta S$.

$\therefore$ The surface integral can also be written as $\iint_S \vec{F} \cdot d\vec{S}$

Note 1. If $\vec{F}$ represents the velocity of a fluid at any point P on a closed surface S. Then the total outward flux of $\vec{F}$ across the surface integral is $\iint_S \vec{F} \cdot d\vec{S}$.

Note 2. When the flux of $\vec{F}$ across every closed surface S in a region E vanishes, then $\vec{F}$ is said to be a **Solenoidal Vector Point Function** in E.

Note 3. Other types of surface integrals are (*i*) $\iint_S \vec{F} \times d\vec{S}$ (*ii*) $\iint_S \phi \, d\vec{S}$ which are both vectors.

Note 4. In order to evaluate surface integrals it is convenient to express them as double integrals taken over the orthogonal projection of S on one of the co-ordinate planes. But this is possible only when any line perpendicular to the co-ordinate plane chosen meets the surface S in not more than one point.

Let R be the orthogonal projection of S on the *xy*-plane and let $\hat{n} = \cos\alpha \hat{i} + \cos\beta \hat{j} + \cos\gamma \hat{k}$ where $\cos\alpha$, $\cos\beta$, $\cos\gamma$ are the direction cosines of $\hat{n}$.

Now, $dx\,dy$ = projection of dS on *xy*-plane = $dS \cos\gamma$

$$\therefore \qquad dS = \frac{dx\,dy}{\cos\gamma}$$

Also,
$$|\hat{k} \cdot \hat{n}| = \cos\gamma \quad \therefore \quad dS = \frac{dx\,dy}{|\hat{k} \cdot \hat{n}|}$$

Hence
$$\iint_S \vec{F} \cdot \hat{n}\, dS = \iint_S \vec{F} \cdot \hat{n} \frac{dx\,dy}{|\hat{k} \cdot \hat{n}|}$$

7.29. VOLUME INTEGRAL

Definition: Any integral which is to be evaluated over a volume is called a Volume Integral. If V is a volume bounded by a surface S, then the triple integrals

$$\iiint_V \phi\, dV \text{ and } \iiint_V \vec{F}\, dV \text{ are called volume integrals.}$$

Obviously, first integral is scalar and second integral is a vector.

Note 1. If we sub-divide the volume V into small cuboids by drawing planes parallel to the coordinate planes, then $dV = dx\,dy\,dz$

$$\iiint_V \phi\, dV = \iiint_V \phi(x, y, z)\, dx\,dy\,dz$$

Note 2. If $\vec{F} = F_1 \hat{i} + F_2 \hat{j} + F_3 \hat{k}$, then

$$\iiint_V \vec{F}\, dV = \hat{i} \iiint_V F_1(x, y, z)\, dx\,dy\,dz + \hat{j} \iiint_V F_2(x, y, z)\, dx\,dy\,dz$$
$$+ \hat{k} \iiint_V F_3(x, y, z)\, dx\,dy\,dz$$

ILLUSTRATIVE EXAMPLES

Example 1. *Show that* $\iint_S \vec{F} \cdot \hat{n}\, dS = \frac{3}{2}$; *where* $\vec{F} = 4xz\,\hat{i} - y^2\hat{j} + yz\hat{k}$ *and S is the surface of the cube bounded by the planes* $x = 0, x = 1, y = 0, y = 1, z = 0, z = 1.$

Sol. The given surface compromised of six surfaces

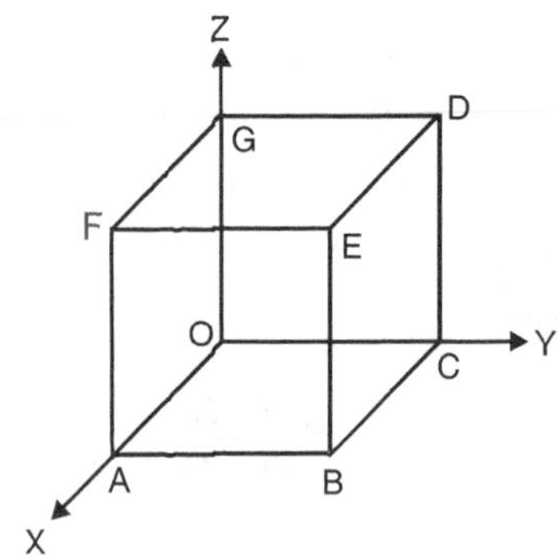

$S_1 \rightarrow$ the square OABC; its equation $z = 0$; $ds = dx\, dy$ outward normal $= -\hat{k}$

$S_2 \rightarrow$ the square DEFG; its equation $z = 1$; $ds = dx\, dy$ outward normal $= \hat{k}$

$S_3 \rightarrow$ the square OAFG; its equation $y = 0$; $ds = dx\, dz$ outward normal $= -\hat{j}$

$S_4 \rightarrow$ the square BCDE; its equation $y = 1$; $ds = dx\, dz$ outward normal $= \hat{j}$

$S_5 \rightarrow$ the square ABEF; its equation $x = 1$; $ds = dy\, dz$ outward normal $= \hat{i}$

$S_6 \rightarrow$ the square OCDG; its equation $x = 0$; $ds = dy\, dz$ outward normal $= -\hat{i}$

$$\iint_S \vec{F} \cdot \hat{n}\, dS = \iint_{S_1} \vec{F} \cdot \hat{n}\, dS + \iint_{S_2} \vec{F} \cdot \hat{n}\, dS + \iint_{S_3} \vec{F} \cdot \hat{n}\, dS$$

$$+ \iint_{S_4} \vec{F} \cdot \hat{n}\, dS + \iint_{S_5} \vec{F} \cdot \hat{n}\, dS + \iint_{S_6} \vec{F} \cdot \hat{n}\, dS \quad \ldots(1)$$

$$\iint_{S_1} \vec{F} \cdot \hat{n}\, dS = \int_0^1\int_0^1 \left(4xz\,\hat{i} - y^2\,\hat{j} + yz\,\hat{k}\right) \cdot \left(-\hat{k}\right) dx\, dy$$

$$= \int_0^1\int_0^1 - yz\, dx\, dy = 0 \qquad (\because z = 0)$$

$$\iint_{S_2} \vec{F} \cdot \hat{n}\, dS = \int_0^1\int_0^1 \left(4xz\,\hat{i} - y^2\,\hat{j} + yz\,\hat{k}\right) \cdot \hat{k}\, dx\, dy = \int_0^1\int_0^1 yz\, dx\, dy = \int_0^1\int_0^1 y\, dy\, dx \qquad (\because z = 1)$$

$$= \int_0^1 \left\{ \frac{y^2}{2} \Big|_0^1 \right\} dx = \int_0^1 \frac{1}{2}\, dx = \frac{x}{2}\Big|_0^1 = \frac{1}{2}$$

$$\iint_{S_3} \vec{F} \bullet \hat{n}\, dS = \int_0^1\int_0^1 \left(4xz\,\hat{i} - y^2\,\hat{j} + yz\,\hat{k}\right) \bullet \left(-\hat{j}\right) dx\, dz = \int_0^1\int_0^1 yz\, dx\, dy = \int_0^1\int_0^1 y^2\, dx\, dz = 0$$

$$(\because y = 0)$$

$$\iint_{S_4} \vec{F} \bullet \hat{n}\, dS = \int_0^1\int_0^1 \left(4xz\,\hat{i} - y^2\,\hat{j} + yz\,\hat{k}\right) \bullet \hat{j}\, dx\, dz = \int_0^1\int_0^1 -y^2\, dx\, dz = \int_0^1\int_0^1 -dz\, dx \qquad (\because y = 1)$$

$$= \int_0^1 \left\{ -z \Big|_0^1 \right\} dx = \int_0^1 -dx = -x \Big|_0^1 = -1$$

$$\iint_{S_5} \vec{F} \bullet \hat{n}\, dS = \int_0^1\int_0^1 \left(4xz\,\hat{i} - y^2\,\hat{j} + yz\,\hat{k}\right) \bullet \hat{i}\, dy\, dz = \int_0^1\int_0^1 4xz\, dy\, dz = \int_0^1\int_0^1 4z\, dy\, dz \qquad (\because x = 1)$$

$$= 4\int_0^1 \left\{ \frac{z^2}{2} \Big|_0^1 \right\} dy = 2\int_0^1 dy = 2$$

$$\iint_{S_6} \vec{F} \bullet \hat{n}\, dS = \int_0^1\int_0^1 \left(4xz\,\hat{i} - y^2\,\hat{j} + yz\,\hat{k}\right) \bullet \left(-\hat{i}\right) dy\, dz = \int_0^1\int_0^1 -4xz\, dy\, dz = 0 \qquad (\because x = 0)$$

$\therefore$ From (1), $\displaystyle\iint_S \vec{F} \bullet \hat{n}\, dS = 0 + \frac{1}{2} + 0 - 1 + 2 + 0 = \frac{3}{2}$.

Example 2. *Compute* $\displaystyle\iint_S \vec{F} \bullet \hat{n}\, dS$ *where* $\vec{F} = 6z\hat{i} - 4\hat{j} + y\hat{k}$ *and S is the portion of the plane* $2x + 3y + 6z = 12$ *in the first octant.* **(P.T.U., Dec. 2012)**

Sol. A vector normal to the surface S is given by $\nabla(2x + 3y + 6z - 12)$

$$= 2\hat{i} + 3\hat{j} + 6\hat{k}$$

$\hat{n}$ = a unit vector normal to surface S

$$= \frac{2\hat{i} + 3\hat{j} + 6\hat{k}}{\sqrt{4 + 9 + 36}} = \frac{1}{7}(2\hat{i} + 3\hat{j} + 6\hat{k})$$

$$\hat{k} \bullet \hat{n} = \frac{6}{7}$$

$$\therefore \qquad \iint_S \vec{F} \bullet \hat{n}\, dS = \iint_R \vec{F} \bullet \hat{N} \frac{dxdy}{|\hat{k} \bullet \hat{n}|}$$

where R is the projection of S (*i.e.*, ΔLMN) on xy-plane

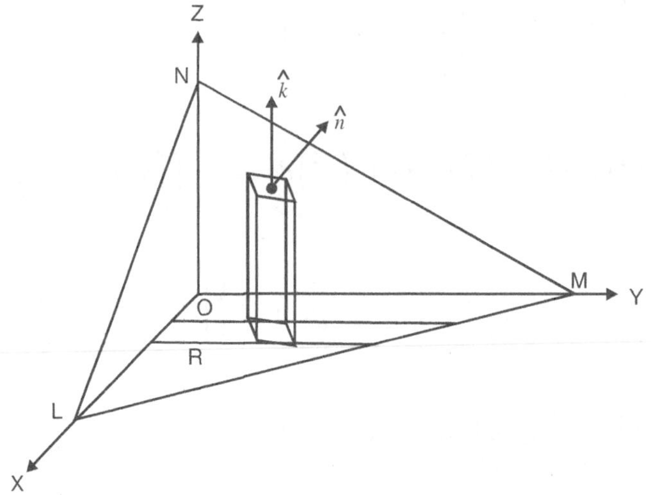

The region R (*i.e.*, ΔOLM) is bounded by x-axis, y-axis, and the line $2x + 3y = 12$, $z = 0$

Now $$\vec{F} \cdot \hat{n} = (6z\hat{i} - 4\hat{j} + y\hat{k}) \cdot \frac{1}{7}(2\hat{i} + 3\hat{j} + 6\hat{k}) = \frac{1}{7}(12z - 12 + 6y)$$

Substituting the value of $z = \dfrac{12 - 2x - 3y}{6}$

$$\vec{F} \cdot \hat{n} = \frac{1}{7}[2(12 - 2x - 3y) - 12 + 6y] = \frac{1}{7}[12 - 4x] = \frac{4}{7}(3 - x)$$

$$\therefore \quad \iint_S \vec{F} \cdot \hat{n}\, dS = \iint_R \frac{4}{7}(3 - x)\frac{dxdy}{6/7}$$

where y varies from 0 to $\dfrac{12 - 2x}{3}$ and x varies from 0 to 6

$$\therefore \quad \iint_S \vec{F} \cdot \hat{n}\, dS = \frac{2}{3}\int_0^6 \int_0^{\frac{12-2x}{3}} (3 - x)\, dy\, dx = \frac{2}{3}\int_0^6 (3 - x)\, y \Big|_0^{\frac{12-2x}{3}} dx$$

$$= \frac{2}{3}\int_0^6 (3 - x)\left(\frac{12 - 2x}{3}\right) dx = \frac{4}{9}\int_0^6 (3 - x)(6 - x)\, dx$$

$$= \frac{4}{9}\int_0^6 (18 - 9x + x^2)\, dx = \frac{4}{9}\left(18x - \frac{9x^2}{2} + \frac{x^3}{3}\right)\Bigg|_0^6$$

$$= \frac{4}{9}(108 - 162 + 72) = 8$$

Example 3. *Evaluate* $\iint_S \vec{A} \cdot \hat{n}\, dS$, *where* $\vec{A} = z\hat{i} + x\hat{j} - 3y^2 z\,\hat{k}$ *and S is the surface of cylinder* $x^2 + y^2 = 16$ *included in the first octant between* $z = 0$ *and* $z = 5$.

Sol. A vector normal to the surface S is given by

$$\nabla (x^2 + y^2) = \hat{i}\, 2x + \hat{j}\, 2y$$

$\hat{n}$, a unit vector normal to S

$$= \frac{2x\,\hat{i} + 2y\,\hat{j}}{\sqrt{4x^2 + 4y^2}} = \frac{x\,\hat{i} + y\,\hat{j}}{\sqrt{x^2 + y^2}} = \frac{x\,\hat{i} + y\,\hat{j}}{4} \qquad \because x^2 + y^2 = 16$$

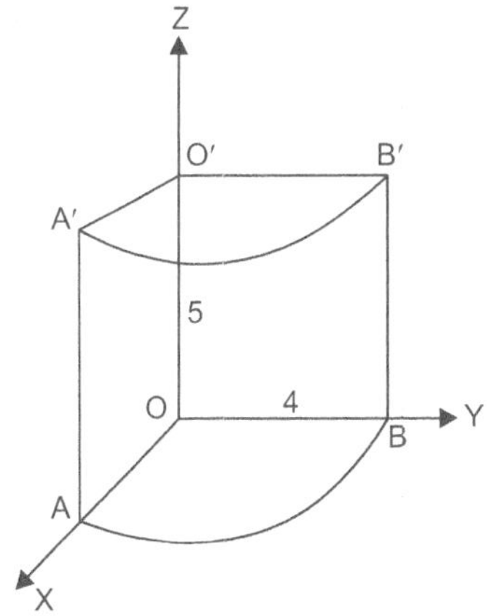

Let R be the projection of S on YZ plane.

[**Note.** We have chosen YZ plane $\because$ $\hat{n}$ is not containing $\hat{k}$]

then $\displaystyle\iint_S \vec{A} \bullet \hat{n}\, dS = \iint_R \vec{A} \bullet \hat{n} \frac{dy\, dz}{|\hat{i} \bullet \hat{n}|}$

where R is OBB′O′ which is enclosed by $y = 0$ to $y = 4$ and $z = 0$ to $z = 5$.

$$\hat{i} \bullet \hat{n} = \hat{i} \bullet \frac{x\,\hat{i} + y\,\hat{j}}{4} = \frac{x}{4}$$

$$\vec{A} \bullet \hat{n} = \left(z\,\hat{i} + x\,\hat{j} - 3y^2 z\,\hat{k}\right) \bullet \frac{x\,\hat{i} + y\,\hat{j}}{4} = \frac{zx + xy}{4} = \frac{x\,(y + z)}{4}$$

$$\therefore \qquad \iint_S \vec{A} \bullet \hat{n}\, dS = \iint_R \frac{x\,(y+z)}{4} \frac{dy\, dz}{\frac{x}{4}} = \int_{z=0}^{5} \int_{y=0}^{4} (y + z)\, dy\, dz$$

$$= \int_0^5 \left\{ \frac{y^2}{2} + zy \Big|_0^4 \right\} dz = \int_0^5 (8 + 4z)\, dz = 8z + 2z^2 \Big|_0^5$$

$$= 40 + 50 = 90.$$

Example 4. *Evaluate* $\displaystyle\iint_S \vec{F} \bullet \hat{n}\, dS$, *where* $\vec{F} = 2x^2 y\hat{i} - y^2\,\hat{j} + 4xz^2\,\hat{k}$ *and S is the closed surface of the region in the first octant bounded by the cylinder* $y^2 + z^2 = 9$ *and the planes* $x = 0$, $x = 2$, $y = 0$ *and* $z = 0$.

Sol. The given surface is comprised of six surfaces as shown below:

Surface	**Equation**	***ds***	**Outward normal**
$S_1 \to$ Rectangle OAEB	$z = 0$	$dx\,dy$	$-\hat{k}$
$S_2 \to$ Rectangle OADC	$y = 0$	$dx\,dz$	$-\hat{j}$
$S_3 \to$ Circular quadrant OBC	$x = 0$	$dy\,dz$	$-\hat{i}$
$S_4 \to$ Circular quadrant ADE	$x = 2$	$dy\,dz$	$\hat{i}$
$S_5 \to$ Curved surface BCDE	$y^2 + z^2 = 9$ varies from 0 to 2	$\dfrac{dx\,dy}{\lvert\hat{n} \bullet \hat{k}\rvert}$	$\hat{n}$

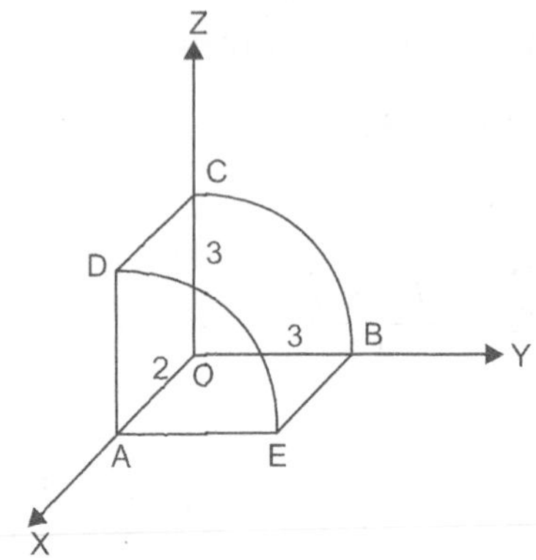

$$\iint_S \vec{F} \bullet \hat{n}\, dS = \iint_{S_1} \vec{F} \bullet \hat{n}\, dS + \iint_{S_2} \vec{F} \bullet \hat{n}\, dS + \iint_{S_3} \vec{F} \bullet \hat{n}\, dS + \iint_{S_4} \vec{F} \bullet \hat{n}\, dS + \iint_{S_5} \vec{F} \bullet \hat{n}\, dS \quad \ldots(1)$$

$$\iint_{S_1} \vec{F} \bullet \hat{n}\, dS = \iint \left(2x^2y\hat{i} - y^2\hat{j} + 4xz^2\hat{k}\right) \bullet \left(-\hat{k}\right) dx\, dy = \iint -4x\, z^2\, dx\, dy = 0 \quad (\because z = 0)$$

$$\iint_{S_2} \vec{F} \bullet \hat{n}\, dS = \iint \left(2x^2y\hat{i} - y^2\hat{j} + 4xz^2\hat{k}\right) \bullet \left(-\hat{j}\right) dx\, dz = \iint y^2\, dx\, dz = 0 \quad (\because y = 0)$$

$$\iint_{S_3} \vec{F} \bullet \hat{n}\, dS = \iint \left(2x^2y\, \hat{i} - y^2\hat{j} + 4xz^2\hat{k}\right) \bullet \left(-\hat{i}\right) dx\, dz = \iint -2x^2y\, dy\, dz = 0 \quad (\because x = 0)$$

$$\iint_{S_4} \left(\vec{F} \bullet \hat{n}\right) dS = \iint \left(2x^2y\hat{i} - y^2\hat{j} + 4xz^2\hat{k}\right) \bullet \hat{i}\, dy\, dz$$

$$= \iint 2x^2y\, dy\, dz$$

$$= \iint 8y\, dy\, dz \quad (\because x = 2)$$

$$= \int_0^3 \left\{ \int_0^{\sqrt{9-z^2}} 8y\, dy \right\} dz = \int_0^3 \left\{ 4y^2 \Big|_0^{\sqrt{9-z^2}} \right\} dz = \int_0^3 4\left(9 - z^2\right) dz = 4 \bullet \left\{ 9z - \frac{z^3}{3} \right\}_0^3$$

$$= 4\,\{27 - 9\} = 72$$

To find $\iint_{S_5} \left(\vec{F} \bullet \hat{n}\right) dS$ we have to first find $\hat{n}$.

A vector normal to $y^2 + z^2 = 9$ is $\nabla\,(y^2 + z^2) = \left(\hat{i}\,\frac{\partial}{\partial x} + \hat{j}\,\frac{\partial}{\partial y} + \hat{k}\,\frac{\partial}{\partial z}\right)\left(y^2 + z^2\right) = 2\left(y\,\hat{j} + z\,\hat{k}\right)$

$$\hat{n} = \text{unit vector normal to } S_5 = 2\,\frac{\left(y\,\hat{j} + z\,\hat{k}\right)}{2\sqrt{y^2 + z^2}} = \frac{y\,\hat{j} + z\,\hat{k}}{3} \quad (\because y^2 + z^2 = 9)$$

$$\left|\hat{n} \bullet \hat{k}\right| = \frac{z}{3} \qquad \therefore \quad dS = \frac{dx\,dy}{\frac{z}{3}} = \frac{3\,dx\,dy}{z}$$

$$\vec{F} \bullet \hat{n} = \frac{1}{3}\left[-y^3 + 4xz^3\right]$$

$$\iint_{S_5} \vec{F} \bullet \hat{n}\, dS = \iint \frac{1}{3}\left(-y^3 + 4xz^3\right)\frac{dx\,dy}{\left|\hat{n} \bullet \hat{k}\right|} = \iint \frac{1}{3}\left(-y^3 + 4xz^3\right)\frac{dx\,dy}{\frac{z}{3}}$$

$$= \iint \left(-\frac{y^3}{z} + 4xz^2\right) dx\,dy$$

Eliminate z; $y^2 + z^2 = 9$

$\therefore$ $z^2 = 9 - y^2$

$$= \int_{y=0}^{3}\int_{x=0}^{2} -\frac{y^3}{\sqrt{9-y^2}} + 4x\left(9-y^2\right)dx\,dy = \int_{y=0}^{3}\left\{-\frac{y^3}{\sqrt{9-y^2}} \bullet x + 2x^2\left(9-y^2\right)\right\}_{x=0}^{2} dy$$

$$= \int_{y=0}^{3}\left\{\frac{-2y^3}{\sqrt{9-y^2}} + 8\left(9-y^2\right)\right\}dy$$

Put $y = 3\sin\theta \Rightarrow dy = 3\cos\theta\, d\theta$

$$= \int_{0}^{\frac{\pi}{2}}\left[\frac{-2\,.\,27\sin^3\theta}{3\cos\theta} + 8 \cdot 9\cos^2\theta\right]3\cos\theta\, d\theta = \int_{0}^{\frac{\pi}{2}}\left\{-54\sin^3\theta + 216\cos^3\theta\right\}d\theta$$

$$= -54\,\frac{2}{3} + 216\,\frac{2}{3} = \frac{2}{3}(162) = 108$$

$\therefore$ From (1), $\iint_S \vec{F} \bullet \hat{n}\, dS = 0 + 0 + 0 + 72 + 108 = 180.$

Example 5. *If* $\vec{F} = \left(2x^2 - 3z\right)\hat{i} - 2xy\,\hat{j} - 4x\,\hat{k}$, *then evaluate* $\iiint_V \nabla \bullet \vec{F}\, dV$, *where V is bounded by the planes x = 0, y = 0, z = 0 and 2x + 2y + z = 4.*

Sol. $\nabla \bullet \vec{F} = \left(\hat{i}\,\frac{\partial}{\partial x} + \hat{j}\,\frac{\partial}{\partial y} + \hat{k}\,\frac{\partial}{\partial z}\right) \bullet \left\{\left(2x^2 - 3z\right)\hat{i} - 2xy\,\hat{j} - 4x\,\hat{k}\right\}$

$$= \frac{\partial}{\partial x}\left(2x^2 - 3z\right) + \frac{\partial}{\partial y}(-2xy) + \frac{\partial}{\partial z}(-4x) = 4x - 2x = 2x$$

$$\iiint_V \nabla \bullet \vec{F}\, dV = \iiint_V 2x\, dx\,dy\,dz = \int_{x=0}^{2}\int_{y=0}^{2-x}\int_{z=0}^{4-2x-2y} 2x\, dz\,dy\,dx$$

$$= \int_0^2 \int_0^{2-x} \left\{ 2xz \Big|_0^{4-2x-2y} \right\} dy\, dx = \int_0^2 \int_0^{2-x} \left(8x - 4x^2 - 4xy\right) dy\, dx$$

$$= \int_0^2 \left[8xy - 4x^2 y - 2xy^2\right]_0^{2-x} dx = \int_0^2 \left\{8x(2-x) - 4x^2(2-x) - 2x(2-x)^2\right\} dx$$

$$= \int_0^2 2x\left\{8 - 4x - 4x + 2x^2 - 4 - x^2 + 4x\right\} dx = \int_0^2 2\left(4x - 4x^2 + x^3\right) dx$$

$$= 2\left\{2x^2 - \frac{4x^3}{3} + \frac{x^4}{4}\right\}_0^2 = 2\left(8 - \frac{32}{3} + 4\right) = \frac{8}{3}.$$

TEST YOUR KNOWLEDGE

1. (*i*) Evaluate $\iint_S \vec{F} \bullet \hat{n}\, dS$, where $\vec{F} = 18z\,\hat{i} - 12\,\hat{j} + 3y\,\hat{k}$ and S is the surface of the plane $2x + 3y + 6z = 12$ in the first octant. ***(P.T.U., May 2010)***
[**Hint:** consult S.E. 2]

(*ii*) Evaluate $\iint_S \vec{A} \bullet \hat{n}\, dS$, where $\vec{A} = (x + y^2)\hat{i} - 2x\hat{j} + 2yz\hat{k}$ and S is the surface of the plane $2x + y + 2z = 6$ is the first octant.

2. Evaluate $\iint_S \vec{A} \bullet \hat{n}\, dS$, where $\vec{A} = 12x^2 y\,\hat{i} - 3yz\,\hat{j} + 2z\,\hat{k}$ and S is the portion of the plane $x + y + z = 1$ included in the first octant.

3. Evaluate $\iint_S \vec{F} \bullet \hat{n}\, dS$, where $\vec{F} = yz\,\hat{i} + zx\,\hat{j} + xy\,\hat{k}$ and S is the part of the surface $x^2 + y^2 + z^2 = 1$ which lies in the first octant. ***(P.T.U. Dec. 2014)***

$$\left[\textbf{Hint: } \hat{n} = \nabla(x^2 + y^2 + z^2 - 1) = x\hat{i} + y\hat{j} + z\hat{k},\ \vec{F} . \hat{n} = 3xyz \right.$$

$$\left. \therefore \iint_S \vec{F} \bullet \hat{n}\, dS = \int_{y=0}^{1} \int_{x=0}^{\sqrt{1-y^2}} 3xyz \frac{dx\, dy}{z} \right]$$

4. Evaluate $\iint_S \vec{F} \bullet \hat{n}\, dS$, where $\vec{F} = 4x\,\hat{i} - 2y^2\,\hat{j} + z^2\,\hat{k}$ and S is the surface bounding the region $x^2 + y^2 = 4$; $z = 0$ and $z = 3$.

5. Evaluate $\iint_S \vec{F} \bullet dS$, where $\vec{F} = x\,\hat{i} + \left(z^2 - zx\right)\hat{j} - xy\,\hat{k}$ and S is the triangular surface with vertices (2, 0, 0), (0, 2, 0) and (0, 0, 4). $\left\{ \textbf{Hint. } \iint_S \vec{F} \bullet d\vec{S} = \iint_S \vec{F} \bullet \hat{n}\, dS \right\}$

6. If velocity vector is $\vec{v} = y\hat{i} + 2\hat{j} + xz\hat{k}$ m/sec show that the flux of water through the parabolic cylinder $y = x^2$, $0 \le x \le 3$, $0 \le z \le 2$ is 69 m^3/sec.
[**Hint:** consult S.E. 2 art 8.2 next chapter]

7. Evaluate $\iiint_V \phi\, dV$, where $\phi = 45\, x^2 y$ and V is the closed region bounded by the planes $4x + 2y + z = 8$, $x = 0$, $y = 0$, $z = 0$.

8. If $\vec{F} = \left(2x^2 - 3z\right)\hat{i} - 2xy\,\hat{j} - 4x\,\hat{k}$, then evaluate $\iiint_V \nabla \times \vec{F}\, dV$, where V is the closed region bounded by planes $x = 0$, $y = 0$, $z = 0$ and $2x + 2y + z = 4$.

Answers

1. (*i*) 24 (*ii*) 81 2. $\dfrac{49}{120}$ 3. $\dfrac{3}{8}$ 4. 4π

5. 10 7. 128 8. $\dfrac{8}{3}\left(\hat{j} - \hat{k}\right)$.

REVIEW OF THE CHAPTER

1. **Vector Function.** If a vector $\vec{r}$ varies continuously with the variation of any scalar quantity t then $\vec{r}$ is said to be vector function of t and is written as $\vec{r} = \vec{f}(t)$.
2. The derivative of a constant vector is always zero.
3. If $\vec{r}$ has constant Magnitude but variable Direction then $\vec{r} \bullet \dfrac{d\vec{r}}{dt} = 0$.
4. **Velocity and Acceleration.** If $\vec{r}$ is the P.V. of a moving point and t, a scalar variable, denoting time then vector $\dfrac{d\vec{r}}{dt}$ represents the velocity vector, its direction is along the tangent at that point. If $\vec{a}$ represents the acceleration at time t then $\vec{a} = \dfrac{d^2\vec{r}}{dt^2}$.
5. **Tangential and Normal Acceleration.** The Tangential component of acceleration at any point of the curve is $\dfrac{dv}{dt}$ or $\dfrac{d^2s}{dt^2}$ of $v\dfrac{dv}{ds}$ where v is the magnitude of velocity at that point. The Normal component of acceleration at that point is $\dfrac{v^2}{\rho}$ where ρ is the radius of curvature at that point.
6. **Radial and Transverse Accelerations.** The radial and transverse components of acceleration at any point of the curve are $\dfrac{d^2r}{dt^2} - r\left(\dfrac{d\theta}{dt}\right)^2$ and $2\dfrac{dr}{dt}\dfrac{d\theta}{dt} + r\dfrac{d^2\theta}{dt^2}$.
7. **Scalar Point Function.** A function $\phi(x, y, z)$ is called a scalar point function if it associates a scalar with every point in region R of a space region. R is called scalar field.
8. **Vector Point Function.** If a function $\vec{V}(x, y, z)$ defined a vector at every point of the region R of a space then $\vec{V}(x, y, z)$ is called a vector point function. Region R is called vector field.

9. **Gradient of a Scalar Field.** Let $\phi(x, y, z)$ be a function defining a scalar field, then the vector $\hat{i}\frac{\partial\phi}{\partial x}+\hat{j}\frac{\partial\phi}{\partial y}+\hat{k}\frac{\partial\phi}{\partial z}$ is called gradient of scalar field ϕ and is denoted by $\nabla\phi$.

10. **Directional Derivative.** Let PQ = δr, then $\underset{\delta r\to 0}{\text{Lt}} = \frac{\delta\phi}{\delta r} = \frac{\partial\phi}{\partial r}$ is called directional derivative of ϕ at P in the direction of PQ. *i.e.*, directional derivative $= \frac{\partial\phi}{\partial r} = \nabla\phi \bullet \hat{N}$, where $\hat{N}$ is a unit vector in the direction of $\overrightarrow{PQ} \cdot \nabla\phi$ gives the maximum rate of change of ϕ and the magnitude of this maximum rate of change is $|\nabla\phi|$.

11. **Properties of Gradient.**

(*i*) If ϕ is a constant scalar point function, then $\nabla\phi = \vec{0}$

(*ii*) If ϕ_1 and ϕ_2 are two scalar point functions
then (*a*) $\nabla(\phi_1 + \phi_2) = \nabla\phi_1 \pm \nabla\phi_2$
(*b*) $\nabla(C_1\phi_1 + C_2\phi_2) = C_1\nabla\phi_1 + C_2\nabla\phi_2$
(*c*) $\nabla(\phi_1\phi_2) = \phi_1\nabla\phi_2 + \phi_2\nabla\phi_1$

12. **Divergance of A vector Point Function.** The divergence of a vector point function is defined as: $\text{div } \vec{V} = \nabla \bullet \vec{V} = \left(\hat{i}\frac{\partial}{\partial x}+\hat{j}\frac{\partial}{\partial y}+\hat{k}\frac{\partial}{\partial z}\right)\bullet \vec{V} = \hat{i}\bullet\frac{\partial\vec{V}}{\partial x}+\hat{j}\bullet\frac{\partial\vec{V}}{\partial y}+\hat{k}\bullet\frac{\partial\vec{V}}{\partial z}$

Clearly div $\vec{V}$ is a scalar quantity.

13. **Curl of a Vecter Point Function.** Curl of a vector point function is defined as:

$$\text{curl } \vec{V} = \nabla\times\vec{V} = \left(\hat{i}\frac{\partial}{\partial x}+\hat{j}\frac{\partial}{\partial y}+\hat{k}\frac{\partial}{\partial z}\right)\times\vec{V} = \hat{i}\times\frac{\partial\vec{V}}{\partial x}+\hat{j}\times\frac{\partial\vec{V}}{\partial y}+\hat{k}\times\frac{\partial\vec{V}}{\partial z}$$

If $\vec{V} = V_1\hat{i} + V_2\hat{j} + V_3\hat{k}$

$$\text{Then curl } \vec{V} = \begin{vmatrix} \hat{i} & \hat{j} & \hat{k} \\ \frac{\partial}{\partial x} & \frac{\partial}{\partial y} & \frac{\partial}{\partial z} \\ V_1 & V_2 & V_3 \end{vmatrix} = \Sigma\hat{i}\left(\frac{\partial V_3}{\partial y}-\frac{\partial V_2}{\partial z}\right)$$

Curl of a vector point function is called rotation of a vector point function.

14. **Properties of Divergence and Curl.**

(*i*) For a constant vector $\vec{a}$, div $\vec{a} = 0$, Curl $\vec{a} = \vec{0}$

(*ii*) $\text{div}(\vec{A} \pm \vec{B}) = \text{div}\vec{A} \pm \text{div}\vec{B}$

(*iii*) $\text{curl}(\vec{A} \pm \vec{B}) = \text{curl}\vec{A} \pm \text{curl}\vec{B}$

(*iv*) If $\vec{A}$ is a vector function and ϕ is a scalar function, then

$\text{div}(\phi\vec{A}) = \phi\,\text{div}\vec{A} + (\text{grad }\phi)\vec{A}$

(*v*) $\text{curl}(\phi\vec{A}) = (\text{grad }\phi)\times\vec{A} + \phi\,\text{curl}\vec{A}$

$\nabla(\vec{A}\bullet\vec{B}) = (\vec{A}\bullet\nabla)\vec{B} + (\vec{B}\bullet\nabla)\vec{A} + \vec{A}\times(\nabla\times\vec{B}) + \vec{B}\times(\nabla\times\vec{A})$

$$\nabla(\vec{A} \bullet \vec{B}) = (\vec{A} \bullet \nabla)\vec{B} + (\vec{B} \bullet \nabla)\vec{A} + \vec{A} \times (\nabla \times \vec{B}) + \vec{B} \times (\nabla \times \vec{A})$$

(*vi*) $\nabla \bullet (\vec{A} \times \vec{B}) = \vec{B} \bullet (\nabla \times \vec{A}) - \vec{A} \bullet (\nabla \times \vec{B})$

(*vii*) $\nabla \times (\vec{A} \times \vec{B}) = (\nabla \bullet \vec{B})\vec{A} - (\nabla \bullet \vec{A})\vec{B} + (\vec{B} \bullet \nabla)\vec{A} - (\vec{A} \bullet \nabla)\vec{B}$

15. (*i*) Div (grad ϕ) = $\nabla^2\phi$ where $\nabla^2 = \dfrac{\partial^2}{\partial x} + \dfrac{\partial^2}{\partial y^2} + \dfrac{\partial^2}{\partial z^2}$

(*ii*) ∇^2 is called **Laplacian Operator** and $\nabla^2\phi = 0$ is called **Laplace equation.**

(*iii*) A function which satisfies Laplace equation is called **Harmonic Function** *i.e.*, ϕ is Harmonic Function.

(*iv*) Div $\bullet$ (curl $\vec{V}$) = 0

(*v*) Curl (curl $\vec{V}$) = grad div $\vec{V} - \nabla^2\vec{V}$ or $\nabla(\nabla \bullet \vec{V}) - (\nabla \bullet \nabla)\vec{V}$

(*vi*) grad (div $\vec{V}$) = curl (curl $\vec{V}$) + $\nabla^2\vec{V}$

$$= \nabla \times (\nabla \times \vec{V}) + (\nabla \bullet \nabla)\vec{V}$$

16. **Line Integral.** Any integral which is evaluated along a curve is called a line integral.

17. **Circulation.** If $\vec{V}$ represents the velocity of a fluid particle and C is a closed curve, then the line integral $\int_C \vec{V} \bullet d\vec{r}$ is called circulation of $\vec{V}$ round the curve C. If the circulation of $\vec{V}$ round every closed curve is a region D vanishes then $\vec{V}$ is called irrotational in D.

18. **Work Done by a Force.** If $\vec{F}$ represents a force acting on a particle moving along an arc AB of a curve C, then the work done by $\vec{F}$ during the displacement from A to B is $\int_A^B \vec{F} \bullet d\vec{r}$.

19. **Surface Integral.** Any integral which is to be evaluated over a surface is called a surface integral. If $\vec{F}$ be a continuous vector point function and S be a surface then surface integrals $\iint_S \vec{F} \bullet d\vec{S}$.

20. **Volume Integral.** Any integral which is to be evaluated over a volume is called volume integral. If V is a volume bounded by a surface S then the triple integral $\iiint_V \phi dV$ or $\iiint_V \vec{F} dV$ are called Volume Integrals.

SHORT ANSWER TYPE QUESTIONS

1. If $\vec{r}$ has a constant magnitude but variable direction then $\vec{r} \bullet \dfrac{d\vec{r}}{dt} = 0$.

(P.T.U., May 2004, Dec. 2011, May 2012)

[**Hint.** Art. 7.5]

2. If $\vec{r}$ has a constant direction but a variable magnitude then $\vec{r} \times \dfrac{d\vec{r}}{dt} = \vec{0}$. [**Hint.** Art 7.6]

(P.T.U., May 2014)

3. Write tangential and normal acceleration of a particle moving in a plane curve.
4. Write radial and transverse acceleration of a particle moving in a plane curve.
5. If $\vec{A} = 5t^2\hat{i} + t\hat{j} - t^3\hat{k}$ and $\vec{B} = \sin t\hat{i} - \cos t\hat{j}$, find $\left(\vec{A}\bullet\vec{B}\right)$.
6. Prove that $\frac{d}{dt}\left[\vec{r}\times\left(\frac{d\vec{r}}{dt}\times\frac{d^2\vec{r}}{dt^2}\right)\right] = \frac{d\vec{r}}{dt}\times\left(\frac{d\vec{r}}{dt}\times\frac{d^2\vec{r}}{dt^2}\right) + \vec{r}\times\left(\frac{d\vec{r}}{dt}\times\frac{d^3\vec{r}}{dt^3}\right)$

   [**Hint.** Art 7.11]
7. If $\hat{r}$ is a unit vector in the direction of $\vec{r}$, then prove that $\hat{r}\times\frac{d\hat{r}}{dt} = \frac{1}{r^2}\vec{r}\times\frac{d\vec{r}}{dt}$.

   [**Hint.** $\vec{r} = r\hat{r} \Rightarrow \hat{r} = \frac{\vec{r}}{r}$; $\frac{d\hat{r}}{dt} = \frac{1}{r}\frac{d\vec{r}}{dt} - \frac{1}{r^2}\frac{dr}{dt}\vec{r}$

   $$\hat{r}\times\frac{d\hat{r}}{dt} = \frac{\vec{r}}{r}\times\left[\frac{1}{r}\frac{d\vec{r}}{dt} - \frac{1}{r^2}\frac{dr}{dt}\vec{r}\right] = \frac{\vec{r}}{r^2}\times\frac{d\vec{r}}{dt} - \frac{1}{r^3}\frac{dr}{dt}\vec{r}\times\vec{r} = \frac{\vec{r}}{r^2}\times\frac{d\vec{r}}{dt}\Bigg]$$
8. If $\vec{r} = \vec{a}\sin 2t + \vec{b}\cos 2t$, where $\vec{a}$ and $\vec{b}$ are constant vectors, prove that $\vec{r}\times\frac{d\vec{r}}{dt} = -2\vec{a}\times\vec{b}$.

   [**Hint.** Consult S.E. 2 Art 7.12]
9. Find the angle between the tangents to the curves $x = t$, $y = t^2$, $z = t^3$ at $t = \pm 1$

   [**Hint.** Consult S.E. 8 Art 7.12]
10. A particle with position vector $\vec{r}$ is moving in a circle with constant angular velocity ω. Show by vector method that acceleration is equal to $-\omega^2\vec{r}$.

    [**Hint.** Let any point on the circle with radius unity be $x = \cos\theta$; $y = \sin\theta$ $\therefore$ $\vec{r} = \cos\theta\,\hat{i} + \sin\theta\,\hat{j}$;

    $\omega = \frac{d\theta}{dt}$ $\therefore$ $\theta = \omega t + c$; $t = 0$, $\omega = 0$ $\therefore$ $c = 0$ $\therefore$ $\theta = \omega t$ $\therefore$ $\vec{r} = \cos\omega t\,\hat{i} + \sin\omega t\,\hat{j}$ $\therefore$ $\vec{a} = \frac{d^2\vec{r}}{dt^2} = -\omega^2\vec{r}$]
11. A particle moves along the curve $x = t^3 + 1$, $y = t^2$, $z = 2t + 5$, where t is the time. Determine its velocity and acceleration vectors at $t = 1$ in the direction $2\hat{i} + 3\hat{j} + 6\hat{k}$. ***(P.T.U., May 2015)***

    [**Hint.** S.E. 9 Art 7.12]
12. The position vector of a point at time t is given by $\vec{r} = e^t\left(\cos t\,\hat{i} + \sin t\,\hat{j}\right)$ show that the angle between the radius vector and acceleration is constant.

    [**Hint.** $\vec{a} = \frac{d^2\vec{r}}{dt^2} = e^t\left[-2\sin t\,\hat{i} + 2\cos t\hat{j}\right]$; $\cos\theta = \frac{\vec{r}\bullet\vec{a}}{|\vec{r}||\vec{a}|} = 0$ $\therefore$ $\theta = \frac{\pi}{2}$; constant]
13. (*i*) Find the unit tangent vector at any point on the curve $x = 2t^2 - 6t$, $y = t^2 + 2$, $z = 4t - 5$. Also, determine unit tangent vector at $t = 2$.

    (*ii*) Find unit tangent vector to any point on the curve $x = a\cos t$, $y = a\sin t$, $z = b\,t$. ***(P.T.U., Dec. 2006)***

    [**Hint.** Consult S.E. 7 Art 7.12]
14. Verify the formula $\frac{d}{dt}\left(\vec{a}\bullet\vec{b}\right) = \vec{a}\bullet\frac{d\vec{b}}{dt} + \frac{d\vec{a}}{dt}\bullet\vec{b}$ for $\vec{a} = 5t^2\,\hat{i} + t\hat{j} - t^3\hat{k}$; $\vec{b} = \sin t\,\hat{i} - \cos t\,\hat{j}$. ***(P.T.U., May 2005)***
15. Distinguish between scalar and vector point functions.
16. (*a*) Define gradient of a scalar point function and hence find gradient of

    (*i*) $x^3 + y^3 + 3xyz$ at $(1, -2, -1)$ (*ii*) $y^2 - 4xy$ at $(1, 2)$ ***(P.T.U., May 2008, Jan. 2009)*** (*b*) Give geometrical interpretation of gradient of a scalar point function.

(*P.T.U., May 2010, May 2012*)

17. Define a level surface.

[**Hint.** Art 7.15]

18. (*i*) Find a unit vector normal to the surface $x^3 + y^3 + 3xyz = 3$ at $(1, 2, -1)$.

(*P.T.U., May 2007, Dec. 2012*)

[**Hint.** Consult S.E. 2 Art 7.17]

(*ii*) Find a unit vector normal to the surface $z = \sqrt{x^2 + y^2}$ at $(1, 1, \sqrt{2})$. **(*P.T.U., Dec. 2013*)**

[**Hint.** Consult S.E. 3 art 7.17]

19. Find the equation of the tangent plane to the surface $x^2 + y^2 = z$ at the point $(2, -1, 5)$.

(*P.T.U., May 2003*)

[**Hint.** Consult S.E. 3 Art 7.17]

20. Find the angle between the surfaces $x^2 + y^2 + z^2 = 9$ and $z = x^2 + y^2 - 3$ at $(2, -1, 2)$.

[**Hint.** S.E. 13 Art 7.17]

21. (*a*) Define directional derivative of a function and find the directional derivatives of the function $\phi\,(x, y, z) = 4xz^3 - 3x^2yz^2$ at $(2, -1, 2)$ along z-axis **(*P.T.U., Jan. 2010*)**

(*b*) Find directional derivative of

$f\,(x, y, z) = xy^2 + yz^3$ at $(2, -1, 1)$ (*i*) in the direction of $\hat{i} + 2\hat{j} + 2\hat{k}$.

(*ii*) in the direction normal to the surface $x \log z - y^2 = -4$ at $(-1, 2, 1)$ **(*P.T.U., May 2014*)**

[**Hint.** Consult S.E. 5 Art 7.17]

22. Find the maximum value of the directional derivatives of $f = x^2 - 2y^2 + 4z^2$ at the point $(1, 1, -1)$.

(*P.T.U., May 2009*)

23. If $\vec{r} = x\hat{i} + y\hat{j} + z\hat{k}$ and $|\vec{r}| = r$, show that

(*i*) $\text{grad } r = \dfrac{\vec{r}}{r}$ (*ii*) $\text{grad}\left(\dfrac{1}{r}\right) = -\dfrac{\vec{r}}{r^3}$ **(*P.T.U., Dec. 2003*)**

(*iii*) $\text{grad } r^n = nr^{n-2}\vec{r}$ (*iv*) $\text{grad}\left(\vec{a}\cdot\vec{r}\right) = \vec{a}$, where $\vec{a}$ is a constant vector

[**Hint.** S.E. 4 Art 7.17]

(*v*) $\text{grad}\,[\,f(r)] = f'(r)\,\nabla r$ (*vi*) $\text{grad}\,(\log r) = \dfrac{\vec{r}}{r^2}$.

24. Define divergence of a vector point function. State its physical interpretation.

(*P.T.U., Dec. 2006, 2011*)

[**Hint.** Art. 7.18, 7.19]

25. Define curl of a vector point function and state its physical interpretation.

[**Hint.** Arts. 7.20, 7.21] **(*P.T.U., Dec. 2003, May 2007, 2010, 2014*)**

26. Define Laplace's equation and Harmonic function of vector calculus.

[**Hint.** Art 7.23(1)]

27. (*i*) Find div $\vec{F}$ and curl $\vec{F}$ where $\vec{F} = \text{grad}(x^3 + y^3 + z^3 - 3xyz)$.

(*P.T.U., Dec. 2010, 2014, May 2015*)

(*ii*) If $\phi(x, y, z) = x - y + 2z^2$ compute the value of $\nabla \times (\nabla\phi)$. **(*P.T.U., Dec. 2013*)**

28. (*i*) Prove that $\nabla \cdot \nabla \times \vec{F} = 0$

[**Hint.** Art 7.23 (3)]

(*ii*) Prove that curl curl $\vec{V}$ = grad div $\vec{V} - \nabla^2\vec{V}$. **(*P.T.U., Dec. 2010*)**

[**Hint.** Art 7.23(4)]

29. If $\vec{r} = x\hat{i} + y\hat{j} + z\hat{k}$, show that (*i*) div $\vec{r}$ = 3; curl $\vec{r} = \vec{0}$

(*ii*) $\text{div}\left(r^n \vec{r}\right) = (n+3)r^n$. [**Hint.** Example 4 Art 7.23] ***(P.T.U., May 2006)***

30. (*i*) Evaluate divergence of $3x^2\hat{i} + 5xy^2\hat{j} + xyz^3\hat{k}$ at (1, 2, 3).

(*ii*) Show that for the vector field $\vec{F} = 2xy\hat{i} + xe^y\hat{j} + 2z\hat{k}$, $\nabla \cdot (\nabla \times \vec{F}) = 0$ ***(P.T.U. Dec. 2013)***

31. If $\vec{A} = x^2z\,\hat{i} - 2y^3z^2\,\hat{j} + xy^2z\,\hat{k}$; find $\nabla \bullet \vec{A}$ at (1, –1, 1).

32. If $\vec{V} = \vec{w} \times \vec{r}$ prove that $\vec{w} = \frac{1}{2}$ curl $\vec{V}$ where $\vec{w}$ is a constant vector.

[**Hint.** Art 7.21]

33. If $\vec{a}$ is a constant vector and $\vec{r} = x\hat{i} + y\hat{j} + z\hat{k}$, prove that curl $(\vec{a} \times \vec{r}) = 2\vec{a}$ where $\vec{a}$ is a constant vector.

[**Hint.** Same as Q.No. 32 here $\vec{w}$ is replaced by $\vec{a}$] ***(P.T.U. May 2010)***

34. If $\vec{F} = (x + y + 1)\,\hat{i} + \hat{j} - (x + y)\hat{k}$ show that $\vec{F} \bullet \text{curl}\,\vec{F} = 0$. ***(P.T.U. Dec. 2012)***

35. If $u = x^2 + y^2 + z^2$ and $\vec{v} = x\hat{i} + y\hat{j} + z\hat{k}$, show that div $(u\vec{v}) = 5u$

[**Hint.** S.E. 3 Art 7.23]

36. If $\vec{v_1}$ and $\vec{v_2}$ be the vectors joining the fixed points (x_1, y_1, z_1), (x_2, y_2, z_2) respectively to a variable point (x, y, z) prove that

(*i*) grad $(\vec{v_1} \bullet \vec{v_2}) = \vec{v_1} \bullet \vec{v_2}$ (*ii*) div $(\vec{v_1} \times \vec{v_2}) = 0$

(*iii*) curl $(\vec{v_1} - \vec{v_2}) = 2\,(\vec{v_1} - \vec{v_2})$.

[**Hint.** (*i*) $\vec{v_1} = (x - x_1)\hat{i} + (y - y_1)\,\hat{j} + (z - z_1)\hat{k}$, $\vec{v_2} = (x - x_2)\hat{i} + (y - y_2)\,\hat{j} + (z - z_2)\hat{k}$

$$\vec{v_1} \bullet \vec{v_2} = (x - x_1)(x - x_2) + (y - y_1)(y - y_2) + (z - z_1)(z - z_2)$$

$$\text{grad}\,(\vec{v_1} \bullet \vec{v_2}) = \sum \hat{i}\frac{\partial}{\partial x}\{(x - x_1)(x - x_2) + (y - y_1)(y - y_2) + (z - z_1)(z - z_2)\}$$

$$= \sum \hat{i}\,\{(x - x_1)\bullet 1 + (x - x_2)\} = \sum \hat{i}\,(x - x_1) + \sum \hat{i}\,(x - x_2) = \vec{v_1} + \vec{v_2}$$

37. Prove that $\nabla^2\left(\frac{1}{r}\right) = 0$ where r is the distance of the variable point P from the origin. ***(P.T.U., June 2003)***

$$\left[\textbf{Hint. } r = \sqrt{x^2 + y^2 + z^2},\ \nabla^2\left(\frac{1}{r}\right) = \left(\frac{\partial^2}{\partial x^2} + \frac{\partial^2}{\partial y^2} + \frac{\partial^2}{\partial z^2}\right)\left(x^2 + y^2 + z^2\right)^{-1/2}\right.$$

$$\left. = \sum \frac{2x^2 - y^2 - z^2}{\left(x^2 + y^2 + z^2\right)^{5/2}} = 0\right]$$

38. Find the value of r satisfying $\frac{d^2\vec{r}}{dt^2} = 6t\,\hat{i} - 24t^2\,\hat{j} + 4\sin t\,\hat{k}$ given that $\vec{r} = 2\hat{i} + \hat{j}$, $\frac{d\vec{r}}{dt} = -\hat{i} - 3\hat{k}$ at $t = 0$.

[**Hint.** Integrate twice w.r.t. t and use given conditions to find the values of two arbitrary constants integration]

39. If $\vec{F} = 3xy\hat{i} - y^2\hat{j}$, evaluate $\int_C \vec{F} \bullet d\vec{r}$ where C is the arc of the parabola $y = 2x^2$ from (0, 0) to (1, 2). [**Hint.** S.E. 8 Art 7.27] ***(P.T.U., May 2004)***

40. If $\vec{r} = t\hat{i} - t^2\hat{j} + (t-1)\hat{k}$ and $\vec{s} = 2t^2\hat{i} + 6t\hat{k}$, evaluate

(*i*) $\int_0^2 \vec{r} \bullet \vec{s}\, dt$ (*ii*) $\int_0^2 (\vec{r} \times \vec{s})\, dt$.

41. (*i*) Find the work done by the force field $\vec{F} = xy\hat{i} + (y-x)\hat{j}$ over the straight line from (1, 1) to (2, 3) ***(P.T.U., Dec. 2014)***

(*ii*) Find work done by the force field $\vec{F} = -xy\hat{i} + y^2\hat{j} + z\hat{k}$ in moving a particle over the circular path $x^2 + y^2 = 4$, $z = 0$ from (2, 0, 0) to (0, 2, 0). ***(P.T.U., June 2008, May 2010)***

42. Find circulation of $\vec{F}$ round the curve C where $\vec{F} = y\hat{i} + z\hat{j} + x\hat{k}$ and C is $x^2 + y^2 = 1$, $z = 0$. ***(P.T.U., Dec. 2003)***

[**Hint.** Parametric equation of the circle is $x = \cos\theta$, $y = \sin\theta$, $z = 0$; $0 \le \theta \le 2\pi$; $\vec{F} = \sin\theta\ \hat{i} + \cos\theta\ \hat{k}$

$$d\vec{r} = -\sin\theta\hat{i} + \cos\theta\hat{j}. \text{ Circulation } \int_C \vec{F} \bullet d\vec{r} = \int_0^{2\pi} -\sin^2\theta\, d\theta = -\pi]$$

43. (*i*) Evaluate $\int_C (x^2 + yz)\, dz$ where C is the curve defined by $x = t$, $y = t^2$, $z = 3t$ for t lying in the interval $1 \le t \le 2$. ***(P.T.U., May 2008)***

(*ii*) Evaluate $\int_C x\, dx - yz\, dy + e^z\, dz$ where C is the curve defined by $x = t^3$, $y = -t$, $z = t^2$ for t lying in the interval $0 \le t \le 2$. ***(P.T.U., Dec. 2013)***

[**Hint.** Consult S.E. 5 art. 7.25]

Answers

3. $\frac{dv}{dt}; \frac{v^2}{\rho}$ **4.** $\frac{d^2r}{dt^2} - r\left(\frac{d\theta}{dt}\right)^2; 2\frac{dr}{dt}\frac{d\theta}{dt} + r\frac{d^2\theta}{dt^2}$

5. $5t^2\cos t + 11t\sin t - \cos t$ **9.** $\cos^{-1}\frac{3}{7}$ **11.** $\frac{24}{7}; \frac{18}{7}$

13. (*i*) $\frac{(2t-3)\hat{i} + t\hat{j} + 2\hat{k}}{\sqrt{5t^2 - 12t + 13}}; \frac{\hat{i} + 2\hat{j} + 2\hat{k}}{3}$ (*ii*) $\frac{(-a\sin t)\hat{i} + (a\cos t)\hat{j} + b\hat{k}}{\sqrt{a^2 + b^2}}$

16. (*a*) (*i*) $3(3\hat{i} + 3\hat{j} - 2\hat{k})$ (*ii*) $-8\hat{i}$

18. (*i*) $\frac{1}{\sqrt{14}}(-\hat{i} + 3\hat{j} + 2\hat{k})$ (*ii*) $\frac{\hat{i} + \hat{j} - \sqrt{2}\hat{k}}{\sqrt{2}}$ **19.** $4x - 2y - z = 5$

20. $\cos^{-1}\frac{8\sqrt{21}}{63}$ **21.** (*a*) 144 (*b*) (*i*) $\frac{11}{3}$ (*ii*) $\frac{15}{\sqrt{17}}$

22. $2\sqrt{21}$ **27.** (*i*) $6(x + y + z)$, $\vec{0}$ (*ii*) $\vec{0}$

30. (*i*) 80 **31.** 14

38. $\vec{r} = (t^3 - t - 2)\ \hat{i} + (1 - 2t^4)\ \hat{j} + (t - 4\sin t)\ \hat{k}$ **39.** $-\frac{7}{6}$

40. (*i*) 12 (*ii*) $-24\hat{i} - \frac{40}{3}\hat{j} + \frac{64}{5}\hat{k}$ **41.** (*i*) $\frac{25}{6}$ (*ii*) $\frac{16}{3}$ **42.** $-\pi$

43. (*i*) 163 (*ii*) $27 + e^4$

8

Application of Vector Calculus

8.1. FLUX

Let $\vec{F}(x, y, z)$ be a continuous vector point function and S be a closed smooth surface. Then Flux of $\vec{F}$ across the surface S is given by the integral $\iint_S \vec{F} \cdot \hat{n}\, dS$ (consult art (7.28).

Let $\overrightarrow{dS}$ be a vector of magnitude dS and direction that of $\hat{n}$ then $\overrightarrow{dS} = \hat{n}\, dS$. Then Flux of $\vec{F}$ can also be written as $\iint_S \vec{F} \cdot \hat{n}\, dS = \iint_S \vec{F} \cdot \overrightarrow{dS}$.

If α, β, γ be the angles which $\hat{n}$ makes with the positive direction of co-ordinate axes and if $<l, m, n>$ are the $d \cdot cs$ of the outward normal then

$$l = \cos\alpha,\ m = \cos\beta,\ n = \cos\gamma$$

$$\therefore \quad \hat{n} = (\cos\alpha)\hat{i} + (\cos\beta)\hat{j} + (\cos\gamma)\hat{k}$$

$$\therefore \quad \vec{F} \cdot \hat{n} = F_1 \cos\alpha + F_2 \cos\beta + F_3 \cos\gamma$$

$$= F_1\, l + F_2\, m + F_3\, n$$

$$\therefore \quad \iint_S \vec{F} \cdot \hat{n}\, dS = \iint (F_1 \cos\alpha + F_2 \cos\beta + F_3 \cos\gamma)\, dS$$

$$= \iint F_1\, dydz + F_2\, dzdx + F_3\, dxdy$$

where $dydz$, $dzdx$, $dxdy$ are orthogonal projections of S on the coordinate planes.

ILLUSTRATIVE EXAMPLES

Example 1. *Find the flux of the vector field* $\vec{A} = (x - 2z)\hat{i} + (x + 3y + z)\hat{j} + (5x + y)\hat{k}$ *through the upper side of the triangle ABC with vertices at the points A(1, 0, 0), B(0, 1, 0), C (0, 0, 1).*

Sol. Equation of the plane in which the triangle ABC lies is $x + y + z = 1$

Unit normal $\hat{n}$ to ABC is

$$\frac{\nabla(x + y + z - 1)}{|\nabla(x + y + z - 1)|} \qquad \text{(Art. 7.15)}$$

$$= \frac{\hat{i} + \hat{j} + \hat{k}}{\sqrt{3}}$$

$$\vec{A} \cdot \hat{n} = \left\{(x - 2z)\hat{i} + (x + 3y + z)\hat{j} + (5x + y)\hat{k}\right\} \cdot \frac{\hat{i} + \hat{j} + \hat{k}}{\sqrt{3}}$$

$$= \frac{1}{\sqrt{3}}\{x - 2z + x + 3y + z + 5x + y\} = \frac{1}{\sqrt{3}}\{7x + 4y - z\}$$

Let AOB be the projection of ABC on xy-plane then

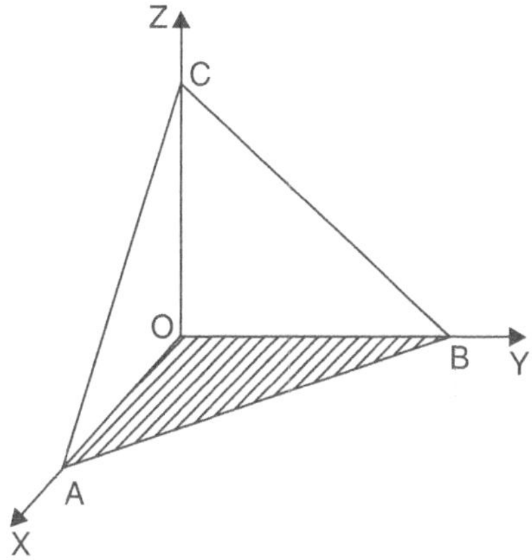

$$dS = \frac{dxdy}{|\hat{n} \cdot \hat{k}|} = \sqrt{3}\, dxdy$$

Flux across the triangle ABC $= \iint_S \vec{A} \cdot \hat{n}\, dS$

$$= \iint_{AOB} \frac{1}{\sqrt{3}}(7x + 4y - z)\sqrt{3}\, dxdy$$

As $\quad x + y + z = 1 \quad \therefore z = 1 - x - y$

y varies from 0 to $1 - x$ and x varies from 0 to 1

$$\therefore \text{ Flux} = \int_0^1 \int_0^{1-x} (7x + 4y - 1 + x + y)\, dydx$$

$$= \int_0^1 \int_0^{1-x} (8x + 5y - 1)\, dydx = \int_0^1 \left. 8xy + 5\frac{y^2}{2} - y \right|_0^{1-x} dx$$

$$= \int_0^1 \left\{8x(1 - x) + \frac{5}{2}(1 - x)^2 - (1 - x)\right\}dx = 8\left(\frac{x^2}{2} - \frac{x^3}{3}\right) + \frac{5}{2}\frac{(1 - x)^3}{-3} - x + \frac{x^2}{2}\Bigg]_0^1$$

$$= 8\left(\frac{1}{2} - \frac{1}{3}\right) - \frac{5}{6} \cdot 0 - 1 + \frac{1}{2} + \frac{5}{6} \cdot 1 = \frac{8}{6} - \frac{1}{2} + \frac{5}{6} = \frac{8 - 3 + 5}{6} = \frac{10}{6} = \frac{5}{3}.$$

Example 2. *Find the flux of the velocity $\vec{V} = y\hat{i} + 2\hat{j} + xz\hat{k}$ through the parabolic cylinder $y = x^2$; $0 \le x \le 3$, $0 \le z \le 2$.*

Sol. Flux of velocity vector $\vec{V} = y\hat{i} + 2\hat{j} + xz\hat{k}$ is $\iint_S \vec{V} \cdot \hat{n}\, ds$ where $\hat{n}$ a unit normal vector to the surface $y = x^2$ in the outward direction.

Let $\quad \phi = x^2 - y$

Normal vector $\quad \nabla\phi = \left(\hat{i}\frac{\partial}{\partial x} + \hat{j}\frac{\partial}{\partial y} + \hat{k}\frac{\partial}{\partial z}\right)(x^2 - y) = \hat{i} \cdot 2x - \hat{j} = 2x\,\hat{i} - \hat{j}$

$$\hat{n} = \frac{\nabla\phi}{|\nabla\phi|} = \frac{2x\,\hat{i} - \hat{j}}{\sqrt{4x^2 + 1}}$$

$$ds = \frac{dx\,dz}{\hat{n} \cdot \hat{j}} \qquad (dx\,dz \;\because\; \text{intervals for } x \text{ and } z \text{ are given})$$

$$= \frac{dx\,dz}{-\dfrac{1}{\sqrt{4x^2 + 1}}} = -\sqrt{4x^2 + 1} \cdot dx\,dz$$

$$\therefore \quad \text{Flux} = \int_{z=0}^{2} \int_{x=0}^{3} (y\hat{i} + 2\hat{j} + xz\hat{k}) \cdot \frac{2x\hat{i} - \hat{j}}{\sqrt{4x^2+1}} \cdot (-\sqrt{4x^2+1})\, dx\, dz$$

$$= -\int_{z=0}^{2} \int_{x=0}^{3} (2xy - 2)\, dxdz = -\int_{z=0}^{2} \int_{x=0}^{3} (2x^3 - 2)\, dxdz \quad \because \quad y = x^2$$

$$= -2 \int_{0}^{2} \left. \frac{x^4}{4} - x \right|_0^3 dz = -2 \int_0^2 \left(\frac{81}{4} - 3\right) dz = -2 \cdot \frac{69}{4} \cdot z \Big|_0^2$$

$$= -2 \cdot \frac{69}{4} \cdot 2 = -69$$

Hence flux of $\vec{V} = 69$.

8.2. SOLENOIDAL VECTOR POINT FUNCTION

Let $\vec{V}(x, y, z)$ be a continuous vector point function. If div $\vec{V} = 0$ everywhere in some region R of the space then $\vec{V}$ is called solenoidal vector. (Consult art. 7.19)

Example 3. *If* $\vec{V} = \dfrac{\vec{r}}{r^3}$ *where* $\vec{r} = x\hat{i} + y\hat{j} + z\hat{k}$, *prove that* $\vec{V}$ *is a solenoidal vector.*

(P.T.U., Dec. 2004, Dec. 2006)

Sol. $\vec{r} = x\hat{i} + y\hat{j} + z\hat{k}$

$$r = |\vec{r}\,| = \sqrt{x^2 + y^2 + z^2} \qquad \therefore \quad \vec{V} = \frac{x\hat{i} + y\hat{j} + z\hat{k}}{(x^2 + y^2 + z^2)^{3/2}}$$

then
$$\text{div } \vec{V} = \left(\hat{i}\frac{\partial}{\partial x} + \hat{j}\frac{\partial}{\partial y} + \hat{k}\frac{\partial}{\partial z}\right) \cdot \frac{x\hat{i} + y\hat{j} + z\hat{k}}{\left(x^2 + y^2 + z^2\right)^{3/2}}$$

$$= \frac{\partial}{\partial x}\left(\frac{x}{\left(x^2+y^2+z^2\right)^{3/2}}\right) + \frac{\partial}{\partial y}\left(\frac{y}{\left(x^2+y^2+z^2\right)^{3/2}}\right) + \frac{\partial}{\partial z}\left(\frac{z}{\left(x^2+y^2+z^2\right)^{3/2}}\right)$$

$$= \sum \frac{\left(x^2+y^2+z^2\right)^{3/2} \cdot 1 - x\,\frac{3}{2}\left(x^2+y^2+z^2\right)^{1/2} \cdot 2x}{\left(x^2+y^2+z^2\right)^3}$$

$$= \sum \frac{\left(x^2+y^2+z^2\right)^{1/2}\left[x^2+y^2+z^2-3x^2\right]}{\left(x^2+y^2+z^2\right)^3} = \sum \frac{y^2+z^2-2x^2}{\left(x^2+y^2+z^2\right)^{5/2}}$$

$$= \frac{y^2+z^2-2x^2+z^2+x^2-2y^2+x^2+y^2-2z^2}{\left(x^2+y^2+z^2\right)^{5/2}} = 0$$

$$\therefore \quad \vec{V} = \frac{\vec{r}}{r^3} \text{ is solenoidal vector.}$$

Example 4. *Show that the vector field given by* $\vec{v} = (-x^2 + yz)\hat{i} + (4y - z^2 x)\hat{j} + (2xz - 4z)\hat{k}$ *is solenoidal.* **(P.T.U., May 2010, May 2011)**

Sol. $\vec{V}$ will be solenoidal if div $\vec{v} = 0$

$$\text{div } \vec{v} = \left(\hat{i}\frac{\partial}{\partial x} + \hat{j}\frac{\partial}{\partial y} + \hat{k}\frac{\partial}{\partial z}\right)\cdot\left[(-x^2 + yz)\hat{i} + (4y - z^2x)\hat{j} + (2xz - 4z)\hat{k}\right]$$

$$= \frac{\partial}{\partial x}(-x^2 + yz) + \frac{\partial}{\partial y}(4y - z^2x) + \frac{\partial}{\partial z}(2xz - 4z) = -2x + 4 + 2x - 4 = 0$$

$\therefore \vec{v}$ is solenoidal vector.

Example 5. *If* $\vec{r} = x\hat{i} + y\hat{j} + z\hat{k}$ *prove that:*

(*i*) *Div* $(r^n\vec{r}) = (n + 3)r^n$ **(P.T.U., May 2006)**

(*ii*) *Find the value of n for which vector* $(r^n\vec{r})$ *is solenoidal vector. Hence deduce that* $\frac{\vec{r}}{r^3}$ *is solenoidal vecctor.* **(P.T.U., May 2006)**

Sol. Let $\vec{F} = r^n\vec{r}$

(*i*) $$\text{div}\vec{F} = \nabla\cdot\vec{F} = \left(\hat{i}\frac{\partial}{\partial x} + \hat{j}\frac{\partial}{\partial y} + \hat{k}\frac{\partial}{\partial z}\right)\cdot\left(r^n\vec{r}\right)$$

Now, $$r^n\vec{r} = \left(x^2 + y^2 + z^2\right)^{\frac{n}{2}}\left(x\hat{i} + y\hat{j} + z\hat{k}\right)$$

$$\left(\because r = |\vec{r}| = \sqrt{x^2 + y^2 + z^2} \quad \therefore r^n = \left(x^2 + y^2 + z^2\right)^{\frac{n}{2}}\right)$$

Now, $$\text{div}\vec{F} = \left(\hat{i}\frac{\partial}{\partial x} + \hat{j}\frac{\partial}{\partial y} + \hat{k}\frac{\partial}{\partial z}\right)\cdot\left[\left(x^2 + y^2 + z^2\right)^{\frac{n}{2}}\left(x\hat{i} + y\hat{j} + z\hat{k}\right)\right]$$

$$= \frac{\partial}{\partial x}\left\{\left(x^2 + y^2 + z^2\right)^{\frac{n}{2}}x\right\} + \frac{\partial}{\partial y}\left\{\left(x^2 + y^2 + z^2\right)^{\frac{n}{2}}y\right\} + \frac{\partial}{\partial z}\left\{\left(x^2 + y^2 + z^2\right)^{\frac{n}{2}}z\right\}$$

$$= \left\{\left(x^2 + y^2 + z^2\right)^{\frac{n}{2}} + \frac{n}{2}\left(x^2 + y^2 + z^2\right)^{\frac{n}{2}-1}\cdot 2x^2\right\}$$

$$+ \left\{\left(x^2 + y^2 + z^2\right)^{\frac{n}{2}} + \frac{n}{2}\left(x^2 + y^2 + z^2\right)^{\frac{n}{2}-1}\cdot 2y^2\right\}$$

$$+ \left\{\left(x^2 + y^2 + z^2\right)^{\frac{n}{2}} + \frac{n}{2}\left(x^2 + y^2 + z^2\right)^{\frac{n}{2}-1}\cdot 2z^2\right\}$$

$$= 3\left(x^2 + y^2 + z^2\right)^{\frac{n}{2}} + \frac{n}{2}\left(x^2 + y^2 + z^2\right)^{\frac{n}{2}-1}\cdot 2\left(x^2 + y^2 + z^2\right)$$

$$= 3\left(x^2 + y^2 + z^2\right)^{\frac{n}{2}} + n\left(x^2 + y^2 + z^2\right)^{\frac{n}{2}} = (3 + n)\left(x^2 + y^2 + z^2\right)^{\frac{n}{2}} = (3 + n)\, r^n$$

$$\therefore \text{div}\left(r^n\vec{r}\right) = (n + 3)r^n$$

(*ii*) Given $\vec{F} = r^n \vec{r}$ is solenoidal $\therefore$ div $\vec{F} = 0$

$$\therefore \quad (3+n)\left(x^2+y^2+z^2\right)^{\frac{n}{2}} = 0$$

$\Rightarrow \quad n + 3 = 0$ or $n = -3 \quad \therefore \quad$ for $n = -3$; $r^n\vec{r}$ is solenoidal.

Put $\quad n = -3$ in $r^n\vec{r}$ we see that div $\left(r^{-3}\vec{r}\right) = 0 \quad \therefore \quad \dfrac{\vec{r}}{r^3}$ is solenoidal vector.

8.3. IRROTATIONAL VECTOR

Let $\vec{V}(x, y, z)$ be a vector point function and if curl $\vec{V} = \vec{0}$, then $\vec{V}$ is called irrotational vector.

(*P.T.U., Dec. 2004*)

e.g., $\vec{r} = x\hat{i} + y\hat{j} + z\hat{k}$ is an irrotational vector.

Proof. $\quad$ Curl $\vec{r} = \begin{vmatrix} \hat{i} & \hat{j} & \hat{k} \\ \dfrac{\partial}{\partial x} & \dfrac{\partial}{\partial y} & \dfrac{\partial}{\partial z} \\ x & y & z \end{vmatrix} = \hat{i}(0-0) + \hat{j}(0-0) + \hat{k}(0-0) = \vec{0}$

$\therefore \quad \vec{r}$ is irrotational.

8.4. CONSERVATIVE FIELD (Irrotational Field)

If a vector point function $\vec{V}(x, y, z)$ can be expressed as gradient of a scalar point function ϕ *i.e.*, if $\vec{V} = \nabla\phi$ then V is called conservative field and ϕ is called potential function of $\vec{V}$.

Obviously if $\vec{V}$ is conservative vector field then curl $\vec{V}$ = curl grad $\phi = \nabla \times \nabla\phi = \vec{0}$
$\therefore$ $\vec{V}$ is irrotational vector and V is irrotational field.

Example 6. *Prove that the vector* $r^n\vec{r}$ *is irrotational, where* $\vec{r} = x\hat{i} + y\hat{j} + z\hat{k}$ *and* $r = |\vec{r}|$.

Sol. Let $\vec{F} = r^n\vec{r}$; $\vec{r} = x\hat{i} + y\hat{j} + z\hat{k} \quad \therefore \quad r = |r| = \sqrt{x^2+y^2+z^2}$

$\vec{F}$ will be irrotational if curl $\vec{F} = \vec{0}$

Now, $\quad$ curl $\vec{F} = \nabla \times \vec{F} = \nabla \times \left(r^n\vec{r}\right)$

$$= \left(\hat{i}\frac{\partial}{\partial x} + \hat{j}\frac{\partial}{\partial y} + \hat{k}\frac{\partial}{\partial z}\right) \times \left(x^2+y^2+z^2\right)^{\frac{n}{2}} \left(x\hat{i} + y\hat{j} + z\hat{k}\right)$$

$$= \begin{vmatrix} \hat{i} & \hat{j} & \hat{k} \\ \dfrac{\partial}{\partial x} & \dfrac{\partial}{\partial y} & \dfrac{\partial}{\partial z} \\ \left(x^2+y^2+z^2\right)^{\frac{n}{2}} x, & \left(x^2+y^2+z^2\right)^{\frac{n}{2}} y, & \left(x^2+y^2+z^2\right)^{\frac{n}{2}} z \end{vmatrix}$$

$$= \Sigma\hat{i}\left\{\frac{\partial}{\partial y}\left(x^2+y^2+z^2\right)^{\frac{n}{2}} z - \frac{\partial}{\partial z}\left(x^2+y^2+z^2\right)^{\frac{n}{2}} y\right\}$$

$$= \Sigma\hat{i}\left\{\frac{n}{2}\left(x^2+y^2+z^2\right)^{\frac{n}{2}-1} z\,.\,2y - \frac{n}{2}\left(x^2+y^2+z^2\right)^{\frac{n}{2}-1} 2zy\right\}$$

$$= \Sigma\hat{i}\left\{nyz\left(x^2+y^2+z^2\right)^{\frac{n}{2}-1} - nyz\left(x^2+y^2+z^2\right)^{\frac{n}{2}-1}\right\}$$

$$\text{curl } \vec{F} = \Sigma\hat{i}\,0 = \vec{0}$$

Hence $r^n \vec{r}$ is irrotational.

Example 7. (*i*) *Show that the vector field* $\vec{F} = 2x\left(y^2+z^3\right)\hat{i} + 2x^2y\,\hat{j} + 3x^2z^2\,\hat{k}$ *is conservative and find a scalar function u (x, y, z) such that* $\vec{F}$ *= grad u.* **(P.T.U., May 2008)**

(*ii*) *Show that* $\vec{F} = \left(y^2\cos x + z^3\right)\hat{i} + \left(2y\sin x - 4\right)\hat{j} + \left(3xz^2 + 2\right)\hat{k}$ *is conservative. Find its scalar potential.* **(P.T.U., May 2009)**

Sol. (*i*) $\vec{F} = 2x\left(y^2+z^3\right)\hat{i} + 2x^2y\,\hat{j} + 3x^2z^2\,\hat{k}$

$$\text{Curl } \vec{F} = \begin{vmatrix} \hat{i} & \hat{j} & \hat{k} \\ \frac{\partial}{\partial x} & \frac{\partial}{\partial y} & \frac{\partial}{\partial z} \\ 2x\left(y^2+z^3\right) & 2x^2y & 3x^2z^2 \end{vmatrix}$$

$$= \hat{i}\left[\frac{\partial}{\partial y}\left(3x^2z^2\right) - \frac{\partial}{\partial z}\left(2x^2y\right)\right] - \hat{j}\left[\frac{\partial}{\partial x}\left(3x^2z^2\right) - \frac{\partial}{\partial z}\left\{2x\left(y^2+z^3\right)\right\}\right]$$
$$+ \hat{k}\left[\frac{\partial}{\partial x}\left(2x^2y\right) - \frac{\partial}{\partial y}\left\{2x\left(y^2+z^3\right)\right\}\right]$$

$$= \hat{i}\,0 - \hat{j}\left(6xz^2 - 6xz^2\right) + \hat{k}\left(4xy - 4xy\right)$$

$$= \hat{i}\,0 - \hat{j}\,0 + \hat{k}\,0 = \vec{0}$$

$\therefore$ $\vec{F}$ is conservative field

Now grad $u = \vec{F}$

i.e., $$\hat{i}\frac{\partial u}{\partial x} + \hat{j}\frac{\partial u}{\partial y} + \hat{k}\frac{\partial u}{\partial z} = 2x\left(y^2+z^3\right)\hat{i} + 2x^2y\,\hat{j} + 3x^2z^2\,\hat{k}$$

$\therefore$ $$\frac{\partial u}{\partial x} = 2x\,(y^2+z^3);\ \frac{\partial u}{\partial y} = 2x^2y;\ \frac{\partial u}{\partial z} = 3x^2z^2.$$

$$du = \frac{\partial u}{\partial x}dx + \frac{\partial u}{\partial y}dy + \frac{\partial u}{\partial z}dz$$

$$= 2x\left(y^2+z^3\right)dx + \left(2x^2y\right)dy + \left(3x^2z^2\right)dz$$

$$= \left(2xy^2\,dx + 2x^2y\,dy\right) + \left(2xz^3\,dx + 3x^2z^2\,dz\right)$$

$= d\,(x^2y^2) + d\,(x^2z^3)$ [By inspection method]

$= d\,(x^2y^2 + x^2z^3)$

Integrating both sides, we get

$$u = x^2y^2 + x^2z^3 + c$$

$$\therefore \quad u = x^2(y^2 + z^3) + c.$$

(*ii*) $\vec{F} = \left(y^2 \cos x + z^3\right)\hat{i} + (2y \sin x - 4)\hat{j} + \left(3xz^2 + 2\right)\hat{k}$

$$\text{Curl } \vec{F} = \begin{vmatrix} \hat{i} & \hat{j} & \hat{k} \\ \dfrac{\partial}{\partial x} & \dfrac{\partial}{\partial y} & \dfrac{\partial}{\partial z} \\ y^2 \cos x + z^3 & 2y \sin x - 4 & 3xz^2 + 2 \end{vmatrix}$$

$$= \hat{i}\left[\frac{\partial}{\partial y}\left(3xz^2 + 2\right) - \frac{\partial}{\partial z}(2y \sin x - 4)\right] - \hat{j}\left[\frac{\partial}{\partial x}\left(3xz^2 + 2\right) - \frac{\partial}{\partial z}\left(y^2 \cos x + z^3\right)\right]$$

$$+ \hat{k}\left[\frac{\partial}{\partial x}(2y \sin x - 4) - \frac{\partial}{\partial y}\left(y^2 \cos x + z^3\right)\right]$$

$$= \hat{i}(0) - \hat{j}\left(3z^2 - 3z^2\right) + \hat{k}(2y \cos x - 2y \cos x)$$

$$= \hat{i}(0) - \hat{j}(0) + \hat{k}(0) = \vec{0}$$

$\therefore$ $\vec{F}$ is conservative field

Now $\qquad \text{grad } u = \vec{F}$

i.e.,

$$\hat{i}\frac{\partial u}{\partial x} + \hat{j}\frac{\partial u}{\partial y} + \hat{k}\frac{\partial u}{\partial z} = \left(y^2 \cos x + z^3\right)\hat{i} + (2y \sin x - 4)\hat{j} + \left(3xz^2 + 2\right)\hat{k}$$

$$\therefore \quad \frac{\partial u}{\partial x} = y^2 \cos x + z^3$$

$$\frac{\partial u}{\partial y} = 2y \sin x - 4$$

$$\frac{\partial u}{\partial z} = 3xz^2 + 2$$

$$du = \frac{\partial u}{\partial x} dx + \frac{\partial u}{\partial y} dy + \frac{\partial u}{\partial z} dz$$

$$= \left(y^2 \cos x + z^3\right)dx + (2y \sin x - 4)dy + \left(3xz^2 + 2\right)dz$$

$$= \left(y^2 \cos x\, dx + 2y \sin x\, dy\right) + \left(z^3 dx + 3xz^2 dz\right) - 4dy + 2dz$$

$$= d\left(y^2 \sin x\right) + d\left(xz^3\right) - 4dy + 2dz$$

$$= d\left(y^2 \sin x + xz^3 - 4y + 2z\right)$$

Integrating both sides, we get

$$u = y^2 \sin x + xz^3 - 4y + 2z + c$$

which is the scalar potential of $\vec{F}$

Example 8. *A fluid motion is given by* $\vec{v} = (y \sin z - \sin x)\hat{i} + (x \sin z + 2yz)\hat{j} + (xy \cos z + y^2)\hat{k}$. *Is the motion irrotational? If so find the velocity potential.*

(P.T.U., Dec. 2005, May 2010)

Sol. $\vec{v} = (y\sin z - \sin x)\hat{i} + (x\sin z + 2yz)\hat{j} + (xy\cos z + y^2)\hat{k}$

$$\text{curl } \vec{v} = \begin{vmatrix} \hat{i} & \hat{j} & \hat{k} \\ \frac{\partial}{\partial x} & \frac{\partial}{\partial y} & \frac{\partial}{\partial z} \\ y\sin z - \sin x & x\sin z + 2yz & xy\cos z + y^2 \end{vmatrix}$$

$$= \hat{i}\left\{\frac{\partial}{\partial y}\left(xy\cos z + y^2\right) - \frac{\partial}{\partial z}\left(x\sin z + 2yz\right)\right\}$$

$$+ \hat{j}\left\{\frac{\partial}{\partial z}(y\sin z - \sin x) - \frac{\partial}{\partial x}\left(xy\cos z + y^2\right)\right\} + \hat{k}\left\{\frac{\partial}{\partial x}(x\sin z + 2yz) - \frac{\partial}{\partial y}(y\sin z - \sin x)\right\}$$

$$= \hat{i}\{x\cos z + 2y - x\cos z - 2y\} + \hat{j}\{y\cos z - y\cos z\} + \hat{k}\{\sin z - \sin z\}$$

$$= \hat{i}\,0 + \hat{j}\,0 + \hat{k}\,0 = \vec{0}$$

$\because$ $\text{curl}\,\vec{v} = \vec{0}$ $\therefore$ Motion is irrotational.

Let u be the velocity potential

$\because$ Motion is irrotational

$\therefore$ $\vec{v} = \text{grad } u.$

$$(y\sin z - \sin x)\,\hat{i} + (x\sin z + 2yz)\,\hat{j} + (xy\cos z + y^2)\hat{k} = \frac{\partial u}{\partial x}\hat{i} + \frac{\partial u}{\partial y}\hat{j} + \frac{\partial u}{\partial z}\hat{k}$$

$$\therefore \quad \frac{\partial u}{\partial x} = y\sin z - \sin x, \frac{\partial u}{\partial y} = x\sin z + 2yz \text{ and } \frac{\partial u}{\partial z} = xy\cos z + y^2$$

$$du = \frac{\partial u}{\partial x}dx + \frac{\partial u}{\partial y}dy + \frac{\partial u}{\partial z}dz = (y\sin z - \sin x)\,dx + (x\sin z + 2yz)\,dy + (xy\cos z + y^2)\,dz$$

$$= (y\sin z\,dx + x\sin z\,dy + xy\cos z\,dz) + (-\sin x\,dx) + (2yz\,dy + y^2\,dz)$$

$$= d\,[xy\sin z] + d\,(\cos x) + d\,(y^2 z) \qquad \text{(by inspection)}$$

$$= d\,[xy\sin z + \cos x + y^2 z]$$

Integrate $u = xy\sin z + \cos x + y^2 z + c$ which is the required velocity potential.

TEST YOUR KNOWLEDGE

1. Find the flux across the surface of the parabolic cylinder $y^2 = 8x$ in the first octant bounded by the planes $y = 4$, $z = 6$ when the velocity vector is $\vec{V} = 2y\,\hat{i} - z\,\hat{j} + x^2\,\hat{k}$.

2. Find the flux of $\vec{A} = \hat{i} - \hat{j} + xyz\,\hat{k}$ through the circular region S bounded by cutting the sphere $x^2 + y^2 + z^2 = a^2$ with the plane $y = x$ (taking the side of S facing positive side of x-axis).

   [**Hint.** S is bounded by $x^2 + x^2 + z^2 = a^2$ (put $y = x$)

   i.e., ellipse $2x^2 + z^2 = a^2$ or $\dfrac{x^2}{\left(\frac{a}{\sqrt{2}}\right)^2} + \dfrac{z^2}{a^2} = 1$, $\hat{n} = \dfrac{\hat{i} - \hat{j}}{\sqrt{2}}$; $dS = \sqrt{2}\,dxdz$, Area of the ellipse $= \dfrac{\pi a^2}{\sqrt{2}}$]

3. Calculate the flux of water through the parabolic cylinder $y = x^2$ between the planes $x = 0$, $z = 0$; $x = 3$, $z = 2$ if the velocity vector is $\vec{A} = y\hat{i} + 2\hat{j} + xz\,\hat{k}$ m/sec. [**Hint:** Consult S.E. 2]

4. Prove that $\vec{A} = 3y^4z^2\,\hat{i} + 4x^3z^2\,\hat{j} - 3x^2y^2\,\hat{k}$ is solenoidal.

5. Prove that $\vec{A} = (2x^2 + 8xy^2z)\hat{i} + (3x^3y - 3xy)\hat{j} - (4y^2z^2 + 2x^3z)\hat{k}$ is not solenoidal but $\vec{B} = xyz^2 \vec{A}$ is solenoidal.

6. Determine the constant b such that $\vec{A} = (bx^2y + yz)\hat{i} + (xy^2 - xz^2)\hat{j} + (2xyz - 2x^2y^2)\hat{k}$ is solenoidal.

7. (*i*) Prove that the vector $\vec{F} = (x + 3y)\hat{i} + (y - 3z)\hat{j} + (x - 2z)\hat{k}$ is solenoidal. **(P.T.U., Dec. 2010)**

(*ii*) If $\vec{a}$ is a constant vector and $\vec{r} = x\hat{i} + y\hat{j} + z\hat{k}$, prove that $\dfrac{\vec{a} \times \vec{r}}{r^n}$ is solenoidal

[**Hint.** S.E. 4 art 7.23]

8. (*i*) Show that the vector field $\vec{A}$, where $\vec{A} = \left(x^2 - y^2 + x\right)\hat{i} - (2xy + y)\hat{j}$ is irrotational. Find the scalar potential u such that $\vec{A}$ = grad u. **(P.T.U., May 2010)**

(*ii*) A vector field is given by $\vec{A} = (x^2 + xy^2)\hat{i} + (y^2 + x^2y)\hat{j}$. Show that the field is conservative and find the scalar potential. **(P.T.U., Dec. 2003, May 2006)**
[**Hint.** Consult solved example 7]

(*iii*) Show that the vector field $\hat{k} = 2xyz^3\hat{i} + x^2z^3\hat{j} + 3x^2yz^2\hat{k}$ is irrotational and find the scalar potential ϕ such that $\hat{k} = \nabla\phi$.

9. Find the constants a, b, c so that $\vec{F} = (x + 2y + az)\hat{i} + (bx - 3y - z)\hat{j} + (4x + cy + 2z)\hat{k}$ is irrotational. If $\vec{F}$ = grad ϕ, show that $\phi = \dfrac{x^2}{2} - \dfrac{3y^2}{2} + z^2 + 2xy + 4xz - yz$. **(P.T.U., Dec. 2011)**

10. If $\vec{E}$ and $\vec{H}$ are irrotational prove that $\vec{E}$ $\vec{H}$ is solenoidal. **(P.T.U., May 2014, Dec. 2014)**

11. For a solenoidal vector $\vec{F}$, prove that Curl $\vec{F} = \nabla^4 \vec{F}$.

12. (*i*) A fluid motion is given by $\vec{V} = (y + z)\hat{i} + (z + x)\hat{j} + (x + y)\hat{k}$. Is this motion irrotational, if so find the velocity potential. **(P.T.U., May 2007, Dec. 2011)**

(*ii*) Test whether the vector $\vec{F} = (y^2 - z^2 + 3yz - 2x)\hat{i} + (3xz + 2xy)\hat{j} + (3xy - 2xz + 2z)\hat{k}$ is irrotational or not. **(P.T.U., May 2012)**

13. Prove that:

(*i*) Curl $\vec{V}$ is a solenoidal vector point function.
[**Hint.** art 7.23 (3)]

(*ii*) Gradient field describes an irrotational motion.
[**Hint.** art 7.23 (2)]

Answers

1. 132 **2.** $\sqrt{2}\,\pi a^2$ **3.** 69 m^3/sec

6. $b = -2$ **8.** (*i*) $\dfrac{1}{3}x^3 - xy^2 + \dfrac{1}{2}(x^2 - y^2)$ (*ii*) $\dfrac{x^3}{3} + \dfrac{y^3}{3} + \dfrac{x^2y^2}{2}$

(*iii*) $\phi = x^2yz^3$

9. $a = 4, b = 2, c = -1$ **12.** (*i*) Yes, vel. pot. $= xy + yz + zx$ (*ii*) Yes.

8.5. DIVERGENCE THEOREM OF GAUSS

(*P.T.U., June 2003, May 2007, Jan. 2010, Dec. 2010, Dec. 2012*)

(Relation between Surface and Volume Integral)

***Statement.** *If* $\vec{F}$ *is a vector point function having continuous first order partial derivatives in the region V bounded by a* **closed surface S,** *then* $\iiint_V \nabla \cdot \vec{F}\, dV = \iint_S \vec{F} \cdot \hat{n}\, dS$, *where* $\hat{n}$ *is the outward drawn unit normal vector to the surface S.*

Or

In other words, the volume integral of divergence of a vector point function $\vec{F}$ *taken over the volume V enclosed by a surface S, is equal to the surface integral of the component of* $\vec{F}$ *taken over a closed surface S.*

Proof. Let $\vec{F} = F_1 \hat{i} + F_2 \hat{j} + F_3 \hat{k}$

$$\nabla \cdot \vec{F} = \frac{\partial F_1}{\partial x} + \frac{\partial F_2}{\partial y} + \frac{\partial F_3}{\partial z}$$

Let α, β, γ be the angles which the outward drawn unit normal vector $\hat{n}$ makes with the positive direction of x, y and z-axes respectively, then $\cos\alpha$, $\cos\beta$ and $\cos\gamma$ are the direction cosines of $\hat{n}$.

$$\therefore \quad \hat{n} = \cos\alpha\,\hat{i} + \cos\beta\,\hat{j} + \cos\gamma\,\hat{k}$$

$$\therefore \quad \vec{F} \cdot \hat{n} = F_1 \cos\alpha + F_2 \cos\beta + F_3 \cos\gamma$$

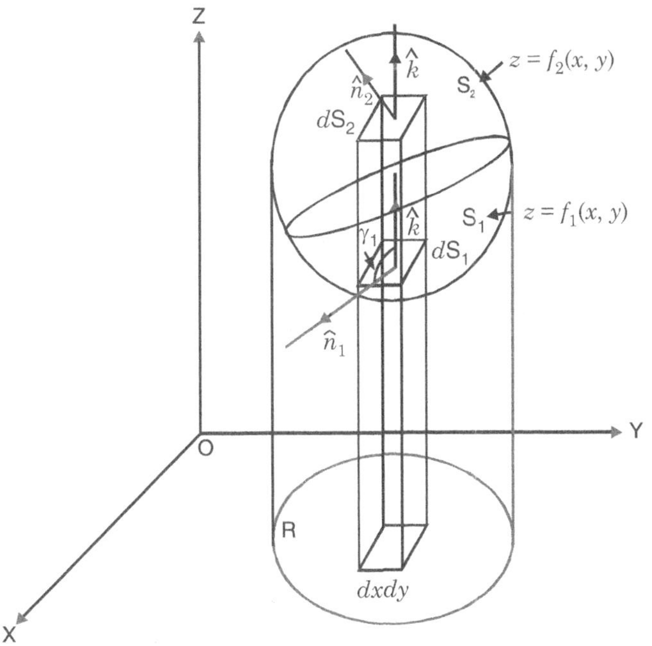

*Proof of this theorem is beyond the syllabus of semester I of P.T.U.

∴ For proving Divergence theorem *i.e.*, for proving $\underset{\text{V}}{} \vec{\text{F}}\, d\text{V} \quad \underset{\text{S}}{\vec{\text{F}}}\ \hat{n}\, d\text{S}$, we have to prove

$$\iiint_{\text{V}} \left(\frac{\partial \text{F}_1}{\partial x} + \frac{\partial \text{F}_2}{\partial y} + \frac{\partial \text{F}_3}{\partial z} \right) dx\, dy\, dz = \iint_{\text{S}} (\text{F}_1 \cos \alpha + \text{F}_2 \cos \beta + \text{F}_3 \cos \gamma)\, d\text{S} \qquad ...(1)$$

Suppose S is such a closed surface that a line parallel to the co-ordinate axes meets it in two points only. Let S_1 and S_2 denote the lower and upper portions of S with equation $z = f_1(x, y)$ and $z = f_2(x, y)$ respectively.

Let R be the projection of S on the xy-plane then we first find the value of $\iiint\limits_{\text{V}} \frac{\partial \text{F}_3}{\partial z}\, dx\, dy\, dz$

$$= \iint\limits_{\text{R}} \left\{ \int_{f_1(x,y)}^{f_2(x,y)} \frac{\partial \text{F}_3}{\partial z}\, dz \right\} dy\, dx = \iint\limits_{\text{R}} \text{F}_3 \Big|_{f_1(x,y)}^{f_2(x,y)} dy\, dx$$

$$= \iint\limits_{\text{R}} \{ \text{F}_3(x, y, f_2) - \text{F}_3(x, y, f_1) \}\, dy\, dx$$

$$= \iint\limits_{\text{R}} \text{F}_3(x, y, f_2)\, dy\, dx - \iint \text{F}_3(x, y, f_1)\, dy\, dx \qquad ...(2)$$

Now, for the upper portion S_2 of S, the normal $\hat{n}_2$ to S_2 makes an angle γ_2 with $\hat{k}$.

$$\therefore \qquad dx\, dy = \cos \gamma_2\, d\text{S}_2 = \hat{k} \cdot \hat{n}_2\, d\text{S}_2$$

For the lower portion S_1 of S, the normal $\hat{n}_1$ to S_1, makes an angle γ_1 with $\hat{k}$.

$$\therefore \qquad dx\, dy = -\cos \gamma_1\, d\text{S}_1 = -\hat{k} \cdot \hat{n}_1\, d\text{S}_1$$

$$\therefore \quad \iint\limits_{\text{R}} \text{F}_3(x, y, f_2)\, dy\, dx = \iint\limits_{\text{S}_2} \text{F}_3\, \hat{k} \cdot \hat{n}_2\, d\text{S}_2 \qquad ...(3)$$

$$\text{and} \quad \iint\limits_{\text{R}} \text{F}_3(x, y, f_1)\, dy\, dx = \iint\limits_{\text{S}_1} -\text{F}_3\, \hat{k} \cdot \hat{n}_1\, d\text{S}_1 \qquad ...(4)$$

Using (3) and (4), (2) becomes

$$\iiint\limits_{\text{V}} \frac{\partial \text{F}_3}{\partial z}\, dx\, dy\, dz = \iint\limits_{\text{S}_2} \text{F}_3\, \hat{k} \cdot \hat{n}_2\, d\text{S}_2 + \iint\limits_{\text{S}_1} \text{F}_3\, \hat{k} \cdot \hat{n}_1\, d\text{S}_1$$

$$= \iint\limits_{\text{S}} \text{F}_3\, \hat{k} \cdot \hat{n}\, d\text{S} = \iint\limits_{\text{S}} \text{F}_3 \cos \gamma\, d\text{S} \qquad ...(5)$$

Similarly, by considering the projection of S on yz and zx-planes, we have

$$\iiint\limits_{\text{V}} \frac{\partial \text{F}_2}{\partial y}\, dx\, dy\, dz = \iint\limits_{\text{S}} \text{F}_2 \cos \beta\, d\text{S} \qquad ...(6)$$

$$\iiint\limits_{\text{V}} \frac{\partial \text{F}_1}{\partial x}\, dx\, dy\, dz = \iint\limits_{\text{S}} \text{F}_1 \cos \alpha\, d\text{S} \qquad ...(7)$$

Adding (5), (6), and (7), we get

$$\iiint_V \left(\frac{\partial F_1}{\partial x} + \frac{\partial F_2}{\partial y} + \frac{\partial F_3}{\partial z}\right) dx\, dy\, dz = \iint_S (F_1 \cos\alpha + F_2 \cos\beta + F_3 \cos\gamma)\, dS$$

or

$$\iiint_V \nabla \cdot \vec{F}\, dV = \iint_S \vec{F} \cdot \hat{n}\, dS \qquad \text{(Using 1)}$$

ILLUSTRATIVE EXAMPLES

Example 1. *Use Divergence theorem to prove the following:*

(*i*) $\iint_S Curl\ \vec{F} \cdot \hat{n}\, dS = 0$ (*ii*) $\iint_S \vec{r} \cdot \hat{n}\, dS = 3V$

(*iii*) $\oint_S \nabla r^2\, \vec{dS}$, *where S is any closed surface enclosing a volume V.*

Sol. (*i*) By Divergence theorem, we have

$$\iint_S \text{Curl}\ \vec{F} \cdot \hat{n}\, dS = \iiint_V \text{div}\left(\text{Curl}\ \vec{F}\right) dV = \iiint_V \left(\nabla \cdot \nabla \times \vec{F}\right) dV$$

Since $\nabla \cdot \nabla \times \vec{F} = 0$

$$\therefore \quad \iint_S \text{Curl}\ \vec{F} \cdot \hat{n}\, dS = 0$$

(*ii*)

$$\iint_S \vec{r} \cdot \hat{n}\, dS = \iiint_V \text{div}\ \vec{r}\, dV$$

Let

$$\vec{r} = x\hat{i} + y\hat{j} + z\hat{k}$$

$$\text{div}\ \vec{r} = \nabla \cdot \vec{r} = \left(\hat{i}\frac{\partial}{\partial x} + \hat{j}\frac{\partial}{\partial y} + \hat{k}\frac{\partial}{\partial z}\right) \cdot \left(x\hat{i} + y\hat{j} + z\hat{k}\right)$$

$$= \frac{\partial x}{\partial x} + \frac{\partial y}{\partial y} + \frac{\partial z}{\partial z} = 1 + 1 + 1 = 3$$

$$\therefore \quad \iint_S \vec{r} \cdot \hat{n}\, dS = \iiint_V 3\, dV = 3V$$

(*iii*)

$$\oint_S \nabla r^2\, d\vec{S} = \iiint_V \text{div}\left(\nabla r^2\right) dV = \iiint_V \nabla^2\left(r^2\right) dV$$

Now,

$$\nabla^2 r^2 = \left(\frac{\partial^2}{\partial x^2} + \frac{\partial^2}{\partial y^2} + \frac{\partial^2}{\partial z^2}\right)\left(x^2 + y^2 + z^2\right)$$

$$= \frac{\partial^2}{\partial x^2}\left(x^2\right) + \frac{\partial^2}{\partial y^2}\left(y^2\right) + \frac{\partial^2}{\partial z^2}\left(z^2\right) = 2 + 2 + 2 = 6.$$

$$\therefore \quad \oint_S \nabla r^2\, d\vec{S} = \iiint_V 6\, dV = 6\,V$$

Example 2. *Use divergence theorem to evaluate* $\iint_S \vec{F}\cdot d\vec{S}$, *where* $\vec{F} = 4x\hat{i} - 2y^2\hat{j} + z^2\hat{k}$ *and S is the surface bounding the region* $x^2 + y^2 = 4$; $z = 0$ *and* $z = 3$. **(P.T.U., Dec. 2002)**

Sol. By Divergence theorem $\iint_S \vec{F}\, d\vec{S} = \iiint_V \text{div}\,\vec{F}\, dV$

$$\text{div}\,\vec{F} = \left(\hat{i}\frac{\partial}{\partial x} + \hat{j}\frac{\partial}{\partial y} + \hat{k}\frac{\partial}{\partial z}\right)\cdot\left(4x\hat{i} - 2y^2\hat{j} + z^2\hat{k}\right) = \frac{\partial}{\partial x}(4x) + \frac{\partial}{\partial y}\left(-2y^2\right) + \frac{\partial}{\partial z}\left(z^2\right)$$

$$= 4 - 4y + 2z$$

$$\therefore \iint_S \vec{F}\cdot d\vec{S} = \iiint (4 - 4y + 2z)\, dx\, dy\, dz$$

$$= \int_{-2}^{2}\int_{y=-\sqrt{4-x^2}}^{y=\sqrt{4-x^2}}\int_{z=0}^{3}(4 - 4y + 2z)\, dz\, dy\, dx = \int_{-2}^{2}\int_{-\sqrt{4-x^2}}^{\sqrt{4-x^2}}\left\{4z - 4yz + z^2\right\}_0^3 dy\, dx$$

$$= \int_{-2}^{2}\int_{-\sqrt{4-x^2}}^{\sqrt{4-x^2}}(12 - 12y + 9)\, dy\, dx = \int_{-2}^{2}\int_{-\sqrt{4-x^2}}^{\sqrt{4-x^2}}(21 - 12y)\, dy\, dx$$

$$= \int_{-2}^{2}\left\{21y - 6y^2\right\}_{-\sqrt{4-x^2}}^{\sqrt{4-x^2}} dx = \int_{-2}^{2}\left\{42\sqrt{4-x^2} - 6\cdot 0\right\} dx$$

$$= 42\int_{-2}^{2}\sqrt{2^2 - x^2}\, dx \qquad \left(\because y^2 \text{ is even, } \therefore\ y^2\Big|_{-a}^{a} = 0\right)$$

$$= 42\left\{\frac{x\sqrt{4-x^2}}{2} + \frac{4}{2}\sin^{-1}\frac{x}{2}\right\}_{-2}^{2} = 42\left\{0 + 2\frac{\pi}{2} - 0 - 2\left(-\frac{\pi}{2}\right)\right\}$$

$$= 42\,(2\pi) = 84\,\pi.$$

Example 3. *Evaluate* $\iint_S \vec{F}\cdot\hat{n}\, dS$ *over the entire surface of the region above the xy-plane bounded by the cone* $z^2 = x^2 + y^2$ *and the plane* $z = 4$ *if* $\vec{F} = 4xz\,\hat{i} + xyz^2\,\hat{j} + 3z\,\hat{k}$.

Sol. By Divergence theorem $\iint_S \vec{F}\cdot\hat{n}\, dS = \iiint_V \text{div}\,\vec{F}\, dV$, where V is the volume enclosed by S

Now, $\operatorname{div} \vec{F} = \left(\hat{i}\frac{\partial}{\partial x} + \hat{j}\frac{\partial}{\partial y} + \hat{k}\frac{\partial}{\partial z}\right) \cdot \left(4xz\,\hat{i} + xyz^2\,\hat{j} + 3z\,\hat{k}\right)$

$$= \frac{\partial}{\partial x}(4xz) + \frac{\partial}{\partial y}\left(xyz^2\right) + \frac{\partial}{\partial z}(3z) = 4z + xz^2 + 3$$

$$\therefore \iint_S \vec{F} \cdot \hat{n}\, dS = \int_0^4 \int_{-z}^{z} \int_{-\sqrt{z^2-y^2}}^{\sqrt{z^2-y^2}} \left(4z + xz^2 + 3\right) dx\, dy\, dz$$

$$= \int_0^4 \int_{-z}^{z} \left\{(4z+3)\,x + \frac{z^2 x^2}{2}\right\}_{-\sqrt{z^2-y^2}}^{\sqrt{z^2-y^2}} dy\, dz = \int_0^4 \int_{-z}^{z} 2\,(4z+3)\sqrt{z^2-y^2}\, dy\, dz$$

$\left(\because \quad x^2 \text{ is even} \quad \therefore \quad x^2 \Big|_{-a}^{a} = 0\right)$

$$= \int_0^4 2\,(4z+3)\left\{\frac{y\sqrt{z^2-y^2}}{2} + \frac{z^2}{2}\sin^{-1}\frac{y}{z}\right\}_{-z}^{z} dz$$

$$= \int_0^4 2\,(4z+3)\left\{0 + \frac{z^2}{2}\cdot 2\,\frac{\pi}{2}\right\} dz$$

$$= \int_0^4 (4z+3)\, z^2 \quad dz = \int_0^4 \left(4z^3\; 3z^2\right) dz = \left\{z^4 + z^3\right\}_0^4 = \pi\,\{4^4 + 4^3\} = 320\,\pi.$$

Example 4. *Evaluate using divergence theorem* $\iint_S (\vec{v} \cdot \hat{n})\, dA$, *where* $\vec{v} = x^2 z\,\hat{i} + y\hat{j} - xz^2\,\hat{k}$ *and S is the boundary of the region bounded by the paraboloid* $z = x^2 + y^2$ *and the plane* $z = 4y$.

(P.T.U., May 2008)

Sol. By Divergence theorem,

$$\iint_S (\vec{v} \cdot \hat{n})\, dA = \iiint_V \operatorname{div} \vec{v}\, dV$$

$$\operatorname{div} \vec{v} = \left(\hat{i}\frac{\partial}{\partial x} + \hat{j}\frac{\partial}{\partial y} + \hat{k}\frac{\partial}{\partial z}\right) \cdot \left(x^2 z\,\hat{i} + y\hat{j} - xz^2\,\hat{k}\right)$$

$$= 2xz + 1 - 2xz = 1$$

$$\therefore \qquad \iint_S (\vec{v} \cdot \hat{n})\, dA = \iiint_V 1 \cdot dV$$

Since the projection of S on xy-plane is $4y = x^2 + y^2$

$\therefore$ z varies from $x^2 + y^2$ to $4y$

x varies from $-\sqrt{4y - y^2}$ to $\sqrt{4y - y^2}$ and y varies from 0 to 4.

$$\therefore \qquad \iint (\vec{v} \cdot \hat{n})\, dA = \int_{y=0}^{4} \int_{x=-\sqrt{4y-y^2}}^{\sqrt{4y-y^2}} \int_{z=x^2+y^2}^{4y} dz\, dx\, dy$$

$$= \int_{y=0}^{4} \int_{x=-\sqrt{4y-y^2}}^{\sqrt{4y-y^2}} \left(4y - x^2 - y^2\right) dx\, dy$$

$$= \int_{y=0}^{4} 4yx - \frac{x^3}{3} - y^2 x \Bigg|_{-\sqrt{4y-y^2}}^{\sqrt{4y-y^2}} dy$$

$$= \int_{0}^{4} \left[\left(4y - y^2\right)\left(2\sqrt{4y-y^2}\right) - \frac{1}{3} . 2\left(4y - y^2\right)^{3/2}\right] dy$$

$$= \frac{4}{3} \int_{0}^{4} \left(4y - y^2\right)^{\frac{3}{2}} dy = \frac{4}{3} \int_{0}^{4} \left[4 - (y-2)^2\right]^{\frac{3}{2}} dy$$

Put $\qquad y - 2 = 2 \sin t \quad \therefore \quad dy = 2 \cos t\, dt$

when $\qquad y = 0,\ t = \frac{-}{2}; \quad$ when $y = 4t = \frac{}{2}$

$$\therefore \qquad \iint (\vec{v} \cdot \vec{n})\, dA = \frac{4}{3} \int_{-\pi/2}^{\pi/2} 8 \cos^3 t \cdot 2 \cos t\, dt$$

$$= \frac{64}{3} \int_{-\pi/2}^{\pi/2} \cos^4 t\, dt$$

$$= \frac{128}{3} \int_{0}^{\pi/2} \cos^4 t\, dt \qquad \text{| Integrand is an even function.}$$

$$= \frac{128}{3} \frac{3 \cdot 1}{4 \cdot 2} \frac{\pi}{2} = 8\pi$$

$$\therefore \qquad \iint_S \vec{v} \cdot \vec{n}\, dA = 8\pi$$

Example 5. *Evaluate surface integral* $\iint \vec{F} \cdot \hat{n}\, dS$, *where* $\vec{F} = \left(x^2 + y^2 + z^2\right)\left(\hat{i} + \hat{j} + \hat{k}\right)$, *S is the surface of tetrahedron* $x = 0$, $y = 0$, $z = 0$, $x + y + z = 2$ *and* $\hat{n}$ *is the unit normal in the outward direction to the closed surface S.*

Sol. By Divergence theorem $\iint \vec{F} \cdot \hat{n}\, dS = \iiint_V \text{div}\, \vec{F}\, dV$

$$\text{div } \vec{F} = \left(\hat{i}\frac{\partial}{\partial x} + \hat{j}\frac{\partial}{\partial y} + \hat{k}\frac{\partial}{\partial z}\right) \cdot \left\{\left(x^2 + y^2 + z^2\right)\left(\hat{i} + \hat{j} + \hat{k}\right)\right\}$$

$$= \frac{\partial}{\partial x}\left(x^2 + y^2 + z^2\right) + \frac{\partial}{\partial y}\left(x^2 + y^2 + z^2\right) + \frac{\partial}{\partial z}\left(x^2 + y^2 + z^2\right)$$

$$\therefore \iint_S \vec{F} \cdot \hat{n}\, dS = \iiint_V \text{div } \vec{F}\, dV = \iiint_V (2x + 2y + 2z)\, dx\, dy\, dz$$

$$= \int_{x=0}^{2} \int_{y=0}^{2-x} \int_{z=0}^{2-x-y} 2(x + y + z)\, dz\, dy\, dx = 2\int_0^2 \int_0^{2-x} \left\{(x + y)z + \frac{z^2}{2}\right\}_0^{2-x-y} dy\, dx$$

$$= 2\int_0^2 \int_0^{2-x} \left\{(x + y)(2 - x - y) + \frac{(2 - x - y)^2}{2}\right\} dy\, dx$$

$$= 2\int_0^2 \int_0^{2-x} \left\{2x + 2y - x^2 - y^2 - 2xy + \frac{(2 - x - y)^2}{2}\right\} dy\, dx$$

$$= 2\int_0^2 \left\{2xy + y^2 - x^2 y - \frac{y^3}{3} - xy^2 - \frac{(2 - x - y)^3}{6}\right\}_0^{2-x} dx$$

$$= 2\int_0^2 \left\{2x(2 - x) + (2 - x)^2 - x^2(2 - x) - \frac{(2 - x)^3}{3} - x(2 - x)^2 + \frac{(2 - x)^3}{6}\right\} dx$$

$$= 2\int_0^2 4x - 2x^2 - 2x^2 + x^3 - 4x - x^3 + 4x^2 + (2 - x)^2 - \frac{(2 - x)^3}{6}\, dx$$

$$= 2\int_0^2 \left\{(2 - x)^2 - \frac{(2 - x)^3}{6}\right\} dx = 2\left\{\frac{(2 - x)^3}{-3} + \frac{(2 - x)^4}{24}\right\}_0^2$$

$$= 2\left\{\frac{8}{3} - \frac{16}{24}\right\} = 2\left(\frac{8}{3} - \frac{2}{3}\right) = 4$$

Example 6. *Use divergence theorem to evaluate* $\iint_S (x\, dy\, dz + y\, dz\, dx + z\, dx\, dy)$, *where S is portion of the plane $x + 2y + 3z = 6$ in the first octant.*

Sol. We know that

$$\iint_S F_1\, dy\, dz + F_2\, dz\, dx + F_3\, dx\, dy = \iiint_V \left(\frac{\partial F_1}{\partial x} + \frac{\partial F_2}{\partial y} + \frac{\partial F_3}{\partial z}\right) dx\, dy\, dz$$

Compare LHS with

$$\iint_S x\, dy\, dz + y\, dz\, dx + z\, dx\, dy$$

We have $F_1 = x$, $F_2 = y$, $F_3 = z$

$$\therefore \quad \iint_S x\, dy\, dz + y\, dz\, dx + z\, dx\, dy$$

$$= \iiint_V (1 + 1 + 1)\, dx\, dy\, dz$$

$$= 3 \iiint_V dx\, dy\, dz$$

= 3 volume of the tetrahedron OABC

$$= 3\left\{\frac{1}{3}\, \Delta\text{OAB height OC}\right\} = 3\left\{\frac{1}{3}\cdot\frac{1}{2}(3)(6)\,2\right\} = 18.$$

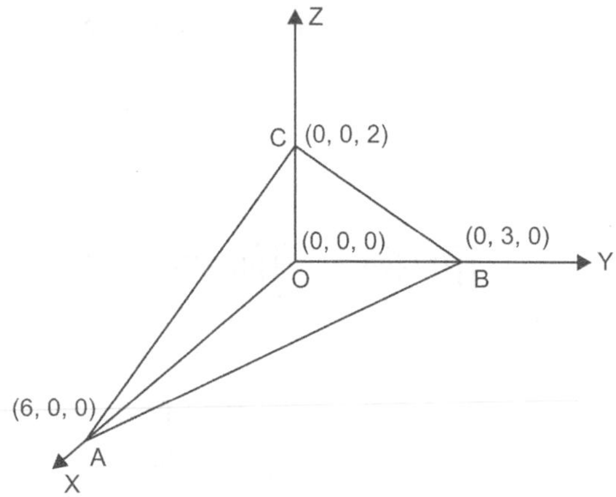

Example 7. *Verify divergence theorem for the function* $\vec{F} = x\hat{i} + y\hat{j} + z\hat{k}$ *over the surface S where S is the piecewise smooth closed surface consisting of the surfaces* S_1 *of the cone.* $z = \sqrt{x^2 + y^2}$ *for* $x^2 + y^2 \le 1$ *together with flat cap* S_2 *consisting of the disc* $x^2 + y^2 \le 1$ *in the plane* $z = 1$. **(P.T.U. Dec. 2013)**

Sol. By divergence theorem

$$\iint_S \vec{F}\cdot\hat{n}\, dS = \iiint_V \text{div}\, \vec{F}\, dV$$

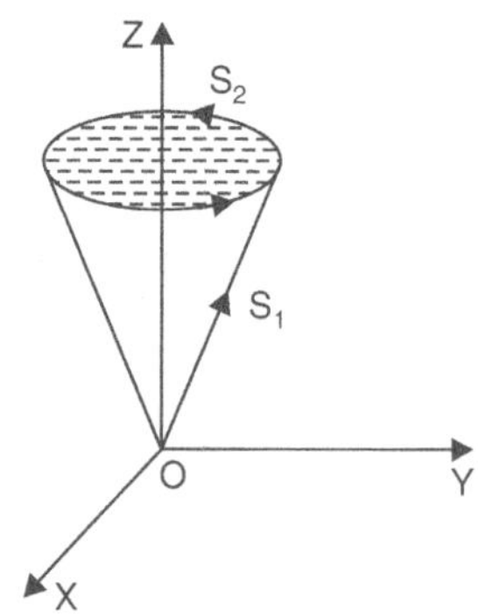

$$\vec{F} = x\hat{i} + y\hat{j} + z\hat{k}$$

$$\text{div}\, \vec{F} = \left(\hat{i}\frac{\partial}{\partial x} + \hat{j}\frac{\partial}{\partial y} + \hat{k}\frac{\partial}{\partial z}\right)\cdot(x\hat{i} + y\hat{j} + z\hat{k})$$

$$= 1 + 1 + 1 = 3$$

$$\iiint_V \text{div}\, \vec{F}\, dV = \iiint_V 3\, dV$$

where V is the volume of the cone $z^2 = x^2 + y^2$.

$z^2 = x^2 + y^2$ is a right circular cone.

$\therefore$ Its semivertical angle is $\frac{\pi}{4}$

and h, the height of the cone = radius of its plane base *i.e.*, $h = r = 1$

$$\therefore \text{ Volume of the cone} = \frac{1}{3}\pi r^2 h = \frac{1}{3}\pi\cdot 1\cdot 1 = \frac{\pi}{3}$$

$$\therefore \quad \iint_S \vec{F}\cdot\hat{n}\, dS = 3\cdot\frac{\pi}{3} = \pi \text{ (or otherwise)}$$

$$\iiint_V 3\, dV = \int_0^1 \int_{-z}^{z} \int_{-\sqrt{z^2-y^2}}^{\sqrt{z^2-y^2}} 3\, dx\, dy\, dz = 3\int_0^1 \int_{-z}^{z} 2\sqrt{z^2 - y^2}\, dy\, dz$$

$$= 6\int_0^1 \left.\frac{y\sqrt{z^2-y^2}}{2} + \frac{z^2}{2}\sin^{-1}\frac{y}{z}\right|_{-z}^{z} dz = 6\int_0^1 \frac{z^2}{2}\,.\,\pi\, dz = 3\pi\,.\left.\frac{z^3}{3}\right|_0^1 = \pi$$

For verification we have to find $\iint_S \vec{F} \cdot \hat{n}\, dS$ along the surfaces S_1 and S_2.

$$\therefore \qquad \iint_{S_1} \vec{F} \cdot \hat{n}\, dS + \iint_{S_2} \vec{F} \cdot \hat{n}\, dS \iint_{S} \vec{F} \cdot \hat{n}\, dS$$

where S_1 is the surface of the cone and S_2 is the surface of the disc.

For $\iint_{S_1} \vec{F} \cdot \hat{n}\, dS$;

$\hat{n}$ is a unit vector normal to the cone which is given by $\vec{n} = \nabla\phi$ where ϕ is given by

$$z = \sqrt{x^2 + y^2} \quad i.e., \quad \phi = z^2 - x^2 - y^2$$

$$\vec{n} = \hat{i}(-2x) + \hat{j}(-2y) + \hat{k}(2z)$$

$$\hat{n} = \frac{-2(x\hat{i} + y\hat{j} - z\hat{k})}{\sqrt{4x^2 + 4y^2 + 4z^2}} = \frac{-2(x\hat{i} + y\hat{j} - z\hat{k})}{2\sqrt{z^2 + z^2}} = -\frac{x\hat{i} + y\hat{j} - z\hat{k}}{\sqrt{2}\, z}$$

and
$$dS = \frac{dx\, dy}{\hat{n} \cdot \hat{k}} = \frac{dx\, dy}{\frac{z}{\sqrt{2}z}} = \sqrt{2}\, dx\, dy$$

$$\therefore \qquad \iint_{S_1} \vec{F} \cdot \hat{n}\, dS = \iint_{S_1} (x\hat{i} + y\hat{j} + z\hat{k}) \cdot \frac{-(x\hat{i} + y\hat{j} - z\hat{k})}{\sqrt{2}\, z} \cdot \sqrt{2}\, dx\, dy$$

$$= \iint_{S_1} \frac{-(x^2 + y^2 - z^2)}{z}\, dx\, dy = 0 \quad \because \quad x^2 + y^2 = z^2$$

Now $\iint_{S_2} \vec{F} \cdot \hat{n}\, dS$

S_2 is the surface of the circular disc $x^2 + y^2 = 1$ $i.e., z = 1$

$$\therefore \qquad \hat{n} = \hat{k} \qquad \text{(outward normal)}$$

and
$$dS = dx\, dy$$

$$\therefore \qquad \iint_{S_2} (x\hat{i} + y\hat{j} + z\hat{k}) \cdot \hat{k}\, dx\, dy$$

$$= \iint_{S_2} z\, dx\, dy = \iint_{S_2} dx\, dy \quad \because \quad z = 1$$

$= \text{area of the disc.} = \pi.$ $\qquad (\because \quad r = 1)$

$$\therefore \qquad \iint_S \vec{F} \cdot \hat{n}\, dS = 0 + \pi = \pi \quad \text{which is same as} \quad \iiint_V \text{div}\, \vec{F}\, dV$$

Hence divergence theorem is verified.

Example 8. *Evaluate* $\iint_S xz^2\, dy\, dz + \left(x^2y - z^3\right) dz\, dx + \left(2xy + y^2z\right) dx\, dy$, *where S is the surface of the hemisphere bounded by* $x^2 + y^2 + z^2 = a^2$ *and* $z = 0$.

Sol. $\iint_S F_1\, dy\, dz + F_2\, dz\, dx + F_3\, dx\, dy = \iint_S \vec{F} \cdot \hat{n}\, dS = \iiint_V \text{div}\, \vec{F}\, dV$

$$= \iiint_V \left(\frac{\partial F_1}{\partial x} + \frac{\partial F_2}{\partial y} + \frac{\partial F_3}{\partial z} \right) dx\, dy\, dz$$

Here $F_1 = xz^2;\ F_2 = x^2 y - z^3;\ F_3 = 2xy + y^2 z$

$$= \iiint_V \left(z^2 + x^2 + y^2\right) dx\, dy\, dz = \iiint_V \left(x^2 + y^2 + z^2\right) dx\, dy\, dz$$

change Cartesian co-ordinates to spherical polar co-ordinates

$$x = r \sin\theta \cos\phi,\ y = r \sin\theta \sin\phi,\ z = r\cos\theta,$$

$$dx\, dy\, dz = r^2 \sin\theta\, dr\, d\theta\, d\phi$$

$$= \int_{=0}^{2} \int_{=0}^{\bar{2}} \int_{r=0}^{a} r^2 \cdot r^2 \sin \quad dr\, d \quad d \quad = \int_0^{2} \int_0^{\bar{2}} \left\{ \frac{a^5}{5} \sin \quad d \right\} d$$

$$= \int_0^{2} \frac{a^5}{5} (-\cos \quad)_0^{\bar{2}}\, d \quad = \frac{a^5}{5} \int_0^{2} d \quad = \frac{a^5}{5} . 2 \quad = \frac{2}{5} a^5.$$

Example 9. *Evaluate* $\int_S \vec{A} \cdot d\vec{S}$, *where* $\vec{A} = x^3\,\hat{i} + y^3\,\hat{j} + z^3\,\hat{k}$ *and S is the surface of the sphere* $x^2 + y^2 + z^2 = a^2$. **(P.T.U., May 2009, Jan. 2010)**

Sol. By divergence theorem, we have $\int_S \vec{A} \cdot d\vec{S} = \iiint_V \text{div}\, \vec{A}\, dV$

$$= \iiint_V \left(\frac{\partial}{\partial x} x^3 + \frac{\partial}{\partial y} y^3 + \frac{\partial}{\partial z} z^3 \right) dx\, dy\, dz = 3 \iiint_V \left(x^2 + y^2 + z^2\right) dx\, dy\, dz$$

To integral it change it into spherical polar co-ordinates by putting

$$x = r \sin\theta \cos\phi,\ y = r\sin\theta \sin\phi,\ z = r\cos\theta,\ dx\, dy\, dz = r^2 \sin\theta\, dr\, d\theta\, d\phi$$

r varies from 0 to a

θ varies from 0 to π and ϕ varies from 0 to 2π

$$\therefore \quad \int_S \vec{A} \cdot d\vec{S} = 3 \int_{=0}^{2} \int_{=0} \int_0^{a} r^2 \cdot r^2 \sin \quad dr\, d \quad d \quad = 3 \int_0^{2} \int_0 \frac{a^5}{5} \sin \quad d \quad d$$

$$= 3 \int_0^{2\pi} \frac{a^5}{5} (-\cos\theta) \Big|_0^{\pi} d\phi = \frac{6\, a^5}{5} \cdot 2\pi = \frac{12\, a^5 \pi}{5}.$$

Example 10. *Verify divergence theorem for* $\vec{F} = \left(x^2 - yz\right)\hat{i} + \left(y^2 - zx\right)\hat{j} + \left(z^2 - xy\right)\hat{k}$ *taken over the rectangular parallelopiped* $0 \le x \le a,\ 0 \le y \le b,\ 0 \le z \le c$.

(P.T.U., May 2007, Dec. 2010, May 2011, Dec. 2012, 2015)

Sol. [For verification of divergence theorem we shall evaluate the volume and surface integrals separately and show that they are equal]

Now, $$\text{div}\, \vec{F} = \frac{\partial}{\partial x}\left(x^2 - yz\right) + \frac{\partial}{\partial y}\left(y^2 - zx\right) + \frac{\partial}{\partial z}\left(z^2 - xy\right) = 2\,(x + y + z)$$

$$\therefore \quad \iiint_V \text{div}\, \vec{F}\, dV = \int_{z=0}^{c} \int_{y=0}^{b} \int_{x=0}^{a} 2\,(x + y + z)\, dx\, dy\, dz$$

$$= 2 \int_0^c \int_0^b \left\{ \frac{x^2}{2} + (y+z)\,x \right\}_0^a dy\,dz = 2 \int_0^c \int_0^b \left\{ \frac{a^2}{2} + (y+z)\,a \right\} dy\,dz$$

$$= 2 \int_0^c \left\{ \frac{a^2}{2}\,y + a\,\frac{y^2}{2} + ayz \right\}_0^b dz = 2 \int_0^c \left\{ \frac{a^2 b}{2} + \frac{ab^2}{2} + abz \right\} dz$$

$$= 2 \left\{ \frac{a^2 b}{2}\,z + \frac{ab^2}{2}\,z + ab\,\frac{z^2}{2} \right\}_0^c = \left\{ a^2bc + ab^2c + abc^2 \right\}$$

$$= abc\,(a + b + c) \qquad \ldots(1)$$

To evaluate surface integral, the surface S consists of six surfaces given below.

	Equation	***d*S**	**Outward normal**
$S_1 \to$ OAC′B	$z = 0$	$dx\,dy$	$-\hat{k}$
$S_2 \to$ CB′PA′	$z = c$	$dx\,dy$	$\hat{k}$
$S_3 \to$ OBA′C	$x = 0$	$dy\,dz$	$-\hat{i}$
$S_4 \to$ AC′PB′	$x = a$	$dy\,dz$	$\hat{i}$
$S_5 \to$ OCB′A	$y = 0$	$dx\,dz$	$-\hat{j}$
$S_6 \to$ BA′PC′	$y = b$	$dx\,dz$	$\hat{j}$

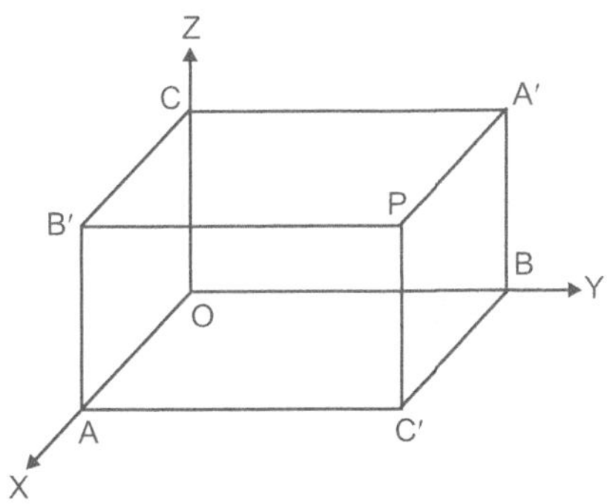

$$\iint_S \vec{F} \cdot \hat{n}\,dS = \iint_{S_1} \vec{F} \cdot \hat{n}\,dS + \iint_{S_2} \vec{F} \cdot \hat{n}\,dS + \iint_{S_3} \vec{F} \cdot \hat{n}\,dS$$

$$+ \iint_{S_4} \vec{F} \cdot \hat{n}\,dS + \iint_{S_5} \vec{F} \cdot \hat{n}\,dS + \iint_{S_6} \vec{F} \cdot \hat{n}\,dS \qquad \ldots(2)$$

$$\iint_{S_1} \vec{F} \cdot \hat{n}\,dS = \iint \left\{ \left(x^2 - yz\right)\hat{i} + \left(y^2 - zx\right)\hat{j} + \left(z^2 - xy\right)\hat{k} \right\} \cdot \left(-\hat{k}\right) dx\,dy$$

$$= \iint \left(xy - z^2\right) dx\,dy = \int_{x=0}^{a} \int_{y=0}^{b} xy\,dx\,dy \qquad (\because z = 0)$$

$$= \int_{x=0}^{a} \frac{xy^2}{2} \Bigg|_0^b dx = \frac{b^2}{2} \cdot \frac{x^2}{2} \Bigg|_0^a = \frac{a^2 b^2}{4}$$

$$\iint_{S_2} \vec{F} \cdot \hat{n}\, dS = \iint \left\{ \left(x^2 - yz\right)\hat{i} + \left(y^2 - zx\right)\hat{j} + \left(z^2 - xy\right)\hat{k} \right\} \cdot \hat{k}\, dx\, dy$$

$$= \iint \left(z^2 - xy\right) dx\, dy = \int_{x=0}^{a} \int_{y=0}^{b} \left(c^2 - xy\right) dy\, dx \qquad (\because z = c)$$

$$= \int_{x=0}^{a} \left(c^2 y - \frac{xy^2}{2}\right)\Bigg|_0^b dx = \int_0^a \left(c^2 b - \frac{xb^2}{2}\right) dx$$

$$= c^2 bx - \frac{x^2 b^2}{4}\Bigg|_0^a = abc^2 - \frac{a^2 b^2}{4}$$

$$\iint_{S_3} \vec{F} \cdot \hat{n}\, dS = \iint \left\{ \left(x^2 - yz\right)\hat{i} + \left(y^2 - zx\right)\hat{j} + \left(z^2 - xy\right)\hat{k} \right\} \cdot \left(-\hat{i}\right) dy\, dz$$

$$= \iint \left(yz - x^2\right) dy\, dz = \iint yz\, dy\, dz \qquad (\because x = 0)$$

$$= \int_{y=0}^{b} \int_{z=0}^{c} yz\, dz\, dy = \int_0^b \frac{yz^2}{2}\Bigg|_0^c dy = \int_0^b \frac{c^2 y}{2} dy = \frac{c^2 y^2}{4}\Bigg|_0^b = \frac{b^2 c^2}{4}$$

$$\iint_{S_4} \vec{F} \cdot \hat{n}\, dS = \iint \left\{ \left(x^2 - yz\right)\hat{i} + \left(y^2 - zx\right)\hat{j} + \left(z^2 - xy\right)\hat{k} \right\} \cdot \hat{i}\, dy\, dz$$

$$= \iint \left(x^2 - yz\right) dy\, dz = \int_{z=0}^{c} \int_{y=0}^{b} \left(a^2 - yz\right) dy\, dz$$

$$= \int_0^c \left\{ a^2 y - \frac{zy^2}{2} \right\}_0^b dz \qquad (\because x = a)$$

$$= \int_0^c \left(a^2 b - \frac{b^2 z}{2}\right) dz = a^2 bz - \frac{b^2}{4} z^2 \Bigg|_0^c = a^2 bc - \frac{b^2 c^2}{4}$$

$$\iint_{S_5} \vec{F} \cdot \hat{n}\, dS = \iint \left\{ \left(x^2 - yz\right)\hat{i} + \left(y^2 - zx\right)\hat{j} + \left(z^2 - xy\right)\hat{k} \right\} \cdot \left(-\hat{j}\right) dx\, dz$$

$$= \iint \left(xz - y^2\right) dx\, dz = \int_{z=0}^{c} \int_{x=0}^{a} xz\, dx\, dz \qquad (\because y = 0)$$

$$= \int_0^c \frac{zx^2}{2}\Bigg|_0^a dz = \int_0^c \frac{a^2}{2} z\, dz = \frac{a^2 z^2}{4}\Bigg|_0^c = \frac{a^2 c^2}{4}$$

$$\iint_{S_6} \vec{F} \cdot \hat{n}\, dS = \iint \left\{ \left(x^2 - yz\right)\hat{i} + \left(y^2 - zx\right)\hat{j} + \left(z^2 - xy\right)\hat{k} \right\} \cdot \hat{j}\, dx\, dz$$

$$= \iint \left(y^2 - zx\right) dx\, dz = \int_{z=0}^{c} \int_{x=0}^{a} \left(b^2 - zx\right) dx\, dz \qquad (\because y = b)$$

$$= \int_0^c \left. b^2x - \frac{zx^2}{2} \right|_0^a dz = \int_0^c b^2a - \frac{za^2}{2} dz = b^2az - \left. \frac{a^2z^2}{4} \right|_0^c$$

$$= b^2ac - \frac{a^2c^2}{4} = ab^2c - \frac{a^2c^2}{4}$$

Substituting the value of S_1, S_2, S_3, S_4, S_5, S_6 in (2)

$$\iint_S \vec{F} \cdot \hat{n}\, dS = \frac{a^2b^2}{4} + abc^2 - \frac{a^2b^2}{4} + \frac{b^2c^2}{4} + a^2bc - \frac{b^2c^2}{4} + ab^2c - \frac{a^2c^2}{4}$$

$$= abc^2 + a^2bc + ab^2c = abc\,(a + b + c) = \iiint_V \text{div}\, \vec{F}\, dV \qquad (\because \text{ of I})$$

Hence $\iint_S \vec{F} \cdot \hat{n}\, dS = \iiint_V \text{div}\, \vec{F}\, dV$

$\therefore$ Divergence theorem is verified.

Example 11. *Verify Divergence theorem for the function* $\vec{F} = 2x^2y\,\hat{i} - y^2\,\hat{j} + 4x\,z^2\,\hat{k}$ *taken over the region in first octant bounded by* $y^2 + z^2 = 9$ *and* $x = 2$.

Sol. $\text{div}\, \vec{F} = \frac{\partial}{\partial x}\left(2x^2y\right) + \frac{\partial}{\partial y}\left(-y^2\right) + \frac{\partial}{\partial z}\left(4xz^2\right) = 4xy - 2y + 8xz$

$$\iiint_V \text{div}\, \vec{F}\, dV = \iiint_V (4xy - 2y + 8xz)\, dx\, dy\, dz$$

$$= \int_{x=0}^{2} \int_{y=0}^{3} \int_{z=0}^{\sqrt{9-y^2}} (4xy - 2y + 8xz)\, dz\, dy\, dx = \int_0^2\int_0^3 \left\{4xyz - 2yz + 4xz^2\right\}_0^{\sqrt{9-y^2}} dy\, dx$$

$$= \int_0^2\int_0^3 \left[(4xy - 2y)\sqrt{9 - y^2} + 4x\left(9 - y^2\right)\right] dy\, dx$$

$$= \int_0^2\int_0^3 \left[(1 - 2x)\left\{(-2y)\left(9 - y^2\right)^{\frac{1}{2}}\right\} + 36x - 4xy^2\right] dy\, dx$$

$$= \int_0^2 \left. (1-2x)\frac{\left(9-y^2\right)^{\frac{3}{2}}}{\frac{3}{2}} + 36xy - \frac{4xy^3}{3} \right|_0^3 dx$$

$$= \int_0^2 \left\{\frac{2\,(1 - 2x)}{3} \cdot 0 + 108x - 36\,x - \frac{2\,(1 - 2x)}{3} \cdot 9^{\frac{3}{2}}\right\} dx$$

$$= \int_0^2 \{72x - 18 + 36\,x\}\, dx = \int_0^2 (108\,x - 18)\, dx = 108\,\frac{x^2}{2} - 18x \Big|_0^2$$

$$= 54 \times 4 - 36 = 180 \qquad \ldots(1)$$

$\iint_S \vec{F}\cdot\hat{n}\,dS$. Here S consists of five surfaces S_1, S_2, S_3, S_4, S_5 shown below:

	Equation	dS	Outward unit normal
$S_1 \to$ OABC	$z = 0$	$dx\,dy$	$-\hat{k}$
$S_2 \to$ OCE	$x = 0$	dS	$-\hat{i}$
$S_3 \to$ OADE	$y = 0$	$dxdz$	$-\hat{j}$
$S_4 \to$ ABD	$\begin{Bmatrix} x = 2 \\ y^2 + z^2 = 9 \end{Bmatrix}$	$dydz$	$\hat{i}$
$S_5 \to$ BDEC	$\begin{Bmatrix} y^2 + z^2 = 9 \\ x \text{ varies from 0 to 2} \end{Bmatrix}$	$\dfrac{dx\,dy}{\lvert\hat{n}\cdot\hat{k}\rvert}$	$\hat{n}$

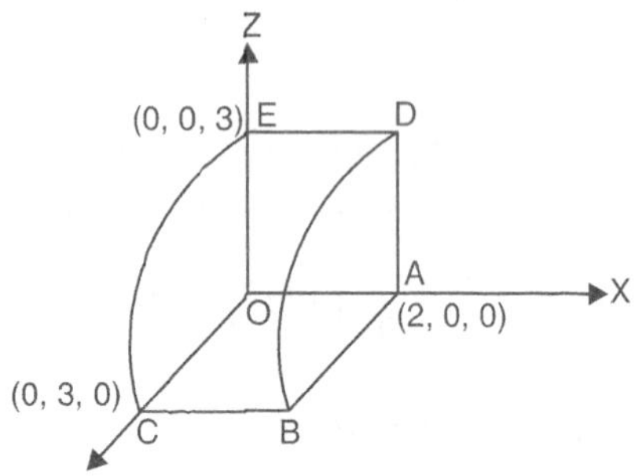

$$\iint_{S_1} \vec{F}\cdot\hat{n}\,dS = \iint\left(2x^2y\,\hat{i} - y^2\hat{j} + 4xz^2\,\hat{k}\right)\cdot\left(-\hat{k}\right)dx\,dy$$

$$= \iint -4x\,z^2\,dx\,dy = 0 \qquad (\because z = 0)$$

$$\iint_{S_2} \vec{F}\cdot\hat{n}\,dS = \iint\left(2x^2y\,\hat{i} - y^2\hat{j} + 4xz^2\,\hat{k}\right)\cdot\left(-\hat{i}\right)dS$$

$$= \iint -2x^2y\,dS = 0 \qquad (\because x = 0)$$

$$\iint_{S_3} \vec{F}\cdot\hat{n}\,dS = \iint\left(2x^2y\hat{i} - y^2\hat{j} + 4xz^2\,\hat{k}\right)\cdot\left(-\hat{j}\right)dS = \iint y^2\,dS = 0 \qquad (\because y = 0)$$

$$\iint_{S_4} \vec{F}\cdot\hat{n}\,dS = \iint\left(2x^2y\,\hat{i} - y^2\hat{j} + 4xz^2\,\hat{k}\right)\cdot\hat{i}\,dS$$

$$= \iint 2x^2y\,dy\,dz = \int_0^3\int_{y=0}^{\sqrt{9-z^2}} 8y\,dy\,dz \qquad (\because x = 2)$$

$$= \int_0^3 \left\{4y^2\right\}_0^{\sqrt{9-z^2}} dz = \int_0^3 4\left(9 - z^2\right)dz = 4\left(9z - \frac{z^3}{3}\right)_0^3 = 4\left(27 - \frac{27}{3}\right)$$

$$= 4\times 27\times\frac{2}{3} = 72$$

$$\iint_{S_5} \vec{F} \cdot \hat{n}\, dS = \iint \left(2x^2 y\, \hat{i} - y^2 \hat{j} + 4xz^2\, \hat{k}\right) \cdot \hat{n}\, dS \qquad \text{[To find } \hat{n}\text{]}$$

$$\nabla \phi = \left(\hat{i}\, \frac{\partial}{\partial x} + \hat{j}\, \frac{\partial}{\partial y} + \hat{k}\, \frac{\partial}{\partial z}\right)\left(y^2 + z^2\right) = 2y\,\hat{j} + 2z\,\hat{k} = 2\left(y\,\hat{j} + z\,\hat{k}\right)$$

$$\hat{n} = \frac{2\left(y\,\hat{j} + z\,\hat{k}\right)}{\sqrt{4y^2 + 4z^2}} = \frac{y\,\hat{j} + z\,\hat{k}}{\sqrt{y^2 + z^2}} = \frac{y\,\hat{j} + z\,\hat{k}}{3} \qquad (\because y^2 + z^2 = 9)$$

$$\iint_{S_5} \vec{F} \cdot \hat{n}\, dS = \iint \left(2x^2 y\, \hat{i} - y^2 \hat{j} + 4xz^2\, \hat{k}\right) \cdot \frac{y\,\hat{j} + z\,\hat{k}}{3}\, \frac{dx\, dy}{\left|\hat{n} \cdot \hat{k}\right|}$$

$$= \iint \frac{-y^3 + 4x\, z^3}{3}\, \frac{dx\, dy}{\frac{z}{3}} = \iint \left(-\frac{y^3}{z} + 4xz^2\right) dx\, dy$$

Eliminate z by putting $z^2 = 9 - y^2$

$$= \int_{y=0}^{3} \int_{x=0}^{2} \left\{-\frac{y^3}{\sqrt{9 - y^2}} + 4x\left(9 - y^2\right)\right\} dx\, dy$$

$$= \int_{y=0}^{3} \left.\frac{-y^3 x}{\sqrt{9 - y^2}} + 2x^2\left(9 - y^2\right)\right|_0^2 dy = \int_0^3 \frac{-2y^3}{\sqrt{9 - y^2}} + 8\left(9 - y^2\right) dy$$

$$= -2\int_0^3 \frac{y^3}{\sqrt{9 - y^2}}\, dy + \left. 72\, y - \frac{8y^3}{3}\right|_0^3$$

Put $y = 3 \sin \theta;\quad dy = 3 \cos \theta\, d\theta$

$$= -2\int_0^{\frac{\pi}{2}} \frac{27 \sin^3 \theta}{3 \cos \theta}\, 3 \cos \theta\, d\theta + 216 - 72$$

$$= -2\int_0^{\frac{\ }{2}} 27 \sin^3 \quad d \quad + 144 = -54\, \frac{2}{3} + 144 = -36 + 144 = 108$$

Substituting the values of S_1, S_2, S_3, S_4, S_5

$$\iint_S \vec{F} \cdot \hat{n}\, dS = 0 + 0 + 0 + 72 + 108 = 180$$

which is same as (1)

Hence $\displaystyle\iint_S \vec{F} \cdot \hat{n}\, dS = \iiint_V \text{div}\, \vec{F}\, dV$

Hence Theorem is verified.

Example 12. *If S is the surface of the sphere* $x^2 + y^2 + z^2 = a^2$, *prove that*

$$\iint_S \left(x^4 + y^4 + z^4\right) dS = 4pa^6.$$ **(P.T.U., May 2004)**

Sol. First of all convert the integral $\iint_S \left(x^4 + y^4 + z^4\right) dS$ into the form $\iint_S \vec{F} \cdot \hat{n}\, dS$

where $\hat{n}$ is a unit vector normal to the surface $\phi(x, y, z) = x^2 + y^2 + z^2 - a^2 = 0$

i.e.,
$$\hat{n} = \frac{\text{grad } \phi}{|\text{grad } \phi|}$$

$$\text{grad } \phi = \left(\hat{i}\frac{\partial}{\partial x} + \hat{j}\frac{\partial}{\partial y} + \hat{k}\frac{\partial}{\partial z}\right)\left(x^2 + y^2 + z^2 - a^2\right) = 2x\hat{i} + 2y\hat{j} + 2z\hat{k}$$

$$|\text{ grad } \phi \text{ }| = 2\sqrt{x^2 + y^2 + z^2} = 2a.$$

$\therefore$
$$\hat{n} = \frac{2\left(x\hat{i} + y\hat{j} + z\hat{k}\right)}{2a} = \frac{x\hat{i} + y\hat{j} + z\hat{k}}{a}$$

Let
$$\vec{F} = f_1\hat{i} + f_2\hat{j} + f_3\hat{k}$$

$$\vec{F} \cdot \hat{n} = x^4 + y^4 + z^4 \qquad \text{(given)}$$

$$\left(f_1\hat{i} + f_2\hat{j} + f_3\hat{k}\right) \cdot \frac{x\hat{i} + y\hat{j} + z\hat{k}}{a} = x^4 + y^4 + z^4$$

$$\frac{1}{a}(f_1 x + f_2 y + f_3 z) = x^4 + y^4 + z^4$$

$\therefore$
$$\frac{f_1 x}{a} = x^4, \frac{f_2 y}{a} = y^4, \frac{f_3 z}{a} = z^4$$

$\therefore$
$$f_1 = ax^3, f_2 = ay^3, f_3 = az^3$$

$$\vec{F} = a\left(x^3\hat{i} + y^3\hat{j} + z^3\hat{k}\right)$$

Now, $\iint_S \left(x^4 + y^4 + z^4\right) dS = \iint_S \vec{F} \cdot \hat{n}\, dS$

$$= \iiint_V \text{div } \vec{F}\, dV = \iiint_V \nabla \cdot a\left(x^3\hat{i} + y^3\hat{j} + z^3\hat{k}\right) dV$$

$$= a\iiint 3\left(x^2 + y^2 + z^2\right) dV = 3a\iiint a^2\, dV = 3a^3 V$$

$$= 3a^3 \frac{4}{3}\pi a^3 = 4\pi a^6.$$

TEST YOUR KNOWLEDGE

1. Use divergence theorem to show that $\oint_C r^n \vec{r} \cdot \overrightarrow{dS} = (n+3) \int_V r^n \, dV$, where $n \neq -3$.

2. If S is any closed surface enclosing a volume V and $\vec{F} = x\hat{i} + 2y\hat{j} + 3z\hat{k}$, prove that $\iint \vec{F} \cdot \hat{n} \, dS = 6V$.

3. Use divergence theorem to evaluate $\iint_S \vec{F} \cdot \hat{n} \, dS$, where $\vec{F} = x^3\hat{i} + x^2y\hat{j} + x^2z\hat{k}$ and S is the surface bounding the region $x^2 + y^2 = a^2$; $z = 0, z = b$. ***(P.T.U., Dec. 2011)***

4. Evaluate $\iint_S \left(x\hat{i} + y\hat{j} + z^2\hat{k}\right) \cdot \hat{n} \, dS = 6V$, where S is the closed surface bounded by the cone $x^2 + y^2 = z^2$ and the plane $z = 1$.

5. Evaluate $\iint_S \left(2xy\hat{i} + yz^2\hat{j} + xz\hat{k}\right) \cdot \overrightarrow{dS}$ over the surface of region bounded by $x = 0, y = 0, y = 3$, $z = 0$ and $x + 2z = 6$.

6. If $\vec{F} = ax\hat{i} + by\hat{j} + cz\hat{k}$; a, b, c are constants show that $\iint_S \vec{F} \cdot \overrightarrow{dS} = \frac{4}{3} p \, (a + b + c)$, where S is the surface of the unit sphere.

7. Evaluate $\iint_S \left(y^2z^2\hat{i} + z^2x^2\hat{j} + z^2y^2\hat{k}\right) \cdot \overrightarrow{dS}$, where S is the part of the sphere $x^2 + y^2 + z^2 = 1$ above xy-plane and bounded by this plane.

8. Use divergence theorem to evaluate $\iint_S \left(y^2z^2\hat{i} + z^2x^2\hat{j} + z^2y^2\hat{k}\right) \cdot \overrightarrow{dS}$, where S is the upper part of the sphere $x^2 + y^2 + z^2 = 9$ above xy-plane.

9. Evaluate $\iint_S \left(z^2 - x\right) dy \, dz - xy \, dx \, dz + 3z \, dx \, dy$, where S is the surface of closed region bounded by $z = 4 - y^2$ and planes $x = 0, x = 3, z = 0$.

10. Find $\iint_S \vec{F} \cdot \hat{n} \, ds$ where $\vec{F} = (2x + 3z)\hat{i} - (xz + y)\hat{j} + (y^2 + 2z)\hat{k}$ and S is the surface of the sphere having centre at $(3, -1, 2)$ and radius 3.

11. Use divergence theorem to compute $\int \vec{F} \cdot \hat{N} \, ds$ where $\vec{F} = x\hat{i} + y\hat{j} + z\hat{k}$ and S is the piecewise smooth closed surface consisting of the surfaces, of the cone $z = \sqrt{x^2 + y^2}$ for $x^2 + y^2 \leq 1$ together with flat cap S_2 consisting of the disc $x^2 + y^2 \leq 1$ in the plane $z = 1$. [**Hint:** Consult S.E. 7]

12. Evaluate $\iint_S (x\hat{i} + y\hat{j} + z^2\hat{k}) \cdot \hat{n} \, ds$ where S is the closed surface bounded by the cone $x^2 + y^2 = z^2$ and the plane $z = 1$.

13. Verify divergence theorem for the following vector point functions

(*i*) $\vec{F} = x^2\hat{i} + + z\hat{j} + + yz\hat{k}$ taken over the cube bounded by $x = 0, x = 1, y = 0, y = 1, z = 0, z = 1$. **(*P.T.U. May 2015*)**

(*ii*) $\vec{F} = \left(x^3 - yz\right)\hat{i} - 2x^2y\hat{j} + 2\hat{k}$ taken over the cube bounded by $x = 0, x = a, y = 0, y = a, z = 0, z = a$.

(*iii*) $\vec{F} = y\hat{i} + x\hat{j} + z^2\hat{k}$ over the cylindrical region bounded by $x^2 + y^2 = a^2$, $z = 0$ and $z = h$.

(*iv*) $\vec{F} = 2x^2y\hat{i} - y^2\hat{j} + 4xz^2\hat{k}$ taken over the region in the first octant bounded by $y^2 + z^2 = 9$ and $x = 2$.

(*v*) $\vec{F} = \left(x + y^2\right)\hat{i} - 2x\hat{j} + 2yz\hat{k}$ taken over the tetrahedron bounded by co-ordinate planes and the plane $2x + y + 2z = 6$.

(*vi*) $\vec{F} = 4xz\hat{i} - xy^2z\hat{j} + 3z\hat{k}$ taken over the region above XOY plane bounded by the cone $z^2 = x^2 + y^2$ and the plane $z = 4$.

Answers

3. $\frac{5}{4}a^4b$ **4.** $\frac{7}{6}$ **5.** $\frac{351}{2}$

7. $\frac{}{12}$ **8.** $\frac{243}{4}$ **9.** 16

10. 108π **11.** π **12.** $\frac{7\pi}{6}$

13. (*i*) $\frac{3}{2}$ (*ii*) $\frac{a^5}{3}$ (*iii*) πa^2h^2 (*iv*) 180 (*v*) 81 (*vi*) 320π.

8.6. GREEN'S THEOREM IN THE PLANE

(*P.T.U., May 2007, Jan. 2009, May 2011*)

***Statement.** *If M (x, y) and N (x, y) be continuous functions of x, y having continuous partial derivative* $\frac{\partial M}{\partial y}$ *and* $\frac{\partial N}{\partial x}$ *in a region R of the xy-plane bounded by a closed curve C, then*

$$\oint_C (M\,dx + N)\,dy = \iint_R \left(\frac{\partial N}{\partial x} - \frac{\partial M}{\partial y}\right) dx\,dy$$

where C is transversed in anticlockwise direction.

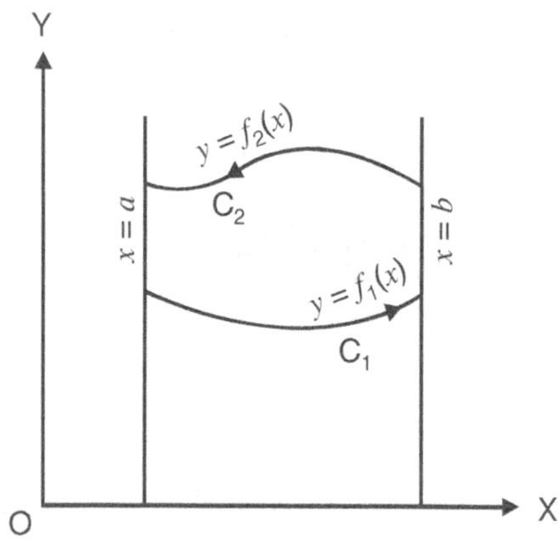

Proof. Consider the region R bounded by a single closed curve C which is cut by any line parallel to either axis at the most in two points. Let R be bounded by $x = a$, $y = f_1(x)$, $x = b$, $y = f_2(x)$ where $f_2(x) \geq f_1(x)$ so that C is divided into two curves C_1 and C_2.

Now,
$$\iint_R \frac{\partial M}{\partial y}\,dx\,dy = \int_{x=a}^{b}\left\{\int_{f_1(x)}^{f_2(x)} \frac{\partial M}{\partial y}\,dy\right\} dx$$

$$= \int_a^b \{M(x, y)\}_{f_1(x)}^{f_2(x)}\,dx = \int_a^b \{M(x, f_2(x)) - M(x, f_1(x))\}\,dx$$

* **Proof of the theorem is beyond the syllabus of semester-I (P.T.U.).**

$$= \int_a^b M(x, f_2)\, dx - \int_a^b M(x, f_1)\, dx = -\int_b^a M(x, f_2)\, dx - \int_a^b M(x, f_1)\, dx$$

$$= -\left[\int_a^b M(x, f_1)\, dx + \int_b^a M(x, f_2)\, dx\right] = -\oint_C M\, dx$$

or
$$\oint_C M\, dx = -\iint_R \frac{\partial M}{\partial y}\, dx\, dy \qquad \ldots(1)$$

Similarly, we can show that $\oint_C N\, dy = \iint_R \frac{\partial N}{\partial x}\, dx\, dy \qquad \ldots(2)$

Adding (1) and (2), we get

$$\oint_C M\, dx + N\, dy = \iint_R \left(\frac{\partial N}{\partial x} - \frac{\partial M}{\partial y}\right) dx\, dy$$

Note 1. This theorem is useful for changing a line integral around a closed curve C into a double integral over the region R enclosed by C.

Note 2. Green's theorem in vector form $\int_C \vec{F} \cdot d\vec{r} = \iint_R (\nabla \times \vec{F}) \cdot \hat{k}\, dR$, where $\vec{F} = f_1 \hat{i} + f_2 \hat{j}$, $\vec{r} = x\hat{i} + y\hat{j}$; $\hat{k}$ is a unit vector along z-axis and $dR = dx\, dy$.

ILLUSTRATIVE EXAMPLES

Example 1. *A vector field $\vec{F}$ is given by $\vec{F} = \sin y \hat{i} + x(1 + \cos y)\hat{j}$. Using Green's theorem evaluate the line integral $\int_C \vec{F} \cdot d\vec{r}$, where C is the circular path given by $x^2 + y^2 = a^2$.*

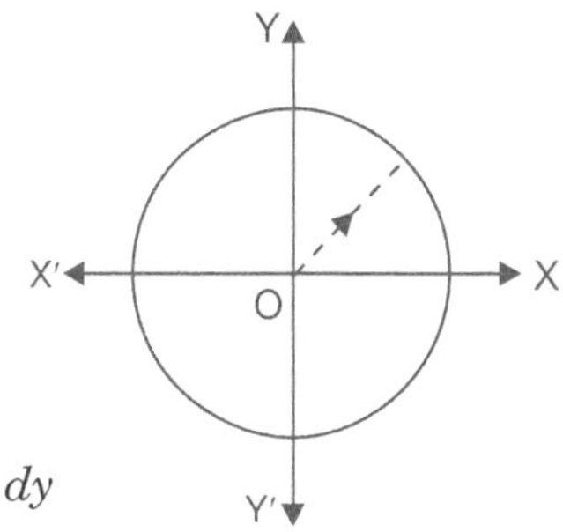

Sol.
$$\vec{F} = \sin y \hat{i} + x(1 + \cos y)\hat{j}$$
$$d\vec{r} = dx\hat{i} + dy\hat{j}$$
$$\vec{F} \cdot d\vec{r} = \sin y\, dx + x(1 + \cos y)\, dy$$

Now,
$$\oint_C \vec{F}\, d\vec{r} = \oint_C \sin y\, dx + x(1 + \cos y)\, dy$$

Apply Green's theorem $\oint_C M\, dx + N\, dy = \iint_R \left(\frac{\partial N}{\partial x} - \frac{\partial M}{\partial y}\right) dx\, dy$

Here, $M = \sin y \qquad N = x(1 + \cos y)$

$$\frac{\partial M}{\partial y} = \cos y \qquad \frac{\partial N}{\partial x} = 1 + \cos y$$

$$\therefore \quad \oint_C \sin y\, dx + x(1 + \cos y)\, dy = \iint_R (1 + \cos y - \cos y)\, dx\, dy = \iint_R dx\, dy \qquad \ldots(1)$$

(where region R is bounded by $x^2 + y^2 = a^2$)

$\therefore$ y varies from $-\sqrt{a^2-x^2}$ to $\sqrt{a^2-x^2}$ and x varies from $-a$ to a.

$$\therefore \oint_C \sin y\,dx + x\,(1+\cos y)\,dy = \int_{-a}^{a}\int_{-\sqrt{a^2-x^2}}^{\sqrt{a^2-x^2}} dy\,dx$$

$$= \int_{-a}^{a}\left(2\sqrt{a^2-x^2}\right)dx = 2\left[\frac{x\sqrt{a^2-x^2}}{2}+\frac{a^2}{2}\sin^{-1}\frac{x}{a}\right]_{-a}^{a} = 2\cdot\frac{a^2}{2}\cdot 2\cdot\frac{\pi}{2} = \pi a^2$$

Or

Second Method to evaluate

$$\int_C \sin y\,dx + x\,(1+\cos y)\,dy = \iint_R dx\,dy$$

= area of the region bounded by the circle $x^2+y^2=a^2$

$= \pi a^2$

Example 2. *Use Green's theorem to evaluate* $\int_C \left(x^2+xy\right)dx + \left(x^2+y^2\right)dy$, *where C is the square formed by the lines* $x = \pm 1,\ y = \pm 1$.

Sol. By Green's theorem $\oint_C \mathrm{M}\,dx + \mathrm{N}\,dy = \iint_R\left(\frac{\partial \mathrm{M}}{\partial y}-\frac{\partial \mathrm{N}}{\partial x}\right)dx\,dy$

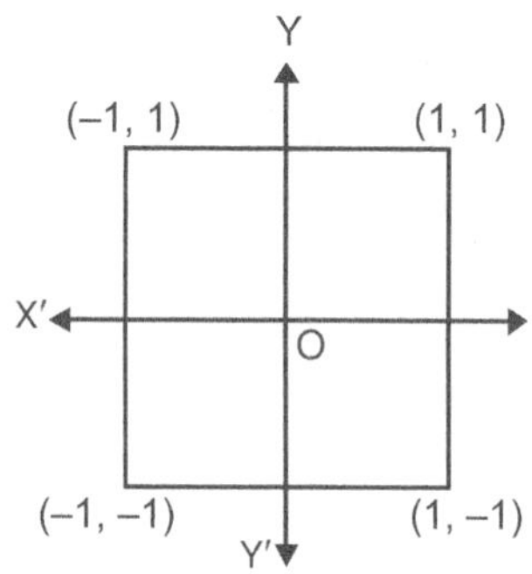

Here, $\oint_C \mathrm{M}dx + \mathrm{N}dy = \int_C \left(x^2+xy\right)dx + \left(x^2+y^2\right)dy$

$$\mathrm{M} = x^2+xy,\quad \mathrm{N} = x^2+y^2$$

$$\frac{\partial \mathrm{M}}{\partial y} = x,\quad \frac{\partial \mathrm{N}}{\partial x} = 2x$$

$$\therefore \int_C \left(x^2+xy\right)dx + \left(x^2+y^2\right)dy$$

$$= \iint_R (2x-x)\,dx\,dy = \iint_R x\,dxdy$$

where R is the region bounded by the square.

$$= \int_{y=-1}^{1}\int_{x=-1}^{1} x\,dx\,dy = \int_{-1}^{1}\frac{x^2}{2}\bigg|_{-1}^{1} dy = \int_{-1}^{1} 0\,.\,dy = 0$$

Hence, $\int_C \left(x^2+xy\right)dx + \left(x^2+y^2\right)dy = 0.$

Example 3. *Using Green's theorem evaluate* $\oint_C (y - \sin x)\, dx + \cos x\, dy$, *where C is the plane triangle enclosed by the lines* $x = \frac{\pi}{2}, y = 0, y = \frac{2}{\pi} x.$

(P.T.U., Dec. 2005, 2011, 2015, May 2015)

Sol. By Green's theorem $\oint_C M dx + N dy = \iint_R \left(\frac{\partial N}{\partial x} - \frac{\partial M}{\partial y} \right) dx\, dy$

Compare $\oint_C M dx + N dy$ with $\oint_C (y - \sin x) dx + \cos x\, dy$

We get $M = y - \sin x, \quad N = \cos x$

$$\frac{\partial M}{\partial y} = 1, \quad \frac{\partial N}{\partial x} = - \sin x$$

$$\therefore \quad \oint_C (y - \sin x) dx + \cos x\, dy = \iint_R (- \sin x - 1) dx\, dy,$$

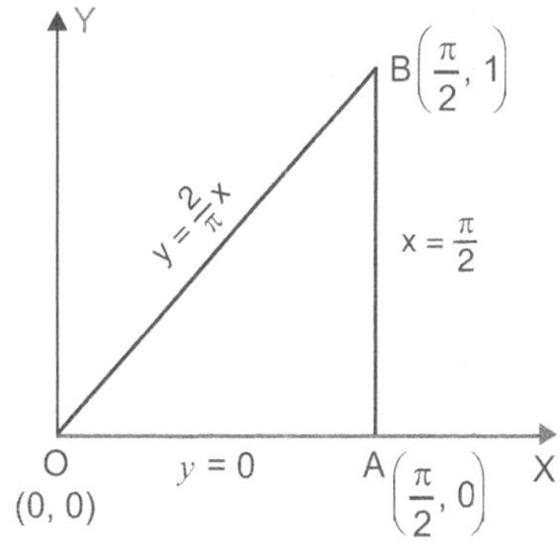

where R is the region enclosed by the triangle OAB (shown in figure)

$$= \int_0^{\pi/2} \int_0^{\frac{2x}{\pi}} (-\sin x - 1) dy\, dx$$

$$= \int_0^{\pi/2} (-\sin x - 1) y \Big|_0^{\frac{2x}{\pi}} dx = \int_0^{\pi/2} -(\sin x + 1) \frac{2}{\pi} x\, dx$$

$$= -\frac{2}{\pi} \int_0^{\pi/2} (x \sin x + x) dx$$

(Integrate by parts)

$$= -\frac{2}{\pi} \left\{ x(-\cos x) \Big|_0^{\pi/2} - \int_0^{\pi/2} 1(-\cos x) dx + \frac{x^2}{2} \Big|_0^{\pi/2} \right\}$$

$$= -\frac{2}{\pi} \left\{ 0 - 0 + \sin x \Big|_0^{\pi/2} + \frac{\pi^2}{8} \right\}$$

$$= -\frac{2}{\pi} \left[1 + \frac{\pi^2}{8} \right] = -\left[\frac{2}{\pi} + \frac{\pi}{4} \right].$$

Example 4. *Use Green's theorem to evaluate* $\oint_C (2x^2 - y^2) dx + (x^2 + y^2) dy$, *where C is the boundary in the xy-plane of the area enclosed by the x-axis and the semi-circle* $x^2 + y^2 = 1$ *in the upper half xy-plane.*

(P.T.U., Jan. 2009)

Sol. By Green's theorem

$$\oint_C M\, dx + N\, dy = \iint_R \left(\frac{\partial N}{\partial x} - \frac{\partial M}{\partial y} \right) dx\, dy$$

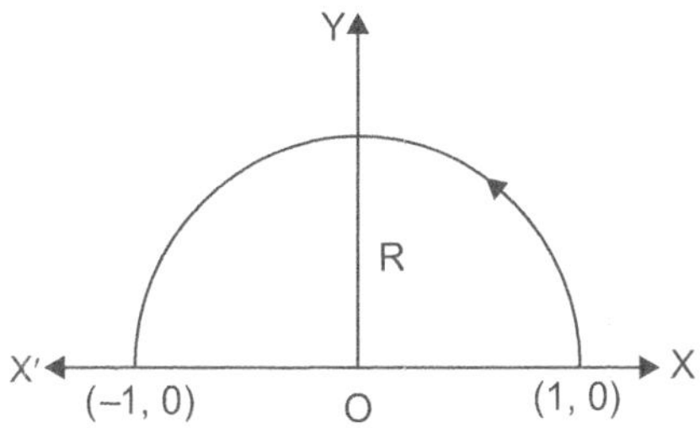

Here, $M = 2x^2 - y^2, \quad N = x^2 + y^2$

$$\frac{\partial M}{\partial y} = -2y, \quad \frac{\partial N}{\partial x} = 2x$$

$$\int_C \left(2x^2 - y^2\right) dx + \left(x^2 + y^2\right) dy = \iint_R (2x + 2y)\, dx\, dy = 2 \int_{-1}^{1} \left\{ \int_0^{\sqrt{1-x^2}} (x + y)\, dy \right\} dx$$

$$= 2 \int_{-1}^{1} \left(xy + \frac{y^2}{2} \right)_0^{\sqrt{1-x^2}} dx = 2 \int_{-1}^{1} \left(x\sqrt{1-x^2} + \frac{1-x^2}{2} \right) dx = \int_{-1}^{1} (2x)\left(1-x^2\right)^{\frac{1}{2}} + x - \frac{x^3}{3}$$

$$= -\frac{2}{3}\left(1-x^2\right)^{\frac{3}{2}} + x - \frac{x^3}{3} \Bigg|_{-1}^{1} = 0 + 1 - \frac{1}{3} + 1 - \frac{1}{3} = 2\left(1 - \frac{1}{3}\right) = \frac{4}{3}.$$

Example 5. *Apply Green's theorem to prove that the area enclosed by a plane curve is* $\frac{1}{2}\int x\,dy - y\,dx$. *Hence find the area of the ellipse whose semi major and minor axes are of lengths a and b.* **(P.T.U., May 2002, Dec. 2003)**

Sol. We know by Green's theorem that $\oint_C M\,dx + N\,dy = \iint_R \left(\frac{\partial N}{\partial x} - \frac{\partial M}{\partial y} \right) dx\,dy$, where R is the plane region bounded by a simple closed curve C.

Compare $\int x\,dy - y\,dx$ with $\int M\,dx + N\,dy$, we have

$$M = -y, \quad N = x$$

$$\therefore \qquad \frac{\partial M}{\partial y} = -1, \quad \frac{\partial N}{\partial x} = 1$$

$$\therefore \qquad \oint_C x\,dy - y\,dx = \iint [1 - (-1)]\, dx\,dy = 2\iint dx\,dy$$

$= 2A$, where A is the area of the region enclosed by C.

$$\therefore \qquad A = \frac{1}{2}\oint_C x\,dy - y\,dx$$

Hence, $\frac{1}{2}\oint_C x\,dy - y\,dx$ is the area enclosed by C.

We know that for an ellipse with semi major axis a and semi minor axis b.

$$x = a\cos\theta,\ y = b\sin\theta$$

$$\therefore \quad \text{Area of the ellipse} = \frac{1}{2}\oint_C x\,dy - y\,dx = \frac{1}{2}\int_0^{2\pi} \left(x\frac{dy}{d\theta} - y\frac{dx}{d\theta} \right) d\theta$$

$$= \frac{1}{2}\int_0^{2\pi} \left[(a\cos\theta)(b\cos\theta) - (b\sin\theta)(-a\sin\theta) \right] d\theta$$

$$= \frac{1}{2}ab\int_0^{2\pi} \left(\cos^2\theta + \sin^2\theta \right) d\theta = \frac{1}{2}ab\int_0^{2\pi} 1\, d\theta$$

$$= \frac{1}{2}ab \cdot \theta \Big|_0^{2\pi} = \pi ab.$$

Example 6. *Verify Green's theorem for* $\int_C (xy + y^2)\,dx + x^2 dy$, *where C is the boundary of the closed region bounded by y = x and y = x^2.* **(P.T.U., May 2005)**

Note. For verification we first evaluate the given integral by Greens theorem and then by taking line integral along *C*.

Sol. Evaluation by Green's theorem,

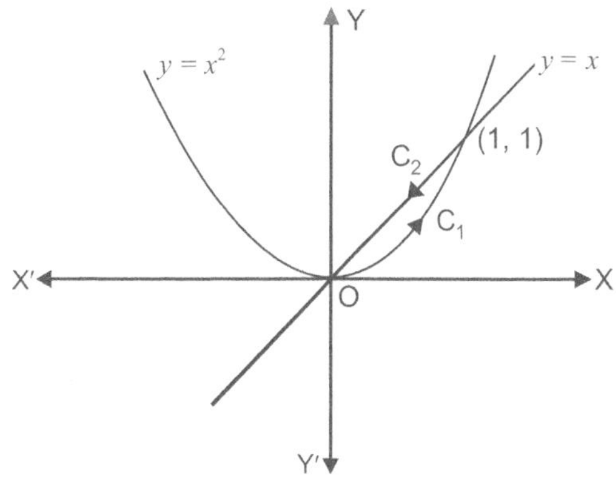

$$\int_C (xy + y^2)\,dx + x^2 dy \text{ with } \int_C \text{M}dx + \text{N}dy$$

Here $\quad \text{M} = xy + y^2, \quad \text{N} = x^2$

$$\frac{\partial \text{M}}{\partial y} = x + 2y, \quad \frac{\partial \text{N}}{\partial x} = 2x$$

By Green's theorem

$$\int_C \text{M}dx + \text{N}dy = \iint_R \left(\frac{\partial \text{N}}{\partial x} - \frac{\partial \text{M}}{\partial y}\right) dxdy$$

$$\therefore \quad \int_C (xy + y^2)\,dx + x^2 dy = \int_{x=0}^{1} \int_{y=x^2}^{x} (2x - x - 2y)\,dy\,dx$$

$$= \int_0^1 \left(xy - y^2\right)\Big|_{x^2}^{x} dx = \int_0^1 (-x^3 + x^4)\,dx$$

$$= -\frac{x^4}{4} + \frac{x^5}{5}\Big|_0^1 = \frac{-1}{4} + \frac{1}{5} = \frac{-1}{20} \qquad \ldots(1)$$

(Evaluation by line integral)

$$\int_C (xy + y^2)\,dx + x^2 dy = \int_{C_1} (xy + y^2)\,dx + x^2 dy + \int_{C_2} (xy + y^2)\,dx + x^2 dy$$

Along C_1; $\quad y = x^2 \quad \therefore \quad dy = 2x\,dx$ and $0 \le x \le 1$

Along C_2; $\quad y = x \quad \therefore \quad dy = dx\ ;\ 1 \le x \le 0$

$$= \int_0^1 \left(x^3 + x^4\right) dx + x^2 \cdot 2xdx + \int_1^0 \left(x^2 + x^2\right) dx + x^2\,dx$$

$$= \int_0^1 \left(3x^3 + x^4\right) dx + \int_1^0 3x^2 dx$$

$$= \frac{3x^4}{4} + \frac{x^5}{5}\Big|_0^1 + x^3\Big|_1^0$$

$$= \frac{3}{4} + \frac{1}{5} - 1 = \frac{1}{5} - \frac{1}{4} = \frac{-1}{20} \text{ which is same as (1)}$$

Hence Green's theorem is verified.

Example 7. *Verify Green's theorem in the plane for* $\oint_C (3x^2 - 8y^2)\, dx + (4y - 6xy)\, dy,$ *where C is the boundary of the region defined by*

(*i*) $y = \sqrt{x},\quad y = x^2$

(*ii*) $x = 0,\ y = 0,\ x + y = 1.$

(P.T.U., May 2014)

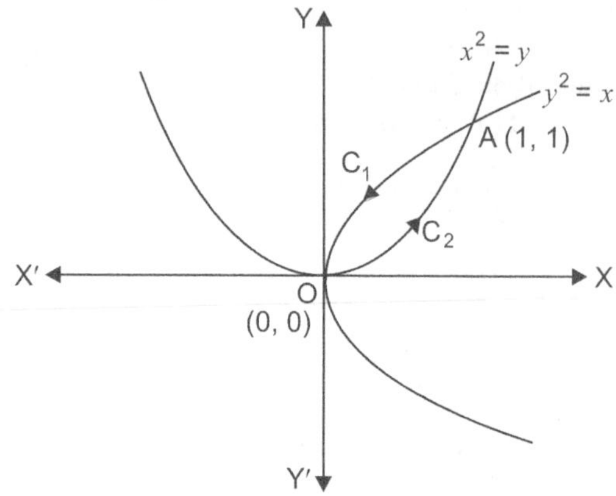

Note. For verification we first evaluate the given integral by Green's theorem and then by taking line integral along C.

Sol. (*i*) The boundary of the region is given by $y = \sqrt{x}$ *i.e.*, $y^2 = x$ which is represented by C_1 and $y = x^2$ which is represented by C_2 both are parabolas (shown in figure). Intersections of parabolas are (0, 0) and (1, 1).

By Green's theorem

$$\int_C M\, dx + N\, dy = \iint_R \left(\frac{\partial N}{\partial x} - \frac{\partial M}{\partial y}\right) dx\, dy$$

Compare $\int_C (3x^2 - 8y^2) dx + (4y - 6xy) dy$ with $\int M dx + N dy$

Here $\quad M = 3x^2 - 8y^2 \qquad N = 4y - 6xy$

$$\frac{\partial M}{\partial y} = -16y \qquad \frac{\partial N}{\partial x} = -6y$$

$$\int_C (3x^2 - 8y^2)\, dx + (4y - 6xy)\, dy = \iint_R (-6y + 16\, y)\, dx\, dy$$

$$= \int_{x=0}^{1} \int_{y=x^2}^{\sqrt{x}} (10\,y)\, dy\, dx = \int_0^1 5y^2 \Big|_{x^2}^{\sqrt{x}} dx = \int_0^1 (5x - 5x^4)\, dx = 5\left(\frac{x^2}{2} - \frac{x^5}{5}\right)_0^1$$

$$= 5\left(\frac{1}{2} - \frac{1}{5}\right) = 5\,\frac{3}{10} = \frac{3}{2}$$

$\therefore\quad$ By Green's theorem $\oint_C (3x^2 - 8y^2)\, dx + (4y - 6xy)\, dy = \frac{3}{2}$ ...(1)

Now, line integral $\oint_C (3x^2 - 8y^2)\, dx + (4y - 6xy)\, dy$

Curve C consists of two parts C_2 and C_1

Along C_2, $\ x^2 = y \qquad$ *i.e.*, $\ y = x^2 \quad \therefore\ dy = 2x\, dx\ $ and $\ x$ varies from 0 to 1

Along C_1, $\ y^2 = x;\ \ x = y^2 \quad \therefore\ dx = 2y\, dy\ $ and $\ y$ varies from 1 to 0.

$\therefore\quad$ Direction of C_1 is from A to O

$$\oint_C \left(3x^2 - 8y^2\right) dx + (4y - 6xy)\, dy$$

$$= \oint_{C_2} \left(3x^2 - 8y^2\right) dx + (4y - 6xy)\, dy + \oint_{C_1} \left(3x^2 - 8y^2\right) dx + (4y - 6xy)\, dy$$

$$= \int_0^1 \left(3x^2 - 8x^4\right) dx + \left(4x^2 - 6x^3\right) 2x\, dx + \int_1^0 \left(3y^4 - 8y^2\right)(2y\, dy) + \left(4y - 6y^3\right) dy$$

$$= \int_0^1 \left(3x^2 - 8x^4 + 8x^3 - 12x^4\right) dx + \int_1^0 \left(6y^5 - 16y^3 + 4y - 6y^3\right) dy$$

$$= x^3 - 20\frac{x^5}{5} + 8\frac{x^4}{4}\Bigg|_0^1 + \left\{\frac{6y^6}{6} - 22\frac{y^4}{4} + \frac{4y^2}{2}\right\}_1^0 = 1 - 4 + 2 - \left[1 - \frac{11}{2} + 2\right]$$

$$= -1 - 3 + \frac{11}{2} = -4 + \frac{11}{2} = \frac{3}{2} \qquad \ldots(2)$$

(1) and (2) are same. So Green's theorem is verified.

(*ii*) Here region is bounded by $x = 0, y = 0, x + y = 1$ which is a triangle OAB as shown in figure.

$\therefore$ By Green's theorem $\int_C \left(3x^2 - 8y^2\right) dx + (4y - 6xy)\, dy = \iint_R 10y\, dy\, dx$

$$= \int_{x=0}^{1} \int_{y=0}^{1-x} 10y\, dy\, dx = \int_0^1 5y^2 \Big|_0^{1-x} dx = \int_0^1 5(1-x)^2\, dx = 5\frac{(1-x)^3}{-3}\Bigg|_0^1 = \frac{5}{3}$$

$\therefore$ By Green's theorem $\int_C \left(3x^2 - 8y^2\right) dx + (4y - 6xy)\, dy = \frac{5}{3}$ ...(1)

Taking line integral of $\int_C \left(3x^2 - 8y^2\right) dx + (4y - 6xy)\, dy$

Line integral consists of three parts

(*i*) $C_1 \to$ integral from 0 to A *i.e.*, $y = 0$, $dy = 0$ and x varies from 0 to 1.

(*ii*) $C_2 \to$ along AB where, $y = 1 - x$, $dy = -\,dx$ and x varies from 1 to 0

(*iii*) $C_3 \to$ along BO, where $x = 0$ $\therefore$ $dx = 0$, y varies from 1 to 0.

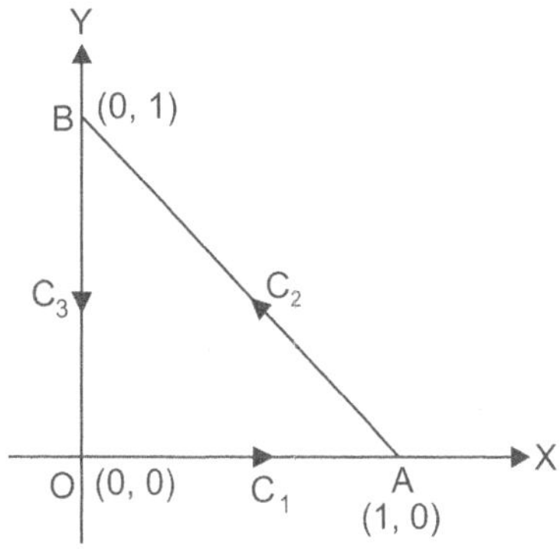

$\therefore$ $\int_C \left(3x^2 - 8y^2\right) dx + (4y - 6xy)\, dy$

$$= \int_{C_1} \left(3x^2 - 8y^2\right) dx + (4y - 6xy)\, dy + \int_{C_2} (3x^2 - 8y^2)\, dx + (4y - 6xy)\, dy$$

$$+ \int_{C_3} (3x^2 - 8y^2)\, dx + (4y - 6xy)\, dy \qquad \ldots(2)$$

$$\int_{C_1}\left(3x^2 - 8y^2\right) dx + (4y - 6xy)\, dy$$

$$= \int_0^1 3x^2\, dx \qquad \because\ y = 0,\quad dy = 0$$

$$= x^3 \Big|_0^1 = 1 \qquad \ldots(3)$$

$$\int_{C_2}\left(3x^2 - 8y^2\right) dx + (4y - 6xy)\, dy$$

$$= \int_1^0 \left\{3x^2 - 8\,(1 - x)^2\right\} dx + \left(4 - 4x - 6x + 6x^2\right)(-\,dx)$$

$$= \int_1^0 \left(3x^2 - 8 + 16x - 8x^2 - 4 + 10x - 6x^2\right) dx$$

$$= \int_1^0 \left(-11x^2 + 26x - 12\right) dx = -11\frac{x^3}{3} + 13x^2 - 12x \Big|_1^0 = \frac{11}{3} - 13 + 12 = \frac{8}{3} \qquad \ldots(4)$$

$$\int_{C_3}\left(3x^2 - 8y^2\right) dx + (4y - 6xy)\, dy$$

$$= \int_1^0 0 + 4y\, dy = 2y^2 \Big|_1^0 = -2 \qquad \ldots(5)$$

Substituting (3), (4), (5) in (2), we get line integral $\int_C \left(3x^2 - 8y^2\right) dx + (4y - 6xy)\, dy$

$$= 1 + \frac{8}{3} - 2 = \frac{8}{3} - 1 = \frac{5}{3} \qquad \ldots(6)$$

(1) and (6) are same.

$\therefore$ Green's theorem is verified.

Example 8. *Verify Green's theorem for* $\oint_C e^{-x}\left(\sin y\, dx + \cos y\, dy\right)$, *where C is the boundary of a rectangle whose vertices are* (0, 0), (π, 0), $\left(\pi, \frac{\pi}{2}\right)$ *and* $\left(0, \frac{\pi}{2}\right)$.

(P.T.U., May 2006, May 2003)

Sol. Boundary of the region *i.e.*, that of a rectangle is given by $y = 0,\ x = \pi,\ y = \frac{\pi}{2},\ x = 0$.

By Green's theorem,

$$\oint \text{M}\, dx + \text{N}\, dy = \iint_R \left(\frac{\partial \text{N}}{\partial x} - \frac{\partial \text{M}}{\partial y}\right) dx\, dy$$

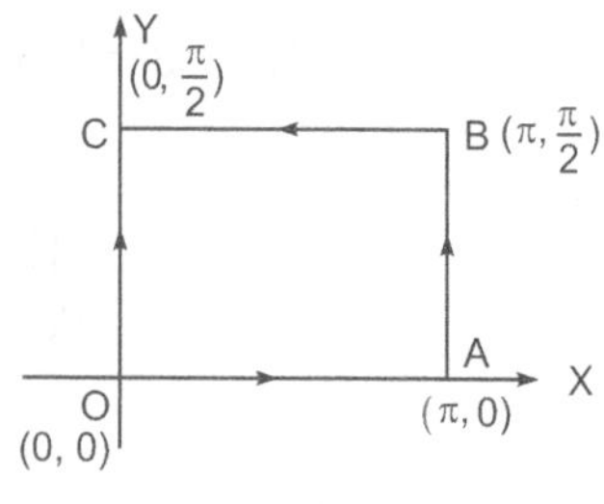

Compare $\oint \mathrm{M}\,dx + \mathrm{N}\,dy$ with $\oint_C e^{-x} \sin y\, dx + e^{-x} \cos y\, dy$, we have

$$\mathrm{M} = e^{-x} \sin y\,; \quad \mathrm{N} = e^{-x} \cos y$$

$$\frac{\partial \mathrm{M}}{\partial y} = e^{-x} \cos y\,; \quad \frac{\partial \mathrm{N}}{\partial x} = -e^{-x} \cos y$$

$\therefore \quad \oint_C e^{-x} (\sin y\, dx + \cos y\, dy) = \iint_R \left(-e^{-x} \cos y - e^{-x} \cos y\right) dx\, dy$, where R is the region OABC (shown in the figure)

$$= -2 \int_{y=0}^{\pi/2} \int_{x=0}^{\pi} e^{-x} \cos y\, dx\, dy = -2 \int_0^{\pi/2} -e^{-x} \cos y \int_0^{\pi} e^{-x}\, dy$$

$$= -2 \int_{y=0}^{\pi/2} (\cos y\, dy) \left(-e^{-x} \Big|_0^{\pi}\right)$$

$$= -2 \int_0^{\pi/2} \left(-e^{-\pi} + 1\right) \cos y\, dy$$

$$= 2 \int_0^{\pi/2} \left(e^{-\pi} - 1\right) \cos y\, dy = 2\left(e^{-\pi} - 1\right) \sin y \Big|_0^{\pi/2} = 2\,(e^{-\pi} - 1) \qquad ...(1)$$

[Now find $\oint_C \mathrm{M}\,dx + \mathrm{N}\,dy = \int_{OABC} e^{-x} (\sin y\, dx + \cos y\, dy)$ by line integral]

$$\therefore \quad \int_{OABC} e^{-x} (\sin y\, dx + \cos y\, dy)$$

$$= \int_{OA} e^{-x} (\sin y\, dx + \cos y\, dy) + \int_{AB} e^{-x} (\sin y\, dx + \cos y\, dy)$$

$$+ \int_{BC} e^{-x} (\sin y\, dx + \cos y\, dy) + \int_{CO} e^{-x} (\sin y\, dx + \cos y\, dy) \qquad ...(2)$$

Along OA; $\quad y = 0 \quad \therefore \quad dy = 0$ and x varies from 0 to π

$$\therefore \quad \int_{OA} e^{-x} (\sin y\, dx + \cos y\, dy) = \int_0^{\pi} e^{-x} (0 + 0) = 0 \qquad ...(3)$$

Along AB; $\quad x = \pi, \quad dx = 0$ and y varies from 0 to $\dfrac{\pi}{2}$

$$\therefore \quad \int_{AB} e^{-x} (\sin y\, dx + \cos y\, dy) = \int_0^{\pi/2} e^{-\pi} \cos y\, dy = e^{-\pi} \sin y \Big|_0^{\pi/2} = e^{-\pi} \qquad ...(4)$$

Along BC; $\quad y = \dfrac{\pi}{2} \quad \therefore \quad dy = 0$ and x varies from π to 0

$$\therefore \quad \int_{BC} e^{-x}(\sin y\, dx + \cos y\, dy) = \int_{\pi}^{0} e^{-x}(1\cdot dx) = -e^{-x}\Big|_{\pi}^{0} = e^{-\pi} - 1 \quad ...(5)$$

Along CO ; $x = 0 \quad \therefore \quad dx = 0$ and y varies from $\frac{\pi}{2}$ to 0

$$\therefore \quad \int_{CO} e^{-x}(\sin y\, dx + \cos y\, dy) = \int_{\pi/2}^{0} (0 + \cos y)\, dy = \sin y\Big|_{\pi/2}^{0} = -1 \quad ...(6)$$

Substituting (3), (4), (5), and (6) in (2), we get

$$\oint_C e^{-x}(\sin y\, dx + \cos y\, dy) = 0 + e^{-\pi} + e^{-\pi} - 1 - 1 = 2(e^{-\pi} - 1)$$

which is same as (1)

$\therefore$ Green's theorem is verified.

Example 9. *Verify Green's theorem for* $\oint_C (x^2 - \cosh y)\, dx + (y + \sin x)\, dy$, *where C is the boundary of a rectangle whose vertices are O (0, 0), A (π, 0), B (π, 1), C (0, 1).*

(P.T.U., Dec. 2006)

Sol. Here boundary of the region is that of a rectangle given by $y = 0$, $x = \pi$, $y = 1$, $x = 0$.

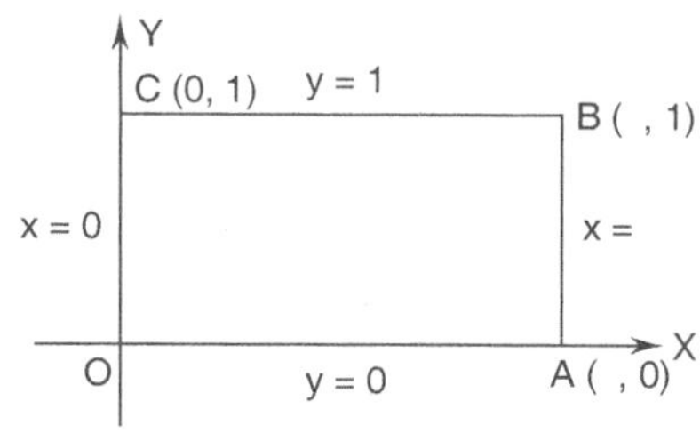

By Green's theorem

$$\oint_C M\, dx + N\, dy = \iint_R \left(\frac{\partial N}{\partial x} - \frac{\partial M}{\partial y}\right) dx\, dy$$

Compare $\oint_C M\, dx + N\, dy$ to $\oint_C (x^2 - \cosh y)\, dx + (y + \sin x)\, dy$

$$M = x^2 - \cosh y; \quad N = y + \sin x$$

$$\therefore \quad \frac{\partial M}{\partial y} = -\sinh y; \quad \frac{\partial N}{\partial x} = \cos x$$

$\therefore \quad \oint_C (x^2 - \cosh y)dx + (y + \sin x)dy = \iint_R (\cos x + \sinh y)\, dx\, dy$ where R is the region enclosed by rectangle shown in the figure

$$\therefore \quad \oint_C (x^2 - \cosh y)dx + (y + \sin x)dy = \int_{y=0}^{1}\int_{x=0}^{\pi} (\cos x + \sinh y)\, dx\, dy$$

$$= \int_0^1 -\sin x + x\sinh y\Big|_0^{\pi} dy = \int_0^1 0 + \pi\sinh y\, dy$$

$$= \pi\cosh y\Big|_0^1 = 0 - \pi = -\pi \quad ...(1)$$

Now, find $\oint_C (x^2 - \cosh y)\, dx + (y + \sin x)\, dy$ by line integral

$$\oint_C \left(x^2 - \cosh y\right) dx + \left(y + \sin x\right) dy$$

$$= \int_{OABC} \left(x^2 - \cosh y\right)dx + \left(y + \sin x\right)dy$$

$$= \int_{OA} \left(x^2 - \cosh y\right)dx + \left(y + \sin x\right)dy + \int_{AB} \left(x^2 - \cosh y\right)dx + \left(y + \sin x\right)dy$$

$$+ \int_{BC} \left(x^2 - \cosh y\right)dx + \left(y + \sin x\right)dy + \int_{CO} \left(x^2 \cosh y\right)dx + \left(y + \sin x\right)dy \quad \ldots(2)$$

Along OA; $y = 0 \quad \because \quad dy = 0 \quad \text{and} \quad 0 \le x \le \pi$

$$\therefore \quad \int_{OA} \left(x^2 - \cosh y\right) dx + \left(y + \sin x\right) dy = \int_0^{\pi} \left(x^2 - 1\right) dx = \frac{x^3}{3} - x \Big|_0^{\pi} = \frac{\pi^3}{3} - \pi \quad \ldots(3)$$

Along AB; $x = \pi \quad \therefore \quad dx = 0; \quad 0 \le y \le 1$

$$\int_{AB} \left(x^2 - \cosh y\right) dx + \left(y + \sin x\right) dy = \int_0^1 y\, dy = \frac{y^2}{2} \Big|_0^1 = \frac{1}{2} \quad \ldots(4)$$

Along BC; $y = 1 \quad \therefore \quad dy = 0; \quad \pi \le x \le 0$

$$\therefore \quad \int_{BC} \left(x^2 - \cosh y\right) dx + \left(y + \sin x\right) dy = \int_{\pi}^0 x^2\, dx = \frac{x^3}{3} = -\frac{\pi^3}{3} \quad \ldots(5)$$

Along CO; $x = 0 \quad \therefore \quad dx = 0; \quad 1 \le y \le 0$

$$\therefore \quad \int_{CO} \left(x^2 - \cosh y\right) dx + \left(y + \sin x\right) dy = \int_1^0 y\, dy = \frac{y^2}{2} \Big|_1^0 = -\frac{1}{2} \quad \ldots(6)$$

Substituting (3), (4), (5) and (6) in (2), we get

$$\oint_C \left(x^2 - \cosh y\right) dx + \left(y + \sin x\right) dy = \frac{\pi^3}{3} - \pi + \frac{1}{2} - \frac{\pi^3}{3} - \frac{1}{2} = -\pi$$

which is same as (1)

$\therefore$ Green's theorem is verified.

TEST YOUR KNOWLEDGE

1. Apply Green's theorem to evaluate $\oint_C x^2 y\, dx + x^2\, dy$, where C is the boundary describing counter clockwise of the triangle with vertices (0, 0), (1, 0), (1, 1).

2. Either directly or by Green's theorem, evaluate the line integral $\int_C e^{-x} (\cos y\, dx - \sin y\, dy)$ where C is the rectangle with vertices (0, 0), (π, 0), $\left(\pi, \frac{\pi}{2}\right)$ and $\left(0, \frac{\pi}{2}\right)$.

3. Use Green's theorem to evaluate the line integrals

(*i*) $\oint_C \left[xy^2\,dx + x^2 dy\right]$ [**Hint.** Consult S.E. 6]

(*ii*) $\int_C 2y^2\,dx + 3x\,dy$, where C is the boundary of the closed region bounded by $y = x$, $y = x^2$.

4. Verify Green's theorem for the following line integrals

(*i*) $\oint_C \left(2xy - x^2\right)dx + \left(x^2 + y^2\right)dy$, where C is the boundary of the region enclosed by $y = x^2$ and $y^2 = x$. [**Hint.** Consult S.E. 7(*i*)]

(*ii*) $\oint_C \left(xy^2 - 2xy\right)dx + \left(x^2 y + 3\right)dy$ around the boundary C of the region enclosed by $y^2 = 8x$ and $x = 2$.

(*iii*) $\int_C \left(x^2 + 2xy\right)dx + \left(y^2 + x^3 y\right)dy$, where C is the square with vertices P (0, 0), Q (1, 0), R (1, 1), S (0, 1).

Answers

1. $\frac{5}{12}$ **2.** $2\left(1 - e^{-\pi}\right)$ **3.** (*i*) $\frac{1}{12}$ (*ii*) $\frac{7}{30}$

4. (*i*) Zero (*ii*) 128/5 (*iii*) 1/2.

8.7. STOKE'S THEOREM (Relation Between Line and Surface Integrals)

(P.T.U., Dec. 2010, 2012, 2014, Jan. 2009, May 2011, 2014, 2015)

Statement.** *If S is an* ***open surface *bounded by a closed curve C and* $\vec{F} = F_1\hat{i} + F_2\hat{j} + F_3\hat{k}$ *be any vector point function having continuous first order partial derivatives, then* $\int_C \vec{F}\cdot d\vec{r} = \iint_S Curl\,\vec{F}\cdot\hat{n}\,dS$ *where* $\hat{n}$ *is a unit normal vector at any point of S drawn in the sense in which a right handed screw would advance when rotated in the sense of discription of C.*

or In other words

[Surface integral of the component of $\vec{F}$ along the normal to the surface S, taken over the surface S bounded by the curve C is equal to the line integral of the vector point function $\vec{F}$ taken along the closed curve C].

Note. When the surface is closed, the boundary of the curve does not exist. In such case

$$\iint_S \text{Curl}\,\vec{F}\cdot\hat{n}\,dS = 0.$$

Proof. Let $\hat{n}$ makes angles α, β, γ with positive direction of x, y, z axes respectively.

$$\therefore \quad \hat{n} = \cos\alpha\,\hat{i} + \cos\beta\,\hat{j} + \cos\gamma\,\hat{k}$$

Also, $\vec{r} = x\,\hat{i} + y\,\hat{j} + z\,\hat{k}$ $\therefore$ $d\vec{r} = dx\,\hat{i} + dy\,\hat{j} + dz\,\hat{k}$

*__Proof of the theorem is beyond the syllabus of sem. I of P.T.U.__

$$\text{Curl } \vec{F} = \nabla \times \vec{F} = \begin{vmatrix} \hat{i} & \hat{j} & \hat{k} \\ \dfrac{\partial}{\partial x} & \dfrac{\partial}{\partial y} & \dfrac{\partial}{\partial z} \\ F_1 & F_2 & F_3 \end{vmatrix} = \left(\frac{\partial F_3}{\partial y} - \frac{\partial F_2}{\partial z}\right)\hat{i} + \left(\frac{\partial F_1}{\partial z} - \frac{\partial F_3}{\partial x}\right)\hat{j} + \left(\frac{\partial F_2}{\partial x} - \frac{\partial F_1}{\partial y}\right)\hat{k}$$

$$\text{Curl } \vec{F} \cdot \hat{n} = \left(\frac{\partial F_3}{\partial y} - \frac{\partial F_2}{\partial z}\right)\cos\alpha + \left(\frac{\partial F_1}{\partial z} - \frac{\partial F_3}{\partial x}\right)\cos\beta + \left(\frac{\partial F_2}{\partial x} - \frac{\partial F_1}{\partial y}\right)\cos\gamma$$

Also $\qquad \vec{F} \cdot d\vec{r} = F_1\, dx + F_2\, dy + F_3\, dz$

$\therefore$ Stoke's theorem can be written as

$$\int_C F_1 dx + F_2 dy + F_3 dz = \iint_S \left\{ \left(\frac{\partial F_3}{\partial y} - \frac{\partial F_2}{\partial z}\right)\cos\alpha + \left(\frac{\partial F_1}{\partial z} - \frac{\partial F_3}{\partial x}\right)\cos\beta + \left(\frac{\partial F_2}{\partial x} - \frac{\partial F_1}{\partial y}\right)\cos\gamma \right\} dS \qquad \ldots(1)$$

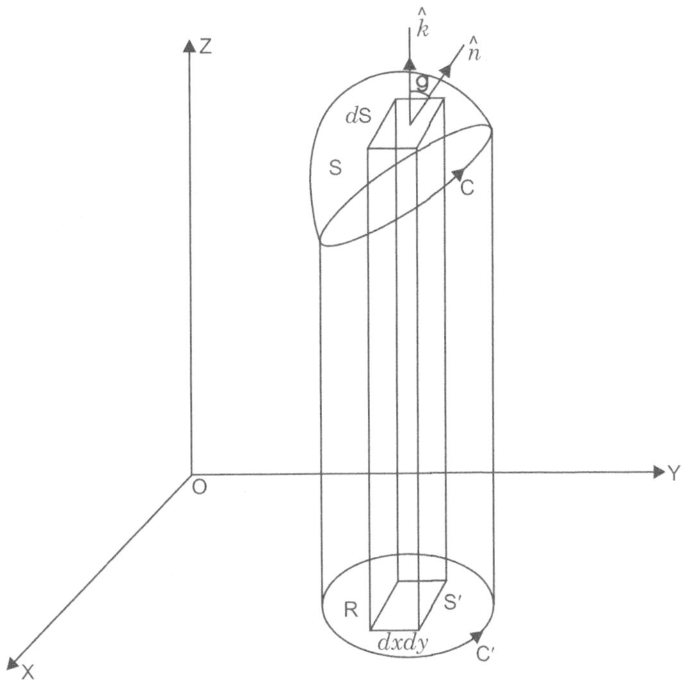

Let $z = \phi(x, y)$ be the equation of the surface S whose projection on the xy-plane is R. Then, the projection of C on the xy-plane is the curve C′ which bounds the region R.

$$\therefore \quad \int_C F_1(x, y, z)\, dx = \int_{C'} F_1(x, y, \;\;)\, dx = \int_{C'} \{F_1(x, y, \;\;)\, dx + 0\, dy\} \qquad \ldots(2)$$

Compare it with Green's theorem in the plane for the region R, we have

$M = F_1(x, y, \ ), \quad N = 0$

$$\therefore \qquad \int_{C'} F_1(x, y, \ )\, dx = -\iint_R \frac{\partial}{\partial y} F_1(x, y, \ )\, dx\, dy$$

$$= -\iint_R \left(\frac{\partial F_1}{\partial y} + \frac{\partial F_1}{\partial z} \bullet \frac{\partial}{\partial y}\right) dx\, dy \qquad \ldots(3) \qquad \left(\because \ \frac{\partial z}{\partial y} = \frac{\partial \phi}{\delta y}\right)$$

Now, the direction ratios of the normal to the surface $z = \phi(x, y)$ are $\frac{\partial}{\partial x}, \frac{\partial}{\partial y}, -1$

$[\because z = \phi(x, y) \quad \text{or} \quad \phi(x, y) - z = 0]$

$$\Rightarrow \qquad \frac{\cos}{\frac{\partial}{\partial x}} = \frac{\cos}{\frac{\partial}{\partial y}} = \frac{\cos}{-1}$$

$$\Rightarrow \qquad \frac{\partial}{\partial y} = -\frac{\cos}{\cos}$$

Since $dx\,dy$ is the projection of dS on xy-plane

$$\therefore \qquad dx\, dy = \cos\gamma\, dS$$

$$\therefore \quad \text{From (3), we have} \quad \int_{C'} F_1(x, y, \phi)\, dx = -\iint_S \left(\frac{\partial F_1}{\partial y} - \frac{\partial F_1}{\partial z}\frac{\cos}{\cos}\right) \cos \quad dS$$

$$= -\iint_S \left\{\frac{\partial F_1}{\partial z} \cos \quad - \frac{\partial F_1}{\partial y} \cos \quad \right\} dS = \iint_S \left\{\frac{\partial F_1}{\partial z} \cos \quad - \frac{\partial F_1}{\partial y} \cos \quad \right\} dS$$

$$\therefore \quad \text{From (2),} \quad \oint_C F_1(x, y, z)\, dx$$

$$= \iint_C \left(\frac{\partial F_1}{\partial z} \cos \quad - \frac{\partial F_1}{\partial y} \cos \quad \right) dS \qquad \ldots(4)$$

Similarly, we can prove that

$$\int_C F_2(x, y, z)\, dy = \iint_S \left(\frac{\partial F_2}{\partial x} \cos \quad - \frac{\partial F_2}{\partial z} \cos \quad \right) dS \qquad \ldots(5)$$

and

$$\int_C F_3(x, y, z)\, dy = \iint_S \left(\frac{\partial F_3}{\partial y} \cos \quad - \frac{\partial F_3}{\partial x} \cos\beta\right) dS \qquad \ldots(6)$$

Adding (4), (5), (6), we get

$$\int_C F_1(x, y, z)\, dx + F_2(x, y, z)\, dy + F_3(x, y, z)\, dz$$

$$= \iint_S \left\{\left(\frac{\partial F_3}{\partial y} - \frac{\partial F_2}{\partial z}\right) \cos \quad \left(\frac{\partial F_1}{\partial z} - \frac{\partial F_3}{\partial x}\right) \cos \quad \left(\frac{\partial F_2}{\partial x} - \frac{\partial F_1}{\partial y}\right) \cos \quad \right\} dS \text{ which is same as (I)}$$

Hence, theorem is proved.

Cor. Green's theorem in a plane as a special case of Stoke's Theorem

Let $\vec{F} = f_1\hat{i} + f_2\hat{j}$ be a vector point function which is continuously differentiable in a region S of the xy-plane bounded by the closed curve C

Then $$\int_C \vec{F}\cdot d\vec{r} = \int_C \left(f_1\hat{i} + f_2\hat{j}\right)\cdot\left(dx\,\hat{i} + dy\,\hat{j}\right) = \int f_1\,dx + f_2\,dy$$

and $$\text{Curl}\,\vec{F}\cdot\hat{n} = \begin{vmatrix} \hat{i} & \hat{j} & \hat{k} \\ \dfrac{\partial}{\partial x} & \dfrac{\partial}{\partial y} & 0 \\ f_1 & f_2 & 0 \end{vmatrix}\cdot\hat{k} = \left\{\left(0\,\hat{i} + 0\,\hat{j}\right) + \left(\frac{\partial f_2}{\partial x} - \frac{\partial f_1}{\partial y}\right)\right\}\cdot\hat{k} = \frac{\partial f_2}{\partial x} - \frac{\partial f_1}{\partial y}$$

Hence Stoke's theorem $\int_C \vec{F}\cdot d\vec{r} = \iint_S \text{Curl}\,\vec{F}\cdot\hat{N}\,dS$ takes the form

$$\int_C f_1\,dx + f_2\,dy = \iint_R \left(\frac{\partial f_2}{\partial x} - \frac{\partial f_1}{\partial y}\right)dx\,dy$$ which is Green's theorem in a plane.

ILLUSTRATIVE EXAMPLES

Example 1. *Evaluate* $\oint_C \vec{F}\cdot d\vec{r}$ *by Stoke's Theorem, where* $\vec{F} = y^2\,\hat{i} + x^2\,\hat{j} - (x+z)\,\hat{k}$ *and C is the boundary of the triangle with vertices at (0, 0, 0), (1, 0, 0), (1, 1, 0).*

(P.T.U., May 2010)

Sol. $\because$ z-coordinate of each vertex of triangle is zero $\therefore$ triangle lies in the xy-plane

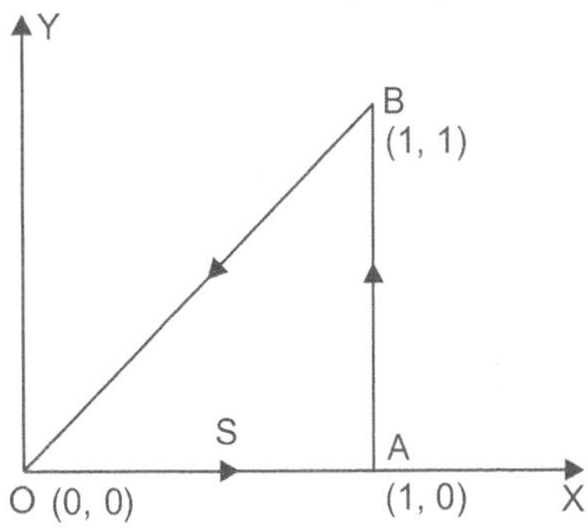

By Stoke's theorem $\oint_C \vec{F}\cdot d\vec{r} = \iint_S \text{Curl}\,\vec{F}\cdot\hat{n}\,dS$

$$\text{Curl}\,\vec{F} = \begin{vmatrix} \hat{i} & \hat{j} & \hat{k} \\ \dfrac{\partial}{\partial x} & \dfrac{\partial}{\partial y} & \dfrac{\partial}{\partial z} \\ y^2 & x^2 & -x-z \end{vmatrix}$$

$$= \hat{i}\,\{0\} + \hat{j}\,\{1\} + \hat{k}\,\{2x - 2y\} = \hat{j} + 2\,(x-y)\,\hat{k}$$

$\hat{n} = \hat{k}$ (unit normal to xy-plane)

$\therefore$ $$\text{Curl}\,\vec{F}\cdot\hat{n} = 2\left\{\hat{j} + 2\,(x-y)\,\hat{k}\right\}\cdot\hat{k} = 2\,(x-y)$$

Now, $$\oint_C \vec{F}\cdot d\vec{r} = \iint_S 2\,(x-y)\,dS, \text{ where } dS = \frac{dx\,dy}{|\hat{n}\cdot\hat{k}|} = \frac{dx\,dy}{|\hat{k}\cdot\hat{k}|} = dx\,dy$$

$= \iint_R 2(x - y)\,dx\,dy$, where R is region of ΔOAB, the projection of the given triangle on xy-plane

$= \int_{x=0}^{1} \int_0^x 2\,(x - y)\,dy\,dx$; equation of OB is $y = x$ $\therefore$ y varies from 0 to x

$$= \int_0^1 \left\{ 2\left(xy - \frac{y^2}{2} \right) \right\}_0^x dx = \int_0^1 2\left(x^2 - \frac{x^2}{2} \right) dx = \int_0^1 x^2\,dx = \left. \frac{x^3}{3} \right|_0^1 = \frac{1}{3}.$$

Example 2. *Using Stoke's theorem or otherwise evaluate* $\int_C \left[(2x - y)dx - yz^2 dy - y^2 z\,dz \right]$, *where C is the circle* $x^2 + y^2 = 1$, *corresponding to the surface of sphere of unit radius.*

Sol. We will use Stoke's theorem to prove the result.

$$\int_C \{(2x - y)\,dx - yz^2\,dy - y^2 z\,dz\}$$

$$= \int_C \left[(2x - y)\hat{i} - yz^2\,\hat{j} - y^2\,z\,\hat{k} \right] \cdot \left[\hat{i}\,dx + \hat{j}\,dy + \hat{k}\,dz \right]$$

$$= \int_C \vec{F} \cdot d\vec{r}, \text{ where } \vec{F} = (2x - y)\,\hat{i} - yz^2\,\hat{j} - y^2 z\,\hat{k}$$

By Stoke's theorem $\int_C \vec{F} \cdot d\vec{r} = \iint_S \text{Curl } \vec{F} \cdot \hat{n}\,dS$

$$\therefore \quad \int_C \{(2x - y)\,dx - yz^2\,dy - y^2 z\,dz\} = \underset{S}{} \text{Curl } \vec{F} \ \hat{n}\,dS$$

$$\text{Curl } \vec{F} = \begin{vmatrix} \hat{i} & \hat{j} & \hat{k} \\ \dfrac{\partial}{\partial x} & \dfrac{\partial}{\partial y} & \dfrac{\partial}{\partial z} \\ 2x - y & -yz^2 & -y^2 z \end{vmatrix} = \hat{i}\,\{-2yz + 2yz\} + \hat{j}\,\{0 - 0\} + \hat{k}\,\{0 + 1\} = \hat{k}$$

$$\text{Curl } \vec{F} \cdot \hat{n} = \hat{k} \cdot \hat{k} = 1$$

$$\iint_S \text{Curl } \vec{F} \cdot \hat{n}\,dS = \iint_S 1 \cdot dS = \iint_S dS = \text{area of the circle of unit radius}$$

$$= \pi\,1^2 = \pi.$$

Example 3. *Using Stoke's theorem for the vector function* $\vec{F} = (2x + y - 2z)\hat{i} + (2x - 4y + z^2)\,\hat{j} + (x - 2y - z^2)\,\hat{k}$ *evaluate the integral* $\oint_C \vec{F} \cdot \vec{dR}$, *where C is the circle with centre at (0, 0, 3) and radius 5 in the plane* $z = 3$.

Sol. $\vec{F} = (2x + y - 2z)\hat{i} + (2x - 4y + z^2)\hat{j} + (x - 2y + z^2)\hat{k}$

and C is the curve given by

$$(x - 0)^2 + (y - 0)^2 + (z - 3)^2 = 25 \; ; \; z = 3$$

i.e., $$x^2 + y^2 + (z - 3)^2 = 25 \; ; \; z = 3$$

(it is a great circle with centre at (0, 0, 3) and radius 5)

By Stokes theorem $\oint_C \vec{F} \cdot d\vec{R} = \iint_S \text{Curl } \vec{F} \cdot \hat{n} \, dS$

$$\text{Curl } \vec{F} = \begin{vmatrix} \hat{i} & \hat{j} & \hat{k} \\ \dfrac{\partial}{\partial x} & \dfrac{\partial}{\partial y} & \dfrac{\partial}{\partial z} \\ 2x + y - 2z & 2x - 4y + z^2 & x - 2y + z^2 \end{vmatrix}$$

$$= (-2 - 2z)\hat{i} - 3\hat{j} + \hat{k}$$

$$\hat{n} = \hat{k} \text{ (unit normal to the plane } z = 3)$$

$$\therefore \quad \oint_C \vec{F} \cdot d\vec{R} = \iint_S \left[(-2 - 2z)\hat{i} - 3\hat{j} + \hat{k}\right] \cdot \hat{k} \, dS$$

$$= \iint_S 1 \, dS = S = \text{(area of the circle)}$$

$$= \pi (5)^2 = 25\pi.$$

Example 4. *Apply Stoke's theorem to find the value of* $\int_C (y\,dx + z\,dy + x\,dz)$*, where C is the curve of intersection of* $x^2 + y^2 + z^2 = a^2$ *and* $x + z = a$. **(P.T.U., May 2014, Dec. 2015)**

Sol. $\int_C (y\,dx + z\,dy + x\,dz) = \int \left(y\hat{i} + z\hat{j} + x\hat{k}\right) \cdot \left(dx\,\hat{i} + dy\,\hat{j} + dz\,\hat{k}\right) = \int \vec{F} \cdot d\vec{r}$

where $\vec{F} = y\hat{i} + z\hat{j} + x\hat{k}$

By Stoke's theorem $\oint_C \vec{F} \cdot d\vec{r} = \iint_S \text{Curl } \vec{F} \cdot \hat{n} \, dS$

$$\text{Curl } \vec{F} = \begin{vmatrix} \hat{i} & \hat{j} & \hat{k} \\ \dfrac{\partial}{\partial x} & \dfrac{\partial}{\partial y} & \dfrac{\partial}{\partial z} \\ y & z & x \end{vmatrix} = \hat{i}\,(-1) + \hat{j}\,(-1) + \hat{k}\,(-1) = -\left(\hat{i} + \hat{j} + \hat{k}\right)$$

Now, $\hat{n}$ is a unit vector normal to the plane of circle whose equation is $x + z - a = 0$ (*i.e.,* in xz-plane)

$\therefore$ Vector normal to $\phi = x + z - a$ is given by

$$\nabla \phi = \left(\hat{i}\frac{\partial}{\partial x} + \hat{j}\frac{\partial}{\partial y} + \hat{k}\frac{\partial}{\partial z}\right)(x + z - a) = \hat{i}\,1 + \hat{k}\,1 = \hat{i} + \hat{k}$$

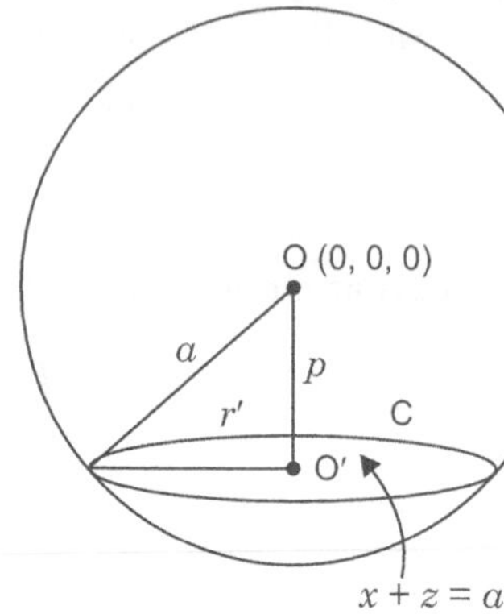

$$\therefore \qquad \hat{n} = \frac{\hat{i} + \hat{k}}{\sqrt{1+1}} = \frac{\hat{i} + \hat{k}}{\sqrt{2}}$$

$\therefore$ By Stoke's theorem $\int_C \{y\,dx + z\,dy + x\,dz\}$

$$= \iint_S -\left(\hat{i} + \hat{j} + \hat{k}\right) \cdot \frac{\hat{i} + \hat{k}}{\sqrt{2}}\,dS$$

$$= -\frac{1}{\sqrt{2}} \iint (1+1)\,dS = -\sqrt{2} \iint_S dS,$$

where S is the surface area of the circle shown in figure. If r' is radius of the circle then $r'^2 = a^2 - p^2$, where $p = \frac{|0 + 0 - a|}{\sqrt{2}} = \frac{a}{\sqrt{2}}$

$$\therefore \qquad r'^2 = a^2 - \frac{a^2}{2} = \frac{a^2}{2}.$$

$$\therefore \int_C \{y\,dx + z\,dy + x\,dz\} = -\sqrt{2}\left(\pi r'^2\right) = -\sqrt{2}\left(\pi \frac{a^2}{2}\right) = -\frac{\pi a^2}{\sqrt{2}}.$$

Example 5. *Apply Stoke's theorem to evaluate* $\int_C 4y\,dx + 2z\,dy + 6y\,dz$, *where C is the curve of intersection of* $x^2 + y^2 + z^2 = 6z$ *and* $z = x + 3$.

Sol. $\int_C 4y\,dx + 2z\,dy + 6y\,dz = \int_C \left\{4y\hat{i} + 2z\hat{j} + 6y\hat{k}\right\} \cdot \left\{\hat{i}\,dx + \hat{j}\,dy + \hat{k}\,dz\right\}$

$$= \int_C \vec{F} \cdot d\vec{r}, \text{ where } \vec{F} = 4y\hat{i} + 2z\hat{j} + 6y\hat{k}$$

$$\vec{r} = x\hat{i} + y\hat{j} + z\hat{k} \quad \therefore\ d\vec{r} = dx\,\hat{i} + dy\,\hat{j} + dz\,\hat{k}$$

By Stoke's theorem $\int_C \vec{F} \cdot d\vec{r} = \iint_S \text{Curl } \vec{F} \cdot \hat{n}\,dS$

$$\text{Curl } \vec{F} = \begin{vmatrix} \hat{i} & \hat{j} & \hat{k} \\ \frac{\partial}{\partial x} & \frac{\partial}{\partial y} & \frac{\partial}{\partial z} \\ 4y & 2z & 6y \end{vmatrix} = \hat{i}\,(6-2) + \hat{j}\,(0) + \hat{k}\,(0-4) = 4\hat{i} - 4\hat{k} = 4\left(\hat{i} - \hat{k}\right)$$

$\hat{n}$ is a unit vector $\perp$ to the plane of the circle given by $z = x + 3$ (in xz-plane)

or $\qquad \phi = x - z + 3$

$$\therefore \qquad \hat{n} = \frac{\nabla}{|\nabla\ |} = \frac{\left(\hat{i}\frac{\partial}{\partial x} + \hat{j}\frac{\partial}{\partial y} + \hat{k}\frac{\partial}{\partial z}\right)(x - z + 3)}{|\nabla\ |} = \frac{\hat{i} - \hat{k}}{\sqrt{2}}$$

$$\text{Curl } \vec{F} \cdot \hat{n} = 4(\hat{i} - \hat{k}) \cdot \frac{\hat{i} - \hat{k}}{\sqrt{2}} = \frac{4}{\sqrt{2}}(1+1) = 4\sqrt{2}$$

$$\therefore \quad \int_C 4y\,dx + 2z\,dy + 6y\,dz = \iint_S 4\sqrt{2}\,dS = 4\sqrt{2}\iint_S dS = 4\sqrt{2} \text{ surface area of the circle}$$

which is intersection of the sphere $x^2 + y^2 + z^2 - 6z = 0$ and plane $x - z + 3 = 0$.

The centre of the sphere is (0, 0, 3) which lies on the plane $x - z + 3 = 0$ which shows that circle is the great circle of the sphere.

$\therefore$ Radius of the circle is same as radius of the sphere = 3

$\therefore$ Area of the circle = $\pi \cdot 3^2 = 9\pi$

$$\therefore \quad \oint_C 4y\,dx + 2z\,dy + 6y\,dz = 4\sqrt{2} \cdot 9\pi = 36\sqrt{2}\pi.$$

Example 6. *Use Stoke's theorem to evaluate* $\int_C \vec{v} \cdot d\vec{r}$, *where* $\vec{v} = y^2\hat{i} + xy\hat{j} + xz\hat{k}$ *and C is the bounding curve of the hemisphere* $x^2 + y^2 + z^2 = 9$, $z > 0$ *oriented in the positive direction.*

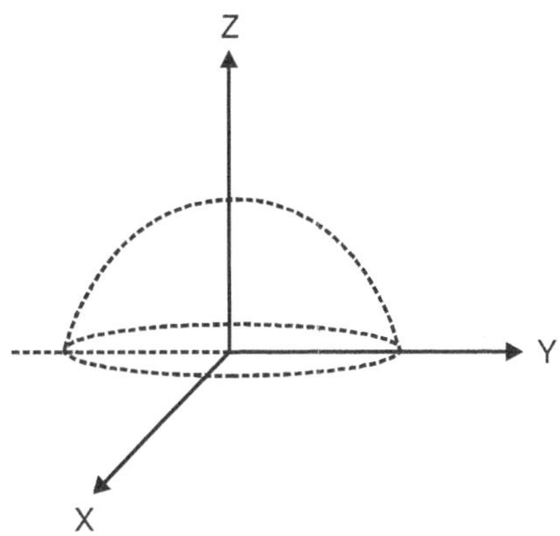

Sol. By Stoke's theorem $\int_C \vec{v} \cdot d\vec{r} = \iint_S \text{Curl } \vec{v} \cdot \hat{n}\,dS$

$$\text{Curl } \vec{v} = \begin{vmatrix} \hat{i} & \hat{j} & \hat{k} \\ \frac{\partial}{\partial x} & \frac{\partial}{\partial y} & \frac{\partial}{\partial z} \\ y^2 & xy & xz \end{vmatrix}$$

$$= \hat{i}\{0 - 0\} + \hat{j}\{0 - z\} + \hat{k}\{y - 2y\} = -z\hat{j} - y\hat{k}$$

$$\hat{n} = \frac{\nabla\phi}{|\nabla\phi|}, \text{ where } \phi(x, y, z) = x^2 + y^2 + z^2 - 9$$

$$\therefore \quad \hat{n} = \frac{\hat{i}(2x) + \hat{j}(2y) + \hat{k}(2z)}{\sqrt{4x^2 + 4y^2 + 4z^2}} = \frac{2(x\hat{i} + y\hat{j} + z\hat{k})}{2 \cdot 3} = \frac{x\hat{i} + y\hat{j} + z\hat{k}}{3}$$

$$(\because x^2 + y^2 + z^2 = 9)$$

$$\oint_C \vec{v} \cdot d\vec{r} = \iint_S (-z\hat{j} - y\hat{k}) \cdot \frac{x\hat{i} + y\hat{j} + z\hat{k}}{3}\,dS = \iint_S \frac{-yz - yz}{3}\,dS$$

$$= -\frac{2}{3}\iint_S yz\,dS = -\frac{2}{3}\iint_R yz\frac{dx\,dy}{|\hat{n} \cdot \hat{k}|} = -\frac{2}{3}\iint_R yz\frac{dx\,dy}{\frac{z}{3}} = -2\iint_R y\,dx\,dy$$

where R is the projection of hemisphere $x^2 + y^2 + z^2 = 9$, $z > 0$ on xy plane.

On xy plane $z = 0$

$\therefore \quad x^2 + y^2 = 9$, where y varies from $-\sqrt{9-x^2}$ to $\sqrt{9-x^2}$ and x varies from -3 to 3.

$$\therefore \quad \int_C \vec{v} \cdot d\vec{r} = \int_{-3}^{3} \left\{ \frac{y^2}{2} \right\}_{-\sqrt{9-x^2}}^{\sqrt{9-x^2}} dx = 0 \qquad \left(\because y^2 \text{ is even} \therefore \left[y^2 \right]_{-a}^{a} = 0 \right)$$

Hence $\quad \int_C \vec{v} \cdot d\vec{r} = 0.$

Note. Here we can directly take $\hat{n} = \hat{k}$, the solution will be simple

as $\int_C \vec{v} \cdot d\vec{r} = \iint_S \left(-z\hat{j} - y\hat{k} \right) \cdot \hat{k}\, dS = \iint_R -y \dfrac{dx\,dy}{|\hat{n} \cdot \hat{k}|}$, where R is the projection of hemisphere on xy-plane

$$= -\int_{-3}^{3} \int_{-\sqrt{9-x^2}}^{\sqrt{9-x^2}} y\,dx\,dy \qquad \because \hat{n} \cdot \hat{k} = \hat{k} \cdot \hat{k} = 1$$

$$= -\int_{-3}^{3} 0\,dx \qquad \because y \text{ is an odd function} \therefore \int_{-a}^{a} y\,dx = 0$$

$$= 0$$

Example 7. *Directly or by Stoke's theorem, evaluate* $\iint_S Curl\,\vec{v} \cdot \hat{n}\, dS$, *where* $\vec{v} = \hat{i}\,y + \hat{j}\,z + \hat{k}\,x$; *S is the surface of the paraboloid* $z = 1 - x^2 - y^2$, $z \geq 0$ *and* $\hat{n}$ *is the unit normal to S.*

Sol.
$$\text{Curl } \vec{v} = \begin{vmatrix} \hat{i} & \hat{j} & \hat{k} \\ \dfrac{\partial}{\partial x} & \dfrac{\partial}{\partial y} & \dfrac{\partial}{\partial z} \\ y^2 & xy & xz \end{vmatrix} = \hat{i}\,(-1) + \hat{j}\,(-1) + \hat{k}\,(-1) = -\left(\hat{i} + \hat{j} + \hat{k} \right)$$

S is given by $\phi\,(x, y, z) = x^2 + y^2 + z - 1 = 0$

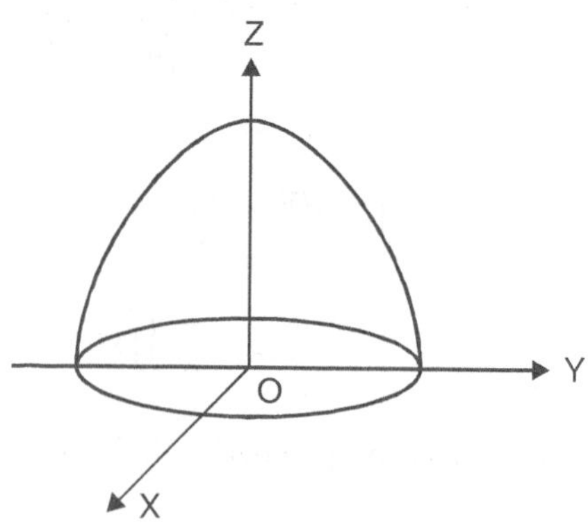

$$\therefore \quad \hat{n} = \frac{\nabla\phi}{|\nabla\phi|} = \frac{2x\hat{i} + 2y\hat{j} + \hat{k}}{\sqrt{4x^2 + 4y^2 + 1}}$$

$$\text{Curl } \vec{v} \cdot \hat{n} = -\frac{2x + 2y + 1}{\sqrt{4x^2 + 4y^2 + 1}}$$

$$\iint_S \text{Curl } \vec{v} \cdot \hat{n}\, dS = \iint_S -\frac{2x + 2y + 1}{\sqrt{4x^2 + 4y^2 + 1}}\, dS$$

$$= \iint_R -\frac{2x+2y+1}{\sqrt{4x^2+4y^2+1}} \frac{dx\,dy}{|\hat{n}\cdot\hat{k}|} = \iint_R -\frac{2x+2y+1}{\sqrt{4x^2+4y^2+1}} \frac{dx\,dy}{\frac{1}{\sqrt{4x^2+4y^2+1}}}$$

$$= \iint_R -(2x+2y+1)\,dx\,dy,$$

where R is projection of the paraboloid on XY plane *i.e.*, $x^2 + y^2 = 1$

$$= -\int_{-1}^{1}\int_{-\sqrt{1-x^2}}^{\sqrt{1-x^2}} (2x+2y+1)\,dy\,dx = -\int_{-1}^{1}\left\{(2x+1)\,y + y^2\right\}_{-\sqrt{1-x^2}}^{\sqrt{1-x^2}} dx$$

$$= -\int_{-1}^{1} 2\,(2x+1)\sqrt{1-x^2}\,dx = 2\int_{-1}^{1}(-2x)\sqrt{1-x^2}\,dx - 2\int_{-1}^{1}\sqrt{1-x^2}\,dx$$

$$= 2\left.\frac{(1-x^2)^{3/2}}{3/2}\right|_{-1}^{1} + 2\left[\frac{x\sqrt{1-x^2}}{2} - \frac{1}{2}\sin^{-1}x\right]_{-1}^{1} = 0 - 2\left[0 + \frac{1}{2}\pi\right] = -\pi$$

Hence $\iint_S \text{Curl}\,\vec{v}\cdot\hat{n}\,dS = -\pi.$

[**Note.** Also we can take directly $\hat{n} = \hat{k}$ the solution becomes easy]

Example 8. *Verify Stoke's theorem for* $\vec{F} = \left(x^2 + y^2\right)\hat{i} - 2xy\,\hat{j}$ *taken around the rectangle bounded by the lines* $x = \pm a$, $y = 0$, $y = b$. **(P.T.U., Dec. 2003)**

Note. *For verification 1st find the value of* $\oint_C \vec{F}\cdot d\vec{r}$ *and then find* $\iint_S \text{Curl}\,\vec{F}\cdot\hat{n}\,dS$ *and then prove both are same.*

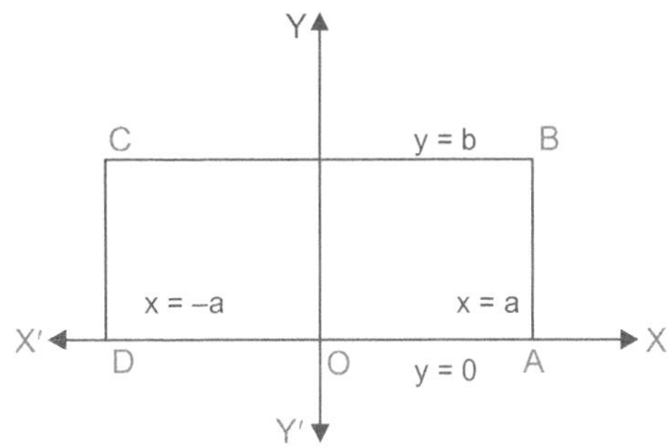

Sol. $\vec{F} = \left(x^2 + y^2\right)\hat{i} - 2xy\,\hat{j}$

$$\oint_C \vec{F}\cdot d\vec{r} = \oint_C \left(\left(x^2+y^2\right)\hat{i} - 2xy\,\hat{j}\right)\cdot\left(dx\,\hat{i} + dy\,\hat{j}\right)$$

$$= \oint_C \left\{\left(x^2+y^2\right)dx - 2xy\,dy\right\}$$

where C consists of AB, BC, CD, DA.

along AB	$x = a$	$\therefore$	$dx = 0$ and	y varies from 0 to b
along BC	$y = b$	$\therefore$	$dy = 0$ and	x varies from a to $-a$
along CD	$x = -a$	$\therefore$	$dx = 0$ and	y varies from b to 0
along DA	$y = 0$	$\therefore$	$dy = 0$ and	x varies from $-a$ to a

$$\therefore \oint_C \left\{\left(x^2+y^2\right)dx - 2xy\,dy\right\} = \int_{AB}\left(x^2+y^2\right)dx - 2xy\,dy$$

$$+ \int_{BC}\left(x^2+y^2\right)dx - 2xy\,dy + \int_{CD}\left(x^2+y^2\right)dx - 2xy\,dy + \int_{DA}\left(x^2+y^2\right)dx - 2xy\,dy$$

$$= \int_0^b 0 - 2ay\,dy + \int_a^{-a} \left(x^2 + b^2\right) dx + \int_b^0 2ay\,dy + \int_{-a}^a x^2\,dx$$

$$= -ay^2 \Big|_0^b + \left\{\frac{x^3}{3} + b^2 x\right\}_a^{-a} + ay^2 \Big|_b^0 + \frac{x^3}{3}\Big|_{-a}^a$$

$$= -ab^2 - \frac{2a^3}{3} - 2ab^2 - ab^2 + \frac{2a^3}{3}$$

$$\therefore \oint_C \left(x^2 + y^2\right) dx - 2xy\,dy = -4\,ab^2 \quad \text{...(1)}$$

Now, find $\iint_S \text{Curl}\,\vec{F} \cdot \hat{n}\,dS$

$$\text{Curl}\,\vec{F} = \begin{vmatrix} \hat{i} & \hat{j} & \hat{k} \\ \frac{\partial}{\partial x} & \frac{\partial}{\partial y} & \frac{\partial}{\partial z} \\ x^2 + y^2 & -2xy & 0 \end{vmatrix} = \hat{i}\,(0) + \hat{j}\,(0) + \hat{k}\,(-2y - 2y) = -4\,y\,\hat{k}$$

$\hat{n} = \hat{k}$ $\quad \because$ Rectangle lies in xy-plane

$$\therefore \quad \text{Curl}\,\vec{F} \cdot \hat{n} = -4y\hat{k} \cdot \hat{k} = -4y$$

$$\iint_S \text{Curl}\,\vec{F} \cdot \hat{n}\,dS = \iint_S -4y\,dS = \iint_R -4y \frac{dx\,dy}{|\hat{n} \cdot \hat{k}|} = \iint_R -4y\,dx\,dy$$

$$= \int_{x=-a}^{a} \int_{y=0}^{b} -4y\,dy\,dx = \int_{-a}^{a} -2y^2 \Big|_0^b dx = -2b^2\,x \Big|_{-a}^{a} = -4\,ab^2$$

$$\therefore \quad \iint_S \text{Curl}\,\vec{F} \cdot \hat{n}\,dS = -4\,ab^2 \quad \text{...(2)}$$

(1) and (2) are same $\quad \therefore$ Stoke's theorem is verified.

Example 9. *Verify Stoke's theorem for* $\vec{F} = (y - z + 2)\hat{i} + (yz + 4)\hat{j} - (xz)\hat{k}$ *over the surface of a cube* $x = 0, y = 0, z = 0, x = 2, y = 2, z = 2$ *above XOY plane.*

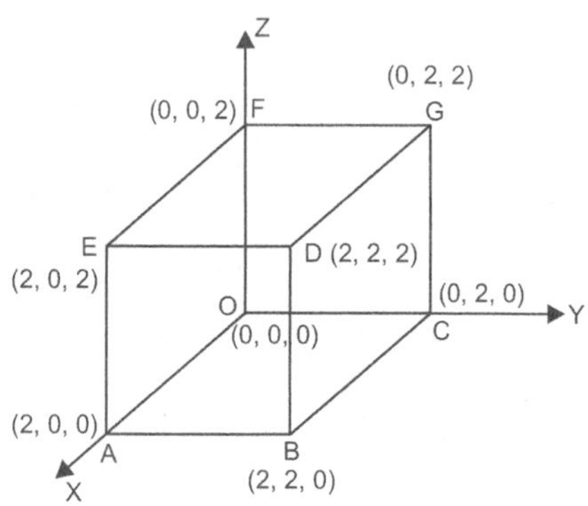

Sol. $\int_C \vec{F} \cdot d\vec{r} = \oint_C (y - z + 2)\,dx + (yz + 4)\,dy - xz\,dz$

where C is the bounding curve of the open surfaces

$\therefore$ C consists of OA, AB, BC, CO

along OA	$y = 0, z = 0,$	$dy = 0, dz = 0$	x varies from	0 to 2
along AB	$x = 2, z = 0,$	$dx = 0, dz = 0$	y varies from	0 to 2
along BC	$y = 2, z = 0,$	$dy = 0, dz = 0$	x varies from	2 to 0
along CO	$x = 0, z = 0,$	$dx = 0, dz = 0$	y varies from	2 to 0

$$\oint_C \vec{F}\cdot d\vec{r} = \oint_C (y-z+2)\,dx + (yz+4)\,dy - xz\,dz$$

$$= \oint_{OA} \vec{F}\cdot d\vec{r} + \oint_{AB} \vec{F}\cdot d\vec{r} + \oint_{BC} \vec{F}\cdot d\vec{r} + \oint_{CO} \vec{F}\cdot d\vec{r} = \int_0^2 2\,dx + \int_0^2 4\,dy + \int_2^0 4\,dx + \int_2^0 4\,dy$$

$$= 4 + 8 - 8 - 8 = -4 \qquad \ldots(1)$$

To obtain surface integral

$$\text{Curl}\,\vec{F} = \begin{vmatrix} \hat{i} & \hat{j} & \hat{k} \\ \dfrac{\partial}{\partial x} & \dfrac{\partial}{\partial y} & \dfrac{\partial}{\partial z} \\ y-z+2 & yz+4 & -xz \end{vmatrix} = \hat{i}\,(0-y) + \hat{j}\,(-1+z) + \hat{k}\,(0-1)$$

$$= -y\,\hat{i} + (z-1)\,\hat{j} - \hat{k}$$

Here we have to integrate over five surfaces ABDE, OCGF, BCGD, OFEA, DGFE, as S is open along the surface OABC

Over the surface ABDE; $x = 2,\quad \hat{n} = \hat{i} \quad dS = dy\,dz$

$$\iint_{ABDE} \text{Curl}\,\vec{F}\cdot\hat{n}\,dS = \iint_R \left\{-y\hat{i} + (z-1)\,\hat{j} - \hat{k}\right\}\cdot\hat{i}\,\frac{dy\,dz}{|\hat{n}\cdot\hat{i}|}$$

$$= \int_0^2\int_0^2 -y\,dy\,dz = \int_0^2 -\frac{y^2}{2}\Big|_0^2\,dz = -2z\Big|_0^2 = -4$$

$$\iint_{OCGF} \text{Curl}\,\vec{F}\cdot\hat{n}\,dS = \iint_R \left\{-y\hat{i} + (z-1)\,\hat{j} - \hat{k}\right\}\cdot\left(-\hat{i}\right)\frac{dy\,dz}{|-\hat{i}\cdot\hat{i}|} = \int_0^2\int_0^2 y\,dy\,dz = 4$$

$$\iint_{BCGD} \text{Curl}\,\vec{F}\cdot\hat{n}\,dS = \iint_R \left\{-y\hat{i} + (z-1)\,\hat{j} - \hat{k}\right\}\cdot\hat{j}\,\frac{dx\,dz}{|\hat{j}\cdot\hat{j}|}$$

$$= \int_0^2\int_0^2 (z-1)\,dx\,dz = \int_0^2 \frac{z^2}{2} - z\Big|_0^2\,dx = \int_0^2 0\,dx = 0$$

$$\iint_{OFEA} \text{Curl}\,\vec{F}\cdot\hat{n}\,dS = \iint_R \left\{-y\hat{i} + (z-1)\,\hat{j} - \hat{k}\right\}\cdot\left(-\hat{j}\right)\frac{dx\,dz}{|-\hat{j}\cdot\hat{j}|}$$

$$= \int_0^2\int_0^2 (-z+1)\,dx\,dz = \int_0^2 \left(\frac{-z^2}{2} + z\right)\Big|_0^2\,dx = \int_0^2 0\,dx = 0$$

$$\iint_{DGFE} \text{Curl}\,\vec{F}\cdot\hat{n}\,dS = \iint_R \left\{-y\hat{i} + (z-1)\,\hat{j} - \hat{k}\right\}\cdot\hat{k}\,\frac{dx\,dy}{|\hat{k}\cdot\hat{k}|} = \int_0^2\int_0^2 -dx\,dy = -4$$

$$\therefore \quad \iint_S \text{Curl}\,\vec{F}\cdot\hat{n}\,dS = -4 + 4 + 0 + 0 - 4 = -4 \qquad \ldots(2)$$

(1) and (2) are same.

$\therefore$ Stoke's theorem is verified.

Example 10. *Verify Stoke's theorem for the vector field* $\vec{F} = (2x - y)\hat{i} - yz^2\,\hat{j} - y^2 z\,\hat{k}$ *over the upper half surface* $x^2 + y^2 + z^2 = 1$ *bounded by its projection on the xy-plane.*

(P.T.U., June 2003)

Sol. $\oint_C \vec{F} \cdot d\vec{r} = \oint_C (2x - y)\,dx - yz^2\,dy - y^2 z\,dz$

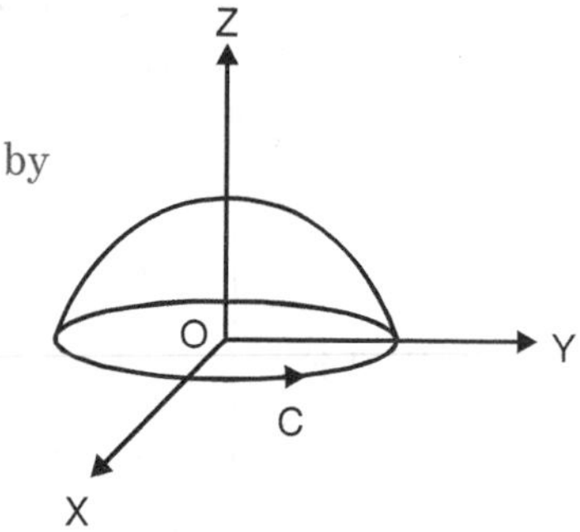

Here path of integration C is a circle in the xy plane given by $x^2 + y^2 = 1$; $z = 0$ $\therefore$ $dz = 0$.

$$\oint_C \vec{F} \cdot d\vec{r} = \int (2x - y)\,dx$$

Any point on circle is $x = \cos t$, $y = \sin t$

$$= \int_0^{2\pi} (2\cos t - \sin t)(-\sin t\,dt)$$

$$= \int_0^{2} (-\sin 2t \quad \sin^2 t)\,dt \quad \frac{\cos 2t}{2}\Bigg|_0^{2} \quad \int_0^{2} \frac{1 \quad \cos 2t}{2}\,dt$$

$$= 0 \quad \frac{1}{2}\, t \quad \frac{\sin 2t}{2}\,\Bigg|_0^{2} = \pi$$

$$\therefore \qquad \oint_C \vec{F} \cdot d\vec{r} = \pi \qquad \ldots(1)$$

Now, to find $\iint_S \text{Curl } \vec{F} \times \hat{n}\, dS$

$$\text{Curl } \vec{F} = \begin{vmatrix} \hat{i} & \hat{j} & \hat{k} \\ \dfrac{\partial}{\partial x} & \dfrac{\partial}{\partial y} & \dfrac{\partial}{\partial z} \\ zx - y & -yz^2 & -y^2 z \end{vmatrix} = \hat{i}\,[-2yz + 2yz] + \hat{j}\,[0 - 0] + \hat{k}\,[0 + 1] = \hat{k}$$

$$\text{Curl } \vec{F} \cdot \hat{n} = \hat{k} \cdot \hat{n}$$

$\iint \hat{k} \cdot \hat{n} \dfrac{dx\,dy}{|\hat{n} \cdot \hat{k}|} = \iint_R dx\,dy$, where R is projection of $x^2 + y^2 + z^2 = 1$ on xy plane given by $x^2 + y^2 = 1$.

$$\therefore \iint_S \text{Curl } \vec{F} \cdot \hat{n}\, dS = \int_{-1}^{1} \int_{-\sqrt{1-x^2}}^{\sqrt{1-x^2}} dy\,dx = \int_{-1}^{1} 2\sqrt{1 - x^2}\,dx = 2\left\{x\sqrt{1 - x^2} + \frac{1}{2}\sin^{-1} x\right\}_{-1}^{1}$$

$$= 2\left[\frac{1}{2}\pi\right] = \pi \qquad \ldots(2)$$

(1) and (2) are same . So Stoke's theorem is satisfied.

Example 11. *Verify Stoke's theorem for* $\vec{F} = -y\hat{i} + x\hat{j} - xyz\hat{k}$ *and S is the surface of the cone* $z = \sqrt{x^2+y^2}$ *for* $x^2 + y^2 \le 9$. **(P.T.U. Dec. 2013)**

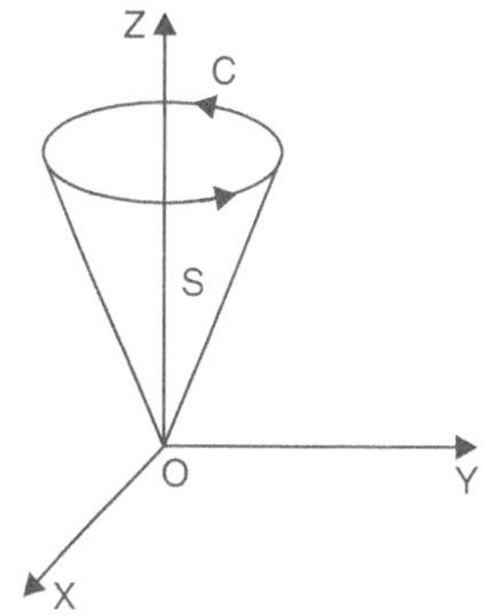

Sol. By Stoke's theorem $\oint_C \vec{F}\cdot d\vec{r} = \iint_S \text{Curl } \vec{F}\cdot\hat{n}\, dS$

Where S is the surface of the cone $z = \sqrt{x^2+y^2}$ for $x^2+y^2 \le 9$, *i.e.*, $z \le 3$

$\therefore$ S is an open surface below the plane $z = 3$ and C is the closed curve given by $x^2 + y^2 = 9$, a circle.

To find $\oint_C \vec{F}\cdot d\vec{r} = \oint_C (-y\hat{i} + x\hat{j} - xyz)\cdot(dx\hat{i} + dy\hat{j} + dz\hat{k})$

$$= \oint_C -y\,dx + x\,dy - xyz\,dz$$

On C; $z = 3$, $x = 3\cos\theta$, $y = 3\sin\theta$

$\therefore$ $dz = 0$, $dx = -3\sin\theta\, d\theta$, $dy = 3\cos\theta\, d\theta$

θ varies from 0 to 2π

$$\therefore \quad \oint_C \vec{F}\cdot d\vec{r} = \int_0^{2\pi} (-3\sin\theta)(-3\sin\theta\, d\theta) + (3\cos\theta)(3\cos\theta)\, d\theta$$

$$= 9\int_0^{2\pi} 1\cdot d\theta = 9\cdot 2\pi = 18\pi$$

Now evaluate $\iint_S \text{Curl } \vec{F}\cdot\hat{n}\, dS$

$$\text{Curl } \vec{F} = \begin{vmatrix} \hat{i} & \hat{j} & \hat{k} \\ \dfrac{\partial}{\partial x} & \dfrac{\partial}{\partial y} & \dfrac{\partial}{\partial z} \\ -y & x & -xyz \end{vmatrix} = \hat{i}(-xz-0) - \hat{j}(-yz-0) + \hat{k}(1+1) = -xz\hat{i} + yz\hat{j} + 2\hat{k}$$

Now $\vec{n}$, the normal vector to surface S *i.e.*, $z^2 = x^2 + y^2$ is

$$\vec{n} = \nabla\phi = \left(\hat{i}\frac{\partial}{\partial x} + \hat{j}\frac{\partial}{\partial y} + \hat{k}\frac{\partial}{\partial z}\right)(z^2 - x^2 - y^2) = -2x\hat{i} - 2y\hat{j} + 2z\hat{k}$$

$$\hat{n} = \frac{\nabla\phi}{|\nabla\phi|} = \frac{-2x\hat{i} - 2y\hat{j} + 2z\hat{k}}{\sqrt{4x^2+4y^2+4z^2}} = \frac{-x\hat{i} - y\hat{j} + z\hat{k}}{\sqrt{x^2+y^2+z^2}} = \frac{-x\hat{i} - y\hat{j} + z\hat{k}}{\sqrt{2z^2}} \qquad [\because\ x^2 + y^2 = z^2]$$

$$= \frac{-x\hat{i} - y\hat{j} + z\hat{k}}{\sqrt{2}z}$$

$$\text{Curl } \vec{F}\cdot\hat{n} = (-xz\hat{i} + yz\hat{j} + 2\hat{k})\cdot\left(\frac{-x\hat{i} - y\hat{j} + z\hat{k}}{\sqrt{2}z}\right)$$

$$= \frac{x^2z - y^2z + 2z}{\sqrt{2}z} = \frac{x^2 - y^2 + 2}{\sqrt{2}}$$

$$dS = \frac{dx\,dy}{\hat{n}\cdot\hat{k}} = \frac{dx\,dy}{\frac{z}{\sqrt{2}z}} = \sqrt{2}\,dx\,dy$$

$$\iint_S \text{Curl}\,\vec{F}\cdot\hat{n}\,dS = \iint_R \frac{x^2 - y^2 + 2}{\sqrt{2}}.\sqrt{2}\,dx\,dy$$

where R is the projection of S on xy-plane

Change it to polar coordinats

$$x = r\cos\theta,\ y = r\sin\theta,\ dxdy = r\,d\theta dr.$$

θ varies from 0 to 2π and r varies from 0 to 3.

$$\iint_S \text{Curl}\,\vec{F}\cdot\hat{n}\,dS = \int_{r=0}^{3}\int_{\theta=0}^{2\pi}(r^2\cos^2\theta - r^2\sin^2\theta + 2)\,r\,d\theta\,dr$$

$$= \int_0^3\int_0^{2\pi}(r^2\cos 2\theta + 2)\,r\,d\theta\,dr = \int_0^3 r^3\frac{\sin 2\theta}{2} + 2r\theta\Big|_0^{2\pi} dr$$

$$= \int_0^3 4\pi r\,dr = 2\pi r^2\Big|_0^3 = 18\pi \qquad ...(2)$$

(1) and (2) are same, so Stoke's theorem is satisfied.

Example 12. *Verify Stoke's Theorem for* $\vec{F} = (x+y)\hat{i} + (2x-z)\hat{j} + (y+z)\hat{k}$ *for the surface of a triangular lamina with vertices (2, 0, 0); (0, 3, 0); (0, 0, 6).* **(P.T.U., May 2010)**

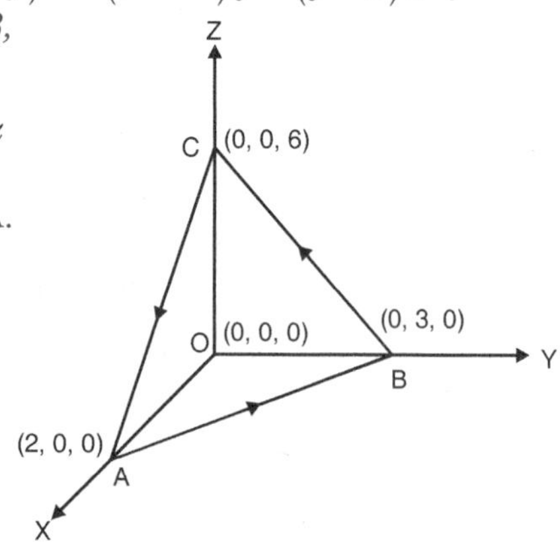

Sol. $\int_C \vec{F}\cdot d\vec{r} = \int_C (x+y)\,dx + (2x-z)\,dy + (y+z)\,dz$

Here path of integration C consists of AB, BC and CA.

Along AB; $z = 0$

$\therefore$ $dz = 0$ and

Equation of AB is $\frac{x}{2} + \frac{y}{3} = 1$ or $y = 3\left(1 - \frac{x}{2}\right)$

$\therefore$ $dy = \frac{-3}{2}dx$

x varies from 2 to 0.

Along BC; $x = 0$ $\therefore$ $dx = 0$

Equation of BC is $\frac{y}{3} + \frac{z}{6} = 1$

i.e., $z = 6\left(1 - \frac{y}{3}\right)$ $\therefore$ $dz = -2dy$ and y varies from 3 to 0.

Along CA; $y = 0$ $\therefore$ $dy = 0$

Equation of CA is $\frac{x}{2} + \frac{z}{6} = 1$ or $z = 6\left(1 - \frac{x}{2}\right)$

$\therefore$ $dz = -3\,dx$ and x varies from 0 to 2.

$$\therefore \quad \oint_C \vec{F}\cdot d\vec{r} = \oint_{AB} \vec{F}\cdot d\vec{r} + \oint_{BC} \vec{F}\cdot d\vec{r} + \oint_{CA} \vec{F}\cdot d\vec{r}$$

$$\oint_{AB} \vec{F}\cdot d\vec{r} = \int_2^0 \left(x + 3 - \frac{3}{2}x\right)dx + 2x\left(-\frac{3}{2}dx\right)$$

$$\left[\because \text{ along AB, } y = 3\left(1 - \frac{x}{2}\right), dy = -\frac{3}{2}dx \text{ and } z = 0\right]$$

$$= \int_2^0 \left(x + 3 - \frac{3}{2}x - 3x\right)dx = \int_2^0 \left(-\frac{7}{2}x + 3\right)dx$$

$$= -\frac{7}{4}x^2 + 3x\Big|_2^0 = 7 - 6 = 1$$

$$\oint_{BC} \vec{F}\cdot d\vec{r} = \int_3^0 -6\left(1 - \frac{y}{3}\right)dy + (y + 6 - 2y)(-2\,dy)$$

$$\left[\because \text{ along BC, } z = 6\left(1 - \frac{y}{3}\right), dz = -2\,dy \text{ and } x = 0\right]$$

$$= \int_3^0 (-6 + 2y - 2y - 12 + 4y)\,dy = \int_3^0 (-18 + 4y)\,dy = -18y + 2y^2\Big|_3^0$$

$$= 54 - 18 = 36$$

$$\oint_{CA} \vec{F}\cdot d\vec{r} = \int_0^2 x\,dx + (6 - 3x)(-3\,dx)$$

$$\left[\because \text{ along CA, } z = 6\left(1 - \frac{x}{2}\right), dz = -3\,dx \text{ and } y = 0\right]$$

$$= \int_0^2 (x - 18 + 9x)\,dx = 5x^2 - 18x\Big|_0^2 = 20 - 36 = -16$$

$$\therefore \quad \oint_C \vec{F}\cdot d\vec{r} = 1 + 36 - 16 = 21 \qquad \ldots(1)$$

Now, $\text{Curl } \vec{F} = \begin{vmatrix} \hat{i} & \hat{j} & \hat{k} \\ \dfrac{\partial}{\partial x} & \dfrac{\partial}{\partial y} & \dfrac{\partial}{\partial z} \\ x + y & 2x - z & y + z \end{vmatrix} = \hat{i}\,(1 + 1) + \hat{j}\,(0 - 0) + \hat{k}\,(2 - 1) = 2\hat{i} + \hat{k}$

Equation of the plane ABC is $\dfrac{x}{2} + \dfrac{y}{3} + \dfrac{z}{6} = 1$

$\therefore$ A vector normal to ABC $= \nabla\left[\dfrac{x}{2} + \dfrac{y}{3} + \dfrac{z}{6} - 1\right]$

$$= \hat{i}\,\frac{1}{2} + \hat{j}\,\frac{1}{3} + \hat{k}\,\frac{1}{6} = \frac{3\hat{i} + 2\hat{j} + \hat{k}}{6}$$

$$\hat{n} = \text{unit vector to ABC} = \frac{3\hat{i} + 2\hat{j} + \hat{k}}{6\,\dfrac{\sqrt{9 + 4 + 1}}{6}} = \frac{3\hat{i} + 2\hat{j} + \hat{k}}{\sqrt{14}}$$

$$\iint_S \text{Curl } \vec{F} \cdot \hat{n}\, dS = \iint_S \left(2\hat{i} + \hat{k}\right) \cdot \frac{3\hat{i} + 2\hat{j} + \hat{k}}{\sqrt{14}}\, dS$$

$$= \iint_S \frac{6+1}{\sqrt{14}}\, dS = \frac{7}{\sqrt{14}} \iint_S dS = \frac{7}{\sqrt{14}} \text{ Area of the triangle ABC}$$

It is difficult to find area of the triangle ABC, so we change dS to $\dfrac{dx\,dy}{\left|\hat{n} \cdot \hat{k}\right|}$

$$= \frac{7}{\sqrt{14}} \iint_R \frac{dx\,dy}{\left|\hat{n} \cdot \hat{k}\right|} = \frac{7}{\sqrt{14}} \iint_R \frac{dx\,dy}{\frac{1}{\sqrt{14}}} = 7 \iint_R dx\,dy,$$

where R is the area of the ΔOAB the projection of ABC on xy-plane.

$$= 7\left(\frac{1}{2}\, 2 \times 3\right) = 21$$

$$\therefore \qquad \iint_S \text{Curl } \vec{F} \cdot \hat{n}\, dS = 21 \qquad \text{...(2)}$$

(1) and (2) are same $\therefore$ Stoke's theorem is verified.

TEST YOUR KNOWLEDGE

1. Evaluate by Stoke's theorem $\oint_C yz\,dx + zx\,dy + xy\,dz$ where C is the curve $x^2 + y^2 = 1$, $z = y^2$.
2. Apply Stoke's theorem to evaluate $\iint_S \nabla \times \vec{F} \cdot \hat{n}\, dS$, where $\vec{F} = y\hat{i} + (x - 2xz)\hat{j} - xy\hat{k}$ and S is the surface of the sphere $x^2 + y^2 + z^2 = a^2$ above xy-plane.
3. Evaluate by Stoke's theorem $\oint_C (\sin z\, dx - \cos x\, dy + \sin y\, dz)$ where C is the boundary of the rectangle $0 \le x \le \pi$, $0 \le y \le 1$, $z = 3$.
4. Evaluate $\oint_C 2y^3\,dx + x^3\,dy + z\,dz$ where C is the trace of the cone $z = \sqrt{x^2 + y^2}$ intersected by the plane $z = 4$ and S is the surface of the cone below $z = 4$. ***(P.T.U. May 2006)***
   (**Hint:** Consult S.E. 11)
5. Use Stoke's theorem to evaluate $\int_C (x + 2y)\,dx + (x - z)\,dy + (y - z)\,dz$, where C is the boundary of the triangle with vertices (2, 0, 0), (0, 3, 0), (0, 0, 6). ***(P.T.U., Dec. 2011)***
   [**Hint.** Consult S.E. 12]
6. Verify Stoke's theorem for the following vector functions:
   (*i*) $\vec{F} = x^2\hat{i} - xy\hat{j}$ integrated round the square in the plane $z = 0$ and bounded by lines $x = 0$, $y = 0$, $x = a$, $y = a$.
   (*ii*) $\vec{F} = \left(x^2 + y - 4\right)\hat{i} + 3xy\hat{j} + \left(2xz + z^2\right)\hat{k}$ over the surface of hemisphere $x^2 + y^2 + z^2 = 16$ above XOY plane. [**Hint:** Consult S.E. 10]

(*iii*) $\vec{F} = (x^2 - y^2)\hat{i} + 2xy\hat{j}$ in the rectangular region of the xy plane bounded by the lines $x = 0$, $x = 1, y = 0, y = 2$. [**Hint.** consult S.E. 8] ***(P.T.U., May 2004)***

(*iv*) $\vec{F} = (x^2 - y^2)\hat{i} + 2xy\hat{j}$ over the box bounded by the planes $x = 0, y = 0, z = 0, x = a, y = b, z = c$; if the face $z = 0$ is cut. [**Hint.** consult S.E. 9] ***(P.T.U. May 2012)***

(*v*) $\vec{F} = y\hat{i} + z\hat{j} + x\hat{k}$, where S is the upper half surface of the sphere $x^2 + y^2 + z^2 = 1$ and C is its boundary.

(*vi*) $\vec{A} = 3y\hat{i} - xz\hat{j} + yz^2\hat{k}$, where S is the surface of the paraboloid $2z = x^2 + y^2$ bounded by $z = 2$ and C is its boundary transversed in counter clockwise direction.

(*vii*) $\vec{F} = -y^3\hat{i} + x^3\hat{j}$ and the closed curve C is the ellipse $\frac{x^2}{a^2} + \frac{y^2}{b^2} = 1$.

(*viii*) $\vec{F} = (2y + z)\hat{i} + (x - z)\hat{j} + (y - x)\hat{k}$ over the portion of the plane $x + y + z = 1$ cut-off by coordinate planes.

Answers

1. 0 **2.** 0 **3.** 2 **4.** 192 π **5.** 15

6. (*i*) $-\frac{a^3}{2}$ (*ii*) -16π (*iii*) 8 (*iv*) $\frac{a^3}{3} + 2ab^2$

(*v*) $-\pi$ (*vi*) -20π (*vii*) $\frac{3\pi ab}{4}(a^2 + b^2)$ (*viii*) 3/2

REVIEW OF THE CHAPTER

1. **Flux.** Let $\vec{F}(x, y, z)$ be a continuous vector point function and S be a closed smooth surface. Then Flux of $\vec{F}$ across the surface S = $\iint_S \vec{F}\cdot\hat{n}\, dS$ or $\iint_S \vec{F}\cdot d\vec{S}$.

2. **Solenoidal Vector.** Let $\vec{V}(x, y, z)$ be a continuous vector point function. $\vec{V}$ will be solenoidal vector if div $\vec{V} = 0$.

3. **Irrotational Vector.** Let $\vec{V}(x, y, z)$ be a continuous vector point function. Then $\vec{V}$ will be irrotational if curl $\vec{V} = \vec{0}$.

4. **Conservative Field.** If a vector point function $\vec{V}(x, y, z)$ can be expressed as gradient of a scalar point function ϕ *i.e.*, $\vec{V} = \nabla\phi$, then V is called conservative field and ϕ is called potential function of $\vec{V}$.
   A conservative vector field is also an irrotational vector field.

5. **Divergence Theorem of Gauss.** If $\vec{F}$ is a vector point function having continuous first order partial derivatives in the region V bounded by a **closed surface S,** then

   $\iiint_V \nabla\cdot\vec{F}\, dV = \iint_S \vec{F}\bullet\hat{n}\, dS$, where $\hat{n}$ is the outward drawn unit normal vector to the surface S.

6. **Green's Theorem.** If M (x, y), N (x, y) be continuous functions of x, y having continuous partial derivatives $\frac{\partial M}{\partial y}$ and $\frac{\partial N}{\partial x}$ in a region R of the xy plane bounded by a closed curve C, then

$$\oint_C (M dx + N dy) = \iint_A \left(\frac{\partial N}{\partial x} - \frac{\partial M}{\partial y}\right) dx\, dy$$ where C is transversed in anticlockwise direction.

7. If S is an **open surface** bounded by a closed curve C and $\vec{F}$ be any vector point function having continuous first order partial derivatives, then $\int_C \vec{F} \cdot d\vec{r} = \iint_S \text{Curl } \vec{F} \cdot \hat{n}\, dS$ where $\hat{n}$ is a unit normal vector at any point of S drawn in the sense in which a right handed screw would advance when rotated in the sense of description of C.

SHORT ANSWER TYPE QUESTIONS

1. Define solenoidal vector and prove that $\frac{\vec{r}}{r^3}$ is solenoidal vector.
[**Hint.** S.E. 2 art 8.2] ***(P.T.U., Dec. 2004, 2006)***
2. Show that $\vec{v} = (-x^2 + yz)\hat{i} + (4y - z^2x)\hat{j} + (2xz - 4z)\hat{k}$ is solenoidal vector. ***(P.T.U., May 2010)***
[**Hint.** S.E. 3 art 8.2]
3. Prove that $\nabla\phi \times \nabla\psi$ is solenoidal. ***(P.T.U., Dec. 2010)***
4. (*i*) Determine the constant b such that $\vec{A} = (bx^2y + yz)\hat{i} + (xy^2 - xz^2)\hat{j} + (2xyz - 2x^2y^2)\hat{k}$ is solenoidal

(*ii*) Find the value of p for which $\vec{V} = (2x + y)\hat{i} + (3x - 2z)\hat{j} + (x + pz)\hat{k}$ is solenoidal.
5. Define irrotational vector and prove that $\frac{\vec{r}}{r}$ is irrotational vector. ***(P.T.U., Dec. 2004, May 2007)***
6. Prove that gradient field describing a motion is irrotational.

OR

Prove that curl grad $\phi = \vec{0}$.
7. Prove that the following vectors are irrotational

(*i*) $\vec{A} = (x^2 + xy^2)\hat{i} + (y^2 + x^2y)\hat{j}$ ***(P.T.U., Dec. 2003, May 2006)***

(*ii*) $\vec{F} = (x^2 - y^2 + x)\hat{i} - (2xy + y)\hat{j}$ ***(P.T.U., Dec. 2012)***

(*iii*) $\vec{F} = (y^2 - z^2 + 3yz - 2x)\hat{i} + (3xz + 2xy)\hat{j} + (3xy - 2xz + 2z)\hat{k}$. ***(P.T.U., Dec. 2012)***

(*iv*) $\vec{v} = -(y + z)\hat{i} + (z + x)\hat{j} + (x + y)\hat{k}$.
8. If $\vec{E}$ and $\vec{H}$ are irrotational vectors prove that $\vec{E} \times \vec{H}$ is solenoidal vector.
(P.T.U., May 2014, Dec. 2014)
[**Hint.** $\nabla \cdot (\vec{E} \times \vec{H}) = \vec{H} \cdot (\nabla \times \vec{E}) - \vec{E} \cdot (\nabla \times \vec{H}) = \vec{H} \cdot \text{curl } \vec{E} - \vec{E} \cdot \text{curl } \vec{H} = 0 - 0 = 0$]
9. Show that the vector field $\vec{v} = (\sin y + z)\hat{i} + (x \cos y - z)\hat{j} + (x - y)\hat{k}$ is a conservative field.
(P.T.U., Dec. 2003)
[**Hint.** Prove curl $\vec{v} = \vec{0}$]
10. For what values of a, b, c the vector function

$\vec{f} = (x + 2y + az)\hat{i} - (bx - 3y - z)\hat{j} + (4x + cy + 2z)\hat{k}$ is irrotational ***(P.T.U., Dec. 2011)***

11. Prove that (*i*) curl $\vec{V}$ is a solenoidal vector
(*ii*) gradient field describes an irrotional motion
(*iii*) If $\phi(x, y, z)$ is any solution of Laplaces equation, then $\nabla\phi$ is both solenoidal and irrotational.

12. If $\vec{a}$ is a constant vector, prove that $\vec{a} \times \vec{r}$ is solenoidal.

13. (*i*) State Gauss Divergence theorem.
(P.T.U., May 2007, Jan. 2010, Dec. 2010, 2012, May 2011, Dec. 2013)
(*ii*) State Green's Theorem in a plane. ***(P.T.U., Dec., 2004, May 2004, 2007)***
(*iii*) State Stoke's Theorem.
(P.T.U., Dec. 2010, 2006, 2014, Jan. 2009, May 2015)

14. Use divergence theorem to prove

(*i*) $\iiint_V \int \frac{dr}{r^2} = \iint_S \frac{\vec{r}\cdot\hat{n}}{r^2}\, dS$ (*ii*) $\iint_S \frac{\vec{r}}{r^3}\cdot\hat{n}\, dS = 0$

(*iii*) $\iint_S r^n\, \vec{r}\cdot d\vec{S} = (n+3)\iiint_V r^n\, dV$ (*iv*) $\iint_S \nabla r^2 \cdot d\vec{S} = 6V$

(*v*) $\iint_S \vec{r}\,.\,d\vec{S} = 3V$, where S is a closed surface enclosing a volume V.

15. Prove that $\int_S \int \vec{F}\cdot\hat{n}\, dS = 6V$ if $\vec{F} = x\hat{i} + 2y\hat{j} + 3z\hat{k}$.

16. Evaluate $\oint_C \vec{F}\cdot d\vec{r}$ where $\vec{F} = y\hat{i} + xz^3\hat{j} - xy^3\hat{k}$ where C is circle $x^2 + y^2 = 4$, $z = 1.5$.

17. Prove that $\iint_S \vec{F}\cdot d\vec{S} = \frac{4}{3}\pi(a+b+c)$ if $\vec{F} = ax\hat{i} + by\hat{j} + cz\hat{k}$ and S is surface of the unit sphere; a, b, c are constants.

Answers

4. (*i*) $b = -2$ (*ii*) $p = -2$

10. $a = 4, b = 2, c = -1$

16. $\frac{19\pi}{2}$

Important Formulae

TRIGONOMETRY

1. Six trigonometric ratios (t-ratios)
   sin θ, cos θ, tan θ, cot θ, sec θ, cosec θ

$$\tan\theta = \frac{\sin\theta}{\cos\theta},\ \cot\theta = \frac{1}{\tan\theta},\ \sec\theta = \frac{1}{\cos\theta},\ \text{cosec}\,\theta = \frac{1}{\sin\theta}.$$

2. $\sin^2\theta + \cos^2\theta = 1$; $\sec^2\theta - \tan^2\theta = 1$; $\text{cosec}^2\theta - \cot^2\theta = 1$
3. $-1 \le \sin\theta \le 1$; $-1 \le \cos\theta \le 1$, tan θ, cot θ can have any value
4.

θ	0°	30°	45°	60°	90°
sin θ =	$\frac{\sqrt{0}}{2}$	$\frac{\sqrt{1}}{2}$	$\frac{\sqrt{2}}{2}$	$\frac{\sqrt{3}}{2}$	$\frac{\sqrt{4}}{2}$
cos θ =	$\frac{\sqrt{4}}{2}$	$\frac{\sqrt{3}}{2}$	$\frac{\sqrt{2}}{2}$	$\frac{\sqrt{1}}{2}$	$\frac{\sqrt{0}}{2}$
tan θ =	$\frac{\sqrt{0}}{\sqrt{4}} = 0$	$\frac{\sqrt{1}}{\sqrt{3}}$	$\frac{\sqrt{2}}{\sqrt{2}}$	$\frac{\sqrt{3}}{\sqrt{1}}$	$\frac{\sqrt{4}}{\sqrt{0}} = \infty$

5. In 1st quadrant all t-ratios are +ve
   In 2nd quadrant sin θ and cosec θ +ve, all other –ve
   In 3rd quadrant tan θ and cot θ +ve, all other –ve
   In 4th quadrant cos θ and sec θ +ve, all other –ve
6. Allied angles

sin (90 – θ) = cos θ;	cos (90 – θ) = sin θ;	tan (90 – θ) = cot θ
sin (90 + θ) = cos θ;	cos (90 + θ) = –sin θ;	tan (90 + θ) = –cot θ
sin (180 – θ) = sin θ;	cos (180 – θ) = –cos θ;	tan (180 – θ) = –tan θ
sin (180 + θ) = –sin θ;	cos (180 + θ) = –cos θ;	tan (180 + θ) = tan θ
sin (360 – θ) = –sin θ;	cos (360 – θ) = cos θ;	tan (360 – θ) = –tan θ
sin (360 + θ) = sin θ;	cos (360 + θ) = cos θ;	tan (360 + θ) = tan θ
sin (–θ) = –sin θ;	cos (– θ) = cos θ;	tan (–θ) = –tan θ

7. $\sin(A \pm B) = \sin A\cos B \pm \cos A\sin B$
   $\cos(A \pm B) = \cos A\cos B \mp \sin A\sin B$

$$\tan(A \pm B) = \frac{\tan A \pm \tan B}{1 \mp \tan A\tan B}$$

8. $2\sin C\cos D = \sin(C + D) + \sin(C - D)$
   $2\cos C\sin D = \sin(C + D) - \sin(C - D)$
   $2\cos C\cos D = \cos(C + D) + \cos(C - D)$
   $2\sin C\sin D = \cos(C - D) - \cos(C + D)$

9. $\cos 2\theta = \cos^2\theta - \sin^2\theta = 2\cos^2\theta - 1 = 1 - 2\sin^2\theta$

$$\therefore \quad \cos^2\theta = \frac{1+\cos 2\theta}{2}, \ \sin^2\theta = \frac{1-\cos 2\theta}{2}$$

$$\tan\theta = \sqrt{\frac{1-\cos 2\theta}{1+\cos 2\theta}}$$

$$\sin 2\theta = 2\sin\theta\cos\theta$$

$$\sin 2\theta = \frac{2\tan\theta}{1+\tan^2\theta}$$

$$\cos 2\theta = \frac{1-\tan^2\theta}{1+\tan^2\theta}; \quad \tan 2\theta = \frac{2\tan\theta}{1-\tan^2\theta}.$$

10. $\sin 3\theta = 3\sin\theta - 4\sin^3\theta$

$$\cos 3\theta = 4\cos^3\theta - 3\cos\theta; \ \tan 3\theta = \frac{3\tan\theta - \tan^3\theta}{1-3\tan^2\theta}.$$

DIFFERENTIAL CALCULUS

1. $\frac{d}{dx}(x^n) = nx^{n-1}$

2. $\frac{d}{dx}(a^x) = a^x \log_e a$

3. $\frac{d}{dx}(e^x) = e^x$

4. $\frac{d}{dx} f(v) = f'(v) = \frac{du}{dx}$ Composite function

5. $\frac{d}{dx}(\sin x) = \cos x$

6. $\frac{d}{dx}(\cos x) = -\sin x$

7. $\frac{d}{dx}(\tan x) = \sec^2 x$

8. $\frac{d}{dx}(\cot x) = -\text{cosec}^2 x$

9. $\frac{d}{dx}(\sec x) = \sec x \tan x$

10. $\frac{d}{dx}(\text{cosec}\, x) = -\text{cosec}\, x \cot x$

11. $\frac{d}{dx}\log x = \frac{1}{x}$

12. $\frac{d}{dx}\log_a x = \frac{1}{x}\log_a e$

13. $\frac{d}{dx}(\sin^{-1} x) = \frac{1}{\sqrt{1-x^2}}$

14. $\frac{d}{dx}(\cos^{-1} x) = \frac{-1}{\sqrt{1-x^2}}$

15. $\frac{d}{dx}(\tan^{-1} x) = \frac{1}{1+x^2}$

16. $\frac{d}{dx}(\cot^{-1} x) = \frac{-1}{1+x^2}$

17. $\frac{d}{dx}(\sec^{-1} x) = \frac{1}{x\sqrt{x^2-1}}$

18. $\frac{d}{dx}(\text{cosec}^{-1} x) = \frac{-1}{x\sqrt{x^2-1}}$

19. $\frac{d}{dx}(uv) = u \cdot \frac{dv}{dx} + v \cdot \frac{du}{dx}$, where u, v are functions of x.

20. $\frac{d}{dx}\left(\frac{u}{v}\right) = \frac{v\frac{du}{dx} - u\cdot\frac{dv}{dx}}{v^2}.$

INTEGRATION

1. $\int x^n\,dx = \dfrac{x^{n+1}}{n+1}\ (n \neq -1);$

2. $\int (ax+b)^n\,dx = \dfrac{(ax+b)^{n+1}}{a(n+1)}\ n \neq -1$

3. $\int \dfrac{1}{x}\,dx = \log|x|;$

4. $\int \dfrac{1}{ax+b}\,dx = \dfrac{\log|ax+b|}{a}$

5. $\int a^x\,dx = \dfrac{a^x}{\log_e a};$

6. $\int a^{mx+c}\,dx = \dfrac{a^{mx+c}}{m\log_e a}$

7. $\int e^x\,dx = e^x;$

8. $\int e^{mx+c}\,dx = \dfrac{e^{mx+c}}{m}$

9. $\int \sin x\,dx = -\cos x;$ and $\int \sin(ax+b)\,dx = -\dfrac{\cos(ax+b)}{a}$

10. $\int \cos x\,dx = \sin x$ and $\int \cos(ax+b)\,dx = \dfrac{\sin(ax+b)}{a}$

11. $\int \tan x\,dx = -\log\cos x$ and $\int \tan(ax+b)\,dx = -\dfrac{\log[\cos(ax+b)]}{a}$

12. $\int \cot x\,dx = \log\sin x$ and $\int \cot(ax+b)\,dx = \dfrac{\log\sin(ax+b)}{a}$

13. $\int \sec x\,dx = \log(\sec x + \tan x)$ or $\log\tan\left(\dfrac{\pi}{4}+\dfrac{x}{2}\right)$

14. $\int \text{cosec}\,x\,dx = \log(\text{cosec}\,x - \cot x)$ or $\log\left(\tan\dfrac{x}{2}\right)$

15. $\int \sec^2 x\,dx = \tan x$

16. $\int \text{cosec}^2 x\,dx = -\cot x$

17. $\int \sec x\tan x\,dx = \sec x$

18. $\int \text{cosec}\,x\cot x\,dx = -\text{cosec}\,x$

19. $\int [f(x)]^n f'(x)\,dx = \dfrac{[f(x)]^{n+1}}{n+1}$, $n \neq -1$ when $n = 1$

20. $\int \dfrac{f'(x)}{f(x)}\,dx = \log|f(x)|$

21. $\int \dfrac{1}{\sqrt{a^2-x^2}}\,dx = \sin^{-1}\dfrac{x}{a}$

22. $\int \dfrac{1}{\sqrt{x^2-a^2}}\,dx = \log|(x+\sqrt{x^2-a^2})|;\ |x| > a$

23. $\int \dfrac{1}{x^2+a^2}\,dx = \dfrac{1}{a}\tan^{-1}\dfrac{x}{a}$

24. $\int \dfrac{1}{\sqrt{x^2+a^2}}\,dx = \log(x+\sqrt{x^2+a^2})$

25. $\int \dfrac{1}{x\sqrt{x^2-a^2}}\,dx = \dfrac{1}{a}\sec^{-1}\dfrac{x}{a};\ x > a$

26. $\int \sqrt{a^2-x^2}\,dx = \dfrac{x\sqrt{a^2-x^2}}{2} + \dfrac{a^2}{2}\sin^{-1}\dfrac{x}{a}$

27. $\int \sqrt{a^2+x^2}\,dx = \dfrac{x\sqrt{a^2+x^2}}{2} + \dfrac{a^2}{2}\log(x+\sqrt{a^2+x^2})$

28. $\int \sqrt{x^2 - a^2}\, dx = \dfrac{x\sqrt{x^2 - a^2}}{2} - \dfrac{a^2}{2} \log (x + \sqrt{x^2 - a^2})$

29. $\int \dfrac{1}{x^2 - a^2}\, dx = \dfrac{1}{2a} \log \left| \dfrac{x - a}{x + a} \right|$

30. $\int \dfrac{1}{a^2 - x^2}\, dx = \dfrac{1}{2a} \log \left| \dfrac{a + x}{a - x} \right|$

31. $\int f(x)\, g(x)\, dx = f(x) \int g(x)\, dx - \int \left[f'(x) \int g(x)\, dx \right] dx$

(1st) (2nd function)

i.e., 1st function × integral of the 2nd function $- \int$ (derivative of 1st × integral of 2nd) dx

32. $\int e^{ax} \sin bx\, dx = \dfrac{e^{ax}}{a^2 + b^2} (a \sin bx - b \cos bx)$

33. $\int e^{ax} \cos bx\, dx = \dfrac{e^{ax}}{a^2 + b^2} (a \cos bx + b \sin bx)$

DEFINITE INTEGRALS

1. $\int_0^{\pi/2} \sin^n x\, dx$ or $\int_0^{\pi/2} \cos^n x\, dx$

$$= \frac{(n-1)(n-3)(n-5)\cdots}{n(n-2)(n-4)(n-6)\cdots} \times \frac{\pi}{2} \text{ if } n \text{ is even otherwise no } \frac{\pi}{2}$$

2. $\int_0^{\pi/2} \sin^p x \cos^q x\, dx$

$$= \frac{[(p-1)(p-3)(p-5)\cdots] \times [(q-1)(q-2)\cdots]}{(p+q)(p+q-2)(p+q-4)} \times \frac{\pi}{2}$$

if p, q both are even integers otherwise no $\dfrac{\pi}{2}$

3. $\int_{-a}^{a} f(x)\, dx = 2 \int_0^a f(x)\, dx$ if $f(x)$ is an even function

$= 0$ if $f(x)$ is an odd function

[If $f(-x) = f(x)$ then $f(x)$ is even if $f(-x) = -f(x)$ then $f(x)$ is odd]

4. $\int_0^{2a} f(x)\, dx = 2 \int_0^a f(x)\, dx$ if $f(2a - x) = f(x)$

$= 0$ if $f(2a - x) = -f(x)$

5. $\int_a^b f(x)\, dx = \int_a^c f(x)\, dx + \int_c^d f(x)\, dx + \int_d^b f(x)\, dx$ (we can continue like thus)

HYPERBOLIC FUNCTIONS

1. (*i*) $\sinh z = \dfrac{e^z - e^{-z}}{2}$ (*ii*) $\cosh z = \dfrac{e^z + e^{-z}}{2}$

(*iii*) $\tanh z = \dfrac{\sinh z}{\cosh z} = \dfrac{e^z - e^{-z}}{e^z + e^{-z}}$

2. $\coth z = \dfrac{1}{\tanh z}$; $\text{sech } z = \dfrac{1}{\cosh z}$; $\text{cosech } z = \dfrac{1}{\sinh z}$

$\cosh^2 z - \sinh^2 z = 1$

3. **Relation between hyperbolic and trigonometric function**

$$\sin iz = i \sinh z; \cos iz = \cosh z$$

or $$\sinh iz = i \sin z; \cosh iz = \cos z$$

4. **Derivatives of hyperbolic functions**

$$\frac{d}{dz}(\sinh z) = \cosh z; \quad \frac{d}{dz}(\cosh z) = \sinh z$$

$$\frac{d}{dz}(\tanh z) = \text{sech}^2 z; \quad \frac{d}{dz}(\coth z) = -\text{cosech}^2 z$$

$$\frac{d}{dz}(\text{sech } z) = -\text{sech} \tanh z; \quad \frac{d}{dz}(\text{cosech } z) = -\text{cosech } z \coth z$$

[First three derivatives +ve last three –ve]

5. **Integration of hyperbolic functions**

$$\int \sinh x \, dz = \cosh z \text{ and } \int \cosh z \, dz = -\sinh z$$

6. **Inverse hyperbolic functions**

$$\sinh^{-1} z = \log (z + \sqrt{z^2 + 1}); \cosh^{-1} z = \log (z + \sqrt{z^2 - 1})$$

$$\tanh^{-1} z = \frac{1}{2} \log \frac{1+z}{1-z}; \ | z | \neq 1$$

[For more details see chapter of complex number]

COORDINATE GEOMETRY

Points and Straight Lines

1. Distance between two points $P(x_1, y_1)$, $Q(x_2, y_2)$ is $PQ = \sqrt{(x_2 - x_1)^2 + (y_2 - y_1)^2}$
2. Coordinates of a point which divides the join of $P(x_1, y_1)$, $Q(x_2, y_2)$ internally in the ratio $m : n$ is

$\left(\dfrac{mx_2 + nx_1}{m+n}, \dfrac{my_2 + ny_1}{m+n}\right)$ and externally in the ratio $m : n$ is $\left(\dfrac{mx_2 - nx_1}{m-n}, \dfrac{my_2 - ny_1}{m-n}\right)$

3. Area of the triangle with vertices $A(x_1, y_1)$, $B(x_2, y_2)$, $C(x_3, y_3)$ is

$$\frac{1}{2}\begin{vmatrix} x_1 & y_1 & 1 \\ x_2 & y_2 & 1 \\ x_3 & y_3 & 1 \end{vmatrix} = \frac{1}{2} \{(x_1y_2 - x_2 y_1) + (x_2y_3 - x_3 y_2) + (x_3y_1 - x_1 y_3)\}$$

4. Angle between two lines $y = m_1x + c_1$, $y = m_2x + c_2$ is given by $\tan\theta = \pm\dfrac{m_1 - m_2}{1 + m_1 m_2}$

(*a*) If the lines are parallel then $m_1 = m_2$

(*b*) If the lines are perpendicular then $m_1m_2 = -1$

(5) Perpendicular distance of a point $P(x_1, y_1)$ from the line $ax + by + c = 0$ is $\dfrac{|ax_1 + by_1 + c|}{\sqrt{a^2 + b^2}}$

Conics:

1. **Parabola:** Four shapes of the parabola

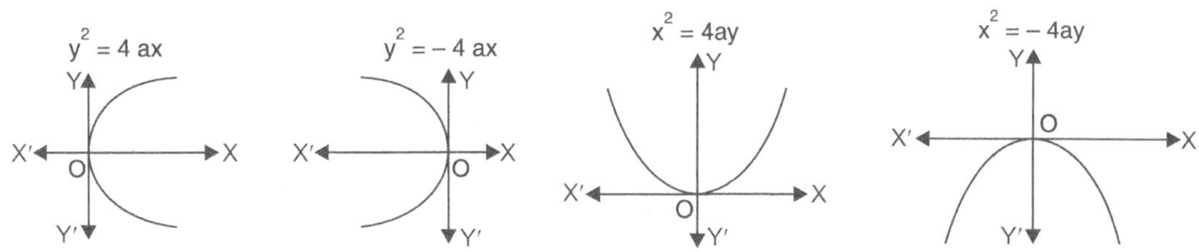

(*i*) (*ii*) (*iii*) (*iv*)

Vertex in each case is (0, 0)

Length of latus rectum in each case is $4a$.

	(*i*)	(*ii*)	(*iii*)	(*iv*)
focus	$(a, 0)$	$(-a, 0)$	$(0, a)$	$(0, -a)$
Axis	$y = 0$	$y = 0$	$x = 0$	$x = 0$
Diretrix	$x = -a$	$x = a$	$y = -a$	$y = a$
Focal distance	$a + x$	$a - x$	$a + y$	$a - y$

Parametric equation of the parabola is $x = at^2$, $y = 2at$ equation of the tangent at (x_1, y_1) to $y^2 = 4ax$ is $yy_1 = 2a\,(x + x_1)$

2. **Ellipse:** Two shapes of the ellipse

$\dfrac{x^2}{a^2} + \dfrac{y^2}{b^2} = 1;\quad a > b > 0$ $\qquad$ $\dfrac{x^2}{b^2} + \dfrac{y^2}{a^2} = 1;\quad a > b > 0$

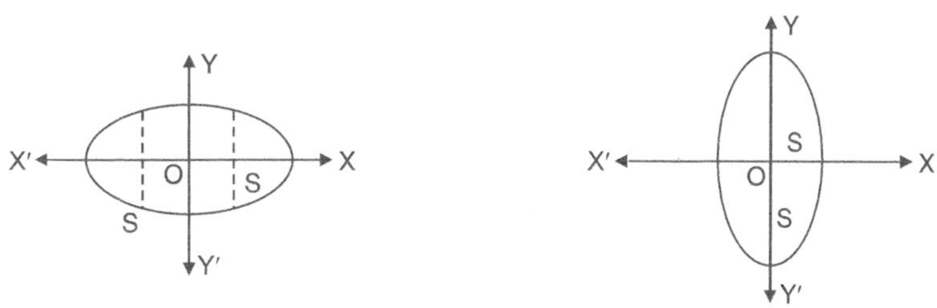

(*i*) (*ii*)

$$b^2 = a^2\,(1 - e^2)$$

	(*i*)	(*ii*)
Vertices	$(a, 0), (-a, 0)$	$(0, a), (0, -a)$
Length of major axis	$2a$	$2a$
Length of minor axis	$2b$	$2b$
Equation of Directrices	$x = \pm a/e$	$y = \pm a/e$

Foci	$(\pm ae, 0)$	$(0, \pm ae)$
Length of latera recta in each case	$= 2b^2/a$	
Focal distances	$a \pm ex$	$a \pm ey$
Parametric form of the ellipse	$x = a \cos \theta$,	$y = b \sin \theta$

Equation of the tangent at (x_1, y_1) is $\dfrac{xx_1}{a^2} + \dfrac{yy_1}{b^2} = 1$

Conjugate diameters of an ellipse are the diameters ⊥ to each other. Extremeties of the conjugate diameters are $P(a \cos \theta, -b \sin \theta)$, $Q\left[a \cos\left(\dfrac{\pi}{2} + \theta\right), b \sin\left(\dfrac{\pi}{2} + \theta\right)\right]$

or $(-a \sin \theta, b \cos \theta)$

$R(-a \cos \theta, -b \sin \theta)$, $S(a \sin \theta, -b \cos \theta)$

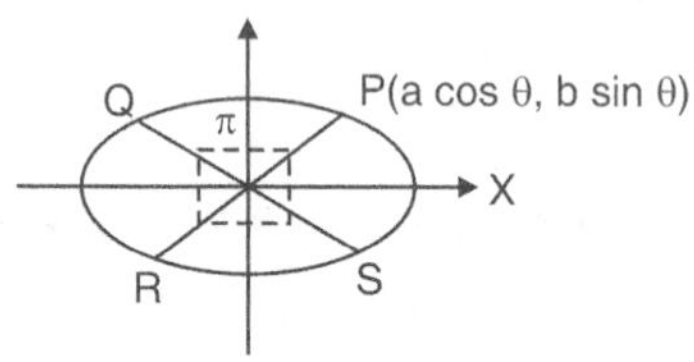

Director circle of the ellipse $\dfrac{x^2}{a^2} + \dfrac{y^2}{b^2} = 1$ is $x^2 + y^2 = a^2 + b^2$

3. Hyperbola: Two shapes of the hyperbola

(*i*) $\dfrac{x^2}{a^2} - \dfrac{y^2}{b^2} = 1, \quad a > b > 0,$ (*ii*) $\dfrac{y^2}{a^2} - \dfrac{x^2}{b^2} = 1, \quad a > b > 0$

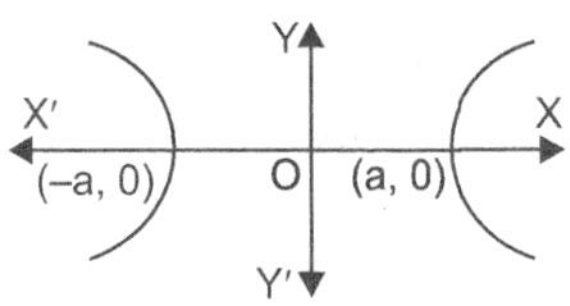

(*i*)

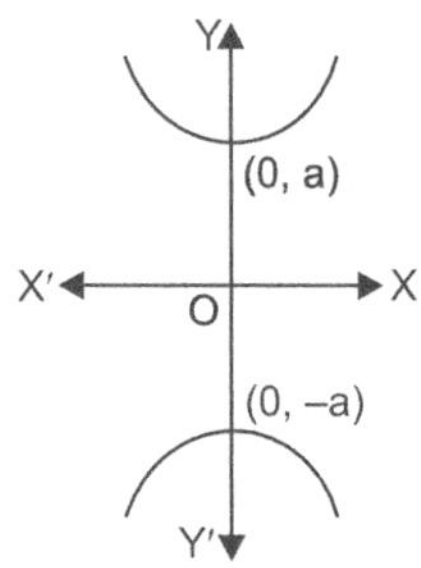

(*ii*)

	(*i*)	(*ii*)
Vertices	$(\pm a, 0)$	$(0, \pm a)$
Length of the transverse axes	$2a$	$2a$
Length of the conjugate axis	$2b$	$2b$
foci	$(\pm ae, 0)$	$(0, \pm ae)$
Equations of the directrices	$x = \pm a/e$	$y = \pm a/e$

Length of the latera recta in each case $= \dfrac{2b^2}{a}$

focal distances	$ex \pm a$	$ey \pm a$

Parametric form of the hyperbola

$$\frac{x^2}{a^2} - \frac{y^2}{b^2} = 1 \text{ is } x = a \sec \theta, y = b \tan \theta$$

Equation of the tangent at (x_1, y_1) to $\dfrac{x^2}{a^2} - \dfrac{y^2}{b^2} = 1$ is $\dfrac{xx_1}{a^2} - \dfrac{yy_1}{b^2} = 1.$